纪蕴芳

Period 3
Christina
Jiyunfang

McGraw-Hill Ryerson

Biology 12

Authors

Trent Carter-Edwards
Upper Canada District School Board

Susanne Gerards
Ottawa Carleton District School Board

Keith Gibbons
London District Catholic School Board

Susan McCallum
York Region District School Board

Robert Noble
Toronto Catholic District School Board

Jennifer Parrington
Durham District School Board

Clyde Ramlochan
Toronto District School Board

Sharon Ramlochan
Toronto District School Board

Contributing Authors

Shahad Abdulnour
University of Toronto

Michelle Anderson
Science Writer and Educator
The Ohio State University

Jonathan Bocknek
Science Writer and Senior Development Editor

David Creasey
University of Victoria

Steven Douglass
York Region District School Board

Lois Edwards
Science Writer and Educator,
Formerly of University of Calgary

Catherine Fan
Science Writer and Educator
Formerly of McMaster University/ Mohawk College

Susan Girvan
Science Writer

Katherine Hamilton
Science Writer and Educator
Formerly of University of Saskatchewan

Christy Hayhoe
Science Writer

Tina Hopper
Science Writer

Glen Hutton
Science Education Consultant

Jody Hyne
Toronto Catholic District School Board

Craig Jackson
Independent Learning Centre, TVO

Adrienne Mason
Science Writer and Educator

Jane McNulty
Science Writer

Chris Schramek
London District Catholic School Board

Sandy Searle
Calgary Board of Education

Ken Stewart
York Region District School Board

Alexandra Venter
Science Writer and Educator
Athabasca University

Athlene Whyte-Smith
Science Writer

Assessment Consultant

Anu Arora
Peel District School Board

Technology and ICT Consultant

Catherine Fan
Science Writer and Educator
Formerly of McMaster University/ Mohawk College

Toronto Montréal Boston Burr Ridge, IL Dubuque, IA Madison, WI New York San Francisco St. Louis Bangkok Bogotá Caracas Kuala Lumpur Lisbon London Madrid Mexico City Milan New Delhi Santiago Seoul Singapore Sydney Taipei

The McGraw·Hill Companies

Copies of this book may be obtained by contacting:

McGraw-Hill Ryerson Limited

e-mail:

orders@mcgrawhill.ca

Toll-free fax:

1-800-463-5885

Toll-free call:

1-800-565-5758

or by mailing your order to:

McGraw-Hill Ryerson Limited
Order Department
300 Water Street
Whitby, ON L1N 9B6

Please quote the ISBN and title when placing your order.

Biology 12U

ISBN-13: 978-0-07-106011-0
ISBN-10: 0-07-106011-1

1 2 3 4 5 6 7 8 9 0 TCP 1 9 8 7 6 5 4 3 2 1

Printed and bound in Canada

EXECUTIVE PUBLISHER: Lenore Brooks
PROJECT MANAGER: Julie Karner
SENIOR PROGRAM CONSULTANT: Jonathan Bocknek
SPECIAL FEATURES COORDINATOR Jane McNulty
DEVELOPMENTAL EDITORS: Michelle Anderson, Lois Edwards, Susan Girvan, Christy Hayhoe, Tina Hopper, Julie Karner, Jane McNulty, Christine Weber
SCIENCE WRITER/EDITOR: Tricia Armstrong
MANAGING EDITOR: Crystal Shortt
SUPERVISING EDITOR: Jaime Smith
COPY EDITORS: Kelli Howey, Wendy Scavuzzo, Linda Jenkins
PHOTO RESEARCH/PERMISSIONS: Maria DeCambra, Monika Schurmann
REVIEW COORDINATOR: Jennifer Keay
EDITORIAL ASSISTANT: Michelle Malda
EDITORIAL INTERN: Daniel McDonald
MANAGER, PRODUCTION SERVICES: Yolanda Pigden
PRODUCTION COORDINATOR: Sheryl MacAdam
SET-UP PHOTOGRAPHY: David Tanaka
COVER DESIGN: Vince Satira
INTERIOR DESIGN: Vince Satira
ELECTRONIC PAGE MAKE-UP: Word & Image Design Studio, Inc.

COVER IMAGES: *ATP molecule:* Public Domain; *mitochondria:* ©Thomas Deerinck, NCMIR/Photo Researchers, Inc.; *neuron and muscle fibres:* Don W. Fawcett/Photo Researchers, Inc.; *elk and wolves:* © NPS photo by Doug Smith

Acknowledgements

Pedagogical Reviewers

Andrea Altenhof
Windsor-Essex Catholic School Board

Samantha Booth
Niagara Catholic District School Board

Kerry D. Dowdell
Greater Essex County District School Board

Alisia D'Silva
York Catholic District School Board

Nikki Giesbrecht
Catholic District School Board of Eastern Ontario

Andrew Jordan
Peel District School Board

Fiona Lawrence-Maki
Rainbow District School Board

Rita Leone
Toronto Catholic District School Board

Dimitrios N. Melegos
Durham District School Board

Ellen Murray
Toronto District School Board

Kamla Kerry-Ann Reid
York Region District School Board

Diana Reis
Dufferin-Peel Catholic District School Board

Catherine Roske
Toronto District School Board

Mary Rupcic
Dufferin-Peel Catholic District School Board

Paris Vasiliou
Ottawa District School Board

Kimberley Walther
Durham Catholic District School Board

Accuracy Reviewers

Doug Bruce, PhD
Brock University

Bhagwati Gupta, PhD
McMaster University

Fiona F. Hunter, PhD
Brock University

Danton H. O'Day, PhD
University of Toronto, Mississauga

Elita Partosoedarso, PhD
University of Ontario Institute of Technology

David Lunn, B. Ed.

Isha DeCoita, PhD
York University, UOIT

Safety Reviewer

Brian Heimbecker
Worker Co-chair, Dufferin-Peel Catholic District School Board

Lab Testers

Jacqueline Deacon
Waterloo Region District School Board

Carey S. Draper
Simcoe County District School Board

Jody Hyne
Toronto Catholic District School Board

Mary-Ann Rupcich
Dufferin-Peel Catholic District School Board

Bias Reviewer

Nancy Christoffer
Markham, Ontario

Catholicity Reviewer

Bernie Smith
York Catholic District School Board

Special Features Writers

Michelle Anderson
Geula Bernstein
Emily Chung
Sharon Oosthoek

Unit Project Writers

Michelle Anderson
Craig Jackson
Tim Lougheed
Craig Saunders

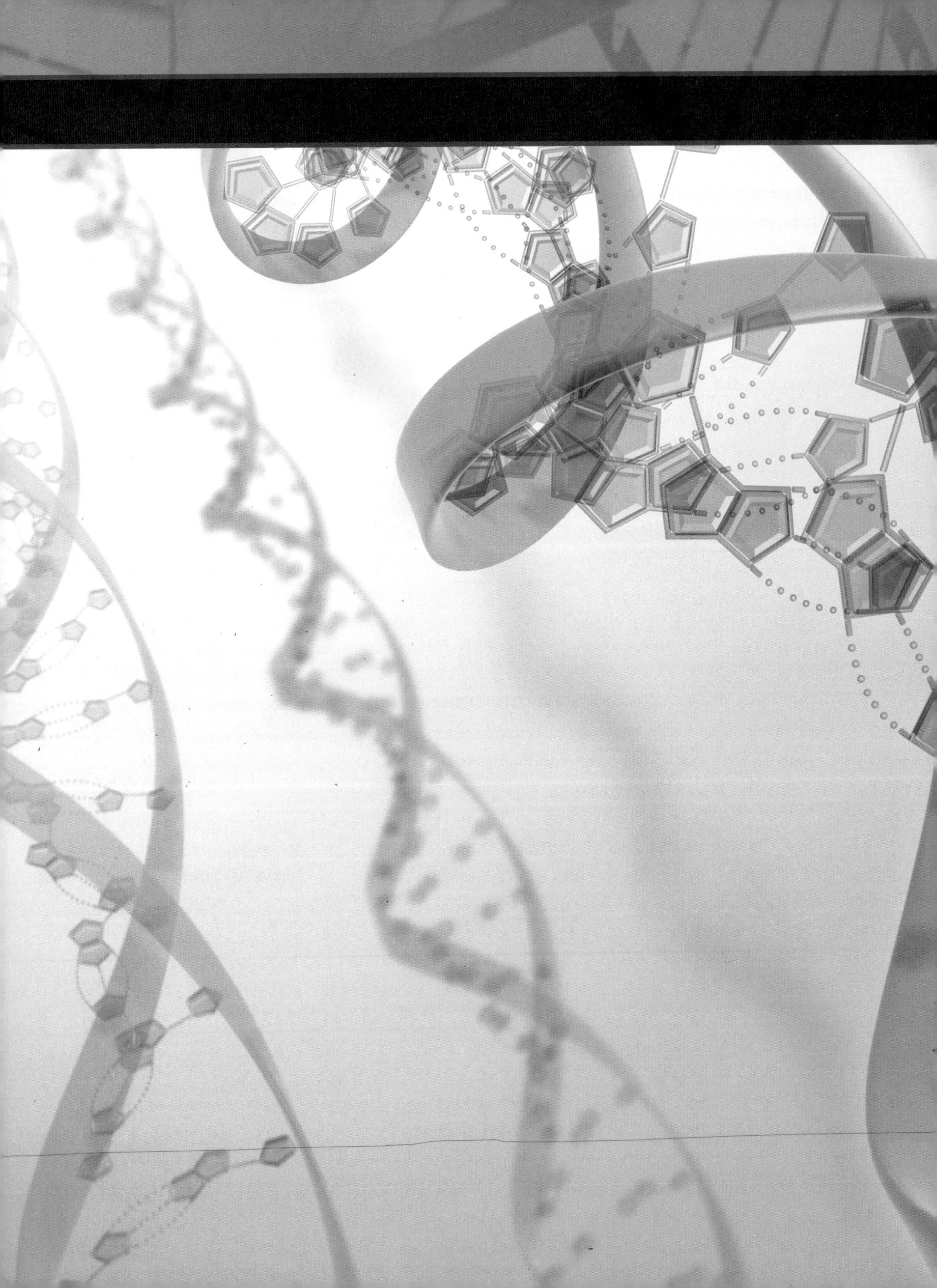

Contents

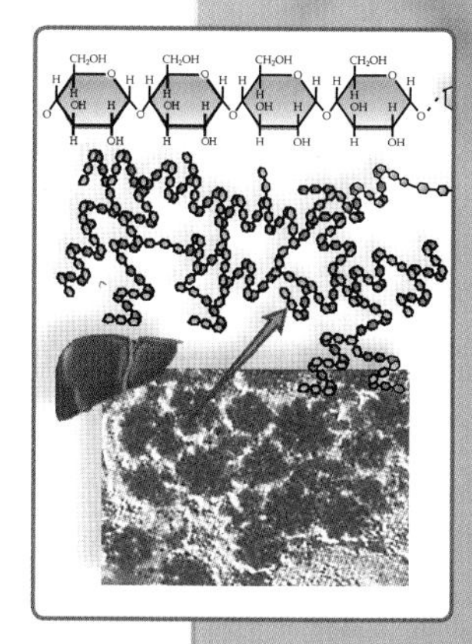

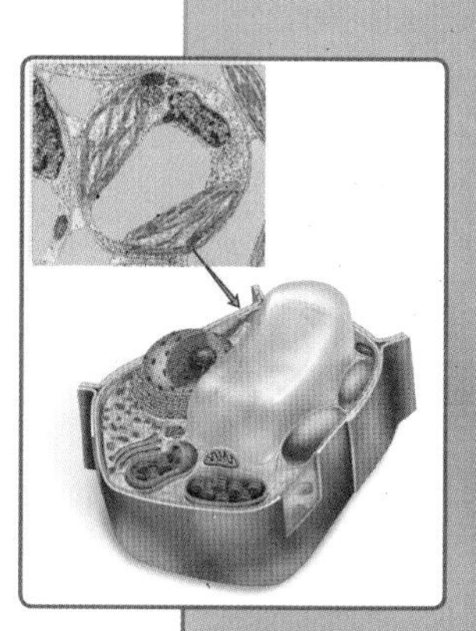

Unit 3 Molecular Genetics

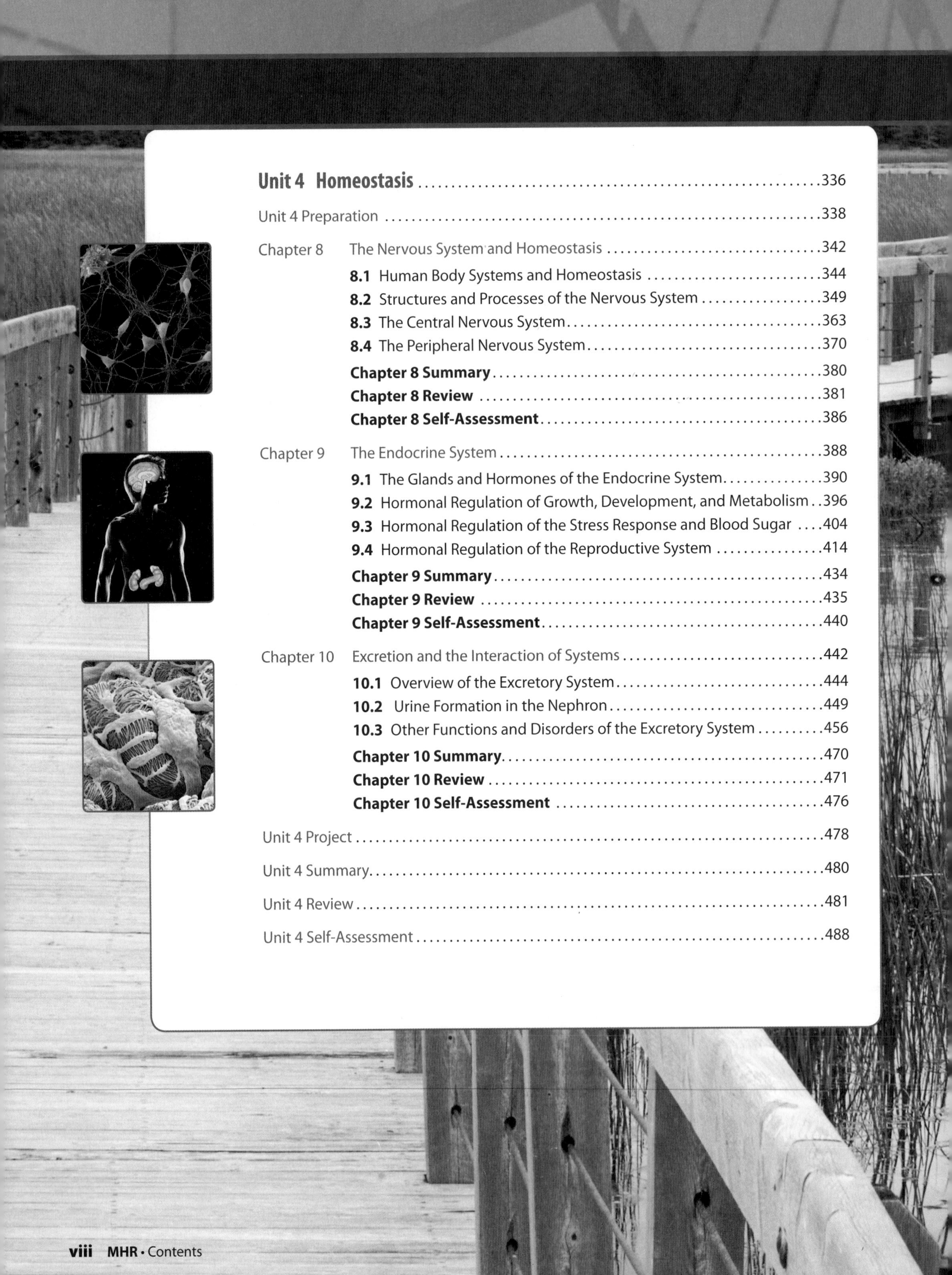

Activities and Investigations

Activities

Investigations

STSE Special Features

Safety in the Biology Lab and Classroom

Keep in mind at all times that working in a biology classroom can involve some risks. *Therefore, become familiar with all facets of laboratory safety, especially for performing investigations safely.* To make the investigations and activities in *Biology 12* safe and enjoyable for you and others who share a common working environment,

- become familiar with and use the following safety rules
- follow any special instructions from your teacher
- *always read* the safety notes before beginning each activity or investigation. Your teacher will tell you about any additional safety rules that are in place at your school.

WHMIS Symbols for Hazardous Materials

Look carefully at the WHMIS (Workplace Hazardous Materials Information System) safety symbols shown here. The WHMIS symbols and the associated material safety data sheets (MSDS) are used throughout Canada to identify dangerous materials. These symbols and the material safety data sheets help you understand all aspects of safe handling of hazardous materials. Your school is required to have these sheets available for all chemicals, and they can also be found by doing an Internet search. Make certain you read the MSDS carefully and understand what these symbols mean.

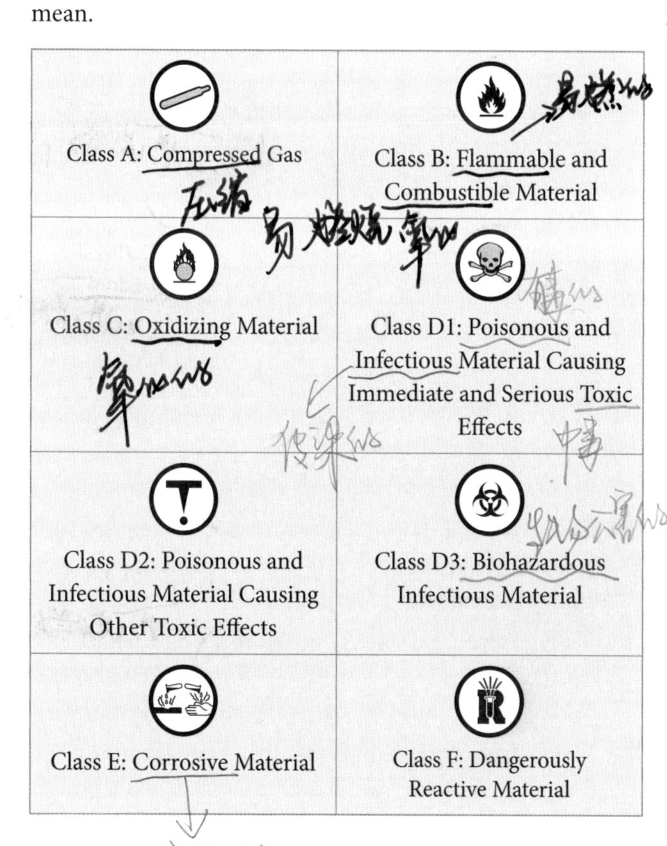

Safety Symbols

Be sure you understand each symbol used in an activity or investigation before you begin.

	**Disposal Alert** This symbol appears when care must be taken to dispose of materials properly.
	**Biological Hazard** This symbol appears when there is danger involving bacteria, fungi, or protists.
	**Thermal Safety** This symbol appears as a reminder to be careful when handling hot objects.
	**Sharp Object Safety** This symbol appears when there is danger of cuts or punctures caused by the use of sharp objects.
	**Fume Safety** This symbol appears when chemicals or chemical reactions could cause dangerous fumes.
 	**Electrical Safety** This symbol appears as a reminder to be careful when using electrical equipment.
	**Skin Protection Safety** This symbol appears when the use of caustic chemicals might irritate the skin or when contact with micro-organisms might transmit infection.
	**Clothing Protection Safety** A lab apron should be worn when this symbol appears.
	**Fire Safety** This symbol appears as a reminder to be careful around open flames.
	**Eye Safety** This symbol appears when there is danger to the eyes and safety glasses should be worn.
	**Poison Safety** This symbol appears when poisonous substances are used.
	**Chemical Safety** This symbol appears when chemicals could cause burns or are poisonous if absorbed through the skin.
	**Animal Safety** This symbol appears when live animals are studied and the safety of the animals and students must be ensured.

General Rules

1. Inform your teacher if you have any allergies, medical conditions, or physical problems (including a hearing impairment) that could affect your classroom work.

2. Inform your teacher if you wear contact lenses. If possible, wear eyeglasses instead of contact lenses, but remember that eyeglasses are not a substitute for proper eye protection.
3. Read through all of the steps in an activity or investigation before beginning. Be sure to read and understand the *Safety Precautions* and safety symbols.
4. Listen carefully to any special instructions your teacher provides. Get your teacher's approval before beginning any investigation that you have designed yourself.
5. *Never* eat, drink, or taste anything in the biology classroom. Never pipette with your mouth. If you are asked to smell a substance, do not hold it directly under your nose. Keep the object at least 20 cm away, and waft the fumes toward your nostrils with your hand.

Safety Equipment and First Aid

6. When you are directed to do so, wear safety goggles and protective equipment in the biology classroom.
7. Know the location and proper use of the nearest fire extinguisher, fire blanket, fire alarm, first-aid kit, spill kit, eyewash station, and drench hose/shower.
8. Never use water to fight an electrical equipment fire. Severe electric shock may result. Use a carbon dioxide or dry chemical fire extinguisher. Report any damaged equipment or frayed cords to your teacher.
9. Cuts, scratches, or any other injuries in the biology classroom should receive immediate medical attention, no matter how minor they seem. If any part of your body comes in contact with a potentially dangerous substance, wash the area immediately and thoroughly with water.
10. If you get any material in your eyes, do not touch them. Wash your eyes immediately and continuously for 15 minutes in an eyewash station, and make sure your teacher is informed. If you wear contact lenses, take your lenses out immediately if you get material in your eyes. Failing to do so may result in material being trapped behind the contact lenses. Flush your eyes continuously with water for 15 minutes, as above.

Lab Precautions

11. Keep your work area clean, dry, and well organized.
12. When using a scalpel or knife, cut away from yourself and others. Always keep the pointed end of any sharp objects directed away from you and others when carrying such objects.
13. Use EXTREME CAUTION when you are near an open flame. Wear heat-resistant safety gloves and any other safety equipment that your teacher or the *Safety Precautions* suggest when heating any item. Be especially careful with a hot plate that may look as though it has cooled down. If you do receive a burn, apply cold water to the burned area immediately. Make sure your teacher is notified.

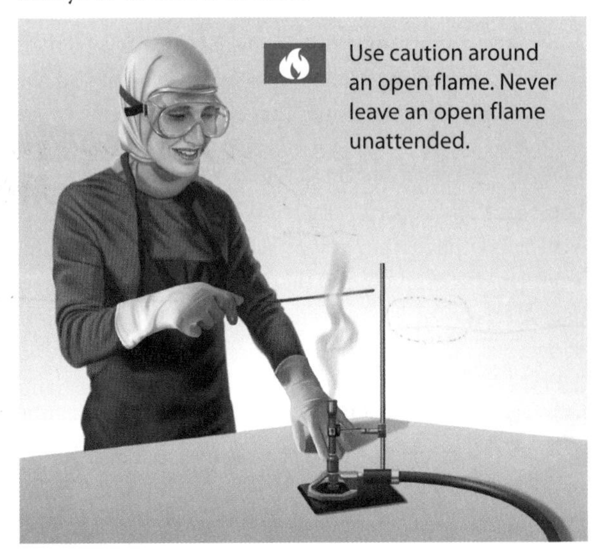

Use caution around an open flame. Never leave an open flame unattended.

14. Keep your hands and work area dry when touching electrical cords, plugs, sockets, or equipment. Ensure cords are placed neatly where they will not be a tripping hazard. Turn OFF all electrical equipment before connecting to or disconnecting from a power supply. When unplugging electrical equipment, do not pull the cord—grasp the plug firmly at the socket and pull gently.
15. When you are heating a test tube, apply heat gently and always slant it so the mouth points away from you and others.

Safety for Animal Dissections

16. Ensure your work area is well ventilated.
17. Always wear appropriate protective equipment for your skin, clothing, and eyes. This will prevent preservatives from harming you.
18. If your scalpel blade breaks, do not replace it yourself. Your teacher will help you dispose of any broken or spent blades in an appropriate sharps container as directed.
19. Make sure you are familiar with the proper use of all dissecting equipment. Whenever possible, use a probe or your gloved fingers to explore a specimen. Scalpels

are not appropriate for this. They can damage the structures you are examining.

Clean-up

20. Wipe up all spills immediately, and always inform your teacher. Acid or base spills on clothing or skin should be diluted and rinsed with water. Small spills of acid solutions can be neutralized with sodium hydrogen carbonate (baking soda). Small spills of basic solutions can be neutralized with sodium hydrogen sulfate or citric acid. For larger spills, an appropriate spill kit should be used.
21. Never use your hands to pick up broken glass. Use a broom and dustpan. Dispose of broken glass and solid substances in the proper containers, as directed by your teacher.
22. Dispose of all specimens, materials, chemicals, and other wastes as instructed by your teacher. Do not dispose of materials in a sink or drain unless directed to do so.
23. Clean equipment before putting it away, according to your teacher's instructions. Turn off the water and gas. Disconnect electrical devices. Wash your hands thoroughly after all activities and investigations.

Working with Living Organisms

24. When in the field, be careful and observant at all times to avoid injury, such as tripping, being poked by branches, etc., or coming into contact with poisonous plants.
25. On a field trip, try not to disturb the area any more than is absolutely necessary. If you must move anything, do so carefully. If you are asked to remove plant material, do so gently. Take as little as possible.
26. In the classroom, remember to treat living organisms with respect. If it is possible, return living organisms to their natural environment when your work is done.
27. When working with micro-organisms, observe your results through the clear lid of the petri dish. Do not open the cover. Make sure that you do not touch your eyes, mouth, or any other part of your face during these investigations.
28. When handling live bacterial cultures, always wear gloves and eye protection. Wash your hands thoroughly with soap immediately after handling any bacterial culture. Culturing bacteria from swabbing areas of the school indiscriminately is not recommended unless closely monitored by your teacher.

Safety in Your On-line Activities

The Internet is like any other resource you use for research—you should confirm the source of the information and the credentials of those supplying it to make sure the information is credible before you use it in your work.

Unlike other resources, however, the Internet has some unique pitfalls you should be aware of, and practices you should follow.

- When you copy or save something from the Internet, you could be saving more than information. Be aware that information you pick up could also include hidden, malicious software code (known as "worms" or "Trojans") that could damage your system or destroy data.
- Avoid sites that contain material that is disturbing, illegal, harmful, and/or was created by exploiting others.
- *Never* give out personal information on-line. Protect your privacy, even if it means not registering to use a site that looks helpful. Discuss ways to use the site while protecting your privacy with your teacher.
- Report any on-line content or activity that you suspect is inappropriate or illegal to your teacher.

Instant Practice

1. One of the materials you plan to use in a Plan Your Own Investigation bears the following symbols:

   Describe the safety precautions you would need to incorporate into your investigation.
2. Describe when you would require an MSDS sheet. What would you do with the information?

UNIT 1 Biochemistry

BIG IDEAS

- Technological applications that affect biological processes and cellular functions are used in the food, pharmaceutical, and medical industries.
- Biological molecules and their chemical properties affect cellular processes and biochemical reactions.
- Biochemical compounds play important structural and functional roles in cells of all living organisms.

Overall Expectations

In this unit, you will learn how to...

- **analyze** technological applications of enzymes in some industrial processes, and **evaluate** technological advances in the field of cellular biology
- **investigate** the chemical structures, functions, and chemical properties of biological molecules involved in some common cellular processes and biochemical reactions
- **demonstrate** an understanding of the structures and functions of biological molecules, and the biochemical reactions required to maintain normal cellular function

Unit Contents

Focussing Questions

1. *What are the structures, functions, and properties of various biological molecules?*
2. *What biochemical reactions are necessary to maintain normal cellular function?*
3. *How is technology related to cellular biology used in industry, medicine, and other applications?*

Go to **scienceontario** to find out more about biochemistry

In 2010, NASA scientists announced they had discovered a new bacterium. This organism could incorporate arsenic, which is normally poisonous to most organisms, into its DNA and other cellular components in place of phosphorus. The structure of DNA, shown here, was determined by scientists in the early 1950s, and it did not include arsenic. The significance of this new discovery could affect our understanding of life as we know it, including new possibilities about environments in which organisms might live. However, shortly after the publication of the NASA scientists' results in the peer-reviewed journal, *Science*, other scientists began to question the methods used in the study and the interpretation of the results. The original research could indeed be valid. But establishing its validity will depend on the results of additional study. In Unit 1, you will learn about biochemistry—the chemistry of the molecules that make life possible. In this unit and throughout the book as a whole, you will also see how the ongoing processes of inquiry and investigation shape and refine scientific knowledge and its many applications.

As you study this unit, look ahead to the Unit 1 Project on pages 96 to 97. Complete the project in stages as you progress through the unit.

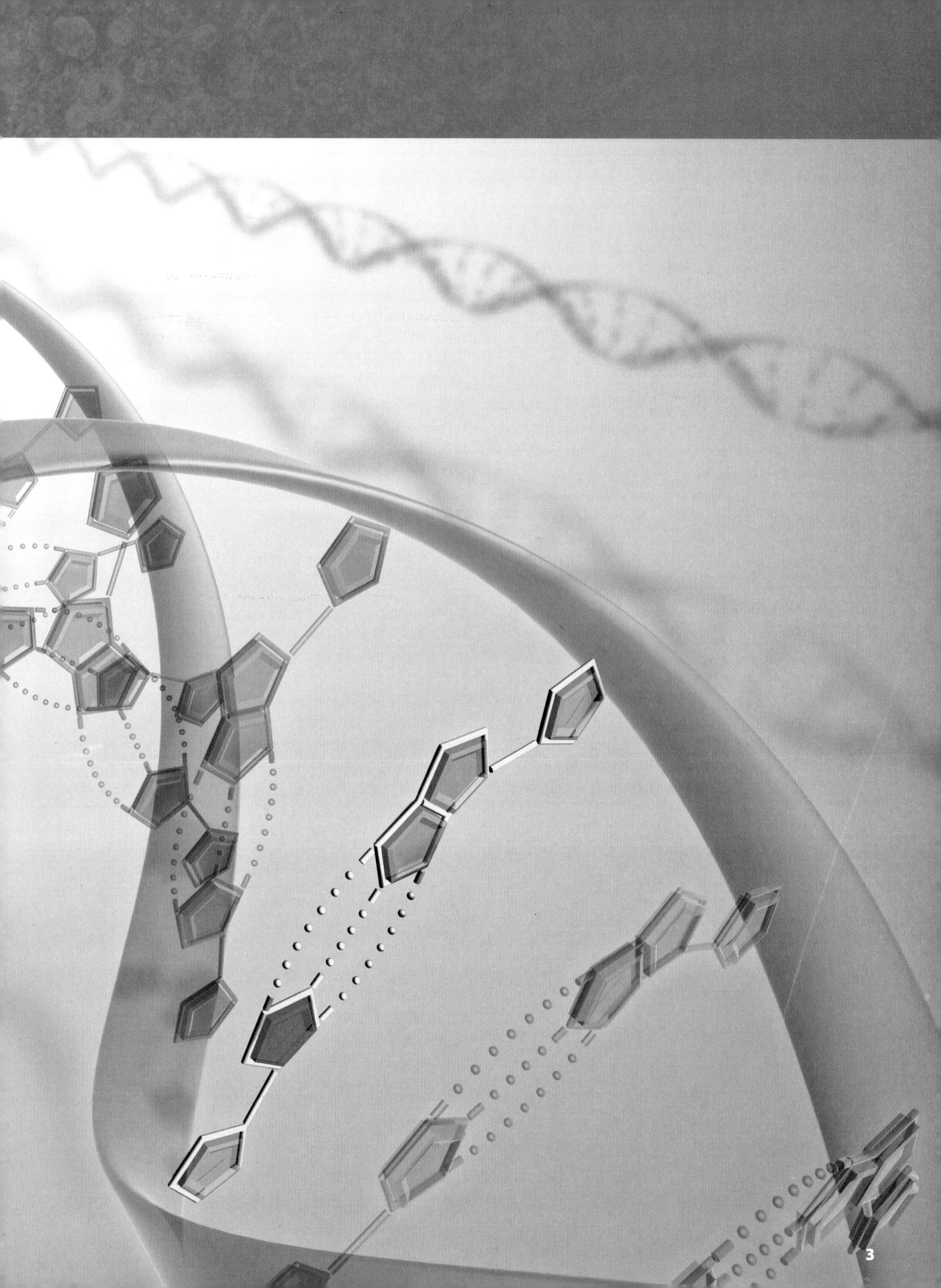

UNIT 1

Preparation

Atoms, Elements, Compounds, and Chemical Bonding

- An element is a pure substance that cannot be broken down into simpler substances through chemical or physical methods. An element consists of only one type of atom. An atom is the smallest component of an element that retains the properties of that element.
- A compound is a pure substance composed of two or more elements that are chemically combined. A compound consists of a specific ratio of two or more types of atoms.
- Atoms are made up of subatomic particles: protons, neutrons, and electrons. The nucleus of an atom is composed of positively charged protons and uncharged neutrons. Negatively charged electrons are present in specific regions, called orbitals, electron shells, or energy levels, which are located at increasing distances from the nucleus.
- Isotopes of an element are atoms that have the same number of protons but different numbers of neutrons.
- Atoms that lose or gain electrons become charged particles called ions. When atoms lose electrons, they form positively charged ions called cations. When atoms gain electrons, they form negatively charged ions called anions. Atoms are at their most stable when their outer electron shell, called the valence shell, is completely filled. For most atoms, this occurs when the valence shell contains eight electrons. For hydrogen and helium, the first two elements in the periodic table, the valence shell is full when it contains two electrons.
- Chemical bonding allows atoms to achieve a stable conformation by sharing electrons with, donating electrons to, or receiving electrons from other atoms.
- In ionic bonding, a chemical bond forms between oppositely charged ions. An ionic compound forms when one or more electrons from one atom are transferred to another atom.
- In covalent bonding, atoms share electrons to achieve stable conformations. Molecular compounds form when atoms share electrons in covalent bonds.
- A molecular formula shows the number of each type of atom in an element or compound.

1. Sketch a simple model of a carbon atom in your notebook. Label each part of the atom.

2. Examine the diagram below that shows two atoms forming a covalent compound. Which statement best describes the formation of a covalent bond?

a. One atom gives up an electron to another atom, forming a chemical bond.

b. Two atoms share one or more valence electrons, forming a chemical bond.

c. One atom gives up two or more electrons to another atom, forming a chemical bond.

d. The protons of one atom attract the neutrons of another atom, and a chemical bond forms.

e. All of the above.

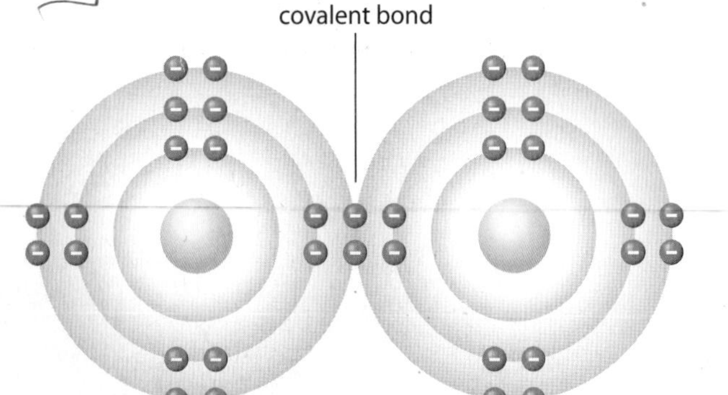

3. Select the correct definition of an ion.

a. An ion is an atom that has gained or lost one or more electrons, and as a result has either a negative charge or a positive charge.

b. An ion is an atom involved in forming a covalent bond.

c. An ion is an atom that has a neutral charge.

d. An ion is one of the subatomic particles making up an atom, along with protons, neutrons, and electrons.

e. An ion is another name for an electron.

4. In your notebook, complete this sentence by filling in the missing words: The name of an ionic compound, for example, sodium chloride, contains the name of the _____ followed by the name of the _____.

5. In a glucose molecule, carbon, hydrogen, and oxygen atoms share electrons so that each carbon and oxygen atom has eight valence electrons, and each hydrogen atom has two valence electrons. Use this information to answer the following questions.

a. What type of bonding is present in glucose, and how do you know?

b. Is glucose a stable molecule? Explain why or why not.

6. Name each compound represented by its molecular formula below. Identify whether the compound is molecular or ionic and explain why.

a. H_2O **c.** $C_6H_{12}O_6$ **e.** $Ca_3(PO_4)_2$ **g.** O_2

b. CO_2 **d.** NaCl **f.** CH_4 **h.** NH_3

7. Write the name of each ion.

a. Cl^- **b.** SO_4^{2-} **c.** Mg^{2+} d. Cu^{2+}

8. Write the symbol for each ion.

a. hydroxide ion **c.** aluminum ion

b. sulfide ion **d.** iron(III) ion

9. The Bohr-Rutherford models below show how covalent bonds between hydrogen and oxygen involve sharing a pair of valence electrons. Identify the compound represented by these diagrams.

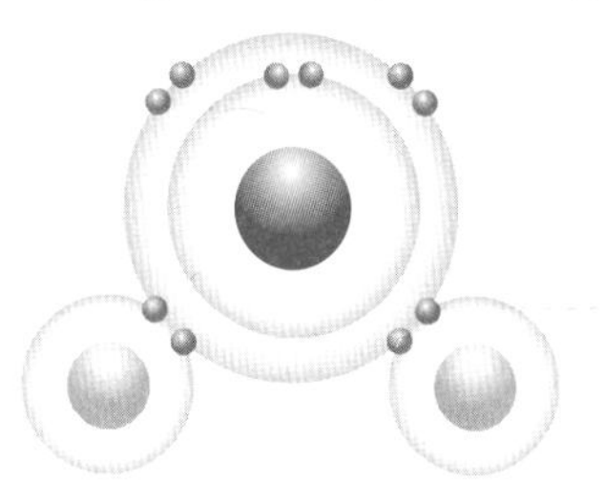

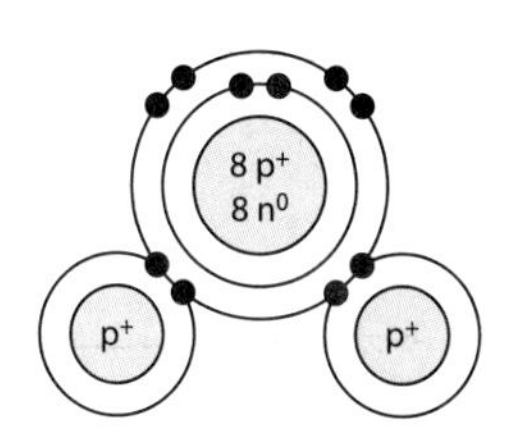

Chemical Reactions

- A chemical reaction takes place when the atoms making up one or more substances, called reactants, rearrange to form one or more new substances, called products.
- A chemical equation is a way of representing the reactants and products in a chemical reaction. In a balanced chemical equation, the total number of each type of atom is the same on both sides of the equation.
- Types of chemical reactions include synthesis, decomposition, and neutralization reactions. In a neutralization reaction, an acid and a base react to form a salt (an ionic compound) and water.
- An acid is a substance that produces hydrogen ions, H^+, when it dissolves in water. A base is a substance that produces hydroxide ions, OH^-, when it dissolves in water. If a solution has a pH lower than 7, it is acidic. If it has a pH higher than 7, it is basic.

10. Write a word equation corresponding to the chemical equation below.

$$2H_2O(\ell) \rightarrow 2H_2(g) + O_2(g)$$

11. Which numbers would balance the chemical equation below?

$$CH_4(g) + \_\_ O_2(g) \rightarrow CO_2(g) + \_\_ H_2O(g)$$

a. 1; 4 **d.** 2; 2

b. 2; 4 **e.** 4; 4

c. 1; 1

12. Identify each substance in the following neutralization reaction.

$$NaOH(aq) + HCl(aq) \rightarrow H_2O(\ell) + NaCl(aq)$$

13. A student measures the pH of a water-based solution to be 8. After the addition of a small amount of a different substance, the pH of the solution increases to 9.

a. Did the student add an acid or a base to the solution? Explain your answer.

b. Describe what happened to the solution in the beaker at the molecular level as the second substance was added.

14. Classify the following reactions as synthesis, decomposition, or neutralization reactions.

a. $CaCO_3(s) \rightarrow CaO(s) + CO_2(g)$

b. $HCl(aq) + NaOH(aq) \rightarrow H_2O(\ell) + NaCl(aq)$

c. $2AgCl(s) \rightarrow 2Ag(s) + Cl_2(g)$

d. $2H_2(g) + O_2(g) \rightarrow 2H_2O(\ell)$

e. $H_2SO_4(aq) + Ca(OH)_2(aq) \rightarrow CaSO_4(aq) + 2H_2O(\ell)$

f. $2NO(g) + O_2(g) \rightarrow 2NO_2(g)$

g. $MgCl_2(\ell) \rightarrow Mg(\ell) + Cl_2(g)$

h. $S_8(s) + 12O_2 \rightarrow 8SO_3(g)$

i. $Cu(s) + 2AgNO_3(aq) \rightarrow 2Ag(s) + Cu(NO_3)_2(aq)$

j. $3Mg(s) + N_2(g) \rightarrow Mg_3N_2(s)$

15. Identify the pH of a solution that is described as being neutral.

Cell Structure and Function

- The cell is the basic organizational unit of life. All cells come from pre-existing cells, and all living things are made of one or more cells.
- There are two basic cell types—prokaryotic and eukaryotic. Prokaryotic cells do not have a membrane-bound nucleus. Eukaryotic cells have a membrane-bound nucleus as well as other membrane-bound organelles. Bacteria and archaea are prokaryotes. Plants, animals, and fungi have eukaryotic cells.
- All cells have a cell membrane that controls what enters and leaves the cell. The membrane separates the inside of the cell from the extracellular environment.
- In plant and animal cells, specific functions to support life are carried out by internal cell parts called organelles. Examples of cellular organelles include mitochondria, Golgi bodies, ribosomes, vesicles, vacuoles, chloroplasts, and a nucleus or multiple nuclei.

16. Which of the following statements is correct?

a. Substances pass through cell membranes in one direction only.
b. Substances move freely in and out of the cell in both directions.
c. Substances other than water cannot move freely in and out of cells.
d. Substances can move freely in and out of red blood cells only.
e. Substances are prevented from entering or leaving cells by the cell membrane.

17. State one important function of proteins in the body.

18. Compare and contrast a molecule and a macromolecule.

19. Write a definition for the term *enzyme*.

20. Which statement best describes the function of a cell membrane?

a. It allows water but nothing else to move freely into and out of the cell.
b. It allows water and other substances to move into the cell, but not out of the cell.
c. It allows water and certain other substances to move freely into and out of the cell, but restricts the passage of other substances into or out of the cell.
d. It always uses energy to transport water and certain other substances into and out of the cell.
e. It never uses energy to transport substances into and out of the cell.

21. Which structure enables some eukaryotic cells to trap light energy from the Sun in the form of energy-rich molecules?

a. cell membrane
b. ribosome
c. nucleus
d. mitochondria
e. chloroplast

22. Which statement best describes the main role of the cytoskeleton?

a. It prevents animal cells from expanding too far and bursting.
b. It provides scaffolding for the cell, determines cell shape, and anchors organelles in place.
c. It provides an internal messaging system for the cell along which chemical messages are sent and received.
d. It permits some substances to enter and leave the cell, but prevents others from doing so.
e. It controls the activities of the cell.

23. The two diagrams below show a prokaryotic cell and a eukaryotic cell. Identify each type of cell. Then use a Venn diagram to compare and contrast these two types of cells.

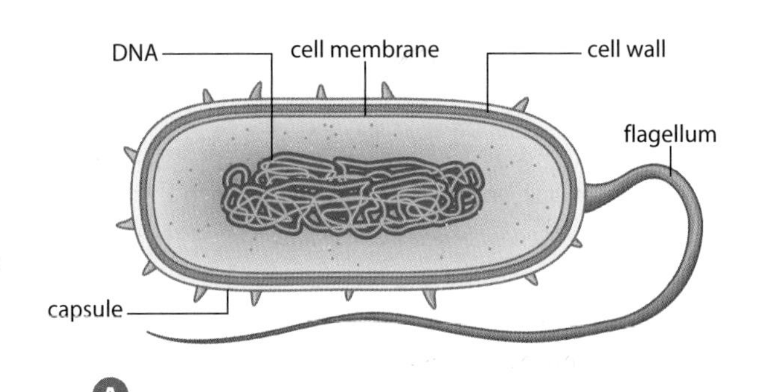

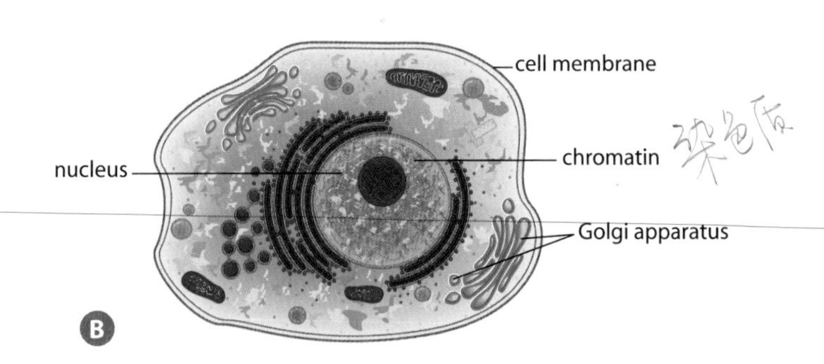

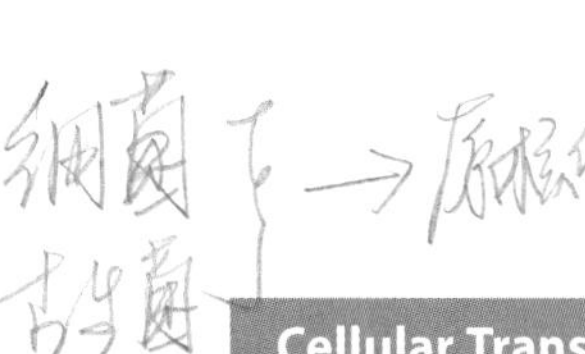

Cellular Transport: Diffusion and Osmosis

- The cell membrane is selectively permeable, which means that some substances can move across it but the passage of other substances is restricted.
- Concentration is the amount of a substance that is dissolved in a solvent.
- Diffusion is the net movement of particles from an area in which their concentration is high to an area in which their concentration is lower.
- Osmosis is the diffusion of water molecules across a semi-permeable membrane from an area of high concentration of water molecules to an area of lower concentration of water molecules.

24. Which is not an example of diffusion?

a. a drop of ink spreading out in a glass of water
b. the movement of oxygen from the lungs into the bloodstream
c. the absorption of water from the soil into plant root cells
d. a teabag steeping in a mug of hot water
e. the movement of water and nutrients rising in a tree

25. Which statement accounts for the ability of water to move into a plant cell by osmosis?

a. The plant cell is a specialized cell that functions as a guard cell.
b. The plant cell is carrying out cellular respiration.
c. The central vacuole is full of water and is pressing against the cell wall.
d. The concentration of water inside the cell is greater than the concentration of water outside the cell.
e. The concentration of water outside the cell is greater than the concentration of water inside the cell.

26. Which of the following statements about osmosis is correct?

a. Water moves from a weaker (less concentrated) solution to a stronger (more concentrated) solution.
b. Water molecules move from an area of higher water concentration to an area of lower water concentration.
c. Water moves from a more dilute solution to a more concentrated solution.
d. All three statements are true.
e. None of the three statements is true.

27. Cell membranes are selectively permeable. Explain the meaning of this term, and describe why this property is important to the cell.

28. Use a Venn diagram to compare and contrast diffusion and osmosis in the context of a cell membrane.

29. The illustration below shows a sugar solution in a U-shaped tube. The tube has a selectively permeable membrane that divides the tube into equal halves. Examine the illustration carefully, and write a caption for it that explains what is occurring and why.

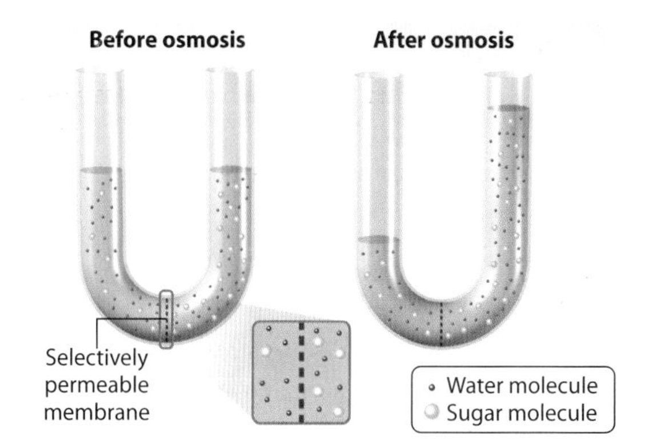

30. An animal cell is placed in a beaker of pure water. Predict what will happen to the cell on a molecular level. Use the terms *osmosis, water molecules, higher concentration,* and *lower concentration* in your explanation.

31. Which process taking place in the cell membrane requires energy?

a. the movement of water from a region of higher concentration inside the cell to a region of lower concentration outside the cell
b. the movement of glucose from a region of higher concentration outside the cell to a region of lower concentration inside the cell
c. the movement of glucose from a region of lower concentration outside the cell to a region of higher concentration inside the cell
d. the movement of oxygen from the bloodstream into the cell
e. the movement of carbon dioxide from the cell into the bloodstream

CHAPTER 1

The Molecules of Life

Specific Expectations

In this chapter, you will learn how to . . .

- B1.1 **analyze** technological applications related to enzyme activity in the food and pharmaceutical industries (1.3)
- B2.1 **use** appropriate terminology related to biochemistry (1.1, 1.2, 1.3)
- B2.3 **construct** and **draw** three-dimensional molecular models of important biochemical compounds, including carbohydrates, proteins, lipids, and nucleic acids (1.2, 1.3)
- B2.4 **conduct** biological tests to identify biochemical compounds found in various food samples, and **compare** the biochemical compounds found in each food to those found in the others (1.2)
- B2.5 **plan** and **conduct** an investigation related to a cellular process, using appropriate laboratory equipment and techniques, and **report** the results in an appropriate format (1.3)
- B3.2 **describe** the structure of important biochemical compounds, including carbohydrates, proteins, lipids, and nucleic acids, and explain their function within cells (1.2)
- B3.3 **identify** common functional groups within biological molecules, and **explain** how they contribute to the function of each molecule (1.2)
- B3.4 **describe** the chemical structures and mechanisms of various enzymes (1.3)
- B3.5 **identify** and **describe** the four main types of biochemical reactions (1.3)

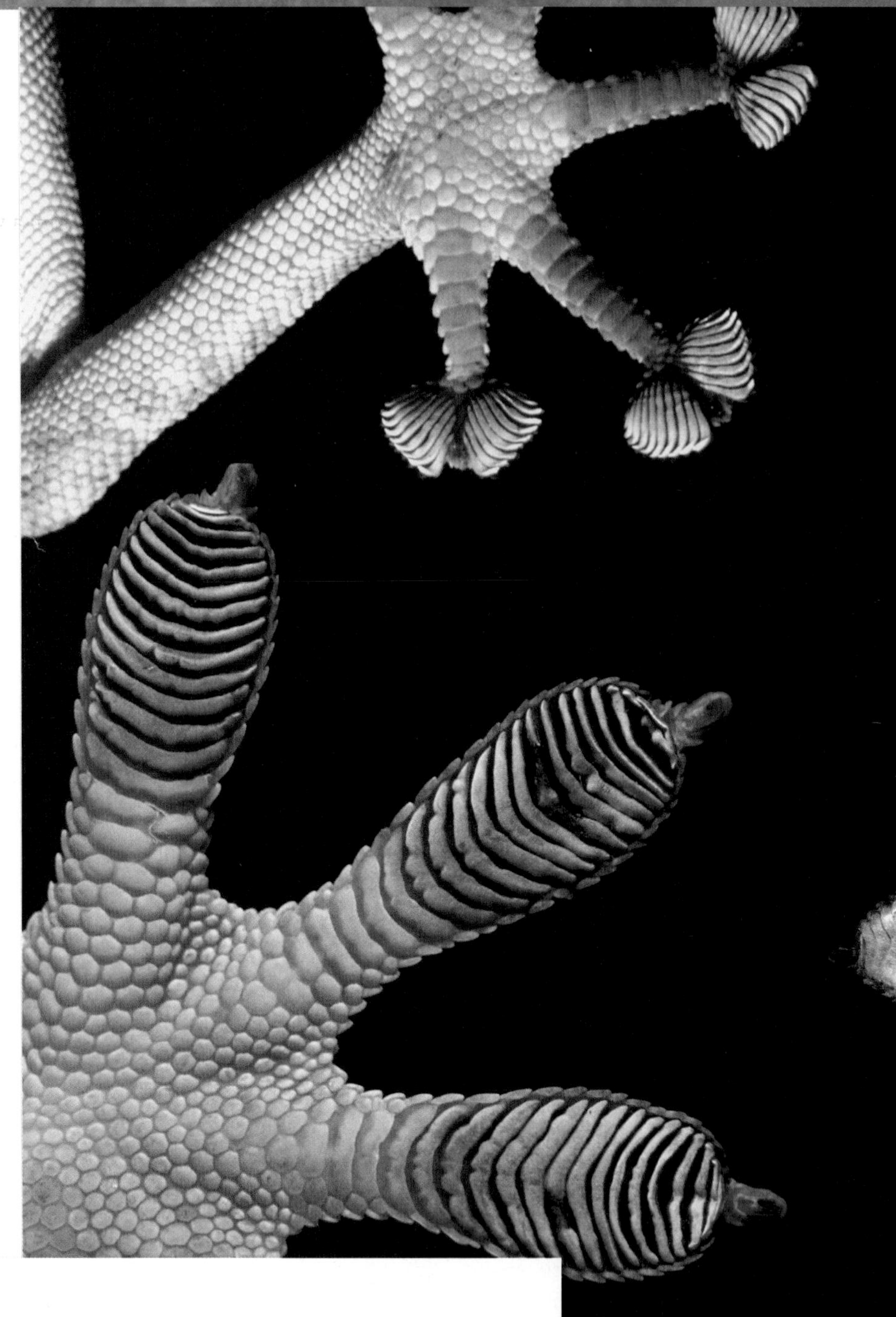

Geckos are small lizards that appear to defy gravity by running up walls and upside down on ceilings as smooth as glass—even on glass itself. Only recently, in 2002, were scientists able to explain this ability. Each gecko toe has about two million densely packed, hair-like structures called setae, and each individual seta is as long as twice the diameter of a human hair. In addition, each seta splits into hundreds of even finer tips, resulting in extremely close contact between the microscopic structures of the feet and any surface. About one million setae—a fraction of the total number a gecko has—could fit onto the surface of a dime, with an adhesive force large enough to lift a 20 kg child! All this is possible due to forces of molecular attraction operating at extremely short distances between the molecules that make up setae and the molecules that make up walls, ceilings, and other surfaces.

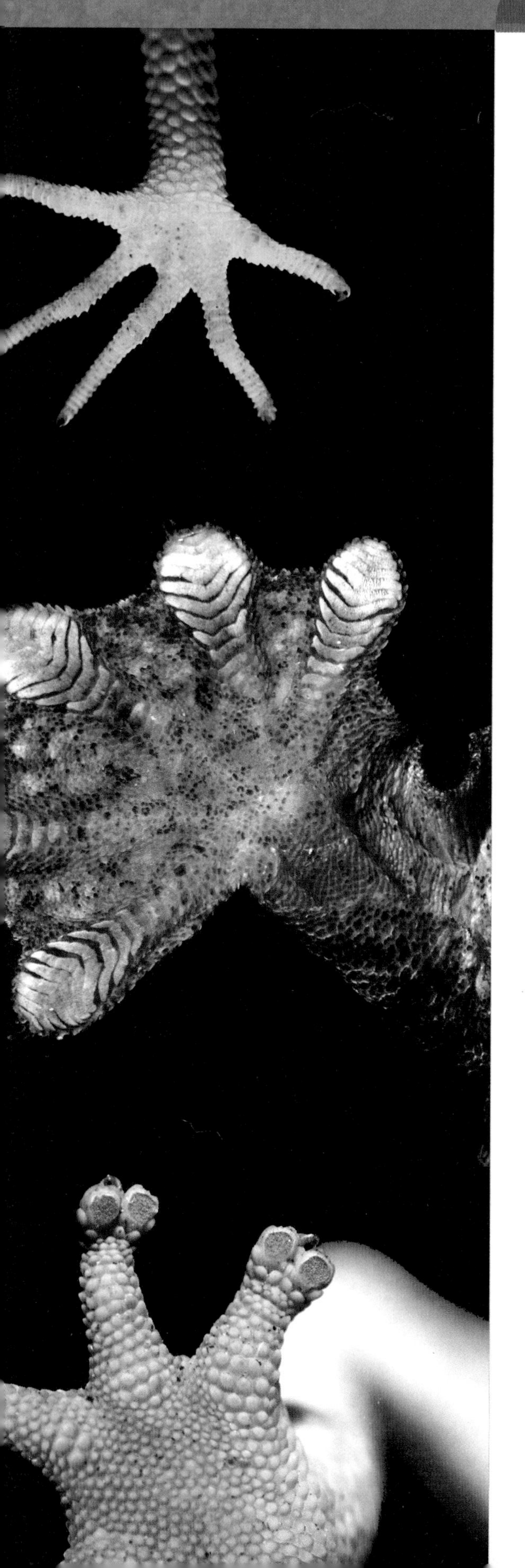

Launch Activity

On the Matter of Gecko Feet

Science is a system for developing knowledge by asking questions and designing ways to answer them. This knowledge does not exist in a vacuum, however. Knowledge cannot be divorced from ways in which people choose to use it. Similarly, knowledge and its application cannot be divorced from possible effects—intended or unintended—on people as well as on the environment. In other words, science, technology, society, and the environment are inseparably linked. In this activity, you will consider this linkage.

Procedure

1. Imagine you are a biologist interested in investigating geckos and their remarkable climbing abilities.
   a. On your own, write two questions that would enable you to begin developing knowledge about geckos and their feet.
   b. Share your questions with a partner, and together write two more questions.
   c. Share your additional questions with another pair of students, and together write at least two more questions.
2. Working the same way as in question 1, identify practical problems for which an understanding of gecko feet could provide a solution. For example, an understanding of gecko feet could lead to the invention of picture frames that can be hung on walls without leaving nail holes or sticky glue residue.
3. Still working as in question 1, identify possible societal and environmental consequences—both intended and unintended—of the solutions from question 2.

Questions

1. Do scientists have any responsibility for how the knowledge they develop might be used by others? Why or why not?
2. Do people who apply scientific knowledge to create solutions to practical problems have any responsibility for how they (or others) use their solutions? Why or why not?
3. Do members of society have any responsibility for the way they use scientific knowledge and technological solutions? Why or why not?

SECTION 1.1

Chemistry in Living Systems

Key Terms

isotope
radioisotope
molecule
organic molecule
biochemistry
intramolecular
intermolecular
hydrogen bond
hydrophobic
hydrophilic
ion
functional group

isotope atoms of the same element that have different numbers of neutrons

radioisotope an unstable isotope that decays over time by emitting radiation

All matter is composed of elements—substances that cannot be broken down into simpler substances by ordinary chemical methods. Only about 92 naturally occurring elements serve as the building blocks of matter, including the matter that comprises you and the millions of species of organisms in the world around you. And yet only six elements—carbon, hydrogen, nitrogen, oxygen, phosphorus, and sulfur—are the chemical foundation for this great diversity of life. Carbon and hydrogen form the underlying structures of biological molecules, with the other four elements providing particular properties to these molecules.

The smallest particle of an element that retains the properties of that element is an atom. Each atom has its own specific *atomic mass*, which is the sum of its protons and neutrons. While all atoms of an element have the same number of protons, the number of neutrons can vary. **Isotopes** are atoms of the same element that differ in the number of their neutrons. For example, carbon has three common isotopes: carbon-12 has six neutrons (the most abundant form), carbon-13 has seven neutrons, and carbon-14 has eight neutrons. Some isotopes are unstable, which means that their nucleus decays (breaks down) by emitting radiation in the form of subatomic particles or electromagnetic waves. Unstable isotopes are radioactive and are referred to as **radioisotopes**. Carbon-14 is an example. Radioisotopes are valuable diagnostic tools in medicine. Using a method called *radioisotope tracing* doctors can inject radioactive material into a patient and trace its movement in the body. For example, cancerous tissues in the body are characterized by a much higher level of activity than healthy tissues. Consequently, cancerous cells take in more glucose—a common cellular energy source—than healthy cells. Injecting a patient with radioactive glucose and then performing a positron emission tomography (PET) scan, such as the one shown in **Figure 1.1**, is one method to diagnose a cancerous tumour

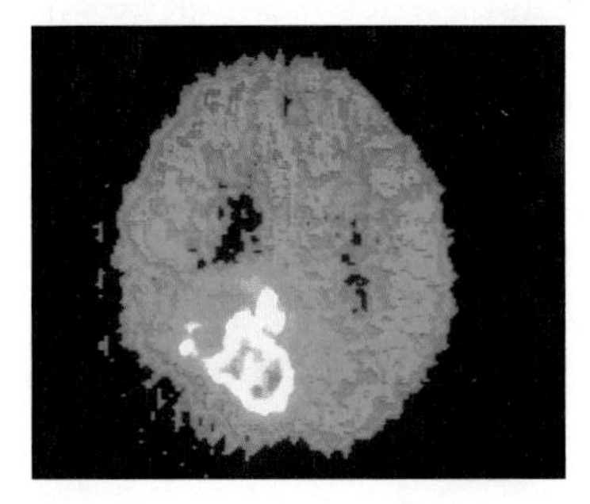

Figure 1.1 This positron emission tomography (PET) scan is of a 62-year-old man's brain. The yellow and orange area represents a tumour, which breaks down the injected radioactive glucose at a faster rate than normal cells do.

molecule a substance composed of two or more non-metal atoms that are covalently bonded together

organic molecule a carbon-containing molecule in which carbon atoms are nearly always bonded to each other and to hydrogen

biochemistry the study of the activity and properties of biologically important molecules

Studying the Interactions of Molecules

For most biological studies, chemical elements are not considered in the form of individual atoms but, rather, as components of **molecules**. Recall that a molecule is composed of two or more atoms and is the smallest unit of a substance that retains the chemical and physical properties of the substance. Many of the molecules of life are organic molecules. **Organic molecules** are carbon-based, and the carbon atoms are usually bonded to each other and to hydrogen. Many organic molecules also include atoms of nitrogen, oxygen, phosphorus, and/or sulfur.

There are major classes of biologically important organic molecules that are the cornerstones of most research in biochemistry. **Biochemistry** is often viewed as a field of study that forms a bridge between chemistry (the study of the properties and interactions of atoms and molecules) and biology (the study of the properties and interactions of cells and organisms). Biochemists are concerned mainly with understanding the properties and interactions of biologically important molecules. Understanding the physical and chemical principles that determine the properties of these molecules is essential to understanding their functions in the cell and in other living systems.

Interactions within Molecules

The forces that hold atoms together within a molecule are **intramolecular** forces ("intra" meaning within). These forces are what are generally thought of as the chemical bonds within a molecule. Bonds within molecules are *covalent bonds*. A covalent bond forms when the electron shells of two non-metal atoms overlap so that valence electrons of each atom are shared between both atoms. Each atom has access to the electrons in the bond, as well as to its other valence electrons. In this way, both atoms obtain a full valence shell. To illustrate this, a molecule of water, H_2O, is shown in **Figure 1.2A**.

intramolecular occurring between atoms within a molecule

Some atoms attract electrons much more strongly than other atoms. This property is referred to as an atom's *electronegativity*. Oxygen, O, nitrogen, N, and chlorine, Cl, are atoms with high electronegativity. Hydrogen, H, carbon, C, and phosphorus, P, are examples of atoms with lower electronegativity. When two atoms with significantly different electronegativities share electrons in a covalent bond, the electrons are more attracted to the atom with the higher electronegativity, so they are more likely to be found near it. Because electrons have a negative charge, this causes that atom to assume a slightly negative charge, called a partial negative charge ($\delta-$). The atom with lower electronegativity assumes a partial positive charge ($\delta+$). This unequal sharing of electrons in a covalent bond creates a *polar* covalent bond. **Figure 1.2B** shows how a water molecule contains two polar covalent O–H bonds. The electrons in each bond are more strongly attracted to the oxygen atom than to the hydrogen atom and are more likely to be found near the oxygen atom. This results in the oxygen atom being partially negative and the hydrogen atoms being partially positive. Molecules such as water, which have regions of partial negative and partial positive charge, are referred to as *polar molecules*.

When covalent bonds are formed between atoms that have similar electronegativities, the electrons are shared fairly equally between the atoms. Therefore, these bonds are considered non-polar. If this type of bond predominates in a molecule, the molecule is considered a non-polar molecule. For example, bonds between carbon and hydrogen atoms are considered non-polar, because carbon and hydrogen have similar electronegativities. As you will see in this unit, the polarity of biological molecules greatly affects their behaviour and functions in a cell.

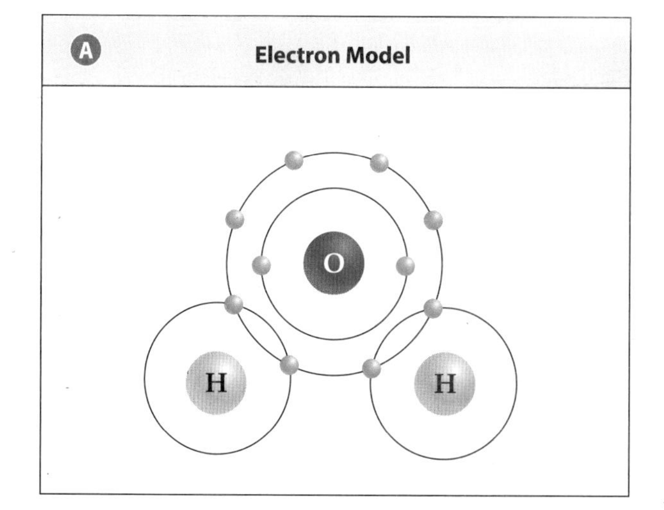

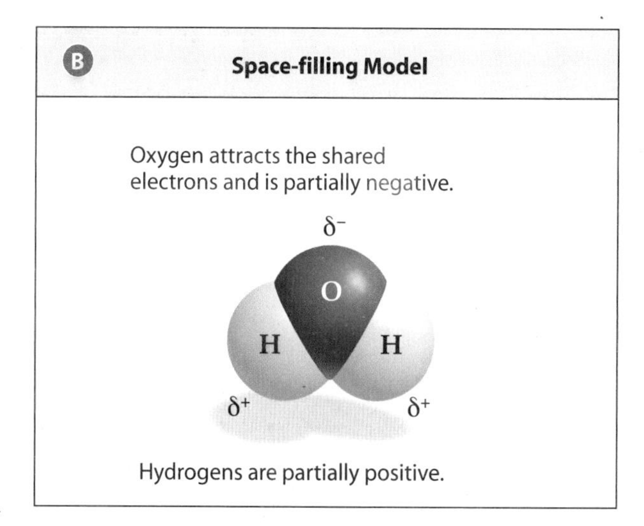

Figure 1.2 As shown in the electron model (**A**), two hydrogen atoms each share a pair of electrons with oxygen to form covalent bonds in a molecule of water, H_2O. Because oxygen is more electronegative than hydrogen, there is a partial negative charge on the oxygen and a partial positive charge on each hydrogen, as shown in the space-filling model (**B**).
***Predict** how two water molecules might interact, based on this diagram.*

intermolecular occurring between atoms of different molecules

hydrogen bond a weak association between an atom with partial negative charge and a hydrogen atom with partial positive charge

Interactions between Molecules

In addition to forces *within* molecules, there are also forces *between* molecules. These **intermolecular** forces ("inter" meaning between) may form between different molecules or between different parts of the same molecule if that molecule is very large. Intermolecular interactions are much weaker than intramolecular interactions. They determine how molecules interact with each other and with different molecules, and therefore they play a vital role in biological systems. Most often, intermolecular interactions are attractive forces, making molecules associate together. However, because they are relatively weak, intermolecular forces can be broken fairly easily if sufficient energy is supplied. As a result, intermolecular forces are responsible for many of the physical properties of substances. Two types of intermolecular interactions are particularly important for biological systems: hydrogen bonding and hydrophobic interactions.

Hydrogen Bonding

With its two polar O–H bonds, a water molecule is a polar molecule, with a slightly positive end and a slightly negative end. The slightly positive hydrogen atoms of one molecule of water are attracted to the slightly negative oxygen atoms of other water molecules. This type of intermolecular attraction is called a **hydrogen bond**, and it is weaker than an ionic or covalent bond. As shown in **Figure 1.3**, a hydrogen bond is represented by a dotted line to distinguish it from the stronger covalent bond. Many biological molecules have polar covalent bonds involving a hydrogen atom and an oxygen or nitrogen atom.

A hydrogen bond can occur between different molecules as well as within the same molecule. Since the cell is an aqueous environment, hydrogen bonding between biological molecules and water is very important. Although a hydrogen bond is more easily broken than a covalent bond, many hydrogen bonds added together can be very strong. Hydrogen bonds between molecules in cells help maintain the proper structure and function of the molecule. For example, the three-dimensional shape of DNA, which stores an organism's genetic information, is maintained by numerous hydrogen bonds. The breaking and reforming of these bonds plays an important role in how DNA functions in the cell.

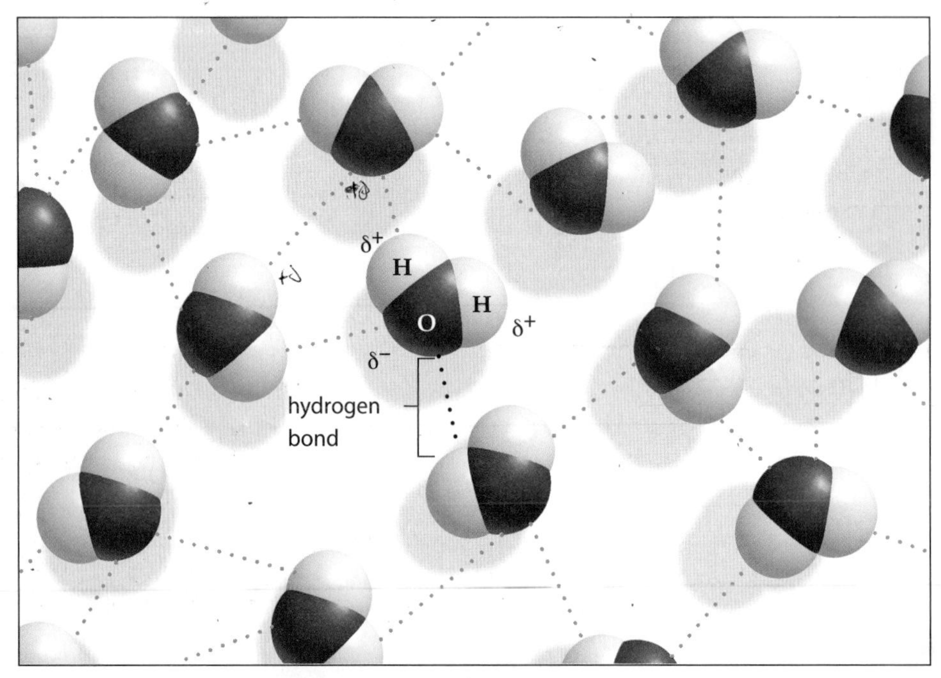

Figure 1.3 In water, hydrogen bonds (dotted lines) form between the partially positive hydrogen atoms of one molecule and the partially negative oxygen atoms on other molecules.

Hydrophobic Interactions

Non-polar molecules such as cooking oil and motor oil do not form hydrogen bonds. When non-polar molecules interact with polar molecules, the non-polar molecules have a natural tendency to clump together, rather than to mix with the polar molecules, as shown in **Figure 1.4**. (Think of the saying, "oil and water don't mix.") If the molecules had human emotions and motivations, it would appear as if the non-polar molecules were drawing or shying away from the polar molecules. Thus, in their interactions with water molecules, non-polar molecules are said to be **hydrophobic** (literally meaning "water-fearing"). Polar molecules, on the other hand, have a natural tendency to form hydrogen bonds with water and are said to be **hydrophilic** (literally meaning "water-loving").

The natural clumping together of non-polar molecules in water is referred to as the *hydrophobic effect.* As you will see in this unit, the hydrophobic effect plays a central role in how cell membranes form and helps to determine the three-dimensional shape of biological molecules such as proteins.

hydrophobic referring to non-polar molecules that do not have attractive interactions with water molecules

hydrophilic referring to polar molecules that have attractive interactions with water molecules

ion an atom or group of atoms that has gained or lost one or more electrons, giving it a positive or negative charge

Ions in Biological Systems

An atom can obtain a stable valence shell by losing or gaining electrons rather than sharing them. For example, the sodium atom, Na, has only one electron in its outer valence shell. Once this electron is given up, the electron shell closer to the sodium nucleus, which already contains eight electrons, becomes the valence shell. When an atom or group of atoms gains or loses electrons, it acquires an electric charge and becomes an **ion**. When an atom or group of atoms loses electrons, the resulting ion is positive and is called a *cation.* When an atom or group of atoms gains electrons, the resulting ion is negative and is called an *anion.* Ions can be composed of only one element, such as the sodium ion, Na^+, or of several elements, such as the bicarbonate ion, HCO_3^-.

Ions are an important part of living systems. For example, hydrogen ions, H^+, are critical to many biological processes, including cellular respiration. Sodium ions, Na^+, are part of transport mechanisms that enable specific molecules to enter cells. For biological processes in the cell, substances that form ions, such as sodium, are almost never considered in the form of *ionic compounds*, such as sodium chloride, NaCl(s). Since the cell is an aqueous environment, almost all ions are considered as free, or dissociated ions (Na^+(aq) and Cl^-(aq)) since they dissolve in water.

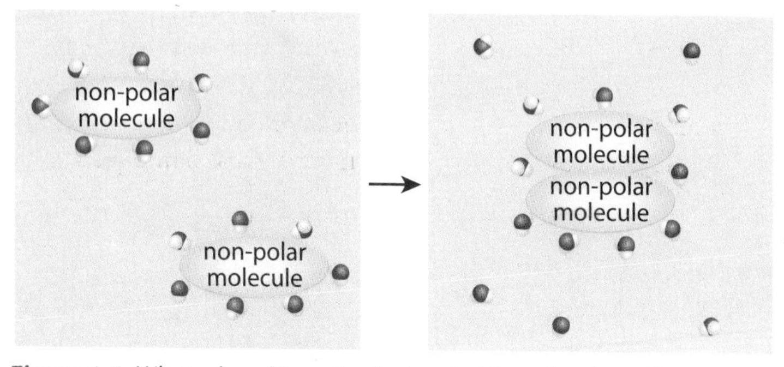

Figure 1.4 When placed in water, hydrophobic molecules will tend to clump together.

Learning Check

1. What is the relationship between elements and atoms?
2. Explain, with reference to subatomic particles and stability, the difference between carbon-12 and carbon-14.
3. Explain how a polar covalent bond is different from an ionic bond.
4. Use a water molecule to describe the relationships among all the following: polar and non-polar molecules, intramolecular and intermolecular forces, hydrophilic and hydrophobic interactions.
5. What is the hydrophobic effect?
6. Biochemistry is one of the many scientific disciplines that bridge the knowledge and understanding of one field of science with another. Identify at least two other "bridging" scientific disciplines, and explain how the knowledge and understanding of one field complements the knowledge and understanding of the other in each case.

Functional Groups Determine the Properties of a Molecule

Organic molecules that are made up of only carbon and hydrogen atoms are called *hydrocarbons*. Hydrocarbons share similar properties—for example, they are non-polar, do not dissolve in water, have relatively low boiling points (depending on size), and are flammable. The covalent bonds between carbon and carbon and between carbon and hydrogen are "energy-rich"; breaking of the bonds releases a great deal of energy. Therefore, hydrocarbons make good fuels. Most of the hydrocarbons you encounter in everyday life, such as acetylene, propane, butane, and octane, are fuels.

functional group an atom or group of atoms attached to a molecule that gives the molecule particular chemical and physical properties

Although hydrocarbons share similar properties, other organic molecules have a wide variety of properties. This is because most organic molecules also have other atoms or groups of other atoms attached to their central carbon-based structure. A cluster of atoms that always behaves in a certain way is called a **functional group**. Functional groups contain atoms such as oxygen (O), nitrogen (N), phosphorus (P), or sulfur (S). Certain chemical properties are always associated with certain functional groups. These functional groups provide the molecules to which they are bonded with those same chemical properties. **Table 1.1** lists the common functional groups of biologically important molecules. For example, the presence of hydroxyl or carbonyl groups on a molecule makes the molecule polar. Also, a carboxyl functional group on a molecule will make it acidic, meaning it will easily release or donate a hydrogen atom to another molecule. Many of these functional groups, and therefore the molecules that contain them, can also participate in hydrogen bonding.

Table 1.1 Important Functional Groups on Biological Molecules

Functional Group	Properties	Structural Formula	Example	Found In
Hydroxyl	polar	—OH	H H H—C—C—OH H H Ethanol	carbohydrates, proteins, nucleic acids, lipids
Carbonyl	polar	O ‖ —C—	H O H—C—C—H H Acetaldehyde	carbohydrates, nucleic acids
Carboxyl	polar, acidic (donates a proton)	—C(=O)—OH	H H—C—C(=O)—OH H Acetic acid	proteins, lipids
Amino	polar, basic (accepts a proton)	—N(—H)—H	O H HO—C—C—N(H)H CH_3 Alanine	proteins, nucleic acids
Sulfhydryl	slightly polar	—S—H	COOH H—C—CH_2—S—H NH_2 Cysteine	proteins
Phosphate	polar, negatively charged	O^- —O—P—O^- ‖ O	OH OH H O H—C—C—C—O—P—O^- H H H O^- Glycerol phosphate	nucleic acids

Structures and Shapes of Molecules

A *molecular formula* shows the number of each type of atom in an element or compound. Examples of molecular formulas include H_2O, $C_3H_7NO_2$, and $C_6H_{12}O_6$. Molecular formulas are useful because they show the number and type of atoms in a molecule. *Structural formulas* show how the different atoms of a molecule are bonded together. When representing molecules using a structural formula, a line is drawn between atoms to indicate a covalent bond. A single line indicates a single covalent bond, double lines indicate a double bond, and triple lines indicate a triple bond. **Figure 1.5** shows some examples of structural formulas. Also shown are simplified diagrams of structures that you will often see when they are written out for biological molecules. In these simplified structures, carbon atoms are sometimes indicated by a bend in a line, so their symbol, C, is not included. Also, hydrogen atoms attached to these carbon atoms are omitted but are assumed to be present.

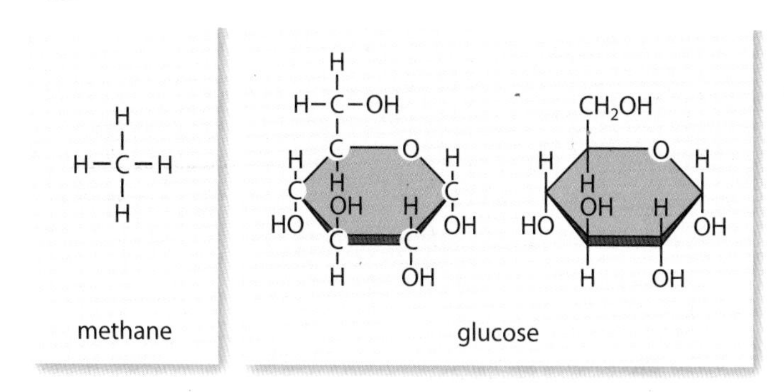

Figure 1.5 Structural formulas show how each atom is bonded together in a molecule. Biological molecules are often drawn using a simplified form, where the intersection of two lines represents a carbon atom and any hydrogen atoms bonded to that carbon are omitted.

***Identify** the polar and non-polar molecules.*

Structural formulas are two-dimensional representations of molecules and the bonds between molecules. However, molecules are not flat—they take up space in three dimensions. In fact, the three-dimensional shape of a molecule influences its behaviour and function. As shown in **Figure 1.6**, a molecule such as methane, CH_4, has a tetrahedral shape. Because they are negatively charged, the electron pairs in covalent bonds repel each other, and move as far apart as possible. If there are four bonds, as in methane, then a tetrahedral shape represents the farthest the electrons in these bonds can be from each other. Depending on the atoms and types of bonds in a molecule, different shapes are possible. Therefore, you will often see large biologically important molecules represented using three-dimensional models, such as space-filling models.

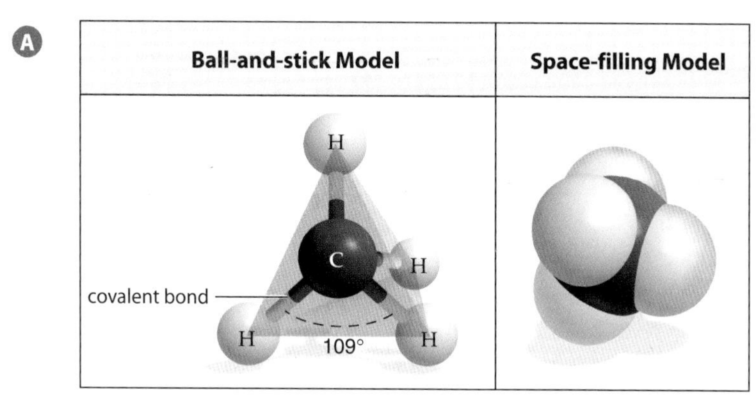

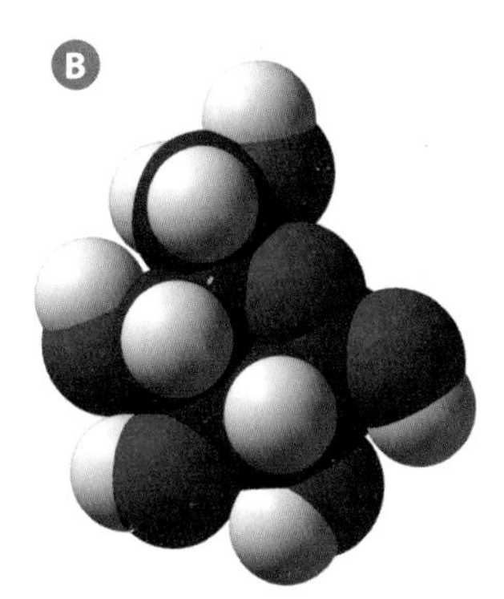

Figure 1.6 Methane (**A**) has four bonds and a particular three-dimensional shape, called tetrahedral. In larger biological molecules such as glucose (**B**), the three-dimensional shape plays a role in the molecule's biological activity. Note that in space-filling models like the ones shown here, each atom is assigned a particular colour by convention. Carbon is black, hydrogen is white, and oxygen is red.

Activity 1.1 Molecular Shapes

In this activity, you will investigate the shapes of the molecules that result when each of carbon, hydrogen, nitrogen, and oxygen form covalent bonds.

Materials

• molecular model kit or other model building supplies

Procedure

1. Copy the following table.

Molecule	Number and Type of Bond	Structural Formula	Three-Dimensional Diagram
methane (CH_4)	four single bonds		
ammonia (NH_3)	three single bonds		
water (H_2O)	two single bonds		
formaldehyde (CH_2O)	one double bond, two single bonds		
ethene (C_2H_4)	one double bond, four single bonds		
propadiene (C_3H_4)	two double bonds, four single bonds		

2. Write the structural formula for the first compound.

3. Using a molecular model kit, or other model-building supplies provided by your teacher, build the first molecule.

4. Draw a three-dimensional diagram for your model using the following method:

1. Position the model so that as many atoms as possible are in the same plane—that is, so the atoms lie within the same flat surface.
2. Draw circles to represent each atom.
3. Use solid lines to represent bonds between atoms that lie in the same plane as the surface of the paper you are drawing on.
4. Use dashed lines to represent bonds between atoms that are directed away from you, into the plane of the paper.
5. Use wedged lines to represent bonds between atoms that are directed toward you, out of the plane of the paper.

5. Repeat steps 2, 3, and 4 for the remaining compounds.

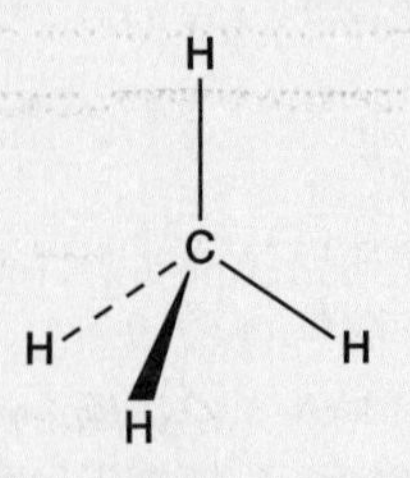

methane

Questions

1. Describe the shape of the molecule when an atom forms each of the following:

a. four single bonds

b. three single bonds

c. two single bonds

d. one double bond, and two single bonds

e. one double bond, and four single bonds

f. two double bonds, and four single bonds

2. Predict the shape of phosphine, PH_3.

3. Phosphorus and sulfur are important elements in biological molecules. Given that phosphorus is in the same chemical family as nitrogen and sulfur is in the same chemical family as oxygen, predict the shapes formed when phosphorus and sulfur form covalent bonds.

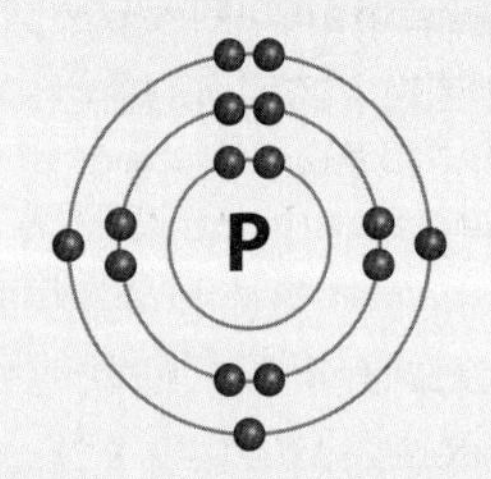

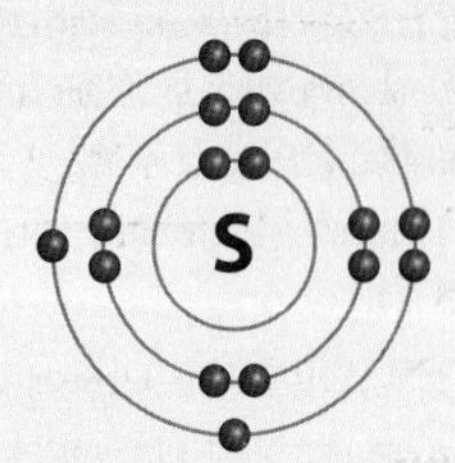

Bohr-Rutherford diagrams of phosphorus, P, (left) and sulfur, S, (right).

Section 1.1 Review

Section Summary

- Organisms are composed primarily of the chemical elements carbon, hydrogen, oxygen, nitrogen, phosphorus, and sulfur.
- Intramolecular interactions occur between atoms within a molecule, forming chemical bonds.
- The polarity of a molecule influences the intermolecular interactions that occur between molecules. Two important types of intermolecular interactions are hydrogen bonding and hydrophobic interactions.
- The functional group or groups on a molecule determine the properties of the molecule. Important functional groups in biological molecules include hydroxyl, carboxyl, carbonyl, amino, sulfhydryl, and phosphate groups.
- The structures of molecules can be represented using a variety of formats. Structural formulas are two-dimensional representations that indicate how the atoms are bonded together. Space-filling models are a common way to represent the three-dimensional structures of molecules.

Review Questions

1. **K/U** Name the four most common elements that make up living systems.
2. **A** **a.** Describe how radioisotopes can be used to locate cancerous tissues in the body.
   b. How might a biologist use the process of radioisotope tracing to study digestion? What type of information could the biologist hope to learn from this investigation?
3. **C** Use a concept map to show the connections between the following terms: *atom*, *functional group*, *isotope*, *molecule*, *radioisotope*, *organic molecule*, and *ion*.
4. **A** Give one example of an intramolecular force, and one example of an intermolecular force. Which type of force is stronger? Use logical reasoning to explain how you know this must be so in order for life to exist.
5. **C** Sketch a structural diagram of a water molecule. Label the covalent bonds, the atoms, and the polarity of each part of the molecule.
6. **T/I** Identify each of the following bonds as polar or non-polar. Provide justification for your answer.
   a. O–O
   b. H–N
   c. C–Cl
   d. P–O
7. **A** Fluorine is never observed in naturally-occurring biological compounds. Since fluorine has a high electronegativity, predict the type of bond that fluorine would likely form with carbon.
8. **K/U** Three students discussed the properties of a water molecule. Ari said, "Water's polarity and water's ability to form hydrogen bonds are actually the same thing. Any polar molecule can form hydrogen bonds with any another polar molecule." Jordan said, "Water's polarity and water's ability to form hydrogen bonds are connected, but they are not the same thing. Some polar molecules cannot form hydrogen bonds, and others, like water, can." Ravi said, "Water's polarity and water's ability to form hydrogen bonds have nothing to do with each other. Even if water had no polar bonds, it would still be able to form hydrogen bonds."
   Which student, if any, do you agree with? Write a paragraph explaining why.
9. **A** Hydrocarbons contain many hydrogen atoms but are not known to carry out hydrogen bonding. Explain this apparent discrepancy.
10. **C** Use a magnet analogy to explain the attractive forces between water molecules. Explain also the limitations of using a magnet analogy for this purpose.
11. **A** Use the hydrophobic effect to explain why the oil in salad dressing will rise to the top of a water-based mixture of vinegar and herbs.
12. **K/U** **a.** How does a cation differ from an atom?
    b. Provide two examples of cations that are important in biological systems.
13. **K/U** What are the advantages of using a molecular formula to depict a molecule, rather than a structural formula? What are the advantages of using a structural formula?
14. **C** Using a table or a Venn diagram, compare the two-dimensional structure of methane with the three-dimensional structure of methane.
15. **T/I** The table compares the percentage (by mass) of three elements in the human body and Earth's crust. Explain why the values in each case are different.

	Percentage in Body	Percentage in Earth's Crust
Hydrogen	9.5	0.14
Carbon	18	0.03
Oxygen	65	47

SECTION 1.2

Biologically Important Molecules

Key Terms

macromolecule
polymer
monomer
carbohydrate
monosaccharide
isomer
disaccharide
polysaccharide
lipid
triglyceride
fatty acid
phospholipid
lipid bilayer
steroid
wax
protein
amino acid
polypeptide
nucleic acid
DNA (deoxyribonucleic acid)
RNA (ribonucleic acid)
nucleotide

macromolecule a large, complex molecule, usually composed of repeating units of smaller molecules covalently linked together

polymer a large molecule composed of repeating units of smaller molecules (monomers)

monomer the smallest repeating unit of a polymer

Many of the molecules of living organisms are composed of thousands of atoms. These molecules are considered **macromolecules**, which are large molecules that often have complex structures. Many macromolecules are **polymers**, which are long chain-like substances composed of many smaller molecules that are linked together by covalent bonds. These smaller molecules are called **monomers**, which can exist individually or as units of a polymer. The monomers in a polymer determine the properties of that polymer. As shown in **Figure 1.7**, there are four main types of biological macromolecules: carbohydrates, lipids, proteins, and nucleic acids. Some occur as polymers that are composed of characteristic monomers. This section will discuss the basic structures of these molecules, as well as some important examples and their functions in the cell.

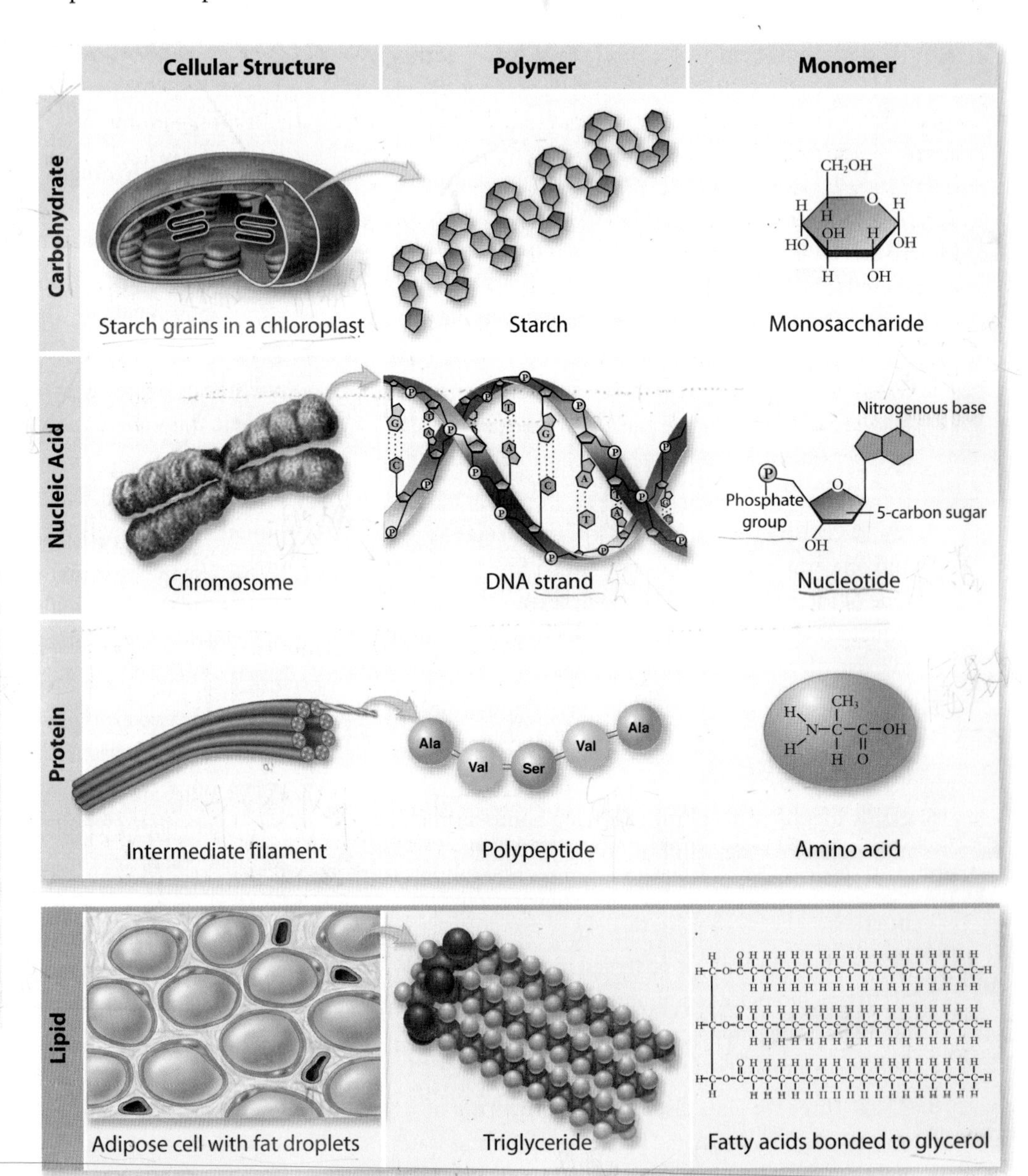

Figure 1.7 Carbohydrates, nucleic acids, proteins, and lipids (shown here as a triglyceride) are biologically important components of larger structures in the cell.

***Explain** why the triglyceride shown here does not fit the monomer-polymer relationship that exists for the other macromolecules.*

Carbohydrates

Sugars and starches are examples of carbohydrates. **Carbohydrates** are a type of biological molecule that always contains carbon, hydrogen, and oxygen, nearly in the ratio of two atoms of hydrogen and one atom of oxygen for every one atom of carbon. This is why a general formula for carbohydrates is often written using the formula $(CH_2O)_n$, where n is the number of carbon atoms. Sugars and starches store energy in a way that is easily accessible by the body. Because they contain a high proportion of hydroxyl functional groups and many contain carbonyl groups, most carbohydrates are polar molecules, and many dissolve in water.

carbohydrate a biological macromolecule that contains carbon, hydrogen, and oxygen in a 1:2:1 ratio

monosaccharide a carbohydrate composed of between three and seven carbon atoms

isomer one of two or more molecules with the same number and type of atoms, but different structural arrangements

disaccharide a carbohydrate composed of two monosaccharides joined by a covalent bond

Monosaccharides and Disaccharides

If the number of carbon atoms in a carbohydrate molecule is between three and seven, the carbohydrate is classified as a simple sugar. These sugars are called **monosaccharides** ("mono" meaning one; "saccharide" meaning sugar) because they are composed of a single carbon-based monomer structure. Examples of common monosaccharides are shown in **Figure 1.8**. Glucose in the body is commonly referred to as blood sugar, because it is the sugar that cells in the body use first for energy. Fructose is often called fruit sugar, since it is a principal sugar in fruits. Galactose is a sugar found in milk. These three simple sugars all have the molecular formula $C_6H_{12}O_6$. The exact three-dimensional shapes of their structures differ, as does the relative arrangement of the hydrogen atoms and hydroxyl groups. Molecules that have the same molecular formula but different structures are called **isomers**. Glucose, fructose, and galactose are isomers of each other. The different structures of these molecules results in them having very different three-dimensional shapes. These differences are enough for the molecules to be treated very differently by your body and in the cell. For example, your taste buds will detect fructose as being much sweeter than glucose, even though they have the same chemical composition.

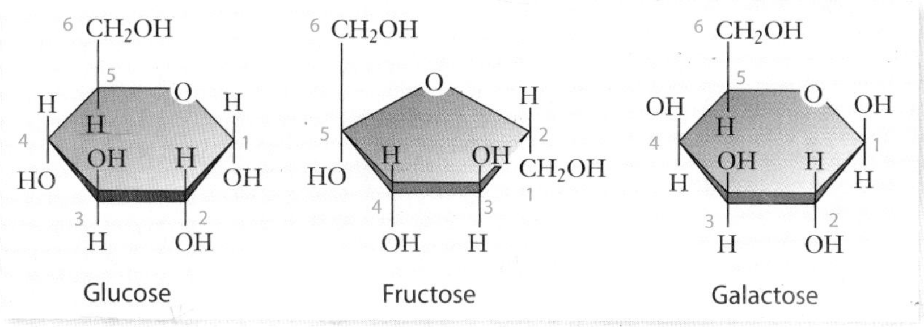

Glucose | Fructose | Galactose

Figure 1.8 Although glucose, fructose, and galactose have the same number and types of atoms, the atoms are arranged to form different molecules. The carbon atoms of monosaccharides with this type of ring structure are numbered by the convention shown here.

Two monosaccharides can join to form a **disaccharide**. The covalent bond between monosaccharides is called a *glycosidic linkage*. It forms between specific hydroxyl groups on each monosaccharide. Common table sugar is the disaccharide sucrose, shown in **Figure 1.9**, which is composed of glucose bonded with fructose. Galactose and glucose bond to form the disaccharide lactose, which is a sugar found in milk and other dairy products. Some people are lactose-intolerant and can experience side effects such as cramping and diarrhea after consuming foods that contain lactose. These side effects result from body cells being unable to digest (break down) lactose into its monosaccharide monomers.

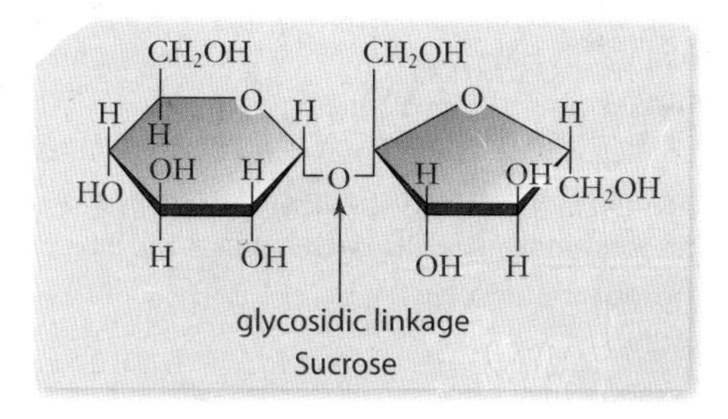

Figure 1.9 Sucrose, commonly called table sugar, is a disaccharide composed of glucose and fructose monomers that are chemically joined by a covalent bond.

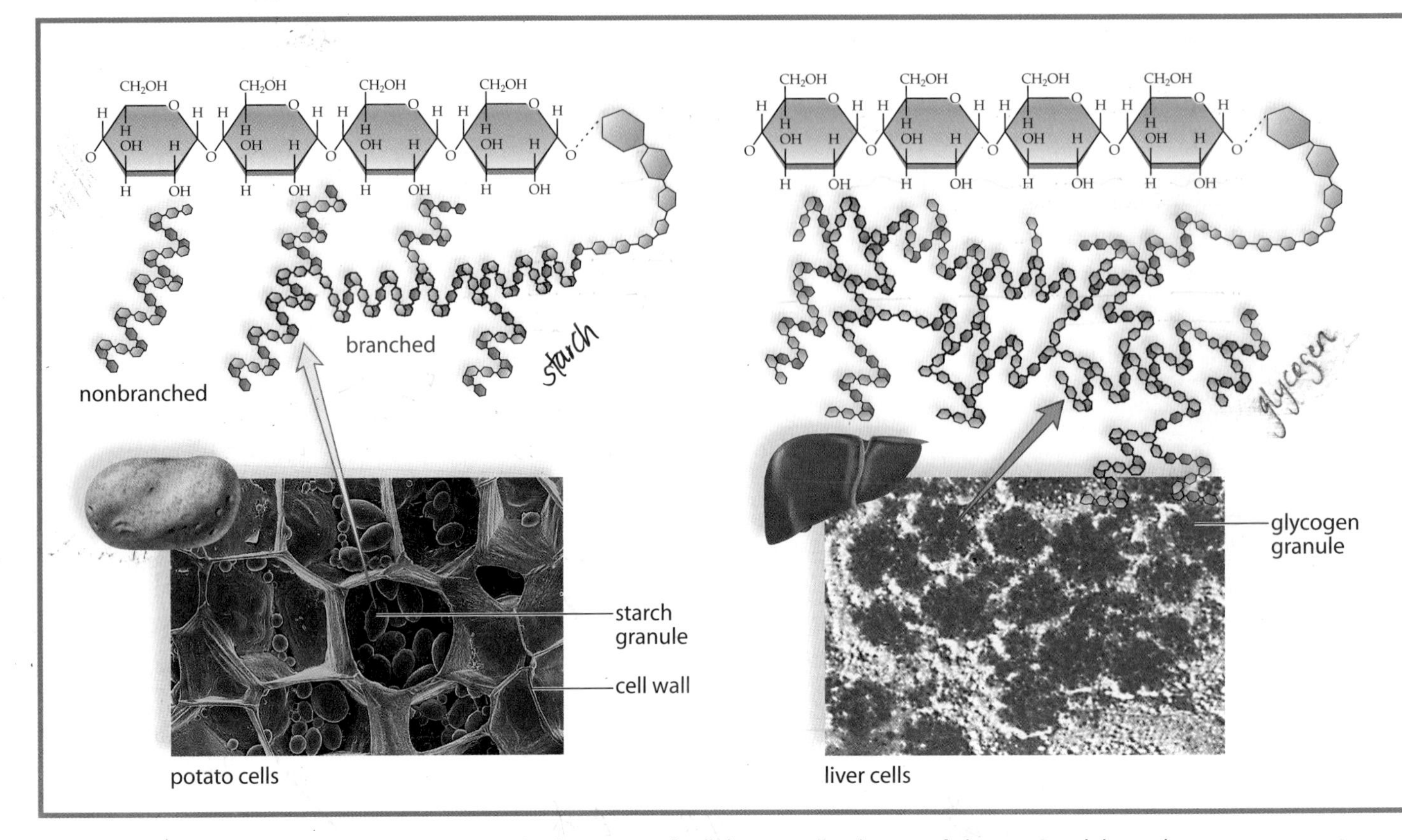

Figure 1.10 Starch, glycogen, and cellulose are all polymers of glucose. Starch has a three-dimensional structure that is much more linear than the highly branched structure of glycogen. Cellulose has a linear structure and has a different type of glycosidic linkage between monomers compared with the other two polysaccharides.

Polysaccharides

polysaccharide a carbohydrate polymer composed of many monosaccharides joined by covalent bonds between particular atoms

Monosaccharides are the monomers for carbohydrates that are more complex. Many monosaccharides can join together by glycosidic linkages to form a **polysaccharide** ("poly" meaning many). Three common polysaccharides are starch, glycogen, and cellulose. All three are composed of monomers of glucose. However, the different ways that the glucose units are linked together results in these molecules having quite different three-dimensional shapes. This, in turn, is reflected in them having very different functions. Plants store glucose in the form of starch, and animals store glucose in the form of glycogen. These molecules provide short-term energy storage, whereby glucose can be easily accessed from their breakdown in the cell. Starch and glycogen differ from each other in the number and type of branching side chains, represented in **Figure 1.10**. Because there are many more branches on the glycogen molecule, it can be broken down for energy much more rapidly than starch can.

Cellulose carries out a completely different function—it provides structural support in plant cell walls. The type of glycosidic linkage between monomers of cellulose is different from the type for starch and glycogen. The different linkages are possible because the hydroxyl group on carbon-1 of glucose can exist in two different positions. These positions are referred to as alpha and beta, as shown in **Figure 1.11**. The alpha form results in starch and glycogen, while the beta form results in cellulose.

Starch and glycogen are digestible by humans and most other animals because we have the enzymes that recognize this glycosidic linkage and catalyze the cleavage into glucose monomers. However, cellulose is indigestible because we lack the enzyme that recognizes the glycosidic linkage in that macromolecule.

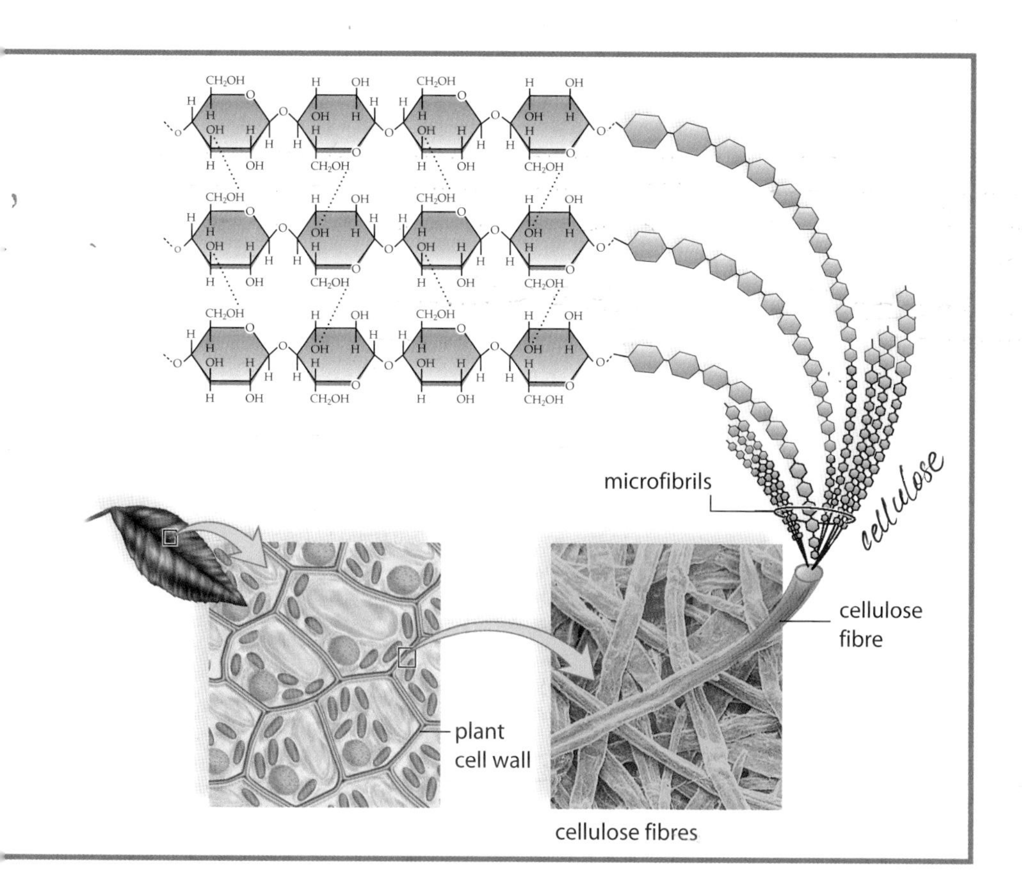

alpha glucose

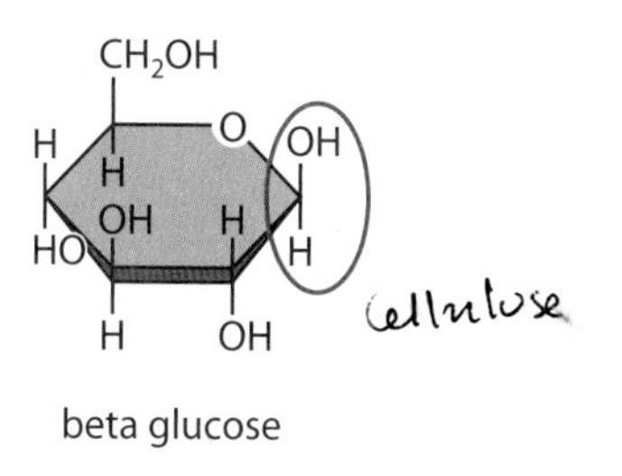

Figure 1.11 The hydroxyl group on carbon-1 of glucose can exist in either the alpha or beta form.

Learning Check

7. Explain how the following terms are related: macromolecule, polymer, monomer.
8. Identify at least three structural and functional characteristics of carbohydrates.
9. Why are glucose, fructose, and galactose isomers?
10. Use a graphic organizer to describe the similarities and differences among monosaccharides, disaccharides, and polysaccharides.
11. Identify two functions of carbohydrates in living systems.
12. Describe the similarities of and differences between starch, glycogen, and cellulose.

Lipids

Like carbohydrates, **lipids** are composed of carbon, hydrogen, and oxygen atoms, but lipids have fewer oxygen atoms and a significantly greater proportion of carbon and hydrogen bonds. As a result, lipids are hydrophobic—they do not dissolve in water. Since the cell is an aqueous environment, the hydrophobic nature of some lipids plays a key role in determining their function.

lipid a biological macromolecule composed of carbon, hydrogen, and oxygen atoms, with a high proportion of non-polar carbon–hydrogen bonds

The presence of many energy-rich C–H bonds makes lipids efficient energy-storage molecules. In fact, lipids yield more than double the energy per gram that carbohydrates do. Because lipids store energy in hydrocarbon chains, their energy is less accessible to cells than energy from carbohydrates. For this reason, lipids provide longer-term energy and are processed by the body after carbohydrate stores have been used up.

Lipids are crucial to life in many ways. For example, lipids insulate against heat loss, they form a protective cushion around major organs, and they are a major component of cell membranes. In non-human organisms, lipids provide water-repelling coatings for fur, feathers, and leaves.

triglyceride a lipid molecule composed of a glycerol molecule and three fatty acids linked by ester bonds

fatty acid a hydrocarbon chain ending in a carboxyl group

Triglycerides: Lipids Used for Energy Storage

As shown in **Figure 1.12**, **triglycerides** are composed of one glycerol molecule and three fatty acid molecules. The bond between the hydroxyl group on a glycerol molecule and the carboxyl group on a fatty acid is called an *ester linkage*, because it results in the formation of an ester functional group. A **fatty acid** is a hydrocarbon chain that ends with an acidic carboxyl group, –COOH. Fatty acids are either saturated or unsaturated. A *saturated fatty acid* has no double bonds between carbon atoms, while an unsaturated fatty acid has one or more double bonds between carbon atoms. If the unsaturated fatty acid has one double bond, it is *monounsaturated*; unsaturated fatty acids with two or more double bonds are *polyunsaturated*. Humans cannot synthesize polyunsaturated fats. Therefore, these essential fats must be consumed in our diet.

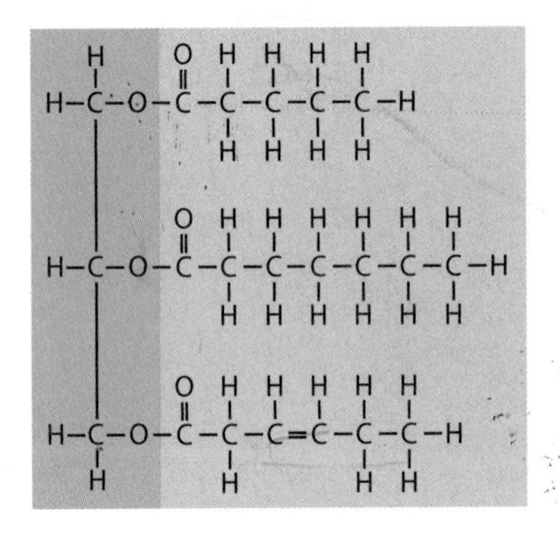

Figure 1.12 A triglyceride is composed of three fatty acids (orange background) and a glycerol molecule (green background). The fatty acids can be saturated (no double bonds) or unsaturated (with one or more double bonds).

As shown in **Figure 1.13**, the presence of double bonds in a triglyceride affects its three-dimensional shape. Naturally occuring unsaturated fats have *cis* double bonds, which cause the long hydrocarbon chain to bend. This alters the physical properties of the triglyceride and the behaviour of the molecule in the body. Triglycerides containing saturated fatty acids, such as lard and butter, are generally solid fats at room temperature. Triglycerides containing unsaturated fatty acids, such as olive oil or canola oil, are generally liquid oils at room temperature. A diet high in saturated fat is linked with heart disease in humans. However, some unsaturated fatty acids, particularly *polyunsaturated fatty acids*, are known to reduce the risk of heart disease.

A food preservation process called *hydrogenation* involves chemical addition of hydrogen to unsaturated fatty acids of triglycerides to produce saturated fats. A by-product of this reaction is the conversion of cis fats to trans fats, whereby remaining double bonds are converted to a trans conformation. Consumption of trans fats is associated with increased risk of heart disease.

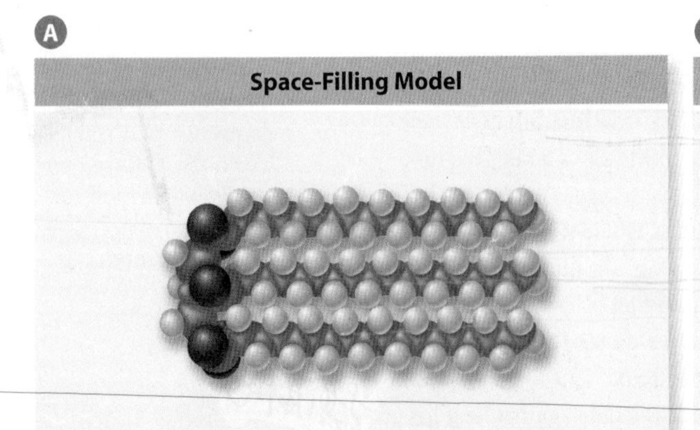

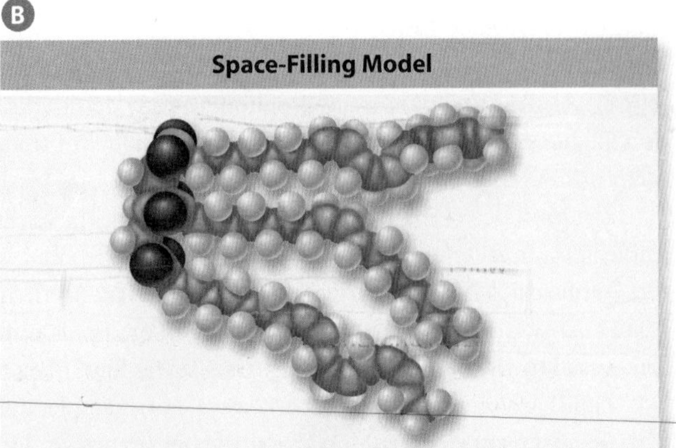

Figure 1.13 A saturated fat (**A**) is a triglyceride that contains only saturated fatty acids. An unsaturated fat (**B**) is a triglyceride containing one or more unsaturated fatty acids.

Phospholipids: Components of Cell Membranes

The main components of cell membranes are **phospholipids**. As shown in **Figure 1.14**, the basic structure of a phospholipid is similar to that of a triglyceride. The difference is that a phosphate group replaces the third fatty acid of a triglyceride. Attached to the phosphate group is an *R* group, which is a group of atoms that varies in composition. It is this *R* group that defines the type of phospholipid. The "head" portion of a phospholipid molecule is polar, while the lower "tail" portion has only non-polar C–C and C–H bonds. Therefore, the "head" of a phospholipid molecule is hydrophilic, while the non-polar "tail" is hydrophobic.

phospholipid a lipid composed of a glycerol molecule bonded to two fatty acids and a phosphate group with an *R* group

lipid bilayer a structure with hydrophilic "heads" of phospholipids directed toward the aqueous environment and hydrophobic "tails" directed toward the centre, interacting with each other

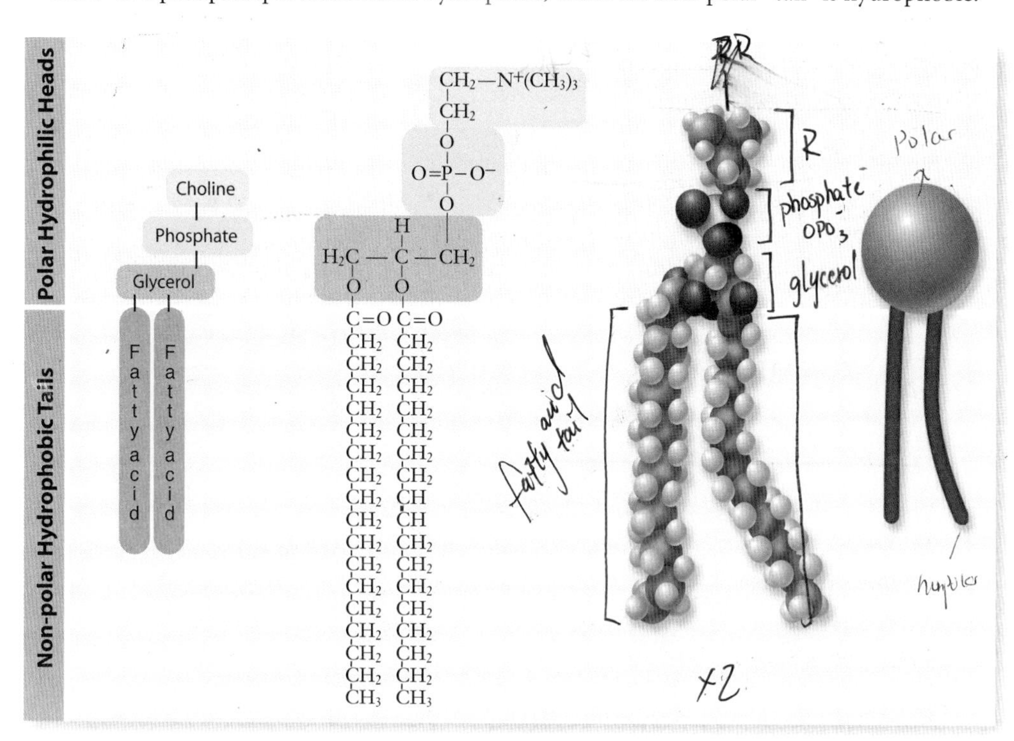

Figure 1.14 Phospholipids have a polar phosphate "head" group and non-polar hydrocarbon "tails" that are fatty acids. The phospholipid shown here has an *R* group called choline and is called phosphatidylcholine.

This dual character of a phospholipid is essential to its function in living organisms. In aqueous environments, phospholipids form a **lipid bilayer**, as shown in **Figure 1.15**. In a phospholipid bilayer, the hydrophilic heads face the aqueous solution on either side of the bilayer. The tails form a hydrophobic interior. In addition to the interior of a cell being an aqueous environment, cells are surrounded by an aqueous extra-cellular fluid. Therefore, membranes of the cell, which are made of phospholipids, adopt this structure. In Chapter 2, you will examine the vital role of lipid bilayers in the cell.

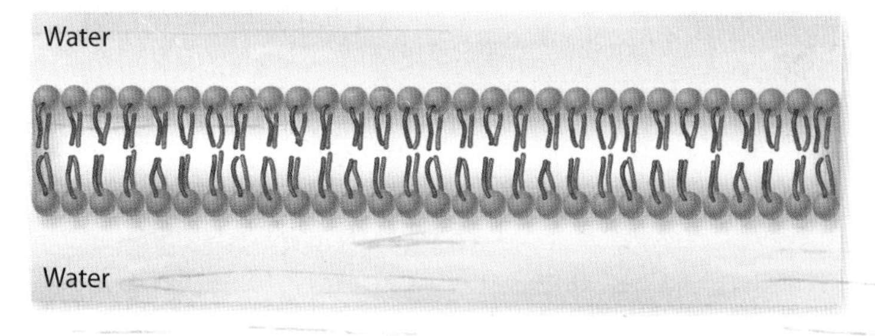

Figure 1.15 In water, phospholipids form a lipid bilayer. They naturally arrange themselves so that the non-polar "tails" are tucked away from the water, and the polar "heads" are directed toward the water.

***Predict** what might happen to a bilayer structure if many of the phospholipids contained unsaturated fatty acids.*

Other Lipids

steroid a lipid composed of four attached carbon-based rings

Steroids are a group of lipids that are composed of four carbon-based rings attached to each other. Each steroid differs depending on the arrangement of the atoms in the rings and the types of functional groups attached to the atoms. Cholesterol, shown in **Figure 1.16**, is a component of cell membranes in animals, is present in the blood of animals, and is the precursor of several other steroids, such as the sex hormones testosterone and estrogen. Testosterone regulates sexual function and aids in building bone and muscle mass. Estrogen regulates sexual function in females, and acts to increase the storage of fat. Cholesterol is made by mammals, and it can also enter the body as part of the diet. Although it performs important functions, high levels of cholesterol in blood can cause fatty material to accumulate inside the lining of blood vessels, reducing blood flow and contributing to heart disease.

In medicine, steroids are used to reduce inflammation. Examples include topical steroid ointments to treat skin conditions and inhalers to treat asthma. *Anabolic steroids* are synthetic compounds that mimic male sex hormones. Used to build muscle mass in people who have cancer and AIDS, anabolic steroids are also frequently misused by athletes and their use is banned in most competitive sports.

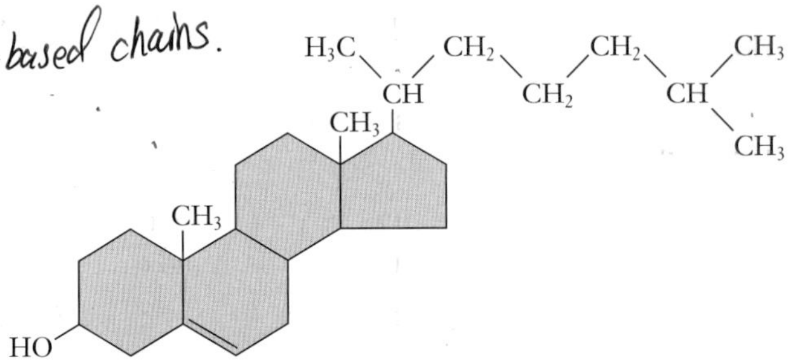

H_3C CH_2 CH_2 CH_3
CH CH_2 CH
CH_3
CH_3
CH_3
HO

Figure 1.16 Cholesterol is a member of a large group of lipids called steroids.

wax lipids composed of long carbon-based chains that are solids at room temperature

Waxes have a diversity of chemical structures, often with long carbon-based chains, and are solid at room temperature. They are produced in both plants (for example, carnauba wax) and animals (for example, earwax, beeswax, and lanolin). In plants, waxes coat the surfaces of leaves, preventing water and solutes from escaping and helping to repel insects. Waxes are present on the skin, fur, and feathers of many animal species and on the exoskeletons of insects. The whale species shown in **Figure 1.17** gets its common name from the wax-filled organ that occupies its head and represents as much as one-third of the total length of the whale and one-quarter of its total mass. Whales can generate sound from this organ, which can be used for sonar to locate prey.

Figure 1.17 The sperm whale gets its name from the spermaceti-filled structure located at its head. Spermaceti is a liquid wax at internal body temperature, but it changes to a milky-white solid with exposure to air. The whale was once hunted for its spermaceti, which was highly prized for its use in candles and lubricant oils.

Learning Check

13. In what ways are lipids similar to and different from carbohydrates?

14. Explain why lipids are efficient energy-storage molecules.

15. Sketch and describe the basic structure of a triglyceride, and explain how the presence of double bonds affects its properties.

16. What is meant by the dual character of a phospholipid molecule, and how is this essential to its function in living systems?

17. Identify two examples of steroids, and explain their significance.

18. What property of waxes is common in both plants and animals? Give an example and its significance in a specific plant and a specific animal.

Proteins

Proteins represent an extremely diverse type of macromolecules that can be classified into groups according to their function, as outlined below. The body contains tens of thousands of different groups of proteins, and each of these groups contains thousands of specific examples.

- **Catalyzing chemical reactions:** There are a large group of proteins that can catalyze, or speed up, specific biological reactions.
- **Providing structural support:** Protein fibres provide structural support in such materials as bones, tendons, skin, hair, nails, claws, and beaks.
- **Transporting substances in the body:** In the blood, proteins transport small molecules such as oxygen. Proteins in cell membranes transport substances in and out of the cell.
- **Enabling organisms to move:** Animals move by muscle contraction, which involves the interaction of the proteins actin and myosin.
- **Regulating cellular processes:** Proteins such as some hormones carry signals between cells that can regulate cell activities. Proteins can also regulate the genetic activity of a cell by altering gene expression.
- **Providing defence from disease:** Antibodies are proteins that combat bacterial and viral infections.

protein a biological macromolecule composed of amino acid monomers linked by covalent bonds

amino acid an organic molecule composed of a central carbon atom bonded to a hydrogen atom, an amino group, a carboxyl group, and a variable *R* group

How is it possible for a single type of macromolecule to carry out such diverse activities in biological systems? The functions of proteins depend on their three-dimensional structures. These structures are highly specific and all begin with the repeating units of which proteins are composed—amino acids.

Amino Acids: Monomers of Proteins

A **protein** is a macromolecule composed of amino acid monomers. As shown in **Figure 1.18**, an **amino acid** contains a central carbon atom that is bonded to the following four atoms or groups of atoms: a hydrogen atom, an amino group, a carboxyl group, and an *R* group. (An *R* group is also called a *side chain*.) Because all amino acids have the same underlying structure, the distinctive shape and properties of an amino acid result from its *R* group. All amino acids are somewhat polar, due to the polar C=O, C–O, C–N, and N–H bonds. Some amino acids are much more polar than others, depending on the polarity of the *R* group.

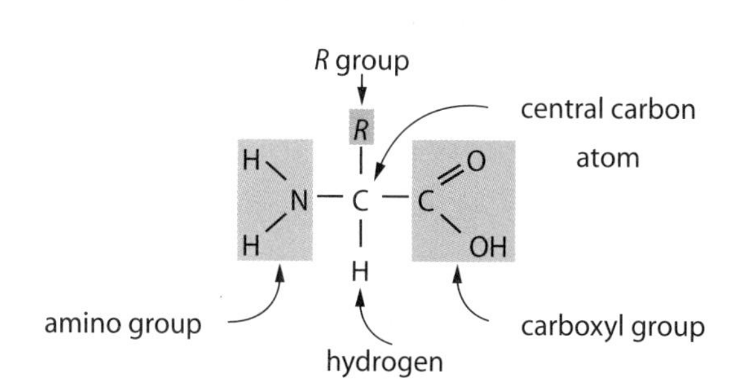

Figure 1.18 An amino acid molecule is composed of a central carbon atom bonded to an amino group, a carboxyl group, and a hydrogen atom. Each amino acid also has an *R* group bonded to the central carbon atom, providing the amino acid with its unique identity.

Figure 1.19 shows the 20 common amino acids that make up most proteins. Eight of these amino acids, called *essential amino acids*, cannot be produced by the human body and must be consumed as part of the diet. These are isoleucine, leucine, lysine, methionine, phenylalanine, threonine, tryptophan, and valine.

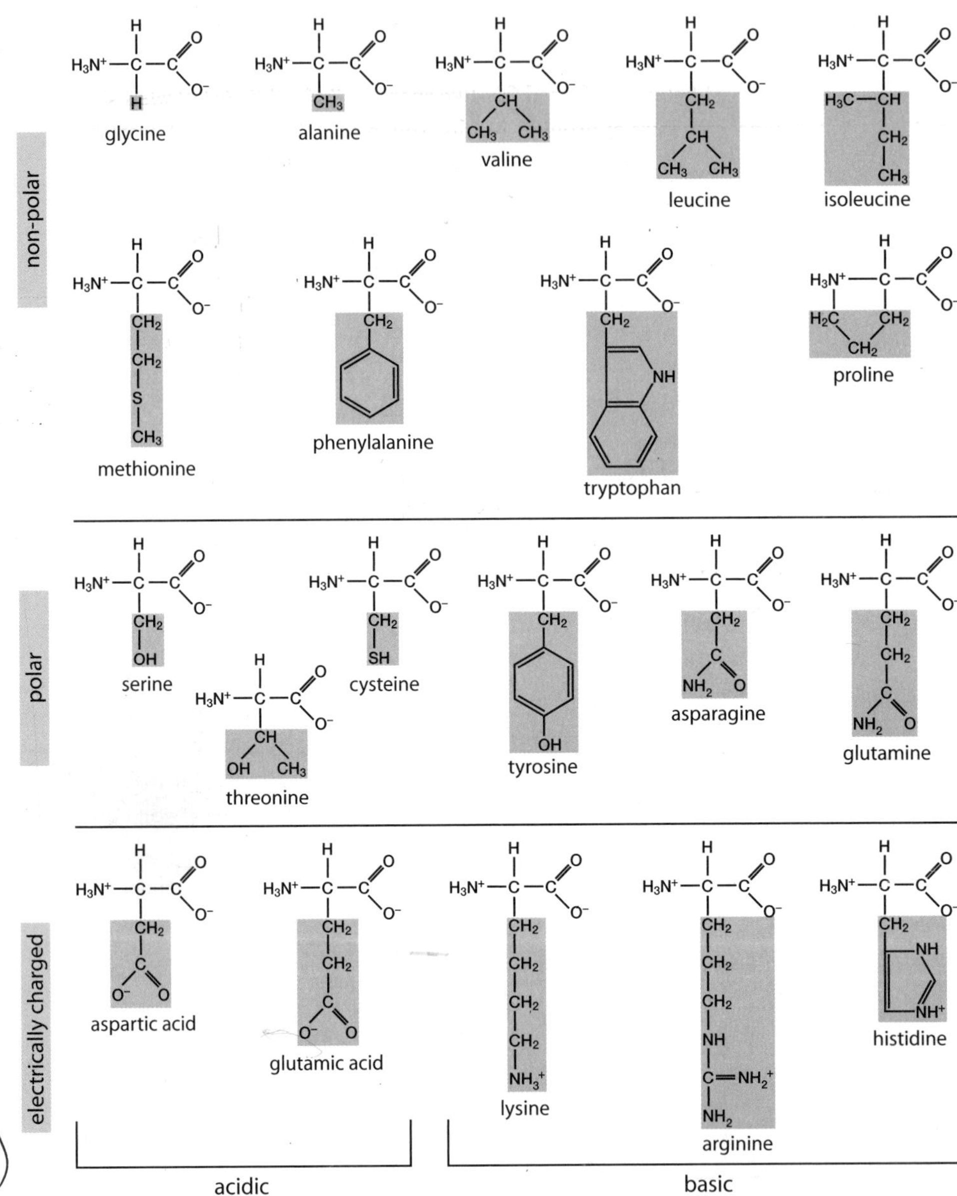

Figure 1.19 The properties of the side chains on the amino acids determine the overall shape of the protein and, therefore, its function. While cysteine is listed with the polar amino acids, it is borderline between polar and non-polar. Note that the carboxyl groups and amino groups of the amino acids are shown in their charged form, which is how they exist in the cell.

polypeptide a polymer composed of many amino acids linked together by covalent bonds

In proteins, amino acids are joined by covalent bonds called *peptide bonds*. As shown in **Figure 1.20**, a peptide bond forms between the carboxyl group on one amino acid and the amino group on another. A polymer composed of amino acid monomers is often called a **polypeptide**. Proteins are composed of one or more polypeptides. Amino acids can occur in any sequence in a polypeptide. Since there are 20 possible amino acids for each position, an enormous variety of proteins are possible. For example, a protein made of 50 amino acids could have 20^{50} different possible sequences. This number is greater than the number of atoms making up the Earth and everything on it.

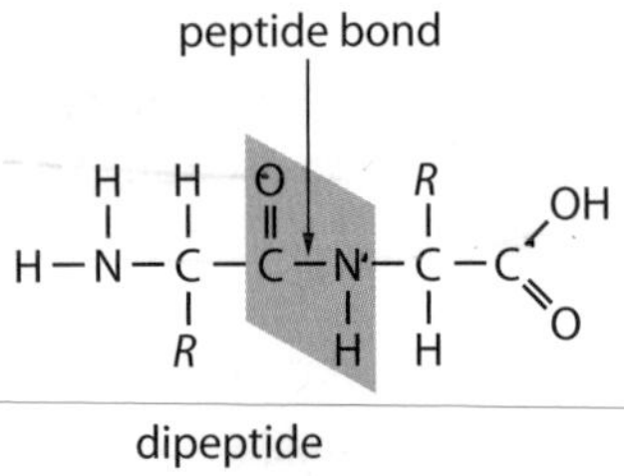

peptide bond

dipeptide

Figure 1.20 Forming a dipeptide

Levels of Protein Organization

The structure of a protein can be divided into four levels of organization, which are shown in **Figure 1.21**. The first level, called the *primary structure*, is the linear sequence of amino acids. The peptide bonds linking the amino acids may be thought of as the backbone of a polypeptide chain. Since the peptide bonds are polar, hydrogen bonding is possible between the C=O of one amino acid and the N–H of another amino acid. This contributes to the next level of organization, called the *secondary structure*. A polypeptide can form a coil-like shape, called an α (alpha) helix, or a folded fan-like shape, called a β (beta) pleated sheet. The three-dimensional structure of proteins results from a complex process of protein folding that produces the *tertiary structure*. Most of the protein folding process occurs naturally as the peptide bonds and the different *R* groups in a polypeptide chain interact with each other and with the aqueous environment of the cell. Most of what determines tertiary structure is the hydrophobic effect. For example, polar hydrophilic groups will be directed toward the aqueous environment, while non-polar hydrophobic groups will tend to be directed toward the interior of the protein's three-dimensional shape. There are also other stabilizing forces, such as hydrogen bonding between *R* groups of different amino acids and electrostatic attractions between oppositely charged *R* groups. Scientists now know that the process of protein folding also involves the assistance of a class of proteins called *molecular chaperones*. These proteins interact with the polypeptide chain and, through a series of steps, produce the final properly folded protein.

Although some proteins are composed of only one polypeptide, others are made of more than one polypeptide, each with its own primary, secondary, and tertiary structures. In proteins with multiple polypeptide chains, these separate polypeptides are arranged into a fourth level of organization, called the *quaternary structure*.

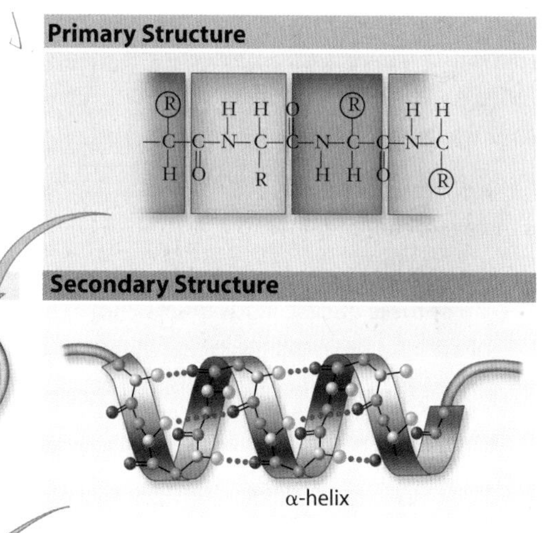

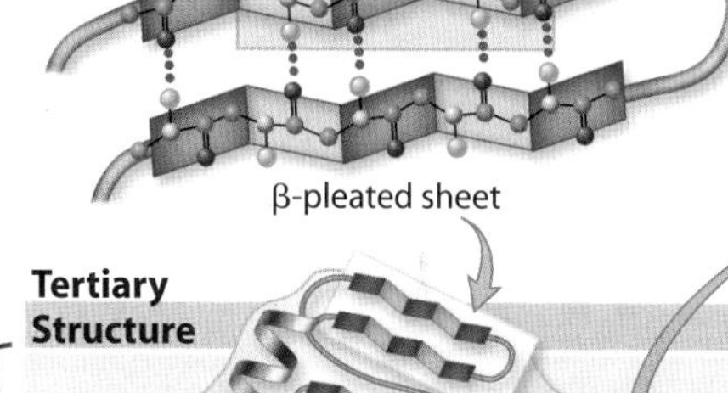

Tertiary Structure

Quaternary Structure

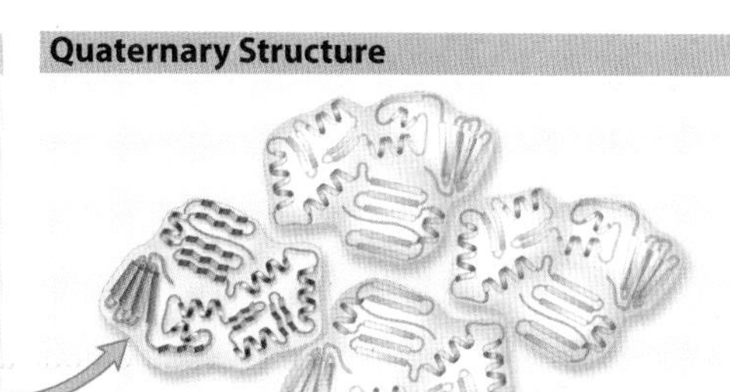

Figure 1.21 The different levels of protein structure are the primary structure (its amino acid sequence), its secondary structure, and its tertiary structure (its overall three-dimensional shape). Proteins composed of more than one polymer chain of amino acids also have quaternary structure.

Under certain conditions, proteins can completely unfold in a process called *denaturation*. Denaturation occurs when the normal bonding between *R* groups is disturbed. Intermolecular bonds break, potentially affecting the secondary, tertiary, and quaternary structures. Conditions that can cause denaturation include extremes of hot and cold temperatures and exposure to certain chemicals. Once a protein loses its normal three-dimensional shape, it is no longer able to perform its usual function.

Learning Check

19. Explain, using at least four functional examples, why proteins represent a diverse type of macromolecule.
20. Describe, using a labelled sketch, the basic structure of an amino acid.
21. What is the significance of an *R* group?
22. Explain why proteins are more structurally and functionally diverse than carbohydrates and lipids.
23. Summarize the four levels of organization in protein structure.
24. Explain how the three-dimensional structure of a protein can be changed (in the absence of gene mutations) and why this can be harmful to an organism.

nucleic acid biological macromolecules composed of nucleotide monomers

DNA (deoxyribonucleic acid) a biological macromolecule composed of nucleotides containing the sugar deoxyribose

RNA (ribonucleic acid) a biological macromolecule composed of nucleotides containing the sugar ribose

nucleotide an organic molecule composed of a sugar bonded to a phosphate group and a nitrogen-containing base

Nucleic Acids

The fourth type of biologically important molecule is a class of macromolecules called **nucleic acids**. The two types of nucleic acids are **DNA (deoxyribonucleic acid)** and **RNA (ribonucleic acid)**. DNA contains the genetic information of an organism. This information is interpreted or decoded into particular amino acid sequences of proteins, which carry out the numerous functions in the cell. The conversion of genetic information that is stored in DNA to the amino acid sequence of proteins is carried out with the assistance of different RNA molecules. The amino acid sequence of a protein is determined by the nucleotide sequences of both DNA and RNA molecules.

DNA and RNA are polymers made of thousands of repeating **nucleotide** monomers. As shown in **Figure 1.22**, a nucleotide is made up of three components that are covalently bonded together:

- a phosphate group
- a sugar with five carbon atoms
- a nitrogen-containing base

While both DNA and RNA are composed of nucleotide monomers, the nucleotide make-up of these nucleic acids differs somewhat. The nucleotides in DNA contain the sugar *deoxyribose*, and the nucleotides in RNA contain the sugar *ribose*. This difference accounts for their respective names. There are four different types of nitrogenous bases in DNA: adenine (A), thymine (T), guanine (G), and cytosine (C). In RNA, all but one of these same bases is used. The exception is that thymine is replaced by the base uracil (U).

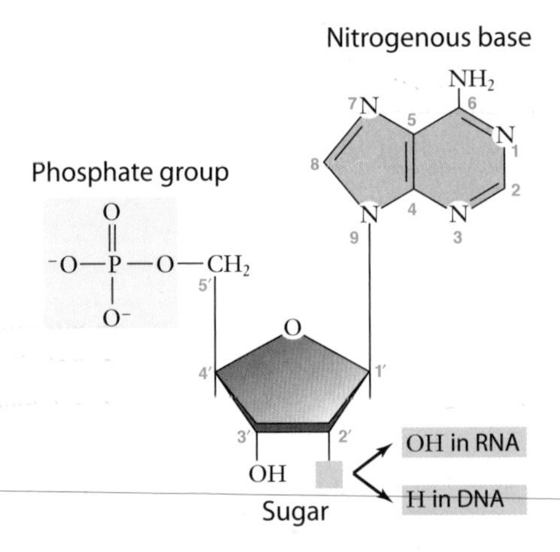

Figure 1.22 Nucleotides contain a phosphate group, a five-carbon sugar, and a nitrogen-containing base. DNA nucleotides contain a deoxyribose sugar, while RNA nucleotides contain ribose as the sugar component. The base shown here is adenine.

A polymer of nucleotides is often referred to as a *strand*. The covalent bond between adjacent nucleotides is called a *phosphodiester bond*, and it occurs between the phosphate group on one nucleotide and a hydroxyl group on the sugar of the next nucleotide in the strand. A nucleic acid strand, therefore, has a backbone made up of alternating phosphates and sugars with the bases projecting to one side of the backbone, as shown in **Figure 1.23**. Just as amino acids in proteins have a specific order, nucleotides and their bases occur in a specific order in a strand of DNA or RNA. DNA is composed of two strands twisted about each other to form a double helix. When unwound, DNA resembles a ladder. The sides of the ladder are made of alternating phosphate and sugar molecules, and the rungs of the ladder are made up of pairs of bases held together with hydrogen bonds. Nucleotide bases always pair together in the same way. Thymine (T) base pairs with adenine (A), and guanine (G) base pairs with cytosine (C). These bases are said to be *complementary* to each other. RNA, on the other hand, is single-stranded. Further details of the structures and functions of DNA and RNA will be discussed in Unit 3.

Suggested Investigation

Inquiry Investigation 1-A, Identifying Biological Macromolecules in Food

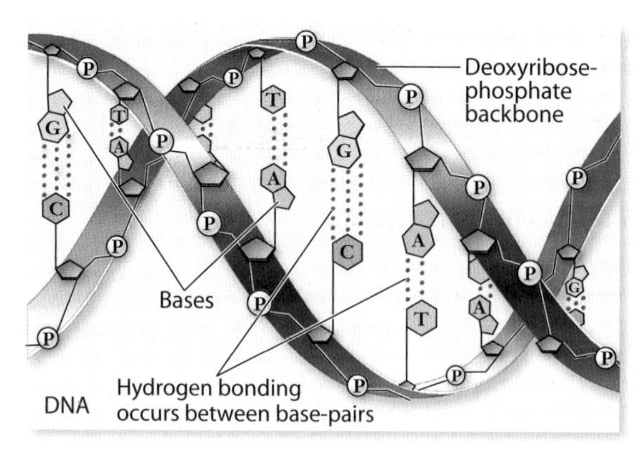

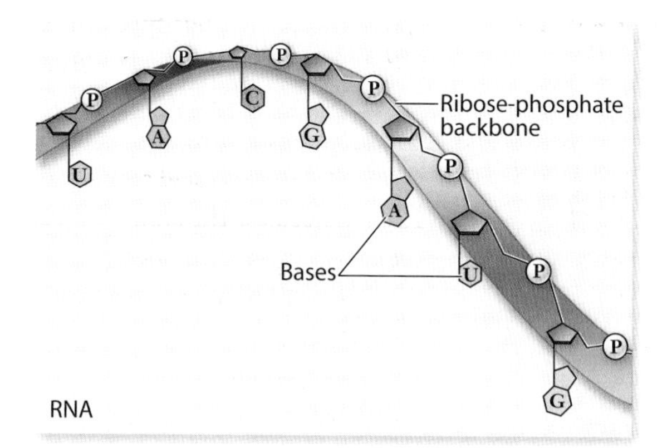

Figure 1.23 DNA and RNA have a backbone of alternating phosphate groups and sugars. DNA is made of two strands that are twisted into the shape of a double helix. Toward the centre of the helix are complementary base pairs. Hydrogen bonds between the bases help hold the two strands together. Three hydrogen bonds occur between G and C base pairs, while two hydrogen bonds occur between A and T base pairs. Since RNA exists as a single-stranded molecule, it does not occur as a double helix.

Activity 1.2 Modelling Biological Molecules

The shape of a biological compound plays a significant role in the ability to carry out a particular function in the body. In this activity, you will build and draw three-dimensional models of carbohydrates, proteins, lipids, nucleic acids, and their monomers in order to better understand their structural similarities and differences.

Materials

- molecular model kit or other model-building supplies, or
- appropriate software package

Procedure

1. Choose one of the following biological molecules to build:

monosaccharide	saturated fatty acid
disaccharide	amino acid
polysaccharide	polypeptide
triglyceride	nucleotide
phospholipid	DNA
unsaturated fatty acid	RNA

2. Using a molecular model kit, model-building supplies, or an appropriate software program, construct a three-dimensional model of your biological compound.
3. Sketch a diagram for each molecule that you build.

Questions

1. On your diagram, identify:
   a. the biological compound by its type
   b. any functional groups, and provide their names
   c. polarity (of the molecule, or parts of the molecule)
   d. solubility in water
2. For each class of biological molecule that you built, identify a structural feature unique to that class.
3. Describe the relationship between the structural features and the function of the molecule you built.

Summary of Biologically Important Molecules

The four types of macromolecules contain mostly carbon, hydrogen, and oxygen, with varying amounts of nitrogen, phosphorus, and sulfur. The presence of certain functional groups on these molecules provides them with particular properties, such as polar groups on phospholipids, or non-polar side chains on amino acids. Functional groups, in turn, influence and determine the functions of macromolecules in the cell. **Table 1.2** outlines important structural features and examples of macromolecules.

Table 1.2 Examples of Biologically Important Molecules

Carbohydrates			
Type	**Structure**	**Examples**	**Some Functions**
Monosaccharide	• Contains a single three- to seven-carbon atom-based structure	Glucose, fructose, galactose	• Glucose is used as a primary energy source
Disaccharide	• Contains two monosaccharides joined by a glycosidic linkage	Sucrose, lactose, maltose	• Sucrose and lactose are dietary sugars that are used for energy
Polysaccharide	• Contains many monosaccharides joined by glycosidic linkages	Starch, glycogen, cellulose	• Glycogen is a form of storing glucose in animals • Cellulose provides structural support in plants
Lipids			
Type	**Structure**	**Examples**	**Some Functions**
Triglyceride	• Contains three fatty acids joined to glycerol by ester linkages	Lard, butter, vegetable oils	• Provides long-term energy storage • Acts to cushion organs and insulate from heat loss
Phospholipid	• Contains two fatty acids and a phosphate group joined to glycerol	Phosphatidylcholine	• Forms the main structure of cell membranes
Steroid	• Contains four carbon-based rings attached to one another	Cholesterol, testosterone, estrogen	• Cholesterol is part of cell membranes • Testosterone and estrogen are sex hormones
Wax	• Contains long carbon-based chains	Earwax, beeswax, spermaceti	• A variety of functions, including protection
Protein			
Type	**Structure**	**Examples**	**Some Functions**
Catalyst	• Contains amino acid monomers joined by peptide bonds • All have primary, secondary, tertiary structure	Amylase, sucrase	• Speeds up chemical reactions
Transport		Hemoglobin, ion channel proteins	• Transports specific substances
Structural		Collagen, keratin	• Provides structure
Movement		Myosin, actin	• Enables movement
Regulatory		Hormones, neurotransmitters	• Carries cellular messages
Defence		Antibodies	• Fights infection
Nucleic Acids			
Type	**Structure**	**Some Functions**	
DNA	• Contains deoxyribonucleotide monomers (A, G, T, C)	Stores genetic information of an organism	
RNA	• Contains ribonucleotide monomers (A, U, G, C)	Participates in protein synthesis	

Section 1.2 Review

Section Summary

- Carbohydrates can act as short-term energy-storage and structural biological molecules. They include polysaccharides, and disaccharides and monosaccharides (sugars). Polysaccharides are macromolecules composed of monosaccharide monomers.
- Lipids are longer-term energy storage and structural biological molecules that contain a higher proportion of non-polar C–C and C–H bonds than carbohydrates. Triglycerides, phospholipids, steroids, and waxes are different types of lipids.
- Proteins are macromolecules composed of amino acid monomers. Proteins have primary, secondary, tertiary, and sometimes quaternary structures, which are essential for their function. Proteins enable chemical reactions in living systems, and also perform a wide variety of structural and regulatory roles.
- Nucleic acids are macromolecules composed of nucleotide monomers. DNA and RNA are both nucleic acids. DNA stores the hereditary information of a cell. RNA plays a central role in the synthesis of proteins.

Review Questions

1. **K/U** A carbon atom can bond in a variety of configurations, which is one reason why carbon atoms form the backbone of all biological molecules. Look through the structural diagrams in this section, and identify at least three different shapes that carbon atoms can form when bonded to each other.

2. **K/U** Identify the monomer for each type of polymer.
   a. protein **b.** glycogen **c.** RNA

3. **K/U** Identify the functional group(s) on each biological molecule.
   a. H_3N^+—C(H)(CH$_2$–SH)—COO^-
   b. (nucleoside structure with NH_2, HO, OH, OH groups)

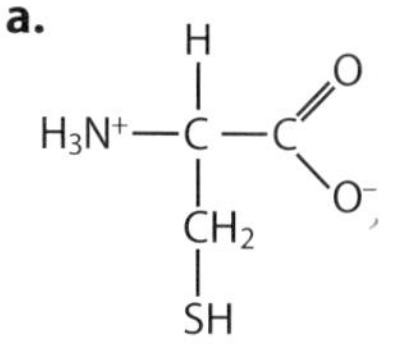

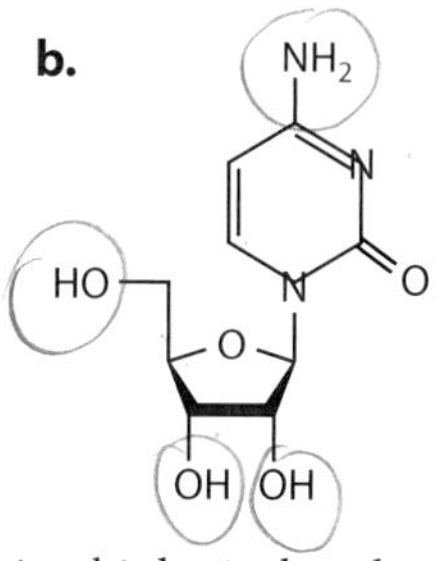

4. **K/U** Arrange the following biological molecules from smallest to largest, and give reasons for your sequence: cellulose, amino acid, ribose, triglyceride.

5. **A** If humans had the ability to digest cellulose, what would a high-fibre diet do to blood glucose levels?

6. **K/U** A sample of fat contains about twice as much energy as the same mass of carbohydrate. Compare the structures of each type of biological molecule to explain why.

7. **C** Use one or more labelled sketches to show the types of intermolecular and intramolecular interactions that determine the final structure of a protein.

8. **K/U** When placed in water, phospholipid molecules form the structures shown here. Explain why.

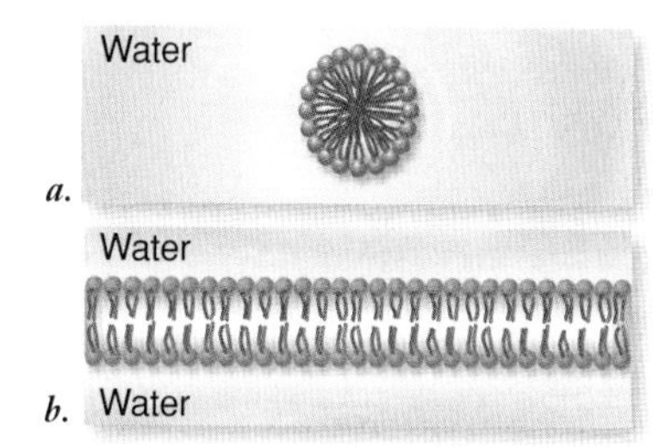

9. **A** Biologists and health science researchers continually conduct research to determine the roles of fatty acids and other macromolecules in health and longevity. Use your knowledge of biochemistry to explain how a balanced diet offsets the need to follow the latest dietary trend.

10. **A** Ninhydrin is a compound that is commonly used in forensic identification, because it turns purple in contact with the amino acids often found in sweat residue. Using the structure provided, answer the questions below.

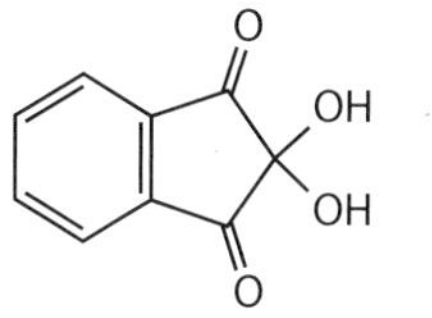

    (ninhydrin structure: O, OH, OH, O)

    a. Determine the number of carbonyl and hydroxyl functional groups present in ninhydrin.
    b. How many carboxylic acid functional groups are present? Explain your answer.

11. **A** Using your understanding of protein structure and protein denaturation, infer how a hair-straightening iron works.

12. **C** Use a Venn diagram to compare and contrast a strand of DNA and a strand of RNA.

SECTION 1.3

Biochemical Reactions

Key Terms

acid
base
pH scale
neutralization reaction
buffer
oxidation
reduction
redox reaction
condensation reaction
hydrolysis reaction
activation energy
catalyst
enzyme
active site
substrate
enzyme-substrate complex
inhibitor
allosteric site
activator

The chemical reactions that are associated with biological processes can be grouped into several types. Often, these biochemical reactions involve a combination of more than one type. The four main types of chemical reactions that biological molecules undergo in the cell are neutralization, oxidation-reduction, condensation, and hydrolysis reactions.

Neutralization (Acid-Base) Reactions

In the context of biological systems, acids and bases are discussed in terms of their behaviour in water. An **acid** can be defined as a substance that produces hydrogen ions, H^+, when it dissolves in water. Thus, acids increase the concentration of hydrogen ions in an aqueous solution. A **base** can be defined as a substance that produces hydroxide ions, OH^-, when it dissolves in water, increasing the concentration of hydroxide ions. As you will see in this section, a base can also be thought of as a substance that accepts or reacts with hydrogen ions.

The **pH scale**, shown in **Figure 1.24**, ranks substances according to the relative concentrations of their hydrogen ions. Substances that have a pH lower than 7 are classified as acids. Substances that have a pH higher than 7 are classified as bases. Because pure water has equal concentrations of hydrogen ions and hydroxide ions, it is neither an acid nor a base. With a pH of 7, water is classified as neutral.

acid a substance that produces hydrogen ions, H^+, when dissolved in water

base a substance that produces hydroxide ions, OH^-, when dissolved in water

pH scale a numerical scale ranging from 0 to 14 that is used to classify aqueous solutions as acidic, basic, or neutral

neutralization reaction a chemical reaction between an acid and a base, producing water and a salt

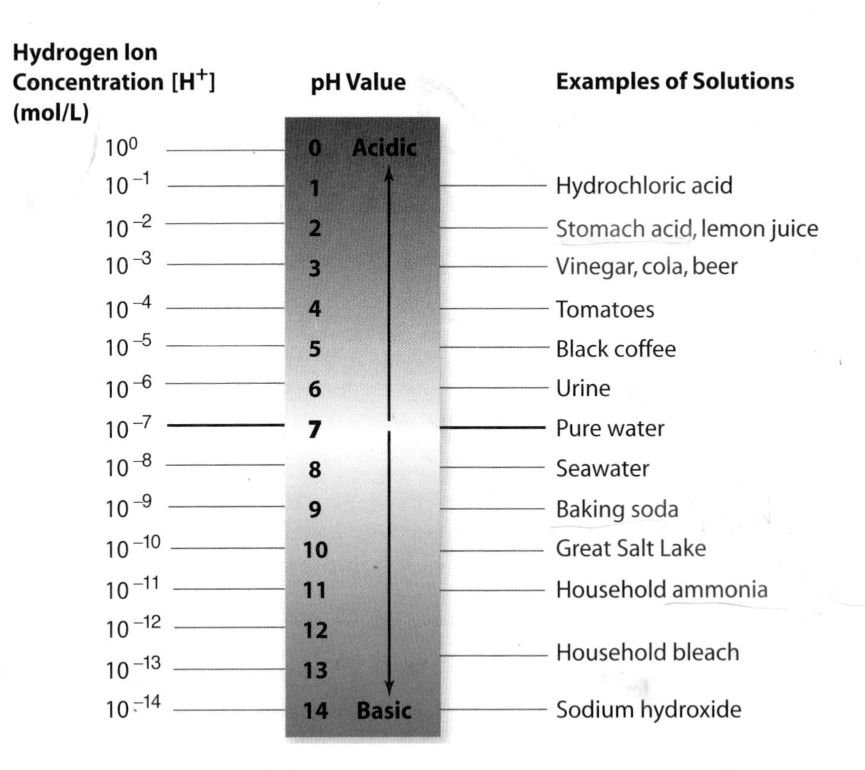

Figure 1.24 Pure water has a pH of 7. Acid solutions have lower pH values, and basic solutions have higher pH values.

When an acid chemically interacts with a base, they undergo a **neutralization reaction** that results in the formation of a salt (an ionic compound) and water. As a result of a neutralization reaction, the acid loses its acidic properties and the base loses its basic properties. In other words, their properties have been cancelled out, or neutralized.

Many of the chemical reactions that occur in the body can only take place within a narrow range of pH values, and illness results when pH changes beyond this range. For example, the normal pH of human blood ranges from 7.35 to 7.45. If blood pH increases to 7.5, as might happen if a person breathes too quickly at high altitudes, feels extremely anxious, or takes too many antacids, the person can feel dizzy and agitated. Such a condition is called alkalosis. On the other hand, acidosis occurs if blood pH falls to within 7.1 and 7.3. Symptoms of acidosis include disorientation and fatigue and can result from severe vomiting, brain damage, and kidney disease. Blood pH that falls below 7.0 and rises beyond 7.8 can be fatal.

To maintain optimum pH ranges, organisms rely on **buffers**—substances that resist changes in pH by releasing hydrogen ions when a fluid is too basic and taking up hydrogen ions when a fluid is too acidic. Most buffers exist as specific pairs of acids and bases. For example, one of the most important buffer systems in human blood involves the pairing of carbonic acid, $H_2CO_3(aq)$, and hydrogen carbonate ion, $HCO_3^-(aq)$. The reaction that takes place between these two substances is shown in **Figure 1.25**. Notice the double arrow in the reaction equations. This arrow means that the reaction can occur in both directions, which is a requirement for buffers.

buffer a substance that minimizes changes in pH by donating or accepting hydrogen ions as needed

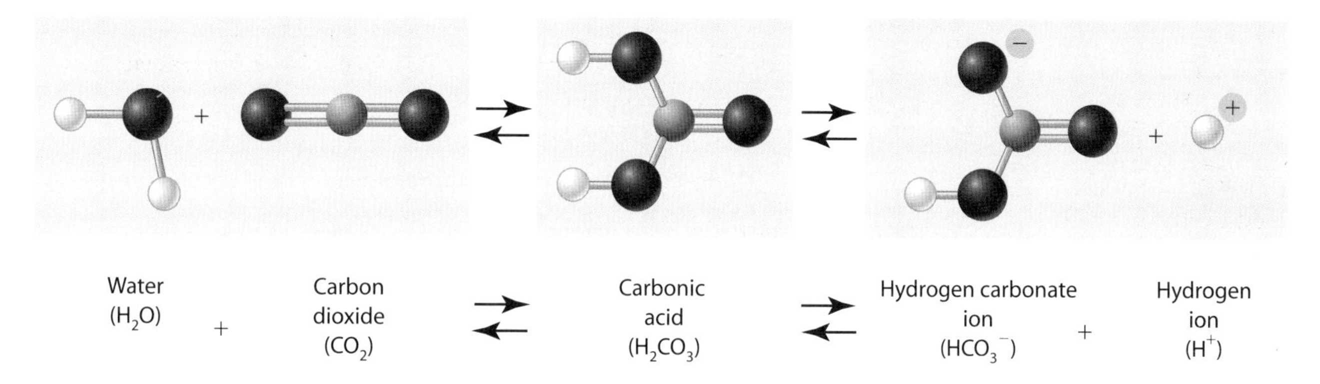

Figure 1.25 The carbonic acid–hydrogen carbonate ion buffer system is a key system in the body. If blood becomes too basic, carbon dioxide and water react to to produce carbonic acid, which dissociates into hydrogen carbonate and hydrogen ions and increases the acidity of the blood.

***Explain** how this buffer system works to reduce the acidity of blood.*

Oxidation-Reduction Reactions

Another key type of chemical reaction in biochemistry is based on the transfer of electrons between molecules. When a molecule loses electrons it becomes *oxidized* and has undergone a process called **oxidation**. Electrons are highly reactive and do not exist on their own or free in the cell. Therefore, when one molecule undergoes oxidation, the reverse process must also occur to another molecule. When a molecule accepts electrons from an oxidized molecule, it becomes *reduced* and has undergone a process called **reduction**. Because oxidations and reductions occur at the same time, the whole reaction is called an oxidation-reduction reaction, which is often abbreviated to the term **redox reaction**.

oxidation a process involving the loss of electrons

reduction a process involving the gain of electrons

redox reaction a chemical reaction that involves the transfer of electrons from one substance to another; also called oxidation-reduction reaction

A common type of redox reaction is a combustion reaction. For example, the combustion of propane, C_3H_8, in a gas barbecue occurs according the following chemical equation:

$$C_3H_8 + 5O_2 \rightarrow 3CO_2 + 4H_2O$$

The propane is oxidized, and the oxygen is reduced. This reaction also releases a large amount of energy—energy that is used to cook food on the barbecue.

Similar types of redox reactions occur in cells. For example, the process of cellular respiration has an overall chemical equation of

$$C_6H_{12}O_6 + 6O_2 \rightarrow 6CO_2 + 6H_2O$$

Sugars such as glucose are oxidized to produce carbon dioxide and water. Note, however, that this chemical equation represents an *overall* reaction. As you will see in Unit 2, cellular respiration involves oxidation of glucose through a series of redox reactions. The energy released at each step in the series is stored in the form of chemical bonds of molecules that are made during the process. This allows the energy to be harnessed by the cell in discrete steps, instead of in one surge that would result in a great deal of energy loss and perhaps harm to the cell.

Condensation and Hydrolysis Reactions

condensation reaction a chemical reaction that results in the formation of a covalent bond between two molecules with the production of a water molecule

hydrolysis reaction a chemical reaction that results in cleavage of a covalent bond with the addition of a water molecule

The assembly of all four types of biological macromolecules involves the same type of chemical reaction—a condensation reaction between the monomers of each polymer. In a **condensation reaction**, an H atom is removed from a functional group on one molecule, and an OH group is removed from another molecule. The two molecules bond to form a larger molecule and water. Because the reaction results in the release of water, condensation reactions are also called *dehydration* reactions. An example of a condensation reaction is shown in **Figure 1.26**.

The breakdown of macromolecules into their monomers also occurs in cells. For example, during the digestion of a polysaccharide, your body breaks down the macromolecule into simpler sugars. This process involves the addition of water to break the bonds between the monomers. In a **hydrolysis reaction**, an H atom from water is added to one monomer, and an OH group is added to the monomer beside that one. The covalent bond between these monomers breaks and the larger molecule is split into two smaller molecules.

Condensation reactions and hydrolysis reactions can be thought of as opposite reactions, and a single chemical equation can be used to represent them. Chemical equations showing the condensation and hydrolysis reactions that are involved in the synthesis and breakdown of each of the four classes of biological molecules are shown in **Figure 1.27** on the opposite page.

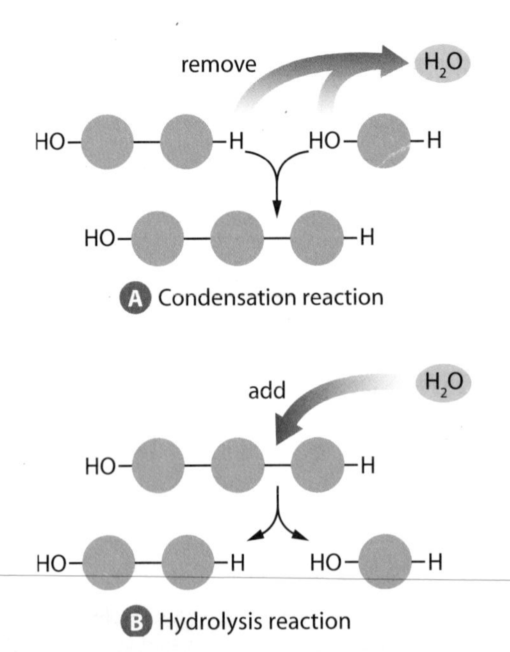

Figure 1.26 In a condensation reaction (**A**), two molecules are joined by a covalent bond, and water is produced as a side product. In a hydrolysis reaction (**B**), a larger molecule is broken apart through the addition of water at each break point.

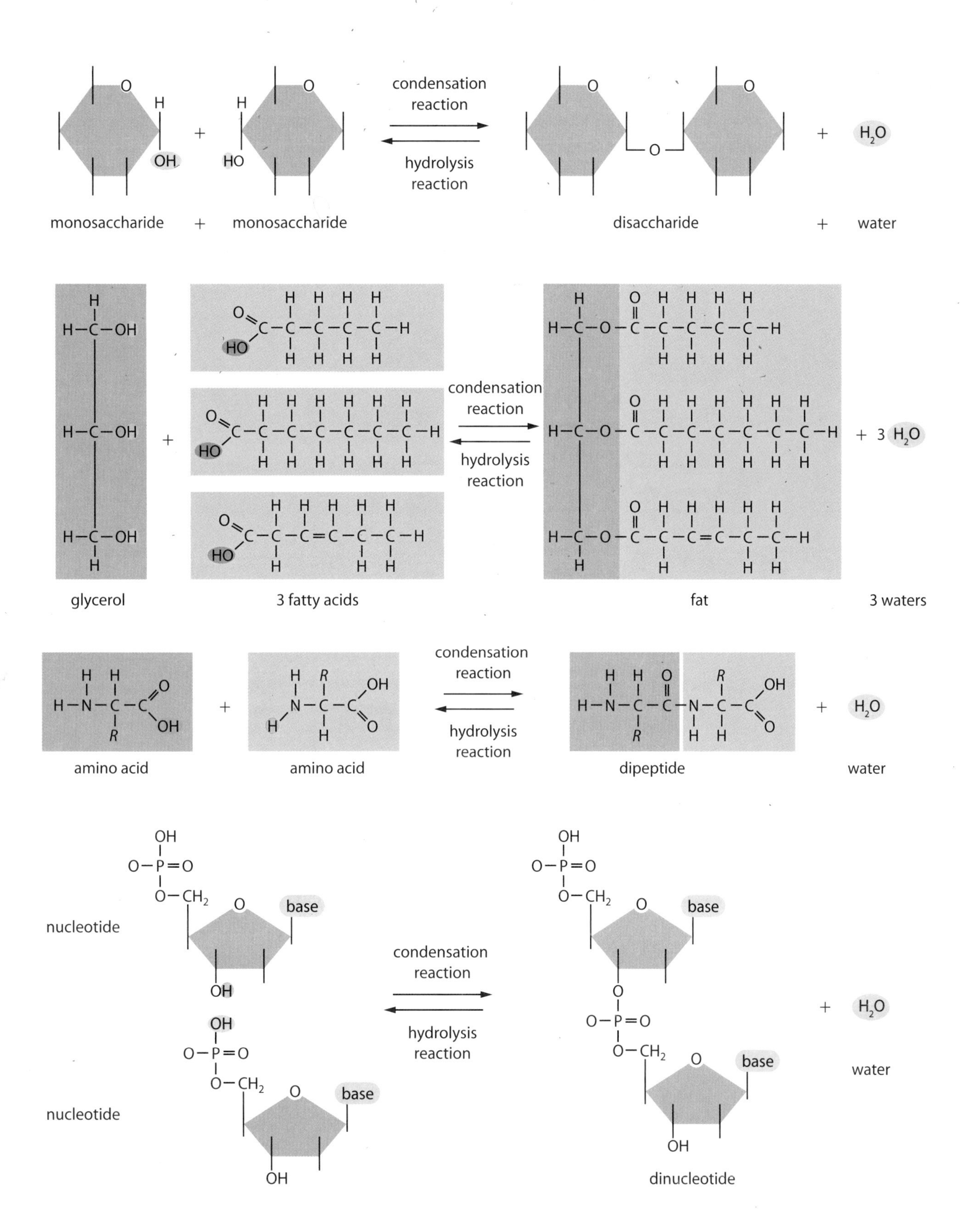

Figure 1.27 Carbohydrates, lipids, proteins, and nucleic acids are assembled by condensation reactions among their monomers, and these polymers are broken down by hydrolysis reactions. The double arrow indicates that a chemical reaction can proceed in a "forward" and a "reverse" direction. As written, the forward is a condensation reaction and the reverse is a hydrolysis reaction. Note that the rings in the carbohydrate and nucleotide structures are drawn in this manner so that a particular molecule is not specified.

Learning Check

25. Identify the products of a neutralization reaction, and explain how the reaction changes the properties of the reacting acid and base.

26. What happens to the electrons that are lost by a compound that is undergoing an oxidation reaction?

27. Explain why a redox reaction must involve changes to two molecules simultaneously.

28. Does a molecule have more energy in its oxidized or reduced form? Explain.

29. Differentiate between a hydrolysis reaction and a condensation reaction.

30. There are four major types of chemical reactions that break apart and build biological molecules. For each type, write a sentence that summarizes a key defining feature that distinguishes it from the others.

activation energy the energy required to initiate a chemical reaction

catalyst a substance that speeds up the rate of a chemical reaction by lowering the activation energy for the reaction; is not consumed in the reaction

enzyme a biological macromolecule that catalyzes, or speeds up, chemical reactions in biological systems

Enzymes Catalyze Biological Reactions

A certain amount of energy is required to begin any reaction. This energy is referred to as the **activation energy** of a reaction. If the activation energy for a reaction is large, it means the reaction will take place very slowly. If the chemical reactions of cells were carried out in a test tube, most would occur too slowly to be of any use to the cell. For example, suppose you performed the reaction of carbon dioxide with water to form carbonic acid, which is part of the buffering system to control blood pH. You would determine that it proceeds quite slowly. Perhaps 200 molecules of carbonic acid would form in about one hour. Reactions this slow are of little use to an organism. One way to increase the rate of any chemical reaction is to increase the temperature of the reactants. In living systems, however, this approach to speeding up reactions has a major drawback. The temperatures at which chemical reactions would need to occur to proceed quickly enough to sustain life are so high that they would permanently denature proteins. This is why a long-lasting fever can be dangerous. If the fever stays too high for too long, major disruptions to cellular reactions occur, and in some cases they can be fatal.

Another way to increase the rate of chemical reactions without increasing temperature is to use a catalyst. A **catalyst** is a substance that speeds up a chemical reaction but is not used up in the reaction. It can be recovered unchanged when the reaction is complete. As shown in **Figure 1.28**, catalysts function by lowering the activation energy of a reaction. Cells manufacture specific proteins that act as catalysts. A protein molecule that acts as a catalyst to increase the rate of a reaction is called an **enzyme**. In red blood cells, for example, an enzyme called carbonic anhydrase enables carbon dioxide and water to react to form about 600 000 molecules of carbonic acid each second!

Almost all chemical reactions in organisms are facilitated by enzymes. The types of chemical reactions discussed in this section are all catalyzed by enzymes. In fact, each example of the different types of reaction is carried out with its own characteristic enzyme. Thus, enzymes not only increase the rate of a chemical reaction, but also do so in a highly specific manner through specific interactions between the enzyme and the reactant(s) for the reaction.

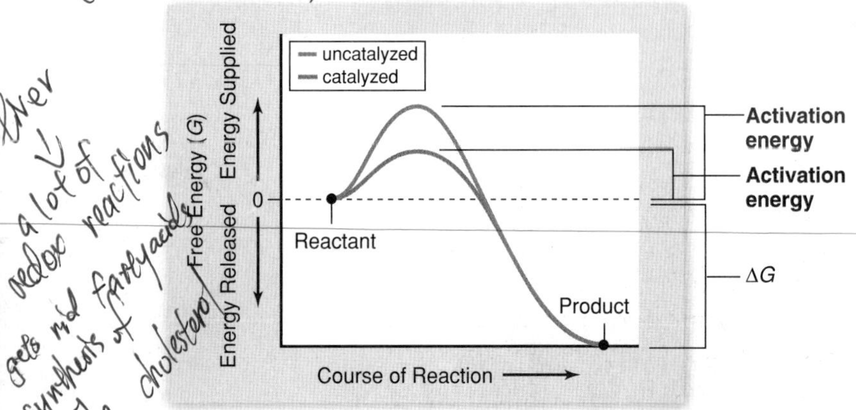

Figure 1.28 Catalysts, such as enzymes, reduce the activation energy required for a reaction to begin.

Enzymes Bind with a Substrate

Like other proteins, enzymes are composed of long chains of amino acids folded into particular three-dimensional shapes, with primary, secondary, tertiary, and often quaternary structures. Most enzymes have globular shapes, with pockets or indentations on their surfaces. These indentations are called **active sites**. The unique shape and function of an active site are determined by the sequence of amino acids in that section of the protein.

As shown in **Figure 1.29**, an active site of an enzyme interacts in a specific manner with the reactant for a reaction, called the **substrate**. During an enzyme-catalyzed reaction, the substrate joins with the enzyme to form an **enzyme-substrate complex**. The substrate fits closely into the active site, because enzymes can adjust their shapes slightly to accommodate substrates. Intermolecular bonds, such as hydrogen bonds, form between the enzyme and the substrate as the enzyme adjusts its shape. The change in shape of the active site to accommodate the substrate is called *induced fit*.

active site the site on an enzyme where the substrate binds; where the chemical reaction that is catalyzed by the enzyme takes place

substrate a reactant that interacts with the enzyme in an enzyme-catalyzed reaction

enzyme-substrate complex the combined structure of an enzyme with a substrate that is bound to the enzyme's active site

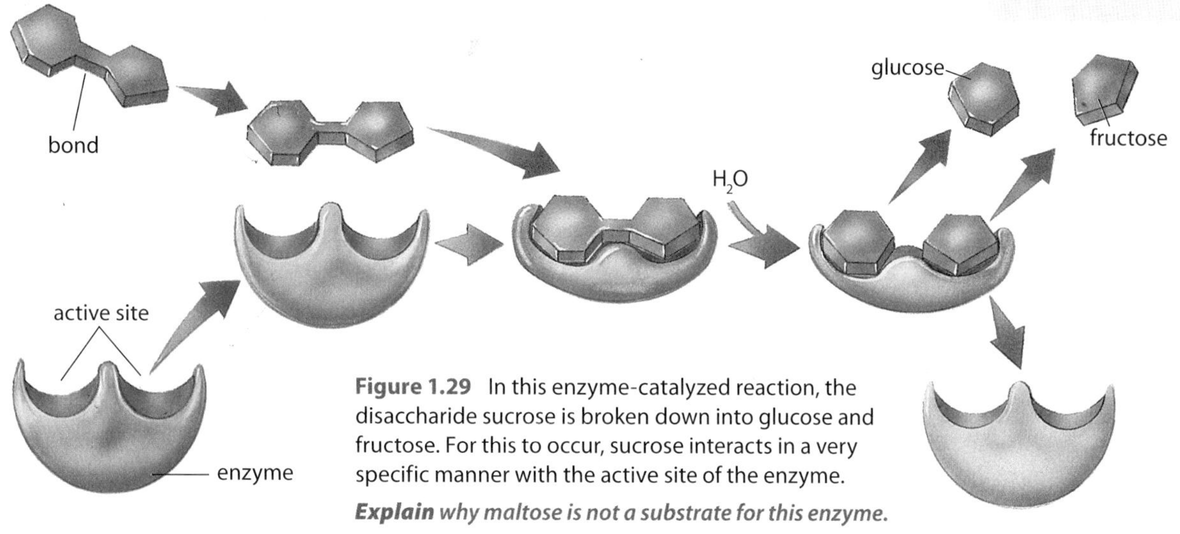

Figure 1.29 In this enzyme-catalyzed reaction, the disaccharide sucrose is broken down into glucose and fructose. For this to occur, sucrose interacts in a very specific manner with the active site of the enzyme.
***Explain** why maltose is not a substrate for this enzyme.*

Enzymes prepare substrates for reaction by changing the substrate, its environment, or both in some way, and thus lowering the activation energy of the reaction. Depending on the enzyme, this process may occur in a variety of ways. For example, the active site may:

- contain amino acid *R* groups that end up close to certain chemical bonds in the substrate, causing these bonds to stretch or bend, which makes the bonds weaker and easier to break;
- bring two substrates together in the correct position for a reaction to occur;
- transfer electrons to or from the substrate (that is, reduce or oxidize it), destabilizing it and making it more likely to react;
- add or remove hydrogen ions to or from the substrate (that is, act as an acid or base), destabilizing it and making it more likely to react.

Once the reaction has taken place, the products of the reaction are released from the enzyme. The enzyme is now able to accept another substrate and begin the process again. This cycle is known as the *catalytic cycle*. Some enzymes require the presence of additional molecules or ions to catalyze a reaction. Organic molecules that assist an enzyme are called *coenzymes*. Some enzymes require the presence of metal ions, such as iron or zinc, which are referred to as *cofactors*. Your body requires small amounts of minerals and vitamins in order to stay healthy. In many cases, this is because those minerals and vitamins are essential to enzyme activity. Without them, enzymes in your body cells cannot catalyze reactions.

Enzyme Classification

Enzymes are classified according to the type of reaction they catalyze. For example, enzymes that catalyze hydrolysis reactions are classified as hydrolases. Because the shape of an enzyme must match its substrate exactly, most enzymes are specific, catalyzing just one specific reaction. Not surprisingly, therefore, thousands of different enzymes exist to catalyze the numerous reactions that take place in organisms. Each enzyme is provided with a unique name, in order to properly identify it. The names of many enzymes consist of the first part of the substrate name, followed by the suffix "*-ase*." For example, the hydrolase enzyme that is responsible for catalyzing the cleavage of the glycosidic linkage in the dissacharide lactose is named lactase.

Activity 1.3 Enzymes in the Food and Pharmaceutical Industries

Enzymes are used various ways in the food and pharmaceutical industries. For example, they may be used:

- as medicines (for example, to aid in digestion and to treat digestive disorders such as pancreatitis);
- to manufacture certain medicines (for example, certain antibiotics, anti-inflammatory products, and clot-dissolving preparations);
- to manufacture and/or process foods and food products (for example, breads, food products that contain eggs, and food products that contain fruits and vegetables).

In this activity, you will do research to find out about the roles and uses of enzymes in the preparation of various foods, food products, and medicines.

Materials

- computer with Internet access
- print resources

Procedure

1. Choose an application of enzymes either in the food and food processing industry or in the pharmaceutical industry. Examples of applications you might choose include the following:
   - enzymes to weaken or strengthen wheat gluten
   - enzymes to derive sweeteners from starch
   - enzymes to produce cheese flavourings
   - enzymes to assist in food allergies and intolerances
   - enzymes to increase yields and improve health benefits of fruit and vegetable juices
   - enzymes to reduce bitterness in protein-containing foods
   - enzymes to improve colour in teas and coffees
   - enzymes to produce penicillin
   - enzymes to treat muscle and joint inflammation
   - enzymes to treat pancreatitis
   - enzyme formulations for use in conjunction with other medicines to reduce required dosage and/or side effects
2. Use print resources, electronic resources, or both to learn which enzyme or enzymes may be used and how they are used. Where possible, investigate the techniques that were used before the adoption of enzymes for the same purpose.
3. Record your findings in the form of a table or some other suitable format of your choice.

Questions

1. Summarize, using a format and reporting style of your choice, the ways in which enzymes are used in the application you researched and the importance of enzymes to that application.

Learning Check

31. Explain the meaning and significance of activation energy as it relates to chemical reactions.
32. Write one or two sentences to explain the meaning of the term enzyme-substrate complex. Your sentence(s) should make clear the meaning of the following additional terms: active site, substrate, catalyst.
33. Why are enzymes important to biological systems if the reactions they facilitate will occur naturally with or without the presence of enzymes?
34. Why would an enzyme be unable to catalyze many different types of reactions?
35. Outline and describe the ways in which an enzyme can prepare a substrate for a reaction.
36. Differentiate between coenzymes and cofactors, and explain their role in biological systems.

Enzyme Activity Is Influenced by Surrounding Conditions

Enzyme activity is affected by any change in condition that alters the enzyme's three-dimensional shape. Temperature and pH are two important factors that can cause this alteration. When the temperature becomes too low, the bonds that determine enzyme shape are not flexible enough to enable substrate molecules to fit properly. At higher temperatures, the bonds are too weak to maintain the enzyme's shape. As shown in **Figure 1.30**, enzymes function best at an optimal temperature range and range of pH values. These ranges are fairly narrow for most enzymes. Most human enzymes work best within the range of pH 6 – 8 to 8. Some enzymes, however, function best in very acidic environments, such as is found in the stomach. For example, because pepsin acts in the acidic environment of the stomach, its optimum pH is much lower than that of trypsin, which acts in the small intestine.

pH 6-8

Although enzymes bind and catalyze substrates very quickly, the formation of an enzyme-substrate complex will take longer if there are few substrates present. For example, in a dilute solution of substrate, enzymes and substrates will encounter each other less frequently than they will in a concentrated solution of substrate. For this reason, enzyme activity generally increases as substrate concentration increases. This is true up to a certain point, when the enzymes in the solution are working at maximum capacity. Past this point, adding more substrate will not affect the rate of the reaction.

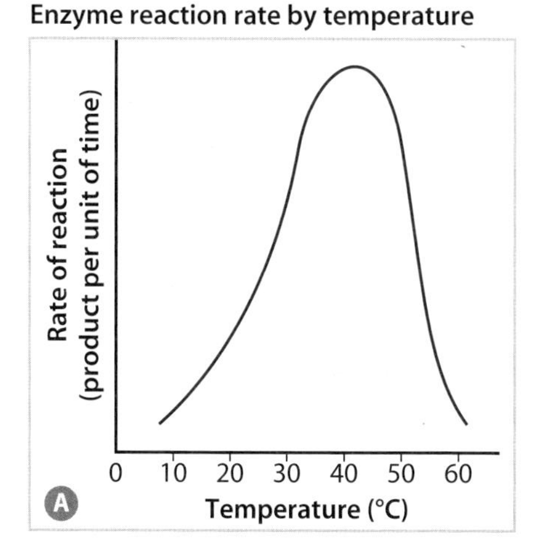

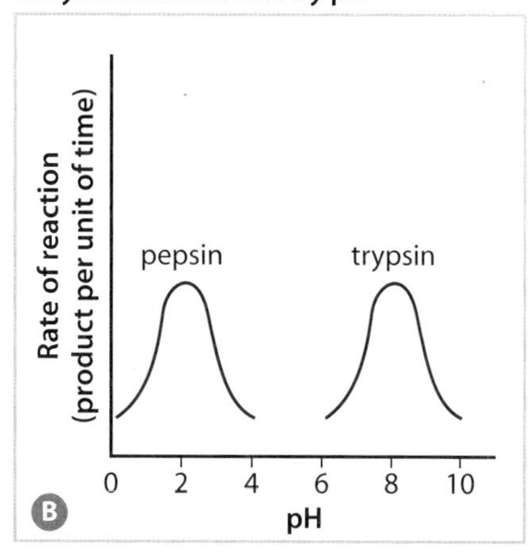

Figure 1.30 The activity of an enzyme is affected by (**A**) temperature and (**B**) pH. Most enzymes in humans, such as trypsin, which helps break down protein in the small intestine, work best at a temperature of about 40°C and a pH of between 6 and 8. Pepsin is an enzyme that acts in the stomach, an acidic environment, so its optimum pH is about 2.

***Predict** the activity of trypsin in the stomach.*

inhibitor a molecule that binds to the allosteric or active site of an enzyme and causes a decrease in the activity of that enzyme

allosteric site a site on an enzyme that is not the active site, where other molecules can interact with and regulate the activity of the enzyme

activator a molecule that binds to the allosteric site of an enzyme and keeps an enzyme active or causes an increase in the activity of that enzyme

Enzyme Activity Is Regulated by Other Molecules

Inhibitors are molecules that interact with an enzyme and reduce the activity of the enzyme. They do this by reducing an enzyme's ability to interact with its substrate. As shown in **Figure 1.31**, this interference in substrate binding can occur by two different mechanisms: *competitive inhibition* and *non-competitive inhibition*. Competitive inhibitors interact with the active site of the enzyme. When both the substrate and an inhibitor are present, the two compete to occupy the active site. When the inhibitor is present in high enough concentration, it will out-compete the substrate for the active site. This blocks the substrate from binding, and the reaction that the enzyme normally catalyzes does not occur.

In addition to an active site, many enzymes have an allosteric site. Other molecules that bind to this site can alter the activity of the enzyme by altering the conformation or three-dimensional shape of the enzyme. Non-competitive inhibitors bind to an **allosteric site** on an enzyme. This causes the conformation of the enzyme to change in such a way as to reduce its ability to interact with the substrate at its active site. As a result, there is a decrease in the activity of the enzyme. As you will learn in Unit 2, many biochemical reactions are grouped together in pathways. The product of one reaction acts as a substrate for the enzyme that catalyzes the next reaction in the pathway. A common way that biochemical pathways are regulated is by a process called feedback inhibition. In feedback inhibition, the product of the last reaction of the pathway is a non-competitive inhibitor of the enzyme that catalyzes a reaction at the beginning of the pathway. This form of regulation is a way that the cell has for ensuring that products of a pathway are not produced unnecessarily. When enough product is available, its synthesis and all the reactions related to its synthesis are turned off or reduced.

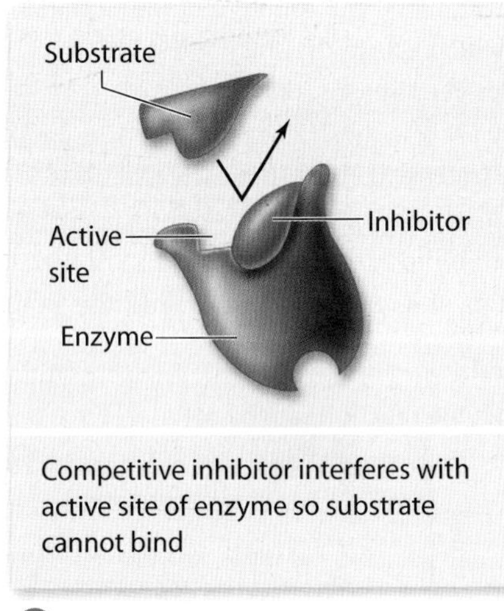

A Competitive inhibition

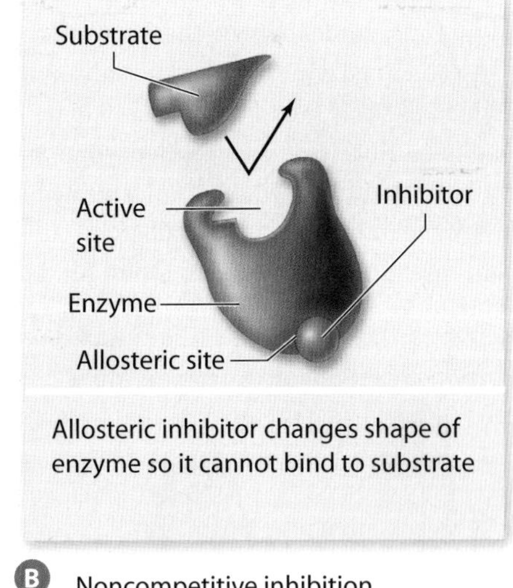

B Noncompetitive inhibition

Figure 1.31 In competitive inhibition, a substance binds to the active site of an enzyme, preventing substrates from binding. In non-competitive inhibition, such as allosteric inhibition, an inhibitor prevents the enzyme from working, but the inhibitor does not bind to the active site.

Suggested Investigation

Plan Your Own Investigation 1-B Investigating Factors Affecting Enzyme Activity

Activator molecules can also bind to an allosteric site. In this case, the conformation of the enzyme alters in such a way as to cause an increase in enzyme activity. The regulation of enzyme activity by activators and inhibitors binding to allosteric sites is called *allosteric regulation*.

STSE

BIOLOGY Connections

Fabry Disease: A Serious Result of Enzyme Deficiency

Fabry disease, a rare condition that affects one in every 40 000 men and one in every 100 000 women, occurs today in about 250 Canadians. It is caused by a deficiency in a lysosomal enzyme called α-galactosidase A. Alpha-galactosidase A is one of many types of enzymes in the lysosomes of cells. These enzymes act on proteins, fats, nucleic acids, and sugars to either break down or transform them. Without the action of lysosomal enzymes, these substances can build up in cells, leading to toxic effects.

Normally, α-galactosidase A will break down its substrate, globotriaosylceramide (Gb3), through a hydrolysis reaction. Gb3 is a glycolipid—a type of lipid molecule with a carbohydrate tail attached. In Fabry disease, a defect in the gene that codes for α-galactosidase A prevents the gene product from folding properly. This prevents the enzyme from carrying out its usual function—breaking down Gb3. The resulting build-up of Gb3 in cells causes widespread damage to tissues and organs. Symptoms of Fabry disease include skin spots, fatigue, and pain, as well as heart problems, kidney damage, and strokes.

ENZYME REPLACEMENT THERAPY The idea of replacing the α-galactosidase A activity that patients with Fabry disease are lacking gave rise to Enzyme Replacement Therapy (ERT) for the disease. This therapy is the result of recombinant DNA technology, in which copies of the human enzyme are produced by inserting DNA into cells cultured *in vitro*. The recombinant α-galactosidase A enzyme is then injected into patients to help replenish the enzyme levels. Although the replacement therapy is not a cure, the treatment can give patients some relief from symptoms.

Canadians advocating for a national orphan disease drug plan demonstrate outside a 2005 meeting held by federal and provincial/territorial heath ministers.

LACK OF FUNDING FOR ORPHAN DISEASES Controversy has accompanied ERT since its approval in 2004. ERT is extremely expensive (about $300 000 per year per person). Because Fabry disease is so rare, scientists are having a hard time enrolling enough people in studies to draw strong conclusions about how well the treatment works. Some patients were given a supply of ERT in manufacturer-run clinical trials and on compassionate grounds, but on a temporary basis only. When access was halted in 2005, patients staged a two-day rally calling for government funding for ongoing treatment. The federal government teamed up with manufacturers and the provincial/territorial governments in 2007 to co-sponsor a $100 million, three-year clinical trial that delivered treatment to patients. But when funding ended, patients' access to treatment was again at risk. Although Ontario has since led negotiations to continue funding for treatment through the drug trial, rare-disease advocacy groups are seeking a more permanent, long-term solution for this and other *orphan diseases*. Like orphans who lack parental support, these diseases do not receive sufficient support in terms of research, funding, and awareness, due to the low incidence of these diseases in the population. Advocacy groups argue that a national orphan disease medication program should be established.

Connect to Society

1. What policies govern the coverage of orphan disease treatments in other countries such as the United States, Australia, and Japan? What is the status of the development of a national policy in Canada? Who are stakeholders in advocating for and implementing such a policy in Canada, and what is the Ontario government's role? List the advantages and disadvantages of implementing a national orphan drug policy from the point of view of each stakeholder.
2. Enzyme replacement therapy is one of many examples of new technological applications for enzymes in medicine. Research the use of enzymes as treatments for other medical conditions. How and why are enzymes used to treat these conditions?

Section 1.3 Review

Section Summary

- A neutralization reaction is a reaction between an acid and a base, producing water and a salt as products. Buffers in biological systems ensure that changes in pH are minimized.
- In a redox reaction, electrons are transferred between substances.
- In a condensation reaction, two or more molecules are linked together and a water molecule is released per bond formed. In a hydrolysis reaction, a large molecule such as a polymer is broken down into smaller molecules, as water is added to bonds between monomers.
- An enzyme is a protein that catalyzes a chemical reaction, enabling it to proceed faster than it otherwise would do. The basic mechanism of action involves a substrate that fits into the active site of the enzyme to form an enzyme-substrate complex.
- Many enzymes have an allosteric site, where inhibitors or activators may bind to regulate enzyme activity. Enzyme activity is also affected by factors such as temperature, pH, and concentration of substrate.

Review Questions

1. K/U Identify each type of reaction.
   a. $2NaOH + H_2SO_4 \rightarrow Na_2SO_4 + 2H_2O$
   b. $C_6H_{12}O_6 + 6O_2 \rightarrow 6CO_2 + 6H_2O$
   c. glycine + glycine $\rightarrow$ dipeptide
   d. sucrose $\rightarrow$ glucose + fructose
2. K/U Record one more different example of each type of reaction that you identified in question 1.
3. C Prepare a sketch that shows how a non-competitive inhibitor affects enzyme activity. Include a short written description of how this type of inhibition works.
4. K/U Give one example of how each type of reaction is used in a biological system.
   a. condensation reaction
   b. hydrolysis reaction
   c. neutralization reaction
   d. redox reaction
5. K/U What is the importance of buffers in biological systems? Provide one detailed example.
6. A Predict the effects of consuming too many antacids on healthy gastric enzyme activity in the human digestive system.
7. C Prepare a sketch showing how a competitive inhibitor affects enzyme activity. Include a short written description of how this type of inhibition works.
8. C Use a Venn diagram to compare and contrast competitive and non-competitive inhibition.
9. K/U Describe how an enzyme acts to change the rate of a chemical reaction in a biological system. Include the term activation energy in your answer.
10. T/I A biochemist is given an unknown biological compound. The biochemist tests the compound and obtains the following results:
    - One molecule of the compound contains between 15 and 20 atoms.
    - The compound dissolves in water.
    - When the compound is dissolved in water, the solution has a pH of 5.
    - When the solution is heated, a reaction takes place that produces larger molecules.

    Based on these results, give one specific suggestion for what this biological compound might be. Explain your answer.
11. K/U An enzyme is isolated from a cell culture but is found to have lost its function. What factors would you examine as you try to restore the ability of the enzyme to catalyze reactions?
12. T/I A biologist heated an enzyme-catalyzed reaction slowly over 20 min at 1°C per minute, starting at 20°C. From 0 to 9 minutes, the rate of reaction increased from 3 mol/s to 7 mol/s. From 9 to 11 min, the rate of reaction dropped quickly to 0.4 mol/s. From 11 to 20 min, the rate of reaction gradually increased from 0.4 mol/s to 0.9 mol/s.
    a. Use the data provided in the description to construct a rough graph of rate of reaction versus temperature for the enzyme-catalyzed reaction.
    b. Analyze your graph, explaining each change in the rate. For example, why did the rate increase from 0 to 9 min? What happened at 9 min? What might have happened between 11 and 20 min?

Inquiry
INVESTIGATION

1-A

Skill Check

Initiating and Planning

✓ Performing and Recording

✓ Analyzing and Interpreting

✓ Communicating

Safety Precautions

- Iodine, Benedict's solution, and Biuret reagent are toxic and can stain skin and surfaces.
- Do not allow the hot water bath to boil vigorously, as test tubes can break.
- Clean up all spills immediately with plenty of water, and inform your teacher if a spill occurs.

Materials

- distilled water
- Biuret reagent
- albumin solution
- pepsin solution
- starch suspension
- iodine solution
- Benedict's solution
- glucose solution
- onion juice
- potato juice
- vegetable oil
- butter or margarine
- 4 food samples
- 7 small squares of brown paper
- 5 test tubes
- test tube rack
- millimetre ruler
- wax pencil
- hot plate
- large beaker (500 mL or larger)
- tongs
- mortar and pestle

Identifying Biological Macromolecules in Food

Biochemists have developed standard tests to determine the presence of the most abundant macromolecules made by cells: carbohydrates (sugars, starches), lipids (fats), and proteins. In this investigation, your group will conduct some of these standard tests to identify the presence of sugar, starch, lipid, and protein in known samples. You will then use the same tests to analyze different food samples for the presence of these macromolecules. The class test results of various foods will then be combined and analyzed.

Pre-Lab Questions

1. Why are you required to wear safety eyewear and protective clothing while conducting this investigation?
2. Why must the test tubes be thoroughly cleaned and dried when one part of the procedure is completed, before starting the next part of the procedure?
3. What colours indicate a positive test and a negative test for proteins, sugar, and starch for the indicators used in this investigation?

Question

What biochemical macromolecules are present in different food samples?

Procedure

In the first four parts of the procedure, you will learn how to perform different tests to identify protein, starch, sugar, and fat. Some of these tests involve the use of an indicator—a chemical that changes colour when it reacts with a specific substance. When performing these tests:

- add the indicator by holding the dropper bottle over the test tube and allow the drops to "free-fall" into the solution. Do not touch the inside wall of the test tube with the dropper bottle.
- give each tube a gentle shake after adding the indicator, in order to make sure that the solution is properly mixed. You can do this by gently tapping the side of the tube with your finger.

Part 1: Test for Proteins

Biuret reagent has a blue colour that changes to violet in the presence of proteins.

1. Use a millimetre ruler and a wax pencil to mark and label four clean test tubes at the 2 cm and 4 cm levels. Fill each test tube as follows:
    - Test tube 1: Fill to the 2 cm mark with distilled water, and then add Biuret reagent to the 4 cm mark.
    - Test tube 2: Fill to the 2 cm mark with albumin solution, and then add Biuret reagent to the 4 cm mark. (Albumin is a protein.)
    - Test tube 3: Fill to the 2 cm mark with pepsin solution, and then add Biuret reagent to the 4 cm mark. (Pepsin is an enzyme.)
    - Test tube 4: Fill to the 2 cm mark with starch suspension, and then add Biuret reagent to the 4 cm mark.

2. Be sure to tap or swirl each tube to mix the solutions. Record your results and conclusions in a table like the one that follows.

Biuret Test for Protein

Test Tube	Contents	Colour Change	Conclusions
1	distilled water		
2	albumin		
3	pepsin		
4	starch		

3. Dispose of the contents of the test tubes as directed by your teacher. Clean and dry the test tubes.

Part 2: Test for Starch

Iodine solution turns from a brownish colour to a blue-black in the presence of starch.

1. Use a millimetre ruler and a wax pencil to mark and label two clean test tubes at the 1 cm level. Fill each test tube as follows:
   - Test tube 1: Fill to the 1 cm mark with starch suspension, and then add five drops of iodine solution. (Be sure to shake the starch suspension well before taking your sample.)
   - Test tube 2: Fill to the 1 cm mark with distilled water, and then add five drops of iodine solution.
2. Note the final colour change. Record your results and conclusions in a table like the one that follows.

Iodine Test for Starch

Test tube	Contents	Colour Change	Conclusions
1	starch suspension		
2	distilled water		

3. Dispose of the contents of the test tubes as directed by your teacher. Clean and dry the test tubes.

Part 3: Test for Sugars

Sugars react with Benedict's solution after being heated in a boiling water bath. Increasing concentrations of sugar give a continuum of colours, as shown in the following table.

Typical Reactions for Benedict's Solution

Chemical	Chemical Category	Benedict's Solution (after heating)
distilled water	inorganic	blue (no change)
glucose	monosaccharide (carbohydrate)	varies with concentration: very low: green low: yellow moderate: yellow-orange high: orange very high: orange-red
maltose	disaccharide (carbohydrate)	varies with concentration (See results for glucose.)
starch	polysaccharide (carbohydrate)	blue (no change)

1. Use a millimetre ruler and a wax pencil to mark and label five clean test tubes at the 1 cm and 3 cm levels. Fill each test tube as follows:
   - Test tube 1: Fill to the 1 cm mark with distilled water, and then add Benedict's solution to the 3 cm mark.
   - Test tube 2: Fill to the 1 cm mark with glucose solution, and then add Benedict's solution to the 3 cm mark.
   - Test tube 3: Put a few drops of onion juice in the test tube. Fill to the 1 cm mark with distilled water, and then add Benedict's solution to the 3 cm mark.
   - Test tube 4: Put a few drops of potato juice in the test tube. Fill to the 1 cm mark with distilled water, and then add Benedict's solution to the 3 cm mark.
   - Test tube 5: Fill to the 1 cm mark with starch suspension; add Benedict's solution to the 3 cm mark.
2. Heat all five test tubes in a boiling water bath for about 5 min.
3. Note the final colour change. Record your results and conclusions in a table like the one that follows.

Benedict's Test for Sugars

Test Tube	Contents	Colour Change	Conclusions
1	distilled water		
2	glucose solution		
3	onion juice		
4	potato juice		
5	starch suspension		

4. Dispose of the contents of the test tubes as directed by your teacher. Clean and dry the test tubes.

Part 4: Test for Fats

Fats leave a translucent, oily spot on paper. Liquid fats penetrate paper, while solid fats rest predominantly on top.

1. Place a small drop of distilled water on a square of brown paper. Describe the immediate effect.
2. Place a small drop of vegetable oil on a square of brown paper. Describe the immediate effect.
3. Place a small quantity of butter or margarine on a square of brown paper. Describe the immediate effect.
4. Wait about 5 min. Examine each piece of paper to determine which test material penetrates the paper. Record your results and conclusions in a table like the one that follows.

Paper Test for Fats

Sample	Results
distilled water	
oil (liquid fat)	
butter or margarine (solid fat)	

Part 5: Testing for the Presence of Biochemical Macromolecules in Foods

1. Your teacher will give you four small samples of the first foodstuff to be tested.
2. Prepare solid samples for testing by mashing them with a mortar and pestle, and adding a few drops of distilled water.
3. Using the procedures followed in parts 1 to 4, test the sample for the presence of each of protein, starch, sugar, and fat.
4. Record your results and conclusions in a table. Suggested headings for the table are provided below.

Food Sample	Test	Results	Conclusion
Sample 1	protein		
	starch		
	sugar		
	fat		

5. Dispose of the contents of the test tubes as directed by your teacher. Clean and dry the test tubes.
6. Repeat steps 1 to 5 for each of the food samples.

Analyze and Interpret

1. Why was distilled water tested?
2. Describe any limitations to the usefulness of these tests for food samples.

Conclude and Communicate

3. Describe a positive test for:
   - **a.** protein
   - **b.** starch
   - **c.** sugars
   - **d.** fats (lipids)
4. The class tested a number of food samples. Did any of the results surprise you? Explain your answer.

Extend Further

5. **INQUIRY** Starch is digested into sugars by the enzyme amylase. Design a procedure that could be used to measure amylase activity.
6. **RESEARCH** In this investigation, you performed qualitative tests for the presence of certain biochemical substances. How are the relative quantities of these biochemical substances determined? Use the Internet or other sources to learn more about quantitative testing methods.

Plan Your Own
INVESTIGATION

1-B

Skill Check

- ✓ Initiating and Planning
- ✓ Performing and Recording
- ✓ Analyzing and Interpreting
- ✓ Communicating

Safety Precautions

- Wear gloves and safety eyewear throughout this investigation.
- Hydrochloric acid and sodium hydroxide are corrosive. Avoid any contact with skin, eyes, or clothes.
- Flush spills on skin immediately with copious amounts of cool water and inform your teacher.
- Exercise care when heating liquids and using a hot plate.
- Take care when using sharp instruments.
- Do not taste any substances in the laboratory.

Suggested Materials

- 0.1 mol/L hydrochloric acid, HCl
- 0.1 mol/L sodium hydroxide, NaOH
- 3% hydrogen peroxide, H_2O_2
- raw peeled potato
- ice
- test tubes, rack, and marker
- beaker for hot water bath
- hot plate
- beaker tongs
- medicine droppers
- distilled water
- scalpel or sharp knife
- ruler
- forceps
- clock or timer
- thermometer or probe
- pH indicator paper or probe

Investigating Factors Affecting Enzyme Activity

The compound hydrogen peroxide, H_2O_2, is a by-product of metabolic reactions in most living organisms. However, hydrogen peroxide is damaging to molecules inside cells. As a result, nearly all organisms produce the enzyme peroxidase, which breaks down hydrogen peroxide as it is formed. Potatoes are one source of peroxidase. Peroxidase facilitates the breakdown of hydrogen peroxide into water and gaseous oxygen. This reaction can be detected by observing the oxygen bubbles generated.

In this investigation, you will design a procedure and, with your teacher's approval, carry out experiments to test the effects of temperature and pH on peroxidase activity.

Pre-Lab Questions

1. What is an enzyme?
2. What is the substrate of the enzyme peroxidase? How do you know?
3. What is the effect of adding sodium hydroxide to an aqueous solution? What is the effect of adding hydrochloric acid?

Question

How effectively does the enzyme peroxidase work at different temperatures and pH values?

Hypothesis

Make a hypothesis about how you think temperature and pH will affect the rate at which the enzyme peroxidase breaks down hydrogen peroxide. Consider both low and high temperatures and pH values.

Plan and Conduct

1. Examine the materials provided by your teacher. As a group, list ways you might test your hypothesis.
2. Agree on a method(s) your group will use to test your hypothesis.
3. Your experimental design should include a control and test one variable at a time. Will you be collecting quantitative or qualitative data?
4. What will be your independent variable? What will be your dependent variable? How will you set up your control?
5. How will you determine peroxidase activity? How will you measure the amount of oxygen produced? Have you designed a table for collecting data?
6. Will you conduct more than one trial? How long will you allow each trial to run?
7. How will you analyze your data?
8. Write a numbered procedure for your investigation that lists each step, and prepare a list of materials that includes the amounts you will require. Before beginning the invesigation, have your teacher check and approve your plan.

Analyze and Interpret

9. Draw graphs showing the relationship between temperature and oxygen produced, and between pH and oxygen produced.

Conclude and Communicate

10. What did this investigation indicate about the activity of peroxidase?

11. At what temperature did peroxidase work best? at what pH?

12. What was the purpose of using control samples?

13. Do your data support or disprove your hypothesis? Explain.

14. Infer how the presence of hydrochloric acid affects the activity of peroxidase.

15. If you have ever used hydrogen peroxide as an antiseptic to treat a cut or scrape, you know that it foams as soon as it touches an open wound. Account for this observation.

Extend Further

16. INQUIRY Design an investigation in which you use hydrogen peroxide to test for the presence of peroxidase in other foods, such as pieces of other vegetables or meat. Determine which food shows the greatest peroxidase activity. Explain the differences in enzyme activity among different foods. With permission from your teacher and with suitable safety precautions in place, carry out your investigation.

17. RESEARCH The human body contains peroxisomes, which are microscopic organelles in cells that contain enzymes to detoxify substances, such as hydrogen peroxide. Research disorders that affect peroxisomes and the symptoms these disorders cause. What are some advances in cellular biology that are being used to treat these disorders?

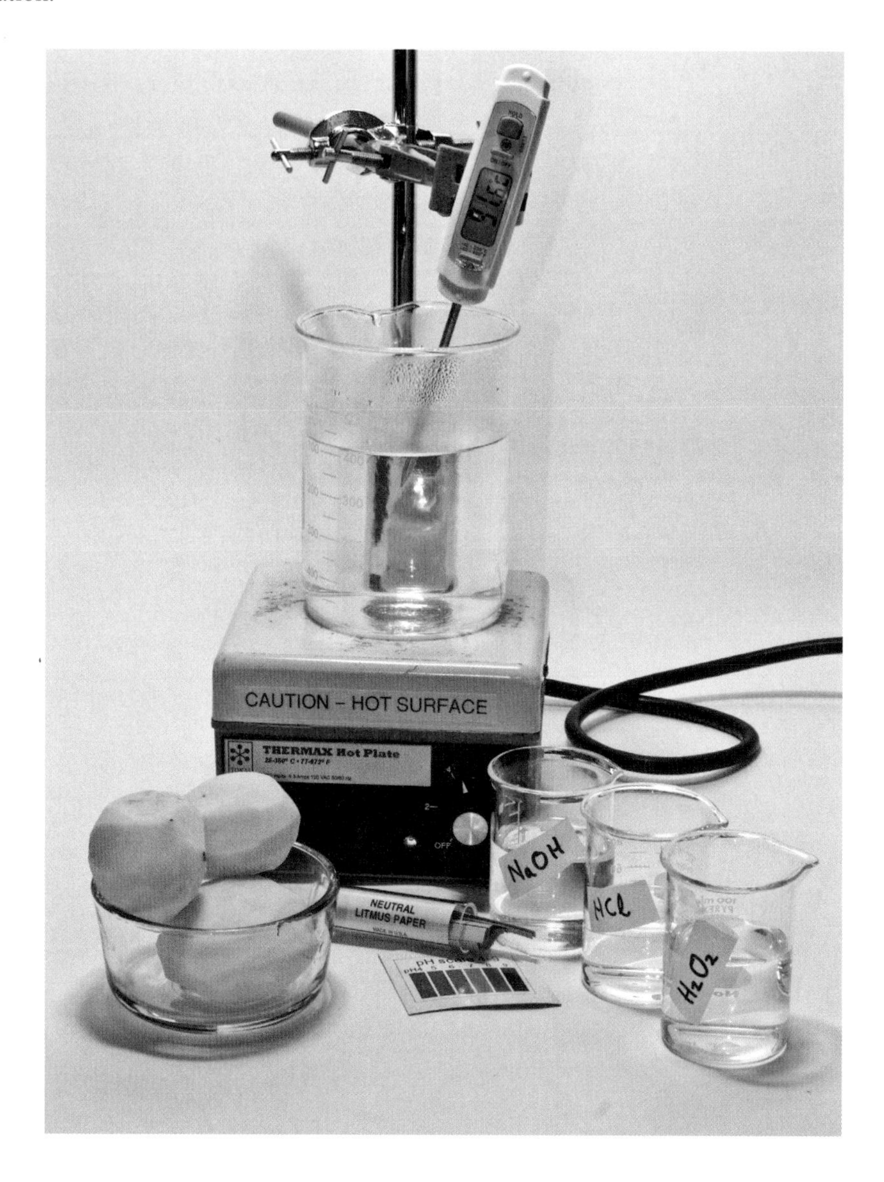

Chapter 1 SUMMARY

Section 1.1 Chemistry in Living Systems

The properties and interactions of biologically important molecules determine their function in living systems.

Key Terms

isotope
radioisotope
molecule
organic molecule
biochemistry
intramolecular
intermolecular
hydrogen bond
hydrophobic
hydrophilic
ion
functional group

Key Concepts

- Organisms are composed primarily of the chemical elements carbon, hydrogen, oxygen, nitrogen, phosphorus, and sulfur.
- Intramolecular interactions occur between atoms within a molecule, forming chemical bonds.
- The polarity of a molecule influences the intermolecular interactions that occur between molecules. Two important types of intermolecular interactions are hydrogen bonding and hydrophobic interactions.
- The functional group or groups on a molecule determine the properties of the molecule. Important functional groups in biological molecules include hydroxyl, carboxyl, carbonyl, amino, sulfhydryl, and phosphate groups.
- The structures of molecules can be represented using a variety of formats. Structural formulas are two-dimensional representations that indicate how the atoms are bonded together. Space-filling models are a common way to represent the three-dimensional structures of molecules.

Section 1.2 Biologically Important Molecules

Carbohydrates, lipids, proteins, and nucleic acids (for example, DNA and RNA) are important macromolecules that are a part of living systems.

Key Terms

macromolecule
polymer
monomer
carbohydrate
monosaccharide
isomer
disaccharide
polysaccharide
lipid
triglyceride
fatty acid
phospholipid
lipid bilayer
steroid
wax
protein
amino acid
polypeptide
nucleic acid
DNA (deoxyribonucleic acid)
RNA (ribonucleic acid)
nucleotide

Key Concepts

- Carbohydrates can act as short-term energy-storage and structural biological molecules. They include polysaccharides, and disaccharides and monosaccharides (sugars). Polysaccharides are macromolecules composed of monosaccharide monomers.
- Lipids are longer-term energy storage and structural biological molecules that contain a higher proportion of non-polar C–C and C–H bonds than carbohydrates. Triglycerides, phospholipids, steroids, and waxes are different types of lipids.
- Proteins are macromolecules composed of amino acid monomers. Proteins have primary, secondary, tertiary, and sometimes quaternary structures, which are essential for their function. Proteins enable chemical reactions in living systems, and also perform a wide variety of structural and regulatory roles.
- Nucleic acids are macromolecules composed of nucleotide monomers. DNA and RNA are both nucleic acids. DNA stores the hereditary information of a cell. RNA plays a central role in the synthesis of proteins.

Chapter 1 REVIEW

Section 1.3 Biochemical Reactions

Chemical reactions that take place in living systems include neutralization, oxidation, reduction, condensation, and hydrolysis reactions, many of which may be catalyzed by enzymes.

Key Terms

acid	activation energy
base	catalyst
pH scale	enzyme
neutralization reaction	active site
buffer	substrate
oxidation	enzyme-substrate complex
reduction	inhibitor
redox reaction	allosteric site
condensation reaction	activator
hydrolysis reaction	

Key Concepts

- A neutralization reaction is a reaction between an acid and a base, producing water and a salt as products. Buffers in biological systems ensure that changes in pH are minimized.
- In a redox reaction, electrons are transferred between substances.
- In a condensation reaction, two or more molecules are linked together and a water molecule is released per bond formed. In a hydrolysis reaction, a large molecule such as a polymer is broken down into smaller molecules, as water is added to bonds between monomers.
- An enzyme is a protein that catalyzes a chemical reaction, enabling it to proceed faster than it otherwise would do. The basic mechanism of action involves a substrate that fits into the active site of the enzyme to form an enzyme-substrate complex.
- Many enzymes have an allosteric site, where inhibitors or activators may bind to regulate enzyme activity. Enzyme activity is also affected by factors such as temperature, pH, and concentration of substrate.

Knowledge and Understanding

Select the letter of the best answer below.

1. How is a polymer formed from multiple monomers?
   a. from the growth of the chain of carbon atoms
   b. by the removal of an –OH group and a hydrogen atom
   c. by the addition of an –OH group and a hydrogen atom
   d. through hydrogen bonding
   e. through intermolecular forces

2. Which carbohydrate would you find as part of a molecule of RNA?
   a. galactose
   b. deoxyribose
   c. ribose
   d. glucose
   e. amylase

3. A triglyceride is a form of_____ composed of_____.
   a. lipid; fatty acids and glucose
   b. lipid; fatty acids and glycerol
   c. lipid; cholesterol
   d. carbohydrate; fatty acids
   e. lipid; fatty acids and glycerase

4. What makes starch different from cellulose?
   a. Starch is produced by plant cells, and cellulose is produced by animal cells.
   b. Cellulose forms long filaments, while starch is highly branched.
   c. Starch is insoluble, and cellulose is soluble.
   d. All of the above.
   e. None of the above.

5. Which of the following does not belong?
   a. methane
   b. cyclohexane
   c. propanol
   d. sodium chloride
   e. water

6. Cholesterol is a precursor for which of the following?
   a. testosterone
   b. fatty acids
   c. trans fat
   d. phospholipids
   e. All of the above are synthesized from cholesterol.

7. Which of the following is not primarily composed of protein?
   a. hair
   b. urine
   c. nails
   d. enzymes
   e. collagen

8. Identify which of the following reactions would result in the production of water.
   a. hydrolysis reaction
   b. formation of a disulfide bond
   c. formation of a peptide bond
   d. breaking a disaccharide
   e. All of the above will produce water.
9. A polypeptide chain is held together by
   a. hydrolysis bonds
   b. glucose linkages
   c. phosphate linkages
   d. peptide bonds
   e. disulfide bridges
10. Which of the following factors do not impact enzyme activity?
    a. changes in temperature
    b. changes in substrate concentration
    c. changes in pH
    d. changes in ionic concentrations
    e. All of the above impact enzyme activity.
11. Allosteric sites are responsible for
    a. regulating enzyme activity
    b. denaturing an enzyme
    c. increasing the substrate concentration
    d. decreasing the substrate concentration
    e. forming hydrogen bonds
12. Which statement regarding enzymes is false?
    a. Enzymes speed up reaction rate.
    b. Enzymes lower the activation energy of a reaction.
    c. Enzyme reaction rate increases with temperature up to a certain point.
    d. Enzyme reaction rate decreases as the concentration of substrate increases.
    e. All of the above are true regarding enzymes.
13. A reaction produces water as a product. Which statement regarding the reaction is false?
    a. A condensation reaction may have occurred.
    b. A dehydration reaction may have occurred.
    c. A neutralization reaction may have occurred.
    d. A hydrolysis reaction may have occurred.
    e. All of the above are true.
14. Which part of an amino acid has the greatest influence on the overall structure of a protein?
    a. the $-NH_2$ amino group
    b. the R group
    c. the –COOH carboxyl group
    d. Both a and c
    e. None of the above.

Answer the questions below.

15. What are two main carbon-containing molecules in organisms?
16. Identify the property of water that allows for hydrogen bonding.
17. Summarize the steps in protein folding.
18. How would the function of an enzyme be affected if human body temperature increased?
19. Identify the differences between a condensation reaction and a hydrolysis reaction.
20. What is the role of an allosteric inhibitor?
21. How does a buffer assist in cellular function?
22. Would enzyme denaturation occur at the primary, secondary, or tertiary structure? Explain.
23. What is the role of an allosteric site in enzyme activity?
24. Differentiate between coenzymes and cofactors.
25. List two changes that could cause an enzyme to denature.
26. What is the difference between an organic molecule and an inorganic molecule? Provide two examples of each.
27. What happens to enzyme activity when slight deviations in pH or temperature are introduced?
28. Identify the monomer for each type of polymer as well as the general type of reaction required to make the polymer.
    a. protein
    b. glycogen
    c. RNA
29. Differentiate between an isotope and a radioactive isotope.
30. In three sentences or fewer, describe how the four main types of biological macromolecules differ and what they have in common.

Thinking and Investigation

31. Hypothesize what would happen to an enzyme if a substrate covalently bonded with the active site.
32. Infer what happens to the function of amylase when it is digested.
33. Genetic mutations that result in an improvement in enzyme function are rare. Draw a conclusion as to why most changes are devastating to an enzyme.

production H_2O (dehydration)

34. Enteric coatings are applied to medication to prevent stomach acid from breaking down the medication prior to delivery. Cellulose, which is indigestible by humans, was initially investigated as a coating material before being modified.

a. Although enzymes do not exist in humans to digest cellulose, a pharmacologist discovers that some medication is entering circulation. Propose a mechanism that explains this.

b. Currently, cellulose is modified with succinate, a naturally occurring compound produced in the metabolism of sugar. Infer why this method is a highly reliable method of releasing medication in the small intestine.

35. Use your understanding of the properties of intermolecular and intramolecular forces to explain the fact that water rises in narrow tubes, against the force of gravity, as shown in the diagram below.

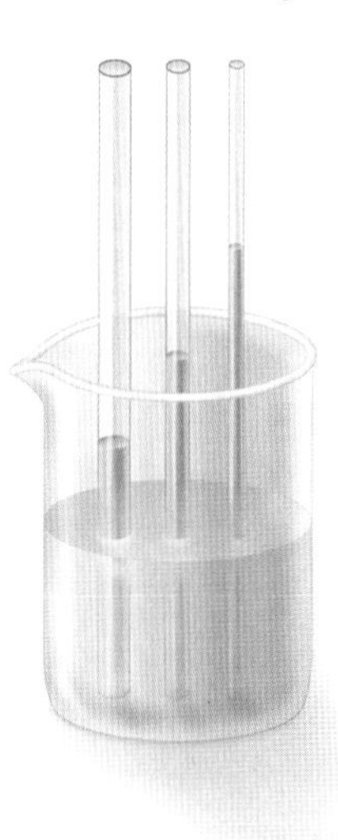

36. In the mid-1990s, bread machines became very popular as they provided optimal conditions (such as a specific temperature) for yeast to produce carbon dioxide, causing the dough to rise.

a. Explain why a specific temperature is optimal for this process.

b. Outline steps you would take to determine the optimal temperature for yeast activity using an apparatus found in the kitchen.

c. What variables would you hold constant to ensure accurate testing?

d. How could you measure carbon dioxide production accurately?

37. A vegetarian has become lactose intolerant but wishes to continue consuming dairy products as a source of protein and nutrients. He plans to take a minimum amount of a lactase digestive supplement to avoid excess exposure to pharmaceuticals, yet still enjoy dairy products. To determine the minimum amount, he breaks each tablet into four pieces each. On the first morning, he takes eight pieces—equivalent to the recommended dosage. Each day he consumes one fewer piece until his lactose intolerance is felt. After evaluating his procedure, make three suggestions that would ensure the most accurate result possible.

38. Artificial enzymes are roughly 1000 times less efficient than naturally occurring enzymes. Make two inferences that could explain this difference.

39. The production of corn syrup from starch involves the hydrolysis of the starch into simple sugars.

a. What are the drawbacks to using acid to catalyze this reaction?

b. Suggest an alternative that would be faster and safer.

40. Manpreet conducts an investigation to determine the rate of enzyme activity. She places a known amount of starch in a carefully measured amylase solution and times the reaction. After one hour she measures the concentration of glucose. Explain the flaw in her procedure that limits the usefulness of her conclusions.

41. Detection of HIV occurs through a process called enzyme-linked immunosorbent assay, or ELISA. In this test, enzymes are linked to proteins in the immune system called antibodies. When the protein-based antibody comes in contact with the protein markers on the virus, the enzyme causes a visible colour change. Why must scientists pay special attention to pH, ionic concentrations, and temperatures during these assays?

42. A starch solution is placed into dialysis tubing and suspended in distilled water. The distilled water contains amylase, the salivary enzyme that breaks down starch, as well as Benedict's solution for detecting glucose. The mixture is heated for 10 minutes, but the colour remains red. Evaluate each of the following statements for accuracy.

a. The enzyme is inactive since the starch was not broken down.

b. The dialysis tubing must not be permeable to the enzyme.

c. Benedict's solution will not detect glucose when it is bound to the enzyme.

43. A biologist is investigating an hypothesis linking decreased brain function to fatty acid levels in the body. How can a baseline measurement of fat be obtained from cellular tissue? Use your knowledge of laboratory practices and common sense to suggest a strategy.

44. Salivary amylase can be collected by having an individual salivate into a test tube for several minutes. A group of three students each perform an investigation in which their own amylase sample is used to break down starch. Results are accurately measured. When complete, one student's sample still contains a large amount of starch and very little glucose.
 a. All three students conclude that this result must be caused by a slower enzyme via a less specific active site. Is this the only explanation? Explain.
 b. How could the accuracy of the investigation be improved?

45. A biologist heats a beaker containing a clear, colourless aqueous solution of an unknown biological substance. After a few hours, the solution begins to thicken. The biologist knows that either a hydrolysis or condensation reaction took place. Based on the observations available, which type of reaction might have occurred, and why?

46. Hydrogen bonds and hydrophobic interactions play important roles in stabilizing and organizing biological macromolecules.
 a. Describe how hydrogen bonds and hydrophobic interactions affect the form and function of proteins.
 b. Infer how a disruption in the hydrogen bonds of nucleic acids would affect form and function.
 c. Infer how a disruption in hydrophobic interactions would affect form and function in lipids.

47. To produce buttons, a piece of plastic is fastened to a steel backing using a high-pressure hand press. To create an analogy for enzyme activity, identify what represents the enzyme and what represents the substrates in this process. How is the "enzyme" here different from a biological enzyme?

48. Use your understanding of enzymes and protein folding to infer how having a fever helps fight bacterial infection.

Communication

49. Examine the diagram below.
 a. Name the chemical reaction that is taking place.
 b. Write a caption for this diagram that explains what is happening.
 c. Use the same style as is used in this diagram to draw the chemical reaction that is effectively the opposite of the one shown.

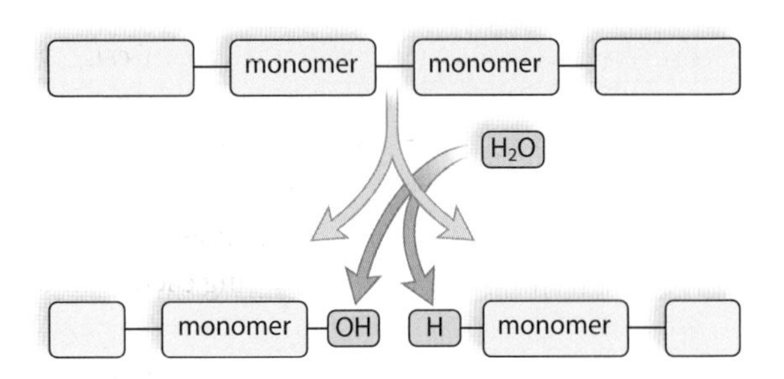

50. Before the structure of DNA was discovered, the proportion of adenine and thymine in DNA was always found to be identical. The same was not true for adenine and uracil in RNA. Based on this, infer the structural differences between DNA and RNA and draw a diagram to show the differences.

51. Lipids, carbohydrates, and proteins all contain different amounts of energy per unit mass. Propose a procedure that would qualitatively show the energy differences among these macromolecules.

52. Construct a graph illustrating the concentrations of reactants and products in an enzyme-catalyzed reaction. How does your graph compare to a non-catalyzed reaction?

53. Draw a graph to show the trend of the data you would expect when comparing enzyme activity and substrate concentration.

54. List the various foods you have eaten today. Use an asterisk to indicate which food has the best representation of all the macromolecules discussed in this chapter.

55. In a table or other graphic organizer, differentiate between RNA and DNA.

56. Make a labelled sketch to show the features of hydroxyl, –OH, that make it a polar functional group.

57. Explain in an email to a classmate why it would be difficult to find an enzyme that functions at a very low pH.

58. Construct a Venn diagram to compare and contrast condensation and hydrolysis reactions.

59. A healthy diet contains more complex carbohydrates than simple sugars. Based on what you have learned about monosaccharides and polysaccharides, draw a simple diagram to illustrate the difference between simple sugars and complex carbohydrates.

60. Explain clearly, with the assistance of a diagram, whether oxidation can occur without the presence of oxygen.

61. Sketch a representation of the condensation and hydrolysis of the molecules listed below. Design symbols to represent the molecules involved. The symbols will be used for a class of grade 7 students, who do not have any understanding of the chemistry concepts related to structural and molecular formulas. Explain your reasoning for designing each symbol as you did.
   - **a.** a disaccharide from two molecules of glucose
   - **b.** a triglyceride from one molecule of glycerol and three fatty acid molecules
   - **c.** a dipeptide from two amino acid molecules

62. Clearly explain how an enzyme can take part in a reaction that involves sucrose but not in a reaction that involves maltose.

63. Summarize your learning in this chapter using a graphic organizer. To help you, the Chapter 1 Summary lists the Key Terms and Key Concepts. Refer to Using Graphic Organizers in Appendix A to help you decide which graphic organizer to use.

Application

64. BIG IDEAS Technological applications that affect biological processes and cellular functions are used in the food, pharmaceutical, and medical industries. Fluorine is often added to medication to prevent the body from breaking the medication down quickly. Since F is highly electronegative, what factor may be responsible for the slow degradation?

65. Would you expect the following amino acid to be soluble in water? After copying the figure into your notebook, clearly label the regions of the molecule that support your solubility decision. These areas should be referenced in your explanation.

H
H_3N^+ — C — C (=O)(—O^-)
H

66. BIG IDEAS Biological molecules and their chemical properties affect cellular processes and biochemical reactions. Arctic fish produce large amounts of antifreeze proteins during cold winter months. These proteins prevent water from forming large ice crystals and damaging soft tissue. Would you expect these proteins to be hydrophobic or hydrophilic? Explain your answer with reference to the role of the proteins.

67. Heat shock proteins are produced when cells are subjected to elevated temperatures. These proteins minimize the need for new enzyme synthesis when temperature conditions return to normal. What do you think is the function of heat shock proteins with respect to other proteins in the cell?

68. Why is an understanding of hydrogen bonding essential to the study of biochemistry?

69. Mucus in the nasal passage is produced by cells that line the airways. Mucus is composed of glycoprotein (protein with carbohydrate side chains) and water. It ensures that airways are kept moist and that drainage occurs regularly. When mucus fails to drain, it often becomes the site of an infection requiring antibiotics.
   - **a.** Where do the bacteria come from that lead to the infection?
   - **b.** Why is a non-draining sinus an ideal location for a bacterial infection?
   - **c.** Under normal circumstances, how would the bacteria be handled by the body?

70. The production of high-fructose corn syrup relies on enzymes for the conversion of starch to simple sugars. What features of the active site would be characteristic of this enzyme?

71. Use your knowledge of polarity to explain why methane, CH_4, has a low boiling point and is a gas at room temperature.

72. Enzymes intended for use in laboratory work are always shipped in a buffer solution. Explain why.

73. Use what you know about the polarities and charge distributions of water and biological molecules to explain how you think the interactions of biological molecules would be different if the liquid of biological systems was a non-polar molecule oil instead of water.

74. In artificial DNA replication, a crucial step involves the heating of DNA to near boiling temperatures prior to enzyme-based replication. What happens to the DNA when it is heated?

Chapter 1 SELF-ASSESSMENT

Select the letter of the best answer below.

1. K/U Which of these provides the most accurate information about the shape of a molecule?

a. electron model
b. structural formula
c. molecular formula
d. space-filling model
e. compound formula

2. K/U What feature could help distinguish between two carbohydrates?

a. ratio of C:H:O
b. boiling point
c. mass
d. internal energy
e. None of the above.

Use the diagram below to answer questions 3 - 5.

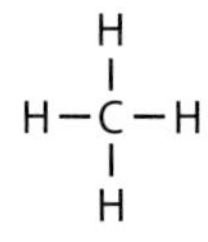

3. K/U The molecule shown is best described as:

a. polar
b. charged
c. ionic
d. hydrophilic
e. non-polar

4. K/U In which part of the molecule is the stored energy that is most important in chemical reactions?

a. in the nucleus of each atom
b. in the covalent bonds
c. in the ionic bond
d. in the electrons
e. in the polar regions

5. K/U Which term best describes the structure of the molecule?

a. linear
b. cross-shaped
c. pyramidal
d. trapezoidal
e. tetrahedral

6. K/U Condensation reactions are known for:

i producing acid
ii producing water
iii occurring in acid
iv consuming water
v attaching two amino acids

a. i and iii
b. iii and iv
c. iii, iv, v
d. ii, v
e. i, ii, iv, v

7. K/U The shape and charge distribution of a protein are important because

a. the shape changes over time as the protein ages, giving it new properties
b. these properties govern the function of each protein
c. the charge and shape are identical for most proteins
d. proteins define life
e. proteins have no chemical bonds

8. K/U A carbon atom can make a maximum of ___ covalent bonds.

a. one
b. two
c. three
d. four
e. six

9. K/U The chemical elements that are most common in living systems are

a. C, S, Fe, H, N
b. C, Cl, Mg, Fe, H
c. C, H O, N, P
d. C, I, Na, K, O
e. C, H, O, F, P

10. K/U An example of a condensation reaction would be

a. building amino acids from proteins
b. building lipids from fatty acids
c. breaking down glucose into starch
d. breaking down starch into glucose
e. none of the above

Use sentences and diagrams as appropriate to answer the questions below.

11. K/U Why does water expand when it freezes?

12. C A classmate asks, "What happens when an allosteric inhibitor enters the active site of the enzyme?" What would you tell your classmate?

13. C Use a diagram to illustrate a peptide bond. Write a caption to go with your illustration, describing the formation of a peptide bond in words.

14. K/U Write a word equation for a general neutralization reaction.

15. K/U Explain what a functional group is, and state two characteristics that make functional groups important to living systems.

16. A Redox reactions occur in non-living systems as well as in living systems.

   a. State a general definition for a redox reaction.

   b. Demonstrate how the reaction between a sodium atom and a chlorine atom to form the ionic compound sodium chloride is a redox reaction.

17. A Infer why cattle chew the same grass for an extended period of time.

18. T/I Identify the following reactions as oxidation, reduction, redox, neutralization, condensation, or hydrolysis. Give reasons for your identification in each case. (You are not required to name any of the compounds in these reactions.)

   a. $CH_3O–H + HO–CH_3 \rightarrow CH_3–O–CH_3 + H_2O$

   b. $CH_3NH_2 + HO–H \rightarrow CH_3NH_3^+ + HO^-$

   c. $CH_4 + O_2 \rightarrow CH_2O + H_2O$

   d. $H_2 \rightarrow 2H^+ + 2e^-$

   e. $O_2 + 4H^+ + 4e^- \rightarrow 2H_2O$

   f. $2H_2 + O_2 \rightarrow 2H_2O$

   g. $C_6H_{12}O_6 + 6O_2 \rightarrow 6CO_2 + 6H_2O$

   h. $CH_3–O–CH_3 + H_2O \rightarrow CH_3O–H + HO–CH_3$

   i. $CH_3COO–H + CH_3NH_2 \rightarrow CH_3COO^- + CH_3NH_3^+$

   j. $Cl_2 + 2e^- \rightarrow 2Cl^-$

19. C Examine the structural diagram below.

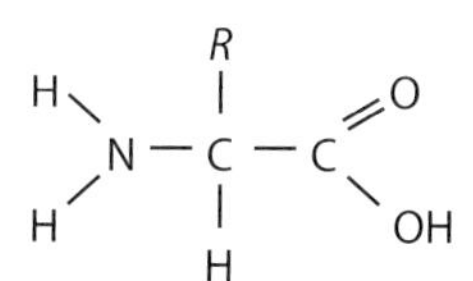

   a. Identify the type of molecule this is.

   b. Copy the structural diagram, and add labels to identify all five features that characterize this type of molecule.

   c. Identify the one feature that is significant for determining the distinctive shape and properties of polymers of this type of molecule.

   d. Explain, with the aid of a diagram, how two different molecules of this type may join together to form a polymer.

20. K/U Explain what activation energy is, and explain the role of enzymes in relation to activation energy and chemical reactions.

21. A The diagrams below show an enzyme called lysozyme that hydrolyzes its substrate, which is a polysaccharide that makes up the cell walls of bacteria. Write a detailed caption that could be used to accompany these two diagrams in a textbook.

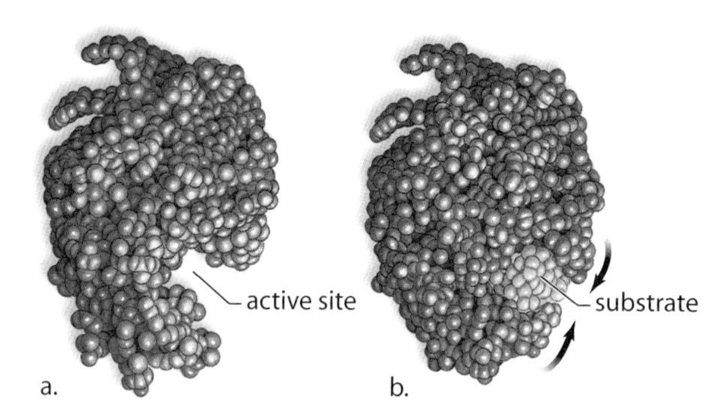

22. C Design a summary chart to record the type, structure, examples, and functions of the four types of macromolecules.

23. A What are the advantages of having a particular biochemical reaction such as the breakdown of a sugar proceed as a set of pathways, rather than as just a single reaction?

24. T/I It used to be common for textbooks to use an analogy of a lock and key to describe the mechanism by which substrates bind to the active site of an enzyme.

   a. Infer how this lock-and-key analogy works, and write a brief description of it.

   b. Name the model that is considered more accurate than the lock-and-key model for enzyme-substrate interaction, and explain why it is more accurate.

25. C Explain to a grade 6 school student the property of water that makes it an excellent transporter of nutrients.

Self-Check

If you missed question...	1	2	3	4	5	6	7	8	9	10	11	12	13	14	15	16	17	18	19	20	21	22	23	24	25
Review section(s)...	1.1	1.2	1.1	1.2	1.1	1.3	1.2	1.1	1.1	1.3	1.1	1.3	1.2	1.3	1.2	1.3	1.3	1.3	1.2 1.3	1.3	1.3	1.2	1.3	1.3	1.1

CHAPTER 2

The Cell and Its Components

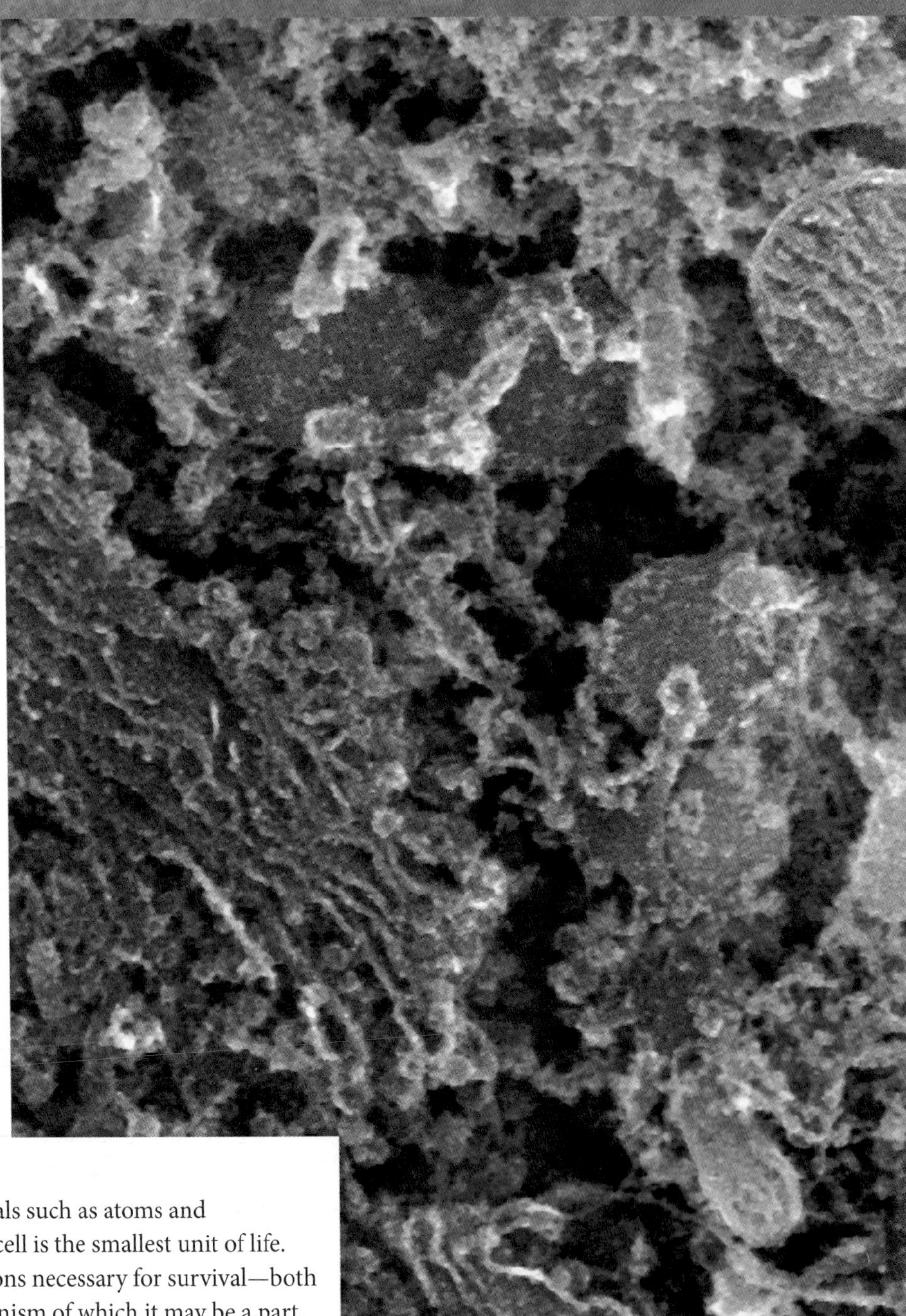

Specific Expectations

In this chapter, you will learn how to . . .

- B1.2 **evaluate**, on the basis of research, some advances in cellular biology and related technological applications (2.2)
- B2.1 **use** appropriate terminology related to biochemistry (2.1, 2.2)
- B2.2 **plan** and **conduct** an investigation to demonstrate the movement of substances across a membrane (2.2)
- B3.1 **explain** the roles of various organelles, such as lysosomes, vacuoles, mitochondria, internal cell membranes, ribosomes, smooth and rough endoplasmic reticulum, and Golgi bodies, in cellular processes (2.1)
- B3.6 **describe** the structure of cell membranes according to the fluid mosaic model, and **explain** the dynamics of passive transport, facilitated diffusion, and the movement of large particles across the cell membrane by the processes of endocytosis and exocytosis (2.2)

The cell is composed of non-living materials such as atoms and molecules, but it is itself a living entity. The cell is the smallest unit of life. Yet it is capable of performing all the functions necessary for survival—both for itself and for an entire multicellular organism of which it may be a part. Scientists have studied the cell with great curiosity and intense interest since its discovery in the 1660s. Modern developments in microscopy now provide the ability to examine the structure and components of cells in far greater detail than ever before. Detailed images of the interior of the cell, such as the one shown here, have provided insight into the inner workings of the cell and have in turn led to the development of new technologies that combat illness and disease, prolong life, and improve its quality.

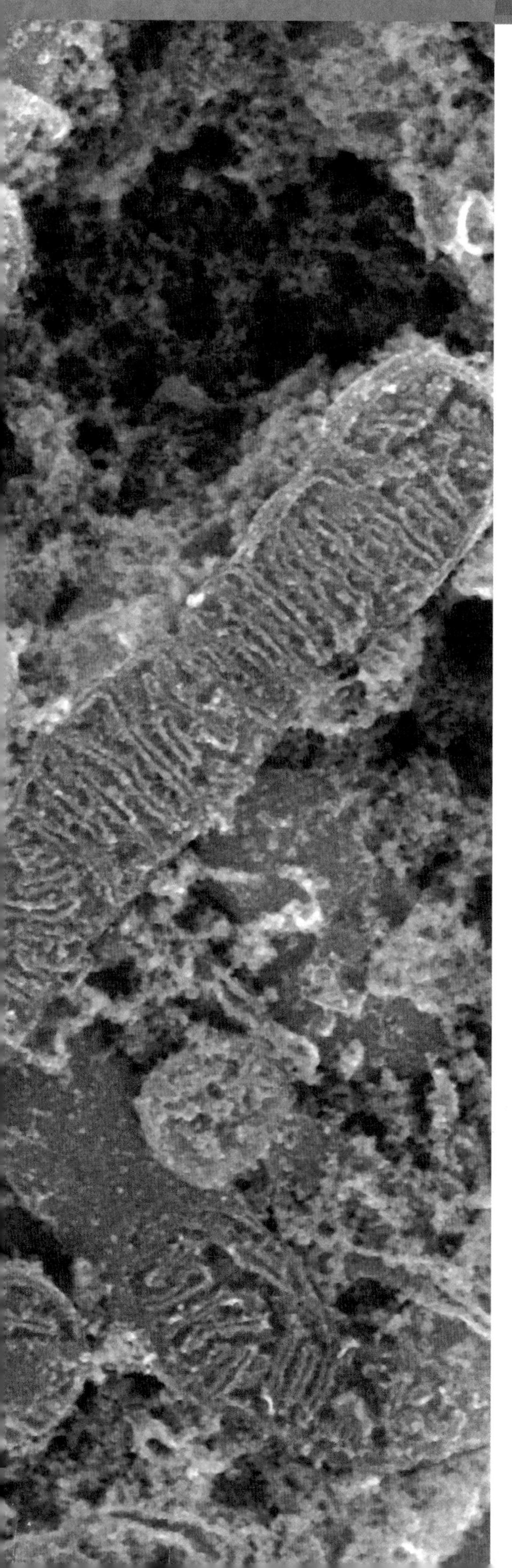

Launch Activity

Look a Little Closer

The bacterium *Staphylococcus aureus* is microscopic, but its effects on the body can be deadly. *S. aureus* can cause food poisoning, boils, rashes, blood infections, and kidney failure, among other problems. In addition, some strains of *S. aureus* are resistant to antibiotics that would normally damage the bacterial cell wall. How can microscopy be used in the fight against this bacterium? What can you learn from different microscope images?

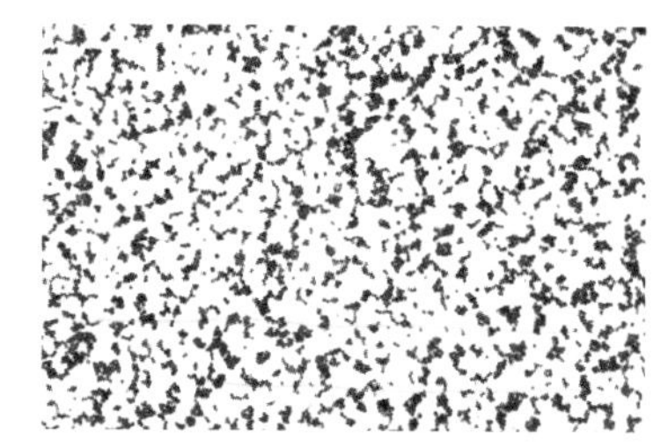

Staphyloccous aureus as viewed under a light microscope.

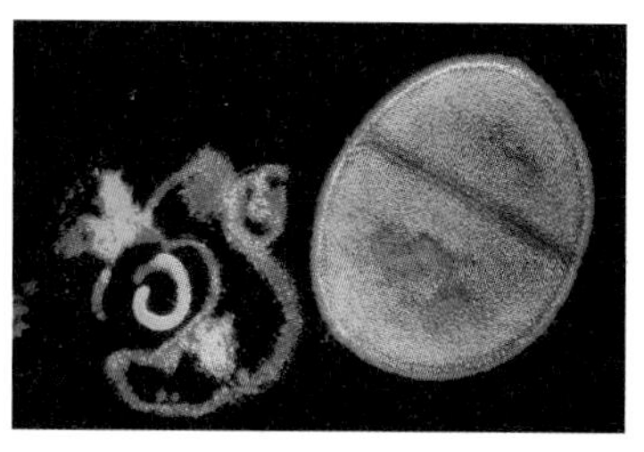

Staphyloccous aureus as viewed under a transmission electron micrograph

The microscope image on the left was made with a light microscope. A technique called a Gram stain was used to stain the bacterial cells. *S. aureus* is described as Gram positive because it stains purple with the Gram stain. The microscope image on the right was made with a transmission electron microscope (TEM). The TEM directs a beam of electrons through thin slices of the specimen in order to produce an image. Colour can be added to the image later, as shown here.

Procedure

1. Your group will be divided into two teams. The clinical microbiology team will examine the light microscope image. The research lab team will examine the TEM image.
2. With your team, examine your assigned microscope image and make a list of all the details you can observe. For example, note shapes, arrangements of shapes, colours, and level of detail.
3. Take turns with the other team to present your findings to one another.
4. As a group, re-examine and compare the microscope images and list some information that each type of microscopy *cannot* provide.

Questions

1. One of the goals of a clinical microbiologist is to find out which bacterial species is making a person ill. How might a light microscope image help a clinical microbiologist to identify a bacterial species?
2. Research microbiologists have the potential to help people around the world by studying disease-causing bacteria. How might a TEM image help researchers to study the effects of an antibiotic on bacteria?
3. Is one type of microscopy more useful than another? Justify your response.

SECTION 2.1

Structures and Functions of Eukaryotic Cells

Key Terms

nucleolus
nuclear envelope
nuclear pore complexes
endoplasmic reticulum (ER)
ribosome
endomembrane system
vesicle
Golgi apparatus
lysosome
peroxisome
vacuole
chloroplast
mitochondrion
cell wall
cytoskeleton
fluid mosaic model

Animals, plants, fungi, and protists are composed of eukaryotic cells. Cellular organization varies among different organisms, but all eukaryotic cells have these features in common.

- The genetic material—DNA—is contained within a membrane-bound nucleus.
- A *cell membrane* comprised of a *phospholipid bilayer* (double layer) and embedded proteins separates the cell's contents from its surroundings. Note: You will study the cell membrane in section 2.2.
- Filling the cell interior is the jelly-like *cytoplasm*, which consists of everything outside the nucleus but within the cell membrane. This includes the organelles, cytosol, and molecules and ions dissolved or suspended in the cytosol. The cytosol is the fluid itself.

Figure 2.1 and **Figure 2.2** identify the components of a generalized animal and plant cell. The structures and organelles shown in these diagrams and described in the remainder of this section are common to most eukaryotic cells, but there are exceptions. For example, erythrocytes (red blood cells) lack nuclei and the genetic material contained in them, so they are not capable of reproduction.

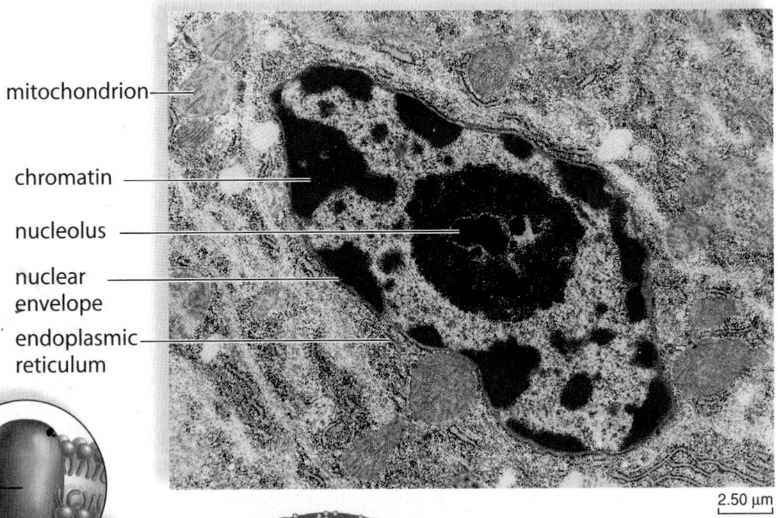

Figure 2.1 Although most animal cells contain the structures shown here, there is tremendous diversity in the form, size, and specialized features of animal cells.

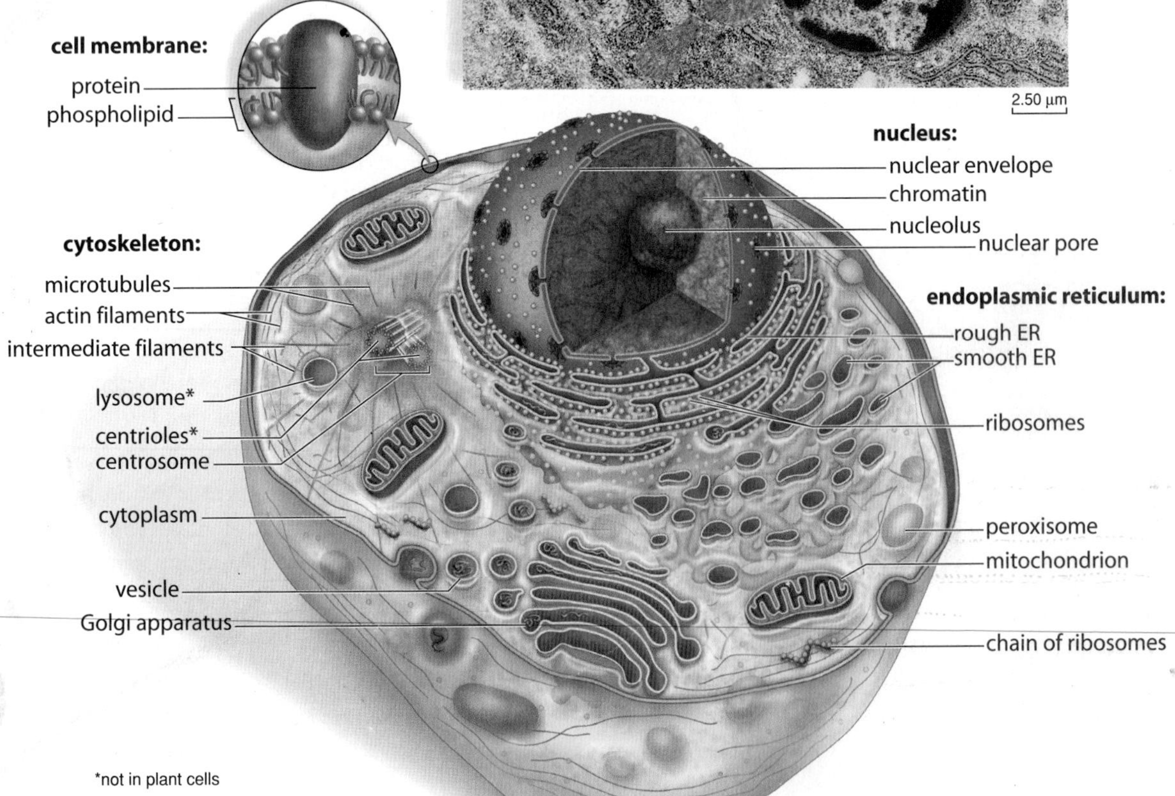

*not in plant cells

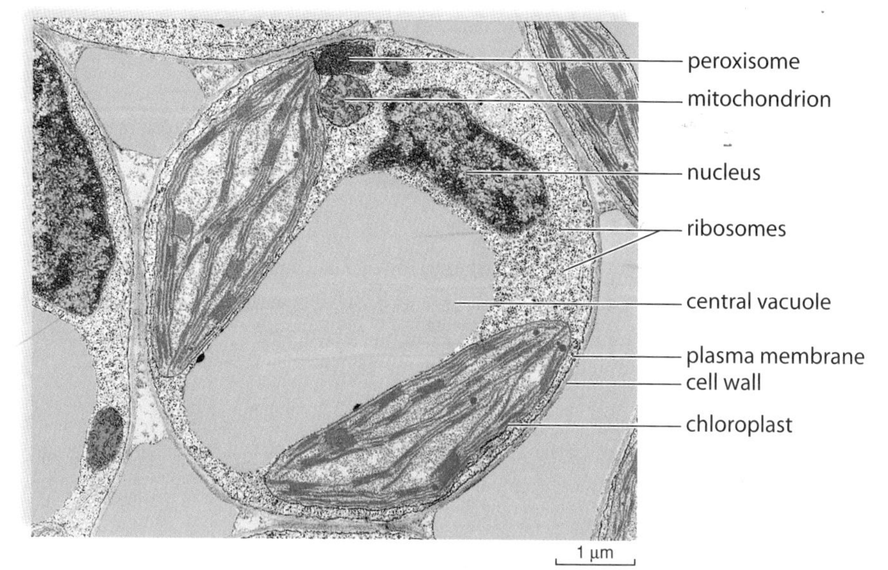

Figure 2.2 Although most plant cells contain the structures shown here, plant cells also exhibit great diversity in their form, size, and specialized features.

***Identify** the organelles that plant cells have but which animal cells lack, and **explain** their significance to plants.*

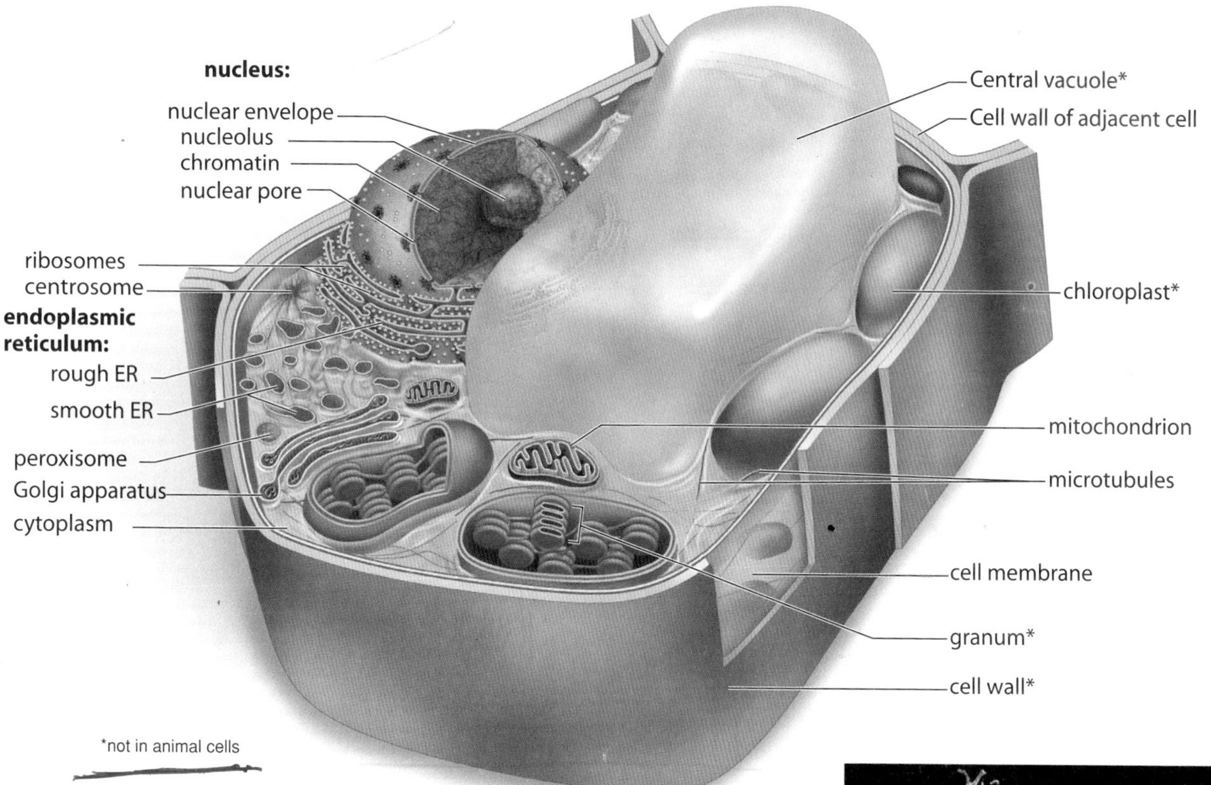

The Nucleus

The cell nucleus, shown in **Figure 2.3**, contains DNA, which stores and replicates the genetic information of the cell. Each molecule of DNA in the nucleus combines with an equal mass of protein to form a *chromosome*. The number of chromosomes in the nucleus varies from species to species. For example, humans have 46 chromosomes, while mosquitoes have only 6 chromosomes. Chromosomes are visible only in dividing cells. In a non-dividing cell, *chromatin*, a complex mixture of DNA and proteins, represents the unfolded state of chromosomes.

Figure 2.3 The nucleus here is outlined with an orange line. The large region, coloured orange, within the nucleus is the nucleolus.

nucleolus a non-membrane-bound structure in the nucleus, which contains RNA and proteins

nuclear envelope a double membrane surrounding the nucleus

nuclear pore complex a group of proteins forming openings in the nuclear envelope

Various structures and regions of the nucleus are shown in greater detail in **Figure 2.4**. A thick fluid called *nucleoplasm* fills the nucleus, and a network of protein fibres called the *nuclear matrix* provides internal structure and support. Within the nucleus is the **nucleolus**, a denser region containing RNA, protein, and chromatin. The nucleus is surrounded by the **nuclear envelope**, a double membrane consisting of two phospholipid bilayers, which separates the nucleus from the rest of the cell. The narrow space between these, or any two, bilayers is called the *lumen*. The nuclear envelope is studded with thousands of **nuclear pore complexes**, groups of proteins that form openings in the nuclear envelope. Small particles such as water and ions travel freely through these openings, but the passage of macromolecules such as RNA is controlled by the nuclear pores.

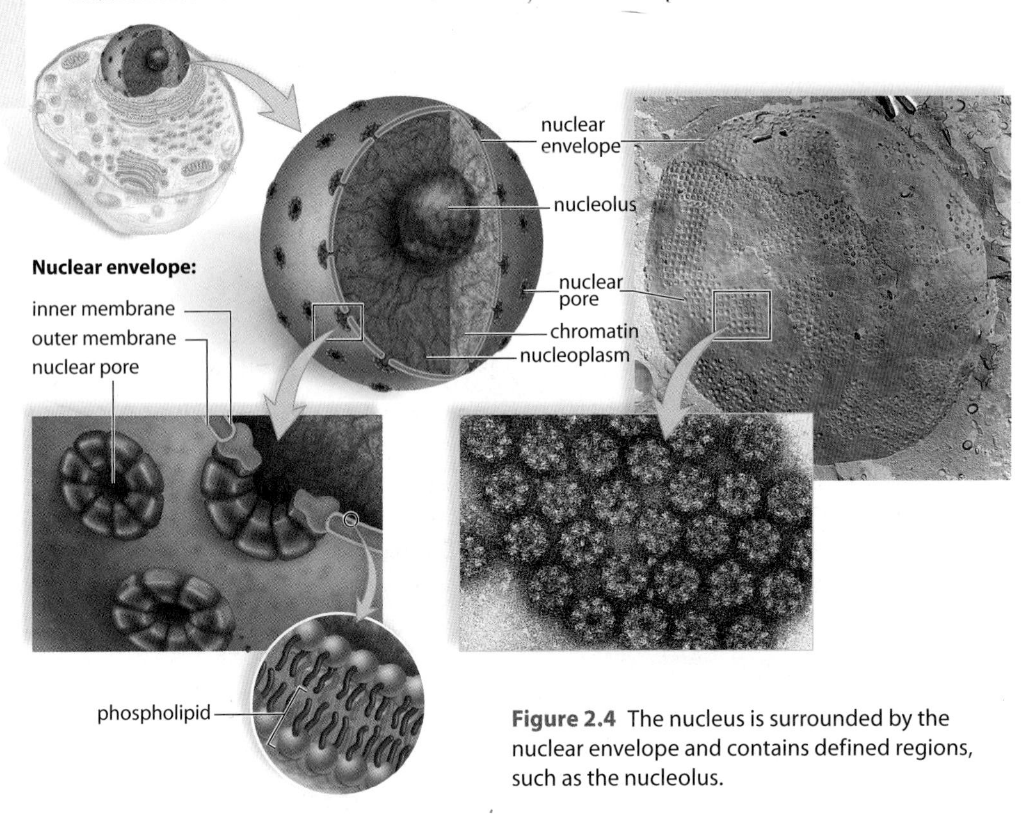

Figure 2.4 The nucleus is surrounded by the nuclear envelope and contains defined regions, such as the nucleolus.

The Endoplasmic Reticulum

endoplasmic reticulum (ER) a complex system of channels and sacs composed of membranes enclosing a lumen; made up of two parts, the rough ER and the smooth ER

ribosome a structure composed of RNA and proteins, and responsible for synthesis of polypeptides in the cytosol and on the surface of the rough endoplasmic reticulum

The nuclear envelope is connected to and part of a complex of membrane-bound tubules and sacs called the **endoplasmic reticulum (ER)**, shown in **Figure 2.5**. The ER surface regions devoted to the synthesis of proteins are studded with **ribosomes**—molecular aggregates of proteins and RNA. Through an electron microscope, ribosome-rich parts of the ER look like sandpaper and are thus called *rough endoplasmic reticulum*. Proteins that are part of membranes or intended for export from the cell are assembled by rough ER ribosomes. Proteins that function in the cytosol are made by ribosomes that are freely suspended there.

Regions of the ER that have no bound ribosomes are called smooth endoplasmic reticulum. The smooth ER synthesizes lipids and lipid-containing molecules such as the phospholipids that make up membranes. Smooth ER performs other functions depending on the type of cell. For example, in the liver, smooth ER helps detoxify drugs and alcohol. In the testes and ovaries, smooth ER produces testosterone and estrogen.

Ribosomes of eukaryotes have different structures and mechanisms compared with those of prokaryotes. This is one reason why antibiotics taken for bacterial infections kill the bacterial cells but not the cells of the body. For example, tetracycline is an antibiotic that inhibits protein synthesis in prokaryotic ribosomes, but it does not affect protein synthesis in human cells.

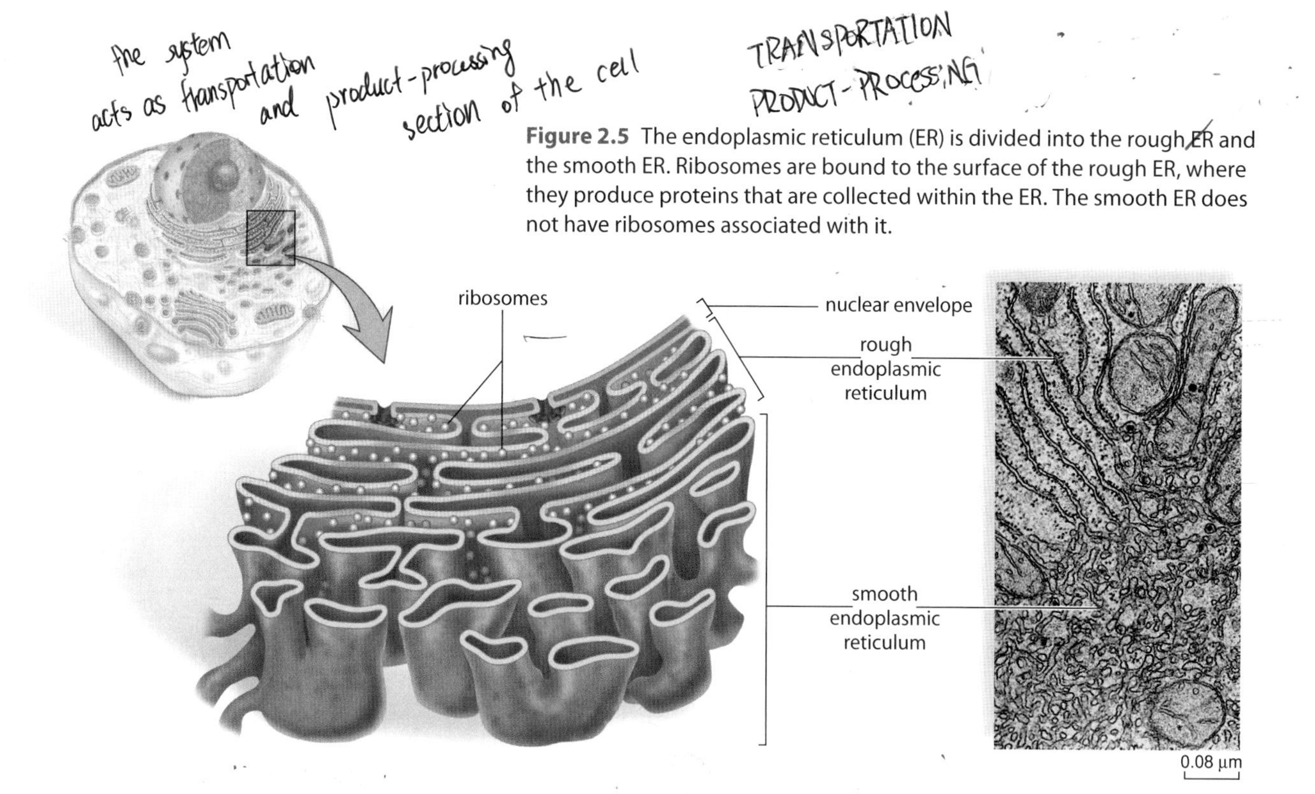
ribosomes
nuclear envelope
rough endoplasmic reticulum
smooth endoplasmic reticulum
0.08 μm

Figure 2.5 The endoplasmic reticulum (ER) is divided into the rough ER and the smooth ER. Ribosomes are bound to the surface of the rough ER, where they produce proteins that are collected within the ER. The smooth ER does not have ribosomes associated with it.

The Endomembrane System: Protein Modification and Transport

The **endomembrane system**, shown in **Figure 2.6**, consists of the nuclear envelope, the endoplasmic reticulum, the Golgi apparatus (described on the next page), and vesicles (also described on the next page). This system acts as the transportation and product-processing section of the cell. The endomembrane system compartmentalizes the cell so that particular functions are restricted to specific regions. The organelles that make up the endomembrane system are connected to one another either directly or by transport vesicles.

endomembrane system the system within the cell that acts to synthesize, modify, and transport proteins and other cell products; includes the endoplasmic reticulum, the Golgi apparatus, vesicles, and the cell membrane, among other structures

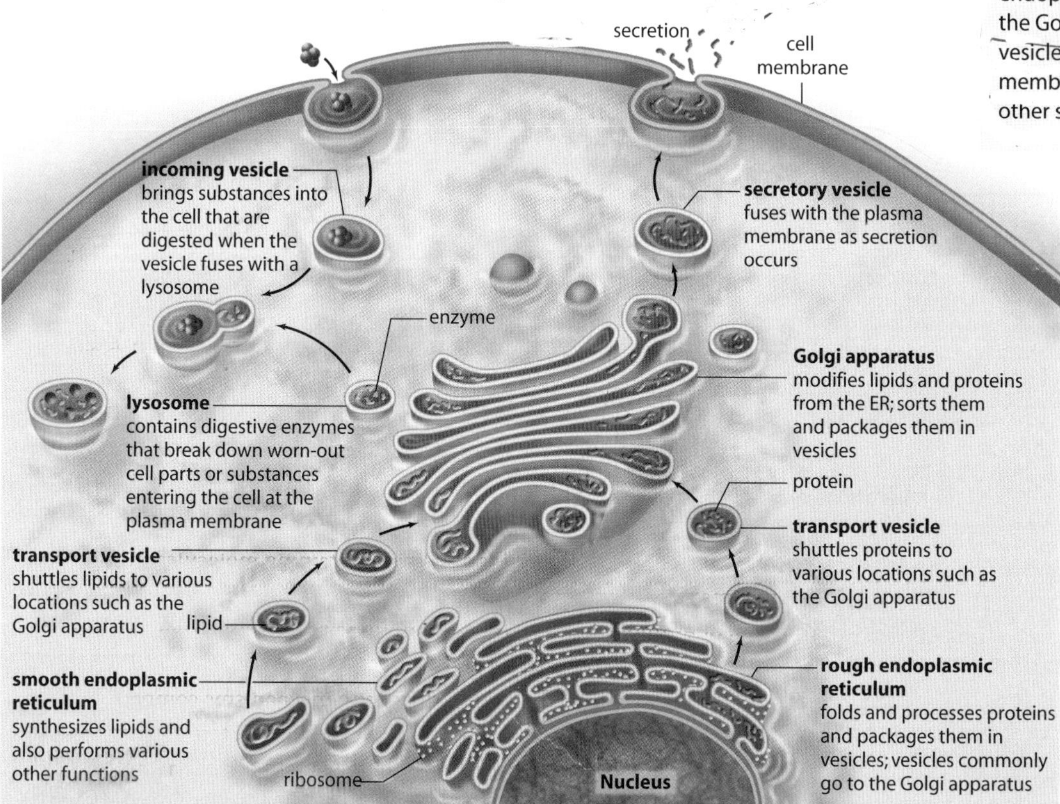
secretion
cell membrane
incoming vesicle brings substances into the cell that are digested when the vesicle fuses with a lysosome
secretory vesicle fuses with the plasma membrane as secretion occurs
enzyme
Golgi apparatus modifies lipids and proteins from the ER; sorts them and packages them in vesicles
lysosome contains digestive enzymes that break down worn-out cell parts or substances entering the cell at the plasma membrane
protein
transport vesicle shuttles proteins to various locations such as the Golgi apparatus
transport vesicle shuttles lipids to various locations such as the Golgi apparatus
lipid
smooth endoplasmic reticulum synthesizes lipids and also performs various other functions
rough endoplasmic reticulum folds and processes proteins and packages them in vesicles; vesicles commonly go to the Golgi apparatus
ribosome
Nucleus

Figure 2.6 The endomembrane system is composed of different organelles that are connected and work together to carry out a number of processes in the cell.

Functions of the Endomembrane System

The endomembrane system modifies and transports proteins, as described below.

1. On the surface of the rough ER, polypeptides are produced by bound ribosomes and extruded into the lumen, rather than being released into the cytosol.
2. These polypeptides travel through the lumen to the smooth ER, where they are stored and processed. When proteins are ready for transport, pieces of smooth ER pinch off to form **vesicles** containing the protein.
3. Vesicles from the smooth ER travel across the cell to the *cis face* of the **Golgi apparatus**, which is a stack of curved membrane sacs, shown in **Figure 2.7**. There, the vesicles merge with the membrane of the Golgi apparatus and release their contents into the interior. In the Golgi apparatus, some proteins are stored and others are modified further. For example, some proteins have carbohydrate chains added to them in the Golgi apparatus or in the ER, converting them into *glycoproteins*, which are important parts of cell membranes. (Note: The Golgi apparatus is called Golgi bodies in some resources.)
4. When the modified proteins are ready for transport, pieces of the Golgi apparatus pinch off from the *trans face* to form vesicles. These vesicles transport the proteins to the cell membrane, or to other destinations within the cell.

vesicle a membrane-enclosed sac used for transport and storage

Golgi apparatus a stack of curved membrane sacs that packages, processes, sorts, and distributes proteins, lipids, and other substances within the cell; acts like a "post office" for the cell

lysosome a membrane-bound vesicle containing enzymes that catalyze hydrolysis reactions, breaking down macromolecules

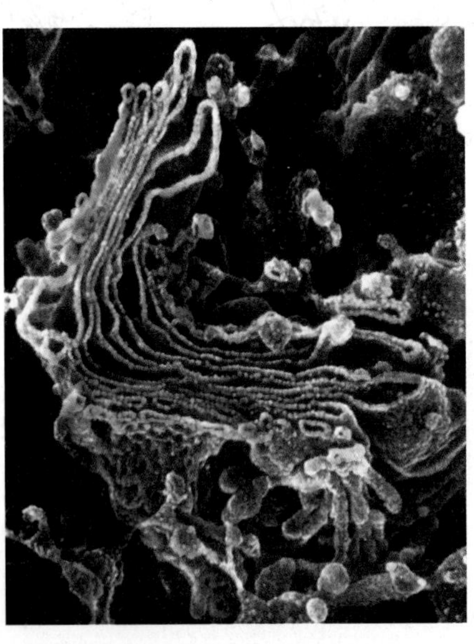

Figure 2.7 Proteins and lipids enter the Golgi apparatus at its *cis* face, or entry face, and leave at the *trans* face, or exit face. The membrane of the Golgi apparatus has a dynamic structure, constantly joining with vesicles at one face, and pinching off to produce vesicles at the other face.

Additional Functions of the Endomembrane System

The endomembrane system has other functions in addition to the modification and transport of proteins. As noted earlier, the smooth ER is responsible for the synthesis and metabolism of lipids, including the steroids and phospholipids that make up cell membranes and organelle membranes. The Golgi apparatus sorts, packages, and distributes these lipids as well as proteins. The Golgi apparatus also manufactures macromolecules, particularly carbohydrates. For example, the Golgi apparatus in many plant cells synthesizes *pectins*, which are non-cellulose structural polysaccharides found in cell walls.

In animal cells, the Golgi apparatus also produces **lysosomes**, which are membrane-enclosed sacs containing digestive enzymes. Lysosomes contain more than 40 enzymes that catalyze hydrolysis reactions, breaking down macromolecules into smaller molecules that can be reused by the cell. Lysosomes break down parts of the cell that are old or no longer needed. They also break down bacteria and other foreign particles that have been ingested by the cell. The enzymes in lysosomes function best at an acidic pH of around 5. Since the cytosol of a cell has a pH of about 7.2, this difference in pH acts as a safeguard for the cell. Even if a lysosome breaks apart, spilling its enzymes into the cell, the enzymes are unlikely to break down the parts of the living cell.

Peroxisomes

Like lysosomes, **peroxisomes** are membrane-enclosed sacs containing enzymes. Peroxisomes form by budding off from the endoplasmic reticulum. Unlike the enzymes in lysosomes, which catalyze hydrolysis reactions, the enzymes in peroxisomes are *oxidases* that catalyze redox reactions. Peroxisomes break down many biological molecules and some toxic molecules. Because toxic substances accumulate in the liver, liver cells contain many peroxisomes. For example, peroxisomes in liver cells oxidize and break down alcohol molecules. Many of the reactions that take place in peroxisomes produce toxic hydrogen peroxide, H_2O_2, so all peroxisomes contain an enzyme known as *catalase* that breaks down hydrogen peroxide into water and oxygen gas. Peroxisomes in some cells synthesize molecules. For example, peroxisomes in liver cells participate in the synthesis of cholesterol and bile acids.

peroxisome membrane-bound sac containing oxidative enzymes that break down excess fatty acids and hydrogen peroxide, and participate in the synthesis of bile acids and cholesterol

vacuole a large, membrane-bound sac in plant cells and some other cells that stores water, ions, macromolecules, sugars, and amino acids

Vesicles and Vacuoles

The term "vesicle" is used to describe membrane-bound sacs used for the transport and storage of substances in the cell. Vesicles form by pinching off from cell membranes and organelle membranes. They can fuse with cell membranes and organelle membranes to release their contents. A typical animal cell contains many small vesicles. Plant cells contain instead a single large central vesicle, called a **vacuole**, shown in **Figure 2.8**. The vacuole stores water, ions, sugars, amino acids, and macromolecules. It also contains enzymes that break down macromolecules and cell wastes. The quantity of water in the central vacuole determines the *turgor pressure*, or internal pressure, of the plant cell. A full vacuole presses against the cell wall, increasing turgor pressure and causing the plant cell to be rigid. This pressure is the source of the rigidity in the flexible stems of herbaceous plants. Without enough water, a vacuole will shrink and pull away from the cell wall. Thus, unwatered plants wilt as the turgor pressure in their cells decreases.

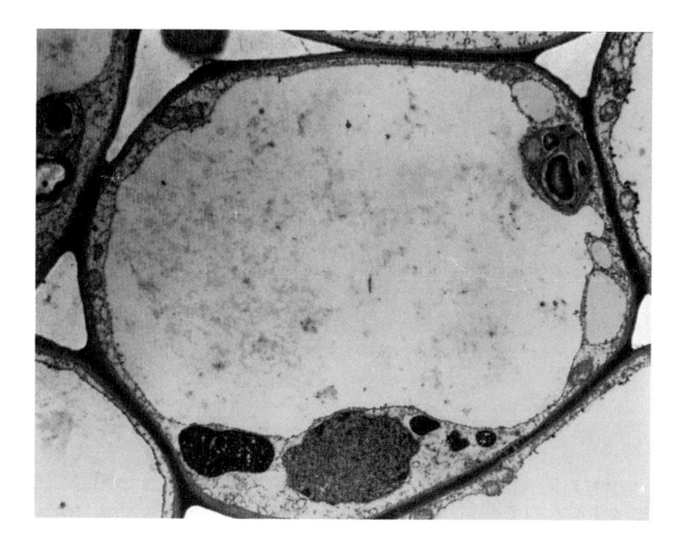

Figure 2.8 The vacuole stores water and other molecules in plant cells.

Learning Check

1. Make a sketch of the nucleus, label its components, and write a detailed caption that describes how the nucleus is organized.
2. Differentiate between the rough and smooth endoplasmic reticulum, and describe the function of each.
3. Explain what the endomembrane system is, including the cell structures that are part of it, and describe its functions.
4. What are peroxisomes?
5. Explain what a vacuole is and how it differs from a vesicle.
6. Explain why the Golgi apparatus is often described as the "post office" of the cell.

chloroplast an organelle in the cells of photosynthetic organisms in which light energy from the Sun is captured and stored in the form of high-energy organic molecules such as glucose

Chloroplasts and Mitochondria

The cells of eukaryotic organisms that carry out photosynthesis typically have one to several hundred **chloroplasts**. These organelles contain the photosynthetic pigment, chlorophyll, which absorbs light energy as part of the process that converts carbon dioxide and water, through redox reactions, into energy-rich organic molecules. As shown in **Figure 2.9**, a thick liquid called *stroma* in the inner membrane surrounds a system of flattened disks called *thylakoids*, which contain chlorophyll in their membranes. A stack of thylakoids is called a *granum* (plural: grana). You will study chloroplasts and photosynthesis in Chapter 4.

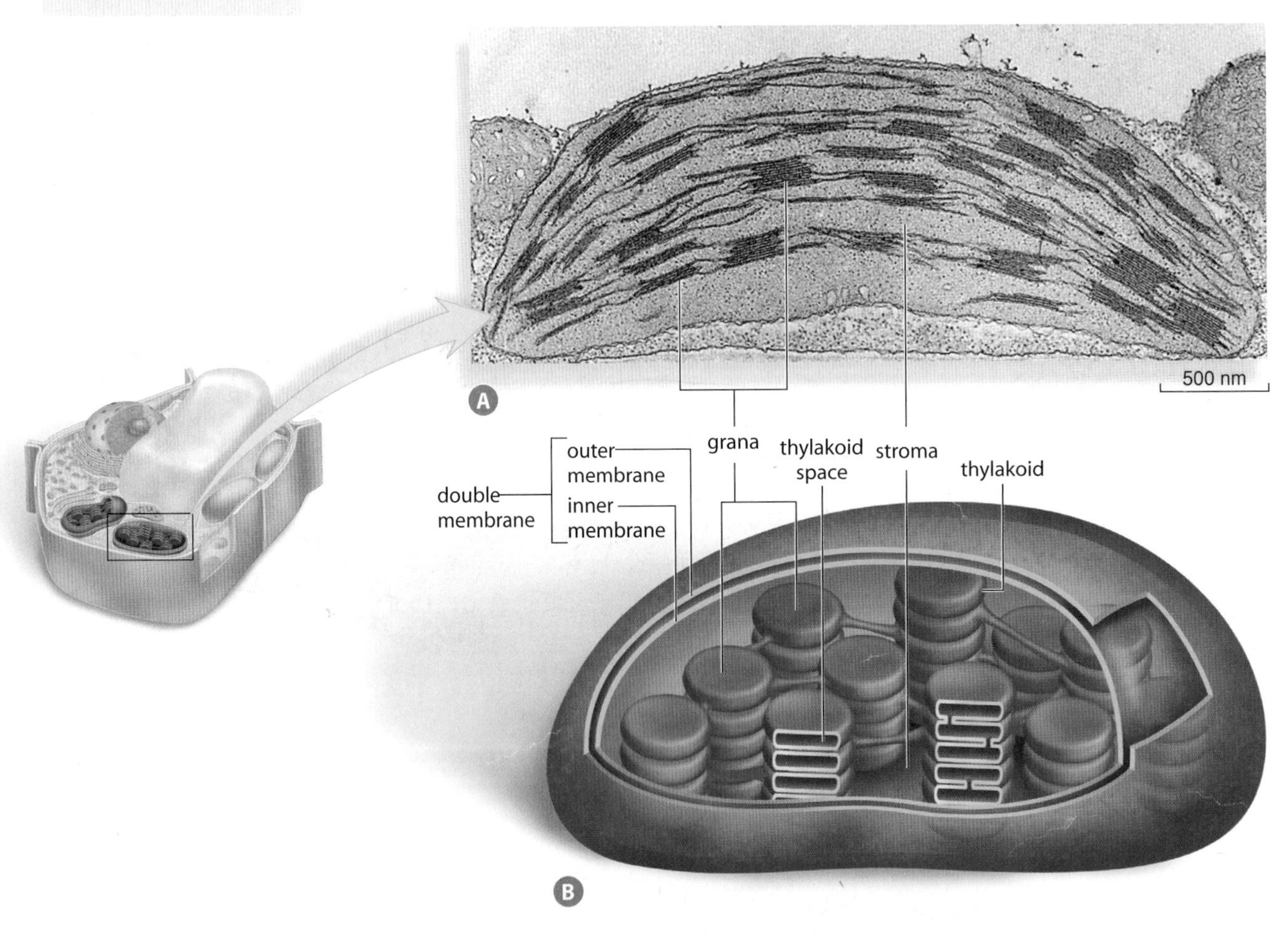

Figure 2.9 Chloroplasts are filled with grana, which are stacks of chloroplyll-containing thylakoids. Chloropyll gives plants their green colour and allows the thylakoids to trap light energy from the Sun.

mitochondrion an organelle in eukaryotic cells in which high-energy organic molecules are oxidized to obtain energy

Activities and chemical reactions in the cell require a steady supply of energy. In eukaryotic cells, **mitochondria** break down high-energy organic molecules to convert stored energy into usable energy. As shown in **Figure 2.10**, mitochondria have a smooth outer membrane and a folded inner membrane. The folds of the inner membrane are called *cristae*, and the fluid-filled space in the inner membrane is called the *matrix*. Both mitochondria and chloroplasts contain some of their own DNA, which encodes some, but not all, of their own proteins. You will study mitochondria and cellular respiration in Chapter 3.

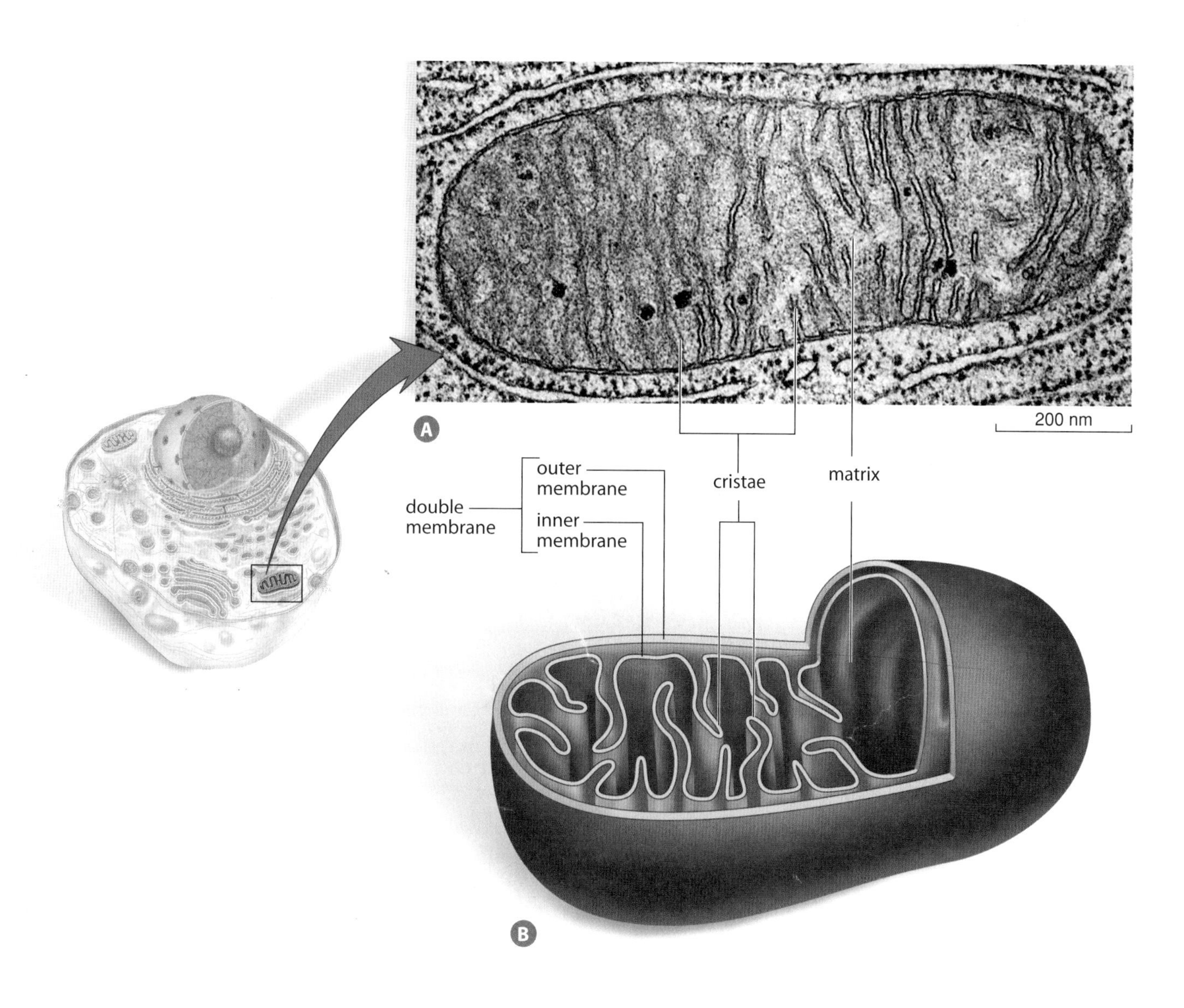

Figure 2.10 Mitochondria are involved in breaking down high-energy organic molecules and storing released energy that can be used by the cell.

The Cell Wall and the Cytoskeleton

Cells of plants, fungi, and many types of protists have a **cell wall**, which provides protection and support. The composition of the cell wall varies with the type of cell, but it is usually a combination of polysaccharides, glycoproteins, or both. For example, cellulose and other substances such as pectins comprise plant cell walls, while chitin comprises fungal cell walls.

All cells contain an internal network of protein fibres called the **cytoskeleton**. The fibres of the cytoskeleton extend throughout the cytoplasm, providing structure and anchoring the cell membrane and organelles in place. Vesicles and other organelles move along these fibres, which act like tracks that lead from one part of the cell to another. In some cells, cytoskeleton fibres form appendages that enable the cell to propel itself through the fluid surrounding it. **Table 2.1** on the next page identifies and compares the functions of the three types of protein fibres in the cytoskeleton.

cell wall a rigid layer surrounding plant, algae, fungal, bacterial, and some archaea cells, composed of proteins and/or carbohydrates; gives the cell its shape and structural support

cytoskeleton a network of protein fibres that extends throughout the cytosol, providing structure, shape, support, and motility

Table 2.1 Functions of Protein Fibres in the Cytoskeleton

Type of Fibre	Size	Structure	Selected Functions
microtubules	Thickest fibres (average of 25 nm in diameter)	Proteins that form hollow tubes	• Maintain cell shape • Facilitate movement of organelles • Assist in cell division (spindle formation)
intermediate filaments	Intermediate thickness (average of 10 nm in diameter)	Proteins coiled together into cables	• Maintain cell shape • Anchor some organelles • Form the internal scaffolding of the nucleus
microfilaments	Thinnest fibres (average of 8 nm in diameter)	Two strands of actin wound together	• Maintain cell shape • Involved in muscle contraction • Assist in cell division (cleavage furrow)

Cilia and Flagella

Cilia and *flagella*, shown in **Figure 2.11**, are appendages that develop on the outside of some eukaryotic cells. If there are just one or two longer appendages, they are called flagella. If many shorter appendages are present, they are referred to as cilia. These structures are composed of an internal shaft made of microtubules, covered with an outer membrane that is a continuation of the cell membrane.

Flagella are like tails, and their whip-like movement propels cells. For example, a human sperm cell has a single flagellum, while a sperm cell of a cycad (a type of tree) has thousands of flagella. In unicellular protists such as paramecia, the wave-like motion of cilia enable the organisms to move. In multicellular organisms such as human, cells that line the upper respiratory tract have cilia that sweep debris trapped within mucus back up into the throat, where it can swallowed or ejected by coughing.

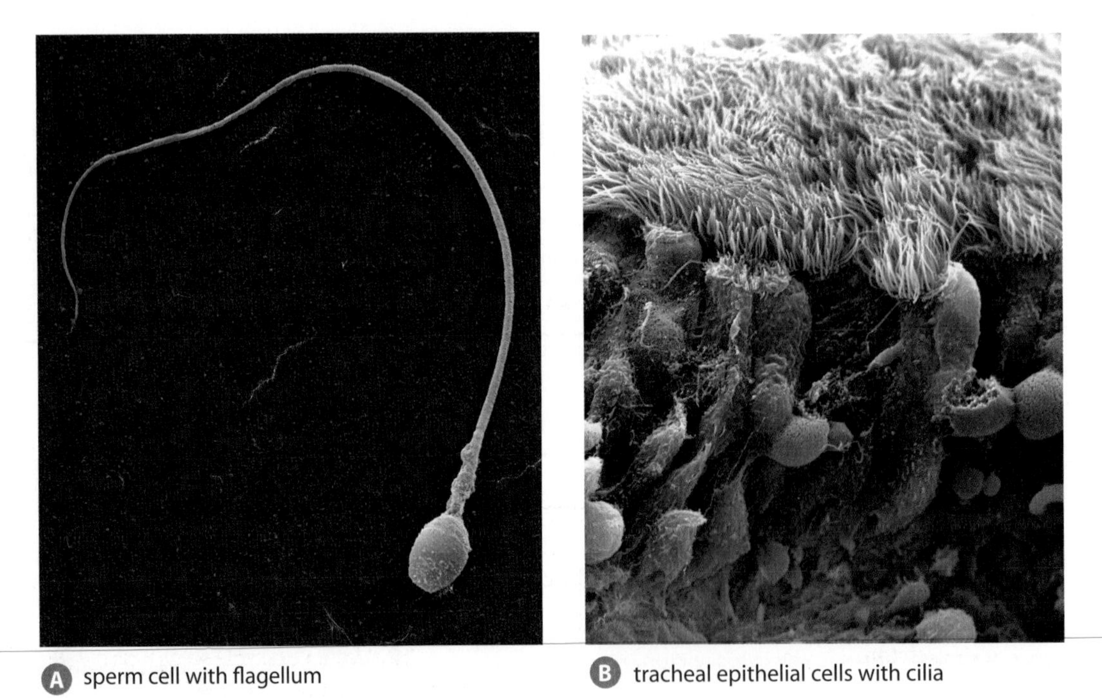

A sperm cell with flagellum

B tracheal epithelial cells with cilia

Figure 2.11 Cilia and flagella are long, thin appendages that allow cells to move themselves, or to move substances over their surface.

Activity 2.1 | A Bright Idea: Fluorescence Microscopy

Fluorescence microscopy is a type of light microscopy that makes it possible to see the cytoskeleton, organelles, proteins, and even ions within cells. Fluorescent compounds are used to stain specimens which are exposed to ultraviolet light. The compounds then emit bright visible light of various colours.

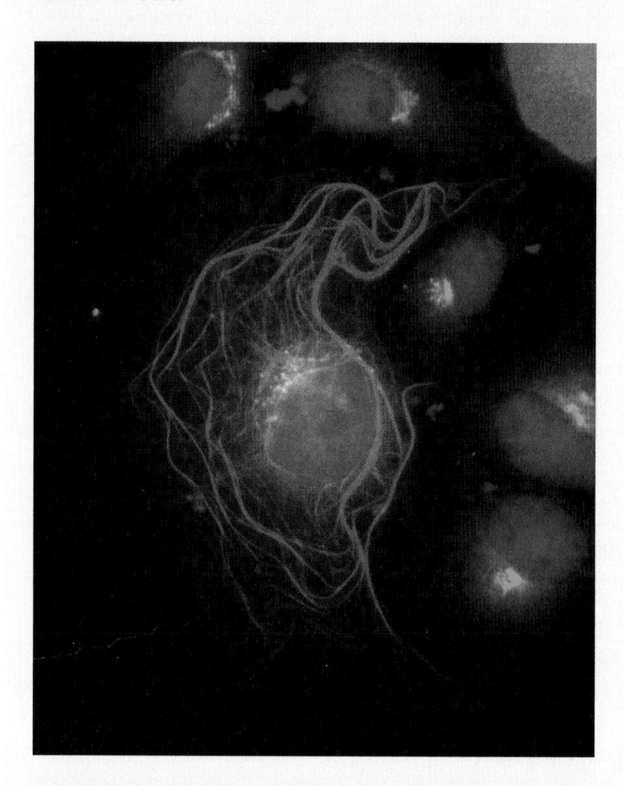

These kidney cells were stained with fluorescent compounds that make it easy to observe different organelles. The nuclei are blue, microtubules are red, and Golgi bodies are green. Fluorescence microscopy can also be used to track the transport of proteins and lipids. In what other ways can fluorescence microscopy enhance understanding of cell organelles and their functions?

Materials

- computer with Internet access

Procedure

1. Read the table of selected techniques.
2. Beginning with this book, search for a fluorescence microscope image that shows one or more organelles or cell structures. Continue your research using the Internet.
3. Once you have located a suitable image, record the source. Then identify which organelles are shown and their colours. Find out which technique was used to produce the image and how it was carried out.

Questions

1. What can you learn about cells and cell functions with fluorescence microscopy that you cannot learn with a compound light microscope?

Selected Fluorescence Microscopy Techniques

Technique	Description
Confocal microscopy	Optical slices of a specimen are assembled into a clear three-dimensional image.
Fluorescent In Situ Hybridization (FISH)	Dye-tagged antibodies that bind to specific DNA sequences are used to stain chromosomes.
Indirect immunofluorescence	A primary antibody binds only to highly specific cell components; a secondary, dye-tagged antibody binds to the primary antibody.
Ion staining	Fluorescent probes are added to cells and, if certain ions are present, the cells will fluoresce.

Learning Check

7. Describe similarities and differences between chloroplasts and mitochondria.
8. Describe the structure and function of the cell wall.
9. Describe the structure and function of the cytoskeleton.
10. Compare the functions of the protein fibres in the cytoskeleton.
11. Use an example to describe the structure and function of cilia.
12. Use an example to describe the structure and function of flagella.

The Cell Membrane

All living cells exist in an aqueous medium. For a unicellular organism such as an alga, this medium might be pond water. For the cells of a multicellular organism such as an animal, the aqueous medium is the extracellular fluid that surrounds all cells. The contents of cells are physically separated from this aqueous environment by the cell membrane, which functions as a selective, dynamic cellular boundary. If this remarkable and remarkably thin membrane does not function properly, cellular processes fail, and cells die. The cell membrane is so thin, in fact, that if the cell were the size of a car, the cell membrane would be as thick as a sheet of paper—a mere 0.006 nm across.

The cell membrane maintains the integrity of the cell of which it is a part by regulating the passage of molecules and ions into and out of the cell. In the early 1900s, researchers noted that lipid-soluble molecules entered cells more rapidly than water-soluble molecules. This prompted the hypothesis that lipids are a component of the cell membrane. By 1925, chemical analysis had demonstrated, however, that phospholipids are a component of cell membranes and that they are arranged around the cell in two layers (a bilayer).

The presence of lipids cannot account for all the properties of the cell membrane, such as the fact that some non-lipid substances can pass through it. Researchers in the 1940s hypothesized that proteins are a part of the membranes and proposed a model in which a phospholipid bilayer is sandwiched between two continuous layers of proteins. By the 1950s, electron microscope views of the cell membrane confirmed a sandwich-like appearance, but a suitable model that could link the structure and properties of membranes to various functions remained elusive.

In 1972, two American biologists, Jonathan Singer and Garth Nicolson, proposed a model for membranes that remains in use today. They visualized proteins inserted into the phospholipid bilayer with their non-polar segments in contact with the non-polar interior of the bilayer and their polar portions protruding from the membrane surface. In this **fluid mosaic model**, shown in **Figure 2.12**, an assortment of proteins and other molecules (in other words, "the mosaic") floats in or on the fluid phospholipid bilayer.

fluid mosaic model the accepted model of the cell membrane, which is a basic framework of a semi-fluid phospholipid bilayer with a mosaic of proteins; carbohydrates may be attached to lipids or proteins

Activity 2.2 The Path to the Fluid Mosaic Model

A neutrotransmitter is a chemical that enables nerve cells to communicate with one another. An abnormal production of certain neurotransmitters has been linked to disorders such as depression, bipolar disorder, anxiety disorders, and schizophrenia. Treatments for these disorders include pharmaceutical medications that affect neurotransmitters in some way.

Materials

- print and Internet resources

Procedure

1. To investigate the many researchers and events in this activity most efficiently, cooperative group work is a good idea. For example, you could work in small teams with each team member responsible for researching several people and their contributions.
2. In a group, investigate the role of each of the following individuals or groups of individuals in developing an understanding of the structure and behaviour of membranes and the interactions of lipids (oils) and water.

Note: Some of these people contributed single ideas and/or techniques, whereas others contributed many more. Be sure to consult a minimum of three information resources for each person to ensure that you have located relevant and reliable information.

- Benjamin Franklin
- Lord Rayleigh (John William Strutt)
- Agnes Pockels
- Charles Ernest Overton
- Irving Langmuir
- Ernest Gorter and F. Grendel
- James Danielli, E. Newton Harvey, and Hugh Davson
- J. David Robertson
- George E. Palade
- Jonathan Singer and Garth Nicolson

Questions

Record and synthesize the information you gather in the form of a summary table, a graphic organizer, a timeline, or another format of your choice.

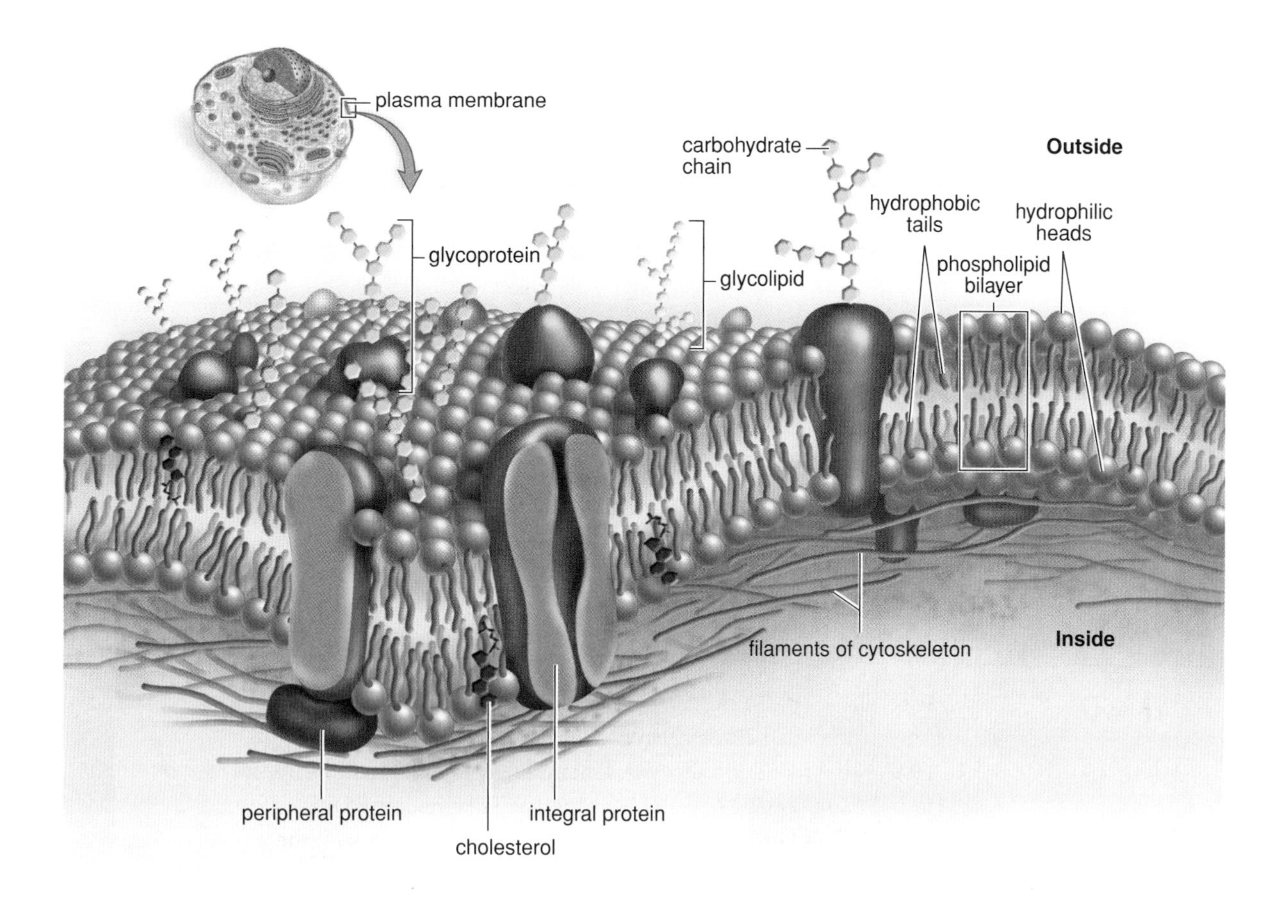

Figure 2.12 In the modern fluid mosaic model, the basic framework of a cell membrane is a phospholipid bilayer into which proteins are inserted. These proteins may be bound on the surface to other proteins or to lipids, including glycoproteins and glycolipids. Glycoproteins and glycolipids are proteins and lipids covalently bonded to carbohydrates.

Features of the Fluid Mosaic Model

According to the fluid mosaic model, each layer—sometimes called a *leaflet*—of a membrane bilayer is composed of various macromolecules. Phospholipids act as the "scaffolding" in which proteins and other macromolecules are embedded. Because membrane lipids are held together by weak intermolecular forces rather than by strong covalent bonds, the molecules in a membrane can move about freely. In fact, phospholipids within the same layer in a membrane exchange places millions of times in a single second, leading to a continual rearrangement of the membrane surfaces. If a puncture or tear occurs in a membrane, molecules will quickly rearrange themselves to seal the rupture.

The lipid bilayer structure of membranes can be explained based on chemical principles and the properties of the phospholipid molecules that form these structures. Recall that a phospholipid molecule has a hydrophilic, polar "head" group and two hydrophobic, non-polar "tails" composed of fatty acids. When placed in water, phospholipids spontaneously form structures in which the polar "heads" cluster together, facing the water molecules, while the non-polar "tails" are shielded from the water. Intermolecular interactions, such as hydrogen bonding, occur between water molecules and between water molecules and the polar "heads" of the phospholipids. The non-polar "tails" cluster together and are held together by hydrophobic interactions. As a result, the polar "heads" end up facing out, and the non-polar tails face inward, away from the aqueous environment.

Learning Check

13. Describe at least two functions of the cell membrane.

14. What kinds of molecules make up a cell membrane?

15. Why are the properties of the cell membrane not adequately explained by the presence of lipids alone in their structure?

16. Use the fluid mosaic model to describe how the components of a cell membrane are organized.

17. Explain why a cell membrane is a dynamic structure, rather than static such as the wall of a building.

18. Describe what happens when phosopholipids are mixed with water, and explain why it happens.

The Fluidity of a Phospholipid Bilayer

At room temperature, a phospholipid bilayer has a viscosity similar to that of vegetable oil. The fluidity of a bilayer is an important property. If it is too fluid, a bilayer permits too many molecules to diffuse in and out of a cell. If it is not fluid enough, a bilayer prevents too many molecules from crossing. The main factors that affect fluidity include the following.

- Temperature: With increasing temperature, the bilayer becomes increasingly fluid until it is unable to act as a barrier. At decreasing temperatures, the bilayer eventually solidifies into a gel-like state.
- Presence of double bonds in the fatty acid "tails": Double bonds form "kinks" in a fatty acid tail. The presence of one or more double bonds causes fatty acids to be less tightly packed and more fluid.
- Fatty acid "tail" length: Longer fatty acid "tails" have more intermolecular attractions and hold together more tightly compared to shorter fatty acid tails, thus reducing fluidity. The most common length of a fatty acid is 16 or 18 carbon atoms.

The presence of cholesterol in cell membranes also affects fluidity. Many eukaryotic cell membranes contain cholesterol molecules. At room temperature and higher, the presence of cholesterol increases the intermolecular forces in the membrane and holds it more tightly together, thus reducing fluidity. For example, cholesterol keeps human cell membranes from being too fluid at body temperature. At lower temperatures, however, cholesterol molecules break up the packing that occurs as phospholipids solidify into a gel. As a result, cholesterol increases the fluidity of the cell membrane at low temperatures.

The Function of Proteins in a Phospholipid Bilayer

Proteins associated with membranes are: integral proteins or peripheral proteins. *Integral proteins* are embedded in the membrane, while *peripheral proteins* are more loosely and temporarily attached to the outer regions of the membrane or to integral proteins.

Peripheral proteins and some integral proteins help to stabilize membranes, and hold them in place by linking them with the cytoskeleton of the cell. Membrane proteins also determine the function of the membrane by performing the following functions.

- Transport: Proteins play an essential role in transporting substances across the cell membranes. This important function of proteins is the subject of the next section.
- Reaction catalysis: Enzymes in cell membranes carry out chemical reactions.
- Cell recognition: The carbohydrate chains that protrude from glycoproteins on the outer layer of the cell membrane enable cells to "recognize" each other. As a result, cells in the body can identify harmful "intruders" such as disease-causing bacteria.
- Signal reception and transduction: Receptor proteins in cell membranes bind to signal molecules, such as hormones, and change shape as a result. This initiates a cellular response to the signal, enabling cells to receive and respond to signals from the brain and other organs.

Section 2.1 Review

Section Summary

- Animals, plants, fungi, and protists are composed of eukaryotic cells, which have DNA, a cell membrane, and cytoplasm. Cytoplasm consists of organelles, the cytosol, and molecules and ions dissolved or suspended in the cytosol. The nucleus includes the nuclear envelope, which is studded with nuclear pore complexes, the nuclear matrix, and the nucleolus.
- The endoplasmic reticulum (ER), consisting of the rough ER and the smooth ER, is a system of channels and membrane-bound-sacs enclosing a narrow space called the lumen.
- The endomembrane system includes the nuclear envelope, the endoplasmic reticulum, the Golgi apparatus, the cell membrane, and vesicles. This system synthesizes, modifies, and transports proteins and other cell products.
- Animal cells contain many small vesicles. Plant cells contain a single large central vesicle called a vacuole.
- Chloroplasts trap light energy from the Sun in the form of high-energy organic molecules. Mitochondria break down high-energy organic molecules to release usable energy.
- Cells of plants, fungi, and many types of protists have a cell wall, which provides protection and support.
- The cytoskeleton provides structure, shape, support, and motility.
- The fluid mosaic model visualizes the cell membrane as a mosaic of proteins and other molecules in a fluid phospholipid bilayer.

Review Questions

1. T/I While researching "eukaryotic cells" online, you and a classmate are surprised to find visuals of various cells that look distinctly different from each other—for example, a bread yeast cell, a pea leaf stoma, and a human liver cell. Why are these highly diverse cells classified together?
2. C In an illustrated table, make labelled sketches to show the functions of the various structures and regions of the nucleus of a cell.
3. T/I What are two important general functions of the organelles in eukaryotic cells?
4. C Using a Venn diagram, compare and contrast rough ER and smooth ER.
5. T/I "The endomembrane system compartmentalizes the cell so that particular functions are restricted to specific regions." Explain how and why a eukaryotic cell could not function or even exist without the endomembrane system.
6. C Use a flowchart to represent the biochemical functions of lysosomes and peroxisomes.
7. K/U Use the following headings to design a summary chart of the structures and organelles of generalized animal and plant cells: Cell Structure or Organelle; Description; Function; Plant, Animal, or Both.
8. A In an animal, which cells would you predict would have the highest concentration of mitochondria? Explain your answer.
9. K/U Name the cells in your body that have many peroxisomes, and explain why it makes sense that they do.
10. K/U Explain the crucial role of the cell membrane in maintaining the integrity of the cell.
11. C In a table, list and describe the features of the fluid mosaic model of the cell membrane.
12. T/I Create an analogy for the structure and function of the fluid mosaic model that would help a younger student understand this model.
13. A Examine the photograph below, which also appeared on the opening page of this chapter. Identify all the cell structures and organelles that you can, and explain how you recognized them.

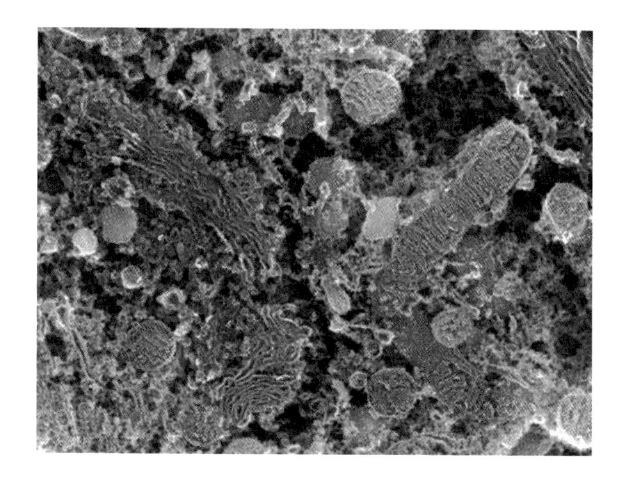

SECTION 2.2

The Transport of Substances Across a Cell Membrane

Key Terms

semi-permeable
passive transport
concentration gradient
diffusion
osmosis
facilitated diffusion
channel protein
carrier protein
active transport
electrochemical gradient
membrane-assisted transport
endocytosis
phagocytosis
pinocytosis
receptor-mediated endocytosis
exocytosis

The cell membrane is able to regulate the passage of substances into and out of the cell, because it is *semi-permeable*. That is, certain substances can move across the membrane while other substances cannot. Processes that enable substances to move in and out of cells without an input of energy from the cell are referred to as **passive transport**. Some ions and molecules can move passively across the cell membrane fairly easily because of a **concentration gradient**—a difference between the concentration on the inside of the membrane and the concentration on the outside of the membrane. Some other substances also move in response to a gradient, but they do so through specific channels formed by proteins in the membrane. Three forms of passive transport are diffusion, osmosis, and facilitated diffusion.

passive transport the movement of ions or molecules across a cell membrane from a region of higher concentration to a region of lower concentration, without the input of energy

concentration gradient a difference in concentration between one side of a membrane and the other

diffusion the net movement of ions or molecules from an area of higher concentration to an area of lower concentration

Passive Transport by Diffusion

Molecules and ions dissolved in the cytoplasm and extracellular fluid are in constant random motion. This random motion causes a net movement of these substances from regions of higher concentration to regions of lower concentration. This process, called **diffusion**, is illustrated in **Figure 2.13**. Net movement driven by diffusion will continue until the concentration is the same in all regions. In the context of cells, diffusion involves differences in the concentration of substances on either side of a cell membrane., Therefore, the relative concentrations both inside and outside the cell, as well as how readily a molecule or ion can cross the membrane, are both factors that affect diffusion.

The major barrier to crossing a biological membrane is the membrane's hydrophobic interior that repels polar molecules but not non-polar molecules. If a concentration difference exists for a non-polar molecule such as oxygen, it will move across the membrane until the concentration is equal on both sides. At that time, movement in both directions still occurs, but there is no net change in either direction. Factors that affect the rate of diffusion include the following.

- Molecule size: The larger a molecule is, the more difficult it is for it to diffuse across a membrane. As a result, the rate of diffusion decreases with molecule size.
- Molecule polarity: Although small polar molecules can cross membranes, their rates of diffusion are generally lower than those of non-polar molecules of the same size.
- Molecule or ion charge: In general, charged molecules and ions cannot diffuse across a cell membrane.

Temperature and pressure also affect the rate of diffusion. At higher temperatures, molecules have more energy and move faster, thus increasing the rate of diffusion. At higher pressures, molecules are forced across the membrane and the rate of diffusion increases.

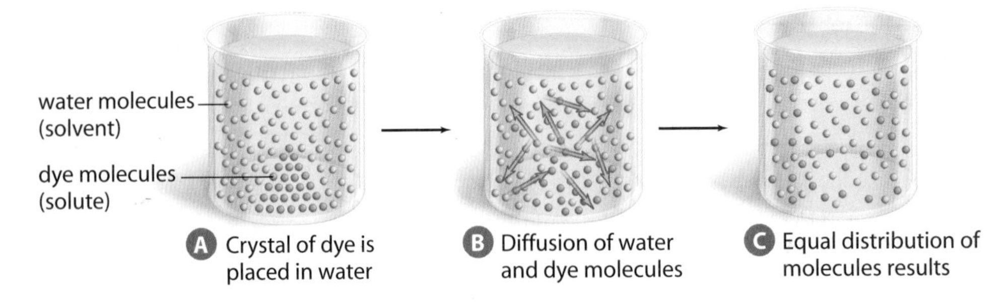

Figure 2.13 When a crystal of dye is dissolved in water, there is a net movement of dye molecules from a higher concentration to a lower concentration. At the same time, there is a net movement of water molecules from a higher to a lower concentration. Eventually, the water and dye molecules are evenly distributed throughout the system.

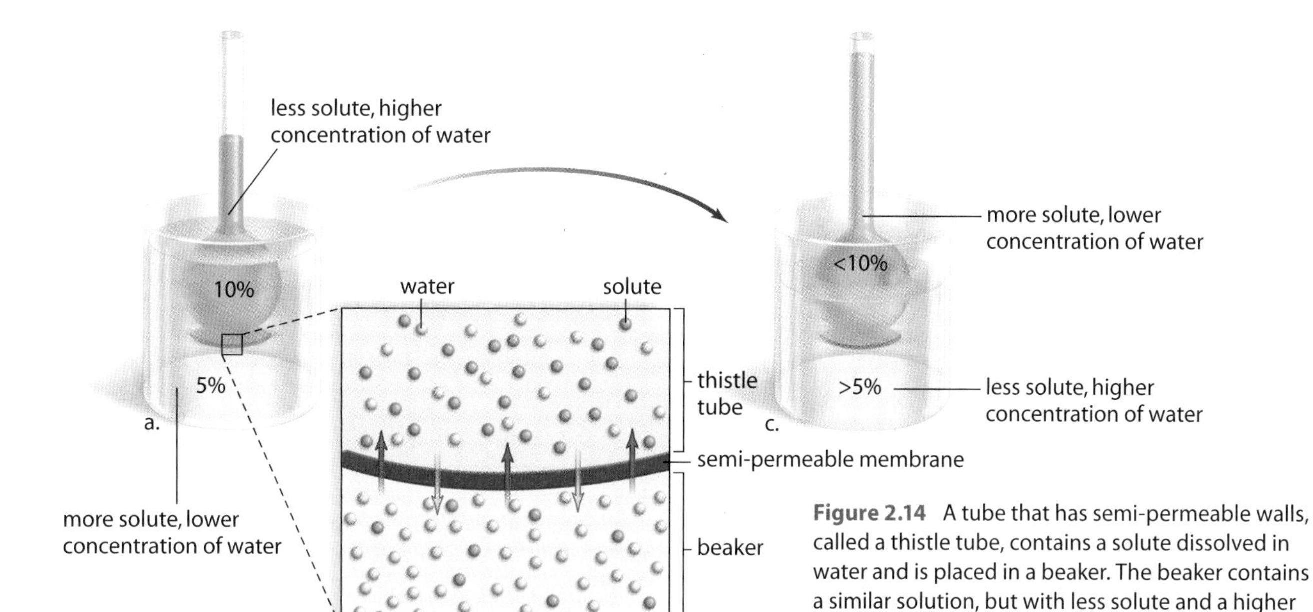

Figure 2.14 A tube that has semi-permeable walls, called a thistle tube, contains a solute dissolved in water and is placed in a beaker. The beaker contains a similar solution, but with less solute and a higher concentration of water.

***Explain** why the water level in the tube rises.*

Passive Transport by Osmosis

The aqueous cytoplasm is a solvent for cellular molecules and ions. Cells must maintain enough water to enable cellular processes. However, cells also interact with extracellular fluid, the composition of which is constantly changing. If too much water enters a cell, it swells. If too much water leaves a cell, it shrinks. Either response can affect the ability of a cell to function. Thus, the regulation of water entry is of crucial importance to a cell.

osmosis the movement of water from an area of higher concentration to an area of lower concentration, across a semi-permeable membrane

Movement of water molecules across biological membranes is called **osmosis**. In osmosis, water molecules move because the membrane is impermeable to the solute, and the solute concentrations may differ on either side of the membrane, as shown in **Figure 2.14**. Water molecules move in or out of a cell, along their concentration gradient, until their concentrations on both sides of the membrane are equal. At that time, water molecules continue to move in and out, but there is no net diffusion of water.

The concentration of all solutes in a solution determines its osmotic concentration. If two solutions have unequal osmotic concentrations, the solution with the higher concentration is *hypertonic* (hyper = "more than"). The solution with the lower concentration is *hypotonic* (hypo = "less than"). When two solutions have the same osmotic concentration, they are *isotonic* (iso = "equal"). **Figure 2.15** shows the effect of osmotic concentration on an animal cell and on a plant cell.

Suggested Investigation

Plan Your Own Investigation 2-A Demonstrating Osmosis

Animal cells

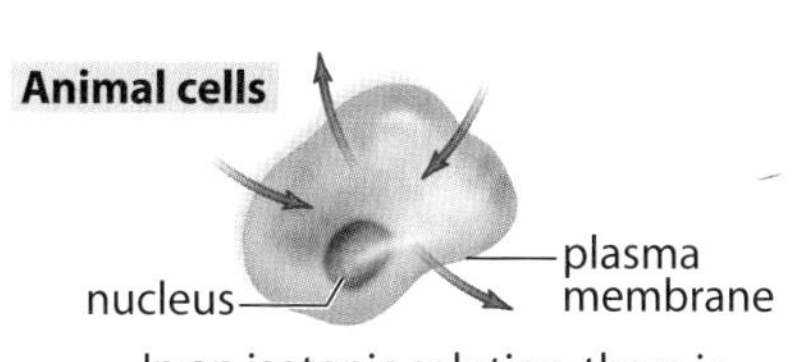

In an isotonic solution, there is no net movement of water.

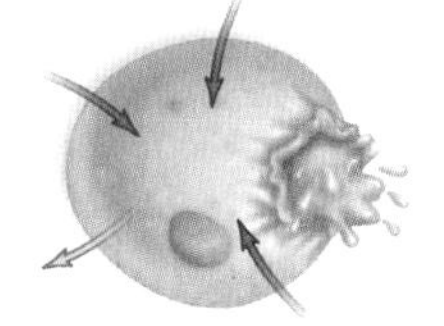

In a hypotonic solution, water enters the cell, which may burst (lysis).

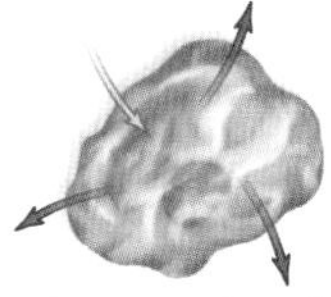
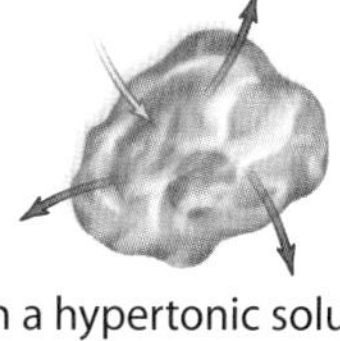

In a hypertonic solution, water leaves the cell, which shrivels (crenation).

Plant cells

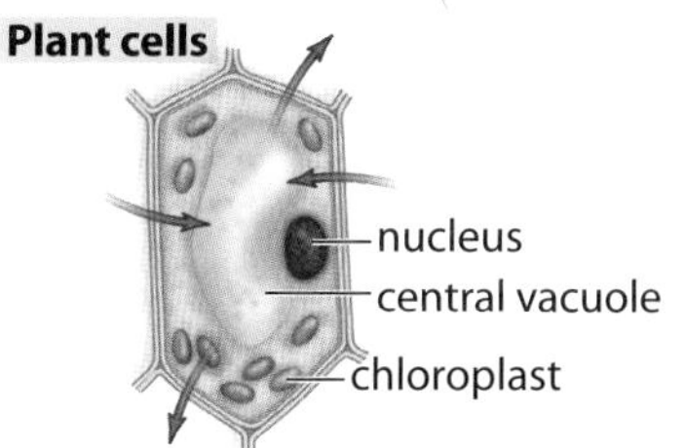

In an isotonic solution, there is no net movement of water.

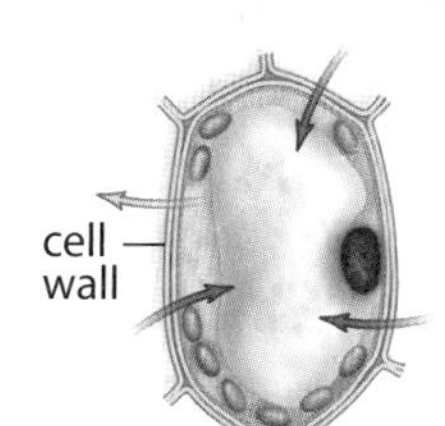

In a hypotonic solution, the central vacuole fills with water, turgor pressure develops, and chloroplasts are seen next to the cell wall.

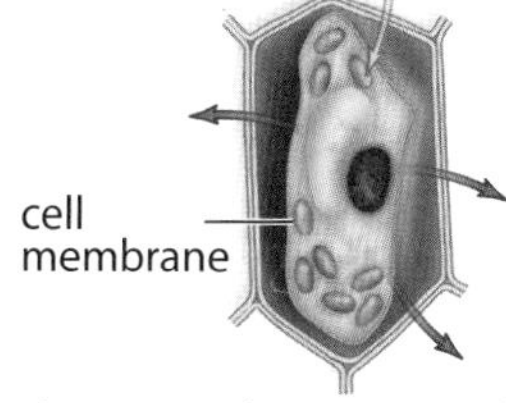

In a hypertonic solution, the central vacuole loses water, the cytoplasm shrinks (plasmolysis), and chloroplast are seen in the center of the cell.

Figure 2.15 Isotonic and hypotonic solutions. Arrows indicate the movement of water molecules.

Learning Check

19. What is a concentration gradient?
20. Describe the process of diffusion, and explain why it occurs.
21. What are three factors that affect the rate of diffusion, and why do they affect it?
22. Explain the similarities and differences between diffusion and osmosis.
23. Would you expect the normal environment of your cells to be typically isotonic, hypertonic, or hypotonic? Explain your reasoning.
34. Would you expect the normal environment of a plant cell to be typically isotonic, hypertonic, or hypotonic? Explain your reasoning

Passive Transport by Facilitated Diffusion

Many important molecules required by cells cannot easily cross the plasma membrane. These molecules can still enter the cell by diffusion through specific channel proteins or carrier proteins embedded in the plasma membrane, as long as there is a higher concentration of the molecule outside the cell than inside. This process of diffusion that is mediated by a membrane protein is called **facilitated diffusion**. Channel proteins have a hydrophilic interior that provides an aqueous channel through which polar molecules can pass when the channel is open. Carrier proteins, in contrast to channels, bind specifically to the molecule they assist, much like an enzyme binds to its substrate.

facilitated diffusion the transport of ions or molecules across a membrane by means of a membrane protein along the concentration gradient for that ion or molecule

channel protein a membrane protein that forms a channel across a cell membrane, which allows specific ions or molecules to cross the membrane along their concentration gradients

carrier protein a membrane protein that binds to and transports one or more particles of a substance from one side of a membrane to the other, along the concentration gradient for that substance

Channel Proteins

Channel proteins form highly specific channels through the cell membrane, as shown in **Figure 2.16A**. The structure of a channel protein determines which particles can travel through it. A channel protein has a tubular shape, like a hollow cylinder. This cylinder is usually composed of one or more helixes, like coiled springs. Recall from Chapter 1 that proteins are composed of linked amino acids, which may have polar, non-polar, or charged side chains. The exterior of a channel protein is usually composed of amino acids with non-polar side chains that interact with the non-polar interior of the cell membrane anchoring the protein in place. The shape and size of the hole through a channel protein determines the shape and size of particles that can pass through it.

Some channel proteins remain open all the time, while others have gates that the cell can open or close to allow or prevent the passage of particles. Different types of gates open or close in response to a variety of signals, such as hormones, electric charge, pressure, or even light.

In general, channel proteins permit the passage of ions or polar molecules. For example, sodium channel proteins allow sodium ions, Na+, to cross the membrane, and potassium channel proteins allow potassium ions, K^+, to cross it. Cystic fibrosis (CF) is a disease that results in the production of very thick mucus in the breathing passages and pancreas. CF is caused by defective chloride ion channel proteins that do not allow the proper movement of chloride ions, Cl^-, across the cell membrane. This, in turn, interrupts the proper balance of water movement into and out of the cell, which causes the formation of a thick layer of mucus.

Carrier Proteins

Carrier proteins bind to specific molecules, transport them across the membrane, and then release them on the other side, as shown in **Figure 2.16B**. Because they bind to the molecules they are carrying, carrier proteins change shape while transporting molecules. While channel proteins usually transport ions or small polar molecules, carrier proteins can also transport larger molecules such as glucose and amino acids. Because they bind to only a few molecules at a time, carrier proteins have lower rates of diffusion compared to channel proteins.

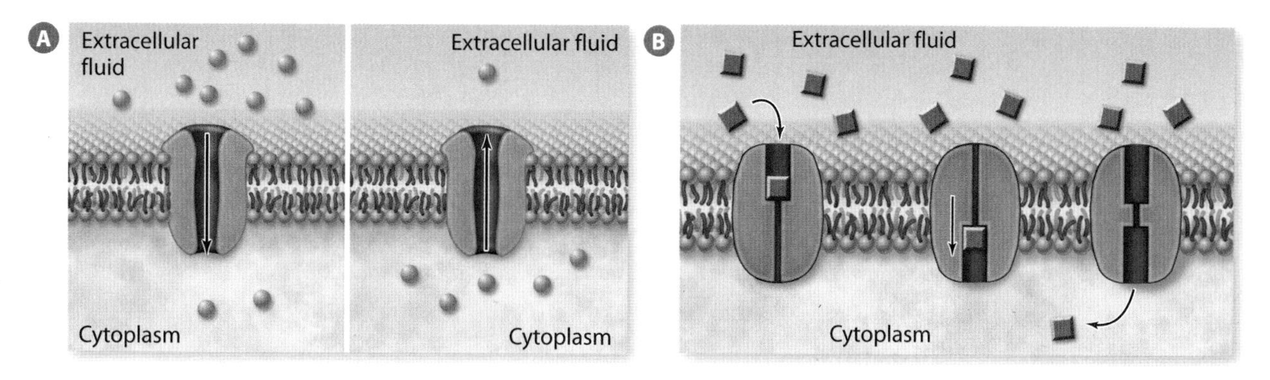

Figure 2.16 Facilitated diffusion involves membrane proteins. **A** Channel proteins form channels through membranes, which allow passage of specific ions and molecules from areas of higher concentration to areas of lower concentration. **B** Carrier proteins bind to molecules and carry them across a membrane from an area of higher concentration to an area of lower concentration.

As with a channel protein, the exterior of a carrier protein is usually composed of non-polar amino acids that interact with the non-polar interior of the membrane. Similarly, the interior of a carrier protein is lined with amino acids that can bind to the particle to be transported. For example, a carrier protein such as Glut1, which transports glucose molecules, is lined with polar or charged amino acids that can form intermolecular bonds with glucose molecules.

Malfunctions in carrier proteins can cause a variety of diseases. For example, cystinurea is a hereditary disease caused by the inability of carrier proteins to remove cystine and some other amino acids from urine. If it is not removed from urine, cystine crystallizes into painful stones, or *calculi*, that can block the flow of urine in the urinary tract.

Active Transport: Movement against a Concentration Gradient

Diffusion, facilitated diffusion, and osmosis are passive transport processes that move substances down their concentration gradients. However, cells also can actively move substances across a cell membrane against, or up, their concentration gradients. This process, called **active transport**, requires the expenditure of energy, usually from ATP.

active transport the transport of a solute across a membrane against its gradient

ATP, or adenosine triphosphate, is the main source of energy in the cell. An ATP molecule is derived from an adenosine nucleotide, but it has a triple phosphate group instead of a single phosphate group. The hydrolysis of the end phosphate group from an ATP molecule releases energy, as shown in **Figure 2.17**. This energy is then used by the cell for other activities. The use of energy from ATP in active transport can be direct or indirect. As you will see below, direct use of ATP is called primary active transport, and indirect use is called secondary active transport. (Although the remainder of the discussion of active transport will occasionally refer to ATP, you will learn more about this important molecule and its role in the process of cellular respiration and metabolism in Chapter 3.)

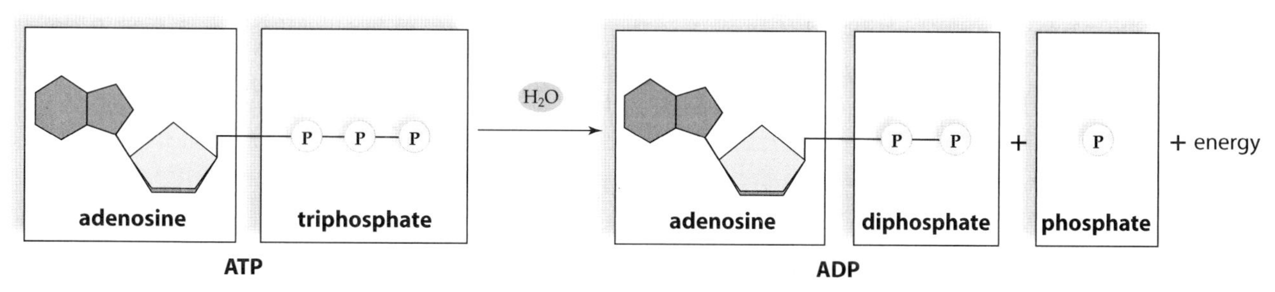

Figure 2.17 ATP undergoes hydrolysis to form ADP and phosphate, with the release of energy. The cell uses this energy for various functions, including the transport of molecules and ions across the cell membrane against their concentration gradients.

Primary Active Transport

A cellular process that uses ATP directly to move molecules or ions from one side of a membrane to the other is called *primary active transport*. For example, ion pumps are carrier proteins that use ATP to "pump" ions from one side of a membrane to the other, against a concentration gradient. One of the most well-studied examples is the *sodium-potassium pump*. This system transports sodium ions out of the cell while transporting potassium ions into the cell. Both processes occur against concentration gradients, so this carrier protein requires ATP to function, as shown in **Figure 2.18**.

At step 1 in the diagram, three sodium ions, Na^+, on the inside of the cell bind to the ion pump in the cell membrane. At step 2, an ATP molecule also binds to the ion pump, and it is hydrolysed to ADP and a phosphate group. The ADP is released, and the phosphate group temporarily attaches to the ion pump. This causes the ion pump to undergo a change in its shape, which releases sodium ions to the outside of the cell. On the outside of the cell, at step 3, two potassium ions, K+, bind to the ion pump. This binding causes the release of the phosphate group from the protein. The protein returns to its original shape, which causes the release of the potassium ions into the cytosol. The ion pump is then available to transport more sodium ions out of the cell.

Figure 2.18 The sodium-potassium ion pump transports ions across the cell membrane.

***Explain** why the action of a sodium-potassium pump will result in a build-up of negative charge inside the cell.*

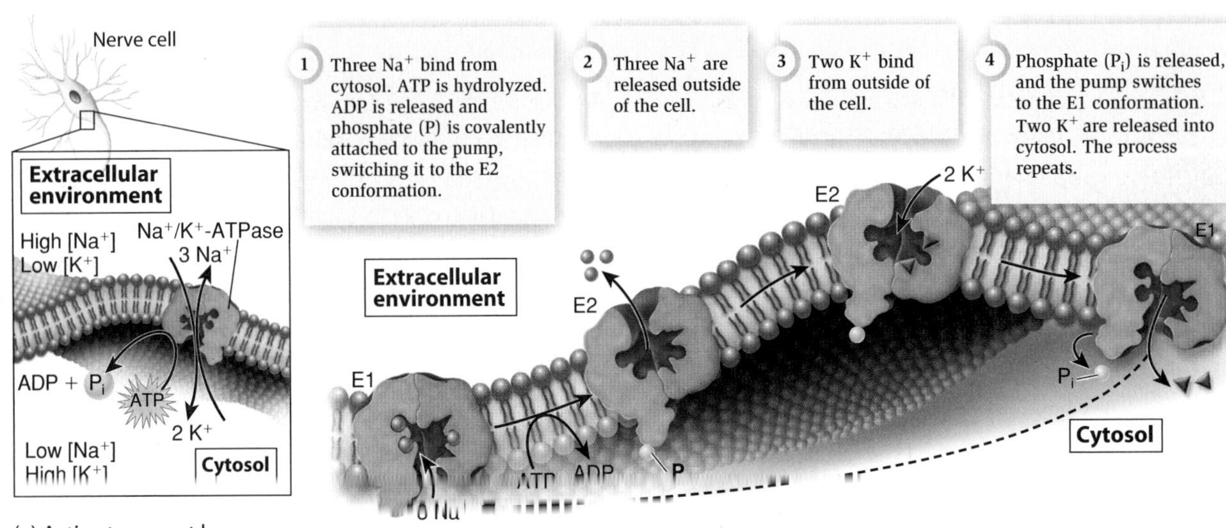

Nerve cell
Extracellular environment
High [Na⁺]
Low [K⁺]
Na⁺/K⁺-ATPase
3 Na⁺
ADP + Pᵢ
ATP
2 K⁺
Low [Na⁺]
High [K⁺]
Cytosol
1 Three Na⁺ bind from cytosol. ATP is hydrolyzed. ADP is released and phosphate (P) is covalently attached to the pump, switching it to the E2 conformation.
2 Three Na⁺ are released outside of the cell.
3 Two K⁺ bind from outside of the cell.
4 Phosphate (Pᵢ) is released, and the pump switches to the E1 conformation. Two K⁺ are released into cytosol. The process repeats.
Extracellular environment
E1
E2
E2
2 K⁺
E1
ATP
ADP
P
Pᵢ
Cytosol

(a) Active transport by the Na^+/ K^+-ATPase

(b) Mechanism of pumping

Secondary Active Transport

As an ion pump functions, a difference in charge, or electric potential, builds up across the membrane. One side of the membrane gains a more positive or negative charge compared to the other side, due to the accumulation of positive or negative ions. At the same time, a concentration gradient builds up across the membrane as the concentration of ions on one side increases compared with the other side. The combination of a concentration gradient and an electric potential across a membrane is called an *electrochemical gradient*. An electrochemical gradient stores potential energy that can be used by the cell.

Secondary active transport uses an electrochemical gradient as a source of energy to transport molecules or ions across a cell membrane. An example of secondary active transport is the hydrogen-sucrose pump. As shown in **Figure 2.19**, hydrogen ions are first pumped out of the cell by a hydrogen ion pump, which uses ATP as an energy source. This process creates an electrochemical gradient, with the area of higher concentration and greater positive charge outside the cell. Sucrose molecules outside the cell bind to a hydrogen-sucrose pump in the cell membrane. As well as binding sucrose molecules, this carrier protein allows hydrogen ions to move into the cell. As they do so, the hydrogen ions provide the energy that transports sucrose against its concentration gradient.

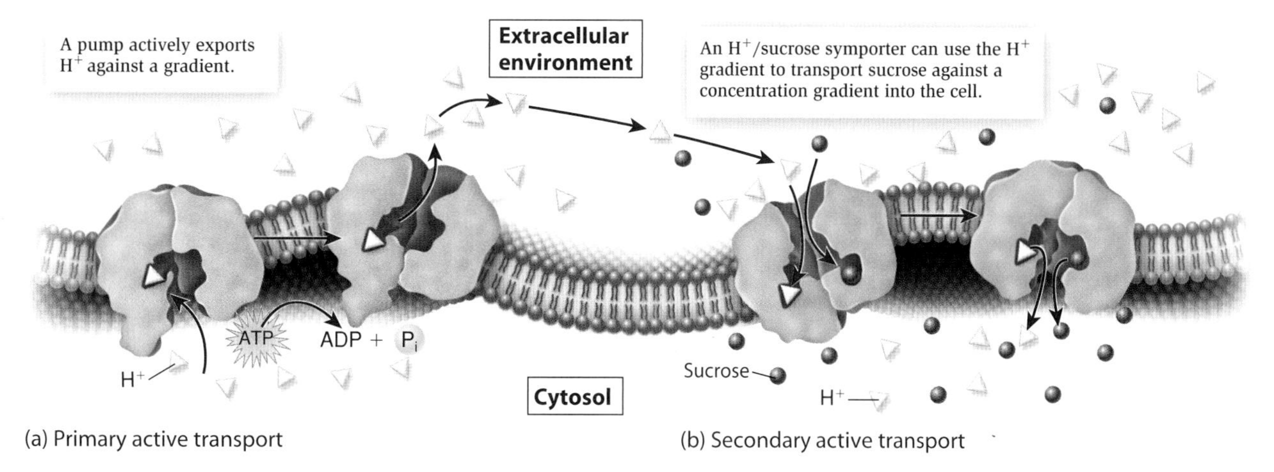
A pump actively exports H^+ against a gradient.
Extracellular environment
An H^+/sucrose symporter can use the H^+ gradient to transport sucrose against a concentration gradient into the cell.
ATP
ADP + P_i
H^+
Cytosol
Sucrose
H^+
(a) Primary active transport
(b) Secondary active transport

Figure 2.19 In secondary active transport, the electrochemical gradient created by primary active transport via an ion pump is used by a different protein to transport other molecules across a cell membrane. This kind of transport is common in bacteria and in plant cells.

Learning Check

25. What do facilitated diffusion and active transport have in common, and how do they differ?
26. Compare and contrast a channel protein and a carrier protein.
27. What is ATP, and what role does it play in active transport?
28. How is an electrochemical gradient similar to and different from a concentration gradient?
29. Distinguish between primary active transport and secondary active transport.
30. What is the sodium-potassium pump, and how does it work?

Activity 2.3 Understanding the Sodium-Potassium Pump

Palytoxin is a deadly compound found in certain marine animals. When scientists first isolated palytoxin from sea corals in the 1970s, they did not know how it affected people exposed to it. In time, they began to suspect that the toxin was interfering with the sodium-potassium pump. Researchers have measured the effect of palytoxin on ion transport through the sodium-potassium pump using the patch-clamp technique. This involves using a fine-tipped microelectrode to measure the electric current across pumps in the cell membrane. In this activity, you will examine some of the researchers' results and conclusions.

Procedure

1. Read the following observations that researchers made after adding palytoxin to a membrane, and then answer the questions.
   - Observation 1: The current across a single pump jumped from 0 picoamperes to 1 picoamperes.
   - Observation 2: When ATP was added to the cytoplasm-facing side of the membrane, the current across a group of pumps increased by a factor of 8 times.
   - Observation 3: A molecule 0.75 nm in diameter was able to pass through the pump. (For comparison purposes, a hydrogen atom measures 0.1 nm in diameter.)

Questions

1. How does the patch-clamp technique help researchers study ion transport across cell membranes?
2. In general, about 10^7–10^8 ions/s pass through an open ion channel. In contrast, only 10^2 ions/s pass through an ion pump. How would you expect the strength of an electric current across an ion channel to compare with the strength across an ion pump?
3. What does Observation 1 suggest about ion flow through the sodium-potassium pump when palytoxin is added?
4. Given that the sodium-potassium pump is a form of active transport, suggest an explanation for Observation 2.
5. What does Observation 3 suggest about the size of the passage through the sodium-potassium pump?

Membrane-Assisted Transport

Although a cell can accumulate and excrete smaller molecules and ions using membrane proteins, macromolecules are too large to cross the cell membrane through a channel or by means of a carrier protein. Instead, the cell forms vesicles to surround incoming or outgoing material and move it across the cell membrane through **membrane-assisted transport**. Like active transport, membrane-assisted transport requires energy from the cell. The two forms of membrane-assisted transport are endocytosis and exocytosis.

membrane-assisted transport transport method used to move materials that are too large to cross the cell membrane through a channel or carrier protein

endocytosis process by which the cell membrane engulfs extracellular material to bring it inside the cell

phagocytosis endocytosis involving solid particles

pinocytosis endocytosis involving liquid particles

Endocytosis

Endocytosis is the process by which a cell engulfs material by folding the cell membrane around it and then pinching off to form a vesicle inside the cell. **Figure 2.20** shows three methods of endocytosis: phagocytosis, pinocytosis, and receptor-mediated endocytosis.

If the material the cell takes in is made up of discrete particles, such as an organism or some other fragment of organic matter, the process is called **phagocytosis** (which literally means "cell-eating"). If the material the cell takes in is liquid, the process is called **pinocytosis** (which literally means "cell-drinking"). Virtually all eukaryotic cells constantly carry out these kinds of endocytotic processes, trapping particles and extracellular fluid in vesicles and ingesting them.

Receptor-mediated endocytosis involves the use of receptor proteins on a portion of a cell membrane that bind with specific molecules outside the cell. The area of the cell membrane containing receptor proteins is called a coated pit, because it is coated with a layer of protein. During this form of endocytosis, the receptor proteins bind with molecules and the pit folds inward to form a vesicle. The contents of the vesicle may be used by the cell or digested by the cell, and the receptor proteins may be recycled to the cell membrane.

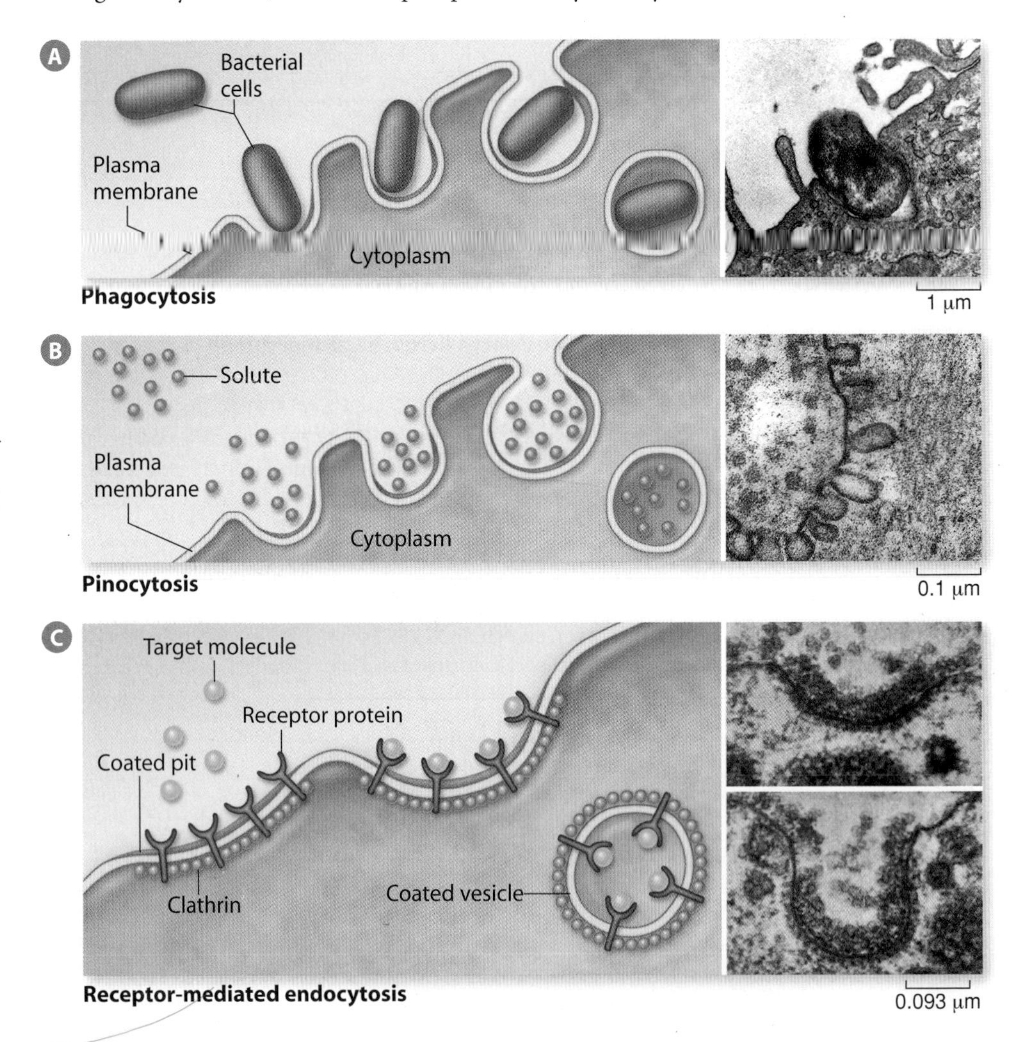

Figure 2.20 In phagocytosis, a cell engulfs a large particle along with some of the liquid surrounding it. In pinocytosis, a cell engulfs a liquid and the small particles dissolved or suspended in it. In receptor-mediated endocytosis, receptor proteins in the cell membrane bind to specific molecules outside the cell. The cell membrane folds inward to create a vesicle containing the bound particles. These vesicles are coated with clathrin, a protein that forms a cage around a vesicle.

Exocytosis

Macromolecules and other large particles can leave a cell by a process called exocytosis, which is shown in **Figure 2.21**. Exocytosis is the opposite of endocytosis. In **exocytosis**, vesicles that contain cell products to be released, or waste material to be excreted, fuse with the cell membrane and empty their contents into the extracellular environment. The vesicle itself becomes part of the cell membrane.

exocytosis transport method in which a vacuole fuses with the cell membrane and releases its contents outside the cell

In plant cells, exocytosis is an important means of exporting through the cell membrane the materials needed to construct the cell wall. In animal cells exocytosis provides a mechanism for secreting (releasing) many hormones, neurotransmitters, digestive enzymes, and other substances. For example, specialized glands secrete sebum, which is an oily substance that lubricates human skin, hair, and eyes. As another example, cells in the intestines of animals secrete enzymes and acid that aid in the digestive process.

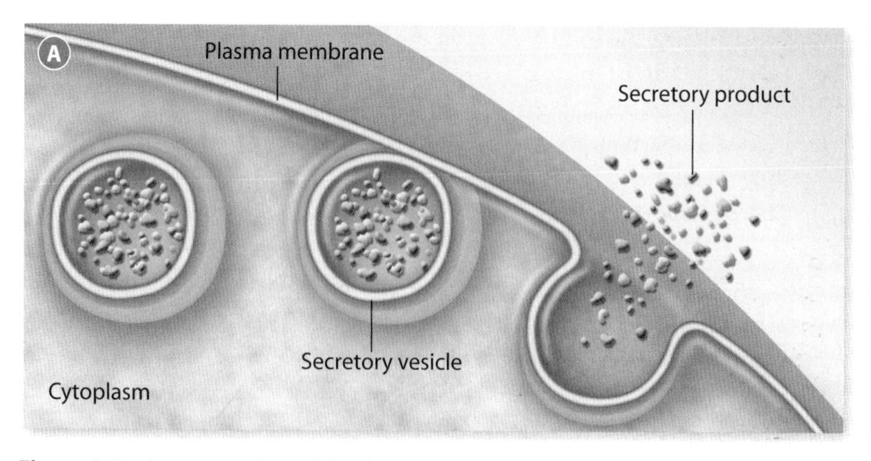

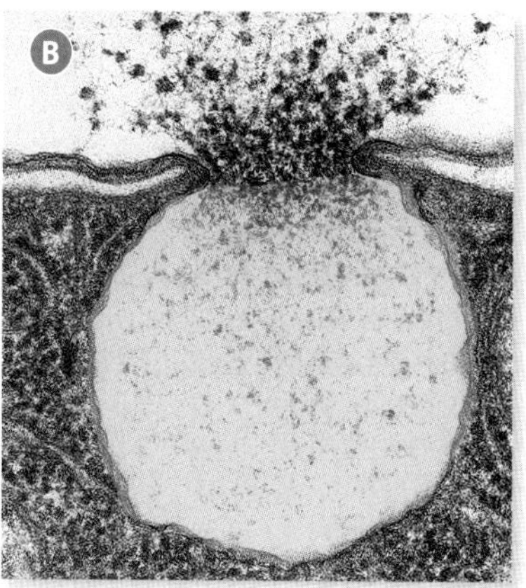

Figure 2.21 In exocytosis, vesicles that contain materials to be released from the cell fuse with the cell membrane and then release their contents into the extracellular environment.

Cellular Transport: A Summary

Table 2.2 summarizes the various mechanisms by which cells transport molecules, ions, and cellular materials or products across a cell membrane.

Suggested Investigation

Inquiry Investigation 2-B Diffusion Across a Semi-permeable Membrane

Table 2.2 Mechanisms for Transport of Substances Across a Cell Membrane

Is Energy Required for the Mechanism to Function?	Type of Cellular Transport Mechanism	Primary Direction of Movement of Substances	Essential Related Factor(s)	Examples of Transported Substances
No	diffusion	toward lower concentration	concentration gradient	lipid-soluble molecules, water, gases
No	facilitated diffusion	toward lower concentration	channel protein or carrier protein and concentration gradient	some sugars and amino acids
Yes	active transport	toward higher concentration	carrier protein and energy	sugars, amino acids, ions
Yes	endocytosis	toward interior of cell	vesicle formation	macromolecules
Yes	exocytosis	toward exterior of cell	fusion of vesicle with cell membrane	macromolecules

Investigating Multi-Drug Resistance in Cancer Cells

The Sharom lab team. Back row, left to right: Pulari Krishnankutty Nair, David Ward, Ashley Parfitt, Dr. Frances Sharom, Adam Clay, Peihua Lu. Front row, left to right: Kevin Courtney, Dr. Miguel Lugo, Jonathan Crawford, Joseph Chu. Not present: Dr. Gavin King.

Related Career

Oncologists are medical doctors who specialize in cancer treatment. An oncologist may be involved in cancer screening and diagnosis. This type of doctor is typically responsible for patient therapy and any patient follow up or palliative care as well. Becoming an oncologist in Canada involves completion of an undergraduate degree and a degree in medicine, followed by a period of further training, or residency.

Chemotherapeutic medications are often used to fight cancerous cells. Sometimes these cells can become resistant to the medications, however. The problem is compounded when there is resistance to several chemotherapeutic medications—a situation called multi-drug resistance.

Dr. Frances Sharom is a professor in the Department of Molecular and Cellular Biology at the University of Guelph in Guelph, Ontario. The Canadian scientist, shown above with the members of her lab team, is also the Canada Research Chair in Membrane Protein Biology. Dr. Sharom and members of her team are especially interested in multi-drug resistance due to the presence of the P-glycoprotein (Pgp) multi-drug transporter in the plasma membrane of cancerous tumour cells. A glycoprotein is a protein that has one or more carbohydrates attached to it. The Pgp transporter causes multi-drug resistance in these cells by pumping out hydrophobic chemotherapeutic medications. Because the cells are able to pump out the medication, they are less responsive to chemotherapeutic treatment. Medications pumped out by the Pgp transporter include TAXOL™, which is developed from the bark and needles of yew trees (*Taxus sp.*). TAXOL™ is used to treat several types of cancers, including breast, lung, bladder, and ovarian cancers. Vinblastine, a chemical that occurs naturally in the Madagascar periwinkle plant (*Catharanthus roseus*), is used to treat various lymphomas, as well as breast, testicular, and bladder cancer. It is also susceptible to transport out of the cell by the Pgp transporter. Thus, cancerous tumour cells that have Pgp-type resistance (that is, have the gene for the Pgp transporter) are less responsive to vinblastine treatment. Fortunately, the ATP-driven pump action that enables the Pgp transporter to transport chemotherapeutic medications from the cell is susceptible to other chemicals known as chemosensitizers, or modulators. These chemicals may reduce Pgp-type multi-drug resistance when administered with chemotherapeutic medications.

Dr. Sharom and her team are especially interested in how the Pgp transporter binds to medications and transports them out of the cell. They are also interested in how these processes are powered by the hydrolysis of ATP. The researchers use a technology known as fluorescence spectroscopy to map the multi-drug binding pocket of the Pgp transporter and to identify the conformational changes in the transporter when it binds to chemotherapeutic medications.

QUESTIONS

1. Draw a diagram or flowchart to illustrate how chemotherapeutic medications interact with cancerous tumour cells possessing the Pgp transporter, as well as how these medications interact with tumour cells lacking this transporter. Write a detailed caption for your diagram, clearly indicating which cells are drug-resistant.
2. Suggest how an understanding of the method by which the Pgp transporter binds to medications and transports them from the cell could be applied to cancer treatments.
3. Brainstorm three other careers that are related to the work described in this feature. Use Internet and/or print resources to research one of these careers. Then write a brief summary explaining the nature of this career.

Section 2.2 Review

Section Summary

- Transport of substances across membranes can occur by diffusion—the passive movement of a substance along a concentration gradient.
- Ions and large hydrophilic molecules cannot cross the cell membrane, but diffusion can still occur through facilitated diffusion, which involves the help of channel proteins and carrier proteins.
- Osmosis is the diffusion of water across membranes. The direction of movement depends on the solute concentration on either side of the membrane.
- Active transport uses energy and specialized protein carriers to move materials against a concentration gradient.
- Primary active transport uses pumps that directly use energy and generate a gradient.
- Secondary active transport involves the use of an existing gradient to actively transport another substance.
- In endocytosis , the cell membrane surrounds material and pinches off to form a vesicle. Phagocytosis is endocytosis involving solid particles. Pinocytosis is endocytosis involving liquid particles. In receptor-mediated endocytosis, specific molecules bind to receptors on the cell membrane.
- In exocytosis, material in a vesicle is secreted when the vesicle fuses with the membrane.

Review Questions

1. K/U Use the following terms to explain the movement of water across a membrane: solute, solvent, concentration.
2. C Draw an animal cell in an isotonic environment. Add labels and a caption to explain clearly the movement of substances in and out of this cell and the effect of this movement on the cell.
3. K/U Describe at least two different mechanisms a cell has to bring in material that otherwise cannot pass through the cell membrane.
4. A A drop of a 5% solution of NaCl is added to a leaf of an aquatic plant. When the leaf is viewed under a microscope, colourless regions appear at the edges of each cell as the cell membranes shrink from the cell walls. Describe what is happening and why.
5. T/I An egg is placed in a jar of household vinegar and left for about one week, after which time the shell has dissolved completely, leaving a thin membrane to contain the contents of the egg. The egg is then carefully removed and the vinegar residue is carefully washed from the membrane. Describe two procedures that could be performed to investigate active and/or passive transport with this membrane system.
6. K/U Explain what a channel protein is and why channel proteins are important to cells.
7. K/U Describe the function of cholesterol as it relates to the cell membrane.
8. C Use a Venn diagram to compare the similarities and differences of endocytosis and active transport.
9. K/U Identify three characteristics of substances that affect their rate of diffusion. Provide an explanation for each.
10. A If a cell membrane were completely permeable to all substances, could the cell continue to live? Explain your answer.
11. K/U Do substances that are moving in and out of a cell by diffusion and osmosis stop moving once the substances are evenly distributed on either side of the membrane? Explain your reasoning.
12. A Explain how the properties of the molecules and macromolecules that comprise biological membranes are important to processes that transport materials in and out of cells.
13. T/I The graph below shows the relative concentrations of five different ions inside and outside the cells of a unicellular pond alga. Interpret and explain the information in the graph.

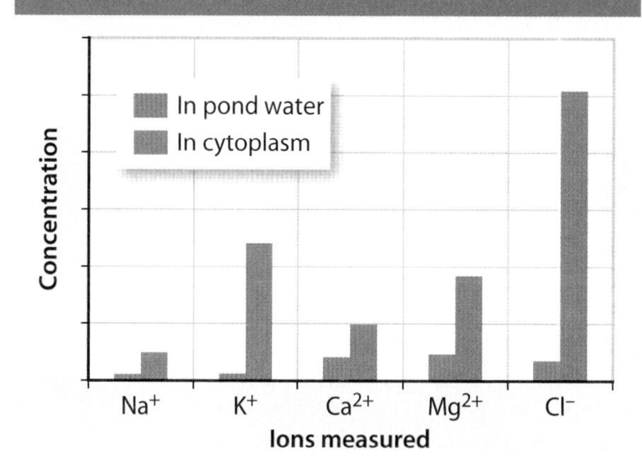

Plan Your Own
INVESTIGATION

2-A

Skill Check

- ✓ Initiating and Planning
- ✓ Performing and Recording
- ✓ Analyzing and Interpreting
- ✓ Communicating

Safety Precautions

- Be sure your hands are dry when you pick up any glassware so that you do not drop it.

Suggested Materials

- 250 mL beakers
- 10 percent NaCl solution
- distilled or filtered water
- potato strips
- scale or balance

*Go to **Scientific Inquiry** in **Appendix A** for information on formulating a hypothesis and prediction, and planning an investigation.*

*Go to **Organizing Data in a Table** in **Appendix A** for help with designing a table for data.*

*Go to **Significant Digits and Rounding** in **Appendix A** for help with reporting measurements.*

Demonstrating Osmosis

How can you observe the effects of osmosis without using a microscope or chemical tests? Working in groups and using the material provided, you will design and conduct an investigation to demonstrate osmosis. Your investigation must enable you to draw conclusions about the following.

- What is the solute concentration of the solutions used relative to the sample of plant material?
- What is the direction of the flow of water between a sample of plant material and the surrounding solution?

Pre-Lab Questions

1. Why is it so important not to eat or drink anything in the lab?
2. Describe the direction of the flow of water when plant cells are placed in a hypertonic solution.
3. What is the difference between a *control* in an investigation and *controlled variables*?
4. Identify a quantitative observation that you could make to determine whether a sample of plant material has gained or lost water.
5. Identify one or more qualitative observations that you could make to determine whether a sample of plant material has gained or lost water.

Question

How does solute concentration influence the direction of osmosis?

Hypothesis

Formulate a hypothesis about the direction of osmosis between a sample of plant material and a surrounding solution. Use this hypothesis as the basis of your experimental design.

Prediction

On the basis of your hypothesis, formulate a prediction about the results that you expect to observe.

Plan and Conduct

1. With your group, brainstorm different methods you could use the materials provided to test your hypothesis. Select one method for your experimental design.
2. As you prepare your procedure, consider the time required for each step. Also be sure to include a step about cleaning up your station.
4. Prepare the data table you will use to record your observations.
5. Review your procedure with your teacher. Do not begin the investigation until your teacher has approved your procedure.
6. Record your observations in your table. Make notes about any findings that do not fit in your data table. Record any questions that come up as you conduct your investigation.

Analyze and Interpret

1. Was your prediction correct? Make reference to specific results to explain what you mean.
2. Using the terms "isotonic," "hypotonic," and "hypertonic" as appropriate, describe the solute concentration of the solutions used relative to the sample of plant material.
3. What did you use as a control in your investigation? Why was this control necessary?

Conclude and Communicate

4. Did your observations support or refute your hypothesis? Explain.
5. State your conclusions about the influence of solute concentration on the direction of osmosis.

Extend Further

6. **INQUIRY** How could you revise your procedure to compare osmosis in different types of samples? What samples would you test? What controlled variables would you use? Design a new procedure to investigate these questions. Review your new procedure with your teacher, then carry it out and report on your findings.
7. **RESEARCH** What is oral rehydration therapy, and how is related to this investigation? When and why is oral rehydration therapy used, and how does it work?

Inquiry
INVESTIGATION

2-B

Skill Check

Initiating and Planning

✓ Performing and Recording

✓ Analyzing and Interpreting

✓ Communicating

Safety Precautions

- Be sure your hands are dry when you pick up any glassware so that you do not drop it.
- Wear gloves, a lab coat, and eye protection when handling Lugol's iodine.

Suggested Materials

- 15 cm length of pre-soaked dialysis tubing
- string or thread
- scissors
- sink or tray
- 10 percent glucose solution in a beaker
- 2 mL syringe or medicine dropper
- 10 percent starch solution in a beaker
- 300 mL beaker
- distilled or filtered water
- 4 glucose test strips
- watch or clock
- 1 percent starch solution in a beaker
- Lugol's iodine in a squeeze bottle

Go to *Writing a Lab Report* in *Appendix A* help with communicating the results of your investigation.

Diffusion Across a Semi-permeable Membrane

Dialysis tubing is a semi-permeable membrane used to separate small dissolved particles from large ones. Ions, water, and small organic molecules can pass through pores in the membrane by simple diffusion. Large molecules, however, cannot fit through the pores.

In this investigation, you will compare the ability of two solutes—starch and glucose—to diffuse across dialysis tubing. This tubing will serve as a model of a cell membrane.

Pre-Lab Questions

1. How many pieces of glassware should you carry at a time? Why?
2. Why is it important not to wear contact lenses in the laboratory?
3. What chemical test indicates the presence of starch?
4. Why is it necessary to wait a specified time after using a glucose test strip before you can read the result?

Question

Will starch or glucose be able to diffuse across the dialysis tubing?

Hypothesis

Make a hypothesis about which molecules will or will not be able to diffuse across the dialysis tubing.

Procedure

1. Obtain a piece of dialysis tubing from the container in which it is soaking. Cut a length of string or thread about 10 cm or 15 cm long. With the help of your lab partner, fold over one end of the dialysis tubing about 5 mm and tightly tie this end to keep it closed.
2. Gently rub the dialysis tubing at the free end so that it opens. While working over a sink or tray, pour the 10 percent glucose solution into the dialysis tubing until it is about half full.
3. Use a 2 mL syringe or medicine dropper to add 2 mL of 10 percent starch solution to the dialysis tubing.
4. Fold over the free end of the dialysis tubing and tie it tightly shut. If possible, rinse the dialysis tubing under running water.
5. The solution-filled dialysis tubing is your model cell. Place the model cell in a 300 mL beaker. Pour enough filtered or distilled water over the model cell to cover it. Note the time.
6. Test for glucose in the water using a glucose test strip. Be sure to follow the product instructions for using and reading the test strip.
7. After 45 min, use fresh glucose test strips to test fresh 10 percent glucose solution, 1 percent starch solution, and the water surrounding your model cell. Record your observations.
8. Put on your protective eyewear and gloves. Add a few drops of Lugol's iodine to the 1 percent starch solution. Keep adding Lugol's iodine until the mixture changes colour.
9. Add 1 mL–2 mL of Lugol's iodine to the water surrounding your model cell. Record your observations.
10. Clean up as directed by your teacher.

Analyze and Interpret

1. How did you check for the diffusion of glucose and starch across the model cell membrane?
2. What was an example of a control in this investigation?
3. If you added 20 mL of 10 percent glucose solution in the dialysis tubing and then added 2 mL of 10 percent starch solution, what was the total concentration of starch in the model cell?
4. Describe the appearance of the model cell when you first covered it with water and 45 min later.

Conclude and Communicate

5. Describe the relative solute concentrations inside and around the model cell when you first covered it with water.
6. Describe the direction of diffusion of solute and water molecules across the model cell membrane over the course of the investigation.
7. Was your hypothesis supported by your observations? Explain.
8. Is dialysis tubing an effective model of a cell membrane? Justify your response.

Extend Further

9. **INQUIRY** Design an investigation to determine the approximate size of molecules of a common household substance, such as vitamin C, beet juice, or vinegar.
10. **RESEARCH** How did Canadian surgeon Gordon Murray's invention help people with failing kidneys? Use your understanding of diffusion to explain how his invention worked.

Case Study

Synthetic Red Blood Cells

Benefit or Detriment to Society?

Scenario

For many years, scientists have sought to design synthetic red blood cells (sRBCs) that can transport oxygen throughout the body. In 1957, Thomas Chang, then an undergraduate student at McGill University in Montreal, experimented with improvised materials in his dormitory to construct a permeable "cell" that carried hemoglobin, the iron-containing molecule in red blood cells responsible for oxygen transport in the body.

Since then, many other scientists have tried to create sRBCs and there have been several successes. Many scientists believe that practical application of sRBCs in medicine will become a reality in the near future. However, this future is by no means assured. Read the following articles, each of which has a different viewpoint regarding the future of sRBCs, and then decide whether this technology is beneficial or detrimental to society.

Synthetic Red Blood Cell Research: A Call for Further Funding

Ivan Mikhailovich, PhD

An artist's conception of theoretical synthetic blood cells, or respirocytes, in a blood vessel. Synthetic cells would have the same function as red blood cells and could be used to treat various blood conditions.

SYNTHETIC BLOOD GIVES LIFE WHERE REAL BLOOD CANNOT. Thanks to sRBC technology, generations to come can be assured a stable and safe blood supply. Access to this blood supply would not depend on refrigeration. The relatively short shelf life of donated blood would no longer be a problem. Nor would this blood carry with it the risks of blood-borne diseases such as HIV and hepatitis C. Adverse reactions due to blood-type incompatibility would be overcome. Such a world is possible, but only with further funding for sRBC research.

Research in sRBC technology has advanced greatly in the last few years. In 2004, Shinji Takeoka published ground-breaking research in the *Journal of the Japan Medical Association*. Takeoka's team encapsulated a high concentration of hemoglobin (isolated from expired donated blood) within an artificial phospholipid bilayer membrane. The stability of this artificial bilayer ensures a long shelf life. In liquid state, the cells have a shelf life of approximately two years at room temperature. They last even longer if dehydrated. In comparison, donated blood lasts only three weeks with refrigeration.

In 2009, Samir Mitragotri and his team at the University of California, Santa Barbara created sRBCs that mimic natural red blood cells even more closely, in both shape and function. Mitragotri's team synthesized these cells by coating a polymer mould, which was later removed, with numerous layers of hemoglobin and other proteins. The sRBCs have the same doughnut-like shape that increases the surface area of human red blood cells. By treating the polymer with a common household chemical, rubbing alcohol, Mitragotri's team produced artificial cells with the same size, shape, and oxygen-carrying ability as human red blood cells. The sRBCs can even manoeuvre through tiny capillaries.

These advances have led to clinical trials of sRBCs and practical use in medicine. In the near future, sRBCs will also be engineered to target pharmaceuticals to specific areas in the body. With adequate funding, other advances will soon follow.

Nanotechnology Reviews...Synthetic Red Blood Cells

Hardeep Hundial, MSc

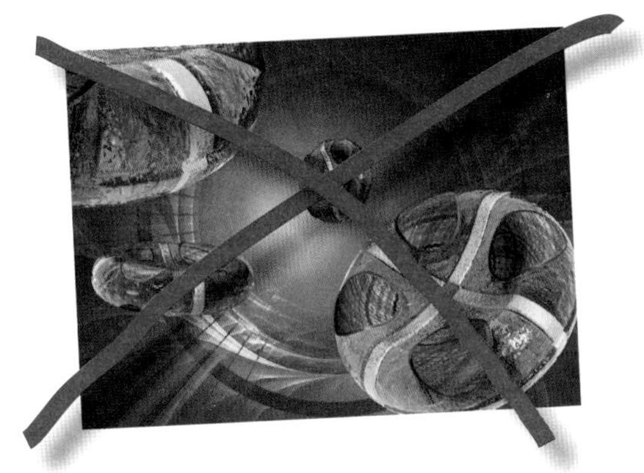

Several practical hurdles make us skeptical that synthetic blood will be used in clinical medicine anytime soon. Two of these hurdles are safety and money.

Disease-free blood that lasts for years and that can be easily transported to accident locations or disaster sites—what could be the downside? A lot, it turns out. Before practical application of this technology becomes possible, sRBCs need to be tested to make sure they are both safe and effective. While animal testing is a starting point, this technology also needs to be tested on its ultimate target, human beings. Human testing, referred to as clinical trials, raises many difficult questions, such as, "Can participant safety be ensured?"

Fortunately, there are trials of other blood substitutes to learn from: clinical trials for non-cellular synthetic blood that began in the 1980s. Non-cellular synthetic blood products are the predecessors of sRBCs. They are based on either modified hemoglobin molecules or fluorocarbons, which contain both fluorine and carbon. Both types of synthetic molecules can carry oxygen and transport it within the body as real blood does. Unlike real blood, however, these blood products contain no cells. Early clinical trials were conducted in the United States and Germany with both trauma victims and volunteers. Several allergic reactions were recorded, as were incidents of kidney failure. In the late 1990s, one company stopped its clinical trials when it learned that nearly half of the patients that received synthetic blood died, compared to 17 percent of patients in the control group. In 2008, the *Journal of the American Medical Association* published a study reporting a 30 percent greater risk of death and a tripling of the heart attack rate in patients that received blood substitutes in clinical trials. As a result, the Ottawa Health Research Institute halted all human blood–substitute testing in Canada.

Will cellular blood substitutes such as sRBCs pose less risk for clinical trial participants than their non-cellular counterparts? There is no way to know until clinical trials take place. However, based on past performance of non-cellular blood substitutes in the human body, such trials would generate serious ethical questions. It makes little sense to continue wasting valuable funding on this research, especially when blood-product screening can now ensure a disease-free product. Funds would be much better spent on initiatives to increase blood donations or research to improve the shelf life of human blood products.

Research and Analyze

1. Find out more about the two main types of non-cellular synthetic blood products. Describe these blood products and the results of clinical trials associated with them, as well as any practical medical applications. Does your research support the claim made in the *Nanotechnology* review that it would be unsafe to test sRBCs on humans? Write a supported opinion piece to express your viewpoint.
2. Respirocytes are theoretical nanomachines that could function as synthetic red blood cells. Scientists claim that such cells could carry much more oxygen than normal red blood cells, enabling humans to stay underwater for hours or sprint at high speeds for several minutes, all on a single breath. Find out more about how respirocytes might influence human performance in the future. Based on your research, decide whether you would prefer to work for a company that has successfully pioneered respirocyte technology or a non-profit group that opposes this technology.
3. Research the benefits and the risks of sRBCs in more detail. Create a table to summarize your findings, as well as the points made in this case study. Using the information in your table, perform a risk-benefit analysis of sRBC technology. Refer to Analyzing STSE Issues in Appendix A if necessary.

Take Action

4. **PLAN** You and a group of other students are volunteering as interns at a science webzine. Your assignment is to research, design, and write an edition about the science and the issues related to synthetic red blood cells (sRBCs). Complete the following questions as a group.
   a. Discuss whether sRBC research is beneficial or detrimental to society. Should sRBC research and associated clinical trials proceed? Refer to your responses to questions 1 to 3 in your discussion.
   b. Based on your discussion, design your webzine edition. Each intern will be responsible for one article in the edition. Your design should include an outline of each proposed article. Articles can have differing viewpoints, but all viewpoints must be supported by research from reputable sources.
5. **ACT** Write and produce the webzine.

Chapter 2 SUMMARY

Section 2.1 Structures and Functions of Eukaryotic Cells

The cell membrane defines the boundary between the internal and external environment of a cell, and organelles compartmentalize its biochemical activities.

Key Terms

nucleolus	lysosome
nuclear envelope	peroxisome
nuclear pore complexes	vacuole
endoplasmic reticulum (ER)	chloroplast
ribosome	mitochondrion
endomembrane system	cell wall
vesicle	cytoskeleton
Golgi apparatus	fluid mosaic model

Key Concepts

- Animals, plants, fungi, and protists are composed of eukaryotic cells, which have DNA, a cell membrane, and cytoplasm. Cytoplasm consists of organelles, the cytosol, and molecules and ions dissolved or suspended in the cytosol. The nucleus includes the nuclear envelope, which is studded with nuclear pore complexes, the nuclear matrix, and the nucleolus.
- The endoplasmic reticulum (ER), consisting of the rough ER and the smooth ER, is a system of channels and membrane-bound-sacs enclosing a narrow space called the lumen.
- The endomembrane system includes the nuclear envelope, the endoplasmic reticulum, the Golgi apparatus, the cell membrane, and vesicles. This system synthesizes, modifies, and transports proteins and other cell products.
- Animal cells contain many small vesicles. Plant cells contain a single large central vesicle called a vacuole.
- Chloroplasts trap light energy from the Sun in the form of high-energy organic molecules. Mitochondria break down high-energy organic molecules to release usable energy.
- Cells of plants, fungi, and many types of protists have a cell wall, which provides protection and support.
- The cytoskeleton provides structure, shape, support, and motility.
- The fluid mosaic model visualizes the cell membrane as a mosaic of proteins and other molecules in a fluid phospholipid bilayer.

Section 2.2 The Transport of Substances Across a Cell Membrane

The passage of substances across a cell membrane takes place by means of passive transport, active transport, and membrane-assisted transport.

Key Terms

passive transport	active transport
concentration gradient	membrane-assisted transport
diffusion	endocytosis
osmosis	phagocytosis
facilitated diffusion	pinocytosis
channel protein	exocytosis
carrier protein	

Key Concepts

- Transport of substances across membranes can occur by diffusion—the passive movement of a substance along a concentration gradient.
- Ions and large hydrophilic molecules cannot cross the cell membrane, but diffusion can still occur through facilitated diffusion, which involves the help of channel proteins and carrier proteins.
- Osmosis is the diffusion of water across membranes. The direction of movement depends on the solute concentration on either side of the membrane.
- Active transport uses energy and specialized protein carriers to move materials against a concentration gradient.
- Primary active transport uses pumps that directly use energy and generate a gradient.
- Secondary active transport involves the use of an existing gradient to actively transport another substance.
- In endocytosis , the cell membrane surrounds material and pinches off to form a vesicle. Phagocytosis is endocytosis involving solid particles. Pinocytosis is endocytosis involving liquid particles. In receptor-mediated endocytosis, specific molecules bind to receptors on the cell membrane.
- In exocytosis, material in a vesicle is secreted when the vesicle fuses with the membrane.

Chapter 2 REVIEW

Knowledge and Understanding

1. How do peroxisomes protect themselves from the reactive hydrogen peroxide they produce?
   a. The hydrogen peroxide is stored within vesicles inside the peroxisomes.
   b. Hydrogen peroxide is passively transported into the cytosol.
   c. Hydrogen peroxide is actively transported out of peroxisomes.
   d. Carbohydrates are combined with hydrogen peroxide to create less reactive substances.
   e. The enzyme catalase breaks down hydrogen peroxide within peroxisomes.
2. Which of the following is common to both mitochondria and chloroplasts?
   a. chlorophyll
   b. grana
   c. thylakoids
   d. double membrane
   e. nucleolus
3. Which cellular process directly uses the energy of ATP hydrolysis to move molecules against their concentration gradient across a membrane?
   a. facilitated diffusion
   b. primary active transport
   c. osmosis
   d. diffusion
   e. secondary active transport
4. Which organelle uses enzyme-catalyzed hydrolysis to break down macromolecules into simpler molecules?
   a. rough endoplasmic reticulum
   b. nucleus
   c. lysosome
   d. peroxisome
   e. Golgi apparatus
5. While observing plant cells, a student noticed that chloroplasts appeared throughout each of the cells. Before the observation she had expected to see the chloroplasts at the periphery of the cell near the cell wall. How can the student's observation be explained?
   a. During preparation the cells may have been exposed to a hypertonic solution that resulted in plasmolysis.
   b. The cells used for the observation were isolated from the roots rather than the shoots of the plant.
   c. Light from the microscope caused the cells to become actively photosynthetic.
   d. During preparation the cells may have lysed as a result of exposure to a hypotonic solution.
   e. Many of the chloroplasts may have contained an excess of thylakoids.
6. Which organelle is involved in both the synthesis of bile acids and redox reactions that break down toxic substances?
   a. lysosome
   b. central vacuole
   c. mitochondrion
   d. peroxisome
   e. Golgi apparatus
7. What do glycoproteins and glycolipids have in common?
   a. Both molecules are modified in lysosomes before being secreted from the cell.
   b. Carbohydrates are covalently bonded to them.
   c. Each is synthesized by ribosomes in the cytosol.
   d. Each is synthesized by ribosomes at the rough ER.
   e. Each is loosely attached to one of the leaflets of a membrane.
8. Chefs often place salad greens in water to make the vegetables crispy. What organelle is principally responsible for this increase in plant cell rigidity?
   a. Golgi apparatus
   b. smooth ER
   c. lysosome
   d. chloroplast
   e. central vacuole
9. Why do phospholipids placed in water form bilayers?
   a. The "heads" of the phospholipids engage in hydrophobic interactions with water molecules.
   b. The fatty acid "tails" engage in hydrogen bonding with water molecules.
   c. Each of the molecules has a polar and a non-polar region.
   d. The water molecules cannot interact with phospholipids since each has a different polarity.
   e. Lipid bilayers are required for the attachment of peripheral proteins.
10. Which event requires a net input of energy?
    a. passage of an ion through a channel protein
    b. passage of an uncharged molecule through a channel protein
    c. the facilitated diffusion of a polar molecule out of a cell by a carrier protein
    d. the unassisted passage of a non-polar solute through the phospholipid bilayer of a membrane as it moves down its concentration gradient
    e. the movement of an ion out of a cell against its electrochemical gradient

11. In the fluid mosaic model of membranes, what does the term "mosaic" tell us about membranes?
  a. Membranes are composed of a mixture of substances.
  b. The components of membranes float in a bilayer of lipids.
  c. Some membrane proteins have polar regions that are exposed at both faces of a membrane.
  d. The two leaflets of any given membrane are identical but may be different when compared to membranes from another source.
  e. Membranes have lipid molecules sandwiched between layers of protein.
12. What transport mechanism exports material from a cell without using a transport protein or the movement of the material directly through the lipid bilayer?
  a. plasmolysis
  b. exocytosis
  c. hydrolysis
  d. signal transduction
  e. pinocytosis
13. Which of the situations below would result in plasmolysis?
  a. placing an animal cell in an isotonic solution
  b. washing a normal plant cell with a hypotonic solution
  c. exposing a typical animal cell to a hypotonic solution
  d. leaving a plant cell in a hypertonic solution
  e. moving a normal, healthy plant cell into an isotonic solution
14. In which of the following does the membrane of a vesicle become part of the cell membrane?
  a. pinocytosis
  b. phagocytosis
  c. exocytosis
  d. active transport
  e. facilitated transport
15. Oxygen enters a cell by which process?
  a. pinocytosis
  b. diffusion
  c. primary active transport
  d. facilitated diffusion
  e. osmosis

Answer the questions below.

16. Identify one example of a protein type made by ribosomes bound to the rough endoplasmic reticulum. Identify one example of a protein type made by ribosomes freely suspended in the cytosol.
17. Describe two structural features of membrane phospholipids that affect membrane fluidity. Use one or two sentences to explain how each feature affects membrane fluidity.
18. Explain why it is beneficial for a cell to use enzymes that work best at an acidic pH inside lysosomes.
19. List three factors that affect the rate of diffusion of a substance through a membrane (a simple lipid bilayer without transport proteins). Use one or two short sentences to describe the effect of each factor on diffusion.
20. Identify the cell structures described in the following statements.
  a. This very long extension of the cell membrane contains many microtubules and is used to propel cells in their environment.
  b. This series of curved membrane sacs is used to modify, sort, and package lipids and proteins.
  c. These flattened, disk-like membranes contain pigment molecules that absorb light energy.
21. Three transport proteins are shown in the diagram below. For each of the labels (a to h) in the diagram, identify the concentration of the transported molecule as either H (high) or L (low). The concentration brackets for one of the transported molecules have been completed for you as an example.

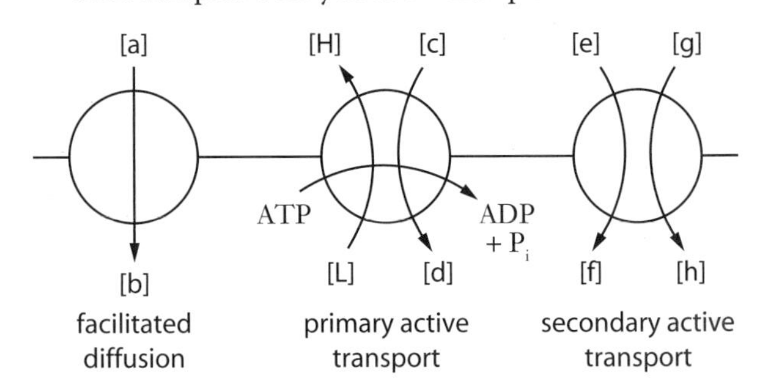

22. Describe the sources of energy used by primary active transport and secondary active transport.
23. Describe both an electrochemical gradient and a concentration gradient across a membrane, and explain the difference between them.
24. Channel and carrier proteins transport molecules at different rates. Explain why this is the case.

25. List the structures that are found in plant cells but not animal cells. Describe one important function of each structure.

26. Explain why most membrane proteins and phospholipids are able to move relative to one another and why this characteristic is important.

Thinking and Investigation

27. Secreted proteins follow a pathway from their synthesis to their release from the cell. Some steps in that pathway are listed below. Arrange and rewrite the steps in the proper order that ends with their release from the cell.
    - **a.** The proteins are packaged into vesicles that form at the *trans* face of the Golgi.
    - **b.** Vesicles merge with the *cis* face of the Golgi apparatus.
    - **c.** Vesicles fuse with the cell membrane and release proteins by exocytosis.
    - **d.** During their synthesis proteins are inserted into the lumen of the rough ER.
    - **e.** Proteins are packaged into vesicles formed at the rough ER.
    - **f.** Proteins are modified in the Golgi apparatus.

28. Using your understanding of osmosis, predict what change(s) would occur in an animal cell suddenly exposed to a solution with a much higher solute concentration than the interior of the cell. Explain your reasoning with one or two sentences using the term(s) isotonic, hypertonic, or hypotonic as appropriate.

29. Which organelle would you expect to find in greater quantity in the cells of an organ specialized for the synthesis of a lipid hormone-smooth ER or rough ER? Explain your reasoning.

30. Explain why the pumping of a different number of sodium and potassium ions across the cell membrane by sodium-potassium pumps is biologically significant.

31. Plants with non-woody stems (herbaceous plants) rely heavily on turgor pressure to help hold them upright.
    - **a.** Given that fact, what do you think the normal, healthy environment is like for those plant cells (isotonic, hypotonic, or hypertonic)?
    - **b.** If the environment of these plant cells had less water but otherwise had the same amount of solutes present as normal (e.g. drought conditions), would the plants be better or less able to stand upright? Explain with one or two sentences.

32. Many polar molecules are able to diffuse directly through the lipid bilayer of a membrane, and yet cells have carrier proteins for the passive transport of many such molecules. Suggest a possible explanation why a cell would have carriers for molecules able to diffuse through a lipid bilayer.

33. Antibodies are proteins of the immune system that are able to recognize and bind specific molecules, even other proteins such as membrane proteins. The binding of an antibody can be visualized by tagging it with a fluorescent dye visible by fluorescence microscopy. This type of microscopy employs an ultraviolet light source that causes the attached dye molecules to fluoresce various colours depending on the type of dye used. During microscopy the ability of the attached dyes to fluoresce can be destroyed at will, even over small areas of a cell surface, using a finely focussed laser beam. These "photobleached" dyes appear dark. Given this technological ability, construct a flowchart that shows the steps you could take to demonstrate that many membrane proteins are able to move laterally when a membrane is in the fluid state.

34. Several cold-sensitive and cold-tolerant varieties of a plant were discovered in the wild. Careful observations revealed that the membrane fluidity of the cold-sensitive plants varied widely with changes in temperature whereas the cold-tolerant plants showed much less fluctuation in membrane fluidity at those same temperatures. Using your knowledge and understanding of membrane fluidity and phospholipid structure, suggest a testable hypothesis for the observations made with these plants.

35. Radioactive amino acids are readily available and can be used by cells to make proteins. The location of radioactive proteins in cells can be detected and imaged microscopically. Given those facts and your understanding of protein synthesis in the endomembrane system, design the procedure for an investigation to show that many proteins synthesized at the rough ER move through compartments of the endomembrane system and emerge at the *trans*-face of the Golgi apparatus.

36. Provide a short point-form explanation for the appearance of carbohydrate chains at the external face but not the cytoplasmic (cytosolic) face of the cell membrane.

37. Two different molecules (A and B) are the same size and both are uncharged polar molecules. Neither molecule diffuses easily through a lipid bilayer, but both have transport proteins for their specific import into cells. Molecule A is transported into cells at a rate that is approximately 1000 times faster than molecule B, even when their concentration gradients are identical. Using your understanding of membrane transport, offer an explanation for the difference in the rates of transport of the two molecules.

Communication

38. Proteins control a wide variety of cellular processes. Create a table like the one shown below, listing three functions of membranes that are directly due to the presence and function of membrane proteins. Use point form to describe each function.

Function	Description

39. Write three or four sentences to explain why phospholipids spontaneously form lipid bilayers when placed in water. Draw and label a diagram of a phospholipid bilayer to support your answer.

40. Explain the following statement in your own words: Unlike pinocytosis, receptor-mediated endocytosis brings specific molecules into a cell.

41. Draw and label a Venn diagram that compares and contrasts pinocytosis and phagocytosis.

42. List five of the several different ways molecules might enter a cell.

43. Construct a concept map for the entry of polar molecules across a lipid bilayer. Be sure to include the major categories of entry as higher-order labels in your map, using the terms "diffusion", "transport protein", and "membrane-assisted transport".

44. Write a "shock and awe" tabloid-type newspaper article about the effects of either hypotonic or hypertonic solutions on red blood cells. Be sure to create a title that will grab and hold your readers. The article should be three to five very short paragraphs in the inverted-pyramid style typical of a newspaper article.

45. Draw and label a Venn diagram that compares and contrasts the channel proteins and carrier proteins used in facilitated diffusion.

46. Construct a Venn diagram that compares and contrasts the carrier proteins used in passive transport and primary active transport.

47. Although some channel proteins are open all the time, others are gated and thus open and close in response to a variety of stimuli. List three types of changes that are known to affect the opening and closing of gated channels.

48. Draw and label a Venn diagram comparing and contrasting the structure, function and location of chloroplasts and mitochondria.

49. Draw a flow chart to explain how the rough ER, smooth ER, Golgi apparatus and vesicles function together as components of the endomembrane system. Begin with the statement:

    Polypeptides are produced by rough ER and put into the lumen.

50. Summarize your learning in this chapter using a graphic organizer. To help you, the Chapter 2 Summary lists the Key Terms and Key Concepts. Refer to Using Graphic Organizers in Appendix A to help you decide which graphic organizer to use.

Application

51. **BIG IDEAS** Biochemical compounds play important structural and functional roles in cells of all organisms. Vesicles were constructed in the lab using only a single type of phospholipid in the presence or absence of cholesterol. Vesicle batch A contained vesicles with lipid bilayers composed only of the phospholipid. Vesicle batch B contained vesicles with both the phospholipid and cholesterol. The membrane fluidity of the two different batches of vesicles was examined at a high temperature and at a low temperature.
    a. Which vesicle batch would you expect to exhibit the greatest membrane fluidity at the high temperature?
    b. Which vesicle batch would you expect to exhibit the lowest membrane fluidity at the low temperature?
    c. Using your understanding of cholesterol and membranes, explain your choices in a short paragraph of a few sentences.

52. The rate of entry of a molecule into cells was examined by a researcher attempting to determine the transport mechanism for the molecule. In several parallel experiments, the cells were exposed to different starting concentrations of the molecule, and the rate of entry was determined at each concentration. As shown

in the graph below, the rate of entry was linear at low starting concentrations, but at high concentrations the rate of entry levelled off.

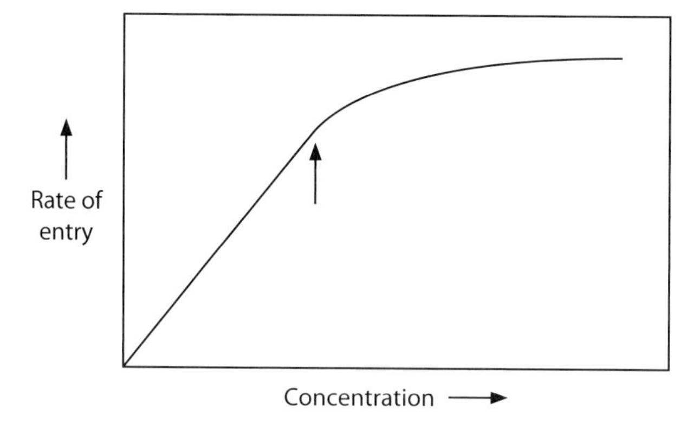

a. What mechanism likely resulted in the entry of the molecules into the cells-diffusion through the lipid bilayer of the membrane or facilitated diffusion?

b. Explain the shape of the line seen in the graph. Why did the rate of diffusion begin to slow at the point indicated by the arrow in the graph?

c. If the plot indicated a linear relationship over the entire range of concentrations (even at very high concentrations), how would this change your answer to part (a)?

53. Predict which of the following cellular processes would be directly affected by the improper functioning of mitochondria. Explain your choice(s).

A. gain or loss of water by a cell

B. facilitated diffusion of glucose into or out of a cell

C. pumping ions by a sodium-potassium pump

D. moving hydrogen ions (protons) against their gradient by a hydrogen ion pump

54. When a large number of animal cells were initially placed into a slightly hypotonic solution, there was a net movement of water across their cell membranes. Once the concentration of water is equal inside and outside the cells, would you expect there to be any movement of water across the cell membranes? Explain your reasoning.

55. Almost all membrane proteins can be easily classified by a biochemist or cell biologist as being either integral or peripheral. What is the principal difference between an integral and a peripheral membrane protein?

56. Predict which type of protein would be released from the phospholipid bilayer by a mild treatment of cells such as the application of heat or a change in the pH of the environment. Explain with one or two sentences.

57. Most cell membrane receptor proteins bind a signal molecule on the exterior side of the cell membrane and then after a change in their shape some other molecule is able to bind the cytosolic side of the receptor protein. Are these types of receptor proteins peripheral or integral? How do you know?

58. While trying to purify a membrane protein a cell biologist discovered that the protein could only be released from the membrane by destroying the phospholipid bilayer with detergent. What type of protein was the cell biologist purifying? Explain.

59. Examine the illustration shown below. Write a detailed caption of at least four sentences to identify and describe the process shown in the illustration.

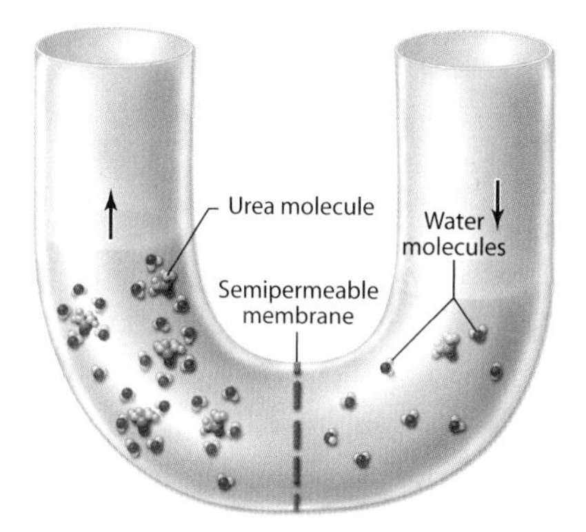

60. **BIG IDEAS** Biological molecules and their chemical properties affect cellular processes and biochemical reactions. Since the enzymes inside a lysosome work best at an acidic pH of ~5 (a higher concentration of protons than in the cytosol), a lysosome needs to establish and maintain a low pH in its interior.

a. What type of gradient is present when two compartments separated by a membrane each have a different pH?

b. Would the maintenance of a special environment like the interior of a lysosome require the expenditure of energy? Explain why (or why not).

c. Using your knowledge of membrane transport suggest a mechanism that could be used by a lysosome to both establish and maintain a low pH in its interior. Draw and label a diagram to help illustrate your written statement(s).

Chapter 2 | SELF-ASSESSMENT

Select the letter of the best answer below.

1. **K/U** Which components does the nucleolus contain?
   a. nucleoplasm
   b. nuclear matrix
   c. RNA and protein
   d. lumen
   e. nuclear core complex

2. **K/U** Which structure is typically found in plant cells but not animal cells?
   a. cell membrane
   b. mitochondria
   c. peroxisome
   d. nuclear envelope
   e. large central vacuole

3. **K/U** Which is a function of rough endoplasmic reticulum?
   a. synthesis of lipids
   b. detoxification of alcohol
   c. production of estrogen
   d. protein synthesis of cell-membrane components
   e. None of the above.

Use the table below to answer question 4.

Results for the F_1 Generation

1	rough endoplasmic reticulum
2	trans face of Golgi apparatus
3	cis face of Golgi apparatus
4	smooth endoplasmic reticulum
5	transport vesicle
6	secretory vesicle

4. **K/U** What is the correct sequence of organelles involved in producing and transporting a glycoprotein that will be a receptor on the cell surface?
   a. 1, 2, 3, 4, 5, 6
   b. 1, 4, 5, 3, 2, 6
   c. 4, 5, 1, 2, 3, 6
   d. 4, 1, 2, 3, 5, 6
   e. 6, 3, 2, 5, 1, 4

5. **K/U** Which cellular component makes up the inner shaft of cilia?
   a. flagella
   b. cell membrane
   c. microfilaments
   d. microtubules
   e. intermediate filaments

6. **K/U** Which statement best describes the movement of molecules that make up a cell membrane?
   a. The membrane molecules are held together by covalent bonds.
   b. Phospholipids leak out of the membrane if it is torn.
   c. The molecules in the membrane can move about fairly freely.
   d. Phospholipids can move about freely, but all proteins are fixed in place.
   e. Proteins can move about freely, but phospholipids are fixed in place.

7. **K/U** Which is an example of passive transport?
   a. A cell takes in particles by pinocytosis.
   b. The sodium-potassium pump transports ions from one side of the membrane to the other, against a concentration gradient.
   c. A pump uses ATP as an energy source to transport hydrogen ions out of a cell.
   d. An ion channel permits sodium ions to diffuse through a cell membrane.
   e. A membrane protein uses a hydrogen ion concentration gradient to transport sucrose against its concentration gradient and into a cell.

8. **A** What form of membrane-assisted transport would a white blood cell use to ingest a bacterial cell?
   a. exocytosis
   b. phagocytosis
   c. pinocytosis
   d. receptor-mediated endocytosis
   e. secretion

Use the diagram below to answer questions 9 and 10.

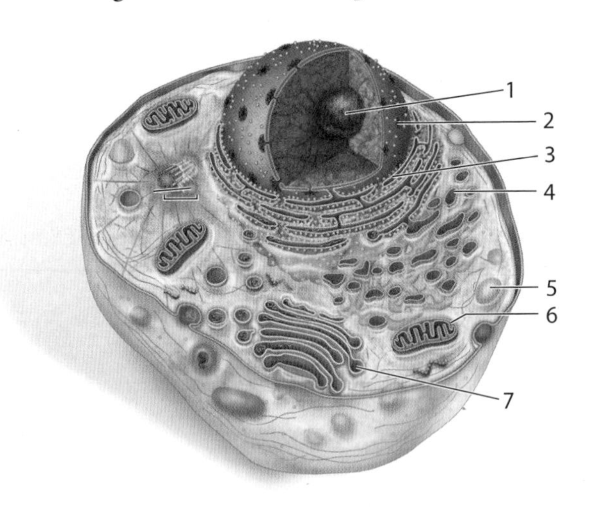

9. K/U Which numbered organelle contains enzymes that break down hydrogen peroxide?
   a. 3
   b. 4
   c. 5
   d. 6
   e. 7

10. K/U Which numbered structure allows RNA to pass through the nuclear envelope?
    a. 1
    b. 2
    c. 3
    d. 4
    e. 5

Use sentences and diagrams as appropriate to answer the questions below.

11. K/U Why are red blood cells incapable of reproduction?

12. K/U In what ways are the roles of cholesterol in the cell membrane contradictory?

13. A Paramecia are unicellular organisms that live in fresh water. They contain contractile vacuoles, which squeeze excess water out of the cell. Explain why paramecia need contractile vacuoles. Use appropriate terminology with respect to osmotic concentrations.

14. A In babies who are born with Tay-Sachs disease, lipids build up in the brain instead of being broken down and their components recycled. Infer why Tay-Sachs disease is known as a lysosomal storage disease.

15. K/U What did the observation that lipid-soluble molecules entered cells more rapidly than water-soluble molecules suggest to researchers who were studying the cell membrane?

16. C Draw and label a Venn diagram that compares and contrasts vesicles in plant cells and animal cells.

17. T/I The prefix *hydro-* refers to water, and the suffix *-lysis* refers to splitting something. Explain how active transport depends on the hydrolysis of a certain molecule.

18. T/I Osmosis occurs readily across the cell membrane by simple diffusion. In some cases, however, water crosses the cell membrane 10 times faster than it would by simple diffusion. What could logically explain this observation?

19. T/I Lizards are sometimes called "cold-blooded" because they depend on the surrounding temperatures to keep themselves warm. Predict how the cell membranes of a lizard might change from winter to summer with respect to the membrane lipids and amount of cholesterol that they contain. Justify your predictions.

20. A On a homework-help website, a student asks the following question: "What is the role of ribosomes on the smooth ER"
    a. Explain why this question, as stated, is scientifically incorrect.
    b. Rewrite the question so that it correctly asks what the student intended to ask.
    c. Answer the rewritten question.

21. C **a.** Draw and label a simple diagram of a section of cell membrane, including the phospholipid bilayer, a channel protein, a carrier protein, and an ion pump.
    b. Indicate on your diagram the types of substances that can cross the cell membrane at various points.

22. T/I What happens to the surface area of a cell during exocytosis? Explain.

23. A Electron microscope images reveal that certain disease-causing bacteria cause the cells lining the intestine to bulge outwards. What does this observation indicate about the effect of the bacteria on the cytoskeleton?

24. T/I How could you test the effect of calcium ions on exocytosis? Assume that you have access to tissue cell cultures, fluorescent labels, and an appropriate microscope, among other materials and equipment.

25. A One of the disease symptoms that results from abnormal mitochondria is extreme muscle weakness. Explain why this symptom occurs.

Self-Check

If you missed question...	1	2	3	4	5	6	7	8	9	10	11	12	13	14	15	16	17	18	19	20	21	22	23	24	25
Review section(s)...	2.1	2.1	2.1	2.1	2.1	1.1	2.2	2.2	2.1	2.1	2.1	2.1	2.2	2.1	2.1	2.1	2.2	2.2	2.1	2.2	2.1 2.2	2.2	2.1 2.2	2.1 2.2	2.1 2.2

Unit 1 Project

An Issue to Analyze

Evaluating Advances in Cellular Biology

Advances in cellular biology have had a major impact on human health worldwide. As one example, when the first cases of acquired immune deficiency syndrome (AIDS) appeared in the late 1970s, they frustrated medical researchers who were trying to find the cause of this wasting disease that could quickly compromise entire body systems. Researchers suspected the cause to be a retrovirus, a type of virus that can transform its RNA into DNA, which is subsequently incorporated into the host cell's genome. If host cells include the crucial *T-lymphocytes* that are part of the immune system, then a person infected with the virus can be left defenceless against infections.

In 1983, a team of researchers at the celebrated Pasteur Institute in Paris finally identified this retrovirus as the human immunodeficiency virus (HIV). Research then focussed on developing a specific class of drugs to counteract HIV. Known as *antiretroviral agents*, these drugs can interfere with the action of HIV in host cells, preventing its spread.

Today, researchers are faced with the reality that anti-retroviral medications can keep HIV in check, but infected individuals remain constantly at risk of developing full-blown AIDs. Therefore, the search continues for an effective medical solution, such as a vaccine that could immunize uninfected individuals against HIV.

What are some issues related to advances in cellular biology and related technological applications?

Initiate and Plan

Select a major advance in cellular biology and related technological applications. Choose a topic of your own or consider one of the following options:

- *Minimizing the impact of stroke.* The interruption of blood flow to the brain can cause massive destruction of brain cells and impair physiological and cognitive functions. How are advances in cellular biology helping to minimize such damage?

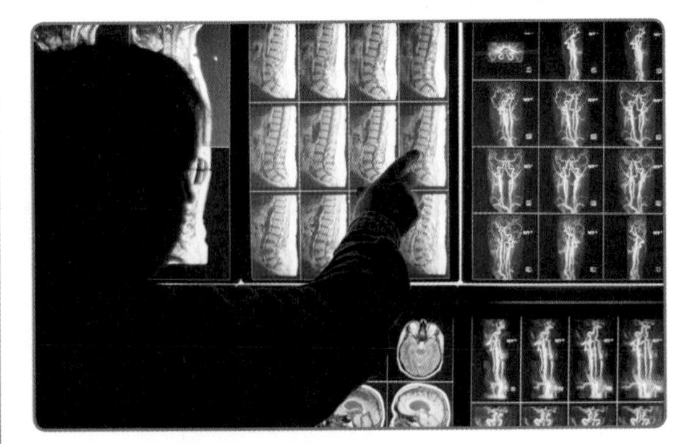

This researcher is examining cellular activity based on visual representations provided by molecular imaging probes—chemical compounds that provide a non-invasive means to diagnose disease at its earliest stages.

- *Treating cancerous tumours.* Cancerous tumours nurture new blood vessels that supply the tumours with nutrients. What advances in cellular biology and related technologies permit doctors to impede the development of cells in these blood vessels in order to inhibit a tumour's growth?
- *Developing medical probes.* Powerful imaging technologies employ detectors that receive radioactive emissions from short-lived isotopes targeted to particular organs or tissues of the body. In each case, the isotopes reach their particular destinations with the help of a probe, such as a protein or an antibody that targets those organs or tissues. How have advances in cellular biology contributed to the identification and refinement of such probes?
- *Combatting viruses.* Vaccinations can stimulate the body's immune system to recognize and reject disease-causing agents, such as the liver-damaging virus that causes hepatitis C. What advances in cellular biology have ensured that hepatitis-C vaccines are effective?

Perform and Record

1. Research one major advance in cellular biology. Describe how new understanding of a particular cellular process, organic molecule, or biochemical interaction has led to the development of a significant medical or technological application.
2. Identify and evaluate significant economic, political, ethical, and/or societal issues related to your chosen advance in cellular biology and related technologies.
3. Consider the following questions to guide your research:
   - Assess the challenges involved in obtaining the cooperation of subjects in medical trials related to advances in cellular biology. What rules should govern the role of study participants? What rules should govern the responsibility of researchers toward these participants?
   - What government agencies or private companies might have an interest in the outcome of the research? Might those interests compete or conflict with each other?
   - What tools and equipment have led to your chosen advance in cellular biology? Is the research largely dependent on this technology, and if so, is the expense or accessibility of this technology a limiting factor? How is the technology funded?
   - Does your chosen example have broad societal implications, comparable to the impact of the discovery of HIV in the 1980s? If so, how might these implications determine the level of public and financial support that such work would receive?
   - What is the potential for this advance in cellular biology to benefit individuals who have a particular medical problem? Who is responsible for determining that potential, and what obligations accompany that decision-making power?

Analyze and Interpret

1. Based on your research, select and organize the information that most clearly highlights the significance of your chosen advance in cellular biology and any related technologies. In addition to benefits associated with the advance in knowledge, decide whether any risks or disadvantages counterbalance the benefits.
2. Offer a recommendation about what you perceive is the most promising direction for research in cellular biology to take. What beneficial results might this research or application yield? What obstacles might stand in the way, and how could stakeholders overcome these obstacles?

Communicate Your Findings

Choose an appropriate form of communication to summarize your findings and express your recommendation. For example, you could create a blog, a podcast, a brief video, or an article for an on-line magazine.

Assessment Criteria

Once you complete your project, ask yourself these questions. Did you…

- ✓ **K/U** describe how new understanding of a particular cellular process, organic molecule, or biochemical interaction has led to the development of a significant medical or technological application?
- ✓ **T/I** identify and evaluate significant economic, political, ethical, and/or societal issues related to your chosen advance in cellular biology and related technologies?
- ✓ **A** offer a recommendation about what you perceive is the most promising direction for research in cellular biology to take?
- ✓ **C** organize and present your work in a format appropriate for your chosen audience and purpose?
- ✓ **C** use scientific vocabulary appropriately?

UNIT 1 SUMMARY

BIG IDEAS

- Technological applications that affect biological processes and cellular functions are used in the food, pharmaceutical, and medical industries.
- Biological molecules and their chemical properties affect cellular processes and biochemical reactions.
- Biochemical compounds play important structural and functional roles in cells of all living organisms.

Overall Expectations

In this unit you learned how to…

- **analyse** technological applications of enzymes in some industrial processes, and **evaluate** technological advances in the field of cellular biology
- **investigate** the chemical structures, functions, and chemical properties of biological molecules involved in some common cellular processes and biochemical reactions
- **demonstrate** an understanding of the structures and functions of biological molecules, and the biochemical reactions required to maintain normal cellular function

Chapter 1 The Molecules of Life

Key Ideas

- Intramolecular interactions occur between atoms within a molecule, forming chemical bonds.
- The polarity of a molecule influences the intermolecular interactions that occur between molecules.
- The functional group or groups on a molecule determine the properties of the molecule. Important functional groups include hydroxyl, carboxyl, carbonyl, amino, sulfhydryl, and phosphate groups.
- Carbohydrates include polysaccharides, and disaccharides and monosaccharides. Polysaccharides are composed of monosaccharide monomers.
- Lipids contain a higher proportion of non-polar C–C and C–H bonds than carbohydrates. Triglycerides, phospholipids, steroids, and waxes are different types of lipids.
- Proteins are composed of amino acid monomers. Proteins have primary, secondary, tertiary, and sometimes quaternary structures, which are essential for their function. Proteins enable chemical reactions in living systems and perform many structural and regulatory roles.

- Nucleic acids are composed of nucleotide monomers. DNA and RNA are both nucleic acids.
- Chemical reactions that take place in biochemical systems include neutralization, redox, condensation, and hydrolysis reactions.
- An enzyme is a protein that catalyzes a chemical reaction, enabling it to proceed faster than it otherwise would do.

Chapter 2 The Cell and Its Components

Key Ideas

- Cytoplasm consists of organelles, the cytosol, and molecules and ions dissolved or suspended in the cytosol.
- The cell nucleus includes the nuclear envelope, nuclear pore complexes, the nuclear matrix, and the nucleolus.
- The endomembrane system includes the nuclear envelope, the endoplasmic reticulum, the Golgi apparatus, the cell membrane, and vesicles. This system synthesizes, modifies, and transports proteins and other cell products.
- Animal cells contain many small vesicles. Plant cells contain a single large central vesicle called a vacuole.
- Chloroplasts trap light energy from the Sun in the form of high-energy organic molecules. Mitochondria break down high-energy organic molecules to release usable energy.
- Cells of plants, fungi, and many types of protists have a cell wall, which provides protection and support.
- The cytoskeleton provides structure, shape, support, and motility.

- The fluid mosaic model visualizes the cell membrane as a mosaic of proteins and other molecules in a fluid phospholipid bilayer.
- Transport of substances across membranes can occur by diffusion, facilitated diffusion, osmosis, and active transport.
- Primary active transport uses pumps that directly use energy and generate a gradient.
- Secondary active transport involves the use of an existing gradient to actively transport another substance.
- Larger particles can move into a cell by endocytosis and out of a cell by exocytosis.

Knowledge and Understanding

Select the letter of the best answer below.

1. Which of the following determines the identity of an atom?
   a. number of protons
   b. number of neutrons
   c. number of electrons
   d. type of chemical bond
   e. conductivity in water
2. The chemical compound CCl_4 is ___ because ___.
   a. polar; the carbon-chlorine bond is polar
   b. polar; the carbon-chlorine bond is non-polar
   c. non-polar; the carbon-chlorine bond is polar
   d. non-polar; the carbon-chlorine bond is non-polar
   e. non-polar; the polar bonds are opposite each other
3. Gate and channel proteins are located in this part of a cell.
   a. nucleus
   b. cell wall
   c. membranes
   d. cytoskeleton
   e. nuclear pores
4. Hydrogen bonding occurs between:
   a. water and water
   b. water and a polar molecule
   c. between two hydrogen-containing polar molecules
   d. between a polar hydrogen and a polar oxygen
   e. all of the above
5. Lysosomes are responsible for:
   a. detoxification in the liver
   b. protein production
   c. transport of materials within the cell
   d. binding to the cell membrane for transport
   e. neutralization of harmful compounds
6. A piece of celery is placed in a bowl of ice-cold water. Which of the following best describes what happens to the cells of the celery?
   a. They shrivel.
   b. Water moves out of the cells by osmosis.
   c. The water vacuoles fill with water and the cells become more plump.
   d. Water moves out of the cells by active transport.
   e. Nothing happens.
7. Which of the following functional groups are commonly associated with amino acids?
   a. amide
   b. hydrates
   c. hydroxide
   d. ether
   e. alcohol
8. The movement of sodium ions across a cell membrane occurs by
   a. active transport
   b. endocytosis
   c. exocytosis
   d. facilitated diffusion
   e. all of the above
9. What does an allosteric activator do?
   a. increase enzyme activity by binding at the active site
   b. decrease enzyme activity by binding at the active site
   c. increase enzyme activity by binding at a remote location
   d. decrease enzyme activity by binding at a remote location
   e. maintains current enzyme activity levels
10. Why do enzyme-catalyzed reactions slow down with time?
    a. enzymes become less efficient
    b. the products become easier to find
    c. after the activation energy has been lowered the reaction catalyzes itself
    d. fewer reactant molecules are available
    e. enzymes compete for the same reactants
11. An isotonic solution has the/a ___ concentration of water as the cell and the/a ___ concentration of solute as the cell.
    a. same, same
    b. same, different
    c. different, different
    d. different, same
    e. partial, partial
12. The mosaic part of the fluid mosaic model of membrane structure refers to
    a. the phospholipids
    b. the proteins
    c. the glycolipids
    d. the cholesterol
    e. all of the above

13. Which of the following structures or organelles could you see most clearly and in reasonable detail with a compound light microscope?
 a. mitochondria
 b. ribosomes
 c. Golgi apparatus
 d. nucleus
 e. smooth endoplasmic reticulum

14. Facilitated diffusion requires:
 a. specific proteins in the cell membrane
 b. ATP
 c. several ions moving down their concentration gradient
 d. sugar or similar energy source
 e. cholesterol

15. Proteins are produced
 a. in the cytoplasm of the endoplasmic reticulum
 b. in the nucleus
 c. in the Golgi apparatus
 d. in the cell membrane
 e. none of the above

Answer the questions below.

16. What is meant by the term passive transport?

17. Describe the two main types of membrane transport proteins.

18. Identify each of the following in the figure below:
 a. the type of biological macromolecule that each reactant is
 b. the type of biological molecule that the product is
 c. the type of biochemical reactions that are represented

19. Many biological molecules are polymers. List three types of polymers commonly found in cells. Identify the monomers for each.

20. Name the types of chemical reactions that occur in biological systems, and give an example for each.

21. Explain the process and causes of plasmolysis.

22. Explain how a channel protein regulates what enters a cell.

23. Compare the structure of glycogen with that of cellulose.

24. Provide two examples of condensation reactions that produce biological macromolecules. Include names of the reactants and products.

25. How are large macromolecules transported into intestinal cells?

26. Describe three features that all eukaryotic cells have in common.

27. What is exocytosis?

28. Distinguish between proteins that are assembled by ribosomes on the surface of the rough endoplasmic reticulum and those that are made by ribosomes freely suspended in the cytoplasm.

29. Describe the two types of endocytosis.

30. Briefly describe the role of molecular forces in determining the three-dimensional shape of a molecule.

31. What information would be required to produce an isotonic solution for a cell?

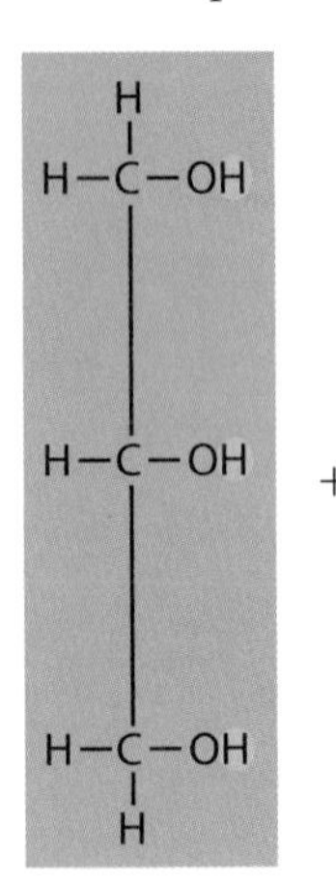

+ → + 3 H_2O

32. Examine the following magnified images of the nuclear envelope. Identify the components of the nuclear envelope outlined by the rectangle and explain how they regulate the passage of particles into and out of the nucleus.

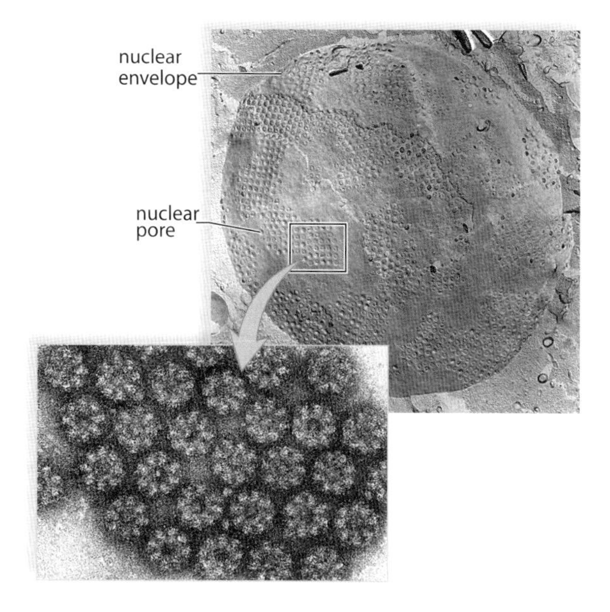

Thinking and Investigation

33. A student uses Benedict's solution and receives the same result for two different carbohydrate solutions. The disaccharide lactose and the monosaccharide glucose both give a positive test. Explain this result.
34. Compare a fat with an oil and explain why both are examples of lipids.
35. How can dialysis tubing be used to demonstrate osmosis?
36. Compare and contrast primary and secondary active transport.
37. Develop an analogy that represents the different levels of protein organization. Include a rationale for why your example is a good representation.
38. Receptor proteins in cell membranes bind to signal molecules. A malfunction in this binding process prevents transport vesicle formation. Explain how such a malfunction would affect the function of the endomembrane system.
39. A person has a genetic disorder that alters the function of certain molecular chaperones. What effect might this have on other proteins in the person's body?
40. How would the structure of DNA differ if hydrogen bonding did not occur?
41. How can bacterial enzymes be used in the development of new antibiotics?
42. Does a diagram showing only the phospholipids in a membrane accurately depict all substances present in the membrane? If not, identify the missing substances and explain their function.
43. A researcher is trying to design an artificial bilayer cell membrane. The researcher's work is currently focussing on membrane proteins. Suggest three functions of membrane proteins the researcher should consider, and explain why.
44. Should a hypertonic, hypotonic or an isotonic solution be used to produce a wet raisin from a grape? Justify your selection and predict any potential problems with this procedure.
45. The diagrams below show four types of membrane proteins. Identify each protein as either a channel protein, a carrier protein, a receptor protein, or an enzymatic protein. Explain which features of each diagram led to your decision.

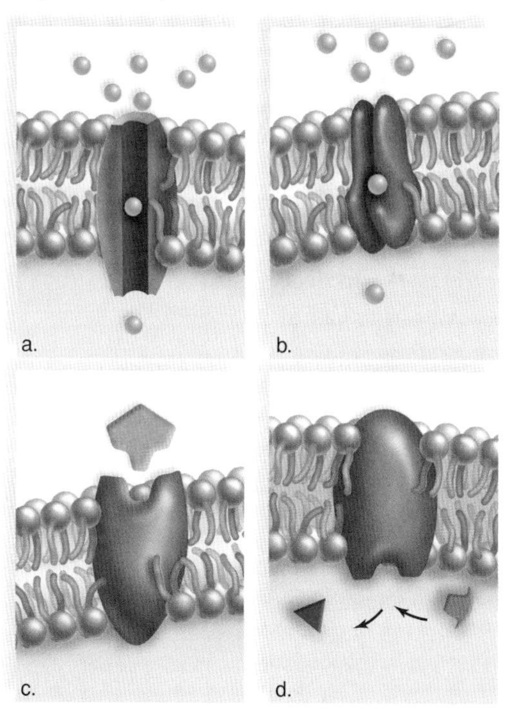

Communication

46. Develop a common, everyday analogy to explain what would happen to a cell under hypotonic conditions.
47. Draw a diagram that shows how active transport can lead to the formation of an electrochemical gradient.
48. Explain, using words and diagrams, how an enzyme would carry out a reaction joining two amino acids.

49. **BIG IDEAS** Technological applications that affect biological processes and cellular functions are used in the food, pharmaceutical, and medical industries. Radioisotopes are dangerous but remain an essential part of medical imaging. Do research to determine what alternatives exist and summarize the chemistry behind them.

What technical problems are presented when imaging relies on radioisotopes? Do research on Ontario's Chalk River plant and learn about difficulties with isotope production.

50. Draw an amino acid that has a hydrophobic R group. Title your drawing with the name of the amino acid. Also, label each of the following components: central carbon atom, carboxyl group, R group, and amino group.

51. Using a Venn diagram or table format, compare and contrast the structures of RNA and DNA.

52. Draw the structure of a phospholipid and indicate the fatty acid tails and head group. Describe and sketch what happens when phospholipids are added to an aqueous environment.

53. Sketch a phospholipid bilayer and explain how intermolecular forces allow the cell membrane to remain intact.

54. Illustrate an enzyme-substrate complex and label key features of the structure. Describe a hypothetical situation that would involve allosteric inhibition of that enzyme. Be sure to include an explanation of how this inhibition occurs and why it is occurring.

55. A living cell is placed in an isotonic solution. Make a sketch to show what would happen to the volume of the cytosol and the concentration of solute.

56. Make a list of common items found in the home that could be used to model the endomembrane system of a cell. Indicate what each represents.

57. Use a bulleted list to outline the functions of the cytoskeleton.

58. Draw a series of sketches to illustrate how exocytosis and endocytosis transport large particles across cell membranes.

59. Design an illustrated graphic organizer that identifies and defines the different types of chemical reactions that occur in living systems.

60. The diagram below is intended to summarize how molecules and ions can (or cannot) passively cross the cell membrane.

a. Interpret the meaning of the two types of arrows used in the diagram.

b. Write a paragraph of one or two sentences that could accompany this diagram as a verbal summary.

c. Imagine that this diagram is the first image in a slideshow presentation about the permeability of the cell membrane and the ways that substances can cross it. Develop an outline for the remaining images in this presentation. Assume that the completed slideshow should have the fewest number of slides possible.

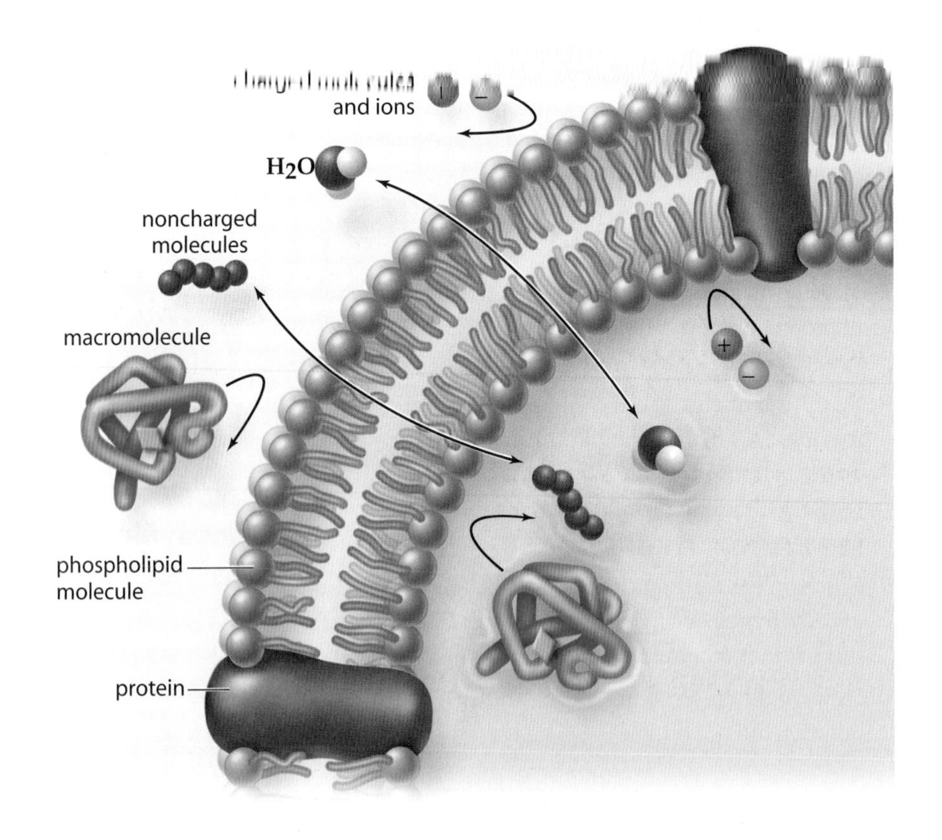

Application

61. Liver cells are packed with glucose. What mechanism could be used to transport more glucose into a liver cell, and why would only this method of transport work?

62. BIG IDEAS Biochemical compounds play important structural and functional roles in cells of all living organisms. How does maintaining the integrity of the cell membrane illustrate the interaction of cell components and the biochemical compounds of which they are composed?

63. Explain why it makes sense that a muscle cell contains many mitochondria, while a white blood cell contains many lysosomes.

64. a. Olestra is an artificial fat substitute that provides zero fat, calories (joules), and cholesterol from foods made with it. This molecule is too large to be metabolized but small enough to behave and taste like fat. Use your knowledge of enzymes to infer why this artificial molecule can provide no food energy (zero calories or joules).

b. Do research to find out which company invented olestra, what problems have been associated with it, and why Health Canada refused to approve olestra as a food additive in 2000.

65. Research the role of enzymes in bioremediation. Although bioremediation has tremendous potential, what problems do you see with its implementation?

66. All enzymes have an optimal temperature at which they work as well as an optimal pH for their activity. Does this mean that enzymes only work at those specific temperature and pH optima? Briefly explain your answer and note the evidence provided in this unit that supports your answer.

67. Infer how manufacturers of timed-release medication rely on enzymes to do the work?

Read the scenario below to answer questions 68 and 69.

A Canadian pharmaceutical company is developing two new antibiotic therapies. The first antibiotic therapy will use the enzymes found in lysosomes to treat bacteraemia, infection of the blood by bacteria. The enzymes will be delivered directly into the bloodstream intravenously. The second antibiotic therapy inhibits protein synthesis in the ribosomes of *E.coli* to treat infection by this bacteria.

68. Lysosomes contain enzymes that break down bacteria that have been ingested by cells. The pharmaceutical company argues that these enzymes could be used to treat bacteraemia. Suggest why this treatment is unlikely to be successful.

69. The second antibiotic therapy inhibits protein synthesis in the ribosomes of bacteria.

a. Would this therapy also affect protein synthesis in human cells? Explain your reasoning.

b. Suggest why this treatment is likely to be successful.

70. BIG IDEAS Biological molecules and their chemical properties affect cellular processes and biochemical reactions. Your cousin mentions that your uncle is on a low cholesterol diet. "He is trying to eradicate all cholesterol from his body," says your cousin. Focusing your argument on cell membranes, explain why the body cannot function without cholesterol.

71. In autoimmune disorders, the body fails to recognize its own cellular components as "self" and views them as disease-causing agents instead. Suggest how glycoproteins in the outer layer of the cell membrane could play a role in such disorders.

72. Examples of cells of the human body that are highly specialized include heart cells, sperm cells, egg (ovum) cells, and red blood cells. Based on prior knowledge, on research, or both, describe the features of these cells that make them so specialized and how these specializations suit their function. Find examples of four other highly specialized cells and provide the same information about them. Prepare a summary chart to record your findings.

73. When broccoli is cooked in boiling water, the brightness of the green colour becomes dull-green. Chefs recommend cooking the broccoli in boiling water that has a small amount of salt added to it in order to maintain the bright green colour of the vegetable. Why would the use of salted water make a difference?

UNIT 1

SELF-ASSESSMENT

Select the letter of the best answer below

1. K/U The number of neutrons determines:
   a. the element
   b. the compound
   c. the type of bond that will form
   d. the isotope
   e. the charge of the atom

2. T/I What feature could help distinguish between two integral proteins?
   a. ATP consumption
   b. binding abilities
   c. polarity
   d. mass
   e. All of the above could help distinguish between two integral proteins.

3. K/U The molecule CH_3OH is:
   a. polar
   b. charged
   c. ionic
   d. hydrophobic
   e. non-polar

4. K/U The association of more than one polypeptide to form a protein is called
   a. primary structure
   b. secondary structure
   c. tertiary structure
   d. quaternary structure
   e. none of the above

5. K/U Ribosomes are known for:
   a. producing proteins
   b. protecting the nucleus
   c. filling the cytosol
   d. being the smallest membrane bound organelle
   e. operating in extreme conditions

6. K/U The tertiary structure of a protein is characterized by
   a. protein folding
   b. the R group of the amino acids
   c. hydrogen bonding
   d. disulphide bridges
   e. All of the above

7. K/U Passive transport does not:
   a. follow the concentration gradient
   b. require energy
   c. preserve ATP
   d. rely on diffusion
   e. occur in animal cells

8. K/U Nucleotides
   a. contain sugar, a nitrogen-containing base, and a phosphate molecule
   b. are the monomers for fats and polysaccharides
   c. join together by covalent bonding between the bases
   d. are found in DNA, RNA, and proteins
   e. none of the above

9. K/U Glycogen provides short-term energy storage. What are the subunits of this polymer?
   a. sucrose
   b. glucose
   c. fatty acids
   d. lactose
   e. ribose

10. K/U Which of the following changes would decrease the activity of an enzyme?
    a. build up of products of the reaction it catalyzes
    b. introduction of acid
    c. introduction of allosteric inhibitor
    d. all of the above
    d. None of the above

Use sentences and diagrams as appropriate to answer the questions below.

11. K/U List three locations where ribosomes can be found in the cell.

12. A How would a cell membrane with trans-fat phospholipids (increased bends in the tails) affect the rate of ion diffusion through the membrane?

13. T/I Winter wheat is planted in the early autumn, grows over the winter when the weather is colder, and is harvested in the spring. As the temperature drops, the composition of the cell membrane of winter wheat changes. Unsaturated fatty acids replace saturated fatty acids in the phospholipids of the membrane. Why is this a suitable adaptation?

14. T/I DNA is a large molecule that is made of two long, thin strands. All of the atoms of each strand are held together by covalent bonds, but the two strands are held together by hydrogen bonds. Knowing that individual hydrogen bonds are weaker than covalent bonds, what would you predict about the stability of the DNA molecule? Explain your answer. (Hint: Think about the overall effect of numerous hydrogen bonds.)

15. T/I A drink claims to be sugar-free. Since it tastes sweet you suspect it is mislabelled. How could you confirm your suspicion?

16. K/U List two functions of peripheral proteins.

17. A You are investigating molecules that inhibit a bacterial enzyme. You discover that the addition of several phosphate groups to an inhibitor improves its effectiveness. Why would knowledge of the three-dimensional structure of the bacterial enzyme help you understand why the phosphate groups improve the inhibitor's effectiveness?

18. C Examine the diagrams of the animal cell below. In each case, identify the cell as being in an isotonic solution, a hypertonic solution, or a hypotonic solution. Then write a caption in your own words for each cell to explain what is happening as a result of being in that solution.

A

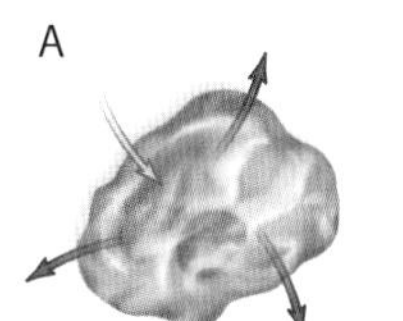

B

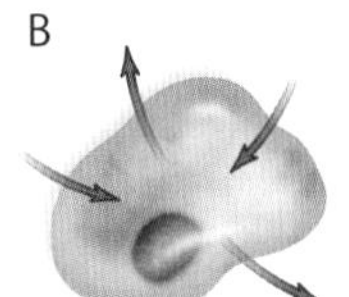

C

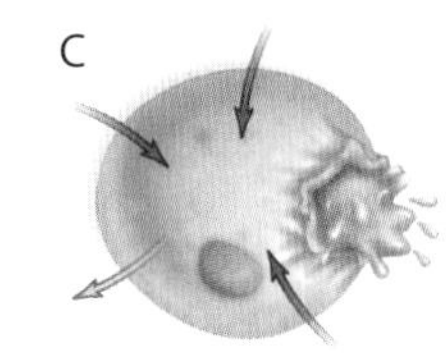

19. K/U Describe how enzyme activity is regulated by other molecules. Use the following terms in your description: inhibitor, allosteric site, competitive inhibition, non-competitive inhibition.

20. K/U Identify the structure or organelle of a cell that performs the following functions.
    a. movement
    b. conversion of energy
    c. disposal of waste
    d. internal transport

21. C The diagrams of plant and animal cells that appear in biology textbooks for the purposes of summarizing the structures and organelles of these cells are usually described as being generalized cell diagrams. Clearly explain why generalized diagrams of plant and animal cells must be used, rather than diagrams that are more specific.

22. C Suggest a title that could be used for diagram A and for diagram B. Then write a caption that explains what is happening in both diagrams and why.

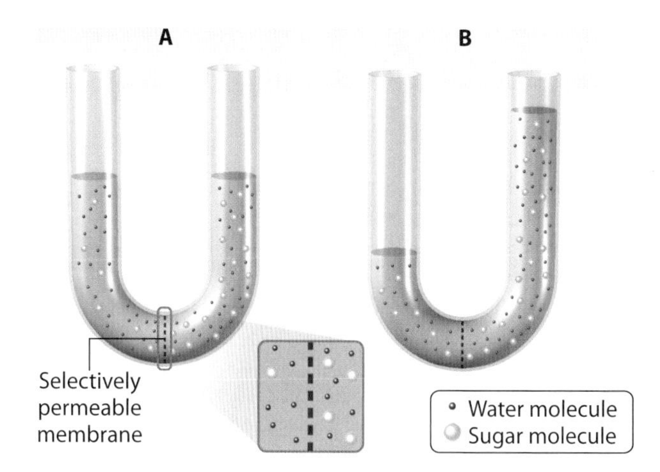

23. A Boiling an egg causes its proteins to denature. Explain what this means and why this change occurs.

24. K/U Explain, with reference to bonding and other atomic properties, why fat stores more energy than an equal mass of carbohydrate.

25. C Outline, in the form of two or three complete and well-constructed sentences, the key features of the fluid mosaic model of a membrane represented below. Explain the significance of these features to the function of a cell.

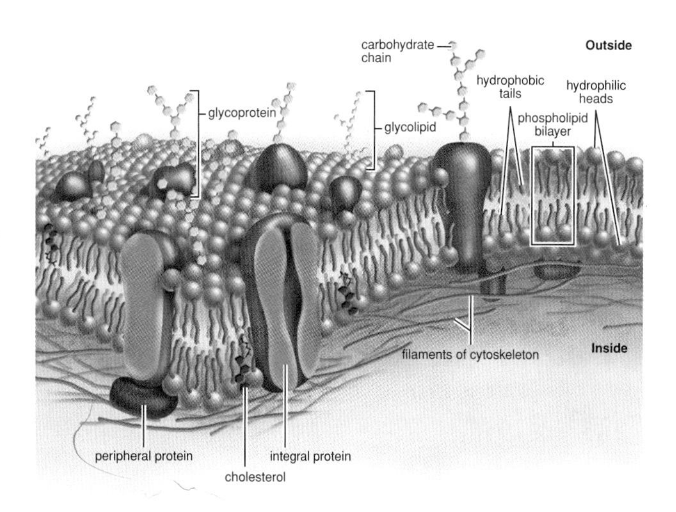

Self-Check

If you missed question...	1	2	3	4	5	6	7	8	9	10	11	12	13	14	15	16	17	18	19	20	21	22	23	24	25
Review section(s)...	1.1	2.1	1.1	1.1	2.1	1.2	2.2	1.2	1.2	2.2	2.1	2.2	1.2	1.1	1.2	2.1	1.3	2.2	1.3	2.1	2.1	2.2	1.3	1.2	2.1

UNIT 2

Metabolic Processes

BIG IDEAS

- All metabolic processes involve chemical changes and energy conversions.
- An understanding of metabolic processes enables people to make informed choices with respect to a range of personal, societal, and environmental issues.

Overall Expectations

In this unit, you will...

- **analyze** the role of metabolic processes in the functioning of biotic and abiotic systems, and **evaluate** the importance of an understanding of these processes and related technologies to personal choices made in everyday life
- **investigate** the products of metabolic processes such as cellular respiration and photosynthesis
- **demonstrate** an understanding of the chemical changes and energy conversions that occur in metabolic processes

Unit Contents

Focussing Questions

1. *What role do metabolic processes play in living and non-living systems on Earth?*
2. *What are the products of photosynthesis and cellular respiration?*
3. *What chemical changes and energy transformations occur in metabolic pathways?*

Go to **scienceontario** to find out more about metabolic processes

Domino Day in Leeuwarden, The Netherlands, is an annual event in which hundreds of participants from all over the world attempt to assemble a world-record-setting number of dominoes into a cascading chain reaction, as shown here. Each domino must be placed precisely so that it is toppled by the one before it, and then transfers this energy as it falls onto the next domino in line. If one domino is taken out of the arrangement, the entire cascade may not proceed as planned.

Like these dominoes, the thousands of chemical reactions that take place in cells must function collectively in order to maintain cellular functions. Many of the chemical reactions that occur in cells, including cellular respiration and photosynthesis, are arranged into pathways made up of many small steps occurring in sequence. Each step is essential for the one following it to proceed, and each step is precisely controlled. In this unit, you will learn about metabolic processes—the chemical reactions that occur within a cell to keep it alive and functioning—and how these processes are regulated.

As you study this unit, look ahead to the Unit 2 Project on pages 186 to 187. Complete the project in stages as you progress through the unit.

UNIT 2 Preparation

Chemical Bonding, Chemical Reactions, and Enzymes

- Intramolecular forces are the forces that hold atoms together in a molecule.
- Intermolecular forces are the attractive forces that exist between molecules.
- In a chemical reaction, chemical bonds in reactants are broken and new chemical bonds form to make the products.
- For every chemical reaction that occurs in living systems, there is a unique enzyme that catalyzes, or speeds up, the reaction.
- Enzymes are proteins and are usually much larger than the reactants and products.
- Activation energy is the amount of energy that must be added to reactants for a chemical reaction to occur.

1. Hydrogen bonds are attractions between
  a. hydrogen atoms on different molecules
  b. hydrogen molecules
  c. hydrogen atoms and oxygen or nitrogen atoms in the same molecule
  d. hydrogen atoms on one molecule and oxygen (or nitrogen) atoms on another molecule
  e. hydrogen atoms and any other atom in the same molecule

2. Hydrophobic interactions are
  a. stronger than hydrogen bonds but weaker than covalent bonds
  b. stronger than covalent bonds
  c. similar in strength to ionic bonds
  d. weaker than hydrogen bonds but stronger than ionic bonds
  e. weaker than hydrogen bonds, ionic bonds, and covalent bonds

3. What types of bonds hold cellular membranes together?

4. The following equation represents a chemical reaction.

$$NO_2(g) + H_2O(\ell) \rightarrow HNO_3(aq)$$

  a. What is the general term for the compounds on the left side of the equation, or in this case, nitrogen dioxide, $NO_2(g)$ and water, $H_2O(\ell)$?
  b. What is the general term for the compounds on the right side of the equation, or in this case, nitric acid, $HNO_3(aq)$?

5. List at least two observations that would indicate that a chemical reaction has occurred.

6. Enzymes are made of
  a. carbohydrates
  b. proteins
  c. nucleic acids
  d. lipids
  e. all of the above

7. When an enzyme catalyzes a chemical reaction, what is the general term for the compounds on the left side of the equation?

8. Describe the structure and function of the active site of an enzyme.

9. Describe one mechanism by which the activity of an enzyme can be regulated.

10. Name the subunits that make up
  a. proteins
  b. carbohydrates
  c. nucleic acids

11. What is the approximate optimal pH for most enzymes?

12. Write a caption to explain the significance of the graph below.

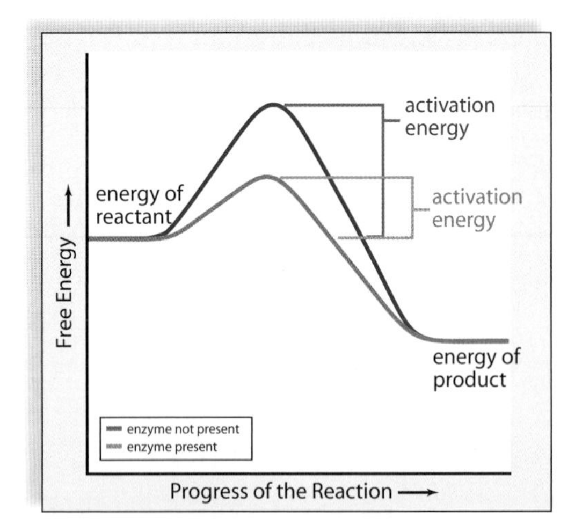

Cellular Organization

- Cells are surrounded by selectively permeable membranes that regulate what enters and leaves the cell.
- Some small, neutral compounds can diffuse through membranes, but ions usually cannot.
- Plant cells have cell walls around the outside of the cell membrane but animal cells do not.
- Eukaryotic cells have many membrane bound organelles such as mitochondria, endoplasmic reticulum, Golgi bodies, lysosomes, peroxisomes, and a nucleus.
- Plant cells have a large central vacuole whereas animal cells have numerous small ones. Plant cells also have chloroplasts.

13. Summarize the fluid mosaic model. Make a sketch of a membrane and label the proteins and phospholipids and explain their functions.

14. Explain the function of the plant cell wall. What functions make the cell wall important in plant cells but not necessary for animal cells?

15. Review the animal cell below. For each numbered structure, supply a label.

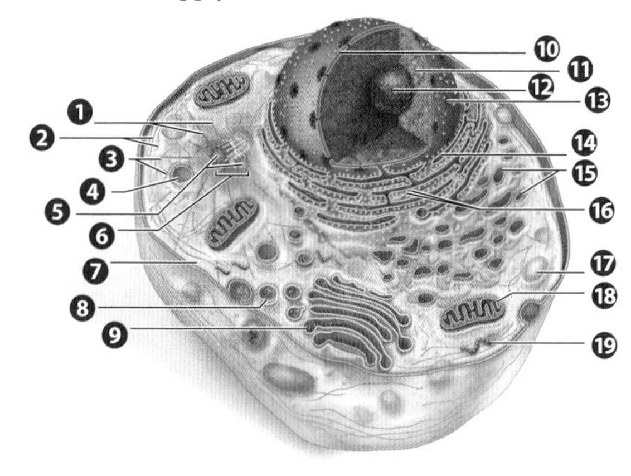

16. Describe the main function of mitochondria.

17. How does oxygen get from the bloodstream to the mitochondria where it is needed?

18. Examine the structure of the leaf cell below. State the name and describe the function of each of the numbered organelles.

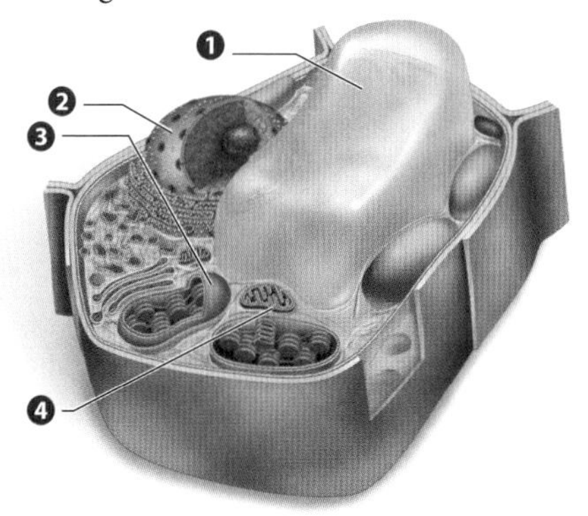

19. Describe the main function of chloroplasts.

20. How are the functions of the chloroplasts and mitochondria related?

Use the diagram below to answer questions 21 to 24.

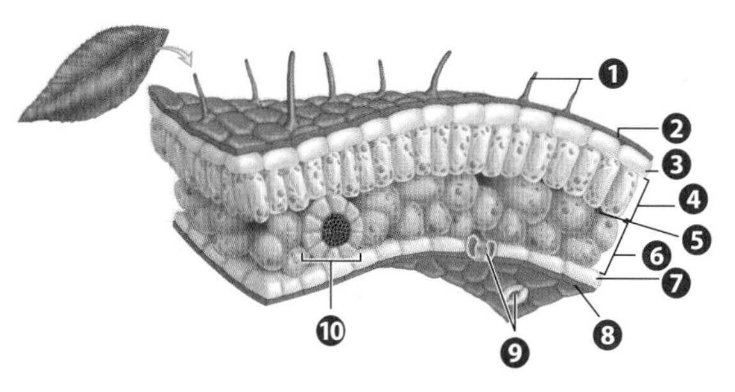

21. For each number in the figure above, provide a label that names the leaf part indicated.

22. The purpose of the air space between the cells in the region labelled as 6 is

a. to cool off the cells
b. to eliminate the carbon dioxide that is formed in the mitochondria
c. to allow oxygen from the air to come in contact with the cells
d. to let the cells dry out
e. to allow carbon dioxide from the air to come in contact with the cells

23. The function of the structure labelled as 9 is to

a. control the air pressure inside the leaf
b. close when the air is dry to prevent water loss
c. close when carbon dioxide outside gets too high
d. close to prevent loss of oxygen when it gets low
e. open when the air moist to let water in

24. Explain the function of the structure labelled as 10.

25. Describe at least three significant differences between animal cells and plant cells.

Cellular transport

- When compounds diffuse across a membrane, they move down their concentration gradient.
- Some compounds cannot diffuse across a membrane, but their movement can be facilitated by specialized proteins that are embedded in the membrane. This process is called passive transport or facilitated diffusion.
- Energy is not required for facilitated diffusion, because compounds or ions travel down their concentration gradient.
- Some compounds can be transported across a membrane against their concentration gradient by using energy. This process is called active transport.
- The source of energy for active transport is usually ATP.
- Secondary active transport is a form of active transport in which one compound or ion is actively transported across a membrane to form a concentration gradient, and a specialized protein in the membrane binds both the compound or ion that was actively transported and a second compound. The second compound can then be transported against its concentration gradient. The energy for transporting the secondary compound is derived from the concentration of the first compound or ion.

26. Explain the concept of diffusion. Use the diagram below to assist you in your explanation.

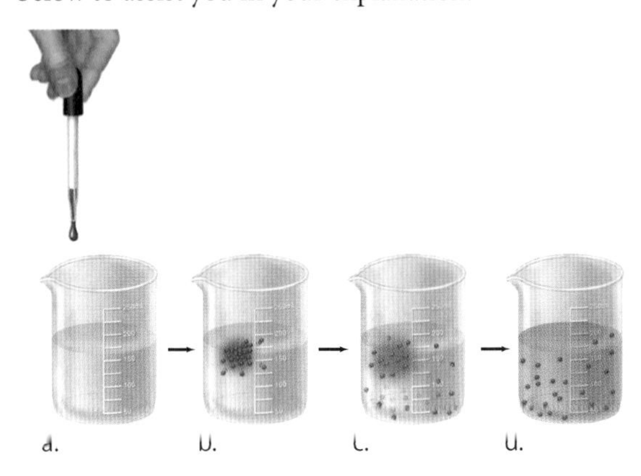

27. When a membrane is described as selectively permeable, what does that communicate to you about the ability of molecules or ions to diffuse across that membrane?

28. The diagram below represents a U tube with a selectively permeable membrane across the bottom. Water is on the right side and a urea solution is on the left side. Water can diffuse across the membrane but urea cannot. Describe what happens in the tube over a period of time. What is the name of this process?

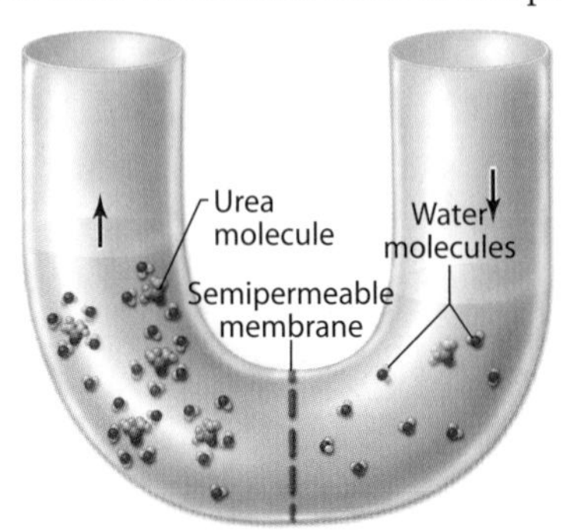

29. Explain how facilitated diffusion differs from simple diffusion as it applies to membranes in living cells.

30. Explain the difference between active transport and passive transport.

31. A protein that recognizes glucose exists in the membranes of red blood cells. This protein provides a channel through which glucose can enter the cell. However, glucose always moves in the direction of its concentration gradient. What type of transport is occuring?

32. In cells that line the kidney tubules, glucose must be moved against its concentration gradient. In this case, sodium ions, Na^+, are actively transported out of the cells, using ATP for energy. A sodium ion gradient is created such that the sodium ion concentration is greater outside the cell than inside. Then a transport protein binds both a sodium ion and a glucose molecule and carries them into the cell.

a. Explain how this transporter can move glucose against its concentration gradient.

b. What is the source of energy for the transport of glucose?

c. What is the name of this type of membrane transport?

33. A classmate is confused about osmosis. He knows that ions and molecules diffuse from a high concentration to a low concentration. He reads that water diffuses into a cell where the concentration of molecules and ions is high, and he thinks that this statement is incorrect. Explain whether the statement is indeed incorrect, and justify your answer.

Energy

- The main form of useful energy for cells is ATP.
- The fundamental source of energy for living systems is the Sun.
- Solar energy is transported as electromagnetic waves, which consist of electric energy and magnetic energy.
- Electromagnetic waves range in wavelength from several metres—or even kilometres—to less than 10^{-12} m, or 0.001 nanometres (nm). (Note: 1 nm = 1×10^{-9} m)
- The electromagnetic spectrum is shown below. Visible light is a very small part of the spectrum, ranging in wavelength from about 400 nm to 700 nm.
- All forms of energy can be transformed from one type to another.
- Energy cannot be created or destroyed but it can be transferred from one object to another.
- In living organisms, energy can be stored in chemical compounds and in concentration gradients.

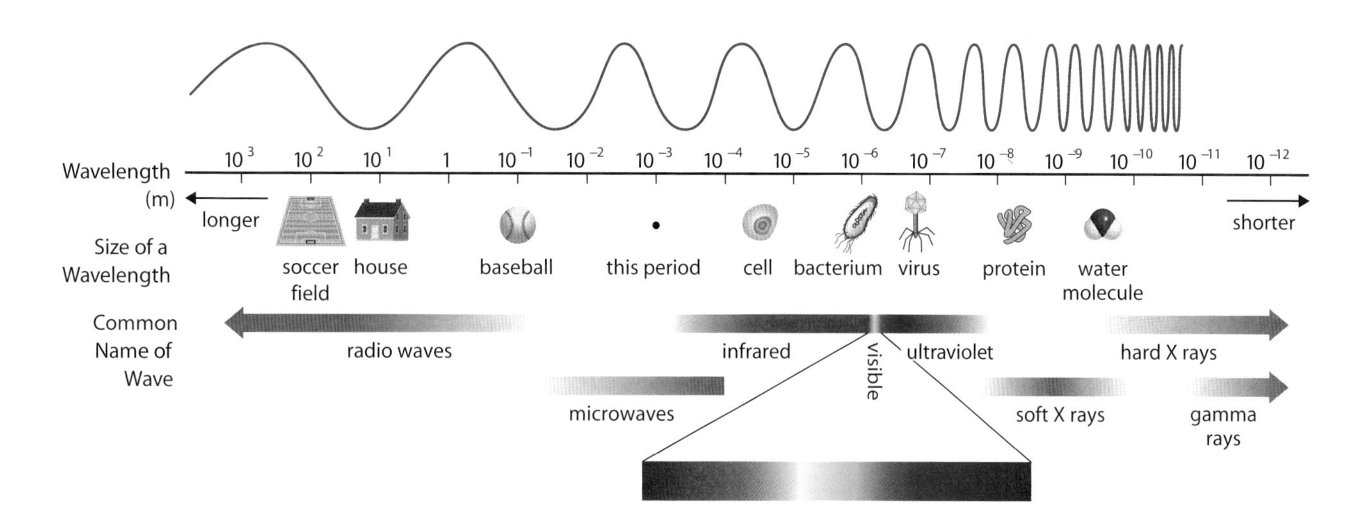

34. Use the diagram of a water wave below to define the wavelength of a wave.

Water Wave

direction of wave motion

wavelength

crest

trough

35. What determines the range of wavelengths that are considered visible light?

36. Describe the different ways in which light can behave when light interacts with matter. For example, one way that light behaves is that it can be absorbed by matter.

37. How can it be said that nearly all of the energy used by living organisms ultimately comes from the Sun when much of the energy humans use for heating, transportation, and electrical energy comes from fossil fuels?

38. People are concerned about conserving energy. However, a basic law of nature states that energy cannot be created or destroyed. In consideration of that law, what is meant by "wasting energy."

39. Give an example of energy transformation in which mechanical energy is transformed into another form of energy.

40. Give an example of transferring energy from one object to another.

41. Explain the meaning of the terms *autotroph* and *heterotroph*. Include in your explanation a discussion of the sources of energy for autotrophs and heterotrophs.

42. In what form do living systems store energy for use in cellular processes?

CHAPTER 3

Energy and Cellular Respiration

Specific Expectations

In this chapter, you will learn how to . . .

- C1.1 **analyze** the role of metabolic processes in the functioning of and interactions between biotic and abiotic systems (3.2, 3.3)
- C1.2 **assess** the relevance, to your personal life and to the community, of an understanding of cell biology and related technologies (3.2, 3.3)
- C2.1 **use** appropriate terminology related to metabolism (3.1, 3.2, 3.3)
- C2.2 **conduct** a laboratory investigation into the process of cellular respiration to **identify** the products of the process, **interpret** the qualitative observations, and **display** them in an appropriate format (3.2, 3.3)
- C3.1 **explain** the chemical changes and energy conversions associated with the processes of aerobic and anaerobic cellular respiration (3.2, 3.3)
- C3.3 **use** the laws of thermodynamics to explain energy transfer in the cell (3.1, 3.2, 3.3)
- C3.4 **describe**, **compare**, and **illustrate** the matter and energy transformations that occur during the process of cellular respiration, including the roles of oxygen and organelles such as the mitochondria (3.2, 3.3)

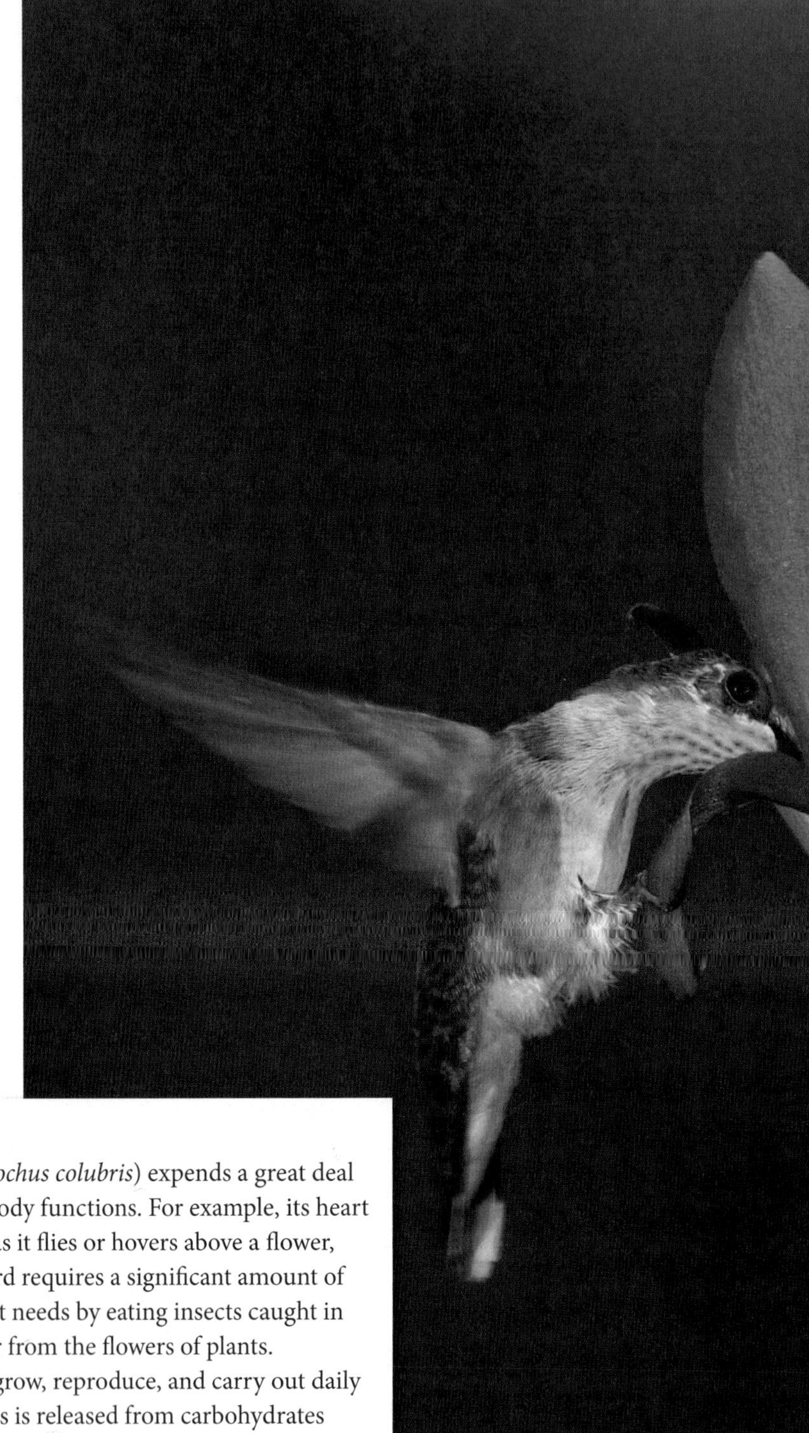

The ruby-throated hummingbird (*Archilochus colubris*) expends a great deal of energy to stay in motion and maintain body functions. For example, its heart rate can reach over 1000 beats per minute as it flies or hovers above a flower, as shown here. As a result, the hummingbird requires a significant amount of energy to stay alive. It acquires the energy it needs by eating insects caught in flight, as well as by consuming sweet nectar from the flowers of plants.

All organisms need energy to survive, grow, reproduce, and carry out daily activities. Energy to support these functions is released from carbohydrates and other energy-rich organic molecules. In animals, plants, and most other organisms, the process that releases this energy is cellular respiration. For a small number of species that live in environments in which there is little or no oxygen, the processes that release this energy are anaerobic respiration and/or fermentation.

Launch Activity

A Flutter of Activity

The ruby-throated hummingbird is found throughout Canada. Ranging in mass from 2.5 g to 4.8 g, it is one of Canada's smallest birds. The hummingbird flaps its wings between 55 and 75 times each second and reaches speeds of 80 km/h or more. Not surprisingly, the ruby-throated hummingbird works up an incredible appetite. It can consume up to three times its body mass in a single day. In this activity, you will compare the basal, or resting, metabolic rate of the hummingbird to that of other animals, as well as yourself. *Metabolic rate* refers to the rate at which the sum total of all chemical reactions necessary to maintain cellular functions occurs.

Materials

- a calculator
- graph paper or computer graphing software

Procedure

1. Study the table below.

Metabolic Rates for Various Animals

	Hummingbird	Cat	Dog	Horse	Elephant
Basal metabolic rate (kJ/day)	42	4.6×10^2	1.5×10^3	5.0×10^4	1.7×10^5
Body mass (kg)	0.003	4	13	1500	5000

2. Graph the data in this table, plotting body mass on the *x*-axis and basal metabolic rate on the *y*-axis.
3. Basal metabolic rate is often reported in terms of per body mass (for example, kJ/day/kg). Calculate the energy (in kJ) needed per day per kg of body mass for each animal. Graph this data, plotting your calculated values on the *y*-axis and body mass on the *x*-axis.
4. Estimate your own body mass, or use a scale to measure it.
5. Using your estimate or measurement of your body mass from step 4, interpolate your basal metabolic rate from your graph.

Questions

1. Based on the graph you plotted in step 1, what conclusion can you make about the relationship between basal metabolic rate and body size?
2. Based on the graph you plotted in step 2, what conclusion can you make about the relationship between basal metabolic rate per kilogram of body weight and body size?
3. Compare your estimated metabolic rate to the metabolic rates of other animals from the table above. What do you notice?
4. Based on the data, infer what factors might influence basal metabolic rate.

SECTION 3.1

Metabolism and Energy

Key Terms

metabolism
metabolic pathway
catabolism
anabolism
energy
kinetic energy
potential energy
bond energy
thermodynamics
entropy
free energy
endergonic
exergonic

All living cells continuously perform thousands of chemical reactions to sustain life. The word metabolism comes from a Greek word that means "change." **Metabolism** refers to all the chemical reactions that change or transform matter and energy in cells. These reactions occur in step-by-step sequences called **metabolic pathways**, in which one substrate or more is changed into a product, and the product becomes a substrate for a subsequent reaction. A unique enzyme catalyzes each of these reactions. In the absence of enzymes, the reactions would not occur fast enough to sustain the life of the cell.

The function of many metabolic pathways is to break down energy-rich compounds such as glucose and convert the energy into a form that the cell can use. The process of breaking down compounds into smaller molecules to release energy is called **catabolism**, and such a process may be referred to as *catabolic*. Active transport is a catabolic process, for example, as is the use of energy by muscle cells to cause motion in the form of muscle contraction. Much of the energy released by catabolic processes, however, is used to synthesize large molecules such as proteins and fats. The process of using energy to build large molecules is called **anabolism**, and such a process may be referred to as *anabolic*. **Figure 3.1** compares representations of catabolic and anabolic reactions.

metabolism the sum of all chemical reactions that occur in the cell

metabolic pathway a sequential series of chemical reactions in living cells; each reaction is catalyzed by an enzyme

catabolism the process of breaking down compounds into smaller molecules to release energy

anabolism the process of using energy build large molecules from smaller molecules

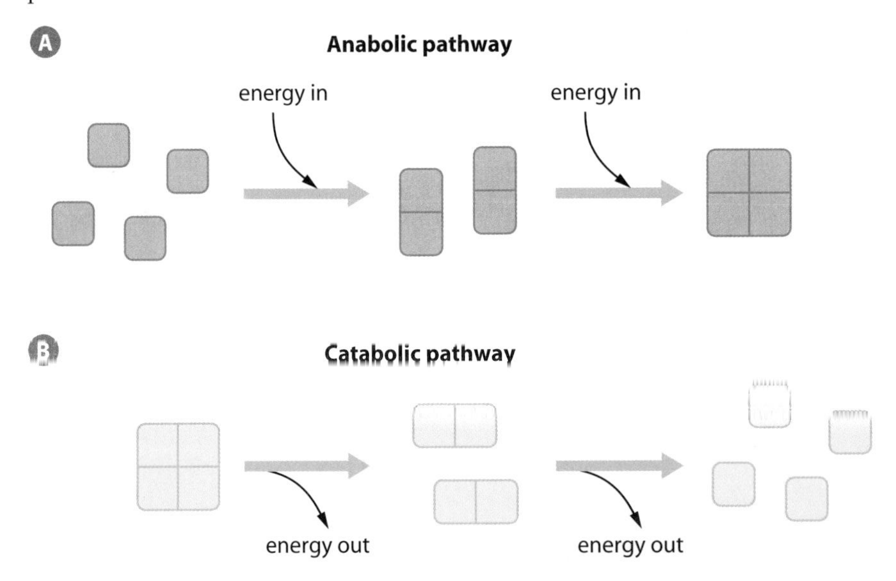

Figure 3.1 Anabolic reactions (**A**) build complex molecules, while catabolic reactions (**B**) reverse that process.

***Explain** why constructing a building with bricks, boards, and cement is analogous to an anabolic reaction, and then describe a similar analogy for a catabolic reaction.*

Energy

The same scientific laws that describe energy in chemistry and physics apply to energy in cells. Therefore, a general understanding of these laws will provide a foundation for the study of metabolism. **Energy** can be defined as the capacity to do work—that is, to change or move matter against an opposing force such as gravity or friction. Energy is often classified as two main types: kinetic and potential. **Kinetic energy** is energy of motion. Moving objects perform work by causing other matter to move. **Potential energy** is stored energy, or energy that is available but not yet released. For example, a boulder perched on a hilltop has potential energy. As it starts to roll downhill, some of its potential energy is transformed into kinetic energy. Much of the work of living cells involves the transformation of potential energy into kinetic energy. For example, the potential energy stored in electrochemical gradients is used to move molecules into and out of cells during active transport.

energy the capacity to do work

kinetic energy the energy of motion

potential energy stored energy

Kinetic energy and potential energy may themselves be classified as different types. For example, the kinetic energy of particles moving in random directions is *thermal energy*. An increase in the kinetic energy of particles of an object increases the temperature of the object. *Heat* is the transfer of thermal energy from one object to another due to a temperature difference between the objects. *Chemical energy* is potential energy stored in the arrangement of the bonds in a compound.

Bond Energy

Whenever a chemical bond forms between two atoms, energy is released. The amount of energy needed to break a bond is the same as the amount of energy released when the bond is formed. This amount of energy is called **bond energy.** Because energy is always released when a bond forms, free (unbonded) atoms can be considered to have more chemical energy than any compound. The relative amounts of chemical energy that compounds possess can be compared by examining the amount of energy released when each compound is formed. **Figure 3.2** shows the relative amounts of chemical energy in different compounds containing one carbon atom, four hydrogen atoms, and four oxygen atoms.

bond energy energy required to break (or form) a chemical bond

When carbon, hydrogen, and oxygen combine to form methane and oxygen, the methane and oxygen molecules have less chemical energy than the individual component atoms. When these same atoms combine to form carbon dioxide and water, they release even more chemical energy. Therefore, methane and oxygen have more chemical energy than carbon dioxide and water. The red arrow in **Figure 3.2** shows how much energy is released when methane and oxygen react to form carbon dioxide and water.

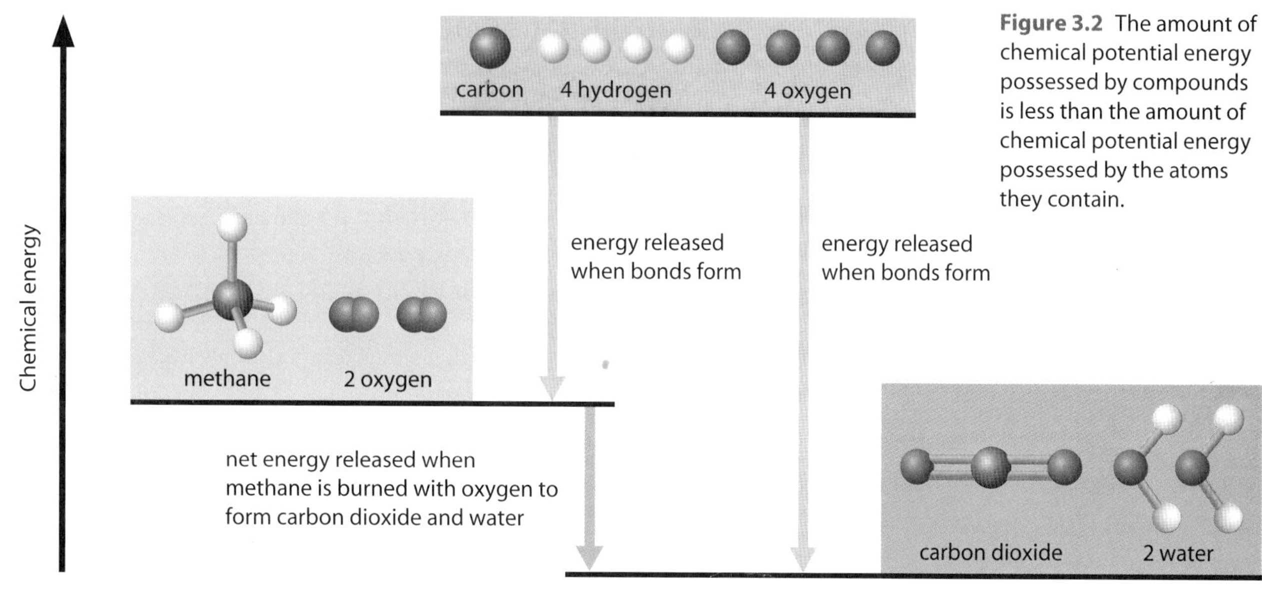

Figure 3.2 The amount of chemical potential energy possessed by compounds is less than the amount of chemical potential energy possessed by the atoms they contain.

The energy released from chemical reactions in a laboratory is usually in the form of thermal energy (heat). The energy released from chemical reactions in living cells can include thermal energy, but it can also be in the form of the movement of compounds across cell membranes, contraction of a muscle, or even the emission of light from compounds within specialized cells in certain organisms, as shown in **Figure 3.3**. In many cases, energy released from one reaction is used to make another reaction occur as part of a metabolic pathway.

Figure 3.3 Chemical reactions within the cells of these jellyfish release energy in the form of light, or phosphorescence.

***Predict** what other form of energy is released by the chemical reactions responsible for phosphorescence.*

The Laws of Thermodynamics

thermodynamics the science that studies the transfer and transformation of thermal energy (heat)

All activities—those that are necessary to enable and sustain life processes as well as those that occur in the non-living world and anywhere else in the universe—involve changes in energy. **Thermodynamics** is the study of these energy changes. Two laws of thermodynamics, called the first and second laws of thermodynamics, describe how energy changes occur. Both laws apply to a system and its surroundings. A *system* can be a whole organism, a group of cells, or a set of substrates and products—whatever object or objects are being studied. *Surroundings* are defined as everything in the universe outside of the system.

In terms of thermodynamics, biological systems are considered to be *open systems*, meaning that the system and its surroundings can exchange matter and energy with each other. The laws of thermodynamics describe how a system can interact with its surroundings and what can, and cannot, occur within a system.

The First Law of Thermodynamics

The first law of thermodynamics concerns the amount of energy in the universe. The first law is also called *the law of the conservation of energy*. In this context, "conservation" refers to maintaining the same amount of energy throughout a process. (This is different from the popular concept of energy conservation as preserving energy or reducing its usage so that it is not wasted or used excessively.)

> **The First Law of Thermodynamics**
> *Energy cannot be created or destroyed, but it can be transformed from one type into another and transferred from one object to another.*

Thus, when a chemical reaction occurs and energy is released, some of the energy can be transformed into mechanical energy, such as the motion of a contracting muscle, and the rest can be transformed into heat or other forms of energy. All of the energy is accounted for. If thermal energy leaves a system such as a living organism, the same amount of thermal energy must enter the surroundings. The energy cannot just disappear or be "lost" to the surroundings.

The first law of thermodynamics also states that energy cannot simply appear. For example, you cannot create the energy you need to go jogging. Chemical energy stored in food molecules must be transformed into kinetic energy in your muscles to enable you to move.

The Second Law of Thermodynamics

entropy a measure of disorder

According to the first thermodynamics law, the total amount of energy in the universe remains constant. Despite this, however, the energy available to do work decreases as more of it is progressively transformed into unusable heat. The second law of thermodynamics concerns the transformation of potential energy into heat, or random molecular motion. It states that the disorder in the universe—more formally called **entropy**—is continuously increasing. Put more simply, disorder is more likely than order. For example, it is much more likely that a stack of books will tumble over than that a pile of books will arrange themselves spontaneously to form a tidy stack. Similarly, it is much more likely that a tidy, orderly room will become more untidy and more disorderly over time than that an untidy room will spontaneously tidy itself.

> **The Second Law of Thermodynamics**
> *During any process, the universe tends toward disorder.*

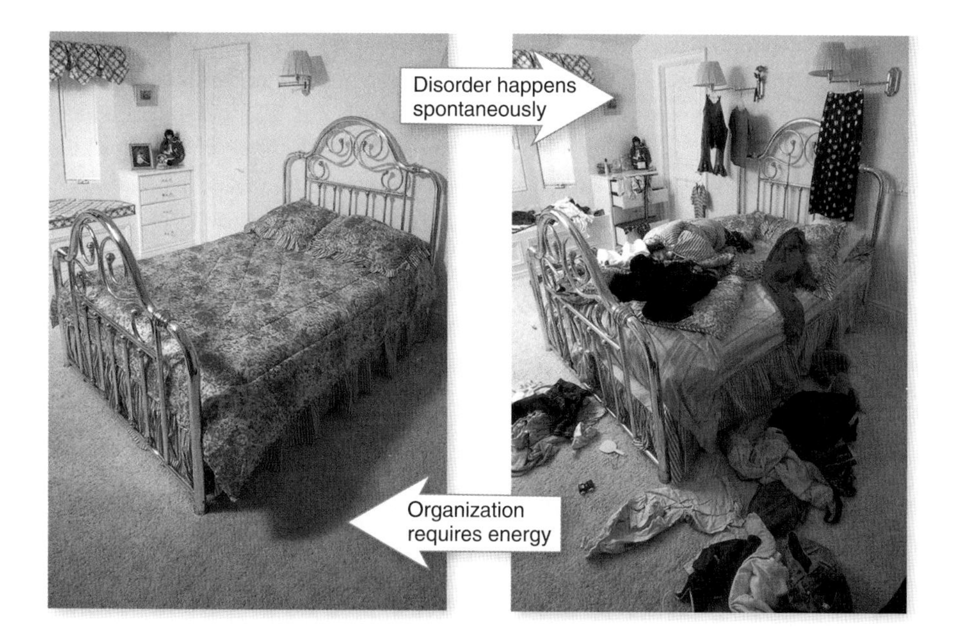
Disorder happens spontaneously
Organization requires energy

Figure 3.4 Entropy in action. Over time, the room becomes more disorganized. Entropy has increased in this room. An input of energy is needed to restore the original ordered condition of the room.

Energy transformations proceed spontaneously to convert matter from a more ordered, less stable condition to a less ordered, more stable condition. For example, in **Figure 3.4**, you could put the pictures into their correct sequence using the information that time had elapsed with only natural processes occurring. Common experience supports this—disorganized rooms do not spontaneously become organized. Thus, the second law of thermodynamics also can be stated simply as "entropy increases." When the universe formed, it held all the potential energy it will ever have. It has become increasingly more disordered ever since, with every energy exchange increasing the amount of entropy.

Because organisms are highly ordered, it might seem that life is an exception to the laws of thermodynamics. However, the second law applies only to closed systems. While they are alive, organisms remain organized because they are not closed systems. They use inputs of matter and energy to reduce randomness (decrease entropy) and thus stay alive. The energy that keeps organisms alive comes ultimately from the Sun. That is, plants transform light energy into the chemical bonds of carbohydrates, which humans and other organisms temporarily store and later use as an energy source.

Predicting Chemical Reactions Based on Changes in Free Energy

It takes energy to break the chemical bonds that hold atoms together. Heat, because it increases the kinetic energy of atoms, makes it easier for the atoms to pull apart. Both chemical bonding and heat have a significant influence on a molecule. Chemical bonding reduces disorder; heat increases it. The net effect—the amount of energy actually available to break and subsequently form other chemical bonds—is referred to as the free energy of that molecule. In a more general sense, **free energy** is defined as the energy available to do work in any system.

free energy energy from a chemical reaction that is available for doing work

For a molecule within a cell, where pressure and volume usually do not change, the free energy is denoted by the symbol G. G is equal to the energy contained in a molecule's chemical bonds, called *enthalpy* and designated H, together with the energy term related to the degree of disorder in the system. This energy term is designated TS, where S is the symbol for entropy and T is temperature. Thus:

$$G = H - TS$$

Chemical reactions break some bonds in the reactants and form new ones in the products. As a result, reactions can produce changes in free energy. When a chemical reaction occurs under conditions of constant temperature, pressure, and volume—as do most biological reactions—the change in free energy (ΔG) is

$$\Delta G = \Delta H - T\Delta S$$

endergonic chemical reaction that requires energy

exergonic chemical reaction that releases energy

Endergonic and Exergonic Reactions

The change in free energy, ΔG, can be used to predict whether a chemical reaction is spontaneous or not. For some reactions, the ΔG is positive, which means that the products of the reaction contain *more* free energy than the reactants. Thus, the bond energy (H) is higher, or the disorder (S) in the system is lower. Such reactions do not proceed spontaneously, because they require an input of energy. Any reaction that requires an input of energy is said to be **endergonic**, which literally means "inward energy."

For other reactions, the ΔG is negative. In this case, the products of the reaction contain less free energy than the reactants. Thus, either the bond energy is lower, or the disorder is higher, or both. Such reactions tend to proceed spontaneously. These reactions release the excess free energy as heat and are said to be **exergonic**, which literally means "outward energy." Any chemical reaction tends to proceed spontaneously if the difference in disorder ($T\Delta S$) is *greater* than the difference in bond energies between reactants and products (ΔH).

Note that spontaneous does not mean the same thing as instantaneous. A spontaneous reaction can proceed very slowly. **Figure 3.5** summarizes the concepts of endergonic and exergonic reactions.

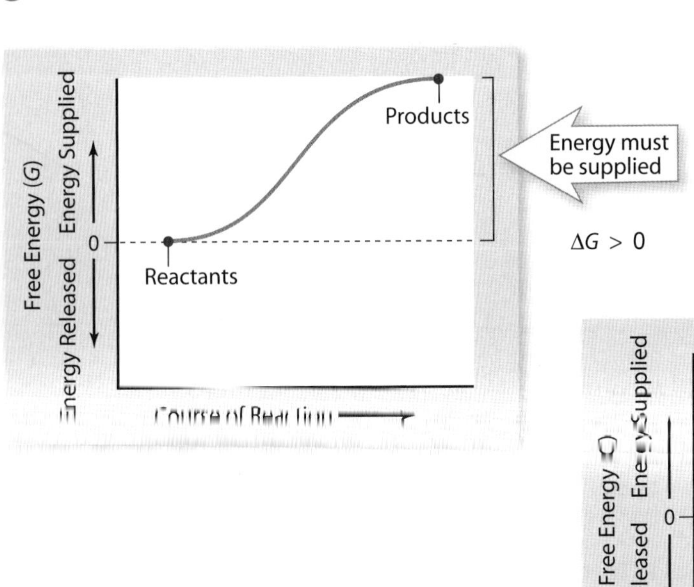

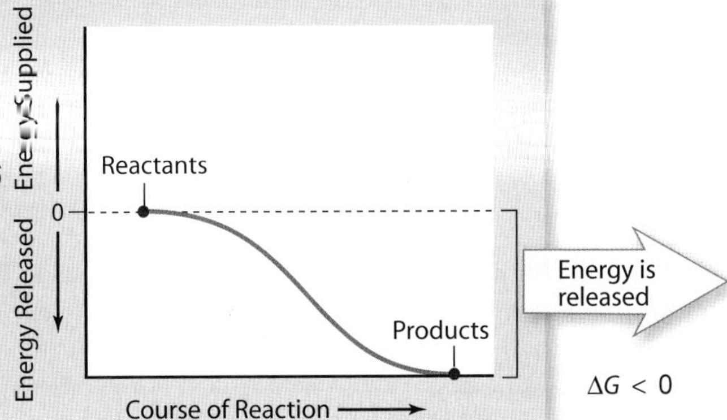

Figure 3.5 In an endergonic reaction (**A**), the products of the reaction contain more energy than the reactants, and energy must be supplied for the reaction to proceed. In an exergonic reaction (**B**), the products contain less energy than the reactants, and excess energy is released.

Identify *the graph that could represent the oxidation of glucose to carbon dioxide and water, and explain why it could.*

Learning Check

1. Write a sentence that not only defines the term metabolism, but also clearly differentiates between the terms anabolism and catabolism.
2. Write a sentence to show the relationship among the following terms: anabolic reactions, catabolic reactions, endergonic reactions, exergonic reactions.
3. Use **Figure 3.2** to explain why the amount of chemical potential energy of a compound is less than the amount of chemical potential energy of the atoms that comprise the compound.
4. What is the significance of the first law of thermodynamics? (In other words, what does it imply about energy transfer and transformations?)
5. What is the significance of the second law of thermodynamics?
6. Describe the relationship among the following concepts: free energy, endergonic reactions, exergonic reactions, entropy.

Thermodynamics and Metabolism

Just because a reaction is spontaneous does not mean that it will necessarily occur. For example, the burning of gasoline will not proceed simply because oxygen and gasoline are mixed. The reaction must be initiated with energy such as from a spark or a flame. Once the reaction has begun, it will proceed in the forward direction until one of the reactants is used up. In fact, most reactions require an input of energy to get started. This energy destabilizes existing chemical bonds and initiates the reaction. In Chapter 1, you learned that this input energy is called *activation energy*. An exergonic reaction may proceed very slowly if the activation energy is quite large. One way that the activation energy of a reaction can be reduced is by using a catalyst. In metabolic pathways, biological catalysts—enzymes—decrease the activation energy of each reaction.

In cells, energy from catabolic reactions is used to power anabolic reactions. The source of energy that links these sets of reactions is the molecule ATP, adenosine triphosphate. ATP is often called the energy currency of the cell, because so many cellular activities depend on ATP. It is the major product of most catabolic pathways, and it is the major source of energy for anabolic pathways.

Figure 3.6 shows the structure of ATP and the hydrolysis of the terminal (last) phosphate group. The red tilde symbols (wavy lines) represent high-energy bonds that, when hydrolyzed, release energy. Each of the phosphate groups in an ATP molecule is negatively charged. The negative charges of the phosphate groups repel each other in such a way that the phosphate groups strain away from each other like opposing teams in a tug-of-war contest. When one of the bonds between the phosphate groups is broken, ATP becomes ADP (adenosine diphosphate) plus an inorganic phosphate (P_i), and a large quantity of energy is released.

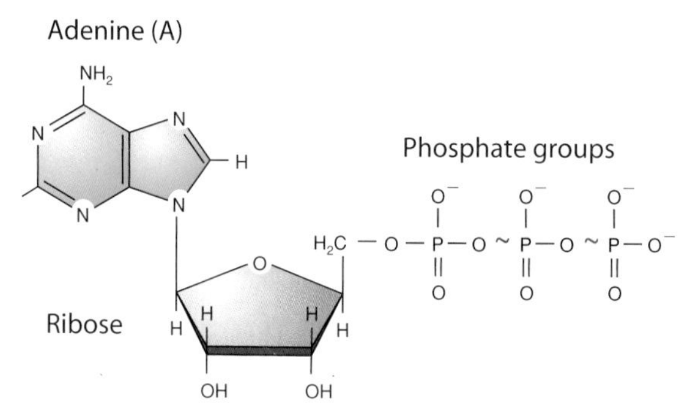

Adenosine triphosphate (ATP)

Hydrolysis of ATP

H_2O

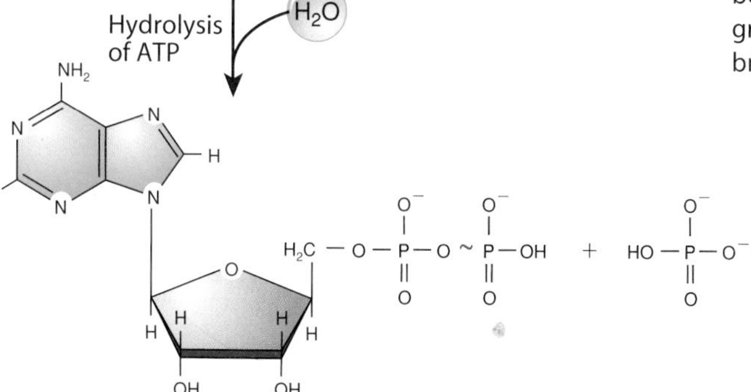

Adenosine diphosphate (ADP)

Phosphate (P_i)

$\Delta G = -30.6$ kJ/mol

Figure 3.6 The hydrolysis of ATP to ADP (adenosine diphosphate) and P_i (inorganic phosphate) is a highly exergonic reaction. Repulsion between negative charges on the neighbouring phosphate groups makes the bonds between the first and second and between the second and third phosphate groups unstable. When these bonds are broken, energy is released.

Coupled Reactions

Cells use ATP to drive endergonic reactions. These reactions do not proceed spontaneously, because their products possess more free energy than their reactants. However, if the cleaving of ATP's terminal high-energy bond releases more energy than the other reaction consumes, the two reactions can be *coupled* so that the energy released by the hydrolysis of ATP can be used to supply the endergonic reaction with energy. Coupled together, these reactions result in a net release of energy ($-\Delta G$) and are therefore exergonic and proceed spontaneously.

The use of ATP can be thought of as a cycle. Cells use exergonic reactions to provide the energy needed to synthesize ATP from ADP + P_i; they then use the hydrolysis of ATP to provide energy for endergonic reactions, as shown in **Figure 3.7**. Most cells typically have only a few seconds' supply of ATP at any given time and continually produce more from ADP and P_i. Even an inactive person turns over an amount of ATP in one day roughly equal to the person's body mass. This statistic makes clear the importance of ATP synthesis.

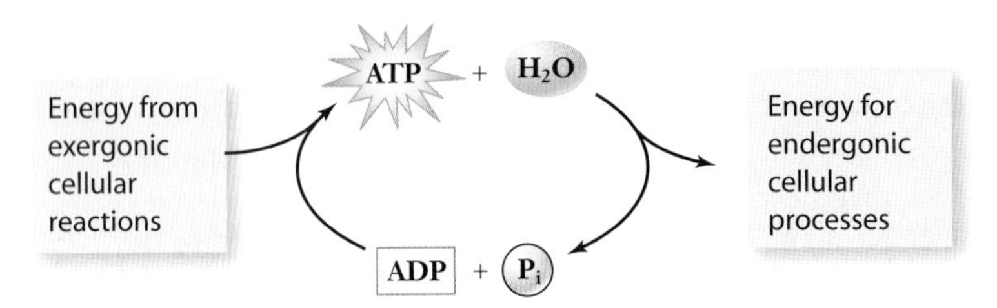

Figure 3.7 ATP is synthesized and hydrolyzed in a cyclic fashion in cells.

Electron Carriers

Redox reactions are coupled reactions that play a key role in the flow of energy through biological systems. When a compound accepts electrons, it becomes reduced, and when it loses electrons, it becomes oxidized. Electrons that pass from one atom to another carry energy with them, so the reduced form of a molecule is always at a higher energy level than the oxidized form. Thus, electrons are said to carry *reducing power*. The amount of energy they carry depends on the energy level they occupy in the atom donating the electrons. *Electron carriers* are compounds that pick up electrons from energy rich compounds and then donate them to low energy compounds. An electron carrier is recycled

Two important electron carriers in metabolic reactions are NAD^+ (nicotinamide adenine dinucleotide) and FAD (flavin adenine dinucleotide). NAD^+ and FAD are the oxidized forms, and NADH and $FADH_2$ are the reduced forms. **Figure 3.8** shows the mechanism by which NAD^+ becomes reduced, forming NADH. You will learn about the role electron carriers play in cellular respiration in the next section.

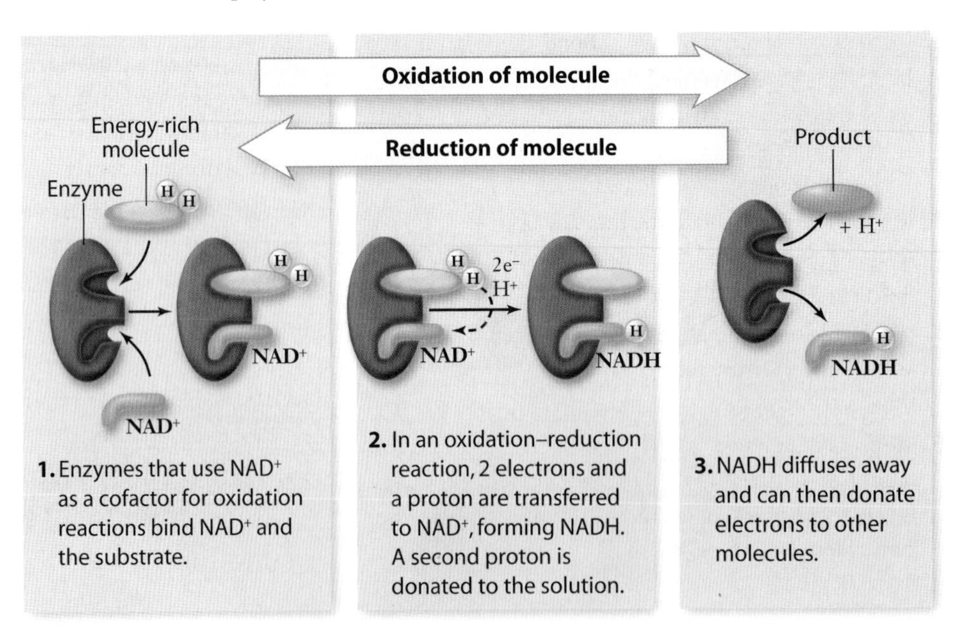

Figure 3.8 While bound to an enzyme, NAD^+ receives electrons from two hydrogen atoms to become reduced to NADH. One proton (H^+ ion) is released into solution. The product of the energy-rich molecule and the NADH leave the enzyme. The reduced NADH then carries the electrons, or reducing power, to another molecule.

Section 3.1 Review

Section Summary

- Metabolism is the sum of all the biochemical reactions in a cell. Anabolic reactions require energy to build up molecules, and catabolic reactions break down molecules and release energy.
- Metabolic pathways are sequences of reactions that use of the product of one reaction as the substrate for the next.
- Energy is the ability to do work, and it can be classified as kinetic or potential.
- Bond energy is the energy needed to break a bond.
- The first law of thermodynamics states that energy cannot be created or destroyed.
- The second law of thermodynamics states that some energy is "lost" as disorder (entropy) increases.
- Free energy is the energy from a chemical reaction that is available to do work.
- Endergonic reactions require energy, while exergonic reactions release energy.
- ATP hydrolysis releases energy to drive endergonic reactions, and it is synthesized using energy from exergonic reactions.
- Electron carriers donate electrons from energy-rich to low-energy compounds.

Review Questions

1. **K/U** What is a metabolic pathway, and why are metabolic pathways advantageous to a cell?
2. **C** Use a labelled diagram to illustrate the differences between anabolism and catabolism.
3. **C** Copy and complete the following table that summarizes different types of energy:

Type of Energy	Description	Example
Kinetic energy		
Potential energy		
Thermal energy		
Chemical energy		
Bond energy		

   a. Bond energy is not a distinct form of energy in the way that thermal energy and chemical energy are considered to be forms of energy. Into what category would you classify bond energy, and why?
   b. At the simplest level of classification, all energy can be classified into two forms: kinetic and potential. Into which of these categories would you classify thermal and chemical energy, and why?
4. **K/U** State and explain the first law of thermodynamics and the second law of thermodynamics. Use examples in your explanations.
5. **K/U** What is the difference between open systems and closed systems? Refer to the laws of thermodynamics in your answer.
6. **A** Some people claim that life's high degree of organization defies the laws of thermodynamics. How is the organization of life actually consistent with the principles of thermodynamics?
7. **A** How is the melting of ice an example of an increase in entropy?
8. **T/I** Use the diagram to answer the questions.

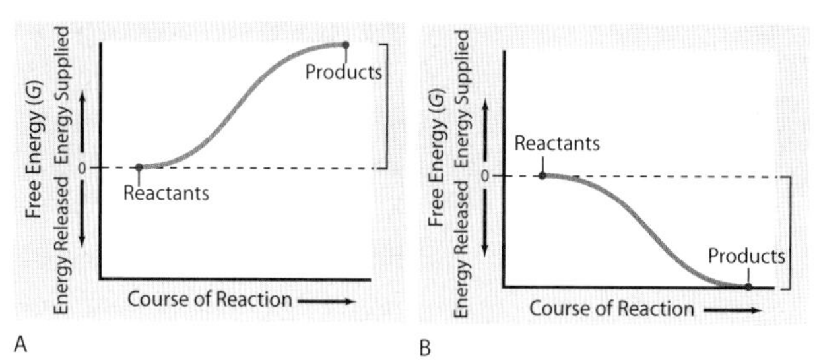

   a. What type of reaction is shown in each graph? Provide an explanation for each answer.
   b. What type of reaction would need energy to be supplied? Which would have energy released?
9. **K/U** What is a spontaneous reaction? How can you predict if a chemical reaction is spontaneous?
10. **T/I** Is an anabolic reaction more likely to be exergonic or endergonic, and would this type of reaction have a positive or negative value for ΔG? Explain your answer.
11. **C** Draw a diagram of a molecule of ATP. Label the different groups that make up this molecule.
12. **C** Use labelled diagrams to illustrate the ATP energy cycle. Use your diagrams to explain how this cycle enables life-sustaining cellular activities.
13. **K/U** Distinguish between oxidation and reduction reactions. Why are these reactions linked?
14. **K/U** Why is the recycling of electron carriers and ATP important for the cell?

SECTION 3.2

Aerobic Respiration

Key Terms

aerobic respiration
substrate level phosphorylation
glycolysis
Krebs cycle
oxidative phosphorylation
electron transport chain
chemiosmosis

aerobic respiration catabolic pathways that require oxygen

substrate level phosphorylation ATP formation from transferring a phosphate group to ADP

Cellular respiration includes the catabolic pathways that break down energy-rich compounds to produce ATP. **Aerobic respiration** refers to those pathways that require oxygen in order to proceed. The following summary reaction for this process represents more than two dozen reactions that take place in different parts of the cell.

$$C_6H_{12}O_6(s) + 6O_2(g) \rightarrow 6CO_2(g) + 6H_2O(\ell) + \text{energy}$$

Overview of Aerobic Respiration

An overview of the reaction pathways that constitute aerobic respiration is shown in **Figure 3.9**. Although part of aerobic respiration, the first pathway, glycolysis, is not truly aerobic. Glycolysis can proceed with or without oxygen. However, the products of glycolysis are the starting materials for the metabolic pathways that follow, and they require oxygen. **Note:** The following description of aerobic respiration is part of the overview discussion accompanying **Figure 3.9**. A more detailed discussion of individual pathways in aerobic respiration follows this overview.

Glycolysis occurs within the cytoplasm of cells. The starting material for glycolysis is glucose, a six-carbon sugar. Each glucose molecule is broken down, through a series of reactions, into two three-carbon compounds. Each of these compounds is converted into a three-carbon pyruvate molecule. Thus, the end products of glycolysis include two molecules of pyruvate for each molecule of glucose that enters glycolysis. The glycolytic pathway also converts two molecules of NAD^+ into two molecules of NADH. The breakdown of glucose into pyruvate includes two reactions that consume ATP, as well as two reactions that produce ATP. Overall, the process consumes two molecules of ATP and produces four molecules of ATP. Thus, this pathway is responsible for a net production of two ATP molecules.

At various steps in the glycolytic pathway, a phosphate group is removed from a substrate molecule and combined with an ADP molecule to form ATP. This process is called **substrate level phosphorylation**. (Phosphorylation refers to any process that involves the combining of phosphate to an organic compound.)

When oxygen is available, pyruvate from glycolysis is transported across the outer and inner mitochondrial membranes into the mitochondrial matrix. There, the three-carbon pyruvate molecule undergoes oxidation. This produces a two-carbon molecule called acetyl-coenzyme A, abbreviated as acetyl-CoA, and releases one carbon atom in the form of carbon dioxide. In the process, one NAD^+ molecule is reduced to form NADH. Since glycolysis produces two pyruvate molecules, pyruvate oxidation actually releases two molecules of carbon dioxide and reduces two molecules of NAD^+ to form two molecules of NADH.

Each acetyl-CoA molecule then enters the Krebs cycle, which is also known as the *citric acid cycle* or the *tricarboxylic acid (TCA) cycle*. Tricarboxylic acid is a general name referring to an organic acid with three carboxyl groups. Citric acid is one example of a tricarboxylic acid. At physiological pH, citric acid exists in its ionized form, called citrate.

Acetyl CoA combines with a four-carbon molecule in the Krebs cycle to form the six-carbon molecule, citrate. In each round of the Krebs cycle, one ATP molecule is formed, three NAD+ molecules are reduced to form three molecules of NADH, and one FAD molecule is reduced to form one molecule of $FADH_2$. Since two molecules of acetyl CoA are produced for every molecule of glucose that enters glycolysis, the total yield from the Krebs cycle is two ATP molecules, six NADH molecules, and two $FADH_2$ molecules.

The substrates, products, and enzymes of the Krebs cycle are in solution in the mitochondrial matrix. The reduced NADH and $FADH_2$ molecules diffuse to specific locations on the inner membrane of the mitochondrion and donate their electrons to a series of electron carriers that are embedded in the inner mitochondrial membrane. As each electron carrier passes an electron to the next carrier, some energy is released. This energy is used to pump protons, H^+, across the inner mitochondrial membrane into the intermembrane space. This creates a proton gradient across the inner mitochondrial membrane, which provides the energy for the enzyme ATP synthase to phosphorylate (add P_i to) ADP molecules to make ATP. Because this energy comes from the flow of electrons from NADH and $FADH_2$ to oxygen, this process is called *oxidative phosphorylation*.

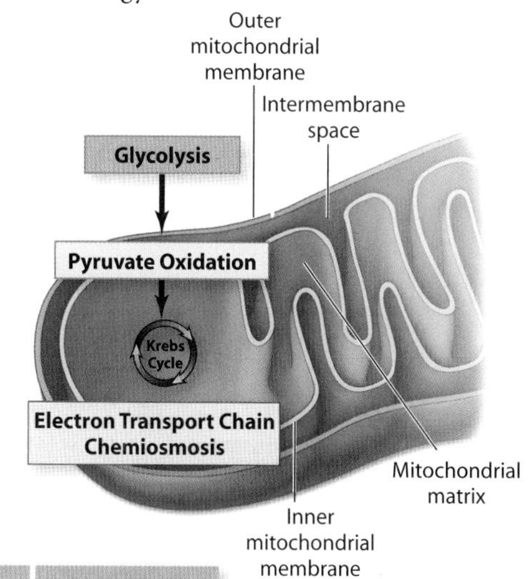

A maximum of 38 molecules of ATP can be generated by the breakdown of one molecule of glucose through the overall process of aerobic respiration. This includes ATP generated directly by glycolysis and the Krebs cycle, ATP generated from the two molecules of NADH produced during glycolysis, and ATP produced through oxidative phosphorylation, from reducing power generated during pyruvate oxidation and the Krebs cycle. A detailed discussion of these processes follows.

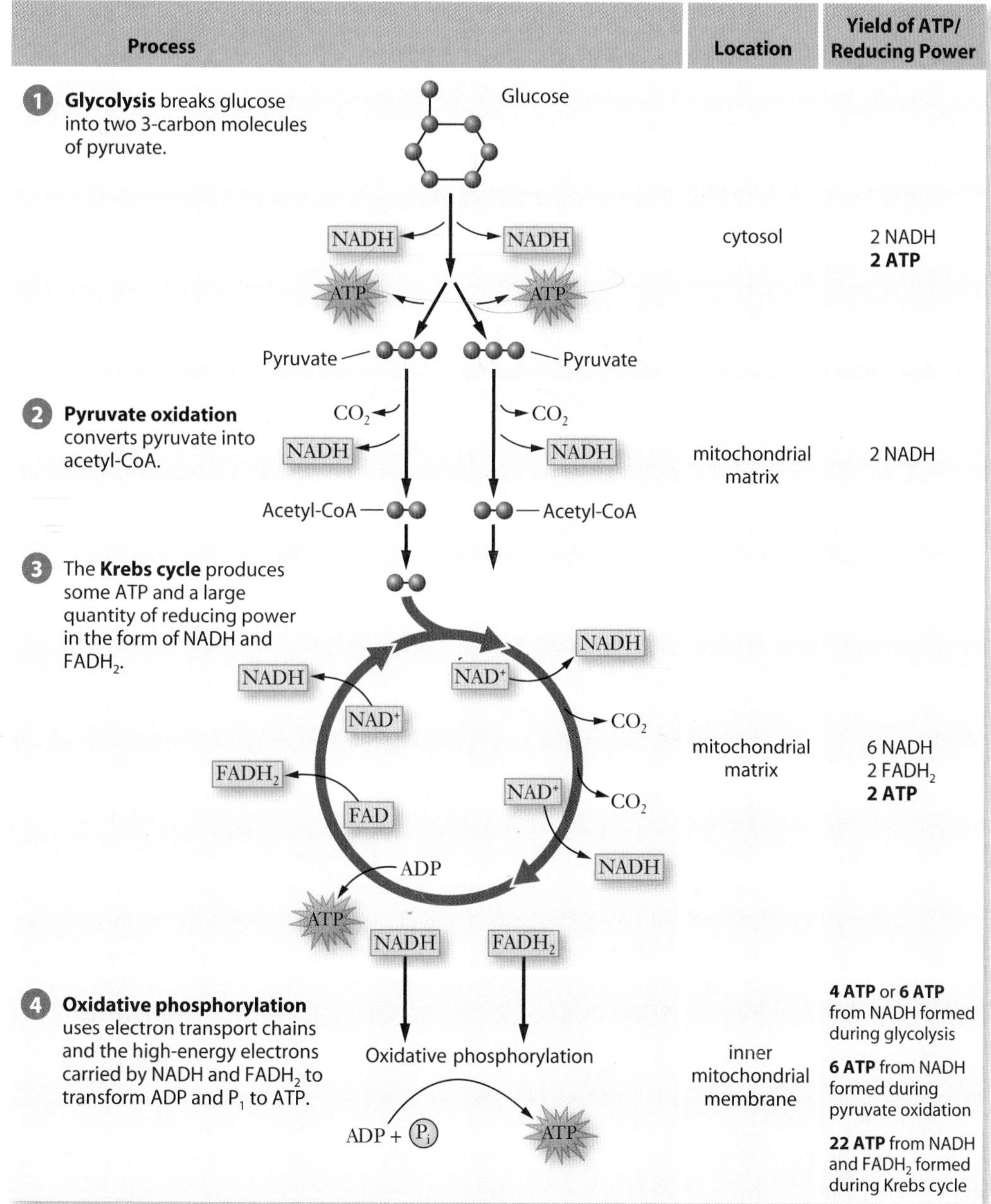

Figure 3.9 This diagram summarizes the metabolism of glucose by aerobic respiration. The maximum total yield of ATP from cellular respiration in eukaryotes is 36 (2 + 2 + 4 + 6 + 22) and in prokaryotes is 38 (2 + 2 + 6 + 6 + 22). These numbers reflect the fact that the yield of ATP from total glycolytic NADH is 4 ATP in eukaryotes and 6 ATP in prokaryotes.

Glycolysis

glycolysis metabolic pathway that breaks glucose down to pyruvate

The term **glycolysis** is derived from two Greek words, *glykos* and *lysis*, which mean sweet and splitting, respectively. As its name implies, glycolysis involves the breakdown ("splitting") of glucose into smaller molecules. The 10 reactions of glycolysis are shown in **Figure 3.10**.

In all cells in which aerobic respiration occurs, the main function of glycolysis is to convert glucose to two molecules of pyruvate. The amount of useful energy in the form of ATP generated by glycolysis is small compared to the ATP produced by the pathways that take place in the mitochondria. Key points about glycolysis appear below.

- In the first part of the glycolytic pathway, in reactions 1 and 3, two molecules of ATP are used to phosphorylate substrate molecules.
- In reaction 4, the six-carbon compound, fructose 1,6-bisphosphate, is split into two different three-carbon compounds: dihydroxyacetone phosphate (DHAP) and glyceraldehyde 3-phosphate (G3P).
- In reaction 5, DHAP is converted into a second G3P molecule. Each of the two G3P molecules proceeds through reactions 6 to 10. In other words, reactions 6 to 10 occur twice for each molecule of glucose that enters glycolysis.
- In reaction 6, an inorganic phosphate group is added to G3P, and an NAD^+ molecule is reduced to form NADH. (Since this reaction occurs for each of the two G3P molecules, two molecules of NAD^+ are reduced to form two molecules of NADH for each molecule of glucose entering glycolysis.)
- In reaction 7, ADP is converted to ATP by substrate level phosphorylation. (Two molecules of ADP are converted to two molecules of ATP for every molecule of glucose entering glycolysis.)
- In reactions 8 and 9, the three-carbon substrate molecules are rearranged and a water molecule is removed. (Two water molecules are removed for each glucose molecule entering glycolysis.)
- Finally, in reaction 10, another molecule of ADP is converted to ATP by substrate level phosphorylation. (Two molecules of ATP are produced in this step for every glucose molecule entering glycolysis.)

To summarize the products of the energy-yielding steps: Two ATP molecules are consumed and four molecules of ATP are produced, resulting in a net gain of two ATP molecules. Two NAD^+ molecules are reduced to form two molecules of NADH.

The net reaction for glycolysis can be written

$$\text{glucose} + 2\,NAD^+ + 2\,ADP + 2\,P_i \rightarrow 2\,\text{pyruvate} + 2H_2O + 2\,NADH + 2\,ATP$$

Learning Check

7. Differentiate between cellular respiration and aerobic cellular respiration.

8. What is the general or summary equation that describes cellular respiration, and must it be a summary of this process, rather than an ordinary chemical equation?

9. What are the stages of energy transfer from glucose to ATP in a general overview of aerobic respiration, and in what parts of the mitochondrion do they occur?

10. What is substrate-level phosphorylation, and when does it generate ATP?

11. What is oxidative phosphorylation, and when does it generate ATP?

12. What is glycolysis, and what are its starting materials and products?

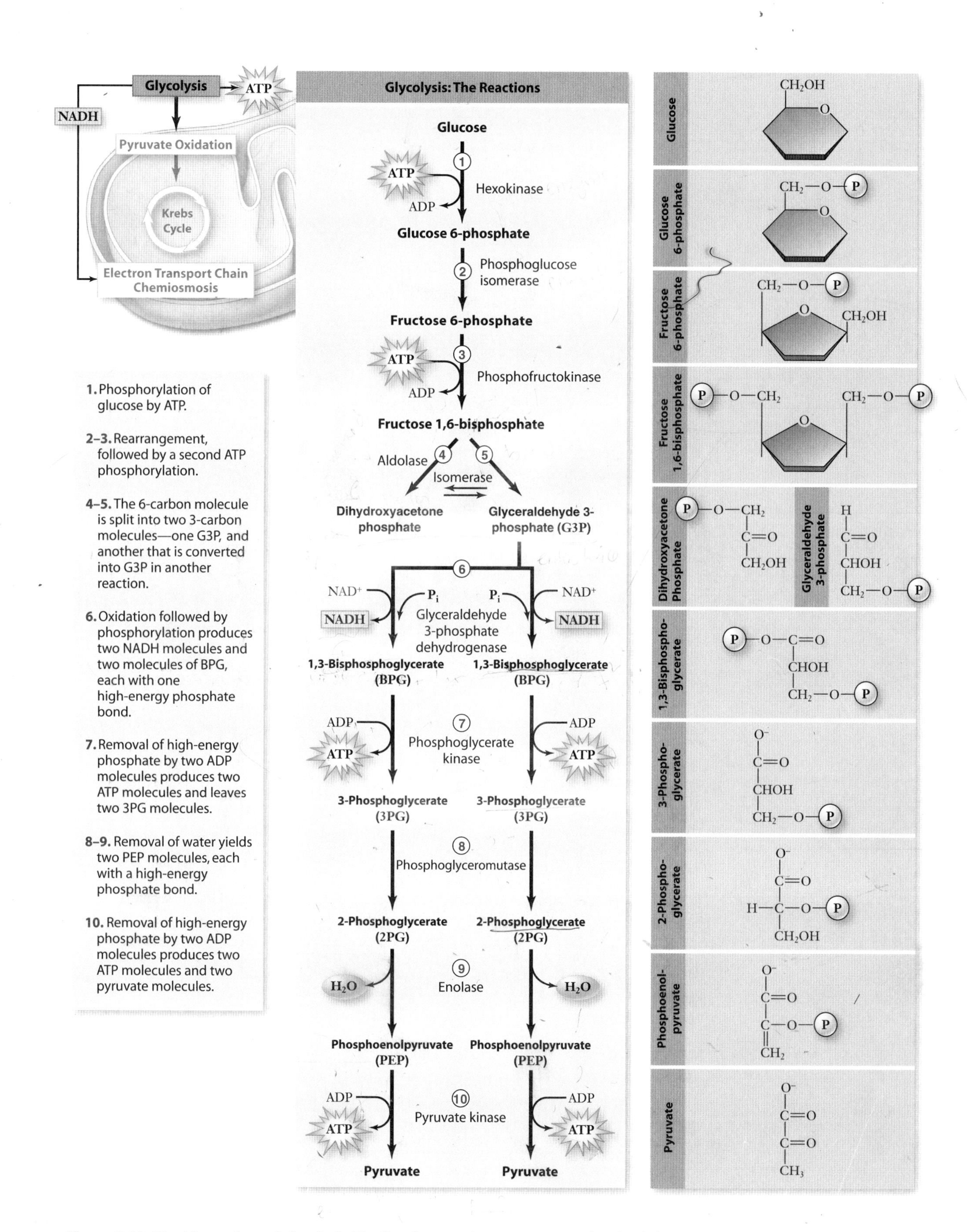

Figure 3.10 The 10 reactions of glycolysis. The first five reactions convert a molecule of glucose into two molecules of G3P. The second five reactions convert G3P into pyruvate.

Pyruvate Oxidation

When oxygen is available, the pyruvate produced in the cytosol is transported into the mitochondrial matrix. There, pyruvate undergoes a reaction, shown in **Figure 3.11**, that converts it into a two-carbon molecule. The reaction takes place on a complex that contains multiple copies of three different enzymes. The reaction occurs in five steps, but the substrate remains bound to the enzyme during all of the steps. One of the three carbons in pyruvate is cleaved off and released as carbon dioxide. The remainder, called an acetyl group, becomes associated with a carrier molecule called coenzyme A (Co-A) to produce acetyl-CoA. This reaction is coupled to the reduction of NAD^+ to produce NADH. For each glucose molecule that enters glycolysis, two pyruvate molecules undergo oxidation, producing two molecules of acetyl-CoA and two NADH molecules. Pyruvate oxidation can be thought of as linking glycolysis to the next major metabolic pathway in aerobic respiration—the Krebs cycle.

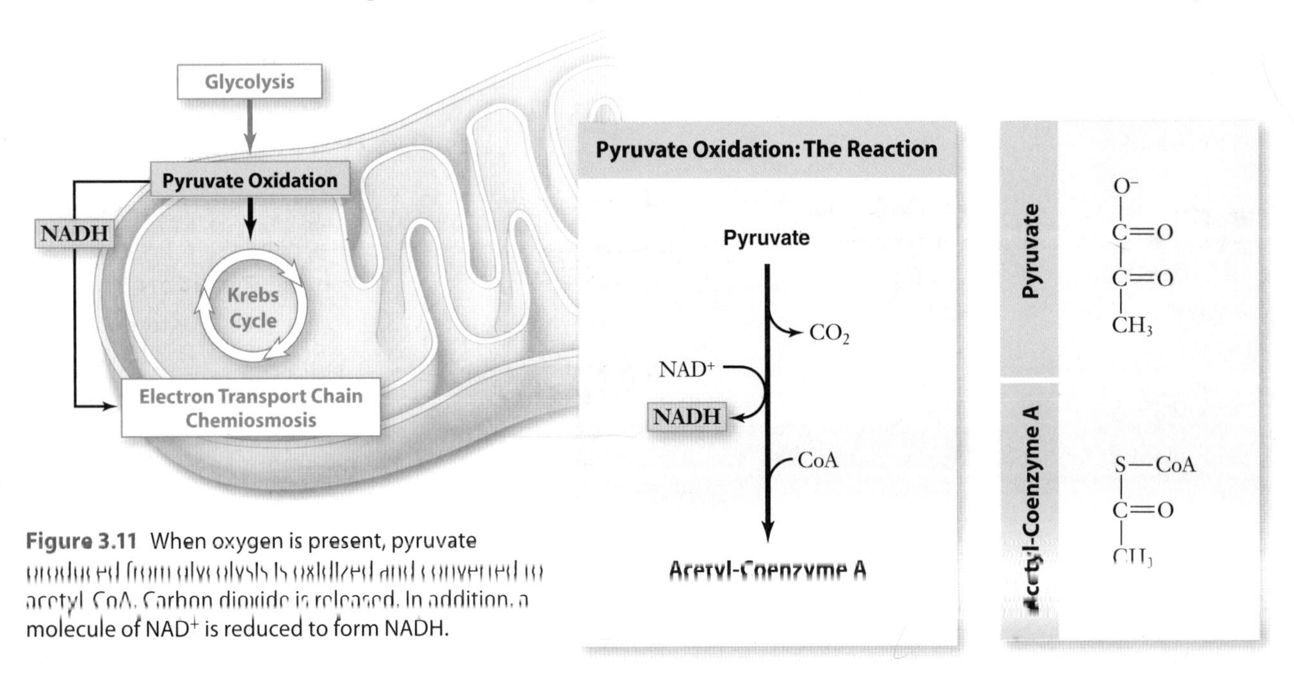

Figure 3.11 When oxygen is present, pyruvate produced from glycolysis is oxidized and converted to acetyl CoA. Carbon dioxide is released. In addition, a molecule of NAD^+ is reduced to form NADH.

The Krebs Cycle

Krebs cycle the cyclic metabolic pathway that acquires acetyl-CoA and oxidizes it to carbon dioxide while regenerating the compound that picks up more acetyl-CoA; converts released energy to ATP, NADH, and $FADH_2$

The Krebs cycle was named after Hans Adolf Krebs, who shared the 1953 Nobel Prize in Physiology or Medicine for discovering this pathway. The **Krebs cycle**, shown in **Figure 3.12**, is a cyclical metabolic pathway that occurs in the mitochondrial matrix. Acetyl CoA is fed into the pathway. Ultimately, the original carbon atoms from glucose that entered glycolysis will be oxidized by the Krebs cycle. The key points to remember about the Krebs cycle are as follows.

- Each molecule of acetyl-CoA carries two carbons from the molecule of glucose that originally entered glycolysis. These two carbons will be released later as carbon dioxide through the activity of the Krebs cycle. For each molecule of glucose that enters glycolysis, two molecules of acetyl-CoA enter the Krebs cycle.
- Acetyl-CoA "delivers" two carbons from glucose to the Krebs cycle by reacting with oxaloacetate to produce a six-carbon molecule called citrate (reaction 1).
- Once citrate is formed, it undergoes a series of five reactions (reactions 2 to 6) that break it down to a four-carbon molecule called succinate. Two of these reactions are oxidation reactions that result in the release of two carbons in the form of carbon dioxide molecules. Coupled to these oxidation reactions are two reduction reactions, each of which reduces a molecule of NAD^+ to produce a molecule of NADH.

- In reaction 6 of the cycle, ATP is produced by substrate-level phosphorylation. This is a complex reaction in which a phosphate group replaces the CoA while the substrate, succinate, is bound to the enzyme. The phosphate group is then added to a molecule of guanosine triphosphate (GTP). The terminal phosphate group from GTP is then transferred to ADP to produce ATP. Although most of the GTP is used to make ATP, a few reactions in the cell use the GTP itself for energy.
- Two of the next three reactions (numbered 7 to 9), are oxidation reactions. These reactions are coupled to the reduction of NAD^+ to form NADH and the reduction of FAD to form $FADH_2$. These electron carriers are used to produce ATP in the oxidative phosphorylation pathway.
- The final product of the last of these three reactions is oxaloacetate. This makes the pathway cyclical—oxoaloacetate is regenerated and ready to react with more acetyl-CoA entering the Krebs cycle.

Ultimately, the carbon atoms from glucose are released in the form of six carbon dioxide molecules, which are considered waste. The energy from the breakdown of glucose is in the form of 4 ATP molecules (2 from glycolysis and 2 from the Krebs cycle), 10 NADH molecules (2 from glycolysis, 2 from the oxidation of pyruvate, and 6 from the Krebs cycle), and 2 $FADH_2$ molecules (from the Krebs cycle).

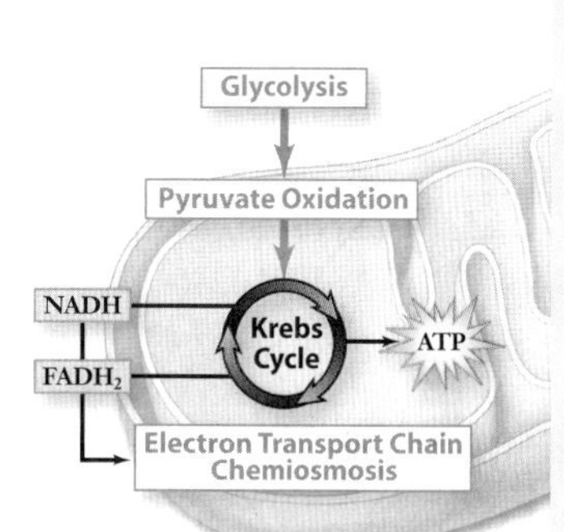

1. Reaction 1: Condensation
2–3. Reactions 2 and 3: Isomerization
4. Reaction 4: The first oxidation
5. Reaction 5: The second oxidation
6. Reaction 6: Substrate-level phosphorylation
7. Reaction 7: The third oxidation
8–9. Reactions 8 and 9: Regeneration of oxaloacetate and the fourth oxidation

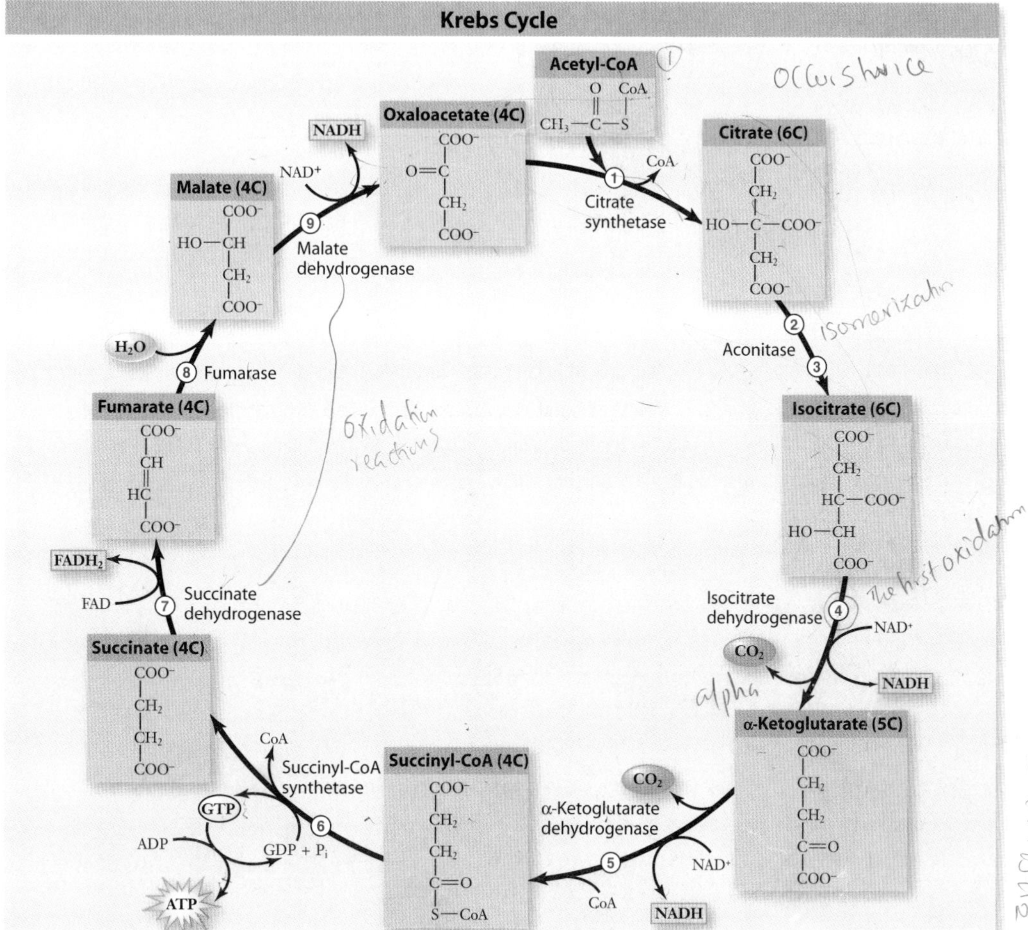

Figure 3.12 The Krebs cycle. This pathway consists of a number of oxidation reactions and completes the breakdown of glucose to carbon dioxide. Reaction numbers 2 and 3 are written on the same arrow because the reaction takes place in two steps but never leaves the surface of the enzyme. The terms written beside the reaction numbers are the names of the enzymes.

Learning Check

13. Explain what NADH is and how it is produced.

14. What is the Krebs cycle, and what is its significance in the process of aerobic respiration?

15. What happens to the pyruvate generated in glycolysis before it enters the Krebs cycle?

16. Explain why for each molecule of glucose that enters glycolysis there is more than one turn of the Krebs cycle.

17. Identify the stages in the Krebs cycle that involve oxidation reactions and the reduction reactions that are coupled to them.

18. Explain how the Krebs cycle generates the following molecules: CO_2, ATP, NADH, and $FADH_2$.

Oxidative Phosphorylation

oxidative phosphorylation a process that couples the oxidation of NADH and $FADH_2$ by the electron transport chain with the synthesis of ATP by phosphorylation of ADP

Through glycolysis and the Krebs cycle, every carbon atom in a molecule of glucose is converted into carbon dioxide. However, very few ATP molecules have been produced. Most of the energy is still in the form of the reduced electron carriers, NADH and $FADH_2$. By this point, no oxygen has been used. The majority of the ATP molecules produced during aerobic respiration come from the process of **oxidative phosphorylation**. It is not until the end of this part of the overall process of aerobic respiration that oxygen acts as an electron acceptor and is converted to water.

The electron transport chain is a series of electron carriers and proteins that are embedded in the inner membrane of the mitochondrion, as shown in **Figure 3.13**. Electrons donated by NADH and $FADH_2$ are transported through this chain, providing the energy needed for oxidative phosphorylation.

When NAD^+ accepts electrons, it accepts two electrons and one hydrogen ion. FAD accepts two electrons and two hydrogen ions. When NADH and $FADH_2$ pass these electrons on to the electron acceptors in the electron transport chain, however, they pass only one at a time. Also, the hydrogen ions do not accompany the electrons, but instead remain in solution in the matrix. The electrons from NADH pass through three major complexes. Each of these complexes uses energy released from the passing electrons to actively transport, or pump, hydrogen ions out of the matrix and into the intermembrane space, thus creating a

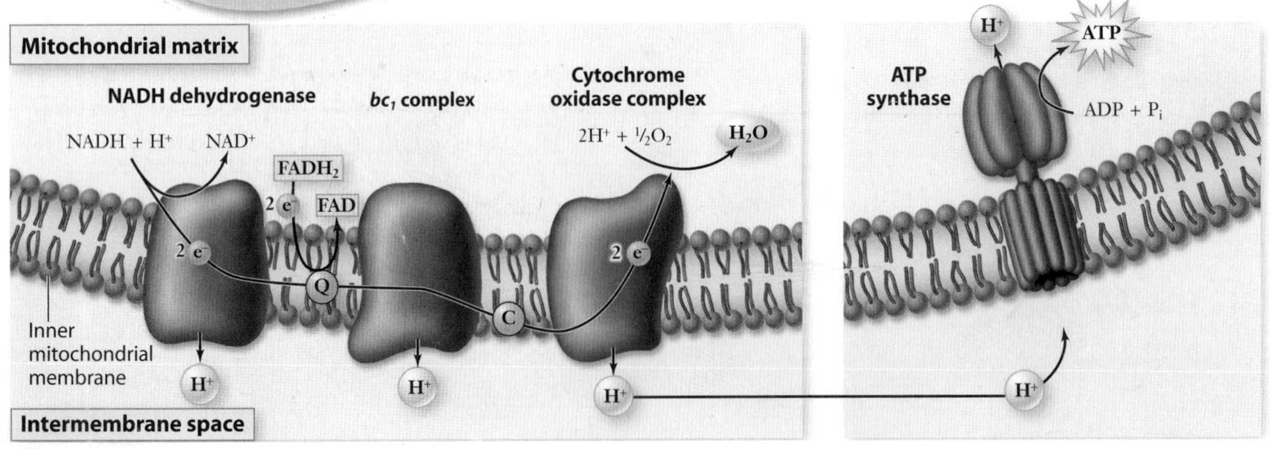

Figure 3.13 The electron carriers continuously cycle between their reduced form and their oxidized form while passing electrons from one to the next and finally to oxygen.

hydrogen ion gradient across the membrane. When $FADH_2$ passes electrons to the **electron transport chain**, these electrons skip the first major complex in the chain. Thus, the energy released by this electron transfer pumps hydrogen ions through only the second and third complex in the chain. The final electron acceptor is oxygen. Each oxygen atom combines with two electrons and two hydrogen ions to form a water molecule. Thus, oxygen is consumed, and water is produced during the process of oxidative phosphorylation.

electron transport chain a series of electron carriers and protein complexes embedded in the inner mitochondrial membrane that accept and donate electrons in a sequential series, resulting in oxidative phosphorylation

chemiosmosis a process that uses energy in a hydrogen ion gradient across the inner mitochondrial membrane to drive phosphorylation of ADP to form ATP

Chemiosmosis

The energy from the reduced NADH and $FADH_2$ is used to establish an electrochemical gradient called a hydrogen ion gradient. The reducing power of NADH and $FADH_2$ is stored as electrical potential energy in this gradient. This gradient can be compared with water behind a hydroelectric dam. The dam restricts the natural flow of a river, preventing water from passing through and flowing downstream. Water builds up behind the dam, increasing the force of the water on the dam. Water is allowed to pass through specific floodgates in the dam in a regulated manner, turning a turbine in the process. A generator converts the mechanical energy of the turning turbine into electrical energy.

Similarly, the inner mitochondrial membrane restricts the passage of hydrogen ions along their concentration gradient. As a result, the concentration of hydrogen ions increases within the intermembrane space, and there is a corresponding increase in positive charge in the intermembrane space relative to the matrix. This electrical potential energy is converted into the chemical potential energy of ATP by complexes called *ATP synthases*. The passage of hydrogen ions through the inner mitochondrial membrane is restricted to facilitated diffusion through the centre of an ATP synthase complex. When electrons are moving down their gradient through an ATP synthase complex, the energy is used to phosphorylate ADP to form ATP. This process is called **chemiosmosis**.

Activity 3.1 Assessing Information about the Effects of Metabolic Toxins

In the 1930s, a compound called dinitrophenol (DNP) was used in diet pills. DNP affects chemiosmosis by disrupting the hydrogen ion concentration gradient. This leads to rapid oxidation of compounds in the Krebs cycle, and encourages the metabolizing of carbohydrates and fats. Since the production of ATP is impaired, energy production in the body is instead given off as significant amounts of heat. People lost a lot of weight quickly, but many people also lost their health and, in some cases, their lives. DNP was later banned, but it is still used by bodybuilders at great risk to their personal safety. DNP is an example of a metabolic toxin—a chemical that impairs or disrupts metabolic pathways. In this activity, you will investigate and report on metabolic toxins that affect the function of mitochondria.

Procedure

1. Choose one of these metabolic toxins to research:
   - antimycin
   - cyanide
   - hydrogen sulfide
   - malonate
   - rotenone
   - arsenic
2. Find at least three sources of information about this toxin. Collect the following information:
   - its physical properties
   - when and why it was first developed and used
   - its effect on metabolism
   - antidotes and/or treatments, if any
3. Create a "Metabolic Toxin Profile" that includes an introductory paragraph summarizing your research findings. Use appropriate headings to organize the information in your profile. Also include the sources of information you have used to construct it.

Questions

1. Classify the metabolic toxins your class investigated on the basis of the metabolic pathway they affect.
2. How did you assess the accuracy of the information sources you used?
3. Examine the profiles and information sources from other students who researched the same toxin you did. Re-examine and re-assess the information sources you used.
4. Are three sources of information sufficient to provide accurate and reliable information when researching a topic such as this? Justify your answer.

Yield of ATP from Aerobic Respiration

Suggested **Investigation**

Inquiry Investigation 3-A, Oxygen Consumption and Heat Production in Germinating Seeds

To determine the maximum number of ATP molecules that can be generated by the energy released when one glucose molecule is fully oxidized, several factors must be considered. First, it is necessary to know how many ATP molecules are produced when the two electrons from NADH and $FADH_2$ are processed through the electron transport chain. Scientists have determined that when two electrons pass through each hydrogen ion pump, one ADP molecule can be phosphorylated to form one ATP. Thus, when the two electrons from one NADH pass through the three hydrogen pumps, three ATP molecules can be formed. When $FADH_2$ passes two electrons through two pumps, two ATP molecules can be formed.

The two NAD^+ molecules reduced by glycolysis yield two NADH molecules in the cytoplasm. However, the mitochondrial membrane is impermeable to NADH. The reducing equivalents, or electrons, must be carried into the mitochondrion by one of two different shuttle mechanisms. In one case, the electrons are delivered to an FAD molecule in the mitochondrion, and two ATP molecules are produced. In the other case, the electrons are delivered to NAD^+, and three ATP molecules are produced.

Figure 3.14 summarizes the yield of ATP at each step in aerobic respiration. The summary accounts for the fact that two pyruvate molecules are produced by one glucose molecule in glycolysis, and thus two acetyl-CoA molecules enter the Krebs cycle. Therefore, the number of ATP molecules generated by substrate-level phosphorylation in the Krebs cycle and the number of NAD^+ and FAD molecules that are reduced in one round of the Krebs cycle are doubled. According to this tally, it is possible to generate as many as 36 or 38 ATP molecules from the energy released from one glucose molecule. (Prokaryotes can generate 38 ATP molecules per molecule of glucose, since they do not have to expend two molecules of ATP to transport the NADH from glycolysis across the mitochondrial membranes.)

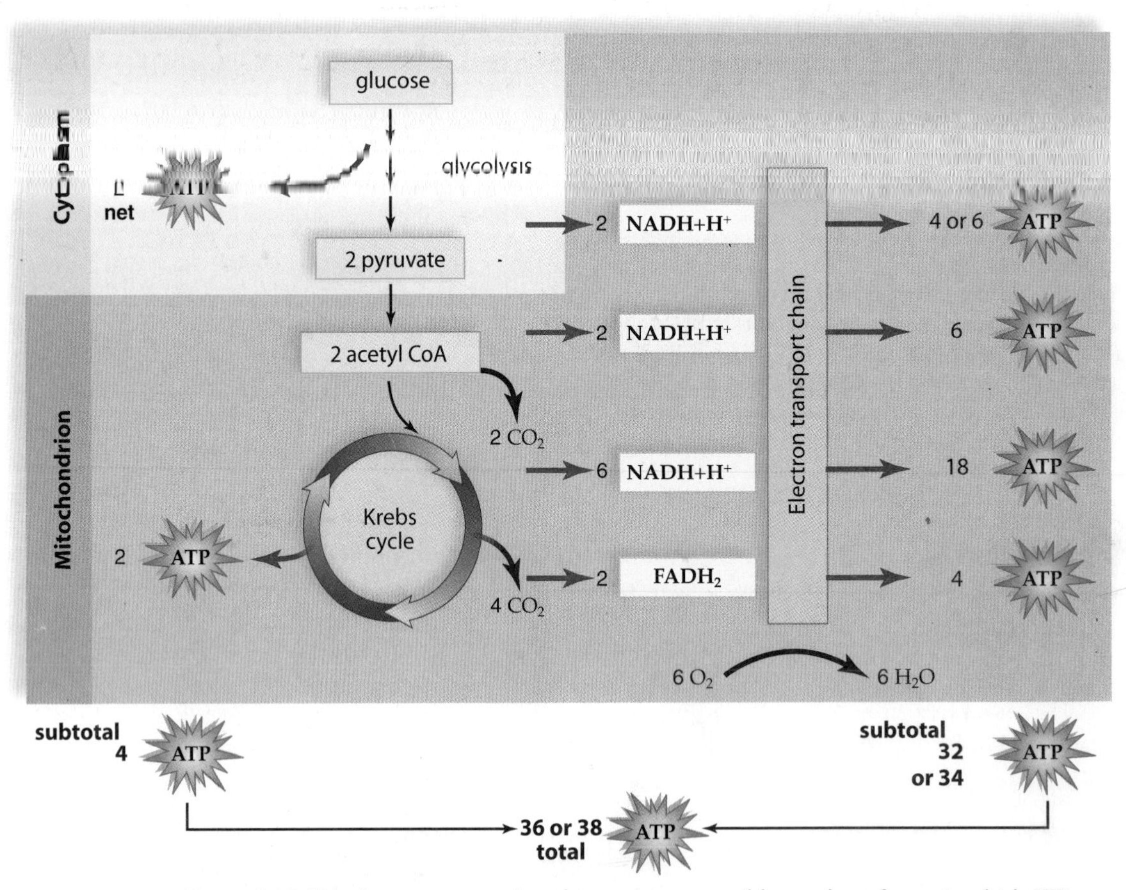

Figure 3.14 This diagram summarizes the maximum possible number of ways in which ATP can be generated in aerobic respiration.

Although 36 or 38 ATP molecules per glucose molecule are theoretically possible, experimental observations of the actual number of ATP molecules per glucose molecule are much lower. There are several reasons to account for these lower values.

- Some protons leak through the inner mitochondrial membrane without passing through an ATP synthase complex.
- Some of the energy from the hydrogen ion gradient in the mitochondria is used to transport pyruvate molecules generated during glycolysis from the cytoplasm into the mitochondria.
- Some energy is used to transport ATP out of the mitochondria for use in the cytoplasm.

Experimentally measured values are closer to 30 to 32 molecules of ATP produced per glucose molecule.

The Interconnections of Metabolic Pathways

A healthy diet consists of far more than just glucose. Some important parts of a diet are vitamins and minerals, which act as coenzymes and cofactors. However, most of a healthy diet is used for energy. Carbohydrates other than glucose, as well as fats and proteins, are used for energy. The pathways by which these other sources of energy enter the metabolic scheme are shown in **Figure 3.15**. Most carbohydrates can be broken down and converted into glucose. After the amino group has been removed from some amino acids, the remainder of the molecule is identical to some intermediate in either glycolysis or the Krebs cycle. For example, the amino acid alanine can be converted directly into pyruvate. The amino acid glutamate can be converted into one of the intermediates in the Krebs cycle. Fat molecules are broken down into glycerol and fatty acids. Glycerol can be converted into G3P, one of the intermediates in glycolysis. Fatty acids are transported into the mitochondria, where carbon atoms are removed two at a time and each two-carbon unit becomes an acetyl-CoA molecule, ready to enter the Krebs cycle.

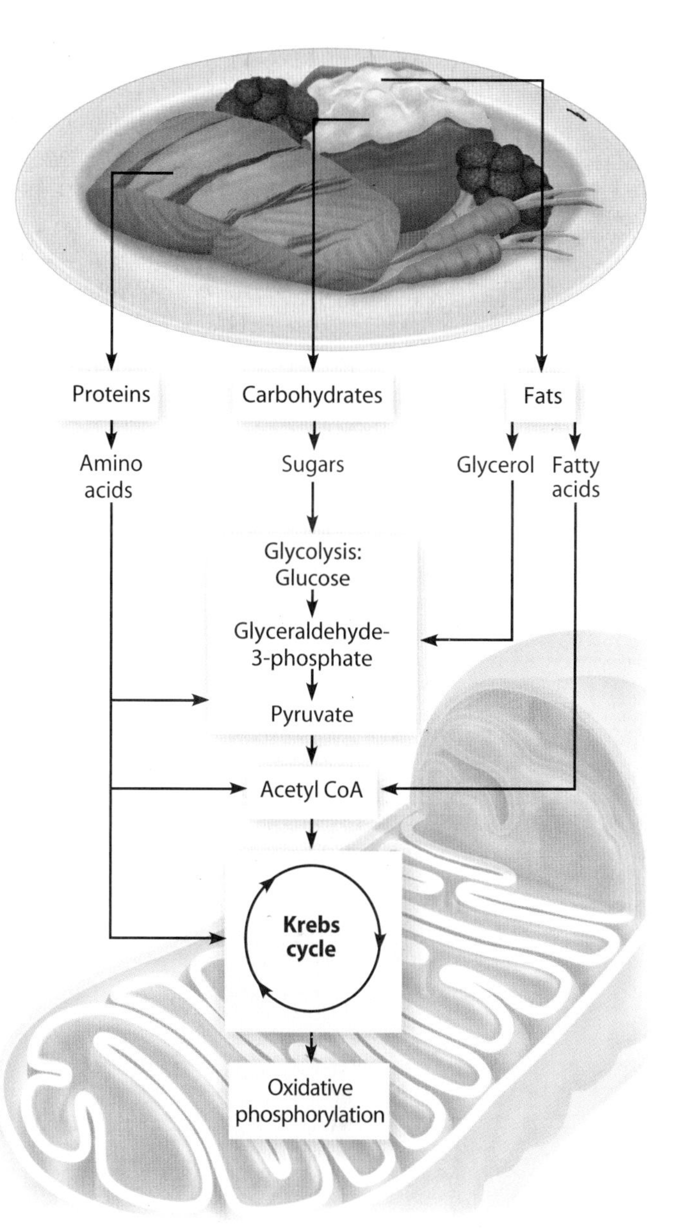

Figure 3.15 The cellular respiratory pathways are not closed. Compounds from the breakdown of all nutrients can be converted into intermediates in glycolysis and the Krebs cycle and can enter and leave at many different stages of the pathways.

***Identify** a common intermediate in the breakdown of fats, carbohydrates, and proteins.*

Regulation of Aerobic Catabolic Pathways

The amount of ATP in living cells remains nearly constant. If it is being used rapidly, it is generated rapidly. If it is being used slowly, it is synthesized slowly. How does the cell determine and control the rate at which to generate ATP? The answer is feedback control. **Figure 3.16** shows the two major enzymes that are controlled by feedback mechanisms.

The enzyme phosphofructokinase is the main control point in glycolysis. Phosphofructokinase has an allosteric binding site for ATP. Thus, when the cell has sufficient ATP to supply energy for the endergonic reactions taking place at a given time, any excess ATP binds to the allosteric site of phosphofructokinase and inhibits the enzyme. Citrate, one of the intermediates in the Krebs cycle, can also inhibit phosphofructokinase. Thus, if there is an accumulation of citrate, pyruvate from glycolysis is not broken down into acetyl-CoA until some of the citrate is used up. In contrast, high levels of ADP activate phosphofructokinase.

A second important control site in aerobic catabolic pathways is the conversion of pyruvate into acetyl-CoA and carbon dioxide. The enzyme that catalyzes this reaction, pyruvate dehydrogenase, is inhibited by excess NADH. Several Krebs cycle enzymes are also inhibited by excess ATP, NADH, and acetyl-CoA. When all of these control mechanisms are working together, the level of ATP remains constant within the cell.

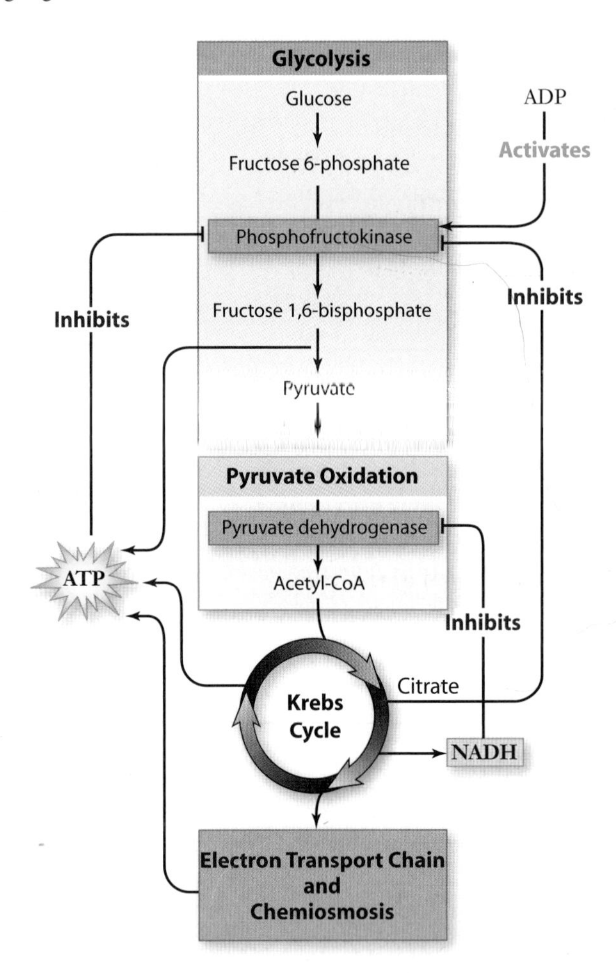

Figure 3.16 Many of the enzymes in glycolysis and the Krebs cycle are controlled by feedback inhibition of products such as ATP and NADH. They can also be activated by ADP. As a result of these control mechanisms, the ratio of ATP to ADP remains constant in living cells.

Suggested Investigation

ThoughtLab Investigation 3-B, Understanding and Treating Mitochondrial Diseases

Section 3.2 Review

Section Summary

- The breakdown of glucose occurs in four stages: glycolysis, pyruvate breakdown, Krebs cycle, and oxidative phosphorylation.
- Glycolysis is the breakdown of glucose to two pyruvates, producing two ATP and two NADH. ATP is made by substrate-level phosphorylation.
- Pyruvate is broken down to CO_2 and an acetyl group that is attached to CoA. NADH is made during this process.
- During the Krebs cycle, the acetyl group attached to CoA is broken down to two CO_2 molecules. Three NADH, one $FADH_2$, and one ATP are made during this process.
- Oxidative phosphorylation involves two events. The electron transport chain oxidizes NADH or $FADH_2$ and generates an H^+ gradient. This gradient is used by ATP synthase to make ATP via chemiosmosis.
- Proteins and fats can enter into glycolysis or Krebs cycle at different points.

Review Questions

1. K/U What is aerobic respiration? Write the overall equation that summarizes this process.
2. C Copy and complete the following table, which summarizes the complete metabolism of glucose.

Process (in order)	Location in the cell	Yield of ATP/ Reducing Power

3. C Write a paragraph that provides an overview of the process of aerobic respiration. The paragraph should communicate essential information without providing too much detail.
4. K/U Distinguish between substrate-level phosphorylation and oxidative phosphorylation.
5. K/U What is the main function of glycolysis? Indicate the reactant and products that result.
6. K/U The following questions refer to the Krebs cycle.
   a. What is the role of the cycle?
   b. What compound that is derived from glucose actually enters the cycle? How is this compound formed?
   c. Ccarbon atoms are oxidized in the cycle. What is reduced?
7. K/U Why is oxidative phosphorylation an important process in aerobic respiration?
8. K/U What is the electron transport chain, and what is its function?
9. C NAD^+ and FAD are sometimes described as being like a "shuttle bus." Is this an accurate analogy? Explain your answer.
10. K/U Describe the process of chemiosmosis. What is the role of chemiosmosis in cellular respiration?
11. T/I Why would lack of oxygen completely inhibit the Krebs cycle and the electron transport chain but not glycolysis?
12. A Rotenone is a broad-spectrum insecticide and that inhibits the electron transport chain. Why might it be toxic to humans?
13. K/U Why is the actual number of ATP molecules per glucose molecule lower than the theoretical amount?
14. K/U How is the cell able to regulate the rate of ATP generation?
15. T/I The five statements below are part of the sequence of events representing the development of the chemiosmotic hypothesis.
    i. ATP is produced; pH drops in the medium containing the vesicles.
    ii. Create vesicles from inner membranes of mitochondria and add O_2 and NADH to simulate conditions in a cell. See what happens as respiration proceeds.
    iii. If the inner mitochondrial membrane is the site of ATP synthesis, then pieces of membrane cultured in the lab and given O_2 and NADH should generate ATP.
    iv. A change in concentration of H^+ occurs on the two sides of the inner mitochondrial membrane as ATP forms from ADP. v. Mitochondria are the sites of ATP formation in a cell.
    a. Place the five statements in the order that best reflects the following sequence of events: Observation, Hypothesis, Experiment, Results, Conclusion.
    b. The conclusion arrived at in part a. led to the development of a new hypothesis. What do you think that hypothesis was? Give reasons for your answer.
16. C Use a diagram to illustrate substrate-level phosphorylation, and write a caption for it that explains the importance of this process to aerobic respiration.

SECTION 3.3

Anaerobic Respiration and Fermentation

Key Terms

anaerobic respiration

fermentation

Many organisms live in anoxic (oxygen-free) environments. In fact, for some organisms, oxygen may be lethal. Different species are adapted to release energy from their food sources in different ways.

Anaerobic Cellular Respiration

anaerobic respiration a metabolic pathway in which an inorganic molecule other than oxygen is used as the final electron acceptor during the chemiosmotic synthesis of ATP

Some organisms live in anoxic environments and thus cannot use oxygen as their final electron acceptor. Instead, they use inorganic compounds such as sulfate, nitrate, or carbon dioxide as electron acceptors. They are said to carry out **anaerobic respiration**.

In some organisms, anaerobic respiration is similar to aerobic respiration. For example, a few bacteria, including *E. coli*, carry out aerobic respiration when oxygen is available. When oxygen is not available but nitrate is, they synthesize an enzyme called nitrate reductase, which can accept electrons from the electron transport chain and pass them to nitrate according to the following equation.

$$NO_3^-(aq) + 2e- + 2H^+ \rightarrow NO_{2}-(aq) + H_2O(\ell)$$

Other organisms, such as methanogens, have different metabolic pathways. However, they use electron transport chains and generate hydrogen ion gradients that provide energy for phosphorylation. They use hydrogen that is synthesized by other organisms as an energy source and carbon dioxide as an electron acceptor. The summary equation for their metabolism is:

$$4H_2(aq) + CO_2(aq) \rightarrow CH_4(g) + 2H_2O(\ell)$$

Some methanogens grow in swamps and marshes and are responsible for marsh gas, which is methane. Some of the prokaryotes that live in the stomachs of cows and other ruminants are methanogens, which are major sources of methane released into the environment.

Fermentation

Many single-celled organisms such as yeasts and some bacteria, use only glycolysis for energy. Multicellular organisms use glycolysis as the first step in aerobic metabolism. During intense exercise, however, oxygen cannot be delivered to muscle cells rapidly enough to supply the energy needs of the cells, so they rely on glycolysis for energy. The NADH that is reduced during glycolysis cannot be reoxidized by electron transport as fast as it is being reduced, and thus muscle cells will run out of oxidized NAD^+ and glycolysis will cease unless another pathway is available. Different organisms and cell types have several different ways to reoxidize the reduced NADH, usually by reducing an organic molecule. These processes are called **fermentation**.

fermentation a cellular respiration pathway that transfers electrons from NADH to an organic acceptor molecule

Fermentation is much less efficient at supplying energy than aerobic respiration, because fermentation only produces the amount of ATP that is generated in glycolysis. Nevertheless, it is commonly used. The name of the pathway is often based on the organic molecule that is reduced. Organisms use many different pathways to reoxidize NADH. Two common pathways are lactate fermentation and ethanol fermentation, outlined in **Figure 3.17** and described below.

Lactate Fermentation

Some single-celled organisms, as well as some animal muscle cells that are temporarily without oxygen, carry out lactate fermentation. The pyruvate generated by glycolysis reacts with NADH to reoxidize it to NAD^+. In the reaction, pyruvate is converted into lactate (also called lactic acid). The reoxidized NAD^+ allows glycolysis to continue. The lactate that is formed in bacteria is secreted into the surrounding medium, causing it to become acidic.

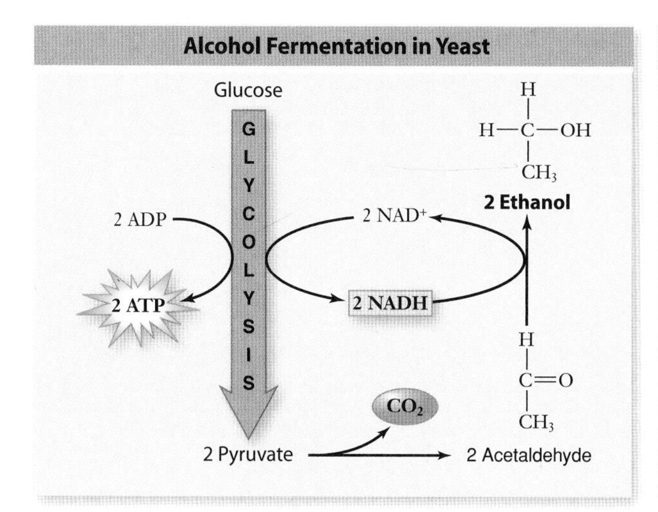

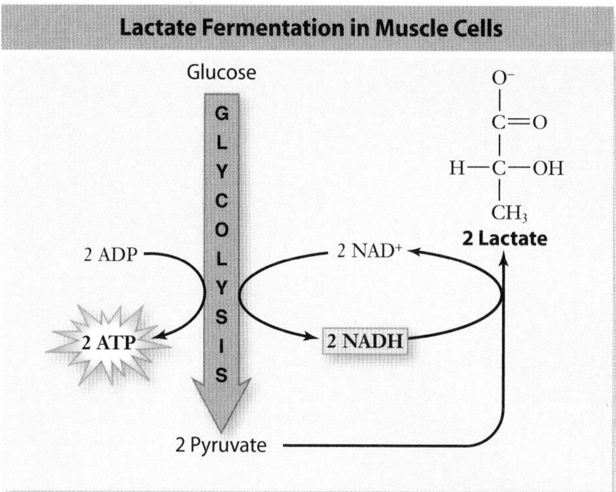

Figure 3.17 Lactate and ethanol fermentation. Muscle cells convert pyruvate into lactate. Yeasts carry out the conversion of pyruvate to ethanol. In each case, the reduction of a metabolite of glucose has oxidized NADH back to NAD^+ to allow glycolysis to continue under anaerobic conditions.

The lactate that is generated in muscles must be reoxidized to protect the tissues from the acidic environment. Oxygen is ultimately needed to allow the lactate to return to the oxidative pathways to be metabolized. The amount of oxygen required to eliminate the lactate is called the *oxygen debt.*

The lactate produced in muscle cells is transported out of the cells into the bloodstream. Researchers thought previously that the lactate was transported to the liver. Now, however, there is evidence that the lactate is taken up by resting muscle cells. Some of the lactate is converted back into pyruvate and oxidized, while some is converted into glycogen, which is stored in the muscle.

Ethanol Fermentation

Yeast and some bacteria are able to function aerobically as well as anaerobically. These organisms are called *facultative anaerobes.* When they function anaerobically, they convert pyruvate to ethanol and carbon dioxide through ethanol fermentation. The process involves two steps, as shown in the B part of **Figure 3.17**.

Fermentation by brewer's yeast (*Saccharomyces cerevisiae*) is used in industry to manufacture baked goods and alcoholic beverages. When used in brewing, a variety of products can be made, depending on the substance being fermented, the variety of yeast used, and whether carbon dioxide is allowed to escape during the process. For example, yeast fermentation may be used to produce wine or champagne from grapes; a syrupy drink, called mead, from honey; or cider from apples. Beer is brewed by fermenting sugars in grain such as barley, rice, or corn.

Fermentation Producing other Organic Compounds

Depending on the organism, fermentation can yield other substances besides lactate and ethanol. Two other examples of fermentation products, acetone and butanol, were essential during World War I. The British needed butanol to make artificial rubber for tires and machinery; acetone was needed to make a smokeless gunpowder called cordite. Prior to the war, acetone was made by heating wood in the absence of oxygen. Up to 100 tonnes of lumber were needed to produce 1 tonne of acetone. When war broke out in 1917, the demand for acetone was great. A swift and efficient means for producing the chemical was needed. In 1915, Chaim Weizmann, a chemist working in Manchester, England, had developed a fermentation process using the anaerobic bacterium *Clostridium acetobutylicum.* Through this process, Weizmann converted 100 tonnes of molasses or grain into 12 tonnes of acetone and 24 tonnes of butanol. For the war effort, Weizmann modified the technique for large-scale production. Today, both acetone and butanol are produced more economically from petrochemicals.

Ethanol Fermentation and Fuel Production

Suggested Investigation

Inquiry Investigation 3-C, Fermentation in Yeast

Glucose is the main fuel for many organisms. However, much of the chemical energy of glucose remains in the compounds that form after glycolysis is complete. The process of fermentation does not remove much of this chemical energy. Therefore, the products of fermentation can still be used for fuel. In organisms that carry out ethanol fermentation, the ethanol they produce is released as a waste product. In fact, the ethanol waste is toxic to yeast. As ethanol concentrations approach 12 percent, the yeast cells begin to die. However, humans learned long ago that this "waste" can be burned. Ethanol was a common lamp fuel during the 1800s, and it was used for early internal combustion engines in cars and other machinery, also starting in the 1800s.

Historically, because gasoline costs less to produce than ethanol, the use of ethanol was limited to small-scale, specialized applications. This situation changed in the late 1970s. At that time, rising oil prices, dwindling petroleum reserves, and environmental concerns caused some governments to invest in alternative energy resources such as ethanol fuels. When gas prices rise, some of these alternative resources become commercially viable sources of fuel. In cars, the use of a gasoline-ethanol fuel mixture (up to 10 percent ethanol and up to 90 percent gasoline) has become common. Cars manufactured after 1980 can use this fuel mixture, called E-10, without any engine modification. Auto companies also design engines that can use fuels with ethanol percentages that are much higher than the 10 percent in gasohol. Brazil is currently leading the world in the production of ethanol and its use as a fuel. Most cars designed and built in Brazil today can burn pure ethanol or various combinations of gasoline and ethanol.

In Canada, the most common source of ethanol is the fermentation of corn and wheat. First the grain is ground into a meal. Then it is mixed with water to form a slurry called "mash." Enzymes added to the mash convert the starches into glucose. The mash is heated to destroy any bacteria, then cooled and placed in fermenters. In the fermenters, yeast is added to the mash. The yeast grows on the glucose under anaerobic conditions and releases the end products, ethanol and carbon dioxide. When the fermentation is complete, the resulting product, called "beer," is approximately 10 percent ethanol and 90 percent water. Distilling the "beer" to eliminate as much of the water as possible yields nearly pure ethanol. A small amount of gasoline is added to make the ethanol unfit for human consumption. The solid residues from the grain and yeast are dried to produce a vitamin- and protein-rich product called Distiller's Dried Grains and Solubles (DDGS) used as livestock feed.

Activity 3.2 Industrial and Domestic Uses of Fermentation Products

Common products of fermentation that are used in industry, as well as in the home, include those that are listed in the table below. In this activity, you will select a fermentation product to investigate.

Materials

- print and/or electronic reference material

Procedure

1. The table to the right lists several end-products of the fermentation process, along with the microorganisms involved in their manufacture. Choose one of the fermentation end-products to research.
2. In the course of your research, identify the use or uses for the fermentation end-product. For example, uses for ethanol include fuel and alcoholic beverages. Also identify the starting material or materials involved. For example, common starting materials for ethanol include malt extract, fruit juices, and plants such as switch grass and sugar cane. Explain the manufacturing pathways that lead from the starting material to the final product.
3. Present your findings in an appropriate format, such as a written report, information brochure, or poster.

End-product(s) of Fermentation	Microorganism(s) Typically Involved
Ethanol	*Saccharomyces cerevisiae*
Acetic acid	*Acetobacter sp.*
Lactic acid	*Lactobacillus delbruckii, Lactobacillus plantarum*; Pediococcus
Acetone and butanol	*Colstridum acetobutylicum*
Glycerol	*Saccharomyces cerevisiae*
Citric acid	*Aspergillus niger*
Sorbose	*Gluconobacter oxydans*

Section 3.3 Review

Section Summary

- Some single-celled organisms that live in conditions of very low oxygen can carry out anaerobic respiration by using an electron acceptor other than oxygen.
- Some single-celled organisms and, during extreme exertion, some muscle cells, use only glycolysis for energy in a process called fermentation.
- In lactate fermentation, pyruvate oxidizes NADH back to NAD^+ and, in the process, is converted into lactate. In single-celled organisms, the lactate is released to the surroundings and in muscle cells, it is released into the bloodstream where it is carried to resting muscle cells and oxidized or converted to glycogen for storage.
- In ethanol fermentation, pyruvate is converted into a two-carbon compound, acetaldehyde, and carbon dioxide. The acetaldehyde reoxidizes the NADH back to NAD^+ for reuse.
- Alcohol fermentation is a useful industrial process, used to generate ethanol for fuel. It is also the same process by which alcoholic beverages are produced.

Review Questions

1. K/U What is anaerobic respiration? Include an example of where this process occurs.
2. C Is anaerobic respiration the same process as fermentation? Use a graphic organizer such as a Venn diagram to present your answer.
3. K/U Why is fermentation less efficient at supplying energy compared to aerobic respiration?
4. T/I Use the diagram below to answer the following questions.
   a. Describe the processes taking place in A and in B.
   b. Where are these processes occurring?
   c. What is the significance of NADH being oxidized back to NAD^+ in both processes?

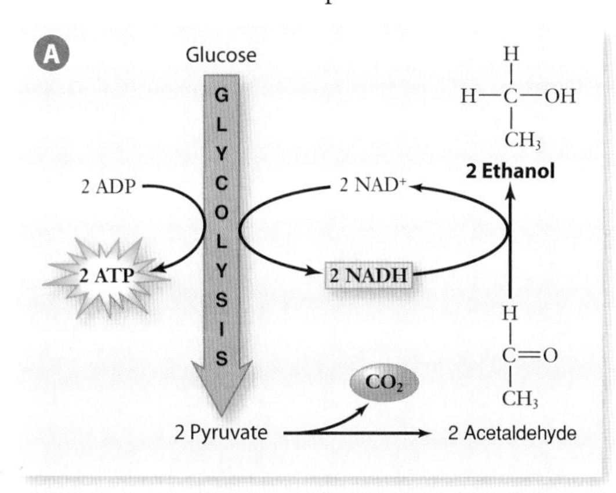

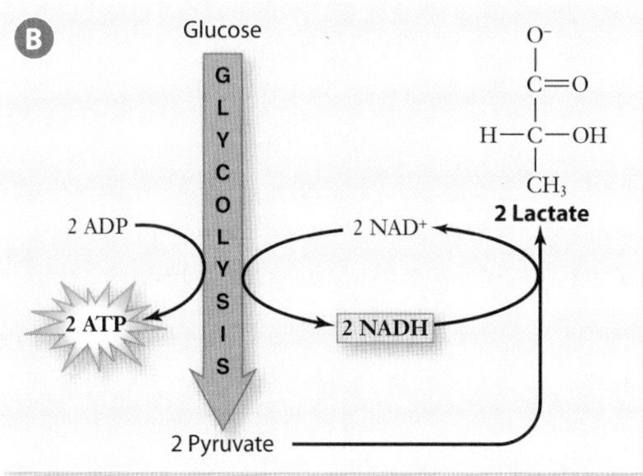

5. K/U Explain why anaerobic respiration produces fewer ATP molecules than aerobic respiration.
6. K/U Using a specific example, explain how anaerobic organisms are able to continue electron transport processes in the absence of oxygen.
7. T/I The anaerobic threshold is the point at which certain muscles do not have enough oxygen to perform aerobic respiration and begin to perform anaerobic respiration. Infer why it is important for competitive long-distance runners to raise their anaerobic threshold.
8. A What are the risks and benefits of promoting grain and corn fermentation as a way to supply society's fuel energy needs?
9. K/U Clearly explain how glycolysis and anaerobic processes are similar and how they are different.
10. A According to Statistics Canada, pig manure is the largest source of greenhouse gas emissions from the hog-production system. Pig manure is also, however, a source of biogas (methane from animal wastes), which can be used as an alternative to traditional fuels.
    a. Using Internet and print resources, research how biogas is produced for commercial consumption.
    b. What are the advantages and disadvantages of using biogas?
    c. The Ontario Ministry of Agriculture, Food and Rural Affairs (OMAFRA) encourages farmers and agricultural businesses to install biogas systems. Construction and implementation of these systems has been approved for many dairy farms in eastern and southwestern Ontario. Use the OMAFRA website to research and write a brief report on how and why these systems would be of benefit not only to farmers and agricultural businesses, but also to society in general.

Inquiry
INVESTIGATION

3-A

Skill Check

- ✓ Initiating and Planning
- ✓ Performing and Recording
- ✓ Analyzing and Interpreting
- ✓ Communicating

Safety Precautions

- If glass tubing is used instead of plastic tubing, handle the tubing very carefully to avoid breakage.
- (Part 2) The buildup of gases in an enclosed container such as a test tube or flask could cause the container to rupture or shatter. Provide a means for venting gases out of any system you use.

Materials

- large test tube
- marker or wax pencil
- ruler
- liquid detergent
- 1 g of seeds of any kind
- balance
- one-hole stopper
- paper towels
- limewater
- spatula
- pipette
- clear adhesive tape
- wad of cotton
- support stand and clamp
- rigid plastic tubing, 20 cm long and bent at right angle
- other materials as needed for Part 2

Oxygen Consumption and Heat Production in Germinating Seeds

When seeds *germinate* (begin to grow and develop), they cannot trap energy from the Sun because they have not yet produced any chlorophyll. In addition, they often germinate under a layer of soil. Therefore, seeds must have enough stored energy to germinate, which includes developing roots and shoots, as well as the chloroplasts and chlorophyll they will need as they mature and are exposed to sunlight.

To germinate, seeds need suitable temperatures, water, and oxygen. The amount of oxygen consumed by the seeds is approximately equal to the amount of carbon dioxide produced as the seeds carry out cellular respiration. In Part 1 of this investigation, you will use an apparatus called a *respirometer* to measure the seeds' consumption of oxygen. The respirometer contains limewater and germinating seeds. As the seeds consume oxygen, they release carbon dioxide. The carbon dioxide is then absorbed by the limewater, creating a slight vacuum in the respirometer. This vacuum will draw a drop of liquid detergent on the end of the tubing inward. The movement of this detergent plug will be measured using a ruler taped to the tubing.

In Part 2, you will design your own investigation to demonstrate and measure the heat produced as germinating seeds respire.

Pre-Lab Questions

1. What metabolic pathways are active in a germinating seed?
2. Read through the procedure for Part 1. What will you be measuring, and what information will you be able to infer from this measurement?
3. Would you expect seeds to germinate in the absence of oxygen? Explain.
4. How could you vent gases from a test tube or flask to prevent the container from rupturing, and how could this affect your results?

Part 1: Oxygen Consumption in Cellular Respiration*

Question

How can you demonstrate quantitatively that germinating seeds consume oxygen?

Prediction

Based on the experimental set-up, predict what will happen to indicate that oxygen is being consumed as the seeds respire.

Procedure

1. Obtain some small plant seeds. If possible, work with a different type of seeds from those used by your classmates. Germinate the seeds by spreading them on wet paper towels a day or two before the lab.
2. Start to make a respirometer by inserting the short end of the tubing into the hole of the stopper. Hold the tubing gently but firmly, close to the end to be inserted into the stopper. Push the tubing gently into the stopper,

being careful to avoid breakage. The long end of the tubing should be sticking out at a right angle as shown in the photograph.

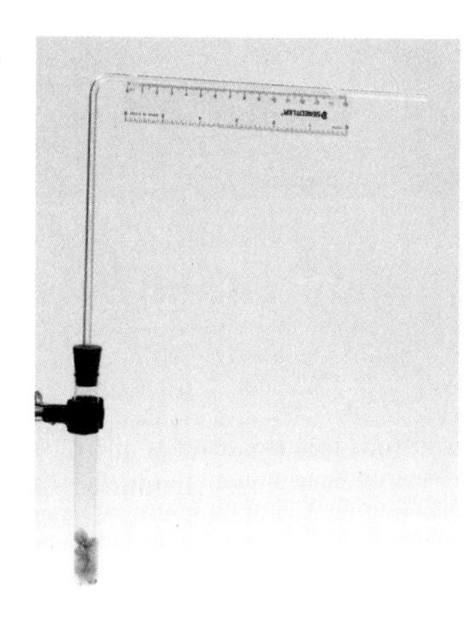

3. Draw a line 0.5 cm above the bottom of the test tube with the marker. Add limewater to the tube up to this mark.
4. Moisten a small wad of cotton and place it on top of the limewater. Now place 1 g of germinating seeds on top of the moistened cotton.
5. Tape the ruler to the tubing as shown in the photograph. Use a pipette to add a drop of detergent just inside the mouth of the tubing at the end that will be outside the test tube.
6. Carefully insert the stopper and tubing into the test tube to form an airtight seal. Use a support stand and clamps to keep the respirometer apparatus in an upright position.
7. Wait 5 min to allow the limewater to absorb any carbon dioxide that was in the respirometer when it was assembled. Mark the position of the detergent drop with a marker or wax pencil. Take an initial reading of where the drop of detergent is with respect to the ruler. Always take the measurement from the same part of the detergent drop. Record the initial reading in a suitable data table.
8. Take readings of the drop's position every minute for 15 min and record them in the data table.

Analyze and Interpret

1. Graph your data, placing time on the x-axis and distance the detergent drop moved on the y-axis.
2. Compare your data with other groups. How did the rate of oxygen consumption compare?

Conclude and Communicate

3. Did your observations indicate that cellular respiration was occurring in the germinating seeds? Explain how you formed this conclusion.
4. Name at least two sources of error that could have affected your observations and data. Explain how significant these sources of error are to the outcome of your investigation.

Part 2: Heat Production in Cellular Respiration

Question

How can you demonstrate that heat is a product of germinating (respiring) seeds?

Hypothesis

State a hypothesis that enables you to obtain quantitative data about the heat given off by germinating seeds.

Plan and Conduct

1. With your group, develop a written plan that outlines the procedure you will follow to test your hypothesis. Be sure to consider the following in your procedure:
   - safety
   - controlled variables
   - data collection and recording
2. Review your hypothesis and procedure with your teacher before you perform it.
3. Decide how you will measure the heat given off by germinating seeds.
4. Decide how you will record and display your data to assist you in analyzing and drawing conclusions about your results.

Analyze and Interpret

1. What variables did you manipulate and control?
2. Compare your results with other groups. Identify possible sources of error and opportunities for improvement in your procedure and/or data collection.

Conclude and Communicate

3. Explain why your results either supported or refuted your hypothesis and prediction.

Extend Further

4. **INQUIRY** Using the same experimental approach you used for Part 1, what would you expect to see if the germinating seeds were able to perform photosynthesis (so that they convert carbon dioxide and water to sugars and oxygen)? Would you need to make any modifications to your procedure or setup to see measurable results? Explain your reasoning.
5. **RESEARCH** Research the link between soil temperature and germination for crop plants. Why is it important for farmers to understand this link?

**Part 1 adapted from Agri-science Resources for High School Sciences, P.E.I. Agriculture Sector Council*

ThoughtLab INVESTIGATION 3-B

Skill Check

- Initiating and Planning
- ✓ Performing and Recording
- ✓ Analyzing and Interpreting
- ✓ Communicating

Materials

- reference books
- computer with Internet access

Understanding and Treating Mitochondrial Diseases

Mitochondria are more than just the energy production centres in a eukaryotic cell. They also function as essential components of most major metabolic pathways used by a cell to build, break down, and recycle organic molecules. For example, mitochondria are required for cholesterol metabolism, for estrogen and testosterone synthesis, for neurotransmitter metabolism, and for hemoglobin production.

Mitochondrial diseases are the result of either inherited or spontaneous mutations in chromosomal DNA or in DNA within the mitochondria themselves. These genetic mutations lead to altered functions of the proteins or RNA molecules that normally reside in mitochondria. Because the functions that mitochondria perform are so diverse, there are many different mitochondrial diseases. There are no cures for mitochondrial diseases, and scientists are just beginning to understand them. Because most mitochondrial diseases are inherited genetically, certain diseases can have an impact on specific communities. For example, one type of mitochondrial disease called Leigh Syndrome is common in communities around Saguenay, Quebec. In this investigation, you will choose and research a mitochondrial disease and assess the possible relevance of this disease to your personal life and to your community. Assess the cost of treatment and accommodation, as well as social costs such as diminished quality of life, against the cost of research dedicated to finding a cure for the mitochondrial disease.

Pre Lab Questions

1. What vital functions do mitochondria carry out in a major organ such as the liver?
2. What do you think might be the level of awareness among Canadians of the occurrence of mitochondrial diseases?
3. Do you think doctors find mitochondrial diseases easy or difficult to diagnose? Why?
4. How might a community be affected by a high incidence of a mitochondrial disease in the population?

Question

What are the effects of a particular mitochondrial disease on energy production and metabolism, and how can you assess the relevance, to your personal life and to the community, of an understanding of mitochondrial disease?

Organize the Data

Conduct research into the causes, symptoms, treatments, and costs associated with the particular mitochondrial disease you have chosen to investigate. (Advocacy groups such as Mitoaction in Canada and the United Mitochondrial Disease Foundation in the United States have posted a wealth of information about many mitochondrial diseases on their websites.) Focus your research on these questions:

- How common is the mitochondrial disease?
- Is the disease an inherited condition or the result of genetic mutation? If it is the latter, what do researchers believe causes the mutation? What part of the mitochondrion is affected, and how does that affect the energy and metabolism, and, more broadly, the organs or biological systems of a person who has the disease?
- What are the symptoms of the disease?
- What treatments are available to manage the symptoms of the disease?
- Are the treatments comparatively inexpensive (for example, vitamin supplements) or comparatively expensive (for example, specialized medications)?
- How much do existing treatment methods cost per patient, per year? Approximately how much would the total treatment cost be for the entire Canadian population that is affected by this disease?
- Are there accommodative devices, such as wheelchairs, that people with the mitochondrial disease will need as their condition advances? How much do such devices cost?
- Does your community have existing infrastructure to accommodate people with mitochondrial disease (for example, wheelchair ramps, accessible washrooms, suitable places for rest)? How might the disease affect a person's quality of life or ability to study or work? Do the treatments themselves have side effects that can result in social or economic costs?
- What is the cost of current research dedicated to finding improved treatments and a possible cure for the mitochondrial disease you have chosen to investigate?

Analyze and Interpret

1. Prepare a graphic organizer to represent the social and economic costs associated with the disease you chose to investigate.
2. Compare these costs to the costs of existing treatments and current research in quest of a cure. Consider any spin-off benefits associated with current research.

Conclude and Communicate

3. Based on analyzing the information you gathered, make a recommendation as to whether or not research into the disease should continue.
4. The United Mitochondrial Disease Foundation states that telling individual stories about mitochondrial disease to a wide audience is important "to make mitochondrial disease more readily recognizable and to create a sense of urgency among policy makers in the medical community." Based on your investigation, do you agree or disagree with this statement? What could be done within your community to generate a sense of urgency?
5. Communicate your findings in a format that you think would be most effective in raising awareness of mitochondrial diseases and the social and economic costs associated with them.

Extend Further

6. **INQUIRY** Based on the most current research published in science journals, how effective are the treatments for the mitochondrial disease you have chosen to investigate? Are there new treatments on the horizon that might work better, cost less, or otherwise improve the lives of people who have the disease?
7. **RESEARCH** In 2009, researchers at McGill University in Montréal, Québec, identified a mutation in a gene that produces the COX1 protein, which is part of the cell's energy production system. This mutation causes Leigh Syndrome. Conduct research to find out what general role the COX1 protein plays in metabolism, and whether this discovery might lead not only to diagnostic tests for this disease, but also to a treatment or a cure.

Inquiry INVESTIGATION 3-C

Skill Check

- Initiating and Planning
- ✓ Performing and Recording
- ✓ Analyzing and Interpreting
- ✓ Communicating

Safety Precautions

- Wash your hands thoroughly with soap before and after the investigation.
- Dispose of waste materials as directed by your teacher.

Materials

- warm tap water
- warm sucrose solutions: 1%, 5%, and 10%
- baker's yeast (dry, active) (Saccharomyces cerevisiae)
- 4 small balloons
- 4 large test tubes (at least 50 mL)
- test tube rack
- [illegible]
- ruler
- marker or grease pencil

Fermentation in Yeast

Without oxygen, yeast use ethanol fermentation for energy. In this investigation, you will observe this process.

Pre-Lab Questions

1. What types of organisms can use fermentation to obtain energy?
2. What do yeast require in order to carry out fermentation?

Question

How does the concentration of sucrose affect the rate of fermentation of yeast?

Procedure

1. Stretch the balloons, and then blow them up and release them several times to soften them and increase their elasticity. Put them aside for step 5.
2. Add 25 mL of warm tap water to the first test tube, and label it control.
3. Add 25 mL of each of the sucrose solutions to the remaining three test tubes. Label each test tube with the percentage of the sucrose solution.
4. Add 1.5 mL of yeast to each test tube.
5. Place a balloon over the top of each test tube.
6. Carefully swirl or gently shake each test tube until the yeast dissolves.
7. Every 5 minutes for 20 minutes, measure the height of the bubbles, and describe the appearance of the balloons for each test tube.

Analyze and Interpret

1. What gas is causing the bubbles to form? How do you know?
2. **a.** Plot time versus height of bubbles for each of the test tubes, plotting the data for all four test tubes on the same graph.
   b. Describe any relationships that you see from the graphs.

Conclude and Communicate

3. Evaluate your predictions on the basis of your observations.

Extend Further

4. **INQUIRY** Design an investigation to determine the effect of temperature on fermentation. Carry out your investigation only if instructed to do so by your teacher.
5. **RESEARCH** Conduct library and/or internet research to find out how the process of fermentation for wine and beer making was first discovered.

STSE

BIOLOGY Connections

Bioaugmentation: Metabolic Processes and Waste Clean-up

Waste generated by human activities often includes water contaminated with chemical mixtures such as petroleum. These chemicals cannot be metabolized by most eukaryotes, and may even be toxic to many living things. Some prokaryotes, however, have specialized biochemical pathways that can metabolize the molecules in these chemicals. Some bacteria can actually feed on substances that pose disposal problems for humans.

How can we take advantage of the metabolic processes habits of these sludge-eating bacteria? The answer lies in bioaugmentation. Bioaugmentation involves the use of specific microbes to break down substances that require disposal, often because they are toxic.

MICROBES ON THE JOB Bioaugmentation is used in various kinds of clean-up operations. Some examples include:

- disposing of grease from restaurant waste to prevent the clogging of plumbing pipes
- treating wastewater from pulp and paper mills
- cleaning up groundwater contaminated with petrochemicals such as diesel fuel

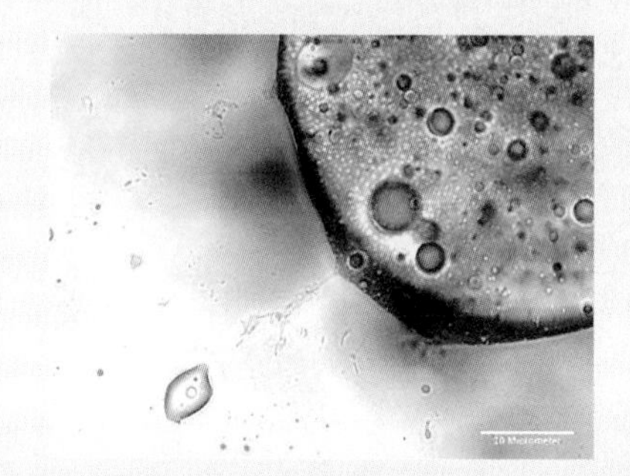

Some prokaryotes, like this petroleum-degrading bacterium, have specialized biochemical pathways that can metabolize molecules that may be toxic to humans.

A number of companies in Canada sell mixtures of microbes that have been specially selected to degrade toxic substances or treat particular waste products. These mixtures, in either powdered or liquid form, may also contain nutrients that help the microbes to grow.

Advantages of bioaugmentation compared to other clean-up methods such as incineration include:

- cheaper cost
- the ability to deploy the process at the contaminated site instead of transporting the waste to another location
- a low impact on the environment, since bioaugmentation involves metabolic processes and organisms that occur in nature

Bioaugmentation is sometimes used to treat wastewater from pulp and paper mills.

WHAT ARE THE LIMITATIONS? Along with its advantages, bioaugmentation also has some disadvantages. Like all living things, microbes need an appropriate environment in order to thrive—an environment that features proper temperature, nutrients, and humidity. These conditions can be challenging to control at a contaminated site. In addition, bacteria cannot metabolize every kind of waste. Synthetic compounds that have been created in a laboratory and are not usually found in nature, such as dioxins, can be difficult for bacteria to break down. Also, bioaugmentation may take longer than some clean-up methods, such as incineration. Finally, in the process of ingesting and metabolizing waste, microbes generate their own waste products. Sometimes these waste products can be compounds that have foul odours or toxic properties. Thus, bioaugmentation is not always the ideal response to every instance of environmental contamination.

Connect to the Environment

1. Bioaugmentation was not used to clean up the huge oil spill in the Gulf of Mexico in 2010. Instead, the oil was skimmed off the water, burned off, or treated with chemical dispersants that broke it up into smaller particles. Conduct research to find out why these methods were used instead of bioaugmentation. Report on the advantages and disadvantages of the methods that were used compared to bioaugmentation.
2. You work for an environmental consulting firm that is trying to decide whether to buy a special formulation of microbes to help clean up wastewater at a local pulp mill. What questions will you ask the company selling the microbes about its product in order to determine whether the advantages of using the microbes outweigh the disadvantages? What criteria would you apply in order to make a decision about whether to use bioaugmentation in this situation?

STSE

Case Study

Fad Diets

Are they healthy and effective?

Scenario

Dr. Health answers readers' questions about nutrition and how to make healthy dietary choices. Part of maintaining a healthy lifestyle includes exercising regularly and consuming the proper amount of calories for your gender, age, and level of physical activity. Dr. Health advises that you always discuss your plans with your primary care physician before beginning any type of diet or exercise plan.

Dr. Health

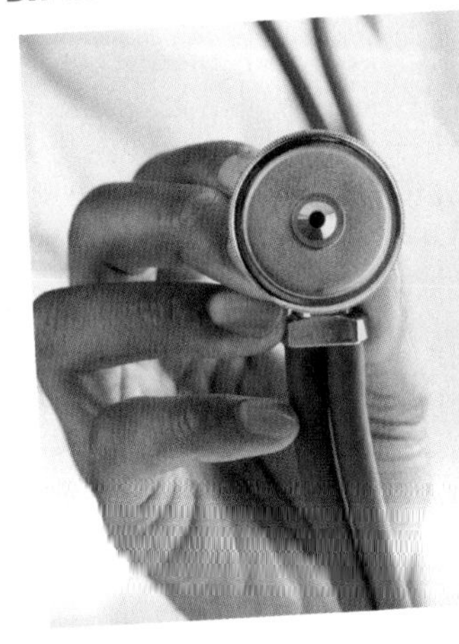

This week's column focuses on fad diets—diets that promise quick weight loss and/or restrict consumption of certain food groups. Dr. Health answers questions about the short-term and long-term effectiveness and nutritional value of these types of diets.

Dear Dr. Health,

My friend is trying a low-carb diet to lose weight. He has stopped eating pasta, bread, rice, and lots of other foods that I learned are good for you. Can you explain how the diet works and if the weight loss is long term?

Thanks,

Craving Carbs in Cornwall

* * *

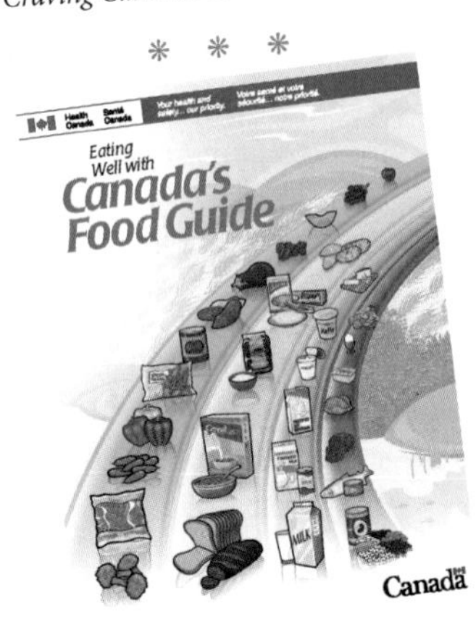

Dear Craving Carbs,

Low-carb diets limit or exclude foods that contain carbohydrates, such as pasta, breads and other baked goods, potatoes and other starchy vegetables, rice, beans, and fruits. Your friend is likely eating foods that are high in or have a relatively high content of proteins and fats, including meat, eggs, dairy products, and fish.

The body uses carbohydrates as a source of energy. They are easily broken down into glucose, which is used by cells to carry out normal metabolic processes. As blood glucose levels rise during digestion, the pancreas releases insulin, which causes cells to take up glucose and subsequently use it for energy. The presence of insulin prevents fat from being broken down and used as an energy source and, instead, uses the glucose from recently consumed carbohydrates. The idea behind a low-carb diet is that by drastically reducing the amount of carbs digested, the amount of insulin released is also reduced, leading the body to burn fat for energy instead. However, despite research in this area, the correlation between reducing insulin secretion and weight loss remains unclear.

Weight loss may occur when adhering to a low-carb diet, partly due to a reduced intake of calories and partly due to water loss. However, if a person returns to his or her previous eating habits, he or she may gain the weight back. One of the health risks associated with low-carb diets is an increase in cholesterol, if the amount of saturated fats ingested increases. High cholesterol is associated with heart disease. Low-carb diets severely limit or exclude vegetables and whole grains and, in some cases, fruits. As a result, these diets are low in fibre, which can lead to constipation. Eliminating one or more food groups from your diet may also result in a lack of proper nutrition. Health Canada recommends eating a variety of foods from the four food groups, including whole grains, fruits, and dark green, leafy vegetables.

Dr. Health

Dear Dr. Health,

I read an advertisement about a quick-weight-loss diet that involved eating mainly cabbage soup, along with some fruits, vegetables, fruit juice, and very little meat for seven days. The ad claimed that a person following the diet could lose 5 kg in a week. Would this really work? Are there any disadvantages to this type of diet?

Pondering in Pembroke

* * *

Dear Pondering,

Quick-weight-loss diets are based on the premise of consuming a low number of calories—usually less than 1200 calories per day. Although this may result in weight loss, much of it may come from water loss rather than a reduction of fat. Because quick-weight-loss diets are not realistic eating plans, people tend to gain back any lost weight after ending the diet.

One problem with quick-weight-loss diets is that they often lack essential nutrients. Also, since the types of foods that can be consumed are often very limited, people may find the diet boring and unsatisfying. Some people report feeling lightheaded and weak while following these types of diets. Quick-weight-loss diets are not a long-term solution to maintaining a healthy weight and incorporating healthy choices into your diet.

* * *

Dear Dr. Health,

My friend wants to eliminate all sources of fat from her diet. I thought we needed to have some fat in our diet to stay healthy. Does she need to be that extreme?

Sincerely,

Fat-Free in Fort Frances

Dear Fat-Free,

Fat, along with carbohydrates and proteins, is one of the macronutrients your body uses for energy. Your body also uses fat to manufacture certain hormones, protect your organs, and absorb certain vitamins such as A, D, and E. Fat is also an essential component of cell membranes. Unfortunately, many people consume too much of the least-desirable types of dietary fat: saturated fats and trans fats. Sources of saturated fat include animal products such as meat, cheese, whole-fat milk, butter, and ice cream. Most trans fats are manufactured during food processing and are sometimes referred to as synthetic trans fats. The consumption of saturated fat and synthetic trans fat is associated with high cholesterol and an increased risk of heart disease.

A third type of fat, unsaturated fat, can be consumed in small amounts as part of a healthy diet. Unsaturated fats are found in oils and plant-based foods such as vegetable oils, nuts, and avocadoes. Fish, such as herring, salmon, and trout, also contain unsaturated fats. Studies show that consuming unsaturated fats, especially polyunsaturated fats, may help decrease the risk of type 2 diabetes and improve cholesterol levels. Health Canada recommends that people include a small amount of unsaturated fat in their diet, about 30–45 mL per day. This is equal to about three tablespoons of oil, such as olive or canola oil, and includes oils used in cooking, salad dressings, and mayonnaise. Intake of saturated fat and trans fat should be limited.

Research and Analyze

1. Conduct research on weight-loss pills that claim to work by speeding up the body's metabolism. Do these pills result in effective short-term and/or long-term weight loss? Why or why not? What health risks are associated with these types of pills? What is Health Canada's position on the use of these types of pills for the purpose of weight loss? Perform a risk/benefit analysis on these types of weight-loss pills. What important facts would you like to make consumers more aware of regarding the use of weight-loss pills?
2. Research more information about very low-calorie diets. What effect do these types of diets have on metabolism? What health risks are associated with extremely low-calorie diets? Plan a school campaign to warn students of the dangers of extremely low-calorie diets and to promote healthy eating and ways to maintain a healthy weight.
3. Infer why, when changing your diet, it is important to know how the cells in your body will react to the introduction of new substances or the removal of other substances.

Take Action

1. **PLAN** Use Canada's Food Guide and the Public Health Agency of Canada's Get Active Tip Sheets to plan five days of meals (breakfast, lunch, dinner, and snacks) and exercise to maintain a healthy metabolism. Make sure your diet includes enough calories to meet the energy needs of your cells, and a variety of foods to meet the nutritional requirements of your body. Make sure you include enough exercise to maintain your cardiovascular health.
2. **ACT** Follow your diet and exercise plan. Record information about how you felt physically and mentally each day. Report any cravings you experienced. How did the plan you followed compare to your usual diet and level of physical activity? How can you incorporate any changes you made into your everyday lifestyle to make them a permanent part of your dietary and physical activity choices?

Chapter 3 SUMMARY

Section 3.1 Metabolism and Energy

Thousands of enzyme-catalyzed chemical reactions occur in metabolic pathways that transform matter and energy in cells to sustain life.

Key Terms

metabolism
metabolic pathway
catabolism
anabolism
energy
kinetic energy
potential energy
bond energy
thermodynamics
entropy
free energy
endergonic
exergonic

Key Concepts

- Metabolism is the sum of all the biochemical reactions in a cell. Anabolic reactions require energy to build up molecules, and catabolic reactions break down molecules and release energy.
- Metabolic pathways are sequences of reactions that use of the product of one reaction as the substrate for the next.
- Energy is the ability to do work, and it can be classified as kinetic or potential.
- Bond energy is the energy needed to break a bond.
- The first law of thermodynamics states that energy cannot be created or destroyed.
- The second law of thermodynamics states that some energy is "lost" as disorder (entropy) increases.
- Free energy is the energy from a chemical reaction that is available to do work.
- Endergonic reactions require energy, while exergonic reactions release energy.
- ATP hydrolysis releases energy to drive endergonic reactions, and it is synthesized using energy from exergonic reactions.
- Electron carriers donate electrons from energy-rich to low-energy compounds.

Section 3.2 Aerobic Respiration

Aerobic respiration is a series of catabolic reactions that result in the complete oxidation of glucose.

Key Terms

aerobic respiration
substrate-level phosphorylation
glycolysis
Krebs cycle
oxidative phosphorylation
electron transport chain
chemiosmosis

Key Concepts

- The breakdown of glucose occurs in four stages: glycolysis, pyruvate breakdown, Krebs cycle, and oxidative phosphorylation.
- Glycolysis is the breakdown of glucose to two pyruvates, producing two ATP and two NADH. ATP is made by substrate-level phosphorylation.
- Pyruvate is broken down to CO_2 and an acetyl group that is attached to CoA. NADH is made during this process.
- During the Krebs cycle, the acetyl group attached to CoA is broken down to two CO_2 molecules. Three NADH, one $FADH_2$, and one ATP are made during this process.
- Oxidative phosphorylation involves two events. The electron transport chain oxidizes NADH or $FADH_2$ and generates an H^+ gradient. This gradient is used by ATP synthase to make ATP via chemiosmosis.
- Proteins and fats can enter into glycolysis or Krebs cycle at different points.

Section 3.3 Anaerobic Respiration and Fermentation

Anaerobic respiration and fermentations are metabolic reactions that do not require oxygen.

Key Terms

anaerobic respiration
fermentation

Key Concepts

- Some single-celled organisms that live in conditions of very low oxygen can carry out anaerobic respiration by using an electron acceptor other than oxygen.
- Some single-celled organisms and, during extreme exertion, some muscle cells, use only glycolysis for energy in a process called fermentation.
- In lactate fermentation, pyruvate oxidizes NADH back to NAD^+ and, in the process, is converted into lactate. In single-celled organisms, the lactate is released to the surroundings and in muscle cells, it is released into the bloodstream where it is carried to resting muscle cells and oxidized or converted to glycogen for storage.
- In ethanol fermentation, pyruvate is converted into a two-carbon compound, acetaldehyde, and carbon dioxide. The acetaldehyde reoxidizes the NADH back to NAD^+ for reuse.
- Alcohol fermentation is a useful industrial process, used to generate ethanol for fuel. It is also the same process by which alcoholic beverages are produced.

Chapter 3 REVIEW

Knowledge and Understanding

Select the letter of the best answer below.

1. Which of the following is an example of an anabolic pathway?
a. alcohol fermentation in yeast
b. lactate fermentation in muscle cells
c. the production of glucose in plants
d. aerobic respiration of glucose
e. anaerobic respiration of glucose

2. A roller coaster has ascended the first hill and is ready for the first drop. As it rolls downhill, what type of energy transformation has occurred?
a. potential energy is transformed to kinetic energy
b. kinetic energy is transformed to potential energy
c. activation energy is transformed to bond energy
d. bond energy is transformed to kinetic energy
e. potential energy is transformed to bond energy

3. Which statement is true?
a. The first law of thermodynamics states that entropy always increases.
b. The first law of thermodynamics applies only to closed systems.
c. The second law of thermodynamics only applies to open systems.
d. The second law of thermodynamics is also called the law of conservation of energy.
e. The second law of thermodynamics applies to closed systems.

4. What are the reactants in the equation for the metabolism of glucose by aerobic respiration?
a. glucose + oxygen + water
b. glucose + oxygen
c. glucose + carbon dioxide + water
d. glucose + carbon dioxide + ATP
e. glucose + oxygen + ATP

5. What is the net production of ATP molecules from two molecules of glucose undergoing glycolysis?
a. 1
b. 2
c. 3
d. 4
e. 8

6. Which compound enters the Krebs cycle for further breakdown?
a. ATP
b. NADH
c. glucose
d. pyruvate
e. acetyl-CoA

7. How is ATP produced in glycolysis and the Krebs cycle?
a. substrate-level phosphorylation
b. oxidative phosphorylation
c. pyruvate oxidation
d. reduction of NAD^+
e. reduction of FAD

8. Which compounds are the electron donors in the electron transport chain?
a. NADH and $FADH_2$
b. NAD^+ and FAD
c. NADH and NAD^+
d. $FADH_2$ and FAD
e. $FADH_2$ and NAD^+

9. What process causes your leg muscles to feel sore after an intense workout?
a. aerobic respiration
b. Krebs cycle
c. glycolysis
d. ethanol fermentation
e. lactate fermentation

10. Bond energy refers to
a. the amount of energy in hydrogen bonds
b. the amount of energy to break a bond
c. the amount of energy in a physical change
d. the sum total of entropy energy in a cell
e. the energy released by ATP

11. Can cellular respiration occur without oxygen?
a. No, oxygen is necessary as the final electron acceptor.
b. No, anaerobic organisms only need glycolysis and fermentation.
c. Yes, because oxygen can be generated by splitting water.
d. Yes, but it requires an alternative to oxygen as a final electron acceptor.
e. None of these answers is correct.

12. A toxin that makes holes in the inner membrane of the mitochondria would
a. stop the movement of electrons down the electron transport chain
b. stop ATP synthesis
c. stop the Krebs cycle
d. all of the above
e. none of the above

13. Which of these processes is/are needed for the complete oxidation of glucose?
a. the Krebs cycle
b. glycolysis
c. pyruvate oxidation
d. all of the above
e. only a and c

Chapter 3 REVIEW

14. The electrons carried by NADH and $FADH_2$ are
   a. pumped into the intermembrane space
   b. transferred to the ATP synthase
   c. moved between proteins in the inner membrane of the mitochondrion
   d. transported into the matrix
   e. considered equivalent to chemiosmosis

Answer the questions below.

15. List the four stages of aerobic respiration. State where each takes place, and what the reactants and products are.
16. How does the presence or absence of oxygen influence the steps pyruvate will undergo?
17. Why is the Krebs cycle also called the citric acid cycle and tricarboxylic acid cycle?
18. Explain the importance of NADH formation during glycolysis, pyruvate oxidation, and the Krebs cycle for ATP formation.
19. Why is it important for the cell to regenerate oxaloacetate? What would be the effect of an oxaloacetate deficiency?
20. Describe how cells use macromolecules other than carbohydrates.
21. How do high levels of ATP and NADH help to regulate aerobic pathways?
22. Explain why energy input is needed for anabolic reactions to occur.
23. What is the importance of catabolic reactions for the cell?
24. Define the term energy, and distinguish between kinetic energy and potential energy.
25. How is bond energy used by cells and living organisms?
26. Using the concepts of the laws of thermodynamics, explain why organisms need a constant input of energy.
27. Why do children need more energy than adults? Explain, using the concept of entropy.
28. Link the following equation $\Delta G = \Delta H - T\Delta S$ to the terms *anabolic*, *catabolic*, *exergonic* and *endergonic*. What is the value of ΔG in each case?
29. Explain why ATP production is considered a coupled reaction.
30. An insecticide blocks a particular enzyme needed to perform a catabolic reaction. Why would such a substance be lethal in sufficient doses?
31. A substance is released into the environment that disrupts the coupled reaction of ATP synthesis. What effect would such a substance have on ecosystems?
32. Why do biologists prefer to use the terms endergonic and exergonic rather than endothermic and exothermic?
33. A person has fallen into a deep coma. Explain why this person's intravenous drip must include glucose.
34. Why is it important that NADH is oxidized before ATP is produced?
35. Do lactic acid and ethanol still contain usable cellular energy? Explain.

Thinking and Investigation

36. Give two examples of energy transformations in living systems in which chemical potential energy is converted into a type of energy other than chemical energy.
37. Which of the following would be a better model to represent the energy transformations in the transport of electrons through the electron transport chain: a ball rolling down a ramp or a ball bouncing down a set of stairs? Explain your reasoning.
38. Explain how diffusion is an example of the second law of thermodynamics in action. Refer to the illustration below as necessary for your answer.

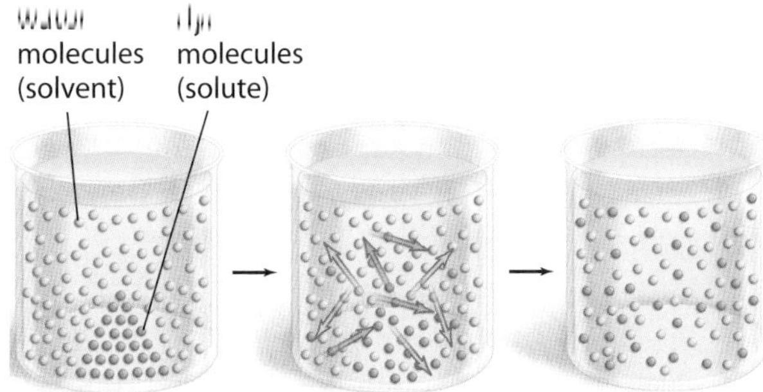

39. The equation for the free energy change of a chemical reaction is $\Delta G = \Delta H - T\Delta S$. Write a word equation that describes the free energy change equation without using the terms free energy, enthalpy, or entropy.
40. Write two chemical equations from glycolysis and two chemical equations from the Krebs cycle that are examples of coupled reactions.
41. Write a chemical reaction from glycolysis that is an example of substrate-level phosphorylation, and indicate which compound is the substrate.

42. A classmate says to you, "I don't get it. You use an ATP to convert glucose to glucose-6-phosphate, and then you use another ATP to convert fructose-6-phosphate to fructose-1,6-bisphosphate. You get one ATP back when 1,3-bisphosphoglycerate is converted to 3-phophoglycerate, and you get another ATP back when you convert phosphoenolpyruvate to pyruvate. That's two ATPs used and two ATPs formed. It looks to me like you break even. Why does everyone say that that you get a net of two ATPs out of glycolysis?" Write a paragraph that clearly explains why this reasoning is incorrect.

43. CoA is sometimes called an acyl group carrier. Examine the two reactions in which CoA is involved and explain what you think that an "acyl group carrier" means.

44. Develop a model or analogy that would help your classmates understand the concept of chemiosmosis.

45. In Chapter 1, you saw the structures of amino acids including the three shown below. Enzymes exist that can remove the amino group (NH_2) and a hydrogen atom from the carbon atom and replace them with a double bonded oxygen atom (=O). After this reaction is carried out on the three amino acids shown below, where would the remaining molecule enter the respiratory reactions?

alanine   aspartic acid   glutamic acid

46. Explain why glycolysis is included with both aerobic respiration and anaerobic respiration.

47. Consider the chemical reactions in alcohol fermentation. Based on these reactions, infer how yeast makes bread dough rise.

48. Design an experiment that could be used to demonstrate the amount of energy that can be released in the combustion of ethanol. An internet search on the concept of calorimetry may help in this question.

49. A classmate tells you that secondary active transport is a type of anabolic reaction. Explain why your classmate might be confused.

50. Is the formation of a starch molecule from simple sugars is an example of an increase or decrease in entropy? Explain your reasoning.

51. How does the level of ATP remain constant within a cell?

52. The hydrolysis of ATP to ADP can be summarized with the following equation:

$$ATP + H_2O \rightarrow ADP + P_i \qquad \Delta G = -30.5 \text{ kJ/mol}$$

a. Which chemical bonds in ATP release large amounts of energy when broken?

b. Based on the ΔG value for the reaction, what can you conclude about the nature of the reaction?

c. What would be the ΔG of the reverse reaction (that is, the formation of ATP from ADP)? Explain your answer.

53. The graph below illustrates an exergonic reaction. What do A and B represent?

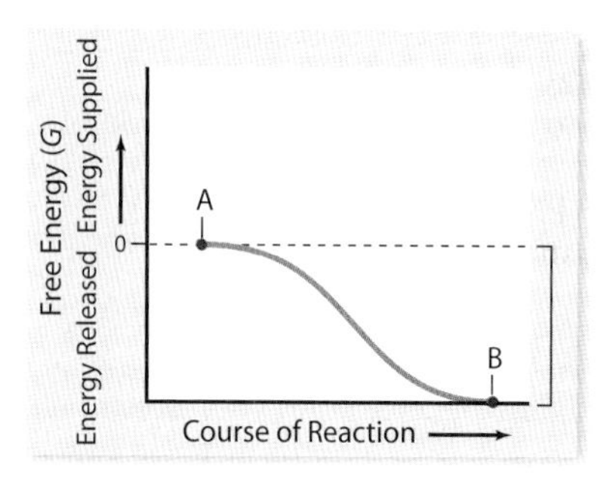

54. The process of glycolysis depends on the correct functioning of many enzymes. How could the effects of a toxin that blocks any of the glycolytic enzymes be overcome by an aerobic organism?

55. Coenzyme A and NAD^+ are vitamin B derivatives. Why would a deficiency of this vitamin complex lead to fatigue?

Communication

56. BIG IDEAS All metabolic processes involve chemical changes and energy conversions. In the process of water (H_2O) and carbon dioxide (CO_2) joining to form methane (CH_4) and oxygen (O_2) molecules, energy is required. Create a fully labelled chemical potential energy diagram for this process. Include in your diagram labels of whether the process is exerogonic or endergonic and whether the change in free energy is positive or negative.

57. Draw a molecule of ADP. Label the different groups that make up this molecule and state the difference between a molecule of ADP and a molecule of ATP.

58. Copy the following chart into your notebook, and complete it with respect to whether the entropy increases or decreases.

Process	Change in Entropy
burning a piece of wood	
table salt dissolves in water	
a toy is assembled	
ice forms	
boxes are unpacked after you move houses	
a librarian shelves books that were returned	
gases are absorbed into the blood from the lungs	

59. Create a graphic organizer to illustrate the oxidation and reduction processes of the formation of NADH and NAD^+.

60. Create a graphic organizer to illustrate the process of gylcolysis. Fully label all materials and energy units in your organizer.

61. Copy the following chart into your notebook, and complete it with respect to the location and yield of ATP and reducing power of each process:

Process	Location	ATP/Reducing Power
glycolysis	cytosol	
pyruvate oxidation	mitrochondrial matrix	
The Krebs cycle	mitrochondrial matrix	
Oxidative phosphorylation	inner mitochondrial membrane	

62. Draw and fully label a diagram to illustrate the Krebs cycle in terms of ATP production and the production of NADH and $FADH_2$. Include the number of each that form in the process.

63. Use a flow chart to summarize the 10 reactions of glycolysis.

64. Use a diagram to summarize the process of alcohol fermentation in yeast.

65. Compare and contrast NADH and $FADH_2$ in terms of their production, processing, and yield in aerobic respiration.

66. Compare and contrast the terms redox chain, oxidative phosphorylation, and chemiosmosis.

67. Use a diagram and a description to explain what reducing power is.

68. The process of chemiosmosis is often compared to being similar to a hydroelectric dam.
    a. Explain this analogy using labelled diagrams and descriptions.
    b. What complex would be the generator in this analogy? Explain your answer.

Application

69. In order to produce large amounts of baker's yeast, the cells are grown in a large vessel, called a bioreactor. The pH, temperature, and amount of nutrients in the media are monitored and adjusted as needed. Oxygen is continuously streamed into the bioreactor and the mixture is stirred. Why is each of these steps critical to the success of the procedure? Explain what would happen if any particular step was left out.

70. Out of all of the potential energy in glucose, only about 40 percent is transferred to ATP through aerobic respiration.
    a. Is the other 60 percent "wasted" or does it serve a useful purpose? Justify your response.
    b. How does the inefficiency of aerobic respiration relate to the second law of thermodynamics?

Use the information below to answer questions 71, 72, and 73.

Slow-oxidative fibres and fast-glycolytic fibres are two types of cells that make up skeletal muscle—the type of muscle that allows you to move your body. Slow-oxidative fibres, which contain many mitochondria, contract slowly and tire out slowly, allowing for sustained postures or movements. Fast-glycolytic fibres, which contain very few mitochondria, contract rapidly to produce powerful, quick movements, but tire out rapidly as well. Most of the ATP produced in fast-glycolytic fibres comes from glycolysis.

71. Suggest why these two types of skeletal muscle fibres have different amounts of mitochondria.

72. It can be much faster to produce ATP by glycolysis than by oxidative phosphorylation. How would it benefit a sprinter to have high numbers of fast-glycolytic fibres?

73. How would it benefit a long distance runner to have high numbers of slow-glycolytic fibres?

74. BIG IDEAS An understanding of metabolic processes enables people to make informed choices with respect to a range of personal, societal, and environmental issues. All plant-eating animals release methane as a result of their digestion, but cattle, by far, emit the most. Methane emitted by cattle is difficult to trap, so it escapes into the atmosphere, where it acts as a greenhouse gas. Research how changing what cattle eat can affect how much methane they emit.

Use the information below to answer questions 75 and 76.

Unlike nitrogen gas, nitrate acts as a fertilizer to plants and many types of algae. When excess nitrate from farmlands or sewage makes its way into natural bodies of water, algae reproduce rapidly, causing an algal bloom. As the algae die, bacteria decompose the dead cells, using up the oxygen in the water. Fish and other aquatic life die off as a result.

75. Identify how algal blooms affect cellular respiration in various types of organisms.

76. Some bacteria can reduce nitrate to nitrogen gas through the process of anaerobic respiration. Explain why these bacteria could be both a problem in agriculture but very useful in sewage treatment.

Use the information below to answer questions 77 and 78.

Along with oxygen, nitrate, and carbon dioxide, there are various electron acceptors that bacteria can use. A microbial fuel cell makes use of this flexibility in order to generate electricity. Bacteria in the fuel cell are provided with wastewater or some other source of nutrients. As they carry out cellular respiration, they transfer electrons to components in the fuel cell, which generates electricity. Although microbial fuel cells are still an experimental technology, their potential is great.

77. Explain how research into microbial fuel cells could advance scientific understanding of electron transport or other processes of cellular respiration.

78. Propose how microbial fuel cells could be used to address two or more environmental issues.

79. Consider the following statement: An understanding of chemistry is essential to a solid understanding of biological processes. State whether or not you agree with this statement and why. Use one or more specific examples from this chapter to justify your response.

80. How might it be beneficial to use a variety of micro-organisms, rather than a single type, to make specialty cheeses with unique flavours, colours, and textures?

81. Although aerobic respiration is a life-sustaining process, it also generates harmful by-products that can remove electrons from proteins, lipids, and DNA. These oxidants—which include hydrogen peroxide and superoxide—must be inactivated for cells to survive. Infer how oxidation-reduction reactions involving antioxidants, such as vitamin C, vitamin E, and coenzyme Q10, can help protect cells.

82. Although rare in humans and most dog breeds, phosphofructokinase deficiency more frequently affects English springer spaniel dogs. The resulting build-up of glucose 6-phosphate and fructose 6-phosphate causes these intermediates to be converted into glycogen, which accumulates in the muscles. How does an understanding of glycolysis help to explain why phosphofructokinase deficiency causes muscle weakness? How could this knowledge be used to pursue possible treatments for this condition?

83. In a newborn baby, the neck and upper back contains brown-coloured fat with specialized mitochondria. Brown-fat mitochondria contain proton channels that divert protons from ATP synthase. Infer how this system could help keep a baby warm.

84. A supplement containing pyruvate is available at the cost of $9/100 g, and promises to give you more energy. Would it make sense to buy it? Explain.

85. Some pills people take to attain weight loss cause the inner mitochondrial membrane to be more permeable to hydrogen ions (H^+). Should the sale and use of such products be regulated? Justify your opinion.

86. Is the formation of a starch molecule from simple sugars is an example of an increase or decrease in entropy? Explain your reasoning.

87. It is known that prolonged exposure to carbon monoxide can cause death. This is due to carbon monoxide molecules binding to the hemoglobin in the blood. This causes a reduced ability to extract oxygen in the lungs. Describe how this can possibly lead to death.

88. Salicylic acid is a metabolite of aspirin (acetylsalicylic acid), and is also commonly used in topical acne medications and face wash lotions. At very high doses, salicylic acid acts an uncoupler of oxidative phosphorylation, since it reduces the proton concentration gradient across the inner mitochondrial membrane. What would be the consequences of very high doses of salicylic acid on cells and the human body? Explain your answer.

Chapter 3 SELF-ASSESSMENT

Select the letter of the best answer below.

1. K/U Which of the following statements is false? The first law of thermodynamics states that

a. energy cannot be created or destroyed
b. energy can be transformed from one type to another
c. the universe tends toward disorder
d. the energy of the universe is constant
e. energy can be transferred from one object to another

2. K/U Which statement is false?

a. Glycolysis occurs in the mitochondrial matrix.
b. Glycolysis breaks down glucose into two three-carbon compounds.
c. Glycolysis uses two ATP molecules and converts them into ADP.
d. Glycolysis reduces two molecules of NAD^+ into NADH.
e. Glycolysis phosphorylates four ADP molecules to make four ATP molecules.

3. K/U Which statement correctly summarizes the function of the Krebs cycle?

a. Glucose is oxidized to carbon dioxide and water to produce usable energy in the form of ATP.
b. Glucose is broken down into two pyruvate molecules which enter the Krebs cycle where three carbon atoms are oxidized to carbon dioxide while generating 36 ATP molecules.
c. Acetyl CoA molecules donate two carbons to oxaloacetate and, in the process of oxidizing two carbon atoms, one turn of the cycle produces four NADH molecules, one $FADH_2$ molecule, and phosphorylates one ADP.
d. A pyruvate molecule from glycolysis loses one carbon atom which is converted to carbon dioxide then NAD^+ is reduced to NADH and coenzyme A picks up the other two carbon atoms.
e. Answers a and c are both correct.

4. K/U What is the location of the enzymes that catalyze the Krebs cycle reactions?

a. embedded in the inner membrane of the mitochondria
b. in the inner membrane space
c. in the cytoplasm
d. embedded in the outer membrane of the mitochondria
e. in the mitochondrial matrix

5. K/U The energy of motion is

a. bond energy
b. potential energy
c. free energy
d. kinetic energy
e. total energy

6. K/U An important function of the electron transport chain is to

a. reduce NADH and $FADH_2$
b. generate a hydrogen ion gradient
c. phosphorylate ATP
d. convert oxygen into water
e. pass electrons to nitrogen

7. T/I Which of the following sequences correctly describes the overall process of metabolizing glucose?

a. pyruvate oxidation, Krebs cycle, glycolysis, electron transport, oxidative phosphorylation
b. oxidative phosphorylation, Krebs cycle, pyruvate oxidation, glycolysis, electron transport
c. Krebs cycle, electron transport, glycolysis, oxidative phosphorylation, pyruvate oxidation
d. glycolysis, pyruvate oxidation, Krebs cycle, electron transport, oxidative phosphorylation
e. Krebs cycle, oxidative phosphorylation, glycolysis, electron transport, pyruvate oxidation

8. K/U Energy to phosphorylate ATP during oxidative phosphorylation is derived from

a. the hydrogen ion gradient across the outer membrane of the mitochondria
b. energy released from a reaction when converting a substrate to a product in glycolysis
c. energy released when electrons are transferred from one carrier to the next
d. energy released from a reaction when converting a substrate to a product in the Krebs cycle
e. the hydrogen ion gradient across the inner membrane of the mitochondria

9. K/U Which of the following statements about the inner mitochondrial membrane is false?

a. It is permeable to positively charged ions.
b. It is impermeable to NADH.
c. The only way hydrogen ions can move into the matrix is through the ATP synthase.
d. It contains hydrogen ion pumps.
e. It contains the electron carriers for electron transport.

10. **K/U** Lactate fermentation occurs in muscle cells when
   a. the cells need lactate
   b. plenty of ATP is available from oxidative phosphorylation
   c. muscles are in poor condition
   d. insufficient oxygen is available
   e. the pH of the cellular fluid is too high

Use sentences and diagrams as appropriate to answer the questions below.

11. **K/U** Explain the difference between anabolic pathways and catabolic pathways.
12. **K/U** Name at least three ways in which the energy from ATP is used in living tissues.
13. **T/I** State the second law of thermodynamics. Explain how the existence of highly ordered living organisms is not a contradiction of the second law.
14. **K/U** What two factors determine the change in free energy of a reaction? Use words and not symbols.
15. **C** Draw graphs that represent endergonic and exergonic reactions. Explain their meaning.
16. **C** Use diagrams to explain how the level of ATP in living cells can remain almost constant while ATP is always being used for energy.
17. **T/I** How does substrate level phosphorylation differ from oxidative phosphorylation?
18. **C** Examine the diagram of a mitochondrion below.
   a. State the name of each of the numbered parts.
   b. Describe one important function of each part.

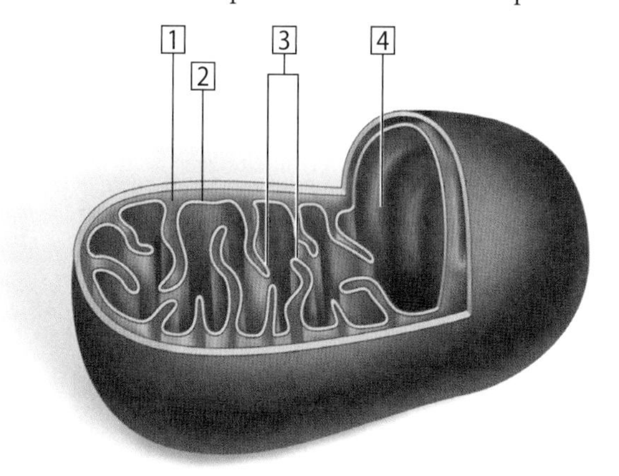

19. **T/I** Molecular oxygen is not involved in any of the reactions of the Krebs cycle. Nevertheless, the Krebs cycle is considered a part of aerobic metabolism. Explain why this is not a contradiction.
20. **T/I** There are four enzymes that catalyze the four reactions in the Krebs cycle in which an electron carrier—NADH or $FADH_2$—is reduced. What term is common to the names of all four enzymes, and why is the term appropriate for the reaction?
21. **C** Discuss the function of molecular oxygen in aerobic metabolism. Think of an analogy for this function in another situation in cellular metabolism or in real life and explain your analogy.
22. **T/I** When determining the amount of ATP that is generated by the complete oxidation of one glucose molecule, the experimental value is always a little lower than the theoretical value. Explain why.
23. **C** For each of the numbers in the diagram, state whether the compound from which the line originates activates or inhibits the enzyme that the line indicates, and explain how it helps to regulate the levels of ATP in the cell.

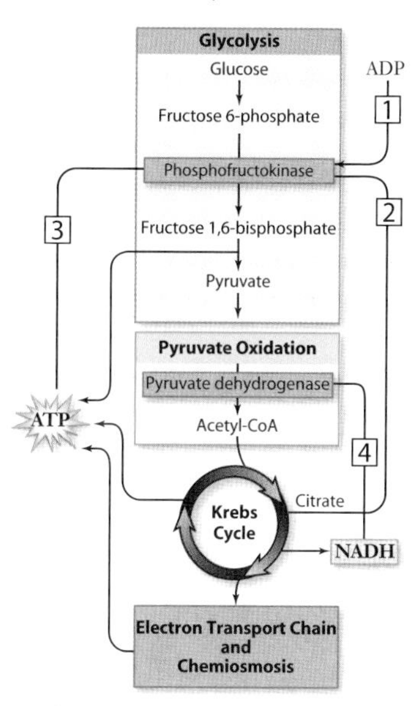

24. **A** Answer the following questions with respect to the digestive system of cattle.
   a. Describe, in detail, why there is so much methane produced by the digestive tract of cattle.
   b. How does this methane cause an environmental problem, and how might the problem be reduced while creating a useful product?
25. **C** Using two or three sentences, describe the circumstances in which lactate is produced in muscles.

Self-Check

If you missed question...	1	2	3	4	5	6	7	8	9	10	11	12	13	14	15	16	17	18	19	20	21	22	23	24	25
Review section(s)...	3.1	3.2	3.2	3.2	3.1	3.2	3.2	3.2	3.2	3.3	3.1	3.2	3.1	3.1	3.1	3.2	3.2	3.2	3.2	3.2	3.2	3.2	3.2	3.3	3.3

CHAPTER 4

Photosynthesis

Specific Expectations

In this chapter, you will learn how to . . .

- C2.1 **use** appropriate terminology related to metabolism (4.1, 4.2)
- C2.3 **conduct** a laboratory investigation into the process of photosynthesis to **identify** the products of the process, **interpret** the qualitative observations, and **display** them in an appropriate format (4.2)
- C3.2 **explain** the chemical changes and energy conversions associated with the processes of photosynthesis (4.1, 4.2)
- C3.4 **describe**, **compare**, and **illustrate** the matter and energy transformations that occur during the process of photosynthesis, including the roles of oxygen and chloroplasts (4.1, 4.2)

In May 2010, members of the forest industry and leading environmental groups signed an agreement for the sustainable management of Canada's boreal forest. This region occupies about three-quarters of the country's total forested area and is home and breeding ground for more than half of Canada's species of plants and animals. As you know from Chapter 3, various metabolic pathways within the cells break down high-energy compounds such as glucose and convert the energy into a usable form—ATP. High-energy compounds do not form spontaneously. They are formed through a series of metabolic pathways known collectively as photosynthesis. Each year, trees and other plants of Canada's boreal region convert about 12.5 million tonnes of carbon, as carbon dioxide from the atmosphere, into the carbon compounds that help sustain their own lives and those of billions of other organisms.

Launch Activity

Seeing Green

In a functioning green plant, light energy is trapped by light-absorbing molecules called *pigments* and is used to synthesize carbohydrates from carbon dioxide and water. The main photosynthetic pigment is chlorophyll. What effect does light have on chlorophyll if it is removed from a living plant?

Materials

- beaker of prepared chlorophyll solution (provided by your teacher)
- strong light source (such as a slide projector)

Procedure

1. In a darkened room, shine a strong beam of light at a sample of chlorophyll solution.
2. Observe the colour of the chlorophyll by viewing the sample at a slight angle.
3. Observe the colour of the chlorophyll by viewing the sample at a right angle to the beam of light.
4. Describe the colours you see in steps 2 and 3.

Analysis

1. Recall from previous studies that visible light is a mixture of different colours (wavelengths). Which colours of light do you think chlorophyll absorbs? Explain your reasoning.
2. When chlorophyll has been removed from a plant and is in solution, it has a property called *fluorescence*. When a pigment molecule absorbs a specific colour (wavelength) of light, its electrons become "excited"—that is, they move to a higher energy level. Almost immediately, the excited electrons return to their original, lower-energy state as they emit (give off) the energy they absorbed. The emitted energy is visible as light of a lower energy and, thus, a longer wavelength. In which step did you observe fluorescence? Suggest a possible explanation for what you observed.

SECTION 4.1

Capturing Solar Energy: The Light-Dependent Reactions

Key Terms

light-dependent reaction
light-independent reaction
thylakoid
pigment
photosystem
photophosphorylation

Photosynthesis transforms the radiant energy of sunlight into the chemical energy of high-energy compounds. Photosynthesizing organisms, such as those shown in **Figure 4.1**, use only an estimated 1 to 2 percent of the Sun's energy that reaches Earth's surface. Nevertheless, the contribution that these organisms make is enormous. For example, they synthesize about 1.4×10^{15} kg of energy-storing glucose and other compounds each year—enough sugar to fill a chain of railway boxcars reaching to the Moon and back 50 times.

Photosynthesis enables plants to produce structural and metabolic substances that aid in their survival. For example, much of the glucose produced by plants is converted to cellulose, which is the main component of plant cell walls. Glucose may also be converted to other sugars as well as storage forms of carbohydrates such as starch. In addition, compounds produced by photosynthesis are involved in the synthesis of other essential cellular substances such as amino acids. In fact, the products of photosynthesis account for nearly 95 percent of the dry weight of green plants. Other organisms, including humans, depend on the molecules, tissues, and substances that plants synthesize for their own use.

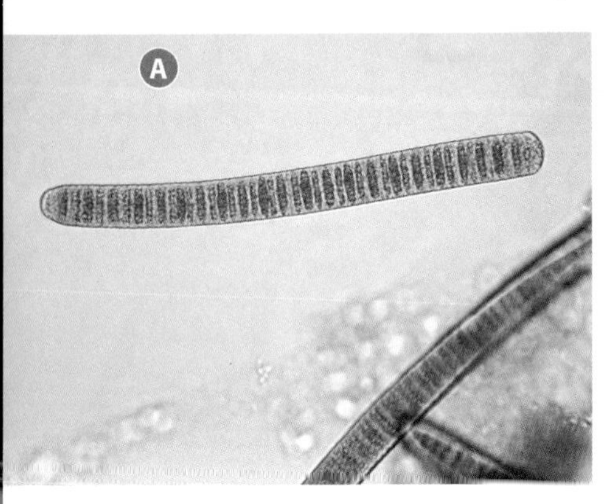

Figure 4.1 Organisms that carry out photosynthesis include **A** cyanobacteria, **B** kelp (a form of algae), and **C** green leaved flowering plants.
Identify *at least two other examples of organisms that carry out photosynthesis.*

light-dependent reaction in photosynthesis, the reaction that traps solar energy and uses it to generate ATP and NADPH

light-independent reaction in photosynthesis, the reaction that assimilates carbon dioxide to produce an organic molecule that can be used to produce biologically important molecules such as carbohydrates

The process of photosynthesis can be summarized by the following equation:

$$6CO_2(g) + 6H_2O(\ell) + \text{energy} \rightarrow C_6H_{12}O_6(s) + 6O_2(g)$$

In other words, carbon dioxide and water, with the addition of energy from the Sun, are used to produce glucose and oxygen. As with cellular respiration, the equation for photosynthesis represents an overall process. In fact, numerous reactions occur between the substrates on the left side of the equation to produce the products on the right side. The arrow in the equation actually represents more than 100 distinct chemical reactions that lead to the end products.

As the term photosynthesis suggests, the process involves two sets of reactions. *Photo*, meaning light, refers to the reactions that capture light energy; *synthesis* refers to the reactions that produce a carbohydrate. The two sets of reactions that make up photosynthesis are called the light-dependent reactions and the light-independent reactions. In the **light-dependent reactions**, light energy is trapped and used to generate two high-energy compounds: ATP and NADPH. NADPH is similar in structure and function to NADH. In the **light-independent reactions**, the energy of ATP and the reducing power of NADPH are used to make a high-energy organic molecule.

For simplicity, most of the discussion of photosynthesis in this chapter will be based on processes that occur in plants. **Figure 4.2** shows the levels of organization in a plant leaf and highlights key structures and materials associated with photosynthesis. Water enters plants through the roots and is transported to the leaves through the veins. Carbon dioxide enters through openings, called stomata, in the leaves. The carbon dioxide and water diffuse into the cells and then enter the *chloroplasts*, where photosynthesis takes place. Most photosynthetic cells contain anywhere from 40 to 200 chloroplasts. A typical leaf may have 500 000 chloroplasts per square millimetre.

A membrane system within the chloroplast forms interconnected disks called **thylakoids** that look like flattened sacs. They are often stacked to form structures called *grana* (singular, *granum*). Thylakoids are central to photosynthesis, because the molecules that absorb the solar energy are embedded in the thylakoid membranes. Surrounding the grana in the chloroplasts is a fluid-filled interior called the *stroma*. The stroma contains the enzymes that catalyze the conversion of the carbon dioxide and water into carbohydrates.

Suggested Investigation

Inquiry Investigation 4-A, Gases Released during Photosynthesis and Cellular Respiration

thylakoid one of many interconnected sac-like membranous disks within the chloroplast, containing the molecules that absorb energy from the Sun

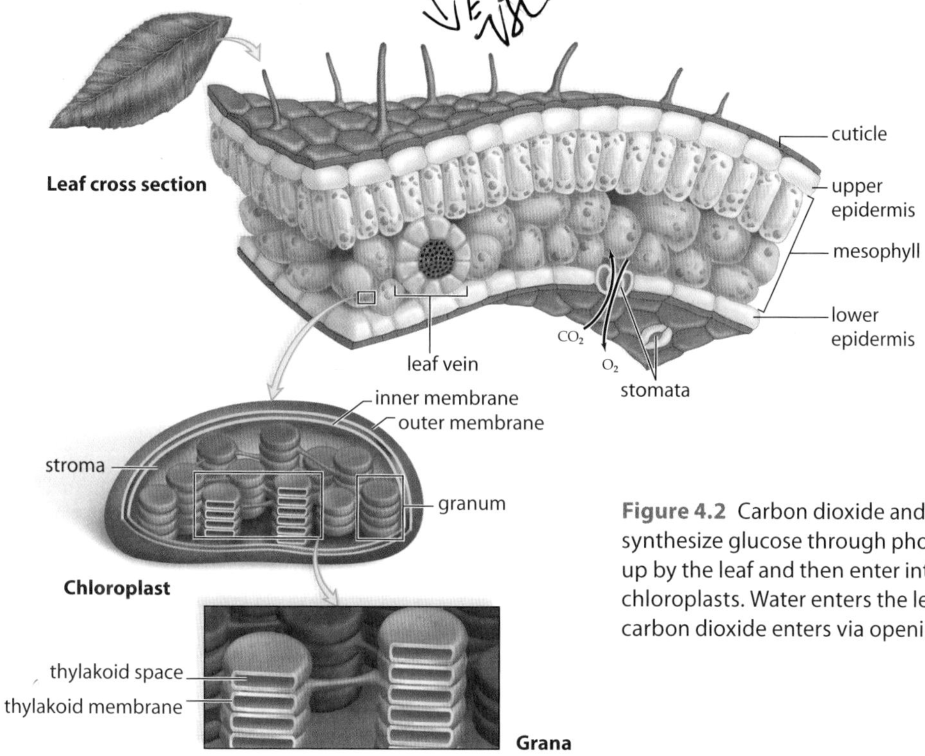

Figure 4.2 Carbon dioxide and water that are used to synthesize glucose through photosynthesis are taken up by the leaf and then enter into plant cells and chloroplasts. Water enters the leaf through veins, and carbon dioxide enters via openings called *stomata*.

Activity 4.1 A Photosynthetic Timeline

In this activity, you will do research to identify key events in the development of an understanding of photosynthesis.

Materials

- print and electronic resources

Procedure

1. Working individually or cooperatively, find out the experimental evidence and conclusions contributed by the following scientists, who were instrumental in constructing an understanding of photosynthesis:
   - Jan Baptista van Helmont
   - John Woodward
   - Stephen Hales
   - Joseph Priestley
   - Jan Ingen-Housz
   - Jean Senebier
   - Antoine Lavoisier
   - Nicolas de Saussure
   - Julius Robert von Mayer
   - Julius von Sachs
2. Design a table or timeline to summarize the names, dates, places, evidence, and findings of the scientists you investigated in step 1.

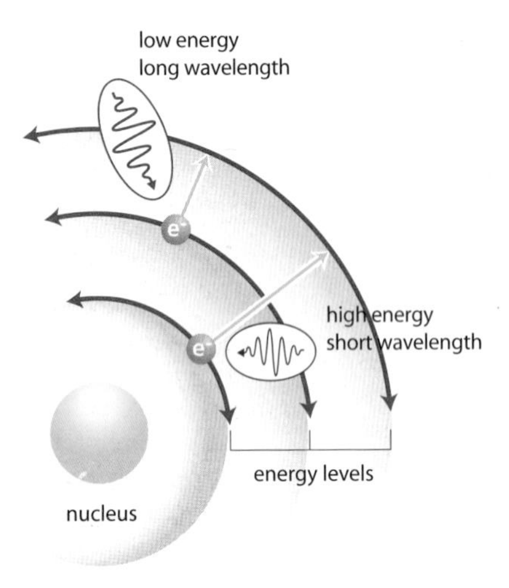

Figure 4.3 The nature of the atom or molecule determines the energy levels that are allowed for its electrons. Electrons can absorb a photon only if it carries exactly enough energy to allow the electron to move up to another allowed energy level.

The Absorption of Light Energy

When any form of matter absorbs light energy, the light is absorbed in the form of packets of energy called *photons*. Photons carry specific amounts of energy. Each wavelength (colour) of visible light is associated with photons of one distinct amount of energy. Longer-wavelength photons have smaller amounts of energy and shorter-wavelength photons have larger amounts of energy. The wavelength of the photon, and thus the colour of light, that an atom or molecule absorbs is determined by the energy levels of the electrons in that atom or molecule. An atom or molecule can absorb photons only if they have an amount of energy that is exactly equal to the difference between two energy levels, as shown in **Figure 4.3**.

A compound that absorbs certain wavelengths of visible light is called a **pigment**. A photosynthetic pigment is a compound that traps light energy and passes it on to other compounds. When sunlight is available, pigments embedded in the thylakoid membranes absorb light energy, initiating the light-dependent reactions. Eventually, the energy is used to synthesize high-energy compounds.

pigment a compound that absorbs certain wavelengths of visible light while reflecting others

Photosynthetic organisms have a variety of photosynthetic pigments, although chlorophyll, which is green, is the main type of photosynthetic pigment in plants. Chloroplasts contain two types of chlorophyll: chlorophyll *a* and chlorophyll *b*. Chlorophyll does not actually absorb green light. To understand why chlorophyll appears green, examine **Figure 4.4A**. When you shine white light through a prism, the prism separates the colours (wavelengths) of light into a spectrum. A chlorophyll solution absorbs red and blue light, and it transmits or reflects green light. Therefore, the light that reaches your eyes is green. An *absorbance spectrum* is a graph that shows the relative amounts of light of different wavelengths that a compound absorbs.

Figure 4.4B shows the absorbance spectra of two of the forms of chlorophyll. Also included is the absorbance spectrum of another pigment called *beta-carotene*. Beta-carotene is responsible for the orange colour of carrots. It can be converted into vitamin A, which can then be converted into retinal, which is the visual pigment in your eyes. Beta-carotene is a member of a very large class of pigments called *carotenoids*. The carotenoids absorb blue and green light, so they are yellow, orange, and red in colour. The coloured leaves of some trees in autumn are due to carotenoids and other pigments.

Suggested Investigation

Inquiry Investigation 4-C, Using Chromatography to Separate Plant Pigments

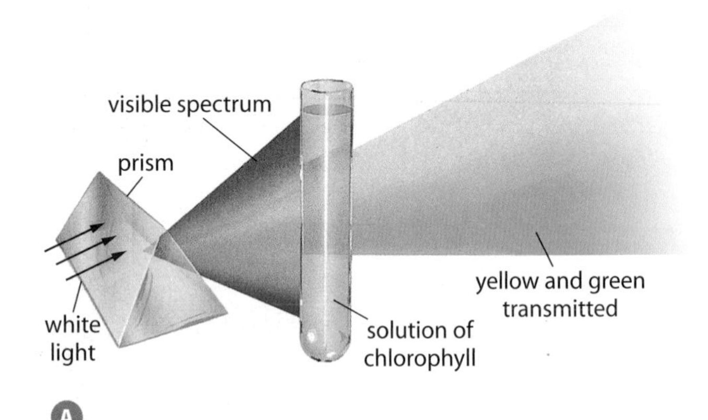

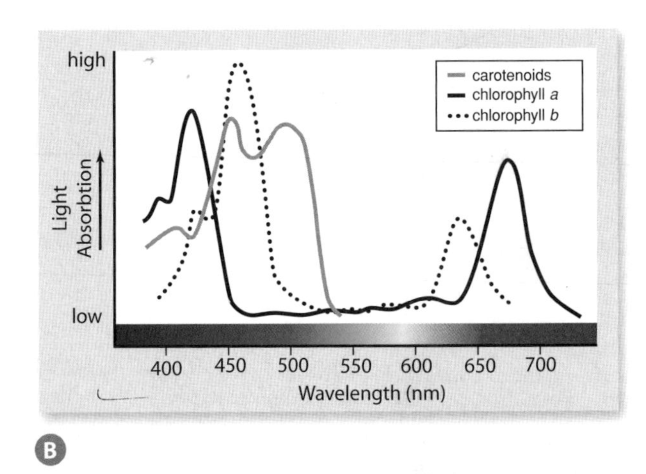

Figure 4.4 A Leaves appear green, because chlorophyll molecules in leaf cells reflect green and yellow wavelengths of light and absorb other wavelengths (red and blue). **B** This absorbance spectrum for three photosynthetic pigments shows that each pigment absorbs a different combination of colours of light.

Figure 4.5 shows another type of spectrum, called an *action spectrum*. An action spectrum shows the relative effectiveness of different wavelengths of light for promoting photosynthesis. It does this by showing the rate at which oxygen is produced by photosynthesis. Notice that the shape of the action spectrum is similar to that of the absorbance spectra in **Figure 4.4B**. This observation links the production of oxygen in photosynthesis with selected wavelengths of light, as well as with the specific pigments that absorb these wavelengths.

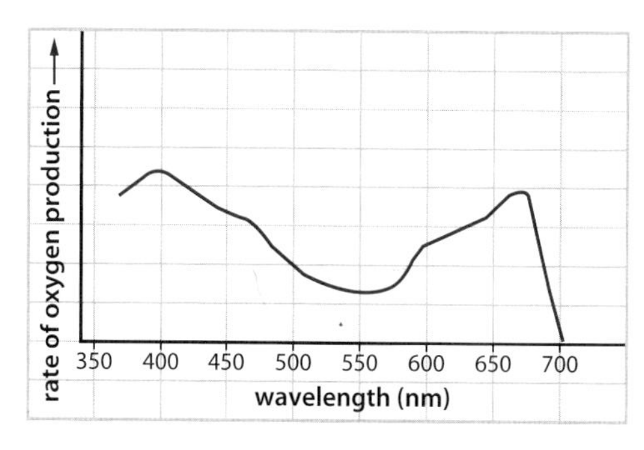

Figure 4.5 This action spectrum for photosynthesis shows the rate at which oxygen is produced when different wavelengths of light are shown on the leaf.

Photosystems Capture the Energy

Pigments do not absorb light independently. Instead, chlorophyll molecules act as clusters of chlorophyll and other pigment molecules that are associated with a specific group of proteins. These clusters are embedded in the thylakoid membranes. The core group of chlorophyll molecules and proteins in the centre of this cluster is called a **photosystem**.

photosystem one of two protein-based complexes composed of clusters of pigments that absorb light energy

Chlorophyll molecules that have been removed from the photosystems and dissolved in solution absorb light of one specific wavelength. However, when chlorophyll molecules are associated with different proteins in a photosystem, they can absorb light energy of various wavelengths. When any pigment molecule absorbs a photon, the molecule passes the energy to a unique pair of chlorophyll *a* molecules associated with a specific group of proteins. This pair of chlorophyll *a* molecules, in combination with these proteins, is called the *reaction centre*, which is shown in **Figure 4.6**.

The antenna complex includes all the surrounding pigment molecules that gather the light energy. The antenna complex transfers light energy to the reaction centre much as a funnel directs liquid into the mouth of a bottle. When a reaction centre has received the energy from the antenna complex, an electron in the reaction centre becomes "excited"—that is, the electron is raised to a higher energy level. The electron then has enough energy to be passed to an electron-accepting molecule. Since this electron acceptor has received an electron, it becomes reduced and is a higher energy level.

The chloroplasts of plants and algae have two photosystems called *photosystem I* and *photosystem II*. The reaction centre pigment molecule of photosystem I is called P700, and the reaction centre pigment molecule of photosystem II is called P680, based on the wavelengths (in nanometres) of light these molecules absorb. The photosystems are named for the order in which they were discovered, not for their sequence in photosynthesis. The photosystems have different roles but work together as part of the light-dependent reactions.

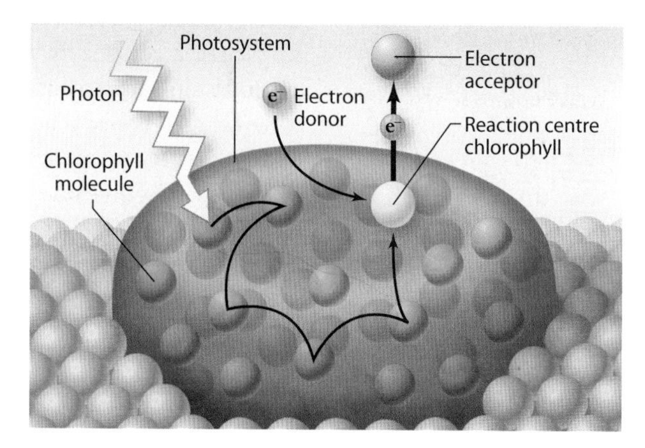

Figure 4.6 How the antenna complex works. The antenna complex is also sometimes referred to as the light-harvesting complex, because it gathers (harvests) energy from light so that the energy can be directed to the P680 molecule in the reaction centre.

Learning Check

1. Why is the balanced chemical equation for photosynthesis referred to as an overall equation, and why is it an oversimplification?
2. What are thylakoids, and why are they important to photosynthesis?
3. Use a Venn diagram to compare the light-dependent and the light-independent reactions of photosynthesis.
4. Plants are green, but a plant provided with only green wavelengths of light will barely survive. Explain why.
5. What is the advantage of plant leaves having a variety of photosynthetic pigments, rather than just chlorophyll?
6. Interpret the shape of the graphs in **Figure 4.4B**. What do the peaks and the trough mean?

The Light-Dependent Reactions

The arrangement of photosystem I and photosystem II and how they work together is shown in **Figure 4.7**. This schematic is often called a Z diagram because of the shape of energy flow through the system. As described in the following steps, the two photosystems work sequentially and have different roles during the light-dependent reactions of photosynthesis.

Step 1: The P680 molecule in the reaction centre of photosystem II absorbs a light photon, exciting an electron. When the excited electron leaves P680 in photosystem II and goes to the electron acceptor, P680 is missing an electron. It is said to have a hole. The $P680^+$, now positively charged, has a powerful attraction for electrons. Although water is very stable molecule, the attraction of $P680^+$ pulls electrons from water. A water-splitting complex holds two water molecules in place as an enzyme strips four electrons from them, one at a time. $P680^+$ accepts these electrons one at a time, and each is passed to another electron carrier. $P680^+$ then absorbs another photon, becomes reduced, and passes on another electron. This process occurs a four times to form one oxygen molecule. The four hydrogen ions from the two water molecules remain in the thylakoid space. The oxygen atoms from the water molecules immediately form an oxygen molecule. This is the oxygen that is released by plants into the environment. Photosystem II can absorb photons, excite P680, and pass electrons to the electron acceptor more than 200 times a second.

Step 2: From the electron acceptor, the energized electrons are transferred, one by one, along a series of electron-carrying molecules. Together, these molecules are referred to as an *electron transport system*. This photosynthetic electron transport system is similar to the electron transport system in mitochondria that is used in cellular respiration. With each transfer of electrons along the system, a small amount of energy is released. The released energy is used by a protein complex called the *b_6-f complex* to pump hydrogen ions from the stroma, across the thylakoid membrane, and into the thylakoid space. Eventually, there are many more hydrogen ions in the thylakoid space than there are in the stroma. This pumping of electrons generates a hydrogen ion concentration gradient across the thylakoid membrane. This is similar to the formation of a hydrogen ion concentration gradient across the inner mitochondrial membrane during cellular respiration.

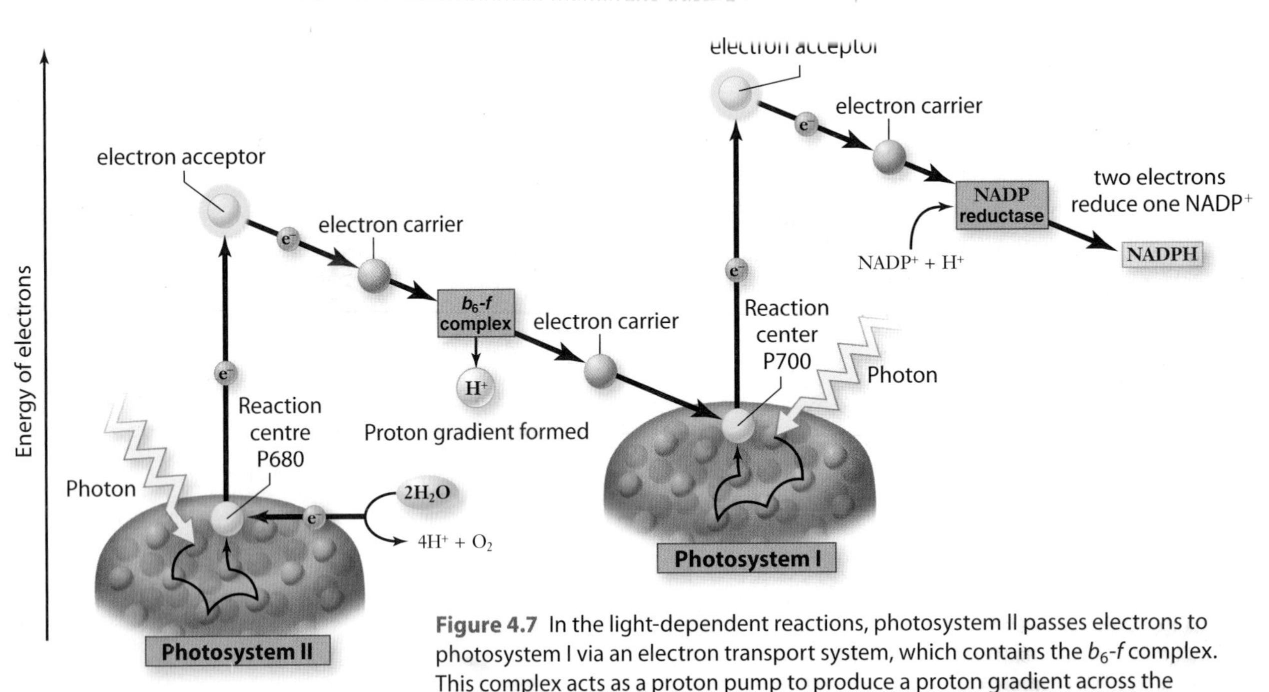

Figure 4.7 In the light-dependent reactions, photosystem II passes electrons to photosystem I via an electron transport system, which contains the b_6-*f* complex. This complex acts as a proton pump to produce a proton gradient across the thylakoid membrane. The electrons lost from the reaction centre of photosystem II are replenished by the oxidation of water. Photosystem I uses the electrons to reduce $NADP^+$ to NADPH.

Step 3: While the events of steps 1 and 2 are taking place, light energy is absorbed by photosystem I. This energy is transferred to the reaction centre P700 molecule, where electrons become excited. Once again, the excited electrons are passed to a high-energy electron acceptor. In photosystem I, the lost electrons are replaced by those that have reached the end of the electron transport system from photosystem II.

Step 4: The electrons that were received by the electron acceptor from photosystem I are used by the enzyme NADP reductase to reduce $NADP^+$ to form NADPH. The reducing power of NADPH will be used in the light-independent reactions.

Making ATP by Chemiosmosis

ATP synthesis in the light-dependent reactions occurs by the same mechanism as ATP synthesis in aerobic respiration. The movement of hydrogen ions is linked to the synthesis of ATP by chemiosmosis. Because the ultimate energy source is light photons, the process is called **photophosphorylation**. **Figure 4.8** shows the arrangement of the two photosystems in the thylakoid membrane, with the electron transport system that connects them, and how they work together to produce the proton gradient that provides the energy to synthesize ATP.

photophosphorylation the use of photons of light to drive the phosphorylation of ADP to produce ATP via chemiosmosis

The hydrogen ions that are pumped from the stroma to the thylakoid space by the b_6-*f* complex of the electron transport chain cannot diffuse back across the membrane, because the membrane is impermeable to these ions. Like mitochondria, chloroplasts also have an ATP synthase enzyme. The chloroplast ATP synthase is embedded in the thylakoid membrane and provides the only pathway for the hydrogen ions to move down the concentration gradient. This pathway is linked to a mechanism that bonds a free phosphate group to an ADP molecule to form ATP. As the hydrogen ions move down their concentration gradient through the ATP synthase molecule, the energy of the gradient is used to generate ATP molecules. Recall that this is the same way ATP is produced by chemiosmosis in mitochondria.

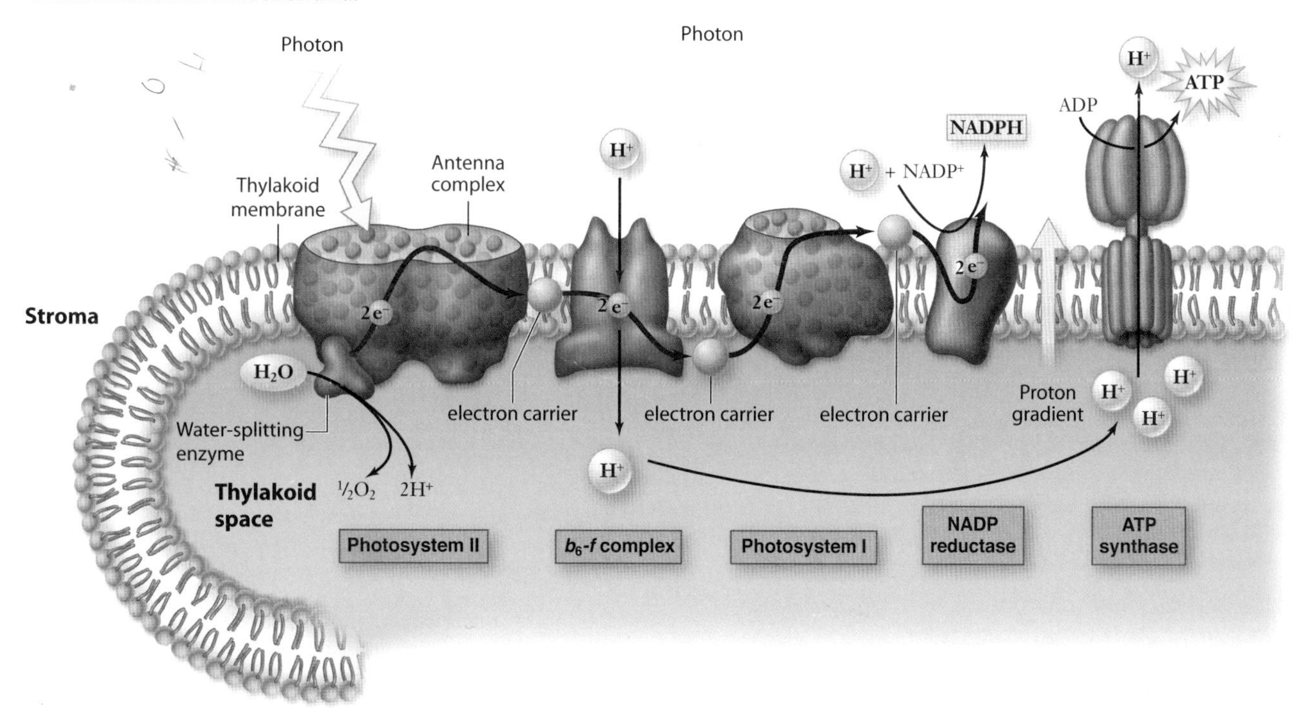

Figure 4.8 Photosystem I, photosystem II, an electron transport system, and the ATP synthase enzyme are embedded in the thylakoid membrane of chloroplasts. ATP synthesis by chemiosmosis in chloroplasts occurs in a way that is very similar to the way it occurs in mitochondria.

***Infer** what would happen to the yield of ATP if a chemical agent that causes membranes to be permeable to protons is added to chloroplasts.*

Noncyclic and Cyclic Photophosphorylation

The production of ATP by the passing of electrons through the Z scheme is often called *noncyclic photophosphorylation*. It is considered noncyclical because the flow of electrons is unidirectional—that is, the electrons are transferred from photosystem II to $NADP^+$ to form NADPH. The passage of one electron pair through this system generates 1 NADPH and slightly more than 1 ATP. However, this ratio of ATP to NADPH is not sufficient for the light-independent reactions. These require three ATP molecules to two NADPH molecules.

Chloroplasts are able to produce more ATP through *cyclic photophosphorylation*. As shown in **Figure 4.9**, excited electrons leave photosystem I and are passed to an electron acceptor. From the electron acceptor, they pass to the b_6-f complex and back to photosystem I. The proton gradient is generated in the same manner as in noncyclic photophosphorylation, and ATP synthesis by chemiosmosis also occurs. Because the same electron that left the P700 chlorophyll molecule in photosystem I returns to fill the hole it left, the process is called cyclical. Notice, however, that neither NADPH nor oxygen is produced in cyclic photophosphorylation. The relative activities of these two photophosphorylation pathways vary depending on the relative amounts of ATP and NADPH that are required by the numerous reactions that are carried out in the stroma.

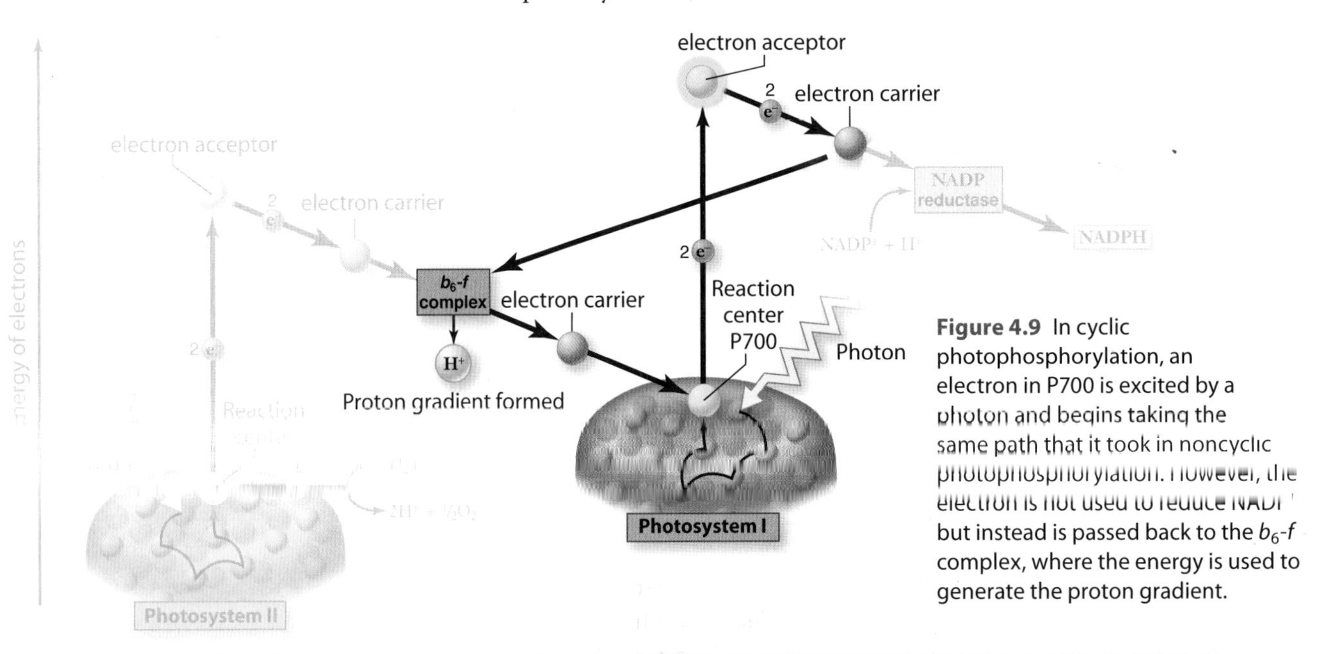

Figure 4.9 In cyclic photophosphorylation, an electron in P700 is excited by a photon and begins taking the same path that it took in noncyclic photophosphorylation. However, the electron is not used to reduce $NADP^+$ but instead is passed back to the b_6-f complex, where the energy is used to generate the proton gradient.

Activity 4.2 Modelling the Source of Oxygen in the Light-Dependent Reactions

In the 1930s, a graduate student at Stanford University, from C.B. van Niel, demonstrated that the source of the oxygen photosynthesis was water molecules, not carbon dioxide as was hypothesized. When radioactive isotopes came into widespread use in the early 1950s, it became possible to test van Niel's findings.

Materials

- construction paper shapes or molecular modelling kits

Procedure

1. Examine the following general equation:

$$CO_2(g) + 2H_2O(\ell) \rightarrow CH_2O(s) + H_2O(\ell) + O_2(g)$$

2. Radioactive water tagged with an isotope of oxygen as a tracer (shown in red) was used. Notice where the tagged oxygen ends up on the right side of the equation.
3. Assume that the investigation was repeated, but this time a radioactive tag was put on the oxygen in $CO_2(g)$.
4. Model what you predict the appearance of the results would be. Your model must include a "tag" to indicate the oxygen isotope on the left side of the arrow as well as where it ends up on the right side of the arrow.
5. Use labels or different colours to indicate what happens to the carbon and hydrogen in this reaction.

Questions

1. Using your model, predict what happens to:
   a. all oxygen molecules from carbon dioxide
   b. all carbon molecules from carbon dioxide
   c. all hydrogen molecules from water

Learning Check

7. What is the function of the splitting of water in the light-dependent reactions?
8. What two events are linked in chemiosmosis?
9. Compare and contrast the thylakoid membrane in plant cells to the cristae found in mitochondria.
10. Sketch a Z diagram and indicate where energy is used to generate high-energy products. Explain what these products are and the mechanism by which they are produced.
11. Distinguish between cyclic and noncyclic photophosphorylation.
12. In both the mitochondria and chloroplast, energy for the phosphorylation of ADP comes from the hydrogen ion gradient. Why is a gradient formed with ions more effective than a gradient of neutral molecules?

The Dream of Mimicking Nature

With ever-growing concerns over the use of fossil fuels for energy and global climate change, engineers and scientists continue to investigate alternative technologies for energy production. The design and construction of technology that traps solar energy and converts it into electrical energy has been successful in devices such as the solar collector shown in **Figure 4.10**. The technology is practical for the space station, with its large solar receptors, and smaller devices such as a handheld calculator. However, the amount of electrical energy that can be produced by solar collector cells at Earth's surface is not sufficient to supply enough of the energy that society needs. As a result, a large percentage of the energy that is used every day comes from the burning of fossil fuels, which adds large amounts of carbon dioxide to the atmosphere. Since carbon dioxide is a greenhouse gas, it contributes to global warming.

Figure 4.10 Solar collectors absorb light energy from the Sun that is usually converted to heat or electrical energy.

In the search for an alternative source of energy, many scientists have been looking closely at hydrogen. Hydrogen is a totally clean fuel. When hydrogen burns in oxygen, the only product is pure, clean water. So why not use hydrogen as a fuel? There is very little hydrogen found free, as a gas, in nature. The optimal source for hydrogen is water. However, to split water into hydrogen and oxygen gases requires more energy than can be produced when the hydrogen is burned. As well, obtaining hydrogen from other compounds usually involves the release of carbon dioxide.

Scientists and engineers see much potential with the idea of developing an artificial system that mimics the splitting of water by photosystem II. This technology would involve using solar energy to split water and generate oxygen and hydrogen gas, which could then be used in devices such as hydrogen fuel cells. Hydrogen fuel cells use hydrogen and oxygen to produce electrical power, heat, and water. Intense focus has been placed on finding ways to produce the hydrogen and oxygen from water. While plants have accomplished the use of solar energy to carry out this reaction, doing the same in a laboratory or engineered device has proven to be a challenge. Some researchers have built artificial reaction centres. Others have devised artificial leaves. To date, practical and economical results remain elusive. It is likely, however, that a practical artificial photosynthetic system will be achieved and marketed in your lifetime.

CANADIAN RESEARCH IN ACTION

Fuel Production through Artificial Photosynthesis

Dr. Ibrahim Dincer supervised students in the Energy Engineering Option of UOIT's Mechanical Engineering Program as they built the first engineering reactor for a solar light-based hydrogen production system, shown in the inset photograph.

Despite international efforts to reduce human consumption of Earth's resources, the global demand for energy and fuel continues to rise. Today, the most commonly used fuels are fossil fuels. This may change in the near future, however, as scientists develop economically feasible, minimally polluting alternatives.

Hydrogen is especially promising as an alternative fuel. Because hydrogen combustion produces only pure water, it generates no pollution. However, the production of hydrogen has a surprising environmental impact. Currently, most hydrogen is generated through reactions in which fossil fuels and water produce hydrogen and carbon dioxide. These reactions release large amounts of carbon dioxide into the atmosphere, contributing to global warming. Fortunately, solar energy, one of the most sustainable and widely available forms of energy, presents an alternative means of generating hydrogen. Through *artificial photosynthesis*, solar energy can be used to split water into hydrogen and oxygen in the laboratory. While this process is completely non-polluting, it has been impractical due to its high cost. Turkish-born Canadian scientist Dr. Ibrahim Dincer hopes that his research into artificial photosynthesis may soon make this technology more practical.

Dr. Dincer is a professor at the University of Ontario Institute of Technology (UOIT) in Oshawa, Ontario. He is currently working on technology that promises to cut the cost of artificial photosynthesis significantly. The cost is reduced by using *supramolecular devices*—complex systems made up of molecular components—to catalyze redox reactions that split water to release hydrogen with the input of solar energy. During the redox reactions, the supramolecular devices are not consumed and therefore have potentially high productivity rates and long lifetimes. Furthermore, they can be recycled.

In artificial photosynthesis, light energy interacts with matter to excite electrons, just as it does in natural photosynthesis. The process involves photo-sensitizers that receive light energy and transfer it to electrons, raising the electrons to a higher energy level. The photo-sensitizers mimic the photosystems involved in the light-dependent reactions in plants. Similarly, artificial photosynthesis involves transferring electrons to electron collection sites, as well as catalysts that facilitate the water-splitting reactions that produce hydrogen and oxygen.

Related Career

Environmental engineers are scientists who apply scientific and technical knowledge to practical problems: the study, maintenance, and restoration of the environment. Environmental engineers may work in a variety of fields. These include waste and water management, remediation, biotechnology, and sustainable development. Completion of an undergraduate degree in environmental engineering is required to become an environmental engineer in Canada. Many universities also offer a co-op program that provides job training as a degree component.

QUESTIONS

1. Compare and contrast how hydrogen is generated during natural and artificial photosynthesis.
2. Suggest at least two different ways in which artificial photosynthesis could have a positive impact on the environment. Include your reasoning in your response.
3. Use Internet and/or print resources to find another career related to the research described in this feature. Briefly describe this career.

Section 4.1 Review

Section Summary

- Photosynthesis includes two distinct sets of reactions: the light-dependent reactions and the light-independent reactions.
- Enzymes and electron carriers responsible for the light-dependent reactions are embedded in the thylakoid membranes of the chloroplasts.
- Pigment molecules absorb photons from the radiant energy of sunlight and use it to energize electrons. The energy of the electrons in noncyclic photophosphorylation is used partially to pump hydrogen ions into the thylakoid space and partially to reduce $NADP^+$. In cyclic photophosphorylation, the energy of the electrons is used only to generate a hydrogen ion gradient across the thylakoid membranes.
- The energy stored in the hydrogen ion gradient is used to phosphorylate ADP.

Review Questions

1. **K/U** Distinguish between light-dependent reactions and light-independent reactions in photosynthesis.
2. **C** Sketch a chloroplast, and label the following parts: outer membrane, inner membrane, stroma, granum, thylakoid membrane, thylakoid space.
3. **C** Describe the type of information that you can obtain from an absorbance spectrum. What, specifically, does the absorbance spectrum of the chlorophylls tell you?
4. **K/U** What is the definition of a pigment? What is the role of pigments in photosynthesis?
5. **C** Sketch the following diagram in your notebook, and use it to answer the following questions.
   a. Label all of the objects in the diagram, including the upward directed arrow on the left.
   b. Circle the two places in the diagram that represent the points where the energy being released is used to generate high energy compounds. For each of the two places, state what high energy compound is generated and how it is formed.
   c. Indicate the point in the diagram at which the oxygen that is released by plants is generated. What is the source of the oxygen?
   d. Indicate the points where "holes" are formed as well as the source of the electrons that fill the holes.

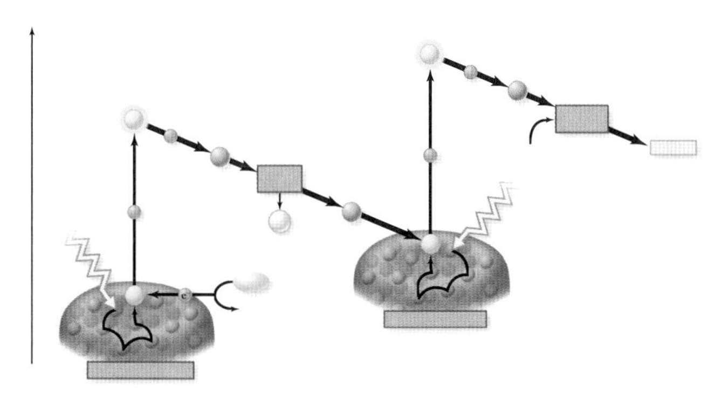

6. **T/I** Could a green plant survive solely on photosystem II if all its photosystem I complexes were damaged by a herbicide? Explain your reasoning.
7. **C** Use a Venn diagram to compare and contrast chloroplasts and mitochondria.
8. **T/I** Predict what would happen to the ability of a plant to trap energy if the only pigment it contained was chlorophyll *a*.
9. **T/I** While walking in part of the rain forest, a student discovers a white-leaved plant and is intrigued. The student concludes that this plant must be maximizing the use of the Sun's energy and probably has quite a large root system where starch is stored. Which conclusions are incorrect? Explain why they are incorrect.
10. **A** Does light cause the oxidation or reduction of chlorophyll *a*? Explain your reasoning.
11. **K/U** Refer to the diagram in question 5. Sketch and label the parts of that diagram that are involved in cyclic photophosphorylation.
12. **K/U** Since cyclic photophosphorylation uses components that are also used in noncyclic photophosphorylation, why does a plant require cyclic photophosphorylation?
13. **A** Develop an analogy that you could use to explain chemiosmosis to a classmate who is having a difficulty understanding it.
14. **K/U** Is the splitting of water as an electron source for photosystem I required so that photosystem I can fulfill all of its tasks? Explain your reasoning.
15. **K/U** Differentiate between phosphorylation and photophosphorylation.

SECTION 4.2

Fixing Carbon Dioxide: The Light-Independent Reactions

Key Terms

Calvin cycle
carbon dioxide fixation
C3 photosynthesis
photorespiration

Chloroplasts in plants contain enzymes in the stroma that, in conjunction with the energy supplied by ATP and NADPH, convert carbon dioxide to carbohydrates. You might see several different names for these reactions. In the past, they were called the "dark reactions" because they do not require light. However, since the reactions can take place in the presence or absence of light, the more accurate terminology is light-independent reactions.

The key initial step in the synthesis of carbohydrates in plants is conversion of carbon dioxide to organic compounds—a process called *CO_2 assimilation*. The assimilation of carbon dioxide is carried out by a cyclical pathway that continually regenerates its intermediates. This pathway is called the Calvin cycle in honour of Melvin Calvin, who identified the pathway in the 1950s. The **Calvin cycle**, shown in **Figure 4.11**, accomplishes the conversion of inorganic carbon, in the form of carbon dioxide from the atmosphere, into organic carbon, in the form of the three-carbon organic molecule glyceraldehyde-3-phosphate (G3P). This product of photosynthesis is then used as a starting substrate in many other metabolic pathways. The reactions of the Calvin cycle can be grouped into three phases, as described on the opposite page.

Calvin cycle in photosynthesis, the reactions that convert carbon dioxide to the three-carbon organic molecule glyceraldehyde-3-phosphate (G3P); can occur in the absence or presence of light; also called the dark reactions and the Calvin-Benson cycle

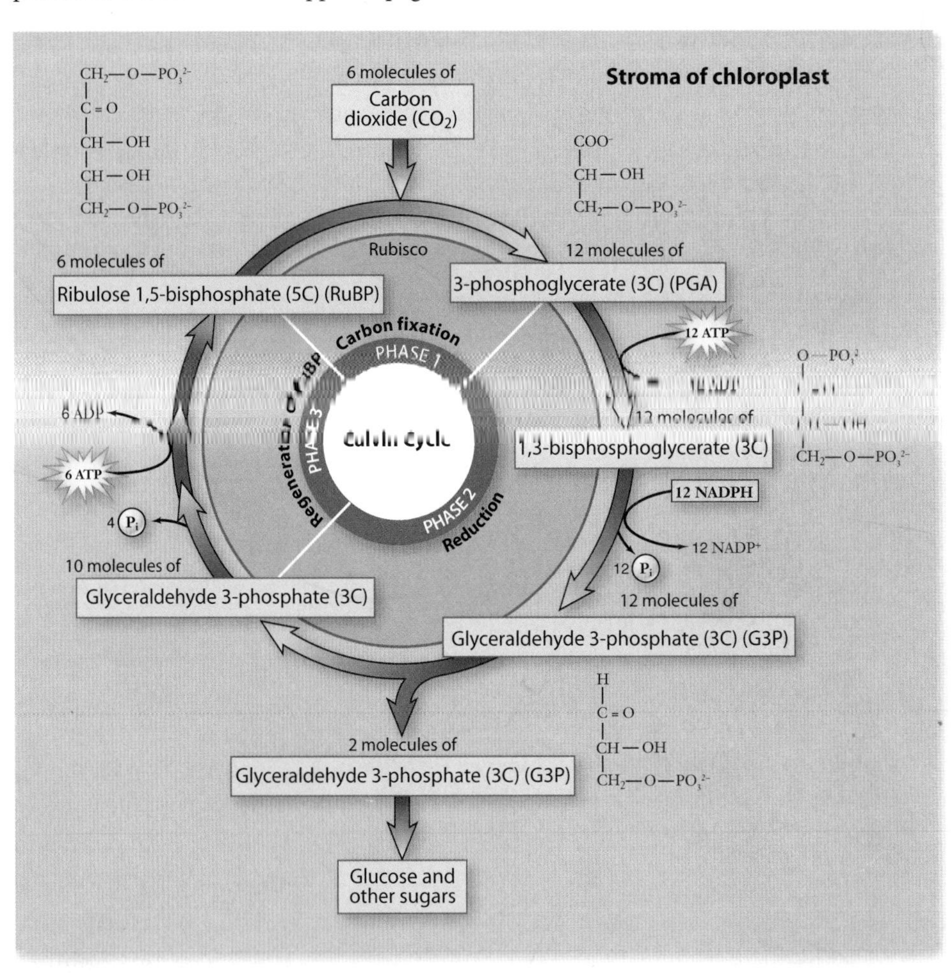

Figure 4.11 The ATP and NADP used in the Calvin cycle are produced in the light-dependent reactions. For every 12 molecules of G3P made in the Calvin cycle, two are used to make glucose and other high-energy compounds.

***Identify** a pathway in cellular respiration that also has G3P as an intermediate.*

1. *Fixing Carbon Dioxide:* The first phase is **carbon dioxide fixation.** The key to this is the chemical bonding of the carbon atom in carbon dioxide to a pre-existing molecule in the stroma. This molecule is a five-carbon compound called ribulose-1,5-bisphosphate, or RuBP for short. The resulting six-carbon compound is unstable and immediately breaks down into two identical three-carbon compounds called 3-phosphoglycerate (PGA). Because these three-carbon compounds are the first stable products of the process, plants that use this method for photosynthesis are called C3 plants, and the process is called **C3 photosynthesis**. The reaction that leads to these three-carbon compounds can be summarized as:

$$CO_2 + \text{RuBP} \rightarrow \text{unstable } C_6 \rightarrow 2 \text{ PGA}$$

This reaction is catalyzed by the enzyme ribulose bisphosphate carboxylase, which is often called *rubisco*. Rubisco is possibly the most abundant protein on Earth.

2. *Reduction:* In the second phase, the newly formed three-carbon compounds are in a low-energy state. To convert them into a higher-energy state, they are first activated by ATP and then reduced by NADPH. The result of these reactions is two molecules of glyceraldehyde-3-phosphate (G3P). In their reduced (higher-energy) state, some of the G3P molecules leave the cycle and may be used to make glucose and other carbohydrates. The remaining G3P molecules move on to the third phase of the cycle, in which RuBP is replenished to keep the cycle going.
3. *Regenerating RuBP:* Most of the reduced G3P molecules are used to make more RuBP. Energy, supplied by ATP, is required to break and reform the chemical bonds to make the five-carbon RuBP from G3P. The Calvin cycle must be completed six times in order to synthesize one molecule of glucose. Of the 12 G3P molecules that are produced in six cycles, 10 are used to regenerate RuBP, and 2 are used to make one glucose molecule.

The net equation for the Calvin cycle is:

$$6CO_2 + 18 \text{ ATP} + 12 \text{ NADPH} + \text{water} \rightarrow 2 \text{ G3P} + 16 \text{ P}_i + 18 \text{ ADP} + 12 \text{ NADP}^+$$

The G3P that is produced can then be used for the synthesis of other molecules that plants require. A great deal of G3P is transported out of the chloroplasts and into the cytoplasm. There, G3P is used to produce a key sugar in plants, sucrose. In times of intensive photosynthesis, when G3P levels can rise quite high, the G3P is used to produce starch. G3P is also the starting substrate for cellulose. Plant oils such as corn oil, safflower oil, and olive oil are derived from G3P. As well, G3P and a source of nitrogen are used to synthesize the amino acids that are used to make proteins. Thus, G3P is a crucial molecule in plant metabolism.

carbon dioxide fixation in the Calvin cycle of photosynthesis, the reaction of carbon dioxide with ribulose bisphosphate to produce two identical three-carbon molecules, 3-phosphoglycerate

C3 photosynthesis a process of converting carbon dioxide to glyceraldehyde-3-phosphate using only the Calvin cycle; involves production of a three-carbon intermediate (PGA)

Suggested **Investigation**

Inquiry Investigation 4-B, The Rate of Photosynthesis

Learning Check

13. What is the function of the Calvin cycle?
14. What product of the Calvin cycle is used in a variety of additional reactions?
15. What role does ribulose-1,5-bisphosphate play in the Calvin cycle?
16. How many G3Ps must be produced in order to make one glucose molecule? What is the function of the other G3P molecules?
17. What processes occur during the reduction phase of the Calvin cycle?
18. List at least three different compounds other than glucose that are made from G3Ps.

Adaptations to Photosynthesis

photorespiration the reaction of oxygen with ribulose-1,5-bisphosphate in a process that reverses carbon fixation and reduces the efficiency of photosynthesis

Rubisco is a critical enzyme in the light-independent reactions. However, it has an undesirable property. Rubisco can use oxygen as a substrate as well as carbon dioxide. In fact, oxygen and carbon dioxide compete with each other for the same active site on the rubisco enzyme. When oxygen reacts with ribulose-1,5-bisphosphate in the process called **photorespiration**, the products are a two-carbon compound called phosphoglycolate and one of the three-carbon compound, 3-phosphoglycerate. As a result of photorespiration, all of the energy used to regenerate the ribulose-1,5-bisphosphate is wasted, thus reducing the efficiency of photosynthesis. Under normal conditions, when the temperature is near 25°C, C3 plants lose 20 percent of the energy used to fix one carbon dioxide molecule. Biologists estimate that the maximum possible efficiency of photosynthesis in C3 plants—assuming each photosystem absorbs the maximum amount of light—is 30 percent. Some laboratory-grown plants, raised under controlled conditions, have reached efficiencies of 25 percent. In nature, however, photosynthetic efficiency ranges from 0.1 percent to 3 percent.

Atmospheric conditions influences the reduction of efficiency due to photorespiration. For example, under hot, dry conditions, leaves begin to lose water through the stomata. In response to these conditions, the stomata close to prevent further loss of water, as shown in **Figure 4.12**. With the stomata closed, the oxygen formed in the light-dependent reactions accumulates inside the leaves and carbon dioxide cannot enter. With the higher ratio of oxygen to carbon dioxide in the leaves, the amount of photorespiration increases significantly.

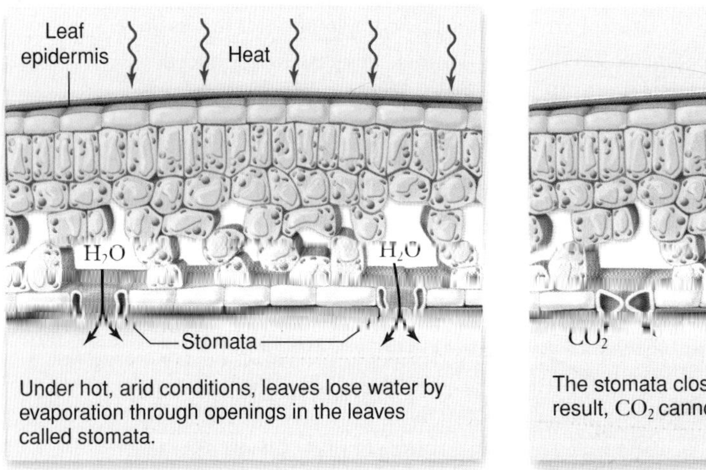

CO_2
CO_2
The stomata close to conserve water but, as a result, CO_2 cannot enter the leaves.

Figure 4.12 Leaves prevent water loss in hot, dry conditions by closing their stomata. Water is conserved but carbon dioxide is prevented from entering.

Some plants that are native to regions in which the climate is normally hot and dry, typically above 28°C, have evolved mechanisms to reduce the amount of photorespiration. These plants fit into two categories: C4 plants and CAM plants. Both use the Calvin cycle; however, they have developed different mechanisms for the uptake and storage of carbon dioxide that increases the ratio of carbon dioxide to oxygen for the Calvin cycle reactions.

C4 Plants

C4 plants have a structure, shown in **Figure 4.13**, that separates the initial uptake of carbon dioxide from the Calvin cycle into different types of cells. In the outer layer of mesophyll cells, carbon dioxide is fixed by addition to a three-carbon compound called *phosphoenolpyruvate* (PEP). The product is the four-carbon compound *oxaloacetate*, giving these plants the name C4. The oxaloacetate is converted to the four-carbon compound *malate* and transported into the bundle-sheath cells. There, the malate is decarboxylated.

The resulting three-carbon compound, *pyruvate*, is transported back into the mesophyll cells and converted into PEP. The bundle-sheath cells are impermeable to carbon dioxide. As a result, carbon dioxide is concentrated in the bundle-sheath cells where the Calvin cycle takes place. This high CO_2 concentration makes the Calvin cycle much more efficient than in C3 plants.

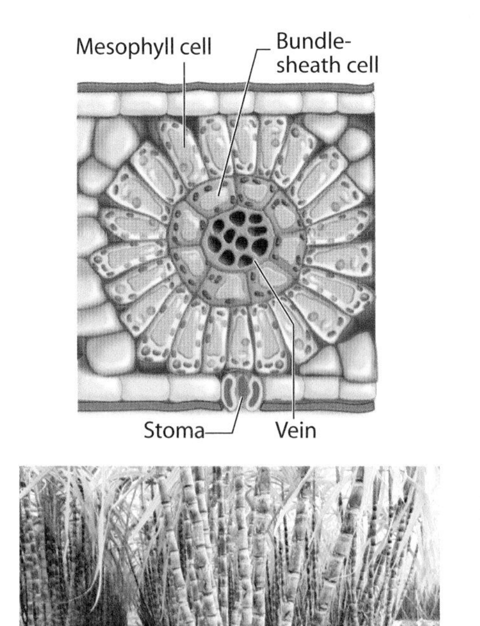

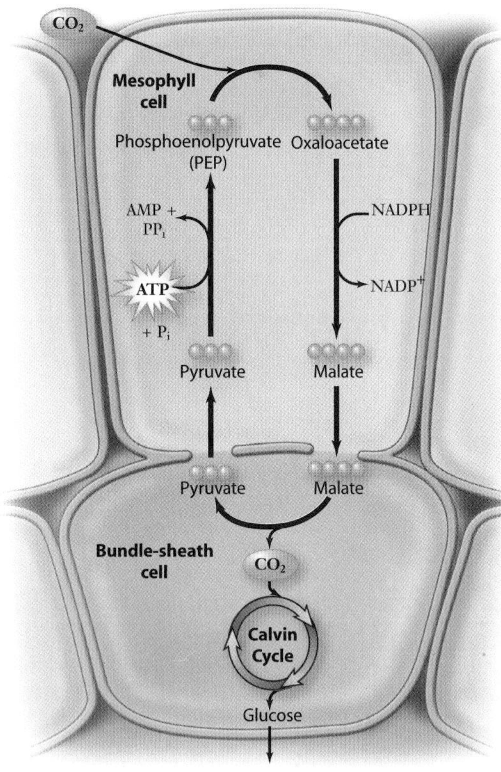

Figure 4.13 C4 plants use energy to "pump" carbon dioxide into the bundle-sheath cells, where it becomes concentrated. Included among the C4 plants are food crops such as corn, sorghum, sugarcane (shown here), and millet. Also included are grasses such as crabgrass and Bermuda grass.

CAM Plants

CAM plants, which include succulent (water-storing) plants such as cacti and pineapples, use a biochemical pathway identical to the C4 plants, but the reactions take place in the same cell. Carbon dioxide fixation is separated from the Calvin cycle by time of day, as shown in **Figure 4.14**, rather than by different cell types. *Crassulaceae* thrive in hot, arid desert conditions. To prevent water loss, their stomata remain closed during the day and open at night. Carbon dioxide is fixed at night while the stomata are open. The reactions proceed until malate is formed. It is then stored in a large vacuole until daytime, when the stomata close. When the light-dependent reactions have produced enough ATP and NADPH to support the Calvin cycle, the malate exists the vacuole and is decarboxylated, freeing the carbon dioxide which is then fixed again by rubisco and enters the Calvin cycle.

Figure 4.14 The opening and closing of stomata in CAM plants, such as pineapple and cacti, are opposite from most plants. The stomata are open at night and closed in the daytime. When the carbon dioxide is removed from the four-carbon compound malate in the daytime, it cannot leave the cell because the stomata are closed. In cool climates, CAM plants are very inefficient, because they use energy to drive the reactions that store carbon dioxide.

The Energy Cycle of Aerobic Respiration and Photosynthesis

The reactions that capture light energy and convert it to organic material are closely related to the reactions of aerobic respiration reactions. As shown in **Figure 4.15**, both of these processes occur in plants and represent a plant cell's energy cycle. The products of aerobic respiration, carbon dioxide and water, are the starting substrates for photosynthesis. The products of photosynthesis, oxygen and glucose, are the starting substrates for aerobic respiration. For plant cells, an outside source of carbon dioxide and water is needed to produce glucose, leaving oxygen as a by-product. For animal cells, which lack chloroplasts, an outside source of glucose and oxygen are needed to generate energy, leaving carbon dioxide and water as by-products. Key features of the metabolic pathways for aerobic cellular respiration and photosynthesis are summarized in **Table 4.1**.

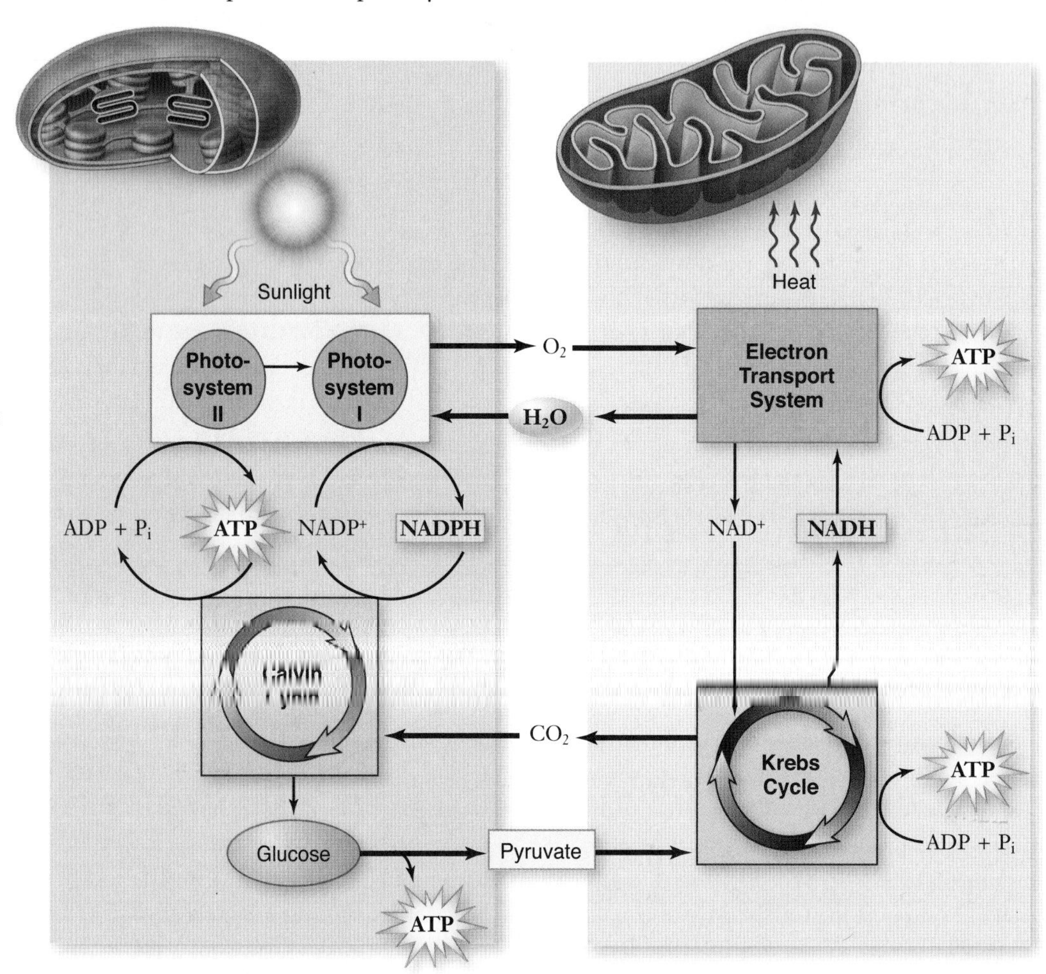

Figure 4.15 Water, carbon dioxide, glucose, and oxygen cycle between chloroplasts and mitochondria in plant cells.

Table 4.1 Aerobic Respiration and Photosynthesis Pathways

	Respiration	Photosynthesis
Overall equation	glucose + $6O_2 \rightarrow$ energy* + $6CO_2$ + $6H_2O$	energy* + $6CO_2$ + $6H_2O \rightarrow$ glucose + $6O_2$
Cell location	Mitochondria	Chloroplasts
Starting substrates	Glucose and oxygen	Carbon dioxide and water
Products	Carbon dioxide and water	Glucose and oxygen
Electron transport chain	Yes	Yes
Electron carriers	NADH, $FADH_2$	NADPH
ATP synthesis by chemiosmosis	Yes	Yes

energy*: for respiration this represents ATP, and for photosynthesis this represents light

Section 4.2 Review

Section Summary

- The light-independent reactions use the energy from ATP and NADPH from the light-dependent reactions to assimilate carbon dioxide and synthesize high-energy organic compounds in a series of reactions called the Calvin cycle.
- Plants that use only the Calvin cycle to assimilate carbon dioxide are called C3 plants.
- Due to photorespiration, C3 plants are extremely inefficient in hot, dry climates.
- Plants that thrive in these environments have evolved two mechanisms to increase their efficiency. C4 plants isolate carbon dioxide fixation from the Calvin cycle reactions by carrying out these reactions in two different cell types. CAM plants isolate dioxide fixation from the Calvin cycle reactions by fixing carbon dioxide at night and carrying out the Calvin cycle reactions in the daytime.

Review Questions

1. K/U In which part of the cell are the enzymes for the Calvin cycle located?
2. K/U Write a reaction equation for the first step in which carbon dioxide is assimilated into an organic compound. Include reactants and products, and label the reaction equation with the name for this phase of the Calvin cycle.
3. K/U Describe the overall function of the reduction phase of the Calvin cycle.
4. T/I From what you learned about glycolysis in Chapter 3, infer what the first two steps are in the conversion of G3P to glucose.
5. C Using a diagram, explain why all of the G3P synthesized in the Calvin cycle cannot be used to make glucose.
6. K/U Name the two types of high-energy molecules that are used to drive the Calvin cycle, and state the number of each needed to synthesize one molecule of glucose.
7. K/U Summarize the uses of the G3P that is synthesized by the Calvin cycle reactions.
8. K/U Define photorespiration, and explain how it affects the Calvin cycle.
9. C Use the diagrams below to explain what happens in C3 plants when the temperature rises above 28°C.

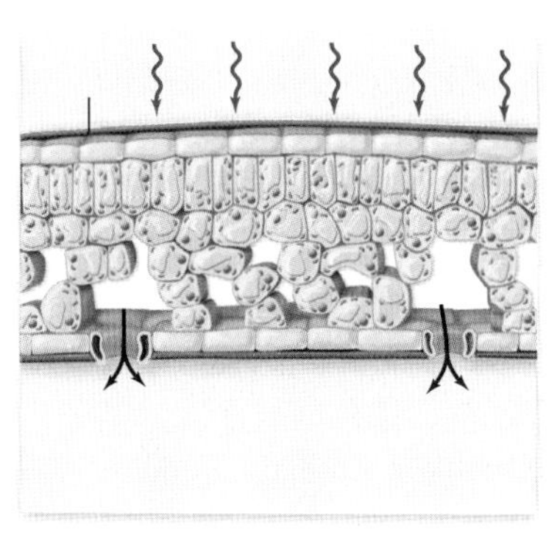

10. K/U Describe the difference in the leaf structure of a C4 plant compared to a C3 plant. Explain how the leaf structure of the C4 plant helps to reduce the amount of photorespiration that occurs in the leaves, even when the air is hot and dry.
11. K/U In the leaves of CAM plants, carbon dioxide fixation and the Calvin cycle occur inside the same cell. How, then, do CAM plants minimize the photorespiration?
12. A Would CAM plants have an advantage over native species in a Boreal forest? Explain why or why not.
13. A In 1771, Joseph Priestley did an experiment with a mint plant and a candle. He put them both in a bell jar and sealed it. He lit the candle by focussing sunlight on the wick with a mirror. The candle soon burned out. He left the jar sealed for 27 days. He lit the candle again and it burned very well. Based on what you know about photosynthesis, explain these observations.
14. T/I Agree or disagree with the following statement, and give reasons for your answer: "Because the light-independent reactions require ATP and NADPH for energy, they are not truly light-independent."

Inquiry
INVESTIGATION

4-A

Skill Check

Initiating and Planning

✓ Performing and Recording

✓ Analyzing and Interpreting

✓ Communicating

Safety Precautions

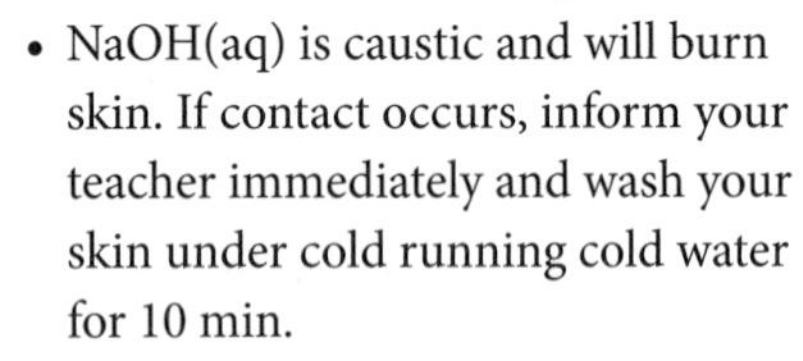

- NaOH(aq) is caustic and will burn skin. If contact occurs, inform your teacher immediately and wash your skin under cold running cold water for 10 min.
- When you blow into any solution, *do not inhale* at any time.

Materials

- 600 mL beaker
- baking soda, $NaHCO_3(s)$
- *Cabomba* (or other aquatic plant)
- bright natural light (or grow light)
- test tube and stopper
- short-stemmed funnel
- wooden splint
- matches
- 0.1 mol/L NaOH(aq) in dropping bottle
- 50 mL Erlenmeyer flask
- bromothymol blue
- straw

Gases Released during Photosynthesis and Cellular Respiration

Photosynthesis and cellular respiration are chemical reactions that produce by-products in the form of gases. How can you detect and identify the gas that plants release? How can you determine what gases you are exhaling?

Pre-Lab Questions

1. Describe the procedure for the proper handling of sodium hydroxide.
2. What are the characteristics of a pH indicator? In what circumstances would it be helpful to use a pH indicator?
3. Read the procedure for Part 1 of the investigation. Explain why it is critical that the test tube that is placed over the stem of the funnel be filled with water so that there is no air in it at all.
4. Why should the plant be placed in a bright light?
5. Why is it necessary to blow out the flame after you light the wooden splint?
6. In Part 2, what is happening when you are blowing into the flask?

Question

How can you identify the gases released by plants and animals?

Procedure

Part 1: Gas Released by Plants

Before starting the procedure, read MSDS sheets for the chemicals used in this investigation. The first 5 steps must be completed the day before the remainder of the investigation is completed.

1. Fill a 600 mL beaker with aquarium water or tap water. Add 2 g of sodium hydrogen carbonate to the water.
2. Hold the small branches of *Cabomba* (or other aquatic plant) under water and clip off the ends of the stems so that air bubbles do not block the vessels.
3. Place a short-stemmed funnel over the branches.
4. Fill a test tube with water and cover the top while you invert it. Hold it under water and position it over the stem of the funnel as shown in the photograph. Be sure there is no air in the test tube.

5. Place the apparatus in bright light and leave it until the next class period.
6. After about 24 h, carefully remove the test tube: keeping it inverted, put your thumb over the mouth of the test tube and stopper it. Follow the directions below to test for the type of gas in the tube. Record your results.

Testing for Gases

Note: For safety purposes, your teacher may conduct some or all of these steps.

1. Wear goggles.
2. Keep the stoppered test tube containing a collected gas inverted and clamp it to a ring stand.
3. Ensure that there are no flammable materials, other than those with which you are working, in the room.
4. Light the wooden splint. Let it burn briefly then blow out the flame. The splint should still be glowing.
5. Remove the stopper from the inverted test tube. Gradually insert the glowing splint up into the inverted test tube. Observe the reaction.
   - If the gas is hydrogen, you will hear a loud pop when the splint reaches the gas.
   - If the gas is carbon dioxide, the splint will go out.
   - If the gas is oxygen, the splint will burst into a small bright flame.

Part 2: Gas Exhaled by Animals

1. Add about 35 mL of water to a 50 mL Erlenmeyer flask. Add a few drops of bromothymol blue to the water and swirl. (Place the flask on a piece of white paper to see the colour more clearly.)
2. Add sodium hydroxide one drop at a time and gently swirl the flask. Stop adding the sodium hydroxide when the water turns a blue colour.
3. Your teacher will give you a fresh, clean straw. Use it to blow gently into the flask. *Do not suck on the straw.* Continue to blow into the solution until you can no longer see any colour change.
4. Add a piece of *Cabomba* to the water and stopper the flask. Place it in a brightly lit place. Leave it in place until you see a change. Record any changes over the next 24 h.
5. Dispose of all materials as instructed by your teacher.

Analyze and Interpret

1. What happened when you inserted the glowing splint into the test tube that had been collecting gas from the Cabomba plant? What is the identity of the gas?
2. Bromothymol blue is called a pH indicator because it changes colour with changes in pH (acidity) of a solution. It is a dark greenish-blue colour in a basic solution and a pale yellow colour in an acidic solution. What happened when you blew into the flask?
3. Read the following points, and then identify the gas in your exhaled breath.
   - Oxygen gas has very low solubility in water and does not react chemically with water.
   - Hydrogen gas is nearly insoluble in water and does not react with water.
   - Carbon dioxide is relatively soluble in water and reacts with water to produce carbonic acid.

Conclude and Communicate

4. What was the gas that you collected in the test tube over the *Cabomba* plant? Explain the source of the gas.
5. What was the gas that you exhaled? Explain the source of the gas.
6. What changes did you observe in the appearance of the solution after Step 4 of Part 2? Explain any changes.

Extend Further

7. **INQUIRY** What did you conclude about the identity of the gas in your breath when you blew out? Suggest another chemical test that you could use to confirm your answer. Design a procedure by which you could perform this test. If time permits, with your teacher's permission, carry out your procedure.
8. **RESEARCH** Do research in print and electronic resources to determine who was the first scientist that discovered the nature of the gas released by plants. Describe his investigations and how he arrived at his conclusions.

Inquiry INVESTIGATION

4-B

Skill Check

Initiating and Planning

✓ Performing and Recording

✓ Analyzing and Interpreting

✓ Communicating

Safety Precautions

Materials

- plant leaf
- single-hole punch
- 10 mL plastic syringe (without the needle)
- 150 W bulb
- liquid dish soap
- 0.25 percent sodium bicarbonate
- medicine dropper
- 200 mL beaker
- lamp with a reflector
- timer

The Rate of Photosynthesis

Oxygen is one of the products of the light-dependent reactions of photosynthesis. The volume of oxygen that is produced by the plant can therefore be used to measure the rate of photosynthesis taking place. However, it is often difficult to accurately measure the volume of gas produced. In this activity, you will use small disks cut from the leaf of a plant to perform a basic floating leaf disk assay. (An assay is a procedure that is used to analyze or determine the amount of substance in a sample.)

Once you have mastered this technique, you will use the floating leaf disk assay to design your own investigation of variables that affect the rate of photosynthesis.

Question

What variables affect the rate of photosynthesis?

Pre-Lab Questions

1. Review the structure of leaves and the chambers through which gases move as shown in the diagram. Why is there air surrounding the mesophyll (spongy) cells inside of the leaf?
2. Read the procedure. What is the purpose of the bicarbonate that is dissolved in the water?
3. Why is there a relationship between the rate of oxygen production and the number of leaf disks that float on top of the water?
4. What is the function of the dish soap in the solution?

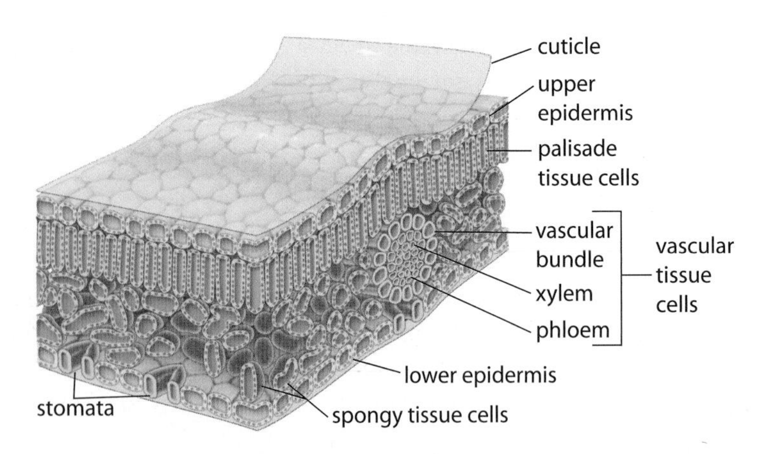

A Palisade tissue cells: long, narrow cells packed with chloroplasts. These cells lie under the upper surface of the leaf and are the sites where most photosynthesis occurs in the leaf.

B Vascular tissue cells: cells that form bundled arrangements of tubes that transport fluids throughout the plant. Xylem tubes carry water and minerals from the roots to the leaves. Phloem tubes carry sugars to various parts of the plant.

C Spongy tissue cells: round and more loosely packed than palisade cells, with many air spaces between them. These cells have chloroplasts, so they perform some photosynthesis. Their structure helps the cells to exchange gases and water with the environment.

D Stomata: small openings in the outer (epidermal) layer that allow carbon dioxide into the leaf and oxygen out of the leaf. Water also diffuses out of the leaf through stomata.

Part 1: Floating Leaf Disk Assay

Procedure

1. Your teacher will give you 100 mL of the 0.25 percent sodium bicarbonate solution. Place it in the beaker.
2. Use the medicine dropper to add 5 drops of liquid dish soap to the bicarbonate solution.
3. Use the single-hole punch to cut 10 uniform leaf disks. Avoid cutting through major leaf veins. Remove the plunger of a plastic syringe and place the leaf disks in the barrel of the syringe. Tap the syringe gently until the leaf disks are near the bottom of the barrel.
4. Replace the plunger in the syringe. Push the plunger down until only a small volume of air remains in the barrel. Be careful not to crush any of the leaf disks.
5. You are going to infiltrate the leaf disks with sodium bicarbonate solution by removing most of the air from the leaf tissue and replacing it with the sodium bicarbonate solution. To do so:
   - Use the plunger to draw 5 mL of solution into the barrel of the syringe.
   - Tap on the syringe to suspend the leaf disks.
   - Hold a finger over the open end of the syringe and draw back on the plunger to create a vacuum.
   - Hold this vacuum for 10 to 15 seconds and then remove your finger from the open end of the syringe. The solution will gradually infiltrate the air spaces inside the leaf disks.
   - Hold the open end of the syringe over the beaker of solution and slowly push the plunger back down, again taking care not to crush the leaf disks.
   - Repeat the infiltration procedure at least 5 times; otherwise your leaf disks may not sink to the bottom of the solution in the beaker.

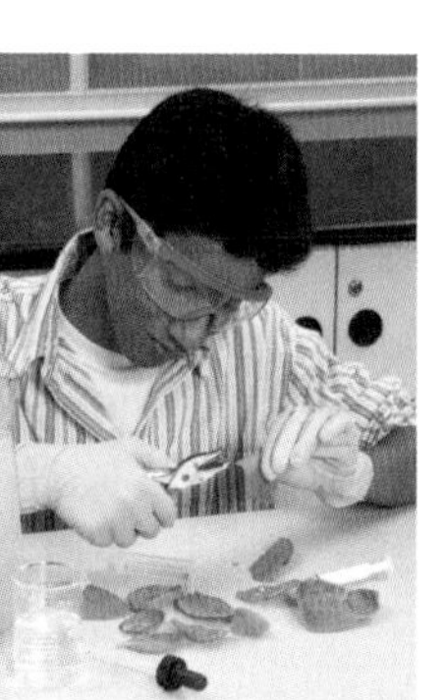

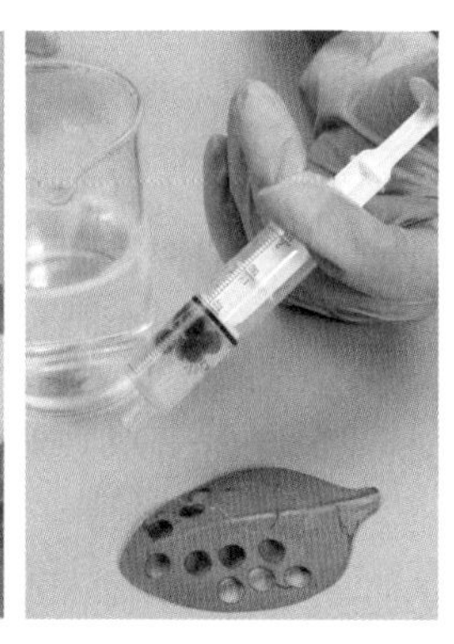

6. Pour the disks and solution from the syringe back into the beaker of sodium bicarbonate and dish soap.
7. If your leaf disks are still floating, carefully add more dish soap—1 drop at a time. You may have to remove the leaf disks and repeat the infiltration procedure if you can't get the disks to sink.
8. Once all of the leaf disks are resting on the bottom, direct white light onto the beaker. Start the timer.
9. At the end of each minute, record the number of disks that have floated to the surface of the solution. Swirl the beaker gently if some disks get stuck to the side, but keep the beaker in the light.
10. Record your results in a suitable data table.

Analyze and Interpret

1. Construct a graph of your data.
2. Using the graph, estimate the time at which 50 percent of the leaf disks were floating on the surface. The point at which 50 percent of the leaf disks are floating will be your point of reference for future investigations.

Conclude and Communicate

3. What variable were you testing in this investigation?
4. Explain why the leaf disks started to float after being exposed to white light.

Part 2: Design Your Own Investigation

A number of variables affect the rate of photosynthesis in a plant leaf. Your challenge is to design an investigation to test the effects of one variable on the rate of photosynthesis. Be sure to:

- state your own question
- make a hypothesis based on this question
- identify the materials that you will require
- write out the experimental procedure you will use
- conduct your investigation
- collect and graph your data
- determine whether your results support or refute your hypothesis
- communicate the results of your investigation in the form of a formal lab report

Extend Further

6. **INQUIRY** In this investigation, you measured the rate of oxygen production. How might you measure the rate of loss of carbon dioxide? Design an investigation for determining the rate of loss of carbon dioxide.
7. **RESEARCH** Brainstorm possible environmental factors that could affect the rate of photosynthesis. Do research on one of these factors and write a short paper on your findings.

Inquiry INVESTIGATION

4-C

Skill Check

- Initiating and Planning
- ✓ Performing and Recording
- ✓ Analyzing and Interpreting
- ✓ Communicating

Safety Precautions

- The solvent is volatile. Ensure there are no flames in the classroom, and avoid breathing the vapours from the solvent. The classroom must be well ventilated.

Materials

- coleus or spinach leaves (or pigment mixture supplied by your teacher)
- isopropanol (solvent)
- chromatography paper
- paper clip
- retort stand
- test-tube clamp
- cork stopper
- watch glass
- large test tube

Using Chromatography to Separate Plant Pigments

Chromatography is a technique that is used to separate and analyze complex mixtures of chemical compounds, such as plant pigments. You will use this technique to examine the pigments in a green leaf.

Pre-Lab Questions

1. Why is isopropanol used as a solvent instead of water?
2. Why does the test tube containing the solvent and paper need to be closed and not disturbed during the formation of the chromatogram?
3. Why is it important to be sure to get as much pigment as possible on the paper before carrying out the chromatography process?

Question

Which pigments can you identify in a green leaf?

Prediction

Predict at least three pigments that you will observe.

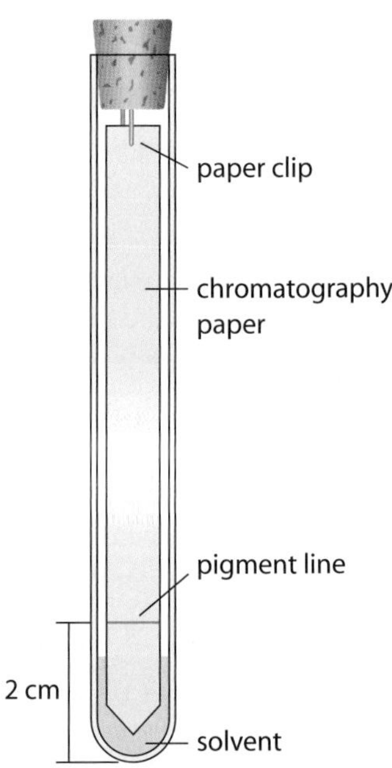

Procedure

1. Attach the large test tube to a retort stand.
2. Set up the cork stopper and the paper clip as shown.
3. Measure a piece of chromatography paper so that it is long enough to hang from the paper clip but not so long that it touches the bottom of the test tube. (Refer to the diagram.) Cut the paper to a point at one end.
4. Place a coleus or spinach leaf over the pointed end of the chromatography paper. Run the edge of a watch glass over the leaf, about 2 cm up from the tip of the paper. Use the watch glass to squeeze out the pigment mixture. Repeat this at least 10 times along the same line to ensure that enough pigment mixture has been deposited onto the paper.
5. Place 5 mL to 10 mL of solvent in the test tube.
6. Hang the chromatography paper from the stopper in the test tube so that the tip of the paper is in the solvent but the pigment mixture is not.
7. Wait until the solvent has travelled up to about 2 cm from the top of the paper.
8. Remove the paper from the test tube. Immediately, before the solvent evaporates, mark the location of the solvent front with a pencil. Also mark the edges of each pigment, as shown in this diagram.

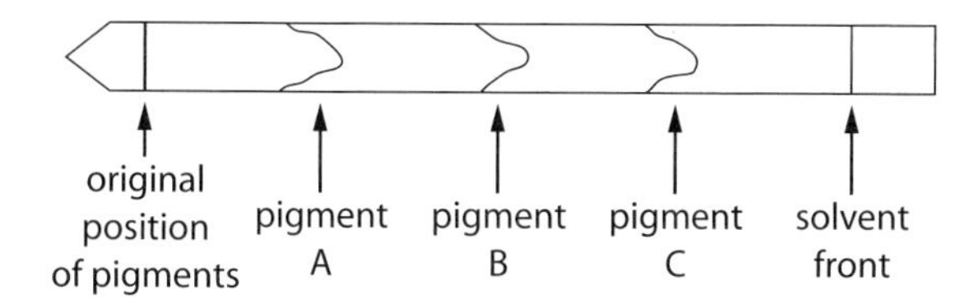

9. Measure and record the distance that each pigment travelled, starting from where you applied the pigment mixture to where each pigment stopped moving up the paper strip.
10. Measure and record the distance that the solvent front travelled.
11. Prepare a data table with the following headings, and record your observations and measurements in the first three columns.

Observations and Data for Chromatography of Plant Pigments

Pigment Colour	Distance Travelled by Pigment (cm)	Distance Travelled by Solvent (cm)	Rf (reference flow) Value	Name of Pigment

12. Calculate the R_f (reference flow) value of each pigment, using the following formula:

$$R_f \text{ value} = \frac{\text{distance travelled by pigment (solute)}}{\text{distance travelled by solvent}}$$

Analyze and Interpret

1. The chromatography paper is now called a chromatogram. Sketch your chromatogram, using different colours to show the different pigments and their positions.
2. Which pigment is (a) most soluble and (b) least soluble in the solvent you used? Explain how you decided.
3. Compare your observations and R_f values with those of your classmates. Identify sources of error in this investigation that might account for any differences.
4. Use the following table as a guide to help you complete the last column in your data table. (You may not have observed all the pigments in this table, or you may have observed other pigments.)

Examples of Plant Pigments and Their Colours

Pigment or Pigment Group	Colour
chlorophyll *a*	bluish-green
chlorophyll *b*	yellowish-green
carotenoids	orange
pheophytin	olive-green
xanthophylls	yellow
phycocyanin	blue
phycoerythrin	red

Conclude and Communicate

5. **a.** Which pigments did you identify in your leaf?
   b. Do you think additional pigments could still be present? Hypothesize how you could find out.

Extend Further

6. **INQUIRY** After performing paper chromatography, it is difficult to recover the pigments from the paper. When scientists want to separate and collect different compounds from a mixture so they can study the separated compounds further, they often use a method called column chromatography. Find a source of information about separating pigments using column chromatography. Write a procedure that you could use to separate and collect pigments using column chromatography. Be sure to include safety precautions.
7. **RESEARCH** Over the years since chromatography was first employed, many new techniques have been developed. Today, scientists use a technique called gas chromatography by which they can separate and identify extremely small amounts of chemical compounds. Carry out research to learn about gas chromatography. Write a paper or design a power point presentation to explain how gas chromatography works.

Chapter 4 SUMMARY

Section 4.1 Capturing Solar Energy: The Light-Dependent Reactions

The main function of the light-dependent reactions of photosynthesis is to use the Sun's light energy, absorbed by chlorophyll, to reduce $NADP^+$ and generate ATP.

Key Terms

light-dependent reaction	pigment
light-independent reaction	photosystem
thylakoid	photophosphorylation

Key Concepts

- Photosynthesis includes two distinct sets of reactions, the light-dependent reactions and the light-independent reactions.
- Enzymes and electron carriers responsible for the light-dependent reactions are embedded in the thylakoid membranes of the chloroplasts.
- Pigment molecules absorb photons from the radiant energy of sunlight and use it to energize electrons. The energy of the electrons in noncyclic photophosphorylation is used partially to pump hydrogen ions into the thylakoid space and partially to reduce $NADP^+$. In cyclic photophosphorylation, the energy of the electrons is used only to generate a hydrogen ion gradient across the thylakoid membranes.
- The energy stored in the hydrogen ion gradient is used to phosphorylate ADP.

Section 4.2 Fixing Carbon Dioxide: The Light-Independent Reactions

During the light-independent reactions, NADPH and ATP are used to reduce carbon dioxide to sugar in the Calvin cycle.

Key Terms

Calvin cycle	C3 photosynthesis
carbon dioxide fixation	photorespiration

Key Concepts

- The light-independent reactions use the energy from ATP and NADPH from the light-dependent reactions to assimilate carbon dioxide and synthesize high-energy organic compounds in a series of reactions called the Calvin cycle.
- Plants that use only the Calvin cycle to assimilate carbon dioxide are called C3 plants.
- Due to photorespiration, C3 plants are extremely inefficient in hot, dry climates.
- Plants that thrive in these environments have evolved two mechanisms to increase their efficiency. C4 plants isolate carbon dioxide fixation from the Calvin cycle reactions by carrying out these reactions in two different cell types. CAM plants isolate dioxide fixation from the Calvin cycle reactions by fixing carbon dioxide at night and carrying out the Calvin cycle reactions in the daytime.

Chapter 4 REVIEW

Knowledge and Understanding

Select the letter of the best answer below.

1. Light that is absorbed by a leaf is
a. not reflected
b. captured by a photosystem
c. used to increase the energy of an electron
d. composed of most colours, other than green and yellow
e. all of the above

2. Which of the following is false regarding light-dependent reactions?
a. produce ATP through chemiosmosis
b. produce NADPH
c. require carbon dioxide as an electron source
d. split water and releases oxygen
e. involve two photosystems

3. The water necessary for photosynthesis
a. is split into H_2 and O_2
b. is directly involved in the synthesis of a carbohydrate
c. provides the electrons to replace lost electrons in photosystem II
d. provides H^+ needed to synthesize G3P
e. none of the above

4. The reaction centre pigment differs from the other pigment molecules of the light-harvesting complex in the following way:
a. The reaction centre pigment is a carotenoid.
b. The reaction centre pigment absorbs light energy and transfers that energy to other molecules without the transfer of electrons.
c. The reaction centre pigment transfers excited electrons to other molecules.
d. The reaction centre pigment does not transfer excited electrons to the primary electron acceptor.
e. The reaction centre acts as an ATP synthase to produce ATP.

5. The electron flow that occurs in photosystem I during noncyclic photophosphorylation produces
a. NADPH
b. oxygen
c. ATP
d. all of the above
e. a) and (c) only

6. During the light-dependent reactions, the high-energy electron from an excited P680
a. eventually moves to $NADP^+$
b. becomes incorporated in water molecules
c. is pumped into the thylakoid space to drive ATP production
d. provides the energy necessary to split water molecules
e. falls back to the low-energy state in photosystem II

7. Which is true for the reactions of the Calvin cycle?
a. occur in the mitochondria
b. occur in the stroma
c. are always light-dependent
d. rely on the presence of oxygen
e. none of the above

8. Starch stored in the __________ of plants is produced from glyceraldehyde-3-phosphate molecules transported from the ____________.
a. roots, chloroplasts
b. chloroplasts, roots
c. roots, cytoplasm
d. cytoplasm, roots
e. cytoplasm, chloroplasts

9. The NADPH produced during the light-dependent reactions is necessary for
a. the carbon fixation phase, which incorporates carbon dioxide into an organic molecule of the Calvin cycle
b. the reduction phase, which produces carbohydrates in the Calvin cycle
c. the regeneration of RuBP of the Calvin cycle
d. all of the above
e. a) and b) only

10. What is the function of RuBP in the Calvin cycle?
a. to accept a proton
b. to generate ATP
c. to accept carbon dioxide
d. to reduce $NADP^+$
e. to oxidize carbon dioxide

11. Which process is considered a redox step?
a. carbon fixation
b. conversion of NADPH to $NADP^+$
c. regeneration of RuBP
d. formation of rubisco
e. none of the above processes

12. During the first phase of the Calvin cycle, carbon dioxide is incorporated into ribulose bisphosphate to form
a. oxaloacetate
b. rubisco
c. RuBP
d. 3-phosphoglycerate
e. G3P

13. The majority of the G3P produced during the reduction and carbohydrate production phase is used to produce

a. glucose
b. ATP
c. RuBP to continue the cycle
d. rubisco
e. all of the above

14. Which of the following is true about C4 plants?

a. They close their stomata during the day and open them at night.
b. They do not use the Calvin cycle for synthesizing carbohydrates.
c. They increase the concentration of carbon dioxide in bundle-sheath cells by carrying it into these cells in the form of malate.
d. Malate leaves bundle-sheath cells to pick up more carbon dioxide.
e. The Calvin cycle reactions occur in the mesophyll cells.

Answer the questions below.

15. What characteristics of a pigment molecule determine the wavelength of light that the pigment will absorb?

16. What capabilities do carotenoids give to a plant that it would not have without these pigments?

17. When plant pigments are extracted from the plants and dissolved in solution, they absorb very specific wavelengths of light. However, pigments in the intact leaf absorb at more wavelengths around the central peak. Why is this true and of what benefit is it to the plant?

18. What are the coloured leaves of some trees in the fall due to?

19. What are the components of a photosystem and how do they work together?

20. List the electron carriers and electron source in the order in which they donate or receive an electron in noncyclic photophosphorylation. Start with the source that donates an electron and then follow the electron to the final electron acceptor.

21. What are the two sources of hydrogen ions that accumulate in the thylakoid space?

22. How do the products of cyclic and noncyclic photophosphorylation differ?

23. Why would scientists be interested in mimicking photosynthesis?

24. Where are the enzymes of the Calvin cycle located?

25. Name the three phases of the Calvin cycle, and state the main function of each phase.

26. What would cause photorespiration to occur in a hot and dry climate?

27. Explain how CAM plants increase the concentration of carbon dioxide in their chloroplasts during the day when their stomata are closed.

28. Explain how photorespiration differs from cellular respiration.

Thinking and Inquiry

29. Why is the H^+ gradient across a thylakoid membrane referred to as a store of energy?

30. Peter Mitchell published his first paper on the chemiosmotic hypothesis in 1961. It was so different from anything biologists had considered previous to that, that they thought his ideas were ridiculous. However, continued testing of the hypothesis supported it until it was accepted. In one experiment, intact grana were separated from chloroplasts and placed in a solution at pH 4. At this low pH, the hydrogen ion concentration is very high. With this high concentration gradient, hydrogen ions were able to diffuse into the thylakoid spaces. Then the grana were removed from the pH 4 solution by centrifugation and placed in a solution at pH 8, which has a very low hydrogen ion concentration gradient. The pH of the solution inside the thylakoid spaces was still 4, for a short time. During this time, the thylakoids synthesized ATP in the dark.

Explain why these results support Mitchell's chemiosmotic hypothesis.

31. In 1803, Thomas Engelman used a combination of filamentous alga and aerobic bacteria to study the effect of various colours of the visible spectrum on the rate of photosynthesis. He passed white light through a prism in order to separate the light into the different colours of the spectrum. Then he exposed different segments of the alga to the various colours. He observed the areas of the spectrum in which the greatest number of bacteria appeared. The following graph shows Engelman's results. Analyze the graph, and explain the results of the experiment.

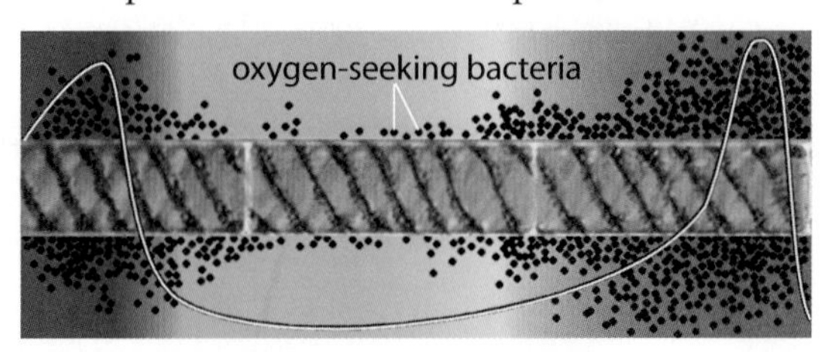

32. Chlorophyll that has been extracted from leaves and dissolved in a solution exhibits fluorescence, meaning that it absorbs light at a short wavelength and then immediately emits light at a longer wavelength. Chlorophyll that is in intact chloroplasts does not fluoresce. Explain why there is a difference in the fluorescent properties of chlorophyll in solution and in the intact chloroplasts.

33. Below are diagrams of an absorbance spectra of the pigments chlorophyll *a*, chlorophyll *b*, and β-carotene and a hypothetical action spectrum for a given plant. Imagine that the hypothetical action spectrum is real. What would the action spectrum indicate about the involvement of the chlorophylls and carotenoids in photosynthesis?

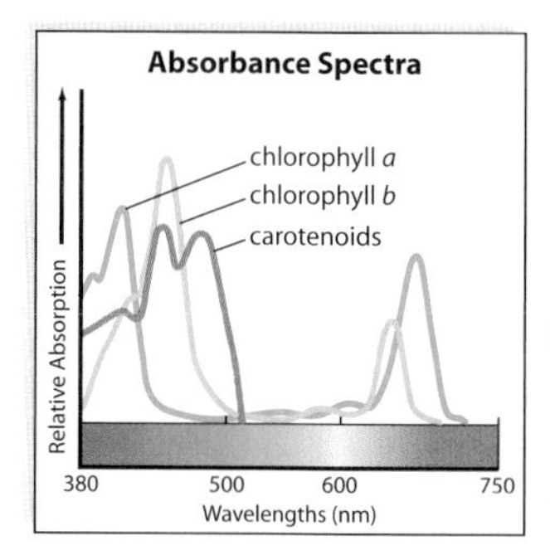

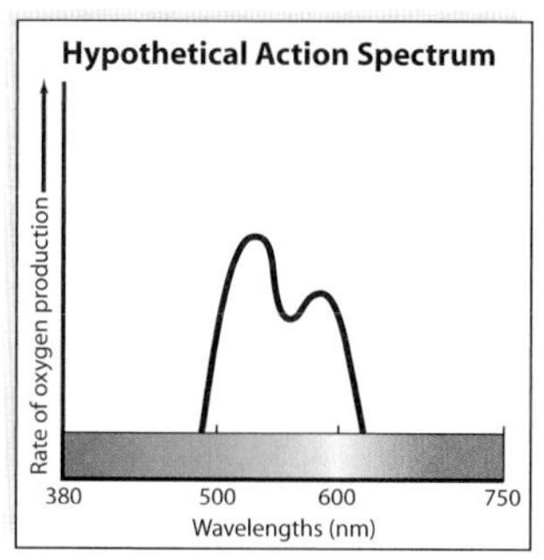

34. Specially designed fluorescent lighting enables people to grow plants indoors. Infer why regular fluorescent lights are not useful for growing indoor plants.

35. In many experiments on isolated chloroplasts, scientists used compounds called "uncouplers" to allow electron transport to continue without phosphorylating any ADP. How could an "uncoupler" act on a chloroplast to "uncouple" electron transport from phosphorylation?

36. If chloroplasts are exposed to light with wavelengths of 700 nm or longer, photosystem I can be excited but photosystem II cannot. What product or products would you expect the light-dependents to generate under these conditions?

37. The Z scheme for light-dependent reactions is always depicted with the primary electron acceptors at a long distance above the P680 or P700 in the photosystems. In reality, the electron acceptors are located very close to the reaction centres in the thylakoid membrane. What is the significance of depicting the Z scheme with the electron acceptors far from the P680 and P700?

38. Some photosynthetic bacteria have only one photosystem. Would you predict that they carry out cyclic or noncyclic photophosphorylation? Explain your reasoning.

39. Why do you think that chloroplasts have such an extensive membrane structure?

40. In the mid 1600s, Jan Baptista van Helmont weighed a small willow tree and dry earth and placed them in an earthen pot. The tree had a mass of 2.2 kg, and the dry soil had a mass of 90 kg. He planted the tree in the soil in the pot and watered it with either rainwater or distilled water. After five years he removed the tree and the soil and dried the soil. The tree had a mass of 77 kg and the soil had lost only 57 g of mass. He concluded that the increase in the mass of the tree must have come from the water. What was the error in van Helmont's conclusion, and why do you think he made that error?

41. Calvin and his colleagues used a radioactive isotope of carbon, ^{14}C, to study the light-independent reactions of photosynthesis. They exposed the green algae, *Chlorella pyrenoidosa*, to ^{14}C labelled carbon dioxide and stopped the reactions at various times by putting a sample of the algae into alcohol. They then analyzed the compounds extracted from the samples for the radioactive label. When the exposure time was very short, only one organic compound contained the label. What compound was labelled? Explain your reasoning.

42. The first law of thermodynamics states that energy cannot be created or destroyed, but it can be transformed into other forms of energy. Starting with the Sun's energy falling on the leaves of a tree, list as many energy transformations and transfers as you can, ending with your sitting in front of a campfire, roasting a marshmallow over the fire, and eating it.

43. In an experiment, a suspension of chloroplasts is carrying out photosynthesis under a bright light. What would happen to the concentrations of (1) ribulose-1,5-bisphosphate and (2) 3-phosphoglycerate if the light were suddenly turned off? Explain your reasoning.

44. Early researchers had a lot of difficulty finding the enzyme that fixed carbon dioxide, because they were looking for an enzyme that would add carbon dioxide to a two-carbon compound. Why might they have been looking for a two-carbon compound?

45. Under cool, moist conditions, why are C4 and CAM plants less efficient than C3 plants?

46. Discuss the importance of the structure of C4 plants as shown in this diagram.

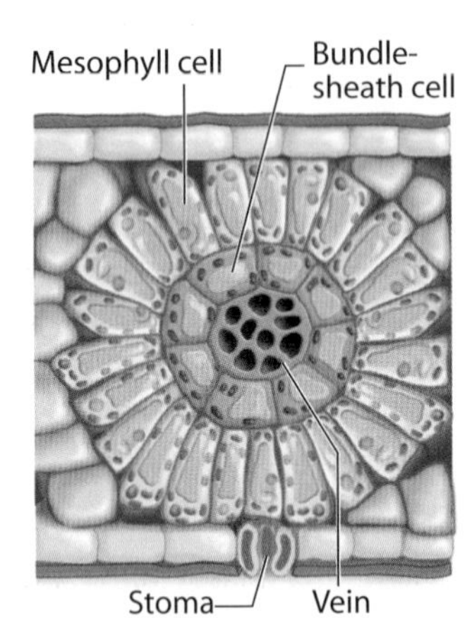

Communication

47. Imagine a hypothetical leaf that is a deep blue colour. Sketch a possible absorbance spectrum for pigments in this hypothetical leaf. Explain the reasoning that you used when drawing the spectrum graph.

48. Make a sketch of a chloroplast. Include and label each of the following: outer membrane, inner membrane, granum, thylakoid membrane, thylakoid space, stroma.

49. Use diagrams to help you explain the relationship between the wavelength of light and the energy of a photon associated with that wavelength.

50. Clearly explain the relationship between absorbance spectra and action spectra.

51. Write a small essay of three or four paragraphs that correctly uses the following terms in ways that make their meaning clear: carotenoid, chemiosmosis, chlorophyll, electron acceptor, electron carrier, hydrogen ion gradient, light-dependent reactions, NADPH, oxygen, P680, photon, photophosphorylation, photosystem, pigment, reaction centre, thylakoid, thylakoid space.

52. Explain why the energy of an electron that becomes excited and leaves photosystem I can either reduce an $NADP^+$ or provide energy to the hydrogen ion pump but cannot do both. Use diagrams to support your explanation.

53. BIG IDEAS All metabolic processes involve chemical changes and energy conversions. Write the net equation for the Calvin cycle.

54. State an undesirable property of rubisco. In no more than two sentences for each example, explain how two types of plants overcome the problems that result from this property.

55. The components of the light-dependent reactions are embedded in a membrane, but the enzymes of the light-independent reactions are not associated with membranes. Write a paragraph that explains these facts.

56. Construct a Venn diagram to compare C3, C4, and CAM plants.

57. Write the summary reaction for photosynthesis. Then create a graphic organizer around this equation to show the origin of each of the components of the reaction, the role of each component in the light-dependent and/or light-independent reactions, and the relationship among the components of the reaction.

58. Summarize your learning in this chapter using a graphic organizer. To help you, the Chapter 4 Summary lists the Key Terms and Key Concepts. Refer to Using Graphic Organizers in Appendix A to help you decide which graphic organizer to use.

Application

59. Do research to find out how β-carotene is related to retinal, the pigment in the retina of the eye. Find out how the function of β-carotene and retinal are related.

60. Write a short essay of three or four paragraphs to express your opinion about the following statement: Photosynthesis and cellular respiration are the two most important chemical reactions on Earth.

61. Biotechnologists have studied ways to genetically engineer plants and animals to synthesize useful compounds, such as pharmaceuticals, that they do not normally synthesize. For reasons other than ethics, why might it be beneficial to develop a plant that produces a desired compound rather than an animal?

62. Develop the basic plan for an educational game involving the Z scheme. Assume that the goal of the game is to help students learn the names of all the electron carriers and their sequence.

63. Scientists and engineers have pursued the dream of inventing an artificial leaf and related technologies that attempt to mimic the process of photosynthesis. Investigate the work of John Turner (U.S. National Renewable Energy Laboratory), Angela Belcher

(MIT—Massachusetts Institute of Technology), Don Nocera (also at MIT) and at least one other researcher of your own discovery. Briefly describe the work these people have done, the successes and failures that have occurred to date, and the current state of this kind of research.

64. Climate monitoring demonstrates a continued increase in carbon dioxide concentration in the atmosphere.
 a. Predict the effect of increasing carbon dioxide concentrations on photorespiration.
 b. Most scientists agree that increasing carbon dioxide concentrations are leading to higher average global temperatures. If temperatures are increasing, does this change your answer to the previous question? Explain why or why not.

65. Choose one of the following topics and write a short essay of two to four paragraphs describing how learning about this concept in photosynthesis helped you understand a related concept in another subject area.
 a. properties of light
 b. absorbance spectra
 c. fall colours
 d. energy conversions
 e. concentration gradients
 f. membrane structure

66. Since the 1990s, researchers have been investigating ways to improve rice production to avoid food shortages in the future. Crop ecologist John Sheehy of the International Rice Research Institute compares the rice plant to a car, suggesting that the aim of researchers it "supercharging the engine. The photosynthetic process is the engine of growth for the rice plant, so, if we can improve that, then the whole plant benefits." A colleague, Robert Zeigler, adds, "This generation must work to assure food security not only for ourselves, but for future generations as well. We must find and develop new ideas to help us further increase rice production while using less land, labour, and water." In aid of this goal, researchers seek to enhance the photosynthetic efficiency of rice plants by converting rice from a C3 to a C4 plant.
 a. What is the benefit of converting rice from a C3 plant to a C4 plant?
 b. Do research to find out how researchers propose to achieve this goal.

67. **BIG IDEAS** An understanding of metabolic processes enables people to make informed choices with respect to a range of personal, societal, and environmental issues. Many pharmaceutical products are derived from compounds found naturally in plants. In many cases, chemists modify the original compound to increase its potency or to reduce side-effects. Biotechnologists are now looking at ways to genetically engineer plants to increase the amount of a pharmaceutical that the organisms produce normally on a smaller scale. Of what benefit might this be to society? What are some possible risks?

68. When plants such as vegetables and flowers are grown in greenhouses in winter, their growth rate greatly increases if the concentration of carbon dioxide is raised two or three times the level in the natural environment. What is the biological basis for the increased rate of growth?

69. In 1941, biologists exposed photosynthesizing cells to water containing a heavy oxygen isotope, designated ^{18}O. The labelled isotope appears in the oxygen gas released in photosynthesis, demonstrating that the oxygen came from water. Explain where the ^{18}O would have ended up if the researchers had used ^{18}O-labelled carbon dioxide instead of water?

70. Use the concept map below to answer the following questions.
 a. Where do electron transport chains fit into the concept map?
 b. What specific event in the light-dependent reactions gives rise to the waste product, O_2?
 c. Redraw the concept map to show how you would incorporate the Calvin cycle, rubisco, C3 plants, C4 plants, and CAM plants into it.

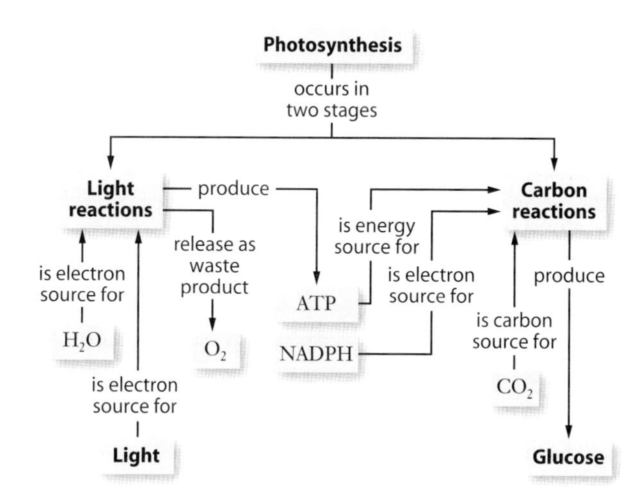

Chapter 4 SELF-ASSESSMENT

Select the letter of the best answer below.

1. K/U Which is the correct summary equation for photosynthesis?

a. $C_6H_{12}O_6(s) + 6O_2(g) \rightarrow 6CO_2(g) + 6H_2O(\ell) +$ energy

b. $6CO_2(g) + 6H_2O(\ell) \rightarrow C_6H_{12}O_6(s) + 6O_2(g) +$ energy

c. $6CO_2(g) + 6O_2(g) +$ energy $\rightarrow C_6H_{12}O_6(s) + 6H_2O(\ell)$

d. $6CO_2(g) + 6H_2O(\ell) +$ energy $\rightarrow C_6H_{12}O_6(s) + 6O_2(g)$

e. $C_6H_{12}O_6(s) + 6H_2O(\ell) \rightarrow 6CO_2(g) + 6O_2(g) +$ energy

2. K/U Photosynthesis and cellular respiration both

a. use oxygen

b. produce carbon dioxide

c. make use of an electron transport chain

d. occur in the chloroplast

e. c and d

3. K/U The noncyclic, but not the cyclic, pathway,

a. generates 3PG

b. generates chlorophyll

c. generates ATP

d. generates NADPH

e. none of the above

4. K/U In the absence of sunlight, plants cannot use the Calvin cycle due to a lack of

a. ATP

b. oxygen

c. NADPH

d. a and c are correct

e. a, b, and c are correct

5. K/U Which wavelengths are absorbed by chlorophyll *a*??

a. Long wavelength light (such as red)

b. Short wavelength light (such as blue)

c. medium wavelength light (such as green)

d. None of the above.

e. a and b are correct

6. K/U Chemiosmosis depends on

a. protein complexes in the thylakoid membrane

b. a difference in H^+ concentration between the thylakoid space and the stroma

c. ATP undergoing a condensation reaction

d. the action spectrum of chlorophyll

e. a and b are correct

7. K/U The O_2 given off by photosynthesis comes from

a. water

b. glucose

c. carbon dioxide

d. rubisco

e. ATP

8. K/U Which of the following components of the light-dependent reactions participates in both cyclic and noncyclic photophosphorylation?

a. NADP reductase

b. P680

c. H_2O

d. b_6-*f* complex

e. photosystem II

9. K/U Which of the following is a goal for scientists who are attempting to mimic certain natural processes in the laboratory?

a. Scientists want to mimic the leaf's ability to reduce $NADP^+$ and use NADH as a source of energy.

b. Scientists want to create artificial membranes and mimic the b_6-f complex to generate a hydrogen ion gradient as a source of energy.

c. Scientists want to study photosystems so they can make them absorb a broader range of wavelengths to make plants more efficient.

d. Scientists want to use artificial thylakoids that can use solar energy to generate ATP to use in food supplements.

e. Scientists want to mimic the splitting of water by photosystem II because hydrogen is a clean source of energy.

10. K/U C3 plants encounter a problem when

a. the environment becomes too cool and the stomata close to retain heat

b. the temperature of the air rises above 28°C and the stomata open to air out the spaces in the leaves

c. the temperature of the air rises above 28°C and the stomata close to conserve water, which prevents carbon dioxide from entering the air spaces in the leaf.

d. the temperature of the air falls below 28°C, which causes chemical reactions to slow down

e. the moisture in the air increases and too much water enters the leaf, preventing the oxygen from escaping

Use sentences and diagrams as appropriate to answer the questions below.

11. K/U For each number in the following diagrams, write a label that states name of the structure.

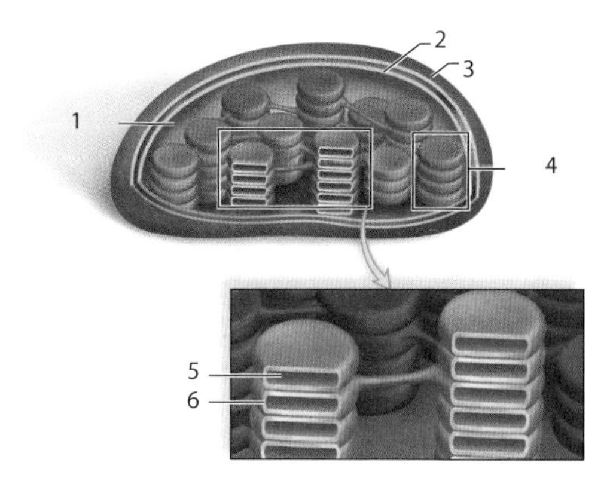

12. T/I You have an object that is not black, but if you shine a green light on the object, it appears black. Explain why it appears black.

13. K/U How does an action spectrum differ from an absorbance spectrum?

14. K/U What is the function of carotenoid pigments?

15. T/I State which, if any, of the high energy compounds normally produced by noncyclic photophosphorylation, would be produced under the following conditions. Explain your answer.

a. Introduction of a certain compound made the thylakoid membrane permeable to hydrogen ions.

b. There was no $NADP^+$ available.

c. The only light that was shining on the chloroplasts had wavelengths of 700 nm and longer.

16. C Make a sketch of a thylakoid membrane and refer to the sketch as you explain the significance of the membrane in photophosphorylation.

17. C In a flowchart, demonstrate the role of water in photophosphorylation. In your flowchart, include the name of the set of reactions in which water participates and show clearly where each part of the water molecule is found after the reaction is complete.

18. A Describe two practical benefits that could come from the study of photosynthesis.

19. K/U What is accomplished during the carbon fixation phase of the Calvin cycle?

20. K/U In the Calvin cycle, 3-phosphoglycerate (PGA) is converted to glyceraldehyde 3-phosphate (G3P) in two steps. What provides the energy that is used in each of those steps?

21. T/I Why is energy required to complete the steps described in question 20?

22. K/U In order to synthesize one glucose molecule, how many turns of the Calvin cycle are required?

23. K/U What forms of energy and how much of each form of energy are required to make one glucose molecule?

24. C Use a graphic organizer to describe the process that occurs in C3 plants when the atmosphere becomes hot and dry. Include these terms: stomata, water, carbon dioxide, oxygen, rubisco, energy.

25. C Use the diagram below to answer the questions.

a. For numbers 1 through 6, state the number of carbon atoms in the compound.

b. For numbers 7 and 8, state the name of the cell type.

c. For number 9, state the name of the process.

d. For number 10, state the name of the compound and explain where it is going next in the process.

e. Explain what is accomplished by the process.

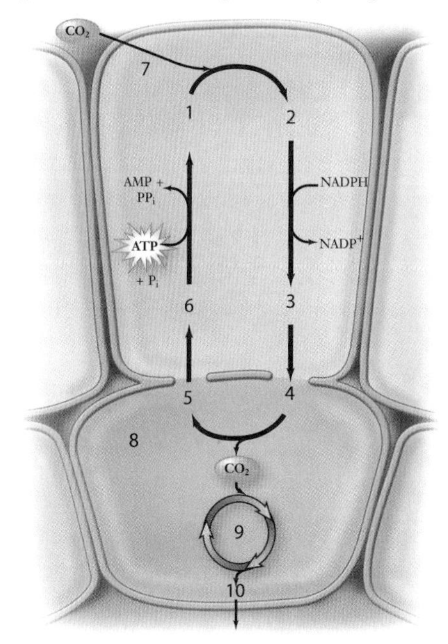

Self-Check

If you missed question...	1	2	3	4	5	6	7	8	9	10	11	12	13	14	15	16	17	18	19	20	21	22	23	24	25
Review section(s)...	4.1	4.1	4.1	4.2	4.1	4.1	4.1	4.1	4.1	4.2	4.1	4.1	4.1	4.1	4.1	4.1	4.1	4.1	4.2	4.2	4.2	4.2	4.1 4.2	4.2	4.2

Unit 2 Project

Conduct an Inquiry

Modelling Metabolic Pathways

The cells of living organisms continuously cycle through numerous metabolic pathways—step-by-step sequences of chemical reactions that transform energy and molecules within cells. These reactions enable cells to live and grow. The cells of all living organisms carry out cellular respiration, either aerobic or anaerobic, which includes the catabolic pathways that break down energy-rich compounds to produce ATP. However, only certain cells carry out both cellular respiration and photosynthesis. This includes the photosynthetic cells in plants and specialized bacteria. In these cells, the anabolic pathways involved in photosynthesis supply the organic molecules that are broken down in cellular respiration.

In this project, you will model the metabolic pathways involved in cellular respiration and photosynthesis and demonstrate how these pathways are linked. Once you have planned and completed your model, you will prepare a presentation of it, choosing a format that is appropriate for your chosen audience and purpose.

How can you model cellular metabolic pathways and their interactions?

Initiate and Plan

1. As a class or in a group, you will model the metabolic pathways involved in cellular respiration and photosynthesis and demonstrate how these pathways are linked. Together, brainstorm a general plan for the design and completion of your model. Assign one student to record notes from your discussion.

   Consider the following questions to guide your brainstorming:

   - How can you design a model of metabolic pathways to clearly illustrate the pathways and their interactions? Be creative. For example, you might ask classmates to act out the interconnected steps involved in metabolic pathways, or you might design a simple computer animation(s) to demonstrate these pathways.
   - How can you ensure that your model demonstrates metabolic pathways both safely and accurately?
   - To what audience will you present your model when it is completed?

2. Once you have brainstormed a general plan for your model in step 1, decide on and record a detailed plan for the model. Be sure to include factual details related to the metabolic pathways you will be modelling. Consider the following guidelines to assist you in your planning:

   - Model cellular respiration and photosynthesis, illustrating how these metabolic pathways are linked and identifying how smaller metabolic pathways, such as glycolysis, fit into these larger pathways.
   - Differentiate between the components in each pathway, such as enzymes, substrates, electron carriers, electrons, and ATP.
   - Show how cellular respiration differs depending on the presence or absence of oxygen.
   - Distinguish between C3 and C4 photosynthesis, and consider whether your model would need to be adjusted to model metabolic pathways in CAM plants.
   - Demonstrate where specific metabolic reactions take place within the cell.

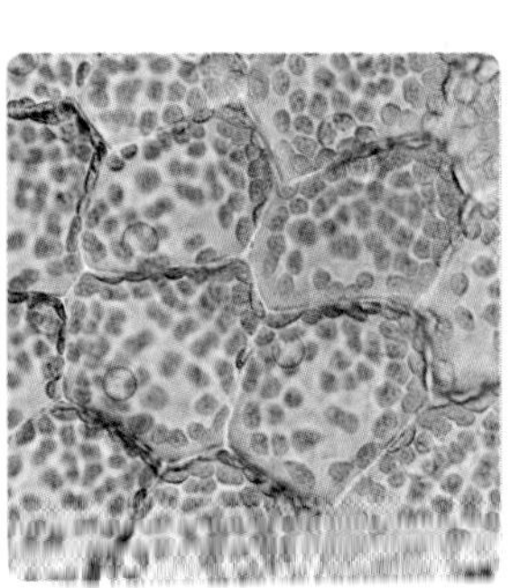

Both cellular respiration and photosynthesis take place in photosynthetic cells, such as these fern leaf cells.

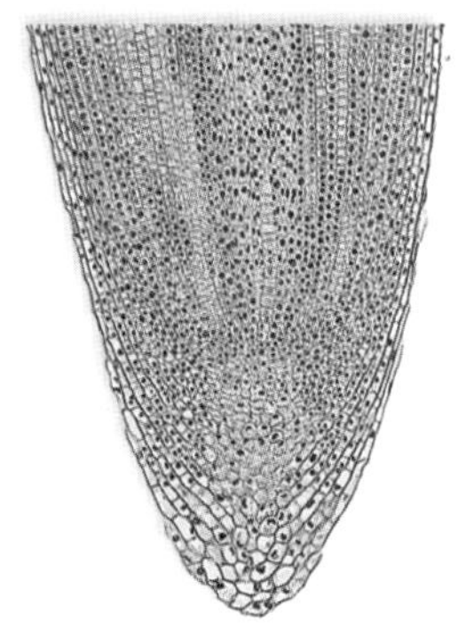

Some cells carry out cellular respiration but not photosynthesis. These include animal cells and non-photosynthetic plant cells, such as these onion root tip cells.

Perform and Record

Following your design and your teacher's instructions, complete your model. Make any necessary adjustments to your design as you proceed.

Analyze and Interpret

1. Hot potato is a children's game in which an object is quickly passed from person to person, with each person attempting to avoid being the one left holding the object, or "hot potato," when play is stopped abruptly. Explain how using an actual hot potato (which cools over time) as the object in this game could represent an effective analogy for electron transport. Would this analogy be more relevant to photosynthetic electron transport or respiratory electron transport? Explain your reasoning.
2. Compare and contrast the flow of energy in cellular respiration and photosynthesis, in terms of the storage/release of energy within the bonds of molecules, as well as ATP production/consumption.
3. Evaluate your model. Explain how it was successful and suggest how it might be improved.
4. Use a graphic organizer or a series of diagrams to summarize the metabolic pathways you illustrated in your model and show how they are linked.

Communicate Your Findings

1. As a class or in a group, decide how you will present your model. Be sure to consider your audience and the purpose of your model when making your decision.
2. Create a narrative summary that describes the steps in your model and explains how your model demonstrates metabolic pathways. Present your summary in a format that fits with your chosen presentation format. For instance, if you present your model in a video or computer animation, you could add descriptive captions or voice-over narration.
   Your narrative summary should include
   - an introduction to the metabolic pathways being modelled
   - a summary of the chemical reactions in each metabolic pathway
   - an explanation of how the metabolic pathways are linked
   - a description of the energy flow in each metabolic pathway

   Additionally, consider the following questions when writing your summary:
   - How will you ensure that the scientific vocabulary is accurate and appropriate to your model and your audience?
   - How can you use your narrative summary to clarify any confusing or challenging concepts in your model?
   - What could you do to make your presentation more interesting or to better engage your audience?
   - If you are using a computer model, how might you use instructional visuals to support your narrative summary?
3. Present your presentation in its final format to its intended audience.

Assessment Criteria

After you complete your project, ask yourself these questions. Did you…

- ✓ T/I brainstorm the general design of your model before you planned more specific details?
- ✓ A decide how your model will represent what you have learned about metabolic pathways?
- ✓ K/U use your model to demonstrate an understanding of how cellular respiration and photosynthesis are linked?
- ✓ T/I make any necessary adjustments to your design while completing your model so that it fulfills its purpose safely and effectively?
- ✓ C use a graphic organizer or a series of diagrams to summarize the metabolic pathways you illustrated in your model and show how they are linked?
- ✓ C decide how you will present your model, considering your audience and the purpose of your model when making your decision?
- ✓ C create a narrative summary that describes and explains your model, adjusting the style of your narrative to match your presentation format?
- ✓ C use scientific vocabulary accurately and appropriately in your presentation?
- ✓ C use instructional visuals, if appropriate, to support your presentation?

UNIT 2

SUMMARY

BIG IDEAS

- All metabolic processes involve chemical changes and energy conversions.
- An understanding of metabolic processes enables people to make informed choices with respect to a range of personal, societal, and environmental issues.

Overall Expectations

In this unit you learned how to…

- **analyze** the role of metabolic processes in the functioning of biotic and abiotic systems, and **evaluate** the importance of an understanding of these processes and related technologies to personal choices made in everyday life
- **investigate** the products of metabolic processes such as cellular respiration and photosynthesis
- **demonstrate** an understanding of the chemical changes and energy conversions that occur in metabolic processes

Chapter 3 Energy and Cellular Respiration

Key Ideas

- Metabolism is the sum of all the biochemical reactions in a cell.
- The first law of thermodynamics states that energy cannot be created or destroyed. The second law of thermodynamics states that some energy is "lost" as disorder (entropy) increases.
- ATP hydrolysis releases energy to drive endergonic reactions, and it is synthesized using energy from exergonic reactions.
- The complete breakdown of glucose occurs in four stages: glycolysis, pyruvate oxidation, Krebs cycle, and oxidative phosphorylation.
- Glycolysis is the breakdown of glucose to two pyruvates, producing two ATP and two NADH. ATP is made by substrate-level phosphorylation.
- Pyruvate is broken down to CO_2 and an acetyl group that is attached to CoA. NADH is made during this process.
- During the Krebs cycle, the acetyl group attached to CoA is broken down to two CO_2 molecules. Three NADH, one $FADH_2$, and one ATP are made during this process.
- Oxidative phosphorylation involves two events. The electron transport chain oxidizes NADH or $FADH_2$ and generates an H^+ gradient. This gradient is used by ATP synthase to make ATP via chemiosmosis.
- Some single-celled organisms that live in conditions of very low oxygen can carry out anaerobic respiration by using an electron acceptor other than oxygen.
- Some single-celled organisms and, during extreme exertion, some muscle cells, use only glycolysis for energy in a process called fermentation. Two types of fermentation are lactate fermentation and ethanol fermentation

Chapter 4 Photosynthesis

Key Ideas

- Photosynthesis includes the light-dependent reactions and the light-independent reactions.
- Enzymes and electron carriers responsible for the light-dependent reactions are embedded in the thylakoid membranes of the chloroplasts.
- Pigment molecules absorb photons from the radiant energy of sunlight and use it to energize electrons. The energy of the electrons in noncyclic photophosphorylation is used partially to pump hydrogen ions into the thylakoid space and partially to reduce $NADP^+$. In cyclic photophosphorylation, the energy of the electrons is used only to generate a hydrogen ion gradient across the thylakoid membranes.
- The energy stored in the hydrogen ion gradient is used to phosphorylate ADP.
- The light-independent reactions use the energy from ATP and NADPH from the light-dependent reactions to assimilate carbon dioxide and synthesize high-energy organic compounds in a series of reactions called the Calvin cycle.
- Plants that use only the Calvin cycle to assimilate carbon dioxide are called C3 plants. Due to photorespiration, C3 plants are extremely inefficient in hot, dry climates.
- Plants that thrive in hot, dry environments have evolved two mechanisms to increase their efficiency. C4 plants isolate carbon dioxide fixation from the Calvin cycle reactions by carrying out these reactions in two different cell types. CAM plants isolate carbon dioxide fixation from the Calvin cycle reactions by fixing carbon dioxide at night and carrying out the Calvin cycle reactions in the daytime.

UNIT 2 REVIEW

Knowledge and Understanding

Select the letter of the best answer below.

1. ATP supplies the energy that drives most cellular activities. The ultimate source of the energy contained in ATP is
   a. glucose
   b. sunlight
   c. ADP
   d. ATP synthase
   e. carbohydrates
2. Which of the following compounds binds to CO_2 during the Calvin cycle?
   a. ribulose bisphosphate
   b. glyceraldehyde 3-phosphate
   c. adenosine triphosphate
   d. ATP synthase
   e. adenosine synthase
3. During glycolysis, high-energy electrons released from the intermediate compounds are first accepted by
   a. ATP
   b. ADP
   c. NAD^+
   d. FAD^+
   e. NADH
4. Aerobic respiration ceases without oxygen because oxygen is a reactant in which of the following pathways?
   a. glycolysis
   b. Krebs cycle preparation
   c. pyruvate oxidation
   d. electron transport
   e. the Calvin cycle
5. Which row correctly identifies the compounds represented by i and ii in the diagram below?

$$C_3H_3O_3 + CoA \rightarrow ii + CO_2$$

Row	i	ii
A.	NADH	C_2H_2OH–CoA
B.	$NADH_2$	C_2H_2OH–CoA
C.	NADH	C_2H_3O–CoA
D.	$NADH_2$	C_2H_3O–CoA
E.	NADH	C_2H_3OH–CoA

   a. A
   b. B
   c. C
   d. D
   e. E
6. Under anaerobic conditions, such as in muscles during strenuous exercise, why is the conversion from pyruvate to lactate required?
   a. to decrease NAD^+ and increase NADH
   b. to decrease NADH and increase NAD^+
   c. to increase NADH and increase NAD^+
   d. to decrease NADH and decrease NAD^+
   e. to keep oxidative phosphorylation functioning
7. In cellular respiration, the energy released by the electron transport chain pumps hydrogen ions across the membrane from
   a. the matrix to the cytoplasm
   b. the cytoplasm to the matrix
   c. the intermembrane space to the matrix
   d. the matrix to the intermembrane space
   e. the cytoplasm to the intermembrane space
8. During chemiosmosis, energy is harvested from the transfer of electrons and converted into usable chemical energy for cell metabolism. Which of the following statements about this energy conversion is correct?
   a. AMP is produced as ADP is dephosphorylated.
   b. ATP is produced as ADP is phosphorylated.
   c. ADP is produced as ATP is dephosphorylated.
   d. ADP is produced as AMP is phosphorylated.
   e. None of the above is correct.
9. Why do leaves appear green?
   a. The green portion of the light that strikes them is converted to heat.
   b. The green portion of the light that strikes them is absorbed.
   c. The green portion of the light that strikes them is destroyed.
   d. The green portion of the light that strikes them is reflected.
   e. The green portion of the light that strikes them is converted to other colours.
10. Comparing ATP to ADP, the compound with more potential energy is
    a. ADP
    b. ATP
    c. Both have the same potential energy.
    d. Neither has any potential energy.
    e. The amount of potential energy of ADP varies.

11. The reducing power of NADH pumps hydrogen ions across the inner mitochondrial membrane, creating a hydrogen-ion gradient. The compound that is produced when the hydrogen ions diffuse back through the inner mitochondrial membrane via ATP synthase is
    a. ADP
    b. ATP
    c. FAD^+
    d. $FADH_2$
    e. AMP
12. As electrons move from photosystem II to electron acceptors, they are replaced by electrons from
    a. the photolysis of H_2O
    b. the electron transport system
    c. chemiosmosis
    d. carbon fixation
    e. absorbance
13. Which of the following statements best describes the relationship between pigments and photosynthesis?
    a. The absorbance spectrum of chlorophyll *a* is similar to the action spectrum for photosynthesis.
    b. The absorbance spectrum of all pigments is similar to the action spectrum for photosynthesis.
    c. Green light is absorbed most effectively by pigments and produces the highest rates of photosynthesis.
    d. The absorbance spectrum of pigments is altered in the autumn when rates of photosynthesis decrease.
    e. Green light is transmitted most effectively by pigments and produces the highest rates of photosynthesis.
14. The transformation of the end product of glycolysis occurs in the
    a. nucleus
    b. chloroplasts
    c. mitochondria
    d. cytoplasm
    e. ribosomes
15. In the Krebs cycle, the oxidation of energy-rich compounds produces the reducing power of
    a. NADH and $FADH_2$
    b. NAD and $FADH_2$
    c. ATP and NADH
    d. $FADH_2$ and ATP
    e. AMP and ATP

Answer the questions below.

16. What three chemical changes occur during pyruvate oxidation?
17. Describe the flow of energy in the electron transport chain.
18. Most cells rely on ATP as a rapid energy source to carry out life processes. How do cells get more ATP to replace what has been used?
19. Describe the type of fermentation that could occur in your cells.
20. Write an equation summarizing the Calvin cycle.
21. How does photorespiration reduce the efficiency of photosynthesis?
22. Define potential energy and kinetic energy in a biological context.
23. What role does ATP synthase play in chemiosmosis in mitochondria?

Use the diagrams of the chloroplast shown below to answer questions 24 to 28. Write your answers in complete sentences using the names of the components labeled A, B, and C.

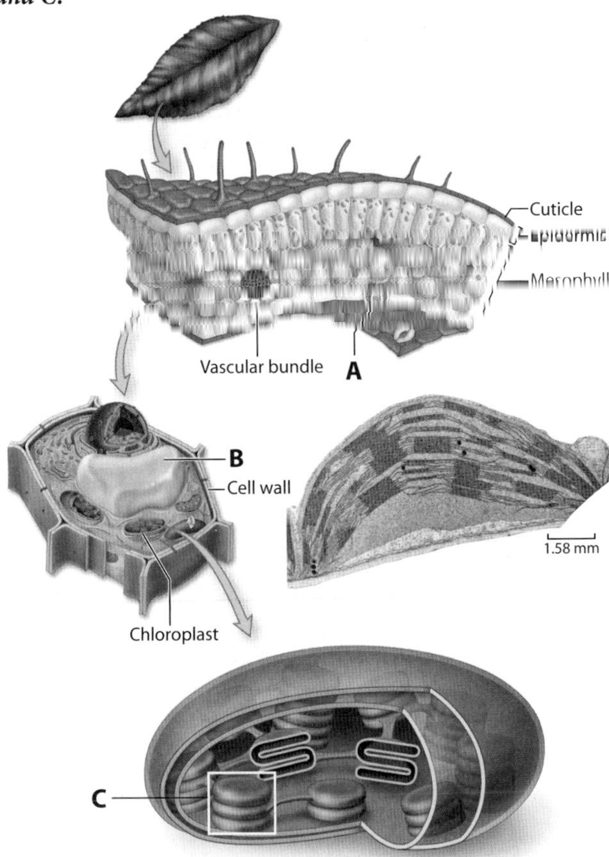

24. Identify the substance needed for photosynthesis that enters through the structure at label A.
25. Name a substance related to photosynthesis that would be carried in B.
26. In which part of the chloroplast would you most likely find 3-carbon sugars? Explain your answer.

27. Name structure C, and state the name of one component of it.

28. Where are the photosynthetic pigments found?

29. Summarize the role of ethanol in the automotive industry.

30. List the end products of the following:
   a. glycolysis
   b. Krebs cycle
   c. Calvin cycle

31. What is the relationship between aerobic cellular respiration, glucose, and oxygen?

32. Discuss the regulation of aerobic catabolic pathways in terms of the main points of regulation and the type of control mechanism that occurs at these points.

33. Describe how the following terms are related: lactic acid fermentation, alcoholic fermentation, anaerobic processes.

Thinking and Investigation

34. Differentiate between the photosystem I reaction centre and the photosystem II reaction centre with respect to their photon requirements, source of electrons, and products.

35. What major advantage does photophosphorylation have over cells that cannot carry out this process?

36. An individual's body mass is always the least in the morning. Account for this loss of mass between evening and morning by providing both metabolic and practical reasons for this consistent observation.

37. In the eighteenth century the experiments of Joseph Priestley demonstrated the oxygen-producing capabilities of plants by supporting the life of a mouse (and then himself!) in a sealed container. It was noted that the mouse survived much longer when there was a plant living in the same container. At the time it was suggested that the plant produced "good air." What variables would have been controlled to ensure that the two experiments were comparable?

38. A friend decides to design a new diet that is characterized by exercise and the removal of carbohydrates to achieve weight loss.
   a. How does a basic understanding of metabolism make this sound promising?
   b. How will the body meet its energy requirements when carbohydrates are reduced and demands for energy increases? What problems will this pose for the body?

39. An experimental apparatus is set up such that the aquatic plant *Cabomba* is in pond water and a pH probe is recording current conditions. The following data are obtained over a 24-hour period:

Time (hours)	pH of water
0	6.58
2	6.88
5	7.04
9	7.02
18	6.76
23	6.63

   a. Based on the pH values of the water, infer whether photosynthesis or cellular respiration is likely the dominant process occurring between each time interval.
   b. Assuming the plant was placed in a window on a sunny day, estimate the time of day at the time point of 0 hours.
   c. Comment on the current weather at the time point of 23 hours.

40. Differentiate between oxidative phosphorylation and chemiosmosis.

41. Suppose that the energy released during cellular respiration was in the form of ADP. Predict the effect on cellular respiration.

42. A biochemist uses her knowledge of cellular structure and function to increase the production of rubisco in the mitochondria.
   a. What is most likely the intended purpose of this modification?
   b. What will likely limit the success of her experiment?

43. Consider the statement, "Under normal conditions, when the temperature is near 25°C, C3 plants lose 20 percent of the energy used to fix one carbon dioxide molecule."
   a. When attempting to verify this statement, how could a control be determined?
   b. What are the independent and dependent variables?
   c. What factors would you hold constant?

44. A scientist decides to measure heat associated with germination using the following procedure. Three seeds are placed in 50 mL of water at the bottom of an Erlenmeyer flask. After the thermometer is placed inside the flask the temperature is recorded each hour. After several days of measurements the seeds germinate but the scientist has not observed a change in temperature.
   a. Summarize problems in the procedure.
   b. Rewrite the procedure to address your concerns and increase the validity of the results.

45. The production of carbon dioxide and water from sugar is an exergonic reaction. To demonstrate this, a student sets up one sugar cube that is exposed to an oxidizing agent. However, a reaction does not occur.
   a. Why did the student think this reaction would occur, and what was the error in reasoning?
   b. Suggest a way to complete this reaction successfully.

46. Herschel conducts a floating leaf assay to determine comparative rates of photosynthesis in plant cells under varied conditions. As photosynthesis occurs, leaf-disks rise to the top as oxygen is formed and gas production increases the buoyancy of the leaf disks. He chooses to use the time it takes for the first disk to rise under white light as a measure of the rate of photosynthesis and labels this data his control.
   a. How can he demonstrate that photosynthesis, not cellular respiration, is responsible for the gas being produced?
   b. In his experiment he illuminates the different samples with three wavelengths of light: blue, green, and red. Predict the rate of leaf-disk rising in each case compared to the control.
   c. How could the accuracy of his experiment be improved? Explain.

Communication

47. **BIG IDEAS** All metabolic processes involve chemical changes and energy conversions. Aerobic cellular respiration includes four main stages: glycolysis, Krebs cycle preparation, the Krebs cycle, and the electron transport system. Using a graphic organizer in your answer, compare these four stages. Include starting and finishing compounds, amount of energy produced, and locations in the cell where each stage occurs.

48. Develop an argument that supports the idea that rubisco is the most essential enzyme for life on Earth.

49. Draw a diagram that shows the roles of the mitochondria and chloroplasts in cellular respiration and photosynthesis in plants, and how the two are linked. In your diagram include the following information: the organelles involved, the reactants and final products of each overall reaction, the electron carriers involved, and where ATP is produced.

50. Construct a graph that summarizes energy available in glucose and its intermediates as sugar is converted into carbon dioxide. In your graph have the y-axis represent free energy and x–axis represent progress. Neither axes require distinct values since they are approximates. On your graph, indicate where each of the following is occurring: glycolysis, pyruvate oxidation, Krebs cycle, release of carbon dioxide.

51. Research scientists are investigating methods for getting photosynthesis to proceed outside of a chloroplast.
   a. Identify a practical use and benefit to society that would come from the successful achievement of this goal.
   b. Perform research to identify one research group that is pursuing this goal. Describe their research goals and general approach. Determine whether the group has estimated or predicted when they believe they will achieve their goal(s).
   c. Summarize your research findings in a presentation, using a format of your choice.

52. Draw and label a diagram of a mitochondrion. In your diagram, include the following labels: matrix, outer membrane, inner membrane, intermembrane space. Also indicate on the diagram where each of the following occurs: glycolysis, pyruvate oxidation, and chemiosmosis.

53. Use a Venn diagram to compare and contrast the products of the type of fermentation that occurs in your cells with the products fermentation that occurs in yeast cells.

54. Summarize the concept of entropy as seen in the second law of thermodynamics. Communicate your understanding by explaining how anabolic processes do not violate the second law of thermodynamics.

55. Creatine-phosphate is used as a source of phosphate in the regulation of ATP levels at the cellular level. Construct a reaction-coupling diagram that represents the transfer of creatine's phosphate to an ADP.

56. Summarize the process of pyruvate production by glycolysis in the form of six main steps.

Application

57. **BIG IDEAS** An understanding of metabolic processes enables people to make informed choices with respect to a range of personal, societal, and environmental issues. A significant amount of time and research dollars have been invested at the public and private levels for a better understanding of metabolic pathways in humans and other organisms of interest. After differentiating between anabolic and catabolic pathways and providing examples, suggest three practical applications of metabolic information.

58. Compounds like dinitrophenol (DNP) disrupt the formation of the hydrogen ion gradient normally occurring in cellular respiration.
   a. What impact would this have on cellular concentrations of ATP?
   b. What physiological occurrences would be representative of DNP consumption?

59. Ethanol is a product of anaerobic respiration and can also be used as a clean fuel in some automobiles. Compare the amount of energy in the molecule of ethanol with the energy in its isolated atoms and the energy in the final combustion products.

60. Phenylketonuria (PKU) is a genetic disorder resulting in the body's inability to process the amino acid phenylalanine.
   a. Conduct research to identify the normal metabolic pathways of this amino acid and the differences that occur in these pathways in PKU.
   b. What difficulties exist for someone with PKU?
   c. What regulations exist in Canada to minimize the risk of PKU?

61. Comment on the following statement, based on your knowledge of cellular respiration: "Some aerobic fitness classes should really be called anaerobic fitness classes because of the short, intense workout they provide."

62. If you run or ride a bike as fast as you can, your leg muscles may begin to feel weak and you may have a burning sensation in your legs. Explain what is happening to your leg muscles.

63. Physical fitness and competitive training is improved through metabolic knowledge and its applications. Summarize the role of lactic acid with respect to physical training and provide suggestions for an athlete looking to improve his or her peak performance.

64. Pyruvate is available as a dietary supplement. Its reported benefits are controversial but include enhanced weight loss and increased endurance levels during physical exercise. Infer how taking pyruvate as a dietary supplement could lead to these effects.

65. Use the following guidelines to develop two analogies.
   a. Use the idea of a rechargeable battery as an analogy for the cycle that "recharges" ATP from ADP and a free phosphate group.
   b. Outline, in the form of a paragraph or a diagram, an analogy to compare the structure and function of chloroplasts with an active solar system. (An active solar system uses a mechanical, chemical, or electrical device to transfer light energy from the Sun to another part of the system for the purpose of doing useful work.)

66. Examine the graphic organizer below. Write a brief essay of three or four paragraphs that describes the processes depicted and explains the significance of these interrelated processes to life on Earth.

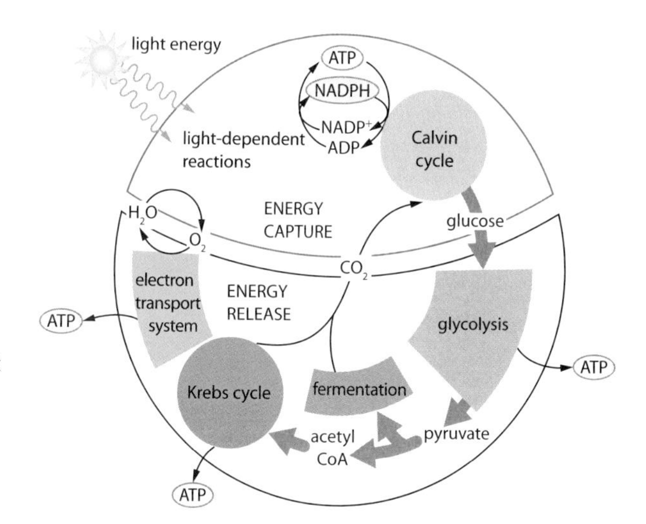

67. The hormone erythropoietin (EPO) is produced by the kidneys. It binds to receptors in bone marrow to stimulate the production of erythrocytes (red blood cells), which increases the oxygen-carrying capacity of the blood. Although EPO is used clinically in cases of anemia, some athletes have adopted its use (illegally) to improve performance—a practice called "blood doping."
   a. Predict how the use of EPO would affect aerobic respiration.
   b. Conduct research into the use and abuse of EPO and its effects on the body.

68. Investigate the experiments that were conducted by Peter Mitchell to test the chemiosmotic hypothesis that he proposed in 1961.

UNIT 2 SELF-ASSESSMENT

Select the letter of the best answer below

1. **K/U** In photosynthesis, ATP functions to provide energy for the
   a. absorption of CO_2
   b. splitting of water molecules into O_2 and H^+
   c. light-dependent reactions
   d. light-independent reactions
   e. absorption of O_2

2. **K/U** Glycolysis, which is the first series of reactions in cellular respiration, occurs in the
   a. cytoplasm
   b. mitochondria
   c. chloroplasts
   d. nucleus
   e. cell membrane

3. **K/U** Which of the following compounds is produced by the Krebs cycle?
   a. pyruvate and CO_2
   b. CO_2 and lactate
   c. CO_2 and ATP
   d. lactate and ATP
   e. lactate and pyruvate

4. **K/U** During cellular respiration, some of the energy stored in the bonds of glucose molecules is transferred to high energy bonds in ATP. The remainder of the energy is transformed into
   a. light
   b. heat
   c. chemical energy
   d. reducing power
   e. mechanical energy

5. **K/U** In the mitochondria, the electron transport system is
   a. associated with the outer membrane
   b. associated with the inner membrane
   c. in the ATP synthase
   d. in the Krebs cycle
   e. none of the above

6. **A** Which of the following pathways is common to both aerobic and anaerobic cellular respiration?
   a. Krebs cycle
   b. glycolysis
   c. electron transport
   d. fermentation
   e. Calvin cycle

7. **K/U** The generation of ATP by the movement of hydrogen ions down their concentration gradient across a membrane occurs because of a process called
   a. chemiosmosis
   b. diffusion
   c. reverse osmosis
   d. passive transport
   e. facilitated transport

8. **K/U** What is the final destination of the electrons released from water in photophosphorylation?
   a. oxygen
   b. NADPH
   c. a hydrogen ion
   d. H_2O
   e. ATP

9. **K/U** In chemiosmosis, the diffusion of hydrogen ions across a membrane in the organelle occurs through passageways created by
   a. NADPH
   b. NADH
   c. ATP synthase
   d. ATP
   e. AMP

10. **T/I** Which of the following molecules is directly shared between the Calvin cycle and the Krebs cycle?
    a. water
    b. oxygen
    c. carbon dioxide
    d. sugar
    e. all of the above

Use sentences and diagrams as appropriate to answer the questions below.

11. **C** Using a graphic organizer, summarize the two laws of thermodynamics and explain their significance to human metabolic processes.

12. **T/I** What advantage does cyclic photophosphorylation hold over noncyclic photophosphorylation? When would a cell favour noncyclic photophosphorylation?

13. **T/I** Differentiate between photophosphorylation and chemiosmosis.

14. **K/U** What is the role of electron carriers in cellular respiration? Provide a specific example.

15. T/I Mariana is concerned that the plants in her bedroom are reducing the available oxygen at night. So, she decides to conduct an experiment to determine their rate of cellular respiration at night. Without access to oxygen sensors, she chooses to set up a glass of limewater to determine if the carbon dioxide levels are increasing in the room. What variables must she control and how can her experiment be improved?

16. T/I Explain, using an analogy, how the Calvin cycle is essential but inefficient.

17. K/U How does the presence of ATP regulate aerobic respiration?

18. T/I What advantage do CAM plants possess that make them ideal for operating in hot and dry climates?

19. T/I Define the roles of kinetic energy and potential energy in metabolism. What is the connection of kinetic and potential energy to entropy and why is an input of energy necessary for the continuation of metabolism?

20. A Solar panels can be thought of as synthetic leaves since they, like plant leaves, capture the Sun's energy.

a. How has an understanding of photosynthesis given researchers a new direction in efforts to harness energy from the Sun.

b. Is it likely that researchers could harness more energy with solar panels than is produced in photosynthesis?

21. A A leading cause of global warming is the presence of additional methane in the atmosphere. Compare and contrast this metabolic waste product with human metabolic waste. Which product has more potential energy—methane or human metabolic waste?

22. C Using a diagram, summarize aerobic cellular respiration. The diagram should include the locations of ATP production and the names of the pathways involved.

23. C Demonstrate an understanding of both the electron transport chain and photophosphorylation by writing a description of these processes that applies to both but specifies neither.

24. A Models such as the one shown in the illustration below are often used to represent the electron transport chain.

a. Explain how this model works.

b. Explain why a metabolic process such as cellular respiration requires a multi-stage release of energy, rather than a one-step exergonic reaction such as the one in which hydrogen combines with oxygen to form water.

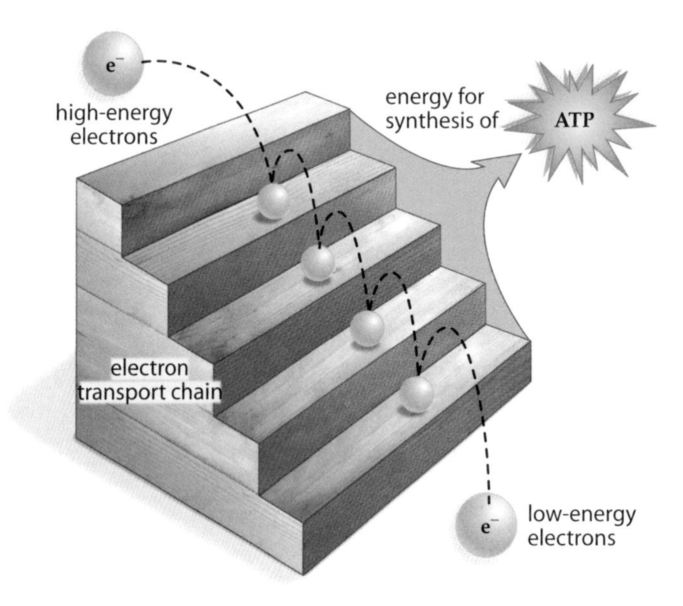

25. A How does the glucose oxidation process support an understanding of other metabolic pathways in human cells?

Self-Check

If you missed question...	1	2	3	4	5	6	7	8	9	10	11	12	13	14	15	16	17	18	19	20	21	22	23	24	25
Review section(s)...	4.2	3.2	3.2	3.1	3.2	3.3	3.2 4.1	4.1	3.2 4.1	3.2 4.1	3.1	4.1	4.1	3.1	3.2	4.2	3.2	4.2	3.1	4.1	3.3	3.2	3.2 4.1	3.2 4.1	3.2

UNIT 3 Molecular Genetics

BIG IDEAS

- DNA contains all the genetic information for any living organism.
- Proteins control a wide variety of cellular processes.
- Genetic research and biotechnology have social, legal, and ethical implications.

Overall Expectations

In this unit, you will learn how to...

- **analyze** some of the social, ethical, and legal issues associated with genetic research and biotechnology
- **investigate**, through laboratory activities, the structures of cell components and their roles in processes that occur within the cell
- **demonstrate** an understanding of concepts related to molecular genetics, and how genetic modification is applied in industry and agriculture

Unit Contents

Focussing Questions

1. *What are the structures and functions of DNA and RNA?*
2. *How are genes expressed?*
3. *What are some of the risks and benefits of biotechnology?*

Go to **scienceontario** to find out more about molecular genetics

This unit discusses some exciting technologies emerging from research in molecular genetics. Scientists can now use genetically modified organisms to produce medications, to profile the DNA sequence of a cancerous tumour, and to treat genetic disorders by introducing the correct form of a disease-related gene into an individual's genome. Another intriguing research focus is the relationship between aging and chromosomes. Scientists have discovered that when chromosomes replicate, they shrink. Each chromosome shortens each time it is replicated, as some of the nucleotides in highly repetitive DNA sequences at the end of the chromosome, known as telomeres, are lost. Eventually, when telomeres have been completely lost, their DNA itself begins to degrade. This affects gene expression. Because the degree of telomere shortening reflects the extent to which an individual's cells are aging, telomere testing will soon be offered as a means of determining relative "physiological age." These tests will also provide people with an indication of their overall health, since telomere length is significantly reduced by other health factors, such as lack of exercise, smoking, stress, obesity, and various diseases.

As you study this unit, look ahead to the Unit 3 Project on pages 326 to 327. Complete the project in stages as you progress through the unit.

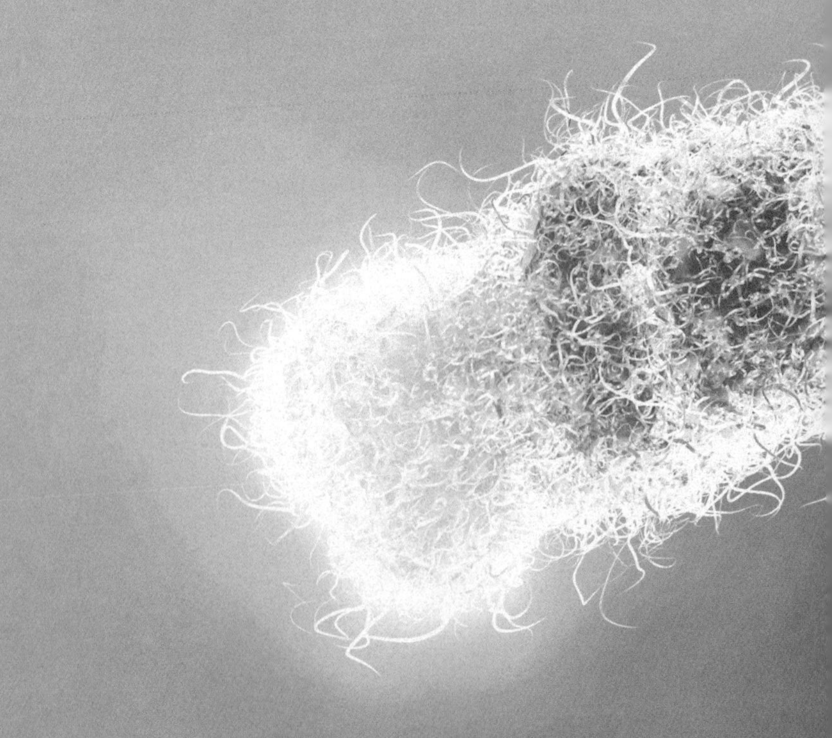

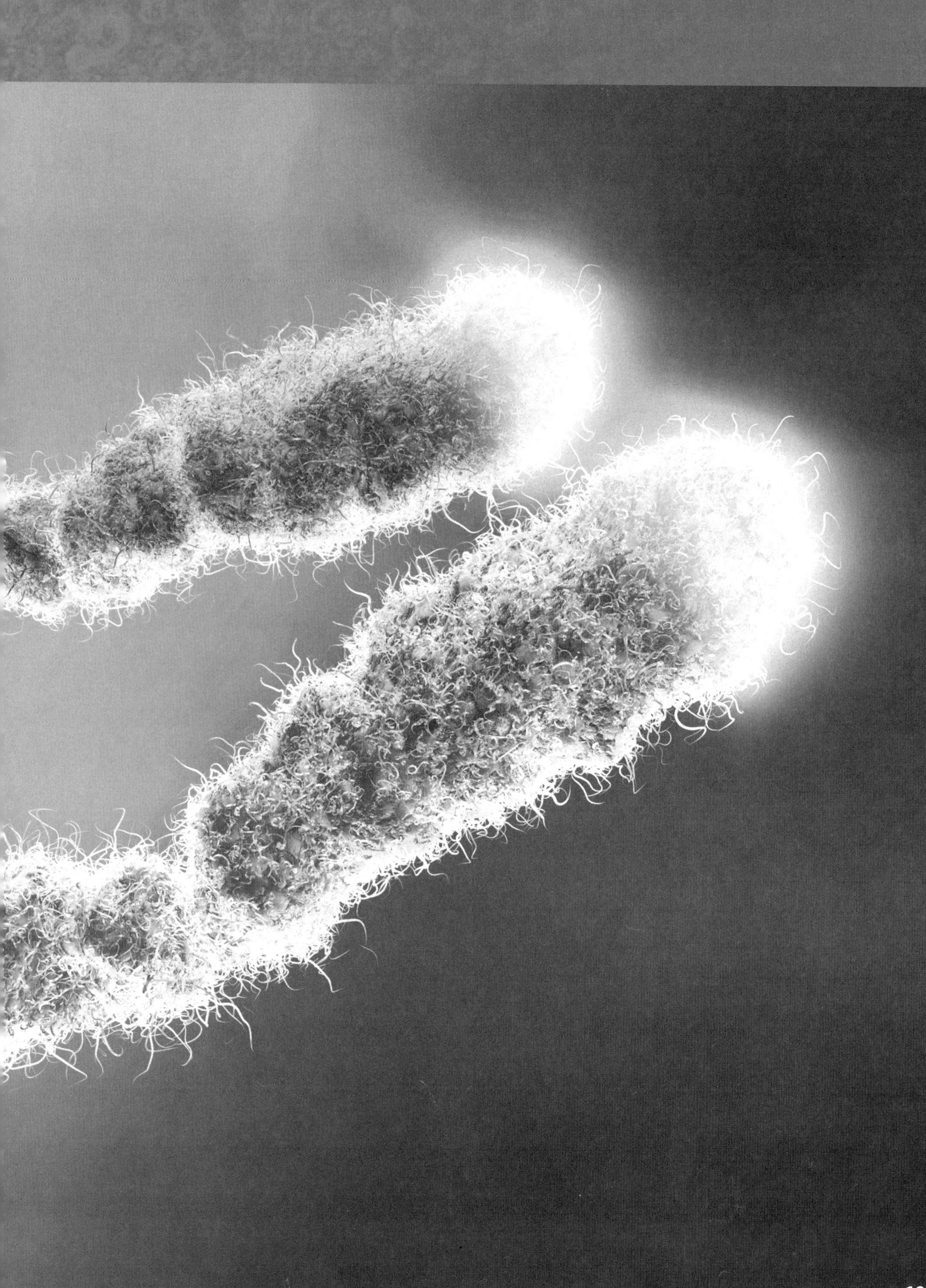

UNIT 3 Preparation

Cells and Microscopic Organisms

- Prokaryotic cells are simpler and smaller than eukaryotic cells. Unlike eukaryotic cells, prokaryotic cells do not have a membrane-bound nucleus or other membrane-bound organelles.
- Viruses consist of DNA or RNA surrounded by a protective protein coat. Viruses are non-cellular. Because they cannot survive independently of cells and must make use of their host's cellular machinery to regenerate, some scientists do not consider viruses to be living organisms.
- Bacteria are cellular organisms that reproduce asexually by binary fission, during which the cell grows and makes a copy of its genetic material.
- In binary fission, once the bacterial cell reaches a certain size, it elongates, and the original chromosome and its copy separate. The cell then builds a partition between the two chromosomes and finally splits into two smaller cells.

1. Distinguish between prokaryotes and eukaryotes by copying and completing the following table.

Characteristic	Prokaryotes	Eukaryotes
Relative cell size		
Cell number in typical organism		
Location of genetic material		
Membrane-bound genetic material		
Number of chromosomes		

2. Prokaryotic organisms include
a. amoebae
b. influenza viruses
c. *E. coli* bacteria
d. mushrooms
e. none of the above

3. Describe three ways in which viruses differ from cells.

4. Viruses make use of their host's cellular machinery to produce multiple copies of themselves through the process of *viral replication*. The host's "cellular machinery" refers primarily to which cellular components?
a. the vacuoles and cell membrane
b. the mitochondria and chloroplasts
c. the ribosomes and nucleus
d. the lysosomes
e. the cytoplasm

5. Bacteria divide through binary fission. During this process
a. mitosis occurs
b. meiosis occurs
c. two genetically unique cells are formed
d. two genetically identical cells are formed
e. conjugation occurs

6. Plasmids are small loops of gene-containing DNA that
a. are found in viruses but not in bacteria
b. are transferred during favourable conditions
c. are transferred between viruses
d. are transferred between bacteria
e. none of the above

7. Examine the cell shown below.

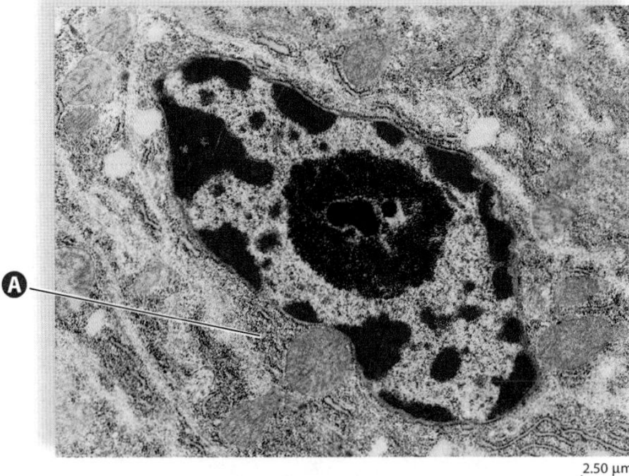

a. Identify the organelle labeled A.
b. Describe two functions it carries out in cells.

Components of Living Systems

- Living systems contain many different molecules. Some molecules are small and simple while others, called macromolecules, are larger and more complex. Macromolecules provide the raw materials necessary for energy production, cellular regulation, and tissue maintenance and repair.
- There are four main categories of macromolecules: carbohydrates, lipids, proteins, and nucleic acids.
- Enzymes are a special class of protein molecule. Enzymes act as catalysts— substances that increase the rate of chemical reactions without being used up in the reaction.
- There are two types of nucleic acids—DNA and RNA. Genes are sections of DNA that contain genetic information for the inheritance of specific traits. Genes contain specific nucleotide sequences that code for polypeptides and certain other molecules.

8. Proteins are made up of smaller sub-units called ________ that are joined to form one or more chains by means of ____________. These chains are referred to as ______________.
   a. amino acids, ionic bonds, polypeptides
   b. nucleic acids, ionic bonds, polypeptides
   c. amino acids, peptide bonds, polypeptides
   d. nucleic acids, peptide bonds, polypeptides
   e. amino acids, peptide bonds, polymers

9. An example of a protein is a(n)
   a. enzyme
   b. insulin molecule
   c. hemoglobin molecule
   d. antibody
   e. all of the above

10. The sites of protein synthesis in cells are the
   a. nuclei
   b. vesicles
   c. lysosomes
   d. ribosomes
   e. peroxisomes

11. Describe the role played by enzymes in living organisms.

12. Most enzymes have globular shapes, with pockets or indentations on their surfaces, as shown below. Describe the function of these indentations.

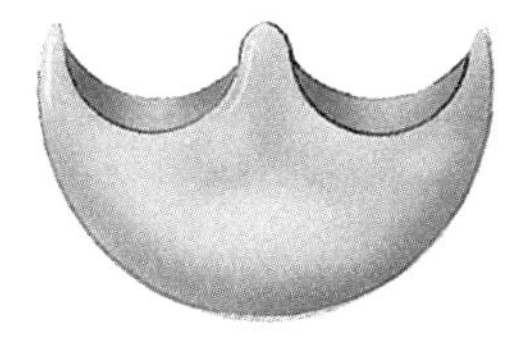

13. Explain how enzymes are classified.

14. Create a flowchart using the following terms: protein, nucleic acid, gene, and nucleotide. Then explain the relationship depicted in your flowchart.

15. The diagram below shows a nucleotide. Nucleotides join together to form nucleic acids.

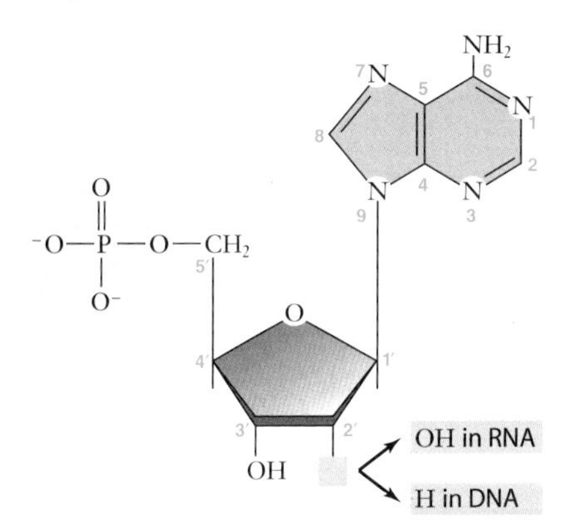

Nucleotides are composed of
   a. a sugar, a phosphate group, and an acid
   b. an alcohol, a phosphate group, and a nitrogenous base
   c. a sugar, a phosphate group, and a protein
   d. a sugar, a phosphate group, and a nitrogenous base
   e. a sugar, a nitrate group, and a nitrogenous base

16. DNA is composed of polymers of nucleotides that are often referred to as *strands*.
   a. Describe the intermolecular forces that link nucleotides together between the two strands that make up a DNA molecule.
   b. Describe the covalent bonds that link adjacent nucleotides together within each strand.

Concepts in Genetics

- Genetics is the field of biology that studies how hereditary information is passed from one generation of organisms or cells to the next.
- In genetics, a trait is a specific feature or characteristic of an organism. Genes carry genetic information for specific traits.
- Homologous chromosomes contain identical gene sequences, but may carry different forms of the same genes.
- Different forms of the same genes are known as alleles.
- Genetic disorders may occur due to errors caused by changes in chromosome structure.

17. A scientist is studying the inheritance of petal colour in wild lupine (*Lupinus perennis*) in Ontario. In genetics, the petal colour of wild lupine is a

a. genome
b. gene
c. genotype
d. mutagen
e. trait

18. Use the concepts of homologous chromosomes and alleles to account for different hair colour in a family.

19. The complete DNA sequence of an organism is that organism's

a. genome
b. gene
c. proteome
d. [illegible]
e. chromosome

20. The combination of alleles for a given trait or the whole genetic composition of an organism is called that organism's

a. genotype
b. genetic profile
c. DNA profile
d. phenotype
e. none of the above

21. The expression of an organism's genotype results in the physical and physiological traits of the organism. These traits are referred to as that organism's

a. genome
b. trait profile
c. genetic profile
d. qualitative traits
e. phenotype

Several changes in chromosome structure are shown in the diagram below. Use this diagram to answer questions 22 to 24.

Error in Chromosone Structure

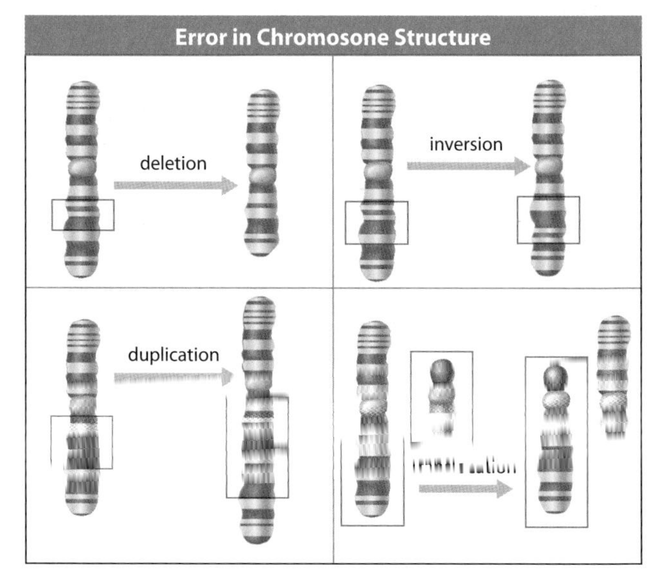

22. Describe the four errors that may occur in chromosome structure, as illustrated in the above diagram.

23. The above changes in chromosome structure occur during meiosis.

a. During what phase of meiosis do the above changes occur?
b. During this phase, chromosomal segments are exchanged in a process called _______________.
c. Explain how chromosome errors may occur during the process described in question b.

24. The chromosome errors described above can, in some instances, result in genetic disorders. In other instances, such errors are not harmful. Suggest why chromosome errors can produce two very different results. (Hint: What types of errors would you expect to be harmful? What types of errors would you expect to be harmless?)

Genetic Research

- Modern technology allows scientists to manipulate DNA and genetically alter organisms.
- The process of producing identical copies of genes, cells, or organisms is called cloning.
- Gene cloning specifically refers to DNA manipulation that produces multiple copies of a DNA segment or gene.
- Inserting foreign DNA into organisms results in transgenic organisms. Transgenic organisms are a type of genetically modified organism (GMO).
- A genetically modified organism is an organism whose genome is altered in some way. This modification is typically for a specific purpose, such as production of a new protein.

The diagram below shows the main steps in creating goats that are genetically modified to produce human protein products for medicinal use. Use this diagram to answer questions 25 and 26.

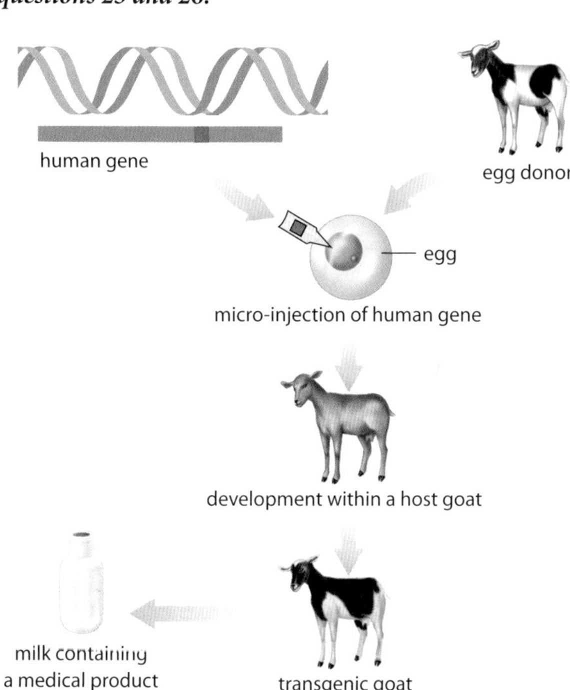

25. The genetically modified organism in this process is the
- **a.** human who donates the gene
- **b.** egg donor goat
- **c.** host goat
- **d.** transgenic goat
- **e.** human who receives the protein product

26. Write a caption that explains what is taking place at each step in the diagram.

27. A scientist is interested in inserting a gene that enables one plant to thrive at low temperatures into another plant that requires warm conditions to grow. To do this, the scientist must first create multiple copies of the gene of interest. This process is referred to as
- **a.** transgenic engineering
- **b.** DNA manipulation
- **c.** gene cloning
- **d.** gene modification
- **e.** reproductive cloning

28. Suggest why viruses are useful tools for scientists who are carrying out gene cloning research.

29. The chromosomes of golden rice have been altered to contain four new genes. The visual below shows a simplification of this process, as the development of golden rice actually involves gene insertions into multiple rice chromosomes.
- **a.** Golden rice is an example of what type of organism?
- **b.** Does golden rice have the same genotype as the original rice plant? Explain.
- **c.** Does golden rice have the same phenotype as the original rice plant? Explain.

Beans

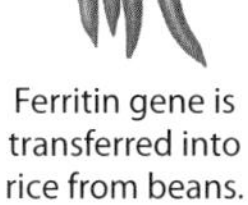

Ferritin gene is transferred into rice from beans.

***Aspergillus* fungus**

Phytase gene is transferred into rice from a fungus.

Wild rice

Metallothionein gene is transferred into rice from wild rice.

Daffodil

Enzymes for β-carotene synthesis are transferred into rice from daffodils.

Fe | *Pt* | rice chromosome | *S* | A_1 A_2 A_3 A_4

Ferritin protein increases iron content of rice.

Phytate, which inhibits iron reabsorption, is destroyed by the phytase enzyme.

Metallothionein protein supplies extra sulfur to increase iron uptake.

β-carotene, a precursor to vitamin A, is synthesized.

CHAPTER 5

The Structure and Function of DNA

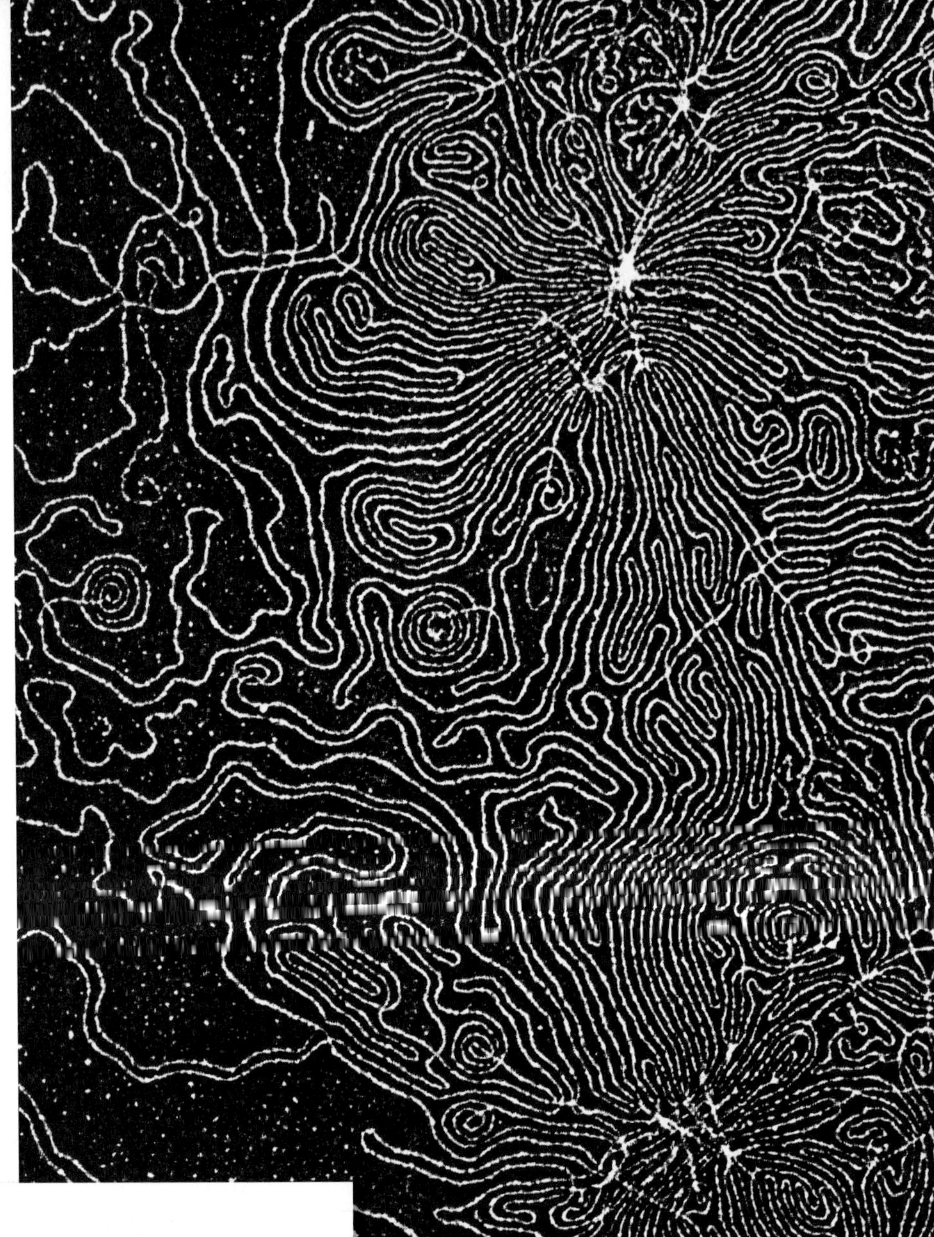

Specific Expectations

In this chapter, you will learn how to . . .

- D1.1 **analyze**, on the basis of research, some of the social, ethical, and legal implications of biotechnology (5.1, 5.2)
- D1.2 **analyze**, on the basis of research, some key aspects of Canadian regulations pertaining to biotechnology and **compare** them to regulations from another jurisdiction (5.2)
- D2.1 **use** appropriate terminology related to molecular genetics (5.1, 5.2)
- D2.3 **conduct** an investigation to extract DNA from a specimen of plant or animal protein (5.2)
- D3.1 **explain** the current model of DNA replication, and **describe** the different repair mechanisms that can correct mistakes in DNA sequencing (5.2)
- D3.7 **describe**, on the basis of research, some of the historical scientific contributions that have advanced our understanding of molecular genetics (5.1, 5.2)

In laboratories around the world, scientists routinely study and alter the genetic makeup of many organisms. However, this ability has only been possible for a relatively short period of time. Before the 1950s, scientists did not even know what the genetic material of cells was. Scientists now know that DNA, shown here in its uncondensed form, carries an organism's genetic information that defines many of its traits, including behaviours and predisposition for certain diseases. The unique properties of DNA provide the stability needed to accurately reproduce and transmit genetic information from one generation to the next, as well as the ability to produce infinite variations that allow a species to adapt and survive. This chapter introduces the fundamental beginnings of modern molecular genetics by exploring the molecular structure and properties of DNA.

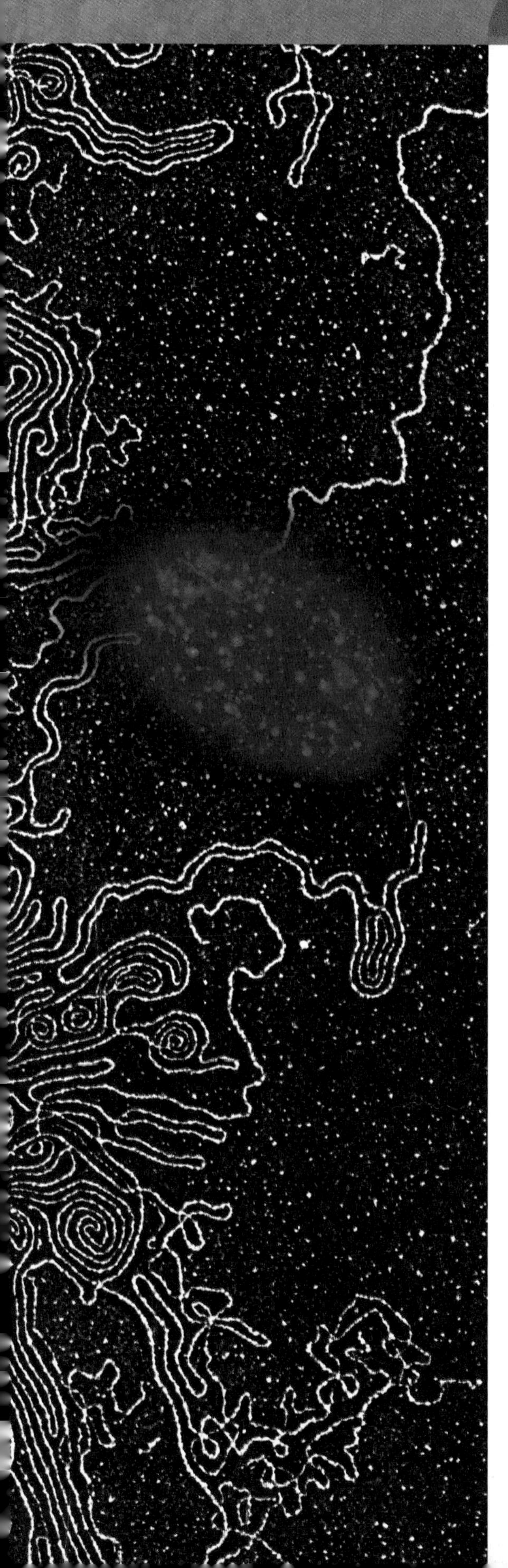

Launch Activity

DNA, Biotechnology, and Society

In this chapter, you will read about the work of many scientists that has helped us gain invaluable knowledge and understanding of the molecular structure and properties of DNA. When this knowledge has been applied with technology, it has led to many processes that affect society today, including DNA evidence being collected at crime scenes, producing genetically modified plants and animals, pinpointing the exact cause of genetic disorders, manufacturing certain medications, and mapping the human genome. As our scientific knowledge and technology continue to advance, more things become possible. However, with each new advancement, questions and controversies may also arise.

Materials

- access to print or Internet resources

Procedure

1. Choose one of the following topics to research:
   - mandatory DNA fingerprinting upon arrest for a felony
   - familial DNA searching to solve crimes
   - cloning deceased pets, such as cats or dogs
   - therapeutic cloning of tissues and organs for transplant
2. Research more information about the topic. Questions that may help guide your research include:
   - How does the process work?
   - Why might it be used?
   - What are the pros and cons involved in its use?
   - What are the ethical, legal, and/or human rights issues associated with the process?
   - What regulations exist or would be needed to oversee the process and protect the individuals or entities involved?

Questions

1. Provide a brief summary of the science behind the process you chose to research.
2. Provide a brief summary of the history of the process.
3. List the pros and cons of the process.
4. Discuss the ethical, legal, and/or human rights controversies surrounding the process.
5. State your opinion on whether the process should be used. Justify your position using facts from your research.

SECTION
5.1 DNA Structure and Organization in the Cell

Key Terms

nucleotide
Chargaff's rule
complementary base pairing
antiparallel
genome
gene
nucleoid
DNA supercoiling
regulatory sequence
histone
nucleosome
chromatin

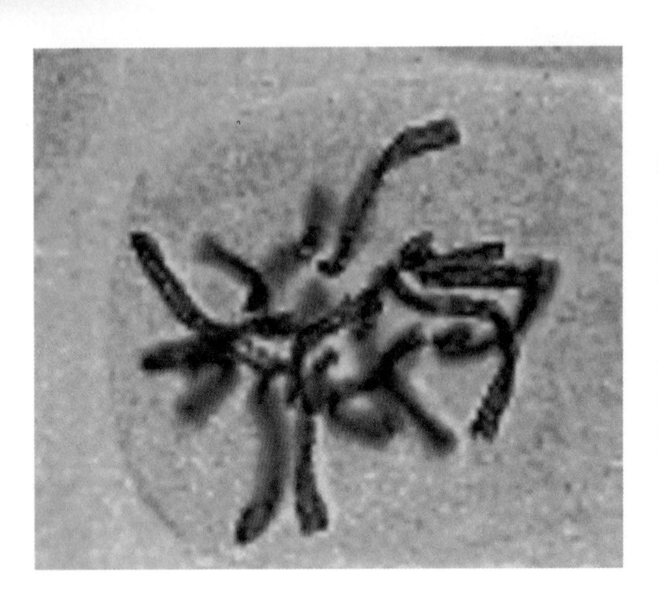

Figure 5.1 During the initial stages of cell division, chromosomes are visible through a compound light microscope. Chromosomes were first observed in 1870 by German biologist Walther Flemming. His scientific paper reporting his observations was published in 1882.

By the start of the 1900s, the connection between chromosomes, shown in **Figure 5.1**, and the inheritance of specific traits was well established. Thus, scientists inferred that the hereditary, or genetic, material was to be found in the chromosomes of cells. Whatever its chemical composition might be, scientists knew that the genetic material had to meet several crucial criteria. It had to contain information that controls the production of enzymes and other proteins. It also had to be able to replicate itself with great accuracy in order to maintain continuity from one generation to the next. In addition, the ability of the genetic material to replicate itself had to allow for occasional mutations as a means for introducing variation within a species. What kind of biological compound had the structural complexity to meet these criteria?

Scientists already [illegible] were composed of two types of macromolecules: proteins and nucleic acids. Proteins were known to be composed of 20 amino acids, and these monomers could be combined in seemingly endless ways to produce thousands, if not millions, of different proteins. Nucleic acids, on the other hand, were composed of just four bases and thus were assumed to have a considerably more limited potential for combinations. In terms of structural complexity, therefore, proteins had a clear advantage over nucleic acids as candidates for the genetic material. It would take about 50 years for scientists to become convinced that deoxyribonucleic acid, DNA, was in fact the molecule of heredity.

Identifying DNA as the Material of Hereditary

One of the first key investigations that would help to establish DNA as the genetic material took place in London, England, in 1928. At that time, bacterial pneumonia was a common, often lethal, disease affecting millions around the world. Because antibiotics had not yet been invented, the bacterium responsible for the disease, *Streptococcus pneumoniae*, was the object of intense study and research. One researcher, microbiologist Frederick Griffith (1879–1941), worked at the British Ministry of Health and was studying the pathology (disease-causing characteristics and effects) of this bacterium. Although Griffith performed many experimental studies with *S. pneumoniae*, one specific approach proved to be pivotal to the history of genetics.

Figure 5.2 shows the experimental design that Griffith used. In this study, two forms of the bacterium were used. One type, called the S-strain, was highly pathogenic but could be made non-pathogenic by heating it. The second type, called the R-strain, was a non-pathogenic form of the bacterium. Griffith discovered that mice died after being injected with a mixture of heat-killed S-strain bacteria and living R-strain bacteria. He concluded that the S-strain had somehow passed on its deadly properties to the live, non-pathogenic R-strain. In a scientific paper published in 1928, Griffith called this phenomenon the transforming principle, because something from the heat-killed pathogenic bacteria must have transformed the living non-pathogenic bacteria to make them deadly.

Griffith's discovery of transformation

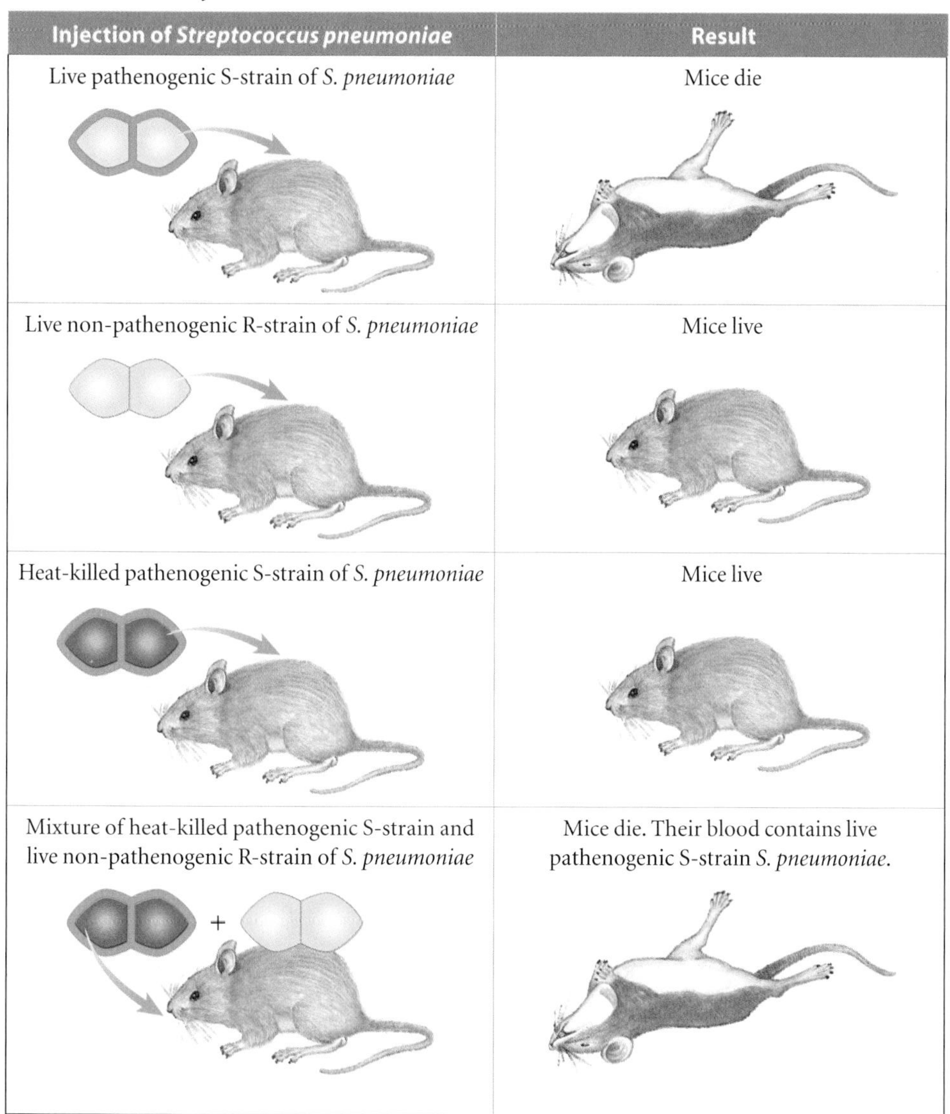

Injection of *Streptococcus pneumoniae*
Result
Live pathenogenic S-strain of *S. pneumoniae*
Mice die
Live non-pathenogenic R-strain of *S. pneumoniae*
Mice live
Heat-killed pathenogenic S-strain of *S. pneumoniae*
Mice live
Mixture of heat-killed pathenogenic S-strain and live non-pathenogenic R-strain of *S. pneumoniae*
+
Mice die. Their blood contains live pathenogenic S-strain *S. pneumoniae.*

Figure 5.2 Fredrick Griffith showed that something was present in the heat-killed S-strain of *S. pneumoniae* that could transform the non-pathogenic R-strain into the pathogenic form.

***Explain** why it was important for Griffith to heat the S-strain bacteria.*

Griffith died of injuries during World War II, but his results spurred the research community to understand the pathology of *S. pneumoniae.* At Rockefeller University in New York, Canadian-American microbiologist Oswald Avery and colleagues were studying the chemical properties of the polysaccharide capsule that surrounds the cells of pathogenic *S. pneumoniae* strains. After reading Griffith's paper, Avery directed his research group's expertise to identifying the molecules in *S. pneumoniae* that cause the transformation.

Identifying Griffith's Transforming Principle

In 1944, Oswald Avery, Colin MacLeod, and Maclyn McCarty published a study that supported the hypothesis that DNA was the hereditary material. By this time, researchers had developed methods to grow bacteria in liquid cultures. In their experiments, Avery and his colleagues prepared identical extracts of the heat-killed S-strain by growing the cells in liquid culture, isolating the bacteria, disrupting the cell membranes, and collecting the cell contents. Then, one of three enzymes was added to each extract. One enzyme specifically destroyed protein, a second enzyme specifically destroyed RNA, and a third enzyme specifically destroyed DNA. Each enzyme-treated extract was then mixed with live R-strain cells. The only extract that did not cause transformation of the R-strain to the pathogenic S-strain was the extract treated with the DNA-destroying enzyme. This result showed that DNA caused the transformation. Griffith's transforming principle was DNA.

Avery and his colleagues also developed methods to isolate and chemically characterize the transforming principle. They prepared and then separated extract of bacterial cultures into different fractions that were tested for their ability to transform R-strain cells into S-strain cells. Eventually, they isolated material with properties consistent with DNA. In spite of this evidence, many scientists at that time held firm to the belief that the more chemically complex proteins were the hereditary material.

Hershey and Chase Demonstrate that DNA Is the Genetic Material

In 1952, the American microbiologist team of Alfred Hershey and Martha Chase designed one of the most famous experiments in the history of genetics to rule out protein in favour of DNA as the hereditary material. Their experiment used bacteriophages, which are viruses that infect bacteria. Bacteriophages have an inner nucleic acid core and an outer protein coat, called a capsid. This structural simplicity made them perfect for the researchers' experiments.

Hershey and Chase used the T2 bacteriophage strain of virus, which consists of a protein coat that surrounds a piece of DNA. When infecting a bacterial cell, the virus attaches to the cell and injects its genetic information into it. The remaining viral structure stays attached to the outside of the bacterium and is referred to as a bacteriophage "ghost." The infected cell manufactures new virus particles using the virus's genetic information. Eventually, the cell bursts and the released viruses infect other cells. Hershey and Chase aimed to determine which part of the virus—the DNA in the viral core or the protein in the capsid—enters bacterial cells and directs the production of more viruses. Once they knew this, they would know whether genetic information was encoded in DNA or protein.

The experimental approach that Hershey and Chase took is shown in **Figure 5.3**. To study the role that the protein and DNA play in T2 infection, Hershey and Chase used radioactive isotopes to trace each type of molecule. Since proteins contain sulfur but DNA does not, they introduced a radioactive source of sulfur (^{35}S) into the protein of the virus by growing infected bacteria in media that contained this isotope. Similarly, since DNA contains phosphorus but T2 protein does not, they were able to produce a virus that had DNA that was specifically labelled with a radioactive isotope of phosphorus (^{32}P).

In one experiment, a virus with radioactively labelled DNA was allowed to infect *Escherichia coli* bacteria. The bacterial cells were then agitated with a blender to remove the bacteriophage ghosts. The material was centrifuged (spun at high speed) to separate the infected bacterial cells, which formed a pellet at the bottom of the centrifuge tube, and the liquid medium, which contained the remnant bacteriophage "ghosts." Hershey and Chase found that most of the radioactively labelled DNA was in the bacteria and not in the liquid. The only way this could happen is if the viral DNA had entered the bacteria.

In a second experiment, a virus with radioactively labelled protein was allowed to infect *E. coli* bacteria. The same procedure was followed to separate the baceriophage ghosts and a pellet of bacteria. For this experiment, Hershey and Chase found most of the radioactively labelled protein in the liquid medium and not in the bacteria. The only way this could happen is if the radioactive protein in the viral capsid remained as part of the bacteriophage ghosts and had not been injected into the bacterial cells.

The Hershey–Chase experiments settled the matter of which molecule—DNA or protein—is the genetic material. The results provided conclusive evidence that viral DNA was transferred to the bacterial cells, and that viral DNA held the genetic information needed for the viruses to reproduce.

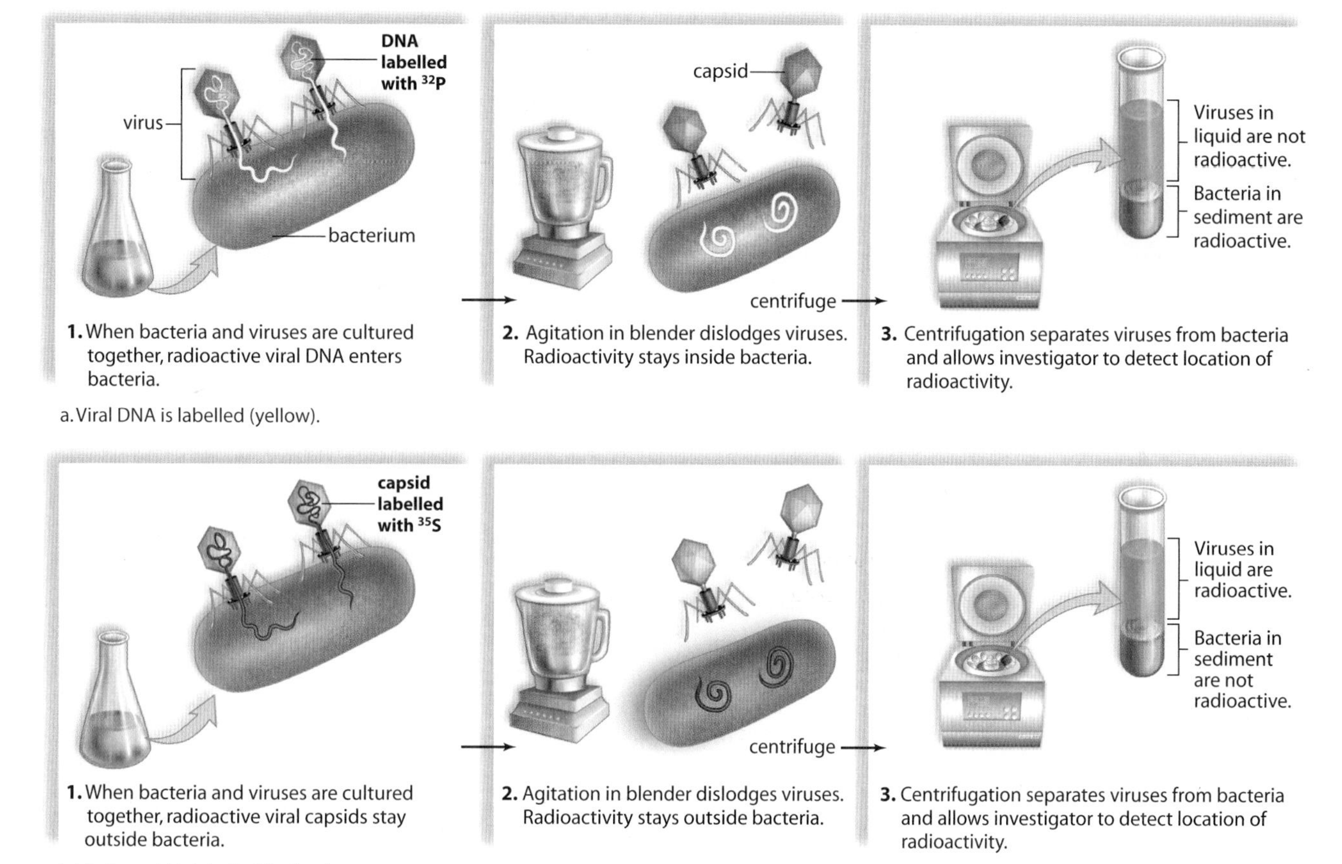

DNA labelled with ³²P
virus
bacterium
capsid
centrifuge
Viruses in liquid are not radioactive.
Bacteria in sediment are radioactive.
1. When bacteria and viruses are cultured together, radioactive viral DNA enters bacteria.
2. Agitation in blender dislodges viruses. Radioactivity stays inside bacteria.
3. Centrifugation separates viruses from bacteria and allows investigator to detect location of radioactivity.
a. Viral DNA is labelled (yellow).
capsid labelled with ³⁵S
centrifuge
Viruses in liquid are radioactive.
Bacteria in sediment are not radioactive.
1. When bacteria and viruses are cultured together, radioactive viral capsids stay outside bacteria.
2. Agitation in blender dislodges viruses. Radioactivity stays outside bacteria.
3. Centrifugation separates viruses from bacteria and allows investigator to detect location of radioactivity.
b. Viral capsid is labelled (yellow).

Figure 5.3 In the Hershey–Chase experiment, radioactive phosphorus in viral DNA and radioactive sulfur in viral proteins were used to trace the transfer of each type of biological molecule into a bacterial host cell.

***Predict** the results if Hershey and Chase had chosen a bacteriophage that had RNA as its genetic material.*

Learning Check

1. Although scientists in the early 1900s had not yet identified the chemical composition of genetic material, what criteria did scientists know genetic material had to meet?
2. How did Griffith test for the existence of a transforming principle?
3. What were the results of the experiments on *S. pneumoniae* done by Avery, MacLeod, and McCarty?
4. Why did Hershey and Chase use two different radioactive isotopes in their experiments?
5. Identify the dependent and independent variables in Hershey and Chase's experiments. What were some controls?
6. If protein were the hereditary material, what would Hershey and Chase have seen in the results of their experiments?

Determining the Chemical Composition and Structure of DNA

Although the role of DNA as the genetic material was not firmly established until 1952, the *existence* of DNA had been discovered more than 80 years earlier. In 1869, a Swiss chemist, Friedrich Miescher, isolated the nuclei of white blood cells that he obtained from pus-soiled hospital bandages. From these nuclei, and later from the nuclei of other cells, he extracted a weakly acidic substance containing nitrogen and phosphorus. Miescher named it *nuclein*, since it was found in the nuclei of cells. Later researchers would rename it nucleic acid. In fact, what Miescher had discovered was DNA.

As some researchers worked to determine whether DNA or protein was the physical agent of heredity, others continued from Miescher's efforts to identify the chemical composition and properties of nucleic acid. In the early 1900s, a Russian-American biochemist, Phoebus Levene, isolated two types of nucleic acid. He named one ribose nucleic acid based on the presence of a five-carbon sugar called ribose. This nucleic acid is now referred to as ribonucleic acid, or RNA. The other nucleic acid contained a five-carbon sugar with a structure similar to ribose but with one less oxygen atom. Levene named it deoxyribose nucleic acid, and it was later renamed deoxyribonucleic acid, or DNA. In 1919, after many years of analysing the results of numerous hydrolysis reactions of nucleic acid from yeast, Levene proposed that RNA and DNA are made up of individual units that he called **nucleotides**. He correctly described each nucleotide as being composed of one of four nitrogen-containing bases, a sugar molecule, and a phosphate group.

nucleotide the repeating unit of nucleic acids; composed of a sugar group, a phosphate group, and a nitrogenous base

The Chemical Composition of the Nucleotides, DNA, and RNA

In later years, other scientists would confirm and extend Levene's work. Today we know that both DNA and RNA are made up of a combination of four different nucleotides. As shown in **Figure 5.4**, each nucleotide in DNA is composed of a five-carbon deoxyribose sugar, a phosphate group, and a nitrogen-containing base, all linked together by covalent bonds. Also, there are four different nitrogenous bases in DNA that can be categorized into two different forms: purines and pyrimidines. The purine bases are adenine (A) and guanine (G). They have two fused rings in their chemical structures. The pyrimidine bases are cytosine (C) and thymine (T). They have a single ring in their chemical structure. In RNA, all but one of the bases are the same as those of DNA. RNA has the pyrimidine base uracil (U) in place of thymine.

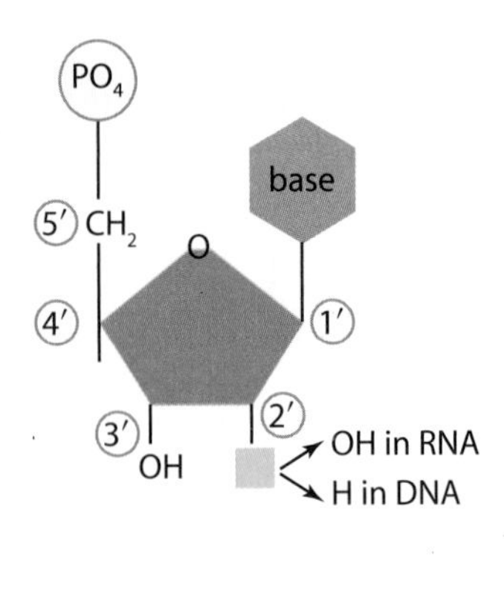

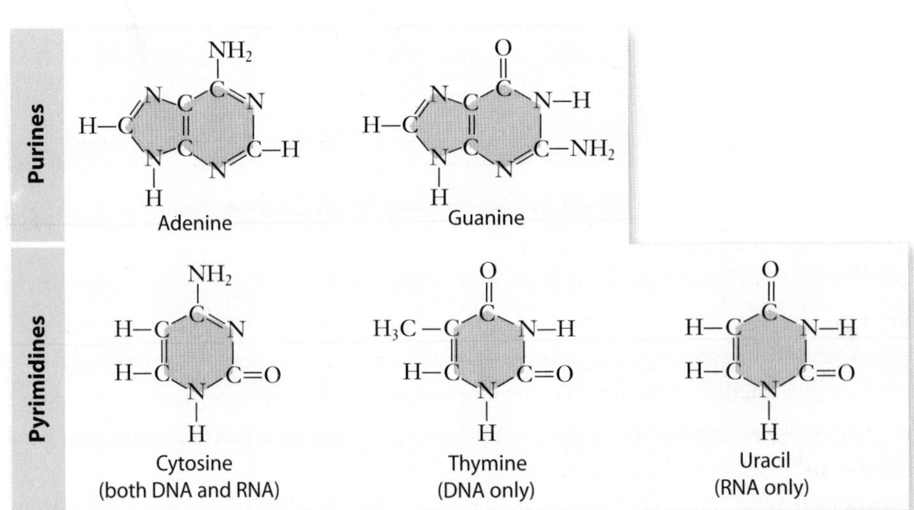

Figure 5.4 The general structure of a DNA nucleotide includes a phosphate group, a deoxyribose sugar group, and a nitrogen-containing base. Nucleotides in RNA have the same basic structure, except a ribose sugar group is used. The sugar groups differ by a hydroxyl group at the 2′ carbon. Both DNA and RNA contain the same purine bases and the cytosine pyrimidine base. However, thymine is only present in DNA, and uracil is only present in RNA.

Nucleotides are often identified by referring to their bases. For DNA these nucleotides are A, G, C, and T. For RNA, these nucleotides are A, G, C, and U.

Levene also proposed that nucleic acids are made up of long chains of nucleotides, in what he referred to as a polynucleotide model of nucleic acids. This model was later shown to be correct. However, Levene incorrectly hypothesized that nucleic acids had a tetranucleotide structure in which four nucleotides were linked together in a repeating sequence, such as ACTGACTGACTG. Since such a simple model could not encode much information, this incorrect model helped to fuel the belief that protein had to be the material of heredity.

Chargaff's Rule: Closing in on the Structure of DNA

An Austrian-American biochemist, Erwin Chargaff, had read the 1944 publication of Avery and his colleagues, which supported DNA as the hereditary material. Chargaff was so inspired by this discovery that he launched a research program to study the chemistry of nucleic acids. His approach was to determine how genetic information could be contained in DNA. Initially, Chargaff looked for differences in DNA among different species.

By the late 1940s, two significant conclusions came from this work. First, Chargaff showed that there is variation in the composition of nucleotides among different species. Thus, he disproved Levene's tetranucleotide hypothesis for the structure of DNA. Second, Chargaff demonstrated that all DNA, regardless of its source, maintains certain properties, even though the composition varies. For example, Chargaff observed that the nucleotides were present in characteristic proportions. As shown in **Table 5.1**, Chargaff found that the amount of adenine in any sample of DNA is always approximately equal to the amount of thymine, and the amount of cytosine is always approximately equal to the amount of guanine. This constant relationship is now known as **Chargaff's rule**. Although Chargaff did not understand its significance, this relationship provided key information for other scientists seeking the structure of DNA.

Chargaff's rule in DNA, the percent composition of adenine is the same as thymine, and the percent composition of cytosine is the same as guanine

Table 5.1 Percent Composition of Each Base from DNA of Several Species

Organism	Adenine	Thymine	Guanine	Cytosine
Mycobacterium tuberculosis	15.1	14.6	34.9	35.4
Escherichia coli	26.0	23.9	24.9	25.2
Yeast	31.3	32.9	18.7	17.1
Drosophila melanogaster	27.3	27.6	22.5	22.5
Mouse	29.2	29.4	21.7	19.7
Human (liver)	30.7	31.2	19.3	18.8

Determining the Three-Dimensional Structure of DNA

By the late 1940s, the scientific community knew

- from Hershey and Chase that DNA is the hereditary material
- from Levene that DNA is a polymer of nucleotides, and that each of the four types of nucleotides contain a phosphate group, a deoxyribose sugar, and one of four possible nitrogen-containing bases (adenine, guanine, cytosine, or thymine)
- from Chargaff that DNA is composed of repeating units of nucleotides in fixed proportions that vary with different species

The answer to a crucial question remained elusive, however. How were the components of DNA arranged to form a three-dimensional structure? In the early 1950s, at the University of Cambridge in England, a partnership formed between James Watson, an American biologist, and Francis Crick, a British physicist who was interested in molecular structures. Working closely together, they decided to determine the structure of DNA by using data that had already been generated by other members of the research community.

They set out to use this information as a basis for building a model of DNA that would explain how this molecule can vary from species to species and even from individual to individual. In addition to the research data that you have already encountered in this section, Watson and Crick relied on crucial discoveries made by two chemists: Linus Pauling and Rosalind Franklin.

Pauling Discovers a Helical Structure for Proteins

American chemist Linus Pauling (1901–1994) developed methods of assembling three-dimensional models based on known distances and bond angles between atoms in molecules. This allowed him to visualize and experiment with how atoms in a molecule might fit together to produce complex structures. Using this technique, Pauling discovered in 1951 that many proteins have helix-shaped structures. Crick later noted that this idea suggested the possibility of a helix shape for the structure of DNA.

Franklin Determines a Helical Structure for DNA

In the early 1950s, British chemist Rosalind Franklin (1920–1958) was working alongside another chemist, Maurice Wilkins, in a London laboratory. Both were using a technique called X-ray diffraction to analyze the structure of biological molecules. In this technique, a purified substance, such as a sample of DNA, is subjected to X rays, which are bent (diffracted) by the molecules that they encounter. The resulting diffraction pattern is captured on photographic film. Mathematical theory applications are then used to interpret the diffraction pattern, and information on the molecular structure is inferred from this.

Franklin, shown in **Figure 5.5A**, in particular had developed fine expertise in X-ray diffraction and was able to obtain the highest resolution photographs at that time. Based on images such as the one in **Figure 5.5B**, Franklin was able to conclude that DNA has a defined helical structure. Franklin also determined that the structure has two regularly repeating patterns—one recurring at intervals of 0.34 nm, and the other recurring at intervals of 3.4 nm. Also, as Franklin prepared her samples, she observed how DNA reacted with water. From her observations, she concluded that the nitrogenous bases were located on the inside of the helical structure, and the sugar-phosphate backbone was located on the outside, facing toward the watery nucleus of the cell.

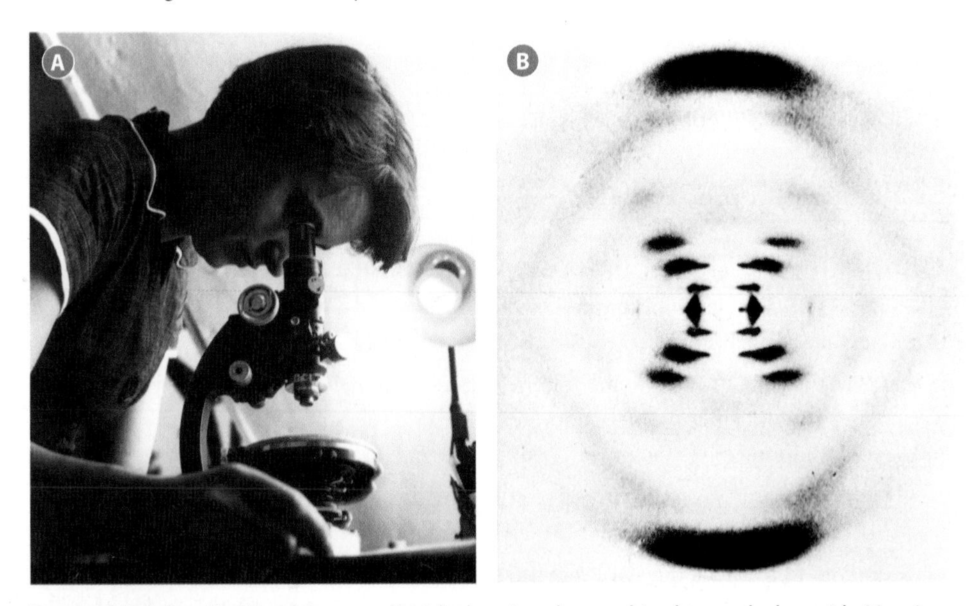

Figure 5.5 **A** Rosalind Franklin was a British chemist who was hired to work alongside Maurice Wilkins at the X-ray diffraction facilities at King's College. **B** In the diffraction image of DNA produced by her, the central x-shaped pattern enabled researchers to infer that DNA has a helical structure.

Watson and Crick Build a Three-Dimensional Model for DNA

Armed with the results, conclusions, and conjectures of their peers, Watson and Crick proceeded to infer and construct a model for DNA. At first, they used basic cardboard cutouts to represent the components of the nucleotides. In a similar way to making a puzzle, Watson and Crick shuffled the pieces around until they came up with possible ways that they could fit together and still reflect the research data. Eventually they built the model of DNA that is shown in **Figure 5.6**.

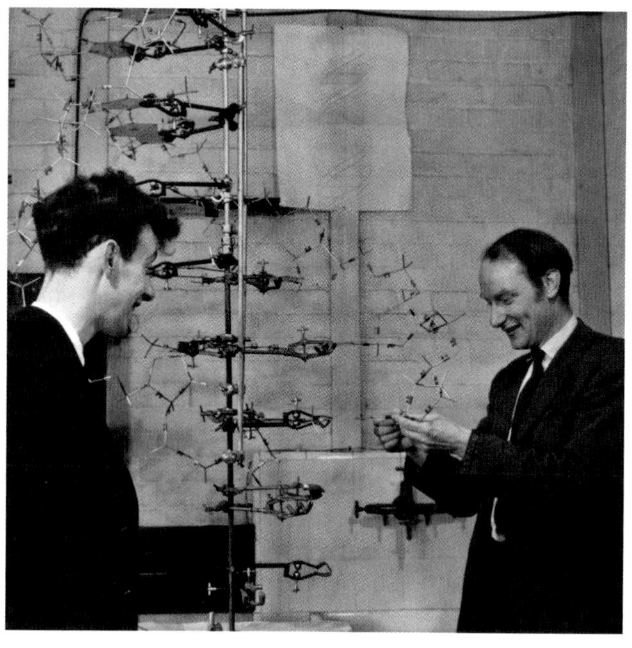

Figure 5.6 James Watson and Francis Crick assembled a model of DNA, which is now recognized as the molecular structure of DNA.

Watson and Crick deduced that DNA has a twisted, ladder-like structure, called a double helix. The sugar-phosphate molecules make up the sides or "handrails" of the ladder, and the bases make up the rungs by protruding inward at regular intervals along each strand. From Franklin's images, Watson and Crick knew that the distance between the sugar-phosphate handrails remained constant over the length of the molecule. However, adenine and guanine have a double-ring structure and thymine and cytosine have a single-ring structure. Chargaff's rule helped them infer that an A nucleotide on one chain always sits across from a T nucleotide on the other chain, while a C nucleotide on one chain always sits across from a G nucleotide on the other chain. This allows for the rungs to be a constant width, which is consistent with Franklin's X-ray photographs. Also, the proposed double-helix structure allows for differences in DNA structure between species, because the base pairs can be in any order. In 1953, Watson and Crick published a two-page paper describing their double-helix model. Soon after, this became the accepted molecular structure for DNA.

Activity 5.1 Science, Ethics, and the Double Helix

Scientific investigation is not free from bias and conflict. In many cases, as well, the line between ethical and unethical behaviour is not so clear. In this activity, you will research a controversial issue concerning the race to determine the structure of DNA.

Procedure

1. Read the following paragraph about Rosalind Franklin's contribution to the discovery of the structure of DNA.

   During her lifetime, the role Franklin played in the discovery of the structure of DNA was, to a great extent, ignored for several reasons. In the 1950s, when she was taking X-ray images of DNA, attitudes toward women in science were sexist. She and Maurice Wilkins, another scientist working on similar research with Franklin at King's College in England, were also at odds with each other. It has been argued that Franklin's images of DNA were shared with Watson and Crick by Wilkins without her agreement. There is a discrepancy between the Franklin version of what happened and the Wilkins version of the same events.

2. Research this issue using various resources in order to answer the questions that follow.

Questions

1. Describe the controversy concerning Franklin's photographs in your own words.
2. Do you believe that Franklin's research was taken without permission? Explain why or why not, making reference to your research findings.
3. If Franklin's work was taken without permission, which scientists, if any, do you believe acted unethically? Explain your reasoning.
4. Research more information about ethical policies that exist at universities, research facilities, and government institutions that would help prevent a situation similar to Franklin's from happening today. Do you think the policies are strong enough? Why or why not? If not, how could they be improved?

complementary base pairing in DNA, the interaction of bases of nucleotides on opposite strands through the formation of hydrogen bonds

antiparallel the directionality of the two strands in a DNA molecule; the strands run in opposite directions, with each end of a DNA molecule containing the 3′ end of one strand and the 5′ end of the other strand

The Modern DNA Model: The DNA Double Helix

Today, scientists can identify the position of every atom in a molecule of DNA. Although some refinements have been made to the structure of DNA since Watson and Crick first proposed their model, many features remain the same. An overview of the important features of the structure of DNA is shown in **Figure 5.7**. They include the following.

- There are two polynucleotide (multiple nucleotide) strands that twist around each other to form a double helix. Each polynucleotide strand has a backbone of alternating phosphate groups and sugars. The bases of each nucleotide are attached to each sugar and protrude inward at regular intervals along each strand. There is a constant total distance between the sugar-phosphate backbones.
- The two strands of a DNA molecule are not identical but are, instead, complementary to each other. This means that a purine molecule is always paired with a pyrimidine molecule. Specifically, adenine (A) always pairs with thymine (T), and guanine (G) always pairs with cytosine (C). This pairing is called **complementary base pairing**. Because the same bases always complement each other, the base sequence of one strand can be determined from the base sequence of the other strand.
- Hydrogen bonds link each complementary base pair. A and T share two hydrogen bonds; C and G share three.
- The two strands of a DNA molecule are **antiparallel**. You can see in **Figure 5.7** that the sugar molecules are oriented differently. Therefore, each strand has directionality, or a specific orientation. (This is similar to how two sports teams are facing in a different direction when they shake hands after a game.) At each end of a DNA molecule, the 5′ end of one strand of DNA lies across from the 3′ end of the complementary strand. The 5′ and 3′ come from the numbering of the carbons on the deoxyribose sugar. The phosphate group is on the 5′ carbon, and the OH group is on the 3′ carbon.
- By convention, the sequence of a DNA strand is always written in the 5′ to 3′ direction.
- The DNA model shown in **Figure 5.7A** and **Figure 5.7B** is called the ribbon model, which clearly shows the components of the DNA molecule. The model for the DNA double helix shown in **Figure 5.7C** is called the space-filling model, in which [illegible] as spheres. This type of structural model emphasizes the surface of DNA. As you can see from this model, the sugar-phosphate backbone, represented by red and white spheres, is on the outermost surface of the double helix. The atoms of the bases, represented by blue spheres, are more internally located within the double-stranded structure.

Learning Check

7. Draw and label a diagram of the general structure of a single nucleotide.

8. How do nucleotides in DNA differ from nucleotides in RNA?

9. Define Chargaff's rule. How did Chargaff's findings overturn earlier beliefs about DNA?

10. How did the work of Franklin and Pauling contribute to the Watson and Crick model of DNA?

11. Use a labelled diagram to illustrate the following features of a DNA molecule:

a. complementary base pairing

b. antiparallel strands

12. In an early model of DNA tested by Watson and Crick, the sugar-phosphate handrails were on the inside of the helix and the nitrogenous bases protruded outward. How is this model inconsistent with experimental evidence about the structure of DNA?

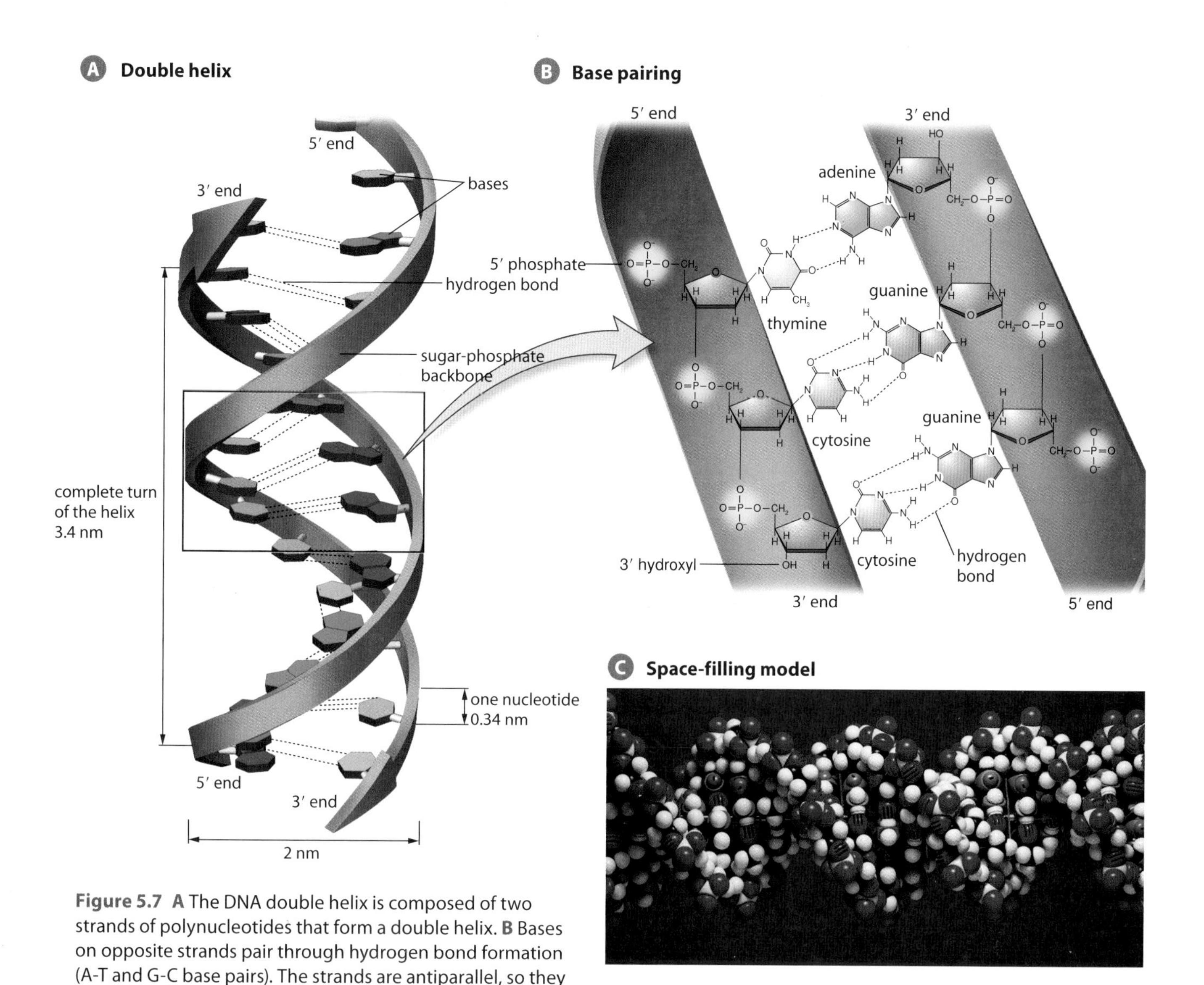

Figure 5.7 A The DNA double helix is composed of two strands of polynucleotides that form a double helix. B Bases on opposite strands pair through hydrogen bond formation (A-T and G-C base pairs). The strands are antiparallel, so they have directionality. C The space-filling model emphasizes the surface of DNA.

Activity 5.2 Modelling DNA Structure

Watson and Crick did not conduct any experiments to determine the structure of DNA. Instead, they worked as synthesizers, examining and interpreting the research and discoveries made by other scientists. Watson and Crick tried different arrangements until they created one model that could account for all the evidence. In this activity, you will work in a group to design and build a model that can be used to simulate the structure of DNA.

Materials

- DNA model building supplies or appropriate software and computer access

Procedure

1. Brainstorm ideas for designing and constructing a three-dimensional model of a short fragment of double stranded DNA (8 to 10base pairs).
2. Consider how you will represent the sugar phosphate backbone, the nitrogenous bases, and hydrogen bonding between complementary base pairs as you design your model.
3. Use your ideas to develop a plan. List the materials and equipment you will need.
4. Create your model.

Questions

1. In what ways is your model useful for explaining the structure of DNA? What are the limitations of your model?
2. A number of scientists contributed to the model proposed by Watson and Crick. Identify the contributions of Levene, Chargaff, Franklin, and Pauling on your model.

The Structure and Organization of Genetic Material in Prokaryotes and Eukaryotes

So far, only the primary and secondary structures of DNA have been considered—that is, how nucleotides are linked together to form a chain and how two chains of nucleotides form a double helix. To relate these structures to DNA function, it is necessary to consider two additional ideas: how DNA is organized in the cell, and how the total genetic material of an organism—its **genome**— is arranged into distinct functional regions on the DNA.

The functional units of DNA are **genes**, because their sequences code for the production of specific proteins or RNA. However, a large amount of the DNA in the genome of an organism does not contain the instructions for producing these molecules. In other words, much of the DNA in an organism's genome is non-coding. In some organisms, non-coding regions account for the majority of the genome.

genome the complete genetic makeup of an organism; an organism's total DNA sequence

gene the basic unit of heredity that determines, in whole or part, a genetic trait; a specific sequence of DNA that encodes for proteins and RNA molecules, and can contain sequences that influence production of these molecules

nucleoid a structure in bacteria that contains the chromosomal DNA

DNA supercoiling the formation of additional coils in the structure of DNA due to twisting forces on the molecule

DNA of Prokaryotic Cells

For most prokaryotes, such as the bacterium *E. coli*, the genetic material is in the form of a circular, double-stranded DNA molecule. Although a single type of chromosome is present in bacteria, they can have more than one copy of that chromosome. Since prokaryotes do not have a nuclear membrane, each bacterial chromosome is packed tightly in a specific region of the cell. This region is called the **nucleoid**, shown in **Figure 5.8**.

The extent of compacting of bacterial chromosomal DNA is roughly one thousand times. How is this possible? Suppose you twist a rubber band like the one shown in **Figure 5.9**. The resulting coiling and compacting is similar to loops and coils formed in the bacterial DNA. Specialized proteins that bind to the bacterial DNA help fold sections of the chromosome into loop-like structures. This folding compacts the DNA ten times more. **DNA supercoiling** achieves further compacting by introducing twists into the DNA structure, as shown in **Figure 5.10**. These twists cause changes in conformation that result in a section of DNA coiling onto itself.

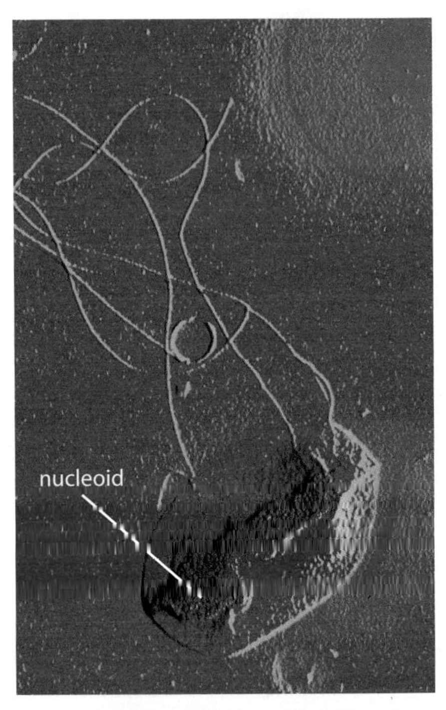

Figure 5.8 The DNA of prokaryotic cells such as *E. coli* is packed near the centre of the cell in the nucleoid, which is not segregated by membranes from the rest of the interior of the cell.

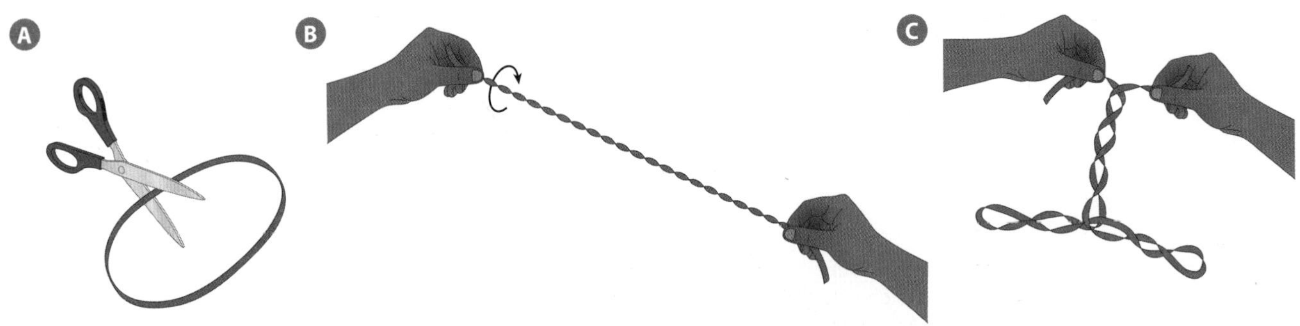

Figure 5.9 Twisting a rubber band can model how bacterial DNA coils and compacts. Supercoiling introduces twists into the DNA that allow it to coil onto itself.

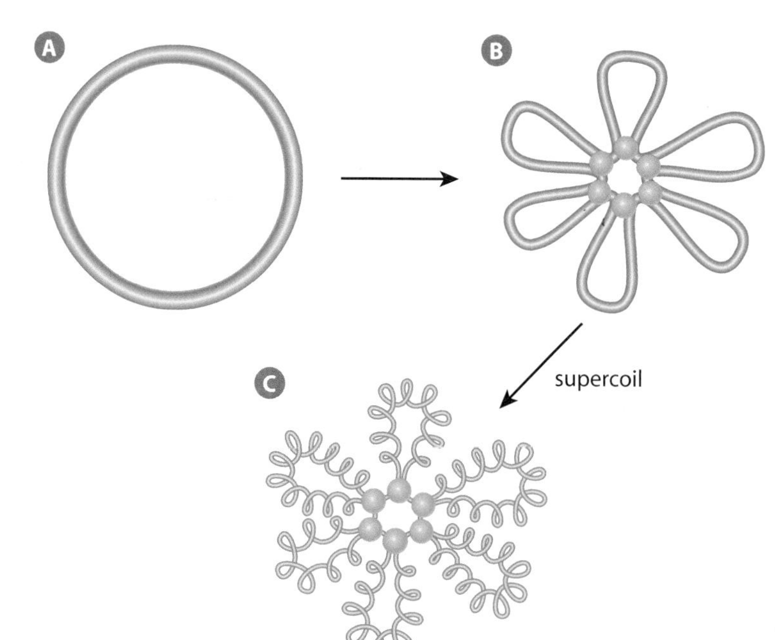

Figure 5.10 **A** The circular chromosomal DNA molecule can be compacted through **(B)** the formation of looped structures. **C** The looped DNA can be further compacted by DNA supercoiling. Note: the coloured balls represent proteins involved in supercoiling ***Explain*** *how using a rubber band helps model supercoiling, and describe any limitations of this model.*

In bacteria, the amount of supercoiling is controlled by two enzymes: topoisomerase I and topoisomerase II. The activities of topoisomerase II are essential for bacterial survival. Therefore, antibacterial drugs have been developed that specifically target and block the activity of this enzyme. Quinolones and coumarins are two examples of classes of drugs that block topoisomerase II activity.

Some prokaryotes have one or more small, circular or linear DNA molecules called plasmids. Plasmids are not part of the nucleoid and often carry non-essential genes. They can be copied and transmitted between cells, or they can be incorporated into the cell's chromosomal DNA and reproduced during cell division. As a result, the hereditary information contained on a plasmid can be transferred from cell to cell.

Most prokaryotes contain only one copy of each gene, so they are haploid organisms. In addition, their genomes contain very little non-essential DNA. The majority of prokaryotic genomes are composed of regions that contain either genes or regulatory sequences. **Regulatory sequences** are sections of DNA sequences that determine when certain genes and the associated cell functions are activated. **Figure 5.11** shows a partial map of the *E. coli* chromosome. The structure of the prokaryotic genome will be looked at in more detail in Chapter 6, when protein synthesis and the regulation of gene expression are discussed.

regulatory sequence a sequence of DNA that regulates the activity of a gene

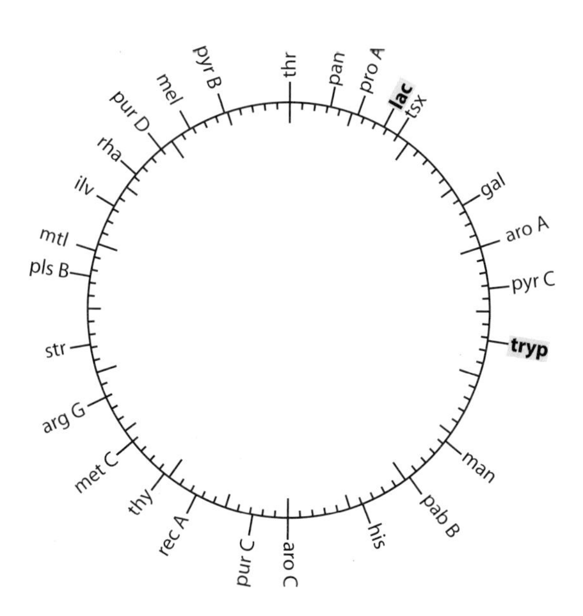

Figure 5.11 This is a partial map of the *E. coli* genome. There are actually thousands of genes in the genome, very few of which are non-essential.

histone a member of a family of proteins that associate with DNA in eukaryotic cells, which acts to help compact the DNA

nucleosome the condensed structure formed when double-stranded DNA wraps around an octamer of histone proteins

DNA of Eukaryotic Cells

For eukaryotic cells, the total amount of DNA is much greater than in prokaryotic cells. Also, the genetic material of eukaryotic cells is contained within the nucleus. Mitochondria and chloroplasts of eukaryotes also contain some of the cellular DNA. If you lined up all the DNA in the nucleus of a single human cell, it would reach about 2 m. The diameter of a nucleus, however, is only about 4 μm. (To visualize this, imagine stuffing a 150 km length of string into a shoe box.) Thus, in eukaryotes, the extent of compacting of the genetic material must be much greater than in prokaryotes. The compacting of DNA in the nucleus is achieved mainly through different levels of organization, which involve interactions between linear sections of DNA and several proteins to form a series of defined structures.

Figure 5.12 shows the levels of organization of genetic material that exist within each chromosome of most eukaryotic cells. At the simplest level, shown in part A, DNA associates with a family of proteins, called **histones**, to form a repeated series of structures called **nucleosomes**. Each nucleosome is composed of double-stranded DNA (146 base pairs in length) that is wrapped around a group of eight histone proteins (two copies each of histones H2A, H2B, H3, and H4). These nucleosome structures are connected by regions called *linker DNA*. The appearance of these repeated structures is often described as "beads on a string."

Further compacting, shown in part B, occurs by the coiling of the nucleosomes, with the aid of H1 histone proteins, into what is often called a 30 nm fibre. At this level, the DNA has been compacted by about 50 times.

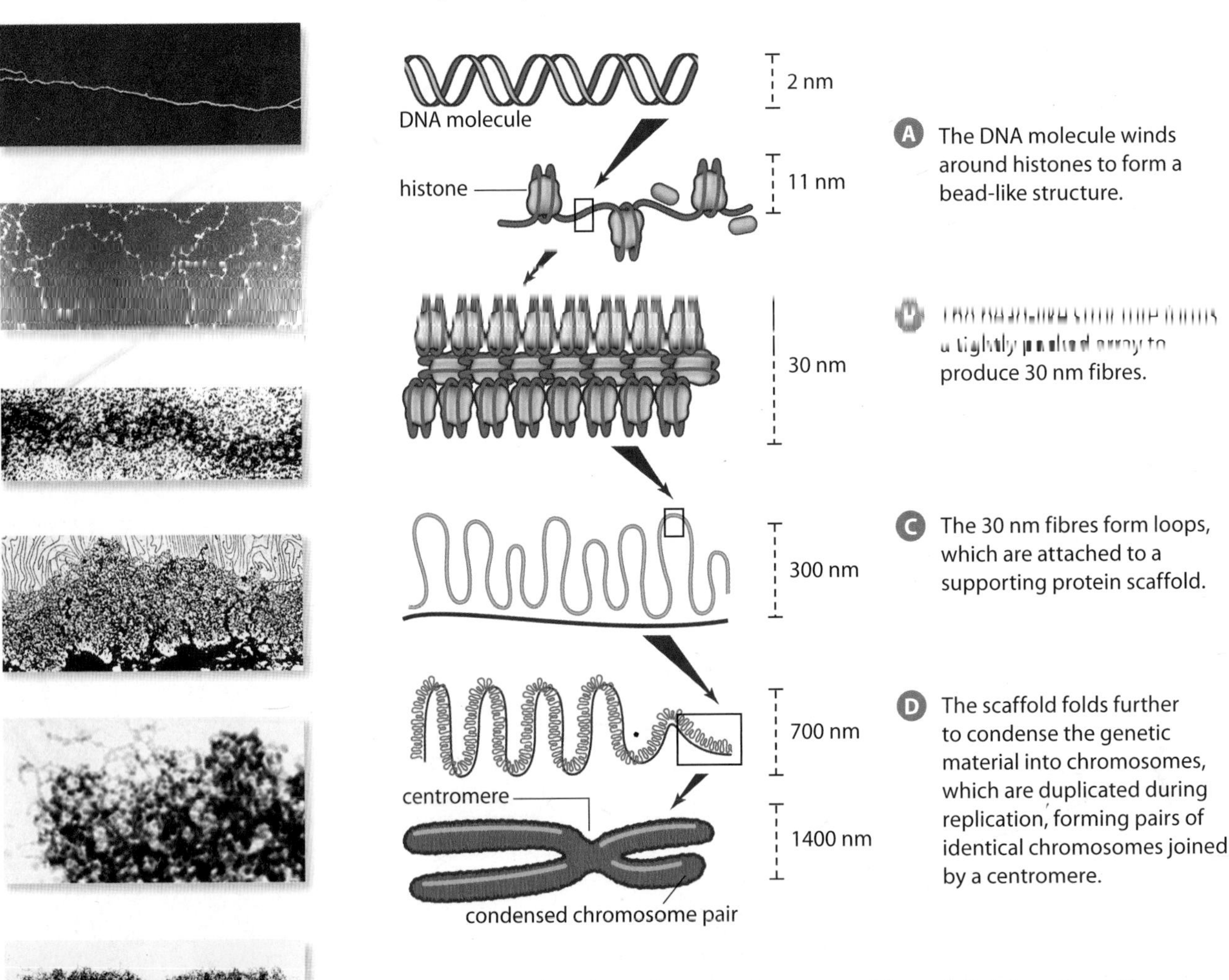

DNA molecule
2 nm
histone
11 nm
A The DNA molecule winds around histones to form a bead-like structure.
30 nm
produce 30 nm fibres.
300 nm
C The 30 nm fibres form loops, which are attached to a supporting protein scaffold.
700 nm
centromere
1400 nm
condensed chromosome pair
D The scaffold folds further to condense the genetic material into chromosomes, which are duplicated during replication, forming pairs of identical chromosomes joined by a centromere.

Figure 5.12 DNA in the eukaryotic nucleus is compacted through its interaction with different proteins, such as histones, to produce defined structures of increasing organization.

Additional compacting of the DNA, shown in part C, involves formation of radial loop domains of the 30 nm fibre. These loops are anchored to a supporting scaffold of proteins in the nucleus.

For most of a cell's life cycle, its genetic material appears as a mass of long, intertwined strands known as **chromatin**. During interphase, the level of chromatin compaction can vary along the length of the chromosome. The 30 nm fibre as looped domains, called *euchromatin*, can undergo further compacting to a more condensed *heterochromatin* structure in some areas. As the genetic material is reorganized during the processes of cellular division, the threads of chromatin condense and become visible under a light microscope as distinct chromosomes, shown in part D.

chromatin the non-condensed form of genetic material that predominates for most of the eukaryotic cell cycle; consists of a complex of DNA and proteins

Variation in the Eukaryotic Genome

The eukaryotic genome can vary a great deal between species. As you have learned, eukaryotic genomes are composed of multiple linear chromosomes that carry a variety of genes. Most eukaryotes are *diploid*—they contain two copies of each gene. Some however, such as ferns and algae, may be haploid and contain one copy of each gene. Some specially bred organisms can have three or more copies of each gene. Seedless watermelons, for instance, are triploid. The organization of genes on each chromosome can differ as well. Genes are not evenly spaced along a chromosome, and they are not equally divided among the chromosomes. For example, in humans, chromosome 19 has 72 million base pairs and about 1450 genes, while chromosome 4 has almost 1.3 billion base pairs but only about 200 genes.

Suggested Investigation

Inquiry Investigation 5-A, Extracting DNA

The size and number of genes in the eukaryote genome vary a great deal. Overall, there is no correlation between an organism's complexity and genome size or number of protein-coding genes. For example, the lungfish (*Protopterus aethiopicus*) shown in **Figure 5.13A** has 40 times more DNA per cell than a human cell. Similarly, the rice (*Oryza sativa*) shown in **Figure 5.13B** has about 51 000 protein-coding genes, while the estimated number of genes in a human being is about between 20 000 and 25 000. Additionally, there is a lack of correlation between genome size and number of protein-coding genes in the genome. Humans have about the same number of protein-coding genes as the small worm-like organism, *Caenorhabditis elegans*, shown in **Figure 5.13C**. However, humans have about 30 times more DNA.

Genes also vary in the molecules they produce. Scientific studies of genome structure now suggest a broader definition of a gene than just a protein-coding sequence. For example, some genes code for RNA molecules that are required for other cellular processes and do not directly result in a protein product. This knowledge has helped scientists understand how the eukaryotic genome has evolved different ways to generate increased biological complexity, without increasing the number of protein-coding genes in the genome.

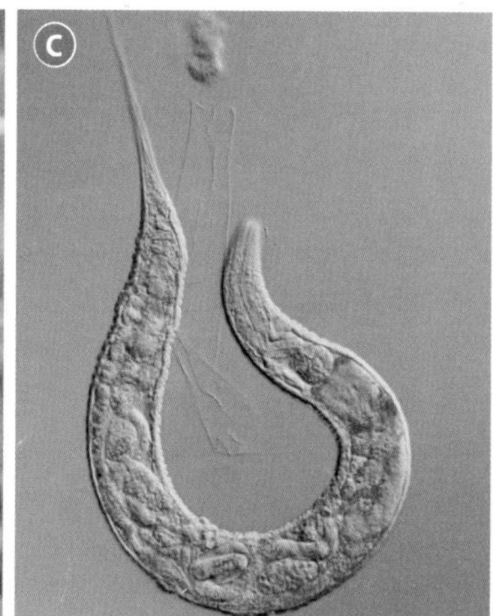

Figure 5.13 **A** The lungfish *Protopterus aethiopicus* has 40 times more DNA per cell than a human cell. **B** Rice, *Oryza sativa*, has about 30 000 more protein-coding genes than a human being. **C** *Caenorhabditis elegans* has about the same number of genes as humans, but much less DNA.

Section 5.1 Review

Section Summary

- DNA was identified as the hereditary material through a series of experiments between 1928 and 1952. Before this identification, many scientists believed that the more chemically complex proteins were the hereditary material of the cell.
- DNA and RNA are nucleic acids made of units called nucleotides. Each nucleotide is composed of one of four nitrogen-containing bases, a sugar, and a phosphate group. The sugar in DNA is deoxyribose, while the sugar in RNA is ribose. For DNA, the nucleotide bases are adenine (A), guanine (G), cytosine (C), and thymine (T). For RNA, they are A, G, C, and uracil (U).
- James Watson and Francis Crick published their three-dimensional structure of DNA in 1953. In this accepted model, DNA has a double-helix structure like a twisted ladder, in which the sides of the ladder are composed of alternating sugars and phosphate groups, while the rungs of the ladder are composed of paired A-T and C-G bases.
- In most prokaryotes, genetic material is in the form of a circular, double-stranded DNA molecule. The nucleoid region of this type of cell contains tightly packed DNA, in part due to DNA supercoiling.
- In eukaryotic cells, genetic material is contained in the nucleus. DNA is compacted through different levels of organization to form a series of defined structures including nucleosomes, chromatin, and chromosomes.

Review Questions

1. K/U How did Griffith's transforming principle experiments on *Streptococcus pneumoniae* contribute to Avery's research?

2. C The Hershey–Chase experiment is often considered to be a landmark or breakthrough study.
  a. Use a graphic organizer to explain the Hershey–Chase experiment.
  b. What was the significance of the results of this experiment?

3. K/U What were Miescher's contributions to the study of hereditary material?

4. C Draw a labelled diagram of the general structure of a nucleotide.

5. A Would the nucleotide composition of a DNA sample from the liver of a human be different from the nucleotide composition of a DNA sample from the liver of a mouse? Explain your answer.

6. T/I Copy the following table into your notebook. Use Chargaff's rule to complete the table. (Assume that the characteristic proportions are exactly equal.)

Nucleotide Composition of DNA in Sample X

Nucleotide	Proportion (%)
A	24
C	
G	
T	

7. C Use a labelled diagram to illustrate how the structure of DNA is "ladder-like."

8. K/U Describe how the contributions of the following scientists influenced the three-dimensional model of DNA proposed by Watson and Crick.
  a. Levene
  b. Chargaff
  c. Franklin
  d. Pauling

9. T/I What would the complementary strand be for each of the following DNA sequences?
  a. 5'-AT[illegible]AAT-3'
  b. 5'-CCGTTAATC-3'
  c. 5'-TTAGCTG[illegible]-3'

10. K/U Explain the difference between a gene and a genome.

11. C Use a Venn diagram to compare and contrast how DNA is organized in prokaryotic cells and eukaryotic cells.

12. C The phrase "beads on a string" is often used to describe how DNA is organized in eukaryotic cells. Use a labelled diagram to explain what this phrase means.

13. A A certain species of invertebrate has a genome that is 6000 times larger than the genome of a particular yeast cell.
  a. What conclusions, if any, could you make about the relative complexities of the two organisms?
  b. What practical applications could result from a study that compared the genomes of the two organisms?

14. T/I Would a mutation in a protein-coding region of DNA be more detrimental than a mutation in a non-coding region? Explain your answer.

SECTION 5.2

DNA Replication

All life depends on the ability of cells to reproduce. Consider humans, for example. Humans are composed of trillions of cells, but each individual starts out as a single fertilized egg cell. Many of our tissues and organs rely on the continual regeneration of new cells, such as red blood cells and skin cells. Also, if a person is injured, new cells are needed to repair the damage. The production of new cells is achieved through mitosis and cytokinesis.

The life cycle of a cell, shown in **Figure 5.14**, is referred to as the cell cycle. For most cells, the *cell cycle* consists of a growth stage, called interphase, followed by a period during which the cell divides to produce two new cells through mitosis and cytoplasmic division. Each new cell, or daughter cell, must contain the same genetic information as the original parent cell so that it can carry on the same functions. Therefore, each new cell requires an exact copy of the parental DNA. A cell copies all of its DNA only once in the cell cycle—during S phase of interphase. This process of copying one DNA molecule into two identical molecules is called **DNA replication**.

Key Terms

DNA replication
semi-conservative replication
origin of replication
helicase
DNA polymerase III
primer
Okazaki fragments
DNA polymerase I
DNA ligase
DNA polymerase II
mismatch repair
telomere

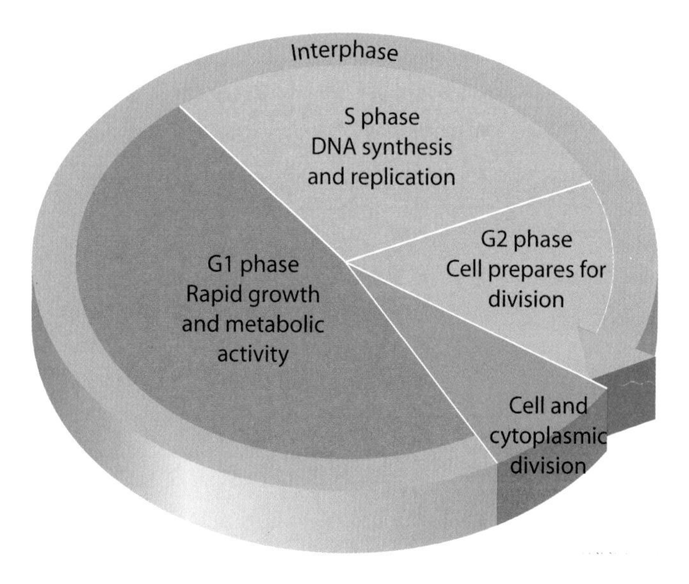

Figure 5.14 DNA is copied during S phase of interphase in the cell cycle.

DNA replication the process of producing two identical DNA molecules from an original, parent DNA molecule

Determining the structure of DNA led Watson and Crick to a greater understanding of how DNA might be replicated. In the concluding remarks of their landmark April 1953 paper on the structure of DNA, Watson and Crick added: "It has not escaped our notice that the specific pairing we have postulated immediately suggests a possible copying mechanism for the genetic material." In other words, due to complementary base pairing, each strand can serve as a template for the production of a new complementary strand.

Approximately one month after publishing their paper, Watson and Crick published a second paper, which focussed on DNA replication. In this paper, Watson and Crick proposed that the two strands of the DNA double-helix molecule unwind and separate, after which each nucleotide chain serves as a template for the formation of a new companion chain. The result would be a pair of daughter DNA molecules, each identical to the parent molecule. The publication of both papers by Watson and Crick inspired many researchers to explore the question of how DNA replicates. Over the course of several years, three competing models emerged from these explorations.

Three Proposed Models for DNA Replication

The three models that were proposed to demonstrate a mechanism for DNA replication are shown in **Figure 5.15**. The *conservative model* proposed that replication involved the formation of two new daughter strands from the parent templates, with the two new strands joining to create a new double helix. The two original strands would then re-form into the parent molecule. In the *semi-conservative model*, each new molecule of DNA would contain one strand of the original complementary DNA molecule and one new parent strand. Thus, each new DNA molecule would conserve half of the strand of the original molecule. (The semi-conservative model was the one proposed by Watson and Crick.) The *dispersive model* proposed that the parental DNA molecules were broken into fragments and that both strands of DNA in each of the daughter molecules were made up of an assortment of parental and new DNA. In 1958, experiments conducted by a pair of American geneticists at the California Institute of Technology were able to distinguish between these hypothesized models and identify the correct mechanism for DNA replication.

Meselson and Stahl Determine the Mechanism of DNA Replication

Matthew Meselson (b. 1930) and Franklin Stahl (b. 1929) reasoned that the proposed models for DNA replication could be tested if they could distinguish between the original parental strand and the newly synthesized daughter strand when DNA is copied. They decided to use two different isotopes of nitrogen to label the DNA in a cell: the common form of nitrogen, ^{14}N, and a rarer form, ^{15}N, which is often called "heavy" nitrogen.

Isotopes of nitrogen were chosen for two main reasons. First, nitrogen is an essential component of DNA, so it would be incorporated into any new daughter strands that were synthesized in the cell. Second, the availability of a "light" form and a "heavy" form of nitrogen meant that the scientists could separate the different DNA strands according to how much of each isotope was present in a newly synthesized DNA molecule. The more "heavy" nitrogen a DNA molecule had, the greater its density. DNA of differing densities can be separated by centrifugation. Afterward, the heavier ^{15}N-containing sample forms a distinct band that is positioned toward the bottom of the centrifugation tube. The lighter ^{14}N-containing sample forms a band of material that is less dense and, therefore, positioned higher in the tube. The details of the experiment are described on the next page and illustrated in **Figure 5.16**.

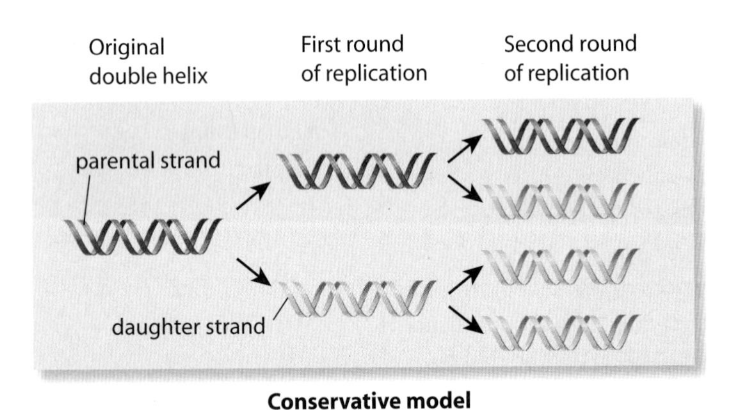

Conservative model

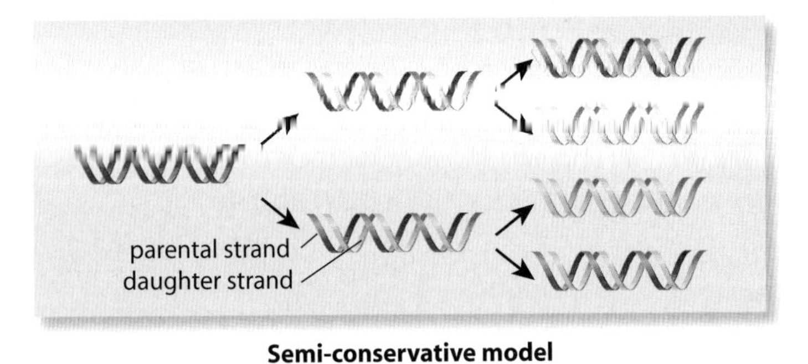

Semi-conservative model

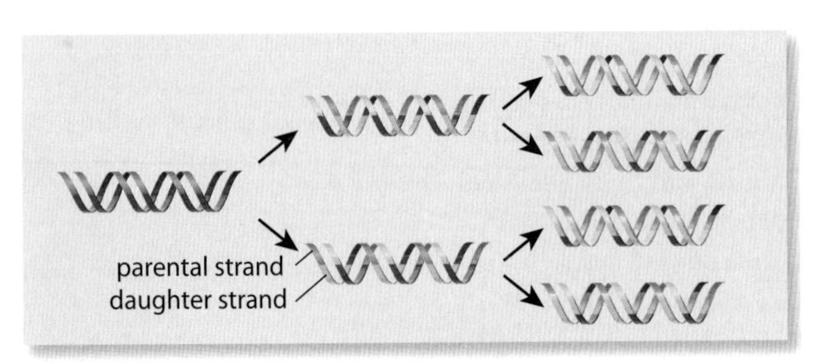

Dispersive model

Figure 5.15 The three models of DNA replication. The conservative model results in one new molecule and conserves the old. The semi-conservative model results in two hybrid molecules of old and new strands. The dispersive model results in hybrid molecules with each strand being a mixture of old and new strands.

First, bacteria were grown in a liquid culture medium containing ^{15}N. This was the only source of nitrogen for the growing and dividing cells and, therefore, the newly synthesized DNA molecules. This culture was grown until the researchers had a population of bacterial cells that contained only the ^{15}N isotope in the DNA. They then transferred that population of bacteria to a culture medium that only had ^{14}N as a source of nitrogen. Samples of cells were removed just before the cells were transferred to ^{14}N-only medium and as the bacteria divided. The DNA was isolated and samples were analyzed by centrifugation. Meselson and Stahl observed the following.

- DNA samples taken just prior to the transfer to ^{14}N media (grown in ^{15}N only) were of uniform density and appeared as a distinct band that corresponded to DNA containing only ^{15}N, and no ^{14}N.
- After one round of replication, the DNA sample formed a single band after centrifugation. However, its position in the tube indicated that its density was midway between DNA with a nitrogen composition of only ^{15}N and DNA with a nitrogen composition of only ^{14}N. The researchers inferred that the DNA was composed of one strand labelled with ^{14}N and one strand labelled with ^{15}N. From this, Meselson and Stahl could rule out the conservative model, since it would have resulted in two bands—one band of ^{15}N-only DNA and another band of ^{14}N-only DNA.
- After a second round of replication, the DNA sample separated into two distinct bands after centrifugation. One band corresponded to ^{14}N-only DNA. The other band corresponded to DNA that had one strand labelled with ^{14}N and the second strand labelled with ^{15}N.
- The same two bands that were observed after the second round of replication continued to be observed after multiple rounds of replication. This discounted the dispersive model, which would have resulted in only one band ever being observed, and supported the semi-conservative model.

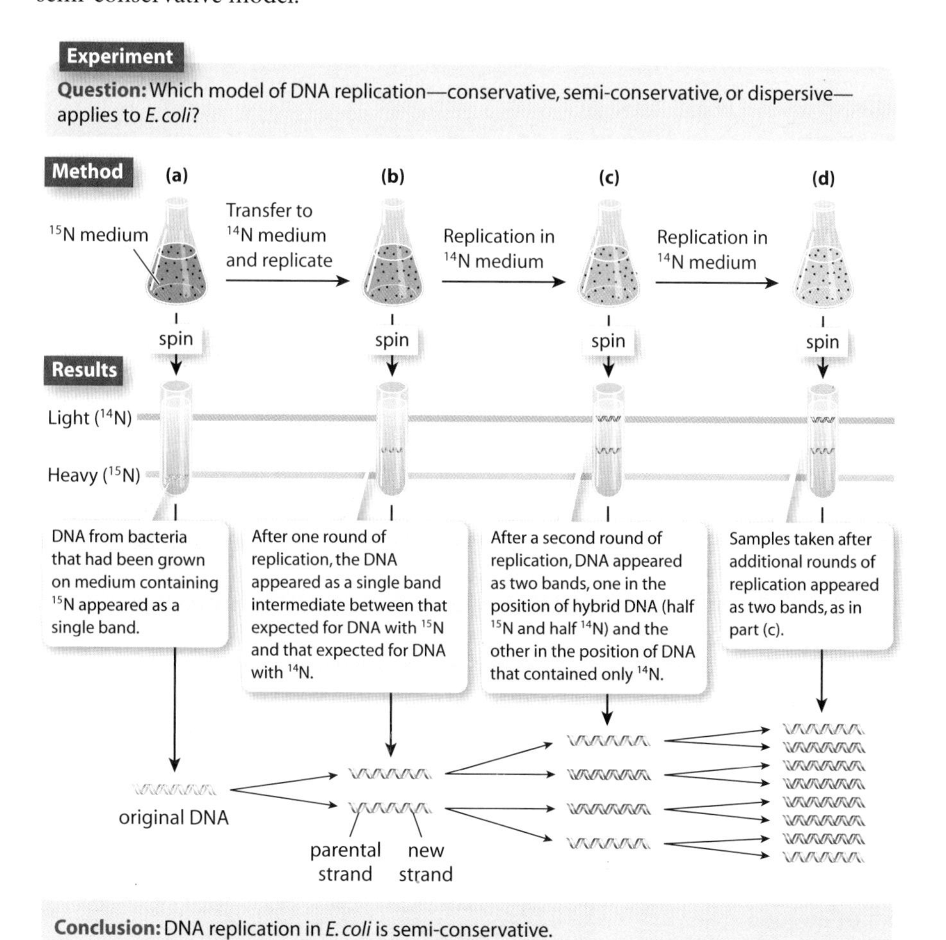

Experiment
Question: Which model of DNA replication—conservative, semi-conservative, or dispersive—applies to E. coli?
Method
(a)
(b)
(c)
(d)
15N medium
Transfer to 14N medium and replicate
Replication in 14N medium
Replication in 14N medium
spin
spin
spin
spin
Results
Light (14N)
Heavy (15N)
DNA from bacteria that had been grown on medium containing 15N appeared as a single band.
After one round of replication, the DNA appeared as a single band intermediate between that expected for DNA with 15N and that expected for DNA with 14N.
After a second round of replication, DNA appeared as two bands, one in the position of hybrid DNA (half 15N and half 14N) and the other in the position of DNA that contained only 14N.
Samples taken after additional rounds of replication appeared as two bands, as in part (c).
original DNA
parental strand
new strand
Conclusion: DNA replication in E. coli is semi-conservative.

Figure 5.16
Meselson and Stahl used two different isotopes of nitrogen to distinguish between parental and daughter DNA strands. These experiments not only proved a particular model, but also disproved the other proposed models.

Infer *How would the results of the Meselson–Stahl experiment be different if they had not started with uniformly labelled ^{15}N-DNA?*

Since the publication of Meselson and Stahl's experiments, many other research studies have supported the semi-conservative model of DNA replication in bacteria as well as in other organisms. Evidence in support of the conservative and dispersive models has never been reported. Thus, Watson and Crick's proposed models of DNA and DNA replication have stood the test of time. Scientists next turned to identifying and studying the individual steps and molecules that are involved in the replication of an organism's genome. Although our knowledge of the replication process has increased in the last several decades, there are still important questions about DNA replication that have yet to be answered.

Learning Check

13. What is the main objective of DNA replication?

14. Why is DNA replication important for cell reproduction?

15. Describe the three proposed models of DNA replication.

16. Why did Meselson and Stahl use two different isotopes of nitrogen in their experiment?

17. What did Meselson and Stahl conclude from their experiment? What results provided convincing evidence for their conclusion?

18. Why is it important for newly replicated daughter strands of DNA to have the same information as the parent strands?

The Molecular Events in DNA Replication

In the process of **semi-conservative replication**, each new molecule of DNA contains one strand of the original parent molecule (parent DNA) and one complementary strand that is newly synthesized (daughter DNA); each resulting DNA molecule conserves half of the original molecule. This process of replication is most often described in three basic phases that rely on the structural features of DNA and a number of specialized proteins.

1. In the *initiation phase*, a portion of the DNA double helix is unwound to expose the bases for new base pairing.
2. In the *elongation phase*, two new strands of DNA are assembled using the parent DNA as a template. The new DNA molecules—each composed of one strand of parent DNA and one strand of daughter DNA—re-form into double helices.
3. In the *termination phase*, the replication process is completed and the two new DNA molecules separate from each another. At that point, the replication machine is dismantled.

The majority of what is known about DNA replication comes from studies in *E. coli*. Therefore, the main events that are described next are those that occur in prokaryotes. Nevertheless, most of the steps and components also occur in eukaryotes. It is also important to keep in mind that, while replication is described as a sequence, all of these events take place simultaneously on the same molecule of DNA.

semi-conservative replication the mechanism of DNA replication in which each newly synthesized DNA molecule is composed of one strand from the original DNA molecule and one new strand

Initiation

Replication starts at a specific nucleotide sequence, called the **origin of replication**. Here several initiator proteins bind to the DNA and begin the process of unwinding the double helix. One group of proteins that is involved in this unwinding process is the **helicase** enzymes. The helicases cleave the hydrogen bonds that link the complementary base pairs between strands together. Other proteins, called *single-strand-binding proteins*, help to stabilize the newly unwound single strands. These strands have a tendency, if unchecked, to re-form into a double helix. These single strand regions serve as the templates that will be used to guide the synthesis of new polynucleotide strands. In addition, the *topoisomerase II enzyme* helps to relieve the strain on the double-helix sections ahead of the replication forks, which results from the unwinding process. As shown in **Figure 5.17**, initiation creates an unwound, oval-shaped area called a *replication bubble*, with two Y-shaped regions at each end of the unwound area. Each Y-shaped area is called a *replication fork*. As replication proceeds, each replication fork moves along the DNA in opposite directions.

origin of replication the DNA sequence where replication begins

helicase a group of enzymes that aid in the unwinding of DNA

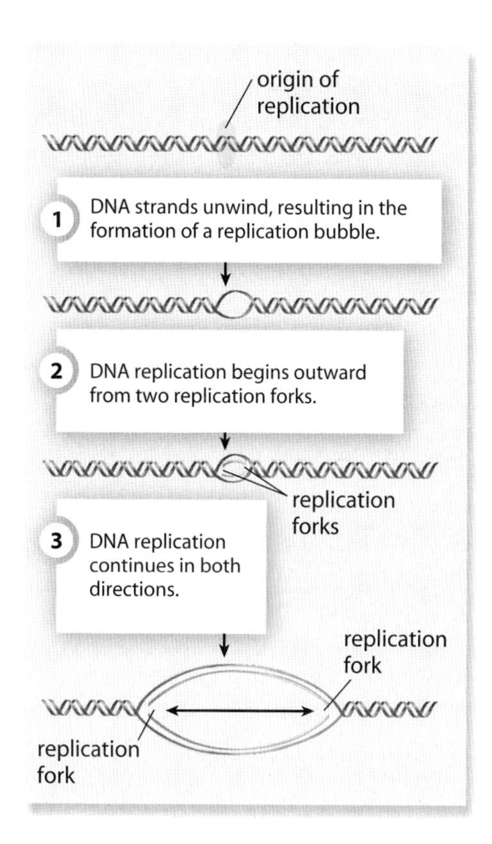
origin of replication
1 DNA strands unwind, resulting in the formation of a replication bubble.
2 DNA replication begins outward from two replication forks.
replication forks
3 DNA replication continues in both directions.
replication fork
replication fork

Figure 5.17 An origin of replication on a DNA molecule marks the start of replication. As the double helix unwinds, a replication bubble forms and produces two replication forks that move along the DNA in opposite directions.

DNA polymerase III an enzyme that adds nucleotides to the 3′ end of a growing polynucleotide strand

Okazaki fragments short DNA fragments that are generated during the synthesis of the lagging strand in DNA replication

Elongation

The elongation phase synthesizes new DNA strands by joining individual nucleotides together. **DNA polymerase III** is the enzyme that catalyses the addition of new nucleotides, one at a time, to create a strand of DNA that is complementary to a parental strand. DNA polymerase III only attaches new nucleotides to the free 3′ hydroxyl end of a pre-existing chain of nucleotides. In addition, DNA polymerase III can only synthesize a new strand from a parent strand in the 5′ to 3′ direction, toward the replication fork.

When double-stranded DNA is separated, both strands are bare templates that do not have free 3' hydroxyl ends for DNA polymerase to begin at. Thus, synthesis of new DNA requires both strands to be started with short fragments of nucleotide sequences complementary to the templates. For one strand, called the *leading strand*, this only needs to happen once. DNA polymerase will keep adding new nucleotides to the 3' end as it moves along in the same direction as the replication fork. However, synthesis of the other strand requires DNA polymerase to move in the opposite direction to the replication fork. As discussed on the next page, this results in synthesis of this new strand, called the *lagging strand*, to occur in short segments and in a discontinuous manner. These short segments of DNA are named **Okazaki fragments** in honour of the scientists who identified them, shown in **Figure 5.18**.

Suggested Investigation

ThoughtLab Investigation 5-B, DNA Replication

Figure 5.18 In 1968, Reiji and Tsuneko Okazaki were the first to identify the short fragments of DNA that are formed in the synthesis of the lagging strand in DNA replication.

primer in DNA replication, a short segment of RNA that is complementary to a part of the 3′ to 5′ DNA template strand and serves as a starting point for addition of nucleotides

DNA polymerase I an enzyme that removes RNA primer and fills gaps between Okazaki fragments on the lagging strand with DNA nucleotides; proofreads newly synthesized DNA

DNA ligase an enzyme that catalyses the joining of Okazaki fragments

Lagging Strand Synthesis

The short fragments of nucleotide sequences that are used to start or "prime" DNA replication are strands of RNA, called RNA **primers**. The RNA primers are synthesized by an enzyme called *primase*. Once a primer is in place, DNA polymerase extends the strand by adding new nucleotides to the free 3' hydroxyl end of the primer. For the synthesis of the lagging strand, the movement away from the replication fork necessitates several primers to be used as replication proceeds. Once each primer is added, a new DNA fragment is generated from the end of each primer. The result is synthesis of the Okazaki fragments, a series of short fragments of DNA that each begin with a section of RNA. Eventually, another DNA polymerase enzyme, **DNA polymerase I**, removes the RNA primer and fills in the space by extending the neighbouring DNA fragment. The Okazaki fragments are then joined together by the enzyme **DNA ligase**.

Termination

As the replication fork progresses along the replicating DNA, only a very short region of DNA is found in a single-stranded form. As soon as the newly formed strands are complete, they rewind automatically into their chemically stable double-helical structure. The protein-DNA complex at each replication fork that carries out replication is often referred to as the *replication machine*. **Figure 5.19** shows the replication machine at work.

The termination phase occurs upon completion of the new DNA strands, and the two new DNA molecules separate from each other. At that point, the replication machine is dismantled. **Table 5.2** summarizes the roles of the key enzymes in DNA replication.

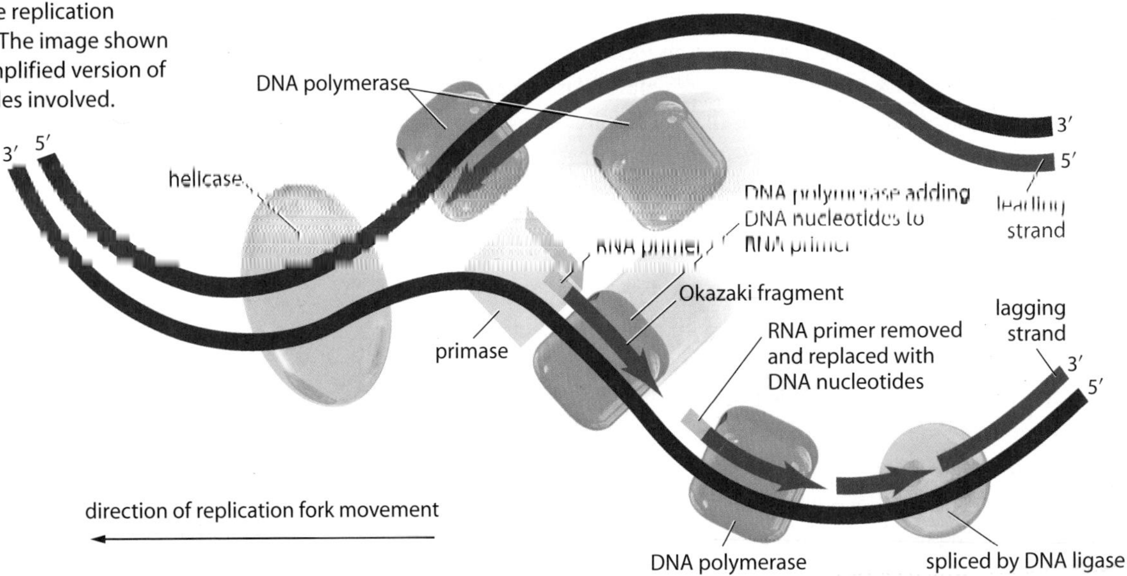

Figure 5.19 At each replication fork lies a complex of proteins and DNA that make up the replication machinery. The image shown here is a simplified version of the molecules involved.

Table 5.2 Important Enzymes in DNA Replication

Proteins and Enzymes	Functions
Helicase	helps to unwind the parent DNA
Primase	synthesizes RNA primer used to generate Okazaki fragments
Single-strand-binding protein	helps to stabilize single-stranded regions of DNA when it unwinds
Topoisomerase II	helps to relieve the strain on the structure of the parent DNA that is generated from the unwinding of the double helix
DNA polymerase I, II, and III	a group of enzymes with differing roles that include • addition of nucleotides to the 3′ end of a growing polynucleotide strand • removal of RNA primer and filling gaps between Okazaki fragments • proofreading newly synthesized DNA
DNA ligase	joins the ends of Okazaki fragments in the lagging strand synthesis

Activity 5.3 Modelling DNA Replication

In this activity, you will use model-building supplies to build a model of DNA replication.

Materials

- DNA model building supplies/software

Procedure

1. Review the steps of DNA replication described in **Figure 15.17** and **Figure 15.19**.
2. Design a model—either a physical model using available materials or a virtual model using appropriate software—that will enable you to simulate the process of DNA replication.
3. Sketch or take a digital photograph of each step in the replication process. Keeping in mind the action of the DNA polymerases, use your model to demonstrate
   a. replication along the leading strand
   b. replication along the lagging strand
   c. the actions of primase, the helicases, the DNA polymerases, and DNA ligase

Questions

1. Label your sketches or digital pictures with the following:
   a. leading strand
   b. lagging strand
   c. primer
   d. Okazaki fragment
2. Identify the action of the key replication enzymes by labelling your sketches or digital pictures.
3. Explain how your model demonstrates semi-conservative replication.

Correcting Errors during DNA Replication

A human cell can copy all of its DNA in a few hours, with an error rate of about one per 1 billion nucleotide pairs. As a comparison, imagine typing one letter for each of the roughly 3 billion base pairs in the human genome. Working non-stop at a rate of one letter per second, this would take you close to 100 years to complete. To match the accuracy of a cell, you could make no more than a single one-letter error every 30 years. Yet, as you have just learned, the replication machine includes many components, all acting simultaneously. It is not surprising, therefore, that errors can occur during replication. In many cases, before they are able to cause a permanent DNA-altering mutation, these errors can be corrected by one of several mechanisms, as you will see below.

One type of error that occurs during replication is a mispairing between a new nucleotide and a nucleotide on the template strand. For example, a T might be paired with a G instead of an A, as shown in **Figure 5.20**. Another type of error during DNA replication can be due to strand slippage, which causes either additions or omissions of nucleotides, as shown in **Figure 5.21**, on the next page. This type of error can result from either the newly synthesizing strand looping out, allowing the addition of an extra nucleotide, or the looping out of the template strand, resulting in a nucleotide not being incorporated where it should.

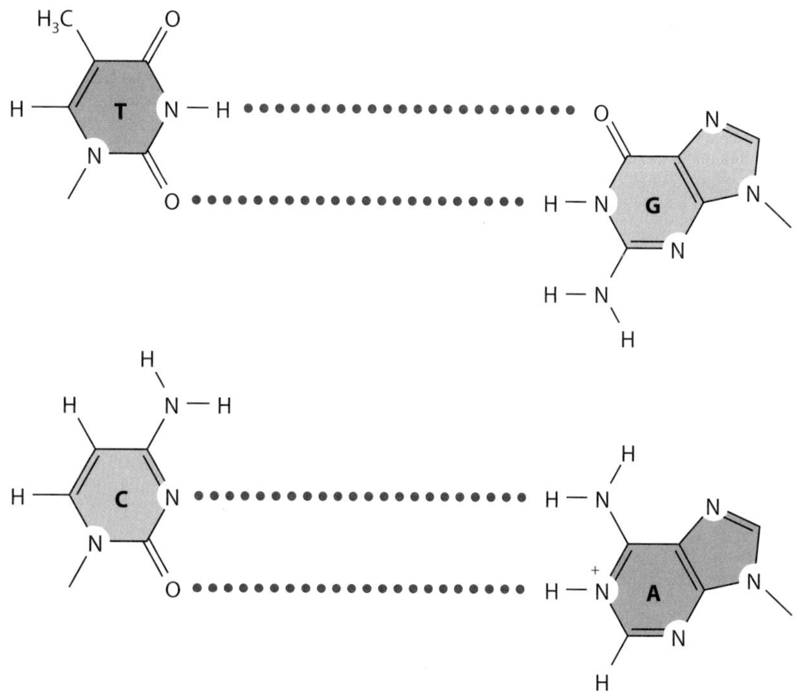

Figure 5.20 Incorrect pairing (mispairing) of bases is thought to occur as a result of flexibility in DNA structure.

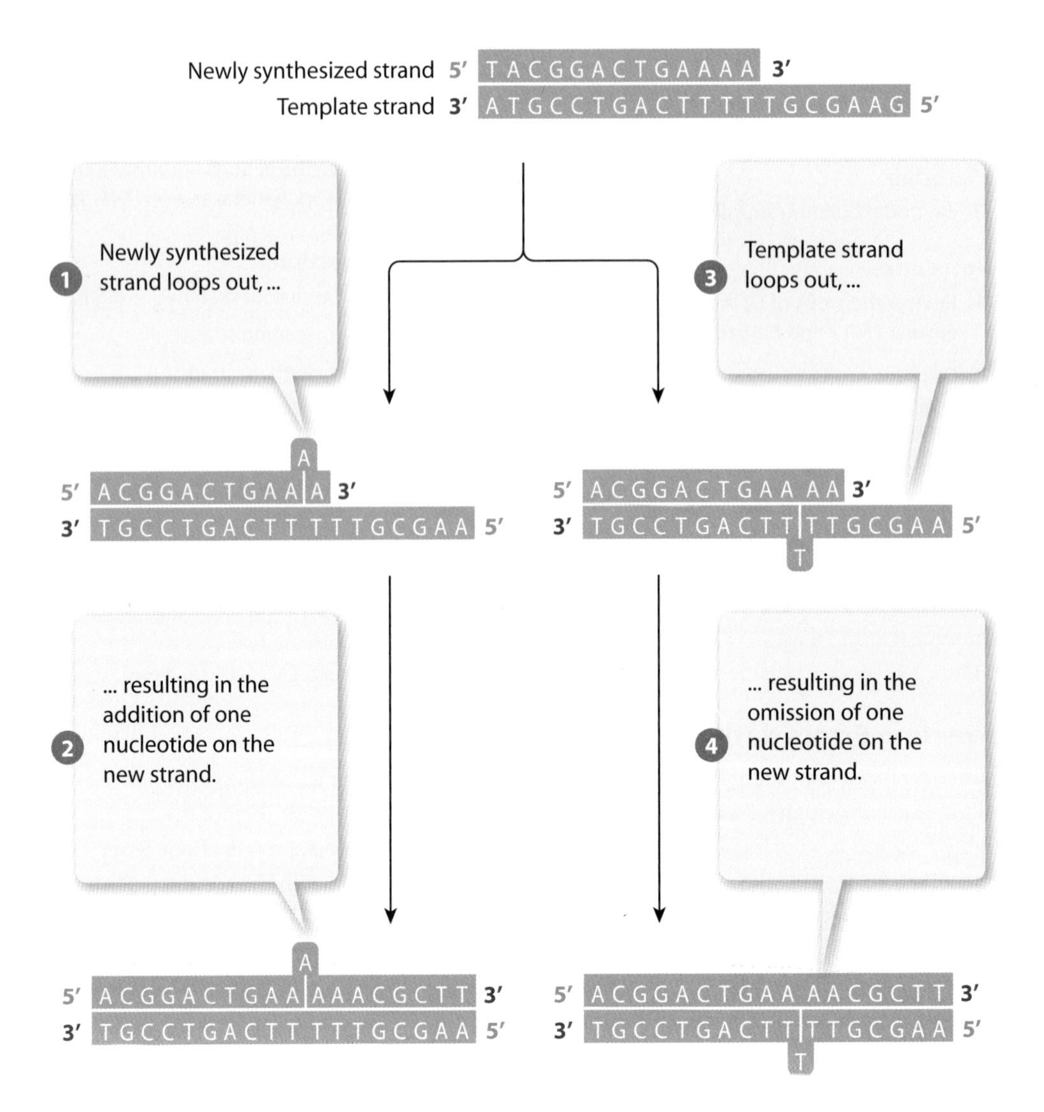

Figure 5.21 Strand slippage during DNA replication can cause the addition or omission of nucleotides in newly synthesized strands, which represent errors.

DNA polymerase II an enzyme that proofreads newly synthesized DNA

mismatch repair a mechanism for repairing errors made during DNA replication, whereby a group of proteins recognize a mispaired nucleotide on the newly synthesized strand and replace it with a correctly paired nucleotide

One mechanism for correcting errors that occur during DNA replication involves DNA polymerases. As well as catalysing the addition of nucleotides and excising primers in DNA replication, DNA polymerase I, along with **DNA polymerase II**, has an important proofreading function. After each nucleotide is added to a new DNA strand, these DNA polymerases can recognize whether or not the correct nucleotide has been added. Replication is stalled when an incorrect nucleotide is added because the 3′ hydroxyl end of the incorrect nucleotide is in the wrong position for the next nucleotide to bond to it. When this occurs, these DNA polymerases excise the incorrect base from the new strand and add the correct base, using the parent strand as a template. This proofreading step repairs about 99% of the mismatch errors that occur during DNA replication.

Another mechanism for correcting DNA replication errors is called **mismatch repair**. This repair mechanism occurs in all species and is similar in both prokaryotes and eukaryotes. The mispairing of bases causes deformities in the newly synthesized molecule. These deformities are recognized by a group of enzymes that bind to the DNA and specifically remove the incorrect base from the daughter strand. Errors that still remain after DNA polymerase proofreading or mismatch repair are then considered mutations in the genome once cell division has occurred. Mutations and the agents that cause them will be explored in the next chapter.

Learning Check

19. List and summarize the three basic phases of DNA replication.
20. Why does elongation take place in a slightly different way on each DNA strand?
21. Explain how the leading and lagging strands of a DNA molecule are replicated.
22. Why does DNA replication require numerous enzymes?
23. Describe one way that errors in replication are corrected.
24. Why is it important that cells are able to replicate DNA quickly and accurately?

Comparing DNA Replication in Eukaryotes and Prokaryotes

DNA replication is carried out in similar ways in prokaryotes and eukaryotes. For example, both

- require origins of replication
- have elongation occur in the 5′ to 3′ direction
- have continuous synthesis of a leading strand and discontinuous synthesis of a lagging strand
- require use of a primer for synthesis of Okazaki fragments in the synthesis of the lagging strand
- use DNA polymerase enzymes

Scientists still have much to learn about DNA replication in eukaryotes. However, important differences in DNA replication between prokaryotes and eukaryotes already have been identified. Several examples are listed below.

- The rate of replication is faster in prokaryotes (about 1000 nucleotides added per second) than in eukaryotes (about 40 nucleotides added per second). Presumably, this is due to more elaborate enzyme complexes that are required in eukaryotic replication and a more stringent proofreading mechanism.
- The DNA polymerase enzymes in eukaryotes differ from those in prokaryotes. They also differ in the number involved. To date, five have been identified in prokaryotes, while 13 have been identified in eukaryotes.
- The smaller circular chromosome of a prokaryote contains a single origin of replication. The larger linear chromosome of a eukaryote may contain thousands of origins of replication.

The linear nature of eukaryotic chromosomes presents an additional problem, when the final RNA primer from the 5′ end of the lagging strand is removed. For linear DNA, there is no adjacent fragment onto which nucleotides can be added and the gap filled, as there is for prokaryotes. Therefore, when this shortened strand is copied in the next round of replication, a shorter chromosome will be produced. To ensure that this loss of DNA does not result in loss of important genetic information, the ends of eukaryotic chromosomes contain highly repetitive sequences, called **telomeres**. Cells have a special enzyme, called telomerase, which synthesizes these telomeric regions and can replace a sequence that has been lost.

telomere a repetitive section of DNA, near each end of a chromosome; the presence of this sequence helps to protect from loss of important genetic information during replication of the linear DNA in eukaryotic cells

Telomerase activity in human cells varies with development in humans. During childhood, when tissues are growing rapidly, telomerase activity is high. As people age and the rate at which tissues grow slows, telomerase activity also slows. This results in a shortening of chromosomes in somatic (body) cells. This shortening means that information from the coding portion of chromosomes may be lost.

BIOLOGY Connections

Biobanks: Violation of Privacy, or Key to Medical Breakthroughs?

Biobanking is the collection and analysis of physical specimens from a wide sampling of individuals, from which DNA can be derived. Biobanks have been established in Sweden, Germany, the United States, Norway, Estonia, Singapore, Saudi Arabia, the United Kingdom, and many other countries. In Canada, millions of dollars have been invested in research related to biobanking.

A primary goal of biobanking is to provide researchers with open access to millions of high-quality samples collected from people around the world. Samples of DNA include tissues collected in hospitals and clinics, for example, when a tissue sample is taken to make a medical diagnosis.

The Benefits and Risks of Biobanking

Researchers hope the genetic data in biobanks, along with access to a person's health records, will provide insights into the interaction between genes and the environment. Making links between genes, environmental factors, and disease can help researchers learn more about disease prevention and lead to an overall improvement in human health.

However, biobanking presents some difficult legal and ethical challenges. For example, how should researchers go about collecting biosamples for genetic research, and how should consent be obtained? How can this sensitive information be protected now and in the future? What safeguards can prevent insurance agencies and employers from using genetic information for inappropriate purposes?

The Beginnings of a Canadian Biobanking Policy

In 2003, a background paper titled *Survey of National Approaches to the Development of Population Biobanks* was prepared for the Canadian Biotechnology Advisory Committee. The purpose of the report was to outline the challenges in developing biobanking policies for Canadians. Nine major challenges identified by the committee are summarized in the table on the right.

Biobanking Challenge	Associated Question or Issue
Consultation	What steps are necessary to obtain and maintain the public trust?
Recruitment	Which individuals, from which segments of the population, will be sampled?
Consent	How will the approval of individuals, and Canadian society as a whole, be obtained?
Governance	Who will oversee and regulate biobanks and their practices?
Commercialization	What rights of ownership do sampled individuals give up—and to whom?
Privacy	How will privacy be ensured, and who will be held accountable?
Communication of research results	Should individuals have access to their personal results or only to those of the aggregate (that is, the population as a whole)?
Contribution to the welfare of the population	What constitutes a benefit in terms of a population, and how is this to be measured?
Contribution to the welfare of humanity	How will all humanity share in and have access to the benefits of biobank research?

Genome Canada and Its Mandate

Genome Canada is an organization that helps to develop national strategies for research in genomics. In January 2010, leading researchers prepared a GE^{3}LS (Genomics and its related Ethical, Economic, Environmental, Legal, and Social aspects) outline of options for developing consent and privacy practices for biobanks. The options included:

- *Option 1:* Specific and fully informed consent is needed for each project (that is, consent is required for each project that uses a person's genetic data).
- *Option 2.* Broad initial consent accompanied by appropriate governance is required (that is, consent is given to use genetic data even though the details of its use in all future research are not provided).
- *Option 3:* Opt-out model (that is, a person's genetic data would be placed in the biobank unless they explicitly request that they do not want their data included).

Connect to Society

1. What Canadian regulations currently exist related to ownership of a genome and use of an individual's genetic data for research? Who makes the decisions related to regulations, and how are regulations enforced and monitored? Are existing regulations acceptable in terms of the social, ethical, and legal implications, or do they need to be strengthened to match the rigour used by other countries? Defend your point of view with specific facts and examples.
2. Which of the options presented in the GE^{3}LS policy brief do you think is preferable, based on the advantages and disadvantages of each one? Defend your position as persuasively as possible.

Section 5.2 Review

Section Summary

- The process of copying one DNA molecule into two identical molecules is called DNA replication, which occurs during the S phase of interphase in the cell cycle.
- Although three models were proposed for DNA replication, the experiments of Meleson and Stahl showed that the semi-conservative was correct. In the semi-conservative model of DNA replication, one double strand of DNA replicates to produce a total of two double strands of DNA, each of which contains one strand of the original DNA molecule and one strand of new DNA.
- In the initiation phase of DNA replication, helicase enzymes help a segment of DNA to unwind from its double-helix structure and separate into two strands. Single-strand binding proteins stabilize the separated strands, resulting in an unwound, oval-shaped area called a replication bubble.
- In the elongation phase of DNA replication, new DNA strands are synthesized by joining individual nucleotides together. This process is catalysed by DNA polymerase III, primase, and DNA ligase enzymes.
- In the termination phase of DNA replication, the two new DNA molecules separate from each other.
- During replication, DNA polymerase II proofreads newly synthesized DNA.
- DNA replication is similar in eukaryotes and prokaryotes in various ways, including the presence of origins of replication, the presence of DNA polymerase enzymes, and elongation in the 5′ to 3′ direction. Replication differs between eukaryotes and prokaryotes in various ways, including a slower replication rate and many more origins of replication in eukaryotes, and different DNA polymerase enzymes in eukaryotes than in prokaryotes.

Review Questions

1. **T/I** Infer why DNA replication must occur before, and not after, cell division occurs.
2. **C** The research of Meselson and Stahl provided convincing evidence for the mechanism of DNA replication.
   a. Using a flowchart, summarize the protocol and results of the Meselson–Stahl experiment.
   b. If DNA replication was conservative, what would Meselson and Stahl have seen after one round of replication? Use a diagram to show what bands would appear in the centrifuged tube.
3. **K/U** List the three basic phases of DNA replication. Write a brief description of what occurs during each phase.
4. **A** Would you expect there to be more replication origins in developing embryo cells or in adult cells? Explain your answer.
5. **C** What is the difference between a replication bubble and a replication fork? Write a short description of each and use labelled diagrams to illustrate your answer.
6. **C** Use a labelled diagram to explain the following statement: "DNA replication is both continuous and discontinuous."
7. **K/U** Why is an RNA primer necessary for DNA replication?
8. **K/U** How are the different DNA polymerase enzymes involved in both DNA replication and DNA repair?
9. **T/I** Temperature-sensitive mutant strains of *E. coli* are useful models for analyzing genes important in DNA replication. Predict what function would be affected in each of the following *E. coli* mutants:
   a. mutation in *dnaG* (primase gene)
   b. mutation in *lig* (ligase gene)
   c. mutation in *dnaB* (helicase gene)
10. **C** Identify and describe the models of replication labelled A, B, and C in the diagram below.

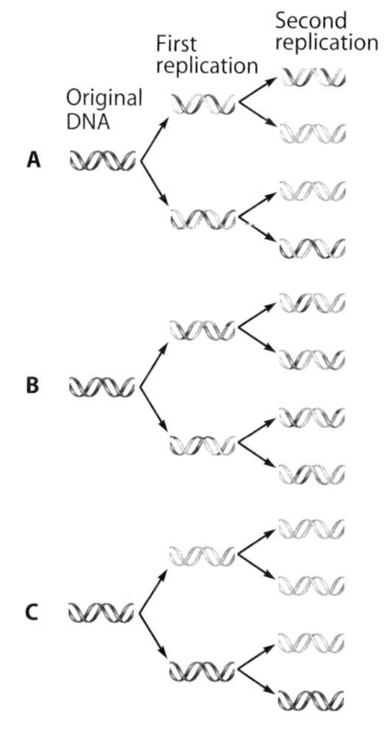

11. **C** Use a graphic organizer to compare and contrast DNA replication in prokaryotes and eukaryotes.

Inquiry
INVESTIGATION

5-A

Skill Check

Initiating and Planning

✓ Performing and Recording

✓ Analyzing and Interpreting

✓ Communicating

Extracting DNA

To study or manipulate DNA, scientists must first isolate it from the rest of the cellular material. To isolate the DNA in bananas, the cells are first separated by squashing them. The cells are then treated with a soap solution that degrades the plasma membranes and nuclear membranes, releasing the cells' contents. The soap also causes proteins and lipids to precipitate out of the solution. Salt is used to stabilize the DNA, which can then be isolated from the solution by adding a layer of ethanol, which causes the DNA to precipitate.

Safety Precautions

- Chemicals used in this investigation may be toxic. Take extra care to avoid getting them in your eyes, on your skin, or on your clothes. If contact does occur, rinse thoroughly with water and inform your teacher.
- Wash your hands thoroughly with soap and water after you have completed this investigation.

Materials

- 1/4 ripe banana
- small resealable plastic bag
- graduated cylinder
- 10 mL 0.9% NaCl solution
- 3 mL dishwashing liquid
- cheesecloth
- 250 mL beaker
- large test tube
- test tube rack
- 95% ethanol (ice-cold)
- stirring rod

Pre-Lab Questions

1. Why is the banana treated with a soap solution?
2. Why is the alcohol poured carefully down the side of the test tube?
3. What safety precautions must be taken when working with ethanol?

Question

How is DNA isolated from plant cells, and what is its appearance?

Procedure

1. Place about one quarter of a ripe banana in a small resealable plastic bag. Seal the bag, and use your fingers to squash the banana until no visible chunks remain

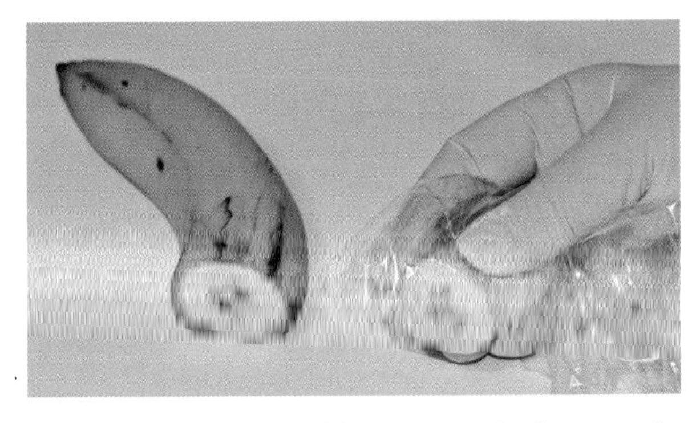

2. Use a graduated cylinder to measure 10 mL of the 0.9% NaCl solution, and add it to the bag. Mix thoroughly.
3. Use a graduated cylinder to measure 3 mL of dishwashing liquid, and add it to the bag. Mix gently as to not create any bubbles.
4. Strain the banana mixture through cheesecloth, and collect the filtrate in a small beaker. Dispose of the banana residue.
5. Transfer 2 mL of the filtrate to a large test tube.
6. Use a graduated cylinder to measure 8 mL of ice-cold ethanol.
7. Tilt the test tube and gently add the ethanol to the test tube by pouring it down the side. The alcohol layer should "float" on the banana filtrate.
8. Let the solution sit for 2–3 minutes. You will see a white substance forming where the two layers meet.
9. Carefully insert a glass stirring rod and swirl it slowly in the alcohol layer to collect the DNA and remove it from the test tube.

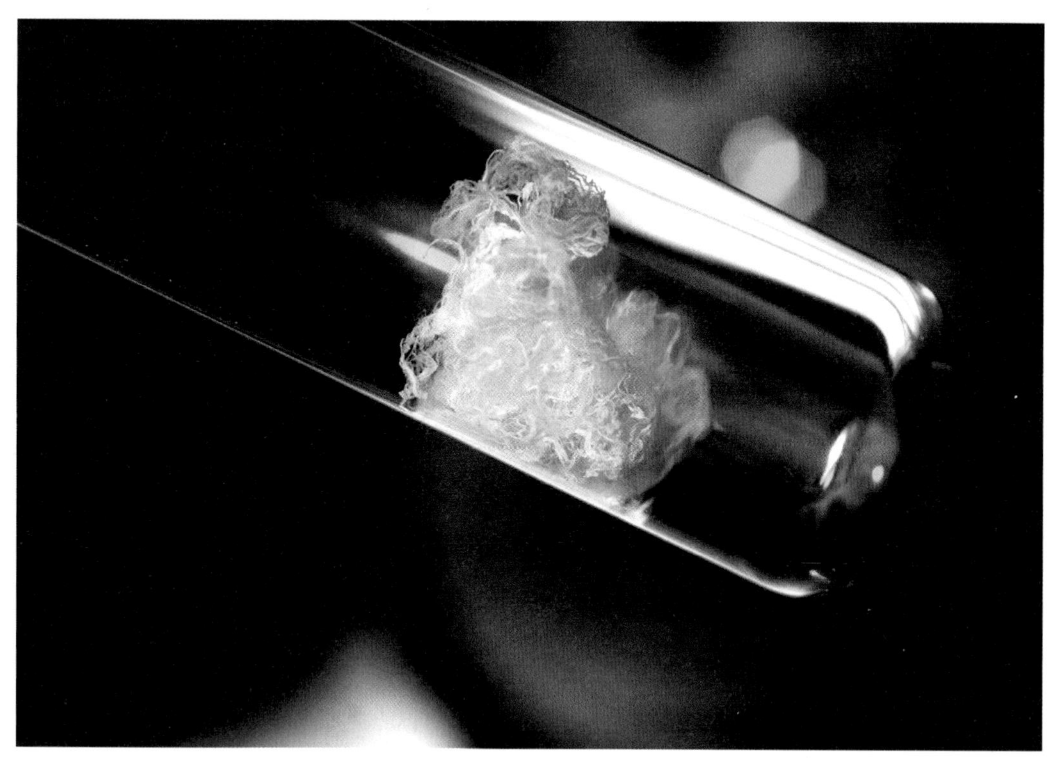

Extracted DNA

Analyze and Interpret

1. A number of steps are required to isolate DNA from cellular contents. Describe what happens at each step, and why it acts to separate the parts of the cell.
2. Do you think the process or results would be different if you were to use a vegetable, animal cells, or a different fruit? Explain.
3. What property of DNA allows it to be "coiled" around a glass stirring rod once precipitated?

Conclude and Communicate

4. Based on your procedure and results, how is DNA isolated from plant or animal cells?
5. Describe the appearance of the DNA collected.

Extend Further

6. **INQUIRY** What other steps could be taken in this procedure to better purify your DNA sample?
7. **RESEARCH** How can you ensure that your sample of DNA does not contain any impurities, including other nucleic acids or proteins?

ThoughtLab INVESTIGATION | **5-B**

Skill Check

Initiating and Planning

✓ Performing and Recording

✓ Analyzing and Interpreting

✓ Communicating

Suggested Materials

- print or Internet resources

DNA Replication

American biochemist Arthur Kornberg was a pioneer of in vitro (in a test tube) studies of DNA replication. This investigation outlines the steps that Kornberg used to determine if nucleotides (deoxyribonucleoside triphosphates) are the substances from which DNA is made.

Pre-Lab Questions

1. Name the three different components for each nucleotide in DNA.
2. If a radiolabelled phosphorus tracer were used to track nucleotides, in which chemical structure would the phosphorus be located?

Question

How can DNA replication be studied in vitro?

Organize the Data

The figure on the opposite page shows the steps Kornberg used to study DNA replication in vitro. Study these steps and answer the questions below.

Analyze and Interpret

1. In step 1, what is the purpose of adding each component to the test tube?
2. What was the purpose of incubating the mixture?
3. Kornberg knew that nucleotides are soluble in an acidic solution, but long DNA strands precipitate out of solution at an acidic pH. What effect did adding the perchloric acid to the test tube have on the mixture?
4. Why was the centrifuge needed in this experiment?
5. The precipitated pellet contained the newly replicated DNA. How would the ^{32}P-radiolabelled nucleotides help in the analysis of the replicated DNA?

Conclude and Communicate

6. If the newly replicated DNA contained radiolabelled nucleotides, what is a reasonable conclusion?
7. Write a paragraph explaining how DNA replication can be studied in vitro. In your paragraph, explain how ^{32}P-radiolabelled nucleotides are used to track the DNA replication process.

Extend Further

8. **INQUIRY** Develop an inquiry question that you could answer by performing a similar experiment. Identify the components that you would use to study DNA in vitro.
9. **RESEARCH** Use print or Internet resources to research some of the historical scientific contributions that have advanced our understanding of molecular genetics. Write a short report explaining the contributions of the scientist.

1. Mix together the extract of *E. coli* proteins, template DNA that is not radiolabelled, and ^{32}P-radiolabelled nucleotides. This is expected to be a complete system that contains everything necessary for DNA synthesis. As a control, a second sample is made in which the template DNA was omitted from the mixture.
2. Incubate the mixture for 30 minutes at 37°C.
3. Add perchloric acid to precipitate DNA. (It does not precipitate free nucleotides.)
4. Centrifuge the tube.
   Note: The radiolabelled nucleotides that have not been incorporated into DNA will remain in the supernatant.
5. Collect the pellet, which contains precipitated DNA and proteins. (The control pellet is not expected to contain DNA.)
6. Count the amount of radioactivity in the pellet using a scintillation counter.

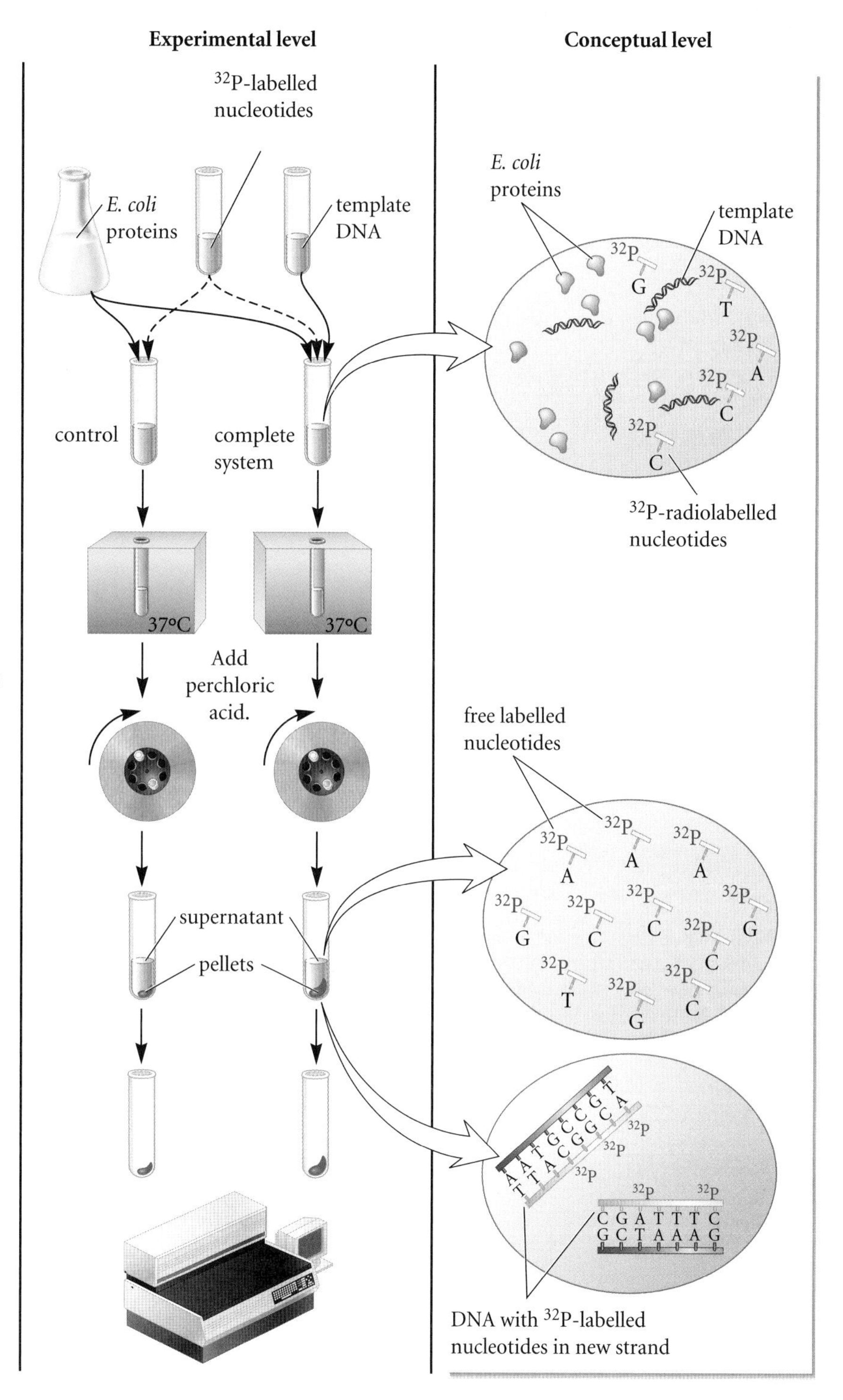

American biochemist Arthur Kornberg was a pioneer of in vitro (in a test tube) studies of DNA replication. His studies in the 1950s provided evidence of the roles that each component played in replication. The steps he used in the laboratory are shown here.

Chapter 5 SUMMARY

Section 5.1 DNA Structure and Organization in the Cell

A molecule of DNA is composed of two strands of linked nucleotides held together by hydrogen bonds and twisted in a double-helix structure.

Key Terms

nucleotide
Chargaff's rule
complementary base pairing
antiparallel
genome
gene
nucleoid
DNA supercoiling
regulatory sequence
histone
nucleosome
chromatin

Key Concepts

- DNA was identified as the hereditary material through a series of experiments between 1928 and 1952. Before this identification, many scientists believed that the more chemically complex proteins were the hereditary material of the cell.
- DNA and RNA are nucleic acids made of units called nucleotides. Each nucleotide is composed of one of four nitrogen-containing bases, a sugar, and a phosphate group. The sugar in DNA is deoxyribose, while the sugar in RNA is ribose. For DNA, the nucleotide bases are adenine (A), guanine (G), cytosine (C), and thymine (T). For RNA, they are A, G, C, and uracil (U).
- James Watson and Francis Crick published their three-dimensional structure of DNA in 1953. In this accepted model, DNA has a double-helix structure like a twisted ladder, in which the sides of the ladder are composed of alternating sugars and phosphate groups, while the rungs of the ladder are composed of paired A-T and C-G bases.
- In most prokaryotes, genetic material is in the form of a circular, double-stranded DNA molecule. The nucleoid region of this type of cell contains tightly packed DNA, in part due to DNA supercoiling.
- In eukaryotic cells, genetic material is contained in the nucleus. DNA is compacted through different levels of organization to form a series of defined structures including nucleosomes, chromatin, and chromosomes.

Section 5.2 DNA Replication

DNA replication occurs in three basic phases: initiation, in which a double strand of DNA unwinds; elongation, in which new DNA strands are synthesized by linking free nucleotides; and termination, in which the two new strands separate.

Key Terms

DNA replication
semi-conservative replication
origin of replication
helicase
DNA polymerase III
primer
Okazaki fragments
DNA polymerase I
DNA ligase
DNA polymerase II
mismatch repair
telomere

Key Concepts

- The process of copying one DNA molecule into two identical molecules is called DNA replication. DNA replication occurs during the S phase of interphase in the cell cycle.
- Although three models were proposed for DNA replication, the experiments of Meleson and Stahl showed that the semi-conservative was correct. In the semi-conservative model of DNA replication, one double strand of DNA replicates to produce a total of two double strands of DNA, each of which contains one strand of the original DNA molecule and one strand of new DNA.
- In the initiation phase of DNA replication, helicase enzymes help a segment of DNA to unwind from its double helix structure and separate into two strands. Single strand binding proteins stabilize the separated strands, resulting in an unwound, oval-shaped area called a replication bubble.
- In the elongation phase of DNA replication, new DNA strands are synthesized by joining free nucleotides together. This process is catalysed by DNA polymerase III, primase, and DNA ligase enzymes.
- In the termination phase of DNA replication, the two new DNA molecules separate from each other.
- During replication, DNA polymerase II proofreads newly synthesized DNA.
- DNA replication is similar in eukaryotes and prokaryotes in various ways, including the presence of origins of replication, the presence of DNA polymerase enzymes, and elongation in the 5′ to 3′ direction. Replication differs between eukaryotes and prokaryotes in various ways, including a slower replication rate and many more origins of replication in eukaryotes, and different DNA polymerase enzymes in eukaryotes than in prokaryotes.

Chapter 5 REVIEW

Knowledge and Understanding

Select the letter of the best answer below.

1. What was the transforming principle in Griffith's experiments on *S. pneumonia*?
   a. protein
   b. lipid
   c. bacteria
   d. DNA
   e. carbohydrate

2. What was the major finding of the Hershey–Chase experiment?
   a. Radioactive phosphorus and radioactive sulfur were both transferred into bacterial cells in equal amounts, proving that DNA was hereditary material.
   b. The majority of radioactive phosphorus was transferred into bacterial cells, proving that DNA was hereditary material.
   c. The majority of radioactive phosphorus was found in the liquid medium, proving that DNA was hereditary material.
   d. The majority of radioactive phosphorus was found in the bacteriophage "ghosts," proving that DNA was hereditary material.
   e. The majority of radioactive sulfur was transferred into bacterial cells, proving that DNA was hereditary material.

3. What are the basic components of a single nucleotide in a DNA molecule?
   a. a phosphate group, a deoxyribose sugar, and a hydrogen
   b. a phosphate group, a ribose sugar, and a nitrogen-containing base
   c. a phosphate group, a deoxyribose sugar, and a pyrimidine base
   d. a phosphate group, a deoxyribose sugar, and a purine base
   e. a phosphate group, a deoxyribose sugar, and a nitrogen-containing base

4. Which of the following types of nitrogen bases are present in a DNA molecule?
   a. adenine, thymine, cytosine, uracil
   b. adenine, thymine, cytosine, guanine
   c. adenine, cytosine, guanine, uracil
   d. thymine, cytosine, guanine, uracil
   e. adenine, guanine, uracil, phosphate

5. If a DNA molecule contains 20 percent of A, approximately what percentage of G is present?
   a. 20 percent
   b. 40 percent
   c. 30 percent
   d. 60 percent
   e. 50 percent

6. What did Rosalind Franklin's X-ray diffraction images reveal about the structure of DNA?
   a. DNA is a polymer of nucleotides.
   b. Each nucleotide has a deoxyribose sugar.
   c. DNA has a helical structure.
   d. DNA is composed of adenine, thymine, cytosine, and guanine.
   e. The amount of adenine is the same as thymine, and the amount of guanine equals the amount of cytosine.

7. In the Watson–Crick model of DNA, what are the "handrails"?
   a. nitrogen bases
   b. sugar-phosphate molecules
   c. hydrogen bonds
   d. purines
   e. pyrimidines

8. In a double-stranded DNA molecule, the complementary strand to 5′-AAACGCTT-3′ is
   a. 5′-TTTGCGAA-3′
   b. 5′-GGGTATCC-3′
   c. 5′-AAACGCTT-3′
   d. 5′-AAGCGTTT-3′
   e. 5′-UUUGCAA-3′

9. In prokaryotic genomes, which regions determine when certain genes and the associated cell functions are activated?
   a. genomic sequences
   b. regulatory sequences
   c. plasmids
   d. chromatin
   e. nucleoids

10. The results of the Meselson–Stahl experiment supported which model of DNA replication?
   a. semi-conservative
   b. conservative
   c. dispersive
   d. antiparallel
   e. complementary base pairing

11. In which phase of DNA replication is the replication bubble created?
    a. synthesis
    b. elongation
    c. termination
    d. initiation
    e. mismatch repair

12. The leading strand of DNA is synthesized
    a. in both 5′ to 3′ and 3′ to 5′ directions.
    b. discontinuously in a 5′ to 3′ direction.
    c. discontinuously in a 3′ to 5′ direction.
    d. continuously in a 5′ to 3′ direction.
    e. continuously in a 3′ to 5′ direction.

13. How would DNA replication be affected if there were a mutation in the gene that codes for DNA ligase?
    a. Okazaki fragments would not be joined.
    b. Errors in DNA replication would be not corrected.
    c. Unwinding of the DNA would be stalled.
    d. Elongation of the leading strand would not occur.
    e. RNA primers would not be synthesized.

14. Which of the following statements about DNA replication in prokaryotes and eukaryotes is true?
    a. Prokaryotes have a single origin of replication, while eukaryotes have multiple origins of replication.
    b. In prokaryotes, elongation occurs in a 3′ to 5′ direction. In eukaryotes, elongation occurs in a 5′ to 3′ direction.
    c. The rate of DNA replication in prokaryotes and eukaryotes is similar.
    d. Prokaryotes and eukaryotes both have telomeres at the ends of their linear chromosomes.
    e. Prokaryotes have continuous synthesis of both leading and lagging strands, while eukaryotes have discontinuous synthesis on both leading and lagging strands.

Answer the questions below.

15. For many years, scientists assumed that proteins rather than DNA made up the material of heredity.
    a. What were some of the factors behind this assumption?
    b. Which experimental results provided strong evidence that genetic information was carried on DNA?

16. How did Rosalind Franklin's X-ray diffraction images contribute to Watson and Crick's research on DNA structure?

17. In building their model of DNA, Watson and Crick built on the work of other scientists who had studied various aspects of the structure of DNA. However, Watson and Crick had to answer several major questions about the structure of DNA in order to develop their model.
    a. Linus Pauling proposed a three-stranded model of DNA. To Watson and Crick, the width of the helix suggested that DNA was two-stranded. Why else did they reason it was two-stranded?
    b. Originally, Watson and Crick placed the sugar-phosphate chain on the inside of the molecule. Why did they move it to the outside?
    c. Why did Watson and Crick call their three-dimensional model of DNA a "double helix"?

18. There is approximately 1 m of DNA in the nucleus of a human cell. How is DNA able to fit inside the nucleus?

19. What is the difference between a purine and a pyrimidine?

20. What is the importance of complementary base pairing in the structure of DNA?

21. Explain how the two strands of a DNA molecule are antiparallel.

22. The Meselson and Stahl experiment provided convincing evidence for a specific model of DNA replication. Why did they use two different isotopes of nitrogen in their experiment?

23. Why is it important that exact copies of DNA are produced during replication?

24. How is RNA involved in DNA replication?

25. What is the replication machine? What are its components?

26. Why are multiple origins of replication advantageous?

27. Describe two mechanisms that correct errors made during DNA replication. Why is it important that these errors are corrected before cell division?

28. Why are telomeres present at the ends of eukaryotic chromosomes?

Thinking and Investigation

29. Where do the components of nucleotides come from?

30. A given organism has many different tissues, but its cells all carry the same genetic information. Explain how this is possible.

31. DNA is sometimes said to be like a language. Explain whether you think this comparison is valid.

32. What would the complementary strand be for the following DNA sequences?
   a. 5′-CTGTACATC-3′
   b. 5′-ATCGCTGAT-3′
   c. 5′-CGGCGTATT-3′

33. A sample of DNA contains A and C nucleotides in the following proportions: A = 34 percent and C = 16 percent. What are the proportions of G and T nucleotides in this sample? (Assume that the characteristic proportions are exactly equal.)

34. Your research team is studying a virus that infects tomato plants. The genetic material of this virus is a single-stranded form of DNA. You extract two samples of DNA from an infected cell: one is the viral DNA and the other is the DNA of the plant cell. The table below shows the results of your analysis of the nucleotide composition of each sample. Which sample is the viral DNA? Explain.

Nucleotide Composition of DNA Samples

Nucleotide	Presence in Sample A (%)	Presence in Sample B (%)
Adenine	30.3	38.5
Cytosine	19.7	10.7
Thymine	30.3	13.3
Guanine	19.7	37.5

35. Your research group isolates a cell line characterized by unusual cell division. In these cells, only small fragments of DNA are found.
   a. Propose a hypothesis about what may be responsible for the small fragments of DNA.
   b. Design an experiment to test your hypothesis. Write a short description or use a flowchart to describe your experimental set-up.

36. Analysis of DNA from two different organisms reveals similar base composition. Based on this information, would the DNA sequences of the two organisms also be similar? Explain your answer.

37. Your research group is studying DNA organization in eukaryotic cells. In one of your investigations, you isolate mouse liver cells and treat them with micrococcal nuclease, an enzyme that preferentially cuts linker DNA. How do you think this enzyme will affect the organization of DNA in your sample? Use a labelled diagram in your explanation.

38. If continuous replication occurred in both strands of DNA, what implications would this have for DNA structure? Use a diagram and a short description to explain your answer.

39. In DNA, weak hydrogen bonds exist between the nitrogen bases of nucleotides on opposite strands, while strong covalent bonds are responsible for mediating the interaction between sugar and phosphate groups on the "handrails." What are the advantages of having these types of chemical bonds?

40. A scientist grows a sample of *E. coli* in liquid growth medium. She then removes a sample of *E. coli* and transfers it to liquid growth medium that contains radioactive phosphorus (^{32}P). What will the distribution of radioactivity be on each DNA strand after one round of replication? after two rounds of replication? Use a labelled diagram to illustrate your answer.

41. Primase synthesizes an RNA primer using the following DNA sequence found on the lagging strand as a template: 5′-AGTCA-3′. What is the sequence of the RNA primer?

42. Would you expect the DNA polymerases to be more active in heart muscle cells (which are mainly in G_1) or in continuously dividing skin cells? Explain.

43. Your research team is working on developing a new anti-bacterial drug that has minimal side effects on eukaryotic cells. What part of the DNA replication machinery would you use as your drug target? Why did you choose this target?

44. Werner's syndrome is a very rare disease that leads to rapid premature ageing after puberty. It is caused by mutations in *WRN*, a gene that codes for a type of DNA helicase. What implications would this have on the DNA of an individual with Werner's syndrome?

45. You have discovered a new type of DNA polymerase that can add nucleotides in the 3′ to 5′ direction. What component or components of the replication machine may no longer be necessary for replication to occur? Explain.

Communication

46. Using a concept map, give an example from the chapter of when different types of investigations were used to address the same hypothesis. Why might using a variety of experimental approaches be necessary?

47. Create a timeline called "The Road to DNA." Include the names and contributions of the scientists who were involved in identifying and characterizing DNA.

48. One strand in a stretch of DNA has the base sequence 5′-CCTGAT-3′. Draw this stretch of DNA, showing both strands. Label the sugar-phosphate backbone, the 5′ and 3′ ends of each strand, and the regions of hydrogen bonding.

49. Compare and contrast the organization of prokaryotic and eukaryotic genomes using a graphic organizer.

50. Using different colours (one for each strand), sketch the DNA molecules that would be present after one, two, and three rounds of semi-conservative replication.

51. Use a flowchart to describe what occurs during three main phases of DNA replication.

52. Create a labelled diagram illustrating the components of a replication bubble. Include arrows to show the direction of unwinding.

53. DNA can only be synthesized in one direction. The replication of DNA strands, however, proceeds in two directions at once. Use a labelled diagram to show how this is possible.

54. Using the table below, list the key enzymes involved in DNA replication. For each enzyme, briefly describe its function and what would happen if it were not present (Assume the absence of any one enzyme does not affect the activity of others.)

Enzyme	Function	Absence of Enzyme

55. Write a one-page journal entry from the point of view of one scientist mentioned in Chapter 5. Describe the current state of knowledge about hereditary material, the challenges you are facing in your research, and your thoughts on the nature and significance of discoveries still to come in your field.

56. Suppose you are a journalist for your local newspaper. You are living in the year 1953, and Watson and Crick just published their paper on the structure of the DNA double-helix. Write a short newspaper article directed at the general public that answers the following questions:

a. What were Watson and Crick's findings?

b. How did Watson and Crick come to their conclusion about the structure of DNA?

c. What are the possible implications of their research?

57. Using a concept map or flowchart, illustrate the relationships among nucleotide, DNA, histone, nucleosome, chromatin, and chromosome.

58. Explain how a defect in the mismatch repair mechanism would affect a cell's DNA.

59. Summarize your learning in this chapter using a graphic organizer. To help you, the Chapter 5 Summary lists the Key Terms and Key Concepts. Refer to Using Graphic Organizers in Appendix A to help you decide which graphic organizer to use.

Application

60. Comparative genomics is the study of similarities and differences among the genomes of different organisms. For example, many human genes have counterparts in the fruit fly. What practical applications could result from comparing genomes of different species?

61. BIG IDEAS Genetic research and biotechnology have social, legal, and ethical implications. To receive funding from the Canadian Institute of Health Research, researchers must fill out a comprehensive grant application that summarizes their project. One of the most important sections on the application requires an explanation of the potential impact that their research will have "within and beyond the academic community."

a. Choose one of the scientists mentioned in this chapter who contributed to the discovery and structure of DNA.

b. Write a paragraph from the perspective of this person explaining how his or her research would have an impact "within and beyond the academic community."

62. As described in the opener for this unit, telomere testing of an individual's chromosomes has the potential to provide information about that person's overall health and the rate at which the person's cells are aging.

a. Explain how telomere length decreases over time.

b. Infer why telomere length is an indicator of physiological age and how it might be affected by lifestyle factors such as exercise, smoking, and diet.

63. The shortening of telomeres that occurs with each round of cell division has been linked to aging. It has been hypothesized that increasing telomerase activity may be the "fountain of youth." What ethical issues surround this type of research?

64. In 1994, a group of researchers led by Dr. Calvin Harley at McMaster University were the first to publish a scientific paper that convincingly showed high levels of telomerase activity in human ovarian cancer tissues. Since then, elevated telomerase activity has been seen in many other human cancers.

a. Why do you think cancer cells have increased telomerase activity?

b. Because of its elevated levels in cancer cells, telomerase has been identified as a possible therapeutic target for cancer treatment. There are, however, concerns about side effects that may occur if telomerase activity is inhibited. What could some of these side effects be?

65. Novel scientific discoveries are usually credited to the scientists whose names appear on the first published paper that describes the findings. In the 1953 *Nature* paper that describes the structure of the DNA double helix, Watson and Crick were listed as the sole authors. However, Wilkins and Franklin are acknowledged in the last paragraph in the paper for their "unpublished experimental results and ideas."

a. Do you think Wilkins and Franklin were given appropriate credit for their contributions? Explain your answer.

b. Do you think that there is an objective way to assess a scientist's contribution to a new discovery? Why or why not?

66. Rosalind Franklin's X-ray diffraction images were shared without her consent. These images contained valuable information that allowed Watson and Crick to propose their breakthrough model of DNA. Do you think that researchers should be obligated to share their results with competing researchers for the "greater good" of scientific progress? Explain your answer.

67. Watson and Crick's discovery of DNA is clearly one of the most important biological discoveries in the last century. What advances in medicine and science can you think of that are built on knowing the structure of DNA?

68. When the results of the sequencing of the entire human genome were first released in 2001, many scientists were surprised to learn that the human genome contains far fewer genes than initially expected. What are some of the implications of this finding for the study of how human genetics relates to human health?

69. Lacking knowledge of Rosalind Franklin's X-ray analysis of the DNA molecule, Linus Pauling proposed a structure for DNA in which the phosphate groups were tightly packed on the inside of the molecule, thus leaving the nitrogenous bases projecting outward. If DNA replication occurred in this structure, how do you think it would differ from what you know is the actual process of DNA replication?

70. The hepatitis B virus (HBV) is a DNA virus that causes inflammation of the liver. Treatment for those infected with HBV includes the anti-viral drug adefovir (Hepsera ®), which acts as an inhibitor of viral DNA polymerase. Why do you think it is important that anti-viral drugs such as adefovir specifically target viral DNA polymerase?

71. BIG IDEAS DNA contains all the genetic information for any living organism. The experiments that were performed to reveal DNA structure and function used a variety of organisms, such as bacteria and mice.

a. How can such diverse organisms demonstrate the same genetic principles?

b. Using Internet and print resources, research an organism used in one of the experiments described in this chapter. Create a graphic organizer to describe the advantages and disadvantages of using this organism in scientific experiments.

c. What are the ethical issues that surround the use of this organism in scientific research?

72. Use the diagram below to answer the following questions.

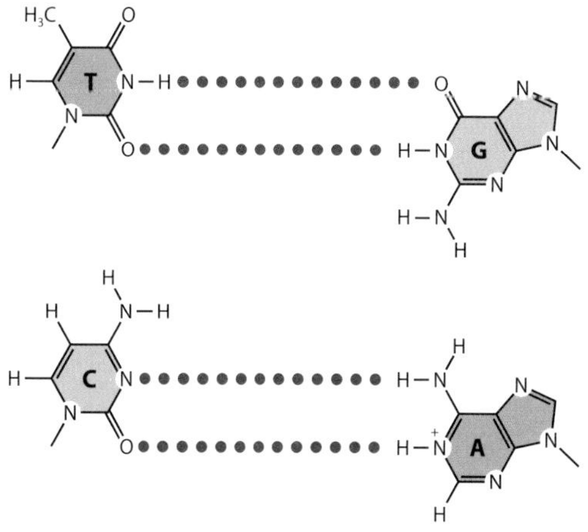

a. What type of DNA replication error is shown?

b. What causes this type of error?

c. How can this error be corrected?

Select the letter of the best answer below.

1. **K/U** Which scientist first proposed the idea of a transforming principle?
   a. Crick
   b. Griffith
   c. Watson
   d. Miescher
   e. Avery

2. **K/U** Identify the following nitrogenous bases.

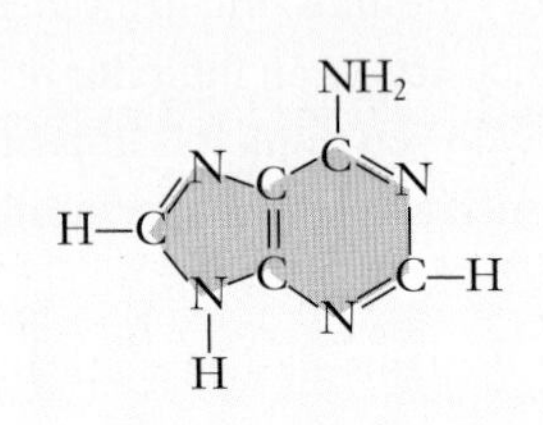

   a. adenine and cytosine
   b. adenine and thymine
   c. adenine and guanine
   d. cytosine and uracil
   e. cytosine and thymine

3. **K/U** Which are the main components of a single nucleotide?
   a. a glucose sugar, a phosphate group, and a nitrogenous base
   b. two sugars, a phosphate group, and a nitrogenous base
   c. a ribose sugar, a deoxyribose sugar, a phosphate group, and a nitrogenous base
   d. a sugar, a phosphate group, and four nitrogenous bases
   e. a sugar, a phosphate group, and a nitrogenous base

4. **K/U** What is the directionality of the two strands in a DNA molecule?
   a. parallel
   b. identical
   c. antiparallel
   d. complementary
   e. double helix

5. **K/U** Which is the complementary strand for the DNA sequence 5′-ATGACG-3′?
   a. 5′-CGTCAT-3′
   b. 5′-TACTGC-3′
   c. 5′-UACUGC-3′
   d. 5′-CGUCAU-3′
   e. 5′-AUCACG-3′

6. **K/U** Which is the functional unit of heredity?
   a. DNA
   b. RNA
   c. protein
   d. genome
   e. gene

7. **K/U** The Meselson–Stahl experiment demonstrated that DNA replication produces two DNA molecules each composed of
   a. two old strands.
   b. two new strands.
   c. one old and one new strand.
   d. two strands with variable proportions of new and old DNA.
   e. a variable number of old and new strands.

8. **K/U** A cell is missing the gene that codes for a helicase enzyme. What affect will this have on replication?
   a. slower unwinding of double helix
   b. faster unwinding of double helix
   c. increased stability of single strands
   d. decreased stability of single strands
   e. decreased strain on double helix

9. **K/U** During which phase(s) of DNA replication does an RNA primer bind to DNA?
   a. termination
   b. initiation
   c. initiation and elongation
   d. initiation and termination
   e. elongation

10. **K/U** The lagging strand of DNA is synthesized
   a. in both 5′ to 3′ and 3′ to 5′ directions.
   b. discontinuously in a 5′ to 3′ direction.
   c. discontinuously in a 3′ to 5′ direction.
   d. continuously in a 5′ to 3′ direction.
   e. continuously in a 3′ to 5′ direction.

Use sentences and diagrams as appropriate to answer the questions below.

11. T/I Imagine that you are analyzing a DNA sample from the liver tissue of a newly discovered species of mouse. Use the information in the table opposite to complete the nucleotide composition of your sample.

Nucleotide Composition of DNA Sample

Nucleotide	Presence in DNA Sample (%)
Adenine	31
Cytosine	
Guanine	
Thymine	

12. K/U Hershey and Chase's experiment with bacteriophages revealed that DNA, and not protein, is the hereditary material.

a. Why did Hershey and Chase use two different radioactive isotopes in their experiments?

b. What experimental result provided evidence for DNA being the hereditary material?

13. K/U Describe the observations made by Franklin that were used by Watson and Crick to determine the molecular structure of DNA.

14. C Use a labelled diagram to explain the "ladder" structure of DNA.

15. K/U How is DNA compacted in prokaryotes and eukaryotes? Why do the differences in the two processes exist?

16. T/I "Both protein-coding and non-coding regions of DNA have equal importance." Do you agree with this statement? Explain your answer.

17. A The emergence of new molecular biology techniques has allowed researchers to determine DNA sequences quickly and efficiently.

a. How could knowledge of a DNA sequence be abused?

b. How could knowing a DNA sequence be helpful?

c. Would you ever consent to having your DNA sequenced? Explain your answer.

18. T/I Copy the following diagram into your notebook, then use it to answer the following questions.

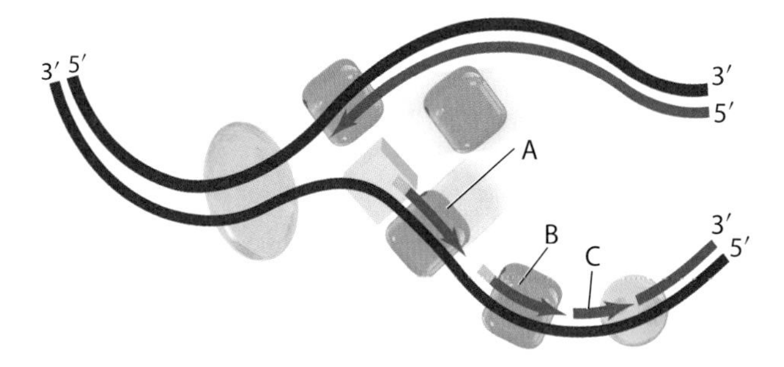

a. Draw an arrow to indicate the direction of replication fork movement.

b. Okazaki fragments are labelled A, B, and C in the diagram. Which Okazaki fragment was made first? How was it made?

c. Why are Okazaki fragments necessary for DNA replication?

19. K/U What structural features of DNA allow replication to occur efficiently and accurately?

20. C Use a labelled diagram to illustrate the components of a replication bubble and a replication fork. Indicate the direction of double-helix unwinding.

21. T/I DNA contains the information that a cell uses to synthesize a particular protein. How do proteins assist in DNA replication?

22. A Do you think individuals could survive without DNA polymerase? Explain your answer.

23. A A subset of colorectal cancers is associated with mutations in the *Mut* genes, which code for proteins involved in repairing DNA mismatches. How would mutations in *Mut* genes affect the DNA of an individual?

24. T/I "Telomeres are like cellular clocks." Explain what this statement means.

25. C Make a Venn diagram like below to compare and contrast replication in prokaryotes and eukaryotes.

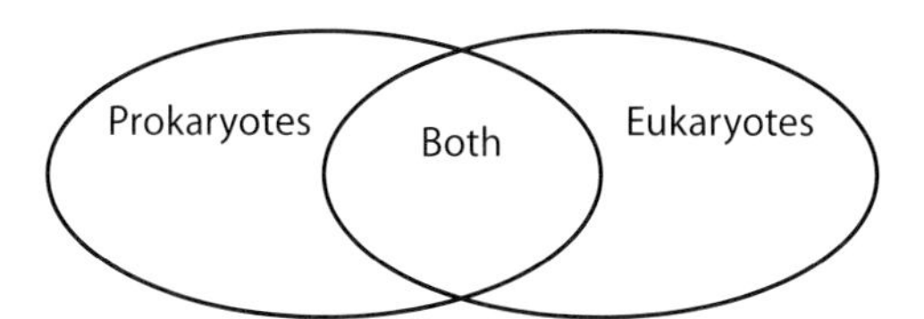

Self-Check

If you missed question...	1	2	3	4	5	6	7	8	9	10	11	12	13	14	15	16	17	18	19	20	21	22	23	24	25
Review section(s)...	5.1	5.1	5.1	5.2	5.1	5.2	5.2	5.2	5.2	5.2	5.1	5.1	5.1	5.1	5.1	5.1	5.2	5.2	5.2	5.2	5.2	5.2	5.2	5.2	5.2

CHAPTER 6

Gene Expression

Specific Expectations

In this chapter, you will learn how to . . .

- D2.1 **use** appropriate terminology related to molecular genetics (6.1, 6.2, 6.3, 6.4)
- D2.2 **analyze** a simulated strand of DNA to determine the genetic code and base pairing of DNA (6.3)
- D2.4 **investigate** and **analyze** the cell components involved in the process of protein synthesis, using appropriate laboratory equipment and techniques, or a computer simulation (6.4)
- D3.1 **explain** the current model of DNA replication, and **describe** the different repair mechanisms that can correct mistakes in DNA sequencing (6.3)
- D3.2 **compare** the structures and functions of RNA and DNA, and **explain** their roles in the process of protein synthesis (6.1, 6.2, 6.3)
- D3.3 **explain** the steps involved in the process of protein synthesis and how genetic expression is controlled in prokaryotes and eukaryotes by regulatory proteins (6.1, 6.2, 6.3, 6.4)
- D3.4 **explain** how mutagens, such as radiation and chemicals, can cause mutations by changing the genetic material in cells (6.3)

A scale played on a piano keyboard and a gene on a chromosome are both a series of simple, linear elements (notes or nucleotide pairs) that produce information. The notes in the scale produce a specific series of sounds ordered by ascending or descending pitch. If one note is changed or skipped, the resulting change to the musical structure of the scale is instantly recognized by the human ear. Similarly, an altered or deleted nucleotide pair will change the structure of a gene. Because the nucleotide pairs in a gene generally code for a specific series of amino acids, the new gene structure may result in an altered or nonfunctional protein.

Launch Activity

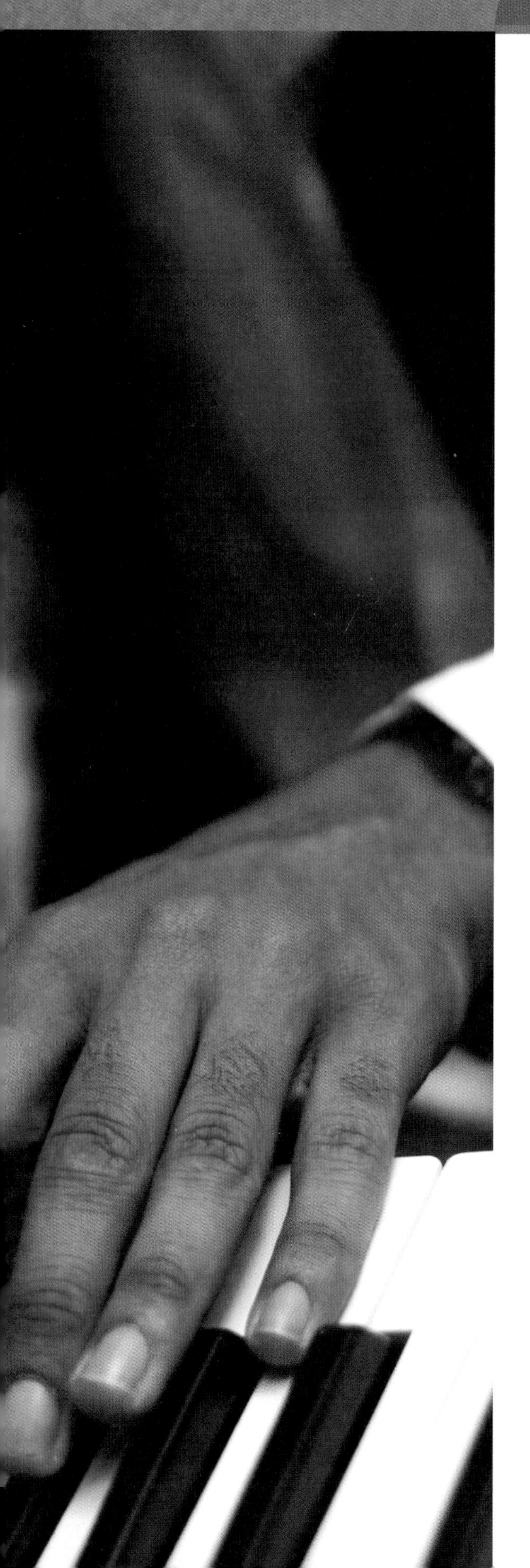

How Can You Read a DNA Sequence?

For this activity, it is helpful to recall that the sequence of bases in a strand of DNA is a code for the production of specific proteins in a cell. Proteins are constructed as amino acids are bonded together in a particular order according to the sequence of bases. In this activity, you will learn more about how the information contained in DNA is interpreted by cell machinery and make inferences about how errors may affect the proteins that are synthesized.

Procedure

1. Study the DNA sequence below.

   AGATGGGGCAGTCGCTAGTGCCCCACAT

2. Separate the sequence of bases into groups of three-letter, or triplet, codes. You should be able to do this in three different ways. (*Hint: Some bases may be left as a single base or in pairs.*)
3. Determine which of the three, newly-grouped sequences would be used to make an amino acid sequence. (*Hint: All amino acid sequences begin with the three-letter sequence ATG.*)
4. An insertion occurs when one or more DNA bases are added to the sequence. A deletion occurs when one or more DNA bases are deleted from a sequence. Using the base sequence you selected in step 3, simulate an insertion. Begin again with the correct base sequence and simulate a deletion.

Questions

1. Assume that each triplet code results in a specific amino acid being brought in to synthesize a protein. Infer how an insertion or a deletion could affect how the sequence is read and which amino acid is brought in.
2. Infer how a change in the amino acid sequence could affect the function of the protein produced.
3. Predict the implications of a protein functioning incorrectly in a unicellular organism and in a multicellular organism.
4. Consider and identify some substances or other agents that could cause an insertion or deletion to occur.

SECTION 6.1

The Transfer of Information from DNA

Key Terms

one-gene/one-polypeptide hypothesis
messenger RNA (mRNA)
genetic code
triplet hypothesis
gene expression
transcription
translation

Early studies in genetics demonstrated a relationship between the inheritance of a certain gene and expression of a particular trait. This formed the basis for the laws of inheritance that you have learned about. A question that naturally followed from this discovery was "*How* does a gene determine a trait?" This began a new wave of research, which focused on identifying the molecules that are involved in the inheritance of a trait. Proteins were known to carry out most of the functions in a cell. Therefore, many researchers began to investigate this question by looking for a relationship between genes and enzymes. In fact, evidence of such a relationship existed in the early 1900s—in the form of black urine.

Establishing a Link between Genes and Protein

The idea of a link between genes and proteins was first introduced by an English chemist and physician, Archibald Garrod (1857–1936). In 1902, Garrod published his studies of patients with a disease called *alcaptonuria*. The disease causes urine to turn black when it is exposed to air. The colour change is caused by increased levels of homogentisic acid in urine.

Garrod's approach was to investigate the disease biochemically, as a series of chemical reactions. He correctly proposed that the build-up of homogentisic acid was due to a defective enzyme in the metabolic pathway that breaks down the amino acid phenylalanine. When Garrod discovered that certain patients were blood relatives, he hypothesized that alcaptonuria was an inherited disease. Garrod suggested that the black urine phenotype was due to what Mendel called a recessive inheritance factor. Having this defective factor resulted in the production of a defective enzyme.

Beadle and Tatum and the One-Gene/One-Enzyme Hypothesis

Garrod's research laid the foundations for demonstrating a relationship between genes and proteins. It also represents the earliest findings that could link inherited human diseases to defective biochemical pathways. However, it was not until 1941 when George Beadle and Edward Tatum from Stanford University in California decided to look for experimental evidence of the relationship between genes and enzymes that Garrod's research was rediscovered. Assuming that a relationship existed, Beadle and Tatum also wanted to determine the quantitative nature of that relationship.

At the time, it was not known whether the production of all enzymes in a biochemical pathway was controlled by one gene, or whether one gene controlled the production of one enzyme. Beadle and Tatum had been working with the fruit fly, *Drosophila melanogaster*. However, they realized that fruit flies were too complex to use for the types of experiments that were needed. They turned, instead, to a bread mold—*Neurospora crassa*.

Normal, or wild-type, *N. crassa* grow on a simple culture medium that contains sugar and inorganic salts. Such a medium is called minimal medium, because it has only the nutritional substances to synthesize all other biochemical compounds *N. crassa* need for growth. Beadle and Tatum created mutant strains of *N. crassa* by exposing them to X-rays, which were known to cause changes in genes. Some of the strains could no longer grow on the minimal medium unless they were provided with additional nutrients. Eventually, Beadle and Tatum isolated some mutant strains that grew only when their growth medium was supplemented with the amino acid arginine.

The synthesis of arginine involves many steps. Beadle and Tatum wanted to identify mutants that affected particular steps of the pathway. They hypothesized that if there is a one-to-one relationship between a gene and an enzyme, then a defective gene will produce a defective enzyme. The appearance of a defective enzyme in one of the steps would mean that the intermediate compound it produces will not be synthesized. As a result, there would be no growth unless the missing intermediate was added to the medium.

As shown in **Figure 6.1**, Beadle and Tatum systematically supplemented the growth media with intermediates of the arginine synthesis pathway. They inferred which enzymes were defective based on whether the strain grew with each addition. Growth would only be possible on media provided with an intermediate that is produced after the step that involves the defective enzyme. Working in this way, Beadle and Tatum were able to isolate mutant strains that were defective at one specific step in the arginine synthesis pathway. This allowed them to pinpoint the step that was affected by each mutation. Eventually, they identified four genes—named E, F, G, and H—that produced a certain enzyme that catalyzed the production of an intermediate.

From their studies, Beadle and Tatum concluded that one gene specifies one enzyme. They called this relationship the *one-gene/one-enzyme hypothesis*. In 1958, Beadle and Tatum shared half of the Nobel Prize in physiology or medicine "for their discovery that genes act by regulating definite chemical events." The original hypothesis was later updated to account for the fact that not all proteins are enzymes, and that many enzymes are composed of more than one polypeptide chain. Thus, it is now more commonly called the **one-gene/one-polypeptide hypothesis**.

one-gene/one-polypeptide hypothesis a proposal that one gene codes for one polypeptide (or protein)

Figure 6.1 Beadle and Tatum's experiment showed that a single gene determines the production of one enzyme.

Predict *whether growth would occur after each addition of the intermediates for a mutant strain that does not produce glutamate.*

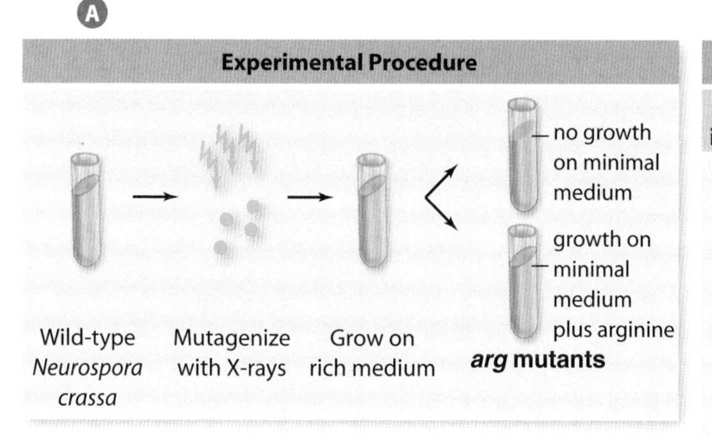

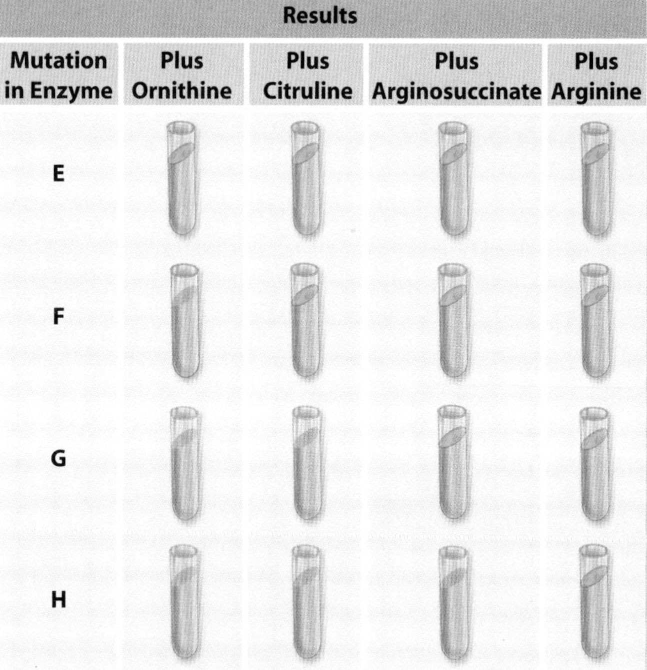

B
Results
Mutation in Enzyme
Plus Ornithine
Plus Citruline
Plus Arginosuccinate
Plus Arginine
E
F
G
H

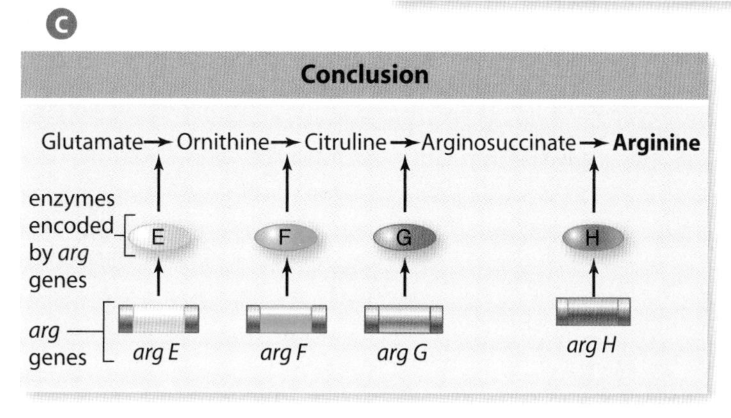

Finding a Messenger between DNA and Proteins

In 1953, the English biochemist Frederick Sanger showed that proteins consist of amino acids covalently linked together. Studying the protein insulin, Sanger established that each insulin molecule was made up of the same sequence of amino acids. Therefore, each protein has a particular amino acid sequence. This discovery came at the same time that Watson and Crick determined that DNA consists of strands of nucleotide sequences. By the early 1960s, a clear link between genes and proteins also had been made, but the nature of this link was still a mystery. How was information in the DNA sequence converted to the amino acid sequence of a protein? And did it happen directly or indirectly, through an intermediary molecule?

In eukaryotes, genes were known to be located on chromosomes that occur only in the nucleus. As well, protein synthesis in eukaryotes was known to occur only in the cytoplasm. Thus, scientists could discount the idea that proteins were directly synthesized from DNA. Some evidence supported the idea that another type of nucleic acid, RNA, was an intermediary between DNA and proteins. This evidence included the fact that RNA could be found in both the nucleus and cytoplasm of eukaryotes. As well, the concentration of RNA in the cytoplasm correlated with the level of protein production. Researchers also studied where RNA is synthesized in the cell, and they followed its transport within the cell. Through these studies, they were able to show that RNA is synthesized in the nucleus and then transported to the cytoplasm. But the question remained—how is RNA acting as an intermediary?

messenger RNA (mRNA) RNA that contains the genetic information of a gene and carries it to the protein synthesis machinery; it provides the information that determines the amino acid sequence of a protein

In 1961, Francois Jacob and Jacques Monod hypothesized the existence of a special type of RNA that acts as a "genetic messenger." They proposed that this RNA, which they called **messenger RNA (mRNA)**, is synthesized from the DNA of genes. The mRNA base sequence would be complementary to the gene DNA sequence. This mRNA nucleotide sequence would provide the amino acid sequence information needed for protein synthesis. In 1964, Jacob, working with Sydney Brenner and Mathew Meselson, confirmed the *messenger RNA hypothesis*. The results of their experiment showed the following:

- When bacteria were infected by a virus, a virus-specific RNA molecule was synthesized and became associated with pre-existing bacterial ribosomes. Recall that ribosomes are the site of protein assembly in the cell.
- The new RNA molecule had a base sequence complimentary to the DNA and carried the genetic information to produce the viral protein.
- This viral RNA molecule was newly synthesized and was not a permanent part of the bacterial ribosomes.

Learning Check

1. How did Garrod's findings demonstrate a link between genes and proteins?
2. Using a flowchart, outline the procedure and results of the experiment performed by Beadle and Tatum on *Neurospora crassa*. What was the significance of their results?
3. What evidence supported the idea that RNA acts as an intermediate molecule between DNA and proteins?
4. How did Jacob and colleagues show that RNA acts as a "genetic messenger"?
5. Explain why it is advantageous for cells to synthesize an mRNA messenger.
6. Predict the sequence of bases in a strand of mRNA that is synthesized from the following strand of DNA.

   5′-CCGTTAATC-3′

The Genetic Code

Once scientists established that mRNA acts as an intermediary, they turned their attention to "cracking" the **genetic code**. This code would help them understand how information is converted from the nucleotide sequence of an mRNA to the amino acid sequence of a protein. Researchers knew that are were only four nucleotides in RNA (A, U, G, and C), but 20 different amino acids. Therefore, there could not be a one-to-one relationship between nucleotide and amino acid. Even using two nucleotides per amino acid would only provide 4 × 4, or 16 possible combinations, which is not enough to code for 20 amino acids. Thus, the minimum combination of the four nucleotides was a triplet code, which could produce 4 × 4 × 4, or 64 possible combinations. From this reasoning came the **triplet hypothesis**, which proposed that the genetic code consists of a combination of three nucleotides, called a codon.

The most convincing support of the triplet hypothesis came from the 1961 studies by Francis Crick and South African biologist Sydney Brenner. These researchers used T4 bacteriophages, which are viruses that infect *E. coli*. They generated a series of viral strains in which the DNA sequence for a specific viral protein needed for *E. coli* infection was altered. Nucleotides were either added to or deleted from the DNA sequence. The results of their studies are shown in **Figure 6.2**.

Crick and Brenner found that when they added or deleted one nucleotide or a pair of nucleotides, the viral protein was not produced. This result could be reversed if additional mutations were able to reinstate the triplet codon. For example, if a nucleotide had been deleted, the addition of a nucleotide resulted in the production of the viral protein and *E. coli* infection. Also, if the mutation added or deleted three nucleotides that were close together, viral protein production and bacterial infection occurred. Crick and Brenner showed that the genetic code is read in triplets, with no spaces in the code—it is read continuously.

genetic code a set of rules for determining how genetic information in the form of a nucleotide sequence is converted to an amino acid sequence of a protein; a code specifying the relationship between a nucleotide codon and an amino acid

triplet hypothesis a proposal that the genetic code is read three nucleotide bases at a time

The nucleotide coding sequence from the wild-type virus is represented by the following hypothetical DNA sequence: CAG—CAG—CAG—CAG—CAG	**Result:** • Normal viral gene • Functional viral protein • Virus kills bacteria
1. The insertion or deletion of a single nucleotide or pair of nucleotides alters all subsequent nucleotide triplets. TCA—GCA—GCA—GCA—GCA ↑ T inserted at the beginning of the sequence.	**Result:** • The entire sequence is shifted to the right, causing each three-base segment to be altered from the original sequence. • Severe mutation in viral gene • Non-functional viral protein • Normal growth of bacteria
2. The insertion or deletion of two nucleotides has the same effect. TTC— AGC—AGC—AGC—AGC ↑ Two T's inserted at the beginning of the sequence.	
3. The insertion or deletion of three nucleotides alters at most two "words" of the code, after which the normal coding resumes. CAG—TTT—CAG—CAG—CAG CAT—TTG—CAG—CAG—CAG CTT—TAG—CAG—CAG—CAG ↑ Three T's inserted in various positions within the sequence.	**Result:** • The virus manufactures a modified version of the polypeptide, which is still partially functional. • Minor mutation in viral gene • Functional viral protein • Virus kills bacteria

Figure 6.2 The results from Crick and Brenner's experiment to test the triplet hypothesis

Determining the Genetic Code

Now that the idea of a genetic code based on multiples of three nucleotides was established, the race was on to determine which codons specified specific amino acids. Between 1961 and 1965, various research groups compared artificially synthesized RNA molecules of known nucleotide sequences with the amino acid sequences of polypeptides. From these studies, the mRNA codons and their corresponding amino acids were determined. These are listed in **Table 6.1**.

By convention, the genetic code is always interpreted in terms of the mRNA codon rather than the nucleotide sequence of the DNA. To read **Table 6.1**, start by finding the first letter of the mRNA codon in the "First Base" column. Then read across the rows in the "Second Base" column to find the second letter of the codon. This will take you to four possible amino acids. Finally, read down the "Third Base" column to find the last letter of the codon. The last letter of the codon, combined with the previous two letters, identifies the amino acid that corresponds to the codon. For example, the three nucleotides UAU code for the amino acid tyrosine. The first letter, U, is in the "First Base" column. The second letter, A, is in the "Second Base" column. The third letter, U, is in the "Third Base" column. Notice that the three nucleotides UAC also code for this same amino acid.

Table 6.1 The Genetic Code

First Base	Second Base				Third Base
	U	C	A	G	
U	UUU phenylalanine UUC phenylalanine UUA leucine UUG leucine	UCU serine UCC serine UCA serine UCG serine	UAU tyrosine UAC tyrosine UAA stop** ✓ UAG stop** ✓	UGU cysteine UGC cysteine UGA stop** ✓ UGG tryptophan	U C A G
C	CUU leucine CUC leucine CUA leucine CUG leucine	CCU proline CCC proline CCA proline CCG proline	CAU histidine CAC histidine CAA glutamine CAG glutamine	CGU arginine CGC arginine CGA arginine CGG arginine	U C A G
A	AUU isoleucine AUC isoleucine AUA isoleucine AUG methionine*	ACU threonine ACC threonine ACA threonine ACG threonine	AAU asparagine AAC asparagine AAA lysine AAG lysine	AGU serine AGC serine AGA arginine AGG arginine	U C A G
G	GUU valine GUC valine GUA valine GUG valine	GCU alanine GCC alanine GCA alanine GCG alanine	GAU aspartate GAC aspartate GAA glutamate GAG glutamate	GGU glycine GGC glycine GGA glycine GGG glycine	U C A G

* AUG is an initiator codon. It also codes for the amino acid methionine.
** UAA, UAG, and UGA are terminator codons.

The genetic code has three important characteristics.

1. The genetic code is redundant. This means that more than one codon can code for the same amino acid. There are only three codons that do not code for any amino acid. As you will learn later in the chapter, these codons serve as "stop" signals to end protein synthesis.
2. The genetic code is continuous. This means that it reads as a series of three-letter codons without spaces, punctuation, or overlap. Therefore, knowing exactly where to start and stop protein synthesis is essential. A shift of one or two nucleotides in either direction can alter the codon groupings and result in an incorrect amino acid sequence.

3. The genetic code is nearly universal. Almost all organisms build proteins with the genetic code shown in **Table 6.1**. (Some rare exceptions are known in some species of protists, for example.) The universality of the genetic code means that a codon in the fruit fly codes for the same amino acid as in a human. As you will learn in the next chapter, this has important implications for genetic techniques, such as cloning. A gene that is taken from one kind of organism and inserted into another kind of organism will produce the same protein.

Gene Expression

Gene expression refers to the synthesis of a protein based on the DNA sequence of a gene. **Figure 6.3** summarizes the path of gene expression. The theory that genetic information flows from DNA to RNA to protein is often called the *central dogma* of genetics.

gene expression the transfer of genetic information from DNA to RNA to protein

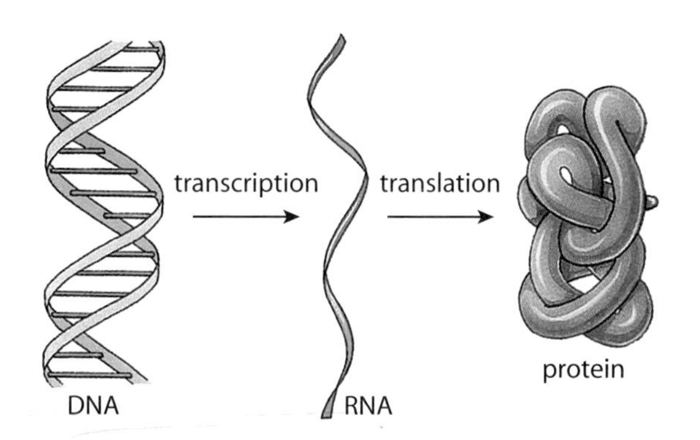

Figure 6.3 The central dogma of genetics describes the transfer of genetic information from DNA, to RNA, and finally to proteins.

***Infer** how drugs that inhibit transcription would affect translation.*

Transcription and Translation

The two steps in the expression of a gene are transcription and translation. In **transcription**, mRNA is synthesized based on the DNA template of a gene. This is followed by **translation**, which involves the production of a protein with an amino acid sequence that is based on the nucleic acid sequence of the mRNA. The translation of nucleotide sequence to amino acid sequence uses the genetic code as outlined in **Figure 6.4**. These two processes, transcription and translation, will be examined in the remainder of this chapter.

transcription the synthesis of RNA from a DNA template

translation the synthesis of protein from an mRNA template

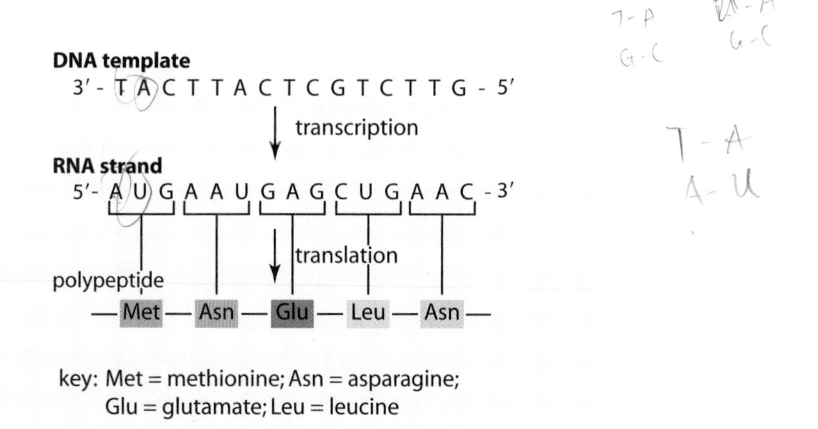

Figure 6.4 Transcription of a DNA template produces an RNA molecule that is a copy of the genetic information. The nucleotide sequence of this RNA molecule is then translated using the genetic code so that the protein coded for by the gene is produced.

Section 6.1 Review

Section Summary

- The idea that a link existed between genes and proteins was first introduced by Garrod. He used his studies of  alcaptonuria to suggest that defective inheritance factors produce defective enzymes.
- From their experiments with the bread mould *Neurospora crassa*, Beadle and Tatum concluded that one gene specifies one enzyme. They called this relationship the one-gene/one-enzyme hypothesis. Later, this was modified to the one-gene/one-polypeptide hypothesis.
- Jacob proposed and demonstrated that mRNA acts as a "genetic messenger." It provides the amino acid sequence information needed for protein synthesis.
- The genetic code specifies how genetic information is converted from a nucleotide sequence to an amino acid sequence of a protein. Experiments by Crick and Brenner showed that the genetic code is read three nucleotides at a time.
- The genetic code is redundant, continuous, and nearly universal.

Review Questions

1. **T/I** "The link between genes and proteins was determined experimentally by Archibald Garrod." Explain why you agree or disagree with this statement.
2. **T/I** Beadle and Tatum wanted to find experimental proof of a relationship between genes and proteins.
   a. How would the results of Beadle and Tatum's experiment have differed if the medium they used also contained the essential amino acids required for growth?
   b. Explain whether their results would still show a relationship between genes and proteins.
3. **C** Create a question for a student you are tutoring to test her knowledge of the one-gene/one-polypeptide hypothesis and its application to modern genetics.
4. **K/U** What was the significance of the experiments performed by Jacob and colleagues?
5. **K/U** Describe the triplet hypothesis.
6. **C** In 1961, Crick and Brenner tested the triplet hypothesis.
   a. Use a flowchart to summarize the Crick and Brenner experiment.
   b. What was the significance of their results?
7. **T/I** Suggest another genetic code system that could code for 20 different amino acids.
8. **K/U** List and explain the three characteristics of the genetic code.
9. **T/I** Infer why growing polypeptides begin with methionine.
10. **K/U** Why are stop codons necessary?
11. **T/I** Suppose that during replication of a gene, one nucleotide is copied incorrectly. However, the protein coded for by this "new" gene is the same as that coded for by the normal gene. How can this be explained?
12. **K/U** Study the illustration below.

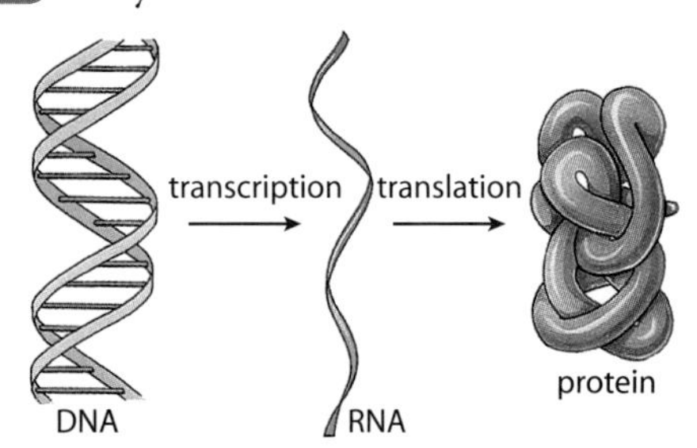

   a. What theory does the illustration represent?
   b. What does this theory explain?
13. **A** "Each person has a unique genetic code." Do you think this is an accurate statement? Explain your answer.
14. **A** Use **Table 6.1** and the information in the diagram below to determine the amino acids that would make up the portion of the polypeptide shown. Include information for a key as well.

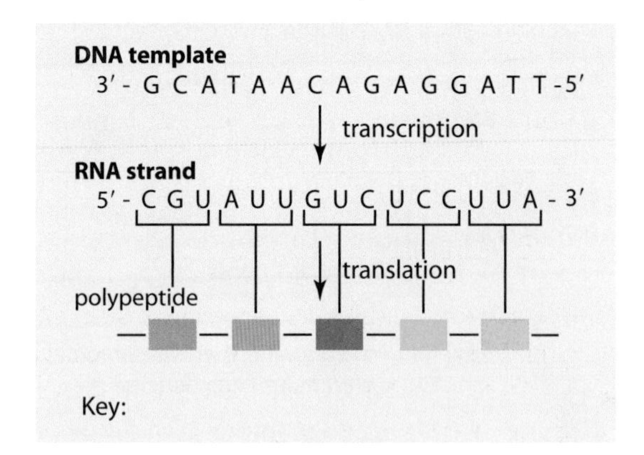

15. **A** Discuss the significance of the universality of the genetic code with respect to the following fields:
   a. evolutionary science
   b. medical science

SECTION 6.2

Transcription: Synthesizing RNA from DNA

DNA is often described as the blueprint of a cell, which stores information needed for survival and reproduction. The first step in how this genetic information is read involves the production of an RNA molecule. The structure of an RNA molecule is shown in **Figure 6.5**. Like DNA, RNA is a polymer of nucleotides. Recall that RNA contains four nucleotides with the bases adenine (A), uracil (U), cytosine (C), and guanine (G). Unlike DNA, RNA is single-stranded. However, it can fold back on itself, and complementary base pairing within the same molecule stabilizes the looped structure. **Table 6.2** summarizes the similarities and differences between RNA and DNA.

Key Terms

RNA polymerase
promoter region
precursor mRNA (pre-mRNA)
mature mRNA
5′ cap
3′ poly-A tail
splicing

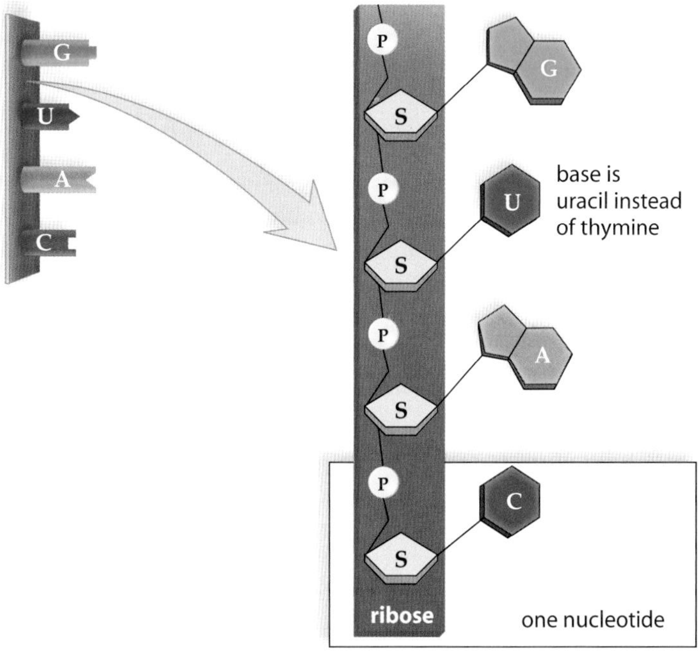

Figure 6.5 RNA is a single-stranded polynucleotide molecule. It contains the bases uracil, adenine, guanine, and cytosine. It also contains ribose as its sugar group.

Table 6.2 Comparison of DNA and RNA

Characteristic	DNA	RNA
Nitrogenous bases	adenine, thymine, guanine, cytosine	adenine, uracil, guanine, cytosine
Sugar	deoxyribose	ribose
Strand structure	double	single

There are several types of RNA molecules. Since all are produced from a DNA template, all are synthesized in the nucleus. However, they are classified according to the different functions they have in the cell. For example, mRNA acts as an intermediary between the DNA sequence of a gene and protein synthesis. The mRNA is used as a template that determines the amino acid sequence of the protein that it codes for. Some of the different RNA molecules and their functions are listed in **Table 6.3**. As you can see, there are many RNAs that do not code for proteins, but that still have essential functions in the cell. Two of these are involved in the process of translation, and their functions will be described in more detail in the last section of this chapter.

Table 6.3 Different RNA Molecules and Their Functions

RNA	Function
Messenger RNA (mRNA)	the template for translation
Transfer RNA (tRNA)	involved in the translation of mRNA
Ribosomal RNA (rRNA)	involved in the translation of mRNA
Small nuclear RNA (snRNA)	involved in modification of mRNA molecules
Micro RNA (miRNA)	involved in regulating gene expression
Small interfering RNA (siRNA)	involved in regulating gene expression
RNA in RNaseP	RNaseP is an enzyme; the RNA is the part of the enzyme
7S RNA	involved in targeting proteins to particular regions in eukaryotic cells
Viral RNA	found in some viral genomes

The Molecular Events of Transcription

The main objective for transcription is to accurately produce a copy of a small section of genomic DNA. In a similar manner to what occurs in DNA replication, there are three defined stages in the transcription process: initiation, elongation, and termination. Note that transcription in eukaryotes and prokaryotes is very similar. The main differences are the proteins involved. More proteins are required for each phase of transcription in eukaryotes, and they form more complex associations.

Initiation

During the initiation phase, the correct position for transcription to start is selected and the transcription machinery, composed of a large protein-DNA complex, is assembled. For each gene, only one strand of the double-stranded DNA molecule is transcribed. This strand is called the *antisense strand* or *template strand*. The other strand, which is not transcribed, is called the *sense strand* or *coding strand*. It has the same sequence as the product mRNA, with thymines instead of uracils. In a single DNA molecule, either strand can serve as the sense strand for different genes.

RNA polymerase the main enzyme that catalyzes the formation of RNA from a DNA template

promoter region a sequence of nucleotides in DNA that indicates where the RNA polymerase complex should bind to initiate transcription

The main enzymes that catalyze the synthesis of RNA are a group of enzymes called **RNA polymerases**. In eukaryotes, each RNA polymerase has a specific function. Once the RNA polymerase complex has bound to the DNA molecule, it unwinds and opens a section of the double helix.

Transcription begins when RNA polymerase binds tightly to a **promoter region** on the DNA. This region contains special sequences of nucleotides. A typical promoter region found in *E. coli* is shown in **Figure 6.6**. Notice how there are two sets of sequences. Both of these are required and need to be correctly positioned relative to each other. This allows the RNA polymerase complex to bind to the correct strand and in the correct orientation, so that the strand is copied in the correct direction.

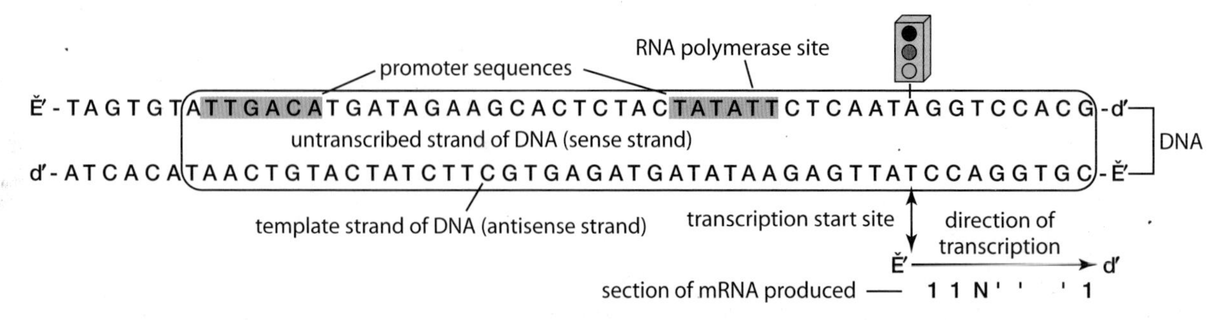

Figure 6.6 The promoter region in *E. coli* is composed of two specifically positioned nucleotide sequence elements. Both of these are required for the RNA polymerase complex to correctly bind to the DNA template.

Elongation

During the elongation phase, shown in **Figure 6.7**, the RNA polymerase complex works its way along the DNA molecule, synthesizing a strand of mRNA that is complementary to the template strand of DNA. In the mRNA strand, however, T is replaced with U. Like DNA polymerase, RNA polymerases work in the 5′ to 3′ direction, adding each new nucleotide to the 3′-OH group of the previous nucleotide. RNA polymerases transcribe only one strand of the template DNA, so there is no need for Okazaki fragments, as in DNA replication.

As soon as the RNA polymerase complex starts to move along the DNA, a second RNA polymerase complex can bind to the promoter region and start to synthesize another mRNA molecule. As shown in **Figure 6.8**, this means that hundreds of copies of mRNA molecules can be made from one gene at one time. Also, RNA polymerase catalyzes the synthesis of mRNA at a much faster rate than DNA polymerase catalyzes the synthesis of DNA. This is mainly because the RNA polymerase complex does not have a proofreading function. An error in transcription would only result in an error in one protein molecule, and not in the genetic make-up of an organism. Being able to synthesize more mRNA in a given amount of time is a greater advantage than minimizing sequence errors.

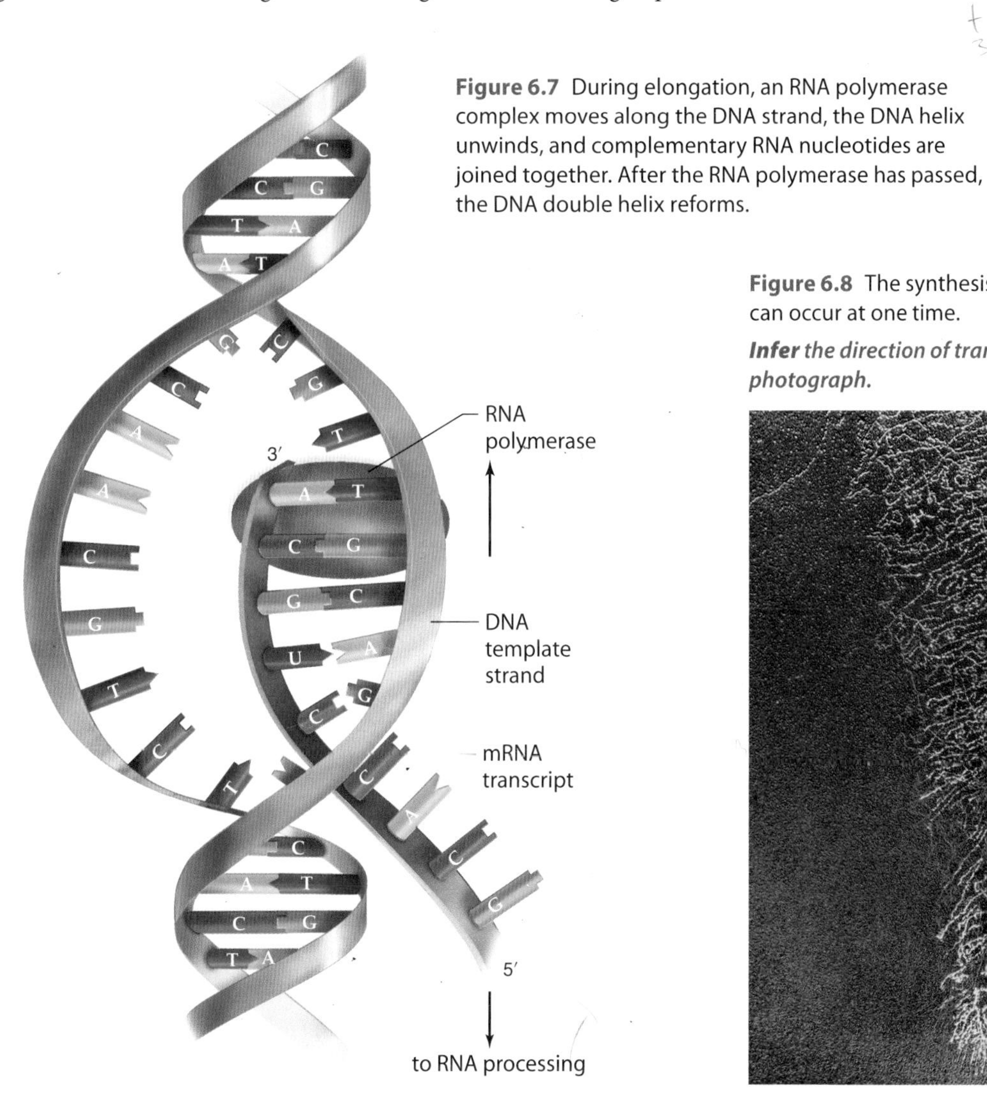

Figure 6.7 During elongation, an RNA polymerase complex moves along the DNA strand, the DNA helix unwinds, and complementary RNA nucleotides are joined together. After the RNA polymerase has passed, the DNA double helix reforms.

Figure 6.8 The synthesis of many mRNA molecules can occur at one time.

***Infer** the direction of transcription in this photograph.*

Termination

For the termination phase, specific nucleotide sequences in the template DNA serve as a signal to stop transcription. When the RNA polymerase complexes reach this signal, they detach from the DNA strand. The new mRNA strand is released from the transcription assembly, and the DNA double helix reforms.

Learning Check

7. List three differences between DNA and RNA.
8. How is RNA involved in protein synthesis?
9. List and describe the three stages of transcription.
10. Predict where mRNA is transported in the cell after transcription is finished.
11. How do cells synthesize multiple strands of mRNA from one gene simultaneously? Why is this advantageous?
12. Would errors in transcription be more or less damaging than errors in DNA replication? Explain your answer.

Activity 6.1 Modelling Transcription

In this activity you will model transcription of a given DNA sequence.

Safety Precautions

• Handle the scissors with care.

Materials

• 6 colours of construction paper
• ruler
• tape
• scissors
• felt marker

Procedure

1. Examine the following DNA sequence:
   5′-ATG CAT GGC TA-3′
   3′-TAC GTA CCG AT-5′
2. Identify the antisense strand (template strand). **Hint:** The coding sequence always begins with AUG on the mRNA strand.
3. Choose a colour of construction paper for each base in mRNA and DNA. Cut out shapes to represent entire nucleotides. Use complementary shapes for bases that pair together, as shown in the photograph. Label the paper nucleotides.
4. Use the paper nucleotides to model the antisense strand of DNA. Draw a line on these nucleotides to distinguish them from mRNA. Tape them together at the back.
5. Cut a circle out of the sixth sheet of construction paper to represent RNA polymerase.

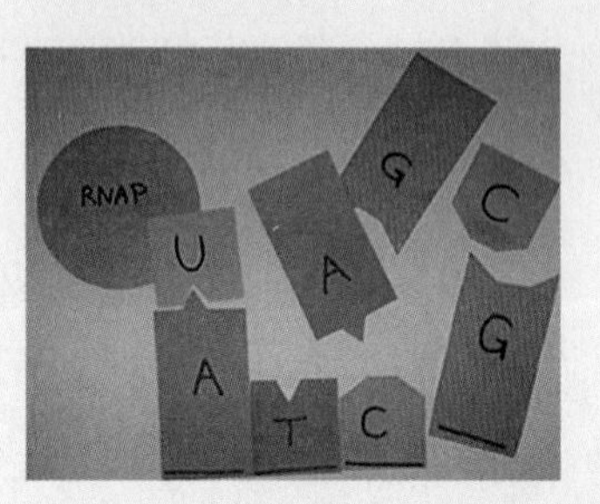

6. Use the additional paper nucleotides and RNA polymerase to simulate transcription of your model DNA strand. Make a sketch or take a digital photograph of your model at the start, middle, and completion of transcription.

Questions

1. Why was it important to identify the antisense strand?
2. Identify the 5′ and 3′ ends on your mRNA model.
3. How does mRNA differ from DNA?
4. How does transcription differ from DNA replication?
5. How would the deletion of one nucleotide in the middle of the antisense strand of DNA affect the sequence of the transcribed mRNA?

mRNA Modifications in Eukaryotes

precursor mRNA (pre-mRNA) mRNA that has not undergone processing

mature mRNA mRNA that has undergone processing

5′ cap modified form of a G nucleotide; added to the 5′ end of an mRNA

3′ poly-A tail a series of A nucleotides added to the 3′ end of mRNA

In prokaryotes, an mRNA molecule can be used in protein synthesis as soon as it is made. In fact, transcription and translation can occur simultaneously. In eukaryotes, the newly synthesized mRNA undergoes modifications before it is transported across the nuclear membrane into the cytoplasm. These modifications convert **precursor mRNA (pre-mRNA)** to **mature mRNA**, as shown in **Figure 6.9** and described below.

- Addition of a **5′ cap**. This involves the covalent linkage of a modified G nucleotide to the 5′ end of the pre-mRNA. The cap is recognized by the protein synthesis machinery.
- Addition of a **3′ poly-A tail**. This involves the covalent linkage of a series of A nucleotides to the 3′ end of the pre-mRNA. The tail makes the mRNA more stable and allows it to exist longer in the cytoplasm.

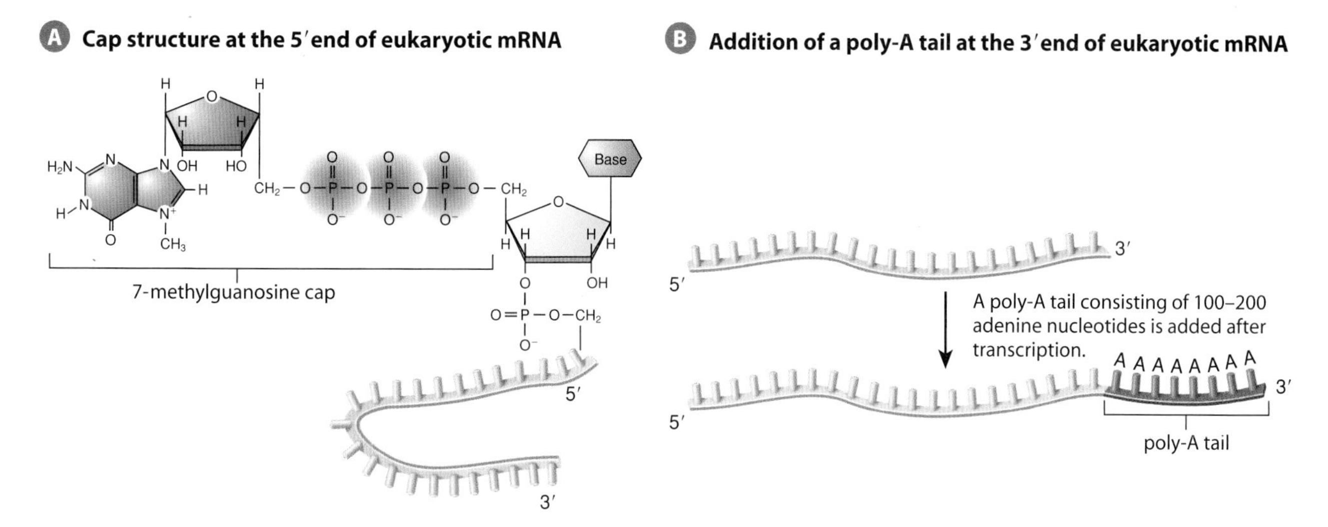
A Cap structure at the 5′ end of eukaryotic mRNA
7-methylguanosine cap
Base
5′
3′
B Addition of a poly-A tail at the 3′ end of eukaryotic mRNA
5′
3′
A poly-A tail consisting of 100–200 adenine nucleotides is added after transcription.
AAAAAAAA
poly-A tail

Figure 6.9 All eukaryotic mRNAs undergo modifications on their ends. **A** A derivative of guanine is added to the 5′ end. **B** A series of nucleotides containing the base adenine are added to the 3′ end.

- Removal of introns. Eukaryotic genes contain non-coding regions called *introns*, which are interspersed among coding regions called *exons*. The intron sequences are removed from pre-mRNA, and the exons are joined together to form the mature mRNA. This process, called **splicing**, is shown in **Figure 6.10**. Particles composed of snRNA and proteins, called snRNPs (pronounced "snurps"), recognize regions where exons and introns meet, and they bind to those areas. The snRNPs interact with other proteins, forming a larger spliceosome complex that removes the introns. For expression of most genes, all of the exons are spliced together. In some cases, however, only certain exons are used to form a mature RNA transcript. This *alternative splicing*, therefore, allows for one gene to code for more than one protein. As a result, certain cell types are able to produce forms of a protein that are specific for that cell.

splicing in mRNA, a process of excising out the introns and combining in the exons

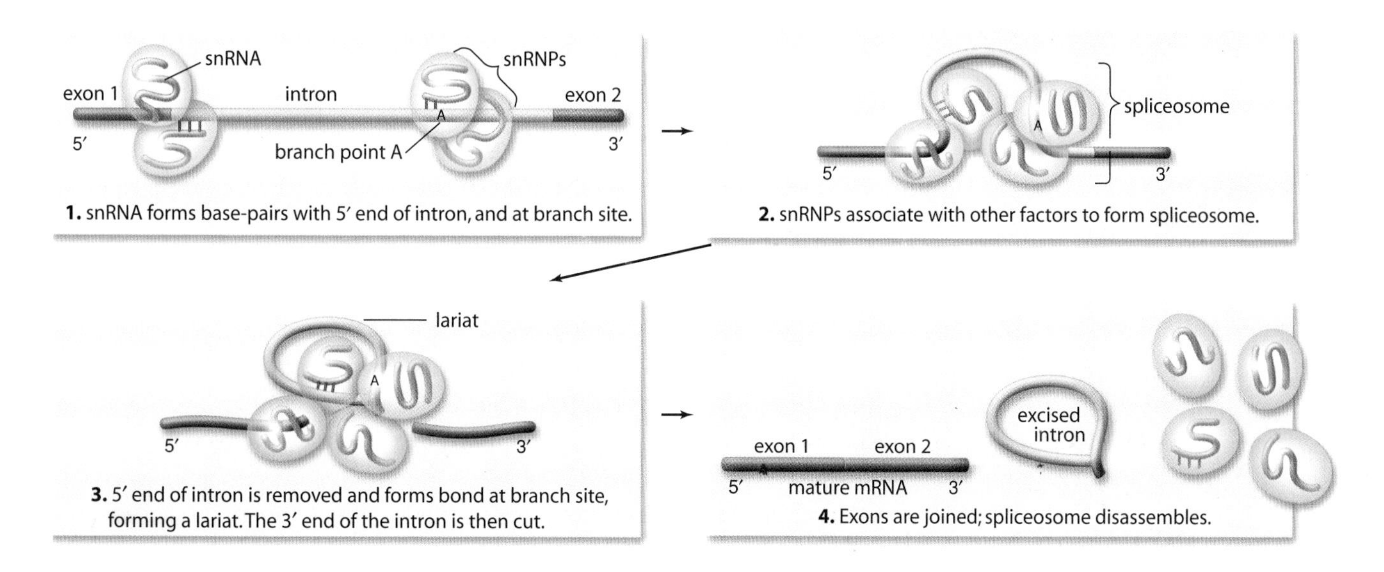
snRNA
snRNPs
exon 1
intron
exon 2
5′
branch point A
3′
1. snRNA forms base-pairs with 5′ end of intron, and at branch site.
spliceosome
5′
3′
2. snRNPs associate with other factors to form spliceosome.
lariat
5′
3′
3. 5′ end of intron is removed and forms bond at branch site, forming a lariat. The 3′ end of the intron is then cut.
excised intron
exon 1
exon 2
5′
mature mRNA
3′
4. Exons are joined; spliceosome disassembles.

Figure 6.10 The process of splicing involves assembly of the spliceosome. An intron forms a loop and the 5′ end is cut and then attached to a nucleotide near the 3′ end, called the branch point. The intron is removed and the exons are spliced together.

Section 6.2 Review

Section Summary

- RNA is a single-stranded polynucleotide. It contains four nucleotides with the bases adenine (A), uracil (U), cytosine (C), and guanine (G).
- There are several types of RNA molecules that are classified according to the different functions they have in a cell.
- There are three defined stages of transcription: initiation, elongation, and termination. During transcription, the antisense strand of a gene is used as a template to synthesize a strand of mRNA.
- In eukaryotes, newly synthesized mRNA undergoes modifications before it is transported across the nuclear membrane into the cytoplasm. As a result, precurser mRNA is converted to mature mRNA.

Review Questions

1. **T/I** Why do cells have several different types of RNA molecules? Explain your answer using examples.
2. **C** Use a flowchart to summarize the three stages of transcription.
3. **K/U** How does transcription differ from DNA replication?
4. **K/U** Distinguish between the following terms: sense strand and antisense strand.
5. **K/U** What is a promoter region? Why are promoter regions necessary for transcription?
6. **K/U** Copy and complete the Venn diagram below to organize information about DNA polymerase and RNA polymerase.

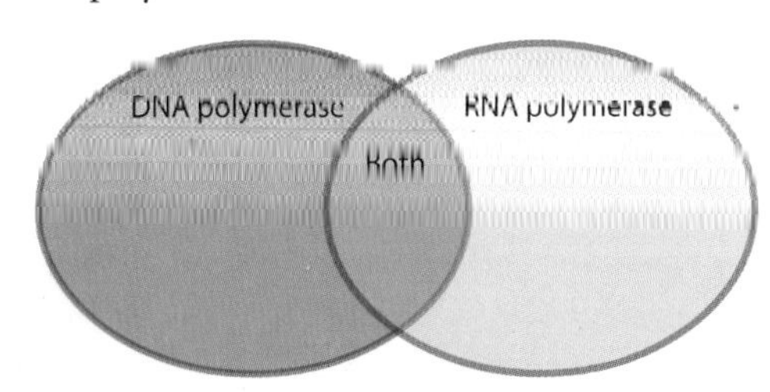

7. **T/I** The following DNA sequence is a fragment from the antisense strand that codes for a protein involved in cell cycle regulation:
   3′- ATACTAGTG - 5′
   a. Write the complementary sense strand.
   b. Which strand will RNA polymerase use in transcription?
   c. What is the nucleotide sequence of the mRNA strand transcribed?
   d. What is the amino acid sequence of the polypeptide produced from the mRNA strand?
8. **A** As you have learned, gene expression involves transcribing information from only one strand of the DNA molecule. What could be some of the biological advantages of double-stranded DNA?
9. **K/U** In prokaryotes, protein synthesis can occur at the same time as transcription. Why is this not the case for eukaryotes?
10. **C** Below is a graphical representation of eukaryotic precursor mRNA.

5′-	Exon	Intron	Exon	Intron	Exon	- 3′

   a. Draw the corresponding mature mRNA.
   b. Label and explain each modification.
11. **A** One of the goals of the Human Genome Project was to identify the number of genes present in the human genome. Before the project, it was estimated that the human genome had 100 000 genes, since it was thought that humans produce approximately 100 000 proteins. The project has since revealed that the human genome contains approximately 25 000 genes, despite estimates for proteins remaining at 100 000–150 000. Use an illustration and a brief paragraph to explain how this is possible.
12. **T/I** How does alternative splicing provide evidence that the one-gene/one-polypeptide hypothesis is not an accurate description of the relationship between genes and proteins?
13. **T/I** Introns are non-coding regions in eukaryotic genes.
   a) Why do introns exist in eukaryotic genes?
   b) How does the absence of introns in prokaryotic genes affect prokaryotic gene expression?

SECTION 6.3

Translation: Synthesizing Proteins from mRNA

The second stage of gene expression involves translating the nucleic acid code of mRNA into the amino acid code of a protein. This process requires the assembly of a complex translation machinery composed of different nucleic acid and protein components. **Table 6.4** provides an overview of the main components and their functions.

Table 6.4 Major Components of the Translation Machinery

Component	Function
mRNA	contains genetic information that determines the amino acid sequence of a protein
tRNA	contains an anticodon that base-pairs with a codon on the mRNA and has the corresponding amino acid attached to it, according to the genetic code
Ribosomes	composed of rRNA and proteins; involved in the process of protein synthesis
Translation factors	proteins that act as accessory factors; needed at each stage of translation

Transfer RNA

Transfer RNA (tRNA) molecules are composed of a single strand of RNA that folds into the characteristic two-dimensional cloverleaf shape shown in **Figure 6.11**. tRNA consists of three stem-loops and a single-stranded region, which in turn fold into a three-dimensional boot-shaped structure. The stem-loops are areas of double-stranded RNA that form through intramolecular base pairing.

Each tRNA molecule has two functional regions. One contains the **anticodon loop**, which is a sequence of three nucleotides that is complementary to a specific mRNA codon. At the opposite end of the molecule, at the 3′ single-stranded region, is the **acceptor stem**, where an amino acid is attached. The **aminoacyl-tRNA synthetase** enzymes are responsible for attaching the appropriate amino acid to a tRNA according to its anticodon. There are 20 different enzymes, one for each of the 20 amino acids. This reaction must be very precise, because tRNAs are responsible for delivering the amino acids to the translation machinery. The correct amino acid must be linked to the appropriate mRNA codon. For example, when the mRNA codon is CGG, the corresponding tRNA has an anticodon that is 3′-GCC-5′, and it should carry the amino acid arginine. Recall that when nucleotide sequences are given, they are written in the 5′ to 3′ direction. Anticodons are written in the 3′ to 5′ direction.

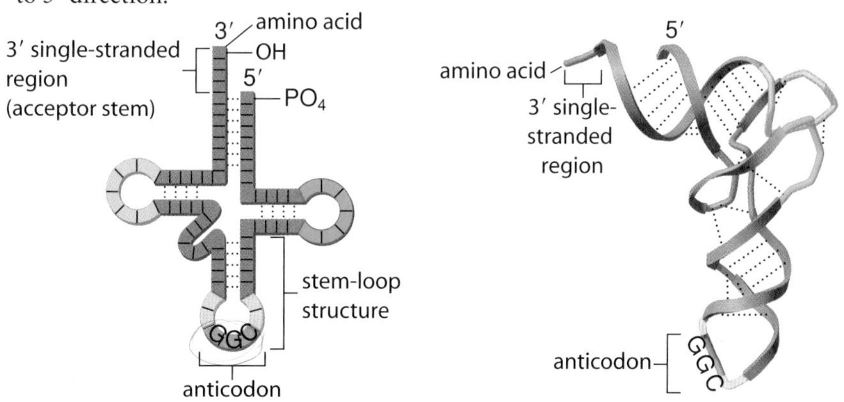

A **Two-dimensional structure of tRNA** **B** **Three-dimensional structure of tRNA**

Figure 6.11 Transfer RNA forms complementary base pairs within each molecule to produce a cloverleaf-shaped two-dimensional structure that folds into a boot-shaped three-dimensional structure. The anticodon is at one end of the molecule, and the point of attachment for the amino acid is at the opposite end.

Apply *Why is the one-gene/one-polypeptide hypothesis inconsistent with the existence of tRNA?*

Key Terms

transfer RNA (tRNA)
anticodon loop
acceptor stem
aminoacyl-tRNA synthetase
ribosome
ribosomal RNA (rRNA)
polyribosome
start codon
reading frame
peptide bond
mutation
single-gene mutation
chromosome mutation
point mutation
frameshift mutation
silent mutation
missense mutation
nonsense mutation
transposon
mutagen

transfer RNA (tRNA) an RNA molecule that links the codons on mRNA to the corresponding amino acid for protein synthesis

anticodon loop a triplet of bases positioned at one end of a tRNA that recognizes and base-pairs with a codon on mRNA during protein synthesis

acceptor stem the 3′ end of a tRNA molecule that is the site of attachment for a particular amino acid, based on the anticodon

aminoacyl-tRNA synthetase an enzyme responsible for attaching an amino acid to a tRNA

ribosome a cell structure composed of proteins and rRNA that provides the site where protein synthesis occurs

ribosomal RNA (rRNA) the RNA that is associated with proteins in the ribosome

polyribosome a structure composed of multiple ribosomes along a strand of mRNA

Ribosomes

Ribosomes are made up of proteins as well as **ribosomal RNAs (rRNAs)** that are bound to the proteins. They are cytoplasmic structures that provide a place where the mRNA, tRNAs
that carry the amino acid molecules, and enzymes involved in protein synthesis can assemble and interact. As shown in **Figure 6.12**, each ribosome is comprised of two sub-units. Each sub-unit is composed of different proteins and rRNA molecules. The ribosome has a binding site for the mRNA and three binding sites for tRNA molecules. These binding sites permit complementary base-pairing between tRNA anticodons and mRNA codons. As soon as the initial portion of mRNA has been translated by one ribosome and the ribosome has begun to move down the mRNA, another ribosome attaches to the same mRNA. Therefore, several ribosomes are often attached to and translating a single mRNA. This whole complex is called a **polyribosome**, and it can synthesize several copies of a polypeptide at the same time.

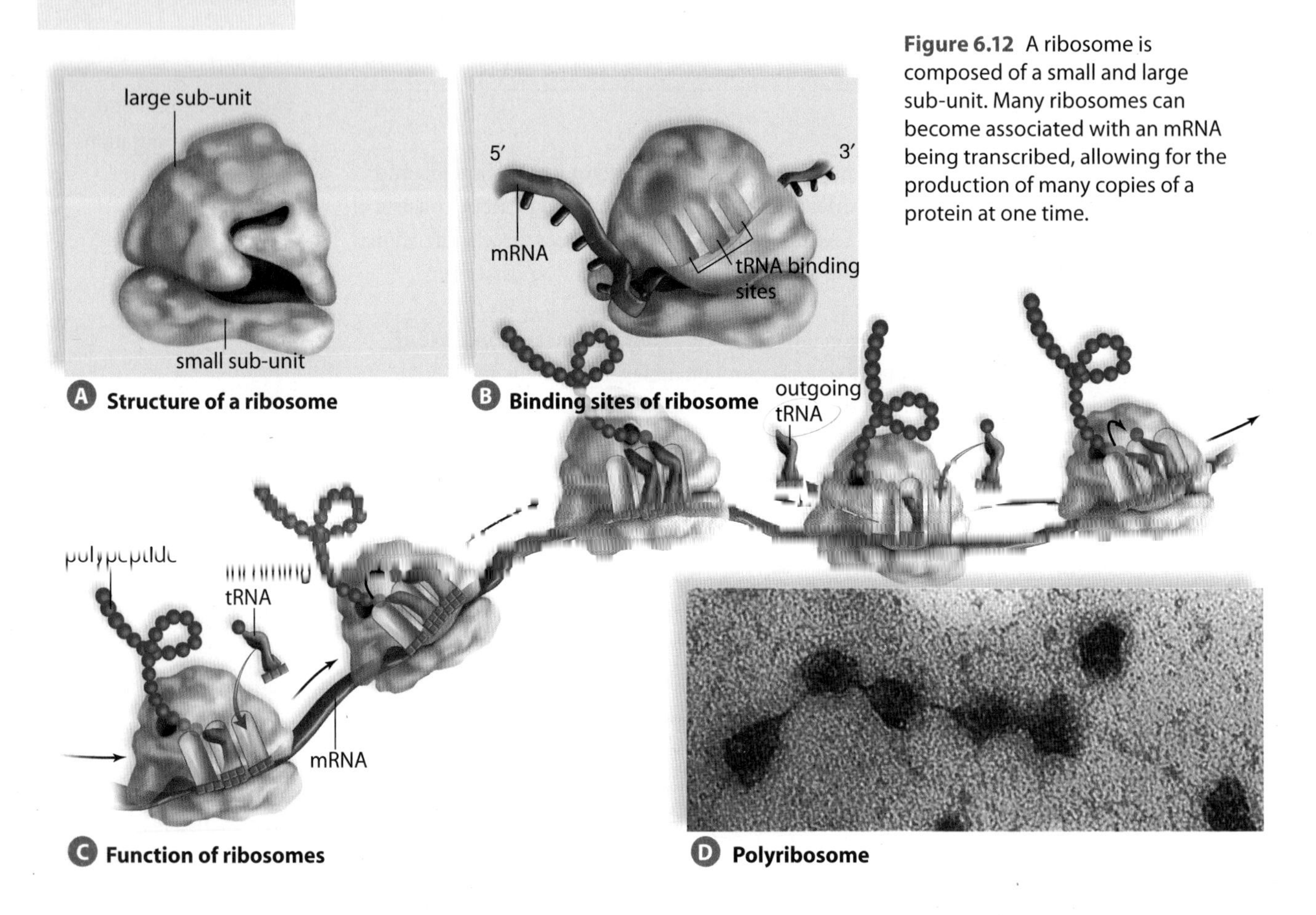

Figure 6.12 A ribosome is composed of a small and large sub-unit. Many ribosomes can become associated with an mRNA being transcribed, allowing for the production of many copies of a protein at one time.

The Molecular Events of Translation

Translation is one of the most energy-consuming processes of the cell. Many protein and nucleic acid components must be synthesized and assembled through a complex array of steps, and energy is needed each time two amino acids are bonded together. Protein synthesis in a bacterium, for instance, is estimated to require 90 percent of the cell's energy.

Translation involves an mRNA being threaded through a ribosome, with the codons of the mRNA base-pairing with the anticodons of tRNA molecules that carry specific amino acids. The order of the codons determines the order of the tRNA molecules at the ribosome and, thus, the sequence of amino acids in a polypeptide. Translation can be divided into three phases: initiation, elongation, and termination.

Initiation

In the initiation phase, all the translation components come together. Proteins called initiation factors assemble the small ribosomal sub-unit, mRNA, initiator tRNA, and the large ribosomal sub-unit for the start of protein synthesis. As shown in **Figure 6.13**, the small ribosomal sub-unit attaches to the mRNA near the **start codon** (AUG). The first tRNA that binds to the codon is the initiator tRNA with its UAC anticodon. In prokaryotes, this tRNA carries a derivative of methionine. In eukaryotes, methionine is used. Then, a large ribosomal sub-unit joins to form the active ribosome. The start codon sets the reading frame for the gene. The **reading frame** establishes how all subsequent codons in the sequence will be read.

The three binding sites for tRNAs are the P (peptide) site, the A (amino acid) site, and the E (exit) site. During protein synthesis, the P site contains the tRNA with the growing polypeptide attached to it; the A site contains the tRNA with the next amino acid to be added to the polypeptide chain; and the uncharged tRNA, which no longer has an amino acid attached to it, is ejected at the E site. At initiation, the initiator tRNA binds to the P site.

Figure 6.13 During initiation of translation, the components assemble. The first tRNA base-pairs with the start codon AUG.

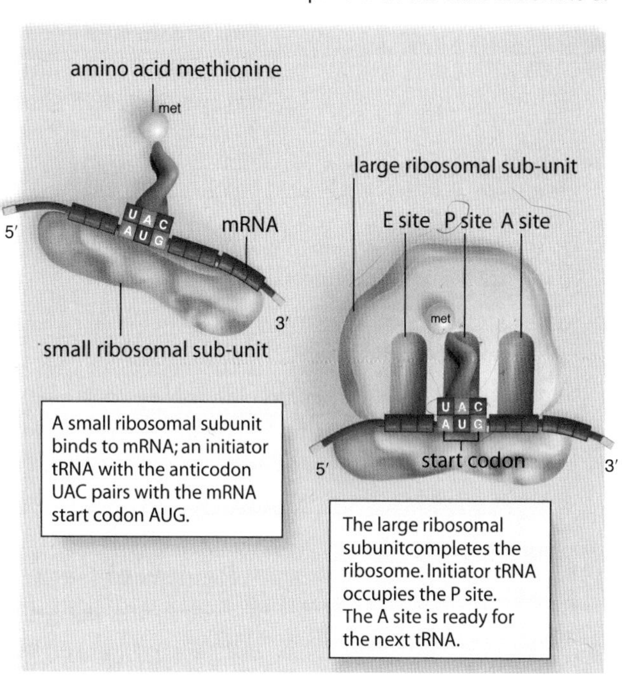

Elongation

In the elongation phase, protein synthesis occurs. The polypeptide becomes longer, one amino acid at a time. In addition to tRNAs, elongation requires elongation factors, which enable tRNA anticodons to bind to mRNA codons. Once the initiator tRNA is bound to the ribosome, the A site is occupied by the tRNA with an anticodon that base-pairs with the second mRNA codon, and that carries the second amino acid of the protein being synthesized. A **peptide bond** forms between the first and second amino acids, and the resulting dipeptide is attached to the tRNA at the A site. The mRNA moves along by one codon, and this complex becomes associated with the P site of the ribosome. The sequential addition of amino acids during elongation, shown in **Figure 6.14**, is a cycle of four steps that is rapidly repeated.

First, a tRNA with an attached polypeptide is in the P site and tRNA carrying the next amino acid enters the A site. Next, the polypeptide chain is transferred to the amino acid of the tRNA in the A site. This makes the polypeptide chain one amino acid longer than before. Last, the mRNA moves forward by one codon, and the polypeptide-bearing tRNA is now at the ribosome P site. The uncharged tRNA exits. The new codon is at the A site and can receive the next complementary tRNA carrying the next amino acid of the polypeptide.

start codon a triple of three bases that specifies the first amino acid of a protein

reading frame collectively, the codons of mRNA that are read to produce an amino acid sequence; it is set by the start codon

peptide bond a covalent bond formed between two amino acids during protein synthesis

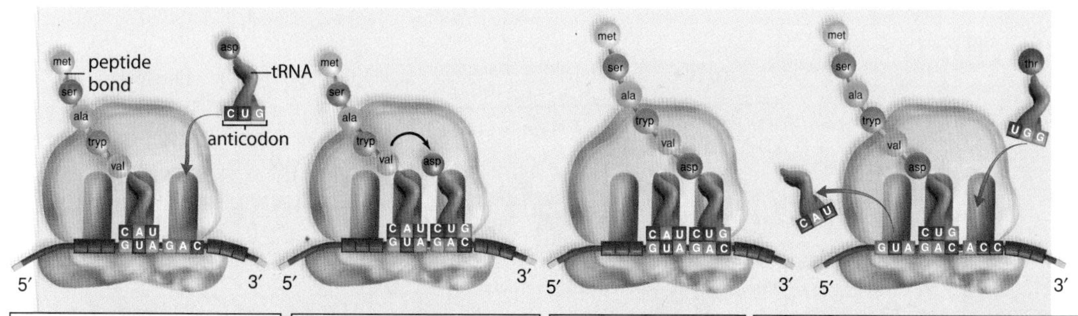

Figure 6.14 During elongation, one amino acid is added at a time onto the growing polypeptide chain.

Termination

The termination phase begins when a stop codon on the mRNA is reached. The polypeptide and the components of the translation machinery are separated. A protein, called a release factor, cleaves (cuts) the polypeptide from the last tRNA. The polypeptide is released and will eventually fold into its three-dimensional shape, ready to carry out its cellular activities.

Learning Check

13. List and explain the major components of the translation machinery.

14. Describe the structure and function of a polyribosome.

15. Explain why translation is an energy-consuming process. Which organelle provides energy for this process?

16. How is translation initiated and terminated?

17. Use labelled diagrams to illustrate the events of the elongation phase of translation.

18. Streptomycin is an antibiotic that works by inhibiting the attachment of tRNA that carries a modified form of methionine. Explain what effect this antibiotic must have on bacteria.

A Review of Gene Expression

Suggested Investigation

Plan Your Own Investigation 6-A, Simulating Protein Synthesis

An overview of the cellular events that are associated with gene expression in eukaryotes is shown in **Figure 6.15**. The expression of a gene refers to the synthesis of a protein that is encoded by that gene. Protein synthesis requires the processes of transcription and translation. During transcription, the DNA sequence of a gene serves as a template for the synthesis of mRNA. In eukaryotes, the mRNA is processed before it leaves the nucleus, which involves removal of introns and addition of a 5′ cap and a 3′ poly-A tail. During translation, the mRNA carries the genetic information to the ribosomes, where protein synthesis occurs. The translation machinery includes tRNAs, which bring amino acids to the ribosomes and act as carriers of the growing polypeptide chain.

The underlying mechanisms for how the DNA sequence information of a gene is decoded to an amino acid sequence of a protein is summarized in **Figure 6.16**. During transcription, the sequence in a region of DNA is copied into a sequence of mRNA codons. Translation of the mRNA codons to an amino acid sequence occurs through tRNA molecules. The amino acid-charged tRNA molecules have anticodons that base-pair with the codons. The order of the codons of the mRNA determines the order that the tRNA-amino acids bind to a ribosome. This determines the order that the amino acids are incorporated into the growing polypeptide chain that will become the complete protein.

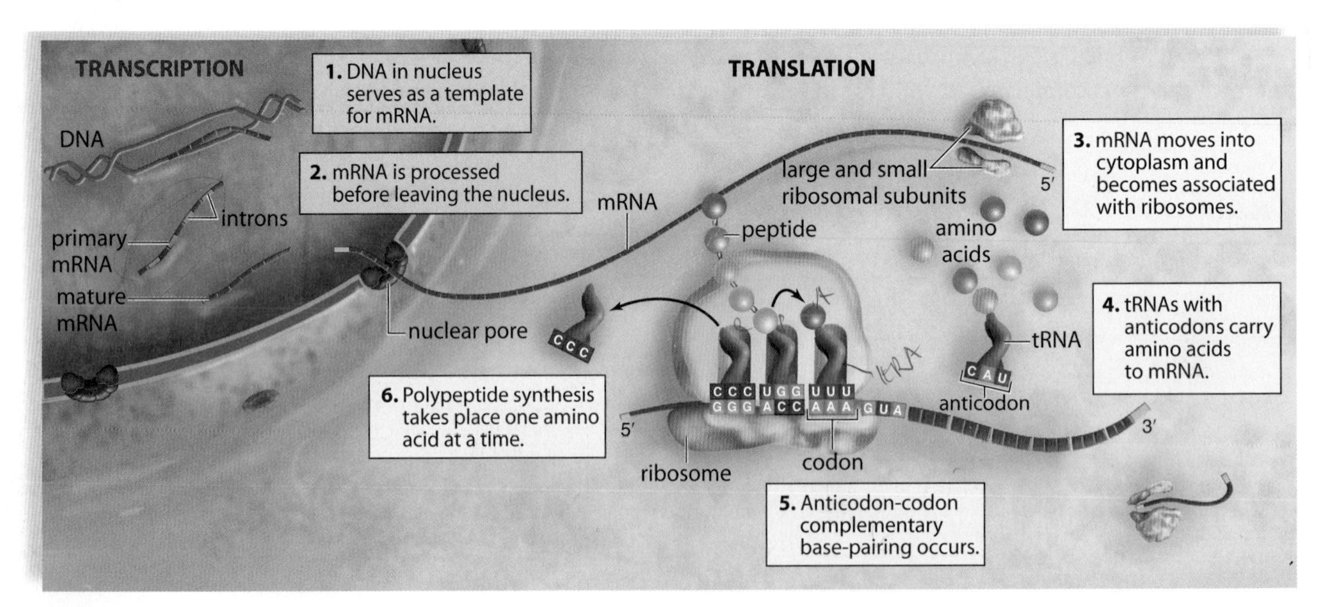

Figure 6.15 Gene expression in eukaryotes involves transcription to form mRNA, which undergoes processing before being exported to the cytoplasm. Once in the cytoplasm, the mRNA is used as the template for protein synthesis by the ribosomes.

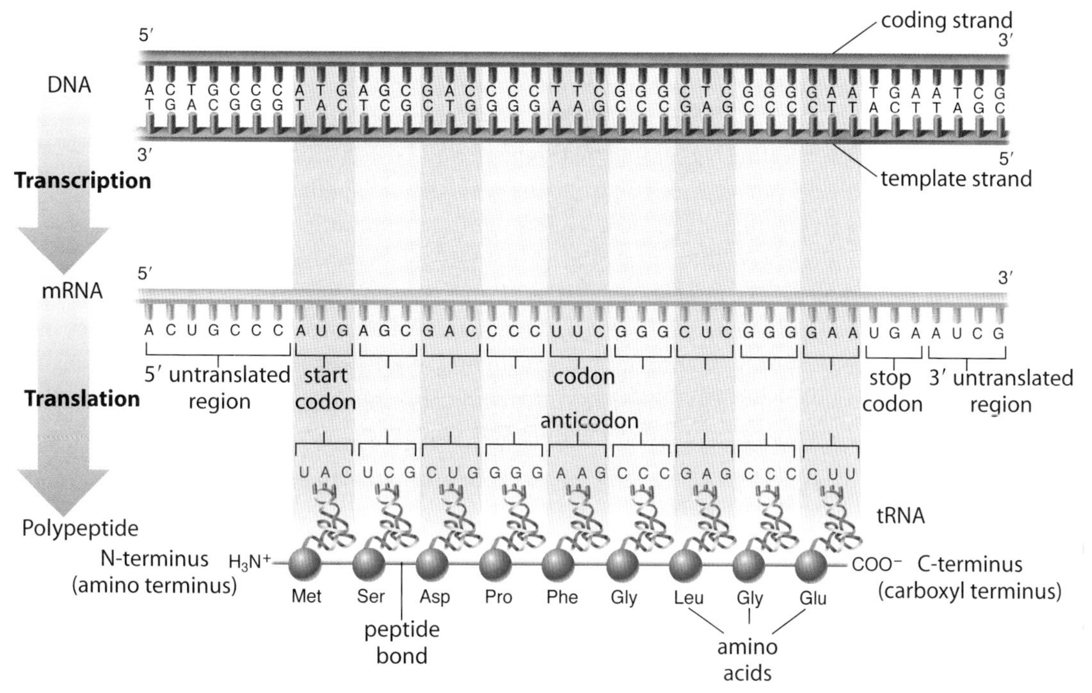

Figure 6.16 When a gene is expressed, the template strand of DNA is transcribed to produce mRNA with a sequence of codons. These codons are then translated to an amino acid sequence.

Apply *An anticodon in a tRNA molecule has the sequence 3′-ACG-5′. Use Table 6.1 to determine which amino acid it carries.*

Activity 6.2 Transcription in Reverse

Amino Acid Abbreviations

Amino Acid	Three-letter Abbreviation
Alanine	Ala
Arginine	Arg
Asparagine	Asn
Aspartate	Asp
Cysteine	Cys
Glutamate	Glu
Glutamine	Gln
Glycine	Gly
Histidine	His
Isoleucine	Ile

Amino Acid	Three-letter Abbreviation
Leucine	Leu
Lysine	Lys
Methionine	Met
Phenylalanine	Phe
Proline	Pro
Serine	Ser
Threonine	Thr
Tryptophan	Trp
Tyrosine	Tyr
Valine	Val

The analysis of DNA can help researchers determine which polypeptides are produced by particular genes. Similarly, the analysis of polypeptides can provide information about the genes that are associated with them. In this activity, you will work backward from a polypeptide chain to construct a stretch of DNA that might code for it.

Procedure

1. Consider the following polypeptide, which might be produced by a bacterial cell:

   —Met—Lys—Asp—Val—Leu—Leu—Phe—Leu—Ala—Glu—

   Referring to **Table 6.1**, and using the information above, draw one possible nucleotide sequence for the DNA molecule that contains the gene for this polypeptide.

2. Draw a labelled diagram to show the mRNA molecule being transcribed from the DNA strand.

Questions

1. Compare your DNA molecule with those of your classmates. How many different sequences could code for the same polypeptide product? What advantage might this give a living cell?

2. The processes of transcription and translation consume a great deal of cellular energy. Why do you think the cell does not simply translate proteins directly from DNA? Brainstorm and discuss some ideas with your classmates.

DNA Mutations and Effects of Mutagens

mutation a permanent change in the nucleotide sequence of a cell's DNA

single-gene mutation a mutation that involves changes in the nucleotide sequence of one gene.

chromosome mutation a mutation that involves changes in chromosomes, and may involve many genes

point mutation a mutation involving a single base pair substitution, insertion, or deletion

In the dynamic environment of a cell, DNA is constantly being replicated, and errors do occur. Enzymes quickly repair some errors. Others can result in a **mutation**—a change in the genetic material of an organism. Because the DNA is changed, all mutations are copied during DNA replication and passed to daughter cells. As well, mutations that occur in reproductive cells can be passed on from one generation to another. Mutations in somatic (body) cells, on the other hand, do not affect future generations.

Mutations are typically neutral or harmful to an organism. In rarer cases, they may be beneficial. Beneficial mutations that affect future generations are important in terms of species change and adaptation over time. Mutations fall into one of two general categories, single-gene mutations and chromosome mutations. **Single-gene mutations** involve changes in the nucleotide sequence of one gene. Conversely, **chromosome mutations** involve changes in chromosomes, and may involve many genes.

Single-Gene Mutations

A single-gene mutation resulting from a change in a single base pair within a DNA sequence is a **point mutation**. A point mutation can involve the substitution of one nucleotide for another. It can also involve the insertion or deletion of a single base pair. A point mutation that involves a nucleotide substitution may have a fairly minor effect on the cell. One reason for this is the redundancy of the genetic code. A change in the coding sequence of a gene does not always result in a change to the polypeptide product of the gene. For example, a change in a DNA sequence from CCT to CCC will not alter the resulting polypeptide, since the associated mRNA codons (GGA and GGG) both code for the same amino acid, glycine.

frameshift mutation a mutation caused by the addition or deletion of a number of nucleotides not divisible by three, resulting in a change in the reading frame

silent mutation a mutation that does not change the amino acid sequence of a protein

missense mutation a mutation that changes the amino acid sequence of a protein

While nucleotide substitutions do not affect neighbouring coding sequences, the insertion or deletion of a number of nucleotides that is not divisible by three can do so. Such an insertion or deletion results in a **frameshift mutation**. A frameshift mutation causes the entire reading frame of the gene to be altered, as shown in **Figure 6.17**. The result is analogous to what happens when you add or delete a letter of a word in a sentence. For example, if the letter C is deleted from the sentence THE CAT ATE THE RAT, the reading frame is altered and it becomes THE ATA TET HER AT, a message that no longer makes sense.

A Normal nucleotide sequence (codons in top row and resulting amino acids beneath them)

GUU–CAU–UUG–ACU–CCC–GAA–GAA
val – his – leu – thr – pro – glu – glu

B Insertion of one nucleotide causes a shift in the reading of all frames beyond the mutation, resulting in a frameshift mutation

GUU–CAU–GUU–GAC–UCC–CGA–AGA A
val – his – val – asp – ser – arg – arg

C Deletion of one nucleotide causes a shift in the reading of all frames beyond the mutation, resulting in a frameshift mutation

A
GUU–CAU–UUG–CUC–CCG–AAG–AA
val – his – leu – leu – pro – lys

Figure 6.17 A frameshift mutation results in alteration of the reading frame due to the insertion or deletion of one (or more) nucleotides.

The examples discussed above explain how mutations affect the nucleotide sequence of DNA. Mutations also may be categorized by how they affect the amino acid sequence of a protein. A mutation that has no effect on the amino acid sequence of a protein is called a **silent mutation**. A mutation that does result an altered amino acid sequence of a protein is called a **missense mutation**. Missense mutations can be harmful. For example, in alcaptonuria, there an accumulation of homogentisic acid is due to a missense mutation in the gene coding for the enzyme that breaks down this compound. The most common mutation results in one amino acid change in the enzyme, which affects the activity of the enzyme.

Missense mutations can also develop new proteins that can help an organism survive. For example, missense mutations may play a role in producing the great variety of antibodies the human body uses to fight new infections. On the other hand, if a change in a gene's coding sequence results in a premature stop codon, then a shortened protein or no protein at all will be made. Such a mutation is called a **nonsense mutation**. These mutations are usually harmful to an organism. **Figure 6.18** shows how a nucleotide substitution in a single coding sequence can result in a silent, missense, or nonsense mutation.

nonsense mutation
a mutation that shortens a protein by introducing a stop codon

A Normal nucleotide sequence (codons in top row and resulting amino acids beneath them)

GUU–CAU–UUG–ACU–CCC–GAA–GAA
val – his – leu – thr – pro – glu – glu

B Silent mutation: The change in nucleotide sequence has no effect on the polypeptide product.

GUU–CAU–UUG–ACC–CCC–GAA–GAA
val – his – leu – thr – pro – glu – glu

C Missense mutation: With valine inserted in place of glutamate, the resulting protein cannot transport oxygen effectively, resulting in sickle cell anemia.

GUU–CAU–UUG–ACU–CCC–GUA–GAA
val – his – leu – thr – pro – val – glu

D Nonsense mutation: Since the codon for the amino acid leucine (UUG) is changed to a premature stop codon, the resulting polypeptide will not function.

GUU–CAU–UAG
val – his – stop

Figure 6.18 A nucleotide substitution can cause silent, missense, or nonsense mutations.
***Identify** a silent mutation and describe its effect on a polypeptide*

Chromosome Mutations

In a previous biology course, you learned that changes to the chromosome number in an organism's genome are always detrimental and often lethal. Mutations can also involve a rearrangement of genetic material, which may affect several genes, including genes located on different chromosomes. Changes to chromosome structure include the *deletion* or *duplication* of portions of chromosomes. There can also be *inversions*, in which a segment of a chromosome is broken and then re-inserted in the opposite direction. Finally, *translocations* involve a section of one chromosome breaking and fusing to another chromosome. **Figure 6.19** reviews these chromosomal mutations.

Causes of Mutations

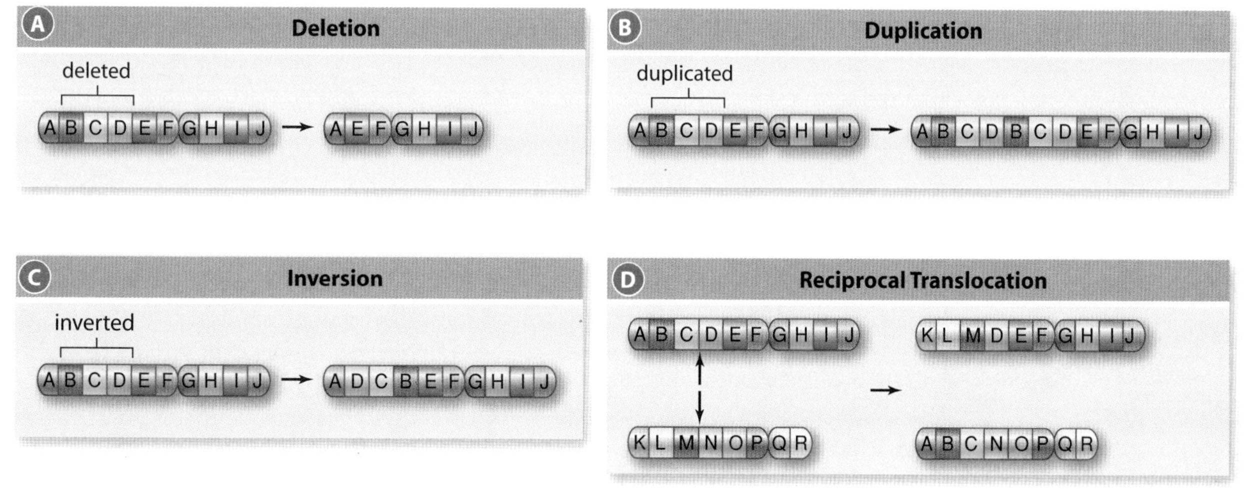

Figure 6.19 Chromosomal mutations include deletions, duplications, inversions, and translocations. If two chromosomes break and exchange material, a reciprocal translocation results.

Many *spontaneous mutations* take place naturally in the cell as a result of normal molecular interactions. The rate of spontaneous mutations varies among organisms and even among different genes in a single cell. Incorrect base pairing by DNA polymerase during DNA replication is one source of spontaneous mutations. Another cellular process, called *DNA transposition*, can disrupt more extensive regions of genetic information. DNA transposition involves the movement of specific DNA sequences within and between chromosomes. Because these sequences effectively swap or exchange places, they are referred to as transposable elements, or transposons for short. They were first described in 1949 by the American geneticist, Barbara McClintock, shown in **Figure 6.20**. While her studies focussed on corn, **transposons** have since been discovered in many species, including humans. For example, in a rare human neurological disorder called Charcot-Marie-Tooth disease, a transposon causes the muscles and nerves of the legs and feet to wither away gradually.

transposon a short segment of DNA capable of moving within the genome of an organism; also called a jumping gene

mutagen an event or substance that increases the rate of changes to the DNA sequence of an organism's genome

All cells can undergo spontaneous mutation, but exposure to certain factors in the environment can increase the rate of mutation. Mutations that are caused by agents outside the cell are referred to as *induced*. A substance or event that increases the rate of mutation in an organism is called a **mutagen**. Mutagens may be physical and chemical.

Figure 6.20 Barbara McClintock was the first individual to study transposons. In 1983, she was awarded the Nobel Prize in physiology or medicine for the discovery of the process of transposition.

Physical Mutagens

X rays are a form of high-energy radiation. They tear through DNA molecules, causing random changes that range from point mutations to the loss of large portions of chromosomes. Because X rays physically change the structure of DNA, they are classified as *physical mutagens*. High-energy radiation is the most damaging form of mutagen known. Ultraviolet (UV) radiation, which is present in sunlight, is less damaging than X rays, but it is still a powerful mutagen. UV radiation can cause a reaction between adjacent pyrimidine (C and T) bases, which distorts the DNA molecule and interferes with replication. Damage from UV radiation, due to exposure to sunlight, is a known cause of melanoma, a form of skin cancer. A single sunburn doubles a light-skinned person's chances of developing skin cancer.

Chemical Mutagens

A *chemical mutagen* is a molecule that can enter the nucleus of a cell and induce mutations by reacting chemically with the DNA. A chemical mutagen may cause a nucleotide substitution or a frame shift mutation. Other chemical mutagens have a structure similar to that of ordinary nucleotides but with different base-pairing properties. When these mutagens are incorporated into a DNA strand, they can cause incorrect nucleotides to be inserted during DNA replication. Examples of chemical mutagens include nitrites (found in small amounts in cured meats), gasoline fumes, and more than 50 different compounds in cigarette smoke. Most chemical mutagens are carcinogenic (associated with one or more forms of cancer).

DNA Repair

Mutations and the phenotypic variations that can result from them may have a positive, neutral, or negative effect on an organism. If the effect is positive, the variation may be passed on to an increasing number of organisms in a population and may lead to the evolution of an existing species or the development of a new species. However, mutations that accumulate too rapidly or are very harmful do not provide a selective advantage. Therefore, cells have a variety of DNA repair mechanisms. Recall from Chapter 5 that DNA polymerase can repair incorrect incorporation of nucleotides during DNA replication. The mismatch repair mechanism also helps to reduce replication errors. In both cases, repairs are made before a mutation occurs.

Suggested Investigation

ThoughtLab Investigation 6-B, The Ames Test for Mutagens

Cells have many other mechanisms that can repair DNA. These all involve a specific set of proteins that act by recognizing and then repairing the damage. These mechanisms are either specific or non-specific.

Specific repair mechanisms fix certain types of damage. For example, UV radiation can cause two adjacent thymines to be covalently linked together to form a structure called a dimer. *Photorepair* is a specific mechanism for repairing the thymine dimer structures, as shown in **Figure 6.21**. An example of a non-specific repair mechanism, an excision repair, is shown in **Figure 6.22**. An excision repair involves removing a damaged region of DNA and replacing it with the correct sequence. This repair is non-specific because it can correct different forms of damage.

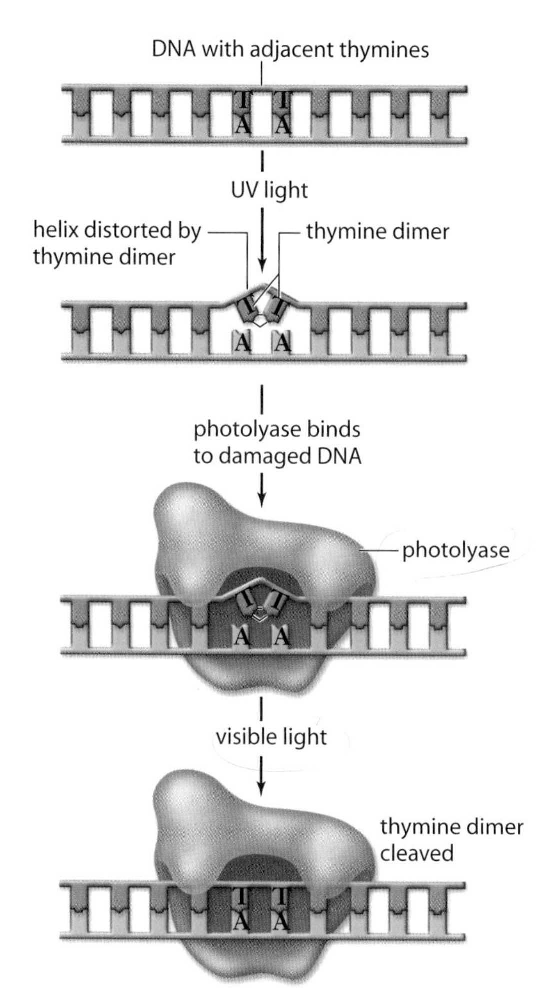

Figure 6.21 Photorepair is a specific mechanism to repair damage to DNA caused by exposure to UV radiation. A photolyase enzyme recognizes the damage, binds to the dimer, and uses visible light to cleave the dimer.

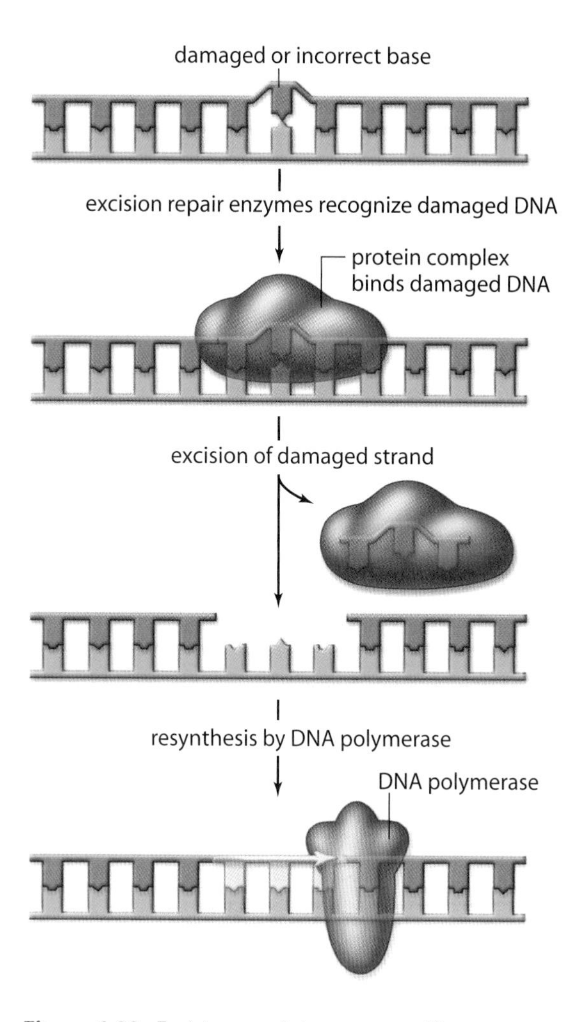

Figure 6.22 Excision repair is a non-specific mechanism of DNA repair because it can fix a variety of damages.

Learning Check

19. How is it possible that some, but not all, mutations get passed from one generation to the next?

20. Would a frameshift mutation occur if six nucleotides were deleted in a sequence of DNA? Explain your answer.

21. Which type of mutation would be least harmful to an organism: silent, missense, or nonsense? Explain your answer.

22. What is the difference between gene mutations and chromosomal mutations?

23. What are two cellular processes that can cause mutations?

24. Use an example to explain the difference between a specific and a non-specific DNA repair mechanism.

Section 6.3 Review

Section Summary

- Translation machinery consists of mRNA, tRNA, ribosomes, and translation factors.
- During translation, the mRNA binds to a ribosome assembly. A tRNA carrying methionine binds with the start codon sequence exposed in the first binding site. Another tRNA molecule then recognizes the codon sequence at the next exposed binding site on the mRNA, and brings the corresponding amino acid to this site. Enzymes bind the amino acids held by adjacent tRNA molecules. As the ribosome progresses along the mRNA, the amino acid chain grows until a stop codon is reached and the new polypeptide is released.
- The processes of gene expression govern the development of living organisms.
- Mutations are permanent changes in DNA. They may involve the insertion, deletion, or substitution of individual nucleotides, or larger-scale rearrangements of portions of chromosomes.
- Mutations may be spontaneous, or they may be induced by exposure to physical or chemical mutagens. Mutations may be harmful to cells. For example, mutations that disrupt gene expression or the cell cycle may result in disorders such as cancer.

Review Questions

1. **K/U** List and describe the major components of the translation machinery.
2. **C** Draw the general structure of a tRNA molecule. Label each component. Write a short caption explaining the function of tRNA.
3. **T/I** Copy the table below into your notebook. The nucleotide sequence in a fragment of an antisense strand of DNA is 3′-CGGAAATTG-5′. Based on this sequence, complete the table.

Codon (mRNA)	Anticodon (tRNA)	Amino Acid

4. **K/U** How does the structure of a ribosome relate to its function?
5. **K/U** Why does translation require a large amount of energy from the cell?
6. **C** Use labelled diagrams to illustrate the four-step cycle in the elongation phase of translation.
7. **T/I** A mutation has occurred in the DNA of an individual's skin cells. Will this mutation be passed onto future offspring?
8. **K/U** Distinguish between gene mutations and chromosomal mutations.
9. **T/I** In what sections of the genome would transposons cause the most damage? Explain your answer.
10. **K/U** List and describe two types of mutagens. Provide an example for each type.
11. **K/U** List and describe two types of DNA repair. Provide an example for each.
12. **T/I** Use the coding sequence below to answer the following questions.

    mRNA codons: UUA-CCU-GUU-AUU
    Amino acids: leu-pro-val-thr

    a. Add a single guanine nucleotide to the beginning of the codon sequence. What type of mutation would this insertion produce?

    b. Write the corresponding amino acids that would result from the addition of the single guanine nucleotide

13. **T/I** Use the DNA sequence template below to answer the following questions.

    3′-TACTTACTCGTCAACCTT - 5′

    a. Write the corresponding mRNA coding sequence and amino acid sequence for the DNA template strand.

    b. A nucleotide substitution occurs in the DNA template at position 14, where adenine is replaced by thymine. Write the mRNA coding sequence and amino acid sequence that would result from this substitution.

    c. What type of mutation would result from this nucleotide substitution?

14. **A** Xeroderma pigmentosum (XP) is a rare genetic disease caused by defective photorepair mechanisms. Individuals with XP have a 1000-fold higher risk of skin cancer than those without XP.

    a. Explain why individuals with XP have a higher risk of skin cancer.

    b. What precautions do you think an individual with XP must take on a daily basis?

SECTION 6.4

Regulation of Gene Expression

So far the discussion of gene expression has focused on the individual events and molecules that are needed for a gene to be expressed. However, to appreciate how gene expression results in an organism's phenotype, the regulation of that expression must be considered. **Gene regulation** refers to control of the level of gene expression. This regulation not only involves whether a gene is active or inactive, but also determines the level of activity and amount of protein that is available in the cell. Some genes are always active and expressed essentially at constant levels, because the proteins are needed for survival of the cell. These are called **constitutive genes**, or *housekeeping genes*. Most genes, however, are regulated so that the protein is expressed only at certain times and in certain amounts.

Key Terms

gene regulation
constitutive gene
operon
operator
repressor
activator
transcription factor
RNA interference

Regulation of Gene Expression in Prokaryotes

Regulation of gene expression in prokaryotes can occur at three different levels: during transcription, during translation, and after the protein is synthesized. By far the most common form of regulation is at the level of transcription, during initiation.

Gene regulation in *E. coli* is one of the most well studied systems. In bacteria, many genes are clustered together in a region that is under the control of a single promoter. (Recall that the promoter is where the RNA polymerase complex binds to DNA and begins transcription.) These regions are called **operons**. Genes that are involved in the same metabolic pathway are often found in the same operon. Since they are all under the control of the same promoter region, these genes are all transcribed together into one continuous mRNA strand, called a *polycistronic* mRNA. Individual proteins are then synthesized from that mRNA.

The *lac* Operon

E. coli can use different sugars as a source of energy and carbon. It does this by quickly adjusting its gene expression according to what sugar is available. For example, when the bacteria are grown in the presence of the sugar lactose, there is an increase in the levels of enzymes that are involved in lactose metabolism. When the lactose is subsequently removed, the levels of these enzymes drastically decline.

The genes that encode the enzymes that are needed to break down lactose are found in the *lac* operon on the *E. coli* chromosome. An illustration of the *lac* operon is shown in **Figure 6.23**. This operon consists of a coding region and a regulatory region. The coding region contains the genes for three enzymes that are required for the breakdown of lactose. The regulatory region contains the promoter for transcription of the lactose-metabolizing enzymes. It also contains sites that regulate that transcription, an operator, and a catabolite activator protein (CAP) binding site. An **operator** is a DNA sequence to which a protein binds to inhibit transcription initiation. This protein is called a **repressor**. The CAP binding site is a DNA sequence to which a specific protein also binds. The binding of CAP increases the rate of transcription of a gene or genes.

gene regulation the control and change to gene expression in response to different conditions in the cell or environment

constitutive gene a gene that is constantly being expressed; it does not undergo regulation of expression

operon a cluster of genes grouped together under the control of one promoter; occurs in prokaryotic genomes

operator a repressor protein that binds to a DNA sequence element; it regulates transcription

repressor a protein that binds to a particular DNA sequence to regulate transcription; inhibits the transcription of a gene or genes

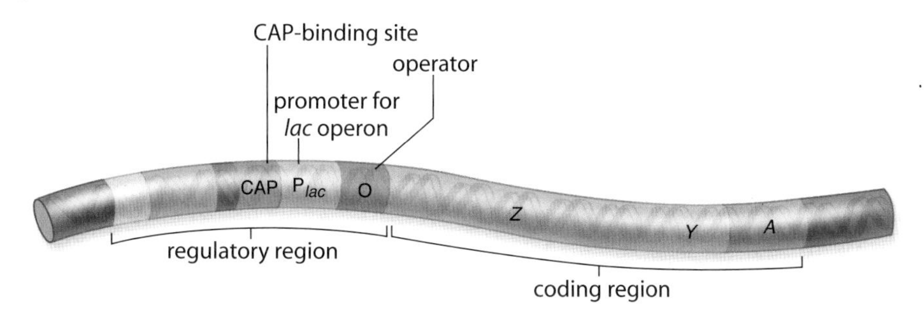

Figure 6.23 The *lac* operon contains three genes (Z, Y, and A) needed for the breakdown of lactose in *E. coli*. All three genes are under the control of one promoter. Therefore, they undergo the same level of regulation. When lactose is not present, the *lac* repressor inhibits transcription.

Activating the *lac* Operon

activator a protein that binds to a particular DNA sequence to regulate transcription; it increases the rate of transcription of a gene or genes

In the absence of lactose, the *lac* repressor protein binds to the operator. This prevents RNA polymerase from binding to the promoter, and transcription cannot occur. When lactose is present, a derivative called allolactose is produced. Allolactose binds to the repressor and the repressor can no longer bind to the operator. This results in the transcription of the genes to produce the required enzymes. The binding of the CAP activator protein can even further enhance levels of transcription. Therefore, the regulation of the *lac* operon is similar to how the ignition and accelerator pedal of a car work. The negative control involves an on/off mechanism that is determined by the repressor, similar to the ignition in a car. The **activator** protein determines how fast the genes are transcribed, just as an accelerator determines the speed of a car.

The *lac* operon is considered an inducible operon. Transcription from it is induced when lactose is present. There are also other operons that are normally active until a repressor turns them off. An example of this type of operon is the *trp* operon.

The *trp* Operon

As shown in **Figure 6.24**, the structure of the *trp* operon is similar to the *lac* operon. The coding region contains five genes for enzymes that are required for the synthesis of the amino acid tryptophan. The regulatory region contains a promoter and an operator region. Under normal conditions, tryptophan must be synthesized, so the repressor does not bind to the operator, and transcription takes place. When tryptophan reaches a certain level in the cell, however, some of it binds to a repressor protein. This binding increases the repressor's ability to bind to the operator, thus reducing transcription activity.

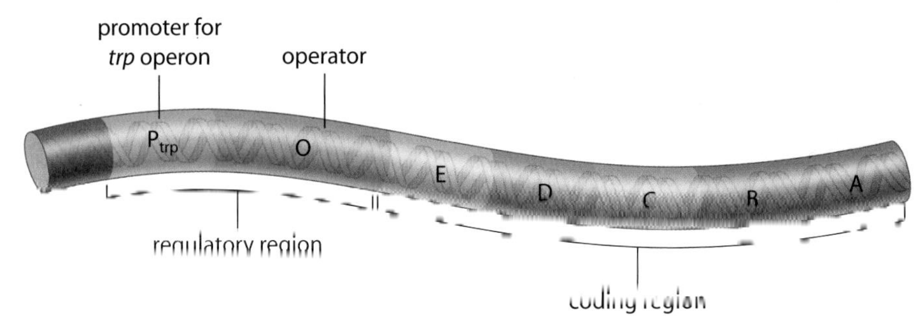

Figure 6.24 The *trp* operon contains five genes that are involved in the synthesis of tryptophan. This operon is normally transcribed, until the cell has sufficient tryptophan. Then the *trp* repressor binds to the promoter and inhibits transcription.

***Compare** and contrast the* trp *operon with the* lac *operon. How are they the same and how do they differ in response to their environment?*

Activity 6.3 Modelling the *lac* Operon

Materials

- variety of materials to build a model of the *lac* operon

Procedure

1. Using the available materials, construct a model for the *lac* operon. Your model must include the regions identified in **Figure 6.23** as well as lactose, repressor protein, activator protein (CAP), allolactose, RNA polymerase, and mRNA.
2. Use your model to illustrate the state of the *lac* operon when
   a. lactose is not present
   b. lactose is present

Questions

1. Make a sketch to show the state of the *lac* operon when lactose is not present. Describe the operon.
2. Make a sketch to show the state of the *lac* operon when lactose is present. Describe the operon.
3. Describe the effect of the following mutations on the *lac* operon and the synthesis of enzymes that digest lactose:
   a. the repressor protein's shape is affected and does not allow it to bind to the operon
   b. the CAP protein is not produced
   c. the repressor protein's shape is affected, inhibiting the binding of allolactose

Learning Check

25. What are constitutive genes and why are they important for cell survival?
26. How is cell specialization possible when all of the cells in an organism contain the same genetic information?
27. Explain why control at the level of transcription saves considerable cellular energy for bacteria.
28. Use a labelled diagram to illustrate the components of an operon.
29. Explain how the *lac* operon is an inducible operon. How does this differ from the *trp* operon?
30. The process of ignition and acceleration in a car is often used as an analogy for regulation of the *lac* operon. What other real-world analogy you could use?

Regulation of Gene Expression in Eukaryotes

As shown in **Figure 6.25**, the regulation of gene expression in eukaryotes is much more complex than it is in prokaryotes, and it occurs at many more levels. There are five levels of gene regulation in eukaryotes: pre-transcriptional, transcriptional, post-transcriptional, translational, and post-translational.

Pre-transcriptional and Transcriptional Control

Regulation of gene expression can take place at the DNA level. Recall that DNA is associated with proteins such as histones to form nucleosomes. These, in turn, assemble into more condensed structures to form chromatin. DNA in highly condensed areas of chromatin is not transcribed, because chromatin acts as a physical barrier to the proteins that are needed to synthesize pre-mRNA. For regions of the genome that need to be expressed, different processes are used by the cell to alter chromatin structure and loosen the nucleosome structures. Once this is done, proteins for initiation of transcription can gain access to the DNA. Methods for accomplishing alterations include chemical modifications of the histone proteins and the use of multi-enzyme structures called chromatin-remodelling complexes.

transcription factor one of a set of proteins required for initiation of transcription; it is required for the RNA polymerase complex to bind to the promoter

Eukaryotic genes are not organized into operons. Rather, each gene has its own promoter, and control of transcription is distinct for that gene. **Transcription factors** must interact with the promoter for RNA polymerase to initiate transcription. Although this interaction is essential, it only allows a certain level of transcription. Certain types of activator proteins also enhance transcription initiation by binding to transcription factors and RNA polymerase, as well as specific sequences of DNA called *enhancers*. There are a vast array and number of activators in eukaryotes, and regulation of a gene usually requires more than one type. Having multiple activators allows gene regulation to be highly tuned to particular conditions or times during the life of the cell. **Figure 6.26** on the following page shows how this enhancement occurs. Notice how some enhancers may bind quite a distance away from the promoter.

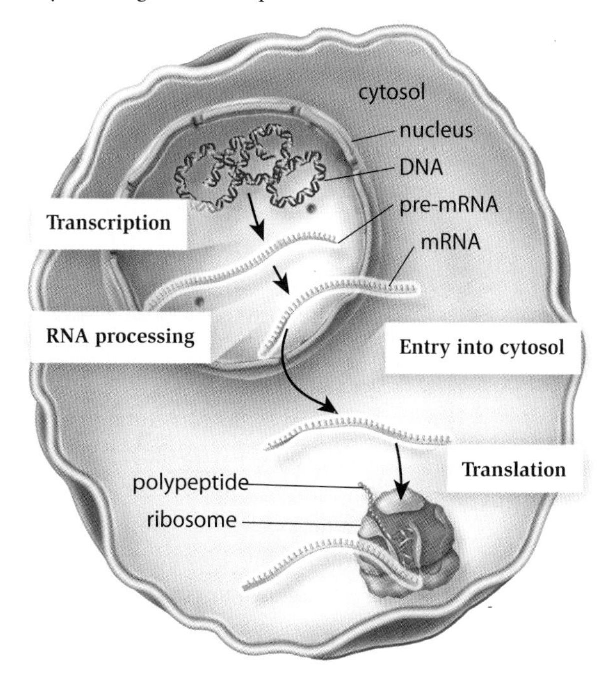

Figure 6.25 Regulation of eukaryotic gene expression occurs at multiple stages of protein production. The regulation points occur in both the nucleus and the cytoplasm.

Figure 6.26 Transcription factor binding is required for transcription by RNA polymerase. The binding of activators enhances the rate of transcription. These activators can bind to enhancer sequences that are hundreds or thousands of base pairs away from the promoter. However, looping of the DNA can allow them to interact with RNA polymerase or certain transcription factors.

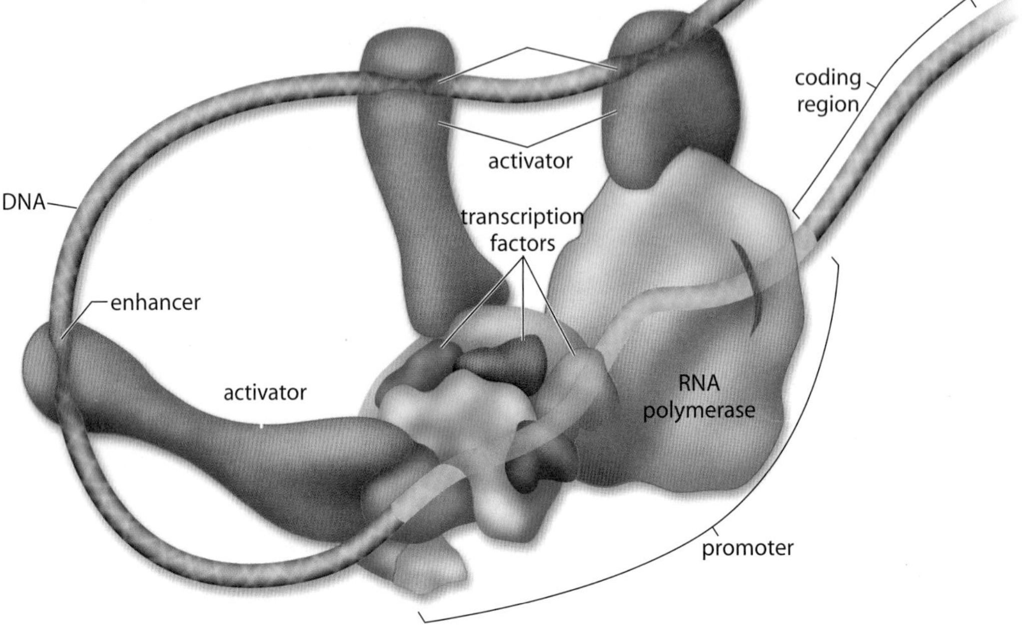

Post-transcriptional and Translational Control

Regulation of gene expression can also occur at the level of the mRNA molecule, once it is synthesized. As discussed earlier, alternative splicing of an mRNA produces different mRNA molecules. These, in turn, provide alternative protein products.

Modifications of the mRNA can be altered so that the 5′ cap and/or 3′ poly-A tail are not added. mRNAs lacking in the modification will either not be transported from the nucleus or undergo rapid degradation in the cell. In either case, the mRNA is no longer available for protein synthesis in the cytoplasm.

RNA interference the regulation of gene expression by small RNAs; it inhibits gene expression by degrading mRNA or inhibiting translation

Small RNA molecules can control gene expression by a mechanism called **RNA interference**. Two of these small RNAs are micro RNA (miRNA) and small interfering RNA (siRNA). These small RNAs can associate with protein complexes and turn off gene expression by either promoting mRNA cleavage or inhibiting translation. The small RNAs can target and interact with specific mRNAs by forming complementary base pairs.

Post-translational Control

Many polypeptides that are synthesized in eukaryotic cells are not active immediately after synthesis. This activation involves one of a number of different modifications. For example, insulin is initially folded into its three-dimensional structure. However, in order to be active several amino acids are removed, which leaves two polypeptide chains. The chains are combined by the formation of a covalent bond between two sulfur atoms that are on each chain. This activates the insulin protein. Other modifications include the covalent linkage of a phosphate group to one or more amino acids in the polypeptide in order for the protein to be functional.

Regulating how long a protein is available in the cell can also be a form of gene regulation. An important pathway that eukaryotic cells have for this involves the attachment of a chain of ubiquitin molecules to a protein. This acts as a signal for the protein to be degraded.

Canadian Research in Action

Expressing Synthetic Antibodies to Treat Cancer

The Sidhu Lab team. Back row, left to right: Dr. Sachdev Sidhu, Alevtina Pavlenco, Dr. Sarav Rajan, Dr. Nish Patel, Esther Lau. Front row, left to right: Dr. Amandeep Gakhal, Dr. Bryce Nelson.

Cancer is one of the leading causes of death in Canada. In the race to develop effective treatments for the disease, a new type of therapy has emerged as a leader. This therapy, which targets specific proteins associated with cancer cells, is called *synthetic antibody technology*. Antibodies are proteins that recognize substances associated with infection and disease and then neutralize the substances or target them for destruction. Although antibodies are produced naturally by the immune system, they can also be engineered and produced in the lab to target specific cancer-related proteins. The proteins these antibodies target are often associated with cell signalling. As a result, antibody binding can stop cancer cells from proliferating, or they can interfere with other signalling events related to these cells.

Dr. Sachdev Sidhu is a prominent scientist in the field of synthetic antibody research. The Sidhu Lab at the Donnelly Centre for Cellular and Biomolecular Research at the University of Toronto is currently collaborating with researchers across Canada to develop large numbers of antibodies that target proteins associated with cancer. In addition to being therapeutic agents, these antibodies will likely be important tools in the field of cancer research. They also play a key role in determining how the products of certain genes function under normal conditions as compared to cancerous conditions.

A great deal is known about the structure of the region of an antibody that binds to the target protein. The chemical groups that are critical for antibody-target binding are also well understood. This knowledge, gained in part by Dr. Sidhu's research, has enabled the Sidhu Lab to produce synthetic antibodies that bind to hundreds of targets. This is accomplished by a quick and cost-effective technology known as *phage display*. *Phages* are viruses that infect bacteria. To synthesize a variety of antibodies, DNA sequences are constructed from different combinations of nucleotides. These sequences are then introduced into regions of the genes that code for the binding site on the antibody. The altered genes for the antibody are then introduced into the phage genome and expressed and displayed on the phage surface as functional antibody proteins. Because many different DNA sequences are possible, a large collection of these sequences, called a phage-displayed antibody library, can be developed. The more DNA sequences such a library holds, the greater the likelihood that some of the sequences will result in the expression of an effective therapeutic antibody.

Related Career

Research technicians are an important part of any lab team. They may carry out a variety of tasks in a lab setting, such as assisting with research, analyzing data, monitoring and maintaining equipment, handling and storing hazardous materials, and ensuring that proper practices are followed in the lab. Research technicians generally have a university undergraduate degree, although they may have completed further studies as well.

Questions

1. What issues would likely need to be addressed to ensure that antibody genes are expressed when inserted into phages?
2. Suggest how synthetic antibody technology could be used to treat a specific disease other than cancer.
3. Use Internet and/or print resources to find another career related to the research covered in this feature. Briefly describe the nature of this career and any required training or education.

Section 6.4 Review

Section Summary

- Gene regulation refers to control of the rate of gene expression in response to different conditions in the cell or environment.
- While constitutive genes are always active and expressed at near constant levels, the majority of genes are regulated so that their product is only expressed at certain times and in certain amounts.
- Regulation of gene expression in prokaryotes can take place during transcription, during translation, and after the protein is synthesized, although it most commonly occurs during initiation.
- Operons may be inactive until transcription is induced or active until a repressor turns them off.
- Regulation of gene expression in eukaryotes may be pre-transcriptional, transcriptional, post-transcriptional, translational, and post-translational.

Review Questions

1. **K/U** What is the relationship between cell specificity and gene regulation?
2. **T/I** What kinds of genes might be included in the category of constitutive genes?
3. **T/I** Why is it advantageous for some genes to be inactive? If these genes are inactive most of the time, why do you think they are still present in the genome?
4. **K/U** Write a short sentence to distinguish between the following pairs.
   a. operon and operator
   b. activator and repressor
5. **C** Draw a labelled diagram of the *lac* operon showing the components of the regulatory and coding regions.
6. **T/I** Describe how each of the following conditions would affect the transcription of lactose-metabolizing enzymes by the *lac* operon.
   a. high levels of lactose
   b. no lactose
   c. high levels of lactose and an absence of activator protein
   d. no lactose and an absence of activator protein
7. **K/U** Describe how the *trp* operon is different from the *lac* operon.
8. **K/U** How does gene regulation differ in prokaryotes and eukaryotes?
9. **T/I** Would you expect DNA sequences for constitutive genes to be in highly condensed areas of chromatin? Explain your answer.
10. **C** Design and create a table with the following headings to summarize the control of gene expression in eukaryotes: Level of Gene Regulation, Location in the Cell, Description, Example.
11. **K/U** What is the role of activators in transcriptional control? Are they necessary for transcription? Explain.
12. **A** Small RNA molecules, such as miRNA and siRNA, can control gene expression by RNA interference. How could this mechanism be useful in the development of new therapeutic drugs and treatments?
13. **K/U** Explain what will happen to a pre-mRNA strand that lacks a 3′ poly-A tail or 5′ cap. How will this affect translation?
14. **A** The *VHL* gene codes for an enzyme responsible for regulating the degradation of specific proteins by "tagging" them with certain molecules. Individuals that have a mutation in the *VHL* gene develop Von-Hippel-Lindau syndrome, which leads to predisposition to many different types of cancer. Speculate on why a mutation in *VHL* may lead to disease.
15. **K/U** Which area labelled on the diagram below represents a transcription factor?

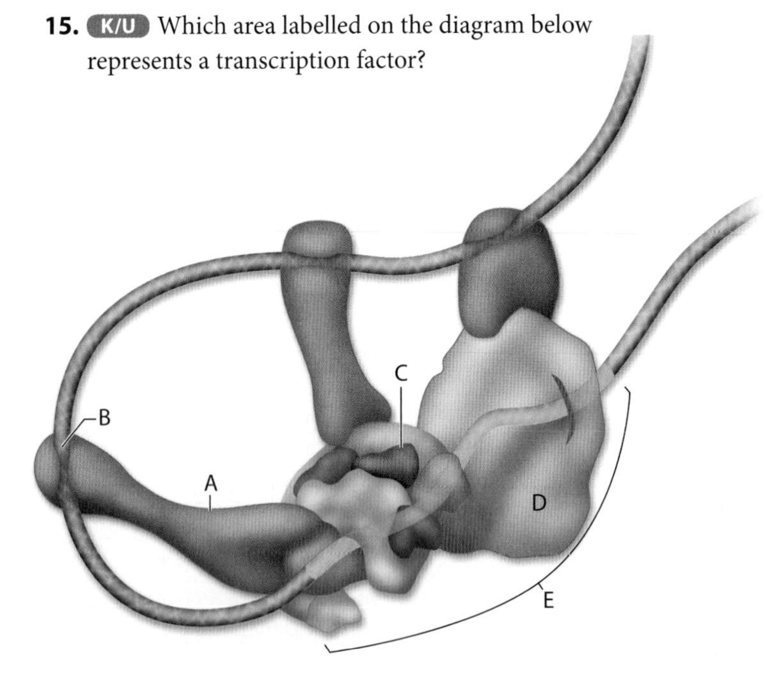

Plan Your Own
INVESTIGATION

6-A

Skill Check

- ✓ Initiating and Planning
- ✓ Performing and Recording
- ✓ Analyzing and Interpreting
- ✓ Communicating

Suggested Materials

- 2 or more protein-synthesis simulations, and/or links to simulations, provided by your teacher
- materials, equipment, and/or software to be determined by your group

Simulating Protein Synthesis

In this investigation, you will work as part of a group to develop a model that simulates the process of protein synthesis.

Pre-Lab Questions

1. What distinct processes are involved in protein synthesis?
2. List the principle molecules in protein synthesis and their functions.

Question

How can you simulate the processes involved in protein synthesis?

Plan and Conduct

1. With your group, develop an overview of protein synthesis by listing the steps in transcription and translation. For each step, jot notes about the structures, molecules, and events involved.
2. Watch the simulations provided by your teacher. Discuss and assess the ways in which these simulations attempt to represent protein synthesis.
3. Compare the details that you compiled in step 1 with the details supplied in the simulations. Discuss any modifications to your step 1 overview.
4. Discuss how you could simulate transcription and translation in your classroom. Your simulation could take any form. For example, you could prepare an interactive computer program, construct a paper flipbook, or design a three-dimensional physical model.
5. When you have agreed on a plan, list the materials and equipment you will need for your simulation.
6. With your teacher's permission, carry out your plan.
7. Present your simulation. Record your classmates' comments.

Analyze and Interpret

1. Explain how your presentation was able to simulate the role of each cell component involved in transcription and translation.

Conclude and Communicate

2. Which part(s) of your presentation seemed to be the most effective at simulating protein synthesis? Which were least effective? Why?

Extend Further

3. **INQUIRY** Given your response to question 2, revise your presentation to better simulate the processes of transcription and translation.
4. **RESEARCH** Use the Internet to find an effective simulation of transcription and translation. Explain what makes this simulation particularly effective.

ThoughtLab INVESTIGATION **6-B**

Skill Check

- Initiating and Planning
- Performing and Recording
- ✓ Analyzing and Interpreting
- ✓ Communicating

Suggested Materials

- print resources and/or computer with Internet access

The Ames Test for Mutagens

It would be an unwelcome task to test every new personal care product, every pollutant, and every food to see if it causes cancer in animals. However, many substances that cause mutations that may lead to the development of cancer in animals are also mutagens in bacteria. The Ames test takes advantage of this relationship by using bacteria to identify mutagens, and, therefore, potential carcinogens.

Bruce Ames developed his testing method while studying mutant strains of *Salmonella* bacteria. These strains have point mutations that make them unable to make histidine, an amino acid the bacteria need in order to survive. As a result, they will not grow unless histidine is provided in their growth medium. These *his–* strains also do a poor job at correcting errors in DNA. When exposed to a mutagen, the strains readily develop new mutations called *back mutations*, which correct errors in the genes for making histidine.

To conduct the Ames test, a suspected mutagen is exposed to rat liver extract, as shown in the diagram below. This step is needed because many carcinogens and mutagens are not directly mutagenic, but become mutagenic when activated by enzymes in the body. A *his– Salmonella* strain is then incubated with the activated test substance. Finally, the bacteria are spread on a growth medium that lacks histidine. If a large number of bacterial colonies grow, this indicates that the test substance caused many back mutations. The more colonies that grow, the more mutagenic the test substance. A control is also used to account for spontaneous back mutations that occur in the absence of any mutagen.

In this investigation, you will examine some Ames test results. You will also look at possible back mutations and their effects on the histidine-synthesizing enzymes of *Salmonella*.

Go to **Constructing Graphs** in Appendix A

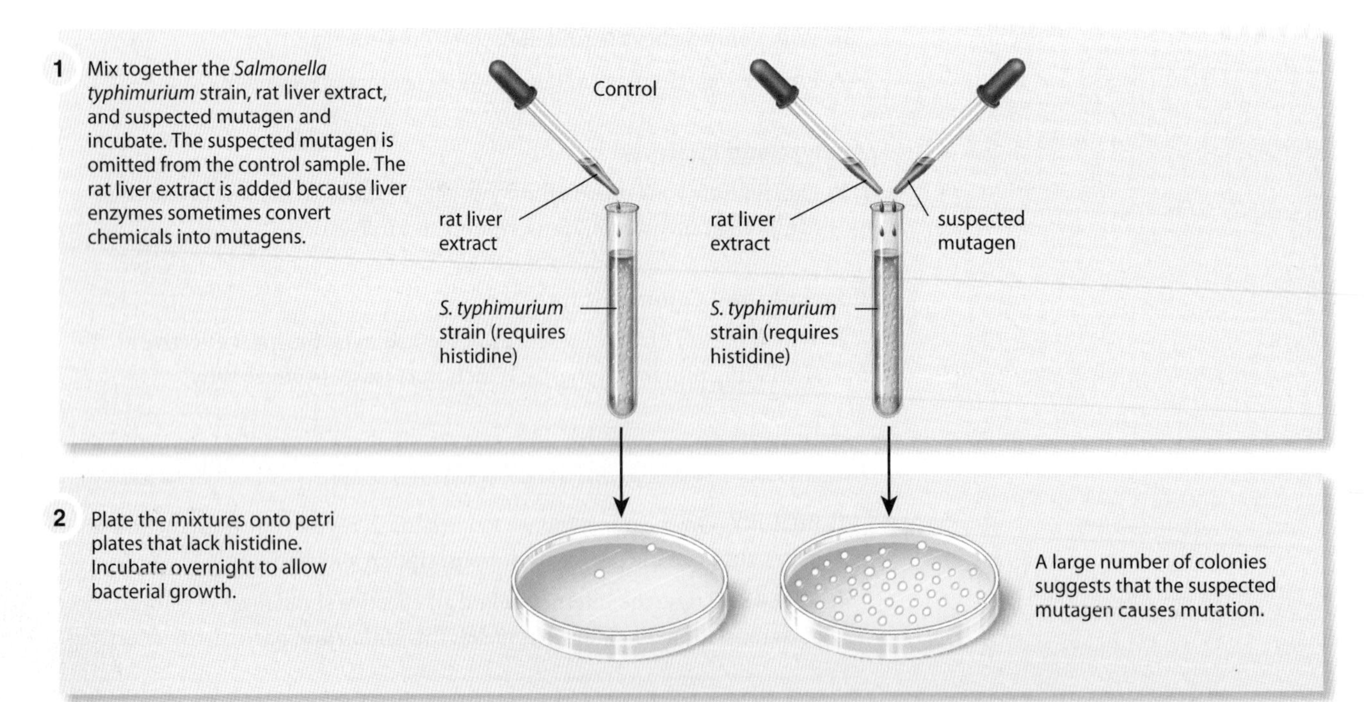

1 Mix together the Salmonella typhimurium strain, rat liver extract, and suspected mutagen and incubate. The suspected mutagen is omitted from the control sample. The rat liver extract is added because liver enzymes sometimes convert chemicals into mutagens.
Control
rat liver extract
S. typhimurium strain (requires histidine)
rat liver extract
suspected mutagen
S. typhimurium strain (requires histidine)
2 Plate the mixtures onto petri plates that lack histidine. Incubate overnight to allow bacterial growth.
A large number of colonies suggests that the suspected mutagen causes mutation.

Pre-Lab Questions

1. What is a point mutation?
2. What causes frameshift mutations?
3. What is meant by "a *his*– mutant"?

Question

How can histidine-requiring strains of bacteria be used to identify mutagens and study mutations?

Organize the Data

Part I: Identifying Mutagens

1. Examine the following data from a study conducted by Bruce Ames' research team. Construct a bar graph to display the data.

	Number of Colonies per Petri Dish	
Test Substance and Common Source	Control with Rat Liver Extract	Test with Rat Liver Extract
1 μg aflatoxin (mouldy peanuts)	26	266
5 μg benzopyrene (wood smoke and cigarette smoke)	44	505
100 μg 2-naphthylamine (once used in dyes)	21	85

Part II: Characterizing the Mutations

2. In *Salmonella*, the biochemical pathway for the synthesis of histidine involves 10 enzymes, which are encoded by 10 genes. Examine the following nucleotide sequences from two *his*– mutants and two possible back mutations that can occur. In each case, describe the type of back mutation.

Gene with Mutation	Nucleotide Sequence in *his*– Strain	Back Mutation
hisG	GGG	GAG
hisD	GAC GGC GCG CGC CTG TGG CGG CCG TCC	GAC GGC GCG CCT GTG GCG GCC GTC

Analyze and Interpret

Part I

1. Refer to your bar graph to answer the following questions.
   a. Which of the test substances are mutagens? Explain.
   b. Do any of the test substances appear to be more mutagenic than the others? Explain.

Part II

2. Determine the mRNA sequences that would be transcribed from the nucleotide sequences from the *his*– mutants and for each possible back mutation.
3. Determine the amino acid or sequence that would be translated from the mRNA sequences from the *his*– mutants and for each possible back mutation.

Conclude and Communicate

Part I

4. On the basis of the test results, which of the test substances is likely carcinogenic? Explain.
5. Suggest why it is harmful to inhale mutagens in smoke or ingest mutagens in food. What could happen to these mutagens once in the body?

Part II

6. How did the back mutation in *hisG* affect the protein produced by this gene?
7. The back mutation in *hisD* occurred in the middle of the gene. Explain how the back mutation affected the protein produced by this gene.

Extend Further

8. **INQUIRY** Many natural substances are being investigated for their potential to prevent mutations or cancer from occurring. Substances that slow the mutation rate or reverse the action of a mutagen are known as *antimutagens*. Design an experiment to test a natural substance for antimutagenic activity.
9. **RESEARCH** Cancer typically involves multiple gene mutations that involve DNA repair genes, oncogenes, or tumour suppressor genes. Find out how mutations in one of these types of genes could lead to cancer.

Chapter 6 SUMMARY

Section 6.1 The Transfer of Information from DNA

After several significant discoveries, scientists showed that the genetic code specifies how genetic information is converted from a nucleotide sequence to an amino acid sequence of a protein.

Key Terms

one-gene/one-polypeptide hypothesis
messenger RNA (mRNA)
genetic code
triplet hypothesis
gene expression
transcription
translation

Key Concepts

- The idea that a link existed between genes and proteins was first introduced by Garrod. He used his studies of alcaptonuria to suggest that defective inheritance factors produce defective enzymes.
- From their experiments with the bread mould *Neurospora crassa*, Beadle and Tatum concluded that one gene specifies one enzyme. They called this relationship the one-gene/one-enzyme hypothesis. Later, this was modified to the one-gene/one-polypeptide hypothesis.
- Jacob proposed and demonstrated that mRNA acts as a "genetic messenger." It provides the amino acid sequence information needed for protein synthesis.
- The genetic code specifies how genetic information is converted from a nucleotide sequence to an amino acid sequence of a protein. Experiments by Crick and Brenner showed that the genetic code is read three nucleotides at a time.
- The genetic code is redundant, continuous, and nearly universal.

Section 6.2 Transcription: Synthesizing RNA from DNA

Transcription is the synthesis of RNA from a DNA template.

Key Terms

RNA polymerase
promoter region
precursor mRNA (pre-mRNA)
mature mRNA
5′ cap
3′ poly-A tail
splicing

Key Concepts

- RNA is a single-stranded polynucleotide. It contains four nucleotides with the bases adenine (A), uracil (U), cytosine (C), and guanine (G).
- There are several types of RNA molecules that are classified according to the different functions they have in a cell.
- There are three defined stages of transcription: initiation, elongation, and termination. During transcription, the antisense strand of a gene is used as a template to synthesize a strand of mRNA.
- In eukaryotes, newly synthesized mRNA undergoes modifications before it is transported across the nuclear membrane into the cytoplasm. As a result, precurser mRNA is converted to mature mRNA.

Section 6.3 Translation: Synthesizing Proteins from mRNA

Translation is the synthesis of protein from a mRNA template.

Key Terms

transfer RNA (tRNA)
anticodon loop
acceptor stem
aminoacyl-tRNA synthetase
ribosome
ribosomal RNA (rRNA)
polyribosome
start codon
reading frame
peptide bond
mutation
single-gene mutation
chromosome mutation
point mutation
frameshift mutation
silent mutation
missense mutation
nonsense mutation
transposon
mutagen

Key Concepts

- Translation machinery consists of mRNA, tRNA, ribosomes, and translation factors.
- During translation, the mRNA binds to a ribosome assembly. A tRNA carrying methionine binds with the start codon sequence exposed in the first binding site.
- Another tRNA molecule then recognizes the codon sequence at the next exposed binding site on the mRNA, and brings the corresponding amino acid to this site. Enzymes bind the amino acids held by adjacent tRNA molecules. As the ribosome progresses along the mRNA, the amino acid chain grows until a stop codon is reached and the new polypeptide is released.
- The processes of gene expression govern the development of living organisms.
- Mutations are permanent changes in DNA. They may involve the insertion, deletion, or substitution of individual nucleotides, or larger-scale rearrangements of portions of chromosomes.
- Mutations may be spontaneous, or they may be induced by exposure to physical or chemical mutagens. Mutations may be harmful to cells. For example, mutations that disrupt gene expression or the cell cycle may result in disorders such as cancer.

Chapter 6 SUMMARY & REVIEW

Section 6.4 Regulation of Gene Expression

Gene regulation involves whether a gene is active or inactive, and determines the level of activity and amount of protein that is available in the cell.

Key Terms

gene regulation	repressor
constitutive gene	activator
operon	transcription factor
operator	RNA interference

Key Concepts

- Gene regulation refers to control of the rate of gene expression in response to different conditions in the cell or environment.
- While constitutive genes are always active and expressed at near constant levels, the majority of genes are regulated so that their product is only expressed at certain times and in certain amounts.
- Regulation of gene expression in prokaryotes can take place during transcription, during translation, and after the protein is synthesized, although it most commonly occurs during initiation.
- Operons may be inactive until transcription is induced or active until a repressor turns them off.
- Regulation of gene expression in eukaryotes may be pre-transcriptional, transcriptional, post-transcriptional, translational, and post-translational.

Knowledge and Understanding

Select the letter of the best answer below.

1. Which statement about RNA is *true*?
   - **a.** RNA is found only in the nucleus.
   - **b.** RNA is found in the nucleus and the cytoplasm.
   - **c.** RNA is synthesized in the cytoplasm and then transported to the nucleus.
   - **d.** RNA contains adenine, thymine, cytosine, and guanine.
   - **e.** RNA contains adenine, thymine, cytosine, and uracil.
2. Who first provided convincing evidence for the triplet hypothesis?
   - **a.** Beadle and Tatum
   - **b.** Jacob and Monod
   - **c.** Watson
   - **d.** Crick and Brenner
   - **e.** Garrod
3. When a gene from one organism is inserted into another kind of organism, it produces the same protein. Which characteristic of a genetic code makes this important implication for gene technology possible?
   - **a.** The genetic code is redundant.
   - **b.** The genetic code is continuous.
   - **c.** The genetic code is universal.
   - **d.** The genetic code has initiation and stop codons.
   - **e.** The genetic code is unique for every individual.
4. A protein is a specific sequence of
   - **a.** amino acids.
   - **b.** nucleotides.
   - **c.** codons.
   - **d.** anticodons.
   - **e.** tRNAs.
5. What is the consequence of changing the codon UGA to UGG?
   - **a.** The protein will be shorter than normal.
   - **b.** The protein will be longer than normal.
   - **c.** The protein will have the same amino acid.
   - **d.** The protein will be normal.
   - **e.** The protein will not be made.
6. Like DNA replication, the first major step in transcription is
   - **a.** binding of the transcription factors.
   - **b.** elongation.
   - **c.** the usage of RNA primers.
   - **d.** recognition of the promoter sites.
   - **e.** unwinding of the double helix.
7. Which statement about RNA polymerase is *true*?
   - **a.** RNA polymerase has a proofreading function.
   - **b.** RNA polymerase works in a 3′ to 5′ direction.
   - **c.** During initiation of transcription, RNA polymerase binds to a promoter region.
   - **d.** RNA polymerase transcribes both strands of DNA at the same time.
   - **e.** RNA polymerase uses the sense strand of DNA as a template.

8. The parts of genes that are transcribed into mRNA but are later removed are called
   a. exons.
   b. spliceosomes.
   c. introns.
   d. peptides.
   e. mutations.

9. To facilitate movement of mRNA from the nucleus into the cytoplasm, what is added to the 3′ end of mRNA?
   a. a cap of modified nucleotides
   b. modified tRNA molecules
   c. ribosomes
   d. a poly-A tail
   e. transcription factors

10. Translation is a process in which a polypeptide is produced from mRNA. What is the role of tRNA in this process?
    a. It carries a particular amino acid to the correct codon site.
    b. It stores genetic information.
    c. It provides information to a ribosome.
    d. It carries genetic information from DNA.
    e. It proofreads the growing polypeptide.

11. The triplet of nucleotides in tRNA that is complementary to a triplet of nucleotides in mRNA is called a(n)
    a. codon.
    b. anticodon.
    c. ribosome.
    d. genctic code.
    e. sequence.

12. If the original sequence is ACGCGT, which represents a point mutation?
    a. ACGGGT
    b. TGCCGT
    c. GGGCCC
    d. ACGCGT
    e. ACCCGU

13. How is transcription directly controlled in eukaryotic cells?
    a. through the use of phosphorylation
    b. through the use of operons
    c. using transcription factors and activators
    d. through condensed chromatin, which allows constant gene activation
    e. through the addition of a 5′cap and a 3′poly-A tail

14. What regions of DNA do eukaryotic activator proteins interact with?
    a. repressors
    b. operators
    c. coding regions
    d. promoters
    e. enhancers

Answer the questions below.

15. Explain why Beadle and Tatum's one-gene/one-enzyme hypothesis is now known as the one-gene/one-polypeptide hypothesis.
16. In what way is the structure of a protein related to the structure of DNA?
17. Why were one-nucleotide and two-nucleotide genetic codes rejected before the proposal of the triplet hypothesis?
18. List and explain the three important characteristics of the genetic code.
19. Define the term "antisense strand." What is its counterpart called, and why?
20. Define the following terms, and explain their significance with respect to gene expression.
    a. codon
    b. anticodon
    c. polyribosome
21. Identify and describe each component's function in translation.
22. List and describe the three binding sites for tRNAs.
23. List three ways in which the process of gene regulation differs between prokaryotic and eukaryotic cells.
24. Using examples, explain the difference between a physical mutagen and a chemical mutagen.
25. Distinguish between specific and non-specific DNA repair mechanisms. Give an example of each.
26. In bacterial cells, transcription and translation can take place at the same time on the same strand of mRNA.
    a. Why is this not possible in a eukaryotic cell?
    b. In what ways could this have both advantages and disadvantages for bacterial cells?
27. Why do different cell types have different rates of transcription and translation?
28. How is complementary base-pairing important in transcription and translation?

Thinking and Investigating

29. An mRNA strand contains the following nucleotide sequence: AUGCCCACUACAUAG. What amino acid sequence does this mRNA code for?
30. A geneticist isolates a gene for a specific trait under study. She also isolates the corresponding mRNA. Upon comparison, the mRNA is found to contain 1000 fewer bases than the DNA sequence. Did the geneticist isolate the wrong DNA? Explain your reasoning.

31. Suppose you are able to remove all of the spliceosomes from a sample of yeast and transfer them to a sample of bacterial cells.

a. How will gene expression be affected after removing all of the spliceosomes from the yeast sample? Explain.

b. How will gene expression be affected after the introduction of spliceosomes into the bacterial sample? Explain.

32. Which amino acid corresponds to each of the following mRNA codons?

a. UCC
b. ACG
c. GUG
d. CAC

33. Write the corresponding anticodon(s) for each of the following amino acids.

a. valine
b. methionine
c. leucine
d. tyrosine

34. A team of researchers has created a series of polypeptide chains for use in an experiment. Most of these polypeptides are made up of long chains of lysine with a small amount of arginine and glutamate.

a. What process might the researchers have used to prepare these polypeptides?

b. What would they have put into the reaction medium to obtain these results?

c. What other amino acids might be found in trace amounts in the polypeptides produced through this process?

35. Use the DNA sequence below to answer the following questions.

3′- GTATAAGCA -5′

a. If the above sequence is an antisense strand, what is the complementary sense strand?

b. What is the nucleotide sequence of the mRNA strand transcribed from the sense strand?

c. What is the amino acid sequence of polypeptide produced from the mRNA strand?

d. What is the sequence of anticodons that would be found on tRNAs delivering amino acids to the mRNA strand?

36. One mutation results in the replacement of a G nucleotide with a T nucleotide in the antisense strand of a DNA molecule. Under what circumstances will this substitution produce each of the following mutations?

a. a silent mutation
b. a missense mutation
c. a nonsense mutation

37. BIG IDEAS Proteins control a wide variety of cellular processes. Many antibiotic drugs work by interfering with protein synthesis in bacteria that cause infection. Explain how each of the following antibiotic mechanisms disrupts gene expression in bacteria.

a. tRNAs misread mRNA codons, binding with the incorrect codon.

b. Methionine is released from the initiation complex before translation can begin.

c. tRNA cannot bind to the ribosome.

d. A tRNA picks up the wrong amino acid.

38. How do proteins assist in their own synthesis?

39. The processes of transcription and translation consume a great deal of cellular energy. Why do you think the cell does not simply translate proteins directly from DNA?

40. Study the diagram below.

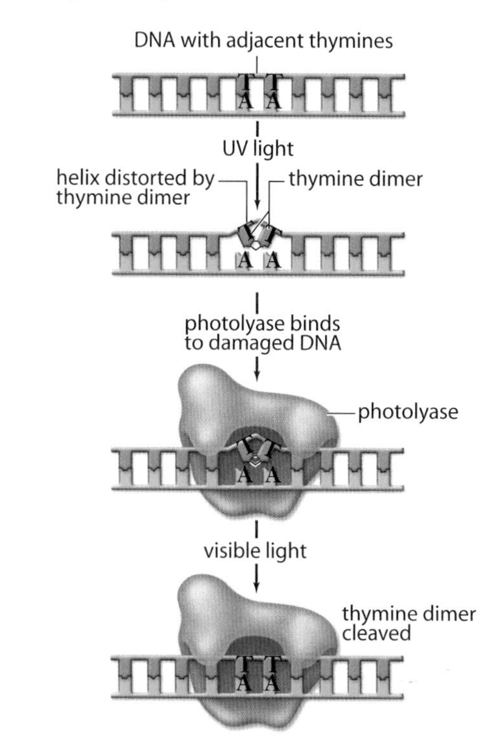

a. What process is occurring in this diagram?

b. What is the significance of this process?

41. If a gene has four exons and three introns, what is the total number of different polypeptide products that the gene could code for? Explain your answer with labelled diagrams.

42. How many bases can be added to or deleted from a sequence of DNA before the reading frame is returned to its original form (with the exception of the affected region of DNA)?

43. A bacterial cell that has been exposed to high levels of X rays soon afterward begins to produce large quantities of many different types of polypeptides, very few of which are the normal, functional proteins it usually produces.
   a. What kind of mutation could account for this effect?
   b. Assume the mutation has no other effect on gene expression. If your hypothesis about the cause of the mutation is correct, what other observation(s) about the polypeptide products would you expect to make?

44. Transposons can "jump" and insert themselves inside coding regions of DNA, causing disruptions in gene expression. Would transposons that "jumped" to non-coding regions be less harmful?

45. Suppose you are studying the regulation of a gene involved in the metabolism of two nutrients, Nutrient X and Nutrient Y, in bacteria. You are trying to determine if these nutrients act as inducers in their operons. The following data were collected from your experiments.

Experimental Data

	Levels of Nutrient in Growth Medium	Transcription of Genes in Operon
Nutrient X	High	High
	Low	Low
Nutrient Y	High	Low
	Low	High

   a. Based on the experimental data, which nutrient operon is most similar to the *lac* operon?
   b. Which nutrient operon is most similar to the *trp* operon? Explain your answer.

46. What types of genes do you think are present in DNA that is in highly condensed areas of chromatin? Explain.

Communication

47. Use a flowchart to outline the stages involved in transcription. Include labelled diagrams for each of the stages.

48. Create a table to show the similarities and differences between RNA and DNA.

49. You have been asked to explain the central dogma to a class of Grade 5 students. Prepare an outline for a five minute lesson about the central dogma. In your lesson plan, consider the following.
   a. What real-world analogies could you use to help students understand the central dogma?
   b. What questions do you think the students would have about the central dogma?
   c. Prepare a set of answers to these anticipated questions.

50. Use labelled diagrams to illustrate how precursor mRNA is modified into mature mRNA in eukaryotes.

51. "Proteins, not DNA, are the key to cell specialization." Write a paragraph explaining what is meant by this statement.

52. Use a labelled diagram to illustrate a ribosome at the initiation of translation.

53. Create a table that compares the main types of RNA involved in gene expression.

54. Your community is hosting a series of public information meetings about health issues. The objective of the series is to teach people about the science behind healthy lifestyle choices. You are asked to make a 10-minute presentation on the topic "DNA and Mutations."
   a. Write a one- or two-sentence key message you would want your audience to remember.
   b. Outline your presentation under five main headings, beginning with "Introduction" and ending with "Conclusion."
   c. Under each heading, list three points you would want to cover, and describe each in a few sentences.

55. The regulation of gene expression in eukaryotes is complex.
   a. Using a flowchart, summarize the five levels of eukaryotic gene regulation in the cell.
   b. Include the locations where each control point occurs.

56. "The *lac* operon is like a thermostat for the cell." Use a graphic organizer to outline arguments for and against this statement.

57. Summarize your learning in this chapter using a graphic organizer. To help you, the Chapter 6 Summary lists the Key Terms and Key Concepts. Refer to Using Graphic Organizers in Appendix A to help you decide which graphic organizer to use.

Application

58. Using Internet and print resources, research how the central dogma has been challenged and modified based on recent evidence about the nature of RNA viruses, such as HIV and other retroviruses.
    a. Summarize your findings by writing a short paper.
    b. Based on your research findings, draw an updated version of **Figure 6.3**, which shows the central dogma. Explain each of the changes that you make to the diagram.

59. BIG IDEAS Genetic research and biotechnology have social, legal, and ethical implications. Almost all living organisms on Earth use the same 20 amino acids. However, molecular biologists have been able to develop a number of artificial amino acids, which can be used to develop synthetic proteins that have many potential applications. In response, some research teams are exploring the possibility of expanding the genetic code to include new nucleotides that could be used to code for artificial amino acids. Use Internet and print resources to find out more about this research.
    a. Write a brief report explaining some of the hurdles that will have to be overcome as scientists try to expand the genetic code.
    b. What are some of the scientific and social implications of this research?

60. The action of the spliceosome is an example of enzymatic activity by a molecule other than a protein. For this reason, some RNA molecules are called ribozymes (ribo{some} + {en}zyme). Some scientists consider the properties of these RNA molecules to be important clues about the origins of life on Earth. Discuss the role such molecules might have played in this scenario. Consider the role of DNA and proteins in DNA replication and gene expression.

61. Some antibiotics work by paralyzing bacterial ribosomes. Using Internet and print resources, research the mechanism of these antibiotics.
    a. Write a short paragraph and draw labelled diagrams to summarize your findings.
    b. How does the mechanism of these antibiotics ensure specificity for bacterial cells?
    c. What is the importance of this specificity?

62. Chemical and physical mutagens, which are known to increase the risk of certain cancers, are very prevalent in our everyday lives. Choose three known mutagens and describe steps you can take to reduce your exposure to them.

63. A molecular biologist creates a form of RNA polymerase that has the same proofreading ability as DNA polymerase. Explain what some of the advantages and disadvantages of this form of RNA polymerase could be
    a. for researchers in a laboratory setting
    b. for living organisms

64. Many major food companies now offer deli meats that are "nitrite-free." These are, however, considerably more expensive than meats preserved with nitrite.
    a. Why do you think there is a price discrepancy between these products?
    b. Would you be more likely to purchase "nitrite-free" products, despite the price difference? Explain.

65. Formaldehyde is a carcinogenic chemical mutagen that is found in many straightening hairsprays. Health Canada recently banned certain brands of hairspray since they exceeded the maximum amount of formaldehyde allowed in household products. Cigarettes, which produce more than 50 known carcinogenic mutagens when smoked, however, are not banned in Canada. Why do you think cigarettes are still allowed to be sold?

66. The study of the structure of genomes, including projects such as the sequencing of the human genome, is often referred to as *genomics*. Many researchers claim that genomics is not nearly as significant as *proteomics*, the study of protein structure and function. Based on the information in this chapter, which field do you think holds the greatest promise for advances in medicine and in understanding human development?

67. BIG IDEAS DNA contains all the genetic information for any living organism. In a June 2010 issue of *Time* magazine, science writer John Cloud wrote an article entitled "Why Your DNA Isn't Your Destiny," which describes the increasing importance of epigenetics in human development. This expanding field of research studies how gene expression can be regulated without making alterations to the DNA sequence. Despite the lack of DNA sequence modification, the changes in gene expression brought on by epigenetic control can affect future generations. Modification of histone proteins and chromatin remodelling are examples of epigenetic control of genes. Using Internet and print resources, write a short report on the implications of epigenetics and "epigenome" research on human health.

Chapter 6 SELF-ASSESSMENT

Select the letter of the best answer below.

1. K/U The codons that code for phenylalanine are UUU an UUC. What characteristic of the genetic code explains this fact?
   - **a.** The genetic code is universal.
   - **b.** The genetic code is continuous.
   - **c.** The genetic code is redundant.
   - **d.** The genetic code has 20 different amino acids.
   - **e.** The genetic code has 64 different codons.
2. K/U Which bases are found in RNA?
   - **a.** adenine, uracil, thymine, and cytosine
   - **b.** adenine, uracil, thymine, and guanine
   - **c.** adenine, thymine, cytosine, and guanine
   - **d.** adenine, thymine, uracil, and cytosine
   - **e.** adenine, uracil, cytosine, and guanine
3. K/U In transcription, what type of strand is transcribed?
   - **a.** DNA sense strand
   - **b.** DNA anti-sense strand
   - **c.** DNA non-coding strand
   - **d.** RNA sense strand
   - **e.** RNA anti-sense strand
4. K/U How would transcription be affected if the promoter region on DNA was mutated?
   - **a.** RNA polymerase would not be able to bind to DNA and transcription would not be initiated.
   - **b.** RNA polymerase would bind to an enhancer region, and transcription would continue.
   - **c.** RNA polymerase would be able to continue initiation and elongation.
   - **d.** RNA polymerase would be able to bind to DNA, but elongation would not occur.
   - **e.** RNA polymerase would be not be able to bind to DNA, and transcription would not be terminated.
5. K/U What are the components of ribosomes?
   - **a.** two sub-units composed of proteins and rRNAs
   - **b.** one sub-unit composed of proteins and rRNAs
   - **c.** two sub-units composed of proteins and tRNAs
   - **d.** one sub-unit composed of proteins and tRNAs
   - **e.** two sub-units composed of proteins and mRNAs
6. K/U What is the corresponding anticodon for the codon CGG?
   - **a.** CGG
   - **b.** CCG
   - **c.** GCC
   - **d.** GCG
   - **e.** CCC
7. K/U Aminoglycosides are a class of antibiotics that specifically bind to bacterial ribosomes to cause inhibition of tRNA from the A site to the P site. Using this information, how are bacteria affected by aminoglycoside usage?
   - **a.** Aminoglycosides prevent bacteria from synthesizing proteins by inhibiting translation.
   - **b.** Aminoglycosides prevent bacteria from synthesizing mRNA by inhibiting transcription.
   - **c.** Aminoglycosides prevent bacteria from synthesizing proteins by inhibiting post-transcriptional modification of mRNA.
   - **d.** Aminoglycosides prevent bacteria from synthesizing proteins by inhibiting transcription.
   - **e.** Aminoglycosides prevent bacteria from synthesizing proteins by inhibiting post-translational modification.
8. K/U A mutation causes a sequence of mRNA codons to change from GUU CAU UUG to GUU CAU UAG. What type of mutation has occurred?
   - **a.** frameshift
   - **b.** silent
   - **c.** missense
   - **d.** nonsense
   - **e.** chemical
9. K/U Which statement about regulation of gene expression in prokaryotes is true?
   - **a.** The most common form of regulation of gene expression is at the level of translation.
   - **b.** The most common form of regulation of gene expression is at the post-transcriptional level.
   - **c.** Regulation of gene expression occurs through operons.
   - **d.** Regulation of gene expression occurs through alternative splicing.
   - **e.** Regulation of gene expression occurs through the addition of a 5′cap and a 3′poly-A tail, and removal of introns.
10. K/U What is the name of the group of proteins that are required for the initiation of transcription in eukaryotes?
    - **a.** RNA polymerases
    - **b.** transcription factors
    - **c.** ribosomes
    - **d.** promoters
    - **e.** enhancers

Use sentences and diagrams as appropriate to answer the questions below.

11. K/U Identify and describe the hypothesis proposed by Beadle and Tatum based on conclusions from their experiments on *Neurospora*.

12. C Use a labelled diagram to illustrate the central dogma.

13. K/U What are the main objectives of transcription and translation?

14. C Create a table to summarize the three major types of RNA involved in gene expression, and describe their functions.

15. T/I The photo below shows transcription in bacteria, with multiple RNA molecules being made from one gene at one time. How is this possible?

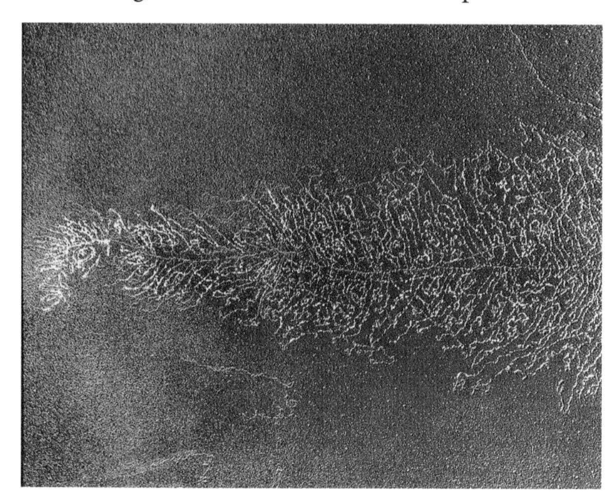

16. K/U List and describe three ways in which eukaryotic mRNA is altered after it is transcribed.

17. T/I List two different mRNA sequences that could correspond to the following amino acid sequence: glycine-leucine-valine-arginine.

18. T/I How is it possible for information on DNA that is confined to the nucleus of a eukaryotic cell to be expressed as protein products outside the nucleus?

19. A A research project called ENCODE (ENCyclopedia of DNA Elements) took 1 percent of the human genome and catalogued 487 genes and 2608 mRNA transcripts. Why are these numbers different?

20. C Draw a labelled diagram of a tRNA molecule that carries the amino acid serine.

21. A Study the diagram below.

normal sequence

GUU–CAU–UUG–ACU–CCC–GAA–GAA
val – his – leu – thr – pro – glu – glu

mutated sequence

GUU–CAU–GUU–GAC–UCC–CGA–AGA A
val – his – val – asp – ser – arg – arg

a. What type of mutation is shown in the diagram?

b. Why do you think this type of mutation is referred to by this term?

22. T/I Use the mRNA sequence below to answer the following questions.

AUG ACA AAA GGU UAC

a. The mRNA sequence can be read in two different directions. Describe how the two interpretations differ.

b. Explain how the amino acid sequence would change if the C in the second codon were deleted.

c. Identify this type of mutation.

23. A A history of sunburn and recreational exposure to the Sun without proper UV radiation protection can lead to an increased risk of skin cancer.

a. Name and describe the mechanism that cells use to repair damage caused by UV exposure.

b. Even though cells in the human body have this repair mechanism, why to do you think skin cancer can still develop in individuals who are exposed to UV radiation?

24. C Create flowcharts to describe how lactose-metabolizing genes are regulated on the *lac* operon in the follow environmental conditions:

a. high amounts of lactose

b. no lactose

25. A What is the importance of the regulation of gene expression? Discuss how the regulation of gene expression has implications on human health and development.

Self-Check

If you missed question...	1	2	3	4	5	6	7	8	9	10	11	12	13	14	15	16	17	18	19	20	21	22	23	24	25
Review section(s)...	6.1	6.1	6.2	6.2	6.2	6.1 6.2	6.2	6.3	6.4	6.4	6.1	6.2	6.2	6.2	6.2	6.2 6.4	6.1 6.2	6.2 6.4	6.2 6.4	6.2	6.3	6.1 6.3	6.3	6.4	6.4

CHAPTER 7

Genetic Research and Biotechnology

Specific Expectations

In this chapter, you will learn how to . . .

- D1.1 **analyze**, on the basis of research, some of the social, ethical, and legal implications of biotechnology (7.1, 7.2)
- D1.2 **analyze**, on the basis of research, some key aspects of Canadian regulations pertaining to biotechnology, and **compare** them to regulations from another jurisdiction (7.1, 7.2)
- D2.1 **use** appropriate terminology related to molecular genetics (7.1, 7.2)
- D3.5 **describe** some examples of genetic modification, and **explain** how it is applied in industry and agriculture (7.2)
- D3.6 **describe** the functions of some of the cell components used in biotechnology (7.1, 7.2)
- D3.7 **describe**, on the basis of research, some of the historical scientific contributions that have advanced our understanding of molecular genetics (7.1, 7.2)

Although we tend to think of molecular genetics when we think of modifying organisms, humans have actually been modifying organisms for thousands of years through the domestication of animals and the selective breeding of animals and plants. Dogs, corn, and roses are just a few examples of organisms that people have modified, long before molecular genetics became a discipline. More recently, however, scientific knowledge has provided the ability to manipulate the genes of organisms—micro-organisms, plants, and animals. We can clone genes, and then use them to alter the genotype and the phenotype of various organisms through genetic engineering. This technology has innumerable uses, from treating cancer and genetic disorders to producing useful proteins. For instance, the goats shown here, Pete and Webster, look like regular goats but they have been genetically modified by Canadian scientists. When these goats mate, their female offspring will secrete spider silk protein in their milk. Spider silk has many possible uses, including as a component of tough, light-weight steel. Because spider silk is not rejected by the human body, this steel could be used to create durable, low-weight artificial limbs, tendons, and other replacement connective tissue for use in humans.

Launch Activity

Biotechnology: Assessing Unintended Consequences

Would you approve of a new technology that could lead to the death of hundreds of thousands of people every year and drastically alter the environment around the world? If not, you would not have welcomed the automobile. We are living during the early years of a developing biotechnology industry. Biotechnology is the use of an organism, or product from an organism, for the benefit of humans. Like the automobile industry, biotechnology is certain to have a widespread impact on society. Few people in the 1920s foresaw how vehicle exhausts would eventually contribute to air pollution and climate change. Similarly, some of the long-term effects of biotechnology, including genetically modifying plants and animals intended for human consumption, DNA fingerprinting, and cloning, may be unpredictable. How can society balance the benefits and risks of new technologies when some of their outcomes are uncertain?

Procedure

1. As a class, list at least three examples of biotechnology that you are familiar with through media or previous science courses.
2. In a small group, discuss how the following viewpoints relate to one of the examples of biotechnology that you listed in step 1.
   - New technologies are sometimes applied before people have had a chance to debate their regulation or to study their implications.
   - The precautionary principle suggests that governments should act in advance to prevent unknown potential harm that could arise from new technologies.
   - The difficulty of predicting the effects of a new technology does not mean that we must reject the technology or give up the effort to control or reduce its risks.
2. With your group members, brainstorm a list of questions that you think would be important to answer before taking a stance on whether or not a new product or procedure available through biotechnology, such as genetically modified plants for human consumption, should be marketed for use by the public.

Questions

1. Do you think government regulations are the best way to protect people from the potential risks of biotechnology? Explain.
2. Do you think the risks and benefits of a new technology affect everybody in society equally? Justify your response.

SECTION 7.1

Techniques for Producing and Analyzing DNA

Key Terms

molecular biology
recombinant DNA
restriction endonuclease
restriction fragment
gene cloning
plasmid
transformation
DNA amplification
polymerase chain reaction (PCR)
gel electrophoresis
DNA fingerprinting
short tandem repeat (STR) profiling
DNA sequencing
dideoxy sequencing
Human Genome Project
site-directed mutagenesis

The development of technologies that enable scientists to isolate and manipulate DNA have led to the formation of new branches of science, including **molecular biology**. Much of molecular biology focuses on using research techniques that investigate the properties of nucleic acids and proteins. In many cases, this research can be done outside the cell, in the controlled environment of a test tube.

Recombinant DNA Technology

An advance in the early 1970s that allowed scientists to construct **recombinant DNA** marked the beginning of a new way to study processes in the cell. Recombinant DNA is DNA that has been prepared in the laboratory by combining fragments of DNA from more than one source. Often these different sources are genomes of different species.

The ability to prepare recombinant DNA became possible due to the isolation of a certain type of enzyme from bacteria. To defend themselves against infection by foreign viruses, most prokaryotic organisms manufacture one or more types of enzymes called *restriction enzymes*. These enzymes were given this name because they restrict the replication of infecting viruses by cleaving (cutting) viral DNA. A specific type of restriction enzyme called a **restriction endonuclease** is of particular importance to molecular biologists. Restriction endonucleases cleave double-stranded DNA within the interior of the DNA strands, rather than at the ends. They do so by recognizing a short sequence of nucleotides, called the *target sequence*, within the DNA. Next, the enzymes cut the strand at a particular point within the sequence. This point is known as a *restriction site*. Each of the different endonucleases that has been isolated by researchers recognizes a different target sequence. An example of a restriction endonuclease reaction is shown in **Figure 7.1**. This reaction illustrates two characteristics of restriction endonucleases that make them especially useful to researchers:

- *Sequence specificity:* The cuts made by restriction endonucleases are specific and predictable. The same enzyme will cut a particular strand of DNA the same way each time, producing an identical set of DNA fragments. These fragments are called **restriction fragments**.
- *Staggered cuts:* Most restriction endonucleases produce a staggered cut that leaves a few unpaired nucleotides on a single strand at each end of the restriction fragment. These short single-stranded regions are often referred to as *sticky ends* or *overhangs*. These sticky ends can form base pairs with other single-stranded regions that have a complementary sequence.

molecular biology the study of the structure and functions of nucleic acids and proteins

recombinant DNA a molecule of DNA composed of genetic material from different sources

restriction endonuclease an enzyme that cleaves (cuts) the interior of double-stranded DNA in a sequence-specific manner

restriction fragment a small segment of DNA generated by cutting a larger piece of DNA with a restriction enzyme

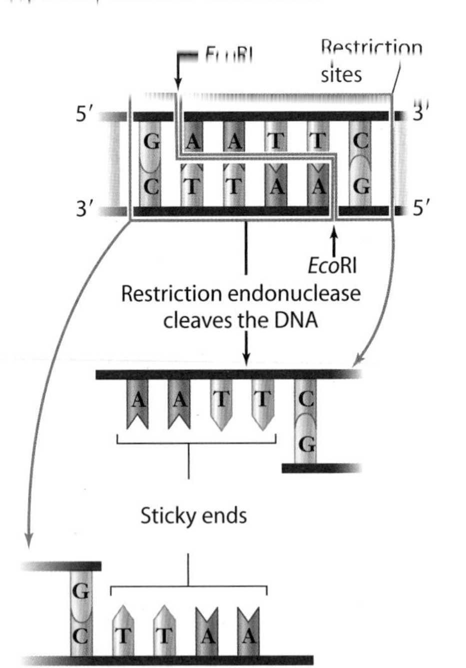

Figure 7.1 *Eco*RI is a restriction endonuclease that recognizes and cleaves the sequence GAATTC. The cut DNA fragment that is produced now has single-stranded sticky ends at each end of the molecule.

***Identify** the sequence of single-stranded DNA that would base pair with the sticky ends shown in the diagram.*

Because base pairing is so specific, cutting DNA fragments with restriction endonucleases that produce sticky ends can limit the fragments that can be combined. To reduce this specificity, restriction endonucleases that produce blunt cuts can be used. The cuts are made so there are no sticky ends that form complementary base pairs. This allows any two fragments of DNA that have blunt ends to be combined. However, these blunt ends also have limitations. The loss of specificity means that any two blunt-end pieces can be combined and many by-products can form, thus making the process less efficient.

The formation of the recombinant molecule is done in the laboratory in very small reaction tubes, like the one shown in **Figure 7.2**. Typically, investigators are working with volumes measured in microlitre quantities. The type of recombinant DNA molecule prepared will depend on the reason for making it. Often a recombinant DNA molecule is produced as a way of making many copies of a specific DNA segment, such as a gene, so it can be isolated and purified for other studies. **Figure 7.3** illustrates how a restriction endonuclease reaction can be used to construct a recombinant DNA molecule.

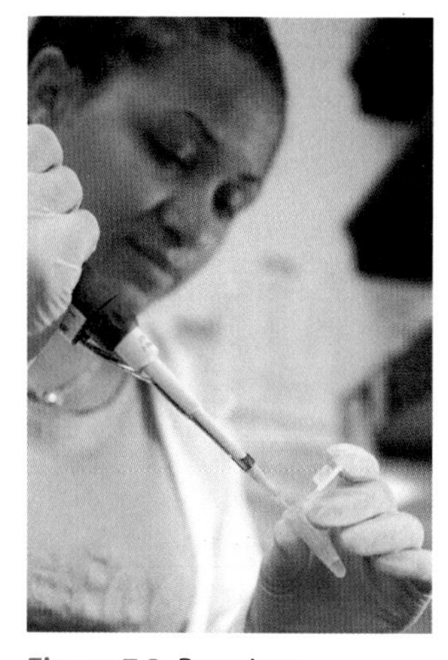

Figure 7.2 Reactions for the preparation of recombinant DNA are done in microlitre quantities. This requires use of special equipment such as this micro-pipette, which can measure and deliver as little as 0.5 μl of a liquid. The reaction is carried out in these microcentrifuge tubes.

Listed below are the basic steps for producing a recombinant DNA molecule.

1. A restriction endonuclease is selected that can cut both DNA fragments to be combined. Ideally, these are enzymes that produce sticky ends, although different enzymes can be used.
2. Each piece of DNA is reacted with the restriction endonuclease enzyme to produce cut DNA fragments.
3. The two cut DNA fragments are incubated with another enzyme, *DNA ligase*. This enzyme seals the breaks in the DNA, forming covalent bonds between the two different fragments. The result is a stable, recombinant DNA molecule. Recall that DNA ligase also joins Okazaki fragments when the lagging strand is synthesized during DNA replication.

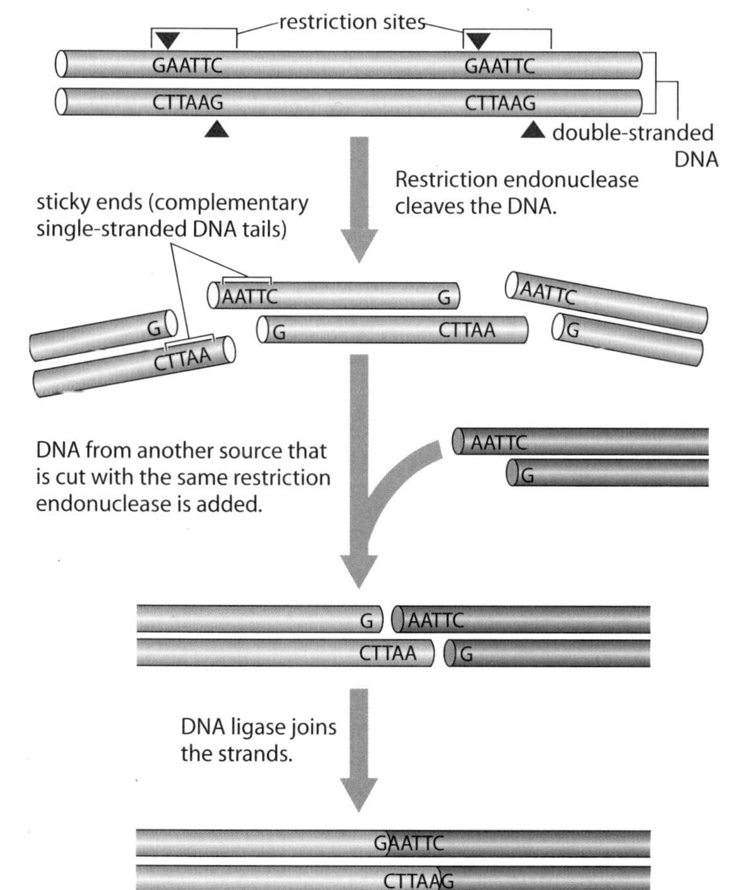

Figure 7.3 To make a recombinant DNA molecule, two fragments from different sources are often cut with the same enzyme to produce complementary single-stranded sticky ends. In this visual, the two fragments are distinguished by colour. The light-blue fragment comes from one source and the dark-blue fragment comes from another source. Base pairing between the complementary sticky ends brings the molecules together. DNA ligase then covalently joins the strands to produce double-stranded recombinant DNA.

Predict *by-products that could be formed in this reaction.*

gene cloning the process of manipulating DNA to produce many identical copies of a gene or another segment of DNA in foreign cells

plasmid in recombinant DNA technology, a self-replicating, closed circular piece of DNA that can act as a carrier of a gene to be cloned in bacteria

transformation in recombinant DNA technology, a process in which a bacterial host takes up a segment of DNA from the environment under particular experimental conditions

Gene Cloning in Bacteria

Recombinant DNA technology has enabled scientists to focus their studies on specific genes and proteins and their particular functions in the cell. Without the ability to work with a specific region of DNA, studying the function of one particular gene or protein within the context of a cell would be extremely difficult. The human cell is very complex, with an estimated 25 000 genes and almost 2 million proteins. Thus, scientists clone the gene or region of DNA that interests them. Recall that, in general, the term *cloning* simply refers to making identical copies of something. **Gene cloning** involves making many identical copies of a gene. There are many reasons for cloning a gene. Typically, scientists wish to either study the gene itself or to use the gene to produce RNA and/or protein in sufficient quantities to study these molecules.

Bacteria are often used as host systems when a researcher is cloning a gene. Bacteria are straightforward and inexpensive to maintain and can be grown easily in large amounts. Also, many reactants used for gene cloning have been optimized for use in an *E. coli* host system. **Figure 7.4** outlines the steps used in cloning a gene in bacteria. While bacteria represent the most common type of host system used in molecular biology, scientists now also routinely use other hosts such as mammalian, yeast, and insect cells.

Generally, gene cloning in bacteria involves the following steps:

1. A recombinant DNA molecule is produced that is composed of the gene to be cloned and an appropriate *vector* for the host system. A vector is a carrier for the gene to be cloned. For gene cloning in bacteria, the vector is a **plasmid**. Recall from Chapter 6 that plasmids are small, circular double-stranded DNA molecules found in some prokaryotes. They are not part of the nucleoid region and often carry non-essential genes, although they may carry genes for antibiotic resistance. During gene cloning, a plasmid acts as a vector for the gene to be cloned. To do so, the plasmid must have an origin of replication that allows it to be copied independently of bacterial chromosomal DNA. It must also carry a gene that makes the bacteria resistant to a certain type of drug. Most commonly, this gene allows for resistance to the antibiotic *ampicillin*. Finally, the plasmid must have one or more restriction endonuclease sites where the gene to be cloned can be inserted using standard recombinant DNA techniques. Often, the DNA to be cloned interrupts the *lacZ* gene. The *lacZ* gene codes for an enzyme that breaks down galactose. The *lacZ* and ampicillin resistance genes are called *selectable markers* because researchers can use them to specifically select for the bacterial colonies that contain the recombinant DNA of interest. This is necessary because some bacteria will also contain plasmids that do not have this DNA inserted into them.
2. The reaction mixture for producing the recombinant DNA is introduced into the bacteria. The process in which foreign DNA is taken up by bacteria is called **transformation**. (Recall that the experiments of Griffith and Avery in Chapter 5 also involved transformation.) To facilitate transformation, bacterial cells are treated with particular chemicals that make the cell membranes permeable to the DNA incubated with the cells. Transformed bacterial cells are those that take up the DNA.
3. The bacterial cells are applied to a Petri dish containing growth media that has been supplemented with (i) the antibiotic ampicillin, and (ii) a derivative of galactose, X-gal, which causes bacterial colonies to turn blue when the bacteria is broken down by the enzyme coded for by the *lacZ* gene.
4. Bacterial colonies containing cells that have the recombinant DNA are identified. This is done by a process of elimination using selectable markers. First, all bacterial colonies that grow on Petri dishes containing ampicillin must contain either the recombinant DNA or the plasmid only. Second, all blue bacterial colonies contain an active *lacZ* gene and, therefore, do not have the gene inserted into them. Third, this means the white colonies produced on these plates must contain the recombinant DNA.

5. Cells from the colonies that contain the recombinant DNA are selected and grown in liquid culture to produce a larger population.
6. The recombinant DNA molecules are isolated and purified from the bacterial cells.
7. A variety of analysis techniques are used to confirm that the correct recombinant DNA molecule has been made.

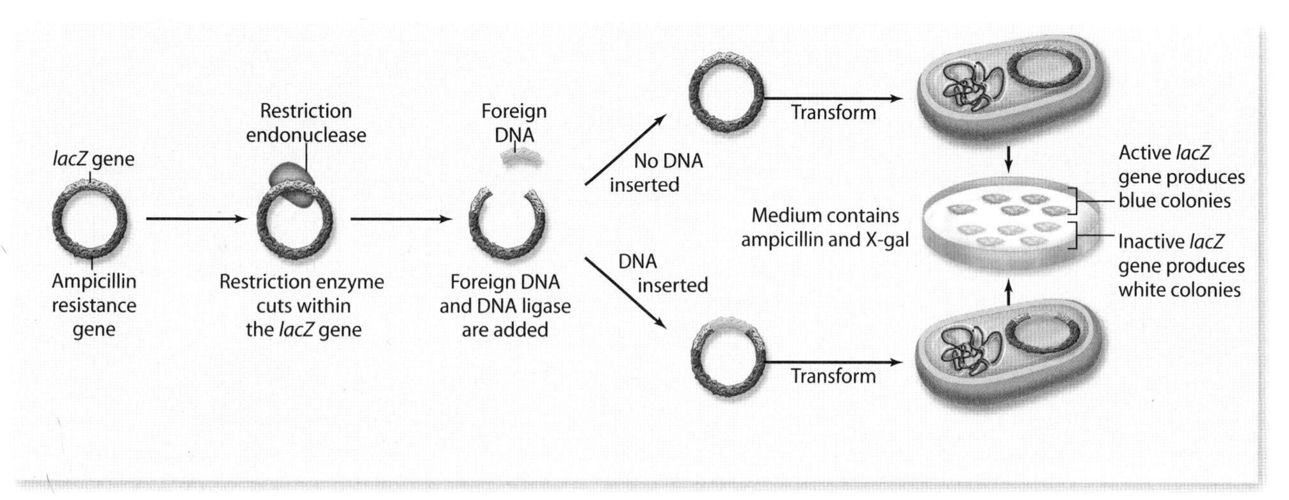

Figure 7.4 The cloning of a gene using a bacterial host involves a recombinant DNA molecule. The recombinant DNA molecule is made up of the gene to be cloned and a plasmid vector. *E. coli* cells take up the plasmid, selectable markers, resistance to ampicillin, and colony colour, thus allowing researchers to identify the bacteria that contain the recombinant DNA.
***Identify** the stage at which the gene has been cloned.*

The Polymerase Chain Reaction

Gene cloning in bacteria allows scientists to produce multiple copies of a gene or another segment of DNA in foreign cells. The process of generating large amounts of DNA for use or analysis is called **DNA amplification**. In gene cloning, the recombinant DNA molecule is transformed into *E. coli* cells. The cells are then grown in liquid media to produce large cultures from which the recombinant DNA molecule can be isolated and purified. At times, however, scientists may want to amplify only a small section of DNA. They may wish to purify this fragment for analysis or use it for cloning purposes instead of isolating sequences from chromosomal DNA. To meet these needs, an alternative DNA amplification method was developed that does not rely on the production of recombinant DNA molecules or host systems. This method is called the **polymerase chain reaction (PCR)**. It was conceptualized in 1983 by an American scientist named Kary Mullis while he was working as a DNA chemist with the Cetus Corporation in California. The development of this technique revolutionized molecular biology. Mullis, shown in **Figure 7.5**, shared the 1993 Nobel Prize in Chemistry "for his invention of the polymerase chain reaction (PCR) method."

DNA amplification the process of producing large quantities of DNA from a sample

polymerase chain reaction (PCR) an automated method for amplifying specific regions of DNA from extremely small quantities

Figure 7.5 Kary Mullis was awarded a Nobel Prize for inventing the polymerase chain reaction method.

Steps in PCR

PCR is highly specific and rapid. The reaction can produce billions of copies of a section of DNA in a test tube within a period of a few hours. The procedure for PCR is outlined in **Figure 7.6**. It involves repeated cycles of the same basic steps, producing copies of the DNA in each cycle. As more cycles are completed, more DNA is produced.

The steps in the polymerase chain reaction are as follows:

1. The DNA sample to be amplified is heated to a high temperature, typically about 95°C, so that the double-stranded DNA is *denatured* (that is, altered structurally by an external stress) into single strands.
2. The DNA sample is cooled in the presence of two nucleotide primers. The primers are complementary to each 3′ end of the DNA fragment to be amplified. The lower temperature allows the primers to *anneal*, or base pair, with the 3′ ends of the single-stranded DNA to be amplified. The temperature used depends on the nucleotide sequence of the primers, but is typically around 55°C.
3. The DNA sample is heated to 72°C, which is the optimal temperature for *Taq* polymerase, a certain type of DNA polymerase used for PCR. The isolation of this DNA polymerase from the heat-loving bacteria *Thermus aquaticus* was one of the key factors that made PCR possible. Like other DNA polymerase enzymes, it synthesizes DNA by the addition of free nucleotides to the ends of the primers via complementary base pairing. However, unlike other DNA polymerases, it can withstand the high temperatures used during the cycling of reactions.
4. The DNA is then taken through several cycles of steps 1 to 3, undergoing repeated rounds of denaturation, primer annealing, and DNA synthesis from extension of the primers. Note that the amount of DNA doubles with each replication cycle. Therefore, one copy of DNA produces two copies after one cycle, four copies after two cycles, eight copies after three cycles, and so on.

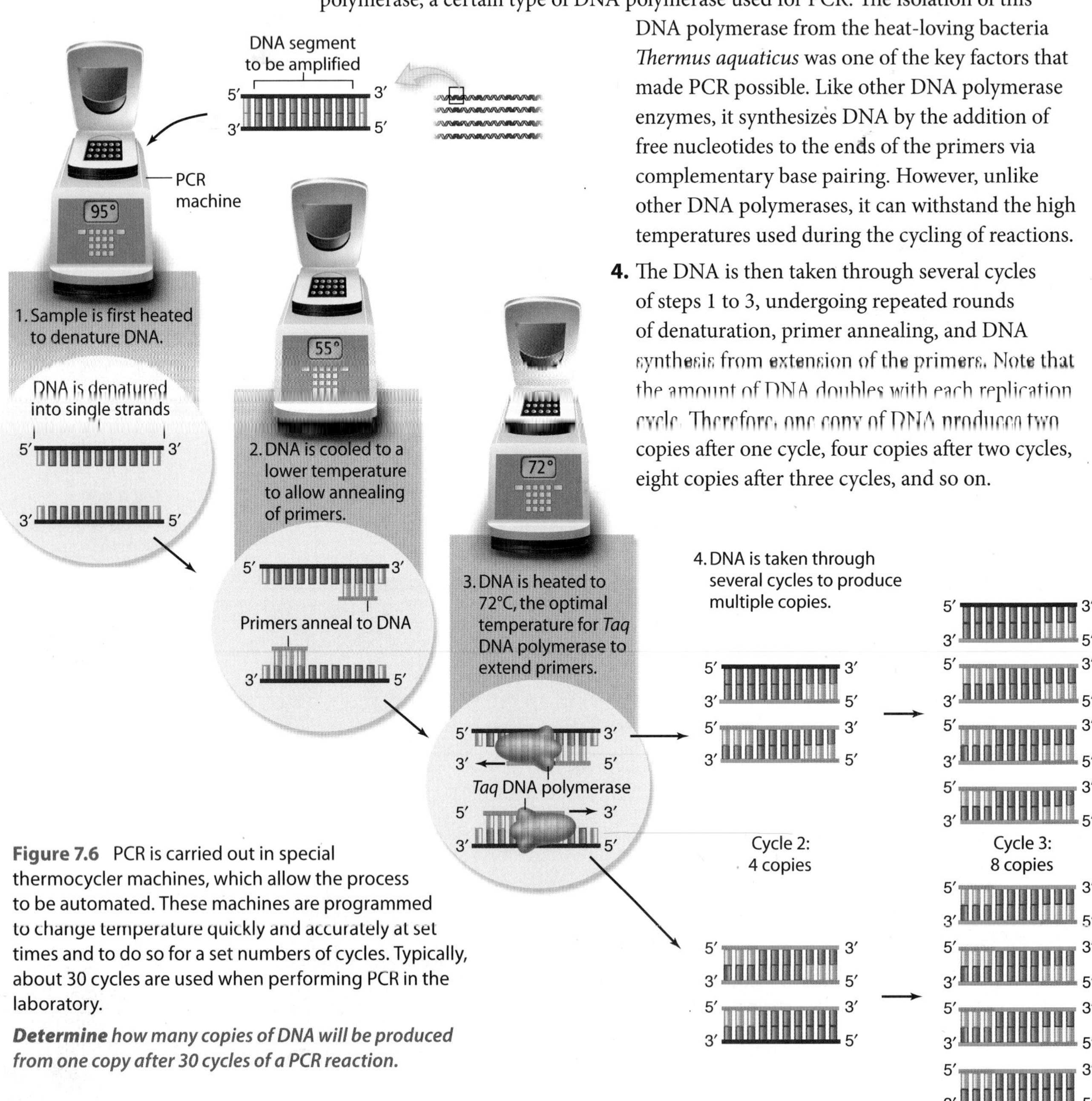

Figure 7.6 PCR is carried out in special thermocycler machines, which allow the process to be automated. These machines are programmed to change temperature quickly and accurately at set times and to do so for a set numbers of cycles. Typically, about 30 cycles are used when performing PCR in the laboratory.

***Determine** how many copies of DNA will be produced from one copy after 30 cycles of a PCR reaction.*

A DNA fragment generated by PCR can be used for a number of applications. It is often the gene of interest to be studied in gene cloning, instead of chromosomal DNA. PCR has also revolutionized a number of fields besides molecular genetics, ranging from medicine to criminal investigations. The method has also shed new light on evolutionary studies by comparing DNA extracted from 76 000-year-old human mummies and from animal fossils millions of years old. Mitochondrial DNA sequences have also been used to study the evolutionary history of human populations. Further, doctors can screen for genetic defects in very early embryos by using a single cell and PCR amplification of its DNA. PCR is now routinely used to amplify DNA from minute quantities found at crime scenes from samples as small as a hair follicle or a skin cell. This amplified DNA can then provide investigators with enough material to be analyzed through various methods to identify both the victim and the criminal.

Learning Check

1. Explain why sequence specificity and staggered cuts make restriction endonucleases useful molecular biology tools.
2. How is complementary base pairing important when constructing a recombinant DNA molecule?
3. What is gene cloning? Why do researchers clone genes?
4. Explain how selectable markers are used in gene cloning.
5. Use a graphic organizer to explain the steps involved in PCR.
6. Describe three practical applications of PCR.

Analyzing DNA Fragment Size

gel electrophoresis tool used to separate molecules according to their mass and charge; can be used to separate fragments of DNA

Once molecular biologists have amplified the DNA of interest, they can choose from a variety of methods to analyze it. One standard method, which is used on a daily basis in the laboratory, is **gel electrophoresis**. Gel electrophoresis uses an electric field to separate negatively charged DNA fragments according to size as they pass through a gel.

As shown in **Figure 7.7**, the gel is submerged in an aqueous solution, called a *buffer*, containing various salts that maintain the pH of the solution. A positively charged anode lies at one end of the buffer bath and a negatively charged cathode lies at the opposite end. The salts in the buffer carry the electric charge. The gel is made of either polyacrylamide or agarose and provides a porous matrix-like support that the DNA molecules weave their way through, repelled by the negative cathode and attracted to the positive anode.

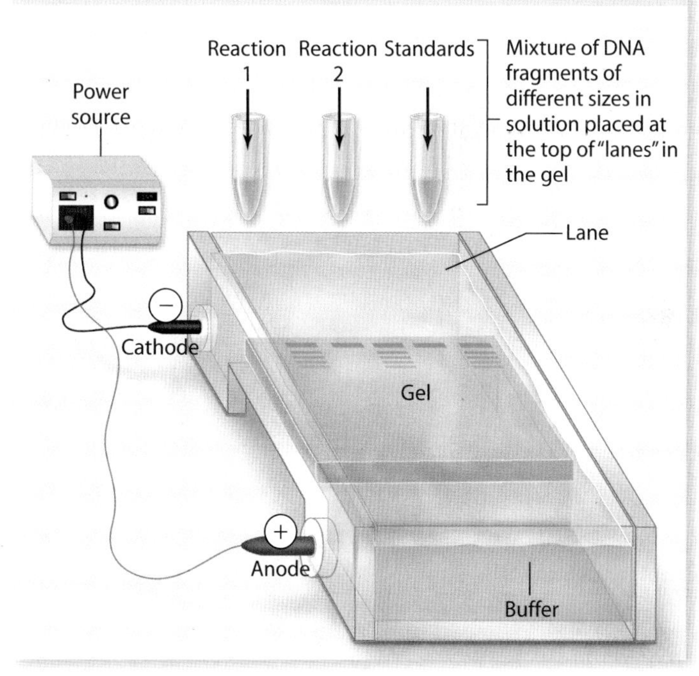

Figure 7.7 In gel electrophoresis, the gel lies in an aqueous buffer solution. A positively charged anode lies at one end and a negatively charged cathode lies at the other end. The salts in the buffer carry the electric charge.

Gel electrophoresis involves the following steps, which are also illustrated in **Figure 7.8**:

1. DNA cannot be seen with the unaided eye. Therefore, before the DNA fragments are added to the gel, scientists use two chemicals. One is a negatively charged dye that turns the samples blue. A chemical called *ethidium bromide* is also added to the gel when it is being prepared. Ethidium bromide associates with DNA and fluoresces under ultraviolet light. Samples of different-sized fragments in solution are applied in preformed wells at one end of the gel.
2. The gel is placed in the buffer solution. A power source is turned on and an electric current runs between the cathode and the anode, through the buffer and the gel. When exposed to this current, DNA fragments move through the gel in their respective lanes. The fragments move toward the positively charged anode because DNA has an overall negative charge due to the presence of phosphate groups along its sugar-phosphate backbone.
3. The smaller the DNA fragment, the more easily it will move through the gel. Therefore, smaller fragments of DNA will travel more quickly and farther through the gel than larger fragments. Over time, the DNA fragments will separate according to size. The negatively charged blue-coloured dye helps scientists follow the progress of the DNA samples.
4. The gel is removed from the buffer and exposed to ultraviolet light. The ethidium bromide in the gel interacts with the fragments of DNA and fluoresces. This allows the different bands of DNA in the gel to be observed and photographed.

In addition to the above steps, a set of standards can be added to one lane of the gel, which includes DNA fragments of known length (refer back to **Figure 7.7** on the previous page). The positions of the fragments in the DNA samples can be compared to these standards. The sizes of the DNA in the samples can then be estimated based on their positions relative to the standards. If needed, a particular band of DNA can be cut out of the gel, extracted from it, and purified for use in further studies.

1 The DNA is treated with chemicals that make it visible. Mixtures of different sized fragments in solution are then added to preformed wells in the gel.

gel

power source

⊖ negative end

⊕ positive end

2 The gel is placed in the buffer solution and an electric current is run between the cathode and anode. The negatively charged DNA fragments move through the gel towards the positive charge.

3 Smaller DNA fragments travel more quickly and farther through the gel than larger fragments. Over time, the fragments separate according to size. Negatively charged dye added before running the gel allows scientists to follow the progress of the DNA.

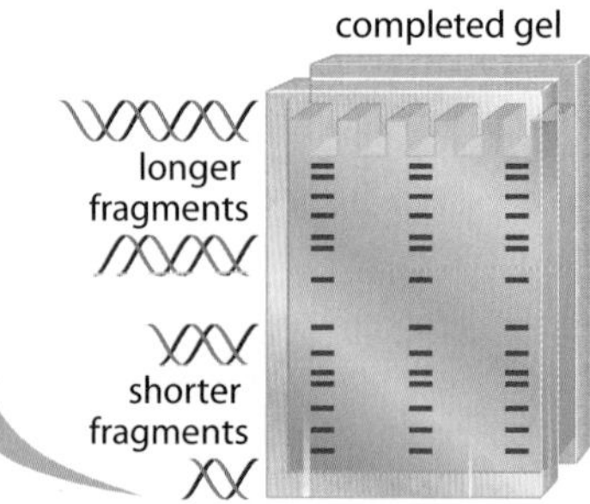

4 The DNA is treated with chemicals that make it visible. Mixtures of different sized fragments in solution are then added to preformed wells in the gel.

Figure 7.8 Gel electrophoresis is a multistep process commonly used in the laboratory to analyze DNA fragment size.

Using a DNA Fingerprint for Identification

Together, restriction enzymes and gel electrophoresis help researchers analyze and compare DNA samples. This type of analytical approach forms the basis for a technique called **DNA fingerprinting** or *DNA profiling*, which involves identifying someone based on their DNA. Like a fingerprint, the DNA of an individual is unique—with the only exception being identical twins. Traditionally, DNA fingerprinting has been done by treating chromosomal DNA with restriction endonucleases and then separating the fragments by gel electrophoresis as shown in **Figure 7.9**. This method uses an approach called *restriction fragment length polymorphism (RFLP)* analysis. The bands on the gel are unique to each individual and the distinct pattern is used as a method of identification by comparing the bands with patterns from an individual of known identity.

DNA fingerprinting a technology used to identify individuals by analyzing the DNA sequence of certain regions of their genome

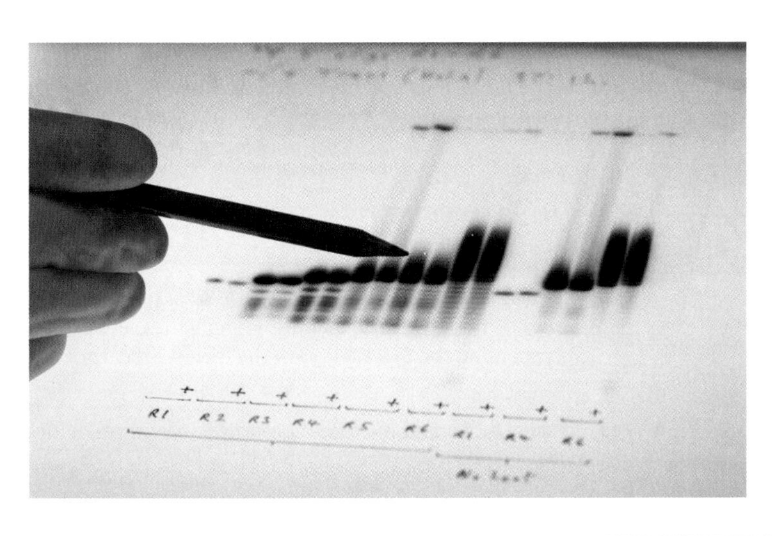

Figure 7.9 Traditional DNA fingerprinting has involved digesting chromosomal DNA and comparing the pattern of bands produced when the fragments are separated by gel electrophoresis.

A more recent method of DNA fingerprinting involves **short tandem repeat (STR) profiling**. STRs are repeating short sequences of DNA in the genome that vary in length between individuals depending on how many copies of a particular STR are present. Numerous different *loci* (locations of DNA sequences or genes) of STRs can be analyzed. The more STR loci that are employed, the more confident scientists are of distinctive results for each person.

short tandem repeat (STR) profiling a technology used to identify individuals based on repeating short sequences of DNA in the genome that vary in length between individuals

Using primers and PCR, STRs of an individual are amplified and then separated by gel electrophoresis. The more repeats at an STR locus, the longer the DNA fragment produced, and so the shorter the distance that the fragment will travel through the gel. In contrast, the fewer repeats there are, the shorter the DNA fragment, and so the farther it will travel through the gel. The different fragments are fluorescently labelled and a detector measures the fluorescence emitted from each STR. The resulting DNA fingerprint is a distinct pattern of peaks with particular molecular masses. An example is shown in **Figure 7.10**.

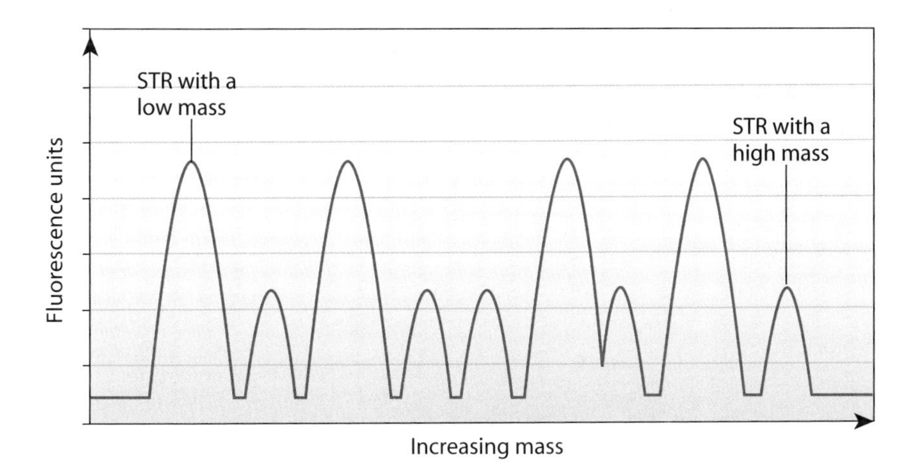

Figure 7.10 DNA fingerprinting using STR profiling produces a series of peaks that represent STRs of differing molecular mass and, therefore, differing lengths. Each individual has a unique series of peaks in their STR profile.

Applications

DNA fingerprinting now has many common applications. It has become invaluable in forensic sciences. For example, investigators at a crime scene might find a small sample of blood or skin tissue. The DNA from this sample is amplified by PCR, and then used to create a DNA fingerprint. This DNA fingerprint can be compared with the DNA fingerprint of a suspect in the crime. A match is very strong evidence that the suspect was present at the crime scene. In 1998, the Royal Canadian Mounted Police converted from RFLP analysis to STR profiling in their forensic laboratories. The Federal Bureau of Investigation is also collecting large numbers of human STR profiles in its databases for current and future crime scene analyses. DNA fingerprinting was extensively used to identify the victims of the September 11, 2001, terrorist attacks in New York City. DNA fingerprints can also be used to solve disputes over parentage. Because DNA is inherited equally from both parents, a child's DNA fingerprint will show some matches with the DNA fingerprint of each parent. Thus, a comparison of the DNA fingerprints of different people can help researchers identify the relationships among them.

Activity 7.1 Reading a DNA Profile

STRs on autosomes are inherited from both parents. The maternal chromosome for a homologous pair carries one version (or *allele*) of an STR. The paternal chromosome may carry the same allele for this STR locus or a different allele with a different number of repeats. An individual with the same alleles for an STR on homologous chromosomes is said to be *homozygous*, while an individual who has different alleles on homologous chromosomes is said to be *heterozygous*.

In this activity, you will analyze partial DNA profiles of individuals with various alleles for an STR locus on chromosome 8. The following diagram shows the results of gel electrophoresis of DNA fragments from a child and three possible sets of parents.

Materials

- computer with Internet access

Procedure

1. The STR with a nucleotide sequence "TCTA" may repeat several times on a chromosome. Identify the bands representing the STR allele with the fewest repeats of the sequence "TCTA."
2. Identify individuals who are homozygous for the STR.
3. Identify the child's biological parents from the three sets of possible parents.
4. Research current Canadian laws regarding mandatory DNA testing.

Questions

1. How did you identify the child's biological parents?
2. Suggest why some bands on the gel were darker (thicker) than others.

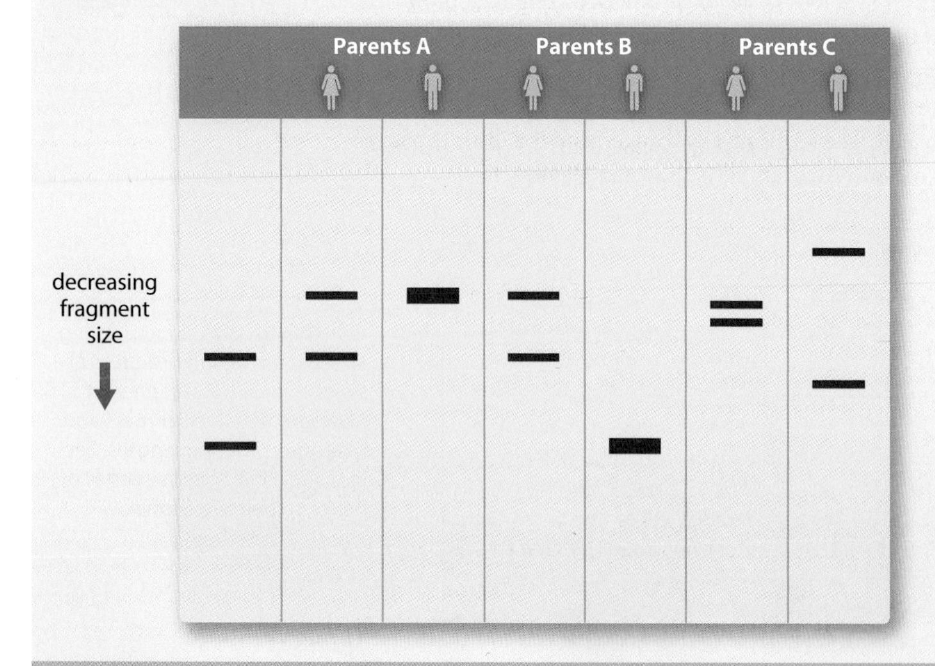

3. In this activity, you examined data for a single STR locus, but in forensic investigations or paternity tests, data for eight or more STR loci are examined. Why is it important to analyze data for several STR loci in such cases?
4. In DNA profiling, it is typical to use fluorescent markers to analyze the results of gel electrophoresis. Describe what a graph of the child's DNA profile for the STR "TCTA" might look like.
5. Do you think that current Canadian laws regarding who must undergo mandatory DNA testing are appropriate? Explain.

Learning Check

7. What characteristic of DNA allows it to migrate in gel electrophoresis?
8. Explain why DNA fragments move through a gel at different rates.
9. Researchers routinely use gel electrophoresis to check whether they have successfully cloned a gene into a plasmid. Why do they use this technique?
10. What is DNA fingerprinting?
11. Explain how STR profiling creates DNA fingerprints.
12. List and explain two practical applications of DNA fingerprinting. What are some ethical issues that surround each application?

Analyzing DNA Sequences

DNA sequencing refers to determining, base by base, the nucleotide sequence of a fragment of DNA. This method represents the ultimate form of DNA analysis and identification. The techniques used to sequence DNA have undergone tremendous changes over time. DNA sequencing was originally developed in the 1970s. The two methods developed at that time involved *manual sequencing* in which researchers themselves performed the processes involved, without the computer-based technology that exists today.

DNA sequencing a method for determining the nucleotide sequence, base by base, of a fragment of DNA

dideoxy sequencing a method for determining the sequence of a DNA fragment using dideoxynucleotides, which cause termination of DNA synthesis during the procedure

Manual DNA Sequencing

In 1976, Allan Maxam and Walter Gilbert at Harvard University developed one of these methods, known as *Maxam-Gilbert sequencing*. Maxam-Gilbert sequencing is a detailed method that relies on radioactive labelling of the single-stranded DNA to be sequenced. Using particular chemicals and reaction conditions, the DNA is then cleaved at specific bases. Ultimately, a series of radioactively labelled fragments are produced and separated by gel electrophoresis. This method was initially popular but, over time, a second method became the preferred technique.

In 1977, Frederick Sanger and colleagues at the University of Cambridge developed **dideoxy sequencing**, also called the *dideoxy chain termination method.* Dideoxy sequencing relies on the principles of DNA replication. DNA polymerase is used to synthesize a series of DNA fragments of differing lengths, using the DNA to be sequenced as the template. The fragments produced all start at the same position, but terminate at different specific bases. The different-sized fragments occur because replication is terminated or stopped due to incorporation of one of four possible dideoxynucleotides (ddA, ddG, ddC, or ddT). As shown in **Figure 7.11**, dideoxynucleotides lack a hydroxyl group (–OH) at the 3′ and 2′ carbons of the ribose sugar. Recall that deoxynucleotides normally found in DNA only lack the hydroxyl group at the 2′ carbon. Since the dideoxynucleotides lack the 3′–OH group that normally reacts with a new nucleotide during DNA synthesis, the reaction stops when a dideoxynucleotide becomes incorporated.

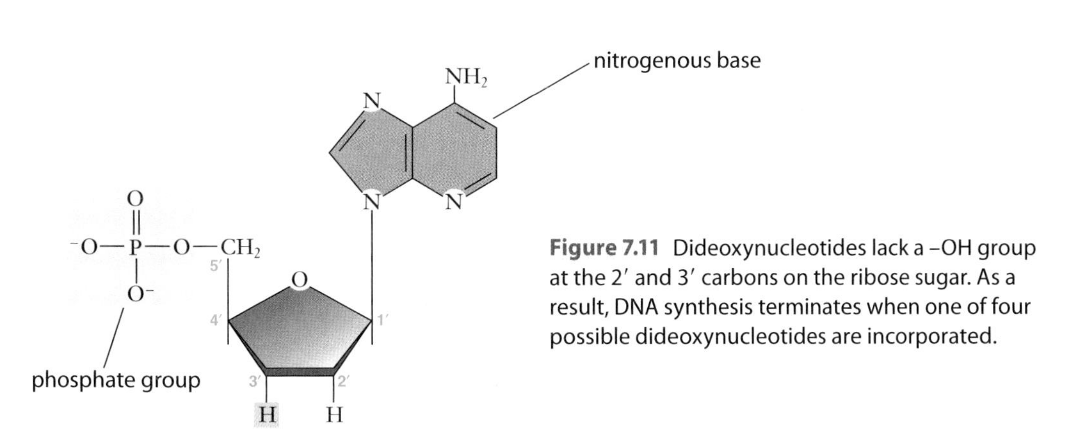

Figure 7.11 Dideoxynucleotides lack a –OH group at the 2′ and 3′ carbons on the ribose sugar. As a result, DNA synthesis terminates when one of four possible dideoxynucleotides are incorporated.

Steps in Dideoxy Sequencing

Dideoxy sequencing is outlined in **Figure 7.12** and involves the following steps:

1. The DNA to be sequenced is denatured to single-stranded DNA and a primer anneals to the 3′ end of the region to be sequenced. The sequence determined from the procedure begins right after the primer has completed its function.
2. Four separate reactions are prepared. Each contains the single-stranded DNA to be sequenced, with primer, deoxynucleotides that will be incorporated into the growing polynucleotide chain, DNA polymerase, and one of the four dideoxynucleotides.
3. The DNA synthesis reaction is allowed to proceed. In each reaction, a series of fragments of differing lengths will be produced. For example, in the reaction containing ddA, a series of fragments will be produced that all end in A. The size of the fragments will depend on where the base T occurs in the DNA template. Keep in mind that the newly synthesized fragments are complementary to the template.
4. Each of the four reactions is separated using gel electrophoresis with a *polyacrylamide* gel. This matrix allows for high resolution of fragments that differ by only one base. Traditionally, a radioactive tag attached to the dideoxynucleotides has been used to visualize the DNA by exposing the gel to X-ray film after electrophoresis, in a method called *autoradiography*.
5. The gel is then "read" from top to bottom, or longest fragment to shortest fragment. Because the type of dideoxynucleotides added to each base is known, scientists can identify which base is located at the end of each strand of DNA separated by the gel. By reading which base is at the end of each strand and comparing the lengths of the strands, the nucleotide sequence of the original DNA molecule can be determined.
6. The sequence read directly from the gel represents the synthesized strand. Therefore, the sequence of the original template is the complement to this sequence.

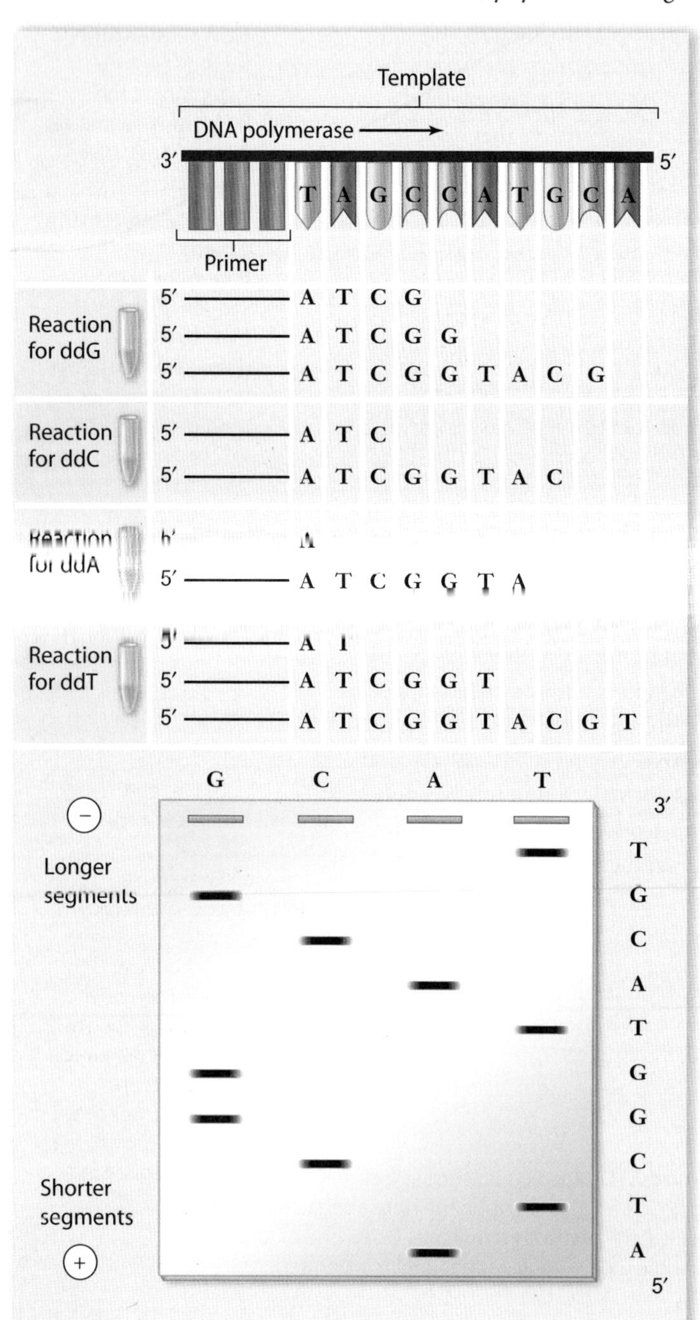

Figure 7.12 The DNA being sequenced by dideoxy sequencing is shown in the top panel, with the primer annealed. Each of the dideoxynucleotides is placed in a separate reaction tube containing the denatured DNA and primer, DNA polymerase, and deoxynucleotides used in the synthesis of DNA. The fragments generated from each reaction show up as bands on the gel. Because the final base of each fragment and the length of each fragment are known, the nucleotide sequence of the original DNA can be determined. Note that the sequence read from the gel is the complement of the DNA fragment being sequenced.

Early Automated DNA Sequencing

The ability to determine the sequence of DNA represented a huge leap forward for scientists in many disciplines. The Nobel committee recognized this great achievement in 1980 when Gilbert and Sanger shared part of the Nobel Prize in Chemistry for "their contributions concerning the determination of base sequences in nucleic acids." DNA sequencing allowed scientists to determine the nucleotide sequence of genes they studied and, consequently, the amino acid sequences of the proteins coded by those genes. As sequence information was amassed, it led to comparisons of gene sequences among populations and between species. One of the most innovative and large-scale projects ever undertaken was the **Human Genome Project**. In addition to determining the DNA sequence of the human genome, another aim of this project was to determine the DNA sequences of genomes for many different organisms. The biggest obstacle to be overcome, however, was handling the enormous amount of DNA-sequence data.

Human Genome Project a project that sequenced the human genome and identified all the genes within it

Although manual sequencing represented an important technological advance, it is laborious and time-consuming. Even skilled researchers were limited to reading only about 300 base pairs of DNA at any one time. Thus, a new era of DNA sequencing began with the development of automated sequencing. The basic reactions remained the same in the initial models for automated sequencing. Nonetheless, automated sequencing has provided numerous advantages, and this method made the Human Genome Project feasible. Early automated sequencing technologies use dideoxynucleotides labelled with dye tags. Each dideoxynucleotide has its own colour of dye tag. Thus, rather than needing four separate reaction tubes, all four dideoxynucleotides are added to one tube. As a result, the products can be run out on a single lane in a gel. Instead of using large polyacrylamide gels, small tubes of gels are used which reduces the time required to separate the fragments. A laser scans the bottom of the tubes and causes the dyes to fluoresce. Because each dideoxynucleotide has its own dye colour, the detector can identify each base according to the colour of the band in the gel. A computer printout of different-coloured peaks is generated, which represents the DNA sequence of the fragment. An example of such a printout is shown in **Figure 7.13**. The development of automated sequencing has increased the size of DNA fragments that can be read from each sequencing reaction. Also, more reactions can be run at once and much less time is required.

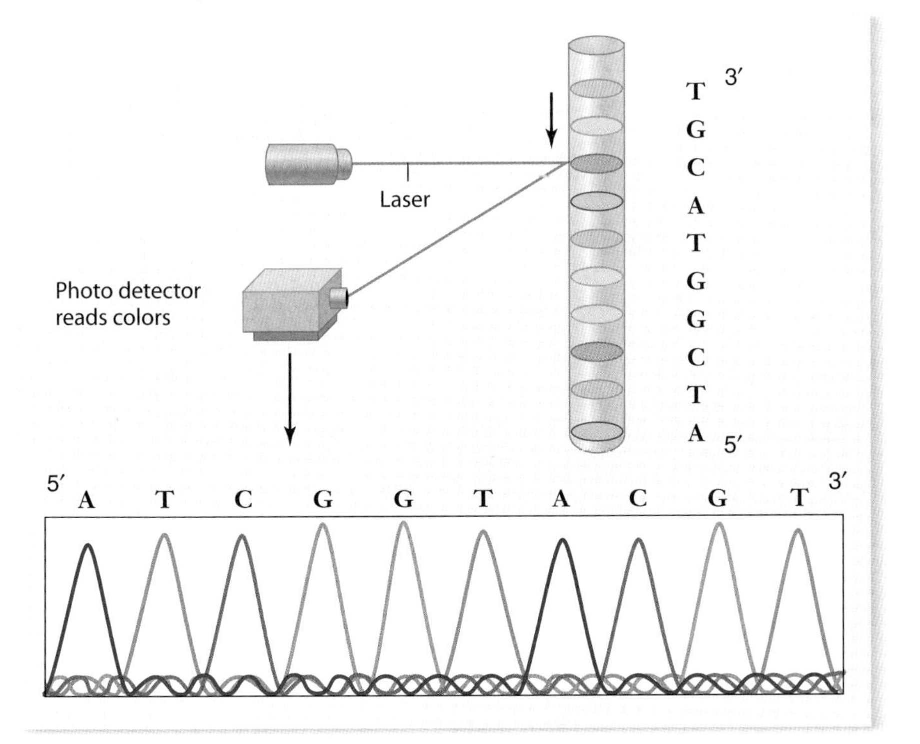

Figure 7.13 Using early automated sequencing technologies, the dideoxynucleotides are fluorescently dyed a particular colour. A detector reads each coloured band that is run through the gel. Since each colour represents a particular base, the printout of coloured peaks from the detector represents the sequence of the fragment that is complementary to the DNA being sequenced.

Next-generation Automated DNA Sequencing

Since 2005, automated sequencing techniques and molecular biology have advanced remarkably. Changing from dideoxynucleotide-based methods, innovative new methods referred to as "next-generation sequencing" are providing an unprecedented 10-fold improvement in data output per year. For example, DNA sequences that would have taken 90 to 100 different reactions to read in 2008 required only about 8 different reactions in 2010.

Why is there so much emphasis on improving this technique? The ability to sequence DNA has numerous applications, particularly in medicine, with the goal of individualized diagnoses and treatments based on a person's genome sequence. For example, a person's genome sequence could become part of their standard medical record. This information could enable a physician to determine a patient's risk of developing a particular disease or their best treatment options for certain diseases. Next-generation sequencing is also being applied to cancer diagnosis and treatment. Cancer involves a wide variety of genomic changes that can range from point mutations to chromosomal rearrangements. The DNA sequence of a cancerous tumour can be unique and is different from an individual's inherited genome. *Tumour profiling* involves determining the DNA sequence of a tumour and, therefore, knowing exactly what type of genetic change has caused a particular cancer. This knowledge will help to better define certain types of cancer. It can also help physicians decide on the best course of treatment out of numerous possible options.

For these applications to be feasible, genome sequencing must be affordable and significantly less time-consuming than it has been. Fortunately, it is becoming so. The Human Genome Project was a decade-long, multibillion-dollar project that involved numerous research labs working across the world. Standard genome sequencing for medical applications would not even be close to feasible based on those standards. Just a few years after the Human Genome Project was completed, however, improvements in automated DNA-sequencing techniques have produced phenomenal results. For example, working at the University of Washington in 2008, Elaine Mardis, shown in **Figure 7.14**, published the sequences of two human genomes. One of these was from a normal cell and the other was from a cancerous cell. This work required only about nine [illegible] using [illegible] the next generation automated sequencing methods. In fact, companies now advertise that they can sequence an entire human genome for less than $100 000.

Figure 7.14 In 2008, Elaine Mardis published the DNA sequences of a normal human cell and a cancerous cell. Next-generation DNA-sequencing machinery is shown behind her.

Making Sequence-Specific Mutations

Once researchers were able to study genes at the detailed level of DNA sequences, a new approach to studying the functions associated with these genes became possible. This approach is based on observing the effects that specific mutations in the gene may have on cellular activity. Before the development of this approach, researchers were limited to studying mutations that were generated naturally or through mutagens. While DNA sequencing allowed researchers to identify what the mutations were, the mutations made were random. Scientists lacked the ability to specifically select the nucleotides they wanted mutated. This problem was resolved, however, thanks to the work of a scientist at the University of British Columbia.

Site-directed Mutagenesis

In 1976, Michael Smith, whose laboratory building is shown in **Figure 7.15**, took a sabbatical to work in Sanger's laboratory at the University of Cambridge. Working with another researcher, Clyde Hutchison of the University of North Carolina, these researchers showed that specific mutations could be made in the DNA sequence of a gene. This technology, now called **site-directed mutagenesis**, has undergone considerable refinement since those initial studies. Nonetheless, the ability to target specific sequences in genes has revolutionized how researchers can study the structure and function of not only DNA and genes, but also the proteins that are coded by those genes. For example, suppose a researcher has hypothesized that a particular amino acid in an enzyme is essential for the activity of the enzyme. That researcher can now alter a *codon* in the corresponding gene to create a form of the enzyme with an altered amino acid. The researcher can then test the hypothesis by studying the activity of the mutant enzyme. For this ground-breaking work, Smith shared the 1993 Nobel Prize in Chemistry "for his fundamental contributions to the establishment of oligonucleotide-based, site-directed mutagenesis and its development for protein studies."

site-directed mutagenesis a method of specifically altering the nucleotide sequence of a region of DNA

Figure 7.15 The Michael Smith Laboratories at the University of British Columbia are named after the Nobel laureate. The building's unique architecture is meant to reflect the innovative molecular genetics research carried out in the labs.

Section 7.1 Review

Section Summary

- Using restriction endonucleases, researchers can combine fragments of DNA from more than one source to form recombinant DNA.
- Bacteria can be used as host systems for DNA amplification, producing multiple copies of a gene for research purposes.
- The polymerase chain reaction (PCR) is an alternative form of DNA amplification in which a sample of DNA is repeatedly denatured through heating, and then replicated using a heat-resistant DNA polymerase.
- In DNA fingerprinting, a sample of DNA from an individual is fragmented, amplified, and then separated by gel electrophoresis to produce a unique "fingerprint" that can be compared to reference samples for identification or matching.
- Dideoxy sequencing enables the nucleotide sequence of a fragment of DNA to be determined by cutting four identical samples of the DNA fragment at specific nucleotide sites and then comparing the sizes of the pieces produced.
- Site-directed mutagenesis is a technology that enables researchers to create specific mutations in the DNA sequence of a gene. As a result, researchers can study the structure and function of DNA, genes, and the proteins coded by these genes.

Review Questions

1. **K/U** Describe the action of a restriction endonuclease. What two features of this action make restriction endonucleases useful to genetic engineers? Explain each feature.
2. **A** Many molecular biology techniques, such as restriction endonuclease reactions, are performed in small reaction tubes due to the small reaction volumes required. What are some possible precautions you would take when working with very small reaction volumes of DNA?
3. The term *cloning* is often used incorrectly in the mainstream media. Explain what the term *gene cloning* actually means. How would you explain gene cloning to someone who has no science background?
4. **K/U** Why are bacteria common hosts used in gene cloning?
5. **C** Use labelled diagrams to describe how plasmids containing selectable markers can be used to screen bacterial colonies for recombinant DNA.
6. **K/U** List and explain two different methods of DNA amplification. Under what circumstances would a researcher use each method?
7. **T/I** A PCR thermocycler in your lab develops a slight malfunction such that it is unable to heat the reaction mixture at the start of the reaction cycle. What effect would this have on the amplification process? Explain.
8. **K/U** Explain why DNA fragments migrate in gel electrophoresis. Which fragments migrate farthest: large or small?
9. **T/I** Suppose you have samples of DNA from two different plants. You want to find out whether these plants are clones. Your laboratory is equipped with restriction endonucleases and gel electrophoresis equipment.
   a. Outline the steps you would take to analyze the DNA samples.
   b. Use a labelled diagram to show the results you would expect if the plants are clones.
10. [illegible] DNA fingerprinting? Describe three of its applications.
11. **T/I** The following diagram shows the results of a gel electrophoresis analysis for a paternity test. Which male is the father of the child? How do you know?

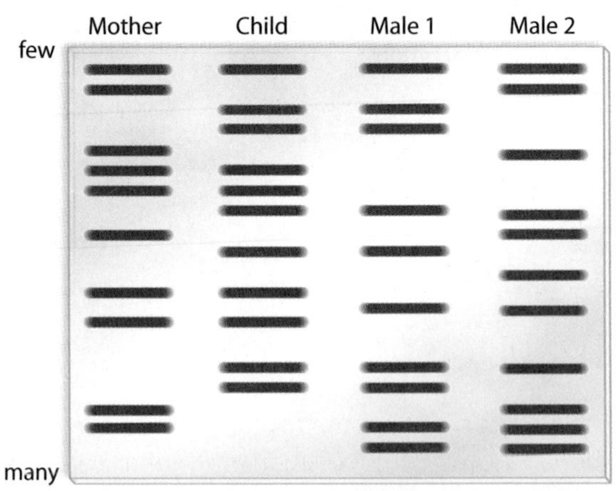

12. **A** What is the Human Genome Project? Describe two of the potential applications of the genetic research conducted through the Human Genome Project.

SECTION 7.2 Production and Regulation of Genetically Engineered Organisms

Scientists' knowledge of the properties of DNA and recombinant DNA techniques has led to breathtaking possibilities. However, this advanced knowledge and technology have also opened the door to controversy. For example, the genetic make-up of an individual can now be altered in an attempt to treat disease associated with a defective gene. In the process of **gene therapy**, the correct form of a gene is introduced into the patient's genome, thereby producing a recombinant human. **Figure 7.16A** shows a dramatic difference in the muscle fibres of a mouse with Duchenne muscular dystrophy before and after gene therapy. Recombinant DNA techniques have also produced recombinant animals that are sold as novelty items. Genomes of the zebrafish (*Danio rerio*) shown in **Figure 7.16B** have been altered by insertion of a gene that causes a fluorescent-coloured phenotype. These examples have stirred up considerable discussion and debate. The ability to alter the human genome has given rise to many serious concerns. For instance, could such technology allow us to one day create "designer" children and, if so, would this action be ethical? Along similar lines, the sale of GloFish® is banned in Canada and across Europe due to ethical issues (for example, the genetic manipulation of animals for sale as pets). Concerns about long-term safety issues regarding the impact these organisms could have on the environment have also been voiced.

Key Terms

gene therapy
genetic engineering
transgenic
genetically modified organism (GMO)
biotechnology
patent
expression vector
bioremediation
biolistic method
Ti plasmid method
gene pharming

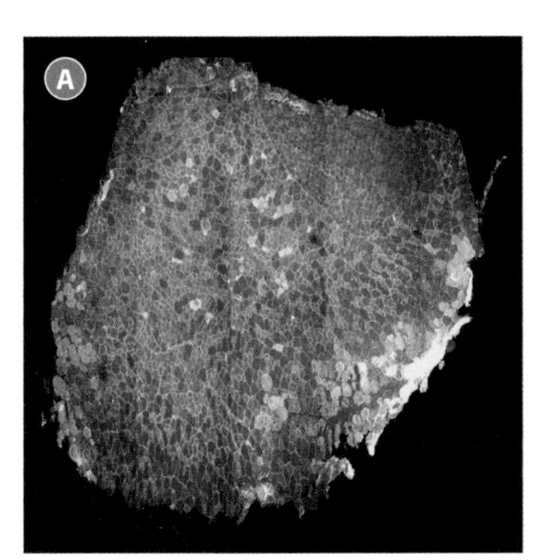

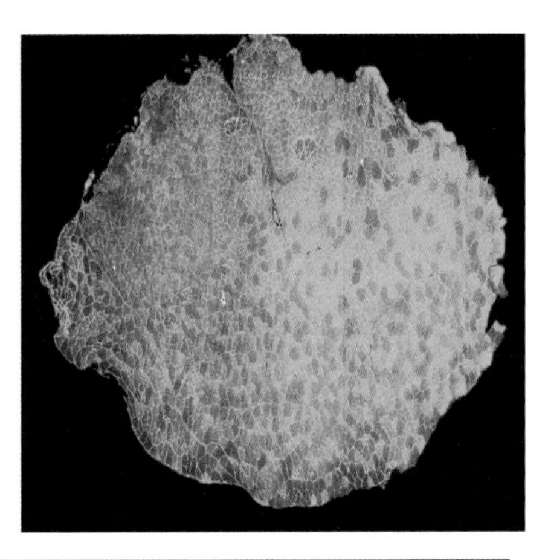

Figure 7.16 **(A)** Before gene therapy (left), the muscle fibres of the mouse do not produce enough dystrophin, a protein required for normal muscle function. The orange and yellow portions of the tissue represent dead cells. After gene therapy (right), dystrophin levels are normal and the muscle fibres are healthy. **(B)** These GloFish® are sold as ornamental fish and novelty items. Both humans and fish are examples of organisms that have been altered by means of recombinant DNA techniques.

The emergence of recombinant DNA technology has given rise to various disciplines that focus on the bioethics and societal impact associated with such advances in genetics. While science has continually pushed the boundaries of what is possible, humanity is now being forced to consider "How far *should* we go?" and, just as importantly, "*Who* should make these decisions?" In many cases, however, technological changes occur so rapidly that discussion of their consequences often follow rather than precede the new technologies.

gene therapy a method for treating genetic disorders by introducing a correct form of the disease-related gene into an individual's genome

genetic engineering alteration of the genetic material of an organism in a specific manner

transgenic describes an organism that is produced from the introduction of foreign DNA into its genome, providing it with a new phenotype

genetically modified organism (GMO) an organism whose genetic material has been modified, often through the insertion of a foreign gene into its genome

biotechnology the use of an organism, or a product from an organism, for the benefit of humans

patent a government ruling giving an individual or organization the sole title or right to make, use, or sell a particular invention

expression vector a plasmid vector that is transformed into a host cell for the purpose of producing a foreign protein

Applications of Genetically Engineered Organisms

Today, researchers can alter the genetic material of numerous organisms including bacteria, plants, and animals. The process of specifically altering the genetic make-up of an organism is called **genetic engineering**. Genetic engineering involves precise changes directed by the researcher. These changes can include making specific changes to the sequence of DNA, such as introducing a mutation into a gene. More elaborate changes are also possible. Genetic engineering that involves the introduction of foreign DNA into an organism's genome, such as a gene from another species, results in a **transgenic** organism. Transgenic bacteria, plants, and animals are commonly referred to as **genetically modified organisms (GMOs)**. Development of GMOs for direct use, or use through a product they produce, represents a major component of a multibillion-dollar biotechnology industry. **Biotechnology** is traditionally described as the application of technologies that involve use of organisms, or products from those organisms, to benefit humans. Today, however, biotechnology is most commonly associated with using the tools of molecular genetics to produce such organisms and their products.

The recombinant DNA techniques and methods of DNA analysis discussed earlier in this chapter are integral to the production of GMOs. While the technologies may have first been discovered and optimized in laboratories for pure research purposes, they are now used in industry in countless ways. This section will consider examples of genetically modified bacteria, plants, and animals to illustrate some of these applications. While reading these examples, consider both the risks and the benefits that may be associated with each example. Also, because the technologies used often produce highly valuable products, give thought to issues surrounding ownership of the GMOs as well. Private companies that use recombinant techniques to produce GMOs generally wish to claim ownership of the organism and its genome. As a result, the development of the biotechnology industry in the 1980s has had a significant impact on the legal profession and courts all over the world. Applications for **patents** for techniques used in the production of GMOs and to claim ownership of modified genomes began to be filed, forcing all stakeholders involved to consider whether a living organism can be patented and the consequences of this decision.

Applications of Transgenic Bacteria in Pharmaceuticals

From a biotechnology viewpoint, the bacteria that contain a recombinant DNA molecule, such as those used to clone genes, are considered GMOs or transgenic organisms. One application of genetically modified bacteria is in the production of medicines. To produce pharmaceuticals, recombinant DNA technology produces transgenic bacteria using a specialized type of plasmid vector called an **expression vector**. This vector is required because the bacteria are not simply being used to make multiple copies of the recombinant DNA. Instead, they are being used as "production factories" for the pharmaceutical for which the gene in the recombinant molecule codes. This requires the vector to have particular sequences that support both transcription and translation of the introduced gene.

A number of pharmaceutical products have now been manufactured using transgenic bacteria. The first successful example was the production of insulin. **Figure 7.17** shows the basic steps that are involved in producing human insulin in bacteria. Human insulin is composed of two polypeptide chains, A and B. These chains are produced in separate batches of bacteria. The expression vectors contain the DNA sequence for the gene for the A chain or the B chain, as well as the gene that codes for a bacterial β-galactosidase enzyme. This results in the production of large amounts of *fusion proteins* comprised of the β-galactosidase enzyme fused to the A chain or B chain. The presence of the β-galactosidase enzyme protects the insulin chains from being degraded by the bacteria, as they would be if expressed on their own. During the protein purification procedure, the β-galactosidase part of the fusion protein is removed and the A and B chains are purified. They are then mixed together under conditions that allow them to properly fold and associate to form a functional human insulin molecule.

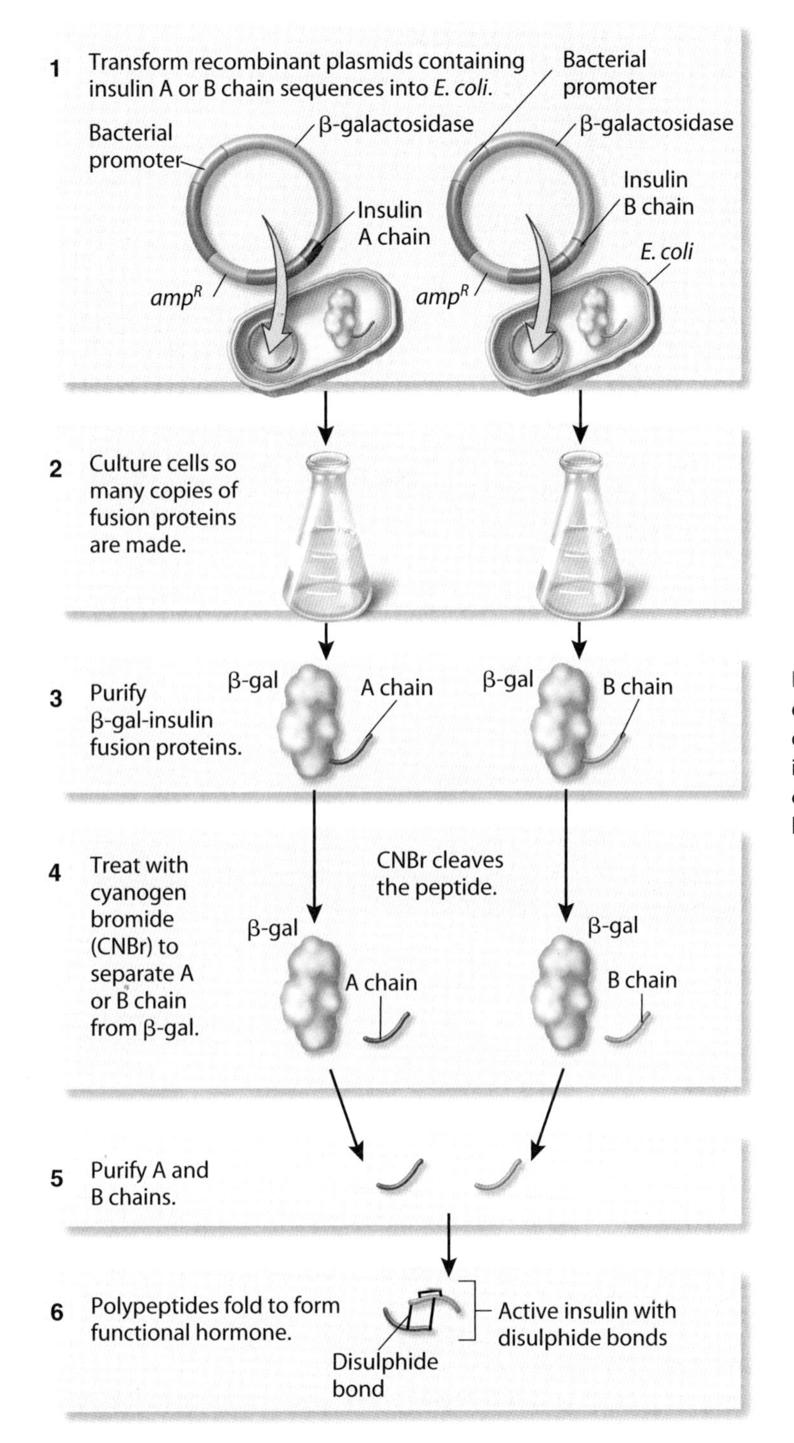
1 Transform recombinant plasmids containing insulin A or B chain sequences into *E. coli*.
Bacterial promoter
β-galactosidase
Insulin A chain
amp^R
Bacterial promoter
β-galactosidase
Insulin B chain
E. coli
amp^R
2 Culture cells so many copies of fusion proteins are made.
3 Purify β-gal-insulin fusion proteins.
β-gal
A chain
β-gal
B chain
4 Treat with cyanogen bromide (CNBr) to separate A or B chain from β-gal.
CNBr cleaves the peptide.
β-gal
A chain
β-gal
B chain
5 Purify A and B chains.
6 Polypeptides fold to form functional hormone.
Active insulin with disulphide bonds
Disulphide bond

Figure 7.17 Bacteria can be engineered to express large quantities of a protein, such as human insulin. For industrial-scale production, large vats called *bioreactors* are used to grow the bacteria.

In 1983, Health Canada approved the sale of human insulin produced by transgenic bacteria. Individuals with insulin-dependent diabetes do not produce sufficient quantities of insulin due to a defect in the beta cells of the pancreas. As a result, they are reliant on injections of insulin. Prior to 1983, these individuals had to use purified insulin from animal pancreas sources in which the hormone is very similar to the human form. This method of isolating insulin is labour-intensive and very expensive. In addition, many people developed allergic reactions to these animal forms of insulin. Using bacteria to produce human insulin has eliminated many of the allergic side effects and has drastically reduced the cost of production. There are now many other successful examples of medicinal proteins that have been produced in bacteria. These include human growth hormone, tissue plasminogen activator (used to treat blood clots), erythropoietin (used to stimulate red blood cell production), and a hepatitis B vaccine.

Suggested **Investigation**

ThoughtLab Investigation 7-A, Constructing the First Genetically Engineered Cells

bioremediation the use of micro-organisms or other living cells for environmental clean-up

Transgenic Bacteria and Bioremediation

Transgenic bacteria have been developed for uses other than medical applications. **Bioremediation** involves reducing environmental pollutants using micro-organisms. Some micro-organisms can convert environmental toxins into non-toxic products. Genetic engineering has made some of these strains even better at this process. One of the earliest examples of such an improvement involved a strain of bacteria that could naturally break down crude oil (petroleum). The bacteria were genetically modified to accomplish this breakdown even more effectively. This particular example is also well known because it is associated with a landmark American Supreme Court case, *Diamond vs. Chakrabarty*. Indian-American microbiologist Ananda Chakrabarty, shown in **Figure 7.18**, developed these transgenic bacteria. He applied to patent the micro-organisms, arguing that they should be considered an invention. In 1980, the court ruled for the first time that a micro-organism could be patented. Thus, this transgenic bacterium became the first recombinant organism to be patented. Although this particular strain did not prove to be very effective at larger-scale oil clean-ups, other transgenic bacteria have been developed and shown to be useful for various environmental applications. For example, some bacteria break down pesticides and herbicides that have been released into water systems. Others were developed to remove sulfur from coal to produce cleaner emissions when the coal is burned.

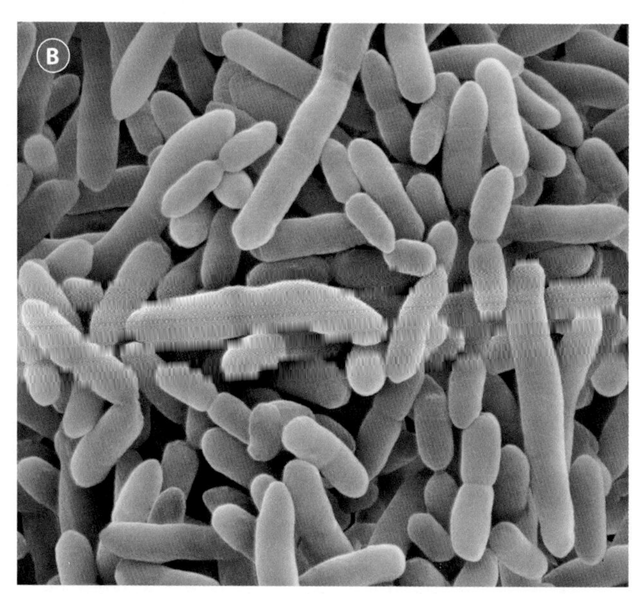

Figure 7.18 In 1971, while working as a scientist in the Research and Development division of General Electric in New York, Dr. Ananda Chakrabarty **(A)** developed a transgenic bacterium **(B)** that could break down oil faster than any naturally occurring bacteria. Dr. Chakrabarty applied for a patent for the transgenic bacteria. Although the application was initially rejected, the patent was eventually granted in 1980.

Learning Check

13. What is genetic engineering? Use an example in your answer.

14. What are the risks and benefits of being able to patent a GMO?

15. How do expression vectors allow transgenic bacteria to become "production factories" for medicines?

16. What are the advantages of using transgenic bacteria to produce human insulin?

17. Explain how transgenic bacteria can be useful for bioremediation.

18. What do you think the possible positive and negative ecological impacts are of using transgenic bacteria for environmental applications?

Transgenic Plants

The manipulation of plants for human benefit has been occurring for thousands of years, since the beginning of agricultural practices. Historically, it has involved the slow and meticulous process of crossbreeding different varieties to produce plants with desired traits. The development of recombinant DNA technology has not only reduced the work and time required to develop new varieties of plants, but it has also increased the scope of potential biotechnological applications.

Agricultural plants are one of the most prominent examples of GMOs. Examples of transgenic crops that have been developed and approved for human consumption include soybeans, corn, canola, tomatoes, and potatoes. The majority of genetic modifications have conferred an increased tolerance to herbicides and/or greater resistance to disease and pest infestations. In Canada, the most prevalent transgenic crop plant is canola. Approximately 80 percent of the canola grown in western Canada is now genetically modified to be resistant to Roundup™, a common herbicide used to kill weeds that compete with canola growth. Other GMO crops grown in Canada include those mentioned above, as well as squash, sugar beets, and flax. Benefits of genetic engineering of crops include increased crop yield, reduction in harvesting cost and pesticide use, and enhanced nutritional value and food quality due to slower spoilage.

Plants are *totipotent*, which means that one cell can grow and divide to produce all the different types of cells in a plant. Since an entire plant can be generated from a somatic cell, a gene can be introduced into somatic tissue such as a plant leaf to produce a transgenic plant. Once the leaf cells become transgenic, the leaf can be treated with plant growth hormones, forming roots and shoots and eventually a complete transgenic plant.

The two major techniques for introducing foreign DNA into plants cells to produce transgenic plants are the **biolistic method** and Ti plasmid transformation. The biolistic method is often called the *gene-gun method* because it involves striking plant cells with tiny particles of gold or platinum that are coated with DNA. This bombardment occurs at a very high speed, which allows the DNA to penetrate the cell wall of plant cells. The once-popular method lacks control with regard to the insertion site of the gene and how many copies are introduced. Many biotechnologists argue against using this method because the introduced gene may insert into a functional gene, potentially altering that gene's function.

The **Ti plasmid method**, outlined in **Figure 7.19** on the following page, uses the same principles as the production of transgenic bacteria. The Ti plasmid, or tumour-inducing plasmid, occurs naturally in the bacterium *Agrobacterium tumefaciens*. These bacteria naturally infect plant cells and cause the formation of a bulbous growth on the plant. Part of the Ti plasmid, called the *T-DNA*, integrates into the plant genome and causes the uncontrolled cell growth that results in a tumour. Researchers have altered the T-DNA of the Ti plasmid so that it no longer causes tumour formation, but still allows for integration of DNA into the plant genome.

biolistic method a method used for producing transgenic plants that involves bombarding plant cells with particles coated in DNA that become integrated into the plant genome

Ti plasmid method a method of producing transgenic plants that uses the tumour-inducing plasmid from *Agrobacterium tumefaciens* as the vector for the insertion of a foreign gene into a plant genome

The steps in the Ti plasmid method are as follows:

1. A recombinant DNA molecule is produced in which the gene of interest is inserted into the altered T-DNA region of the Ti plasmid. The recombinant DNA has a selectable marker. It provides cells that have taken up the plasmid with resistance to the antibiotic *kanamycin*.
2. The recombinant Ti plasmid is taken up by the bacterium *Agrobacterium tumefaciens*.
3. Plant cells are infected with the bacterium. The recombinant DNA carrying the gene of interest integrates into the plant cells.
4. The selectable marker is used to determine which cells have taken up the recombinant DNA. Those that survive when exposed to the antibiotic kanamycin have taken up the DNA. An antibiotic is also used to kill any cells of *Agrobacterium tumefaciens* so only plant cells remain. A transgenic plant is grown from these cells.

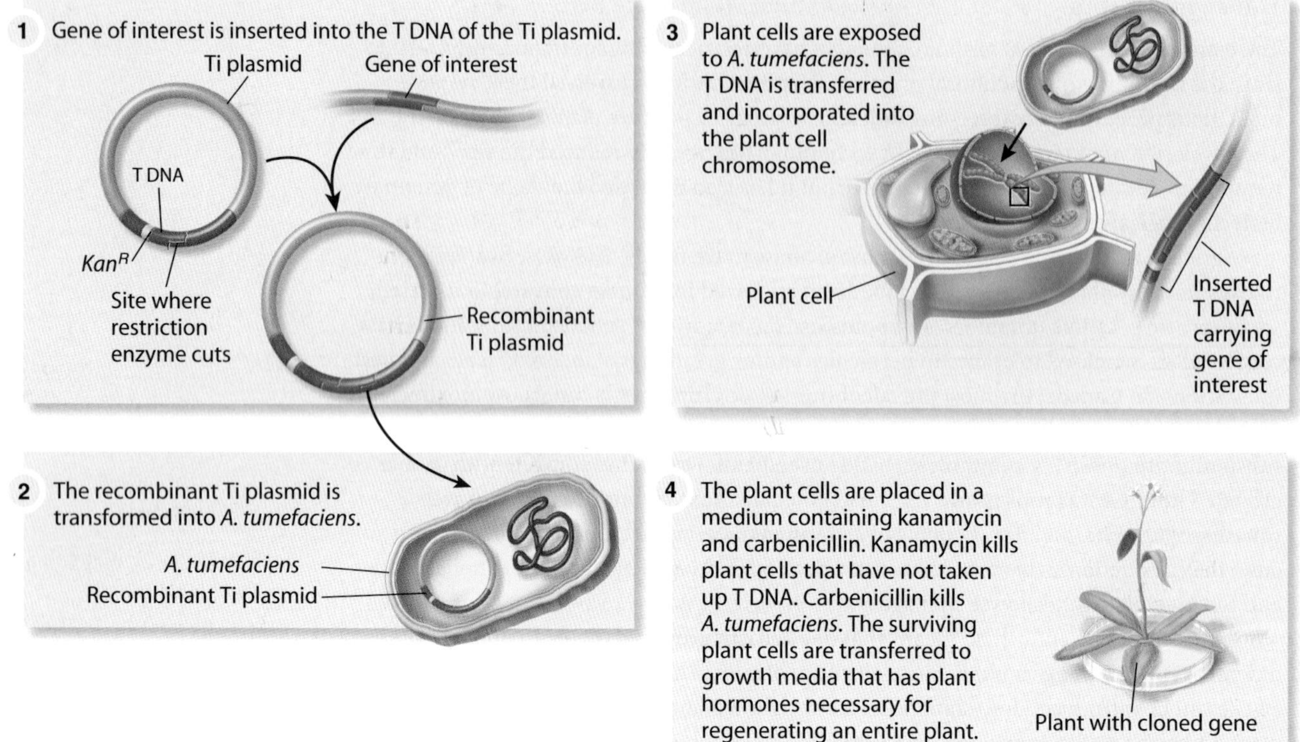
1 Gene of interest is inserted into the T DNA of the Ti plasmid.
Ti plasmid
Gene of interest
T DNA
Kan^R
Site where restriction enzyme cuts
Recombinant Ti plasmid
2 The recombinant Ti plasmid is transformed into A. tumefaciens.
A. tumefaciens
Recombinant Ti plasmid
3 Plant cells are exposed to A. tumefaciens. The T DNA is transferred and incorporated into the plant cell chromosome.
Plant cell
Inserted T DNA carrying gene of interest
4 The plant cells are placed in a medium containing kanamycin and carbenicillin. Kanamycin kills plant cells that have not taken up T DNA. Carbenicillin kills A. tumefaciens. The surviving plant cells are transferred to growth media that has plant hormones necessary for regenerating an entire plant.
Plant with cloned gene

Figure 7.19 The same basic principles that are used to make transgenic bacteria can be applied to plants. In the case of transgenic plants, the vector used to carry the new gene is the Ti plasmid from *Agrobacterium tumefaciens*.

In addition to producing transgenic crops, plants are also being engineered to produce medicinal products. Since the mid-1980s, products such as human growth hormone, clotting factors, and antibodies have been produced in transgenic plants. For instance, one type of antibody made by corn can deliver radioisotopes to tumour cells. Another made by soybeans can be used to treat genital herpes.

The Controversy Surrounding Transgenic Plants

There is a great deal of controversy concerning the growth and consumption of transgenic plants. Much of the controversy is focused on two main questions:

1. Are genetically modified crops safe for human consumption?

Safety of GMO crops for human consumption is being continually monitored. In Canada, each new crop requires regulatory approval. Plants and all other genetically modified organisms intended for human consumption undergo seven to 10 years of health-and-safety research by Health Canada before they can be consumed by Canadians. This research includes a scientific evaluation of how the organism was developed, including how it was genetically altered, and a comparison of the nutritional content of the genetically modified food and its non-modified counterpart. Food safety issues are considered, such as potential for introducing new toxins, allergens, or other secondary effects. To date, adverse reactions to GMO crops have included some reports of increased allergic reactions. No other adverse responses have been observed since GMO crops were introduced into the marketplace. However, some people also argue that sufficient time has not yet passed in which to properly assess any potential long-term effects.

2. Will genetically modified crops have a negative impact on the environment?

One concern regarding possible environmental impacts is that a gene may undergo gene transfer after it has been introduced into a crop. There are two types of potential gene transfer. In *horizontal gene transfer*, an introduced trait—such as resistance to an antibiotic or a pesticide—is transferred to other organisms. These organisms could be other plants and animals, or even bacteria or fungi. Although the possibility of horizontal gene transfer exists, such transfer occurs at a very low rate in nature. Also a concern, *vertical gene transfer* involves the transfer of the gene or trait into the genomes of the natural or wild versions of the same crop. Although preliminary studies done in Mexico have suggested the possibility that this has occurred, other studies show no such evidence. Nonetheless, possible vertical gene transfer continues to be monitored.

A second environmental concern involves the potential harm caused by transgenic crops that express a toxin as a form of insecticidal protection. Could this toxin affect non-target organisms? This concern has been voiced for a common GMO, Bt corn. This form of transgenic corn contains a gene for a protein from a type of bacterium called *Bacillus thuringiensis*. This protein is toxic to many insects. Introduction of the gene into the corn genome provides the transgenic corn with protection from these insects. Thus, Bt crops reduce pesticide use and costs associated with farming the crop. However, controversy ensued for many years with opponents arguing that there were inadvertent negative effects on the monarch butterfly population due to exposure of monarch larvae to pollen from the Bt corn. The numerous studies arising from this controversy concluded that the effect on monarch butterflies was minimal. The fact that insects may become resistant to the Bt protein and related microbial insecticides is also an issue. However, regulations are in place to deal with this possibility. For instance, in Canada, farmers who grow Bt corn are also required to plant a certain amount of non-Bt corn. The presence of the non-Bt corn reduces the likelihood that insects will become resistant to the Bt protein.

Suggested Investigation

ThoughtLab Investigation 7-B, Regulation and Ownership of Genetically Modified Organisms

Learning Check

19. What are the benefits of transgenic crop plants?

20. Explain how the biolistic method introduces foreign DNA into plant cells.

21. Use a flow chart to describe how the Ti plasmid method is used to produce a transgenic plant.

22. Why is gene transfer involving transgenic plants an environmental concern?

23. What are the Canadian government regulations for GMOs that are intended for human consumption? Despite these regulations, why are there still concerns about safety?

24. In what ways could a transgenic plant that has been engineered to secrete an insecticide have an impact on its environment?

Transgenic Animals and Related Controversies

The production of transgenic animals is a much more complex process than the production of transgenic plants. Not surprisingly, it is also highly controversial. To produce transgenic animals, a foreign gene is inserted into the genome of an animal oocyte (egg) that is then fertilized. The fertilized egg is implanted in a host female and allowed to develop. The resulting offspring are the transgenic form of the animal. The procedure has been used to produce transgenic fish, pigs, cows, rabbits, and sheep. Genetically engineering animals with "improved" traits has seen recent success in Canada. At the University of Guelph, pigs were genetically engineered so they can more efficiently break down phosphate in their feed. This means the pigs excrete less phosphate, which reduces phosphate contamination of water sources by runoff from pig farms. As discussed in more detail in the Case Study that follows, these pigs, as well as an over-sized transgenic form of salmon, are now being considered for human consumption of their meat.

gene pharming the process of using transgenic livestock to produce therapeutic proteins (pharmaceuticals)

As with bacteria and plants, transgenic animals are also being produced for medically important human proteins. The use of transgenic animals to produce human therapeutic proteins has been called **gene pharming**, or *molecular pharming*, in reference to the use of farm animals to make pharmaceuticals. As shown in **Figure 7.20**, the approach involves protein production in the mammary glands of the transgenic animal. The human protein is excreted into the animal's milk, the milk is collected, and the protein is purified from it. A foreign protein is targeted to mammary cells through the use of a β-lactoglobulin promoter in the recombinant DNA containing the gene that encodes the protein. β-lactoglobulin is a protein expressed in mammary cells of many animals, such as cows and sheep. Recall from Chapter 6 that promoter regions of genes are sites of regulation of gene expression, which can influence in what type of cell a gene is expressed. The β-lactoglobulin promoter is only functional in mammary cells. Therefore, any gene sequence that follows the promoter sequence in the recombinant DNA will only be expressed in mammary cells.

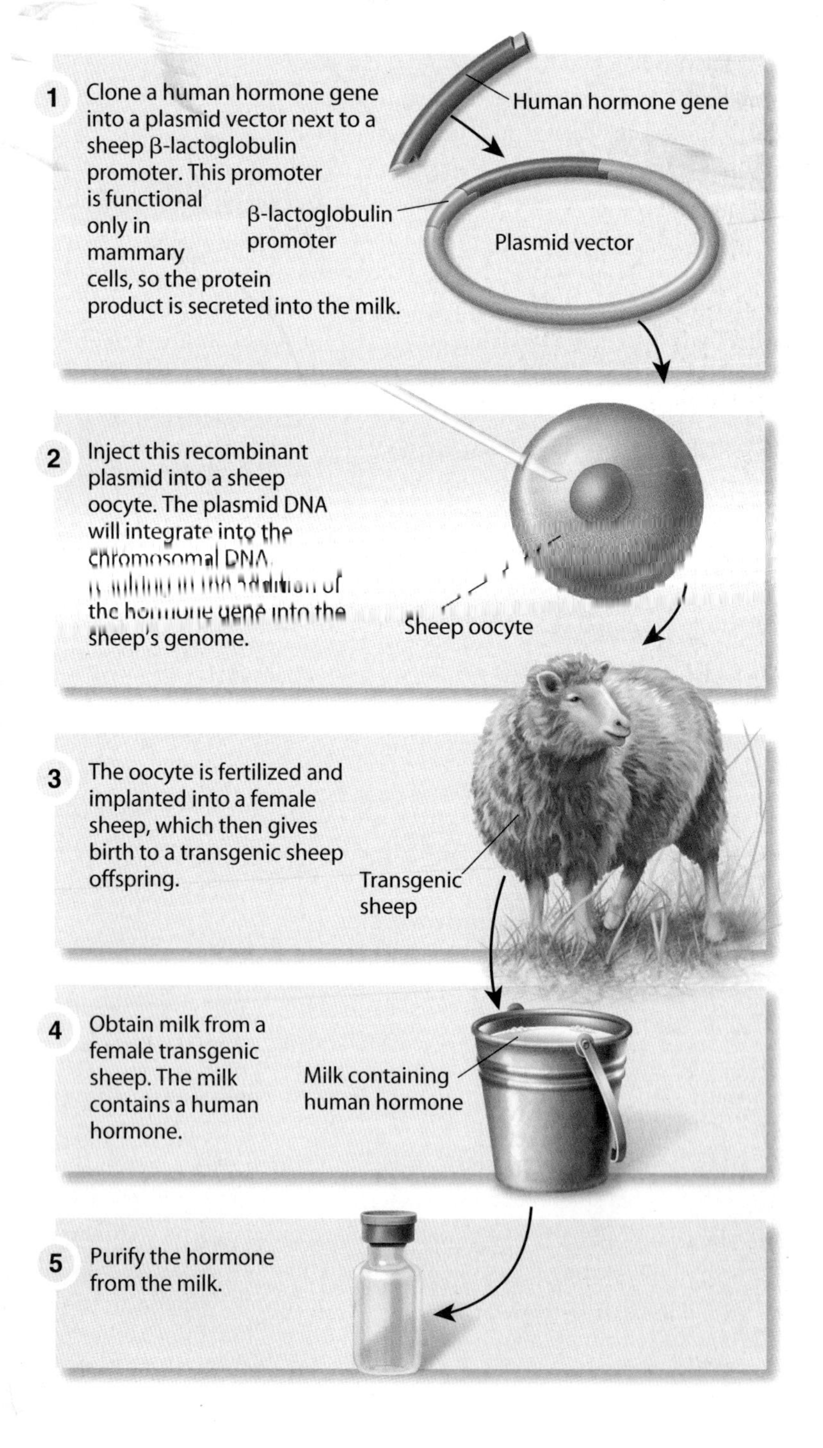
1 Clone a human hormone gene into a plasmid vector next to a sheep β-lactoglobulin promoter. This promoter is functional only in mammary cells, so the protein product is secreted into the milk.
Human hormone gene
β-lactoglobulin promoter
Plasmid vector
2 Inject this recombinant plasmid into a sheep oocyte. The plasmid DNA will integrate into the chromosomal DNA, resulting in the addition of the hormone gene into the sheep's genome.
Sheep oocyte
3 The oocyte is fertilized and implanted into a female sheep, which then gives birth to a transgenic sheep offspring.
Transgenic sheep
4 Obtain milk from a female transgenic sheep. The milk contains a human hormone.
Milk containing human hormone
5 Purify the hormone from the milk.

Figure 7.20 Production of transgenic livestock that secrete a therapeutic human protein involves using a recombinant plasmid vector. This vector contains the gene for the protein as well as a promoter that directs expression of the gene in mammary cells. The recombinant DNA is injected into an oocyte. The oocyte is fertilized and implanted into a host female. The transgenic animals are the offspring produced from that implantation. The milk of female transgenic offspring contains the human protein.

Production of human proteins in livestock such as goats, sheep, and cattle has been highly successful. The production of tissue plasminogen activator Factor IX for treatment of hemophilia (a blood-clotting disorder) and α-1-antitrypsin for treatment of emphysema (a respiratory disorder) are two more successes. Production of therapeutic proteins in transgenic animals is more difficult and expensive than in bacteria. However, proteins are not always produced in their proper form when expressed in bacteria. For example, many human proteins undergo different modifications after they are expressed in human cells which are essential for their function. These modifications cannot occur in bacterial cells in the same way they do in mammalian cells. Also, many foreign proteins are rapidly degraded when expressed in bacteria. For those proteins that do not express well in bacterial systems, gene pharming represents an alternative.

Mammalian Cloning from Somatic Cells

Cloning mammals is another way of genetically modifying organisms. Of course, whole organism cloning does occur in nature. Identical twins are examples of natural human clones—they have the same genome and came from the same fertilized egg. The basic procedure used to clone a mammal is outlined in **Figure 7.21**.

This procedure was first successful with the cloning of Dolly the sheep in 1996 by Ian Wilmut and his colleagues at the Roslin Institute in Edinburgh, Scotland. Wilmut and Dolly are shown together in **Figure 7.22**. Since that time, other species such as cows, pigs, mice, dogs, and cats have been cloned. In 2002, the first pet was cloned, shown in **Figure 7.23** on the following page.

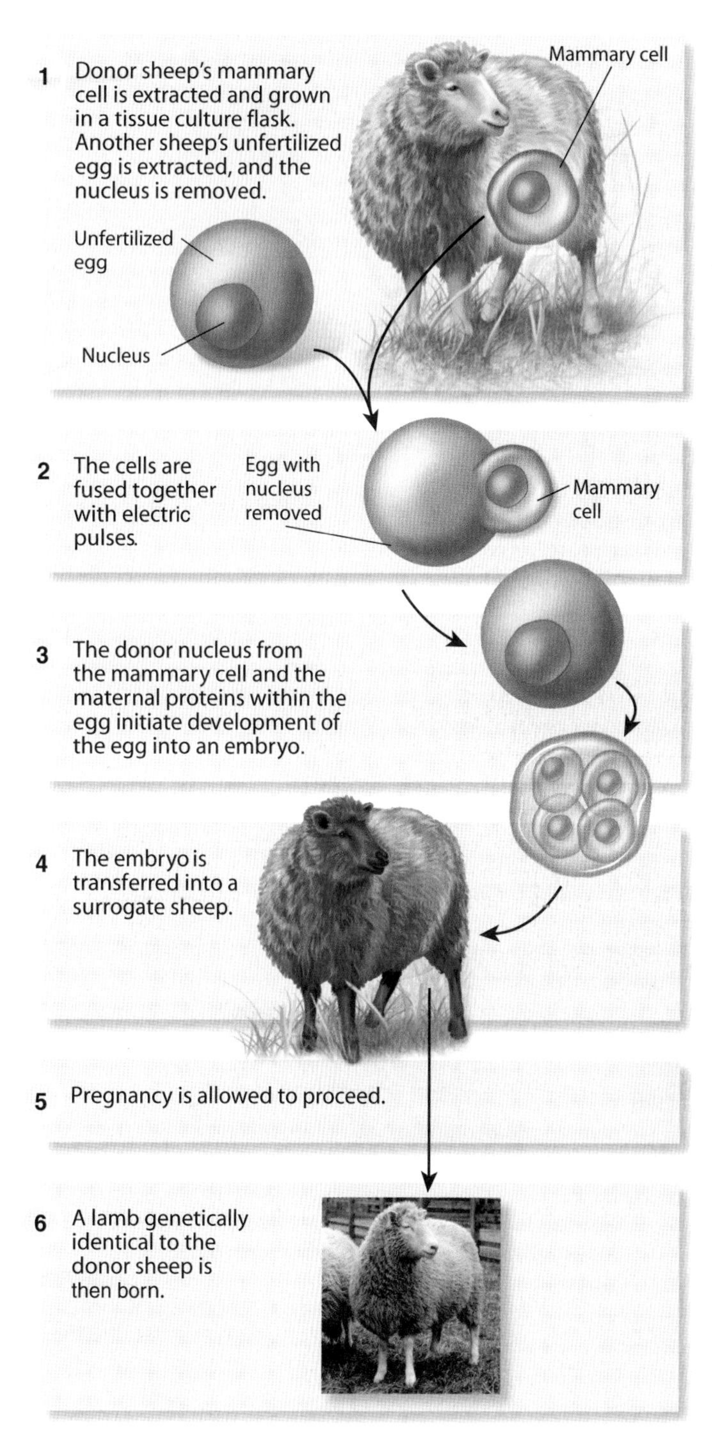

1 Donor sheep's mammary cell is extracted and grown in a tissue culture flask. Another sheep's unfertilized egg is extracted, and the nucleus is removed.
Mammary cell
Unfertilized egg
Nucleus
2 The cells are fused together with electric pulses.
Egg with nucleus removed
Mammary cell
3 The donor nucleus from the mammary cell and the maternal proteins within the egg initiate development of the egg into an embryo.
4 The embryo is transferred into a surrogate sheep.
5 Pregnancy is allowed to proceed.
6 A lamb genetically identical to the donor sheep is then born.

Figure 7.21 In mammalian cloning, the genetic material from a somatic cell is used. The somatic cell is fused with an oocyte that has had the nucleus removed. The resulting embryo is implanted in a host. The offspring is (nearly) identical to the animal that was the source of the somatic cells.

Figure 7.22 Scientist Ian Wilmut and his team successfully cloned Dolly the sheep in 1996.

Figure 7.23 The first pet to be cloned was a cat named CC (for carbon copy).

Despite these advances, mammalian cloning is still considered to be in its preliminary stages. While cloning of livestock and pets for the benefit of humans is controversial, its applications are still being pursued. Human cloning, however, remains a seriously debated issue involving numerous viewpoints. These opinions range from the idea that human cloning is morally wrong and threatens family life as we know it to the idea that it may be a viable reproduction alternative for infertile couples.

Activity 7.2 Human Cloning

Cats, dogs, sheep, and pigs are among the mammals that scientists have successfully cloned. Some companies catering to pet owners even offer cloning services so that the traits of a beloved dog or cat can live on in future generations of clones. The possibility of human cloning may seem just as enticing. A cloned child could donate blood to a sibling suffering from leukemia. A couple who cannot conceive might gain the chance to have babies. What is Canada's law on human cloning? Should this law be amended?

Materials

- reference books/computer with Internet access

Procedure

1. Research the differences between therapeutic cloning and reproductive cloning. Be sure to keep track of all of your sources.
2. Find out what laws govern human cloning in Canada and what that law states.
3. Research some of the arguments in support of and against human cloning.

Questions

1. Which argument or arguments for or against human cloning do you find most persuasive?
2. Are there any forms of human cloning that you think should be permitted despite arguments against them? Explain.
3. Do you think Canadian laws governing human cloning might have to be amended in future? Explain.

Section 7.2 Review

Section Summary

- Transgenic, or genetically modified, bacteria can be used to produce pharmaceutical products such as insulin. GM bacteria are also used in bioremediation of environmental pollutants.
- Transgenic crops have increased tolerance to herbicides and increased resistance to disease and pests, and many have been approved for human consumption.
- Transgenic animals can be used to produce pharmaceutical products such as human proteins, and some, such as pigs and salmon, are being considered for human consumption.
- The production and use of transgenic plants and animals are controversial, in terms of the safety of human consumption and the possibility of negative environmental effects. Health Canada requires that all GMOs intended for human consumption undergo seven to 10 years of health-and-safety research.
- Mammals can be cloned by fusing a somatic cell, including its genetic material, with an oocyte from which the nucleus has been removed, and then fertilizing the resulting embryo and implanting it in a host. Although sheep, cows, and other mammals have been cloned, human cloning remains highly controversial.

Review Questions

1. **A** Is gene therapy a viable method for treating most diseases? Explain your answer.
2. **K/U** Distinguish between the following terms: *transgenic* and *genetically modified organism* (GMO).
3. **A** What do you think are the benefits and risks of researchers' ability to patent organisms? Justify your response.
4. **K/U** What is an expression vector and why is it an important biotechnology tool?
5. **C** Using labelled diagrams, describe how human insulin is produced in bacterial cells. What are the advantages of producing insulin in bacteria?
6. **K/U** What is bioremediation? List and discuss the potential risks and benefits of bioremediation.
7. **K/U** List and describe two techniques that are used to produce transgenic plants.
8. **K/U** **a.** What is gene pharming? Use an example to describe this process.
   b. Identify and describe two benefits that are provided through gene pharming.
   c. Identify and describe two risks that could be associated with gene pharming. (You may need to do research to answer this question.)
   d. Identify and describe two ethical or philosophical positions that object to gene pharming. Do research to answer this question.
   e. Is there any way to find a consensus of opinion on gene pharming? Give reasons to justify your point of view on this question.
9. **A** An aquaculture research corporation based in Prince Edward Island has created a variety of transgenic salmon that grows 10 times faster than wild salmon.
   a. What are the potential advantages of this transgenic fish?
   b. What are the potential risks associated with this fish?
   c. What regulations would you recommend to govern the use of this fish in commercial fish-farming operations?
10. **T/I** You are a member of a genetics research team searching for a cure for male pattern baldness. A member of your research team hands you a culture of bacteria and says, "These bacteria are supposed to express human keratin (the main protein component of human hair) but, so far, they have not expressed any."
    a. Make a list of the possible problems that could be preventing the bacteria from expressing the human protein.
    b. What steps could you take to identify the problem?
    c. What steps could you take to correct the problem?
11. **C** Using a flow chart, describe the cloning of Dolly the sheep.
12. **A** Some wildlife conservation organizations claim that bioengineering and cloning can be used to save endangered species. In addition, they suggest that these technologies could be used to bring back extinct species, using DNA from preserved specimens in museums. Discuss some implications of carrying out these procedures.

ThoughtLab INVESTIGATION

7-A

Skill Check

- Initiating and Planning
- Performing and Recording
- ✓ Analyzing and Interpreting
- ✓ Communicating

Materials

- computer with Internet access

Constructing the First Genetically Engineered Cells

Together, the discovery of restriction endonucleases, research on bacterial plasmids, and the development of gel electrophoresis paved the way for genetic engineering. In 1973, American scientists Stanley Cohen and Herbert Boyer reported that their research team had produced living, replicating bacterial cells that contained recombinant plasmids. These bacterial strains were known as *transformants*. A few months later, the researchers reported that they had developed *Escherichia coli* strains that could transcribe genes from a frog (*Xenopus laevis*).

In this investigation, you will follow some of the key steps taken by Cohen and Boyer to make and select genetically engineered bacterial cells and to confirm their results.

Pre-Lab Questions

1. What do restriction endonucleases do to double-stranded DNA?
2. What is meant by a "sticky end" on a DNA fragment?
3. When gel electrophoresis is conducted, which DNA molecules will move the fastest through an agarose gel?
4. In order to view the DNA in an agarose gel, why is it necessary to add ethidium bromide?

Question

How are genetically engineered bacterial cells made, selected, and identified?

Organize the Data

Go to Organizing Data in a Table in Appendix A for help with designing tables for data.

Part I: Constructing a New Plasmid for *E. coli*

1. Study the diagram below, which shows the two plasmids that Cohen and Boyer's research team began with. Each plasmid contained a gene for resistance to a different antibiotic: pSC101 carried resistance to tetracycline and pSC102 carried resistance to kanamycin. The researchers treated both plasmids with the restriction endonuclease *Eco*RI. They then used DNA ligase to join the fragments of the plasmids together in order to make a new plasmid, pSC105. *E. coli* transformants carrying pSC105 were resistant to both tetracycline and kanamycin.

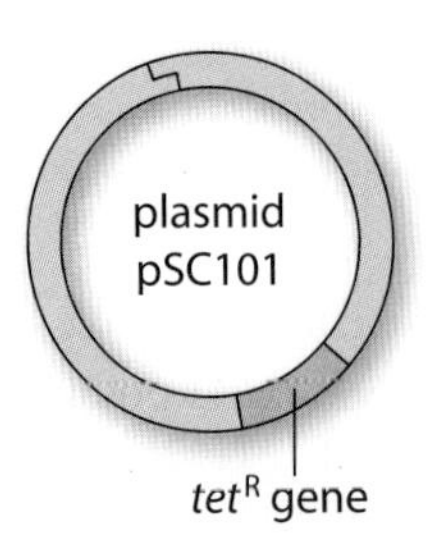

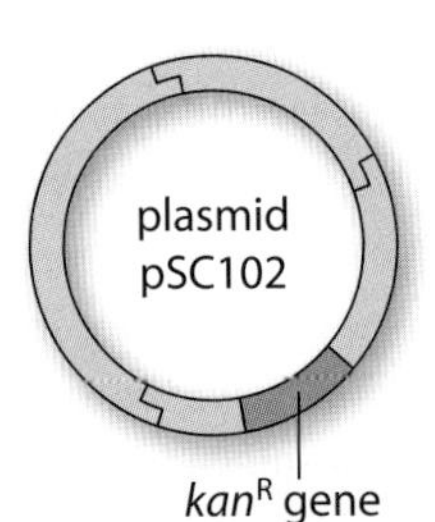

2. Examine the diagram on the left, which shows the results of gel electrophoresis of the plasmid DNA.

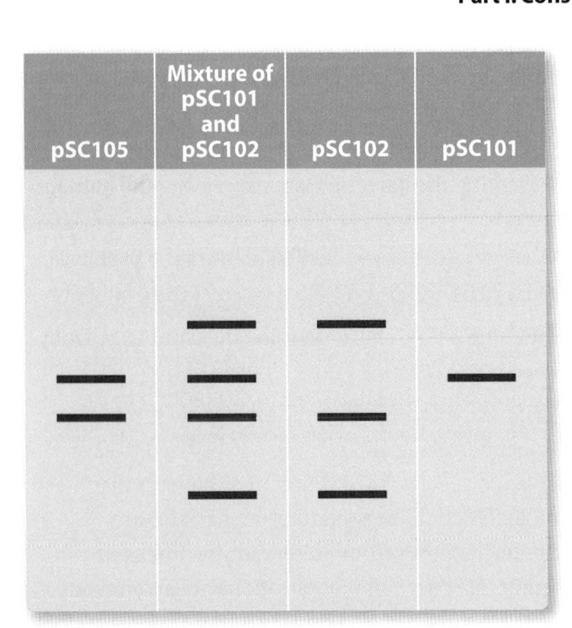

Agarose gel electrophoresis of *Eco*RI digests of plasmids

3. Create a table stating the number of bands in each lane and the relative sizes of the DNA fragments.

Part II: Cloning Eukaryotic Genes in *E. coli*

4. The researchers treated *X. laevis* DNA and pSC101 from *E. coli* with *Eco*RI. They then used DNA ligase to join the *X. laevis* DNA with pSC101 to construct various recombinant plasmids as shown below. The plasmids from the different *E. coli* transformants were numbered CD1 and so on.

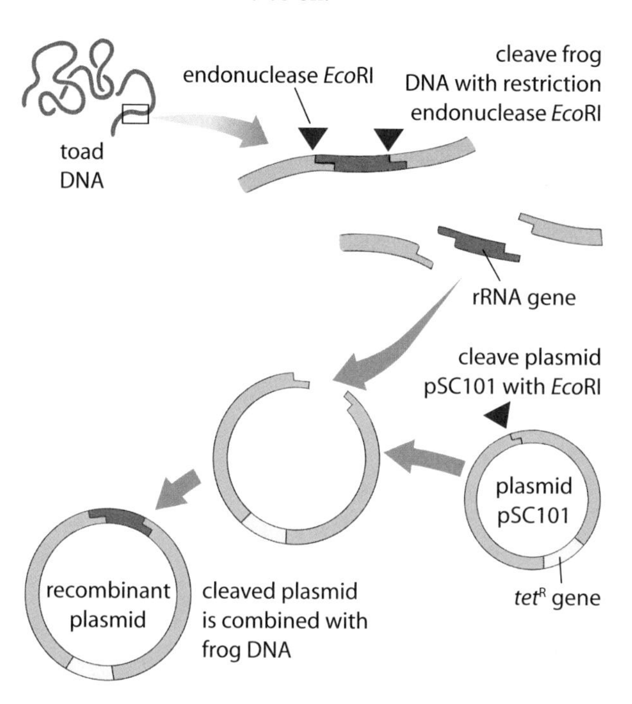

5. Examine the diagram showing some of the results of gel electrophoresis of *X. laevis* and plasmid DNA. Create a table stating the number of bands in each lane and the relative sizes of the DNA fragments.

X. laevis | CD42 | CD18 | pSC101

Agarose gel electrophoresis of *Eco*RI digests of DNA from *X. laevis* and plasmids

Analyze and Interpret

Part I

1. a) How many *Eco*RI restriction sites did pSC101 contain? How do you know? (Hint: Draw a diagram of pSC101 with any *Eco*RI restriction sites.)

b) How many *Eco*RI restriction sites did pSC102 contain? How do you know?

2. Why did the researchers select transformants that could grow on a medium containing both tetracycline and kanamycin?

Part II

3. Describe any similarities in fragment sizes among the recombinant plasmids CD42 and CD18, pSC101, and *X. laevis* DNA.

4. Were transformants carrying plasmids CD42 or CD18 able to grow on a medium containing tetracycline? Explain.

Conclude and Communicate

Part I

5. Which fragments from the other plasmids did pSC105 contain? Explain.

6. Which fragment from pSC102 contained the kanamycin resistance gene? How do you know?

7. How did the researchers know that the *Eco*RI restriction site in pSC101 did not occur within the gene for tetracycline resistance?

Part II

8. Did the plasmids CD42 and CD18 contain *X. laevis* DNA? How do you know?

Extend Further

9. INQUIRY Design a possible procedure for constructing recombinant plasmids containing *lacZ*, *X. laevis* DNA, and genes for resistance to tetracycline and kanamycin. Explain how you could use the selectable markers to select the transformants that you want.

10. RESEARCH Stanley Cohen and a business partner founded the biotechnology company Genentech in 1976. What were the first two products that Genentech made and how did they do it?

ThoughtLab INVESTIGATION

7-B

Skill Check

- Initiating and Planning
- Performing and Recording
- ✓ Analyzing and Interpreting
- ✓ Communicating

Materials

- computer with Internet access
- library resources

Regulation and Ownership of Genetically Modified Organisms

When Saskatchewan farmer Percy Schmeiser met the international biotechnology corporation Monsanto in court, the case revolved around Monsanto's right to control how farmers use its products. Monsanto is the developer of Roundup-Ready™ Canola, a genetically-engineered form of canola that is resistant to the herbicide Roundup™ (also produced by Monsanto). This genetically modified plant has helped farmers increase their crop yields.

Farmers who buy seeds from Monsanto must agree not to save any seeds from their crop, but to buy fresh seeds every year. The farmers are not permitted to exchange Monsanto seeds with other farmers. If Roundup-Ready™ Canola appears by accident in their fields, they must remove and destroy the plants. These regulations provide a way for Monsanto to recover its costs and earn a profit from its research and development.

Many people argue that genetic information is a natural resource that belongs to everyone. On the other hand, if companies cannot earn a profit from their research, there is little incentive for them to invest in genetic studies. A patent is a government license that gives someone ownership over an invention. Only the patent holder has the right to make, use, or sell the invention. In the end, the Canadian courts upheld Monsanto's right to patent the Roundup-Ready™ gene and to control the use and distribution of its seeds.

Should private companies be allowed to patent GMOs or genes? What agencies regulate the control of genetic information? In this investigation, you will research some of the arguments made in the *Monsanto vs. Schmeiser* case and similar scenarios. You and your classmates will then take sides in a debate on the ownership and regulation of genetic information.

Pre-Lab Questions

1. List three practices that you should follow when conducting Internet research.
2. What information should you include in your list of references?
3. What are some of the other terms used to mean "genetically modified"?

Question

What are the rights and responsibilities of individuals or companies who have ownership of genetic information?

Procedure

1. Work with the other members of your group to research the following. Be sure to keep a record of your sources.
   - What arguments and evidence did Monsanto give for suing Percy Schmeiser?
   - How did Schmeiser account for the Roundup-Ready™ Canola plants in his fields?
   - Was there a clear winner in the *Monsanto vs. Schmeiser* court case? Explain.
   - According to the Canadian *Patent Act and Patent Rules*, what types of GMOs can and cannot be patented?
   - What role does the Canadian Environmental Protection Act play in regulating biotechnology?
   - What responsibilities does the *Food and Drugs Act* give to Health Canada in regard to GMOs?
   - What is the role and stance of the Canadian Food Inspection Agency with respect to GMOs?
2. Identify and conduct research on another controversy related to the regulation of GM plants or micro-organisms outside Canada. Some possible topics are listed here:
   - Canadian exports of flax and GM flax to Europe
   - GM papaya grown in Hawaii
   - GM corn grown in Mexico
3. As a class, decide on a resolution to debate that addresses the investigation's overall question. Your group will defend one side of the debate, as assigned by your teacher.

Analyze and Interpret

1. On the basis of the research you conducted, brainstorm possible points for and against the resolution. Be sure to consider legal, ethical, social, environmental, and health implications of the resolution. Refer to the specific scenarios that you investigated.

Conclude and Communicate

2. Engage your opponents in a debate. You should be prepared to do the following:
   - Define the resolution or accept or reject the definition.
   - Present your arguments.
   - Rebut your opponents' arguments.
   - Give a concluding statement.
3. Given what you have heard, reconsider your position on the debate. Decide which arguments you find most compelling.
4. Devise a list of guidelines for assessing whether a particular GM product should be approved for use. Decide what regulations should be placed on its use. Justify your response.

Extend Further

5. **INQUIRY** Design a protocol for testing the safety of a new GMO. In your study consider environmental effects as well as health effects.
6. **RESEARCH** The use of GMOs in medicine is becoming increasingly common. GM plants, for example, can be used as protein factories. Research a specific medical application of a GMO that has not already been discussed in this chapter. Identify the key pros and cons of this application.

STSE

Case Study

Genetically Modified Animals

Do the benefits outweigh the possible risks?

Scenario

Genetically modified plants, such as canola that is resistant to a certain type of herbicide, were first introduced to the world marketplace in the 1990s. Although there has been controversy regarding their health and environmental safety, they remain a significant part of the food industry.

Since then, two new genetically modified (GM) organisms have been hot topics of discussion among those in the biotech field and food safety advocates—salmon that has been genetically modified to grow twice as fast as its non-GM relatives and GM pigs that have a reduced amount of phosphorus in their excrement. Meat from both of these animals could eventually end up on our dinner plates.

The salmon could be the first genetically modified animal to be approved for human consumption but, like GM plant crops, this issue is not without controversy. In this case study, you will read a pamphlet from a non-profit group that advocates for the use of GM animals for human consumption and a web page from a non-profit group that argues against their use. You will then use your research and critical thinking skills to form your opinion on this issue.

Important Benefits of Genetically Modified Animals

Our organization feels that genetic modification of animals presents society with exciting new possibilities, including fast-growing fish and environmentally friendly pigs. Humans can benefit from the use and consumption of these animals in several ways.

Benefits to Society

- **Increased Food Security** Scientists estimate that by 2050, the world population will be 9 billion. As the world continues to grow, many people are concerned with food security (ensuring that all people, at all times, have access to safe and nutritious food to meet their dietary needs and preferences). The AquAdvantage® Salmon (AAS), produced by AquaBounty Technologies, Inc., is a farm-raised fish that has been modified using a gene from a Chinook salmon that accelerates the growth rate of the fish. AquaBounty's salmon grows twice as fast as conventional farm-raised salmon, providing a much needed and popular protein source for people worldwide. Although not officially approved as of April 2011, the United States Food and Drug Administration (USFDA) stated in September 2010 that the meat from the GM salmon was biologically equivalent to meat from non-GM Atlantic salmon and that it was safe for humans to eat.

AquaBounty's salmon are genetically modified to grow faster than wild salmon. These two fish, the larger of which is an AquaBounty salmon, are the same age. The smaller fish is the same species, but has not been genetically modified.

Environmental Benefits

- **Reduced Pollution** The genetically modified "Enviropig™", developed by scientists at the University of Guelph, is able to use an indigestible form of phosphorus found in the grains it is fed. As a result, its urine and fecal matter contain 30 to 65 percent less phosphorus than that of non-GM pigs. This is significant because pig waste material is a source of phosphorus pollution in the environment. Excess phosphorus contained in runoff is used as a nutrient by algae in lakes and oceans. Large algal blooms occur and when the algae die, bacteria decompose the organic material, using oxygen in the process. As a result, the water can become so depleted of oxygen, fish and other organisms cannot survive. Scientists have identified large areas of oceans, called *dead zones* that have been repeatedly affected by this cascade of events.

Although these Enviropigs™ look the same as non-GM pigs, their excrement contains less phosphorus—a substance that pollutes aquatic ecosystems.

In February 2010, Environment Canada announced that the University of Guelph could begin producing the pigs under approved containment conditions, having met the requirements of the Canadian Environmental Protection Act. The University has submitted applications to Health Canada and American regulatory agencies to assess the safety of the pigs for human consumption.

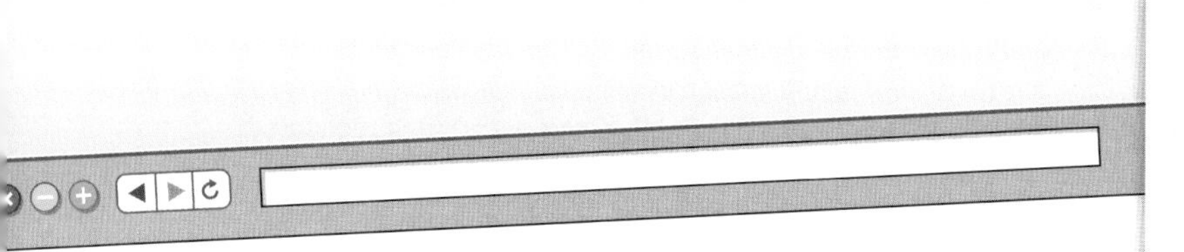

Risks of Genetically Modified Animals

Although it would seem that there are many benefits to the use of genetically modified animals in a commercial agricultural or aquaculture setting, our organization is not quite ready to jump on the bandwagon. We feel the following points need to be carefully considered before any genetically modified animals are approved for human consumption.

Possible Health Risks

- GM foods could contain allergens. For example, about 1 percent of the world's population has a severe allergy to peanuts, tree nuts, or both. If a gene from a nut is incorporated into a GM organism, could that cause an allergic reaction upon consumption? At least one investigation has shown that a gene from a Brazil nut inserted into soybeans caused allergic reactions in humans. As a result, this particular product was never marketed. However, the scenario shows the possible risks of allergens being introduced into plants or animals through genetic modification.

- Antibiotic-resistant genes are used in the process of creating a genetically modified organism. Some scientists are concerned that the antibiotic-resistant genes in genetically modified food products could be conferred to the bacteria that naturally live in the human digestive system.

Environmental Risks

- Some scientists are concerned about the possibility of GM animals escaping from a containment area, such as a farm or a large aquaculture pen along the coast. The potential consequences of GM organisms mixing with wild populations include the GM organisms out-competing the wild organisms for food or changing the habitat in a way that would negatively affect the wild population.

- Scientists are also asking what might happen if a GM animal, such as a salmon, escapes and eventually reproduces with the wild population. The incorporated genes would then enter the gene pool of the wild population and might no longer be controllable.* Since the introduction of crops that are genetically modified to resist certain herbicides, there have been cases in which the GM plants have bred with nearby wild populations of weeds. This confers the herbicide resistance to the weed relative, creating what scientists have dubbed a superweed.

* Editor's Note: The creators of the AquaBounty salmon have responded to this concern by explaining that the GM salmon they produce are all female and are sterile.

Research and Analyze

1. Research more information about how genetically modified animals used for human consumption would be regulated by government agencies such as Environment Canada and the Canadian Food Inspection Agency. What regulations assure that food products from GM animals would be safe? How would the products be tracked in the food system? Would the food products be labelled? Do the regulations in place make you feel as though you are protected as a consumer? Explain your opinion. If you do not feel protected, what changes do you feel would strengthen the system?
2. Some people who oppose using GM organisms in general do so on the ethical grounds that we are tampering with nature. Research more about this point of view. Do you agree or disagree with this point of view? Justify your position.
3. Research more information about the legal implications of genetically modified animals used for human consumption. What entity, such as a corporation or government agency, would be responsible if people were to experience negative health effects as a result of consuming meat or other products from genetically modified animals? What entity would be responsible for any large-scale environmental effects if the GM animals were to mix with a wild population? Are current laws equipped to deal with advances in scientific knowledge and technology in this area? Summarize the results of your research in an essay.

Take Action

1. **PLAN** Suppose you are going to attend a rally about the issue of genetically modified animals for human consumption. Use information in this case study, your research from the previous questions, and any additional research needed, to write a speech that you could give at the rally either for or against the use of these animals for human consumption.
2. **ACTION** Summarize five key points from your speech. Compose an email to your Member of Parliament, stating your position on this issue.

Chapter 7 SUMMARY

Section 7.1 Techniques for Producing and Analyzing DNA

Due to technological developments, researchers are able to combine fragments of DNA from different sources to produce new DNA molecules, amplify and analyze tiny samples of DNA, identify the nucleotide sequence in a fragment of DNA, and cause specific mutations in the DNA sequence of a gene.

Key Terms

molecular biology
recombinant DNA
restriction endonuclease
restriction fragment
gene cloning
plasmid
transformation
DNA amplification
polymerase chain reaction (PCR)
gel electrophoresis
DNA fingerprinting
short tandem repeat (STR) profiling
DNA sequencing
dideoxy sequencing
Human Genome Project
site-directed mutagenesis

Key Concepts

- Using restriction endonucleases, researchers can combine fragments of DNA from more than one source to form recombinant DNA.
- Bacteria can be used as host systems for DNA amplification, producing multiple copies of a gene for research purposes.
- The polymerase chain reaction (PCR) is an alternative form of DNA amplification in which a sample of DNA is repeatedly denatured through heating, and then replicated using a heat-resistant DNA polymerase.
- In DNA fingerprinting, a sample of DNA from an individual is fragmented, amplified, and then separated by gel electrophoresis to produce a unique "fingerprint" that can be compared to reference samples for identification or matching.
- Dideoxy sequencing enables the nucleotide sequence of a fragment of DNA to be determined by cutting four identical samples of the DNA fragment at specific nucleotide sites and then comparing the sizes of the pieces produced.
- Site-directed mutagenesis is a technology that enables researchers to create specific mutations in the DNA sequence of a gene. As a result, researchers can study the structure and function of DNA, genes, and the proteins coded by these genes.

Section 7.2 Production and Regulation of Genetically Engineered Organisms

Genetically modified organisms have significant benefits, including increased resistance to disease and pests, but controversy exists regarding the possibility of negative health and environmental effects from the use of GMOs.

Key Terms

gene therapy
genetic engineering
transgenic
genetically modified organism (GMO)
biotechnology
patent
expression vector
bioremediation
biolistic method
Ti plasmid method
gene pharming

Key Concepts

- Transgenic, or genetically modified, bacteria can be used to produce pharmaceutical products such as insulin. GM bacteria are also used in bioremediation of environmental pollutants.
- Transgenic crops have increased tolerance to herbicides and increased resistance to disease and pests, and many have been approved for human consumption.
- Transgenic animals can be used to produce pharmaceutical products such as human proteins. Some, such as pigs and salmon, are being considered for human consumption.
- The production and use of transgenic plants and animals are controversial, in terms of their safety for human consumption and the possibility of negative environmental effects. Health Canada requires that all GMOs intended for human consumption undergo seven to 10 years of health-and-safety research.
- Mammals can be cloned by fusing a somatic cell, including its genetic material, with an oocyte from which the nucleus has been removed, and then fertilizing the resulting embryo and implanting it in a host. Although sheep, cows, and other mammals have been cloned, human cloning remains highly controversial.

Chapter 7 REVIEW

Knowledge and Understanding

Select the letter of the best answer below.

1. What is the main function of restriction enzyme shown below?

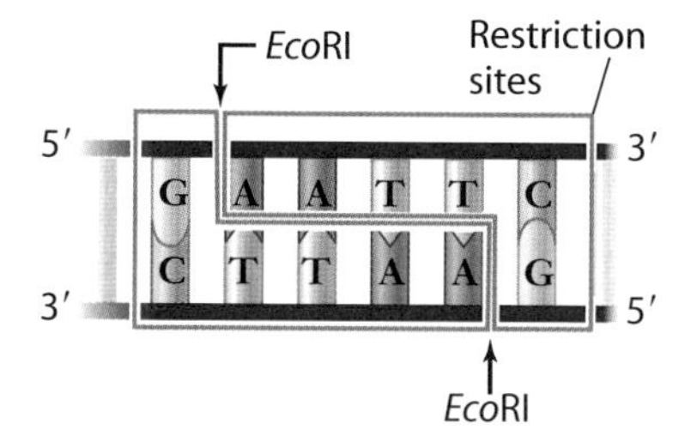

a. to join DNA fragments at the 3′ ends
b. to join DNA fragments at the 5′ ends
c. to cut DNA at specific sites within the DNA
d. to cut DNA from the 5′ or 3′ ends
e. to cut DNA at the 5′ ends

2. Which enzyme catalyzes the joining of fragments to form recombinant DNA?

a. DNA polymerase
b. helicase
c. DNA ligase
d. restriction endonuclease
e. RNA polymerase

3. What is the name of the process that introduces DNA into bacterial cells?

a. gene cloning
b. translocation
c. translation
d. transcription
e. transformation

4. Why are bacteria the most common hosts used in gene cloning?

a. Bacteria can produce selectable markers.
b. Bacteria have the capacity to perform all post-translational modifications on human proteins.
c. Bacteria have mechanisms to prevent the degradation of foreign proteins.
d. Bacteria are resistant to many antibiotics.
e. Bacteria grow quickly in large amounts and are inexpensive to maintain.

5. During PCR, why is the DNA sample initially heated to a high temperature of approximately 95°C?

a. Heating causes the DNA to be denatured into two separate strands.
b. Heating allows DNA polymerase to bind to DNA.
c. Heating allows RNA polymerase to bind to DNA.
d. Heating allows the annealing of primers to DNA.
e. Heating allows the extension of DNA.

6. You are preparing a reaction tube that will undergo PCR in a thermocycler. What components would you include in the reaction tube?

a. DNA, DNA polymerase, primers, nucleotides
b. DNA, DNA polymerase, restriction endonucleases, nucleotides
c. DNA, DNA polymerase, restriction endonucleases, DNA ligase
d. DNA, DNA polymerase, primers, restriction endonucleases, DNA ligase
e. RNA, DNA polymerase, primers, nucleotides

7. Gel electrophoresis separates DNA fragments according to

a. the sequence of their bases.
b. differences in electrical charge.
c. their size.
d. the number of mutations they carry.
e. the amount of DNA loaded in each well.

8. DNA fingerprinting can be done using a very small amount of DNA because

a. gel electrophoresis is very sensitive.
b. the polymerase chain reaction (PCR) allows researchers to amplify samples of DNA.
c. even a very small amount of DNA is likely to contain some restriction sites.
d. no two people have the same set of restriction sites on their DNA.
e. DNA sequencing is automated.

9. What are the dideoxynucleotides used in dideoxy sequencing of DNA?

a. ddA, ddT, ddG, ddC
b. ddA, ddU, ddG, ddC
c. ddU, ddT, ddG, ddC
d. ddA, ddT, G, C
e. A, T, ddG, ddC

10. Because of the Human Genome Project, we now know

a. the sequence of the base pairs of the human genome.
b. the sequence of all genes along the human chromosomes.
c. all the mutations that lead to genetic disorders.
d. all the genes that lead to genetic disorders.
e. all the proteins in the human genome.

11. What type of vector is transformed into host bacterial cells to allow the production of foreign proteins?

a. plasmid
b. transgenic
c. expression
d. recombinant
e. Ti plasmid

12. Some types of bacteria have been genetically engineered to efficiently break down oil. These transgenic bacteria are often used to clean up oil spills. What is this process called?
    a. bioremediation
    b. recombination
    c. transformation
    d. genetic engineering
    e. cloning
13. Which of the following biotechnology products is a result of gene pharming?
    a. herbicide-resistant corn
    b. transgenic sheep that produce human growth hormone in their milk
    c. insulin production in bacteria
    d. pest-resistant wheat
    e. transgenic bacteria that remove pollutants from water
14. In the cloning of Dolly the sheep, what was the source of the genetic material?
    a. germ cell
    b. recombinant DNA plasmid
    c. somatic cell
    d. stem cell
    e. oocyte

Answer the questions below.

15. Why would recombinant DNA technology be restricted if the genetic code were not universal?
16. List and explain the two characteristics of restriction endonucleases that make them useful for researchers.
17. What is an advantage of a drug produced using recombinant DNA technology compared to a drug extracted from natural sources?
18. How does the structure of DNA influence its movement through a gel used in gel electrophoresis?
19. Describe how short tandem repeat (STR) profiling produces a DNA fingerprint.
20. Why does DNA synthesis stop when a dideoxynucleotide is incorporated into DNA?
21. What was the significance of automated DNA sequencing to the Human Genome Project?
22. Describe site-directed mutagenesis. What is the importance of this technique?
23. Distinguish between the following two terms: genetic *engineering* and *biotechnology*. Use examples in your answer.
24. What difficulties can arise in creating transgenic animals versus transgenic bacteria?
25. What are three practical uses of transgenic organisms?
26. Describe two techniques you could use to create pest-resistant transgenic corn.
27. What are the advantages and disadvantages of producing therapeutic human proteins in transgenic animals versus in bacteria?
28. Three different adult sheep were involved in the cloning process that led to the birth of the lamb Dolly.
    a. What were their roles?
    b. Which one was Dolly's clone?

Thinking and Investigation

29. **BIG IDEAS** DNA contains all the genetic information for any living organism. One of the aims of the Human Genome Project was to determine the DNA sequence of genomes from different organisms. Why would this information be useful to researchers?
30. A human gene called *ras* is inserted into mice, creating transgenic animals that develop a variety of tumours. Why are mouse cells able to transcribe and translate human genes?
31. To create a recombinant DNA molecule, you cut a plasmid and your gene of interest with *Eco*RI. Why is the same restriction enzyme used to cut both pieces of DNA?
32. You are given two plasmids, one from each of two bacterial species. One contains gene A and the other contains gene B. You wish to create a single plasmid containing both genes.
    a. Use labelled diagrams to describe the steps you would take to produce this recombinant plasmid.
    b. What other plasmids will result from this procedure? Which of them will be recombinant?
33. Assuming you start with one copy, how many copies of DNA will be produced after 15 cycles of a PCR reaction? After 30 cycles? Show your calculations and provide an explanation.
34. One challenge of working with DNA from ancient (extinct) species is that it is usually present in very small amounts. How could you increase the amount of ancient DNA available for analysis?

35. Complete the diagram below by drawing the number of copies of DNA produced from one copy after each cycle of PCR.

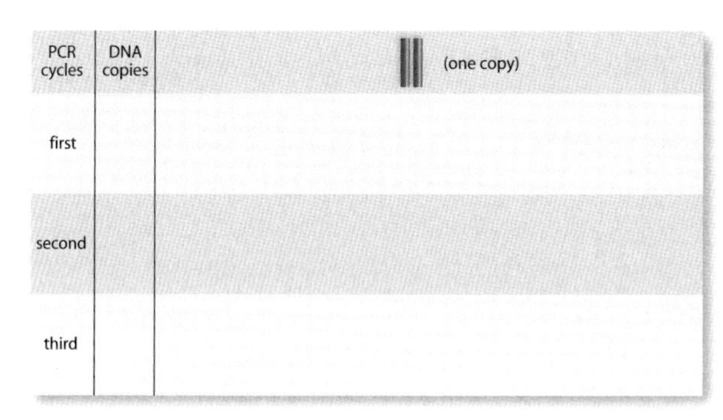

36. You have cloned a gene 500 base pairs in length into a vector 2000 base pairs in length. To ensure the gene was properly inserted into the vector, you cut the purified recombinant DNA into fragments using the same restriction endonuclease initially used to cut the gene and vector. You analyze the resulting DNA fragments using gel electrophoresis.

  a. Sketch the relative positions of the DNA fragments on a gel if the gene was successfully cloned into the vector.

  b. Sketch the relative positions of the DNA fragments if the gene was not succesfully cloned.

37. The DNA fingerprint below shows the results of DNA analysis performed on a woman and a man. If the woman and the man were to have a biological child, what would the results of the child's DNA fingerprint analysis look like?

Woman	Man	Child

38. You are a scientist working in a forensics lab. You have been instructed to perform a DNA fingerprint analysis of a sample of DNA purified from blood found at a crime scene. Design an experimental procedure you would follow to analyze this sample of DNA.

39. Design a primer 10 nucleotides in length that would anneal to the following DNA sequence in a PCR reaction:

5′-GATGGACTGACTGGACTTTTCTC-3′

40. The reaction medium used in the dideoxy sequencing method normally contains only a very low concentration of dideoxynucleotides. What would be the effect of increasing the concentration of these molecules? Explain.

41. This illustration shows the electrophoretic gel pattern that resulted from a dideoxy sequencing process. What is the nucleotide sequence of the original DNA sample?

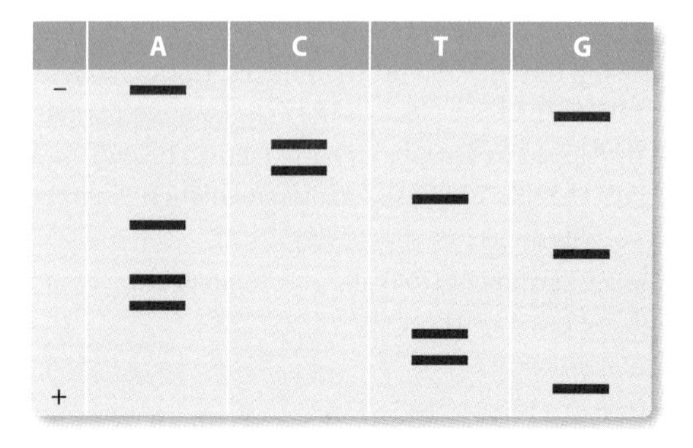

42. A researcher wants to develop a breed of dog with fur colour that changes according to the season. She establishes a partnership with a scientist who breeds stoats (animals with a coat that is white in winter and brown in summer).

  a. Develop an experimental plan that would enable the two researchers to develop a breed of transgenic dog.

  b. Briefly describe one social or ethical issue the researchers should consider before they begin their experiment.

43. BIG IDEAS Proteins control a wide variety of cellular processes. You successfully insert into a bacterial cell a plant gene coding for a growth hormone. When you extract and test the protein product, however, you find it does not function as it should.

  a. Identify and explain why a bacterial cell may produce a non-functioning protein.

  b. What could you do to correct this problem?

Communication

44. Using labelled diagrams, illustrate the differences between sticky ends and blunt ends produced by restriction endonucleases.

45. Using labelled diagrams and written descriptions, explain how to clone a gene in a bacterial host.

46. Using a flow chart, describe how you would amplify a specific sequence of DNA using PCR.

47. You have been asked to teach a Grade 7 class about DNA fingerprinting.
   a. Write a paragraph detailing how you would explain DNA fingerprinting to the class. Include examples and/or analogies in your explanation.
   b. What social and ethical issues would you discuss with the class?
   c. What questions do you anticipate your class would ask? How would you answer those questions?

48. Using Internet and print resources, research one micro-organism, such as the *Pseudomonas putida* developed by Dr. Chakrabarty (shown below), or *E. coli*, that has been used in bioremediation. Summarize your findings in a short essay.

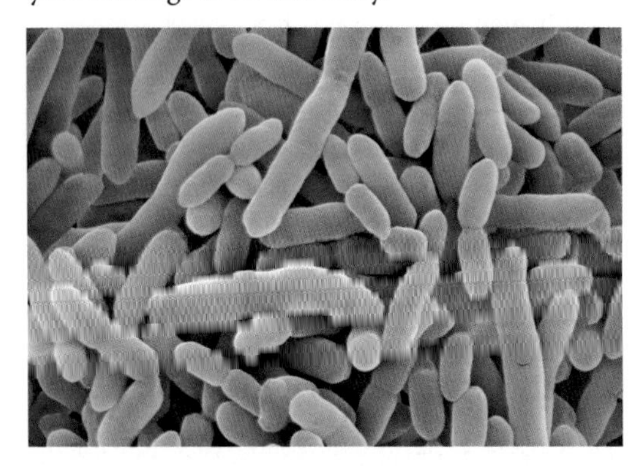

49. Using Internet and print resources, create for the general public a brochure entitled "The Myths and Facts about Genetically Modified Organisms."

50. "Biotechnology is essential to food productivity and security." Do you agree or disagree with this statement? Explain your answer in an opinion piece that would be published in your local newspaper.

51. Use labelled diagrams to illustrate the steps involved in creating a bacterial cell that can produce human insulin.

52. Using a graphic organizer of your choice, discuss the benefits and risks of gene pharming.

53. Working with a partner, debate the following statement: "Human cloning should never be permitted." Then list what you think are the best arguments on each side of the debate. Do research to support your viewpoint with information from credible sources.

54. Summarize your learning in this chapter using a graphic organizer. To help you, the Chapter 7 Summary lists the Key Terms and Key Concepts. Refer to Using Graphic Organizers in Appendix A to help you decide which graphic organizer to use.

Application

55. Do you think government regulations are the best way to protect people from the potential risks of biotechnology? Explain why or why not.

56. Consider the kinds of arguments that a defence lawyer could advance to prevent DNA evidence from being used in court. Suppose you are a forensic scientist called to testify in a case. Detail some of the points you would expect to use in your rebuttal, and why the jury would be wise to consider them.

57. Currently, couples who are expecting a baby often have routine tests performed on the fetus, such as the test for Down syndrome. In the near future, we may be able to test the whole genome—potentially for thousands of genetic disorders.
   a. If you were concerned about a possible genetic disorder in your unborn child, would you undergo this type of screening? Why or why not?
   b. Discuss what you believe are the ethical limits to the use of such technology

58. Health Canada and the Canadian Food Inspection Agency are responsible for developing and enforcing food labelling policies in Canada.
   a. Research what the Canadian government's policies are on the labelling of genetically modified foods (GMFs). Write a brief description that summarizes these policies.
   b. Do you agree with the Canadian government's labelling policies for GMFs? Explain your answer.

59. A farmer plants a strain of transgenic corn in her fields. The corn carries a recombinant gene that confers resistance to a common herbicide. The next year, a species of weed growing near the corn fields is found to be herbicide-resistant. A study shows that the weed is expressing the recombinant gene. You are a journalist assigned to report on the story.
   a. What is the significance of the discovery of the herbicide-resistant weed?
   b. Write two main points you would expect to hear from each of the following individuals you interview: the farmer, an official from the genetic engineering corporation that created the transgenic

corn, the owner of a nearby organic farm, a consumer organization opposed to the development of genetically modified organisms, and a genetics researcher.

60. Imagine you have been hired as an advisor to an international body that establishes conventions for genetic research. Your job is to develop a policy on the collection and ownership of genetic information.

a. What are some of the issues you will consider?

b. Briefly summarize how your policy will balance public and private interests.

61. People did not object to the production of human insulin in bacterial cells used to treat diabetes. Yet some people object to mixing DNA from different animal and plant species in agricultural biotechnology. Why do you think the same general technique is perceived as beneficial in one situation, yet a threat in another?

62. Rice is a staple in the diet of millions of people worldwide, many of them living in less-developed countries. In some of those same countries, vitamin A deficiency is a major cause of blindness in small, malnourished children. Scientists have developed a new form of rice called "golden rice" (shown below), which is genetically modified to assist with the metabolism of vitamin A and might prevent millions of cases of blindness. The scientists who created golden rice claim it is safe for consumption, although critics say the health effects are not yet fully understood.

a. Given its nutritional potential, should golden rice be planted and distributed on a wide scale?

b. What information would you need to make your decision?

63. Health Canada requires and ensures that all food is safe before it enters the Canadian food system. Health Canada's jurisdiction includes what are termed "novel foods." Find out how Health Canada defines novel foods, how they are regulated, and the departments of government that are responsible for regulating products derived using techniques of genetic modification.

64. Recombinant bovine growth hormone (rBGH) has been approved for use in dairy cattle in the United States, but not in Canada.

a. Using Internet and print resources, research why the difference exists between these two countries.

b. Do you agree with Canada's decision to ban rBGH? Explain your answer.

65. What are the advantages of automated DNA sequencing compared to manual DNA sequencing?

66. Using Internet and print resources, research Canada's current legislation regarding human cloning. How does this compare to the human cloning legislation in other countries?

67. **BIG IDEAS** Genetic research and biotechnology have social, legal, and ethical implications.

a. Which biotechnology tool or technique discussed in this chapter do you think will have the greatest impact on your life and the lives of your friends and family members ten years from now?

b. Twenty years from now?

c. Write a brief report that includes as many specific details from this chapter as possible to support your ideas.

68. Several companies in the United States offer "personal DNA analysis." The analysis is based on identifying variants in DNA called *single-nucleotide polymorphisms* (*SNPs*). These companies claim they can offer customers a DNA profile that shows ancestral origins, risk factors for certain diseases, and predicted side-effects of certain drugs. To take advantage of this service, customers mail to the company a small saliva sample for analysis. The cost of this service currently ranges from \$200 to \$1000. Would you consider using this type of service? Why or why not?

69. Copy the following table into your notebook. Using the example of transgenic pigs that have been engineered to serve as donors for organ transplants in humans, fill in the benefits and risks that you foresee.

Using transgenic pigs as organ donors	Benefits	Risks
To individual people		
To society		
To the economy		
To other species		
To the environment		

Chapter 7 SELF-ASSESSMENT

Select the letter of the best answer below.

1. K/U Segments of DNA from two different species can be joined in the lab to form a single molecule of DNA. What is the new molecule called?
   a. recombinant DNA
   b. expression vector
   c. plasmid
   d. selectable marker
   e. clone

2. K/U Bacterial cells that take up DNA under certain experimental conditions are said to be
   a. transferred
   b. transcripts
   c. translations
   d. replications
   e. transformed

3. K/U What are two different methods of DNA amplification?
   a. gene cloning in bacteria and PCR
   b. gene cloning in bacteria and dideoxy sequencing
   c. PCR and dideoxy sequencing
   d. PCR and biotechnology
   e. PCR and bioremediation

4. T/I How many copies of DNA will be produced from one copy after 20 cycles of a PCR reaction?
   a. 100
   b. 1 048 576
   c. [illegible]
   d. 1 000
   e. 524 288

5. K/U What process enables fragments of DNA to be separated for analysis using size and charge differences?
   a. polymerase chain reaction (PCR)
   b. DNA sequencing
   c. cloning
   d. gel electrophoresis
   e. DNA fingerprinting

6. K/U What are the components of a small reaction tube prepared for dideoxy sequencing?
   a. Single-stranded DNA, DNA polymerase, primer, deoxynucleotides, one dideoxynucleotide
   b. Single-stranded DNA, DNA polymerase, primer, one dideoxynucleotide
   c. Double-stranded DNA, DNA polymerase, primer, one dideoxynucleotide
   d. Double-stranded DNA, DNA polymerase, primer, deoxynucleotides, one dideoxynucleotide
   e. Single-stranded DNA, DNA ligase, primer, deoxynucleotides, four dideoxynucleotides

7. T/I The DNA fingerprint below shows the results of DNA analysis performed on a man, a woman, and two children. Based on this evidence, what can you conclude about the relationship between the children and each of the two adults?

Woman	Man	Child 1	Child 2

   a. The woman and man are the biological parents of both child 1 and child 2.
   b. The woman and man are the biological parents of child 1 only.
   c. The woman and man are the biological parents of child 2 only.
   d. The woman and man are not the biological parents of child 1 and child 2.
   e. The woman is the biological mother of child 2 and the man is the biological father of child 1.

8. T/I Your research group wishes to study which regions of DNA are crucial for the proper transcription of a gene. To achieve this, you specifically mutate nucleotides in the promoter regions of the gene. What method do you use to produce these mutations?
   a. restriction endonuclease reaction
   b. high through-put sequencing
   c. site-directed mutagenesis
   d. gene therapy
   e. dideoxy chain termination method

9. K/U What were the main goals of the Human Genome Project?
   a. Sequence the human genome and identify the genes in it.
   b. Sequence the human genome and determine all the genes which cause disease.
   c. Identify the genes in the genome and determine which genes cause disease.
   d. Sequence the human genome and clone mammalian cells.
   e. Identify the genes in the genome and determine which mutations cause disease.

10. K/U Gene pharming is

a. a method of treating genetic disorders by introducing the correct form of a disease-related gene.

b. a method of cloning mammals from somatic cells.

c. the process of using transgenic bacteria to produce therapeutic proteins.

d. the process of using transgenic animals to produce therapeutic proteins.

e. the process of using transgenic plants to produce therapeutic proteins.

Use sentences and diagrams as appropriate to answer the questions below.

11. K/U Explain why these characteristics of restriction endonucleases make the enzymes useful to genetic engineers.

a. specificity

b. staggered cuts

12. K/U Why is DNA ligase an important tool for recombinant DNA technology?

13. T/I You have two plasmids: one plasmid is an expression vector, while the other plasmid contains the gene for human growth hormone (*GH1*).

a. Design an experimental outline you would follow to clone *GH1* into the expression vector. Use labelled diagrams and descriptions in your outline.

b. How would you check that you had successfully cloned *GH1* into the expression vector?

14. T/I A PCR thermocycler was programmed incorrectly and remains at a constant 95°C for every PCR cycle. What are the consequences of this error?

15. K/U Dideoxynucleotides are used in DNA sequencing.

a. What structural feature of dideoxynucleotides makes them useful for DNA sequencing?

b. Why is this structural feature important?

16. A Will cells from your liver and your brain have the same DNA fingerprint? Explain.

17. A Would you be willing to provide your DNA for a national DNA databank? Why or why not?

18. C How would you explain bioremediation to an individual who has no science background? Include examples in your explanation.

19. T/I Your research group is working on producing a new type of pest-resistant tomato. You plan on inserting into the tomato cells genes that produce insecticidal proteins. How would you introduce the insecticide genes into the tomato cells?

20. A Genetic information from the Human Genome Project may be used to develop screening procedures, which could be used by insurance companies or employers. Discuss the possible implications of the use of genetic information for this type of screening.

21. C "The benefits of farming genetically engineered crops outweigh the risks." Do you agree or disagree with this statement? Write a short opinion piece that would be published on your local newspaper's website.

22. A Malaria is an infectious disease caused by parasites of the genus *Plasmodium*. It is passed onto humans through bites from the female *Anopheles* mosquito, which carries the parasite. Malaria causes approximately one million deaths a year, with 90 percent of malaria-related fatalities occurring in sub-Saharan Africa. Researchers at Johns Hopkins University have been working on creating a transgenic mosquito that is resistant to the malaria parasite. The researchers hope that a malaria-resistant transgenic mosquito could eventually be released into nature.

a. What are the risks and benefits of producing this transgenic organism?

b. Do you think transgenic animals should ever be released "into the wild"? Explain your answer.

23. C Use a graphic organizer to compare and contrast the advantages and disadvantages of producing therapeutic proteins in transgenic animals versus in transgenic bacteria.

24. C Using labelled diagrams, illustrate the procedure used in the cloning of Dolly the sheep.

25. A Do you think there are any circumstances in which cloning of humans should be permissible? Explain your answer using examples.

Self-Check

If you missed question...	1	2	3	4	5	6	7	8	9	10	11	12	13	14	15	16	17	18	19	20	21	22	23	24	25
Review section(s)...	7.1	7.1	7.1	7.1	7.1	7.1	7.1	7.1	7.1	7.2	7.1	7.1	7.1	7.1	7.1	7.1	7.2	7.2	7.2	7.2	7.2	7.2	7.2	7.2	7.2

Unit 3 Project

Conduct an Inquiry

Investigating Functions and Applications of RNA Interference

RNA interference (RNAi) is a means of regulating gene expression. In RNAi, processes that involve certain types of RNA control whether genes are expressed and to what extent they are expressed. During RNAi, certain RNA molecules and other proteins bind to messenger RNA (mRNA). This process degrades the mRNA and affects gene expression by preventing or reducing translation. RNAi has important roles and applications in biology, medicine, and biotechnology, some of which are depicted below. In this project, you will investigate RNA interference, focussing on a specific function or application of RNAi. When conducting your research, you may use a variety of online and print resources, as advised by your teacher. You will use the results of your research to prepare a presentation that highlights various aspects of RNAi, including the history of its discovery, how it influences gene regulation, and the role it plays in the specific function or application you have chosen.

What is RNA interference and what is its role in a particular function or application?

Initiate and Plan

1. Examine the functions and applications of RNAi listed below and choose one to investigate further.
- medical therapy to combat a specific virus (for example, influenza, measles, herpes simplex, hepatitis, or HIV)
- medical therapy to treat a disease or condition such as cancer, Huntington's disease, or macular degeneration
- overall regulation of gene expression in cells
- cell type differentiation in organisms
- direction of embryological development in organisms
- plant biotechnology (for example, genetic modifications in cotton, corn, and tomato crops to improve quality and yield)

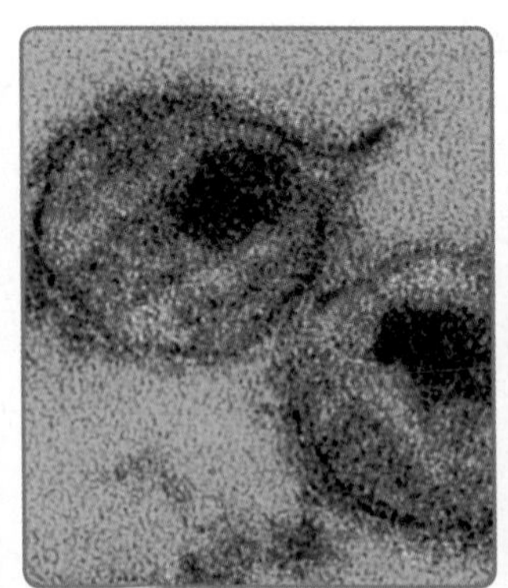

RNAi is used in antiviral therapy to combat viruses such as the human immunodeficiency virus (HIV) shown here (left). RNAi is also used as a biotechnological tool to improve the quality and yield of crops such as tomatoes (right).

2. Review available sources of information that are relevant to the function or application you have chosen to investigate. Refer to Developing Research Skills in Appendix A for tips on evaluating the reliability of your sources.

Perform and Record

1. Conduct some basic background research on RNAi to familiarize yourself with the concept before you begin any in-depth research. As part of this process, research the history of the discovery of RNAi. Concepts or names that may aid you in your research by serving as key search terms include petunias, *C. elegans*, Andrew Fire, and Craig Mello.

2. Research how RNAi controls whether genes are expressed and/or to what degree genes are expressed. Consider the following questions to guide your research:
- How does RNAi affect gene expression?
- What terms, such as *post-transcriptional gene regulation* and *gene silencing*, do you need to define to better understand the scientific material you are reading?
- What roles do the enzyme *dicer* and small interfering RNA carry out in RNAi?
- What is an *RNA-induced silencing complex* and what role does it perform in RNAi?
- What role does complementary base pairing play in RNAi?

- What sources have you used to gather your information? How have you determined that they are credible scientific sources?

3. Research RNAi with respect to the particular function or application you chose in Plan and Initiate step 1 above. Consider the following questions to guide your research, as applicable:
   - Describe the role of RNAi in the function or application you chose.
   - What safety concerns are associated with RNAi-based treatment applications?
   - What ethical issues may arise from certain applications of RNAi technology? How might these issues be resolved?
   - Find out more about past and current research related to your chosen function or application, focussing on Canadian research.

Analyze and Interpret

1. Create a timeline that shows important events in the discovery of RNAi.
2. Explain the basic principles of RNAi. Draw a labelled diagram that shows how RNAi interferes with gene expression.
3. Create a summary of your findings with regard to the RNAi function or application you chose. If you chose to investigate an application, evaluate the contribution of this application to its related field of research.

Communicate Your Findings

4. Decide on the best way to present your findings, such as a poster or an interactive webpage. Be sure to keep your intended purpose and audience in mind when making your decision.
5. Prepare a presentation that focusses on the particular function or application of RNAi that you chose to investigate. Your presentation should include:
   - an explanation of RNAi, including a description and a timeline of historical highlights related to its discovery. Your explanation should also include a diagram showing how RNAi interferes with gene expression
   - a description of the function or application that you chose to investigate. Your description should include a summary of research focussed on this function or application, highlighting any work done by Canadian scientists. If you chose to investigate an RNAi application, you should also discuss the contribution that this application has made to its related field of research
   - a literature citation section that documents the sources you used to complete your research

Assessment Criteria

After you complete your project, ask yourself these questions. Did you…

- ✓ **K/U** choose an appropriate function or application of RNAi to investigate further?
- ✓ **T/I** complete a survey of available information sources and critically evaluate these sources?
- ✓ **C** document your sources using an appropriate academic format?
- ✓ **K/U** complete background research on RNAi and important events related to its discovery?
- ✓ **A** assess the impact that RNAi technology has on science with a particular focus on Canadian research and technology?
- ✓ **C** visually depict the history of RNAi and how RNAi interferes with gene expression?
- ✓ **K/U** summarize your findings related to the RNAi function or application you chose to investigate?
- ✓ **T/I** evaluate the contribution of a particular application to its field of research (if applicable)?
- ✓ **C** create an instructional visual tool to present your findings that is appropriate for your purpose and your audience?
- ✓ **C** use scientific vocabulary that is appropriate to your research and your audience?

UNIT 3 SUMMARY

BIG IDEAS

- DNA contains all the genetic information for any living organism.
- Proteins control a wide variety of cellular processes.
- Genetic research and biotechnology have social, legal, and ethical implications.

Overall Expectations

In this unit you learned how to...

- **analyze** some of the social, ethical, and legal issues associated with genetic research and biotechnology
- **investigate**, through laboratory activities, the structures of cell components and their roles in processes that occur within the cell
- **demonstrate** an understanding of concepts related to molecular genetics, and how genetic modification is applied in industry and agriculture

Chapter 5 The Structure and Function of DNA

Key Ideas

- DNA and RNA are nucleic acids made of units called nucleotides. Each nucleotide is composed of one of four nitrogen-containing bases, a sugar, and a phosphate group.
- The accepted model of DNA, presented by Watson and Crick in 1953, has a double helix structure.
- The process of copying one DNA molecule into two identical molecules is called DNA replication. DNA replication consists of three phases: initiation, elongation, and termination.
- DNA replication is similar in eukaryotes and prokaryotes, including the presence of origins of replication, the presence of DNA polymerase enzymes, and elongation in the 5′ to 3′ direction.

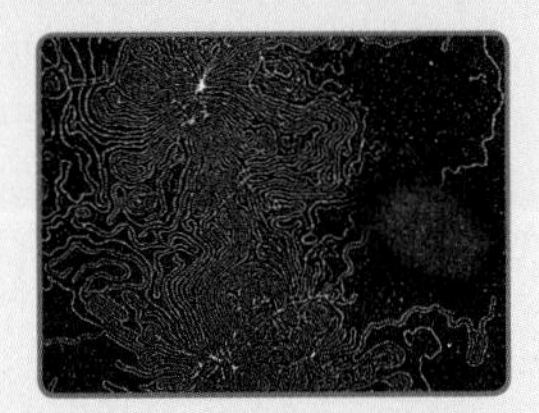

- Replication differs between eukaryotes and prokaryotes, including a slower replication rate and many more origins of replication in eukaryotes, and different DNA polymerase enzymes in eukaryotes than prokaryotes.

Chapter 6 Gene Expression

Key Ideas

- [illegible] universal and [illegible] how genetic information is converted from a nucleotide sequence to an amino acid sequence of a protein.
- RNA is a single-stranded polymer that contains four nucleotides with the bases adenine (A), uracil (U), cytosine (C), and guanine (G).
- During transcription, the antisense strand of a gene is used as a template to synthesize a strand of mRNA.
- During translation, proteins are synthesized from an mRNA template.
- Mutations are permanent [illegible] may be spontaneous, or they may be induced by exposure to physical or chemical mutagens.

- Gene regulation refers to control of the level of gene expression in response to different conditions in the cell or environment. The majority of genes are regulated so that their product is only expressed at certain times and in certain amounts.

Chapter 7 Genetic Research and Biotechnology

Key Ideas

- Using restriction endonucleases, researchers can combine fragments of DNA from more than one source to form recombinant DNA.
- Bacteria can be used as host systems for DNA amplification, producing multiple copies of a gene for research purposes. The polymerase chain reaction (PCR) is an alternative form of DNA amplification.
- Dideoxy sequencing enables the nucleotide sequence of a fragment of DNA to be determined by cutting four identical samples of the DNA fragment at specific nucleotide sites and then comparing the sizes of the pieces produced.
- The production and use of transgenic plants and animals are controversial, in terms of the safety of human consumption and the possibility of negative environmental effects.

- Mammals can be cloned by fusing a somatic cell, including its genetic material, with an oocyte from which the nucleus has been removed, and then implanting the embryo in a host. Although sheep, cows, and other mammals have been cloned, human cloning remains highly controversial.

Knowledge and Understanding

Select the letter of the best answer below.

1. What phase of DNA replication involves helicase activity?
   a. termination
   b. elongation
   c. initiation
   d. transcription
   e. synthesis
2. Who first showed convincing evidence that DNA replication is semi-conservative?
   a. Meselson and Stahl
   b. Watson and Crick
   c. Hershey and Chase
   d. Avery
   e. Franklin
3. Which enzyme that is involved in DNA replication has proofreading capabilities?
   a. helicase
   b. topoisomerase I and II
   c. ligase
   d. primase
   e. DNA polymerase II
4. Spliceosomes are responsible for
   a. removing exons
   b. removing introns
   c. attaching introns together
   d. the synthesis of mRNA
   e. the synthesis of peptides
5. Identify the correct order of organization of genetic material, from largest to smallest.
   a. genome, gene, chromosome, nucleotide
   b. genome, chromosome, gene, nucleotide
   c. gene, chromosome, nucleotide, genome
   d. chromosome, gene, genome, nucleotide
   e. chromosome, genome, nucleotide, gene
6. What is the name of the process that transfers information from DNA to RNA?
   a. translation
   b. synthesis
   c. replication
   d. transcription
   e. elongation
7. What type of DNA replication error is shown in the following diagram?

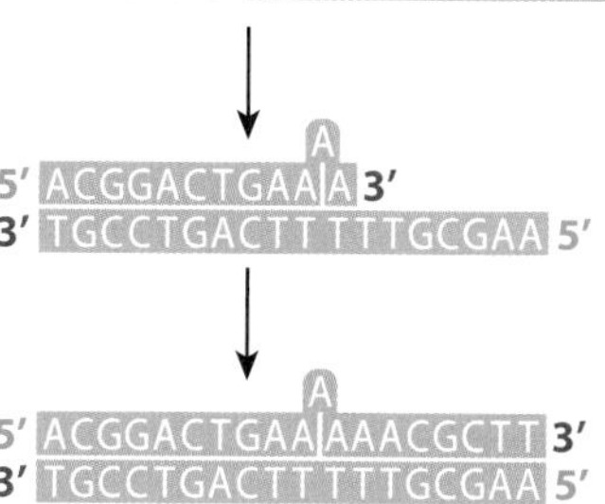

   a. strand slippage causing the addition of a nucleotide
   b. strand slippage causing the deletion of a nucleotide
   c. mispairing causing the addition of a nucleotide
   d. mispairing causing the deletion of a nucleotide
   e. mispairing causing the deletion of a gene
8. What are the non-coding regions of eukaryotic genes called?
   a. chromatin
   b. repetitive DNA
   c. microsatellite DNA
   d. exons
   e. introns
9. What feature of the genetic code helps to protect a cell from the effects of nucleotide substitution?
   a. The genetic code has 64 codons.
   b. The genetic code consists of a combination of three nucleotides.
   c. The genetic code is redundant.
   d. The genetic code is continuous.
   e. The genetic code is universal.
10. What is the corresponding anti-codon for the mRNA codon CGG?
    a. 5′-GCC-3′
    b. 5′-GUU-3′
    c. 3′-CGG-5′
    d. 3′-GCC-5′
    e. 3′-GUU-5′
11. Ultraviolet (UV) radiation is an example of
    a. a transposon
    b. a chemical mutagen
    c. a physical mutagen
    d. a spontaneous mutation
    e. photorepair

12. Which type of mutation would result if the sequence GUU-CAU-UUG had a nucleotide substitution that caused UUG to become UAG?
    a. missense
    b. silent
    c. nonsense
    d. frameshift
    e. inversion
13. In PCR, why is DNA subjected to a cooling phase of approximately 55°C after initially being exposed to a high temperature?
    a. Cooling allows DNA to be denatured.
    b. Cooling allows primers to anneal to DNA.
    c. Cooling causes *Taq* polymerase to synthesize DNA.
    d. Cooling causes *Taq* polymerase to bind to DNA.
    e. Cooling allows RNA polymerase to bind to DNA.
14. What molecular biology technique would be the most appropriate for analyzing the sizes of DNA fragments?
    a. PCR
    b. gel electrophoresis
    c. gene cloning
    d. DNA sequencing
    e. automated DNA sequencing
15. Which is an example of bioremediation?
    a. producing human insulin in bacteria
    b. creating pest-resistant food crops
    c. using microorganisms to break down pollutants
    d. creating transgenic Glo-Fish
    e. creating drought-resistant wheat

Answer the questions below.

16. Why does replication occur in opposite directions?
17. List and explain the similarities and differences between DNA polymerases and RNA polymerases.
18. How do proteins determine the characteristics of cells?
19. How is gene expression a two-step process? What are the benefits of this process?
20. Why does the cell use the majority of its energy for translation?
21. Why is accuracy less important in transcription than in DNA replication?
22. List and explain the similarities and differences in DNA replication between prokaryotes and eukaryotes.
23. Use examples to describe specific and non-specific DNA repair.
24. What is the start codon for the initiation of translation? What are the anticodon and amino acid of the tRNA that pairs with this codon?
25. List and describe the molecular events of translation.
26. Explain how multiple copies of a protein can be produced at the same time. Why would this be advantageous for a cell?
27. List and explain two different methods for amplifying DNA. Under what circumstances would you use each method?
28. The development of quicker and more efficient methods of DNA sequencing has led to many applications in medical research. List and describe two applications of "next-generation" automated DNA sequencing.
29. List and describe two methods for producing transgenic plants.
30. Explain how transgenic animals are produced. What are two applications of transgenic animals?

Thinking and Investigation

31. If the bases on the opposite strands of DNA paired through covalent bonds, what effect would this have on DNA replication?
32. How is complementary base pairing important in the following processes?
    a. DNA replication
    b. transcription
    c. translation
33. "A mutation in a gene's exons is more damaging than a mutation in a gene's introns." Do you agree or disagree with this statement? Explain your answer.
34. The following micrograph shows the "beads on a string" appearance of nucleosomes. Why do nucleosomes appear in this way?

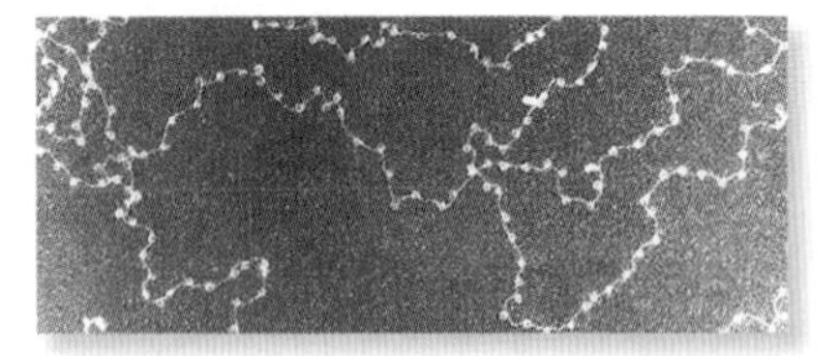

35. A sample of DNA has 26% thymine. What do you expect would be the approximate proportion of adenine, cytosine, and guanine in the sample?
36. Why would it be beneficial for cells to synthesize more than one mRNA from one gene at one time?

37. The following is a graphical representation of the promoter regions in *E. coli:*

a. What would be the consequence of a mutation in Promoter 1?

b. What would be the consequence of a mutation in Promoter 2?

38. Beadle and Tatum's "one gene/one enzyme" hypothesis has been updated to the "one gene/one polypeptide" hypothesis.

a. Why was the hypothesis updated?

b. Why is the "one gene/one polypeptide" hypothesis also an inaccurate description of the relationship between genes and proteins? What do you think would be a more accurate hypothesis?

39. Splicing involves snRNPs ("snurps"), which are composed of snRNA and proteins. How do you think snRNPs are able to recognize the regions where exons and introns meet, and subsequently bind to these areas?

40. Copy the following sentence into your notebook: THE DOG WAS MAD

a. How would the sentence change if it were subjected to a frameshift mutation?

b. What are the possible consequences of a frameshift mutation on gene expression?

41. How is protein synthesis similar to the catalysis of a reaction by an enzyme?

42. The antisense strand of a segment of DNA has the following sequence: 5′-GTTAAAGGC-3′. Use this sequence to answer the following questions:

a. What is the corresponding mRNA codon sequence?

b. What is the anti-codon sequence?

c. What is the corresponding peptide sequence?

43. How are proteins involved in their own synthesis?

44. In bacteria, it is common for two or more structural genes to be arranged together in an operon. Discuss the arrangement of genetic sequences within an operon. What is the biological advantage of an operon organization?

45. Suppose you are a member of a research group who wishes to study the function of a newly identified gene and its protein product. What molecular biology tools would you use in your study? Describe each tool and explain why it would be useful for your research.

46. If there was a mutation in the *lac* repressor protein which prevents it from binding to the operator, how would the following conditions affect the transcription of lactose-metabolizing enzymes?

a. presence of lactose

b. absence of lactose

47. The following diagrams are maps showing the restriction endonuclease sites in a bacterial plasmid (Plasmid A) and in a eukaryotic gene (Gene B) which you wish to clone into the plasmid.

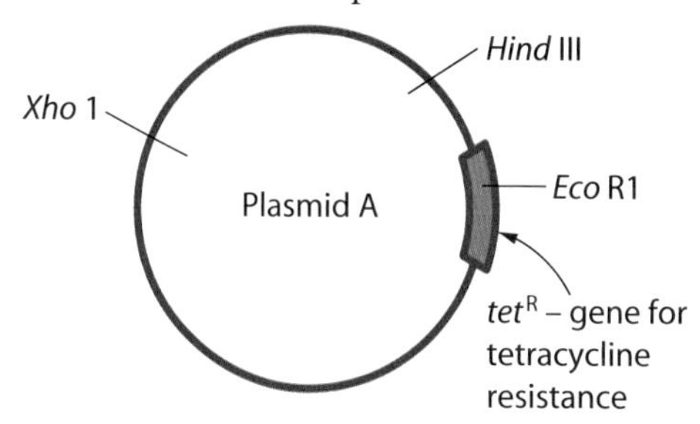

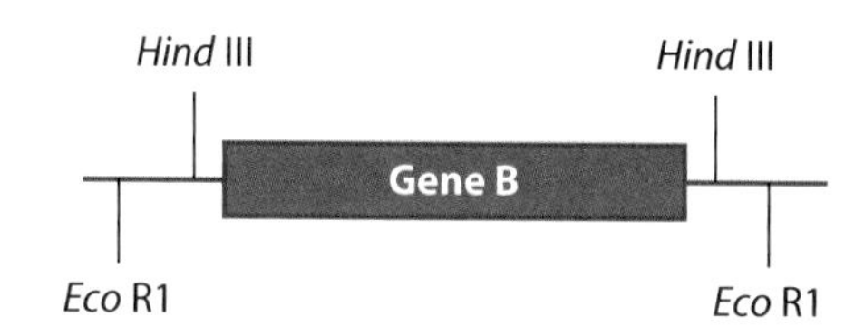

a. Which restriction endonuclease(s) would you use to clone Gene B into Plasmid A? Explain your answer.

b. Explain why you did not choose the other restriction endonuclease(s) to use in your cloning experiment.

c. How would you confirm that you successfully cloned Gene B into Plasmid A?

48. A hair sample found at a crime scene has been submitted to a forensics lab for DNA fingerprinting. Below is the gel electrophoresis analysis of the sample.

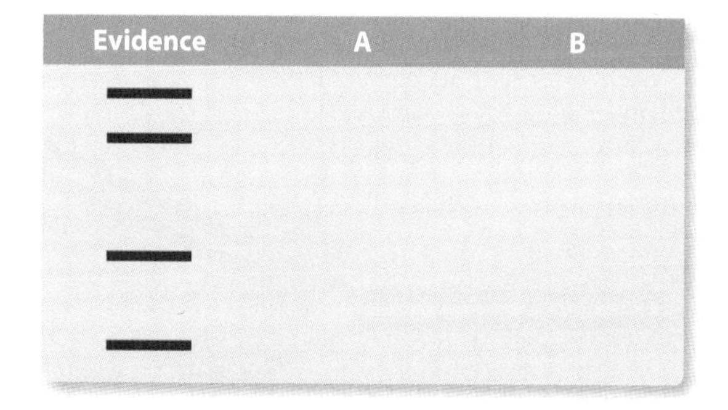

a. Copy the diagram into your notebook.

b. There are two suspects in the case, suspect A and suspect B. Sketch what you would see on the gel for both suspects if the hair sample came from suspect A.

49. A researcher wishes to create a transgenic fruit that ripens more slowly than a natural form does. As a potential aid she has isolated a strain of bacteria that secretes a protein that slows the ripening process in the fruit.
   a. What main steps might she follow to develop the new strain of fruit?
   b. What are the main challenges she will have to overcome?

50. A researcher wishes to produce a therapeutic human anticoagulant protein in cow's milk.
   a. Outline the steps the researcher would take to create a transgenic animal that produced this therapeutic protein.
   b. Draw and label the main features of the recombinant plasmid vector that would be used to create the transgenic animal.
   c. What are the advantages and disadvantages of producing therapeutic proteins in transgenic animals instead of transgenic bacteria?

Communication

51. Describe the two Hershey and Chase experiments using labelled diagrams. What was the significance of their results?

52. Using symbols to represent the different nucleotides, illustrate the molecular structure of a portion of a double-stranded DNA molecule.

53. If DNA replication were to be conservative, how would the pattern of bands have appeared in the Meselson-Stahl experiment after centrifugation? Draw a diagram to illustrate your answer. Additionally, include a sketch of what the pattern of bands Meselson and Stahl actually observed. Explain the significance of their results.

54. Copy the diagram below into your notebook. Replace each letter with its correct label.

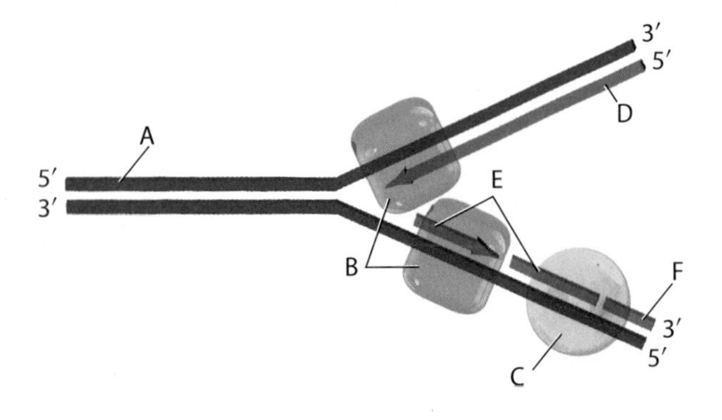

55. Create a study tool, such as a mnemonic, to help you remember the different purines and pyrimidines in DNA and RNA.

56. Use labelled diagrams to illustrate the differences between eukaryotic precursor mRNA and prokaryotic mRNA.

57. A eukaryotic gene has 3 introns and 2 exons. Use labelled diagrams to illustrate the mRNA modifications of this gene that occur after transcription.

58. **BIG IDEAS** DNA contains all the genetic information for any living organism. Use labelled diagrams to summarize gene expression in a eukaryotic cell. Indicate the locations of all the processes involved.

59. Copy and complete the following table on single-gene mutations.

Single-gene Mutations

Type of Mutation	Description	Example	Effect on Cell and Organism

60. Use a table to summarize the functions of the following types of RNA: mRNA, tRNA, and rRNA.

61. Create a graphic organizer that summarizes the five levels of gene regulation in eukaryotes.

62. Use labelled diagrams to illustrate the difference between "sticky ends" and "blunt ends" that are produced by restriction endonucleases.

63. Illustrate how the *trp* operon is different from the *lac* operon by using labelled diagrams.

64. Using Internet and print resources, research and prepare an information pamphlet for the general public on therapeutic cloning. Include the following sections.
   a. What is therapeutic cloning? Explain using diagrams and descriptions of examples.
   b. What is the Canadian government's position on therapeutic cloning?
   c. What are the benefits and risks of therapeutic cloning?

65. Use a flowchart to illustrate an example of gene pharming.

66. Write a Point/Counterpoint editorial article to discuss the following statement: "The cloning of humans should never be permitted. However, the cloning of certain animals and bacteria should continue to be permitted."
  a. In the "Point" article, describe what arguments you would use to defend this statement.
  b. In the "Counterpoint" article, describe what arguments you would use to refute this statement.

Application

67. The use of antibiotics has been of great importance in the battle against many infectious diseases that are caused by microorganisms. Certain antibiotics selectively bind to bacterial (70S) ribosomes but do not inhibit eukaryotic (80S) ribosomes. Their ability to inhibit translation can occur at different steps in the translation process. For example, tetracycline prevents the attachment of tRNA to the ribosome.
  a. How does inhibition of translation antibiotics, such as tetracycline, prevent bacterial growth?
  b. What are the advantages of the antibiotics selectively binding to bacterial ribosomes?

68. BIG IDEAS Proteins control a wide variety of cellular processes. Early-onset Parkinson disease causes rigidity, tremors, and other motor symptoms. Only 2 percent of cases of Parkinson disease are inherited. A subset of these cases is due to mutations in the *parkin* gene. The following three patients all have mutations in the gene *parkin*, which codes for a protein that has 12 exons. For each patient, indicate whether the mutation in their *parkin* gene shortens, lengthens, or does not change the size of the protein.
  I. Patient A's *parkin* gene is deleted.
  II. Patient B's *parkin* gene has a duplication in exon 4.
  III. Patient C's *parkin* gene is missing intron 2.

69. Individuals with Bloom's syndrome have Sun sensitivity, distinct facial rashes, and a higher risk cancer due to a high frequency of chromosomal breaks and rearrangements. Mutations in the *BLM* gene, which is a member of the DNA helicase family, is associated with Bloom's syndrome.
  a. What would be the effects of a mutation in the *BLM* gene on DNA replication?
  b. Gene therapy involving the delivery of a functional *BLM* gene was once considered, but has since been deemed to be an ineffective treatment approach. What are some possible reasons why gene therapy would not be appropriate for Bloom's syndrome?

70. BIG IDEAS Genetic research and biotechnology have social, legal, and ethical implications. In 2000, the Canadian government passed legislation to create a National DNA Data Bank to assist in criminal investigations. The DNA samples and profiles in the bank are from convicted offenders, and currently collect 600-700 new samples per week.
  a. Research the National DNA Data Bank's Privacy and Security policy. How does it ensure that the DNA samples and profiles are used only for law enforcement purposes? Would you change anything about these policies? Why or why not?
  b. Suppose the National DNA Data Bank decides to collect DNA samples from every person who resides in Canada. These samples will only be used to help police investigations. Would you support this endeavour? Explain your answer.

71. The pharmaceutical company Hoffman-La Roche owned the patent for the PCR technique and *Taq* polymerase until it expired in 2005. The patent expiration led to the emergence of lower-cost PCR reagents from competing biotechnology companies, which not only offered researchers more choice, but also allowed PCR to be used in research areas where it previously too expensive to pursue. Do you think scientific techniques and tools should continue to be patentable? Explain your answer.

72. RNA interference (RNAi), which is also known as "gene silencing", is a useful molecular biology technique which has allowed researchers to study the function of genes. RNAi has been proposed as being a possible therapeutic approach for diseases where overexpression of a gene causes deleterious effects. For example, many cancers have been associated with unregulated overexpression of certain genes in the Ras family, which are involved in cell growth, differentiation, and survival.
  a. Why would the Ras family of genes be an appropriate target for RNAi therapy?
  b. What are the drawbacks of RNAi therapy that targeted the Ras family of genes?

73. The idea of having "personalized" medical treatment has been discussed as being a potential application of automated DNA sequencing. It has been proposed that having knowledge of a patient's genome sequence could allow health care professionals to determine treatments that would be most effective for that patient. Would you consent to having your genome as a part of your permanent medical record? Explain.

UNIT 3 SELF-ASSESSMENT

Select the letter of the best answer below

1. K/U In the experiments by Avery and colleagues, what was identified as being the transforming principle?
   a. bacteria
   b. enzymes
   c. DNA
   d. RNA
   e. protein

2. K/U Which of the following sequences is complementary to the sequence 5′ CGCTTAGCA 3′ in double stranded DNA?
   a. 5′ GCGAATCGT 3′
   b. 5′ TGCTAAGCG 3′
   c. 5′ GCGAAUCGU 3′
   d. 5′ TGCTUUGCG 3′
   e. 3′ CGCTTAGCA 5′

3. K/U Which type of bonds exists between bases on opposite strands of DNA?
   a. peptide bonds
   b. covalent bonds
   c. hydrogen bonds
   d. ionic bonds
   e. polar covalent bonds

4. K/U Which post-transcriptional modifications are added to precursor eukaryotic mRNA prior to leaving the nucleus?
   a. 5′ cap, 3′ poly-A tail, and removal of introns
   b. 3′ cap, 5′ poly-A tail, and removal of introns
   c. 5′ cap, 3′ poly-A tail, and removal of exons
   d. 3′ cap, 5′ poly-A tail, and removal of exons
   e. 5′ cap, 3′ poly-A tail, and joining of introns

5. T/I Which are the corresponding tRNA anti-codons for the polypeptide sequence: Ala-Pro-Trp-Arg?
   a. CGA GGA ACC GCA
   b. GCU CCU UGG CGU
   c. CGT GGA ACC GCT
   d. CGT GGT ACC GCT
   e. GCT ACC GGA CGT

6. T/I A mutation causes a sequence of mRNA codons to change from GUU CAU UUG to GUU CAU UAG. Which type of mutation has occurred?
   a. frameshift
   b. silent
   c. missense
   d. nonsense
   e. chemical

7. K/U Which statement about the compounds contained in the smoke shown in the photo below is *false*?

   a. They are chemical mutagens.
   b. They are physical mutagens.
   c. They can induce mutations by reacting chemically with DNA within the nucleus of a cell.
   d. Exposure to these compounds can increase the rate at which mutations occur in an organism.
   e. They can cause a nucleotide substitution or a frameshift mutation to occur.

8. K/U In the absence of lactose, what is the transcriptional state of the lactose-metabolizing genes on the *lac* operon?
   a. Transcription is inhibited due to the *lac* repressor protein binding to the operator.
   b. Transcription is activated due to the *lac* repressor protein binding to the operator.
   c. Transcription is activated since the *lac* repressor protein cannot bind to the operator.
   d. Transcription is inhibited since the *lac* repressor protein cannot bind to the operator.
   e. Transcription is activated since allolactose binds to the repressor, and prevents binding to the operator.

9. K/U In gel electrophoresis, DNA fragments migrate and separate based on
   a. charge and size
   b. size
   c. charge
   d. protein size
   e. number of covalent bonds

10. T/I How many copies of DNA will be produced from one copy of DNA after 35 cycles of a PCR reaction?
   a. 35
   b. 70
   c. 140
   d. 1.25×10^3
   e. 3.43×10^{10}

Use sentences and diagrams as appropriate to answer the questions below.

11. K/U What results from the Hershey and Chase experiment prove that DNA was the heredity material?

12. K/U How did Franklin's image, shown below, contribute to Watson and Crick's model of DNA?

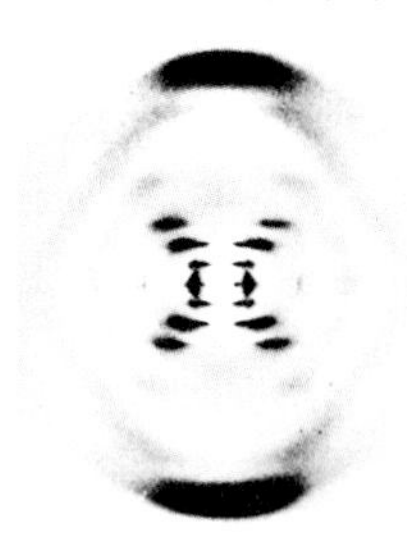

13. K/U Differentiate between the "one gene/one enzyme" hypothesis and the "one gene/one polypeptide" hypothesis.

14. C Use a table to compare the characteristics of DNA and RNA.

15. K/U In what way is the structure of a protein related to the structure of DNA?

16. T/I Your research group has identified a new species of gecko. You find that it contains 18% adenine, but the other three bands are indecipherable. How much each of thymine, guanine, and cytosine would you expect to find?

17. C Copy and complete the following table, which summarizes the important enzymes in DNA replication.

Enzymes in DNA Replication

Enzyme/Protein	Functions
helicases	
	Stabilizes single-stranded regions of DNA during unwinding
	Relieves strain on DNA caused by unwinding
Primase	
DNA polymerase III	
DNA polymerase I	
DNA polymerase II	
	Joins the ends of Okazaki fragments

18. C Use a graphic organizer to illustrate the levels of organization of DNA in eukaryotic cells.

19. T/I The antisense strand of a segment of DNA in a bacterial chromosome has the following base sequence: 5′-TACACATGCATC-3′.

a. Draw a section of the double-stranded DNA molecule that includes this segment.

b. Which end of the segment has a free –OH group?

c. What is the amino acid sequence of the polypeptide product if this is part of a gene?

d. Show how a nucleotide substitution could result in a silent mutation of this gene.

20. C Use labelled diagrams to outline the main stages of translation.

21. T/I How is complementary base pairing important in gene expression?

22. T/I Develop an experimental outline on how you would create transgenic corn that expresses a herbicide-resistant gene called *Bt*.

23. C Use a flowchart to explain the basic steps involved in gel electrophoresis.

24. A A private company has developed a transgenic carrot that secretes its own pesticide. This carrot is therefore resistant to the insects and worms that often damage root crops.

a. What are some of the risks and benefits that the Canadian government will consider when deciding whether to approve this plant for agricultural use?

b. If approved, what advantages will this transgenic carrot offer to farmers? What are some potential drawbacks for farmers?

25. A Suppose that you are an official in the government of Canada. Your job is to support research that will contribute to human health. You must decide how to allocate $100 million in research funding among the following three areas: development of transgenic crops; development of techniques for therapeutic cloning; research into the molecular processes involved in regulation of the cell cycle. How much funding will you allocate to each area? Justify your decision.

Self-Check

If you missed question...	1	2	3	4	5	6	7	8	9	10	11	12	13	14	15	16	17	18	19	20	21	22	23	24	25
Review section(s)...	5.1	5.1	5.1	6.2	6.3	6.3	6.3	6.4	7.1	7.1	5.1	5.1	6.1	5.1 6.2	6.1	5.1	5.2	5.1	5.1 5.2 6.1 6.2 6.3	6.3	6.1 6.2	7.2	7.1	7.2	7.2

UNIT 4 Homeostasis

BIG IDEAS

- Organisms have strict limits on the internal conditions that they can tolerate.
- Systems that maintain homeostasis rely on feedback mechanisms.
- Environmental factors can affect homeostasis.

Overall Expectations

In this unit, you will...

- **evaluate** the impact on the human body of selected chemical substances and of environmental factors related to human activity
- **investigate** the feedback mechanisms that maintain homeostasis in living organisms
- **demonstrate** an understanding of the anatomy and physiology of human body systems, and **explain** the mechanisms that enable the body to maintain homeostasis

Unit Contents

Focussing Questions

1. *Why is it important for homeostasis to be maintained?*
2. *How do body systems including the nervous system, the endocrine system, and the excretory system interact to maintain homeostasis?*
3. *What are some ways in which homeostasis can be disrupted?*

Go to **scienceontario** to find out more about homeostasis

Each of these Cirque du Soleil performers is working hard to maintain her balance, constantly making small adjustments to muscle tension throughout her body to stay in position. The performers in each column also are working together to maintain the balance of the group as each new person is added. If any one of them starts to wobble or become unsteady it could have a ripple effect, causing others to lose their balance and the entire group to fall. This situation is comparable to the interactions among the systems in your body. Body systems including the nervous system, the endocrine system, and the excretory system work together to constantly monitor and maintain a stable and balanced environment within the body. If any one system cannot maintain this balance, other systems may be affected. If one or more systems become too unbalanced, the whole body can be affected.

As you study this unit, look ahead to the Unit 4 Project on pages 478 to 479. Complete the project in stages as you progress through the unit.

UNIT 4 Preparation

Safe and Effective Microscope Techniques

- A compound light microscope uses a series of lenses and a light source to view an object.
- A compound light microscope has two optical systems: the eyepiece lens and the objective lens.
- A microscope is a precision instrument that must be handled carefully to avoid damage to the lenses.
- Proper procedures must be followed to ensure safe and effective use of a microscope.

1. Identify the name and function of each lettered part on the diagram of the compound light microscope shown below.

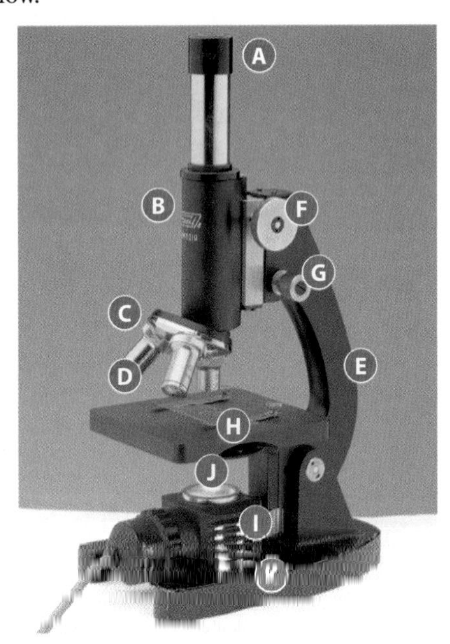

2. Identify which part(s) you must hold when carrying the microscope.

3. When viewing a slide under a microscope, what is the correct procedure?

a. Rotate the highest-power objective lens into place, set the slide on the microscope stage, secure the stage clips in place, and then look through the eyepiece lens.

b. Rotate the highest-power eyepiece lens into place, set the slide on the microscope stage, secure the stage clips in place on the slide, and then look through the objective lens.

c. Set the slide on the microscope stage, secure the stage clips in place on the slide, rotate the lowest-power eyepiece lens into place, and then look through the objective lens.

d. Set the slide on the microscope stage, secure the stage clips in place on the slide, rotate the lowest-power objective lens into place, and then look through the eyepiece lens.

e. Rotate the medium-power eyepiece lens into place, set the slide on the microscope stage, secure the stage clips in place on the slide, and then look through the objective lens.

Safe and Effective Dissection Techniques

- Dissection involves the careful and systematic examination of the internal structures of an organism.
- A well-conducted dissection will reveal not only the location and structure of individual organs, but also how different organs relate to one another in the various systems of an organism's body.
- Extreme care must be taken when using dissecting instruments, particularly scalpels. To every extent possible, you should make cuts away from your body.
- Specimens are preserved in chemical solutions. It is vital to wear plastic gloves, eye protection, and a lab apron at all times, and to work in a well-ventilated area.

4. During a dissection, cuts should be made away from your body? Explain why this is important.

5. You should always consult your teacher for instructions regarding disposal at the end of your dissection. What safety concerns are associated with the disposal of your specimen, and what information should your teacher consult before disposing of dissection materials?

6. Describe the procedure to follow if chemicals come into contact with your skin or eyes.

7. Why must you always wear safety glasses and protective clothing during a dissection?

8. In a dissection, *dorsal* refers to the upper, or back surface, of a specimen, while *ventral* refers to the underside, or front surface. Explain what the following terms mean: *anterior, posterior, proximal, distal.*

Cells, Tissues, Organs, and Systems

- Cell specialization is influenced by the contents of an individual cell's cytoplasm, by environmental factors such as temperature, and by secretions from neighbouring cells.
- The human body is organized in a hierarchy: cells, tissues, organs, systems.
- Groups of similarly specialized cells form tissues.
- The structure of specialized cells is related to their function.
- Animals have four major tissue types: muscle, epithelial, connective, and nervous.
- Organs consist of several different tissue types working together to perform specific functions.
- Each of the human body's 11 organ systems consists of different organs working together to carry out a specific set of functions. The systems interact with one another to allow the body to survive and reproduce.

9. An example of a specialized cell is
  a. a nerve cell
  b. an embryonic stem cell
  c. epithelial tissue
  d. glucose
  e. the skin

10. Give an example of a cell type that would require a large surface area, and explain why this requirement would be necessary.

11. Describe one tissue and one organ that are involved in receiving and responding to an external stimulus, such as pain.

12. Which organ system is responsible for removing liquid wastes from the body?
  a. the circulatory system
  b. the excretory system
  c. the immune system
  d. the integumentary system
  e. the nervous system

13. Give an example of two organ systems that interconnect or rely on each other to function.

14. Nerve cells have long, fibre-like projections. Explain how this structure is related to the function of nerve cells.

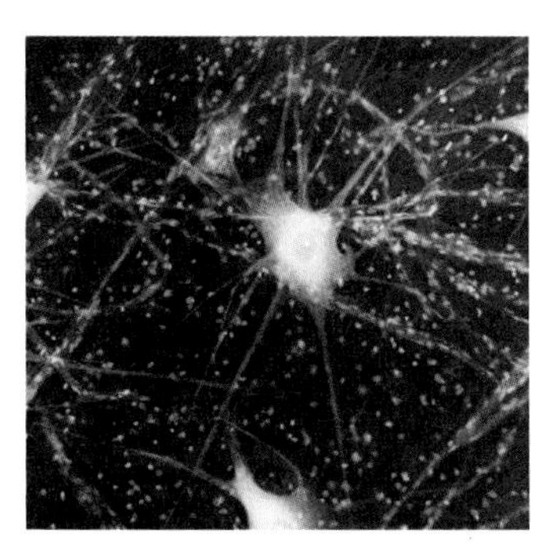

15. Nervous tissue is made of cells called *neurons*. The main function of a neuron is to
  a. remove toxins from the body
  b. act as a barrier to prevent water loss
  c. transport nutrients to body tissues
  d. recognize and destroy foreign cells
  e. receive and transmit electrical signals

16. Which function is associated with the endocrine system?
  a. taking in food and breaking it down
  b. exchanging gases in the lungs
  c. detecting changes in the external environment
  d. manufacturing and releasing hormones
  e. defending the body against infections

17. Which system is shown in the following diagram?

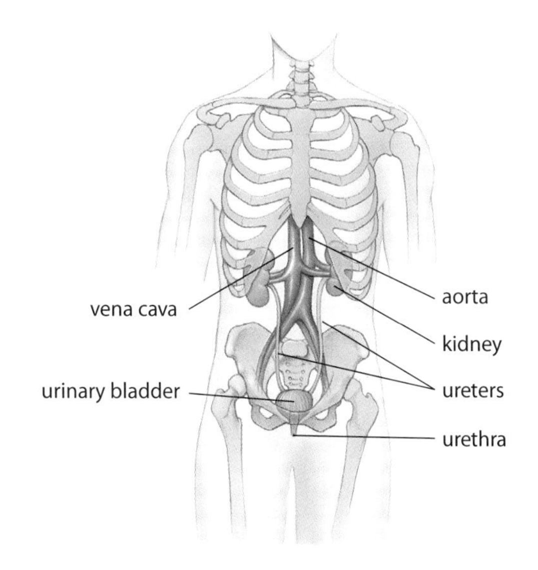

Diffusion, Osmosis, and Transport Across Cellular Membranes

- Eukaryotic cells consist of cytosol and a number of organelles, all enclosed within a cell membrane.
- The cell membrane is a selectively permeable barrier through which all substances, including nutrients, respiratory gases, and wastes, must pass in order to enter or exit the cell. Other cellular membranes are also selectively permeable.
- Water molecules and other molecules pass through cellular membranes by means of different processes.
- Diffusion is the movement of particles from an area of high concentration to an area of lower concentration.
- The difference in concentration between an area of high concentration and an area of lower concentration of a substance is called a *concentration gradient.*
- Oxygen and carbon dioxide can diffuse freely across cell membranes according to their concentration gradients.
- Osmosis is the movement of water molecules through a selectively permeable membrane, from an area of high concentration of water molecules to an area of lower concentration of water molecules.
- Molecules that are larger than oxygen and carbon dioxide, such as glucose, cannot pass through cellular membranes by diffusion. They must cross membranes through active transport or passive transport. The same is true for charged particles, such as calcium ions, Ca^{2+}, or sodium ions, Na^{+}.

18. Muscle cells require a great deal of energy to power muscle contraction and to perform other cellular functions as well. Which organelle is most directly responsible for the production of this energy?
  a. the nucleus
  b. the endoplasmic reticulum
  c. the ribosome
  d. the mitochondrion
  e. the chloroplast

19. Which is an example of osmosis?
  a. oxygen moving into a red blood cell
  b. water moving through the esophagus
  c. glucose moving into the mitochondrion
  d. oxygen moving into the mitochondrion
  e. water moving into a vacuole

20. What would happen to a cell if its membranes were completely permeable rather than selectively permeable?

21. Explain the relationship between the terms *solute, solvent,* and *concentration.*

22. Active transport includes
  a. diffusion
  b. osmosis
  c. diffusion against a concentration gradient
  d. diffusion along a concentration gradient
  e. facilitated diffusion

23. Compare and contrast active transport and passive transport.

Refer to the following information to answer questions 24 to 27.

The diagram below represents a blood cell in water. The green circles represent oxygen molecules.

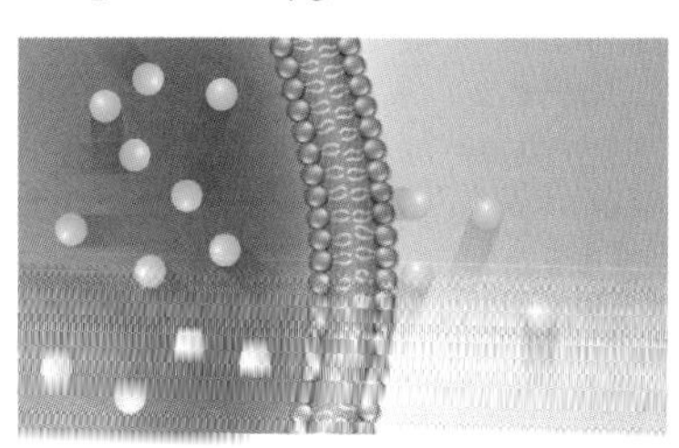

24. Sketch another diagram to show the distribution of oxygen molecules that you would expect to observe several minutes later.

25. Does the situation in question 24 involve active transport, passive transport, or no movement of oxygen molecules?

26. Describe what you would expect to see several minutes later if the green circles represented glucose molecules instead of oxygen molecules, assuming there are no channel proteins or carrier proteins present.

27. Would the situation in question 26 involve active transport, passive transport, or no movement of glucose molecules? Explain.

28. State whether each of the following terms is more closely related to active transport (A) or passive transport (P).
  i. osmosis
  ii. facilitated diffusion
  iii. carrier protein
  iv. diffusion

29. Which of the following pairs of phrases or terms have the same meaning?

a. moving down a concentration gradient; moving against a concentration gradient
b. moving along a concentration gradient; moving from an area of low concentration to an area of high concentration of a substance
c. moving by osmosis; moving by active transport
d. moving down a concentration gradient; moving from an area of high concentration to an area of low concentration of a substance
e. moving against a concentration gradient; moving by passive transport

30. Which of the following molecules is most directly responsible for providing the energy needed for active transport?

a. DNA
b. ATP
c. glucose
d. RNA
e. adenosine

31. What factors determine whether a molecule crosses cellular membranes via active transport versus passive transport?

32. Give three examples of organelles that possess cellular membranes.

Water, Minerals, and pH

- Water is necessary for the functioning of all cells and organs, in which it plays multiple roles.
- Calcium, sodium, and potassium are important minerals that are required for activities such as conducting nerve signals, balancing body fluids, and contracting muscles.
- Biological processes take place within specific limits of acidity and basicity.
- An acid is a compound that produces hydrogen ions (H^+) when it dissolves in water. A base is a compound that forms hydroxide ions (OH^-) when it dissolves in water.
- Hydrogen ion concentration, or pH, is measured on the pH scale, which ranges from 0 to 14. The midpoint of the scale, 7, represents a neutral solution—one that is neither acidic nor basic. Values below 7 represent acidic solutions, while values above 7 represent basic solutions.

33. Some of the functions that water performs in cells include

a. flushing toxins from cells and eliminating waste materials
b. lubricating tissues and joints
c. forming essential body fluids
d. regulating body temperature
e. all of the above

34. Define the term *ion* and give three examples of ions that are biologically important.

35. Describe how the concentration of H^+ ions changes as the pH of a solution decreases.

36. If the pH of a solution changes from 4 to 5, the H^+ ion concentration

a. increases by a factor of 1
b. decreases by a factor of 1
c. increases by a factor of 10
d. decreases by a factor of 10
e. does not change

37. Draw a pH scale and add the following labels: range of acids, range of bases, neutral, stomach acid (pH 1), pure water, ammonia (pH 11).

38. The pH of most cells and the fluid surrounding cells in tissues ranges from

a. 5.8 to 6.5
b. 6.5 to 7.8
c. 7.8 to 8.3
d. 8.3 to 8.9
e. 8.9 to 9.5

39. The reactions that take place in a certain cellular organelle require a low pH. To produce this low pH, proteins in the cell membrane

a. pump Na^+ ions out of the organelle
b. pump Na^+ ions into the organelle
c. pump H^+ ions out of the organelle
d. pump H^+ ions into the organelle
e. break down

CHAPTER 8

The Nervous System and Homeostasis

Specific Expectations

In this chapter, you will learn how to . . .

- E1.1 **assess**, on the basis of findings from a case study, the effects on the human body of taking chemical substances to enhance performance or improve health (8.2, 8.4)
- E2.1 **use** appropriate terminology related to homeostasis (8.1, 8.2, 8.3, 8.4)
- E2.4 **plan** and **conduct** an investigation to study the response mechanism of an invertebrate to external stimuli (8.4)
- E3.1 **describe** the anatomy and physiology of the endocrine, excretory, and nervous systems, and **explain** how these systems interact to maintain homeostasis (8.1, 8.2, 8.3, 8.4)
- E3.3 **describe** the homeostatic processes involved in maintaining water, ionic, thermal, and acid-base equilibrium, and **explain** how these processes help body systems respond to both a change in environment and the effects of medical treatments (8.1)

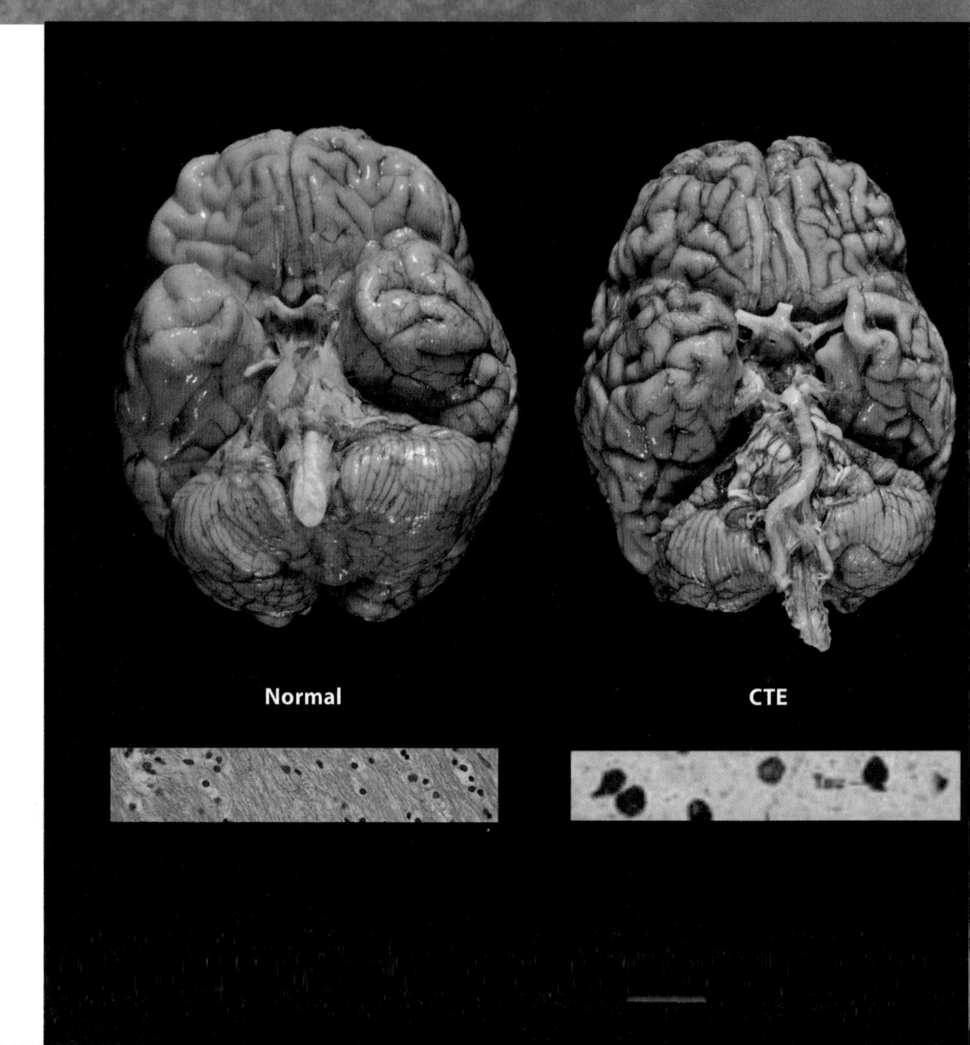

Chronic traumatic encephalopathy (CTE) is a degenerative brain disease that can result from sustaining repeated blows to the head. One of the key markers of CTE is the build-up of an abnormal version of a protein called *tau* throughout the brain, shown in the micrographs above. This is the same protein that appears in the brains of Alzheimer's disease patients. The abnormal version of tau interferes with the normal functioning of the brain. It eventually kills brain cells and disrupts the body's ability to maintain a stable internal environment. Symptoms of CTE are similar to those of Alzheimer's disease, such as memory loss, disorientation and confusion, and erratic behaviour. Other symptoms include tremors, slowed muscular movements, and impaired speech. Emerging research is focussing on the risk of CTE in football and hockey players, boxers, and soldiers exposed to explosions during combat; they all may face an increased risk of long-term brain injury from repeated small impacts accumulated over time.

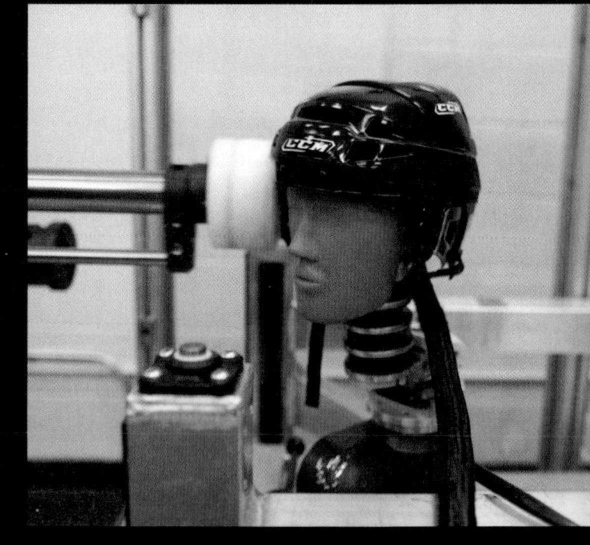

Launch Activity

Thin as an Eggshell

The average thickness for a female human skull is 7.1 mm and for a male is 6.5 mm. Its main purpose is to protect the brain. A 1-m drop will usually break an eggshell, and a 2-m drop can fracture the skull if not protected. In this activity, you will investigate how the design of a model helmet can help reduce injury to the brain due to impact.

Safety Precautions

- Wash egg from your hands and any other surfaces it contacts.
- Wash your hands upon completion of the lab activity.

Materials

- plastic garbage bag
- tape
- tape measure
- 1 egg
- paper towel
- 12 wooden craft sticks
- 1 sealable sandwich bag
- water
- foam peanuts or a foam block
- glue
- scissors

Procedure

1. Secure a garbage bag to the floor to prevent slipping and assist in clean-up.
2. Your teacher will drop an egg from a height of 1.5 m to get a baseline reading of the damage that would occur as a result of an unprotected collision. Observe the condition of the egg after the collision. Record your observations in a data table.
3. Use the materials provided to construct a model "helmet" to protect your egg during a fall.
4. Test your model by dropping it (with the egg in it) from a height of 1.5 m. Be sure to drop it over the plastic you laid down in step 1.
5. Observe the condition of the egg and the model. Record your observations in your data table.
6. Compare your model and observations to those of other groups.

Questions

1. Was your helmet successful at protecting the egg? Explain why or why not.
2. If you were to design another helmet, what improvements would you make? Explain your reasoning.
3. How well does this experiment model the real-world situation of a person wearing a helmet while participating in a sports activity, such as football, cycling, or ice hockey?
4. What factors would you consider if you were designing a helmet for an athlete to wear during competition?
5. How might the design of a helmet be related to CTE?

Alzheimers

Researchers at the University of Ottawa's Neurotrauma Impact Science Laboratory are conducting tests to learn more about how helmets can help protect ice hockey and football players from brain injuries due to impacts.

SECTION 8.1

Human Body Systems and Homeostasis

Key Terms

homeostasis
sensor
control centre
effector
negative feedback system
positive feedback system

The trillions of cells that make up your body can be organized into about 100 different types. Similarly specialized types of cells that perform a common function make up a tissue. Tissues of different types are organized as organs, which themselves are organized structurally and functionally as systems, as shown in **Figure 8.1**. Organ systems work together to perform functions necessary to sustain and maintain the human organism. The systems can be organized into groups based on functions with a common purpose.

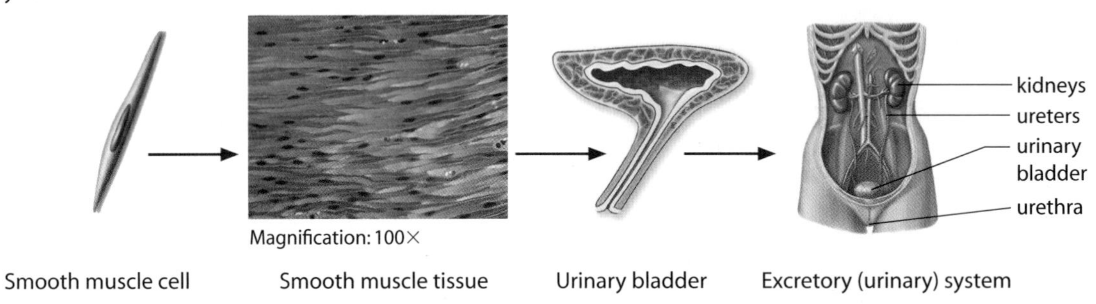

Figure 8.1 Smooth muscle cells make up smooth muscle tissue. Smooth muscle tissue lines the inside of the urinary bladder, an organ that is part of the excretory (urinary) system.
***Describe** another human body system in this same manner.*

Human Body Systems

The circulatory system and lymphatic system both transport materials throughout the body. The heart and blood vessels pump and carry blood through the body. Blood transports nutrients and oxygen to cells and removes waste molecules excreted by cells. In the lymphatic system, vessels absorb fat from the digestive system and collect excess tissue fluid, which is returned to the blood and, thus, the circulatory system. The circulatory system and the lymphatic and immune systems are also involved in protecting the body against disease and substances that are foreign to the body.

Three systems—the digestive, respiratory, and excretory systems—add and/or remove substances from the blood. The digestive system processes food into nutrient molecules that are absorbed by the small intestine and enter the blood. The respiratory system brings oxygen into the body and removes carbon dioxide from the body. It also exchanges gases with the blood. The excretory (urinary) system rids the body of wastes and helps regulate the fluid level and chemical content of the blood.

The sensory receptors in the integumentary system, which consists of the skin, nails, hair, and glands, communicate with the brain and spinal cord via nerve fibres. The muscular and the skeletal systems enable the body and its parts to move. These two systems, along with the integumentary system, also protect and support the internal environment of the body.

The nervous system allows the body to respond to both external and internal stimuli. The endocrine system consists of the hormonal glands that secrete chemicals that serve as messengers between body cells. Both the nervous and the endocrine systems coordinate and regulate the functions of the body's other systems. You will learn more about these two systems and how they work together in this chapter and in Chapter 9.

The reproductive system involves different organs in the male and the female body. However, in both genders, the reproductive system produces and transports gametes and produces sex hormones.

Homeostasis

Whether you are sleeping, studying, enjoying a nice meal, exercising, working outside on a hot day, or hiking in below freezing temperatures, your body is working to maintain your internal temperature near a set point of 37°C. During these activities, your body works to maintain your blood glucose level around 100 mg/mL. Several processes in your body will help keep the pH of your blood near 7.4. Regardless of external conditions, the internal environment of your body remains stable or relatively constant. The tendency of the body to maintain a relatively constant internal environment is known as **homeostasis**. Homeostasis is critical for survival because, like other vertebrates, the human body can survive only within a narrow range of conditions.

homeostasis the tendency of the body to maintain a relatively constant internal environment

Homeostasis is a dynamic process. What this means is that any given variable, such as body temperature, blood glucose levels, or blood oxygen levels, may rise and fall around an average value throughout the course of a day, but still be considered to be in balance. For example, blood glucose levels change in response to consuming food or going long periods without eating. **Figure 8.2** shows how blood glucose levels may change throughout the day in a healthy individual. After a meal, blood glucose levels can rise quickly, especially if you've eaten something with lots of carbohydrates, such as pasta or potatoes. The endocrine system then reacts to bring glucose levels back to a normal value. If you were to go a prolonged period of time without eating, such as skipping a meal, blood glucose levels would start to fall. When that happens, the endocrine and nervous systems work to keep glucose levels within a normal range.

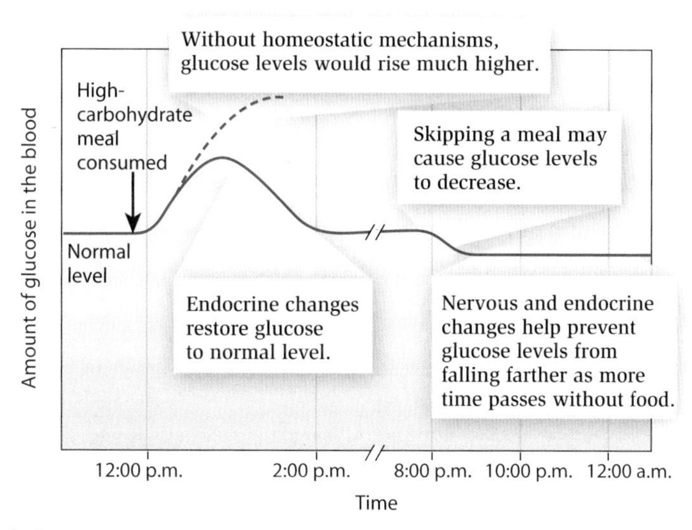

Figure 8.2 Blood glucose levels may rise and fall depending on whether a person has eaten recently. However, even after a sugary meal or skipping a meal, homeostatic mechanisms work to keep glucose levels within a normal range.

Feedback Systems

Whether it is from changes in the external environment or changes within the body, homeostasis is disturbed continually. Body systems respond by constantly monitoring any internal changes and maintaining homeostasis through feedback systems. A feedback system is a cycle of events in which a variable, such as body temperature, blood glucose level, or blood pH, is continually monitored, assessed, and adjusted. A feedback system consists of three components:

- a **sensor**, which detects a change in the internal environment and sends a signal to a control centre
- a **control centre**, which sets the range of values within which a variable should be maintained, receives information from the sensor, and sends signals to effectors when needed
- an **effector**, which receives signals from a control centre and responds, resulting in a change to an internal variable

The body uses two types of feedback systems to regulate its internal environment: negative feedback systems and positive feedback systems.

sensor a body structure that monitors and detects changes in the internal environment

control centre a body structure that sets the range of values within which a variable should be maintained, receives information from the sensor, and sends signals to effectors when needed

effector a body structure that responds to signals from a control centre to effect change in a variable

Learning Check

1. List the four biological levels of organization in a human in order from most simple to most complex.
2. Define the term "homeostasis."
3. What does the term "feedback system" refer to with respect to homeostasis?
4. List the three components of a feedback system and describe each briefly.
5. Summarize the connection between the integumentary system and the nervous system.
6. Using the stomach as an example, illustrate the four biological levels of organization in a human.

negative feedback system mechanism of homeostatic response by which the output of a system reverses a change in a variable, bringing the variable back to within normal range

Negative Feedback Systems

In a **negative feedback system**, the body works to reverse a change detected in a variable so that the variable is brought back to within a normal range. **Figure 8.3A** compares negative feedback to the way a seesaw moves. A seesaw is level when the forces acting on it are balanced. If a change occurs to disrupt this balance, the seesaw can be made level again by applying a force to reverse the change. In terms of negative feedback, a sensor detects a change that disrupts a balanced state and signals a control centre. The control centre then activates an effector, which reverses the change and restores the balanced state.

You know that when you engage in moderate to vigorous exercise for more than a few minutes, you start to sweat. But do you know why? As shown in **Figure 8.3B**, sweating is part of a negative feedback system your body uses to keep your internal temperature as close to 37°C as possible. As you exercise, your muscles produce heat, which raises the temperature of the blood. As a result, signals are sent to the control centre. The control centre directs a response to several effectors, including blood vessels and sweat glands. The blood vessels dilate, resulting in heat loss through radiation and conduction. The sweat glands release sweat. As sweat evaporates from the skin, heat is released from the body. These responses continue until body temperature returns to normal.

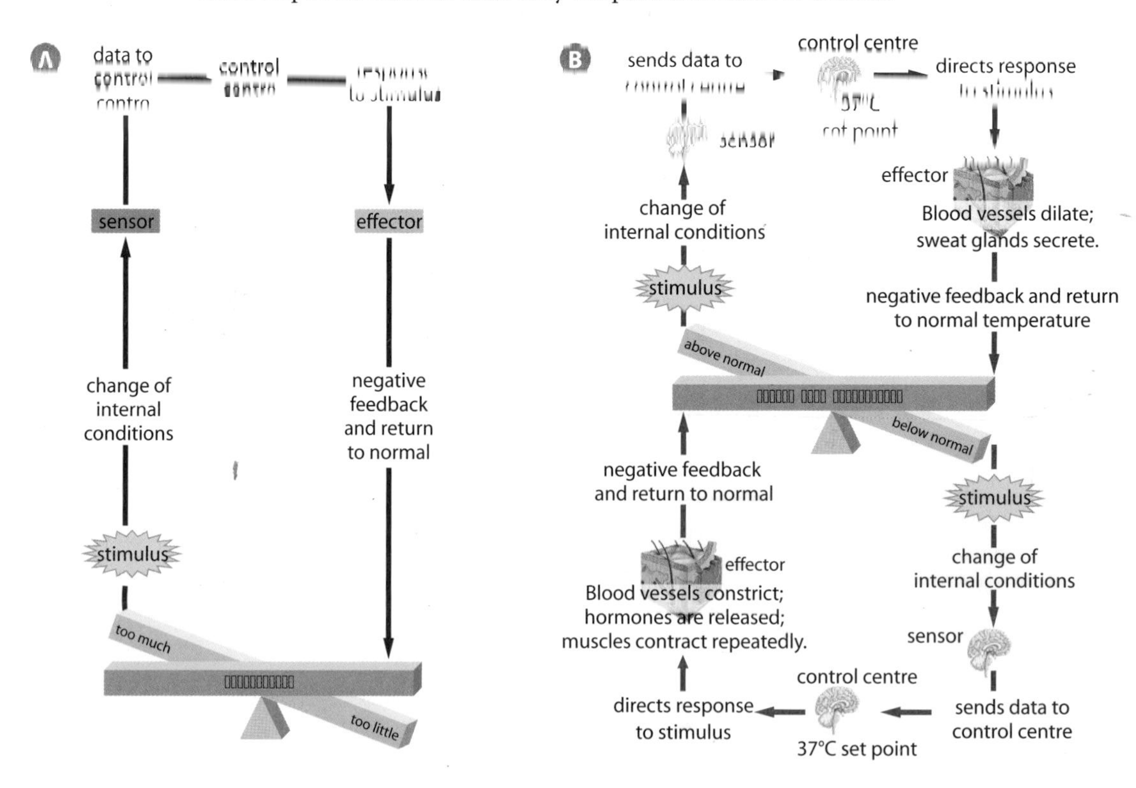

Figure 8.3 A In a negative feedback system, homeostasis is returned to normal. **B** Body temperature is maintained by a negative feedback system.

Identify *the effectors involved in raising body temperature.*

When you are outside on a cold day, such as when you are waiting for the bus to pick you up, you may start to shiver. Like sweating, shivering occurs in response to a change in body temperature; in this case, body temperature has begun to fall below normal. As shown in **Figure 8.3B**, sensors in the skin and brain send messages to the control centre. The control centre sends messages to several effectors. Blood vessels in the skin constrict, decreasing heat loss through the skin. Hormones are released that lead to an increase in body metabolism, which generates heat. Muscles begin to contract repeatedly, which results in shivering, which increases heat production. When body temperature is restored to normal, the feedback cycle and these responses stop.

Positive Feedback Systems

Unlike a negative feedback system, a **positive feedback system** tends to strengthen or increase a change in a variable. One example of a body process that is controlled by a positive feedback system is blood clotting. After an injury occurs, the affected tissues release chemicals that activate platelets. The platelets begin the clotting process. As well, they release chemicals that stimulate further clotting until the bleeding stops.

positive feedback system mechanism of homeostatic response by which the output of a system strengthens or increases a change in a variable

As shown in **Figure 8.4**, a positive feedback system also regulates contractions during childbirth. When a woman is giving birth, the uterus contracts, forcing the baby's head or body into the cervix. The head of the baby presses against the cervix. This stimulates sensors in the cervix. Impulses are sent to the brain (control centre), which causes the pituitary gland (a gland in the endocrine system) to release oxytocin. Oxytocin is a hormone that causes muscles in the wall of the uterus (effectors) to contract. As labour continues, sensors in the cervix continue to send impulses to the brain, which leads to the release of more oxytocin. The release of more oxytocin leads to ever-stronger contractions until birth occurs.

Positive feedback systems are far less common than negative feedback systems. As exemplified above, positive feedback systems tend to be involved in processes that have a definitive cut-off point. In the case of clotting, the feedback cycle stops when bleeding stops. In the case of childbirth, the feedback cycle stops when the birth of the baby occurs. As you read this chapter and the two that follow, you will learn about more examples of negative and positive feedback systems in the human body.

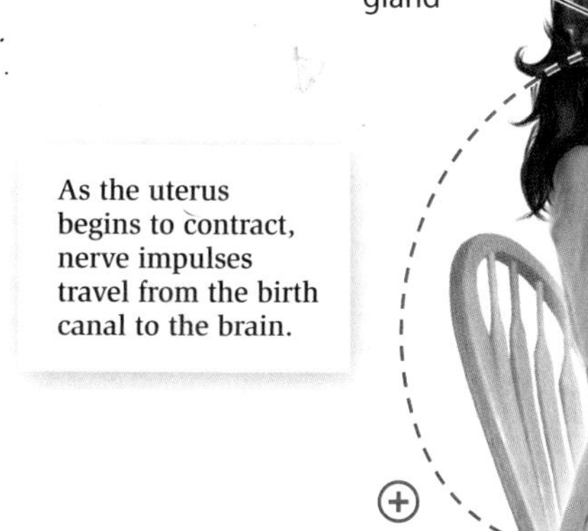
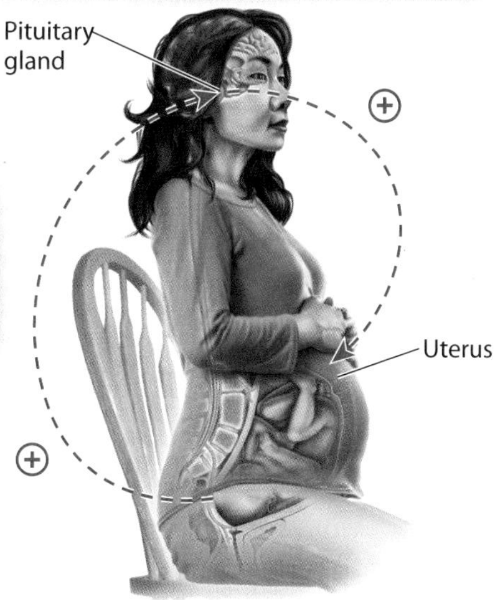

The brain triggers the release of hormones from the pituitary gland, which enter the blood and increase the strength of uterine contractions. The strong contractions stimulate the brain to make even more hormones.
Pituitary gland
As the uterus begins to contract, nerve impulses travel from the birth canal to the brain.
Uterus

Figure 8.4 Contractions during childbirth are regulated by a positive feedback system.

Section 8.1 Review

Section Summary

- The human body is organized in a hierarchy of levels. Cells are the smallest unit in the body. Tissues are groups of similar cells, and organs are tissues that perform a particular function. Organs work together in an organ system.
- The human body systems transport blood and lymph, digest food, excrete wastes, move and protect the body, and maintain homeostasis.
- Homeostasis is the tendency of the body to maintain a relatively constant internal environment and is critical for survival.
- Homeostasis is maintained through feedback systems that continually monitor, assess, and adjust variables in the body's internal environment.
- The two types of feedback systems are negative feedback systems and positive feedback systems.

Review Questions

1. **K/U** The four biological levels of organization are tissues, organ systems, cells, and organs. Arrange these four levels in order from simplest to most complex.
2. **C** Write a paragraph identifying three body systems that are directly involved in kicking a ball. Include how these three systems work together to perform that task.
3. **K/U** Identify a, b, and c in the diagram below. Record your answer in your notebook.

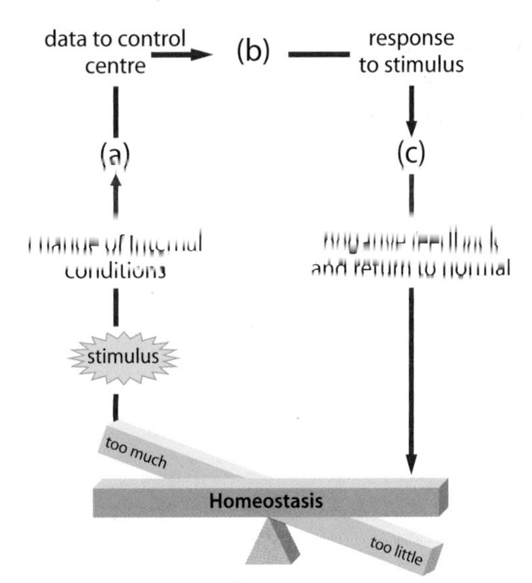

4. **C** Use **Figure 8.3B** to draw a flowchart summarizing the events that occur in your body as it works to maintain homeostasis in response to internal body temperature dropping below normal.
5. **C** Use a Venn diagram or other graphic organizer to compare a negative feedback system to a positive feedback system.
6. **K/U** Describe two examples of where a positive feedback system is used in the body.
7. **T/I** Explain the following statement. "All systems of the body contribute to homeostasis."
8. **C** Human growth hormone (hGH) helps to regulate blood glucose levels. If low blood glucose levels are detected, the brain stimulates the pituitary gland to secrete hGH. Human growth hormone travels through the blood to the liver where it stimulates the liver to convert glycogen to glucose, which is released into the bloodstream raising blood sugar levels. Using a diagram resembling the seesaw model in **Figure 8.3A**, illustrate the negative feedback loop described above.
9. **A** A fever is an increase in body temperature above the normal range. A fever causes metabolic changes that push the person's body temperature still higher. If body temperature rises above 45°C death occurs because cellular proteins change shape and metabolism stops. Is a fever an example of a positive or negative feedback system? Explain your answer.
10. **T/I** The diagram below shows the range in which body temperature is maintained by homeostasis.

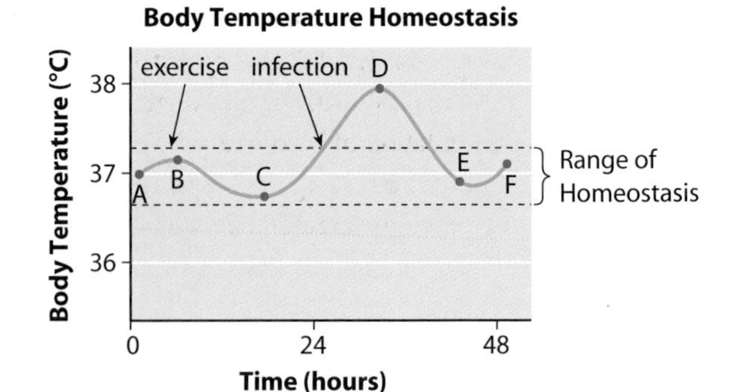

   a. Identify what type of homeostatic feedback system is working between points A and C on this graph. Explain your answer.
   b. Identify what type of homeostatic feedback system is working between points C and D on this graph. Explain your answer.
   c. Infer what occurred between points D and E on the graph during which body temperature returned to normal.

SECTION 8.2

Structures and Processes of the Nervous System

Figure 8.5 This child's nervous system is continuously monitoring and responding to changes in the external and internal environment. In this example, homeostasis is maintained despite the frigid outdoor temperatures.

Key Terms

central nervous system
peripheral nervous system
neuron
nerve
glial cell
myelin sheath
reflex arc
membrane potential
resting membrane potential
polarization
sodium-potassium pump
action potential
synapse
neurotransmitter
acetylcholine

The human nervous system is equipped to sense and respond to continuous change within both the body and the external environment. The nervous system performs the vital function of regulating body structures and processes to maintain homeostasis despite fluctuations in both the internal and the external environment.

For example, the child in **Figure 8.5** lives in what many people consider to be an inhospitable environment, with winter temperatures often falling to –50°C and lower. For the Arctic Inuit, maintaining a constant internal temperature while keeping blood and heat flowing to the extremities is crucial. Researchers have discovered that the nervous systems of people living in colder climates act to constrict blood flow to an extremity (and thus conserve body heat) when the extremity is cooled. However, the constant constriction of blood vessels can lead to frostbite when outside temperatures are low enough. In Inuit who have lived for generations in the far North, the nervous system fluctuates between the constriction and dilation of blood vessels to cooled extremities. The constriction of the vessels has the effect of conserving body heat, but the dilation allows for continued blood flow to prevent frostbite. The careful balance between constriction and dilation helps to ensure that neither too much heat loss nor frostbite occurs. In this way, homeostasis is maintained.

To maintain homeostasis, the human body must react to differences in temperature as well as respond to various internal and external stimuli, and it must regulate these responses. The human nervous system can regulate tens of thousands of activities simultaneously. The nervous system monitors and controls body processes, from automatic functions (such as breathing) to activities that involve fine motor coordination, learning, and thought (such as playing a musical instrument). The brain and spinal cord, and the nerves that emerge from them and connect them to the rest of the body, make up the nervous system. The human nervous system is perhaps the most complex system of any organism. The human brain alone contains more than 100 billion nerve cells, and each nerve cell can have up to 10 000 connections to other nerve cells. This means that a nerve impulse—an electrochemical signal—to or from the brain could travel along 10^{15} possible routes.

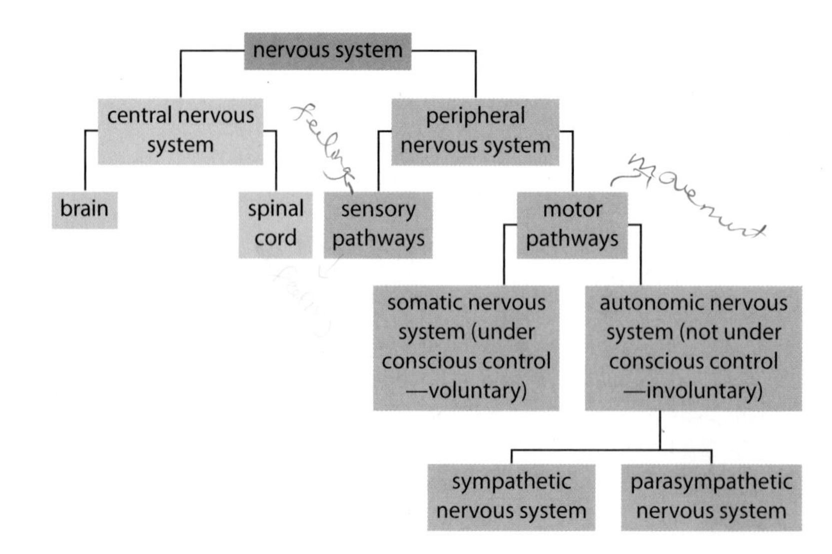

Figure 8.6 The human nervous system is divided into the central nervous system and the peripheral nervous system. The peripheral nervous system is further divided into the somatic and autonomic nervous systems.

central nervous system network of nerves that includes the brain and spinal cord; integrates and processes information sent by nerves

peripheral nervous system network of nerves that carry sensory messages to the central nervous system (CNS) and send information from the CNS to the muscles and glands; consists of the autonomic and somatic system

An Overview of the Nervous System

As shown in **Figure 8.6**, the nervous system has two major divisions: the central nervous system (CNS) and the peripheral nervous system (PNS). Early researchers made this distinction based on where nervous tissue was located in the body—centrally or peripherally (away from the centre). Together, the central nervous system and the peripheral nervous system control sensory input, integration, and motor output.

The **central nervous system**, which consists of the brain and spinal cord, integrates and processes information sent by nerves. The peripheral nervous system includes nerves that carry sensory messages to the central nervous system and nerves that send information from the CNS to the muscles and glands.

The **peripheral nervous system** is further divided into the somatic system and the autonomic system, also shown in Figure 8.6. The somatic system consists of sensory receptors in the head and extremities, nerves that carry sensory information to the central nervous system, and nerves that carry instructions from the central nervous system to the skeletal muscles. The somatic system is under voluntary control. The autonomic system controls glandular secretions and the functioning of the smooth and cardiac muscles. These processes are involuntary. Involuntary processes, such as heartbeat and peristalsis, are those that do not require or involve conscious control. The sympathetic and parasympathetic divisions of the autonomic system often work in opposition to each other to regulate the involuntary processes of the body. You will learn more about the peripheral nervous system in the last section of this chapter.

neuron nerve cell; the structural and functional unit of the nervous system, consisting of a nucleus, cell body, dendrites, and axons; specialized to respond to physical and chemical stimuli, to conduct electrochemical signals, and to release chemicals that regulate various body processes

nerve message pathway of the nervous system; made up of many neurons grouped into bundles and surrounded by protective connective tissue

glial cell support cell of the nervous system that nourishes neurons (nerve-impulse conducting cells), removes their wastes, defends against infection, and provides a supporting framework for all the nervous-system tissue

Cells of the Nervous System

The nervous system is composed of only two main types of cells: neurons and cells that support the neurons, which are called glial cells. **Neurons** are the basic structural and functional units of the nervous system. They are specialized to respond to physical and chemical stimuli, to conduct electrochemical signals, and to release chemicals that regulate various body processes. Individual neurons are organized into tissues called **nerves**.

The activity of neurons is supported by another type of cells called **glial cells**. The word *glial* comes from a Greek word that means "glue." Collectively, glial cells nourish the neurons, remove their wastes, and defend against infection. Glial cells also provide a supporting framework for all the nervous-system tissue. **Figure 8.7** shows a small sample of this tissue.

The Structure of a Neuron

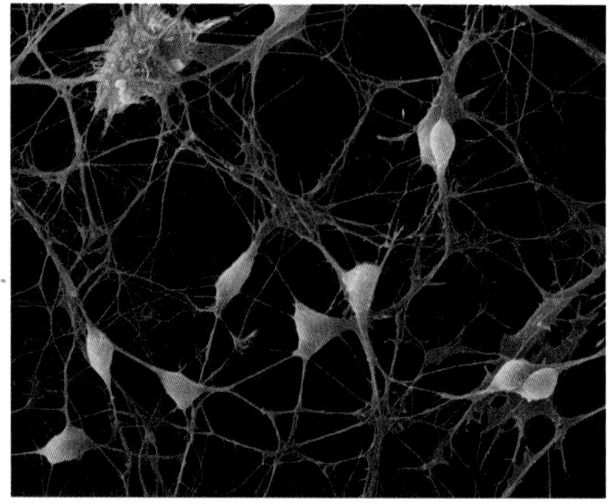
Magnification: 600×

Figure 8.7 Glial cells, shown in green in this micrograph, support neurons (shown in orange).

Neurons have many of the same features as other body cells, such as a cell membrane, cytoplasm, mitochondria, and a nucleus. In addition, neurons have specialized cell structures that enable them to transmit nerve impulses. Different types of neurons are different shapes and sizes. In general, however, they share four common features: dendrites, a cell body, an axon, and branching ends, all shown in **Figure 8.8**.

Dendrites are short, branching terminals that receive nerve impulses from other neurons or sensory receptors, and relay the impulse to the cell body. The dendrites are numerous and highly branched, which increases the surface area available to receive information. The *cell body* contains the nucleus and is the site of the cell's metabolic reactions. The cell body also processes input from the dendrites. If the input received is large enough, the cell body relays it to the axon, where an impulse is initiated.

The *axon* conducts impulses away from the cell body. Axons range in length from 1 mm to 1 m, depending on the neuron's location in the body. For example, the sciatic nerve in the leg contains neuronal axons that extend from the spinal cord all the way to the muscles in the foot, a distance of over 1 m. The terminal end of an axon branches into many fibres, as shown in **Figure 8.8**. To communicate with adjacent neurons, glands, or muscles, the axon terminal releases chemical signals into the space between it and the receptors or dendrites of neighbouring cells.

The axons of some neurons are enclosed in a fatty, insulating layer called the **myelin sheath**, which gives the axons a glistening white appearance. The myelin sheath protects neurons and speeds the rate of nerve impulse transmission. *Schwann cells*, a type of glial cell, form myelin by wrapping themselves around the axon. You will learn more about the importance of myelin sheaths later in this section.

myelin sheath the fatty, insulating layer around the axon of a nerve cell, composed of Schwann cells; protects myelinated axons and speeds the rate of nerve impulse transmission

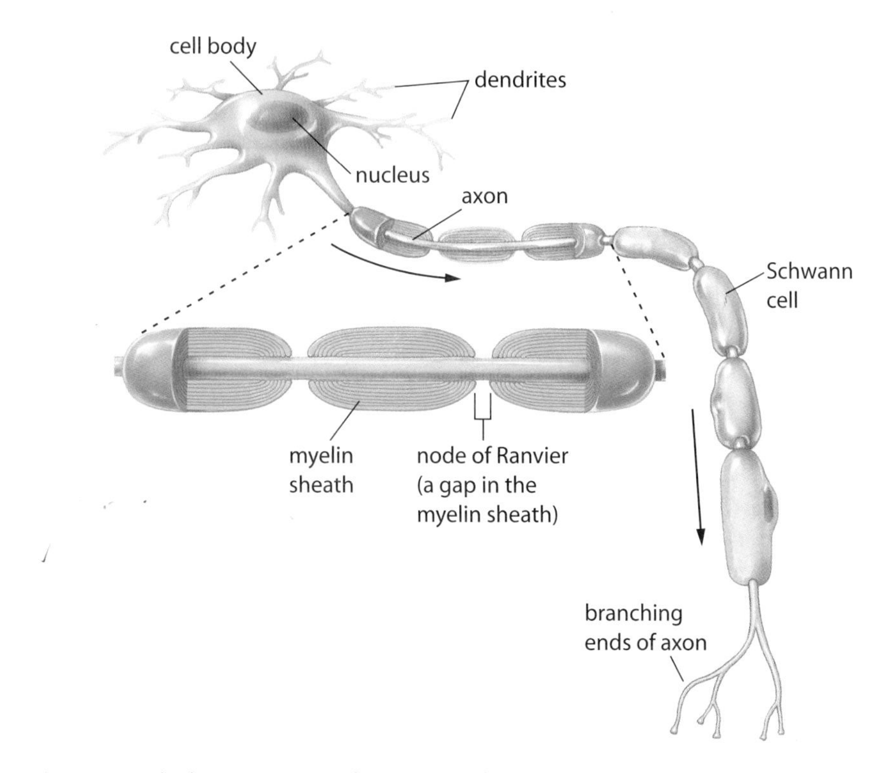

Figure 8.8 The basic structure of a neuron includes dendrites, the cell body, and an axon. The neuron shown here is a myelinated neuron, which means it is covered by a myelin sheath.

Classifying Neurons

Neurons can be classified based on their structure as well as their function. Structurally, neurons are classified based on the number of processes that extend from the cell body. **Table 8.1** describes three types of neurons based on structure: multipolar, bipolar, and unipolar neurons.

Table 8.1 Types of Neurons

Multipolar Neuron	Bipolar Neuron	Unipolar Neuron
Dendrites Cell body Axon	Dendrite Cell body Axon	Dendrites Cell body Axon Axon branches function as a single axon.
• Has several dendrites • Has a single axon • Found in the brain and spinal cord	• Has a single main dendrite • Has a single axon • Found in the inner ear, the retina of the eye, and the olfactory area of the brain	• Has a single process that extends from the cell body • Dendrite and axon are fused • Found in the peripheral nervous system

Functionally, neurons are classified as one of three main types: sensory neurons, interneurons, or motor neurons. These three main types of neurons form the basic impulse-transmission pathway of the entire nervous system. This pathway, shown in **Figure 8.9**, depends on three overlapping functions: sensory input, integration, and motor output.

1. Sensory input: Sensory receptors, such as those in the skin, receive stimuli and form a nerve impulse. Sensory neurons transmit impulses from the sensory receptors to the central nervous system (brain and spinal cord).
2. Integration: Interneurons are found entirely within the central nervous system. They act as a link between the sensory and motor neurons. They process and integrate incoming sensory information, and relay outgoing motor information.
3. Motor output: Motor neurons transmit information from the central nervous system to effectors. Effectors include muscles, glands, and other organs that respond to impulses from motor neurons.

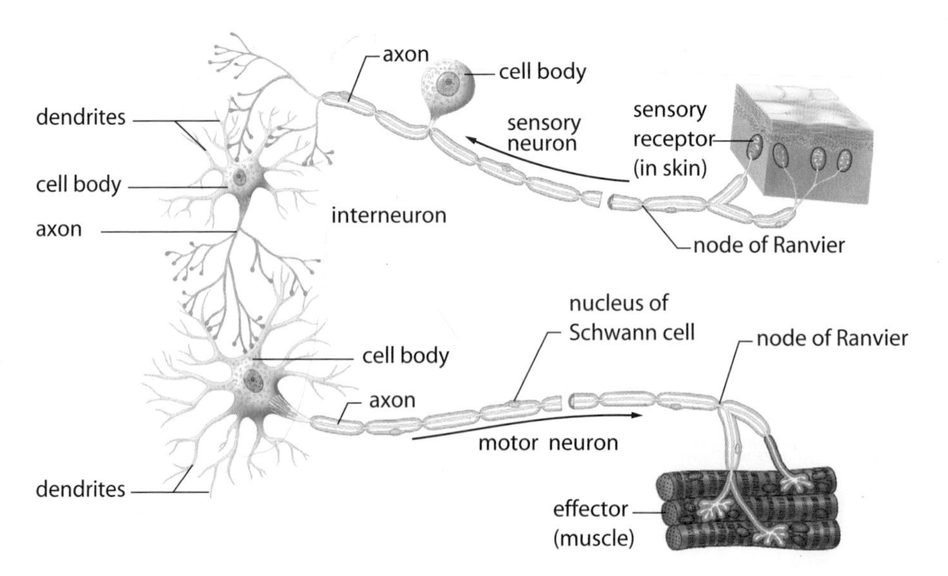

Figure 8.9 This diagram shows how a sensory neuron, an interneuron, and a motor neuron are arranged in the nervous system. (The breaks indicate that the axons are longer than shown.)

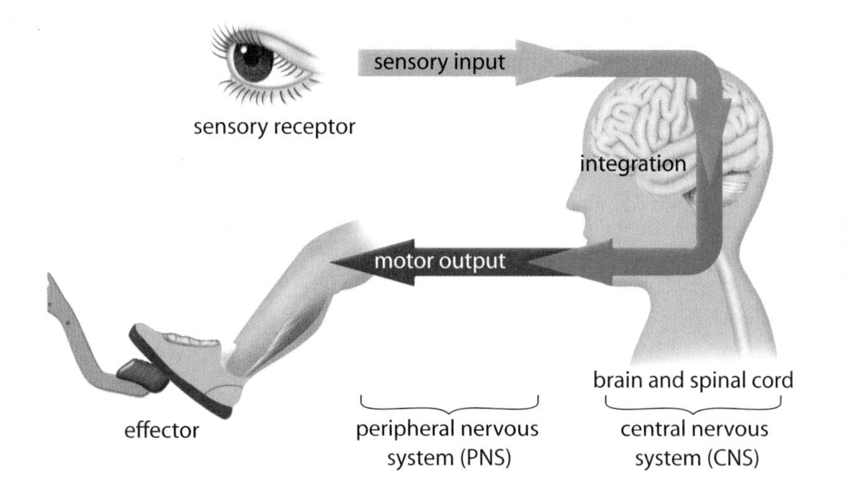

Figure 8.10 Your nervous system reacts to sensory input through a basic impulse pathway.

Figure 8.10 shows another example of a basic neural transmission pathway. Suppose that you are driving a car, and a cat darts onto the road in front of you. Sensory receptors in your eyes collect the information (the sight of the cat on the road), and sensory neurons transmit this information by conducting electrochemical signals to the brain and spinal cord. Here the information is integrated by interneurons. Motor neurons then carry motor output signals to the muscles (effectors), causing you to extend your foot and press the brake.

Suggested Investigation

Inquiry Investigation 8-A, Move Fast! Reflex Responses

The Reflex Arc

Some neurons are organized to enable your body to react rapidly in times of danger, even before you are consciously aware of the threat. These sudden, involuntary responses to certain stimuli are called *reflexes*. Examples of reflexes are jerking your hand away from a hot or sharp object, blinking when an object moves toward your eye, or vomiting in response to food that irritates your stomach. **Reflex arcs** are simple connections of neurons that explain reflexive behaviours. They can be used to model the basic organization of the nervous system.

reflex arc simple connection of neurons that results in a reflex action in response to a stimulus

Reflex arcs usually involve only three neurons to transmit messages. As a result, reflexes can be very rapid, occurring in about 50 ms (milliseconds). Withdrawal reflexes, for example, depend on only three neurons. **Figure 8.11** illustrates a typical neural circuit, as well as a withdrawal reflex from a potentially painful situation. Receptors in the skin sense the pressure of the cactus needle and initiate an impulse in a sensory neuron. The impulse carried by the sensory neuron then activates the interneuron in the spinal cord. The interneuron signals the motor neuron to instruct the muscle to contract and withdraw the hand.

A reflex arc moves directly to and from the brain or spinal cord, before the brain centres involved with voluntary control have time to process the sensory information. This is why, after stepping on a stone, you would not feel pain or cry out until after your foot was withdrawn, once the brain has had time to process the information.

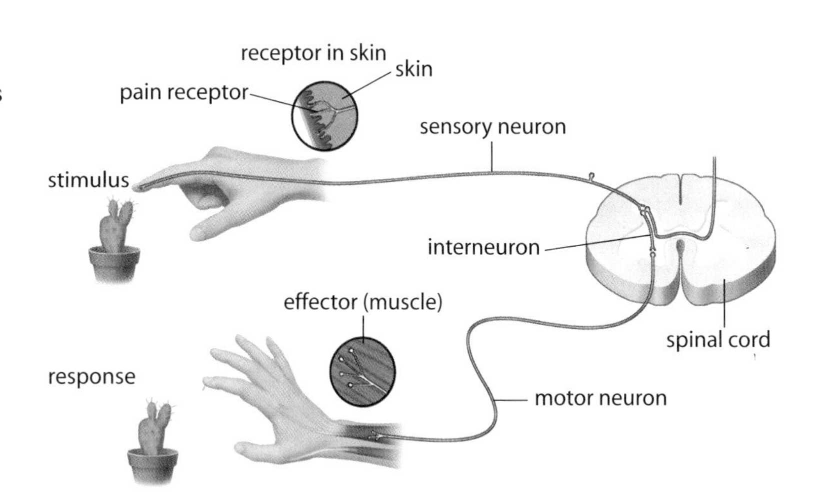

Figure 8.11 In this example of a withdrawal reflex, receptors in the skin perceive the stimulus. Sensory information is conducted from the senses into the spinal cord. Motor information is then conducted away from the spinal cord to the muscles and glands.

Learning Check

7. Explain why the nervous system is critical for maintaining homeostasis.
8. Create a table to identify the different systems in the nervous system and explain the structure and function of each.
9. Compare and contrast the basic function of neurons and glial cells.
10. Draw a neuron, label its basic structures, and identify their functions.
11. List three types of neurons based on structure. Explain the differences among these three neurons.
12. Identify the basic neural pathway that is involved as you dodge a tennis ball. Compare this pathway with a withdrawal reflex.

The Electrical Nature of Nerves

In the late 1700s, Italian scientists Luigi Galvani and Alessandro Volta experimented with ways to stimulate the contraction of frog leg muscles placed in salt solutions. They discovered that stimulating either the nerve that led to the muscle or the muscle directly with electrical current caused the muscle to contract. Galvani hypothesized that electric current could be generated by the tissue itself, a concept he referred to as "animal electricity." Experimental evidence gathered by other scientists over the centuries has supported Galvani's hypothesis. Today, we know that neurons use electrical signals to communicate with other neurons, muscles, and glands. These signals, called nerve impulses, involve changes in the amount of electric charge across a cell's plasma membrane.

membrane potential electrical charge separation across a cell membrane; a form of potential energy

resting membrane potential potential difference across the membrane in a resting neuron

Resting Membrane Potential

When microelectrodes are inserted in an inactive, or resting, neuron, measurements from a voltmeter indicate an electrical potential difference (voltage) across the neural membrane. In a resting neuron, the cytoplasmic side of the membrane (inside of the cell) is negative, relative to the extracellular side (outside of the cell). The charge separation across the membrane is a form of potential energy, or **membrane potential**.

The **resting membrane potential** is the potential difference across the membrane in a resting neuron. The resting membrane potential of most unstimulated neurons is about −70 mV (millivolts). The electrical charge is negative on the inside of the cell, relative to the outside of the cell, as shown in **Figure 8.12**. The resting membrane potential provides energy for the generation of a nerve impulse in response to an appropriate stimulus.

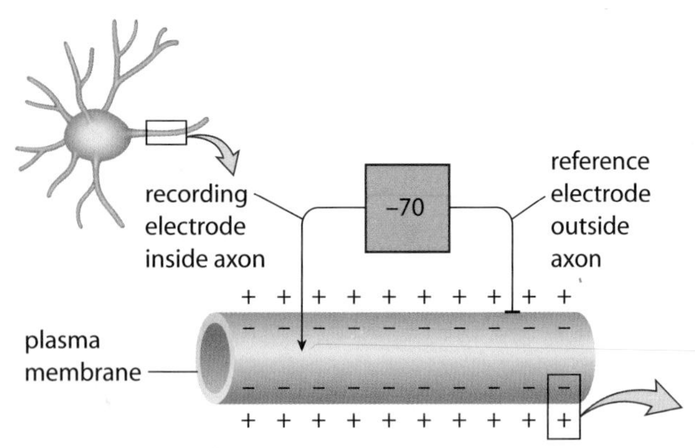

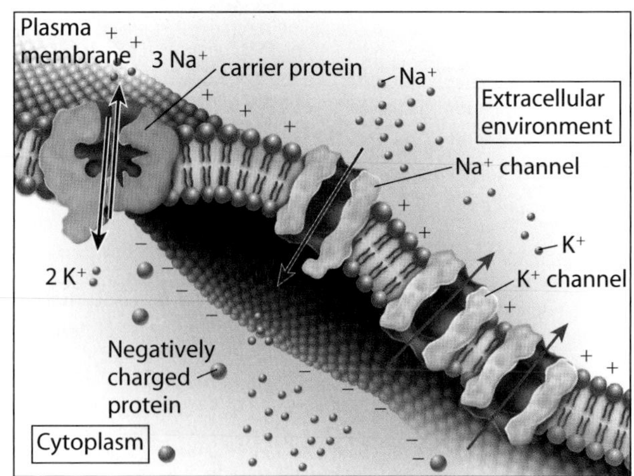

Figure 8.12 Three main factors influence resting membrane potential. Large, negatively charged proteins are located inside the cell. Channels in the membrane allow K^+ to diffuse out of the cell more easily than Na^+ can move into the cell. The sodium-potassium pump moves sodium and potassium ions across the cell membrane in different ratios.

Three factors contribute to maintaining resting membrane potential. First, large protein molecules that are negatively charged are present in the intracellular fluid but not outside of the cell. These proteins are so large that they cannot pass through the cell membrane, contributing to the negative charge in the interior of the cell. Second, the plasma membrane contains ion-specific channels that allow for the passive movement of ions, such as sodium (Na^+) and potassium (K^+), across the membrane. In particular, K^+ channels tend to be open more often at resting potential. This means that potassium can move out of the cell more readily, whereas sodium cannot move into the cell as easily, making the interior of the cell more negative relative to the exterior. Third, and most important, is the sodium-potassium pump, which actively transports Na^+ and K^+ in ratios that leave the inside of the cell negatively charged compared to the outside of the cell. The process of generating a resting membrane potential of −70 mV is called **polarization**.

polarization lowering the membrane potential of the cell below its equilibrium value; in nerves, the process of generating a resting membrane potential of −70 mV

sodium-potassium pump system involving a carrier protein in the plasma membrane that uses the energy of ATP to transport sodium ions out of and potassium ions into animal cells; important in nerve and muscle cells

Sodium-Potassium Pump

The most important contributor to the separation of charge and the resulting electrical potential difference across the membrane is the **sodium-potassium pump**. This system uses the energy of ATP to transport sodium ions out of the cells and potassium ions into the cells. The process is shown in **Figure 8.13**.

Notice that for every three sodium ions transported out of the cell, two potassium ions are transported into the cell. As a result, an excess of positive charge accumulates outside of the cell. Recall that the cell membrane is not totally impermeable to sodium and potassium ions, so they also leak slowly by diffusion across the membrane in the direction of their concentration gradient. However, potassium ions are able to diffuse out of the cell more easily than sodium ions can diffuse into the cell. The overall result of the active transport of sodium and potassium ions across the membrane, and their subsequent diffusion back across the membrane, is a constant membrane potential of −70 mV.

You might wonder why the −70 mV potential difference across the neuronal membrane is called the *resting* membrane potential when the sodium-potassium pump is constantly using energy to transport these ions. The term "resting" means that no nerve impulses are being transmitted along the axon. The resting potential maintains the axon membrane in a condition of readiness for an impulse to occur. The energy for any eventual impulses is stored in the electrochemical gradient across the membrane.

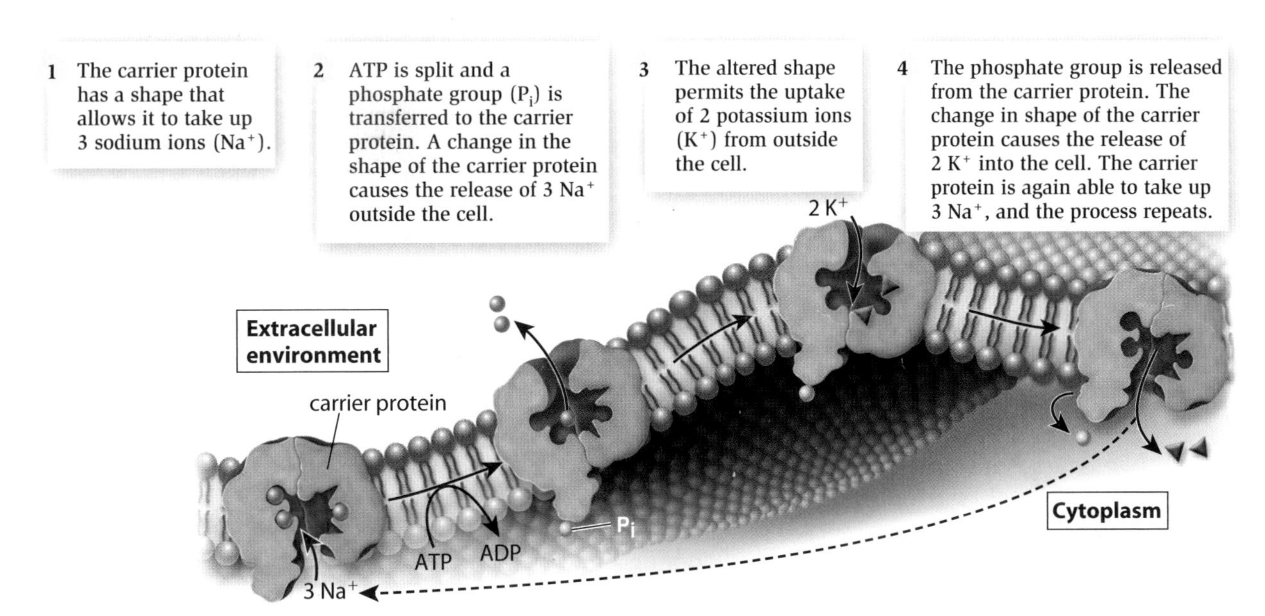

1 The carrier protein has a shape that allows it to take up 3 sodium ions (Na⁺).
2 ATP is split and a phosphate group (Pi) is transferred to the carrier protein. A change in the shape of the carrier protein causes the release of 3 Na⁺ outside the cell.
3 The altered shape permits the uptake of 2 potassium ions (K⁺) from outside the cell.
4 The phosphate group is released from the carrier protein. The change in shape of the carrier protein causes the release of 2 K⁺ into the cell. The carrier protein is again able to take up 3 Na⁺, and the process repeats.
2 K⁺
Extracellular environment
carrier protein
Pi
ATP
ADP
3 Na⁺
Cytoplasm

Figure 8.13 The sodium-potassium pump actively transports three sodium ions (Na^+) outside of the cell for every two potassium ions (K^+) moved inside the cell. Small amounts of Na^+ and K^+ also diffuse ("leak") slowly across the cell membrane, following their concentration gradient.

Suggested **Investigation**

Inquiry Investigation 8-B, Modelling Resting Membrane Potential

Action Potential

A nerve cell is polarized because of the difference in charge across the membrane—specifically that the inside of the cell is more negative than the outside of the cell. Changes in membrane potential are changes in the degree of polarization. *Depolarization* occurs when the cell becomes less polarized, meaning that the membrane potential is reduced to less than the resting potential of −70 mV. During depolarization, the inside of the cell becomes less negative relative to the outside of the cell. What causes a nerve cell to become depolarized, and how does the change in charge occur?

action potential in an axon, the change in charge that occurs when the gates of the K^+ channels close and the gates of the Na^+ channels open after a wave of depolarization is triggered

An action potential causes depolarization to occur. An **action potential** is the movement of an electrical impulse along the plasma membrane of an axon. It results in a rapid change in polarity across the axon membrane as the nerve impulse occurs. An action potential is an *all-or-none* phenomenon. If a stimulus causes the axon membrane to depolarize to a certain level, referred to as the *threshold potential*, an action potential occurs. Threshold potentials can vary slightly, depending on the type of neuron, but they are usually close to −50 mV. In an all-or-none response, the strength of an action potential does not change based on the strength of the stimulus. However, a strong stimulus can cause an axon to start an action potential more often in a given time period than a weak stimulus.

Figure 8.14 illustrates the changes in the membrane potential that occur during an action potential. The events are also sequenced below. Notice that all of these events occur within a period of a few milliseconds. As well, they occur in one small region of the axon membrane.

- An action potential is triggered when the threshold potential is reached.
- When the membrane potential reaches threshold, special structures in the membrane called voltage-gated sodium channels open and make the membrane very permeable to sodium ions. The sodium ions on the outside of the axon suddenly move down their concentration gradient and rush into the axon. Within a millisecond or less, enough positively charged sodium ions have crossed the membrane to make the potential difference across the membrane in that tiny region of the axon +40 mV.
- As a result of the change in membrane potential, the sodium channels close and voltage-gated potassium channels open. The potassium ions now move down their concentration gradient toward the outside of the axon, carrying positive charge out of the neuron. As a result, the membrane potential becomes more negative again. In fact, the membrane potential becomes slightly more negative than its original resting potential, becoming *hyperpolarized* to about −90 mV. At this point, the potassium channels close.
- The sodium-potassium pump and the small amount of naturally occurring diffusion quickly bring the membrane back to its normal resting potential of −70 mV. The membrane is now *repolarized*—that is, returned to its previous polarization.
- For the next few milliseconds after an action potential, the membrane cannot be stimulated to undergo another action potential. This brief period of time is called the *refractory period* of the membrane.

The entire process described above continues down the length of an axon until it reaches the end, where it initiates a response at the junction with the next cell.

Suggested **Investigation**

Plan Your Own Investigation 8-C, Invertebrate Responses to External Stimuli

Our present understanding of action potentials began with experiments carried out in the 1940s by two British scientists, Alan Hodgkin and Andrew Huxley. Using the axon from a giant squid and measuring ion movements into and out of the cell, they showed that the action potential depends on voltage-gated Na^+ and K^+ channels. They noted that Na^+ channels open and close rapidly, while voltage-gated K^+ channels open later than Na^+ channels and close only toward the end of the action potential. This results in Na^+ channels being involved in the initiation and early phase of the action potential, while K^+ channels control the duration and termination of the action potential.

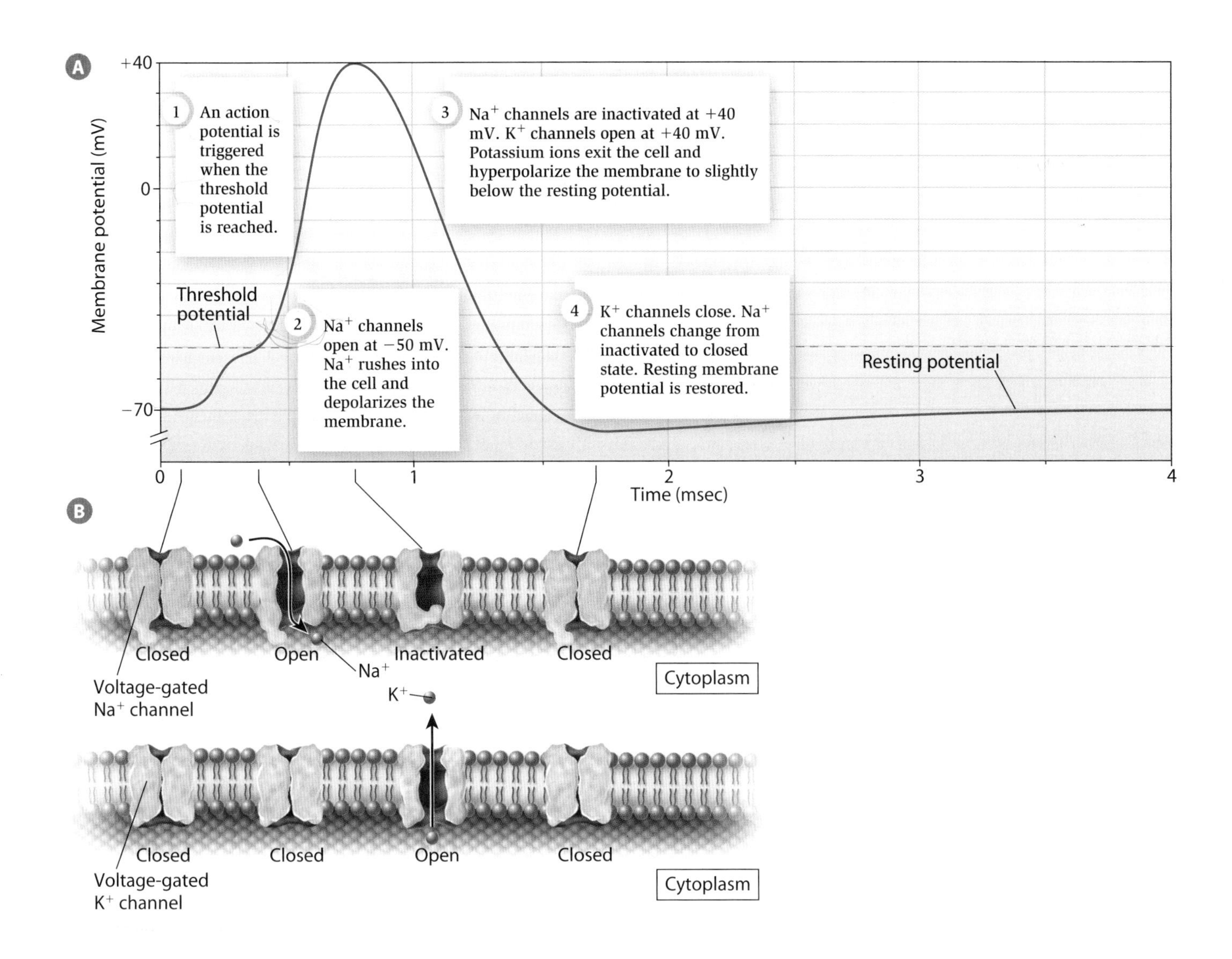
A
+40
0
−70
Membrane potential (mV)
1 An action potential is triggered when the threshold potential is reached.
Threshold potential
2 Na+ channels open at −50 mV. Na+ rushes into the cell and depolarizes the membrane.
3 Na+ channels are inactivated at +40 mV. K+ channels open at +40 mV. Potassium ions exit the cell and hyperpolarize the membrane to slightly below the resting potential.
4 K+ channels close. Na+ channels change from inactivated to closed state. Resting membrane potential is restored.
Resting potential
0
1
2
3
4
Time (msec)
B
Closed
Open
Inactivated
Closed
Na+
Cytoplasm
Voltage-gated Na+ channel
K+
Closed
Closed
Open
Closed
Cytoplasm
Voltage-gated K+ channel

Figure 8.14 **A** The graph shows the changes that occur to membrane potential as an action potential travels down an axon. **B** The changes in membrane potential are caused by the opening and closing of sodium channels and potassium channels.

Predict *How would the curve in the graph change if the sodium channels were unable to close?*

Learning Check

13. Explain what the resting membrane potential is, and why it is significant to the functioning of neurons.

14. Identify three factors that contribute to the resting membrane potential of a neuron.

15. Refer to **Figure 8.13** to summarize, in point form, how the sodium-potassium pump contributes to the separation of charge and the resulting electrical potential difference across the membrane of a neuron.

16. Summarize the changes that occur in an axon as a nerve impulse is transmitted.

17. Explain the importance of repolarization in the transmission of a nerve impulse.

18. Tetrodotoxin is a neurotoxin found in puffer fish. This large molecule blocks the sodium channels in neurons. Infer the effect tetrodotoxin would have on the propagation of an action potential in a neuron.

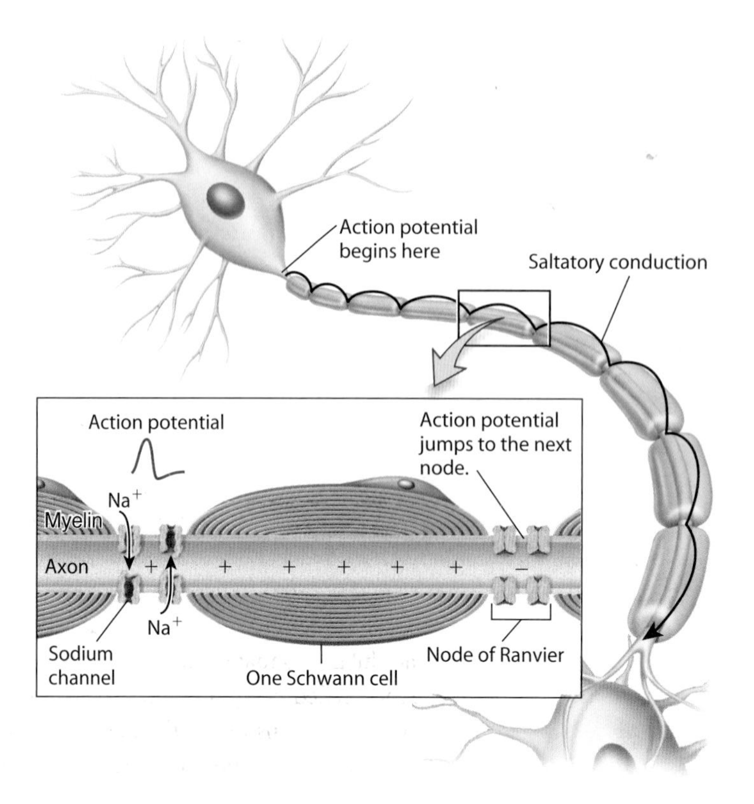

Figure 8.15 In myelinated neurons, action potentials are generated only at the nodes of Ranvier, which lack a surrounding sheath of myelin.
Explain *What is the main advantage of saltatory conduction?*

Myelinated Nerve Impulse

A nerve impulse consists of a series of action potentials. How does one action potential stimulate another? Recall from earlier in this section that the axons of some neurons are enclosed in a fatty, insulating layer called a myelin sheath. At regular intervals, the axons of myelinated neurons have exposed areas known as *nodes of Ranvier*, shown in Figure 8.15.

Nodes of Ranvier contain many voltage-gated sodium channels. The nodes of Ranvier are the only areas of myelinated axons that have enough sodium channels to depolarize the membrane and elicit an action potential. When the sodium ions move into the cell, the charge moves quickly through the cytoplasm to the next node. When the sodium ions reach the neighbouring node of Ranvier, the positive charges reduce the net negative charge inside the axonal membrane. The presence of the positively charged sodium ions causes the membrane at the node to become depolarized to threshold. Since an action potential just occurred at the node to the left, that membrane is refractory, which means that it cannot yet be stimulated to undergo another action potential. This mechanism prevents impulses from going backward. The membrane of the node of Ranvier to the right is not refractory, so the depolarization initiates an action potential at this node. The same process occurs at each node until it reaches the end of the neuron. This process of one action potential stimulating the production of another one at the next node constitutes the nerve impulse.

Because action potentials are forced to "jump" from one node of Ranvier to the next due to the myelin sheath, the conduction of an impulse along a myelinated neuron is called saltatory conduction. The word *saltatory* comes from a Latin word that means to jump or leap. In unmyelinated neurons, conduction of a nerve impulse is continuous. Rather than jumping from one section of an axon to another, action potentials in unmyelinated neurons cause the release of sodium along each adjacent portion of a membrane. As a result of this step-by-step conduction along the axon, the transmission of an impulse along an unmyelinated axon is much slower than the saltatory conduction along a myelinated axon—about 0.5 m/s, compared with as much as 120 m/s in a myelinated axon.

Signal Transmission across a Synapse

The simplest neural pathways have at least two neurons and one connection between the neurons. Other neural pathways can involve thousands of neurons and their connections as an impulse travels from the origin of the stimulus, through the sensory neurons to the brain, and back through motor neurons to the muscles or glands. The connection between two neurons, or a neuron and an effector, is called a **synapse**. A *neuromuscular junction* is a synapse between a motor neuron and a muscle cell.

An impulse travels the length of the axon until it reaches the far end, called the synaptic terminal. Neurons are not directly connected, but have a small gap between them called the synaptic cleft. Although the synaptic cleft is only about 0.02 μm wide, neurons are not close enough for the impulse to jump from one to the other. How, then, does the impulse proceed from the presynaptic neuron, which sends out information, to the postsynaptic neuron, which receives the information?

Chemical messengers called **neurotransmitters** carry the neural signal from one neuron to another. Neurotransmitters can also carry the neural signal from a neuron to an effector, such as a gland or muscle fibre. **Figure 8.16** shows the sequence of events in the movement of an impulse across a synapse. When an action potential arrives at the end of a presynaptic neuron, the impulse causes intracellular sacs that contain neurotransmitters to fuse with the membrane of the axon. These sacs, called synaptic vesicles, release their contents into the synaptic cleft by exocytosis. The neurotransmitters then diffuse across the synapse, taking about 0.5 to 1 ms to reach the dendrites of the postsynaptic neuron, or cell membrane of the effector.

Upon reaching the postsynaptic membrane, the neurotransmitters bind to specific receptor proteins in this membrane. As **Figure 8.16** illustrates, the receptor proteins trigger ion-specific channels to open. This depolarizes the postsynaptic membrane and, if the threshold potential is reached, initiates an action potential. The impulse will travel along the postsynaptic neuron to its terminal and to the next neuron or an effector.

synapse junct between two neu, or between a neuron and an effector (muscle or gland)

neurotransmitter chemical messenger secreted by neurons to carry a neural signal from one neuron to another, or from a neuron to an effector, such as a gland or muscle fibre

Suggested Investigation

Thought Lab Investigation 8-D, The Effect of Drugs on Neurons and Synapses

Figure 8.16 Neurotransmitters carry a nerve impulse across the synapse from one neuron to another. The pink arrows show the direction of nerve impulse transmission.

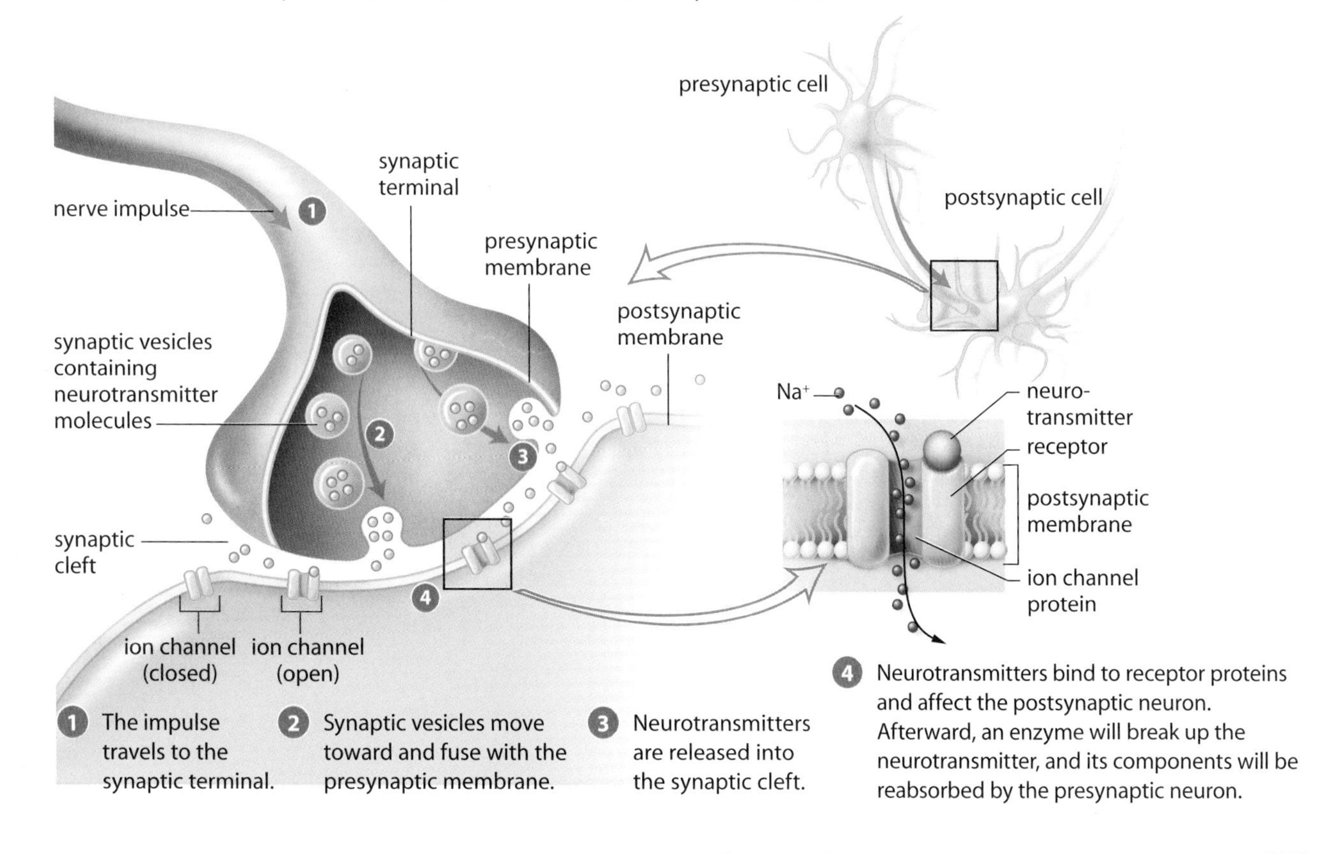
presynaptic cell
postsynaptic cell
synaptic terminal
nerve impulse
presynaptic membrane
postsynaptic membrane
synaptic vesicles containing neurotransmitter molecules
synaptic cleft
ion channel (closed)
ion channel (open)
Na+
neuro-transmitter
receptor
postsynaptic membrane
ion channel protein
1 The impulse travels to the synaptic terminal.
2 Synaptic vesicles move toward and fuse with the presynaptic membrane.
3 Neurotransmitters are released into the synaptic cleft.
4 Neurotransmitters bind to receptor proteins and affect the postsynaptic neuron. Afterward, an enzyme will break up the neurotransmitter, and its components will be reabsorbed by the presynaptic neuron.

acetylcholine the primary neurotransmitter of both the somatic nervous system and the parasympathetic nervous system

Neurotransmitters

Neurotransmitters have either excitatory or inhibitory effects on the postsynaptic membrane. If the effect is *excitatory*, the receptor proteins will trigger ion channels that open to allow positive ions, such as sodium, to flow into the postsynaptic neuron. As a result, the membrane becomes slightly depolarized. If the neurotransmitter is *inhibitory*, the receptor will trigger potassium channels to open, allowing potassium ions to flow out. This results in a more negative membrane potential, resulting in hyperpolarization.

One example of an excitatory neurotransmitter is acetylcholine. **Acetylcholine** is a neurotransmitter that crosses a neuromuscular junction. Acetylcholine excites the muscle cell membrane, causing depolarization and contraction of the muscle fibre. There are more than 50 substances in the human body that can act as neurotransmitters. **Table 8.2** lists some common neurotransmitters and their functions.

Table 8.2 Selected Neurotransmitters and Their Functions

Neurotransmitter	Function	Effects of Abnormal Production
Dopamine	Affects the brain synapses in the control of body movements; is linked to sensations of pleasure, such as eating	Excessive production linked to schizophrenia, a disorder in which the individual's perception of reality is greatly distorted; inadequate production linked to Parkinson's disease, a progressive disorder that destroys neurons, causing tremors, slurred speech, and coordination problems
Serotonin	Regulates temperature and sensory perception; is involved in mood control	Inadequate amounts in the brain synapses linked to depression
Endorphins	Act as natural painkillers in synapses in the brain; also affect emotional areas of the brain	Deficiency linked to an increased risk of alcoholism
Norepinephrine	Used by the brain and some autonomic neurons; complements the actions of the hormone epinephrine, which readies the body to respond to danger or other stressful situations	Overproduction linked to high blood pressure, anxiety, and insomnia; deficiency linked to hunger cravings and exhaustion

Activity 8.1 How Do Certain Medications Help Neurotransmission in the Brain?

An abnormal production of certain neurotransmitters has been linked to disorders such as depression, bipolar disorder, anxiety disorders, and schizophrenia. Treatments for these disorders include pharmaceutical medications that affect neurotransmitters in some way.

Materials

- reference books
- computer with Internet access

Procedure

1. Use reference books or the Internet to research information about the following medications: selective serotonin reuptake inhibitors (SSRIs), monoamine oxidase inhibitors (MAOIs), tricyclic antidepressants (TCAs), antipsychotics, and lithium.
2. Create a table to organize the results of your research. Column headings should include "Class of Medications," "Neurotransmitter Acted On," "How It Works," "Intended Effects on Patient," "Unintended (Side) Effects on Patient." Record your results in the table.

Questions

1. Create a diagram to explain how antidepressants, such as SSRIs and MAOIs, work. Include detailed captions to explain the key points or steps in the process.
2. Why is it important that patients taking antidepressants, especially patients under 18 years of age, be monitored by a health-care professional?

CANADIAN RESEARCH IN ACTION

Examining the Nervous System of a Snail

Gaynor Spencer investigates the signals that guide the growth processes of neurons..

Related Career

Registered dieticians help people manage their diets to promote good health and avoid nutrient deficiencies. For example, a woman might consult a registered dietician during pregnancy to ensure that her baby receives the vitamins and minerals required for normal development, including the vitamin A necessary for healthy neural development. Registered dieticians in Canada require a four-year bachelor's degree in food and nutrition, a period of supervised training, and certification by a provincial dietetics organization.

Gaynor Spencer is an associate professor of biological sciences at Brock University in St. Catharines, ON. Her research focusses on the nervous system of *Lymnaea stagnalis*, or the freshwater snail. Spencer began to study snails during her undergraduate years at Leeds University in the United Kingdom. The brains of these invertebrates offered some opportunities that didn't exist when studying vertebrates such as mice or rats. Snail brains are relatively simple compared to the brains of vertebrates. While human brains may have up to an estimated 100 billion neurons, snail brains may have only 30 to 40 thousand. This means the snail brain can be "mapped", and the functions of some of the larger neurons determined.

During a post-doctorate fellowship at the University of Calgary, Spencer developed an interest in growth cones, which are structures that form at the end of the growing axons of neurons. Growth cones guide the growth of these axons during brain development, and help the cells find their correct targets. Growth cones are also important in the process of nerve regeneration.

Spencer is currently investigating the effect of retinoic acid on the growth cones of neurons taken from the snail brain. Scientists have found evidence to suggest that retinoic acid may guide growing axons and assist in regeneration of some nerve tissues. The Brock research group has shown that growth cones, growing in a dish, will turn towards retinoic acid. This suggests that retinoic acid may be used by the brain as an attractive signal for growing axons.

Retinoic acid originates from vitamin A obtained from the diet. Vitamin A is involved in the formation of light-sensitive chemicals in the retina and in the growth of bones and teeth. It is also necessary for healthy neural development in fetuses. For example, vitamin A deficiency during pregnancy might adversely affect fetal nervous system development.

Spencer says, "Understanding the role of retinoic acid, which comes from vitamin A, might eventually help us better understand some of the problems associated with vitamin A deficiency. My goal is to help build the knowledge base of how retinoic acid is acting at the cellular level, specifically on how it might guide growing axons to their correct targets during development or regeneration. Right now we are looking at single cells in a dish. Later, we hope that our research will involve the intact animal."

QUESTIONS

1. Explain why *Lymnaea stagnalis* is a good research organism for studying the nervous system.
2. Suggest two potential applications for Dr. Spencer's work.
3. Suggest three careers, other than the one listed above, that are related to the research described here. Use the Internet or print resources to find out more about one of these careers.

Section 8.2 Review

Section Summary

- The human nervous system is a complex system composed of many subsystems that all work together to maintain homeostasis in the body.
- The neuron is the functional unit of the nervous system. Neurons can be classified based on structure (multipolar, bipolar, or unipolar) or function (sensory neurons, interneurons, and motor neurons).
- All cells have a membrane potential, but the neuron is unique in that it can change the potential of its membrane to generate an impulse. An impulse is transmitted from one neuron to the next by neurotransmitters at a synapse.

Review Questions

1. K/U Copy the letters A through H into your notebook. Identify the division of the nervous system that is represented by each letter.

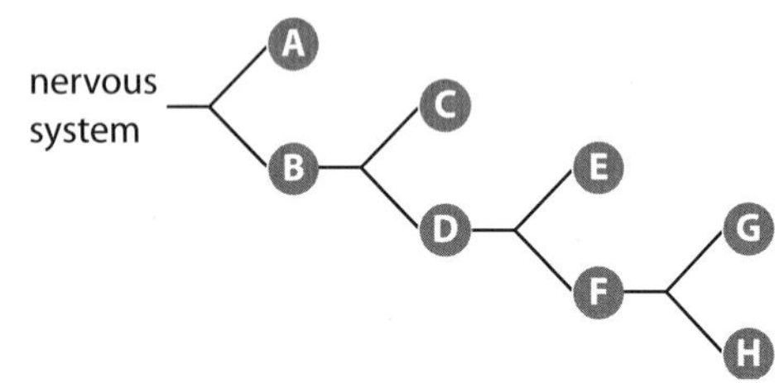

2. K/U What is a nerve?
3. K/U What are Schwann cells and what is their relationship to myelinated neurons?
4. C Using a table, identify and describe the three types of neurons classified based on function.
5. C Use a flowchart to summarize a basic reflex arc. Start your flowchart with a specific stimulus.
6. K/U The sodium potassium pump uses ATP to establish a resting membrane potential of –70 mV. Explain how a membrane that is said to be "resting" can continually use energy to transport ions.
7. K/U Summarize the events involved in impulse transmission from the presynaptic neuron to the postsynaptic neuron.
8. C Construct a Venn diagram to compare and contrast an excitatory response with an inhibitory response.
9. T/I Cocaine affects a synapse by blocking the reuptake of the neurotransmitter dopamine by the presynaptic neurons. Therefore, the levels of dopamine continue to build in the synapse, causing certain effects on the body.
   a. Explain how cocaine interferes with neural transmission across the synapse.
   b. Describe the natural role of dopamine in the brain.
   c. Formulate a hypothesis about how cocaine could be addictive after only one use.
10. A Examine the graph below, and answer the questions that follow.

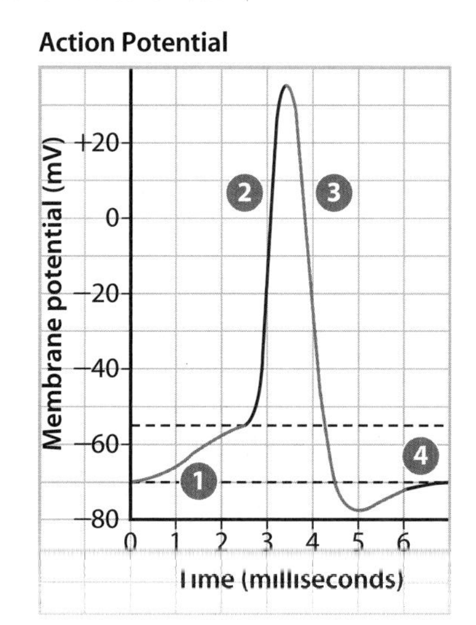

   a. In your notebook, indicate the specific events that are occurring at 1, 2, 3, and 4.
   b. At which area of the graph are sodium ions rapidly entering the neuron?
   c. At which area of the graph are potassium ions rapidly leaving the neuron?
   d. At which area of the graph is the sodium ion concentration higher outside than inside the neuron?
11. T/I Cholinesterase is an enzyme released into a neuromuscular junction, where it breaks down acetylcholine. The acetylcholine is then removed from the protein receptors, allowing the ion channels to close and the membrane to repolarize in a fraction of a second. A nerve gas called sarin blocks the release of cholinesterase into the neuromuscular junction. Hypothesize what would happen to critical muscles, such as the heart and diaphragm, as a result of exposure to sarin.

SECTION 8.3

The Central Nervous System

Key Terms

meninges
cerebellum
medulla oblongata
pons
midbrain
thalamus
hypothalamus
cerebrum
blood-brain barrier
cerebrospinal fluid
cerebral cortex
corpus callosum
occipital lobes
temporal lobes
parietal lobes
frontal lobes

The central nervous system, shown in **Figure 8.17**, is the structural and functional centre for the entire nervous system. As the site of neural integration and processing, the central nervous system receives information from the sensory neurons, evaluates this information, and initiates outgoing responses to the body. Damage to the central nervous system can therefore affect temperament, motor control, and homeostasis. Recall the example of chronic traumatic encephalopathy (CTE) from the beginning of the chapter. CTE results in the degeneration of brain cells and leads to erratic behaviour, loss of motor control, and the inability to maintain homeostasis. CTE can only be officially diagnosed after death, when the brain is examined during an autopsy. Many former football and ice hockey players have made arrangements to donate their brains after death to research institutes that are studying CTE, so scientists can learn more about the disease.

Recall from Section 8.2 that neurons can be myelinated or unmyelinated. In the central nervous system, myelinated neurons form what is known as white matter, and unmyelinated neurons form the grey matter. *Grey matter* is grey because it contains mostly cell bodies, dendrites, and short, unmyelinated axons. Grey matter is found around the outside areas of the brain and forms the H-shaped core of the spinal cord, as shown in **Figure 8.17**. *White matter* is white because it contains myelinated axons that run together in tracts. White matter forms the inner region of some areas of the brain, and the outer area of the spinal cord.

The Spinal Cord

The spinal cord is a column of nerve tissue that extends out of the skull from the brain, and downward through a canal within the backbone. The spinal cord is a vital communication link between the brain and the peripheral nervous system. Within the spinal cord, sensory nerves carry messages from the body to the brain for interpretation, and motor nerves relay messages from the brain to the effectors. The spinal cord is also the primary reflex centre, coordinating rapidly incoming and outgoing neural information.

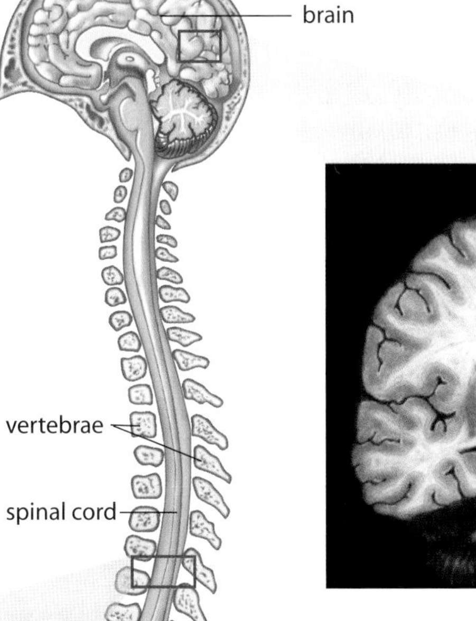

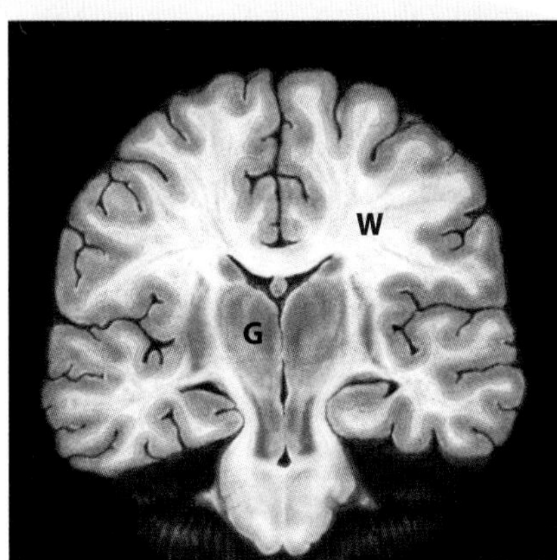

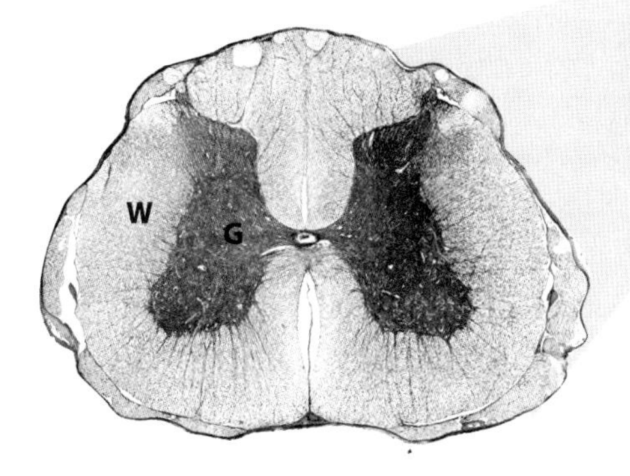

Figure 8.17 The central nervous system consists of the brain and spinal cord. Cross sections of the brain and spinal cord show the locations of grey matter (G) and white matter (W).

Spinal Cord Structure

A cross section of the spinal cord, shown in **Figure 8.18**, reveals both white matter and grey matter. The outer white matter consists of myelinated nerve fibres. The butterfly-shaped core is made up of grey matter, which contains unmyelinated neurons as well as the cell bodies and dendrites of many spinal neurons.

The delicate tissues of the spinal cord are protected by cerebrospinal fluid, soft tissue layers, and the spinal column. The spinal column consists of a series of backbones (vertebrae). Injury to the spinal column can also damage the spinal cord, resulting in paralysis.

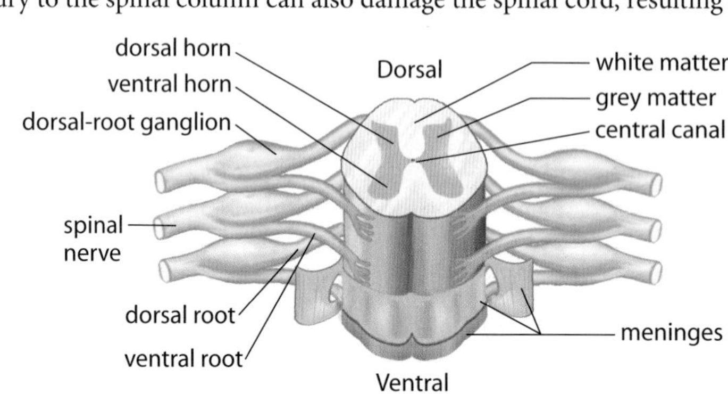

Figure 8.18 The spinal cord is protected by the meninges and the bony vertebrae.

The Brain

It has been only in the last two centuries that researchers have begun to unravel the intricate workings of the human brain. Scientists have discovered the brain's central role in maintaining homeostasis. They have identified the brain as the centre for intelligence, consciousness, and emotion. Yet, in many ways, researchers are just beginning to understand the relationships among the brain's structures and functions. Despite its relatively small size, scientists estimate that there are more neurons in the human brain than stars in the Milky Way Galaxy—over 100 billion.

Figure 8.19 shows that the brain can be subdivided into three general regions: the hindbrain, the midbrain, and the forebrain. Despite its central importance, the brain is fragile and has a gelatin like consistency. The skull, however, forms a protective bony armour around the brain. In addition, the meninges, three layers of tough, elastic tissue within the skull and spinal column, directly enclose the brain and spinal cord. The meninges are shown in detail in **Figure 8.20** on page 366. To visualize the brain, meninges, and skull, think of a peanut wrapped in its red skin and inside its shell.

meninges three layers of tough, elastic tissue within the skull and spinal column that directly enclose the brain and spinal cord

Activity 8.2 Examining Spinal Cord Tissue

In this activity, you will examine a microscope slide of the spinal cord along with a corresponding photograph.

Materials

- light microscope
- prepared slide of spinal cord tissue

Procedure

1. Observe the slide under the microscope or examine the photograph shown on the right.
2. Sketch your observations. Use **Figure 8.18** to label the following structures on your sketch: spinal cord, spinal nerve, white matter, grey matter, central canal, dorsal horn, ventral horn.

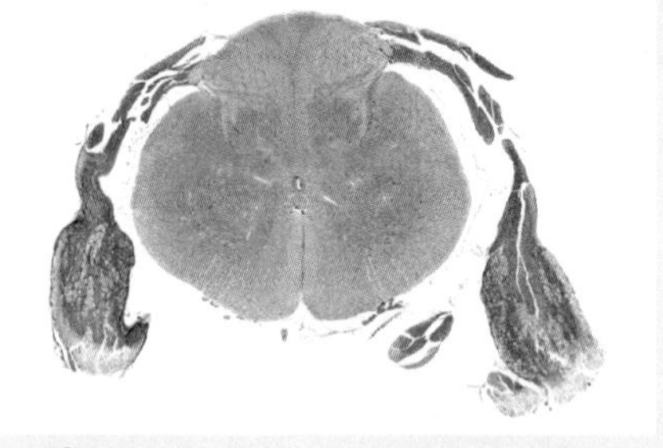

Magnification: 4×

Questions

1. What is the function of the spinal cord?
2. What is the difference between the white matter and the grey matter in the spinal cord?

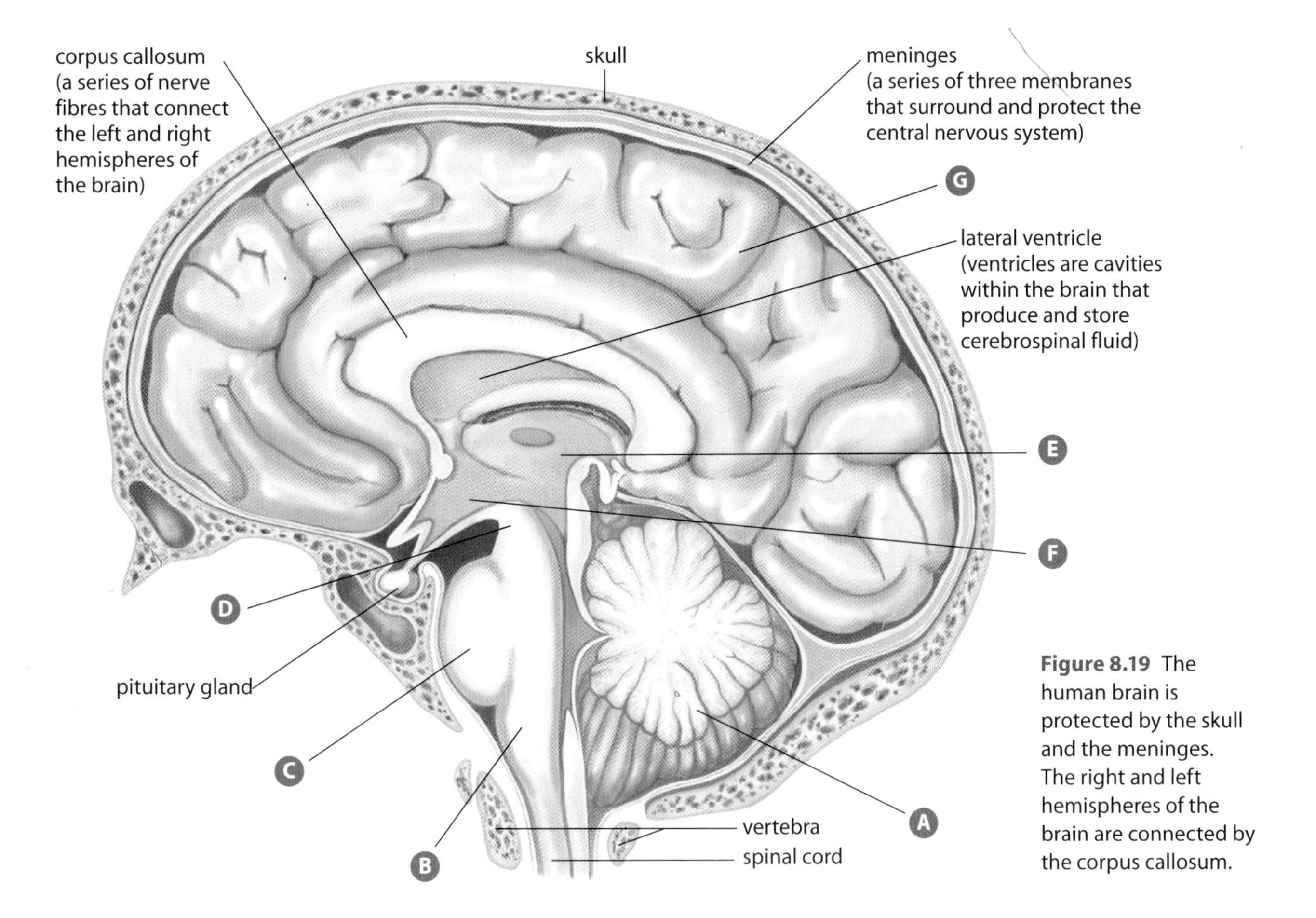

Figure 8.19 The human brain is protected by the skull and the meninges. The right and left hemispheres of the brain are connected by the corpus callosum.

THE HINDBRAIN: COORDINATION AND HOMEOSTASIS

A The **cerebellum** is a walnut-shaped structure located below and largely behind the cerebrum, which is described below. The word *cerebellum* comes from the Latin word for "little brain." This part of the brain is involved in the unconscious coordination of posture, reflexes, and body movements, as well as fine, voluntary motor skills, such as those used to hit a tennis ball, ride a bicycle, or write. The cerebellum receives information from specialized sensors, called proprioceptors, located within skeletal muscles and joints.

B The **medulla oblongata** sits at the base of the brainstem, where it connects the brain with the spinal cord. It coordinates many reflexes and automatic bodily functions that maintain homeostasis, including heart rate, constriction or dilation of blood vessels, and the rate and depth of breathing, swallowing, and coughing.

C The **pons** is found above and in front of the medulla oblongata in the brainstem. The pons serves as a relay centre between the neurons of the right and left halves of the cerebrum, the cerebellum, and the rest of the brain.

THE MIDBRAIN: PROCESSING SENSORY INPUT

D The **midbrain** is found above the pons in the brainstem and is involved in processing information from sensory neurons in the eyes, ears, and nose. It relays visual and auditory information between areas of the hindbrain and forebrain. As well, it plays an important role in eye movement and control of skeletal muscles.

THE FOREBRAIN: THOUGHT, LEARNING, AND EMOTION

E The **thalamus** sits at the base of the forebrain. It consists of neurons that provide connections between various parts of the brain. These connections are mainly between the forebrain and hindbrain, and between areas of the sensory system (except for the sense of smell) and cerebellum. The thalamus is often referred to as "the great relay station" of the brain.

F The **hypothalamus**, which lies just below the thalamus, helps to regulate the body's internal environment, as well as certain aspects of behaviour. The hypothalamus contains neurons that control blood pressure, heart rate, body temperature, and basic drives (such as thirst and hunger) and emotions (such as fear, rage, and pleasure). Brain damage or a tumour that affects the hypothalamus can cause a person to display unusual, even violent behaviour. The hypothalamus is also a major link between the nervous and endocrine (hormone) systems (which you will study in Chapter 9). The hypothalamus coordinates the actions of the pituitary gland, by producing and regulating the release of certain hormones.

G The **cerebrum** is the largest part of the brain and accounts for more than four fifths of the total weight of the brain. The cerebrum is divided into right and left cerebral hemispheres, which contain the centres for intellect, learning and memory, consciousness, and language; it interprets and controls the response to sensory information.

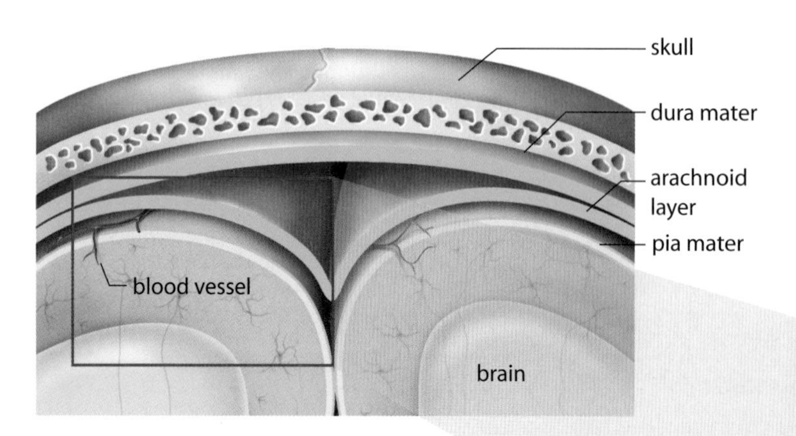

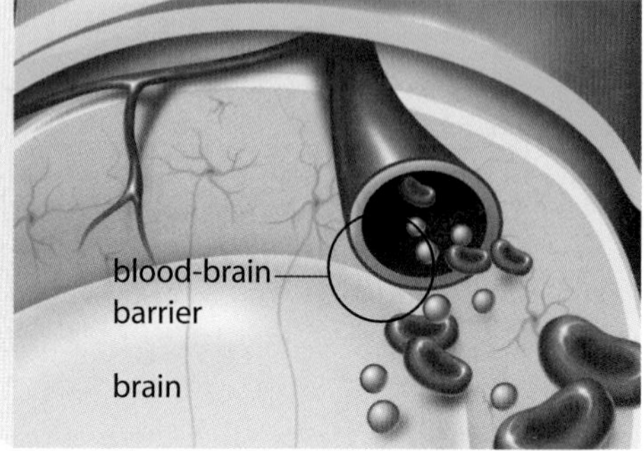

Figure 8.20 **A** Three layers of tissue, called the meninges, surround and protect the brain and spinal cord. **B** The blood-brain barrier. Only certain substances can pass through the tight seal formed by the blood-brain barrier.

The Blood-Brain Barrier

The meninges protect the central nervous system by preventing the direct circulation of blood through the cells of the brain and spinal cord. This separation of the blood and central nervous system is called the **blood-brain barrier**, shown in **Figure 8.20**. Scientists discovered this barrier in the early 1900s when they injected blue dye into the bloodstream of a laboratory rodent and all the body tissues turned blue except for the brain and spinal cord.

The blood capillaries that lead to the brain are made up of tightly fused epithelial cells. Large numbers of certain types of glial cells, called astrocytes, surround the capillaries as well. Scientists think that the astrocytes selectively allow some substances to pass from the blood into brain tissues, while inhibiting the passage of other substances. For example, the brain requires a constant supply of oxygen and nutrients, such as glucose, in order to function properly. If the oxygen supply to the brain is disrupted for even a few minutes, massive damage can occur in the brain. Without a proper supply of glucose to the brain, mental confusion, dizziness, and loss of consciousness can occur. Oxygen and glucose are examples of substances that can cross the blood-brain barrier by special transport mechanisms. Other, lipid-soluble substances, including caffeine, nicotine, and alcohol, are able to pass directly through the lipid bilayer of the cell membrane of the capillaries in the brain. This is why caffeine, nicotine, alcohol, and other lipid-soluble substances have such rapid effects on brain function.

The blood-brain barrier supplies the brain with nutrients and oxygen; however, it also protects the brain by blocking the entrance of potentially harmful substances, such as toxins and infectious agents.

blood-brain barrier protective barrier formed by glial cells and blood vessels that separates the blood from the central nervous system; selectively controls the entrance of substances into the brain from the blood

cerebrospinal fluid dense, clear liquid derived from blood plasma, found in the ventricles of the brain, in the central canal of the spinal cord, and in association with the meninges; transports hormones, white blood cells, and nutrients across the blood-brain barrier to the cells of the brain and spinal cord; acts as a shock absorber to cushion the brain

Cerebrospinal Fluid

Circulating throughout the spaces within the brain and spinal cord is the cerebrospinal fluid. The total volume of **cerebrospinal fluid** in an adult human is about 150 mL at any one time. The fluid is replaced about four times each day, and the total amount of fluid produced each day is about 500 mL. The cerebrospinal fluid transports hormones, white blood cells, and nutrients across the blood-brain barrier for cells of the brain and spinal cord. It also circulates between two layers of the meninges, the arachnoid and pia mater, acting as a shock absorber to cushion the brain.

Learning Check

19. Identify the main structures of the central nervous system, and describe the general functions of each structure.
20. Identify and describe the two types of nervous tissue found in the central nervous system.
21. How is the spinal cord protected?
22. Using a diagram, identify the major structures of the human brain.
23. Summarize the functions of the blood-brain barrier.
24. Identify five homeostatic functions that your brain is carrying out as you are studying.

The Structure and Function of the Cerebrum

Each half of the cerebrum consists of an internal mass of white matter and a thin, outer covering of grey matter, called the **cerebral cortex**. The cerebral cortex is responsible for language, memory, personality, vision, conscious thought, and other activities that are associated with thinking and feeling. The cerebral cortex is about 5 mm thick and, as shown in **Figure 8.21**, is highly convoluted, or folded. This allows it to fit a high concentration of grey matter within the confines of the skull. Relative to a smooth surface, the convolutions and fissures greatly increase the surface area, so that the cerebral cortex covers about 0.5 m^2, or about the area of an open newspaper.

The right and left halves of the cerebrum are called the cerebral hemispheres. They are linked by a bundle of white matter called the **corpus callosum**. The corpus callosum sends messages from one cerebral hemisphere to the other, telling each half of the brain what the other half is doing. Surgical isolation of the hemispheres is sometimes used to treat epilepsy, a condition that causes uncontrollable seizures. Scientists think that epilepsy can be caused by an overload of neurological electrical activities, so the corpus callosum is cut to prevent the spread of the epileptic seizures from one hemisphere to the other.

cerebral cortex thin outer covering of grey matter that covers each cerebral hemisphere of the brain; responsible for language, memory, personality, conscious thought, and other activities that are associated with thinking and feeling

corpus callosum bundle of white matter that joins the two cerebral hemispheres of the cerebrum of the brain; sends messages from one cerebral hemisphere to the other, telling each half of the brain what the other half is doing

Research indicates that, while every cognitive function contains right-brain and left-brain components, some functions seem to have a dominant hemisphere. In general, the right-brain, or right cerebral hemisphere, is associated with holistic and intuitive thinking, visual-spatial skills, and artistic abilities. The left-brain, or left cerebral hemisphere, is linked to segmental, sequential, and logical ways of thinking, and to linguistic and mathematical skills.

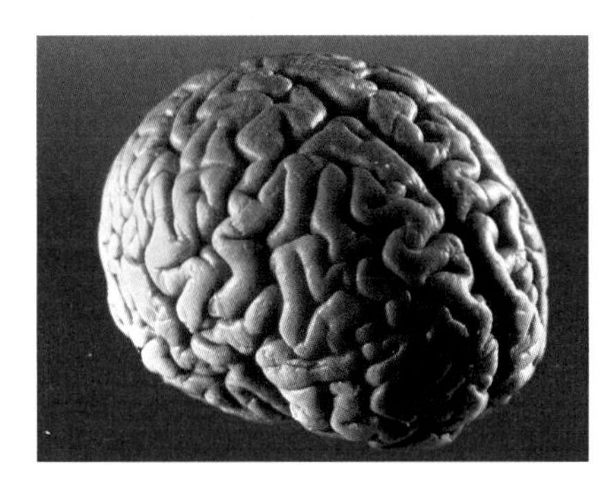

Figure 8.21 The folds on the cerebral cortex increase the surface area of grey matter in the cerebrum.

Activity 8.3 The Stroop Effect

One way to illustrate the difference between right-brain and left-brain processing is the Stroop effect.

Procedure

Try to say aloud the colour in which the word is written in the figure on the right, rather than the word itself.

YELLOW	BLUE	ORANGE
BLACK	RED	GREEN
PURPLE	YELLOW	RED
ORANGE	GREEN	BLACK
BLUE	RED	PURPLE
GREEN	BLUE	ORANGE

Questions

1. What happened when you carried out the procedure?
2. One theory suggests that one side of the brain may dominate in word recognition, while the other side may dominate in colour recognition. Why might this make it difficult to carry out the procedure?

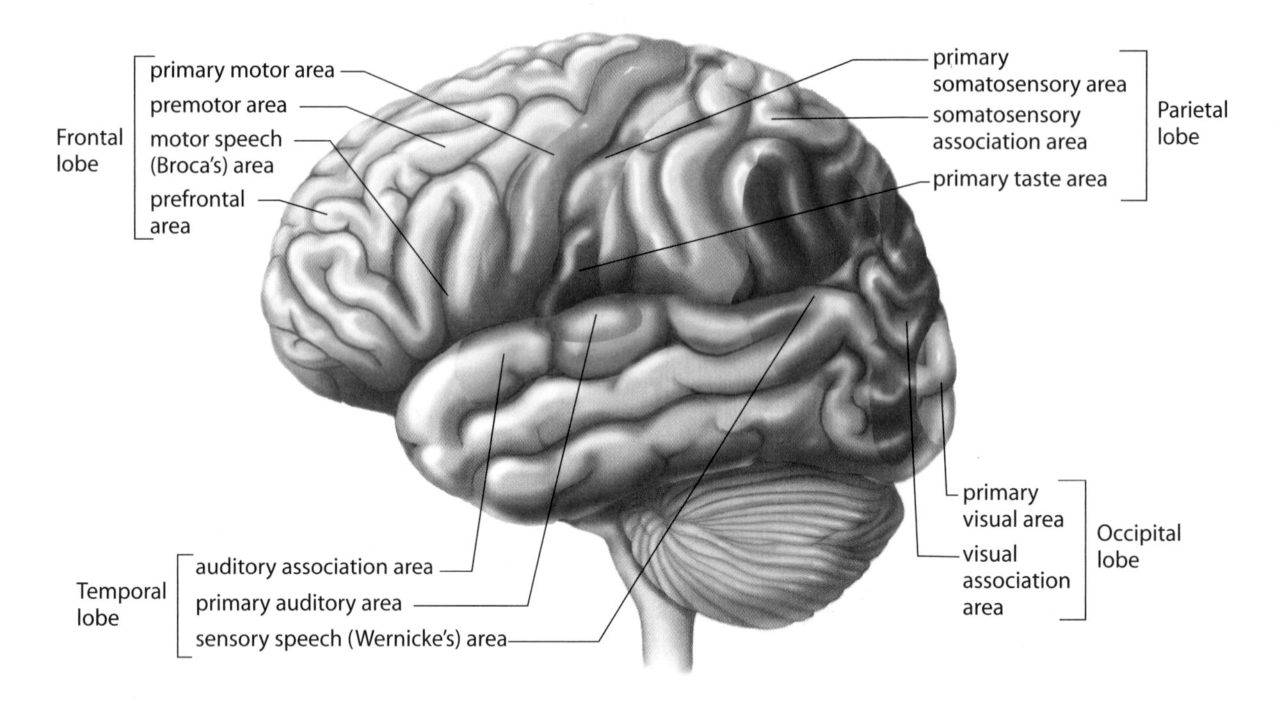

Figure 8.22 The cerebral cortex is divided into four pairs of lobes: frontal, temporal, parietal, and occipital.

***Identify** which lobes of the cerebrum would be used if you were to read a book aloud.*

occipital lobe lobe of the cerebral cortex that receives and analyzes visual information, and is needed for recognition of what is being seen

temporal lobe lobe of the cerebral cortex that shares in the processing of visual information but its main function is auditory reception

parietal lobe lobe of the cerebral cortex that receives and processes sensory information from the skin, and helps to process information about the body's position and orientation

frontal lobe lobe of the cerebral cortex that integrates information from other parts of the brain and controls reasoning, critical thinking, memory, and personality

The Cerebral Cortex

Figure 8.22 shows that each hemisphere of the cerebral cortex can be divided into four pairs of lobes. Each pair of lobes is associated with a different function. The **occipital lobes** receive and analyze visual information. If the occipital lobes are stimulated by surgery or trauma, the individual will see light. The occipital lobes are also needed for recognition of what is being seen. Damage to the occipital lobes can result in a person being able to see objects, but not able to recognize them.

The **temporal lobes** share in the processing of visual information, although their main function is auditory reception. These lobes are also linked to understanding speech and retrieving visual and verbal memories.

The **parietal lobes** receive and process sensory information from the skin. The primary sensory areas extend in a band from the right to left side of the cerebrum. The proportion of a parietal lobe devoted to a particular part of the body is related to the importance of sensory information for this part of the body. The highest concentrations of sensory receptors occur in the face, hands, and genitals, making these areas of the body highly sensitive. The parietal lobes also help to process information about the body's position and orientation.

The **frontal lobes** are named for their location at the front of the cerebrum. The frontal lobes integrate information from other parts of the brain and control reasoning, critical thinking, memory, and personality. One area of the frontal lobes is associated with language use.

The frontal lobes also contain motor areas that control various aspects of precise, voluntary motor movement, such as playing a piano. Similar to the sensory areas in the parietal lobes, the proportion of motor area in the frontal lobes devoted to a particular part of the body correlates with the degree of complexity of movement that body structure can make. The nerves leading from the right and left frontal lobes cross over in the brainstem, so that each side of the brain controls muscles on the opposite side of the body.

Suggested Investigation

Inquiry Investigation 8-E, The Brain

Section 8.3 Review

Section Summary

- The central nervous system, which consists of the brain and spinal cord, is the control centre of the nervous system.
- The brain can be subdivided into three general regions: the hindbrain, the midbrain, and the forebrain.
- The outer layer of the cerebrum, called the cerebral cortex, is composed of grey matter, and is thought to be the source of human intellect.
- The right and left halves of the cerebral cortex are made of four pairs of lobes, each of which is associated with particular functions.

Review Questions

1. K/U What are the structures and functions of the central nervous system?
2. C Use a Venn diagram to compare and contrast grey matter and white matter.
3. K/U Explain the following statement: "The spinal cord is a vital communication link between the brain and the peripheral nervous system."
4. A Explain why damage to the spinal cord could result in the loss of muscle function.
5. K/U Describe the three main tissues that support and protect the central nervous system.
6. A Meningitis is an inflammation of the meninges, which cover the brain and spinal cord. It is diagnosed by a spinal tap, which involves analyzing a sample of the cerebrospinal fluid. Explain why the same information could not be determined from a regular blood sample.
7. K/U Identify the three general regions of the brain.
8. T/I Infer why the brain, which comprises only 2 percent of the body's total weight, uses at least 20 percent of the body's oxygen and energy supplies. [**Hint:** Think about the role that ATP plays in the transmission of nervous impulses.]
9. C Using a table, identify the major structures in the hindbrain, indicate their location in the brain, and summarize the function of each.
10. C Using a table, compare the functions of the thalamus to the functions of the hypothalamus.
11. C Make a sketch of the cerebral cortex. Use it to identify the lobe that would be stimulated in each situation, and explain why.
    - **a.** Seeing this question.
    - **b.** Thinking about this question.
    - **c.** Hearing this question read to you by someone else.
    - **d.** Reading this question to someone else.
12. C The letters on the diagram below indicate possible areas of brain damage. In table format, list the possible areas of brain damage (A to G), and describe the functional problems that might result from damage in each area.

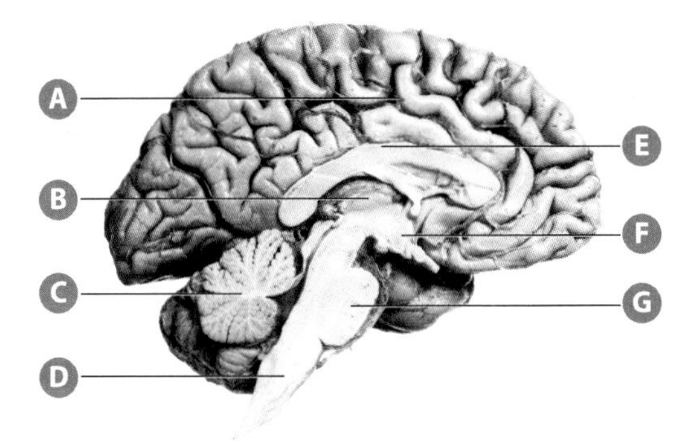

13. A A stroke has caused damage to certain areas of a person's brain. Upon examination, a doctor notices that the person has difficulty understanding speech and the left side of the body is paralyzed. Identify the specific areas of the brain that are damaged, and explain how this damage might cause the symptoms.
14. K/U Describe the blood-brain barrier.
15. K/U Describe the functions of cerebrospinal fluid.
16. T/I Hypothesize why researchers might have difficulty treating neurological disorders with medications that are not lipid soluble.
17. A Identify which area of the brain is used when you are fastening a button or tying your shoelaces.
18. T/I Suppose that a person's eyes and optic nerve are functioning normally, yet the individual cannot see. Provide a possible explanation for how this could occur.
19. K/U Which part of the brain is primarily responsible for controlling balance, posture, and coordination?
    - **a.** pons
    - **b.** medulla oblongata
    - **c.** cerebral cortex
    - **d.** cerebellum

SECTION 8.4

The Peripheral Nervous System

Key Terms

somatic system
autonomic system
sympathetic nervous system
norepinephrine
parasympathetic nervous system

Figure 8.23 The peripheral nervous system is essential for various activities, such as catching a football.

Identify *What types of motor neurons might be activated in the football player shown here?*

As the football player in **Figure 8.23** lunges for the ball, sensory nerves enable him to see the ball, feel its texture, hear the roar of the crowd, and gather information about the positions [illegible] free from the defender, and they increase his heart and breathing rates. The peripheral nervous system consists of nerves that link the brain and spinal cord to the rest of the body, including the senses, muscles, glands, and internal organs. Sensory neurons carry information from all parts of the body to the central nervous system, and motor neurons carry information from the central nervous system to the effectors. The two main divisions of the peripheral nervous system are the somatic system and the autonomic system.

The Somatic System

somatic system in vertebrates, division of the peripheral nervous system that controls voluntary movement of skeletal muscles

The **somatic system** is largely under voluntary control, and its neurons service the head, trunk, and limbs. Its sensory neurons carry information about the external environment inward, from the receptors in the skin, tendons, and skeletal muscles. Its motor neurons carry information to the skeletal muscles. Your decision to turn this page in order to continue reading exemplifies the action of the somatic motor nerves.

The somatic system includes 12 pairs of cranial nerves and 31 pairs of spinal nerves, all of which are myelinated. The cranial nerves are largely associated with functions in the head, neck, and face. An exception is the vagus nerve, which has branches to the throat and larynx, but also connects to many internal organs, including the heart, lungs, bronchi, digestive tract, liver, and pancreas.

Figure 8.24 shows the basic divisions of the spinal nerves that emerge from each side of the spinal cord. Each spinal nerve contains both sensory and motor neurons, which service the area of the body where they are found. For example, thoracic nerves control the muscles of the rib cage.

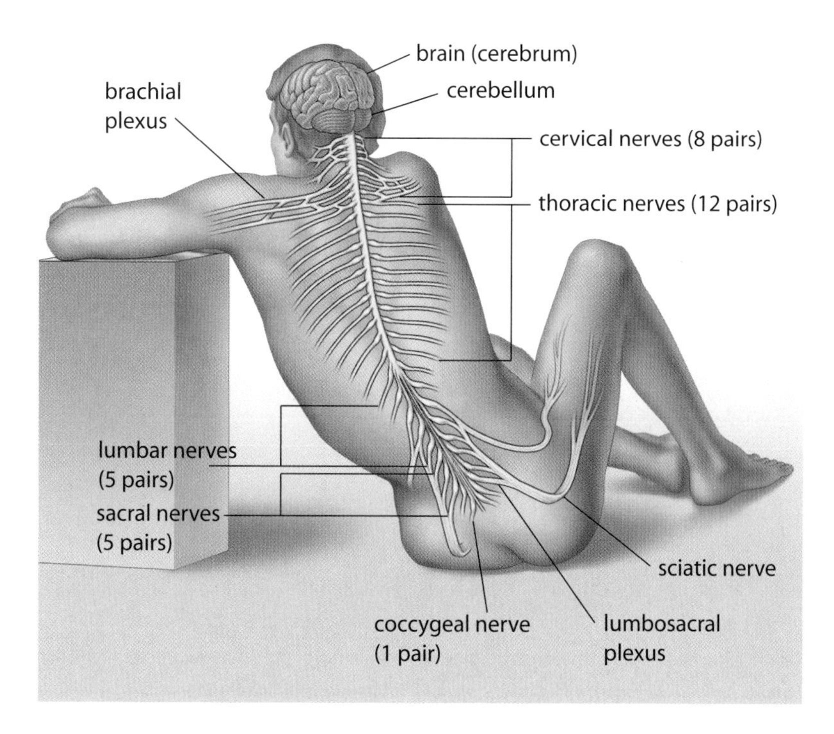

Figure 8.24 The spinal nerves are named for the region of the body where they are located: cervical, thoracic, lumbar, and sacral.

autonomic system in vertebrates, the division of the peripheral nervous system that controls involuntary glandular secretions and the functions of smooth and cardiac muscle

sympathetic nervous system division of the autonomic system that regulates involuntary processes in the body; works in opposition to the parasympathetic nervous system; typically activated in stress-related situations

norepinephrine neurotransmitter released by sympathetic neurons of the autonomic system to produce an excitatory effect on target muscles; also a hormone produced by the adrenal medulla

The Autonomic System

Imagine yourself in a stressful situation. Which systems of your body might be stimulated? Which systems might be suppressed? How would these systems return to their initial states? Your internal reactions to the situation would be controlled by the **autonomic system**. In contrast to the somatic system, the autonomic system is under automatic, or involuntary control. Its nerves either stimulate or inhibit the glands or the cardiac or smooth muscle. The autonomic system maintains homeostasis by adjusting the body to variations in the external and internal environments without an individual having to think about it and control it consciously.

The hypothalamus and medulla oblongata control the autonomic system, which has neurons that are bundled together with somatic system neurons in the cranial and spinal nerves. The sympathetic and parasympathetic divisions of the autonomic system carry information to the effectors. In general, these two divisions have opposing functions.

The Sympathetic Nervous System

The **sympathetic nervous system** is typically activated in stressful situations, a response often referred to as the fight-or-flight response. The sympathetic neurons release a neurotransmitter called **norepinephrine**, which has an excitatory effect on its target muscles. As well, the sympathetic nerves trigger the adrenal glands to release epinephrine and additional norepinephrine, both of which also function as hormones that activate the stress response. At the same time, the sympathetic nervous system inhibits some areas of the body. For example, in order to run from danger, the skeletal muscles need a boost of energy. Therefore, blood pressure increases and the heart beats faster, while digestion slows down and the sphincter controlling the bladder constricts. Although modern stressors are often not life threatening, the sympathetic nervous system may react as if they are. This can lead to feelings of anxiety. As shown in **Figure 8.25**, some people meditate to help reduce anxiety. Scientists are collecting evidence which shows that meditation reduces the activity of the sympathetic nervous system and increases the activity of the parasympathetic nervous system, leading to a sense of calm and well-being.

Figure 8.25 Meditation may reduce the activity of the sympathetic nervous system.

Learning Check

25. What is the peripheral nervous system?
26. What are the major structures involved in the somatic system?
27. What role does the somatic system play in the functioning of the body?
28. What type of motor neurons are involved in the autonomic nervous system?
29. What generally triggers the sympathetic nervous system, and what are its effects on the body?
30. Why is the response of the sympathetic nervous system to stress called a flight-or-flight response?

The Parasympathetic Nervous System

parasympathetic nervous system division of the autonomic system that regulates involuntary processes in the body; works in opposition to the sympathetic nervous system; typically activated when the body is calm and at rest

The **parasympathetic nervous system** is activated when the body is calm and at rest. It acts to restore and conserve energy. Sometimes referred to as the rest-and-digest response, the parasympathetic nervous system slows the heart rate, reduces the blood pressure, promotes the digestion of food, and stimulates the reproductive organs by dilating blood vessels to the genitals. The parasympathetic system uses the neurotransmitter acetylcholine to control organ responses.

The two branches of the autonomic system are much like the gas pedal and brake pedal of a car. At a given instant, high levels of sympathetic stimulation might cause the heart to beat faster, while parasympathetic signals would counter this effect and bring the heart rate back down. Depending on the situation and organs involved, the sympathetic and parasympathetic systems work in opposition to each other in order to maintain homeostasis. **Table 8.3** summarizes the functions of the sympathetic and parasympathetic nervous systems.

Certain drugs can act as either stimulants or depressants by directly affecting the sympathetic and parasympathetic nervous systems. Caffeine, for example, is a commonly used stimulant that causes the sympathetic nervous system to increase the heart rate and blood pressure.

Table 8.3 Functions of the Sympathetic and Parasympathetic Nervous Systems

Effector	Effect of Sympathetic Nervous System	Effect of Parasympathetic Nervous System
Tear ducts	Inhibits tears	Stimulates tears
Pupils	Dilates pupils	Constricts pupils
Salivary glands	Inhibits salivation	Stimulates salivation
Lungs	Dilates air passages	Constricts air passages
Heart	Speeds heart rate	Slows heart rate
Liver	Stimulates liver to release glucose	Stimulates gall bladder to release bile
Kidneys, stomach, pancreas	Inhibits activity of kidneys, stomach, and pancreas	Increases activity of stomach and pancreas
Adrenal glands	Stimulates adrenal secretion	No known effect
Small and large intestine	Decreases intestinal activity	Increases intestinal activity
Urinary bladder	Inhibits urination	Stimulates urination

Section 8.4 Review

Section Summary

- The peripheral nervous system consists of the somatic and autonomic nervous systems.
- Homeostasis is maintained in the body by the actions of the sympathetic and parasympathetic nervous systems.
- In general, the sympathetic nervous system prepares the body for fight-or-flight, while the parasympathetic system returns the organs to a resting state.

Review Questions

1. K/U Compare the general functions of the central nervous system with the functions of the peripheral nervous system.
2. K/U Identify which division of the peripheral nervous system is under voluntary control and which division is under involuntary control. Compare the functions of these two divisions.
3. K/U What is the vagus nerve? List five organs regulated in part by this nerve.
4. A Imagine that you are hiking in the mountains one afternoon with friends. As you turn a corner, you come across a mother bear and her cubs standing in the middle of the trail.
   a. Identify the specific division of the nervous system that is responsible for the body's response to this situation.
   b. Describe at least six physiological responses you might have upon seeing the bears.
   c. Indicate the division of the nervous system that is responsible for returning the body back to equilibrium after the event is over.
5. K/U Identify the neurotransmitters and hormones released by the sympathetic nervous system.
6. C When frightened, your sympathetic nervous system prepares you to run away from the danger or fight. In order to run faster, your skeletal muscles need a boost of energy. Use a table to identify three specific physiological changes that provide this extra energy to the muscles, and explain each change.
7. A Which division of the autonomic nervous system is controlling the response to sad news shown in the photograph to the right?

8. K/U Why does stimulation of the sympathetic nervous system slow down the activity of the organs of the digestive system?
9. K/U Why is the parasympathetic nervous system response sometimes referred to as a "rest-and-digest" response?
10. C Using two examples, write a paragraph explaining the following statement. "The sympathetic nervous system and the parasympathetic nervous system typically function in opposition to each other. However, instead of stating that these two systems are antagonistic, it might be better to say that they are complementary systems."
11. T/I It is often recommended that people wait an hour or so after eating a big meal before exercising vigorously. Infer the reasoning behind this advice.
12. T/I In a healthy individual, the sympathetic and parasympathetic nervous systems work together to adapt to ongoing, minute changes in the external and internal environments. These oscillations can be diagrammed as shown in the graph below.
    a. Construct your own graph to illustrate what these oscillations might look like when a person is running and then recovering from running.
    b. Include appropriate labels and give your graph a title.
    c. Write a caption for your diagram.

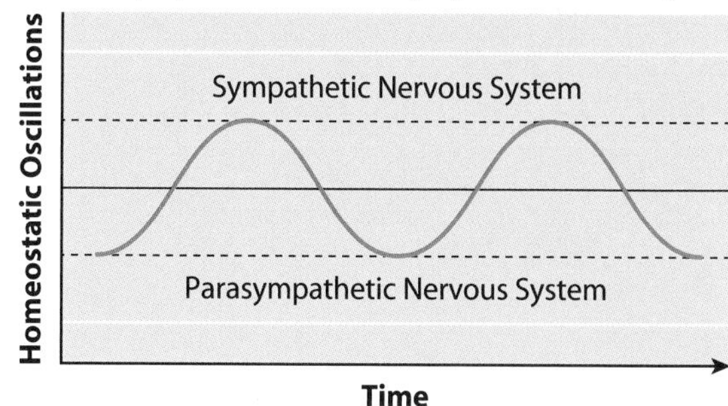

Inquiry INVESTIGATION

8-A

Skill Check

- Initiating and Planning
- ✓ Performing and Recording
- ✓ Analyzing and Interpreting
- ✓ Communicating

Safety Precautions

- Do not use excessive force when testing the knee-jerk reflex.

Materials

- room light
- 20 cm × 20 cm clear plastic sheet
- cotton balls
- chair

Move Fast! Reflex Responses

In this investigation, you will try to initiate three common reflexes.

Pre-Lab Questions

1. What is a reflex? Identify the reflexes you will test in this investigation.
2. Describe the basic stimulus-response loop involved in a reflex.
3. Why is it important to adhere to safety precautions in this investigation?

Question

How can reflexes be tested?

Procedure

Part 1: Pupillary Reflex

1. Work with a partner. Dim the lights in the room for a few minutes. Observe the size of the pupils in your partner's eyes.
2. Turn on the lights. Observe the size of the pupils.

Part 2: Blink Reflex

1. Have your partner hold a piece of clear plastic in front of his or her face.
2. Without warning, quickly throw a cotton ball at your partner's eyes. Observe your partner's eyes as he or she reacts.

Part 3: Knee-Jerk (Patellar) Reflex

1. Have your partner sit in a chair with legs crossed, so the top leg swings freely.
2. Tap the top leg softly, just below the knee, with the side of your hand. The leg should kick out immediately, demonstrating the patellar reflex.

Analyze and Interpret

1. Copy the diagram on the left into your notebook. Use it as a model to summarize each reflex response that you tested in this investigation.

Conclude and Communicate

2. Describe how the reflexes tested in this investigation might protect the body.

Extend Further

3. **INQUIRY** Design a procedure to test one of the other reflexes of the body. Hypothesize what the reflex arc might look like for the reflex you test. Suggest a way in which this reflex would help to protect the body.
4. **RESEARCH** Health-care professionals observe reflex responses to evaluate nervous system function. Research more information about the plantar flexion reflex and the Babinski sign. Describe the reflex. What is the normal response to stimulation? What is the significance of the Babinski sign?

Inquiry INVESTIGATION

8-B

Skill Check

Initiating and Planning
✓ Performing and Recording
✓ Analyzing and Interpreting
✓ Communicating

Safety Precautions

- Wash your hands after completing this investigation, or immediately if your skin is exposed to the solutions used.
- Wear goggles to protect your eyes against accidental splashes. The solutions used are irritating to the eyes.
- Be careful when working with the copper wire as the ends may be sharp.

Materials

- 2 strips uninsulated copper wire, each 40 cm long
- pen
- DC millivolt meter
- 400 mL beaker
- 3 mol/L sodium chloride solution
- 22 cm moistened dialysis tubing
- string
- 3 mol/L potassium chloride solution
- elastic band

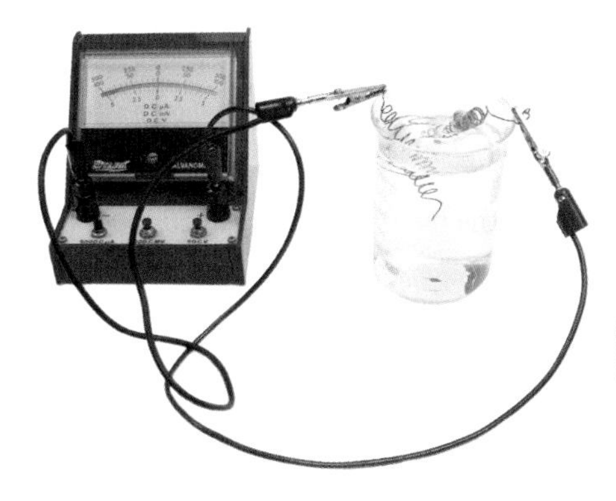

Modelling Resting Membrane Potential

In this investigation, you will build a simple model of the neural membrane to demonstrate how the resting membrane potential is established.

Pre-Lab Questions

1. Describe the membrane potential of a resting neuron.
2. How will you protect your eyes against accidental splashes?

Question

How can you model resting membrane potential?

Procedure

1. Create a table to record the time and voltage data collected.
2. Take the two 40 cm strips of copper wire and tightly wind one end of each around a pen in order to form a coil of about 8 cm. Remove the pen from the coil.
3. Attach the uncoiled ends to the millivolt meter, as shown in the photograph. Each wire will serve as an electrode.
4. Pour about 300 mL of the sodium chloride solution into the 400 mL beaker. Take the copper electrode attached to the positive (red) terminal of the meter, and immerse the free end in the solution.
5. Tie off one end of moistened dialysis tubing. Fill two thirds of the tubing with potassium chloride solution. Place the free end of the other copper electrode in the solution in the tubing. Secure the end of the tubing around the wire with an elastic band.
6. While one group member observes the needle on the millivolt meter, put the dialysis tubing in the beaker holding the solution of sodium chloride.
7. Leave the dialysis tubing in the beaker, and continue to monitor the electric potential every 5 min, until a trend is established. Record each value.

Analyze and Interpret

1. Graph the data from your data table.
2. Explain what the dialysis tubing, potassium chloride solution, and sodium chloride represent in this model of a resting neuron.

Conclude and Communicate

3. Describe and explain what happened to the magnitude of the electric potential over time. If this occurred in a neuron what would happen?
4. How did your model illustrate the mechanism of ion channel diffusion?

Extend Further

5. **INQUIRY** Hypothesize how you might be able to increase the electric potential across the dialysis tubing.
6. **RESEARCH** Find out more information about how local anesthetics, such as lidocaine, work. How do these drugs block pain and other sensations?

Plan Your Own INVESTIGATION 8-C

Skill Check

- ✓ Initiating and Planning
- ✓ Performing and Recording
- ✓ Analyzing and Interpreting
- ✓ Communicating

Safety Precautions

Wash your hands with soap and water upon completion of the investigation.

Suggested Materials

- invertebrate organism(s), such as sowbugs (*Porcellio scaber*), earthworms (*Lumbricus terrestris*), or water fleas (*Daphnia magna*)
- materials for test chamber, such as a Petri dish, lab tray, beaker, paper towels, water, and/or soil
- equipment to provide a stimulus, such as a light source, a heat source, ice cubes, music, or water
- timer
- ruler
- thermometer

Invertebrate Responses to External Stimuli

Different species have developed diverse homeostatic systems in order to survive. Therefore, a particular stimulus may result in different responses in different species. To be able to conduct an experiment on the behaviour of an organism, you should know about its natural environment and physiology. For example, some invertebrate species are attracted to light, while others avoid it. The same applies to water, heat, and a variety of other stimuli. This activity can be carried out using any one of a number of invertebrate species, but one possibility is the sowbug (*Porcellio scaber*), shown below.

The sowbug, also called the woodlouse, is often found under logs or rocks in gardens or the forest floor. It prefers moist conditions and feeds on decaying plant matter.

Ethical Guidelines

Handle the invertebrate organisms with care. Treat them with respect. Animals should be returned to their natural environment or a suitable holding tank as soon as the investigation is complete. These actions support ethical approaches in fieldwork where animals are returned to their habitat after observations have been made.

Pre-Lab Questions

1. What is the difference between qualitative and quantitative data?
2. Identify the independent and dependent variables in your experiment.
3. What actions will you take to follow ethical guidelines in this investigation?
4. What safety precautions will you take in this investigation?

Question

How can you demonstrate a particular invertebrate's responses to a chosen stimulus?

Plan and Conduct

1. Select an invertebrate species for your study. After you have made your choice, gather information on the species to provide the background you will need to design your experiment.
2. Choose a particular stimulus, such as light, heat, moisture content of soil, or noise (music), that you predict will produce a consistent response by the organism.
3. Based on your research, formulate a hypothesis stating how the invertebrate will respond to the stimulus. The hypothesis will form the basis of your experimental design.
4. Design an experiment that will show a positive or negative response to the chosen stimulus by the organism. Be sure that your design has only one independent variable and that you keep as many other variables as possible constant. Consider building an apparatus to provide the necessary controls.
5. As you design your experiment keep in mind that it is important that you have a setup that allows the organism a clear choice. For example, you could have a choice chamber with one side cooled using ice cubes, while the other side is at room temperature. If you have one side cooled with ice and the other under a heat lamp, you may have difficulty telling the difference between an organism attracted to one extreme or repelled from the other.
6. Determine how you will measure the response of the organism, or the dependent variable, in this investigation. Will you measure it using quantitative data (data that consist of numbers) or qualitative data (data that are descriptive)?
7. Consider how many trials you will carry out. Be sure you have enough repetitions to make your results reliable.
8. Construct a data table in which to record your observations.
9. Have your teacher review and approve your plan.
10. Carry out your experiment.

Analyze and Interpret

1. Did the animals in your experiment show a consistent response to the stimulus? Explain your reasoning.
2. Justify whether your observations supported your hypothesis.
3. Were there any anomalous findings? What could account for them?

Conclude and Communicate

4. Very few experiments are free from possible sources of error. What are some other factors that may have affected the behaviour of the animals?
5. What changes would you make to your design if you were to repeat this experiment? Explain your reasoning.
6. Write a paragraph summarizing the results of your investigation. Include information about the conclusions you reached about how your organism responded to the stimulus.
7. Find another group that used the same organism, but a different stimulus than your group or that used a different organism and the same stimulus as your group. Compare your results. What did you learn about your organism or stimulus based on the comparison?

Extend Further

8. **INQUIRY** Assume that next year's class has access to your lab report. What would be an appropriate follow-up question for next year's class to base their lab on, assuming that they had the same materials and species to choose from?
9. **RESEARCH** Octopods are considered to have the most complex brains of all invertebrates. Research more information about the nervous system of octopods. How do they respond to different stimuli? What experiments have been done to measure their intelligence and their ability to learn?

ThoughtLab INVESTIGATION

8-D

Skill Check

- Initiating and Planning
- ✓ Performing and Recording
- ✓ Analyzing and Interpreting
- ✓ Communicating

Materials

- reference books
- computer with Internet access
- handout from your teacher

The Effect of Drugs on Neurons and Synapses

A drug is a non-food substance that changes the way the body functions. Most drugs, legal or illegal, affect neurons and synapses by either promoting or decreasing the action of a neurotransmitter.

Pre-Lab Questions

1. What is a drug?
2. What are some effects of drugs on the nervous system?

Question

How do drugs affect the nervous system?

Procedure

1. With a partner, use the information in the handout, as well as library and/or Internet resources, to create a drug information pamphlet about one of the drugs presented. Your information pamphlet should include
   - a detailed explanation of how the drug affects the nervous system and other body functions
   - any hazards to the nervous system and the entire body from short-term and long-term use of the drug
   - illustrations that help explain concepts in detail
2. Present your pamphlet to another group or to the rest of the class.

Analyze and Interpret

1. Explain the role of the neurotransmitter gamma-aminobutyric acid (GABA) in producing a sedative effect on the nervous system.
2. Compare and contrast valerian root and benzodiazepines.
3. Explain how dopamine affects the brain.

Conclude and Communicate

4. Is the drug you chose to research legal or illegal? Explain the reasoning for its classification.

Extend Further

5. **INQUIRY** Hypothesize what might make the drug you investigated addictive.
6. **RESEARCH** Valerian root, St. John's Wort, and L-tryptophan are all examples of natural health products used to help treat anxiety disorders and/or depression. The federal government regulates natural health products under the Natural Health Products Regulations of the Food and Drugs Act. Research more information about the regulations and explain why it is important that they are in place.

Inquiry INVESTIGATION

8-E

Skill Check

Initiating and Planning

✓ Performing and Recording

✓ Analyzing and Interpreting

✓ Communicating

The Brain

In this investigation, you will use models, photographs, and a mammalian brain to learn about the principal structural areas of the brain and their functions.

Materials

- preserved sheep brain
- paper towel
- dissecting tray
- dissecting kit

Safety Precautions

- Use caution when handling sharp instruments.
- Wash your hands well when finished the dissection.
- Disinfect the equipment and area when finished.

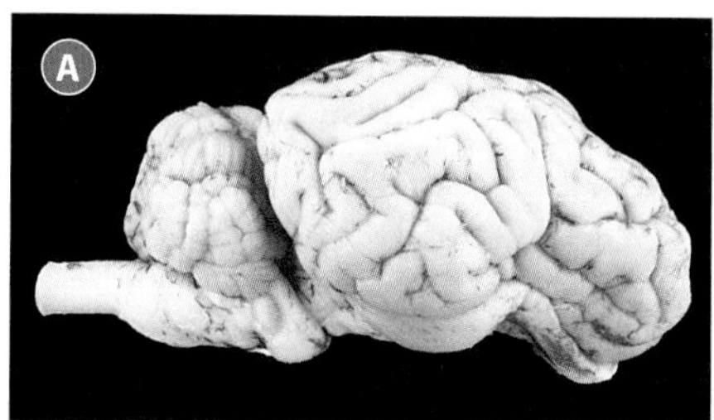

lateral view—whole brain

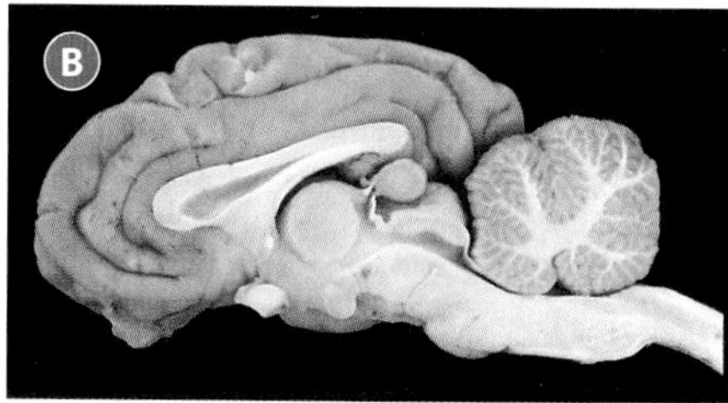

lateral view—cross section

Pre-Lab Questions

1. What structure separates the right and left hemispheres of the brain?
2. Where are the olfactory bulbs located?

Question

What are the principal structures of the brain?

Procedure

1. Your teacher will give you a sheep brain. Follow your teacher's instructions for rinsing the brain. Then place the brain in the dissecting tray.
2. Examine photograph A. Identify each lobe of the cerebrum.
3. Examine the outer surface of the brain. Notice the convolutions and fissures and that the cerebrum is divided into a right side and a left side.
4. Sketch and label the outer surface of the brain.
5. Examine photograph B, showing a cross section of the sheep brain.
6. Make a gentle incision through the corpus callosum of the sheep brain to separate the right and left hemispheres. Then separate the rest of the brain by cutting through the centre of the mid and hind parts.
7. Using photograph B as a guide, identify, sketch, and label the following structures: spinal cord, cerebellum, medulla oblongata, pons, midbrain, thalamus, hypothalamus, pituitary gland, corpus callosum, and cerebrum.
8. Follow your teacher's instructions to dispose of the sheep brain and wash the dissecting tray.

Analyze and Interpret

1. Make a table to record the function of each structure you identified.

Conclude and Communicate

2. Compare the structure of the sheep brain with that of the human brain shown in **Figure 8.19**. What similarities and differences can you identify?

Extend Further

3. **INQUIRY** With a partner, build a model of the human brain and present it to the class. Include all the key structures and functions on your model.
4. **RESEARCH** Find out more information about concussions. How do they occur? What are the symptoms of a concussion? How are concussions linked to chronic traumatic encephalopathy (CTE)?

Chapter 8 SUMMARY

Section 8.1 Human Body Systems and Homeostasis

Human body systems work to maintain homeostasis, which is critical for survival.

Key Terms

homeostasis
sensor
control centre
effector
negative feedback system
positive feedback system

Key Concepts

- The human body is organized in a hierarchy of levels. Cells are the smallest unit of the body. Tissues are groups of similar cells, and organs are tissues that perform a particular function. Organs work together in an organ system.
- The human body systems transport blood and lymph, digest food, excrete wastes, move and protect the body, and maintain homeostasis.
- Homeostasis is the tendency of the body to maintain a relatively constant internal environment and is critical for survival.
- Homeostasis is maintained through feedback systems that continually monitor, assess, and adjust variables in the body's internal environment.
- The two types of feedback systems are negative feedback systems and positive feedback systems.

Section 8.2 Structures and Processes of the Nervous System

The nervous system gathers and interprets information from the body and coordinates responses to both external and internal stimuli.

Key Terms

central nervous system
peripheral nervous system
neuron
nerve
glial cell
myelin sheath
reflex arc
membrane potential
resting membrane potential
polarization
sodium-potassium pump
action potential
synapse
neurotransmitter
acetylcholine

Key Concepts

- The human nervous system is a complex system composed of many subsystems that all work together to maintain homeostasis in the body.
- The neuron is the functional unit of the nervous system. Neurons can be classified based on structure (multipolar, bipolar, or unipolar) or function (sensory neurons, interneurons, motor neurons).
- All cells have a membrane potential, but the neuron is unique in that it can change the potential of its membrane to generate an impulse. An impulse is transmitted from one neuron to the next by neurotransmitters at a synapse.

Section 8.3 The Central Nervous System

The central nervous system is the control centre of the body and consists of the brain and spinal cord.

Key Terms

meninges
cerebellum
medulla oblongata
pons
midbrain
thalamus
hypothalamus
cerebrum
blood-brain barrier
cerebrospinal fluid
cerebral cortex
corpus callosum
occipital lobe
temporal lobe
parietal lobe
frontal lobe

Key Concepts

- The central nervous system, which consists of the brain and spinal cord, is the control centre of the nervous system.
- The brain can be subdivided into three general regions: the hindbrain, the midbrain, and the forebrain.
- The outer layer of the cerebrum, called the cerebral cortex, is composed of grey matter, and is thought to be the source of human intellect.
- The right and left halves of the cerebral cortex are made of four pairs of lobes, each of which is associated with particular functions.

Section 8.4 The Peripheral Nervous System

The peripheral nervous system gathers and relays sensory information to the muscles and glands for response.

Key Terms

somatic system
autonomic system
sympathetic nervous system
norepinephrine
parasympathetic nervous system

Key Concepts

- The peripheral nervous system consists of the somatic and autonomic nervous systems.
- Homeostasis is maintained in the body by the actions of the sympathetic and parasympathetic nervous systems.
- In general, the sympathetic nervous system prepares the body for fight-or-flight, while the parasympathetic system returns the organs to a resting state.

Chapter 8 REVIEW

Knowledge and Understanding

Select the letter of the best answer below.

1. Identify the two systems that regulate and coordinate the function of the body's other systems.
   a. digestive and circulatory systems
   b. respiratory and excretory systems
   c. immune and lymphatic systems
   d. endocrine and nervous systems
   e. reproductive and skeletal systems
2. Which system reverses a change in a variable, bringing it back to within a normal range?
   a. a negative feedback system
   b. a positive feedback system
   c. the nervous system
   d. the endocrine system
   e. a regulatory system
3. Which division of the nervous system controls the senses of touch, taste, sight, sound, and smell?
   a. motor division of the central nervous system
   b. motor division of the peripheral nervous system
   c. parasympathetic division of the autonomic system
   d. somatic division of the central nervous system
   e. somatic division of the peripheral nervous system
4. Which statement about the neurons below could be true?

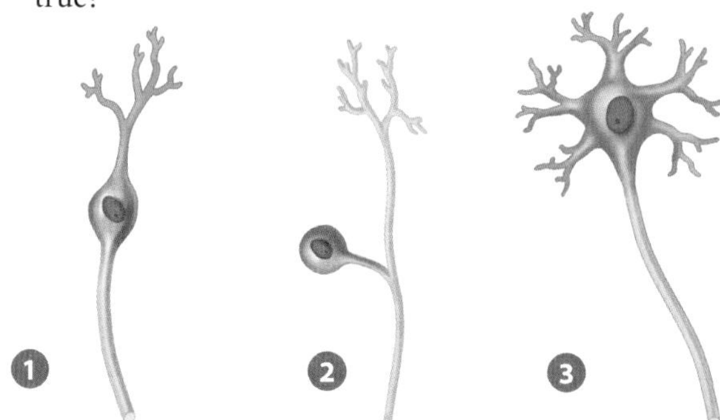

   a. Neuron 1 is a bipolar neuron in the retina.
   b. Neuron 1 is a bipolar neuron in the brain.
   c. Neuron 3 is a bipolar neuron in the inner ear.
   d. Neuron 3 is a bipolar neuron in the brain.
   e. Neuron 2 is a bipolar neuron in the PNS.
5. Which is the barrier that selectively controls the entrance of substances into the brain from the blood?
   a. meninges
   b. ventricles of the brain
   c. blood-brain barrier
   d. medulla oblongata
   e. midbrain
6. You touch a hot stove and withdraw your hand before you perceive the pain. Identify the neural pathway involved in this response.
   a. stimulus → motor neuron → interneuron → brain → sensory neuron → effector
   b. stimulus → sensory neuron → interneuron → brain → motor neuron → effector
   c. effector → sensory neuron → interneuron → brain → motor neuron → stimulus
   d. effector → motor neuron → interneuron → sensory neuron → stimulus
   e. stimulus → sensory neuron → interneuron → motor neuron → effector
7. Which refers to the time during which a neuron cannot be stimulated to undergo another action potential?
   a. the resting membrane potential
   b. the refractory period
   c. polarization of the neuron
   d. the membrane potential
   e. saltatory conduction of the action potential
8. Which statement about the sympathetic nervous system (SNS) in a stressful situation is *false*?
   a. The SNS releases the neurotransmitter norepinephrine.
   b. The SNS increases blood pressure.
   c. The SNS increases breathing rate.
   d. The SNS increases heart rate.
   e. The SNS increases the rate of digestion.

Answer the questions below.

9. Define homeostasis and give two examples of variables that are maintained within a specific range in the human body.
10. Describe a negative feedback system.
11. Create a table comparing the structures and functions of the central nervous system to those of the peripheral nervous system.
12. Identify the four features common to all neurons and describe the function of each.
13. Identify two functions of the myelin sheath.
14. What is a reflex and why are they important?
15. While nailing boards onto a fence, you accidentally hit your hand with a hammer. Sequence the path of neural transmission from the original stimulus to your response as you drop the hammer. Include the types of neurons and their functions.

16. Describe the term "polarization" as it applies to a neuron.
17. What is an action potential in an axon?
18. Explain why saltatory conduction occurs in myelinated neurons but not unmyelinated neurons.
19. Compare white matter with grey matter. Identify the location of each, and describe its function.
20. If the motor area of the right cerebral cortex was damaged in an automobile accident, which side of the body would be affected? Why?
21. In a snowmobile accident, a person receives a severe spinal cord injury. Explain why the person loses all sensation below the injured area.
22. A person complains of a noticeable decrease in muscle coordination after an injury to the brain. Which area of the brain is most likely affected? Explain.
23. A person with epilepsy can have severe epileptic seizures. Explain why severing the corpus callosum is used to treat some cases of epilepsy.
24. If the blood supply to an area of the brain is interrupted, as in a stroke, this part of the brain can be damaged, resulting in a loss of function. In the diagram below, the letters A to D indicate specific lobes of the cerebral cortex that have been damaged due to a stroke. Name the structures that correspond to each letter and describe which brain function would be affected in each case.

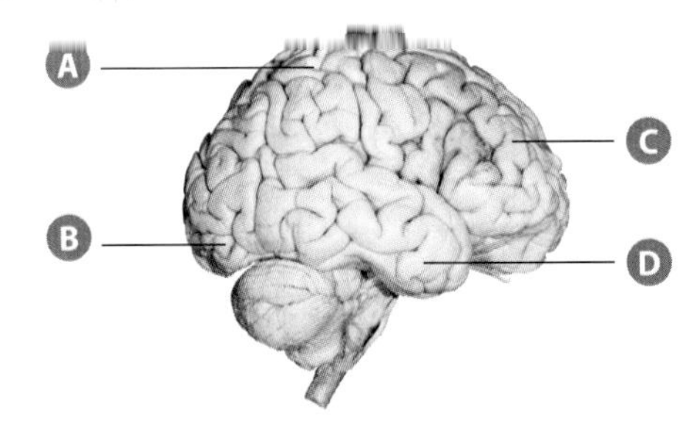

25. Compare the functions of the sympathetic and parasympathetic divisions of the autonomic system. Give specific examples of their physiological effects in the body.

Thinking and Investigation

26. BIG IDEAS Environmental factors can affect homeostasis. Neonicotinoid insecticides mimic the action of the neurotransmitter acetylcholine (ACh) in insects. Cholinesterase, the enzyme that breaks down ACh, is not affected by these insecticides. However, exposure to this class of insecticides leads to continual stimulation of a nerve. Hypothesize how neonicotinoid insecticides actually kill the insects.

Use the information below to answer questions 27—29.

Fibromyalgia is a chronic illness that causes physical pain and debilitating fatigue. In clinical studies, researchers found that patients with fibromyalgia have hyperactivity of the sympathetic nervous system. This hyperactivity induces excessive neurotransmitter secretions. Scientists hypothesize that the excessive neurotransmitter secretions could sensitize pain receptors and thus induce widespread pain and widespread tenderness.

27. Identify the neurotransmitter that would be continually released in people with "relentless hyperactivity."
28. Design an investigation to test the hypothesis stated above.
29. Suppose the hypothesis is supported by the results of the investigation. Infer a possible treatment for fibromyalgia.

Use the information below to answer questions 30–32.

In 1921, an Austrian scientist named Otto Loewi conducted an investigation involving two frog hearts that had been dissected out of living frogs. His procedure and observations are summarized below.

Procedure

- Heart #1, still connected to a functioning vagus nerve, was placed in a saline-filled chamber labelled Chamber #1.
- Heart #2 had been stripped of its vagal and sympathetic nerves. This heart was placed in a second saline-filled chamber labelled Chamber #2.
- Chamber #1 was connected to Chamber #2. This connection allowed the fluid from Chamber #1 to flow into Chamber #2.
- Loewi then electrically stimulated the vagus nerve attached to Heart #1.

Observations

- Loewi observed that Heart #1 slowed down.
- He then observed that, after a short delay, Heart #2 also slowed down.

30. Based on his observations, predict to which component of the autonomic nervous system the vagus nerve belongs. Justify your prediction.
31. Propose an inference explaining his second observation.
32. What did this investigation demonstrate?

33. Describe a safeguard that prevents a neuron from carrying an impulse in the wrong direction. Make an inference as to the effect of an impulse travelling in both directions in a neuron.

34. Scientists know that humans and other animals need to sleep. Sleep deprivation can affect us mentally and physically. Research information about why we sleep. Consider the following questions as you conduct your research.
   a. How is sleep related to brain function?
   b. How can sleep deprivation affect homeostasis? Identify the different body systems affected as part of your explanation.
   c. List three hypotheses scientists have presented regarding why we sleep.
   d. Choose one hypothesis from above and explain the investigation performed to test the hypothesis. Did the results support the hypothesis? Why or why not?

Use the information below to answer questions 35–38.

Some researchers think that the blood-brain barrier (BBB) is defective in patients with Alzheimer's disease (AD). The defective BBB would allow a certain protein to enter the brain from the systemic circulation. Alzheimer's disease (AD) may be caused by deposition of these protein plaques in brain tissue. In one study researchers compared BBB function in AD subjects with that in elderly control subjects. AD subjects and control subjects were residents of a continuing-care retirement community that had a 99-bed Alzheimer's unit. The control group was matched in age and education and subjects were independent, with no cognitive impairment. The 14 AD subjects were from a "mildly to moderately" cognitively impaired group with a diagnosis of probable AD. AD and control subjects were screened medically, neurologically, and cognitively prior to the study and were excluded if they demonstrated evidence of neurological disease other than AD.

35. Identify the hypothesis being tested in this investigation.

36. Describe the variables that were controlled in this investigation.

37. Explain the possible link between a defective blood-brain barrier and Alzheimer's disease.

38. Describe the results the researchers would need to obtain in order to support the hypothesis in this investigation.

Communication

39. Draw a flowchart summarizing the regulation of contractions during childbirth.

40. **BIG IDEAS** Systems that maintain homeostasis rely on feedback mechanisms. Create a 10-minute lesson that you would present to a Grade 6 class about how negative feedback systems help maintain homeostasis. Include the following information.
   a. definition of homeostasis
   b. example of homeostasis in the human body
   c. explanation and example of a negative feedback system
   d. diagram of a negative feedback system
   e. assignment to help students process the information learned from the lesson

41. Create a diagram showing the main divisions of the nervous system. Describe the key features of each division.

42. Examine the diagram below. Create a table to record
   - the structures and types of neurons indicated by the letters in the diagram
   - the functions of these structures and neurons

   Under your table, indicate the direction of neuron transmission.

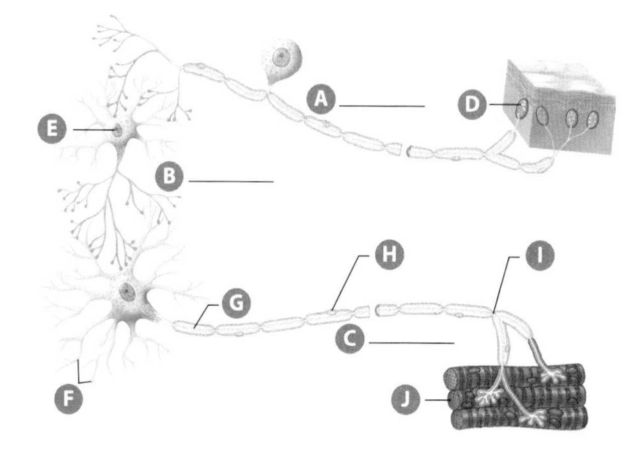

43. Using a diagram, explain depolarization, action potential, and repolarization of the neuron.

44. Use a graphic organizer to organize information about the forebrain. Be sure to identify the major structures in the forebrain and their location in the brain, and summarize the function of each.

45. Write a paragraph explaining the effects of sympathetic nervous system stimulation on the body.

46. The diagram below indicates different ion concentrations from the inside to the outside of a neuron while the neuron is at rest.

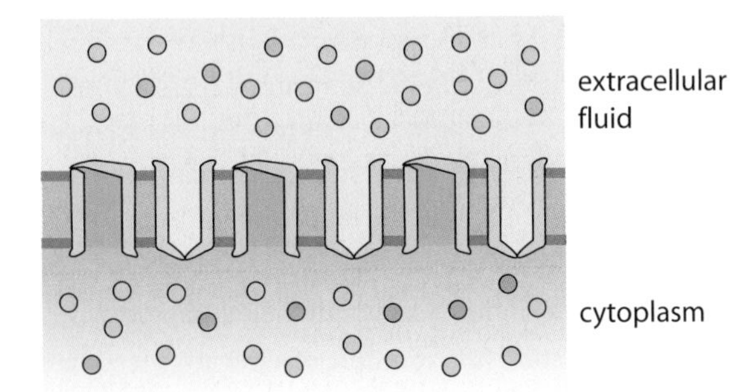

a. Draw this diagram in your notebook, and indicate the sodium ions, potassium ions, sodium ion channels, and potassium ion channels.

b. Also indicate the charge inside and outside the neuron.

c. Explain how the different ion concentrations are established.

47. Summarize your learning in this chapter using a graphic organizer. To help you, the Chapter 8 Summary lists the Key Terms and Key Concepts. Refer to Using Graphic Organizers in Appendix A to help you decide which graphic organizer to use.

Application

48. BIG IDEA Organisms have [illegible] limits on the internal conditions that they can tolerate. Research more information about heat-related illnesses, including heat exhaustion and heat stroke. Consider the following questions as you conduct your research.

a. What conditions, both external and internal, can lead to the development of these illnesses?

b. How are these illnesses related to homeostasis?

c. What are the symptoms of these illnesses?

d. What are the treatments for these illnesses?

e. What happens if homeostasis is not restored?

49. Based on what you know about threshold potential, explain why some people seem to be more tolerant of pain than others.

50. If food is not preserved properly, *Clostridium botulinum* bacteria can start to reproduce and release a neurotoxin called botulinum toxin. Botulinum toxin inhibits the action of acetylcholine, causing botulism. What symptoms would you expect to observe in someone with botulism? Provide an explanation.

Use the information below to answer questions 51-53.

In a classic investigation, the strength of a neural stimulus and the resulting muscle contraction are compared. A single motor neuron that synapses with a muscle fibre is suspended. The other end of the muscle fibre is attached to a mass. If an electrical stimulus is sufficient to cause an impulse in the neuron, the muscle will contract and lift the mass. The following data were obtained from the investigation. Analyze the data, and answer the following questions.

Strength of Stimulus and Muscle Contraction

Strength of Stimulus (mV)	Mass Lifted by Muscle Contraction (g)
1	0
2	10
3	?
4	?

51. Define "threshold potential." What is the minimum size of the stimulus required to reach the threshold potential for this motor neuron?

52. Explain the all-or-none response. Then predict the mass that could be lifted at 3 mV of stimuli and at 4 mV of stimuli.

53. Choose a specific example of a sensory neuron, and explain neural stimulation and impulses in terms of the [illegible] potential and the all or none response.

Use the information below to answer questions 54-58.

One way to model the action potential is to line up several dominoes and initiate a cascade event, in which each successive domino knocks down the next domino.

54. In this model, the hand provides the initial energy. What provides the initial energy in a neural impulse?

55. The finger has to contact the first domino just hard enough to get it to fall. Which response does this represent in a real neuron?

56. Once the dominoes start to fall, they all fall in succession. What does this action represent in the real neuron?

57. The dominoes always fall in one direction. Contrast this with the direction of impulse transmission in a real neuron.

58. No matter how many times the dominoes fall, they always move at the same speed and intensity. What principle does this represent in the real neuron?

59. Researchers are experimenting with new technologies that could help people with missing limbs. In one investigation, electrodes implanted in the nervous tissue of a monkey were connected to an artificial hand. The monkey's nervous system was able to direct the artificial hand to move. This photograph shows the monkey raising a piece of zucchini to its mouth using the thought-controlled robotic arm.

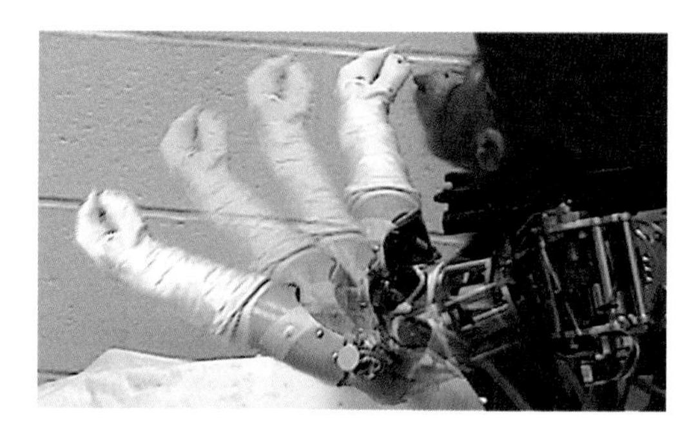

- **a.** What area of the brain directs the movement of the robotic arm? (Assume that the structure of a monkey brain is similar to the structure of a human brain.)
- **b.** Using a flowchart, illustrate the basic neural pathway from the sensory stimulus to the motor output.
- **c.** Compare this artificial pathway with the actual neural pathway to a biologically functional limb.
- **d.** What are some other potential applications for this technology?
- **e.** Do the benefits to human life justify this form of animal research?

60. Trigeminal neuralgia (TN) is a chronic pain condition that affects the trigeminal nerve, one of the largest nerves in the head. The disorder causes extreme, sporadic burning face pain that lasts anywhere from a few seconds to as long as two minutes per episode. The intensity of pain can be physically and mentally incapacitating. Based on this information, predict the type of nerve affected by this condition. Justify your prediction.

61. Research information about amnesia, the lack of memory. Consider the following questions as you conduct your research.

- **a.** What causes amnesia?
- **b.** What are the different types of amnesia? Which area of the brain is associated with each type?
- **c.** How does studying patients with amnesia help scientists learn more about brain function and how memories are formed?

Use the information below to answer questions 62-64.

There is a thermostat, which contains a thermometer, on the wall of a classroom, and there is a furnace in the basement of the school. The thermostat is set to a temperature of 20°C. The temperature in the room drops to 18°C. The thermometer inside the thermostat detects this change and sends an electric signal to the furnace, turning it on. Gradually, the classroom warms up. The temperature in the room increases to 22°C and the furnace shuts down. The room slowly returns to 20°C and then gets slightly cooler as heat flows from the room.

62. List the three components of a feedback system and identify these components in the example of a room heating system.

63. Sketch a graph showing how the temperature in the classroom fluctuates around the set point of 20°C.

64. Explain why this heating system is considered to be an example of a negative feedback system.

Use the information below to answer questions 65 and 66.

Multiple sclerosis (MS) is a disease of the central nervous system (CNS) that attacks the myelin sheath of myelinated axons. The early stages of this disease are characterized by inflammation of the myelin along neurons. When this happens, the usual flow of nerve impulses along axons is interrupted or distorted.

65. Symptoms in some people with MS may include tingling, a burning sensation, or loss of sensation. Identify the type of neuron affected in these individuals. Explain your answer.

66. Predict what happens to the action potential in a neuron affected by MS.

67. Suppose you are at the movie theatre enjoying the latest popular horror film. The table below lists several situations that occur as you watch the film. Copy the table into your notebook. Give it a title and identify the automatic response that occurs in your body and which system (sympathetic or parasympathetic) would control the response for each situation.

Situation	Automatic Response	System That Controls Response
Your eyes adjust to the darkened theatre.		
You are startled when the villain jumps out from behind the sofa in the movie.		
You eat nacho chips and melted cheese.		

Chapter 8 SELF-ASSESSMENT

Select the letter of the best answer below.

1. K/U The generation of a nerve impulse begins with a slight leakage of sodium ions through sodium channels in the neuron. This results in a change in the membrane potential, which, in turn, causes more sodium channels to open which creates the action potential. Which is this an example of?
a. a negative feedback system
b. a positive feedback system
c. the integumentary system
d. the endocrine system
e. a regulatory system

Use the graph below to answer questions 2 and 3.

Action Potential

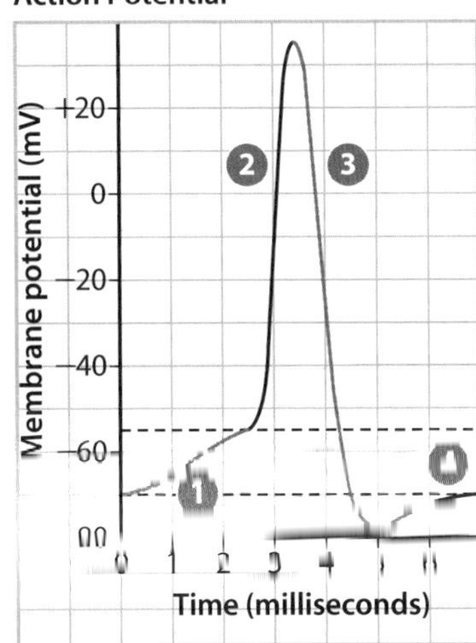

2. K/U What is the resting membrane potential of this neuron?
a. −55 mV
b. +35 mV
c. −80 mV
d. −70 mV
e. 0 mV

3. K/U Which would happen to the action potential if a much larger stimulus were applied to this neuron?
a. More sodium ion channels would open.
b. Fewer sodium ion channels would open.
c. More potassium ion channels would open.
d. Fewer potassium ion channels would open.
e. No changes to the action potential would occur.

4. K/U Transmission of the nerve impulse across the synapse is accomplished by which of the following?
a. sodium-potassium pump
b. release of a neurotransmitter by a dendrite
c. release of a neurotransmitter by an axon
d. release of a neurotransmitter by the cell body
e. saltatory conduction

5. K/U Identify the area of the brain that is responsible for maintaining homeostasis by regulating processes such as hunger, thirst, body temperature, and water balance.
a. thalamus
b. frontal cortex
c. pons
d. corpus callosum
e. hypothalamus

6. K/U Which is responsible for carrying impulses from the external sensory receptors to the CNS as well as carrying motor commands to the skeletal muscles?
a. parasympathetic system
b. somatic nervous system
c. sympathetic system
d. central nervous system
e. autonomic nervous system

7. K/U Which choice below correctly matches the effector to the effects of sympathetic stimulation?
a. pupils of the eye → constricts pupils
b. lungs → constricts air passages
c. heart → increases heart rate
d. stomach → increases activity of stomach
e. urinary bladder → stimulates urination

8. K/U Which are the main neurotransmitters of the sympathetic nervous system and the parasympathetic nervous system, respectively?
a. dopamine and serotonin
b. dopamine and norepinephrine
c. serotonin and acetylcholine
d. endorphins and norepinephrine
e. norepinephrine and acetylcholine

9. K/U Which lobe of the cerebral cortex is responsible for understanding speech and retrieving visual and verbal memories?
a. temporal lobe
b. parietal lobe
c. occipital lobe
d. frontal lobe
e. motor lobe

10. K/U Which of the following statements about the parasympathetic nervous system (PNS) is *false*?
a. The PNS releases acetylcholine.
b. The PNS lowers blood pressure.
c. The PNS decreases breathing rate.
d. The PNS slows heart rate.
e. The PNS decreases intestinal activity.

Use sentences and diagrams as appropriate to answer the questions below.

11. **K/U** Identify the body systems involved in the homeostatic regulation of the human body.

12. **C** Create a table that identifies the organ systems and their functions in the human body.

13. **C** Based on the information presented in this graph, write a paragraph explaining homeostasis.

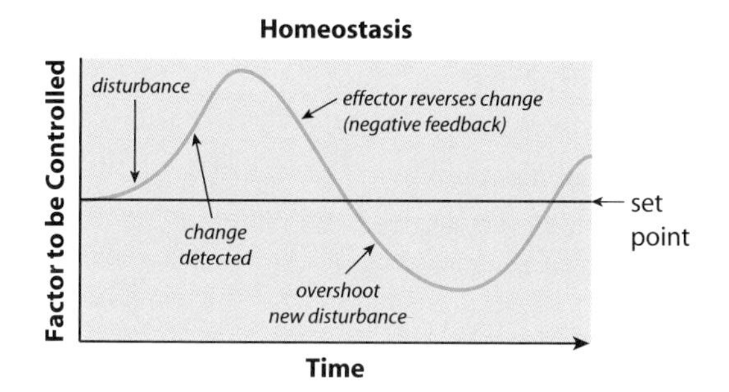

14. **A** Hormones control the secretion and formation of breast milk in a mother. The more the baby suckles, the more milk is produced. What type of feedback system is described? Explain.

15. **K/U** Distinguish between neurons and glial cells and give an example of a glial cell.

16. **T/I** The diagram below shows two neurons.

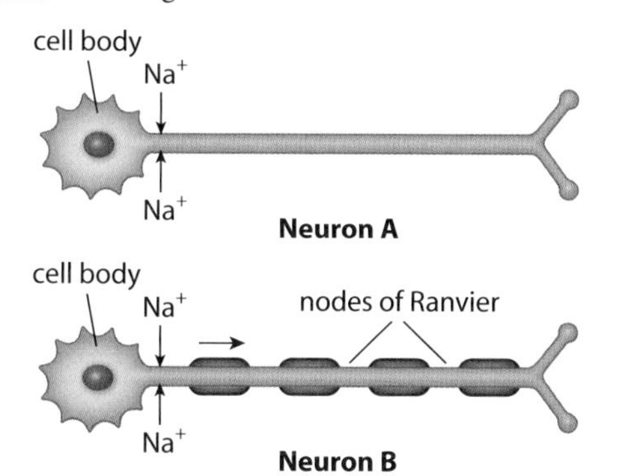

   a. Predict the pathway of an action potential as it moves down neuron A. Do the same for neuron B.

   b. Design an investigation to test your prediction.

17. **C** Write a paragraph comparing a reflex arc to a response involving more conscious thought, such as stopping a car when a cat jumps in front of it.

18. **T/I** A neuron is connected to a voltmeter. It has a resting membrane potential of –70 mV. The threshold potential of this neuron is –55 mV. Predict the response if a researcher applies a stimulus to this neuron that results in a membrane potential of –40 mV. Justify your prediction.

19. **C** Summarize, in point form, the steps in the repolarization of a neuron.

20. **A** The earliest symptoms of amyotrophic lateral sclerosis (ALS) are weakness and/or muscle atrophy. Other symptoms include twitching, cramping, or stiffness of affected muscles, and/or slurred and nasal speech. People living with ALS become progressively paralyzed. Infer the type of neuron affected by ALS. Explain your answer.

21. **C** Summarize the events that occur at a synapse as an action potential reaches the end of an axon.

22. **T/I** A patient with uncontrolled seizures had an area of his brain partially severed by a surgeon in an attempt to control his illness. The surgery did not prevent the patient's ability to walk, talk, or eat. When the patient holds an object in his right hand, he can state the name of the object. However, when the patient holds the same object in his left hand, he is unable to state the name of the object.

   a. Identify the area of the brain that was removed by the surgery in this patient and describe its function.

   b. Propose a hypothesis to explain why this patient would develop these symptoms.

23. **A** Identify symptoms a person who has sustained damage to their cerebellum might experience.

24. **A** Neuritis is a medical condition characterized by an inflamed nerve or an inflamed portion of the nervous system.

   a. Identify the types of neurons affected if a person experiences tingling, burning, pins-and-needles sensations, stabbing, or even loss of sensation.

   b. Identify the types of neurons affected if an individual experiences a slight loss of muscle tone or paralysis with muscle wasting.

25. **K/U** Identify the two areas of the brain that control the autonomic nervous system.

Self-Check

If you missed question...	1	2	3	4	5	6	7	8	9	10	11	12	13	14	15	16	17	18	19	20	21	22	23	24	25
Review section(s)...	8.1	8.2	8.2	8.2	8.3	8.4	8.4	8.4	8.3	8.4	8.1	8.1	8.1	8.1	8.2	8.2	8.2	8.2	8.2	8.2	8.2	8.3	8.3	8.2	8.4

CHAPTER

9 The Endocrine System

Specific Expectations

In this chapter, you will learn how to . . .

- E1.1 **assess**, on the basis of findings from a case study, the effects on the human body of taking chemical substances to enhance performance or improve health (9.3, 9.4)
- E1.2 **evaluate**, on the basis of research, some of the human health issues that arise from the impact of human activities on the environment (9.4)
- E2.1 **use** appropriate terminology related to homeostasis (9.1, 9.2, 9.3, 9.4)
- E2.2 **plan** and **construct** a model to illustrate the essential components of the homeostatic process (9.4)
- E2.3 **plan** and **conduct** an investigation to study a feedback system (9.4)
- E3.1 **describe** the anatomy and physiology of the endocrine, excretory, and nervous systems, and **explain** how these systems interact to maintain homeostasis (9.1, 9.2, 9.3, 9.4)
- E3.2 **explain** how reproductive hormones act in human feedback [illegible] maintain homeostasis (9.4)

Imagine yourself standing 150 metres above the water on the edge of a bridge. Your heart is racing, your breathing rate has increased, and you have butterflies in your stomach. You're about to bungee jump off the bridge. The hormones epinephrine and norepinephrine are coursing through your body. They were released in response to the stimulus of doing something potentially life-threatening. Your body is preparing you for action—blood flow to your heart and muscles has increased, your liver is converting glycogen to glucose, and the rate of cellular metabolism has increased. All of this occurs so that your muscles will have more energy to react to the situation.

Launch Activity

Don't Sweat It

Many modern stressors, such as exams, deadlines, and bills, are not life-threatening. However, the body often responds as if they were, and produces the same physiological changes that would occur in a truly dangerous situation. In this activity you will monitor some of the changes that occur in the body in response to a stressful, though not life-threatening, situation.

Safety Precautions

Do not take the role of subject in this activity if you have a medical condition, such as high or low blood pressure, that is made worse by stress.

Suggested Materials

- timer
- test questions

Procedure

1. Choose a stressful activity that does not involve physical activity, such as giving a speech in front of the rest of the class or taking a test.
2. Decide which variables you will measure to determine the physiological reactions to the stressor. The variables could include changes in respiration rates, pulse, blood pressure, and pupil dilation, as well as reports from participants on how they feel (sweat levels, butterflies in stomach, muscle tension, hands shaking, etc.).
3. Construct a data table in which to record your data.
4. To record control data, have the participants sit at a desk for 1 min with eyes closed, taking deep relaxing breaths. At the end of the relaxation period, measure the variables you've chosen to track.
5. Have participants perform the activity for three minutes, then stop. Immediately begin recording data. Continue to record data every minute until the readings have returned to their initial measurements.

Questions

1. Summarize the main physiological changes that occurred in the subject as a result of the stressor.
2. How would the physiological changes that you listed in question 1 be useful responses to a life-threatening situation?
3. The reactions you measured are under the control of the nervous system, as well as the endocrine system, which releases chemical messengers called hormones into the blood. Infer why it took longer for some variables to return to initial levels than others.
4. List 10 or more stressful situations that people in modern societies experience. Which are truly dangerous? Which, if any, might be considered positive?

SECTION 9.1 The Glands and Hormones of the Endocrine System

Key Terms

endocrine gland
hormone
endocrine system
tropic hormone

Figure 9.1 Each of these speed skaters relies on various body systems in order to make it across the finish line. The organs of their body systems are composed of trillions of cells, some of which help to carry oxygen, remove wastes, move muscles, and hear sounds. Some cells communicate critical information from one area of the body to another.

endocrine gland ductless gland that secretes hormones directly into the bloodstream

hormone chemical messenger sent to many parts of the body to produce a specific effect on a target cell or organ

endocrine system in vertebrates, system that works in parallel with the nervous system to maintain homeostasis by releasing chemical hormones from various glands; composed of the hormone-producing glands and tissues of the body

Each of the speed skaters in **Figure 9.1** relies on complex internal systems in order to reach speeds of 50 km/h or more, to balance [illegible] position of the other skaters, all while continuing to maintain homeostasis. For example, the respiratory, circulatory, and muscular-skeletal systems of each skater are regulated and coordinated. These systems, along with the nervous system, are working together to bring oxygen to muscle cells, remove carbon dioxide from the body, and move each skater over the ice. Another system, the endocrine system that you will read about in this chapter, helps control the availability of glucose, which their cells are using for energy. All of this means that the functioning of the over 100 trillion diverse cells making up the tissues and organs of each of their bodies must also be regulated and coordinated. In order for this to occur, the cells must be able to communicate with one another. The body systems that facilitate cellular communication and control are the nervous and endocrine systems.

Recall from Chapter 8 that nervous system messages tend to be transmitted rapidly to precise locations in the body, such as the reflex arc that causes you to withdraw your hand from a hot stove. In addition to cellular communication through neurons, the body secretes chemical messengers from glands. **Endocrine glands** secrete chemical messengers called **hormones** directly into the bloodstream, which transports the hormones throughout the body. The original Greek meaning of the word hormone is to "excite" or "set in motion."

The endocrine glands and the hormones that they secrete make up the **endocrine system**. Compared to the rapid actions of the nervous system, the endocrine system typically has slower and longer acting effects, and affects a broader range of cell types.

The Endocrine Glands

Scientists have identified over 200 hormones or hormone-like chemicals in the human body. Some regulate growth and development, some speed up or slow down the body's metabolism, and others regulate blood pressure or the immune response. **Figure 9.2** shows the location in the body of glands that function exclusively as endocrine glands: the pituitary, pineal, thyroid, parathyroid, and adrenal glands, as well as tissues and organs that secrete hormones (but do not function exclusively as endocrine glands): the hypothalamus, thymus, pancreas, testes, and ovaries. **Table 9.1** on the next page provides an overview of some of these glands, the hormones they release, and their effects on tissues and organs. When hormones are released, they act on target cells. Target cells are cells whose activity is affected by a particular hormone.

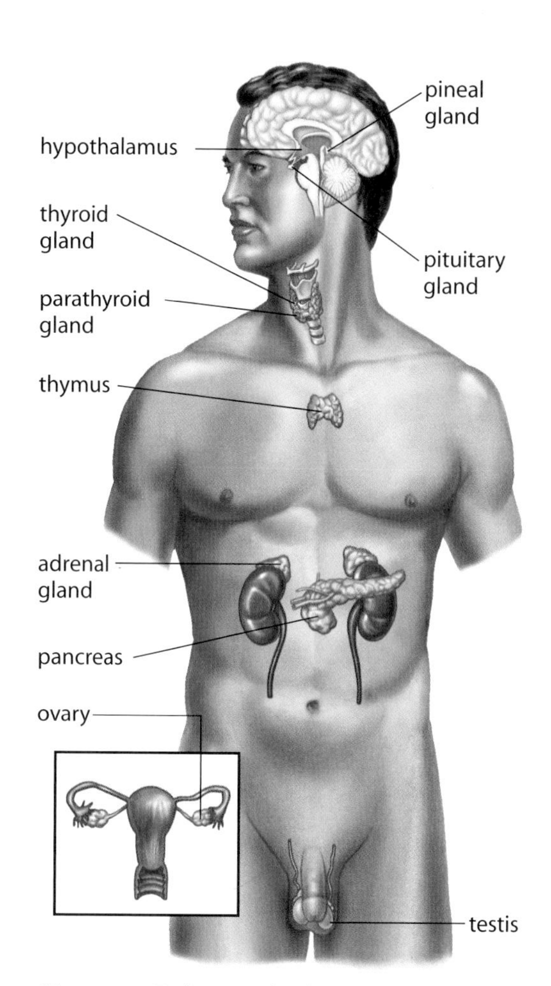

Figure 9.2 Endocrine glands include the pituitary gland, the thyroid gland, and the adrenal glands.

Hormone Action on Target Cells

When hormones encounter their target cells, how do they affect them? Each target cell contains receptor proteins. Circulating hormones bind to their specific receptor proteins, much like a key fits into a lock. A general example of target cells and their receptors is shown in **Figure 9.3**. Human growth hormone (hGH) can be used as a specific example. hGH circulates in the bloodstream and interacts with liver, muscle, and bone cells. Each of these cell types contains receptor proteins specifically shaped to bind with hGH. When the hGH binds to its receptor, this triggers other reactions in the target cell. In other words, the target cell receives and responds to the chemical message sent by the hormone.

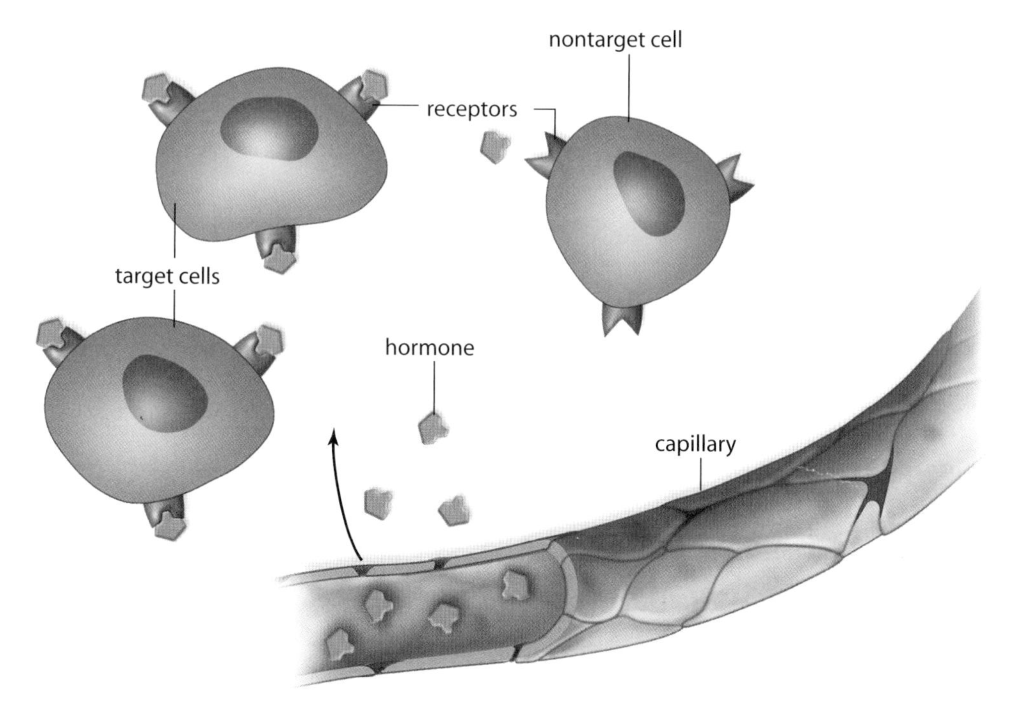

Figure 9.3 Most hormones are distributed by the bloodstream to target cells. Target cells have receptors for the hormone, and the hormone combines with the receptor as a key fits a lock.

***Explain** why the lock and key analogy is appropriate when describing hormones and target cells.*

Table 9.1 The Principal Endocrine Glands and Some of Their Hormones

Endocrine Gland	Hormone Secreted	Effects of Hormone on Target Tissues/Organs
Hypothalamus	Hypothalamic releasing and inhibiting hormones	Regulates anterior pituitary hormones
Anterior pituitary	Human growth hormone (hGH)	Stimulates cell division, bone and muscle growth, and metabolic functions
	Thyroid-stimulating hormone (TSH)	Stimulates the thyroid gland
	Adrenocorticotropic hormone (ACTH)	Stimulates the adrenal cortex to secrete glucocorticoids
	Follicle-stimulating hormone (FSH)	Stimulates production of ova and sperm from the ovaries and testes
	Luteinizing hormone (LH)	Stimulates sex hormone production from the ovaries and testes
	Prolactin (PRL)	Stimulates milk production from the mammary glands
Posterior pituitary	Antidiuretic hormone (ADH)	Promotes the retention of water by the kidneys
	Oxytocin (OCT)	Stimulates uterine muscle contractions and release of milk by the mammary glands
Thyroid	Thyroxine (T_4)	Affects all tissues; increases metabolic rate and regulates growth and development
	Calcitonin	Targets bones and kidneys to lower blood calcium by inhibiting release of calcium from bone and reabsorption of calcium by kidneys
Parathyroid	Parathyroid hormone (PTH)	Raises blood calcium levels by stimulating the bone cells to release calcium, the intestine to absorb calcium from food, and the kidneys to reabsorb calcium
Adrenal cortex	Glucocorticoids (for example, cortisol)	Stimulates tissues to raise blood glucose and break down protein
	Mineralocorticoids (for example, aldosterone)	Promotes reabsorption of sodium and water by the kidneys
	Gonadocorticoids	Promotes secondary sexual characteristics
Adrenal medulla	Epinephrine and norepinephrine	Fight-or-flight hormones
Pancreas	Insulin	Lowers blood glucose levels and promotes the formation of glycogen in the liver
	Glucagon	Raises blood glucose levels by converting glycogen in the liver to glucose
Ovaries	Estrogen	Stimulates uterine lining growth and promotes development of the female secondary sexual characteristics
	Progesterone	Promotes growth of the uterine lining and prevents uterine muscle contractions
Testes	Testosterone	Promotes sperm formation and development of the male secondary sexual characteristics

Learning Check

1. Define the term "hormone" using an example.
2. Describe the endocrine system and identify the general function of the system.
3. Using a table, compare the structures and the mechanisms of action of the nervous system to those of the endocrine system.
4. Using a diagram, explain the relationship between a hormone and its target cells.
5. Suppose you are about to skydive out of a plane that is 3000 m above the ground. Which endocrine gland is likely releasing hormones at the moment you step to the plane door? Explain your reasoning.
6. How does the situation in question 5 involve both the nervous system and the endocrine system?

Steroid Hormones and Water-Soluble Hormones

Steroid hormones, such as testosterone, estrogen, and cortisol, are lipid-based. As shown in **Figure 9.4A**, steroid hormones, also called lipid-soluble hormones, can easily diffuse through the lipid bilayer of cell membranes. Inside the target cell, steroid hormones bind to their receptor proteins. This interaction activates specific genes, causing changes in the cell. For example, estrogen can trigger cell growth.

Epinephrine, human growth hormone (hGH), thyroxine (T_4), and insulin are water-soluble hormones. *Water-soluble hormones*, such as amino acid–based hormones, cannot diffuse across the cell membrane. Typically, a water-soluble hormone will bind to a receptor protein on the surface of the target cell, as shown in **Figure 9.4B**. This starts a cascade of reactions inside the target cell. Much like a telephone tree, in which everyone on a list calls several other people, each reaction that occurs in the target cell triggers many other reactions. As a result of this process, the impact of the hormone is greatly amplified.

For example, a single molecule of epinephrine in the liver can trigger the conversion of glycogen into about one million molecules of glucose. As shown in **Figure 9.4B**, when epinephrine reaches the liver, it stimulates the conversion of ATP to cyclic adenosine monophosphate (cAMP). The cAMP triggers an enzyme cascade that results in many molecules of glycogen being broken down into glucose. The glucose enters the bloodstream and will eventually be used by cells for energy. Once a hormone's message has been delivered, enzymes inactivate the hormone, since any lingering effect could potentially be very disruptive.

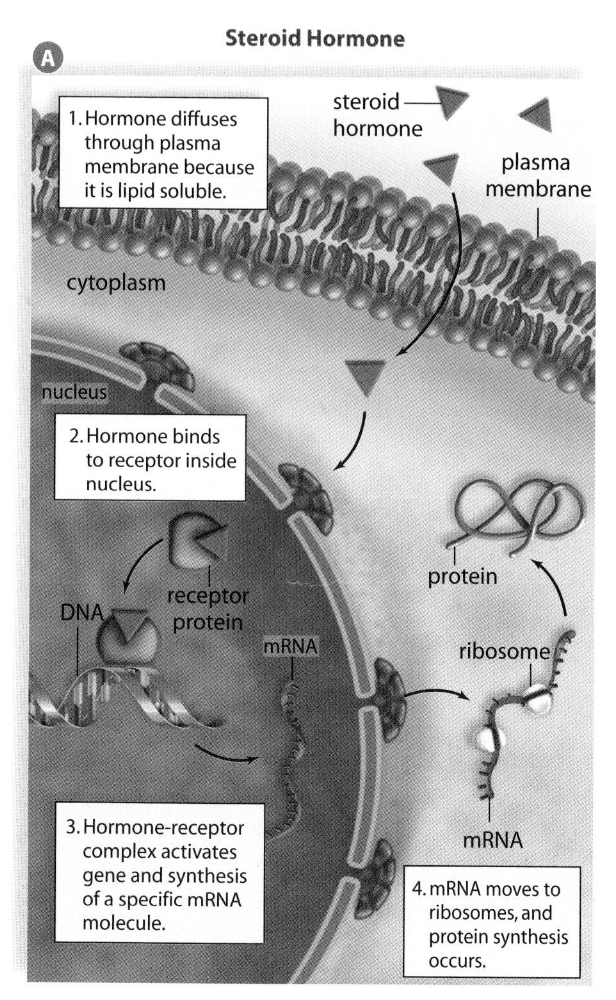

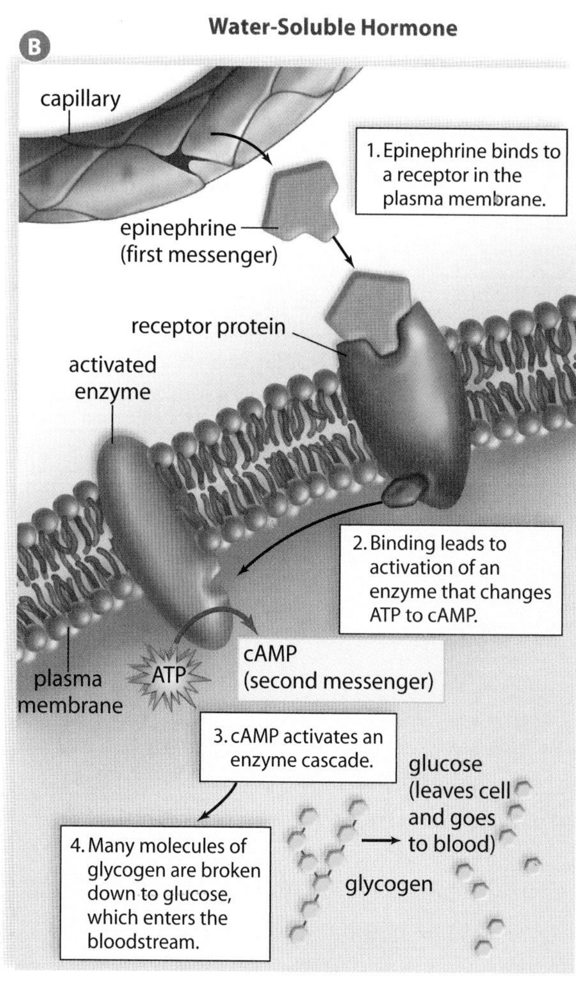

Figure 9.4 A A steroid hormone passes directly through a target cell's membrane before binding to a receptor in the nucleus or cytoplasm. **B** A water-soluble hormone binds to a receptor in the plasma membrane. This sets off a cascade of reactions inside the target cell.

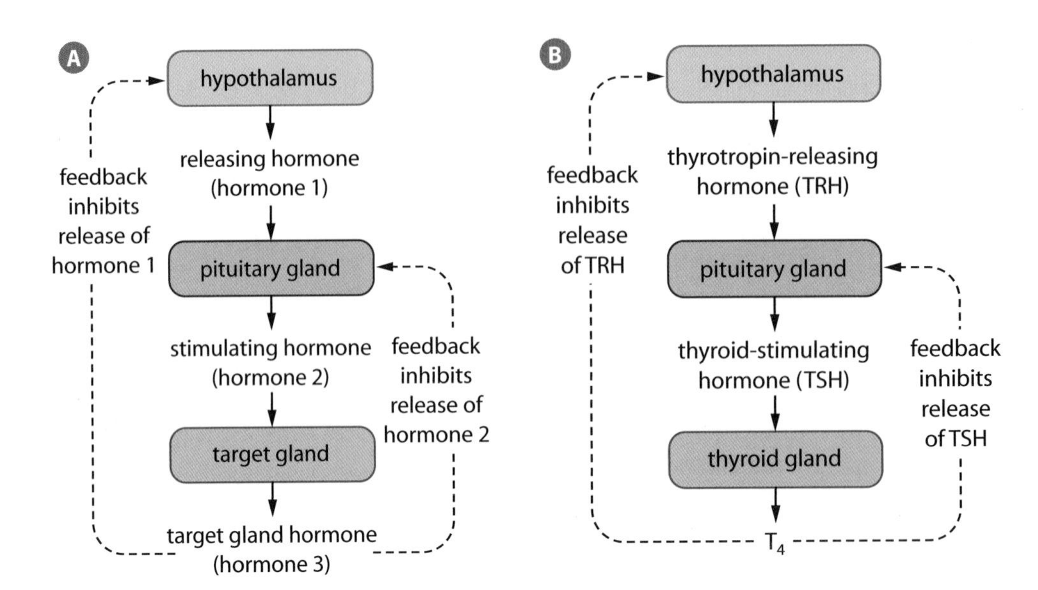

Figure 9.5 Tropic hormones act on other endocrine glands. In these loops, the hormone secreted by the target gland will affect other tissues in the body, such as the bones and muscles.

Explain *How are tropic hormones regulated by negative feedback?*

Regulating the Regulators

For many years, scientists referred to the pituitary gland as the master gland, because many of the hormones it secretes stimulate other endocrine glands. However, further research has shown that the pituitary gland is in fact controlled by the hypothalamus. After receiving signals from various sensors in the body, the hypothalamus secretes what are referred to as releasing hormones, which often travel to the pituitary gland. The *releasing hormones* stimulate the pituitary gland to secrete hormones that act on other endocrine glands. Hormones that stimulate endocrine glands to release other hormones are called **tropic hormones**. Many of the hormones released from the hypothalamus and anterior pituitary are tropic hormones. Together, the hypothalamus and pituitary gland control many physiological processes that maintain homeostasis.

tropic hormone hormone that targets endocrine glands and stimulates them to release other hormones

Figure 9.5A shows the general mechanism of action of tropic hormones. Typically, the hypothalamus secretes a releasing hormone into the anterior pituitary. This causes the anterior pituitary to release a second tropic hormone into the bloodstream. The second tropic hormone then stimulates the target gland to release a third hormone into the blood. This hormone travels to another target tissue and produces an effect. Like many hormones, this system is controlled by a negative feedback loop. In this case, the third hormone prevents further release of the first two hormones in the pathway. A specific example is the feedback system that controls thyroid-stimulating hormone (TSH), as shown in **Figure 9.5B**. Low blood levels of the thyroid hormone T_4 initiate the response from the hypothalamus. When blood levels of T_4 increase, the release of TRH and TSH is inhibited.

Working Together to Maintain Homeostasis

Homeostasis depends on the close relationship between the nervous system and the endocrine system. The functions of these two systems often overlap.

- Some nervous system structures, such as cells in the hypothalamus, secrete hormones.
- Several chemicals function as both neurotransmitters and hormones. Epinephrine acts as a neurotransmitter in the nervous system, and as a hormone in the fight-or-flight response.
- The endocrine and nervous systems are regulated by feedback loops.
- The regulation of several physiological processes involves the nervous and endocrine systems acting together. For example, when a mother breastfeeds her baby, the baby's suckling initiates a sensory message in the mother's neurons that travels to the hypothalamus. This triggers the pituitary to release the hormone oxytocin. Oxytocin travels to the mammary glands of the breast, causing the secretion of milk.

Suggested Investigation

ThoughtLab Investigation 9-A, Regulation of Melatonin

Section 9.1 Review

Section Summary

- The nervous system rapidly affects specific tissues, to which it is directly connected by neurons. The endocrine system relies on chemical messengers called hormones, which circulate in the blood and have broad, long-lasting effects.
- Steroid hormones can easily diffuse through the lipid bilayer of cell membranes. Water-soluble hormones, such as amino acid–based hormones, cannot diffuse across the cell membrane.
- Many hormones are regulated by negative feedback mechanisms.
- The nervous and endocrine systems are self-regulating and help regulate other body systems, thereby maintaining homeostasis.

Review Questions

1. K/U Define homeostasis and explain how the endocrine system helps to maintain homeostasis.
2. K/U Identify which glands secrete the following hormones:
   a. thyroxine
   b. cortisol
   c. human growth hormone
   d. insulin
   e. glucagon
3. T/I Sex hormones are easily absorbed through the skin and resist degradation better than many other hormones. In a 1998 paper in the journal *Clinical Pediatrics,* Dr. Chandra Tiwary reported an outbreak of early breast development in four young African-American girls who used shampoos containing estrogen.
   a. What characteristics of estrogen may have led Dr. Tiwary to infer that it was the cause of the early onset of puberty in these girls?
   b. Assuming the cause of the early puberty was exposure to estrogen, predict what happened when the girls stopped using the shampoo.
4. C Use a Venn diagram to compare and contrast steroid hormones and water-soluble hormones.
5. T/I Infer why steroid hormones usually take longer to have an effect than water-soluble hormones.
6. K/U How are many endocrine glands regulated?
7. A Which is an example of a tropic hormone? Explain your reasoning.
   a. calcitonin
   b. follicle-stimulating hormone (FHS)
   c. glucagon
   d. oxytocin
   e. thyroxine (T_4)
8. A Use an example to explain how the nervous system and endocrine systems work together to regulate a response in the body.
9. C Low blood levels of cortisol stimulate the hypothalamus to secrete corticotropin-releasing hormone (CRH). CRH travels to the pituitary gland and stimulates the secretion of adrenocorticotropin hormone (ACTH). ACTH travels to the adrenal glands, which then secrete cortisol. The increased levels of cortisol in the bloodstream subsequently inhibit the release of CRH and ACTH. Using **Figure 9.5** as a guide, make a diagram that shows the feedback system that controls ACTH.
10. A Use the diagram below to identify the feedback system that controls the release of insulin.

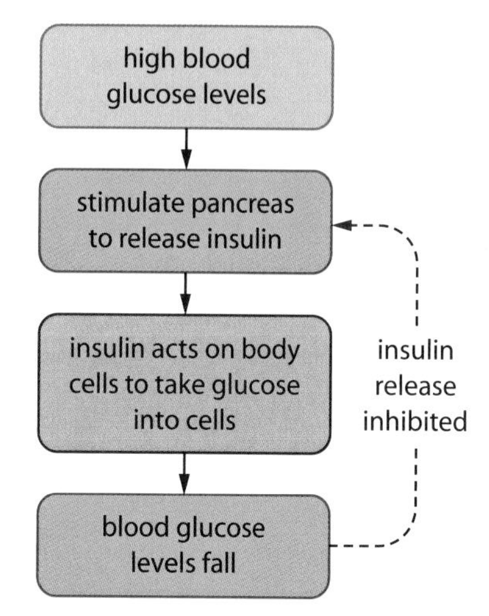

11. T/I Suppose a scientist has discovered a new hormone. It is not clear what gland produces the hormone, but people who produce above average amounts of this hormone also produce very high levels of insulin. Based on your knowledge of how tropic hormones function, provide a possible explanation for the observation.
12. K/U Provide four reasons why the distinction between the nervous and endocrine systems is sometimes blurred.

SECTION
9.2

Hormonal Regulation of Growth, Development, and Metabolism

Key Terms

posterior pituitary
anterior pituitary
human growth hormone (hGH)
thyroid gland
thyroxine (T_4)
hypothyroidism
hyperthyroidism
thyroid-stimulating hormone (TSH)
goitre

You may have heard the expression "growing like a weed" used to refer to an adolescent who has grown several centimetres in just a few months. And you may have heard people say they have a "fast metabolism"—meaning that they can eat whatever they want and not gain any weight. What controls the growth and development of your muscles and bones? What controls the rate of your metabolism? These things are controlled by hormones released by the pituitary gland and thyroid gland, respectively. Recall from Section 9.1 that the pituitary gland is controlled by the hypothalamus via releasing hormones and it secretes tropic hormones. The hormones secreted by the thyroid gland help regulate the metabolic rate of the body. They are involved in growth and development as well. The thyroid secretes thyroxine (T_4) and calcitonin.

The Pituitary Gland

The pituitary gland has two lobes and is about one centimetre in diameter—about the size of a pea. As shown in **Figure 9.6**, it sits in a bony cavity attached by a thin stalk to the hypothalamus at the base of the brain. If you point a finger right between your eyes, and point another finger toward your auditory canal, you will be pointing at your pituitary gland, which is located at the spot where the imaginary lines cross. Despite its small size, this gland releases six main hormones involved in the body's metabolism, growth, development, reproduction, and other critical life functions.

The anterior pituitary and posterior pituitary make up the two lobes of the pituitary gland. Each lobe is really a separate gland, and they release different hormones. The **posterior pituitary** is considered part of the nervous system. The posterior pituitary does not produce any hormones; instead, it stores and releases the hormones ADH and oxytocin, which are produced in the hypothalamus and transferred to the posterior pituitary by neurons.

The **anterior pituitary** is a true hormone-synthesizing gland. As shown in **Figure 9.7**, [illegible] major hormones: thyroid-stimulating hormone (TSH), [illegible] (ACTH), prolactin (PRL), human growth hormone (hGH), follicle-stimulating hormone (FSH), and luteinizing hormone (LH). A series of blood vessels called a portal system carries releasing hormones from the hypothalamus to the anterior pituitary, and these hormones either stimulate or inhibit the release of hormones from this gland.

The hormones of the pituitary will be studied in detail in the remainder of this chapter.

posterior pituitary posterior lobe of the pituitary gland; an endocrine gland that stores and releases antidiuretic hormone (ADH) and oxytocin, which are produced in the hypothalamus and transferred to the posterior pituitary by neuronal axons

anterior pituitary anterior lobe of the pituitary gland; an endocrine gland that synthesizes and secretes six major hormones: human growth hormone (hGH), prolactin (PRL), thyroid-stimulating hormone (TSH), adrenocorticotropic hormone (ACTH), follicle-stimulating hormone (FSH), and luteinizing hormone (LH)

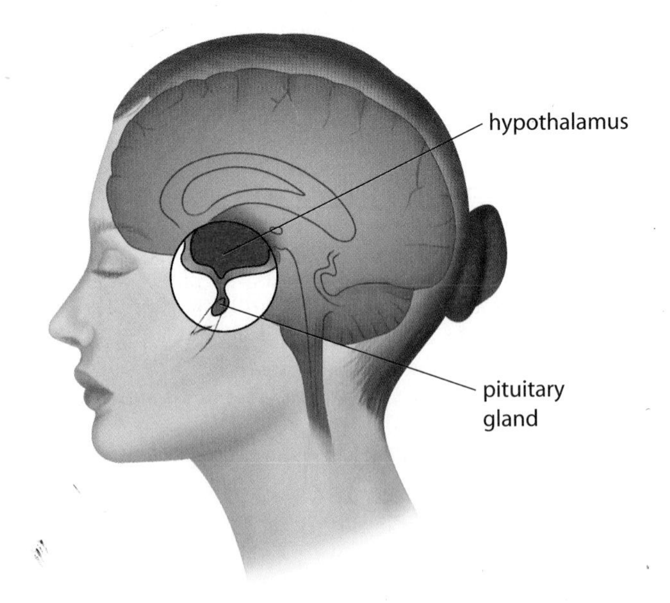

Figure 9.6 The pituitary gland is located below the hypothalamus at the base of the brain.

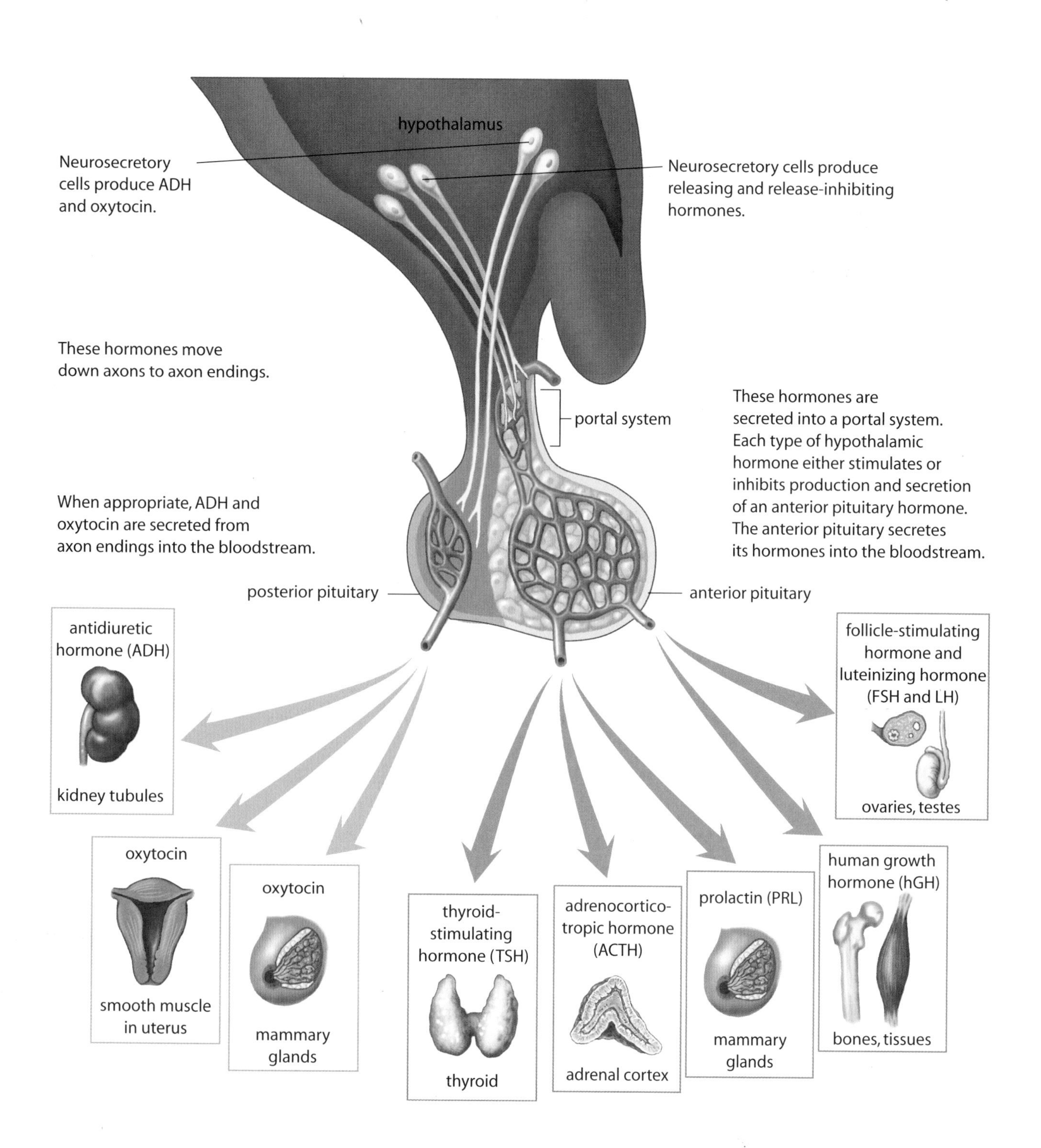
hypothalamus
Neurosecretory cells produce ADH and oxytocin.
Neurosecretory cells produce releasing and release-inhibiting hormones.
These hormones move down axons to axon endings.
portal system
These hormones are secreted into a portal system. Each type of hypothalamic hormone either stimulates or inhibits production and secretion of an anterior pituitary hormone. The anterior pituitary secretes its hormones into the bloodstream.
When appropriate, ADH and oxytocin are secreted from axon endings into the bloodstream.
posterior pituitary
anterior pituitary
antidiuretic hormone (ADH)
kidney tubules
follicle-stimulating hormone and luteinizing hormone (FSH and LH)
ovaries, testes
oxytocin
smooth muscle in uterus
oxytocin
mammary glands
thyroid-stimulating hormone (TSH)
thyroid
adrenocortico-tropic hormone (ACTH)
adrenal cortex
prolactin (PRL)
mammary glands
human growth hormone (hGH)
bones, tissues

Figure 9.7 The hypothalamus produces two hormones, ADH and oxytocin, which are stored and released by the posterior pituitary (left). The hypothalamus controls the secretions of the anterior pituitary (right). The anterior pituitary controls the secretions of the thyroid gland, the adrenal cortex, mammary glands, ovaries, and testes, all of which are also endocrine glands. These activities of the hypothalamus illustrate that the nervous system exerts control over the endocrine system and how both systems interact to maintain homeostasis.

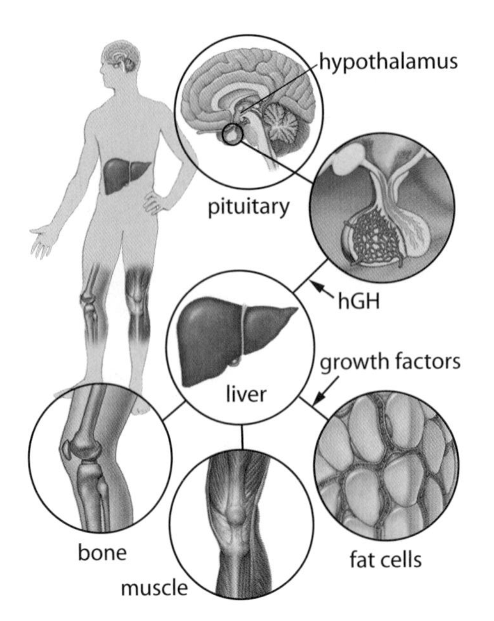

Figure 9.8 The targets of hGH include the liver, muscle cells, and bone cells.

Infer *Why is hGH considered a tropic hormone?*

human growth hormone (hGH) hormone that ultimately affects almost every body tissue, by direct stimulation or via tropic effects; stimulates the liver to secrete hormones called growth factors, which, along with hGH, influence many physiological processes

Human Growth Hormone

The anterior pituitary regulates growth, development, and metabolism through the production and secretion of **human growth hormone (hGH)**. This hormone ultimately affects almost every body tissue. It can affect some tissues by direct stimulation, but the majority of the effects are tropic. **Figure 9.8** shows how hGH stimulates the liver to secrete hormones called growth factors. Together, hGH and the growth factors influence many physiological processes. For example, they increase:

- protein synthesis
- cell division and growth, especially the growth of cartilage, bone, and muscle
- metabolic breakdown and release of fats stored in adipose (fat) tissue.

hGH stimulates the growth of muscles, connective tissue, and the growth plates at the end of the long bones, which causes elongation of these bones. If the pituitary gland [illegible] during childhood, it can result in a condition called gigantism, shown in **Figure 9.9A**. Insufficient hGH production during childhood results in pituitary dwarfism. In this case, an affected person will be of extremely small stature as an adult but will have typical body proportions, as shown in **Figure 9.9B**.

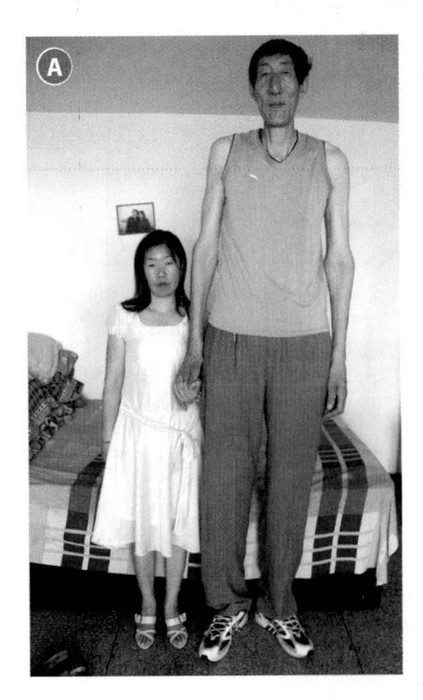

Figure 9.9 A The world's tallest man stands 2.36 m in height. His wife is 1.68 m tall. **B** People with pituitary dwarfism have typical body proportions.

When someone reaches adulthood and skeletal growth is completed, overproduction of hGH can lead to a condition called acromegaly. The excess hGH can no longer cause an increase in height, and so the bones and soft tissues of the body widen. Thus, over time, the face widens, the ribs thicken, and the feet and hands enlarge. However, the condition affects more than just a person's appearance. As **Figure 9.10** shows, some of the effects of untreated acromegaly include debilitating headaches, an enlarged heart, liver, and kidneys, fatigue, breathing problems, cardiovascular diseases, sugar intolerance leading to diabetes, muscle weakness, and colon cancer.

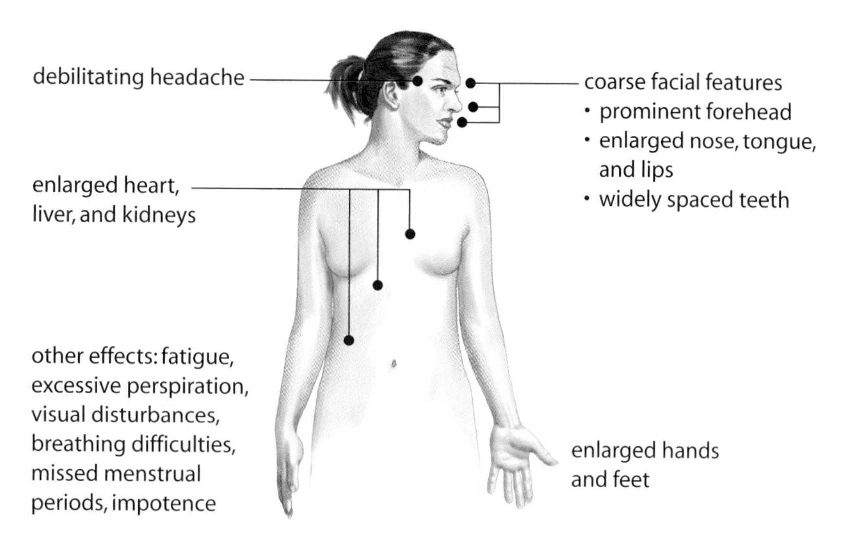

Figure 9.10 Acromegaly results from excessive production of hGH during adulthood. It may be difficult to diagnose the condition in the early stages before a person's appearance noticeably changes.

Scientists first began to understand the function of hGH by studying and treating children with insufficient hGH production, which leads to dwarfism. Researchers found that by injecting the children with material from pituitary gland tissue from human cadavers, the children often grew taller. Sadly, some of the children who received hGH treatment were infected by a form of Creutzfeldt-Jakob disease and died. In addition to the risk of infection with this procedure, it is difficult to obtain sufficient quantities of hGH from organ donations. Since 1985, however, genetic engineering has been used to produce synthetic hGH. The gene that codes for hGH is inserted into bacteria. The altered and rapidly reproducing bacteria are biological factories that make hGH.

Learning Check

7. Explain the relationship between the hypothalamus and the pituitary gland.
8. In point form, summarize how the hypothalamus controls the production and secretion of hormones from the anterior pituitary gland.
9. Use a Venn diagram to compare and contrast the posterior pituitary gland and anterior pituitary gland.
10. Thyroid-stimulating hormone (TSH) is secreted by the anterior pituitary gland. It travels to the thyroid gland, causing the thyroid to secrete thyroxin. Infer what type of hormone TSH is.
11. List three effects that human growth hormone (hGH) has on the body.
12. Explain why a lower than normal secretion of hGH during childhood can result in a condition known as pituitary dwarfism.

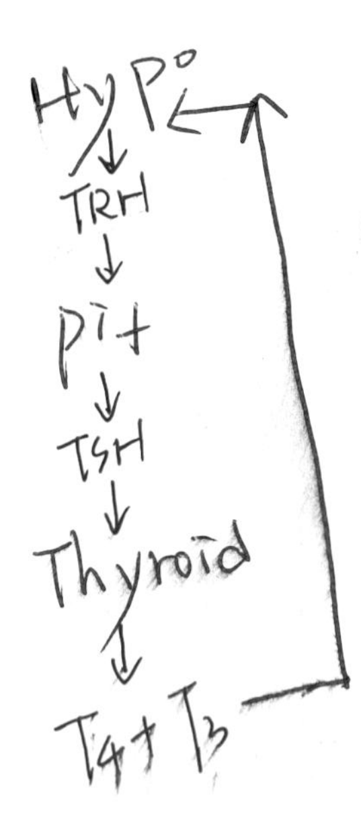

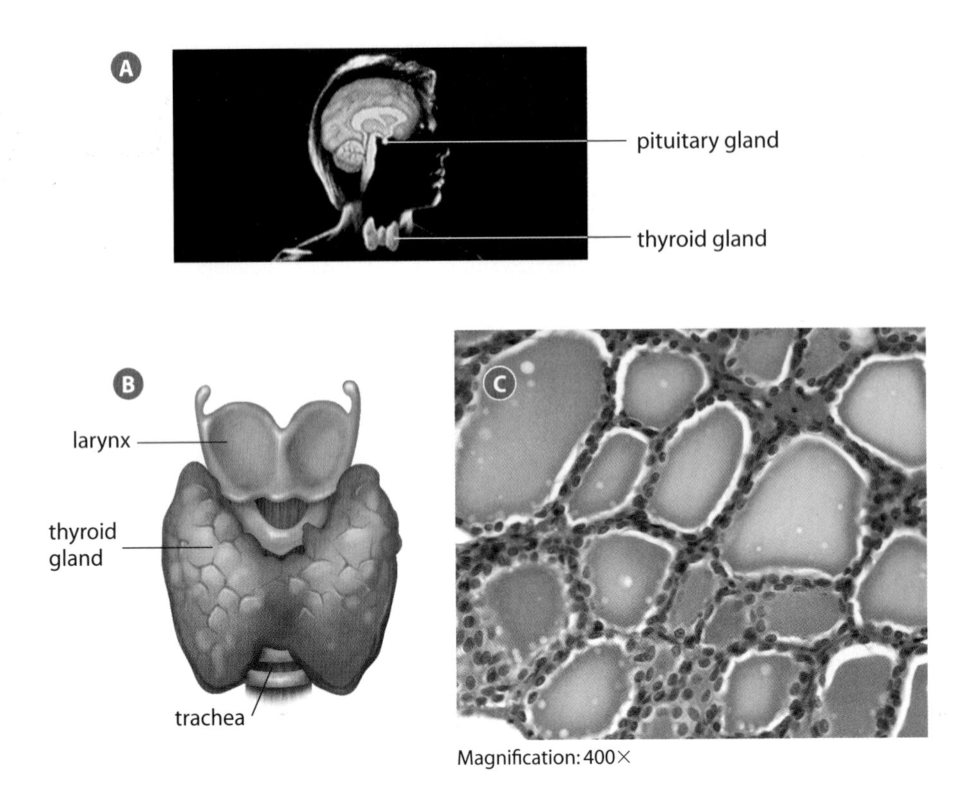

Figure 9.11 A This silhouette of the human body shows the location of the thyroid gland. **B** A close-up illustration of the thyroid gland shows it in front of the trachea. **C** This micrograph shows the cells that make up the thyroid, which produce the hormone thyroxine, also called T_4.

The Thyroid Gland: A Metabolic Thermostat

thyroid gland butterfly-shaped gland located below the larynx in the neck, produces the hormone thyroxine that helps regulate metabolism and growth

thyroxine (T_4) hormone produced by the thyroid and released into the bloodstream; controls the rate at which the body metabolizes fats, proteins, and carbohydrates for energy

hypothyroidism condition resulting when the thyroid produces extremely low levels of thyroxine

Do you know someone who can eat anything, without ever gaining weight? You might have heard that this person has a high metabolism, which implies that they burn energy very quickly. Is it biologically possible to have a high or low metabolism? How do the thyroid hormones influence the metabolic rate? What happens when there is an imbalance of thyroid hormones or a nutrient deficiency?

As shown in **Figure 9.11**, the **thyroid gland** lies directly below the larynx (voice box). It has two lobes, one on either side of the trachea (windpipe), which are joined by a narrow band of tissue. Millions of cells within the thyroid secrete immature thyroid hormones into the spaces between the cells. Here, one of these hormones, **thyroxine (T_4)**, will become functional and be released into the bloodstream. The primary effect of thyroxine is to increase the rate at which the body metabolizes fats, proteins, and carbohydrates for energy. Thyroxine does not have one specific target organ, but instead stimulates the cells of the heart, skeletal muscles, liver, and kidneys to increase the rate of cellular respiration. Thyroxine also plays an important role in the growth and development of children by influencing the organization of various cells into tissues and organs.

If the thyroid fails to develop properly during childhood, a condition called cretinism can result. In this case, the thyroid produces extremely low quantities of thyroxine, and the person is said to have severe **hypothyroidism**. Individuals with cretinism are stocky and shorter than average, and without hormonal injections early on in life they will have mental developmental delays.

Adults with hypothyroidism tend to feel tired much of the time, have a slow pulse rate and puffy skin, and experience hair loss and weight gain. This explains why someone with a slow metabolism due to an underactive thyroid may eat very little, but still gain weight. Hypothyroidism is rare, however. For most people, diet and activity are the main factors in weight gain.

Overproduction of thyroxine is called **hyperthyroidism**. Since thyroxine stimulates metabolism, which releases stored energy as ATP, the symptoms of hyperthyroidism include anxiety, insomnia, heat intolerance, an irregular heartbeat, and weight loss. Graves' disease is a severe state of hyperthyroidism that results when the body's immune system attacks the thyroid. In addition to the other symptoms of hyperthyroidism, Graves' disease produces swelling of the muscles around the eyes, which causes them to protrude and interferes with vision. Hyperthyroidism can be treated by medications, or removal or irradiation of part of the thyroid.

Thyroxine secretion is controlled by negative feedback. The anterior pituitary releases a hormone called **thyroid-stimulating hormone (TSH)**, which causes the thyroid gland to secrete thyroxine. As thyroxine levels rise in the blood, thyroxine itself feeds back to the hypothalamus and anterior pituitary, which suppresses the secretion of TSH and, therefore, thyroxine, as shown in **Figure 9.12**. When the body is healthy and homeostasis is being maintained, the amount of thyroxine in the bloodstream remains relatively constant.

The thyroid requires iodine in order to make the thyroid hormones. (The short form for thyroxine, T_4, refers to the four iodine molecules in the hormone.) If there is insufficient iodine in the diet, thyroxine cannot be made, and there will be no signal to stop the secretion of TSH by the anterior pituitary. The relentless stimulation of the thyroid gland by TSH causes a **goitre**, an enlargement of the thyroid gland. Aside from visible swelling in the neck, additional symptoms of a goitre include difficulty breathing and/or swallowing, and coughing.

In some places, such as the Great Lakes region in Canada, iodine is lacking in the soil and, therefore, in the drinking water. In Canada it is uncommon for people to have goitres, however, because salt refiners add iodine to salt, making it iodized. Other dietary sources of iodine include seafood, fish (such as cod, haddock, and perch), kelp, and dairy products.

hyperthyroidism condition resulting when the thyroid produces extremely high levels of thyroxine

thyroid-stimulating hormone (TSH) a hormone released by the anterior pituitary which causes the thyroid gland to secrete thyroxine; controlled by a negative feedback mechanism: rising thyroxine levels in the blood detected by the hypothalamus and anterior pituitary suppress the secretion of TSH and, therefore, thyroxine

goitre enlargement of the thyroid gland characterized by a large swelling in the throat, often associated with a deficiency of iodine; occurs when the thyroid gland is constantly stimulated by thyroxine-stimulating hormone (TSH), but is unable to synthesize thyroxine to create a negative feedback loop

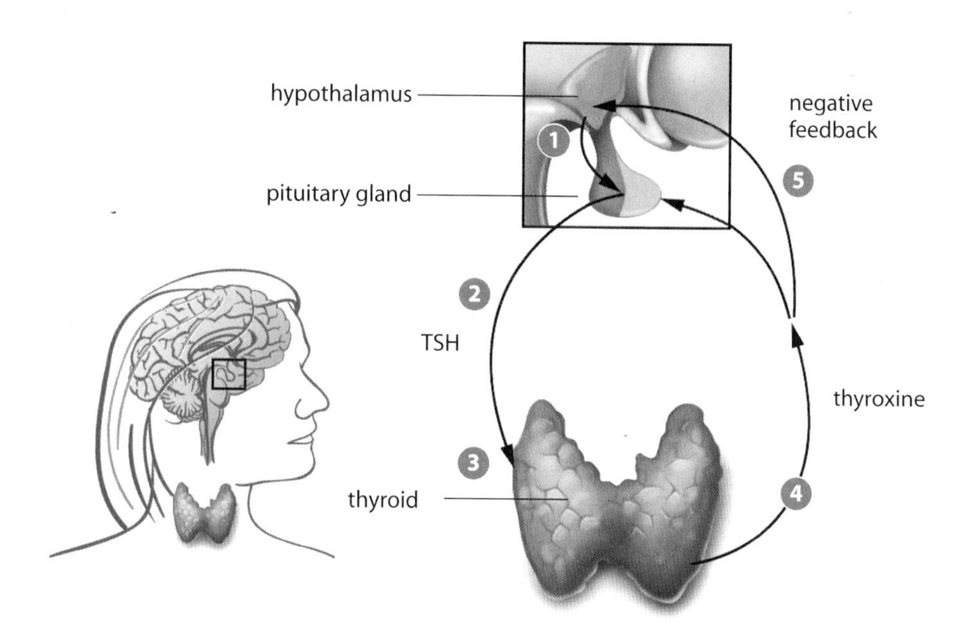

Figure 9.12 This illustration shows the regulation of the thyroid gland by negative feedback.
(1) The hypothalamus secretes a releasing hormone that stimulates the anterior pituitary gland.
(2) The anterior pituitary releases TSH into the bloodstream.
(3) TSH targets the thyroid gland.
(4) TSH causes the thyroid to secrete thyroxine into the bloodstream. Thyroxine stimulates increased cellular respiration in target cells throughout the body.
(5) High levels of thyroxine cause negative feedback on the pituitary and hypothalamus, shutting down production of TSH.

Learning Check

13. Describe the thyroid gland and identify the general functions of this endocrine gland.

14. List the symptoms an adult with hyperthyroidism might experience.

15. Use a flowchart or other graphic organizer to summarize the regulation of the thyroid gland by negative feedback.

16. Explain why hypothyroidism can cause a goitre to develop.

17. Make a Venn diagram to compare and contrast Graves' disease and cretinism.

18. What type of condition can result if an adult is iodine deficient? Describe the symptoms of this condition.

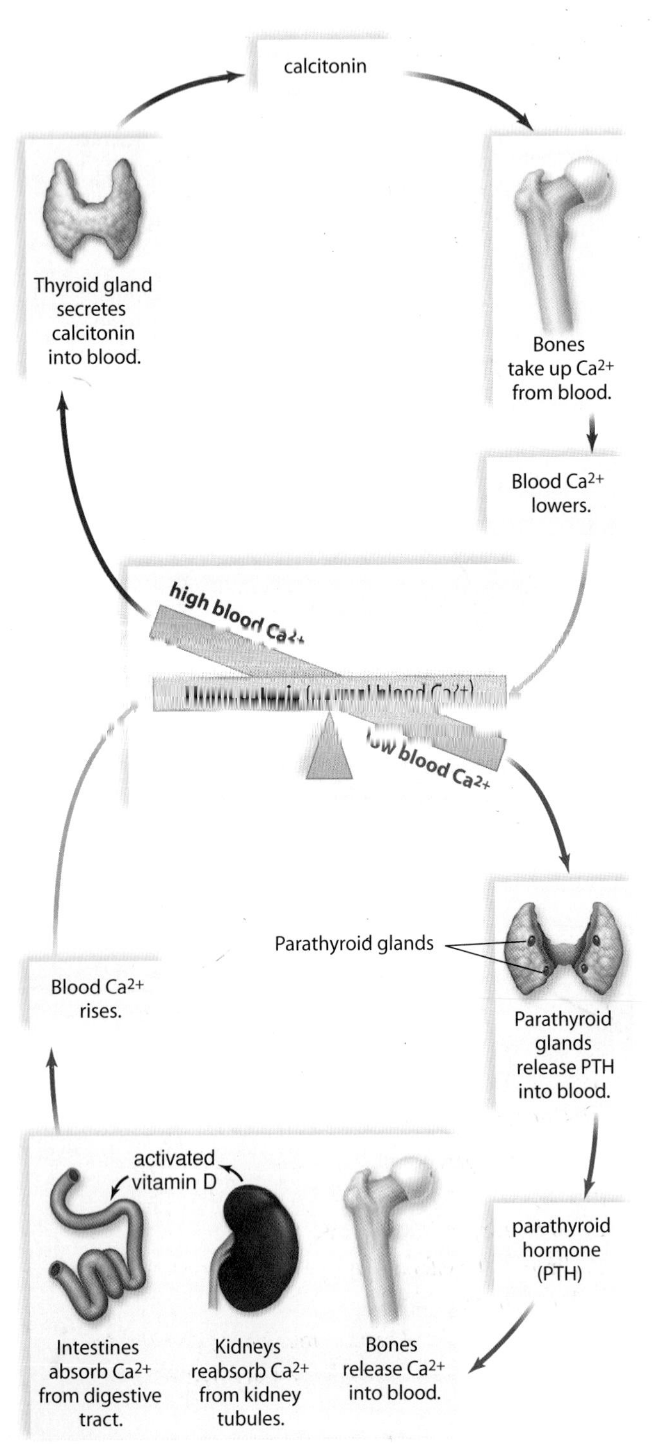

The Thyroid Gland and Calcitonin

Calcium (Ca^{2+}) is essential for healthy teeth and skeletal development. This mineral also plays a crucial role in blood clotting, nerve conduction, and muscle contraction. Calcium levels in the blood are regulated, in part, by a hormone called *calcitonin.* When the concentration of calcium in the blood rises too high, calcitonin stimulates the uptake of calcium into bones, which lowers its concentration in the blood, as shown in **Figure 9.13**. A different hormone, secreted by the parathyroid glands, is released if blood calcium levels get too low.

The Parathyroid Glands and Calcium Homeostasis

The parathyroid glands are four small glands attached to the thyroid. The parathyroid glands produce a hormone called *parathyroid hormone (PTH).* The body synthesizes and releases PTH in response to falling concentrations of calcium in the blood. PTH stimulates bone cells to break down bone material (calcium phosphate) and secrete calcium into the blood. PTH also stimulates the kidneys to reabsorb calcium from the urine, activating vitamin D in the process. Vitamin D, in turn, stimulates the absorption of calcium from food in the intestine. These effects, outlined in **Figure 9.13**, bring the concentration of calcium in the blood back within a normal range so that the parathyroid glands no longer secrete PTH.

Figure 9.13 Negative feedback mechanisms regulate the concentration of calcium in the blood. When blood calcium level is high, the thyroid gland releases calcitonin. Calcitonin promotes the uptake of calcium by the bones and blood calcium level returns to normal. When blood concentration of calcium (Ca^{2+}) is low, PTH is released by the parathyroid glands. PTH directly stimulates the breakdown of bone and the reabsorption of Ca^{2+} by the kidneys. It also indirectly promotes the absorption of Ca^{2+} in the intestine by stimulating the production of vitamin D. The combination of these actions brings blood calcium levels back to normal.

Section 9.2 Review

Section Summary

- The hypothalamus controls the pituitary gland. The pituitary gland has two lobes that store and release tropic hormones.
- The anterior pituitary gland releases human growth hormone (hGH), which stimulates fat metabolism and targets the liver to release hormones that stimulate protein synthesis and muscle and bone growth.
- The thyroid gland secretes hormones that regulate cell metabolism, growth, and development.
- Thyroxine secretion is regulated by the release of thyroid-stimulating hormone (TSH) from the anterior pituitary. TSH is regulated by negative feedback by thyroxine on the hypothalamus and pituitary.
- The parathyroid glands secrete parathyroid hormone (PTH), which raises blood calcium levels.

Review Questions

1. **C** Using a graphic organizer, organize information about the hormones of the posterior pituitary gland and those of the anterior pituitary gland. Include information about the target cells or glands of the hormones and what effect they produce.
2. **K/U** Why could hGH be called a tropic hormone?
3. **K/U** Compare and contrast gigantism and acromegaly.
4. **C** Growth hormone–releasing hormone (GHRH) is produced by the hypothalamus when levels of human growth hormone get too low. GHRH stimulates hGH production and release by cells in the anterior pituitary. hGH stimulates the liver to secrete hormones called growth factors. Somatostatin, also known as growth hormone–inhibiting hormone, is produced by the hypothalamus when levels of hGH or levels of the growth factors released from the liver also get too high.
   a. Using a flowchart similar to **Figure 9.5**, summarize the regulation of hGH levels in the body.
   b. In the general negative feedback system for tropic hormones, the third hormone inhibits the release of the first hormone. Explain how the feedback mechanism for hGH is different.
5. **A** In adults, the production of natural hGH declines with age. There has been research on the use of synthetic hGH to counteract the effects of aging. Some scientists claim that use of hGH has effects such as an increase in muscle mass, a decrease in body fat, and an increase in energy levels. Some of the known side effects of using hGH include heart problems, organ failure, and overgrowth of muscle and bone. Evaluate the advantages and disadvantages of using hGH for this purpose.
6. **A** Explain how the thyroid gland is like a metabolic thermostat.
7. **A** A young child with cretinism is not growing and her parents would like her to take synthetic hGH. Will this treatment help the child? Explain your answer.
8. **K/U** The diagram below shows feedback mechanisms associated with the thyroid gland. In your notebook, identify what is occurring at the labels (A) through (F), and name the hormones involved at each point.

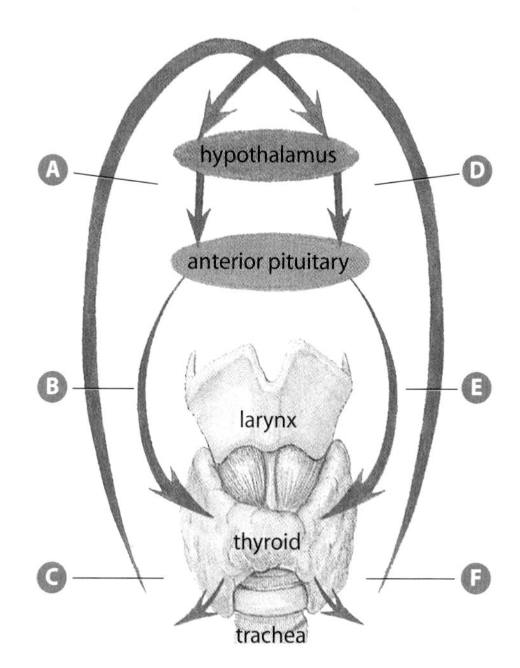

9. **C** Construct a table or Venn diagram to compare and contrast hyperthyroidism with hypothyroidism.
10. **T/I** Osteoporosis is a disease that results from a loss of bone mass, causing bones to become brittle and break more easily. Infer why calcitonin is given as a treatment for people with osteoporosis.
11. **T/I** If the parathyroid glands secrete too much parathyroid hormone (PTH), a condition called hyperparathyroidism results.
   a. Predict the effects that hyperparathyroidism would have on the body.
   b. Infer the types of tests a doctor may perform to diagnose a patient with hyperparathyroidism.

SECTION 9.3

Hormonal Regulation of the Stress Response and Blood Sugar

Key Terms

adrenal gland
adrenal medulla
epinephrine
short-term stress response
adrenal cortex
long-term stress response
cortisol
adrenocorticotropic hormone (ACTH)
aldosterone
pancreas
islets of Langerhans
beta cell
alpha cell
insulin
glucagon
diabetes mellitus
hyperglycemia

Figure 9.14 What happens to your body when you experience stress? How does the endocrine system help you cope with stressful situations?

Imagine yourself on the roller coaster in **Figure 9.14**. You will climb into the sky at a 90° angle, you will reach speeds of 90 km/h, and you will hang upside down in your seat. What physiological changes will be occurring in your body as you fly through the twists and turns of the ride? The stress response involves many interacting hormone pathways, including those that regulate metabolism, heart rate, and breathing. In this section you will focus on the hormones of the adrenal glands and their effects on the body.

The human body has two **adrenal glands**, which are located on top of the kidneys, as shown in **Figure 9.15**. The adrenal glands are named for two Latin words that mean "near the kidney." Each gland is composed of an inner layer, the adrenal medulla, and an outer layer, the adrenal cortex. The adrenal cortex produces hormones that are different in structure and function from the hormones produced by the adrenal medulla.

adrenal gland one of a pair of organs located on top of the kidneys; composed of two layers: an outer cortex and an inner medulla; each layer produces different hormones and functions as an independent organ

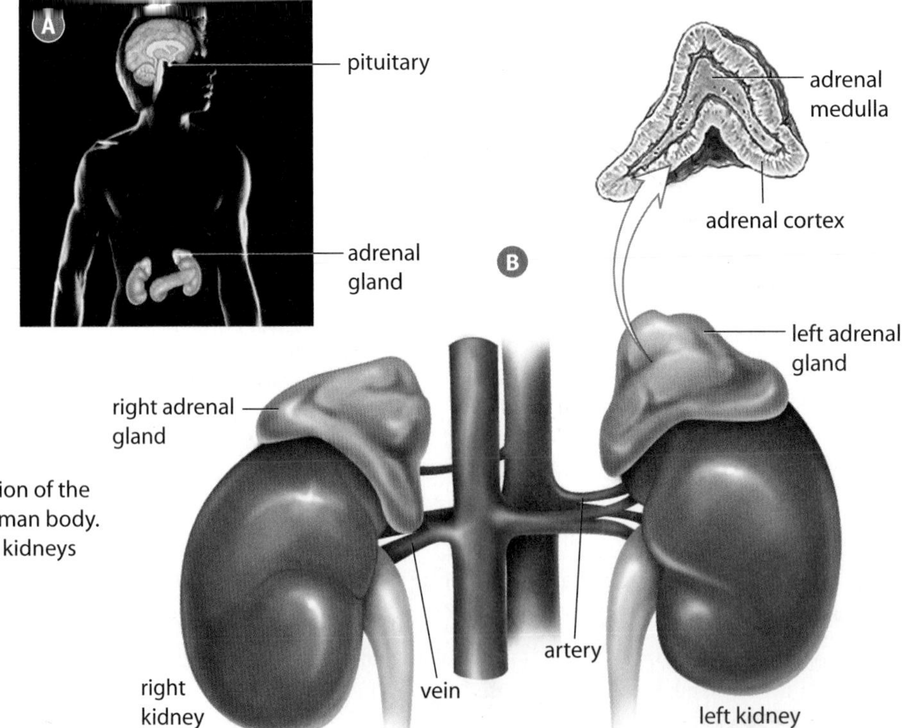

Figure 9.15 **A** The location of the adrenal glands in the human body. **B** A close-up view of the kidneys and adrenal glands.

The Adrenal Medulla: Regulating the Short-Term Stress Response

The **adrenal medulla** produces two closely related hormones: **epinephrine** and norepinephrine. (These hormones are also called adrenaline and noradrenaline, respectively.) These hormones regulate a **short-term stress response** that is commonly referred to as the *fight-or-flight response*. The effects of these hormones are similar to those caused by stimulation of the sympathetic nervous system. In fact, in the developing embryo, sympathetic neurons and adrenal medulla cells are both formed from nervous system tissue, which is why the adrenal medulla is considered a neuroendocrine structure.

Like the sympathetic nervous system, the hormones of the adrenal medulla prepare the body for fight-or-flight by increasing metabolism. In response to a stressor, neurons of the sympathetic nervous system carry a signal from the hypothalamus directly to the adrenal medulla. These neurons (rather than hormones) stimulate the adrenal medulla to secrete epinephrine and a small amount of norepinephrine. Recall from Chapter 8 that norepinephrine also functions as a neurotransmitter, which has an excitatory effect on its target muscles. When released by the adrenal medulla, it functions as a hormone. As shown in **Figure 9.16**, epinephrine and norepinephrine trigger an increase in breathing rate, heart rate, blood pressure, blood flow to the heart and muscles, and the conversion of glycogen to glucose in the liver. At the same time, the pupils of the eyes dilate, and blood flow to the extremities decreases. Epinephrine acts quickly. This is why epinephrine injections can be used to treat different life-threatening conditions. For example, they can be used to stimulate the heart to start beating in someone with cardiac arrest. In cases of anaphylactic shock caused by severe allergies to substances such as nuts, bee sting venom, or certain medications, injected epinephrine will open up the air passages and restore breathing.

The release of epinephrine and norepinephrine is rapid because it is under nervous system control. Although the hormonal effects are similar to those of the sympathetic nervous system, their influence on the body lasts about 10 times longer.

adrenal medulla the inner layer of the adrenal glands that produces epinephrine and norepinephrine, hormones that regulate the short-term stress response

epinephrine hormone produced by the adrenal cortex that helps regulate the short-term stress response; also known as adrenaline

short-term stress response the body's acute reaction to stress in which the sympathetic nervous system is stimulated; also known as fight-or-flight response

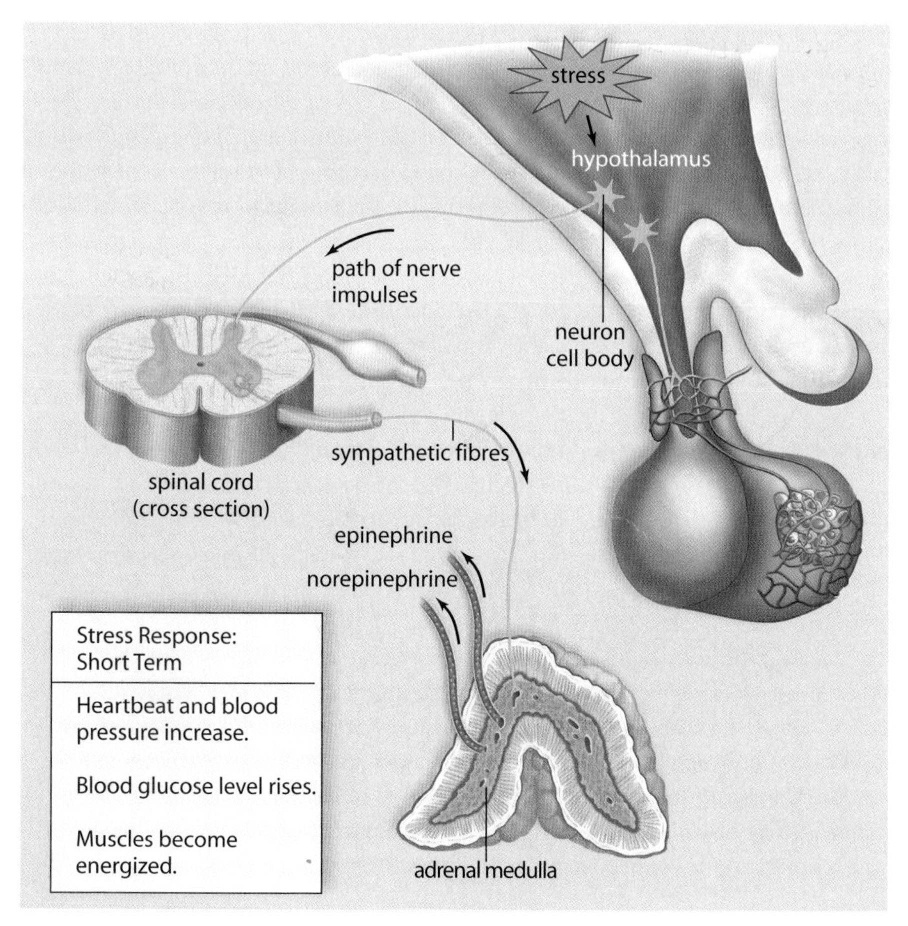

Figure 9.16 Nervous stimulation causes the adrenal medulla to provide a rapid, but short-term, stress response.

adrenal cortex the outer layer of the adrenal glands that produces glucocorticoids and mineralocorticoids, hormones that regulate the long-term stress response; also secretes a small amount of gonadocorticoids, female and male sex hormones that supplement the hormones produced by the gonads (testes and ovaries)

long-term stress response sustained physiological response to stressors, characterized by increases in blood glucose and blood pressure, and decrease in inflammatory response; regulated by hormones produced by the adrenal cortex

cortisol a type of glucocorticoid hormone released by the adrenal cortex of the adrenal gland in a long-term stress response; triggers an increase in blood glucose levels and reduces inflammation

adrenocorticotropic hormone (ACTH) hormone synthesized by the anterior pituitary gland to target the adrenal cortex and regulate the production of glucocorticoids

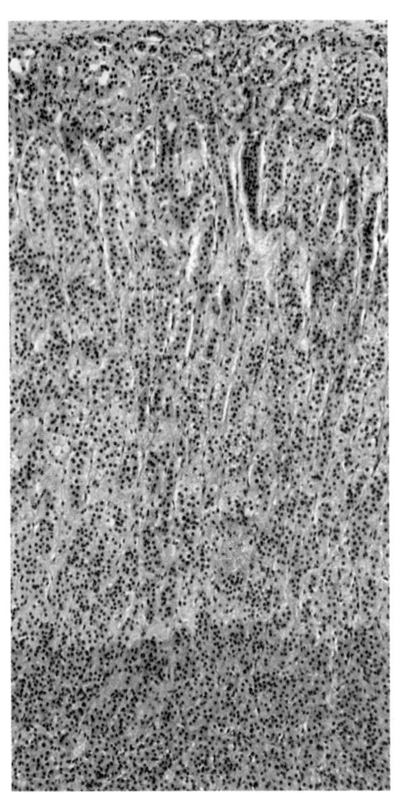
Magnification: 35×

Figure 9.17 A micrograph of the cells of the adrenal cortex. Notice the bands of cells. The cells in the lower and central bands produce glucocorticoids and gonadocorticoids. The cells in the upper band secrete mineralocorticoids.

The Adrenal Cortex: Regulating the Long-Term Stress Response

Figure 9.17 shows the three bands of cells contained within the tissue of the **adrenal cortex**. These cells produce the stress hormones that trigger the sustained physiological responses that make up the **long-term stress response**. These hormones include the glucocorticoids, the mineralocorticoids, and the gonadocorticoids. The glucocorticoids increase blood sugar, and the mineralocorticoids increase blood pressure. The adrenal cortex also secretes a small amount of female and male sex hormones, called gonadocorticoids, which supplement the hormones produced by the gonads (testes and ovaries). A comparison of the short-term stress response and long-term stress response is shown in **Figure 9.18**.

Cortisol

Cortisol is the most abundant glucocorticoid. Like the other hormones produced by the adrenal cortex, cortisol is a steroid hormone synthesized from cholesterol. As shown on the right side of **Figure 9.18**, when the brain detects danger, it directs the hypothalamus to secrete a releasing hormone. The releasing hormone stimulates the anterior pituitary gland to secrete **adrenocorticotropic hormone (ACTH)**. ACTH targets the adrenal cortex, which causes the release of the stress hormone cortisol. Cortisol works often in conjunction with epinephrine, but is longer lasting. The main function of cortisol in the body is to raise the blood glucose levels. Cortisol does this by promoting the breakdown of muscle protein into amino acids. The amino acids are taken out of the blood by the liver, where they are used to make glucose, which is then released back into the blood. Cortisol also prompts the breakdown of fat cells, which also releases glucose. Increased cortisol levels in the blood cause negative feedback on the hypothalamus and anterior pituitary, which suppresses ACTH production, and stops the release of cortisol.

When faced with immediate danger, or playing a vigorous sport, epinephrine and cortisol are just what the body needs. In the long term, however, the sustained high levels of cortisol in chronic stress can impair thinking, damage the heart, cause high blood pressure, lead to diabetes, increase susceptibility to infection, and even cause early death. In Japan, where long work hours and high-stress jobs are common, so many business people have died from heart attacks and strokes that the phenomenon has been named *karoshi*, which means "death from overwork."

One of the ways the body fights disease is by inflammation, in which cells of the immune system attack foreign material, such as invading bacteria. Cortisol is a natural anti-inflammatory in the body and suppresses the actions of the immune system. This is probably why sustained high levels of cortisol make people more susceptible to infections. Synthesized cortisol is commonly used as a medication to reduce the undesirable inflammation associated with asthma, arthritis, or joint injuries. Unfortunately, cortisol inhibits the regeneration of connective tissue, and should therefore be used only when necessary.

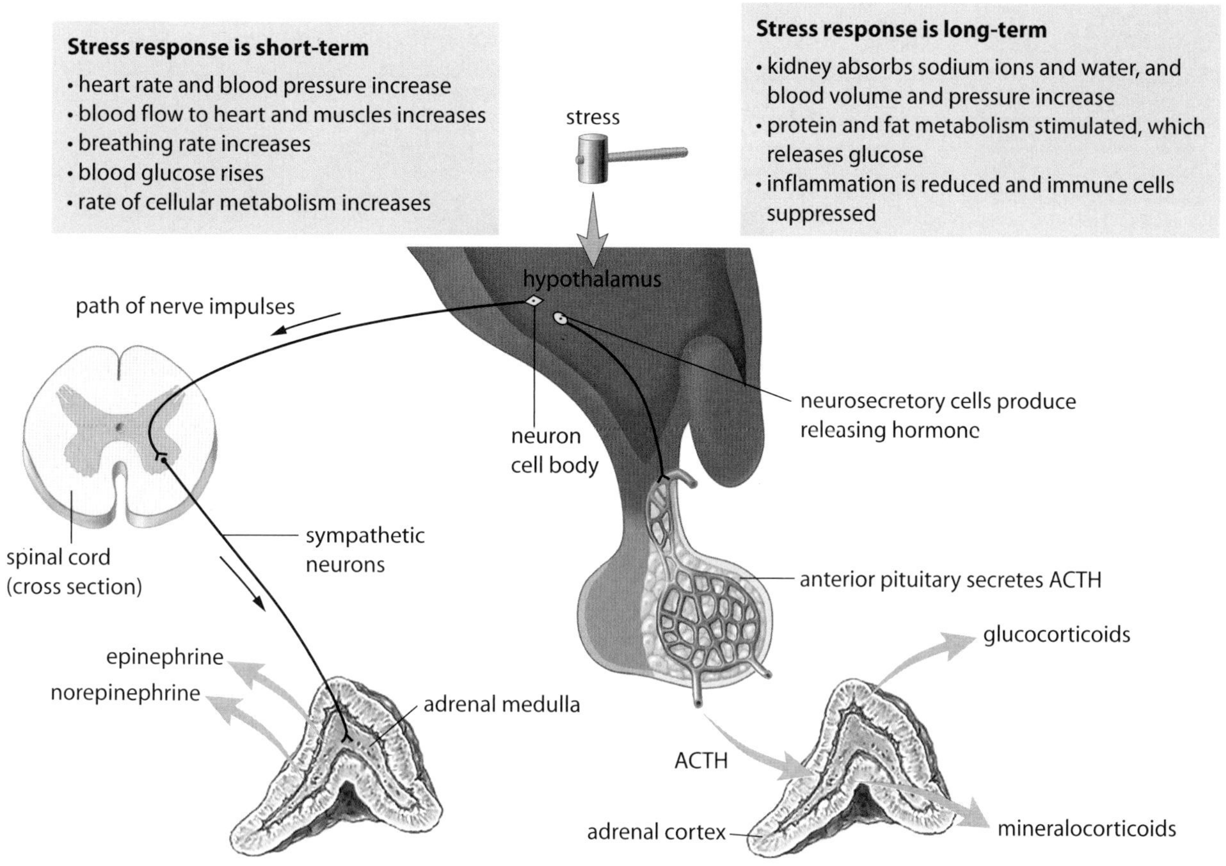

Stress response is short-term
- heart rate and blood pressure increase
- blood flow to heart and muscles increases
- breathing rate increases
- blood glucose rises
- rate of cellular metabolism increases

Stress response is long-term
- kidney absorbs sodium ions and water, and blood volume and pressure increase
- protein and fat metabolism stimulated, which releases glucose
- inflammation is reduced and immune cells suppressed

stress
hypothalamus
path of nerve impulses
neuron cell body
neurosecretory cells produce releasing hormone
spinal cord (cross section)
sympathetic neurons
anterior pituitary secretes ACTH
glucocorticoids
epinephrine
norepinephrine
adrenal medulla
ACTH
adrenal cortex
mineralocorticoids

Figure 9.18 The adrenal medulla and adrenal cortex are under the control of the hypothalamus. The adrenal medulla provides a rapid and short-lived stress response, while the adrenal cortex provides a sustained stress response.

Aldosterone

The main mineralocorticoid is the hormone **aldosterone**. Aldosterone stimulates the kidneys to increase the absorption of sodium into the blood. This increases the concentration of solutes in the blood, which draws in more water from the kidneys, raising blood pressure.

If the adrenal cortex is damaged, Addison's disease can result. In this case, the body secretes inadequate amounts of mineralocorticoids and glucocorticoids. The symptoms of Addison's disease include hypoglycemia (low blood sugar), sodium and potassium imbalances, and rapid weight loss. Low aldosterone results in a loss of sodium and water from the blood due to increased urine output. As a result, blood pressure drops. A person with this condition needs to be treated within days, or the severe electrolyte imbalance will be fatal. Former United States President John F. Kennedy had this condition. Doctors controlled his symptoms with injections of glucocorticoids and mineralocorticoids.

aldosterone a type of mineralocorticoid hormone secreted by the adrenal cortex

Learning Check

19. Describe the adrenal glands.
20. Explain why the adrenal medulla is considered to be a neuroendocrine structure.
21. List five physiological responses that epinephrine and norepinephrine have on the human body and explain these responses in terms of the fight-or-flight response.
22. Construct a flowchart to show the feedback system that controls the release of cortisol.
23. Explain why a person having a severe allergic reaction would be given epinephrine.
24. Many athletes receive synthetic cortisol injections to reduce the inflammation of sprained joints. Infer why physicians limit the number of these injections.

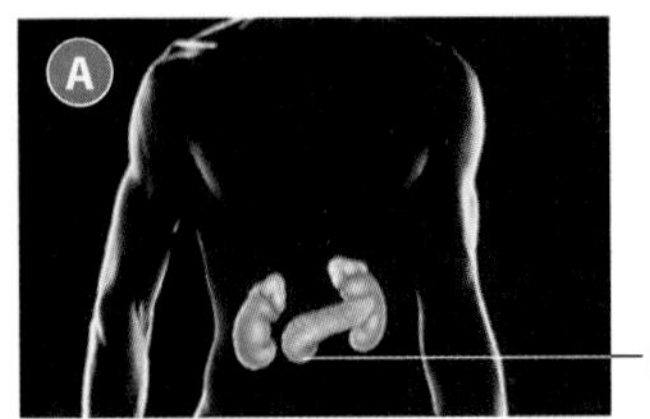

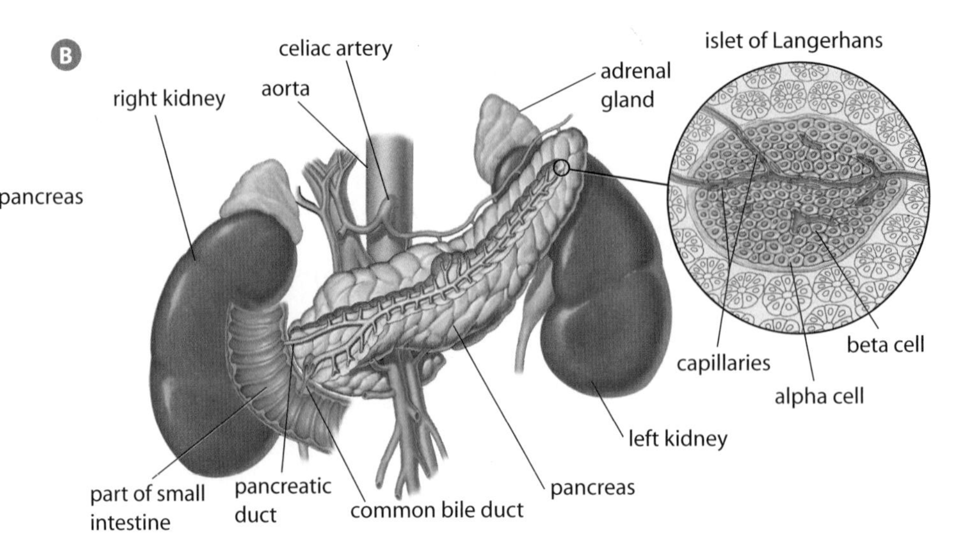

Figure 9.19 A The silhouette shows the location of the pancreas in the human body. **B** A close-up view of the pancreas shows one of the many islets of Langerhans on the pancreas's surface.

pancreas small gland in the abdomen that secretes digestive enzymes into the small intestine; also secretes the hormone insulin

islets of Langerhans cluster of endocrine cells found throughout the pancreas, consisting of glucagon-producing alpha cells and insulin-producing beta cells

beta cell cells of the pancreas that secrete insulin to decrease the level of blood glucose

alpha cell cells of the pancreas that secrete glucagon to increase the level of blood glucose

The Hormones of the Pancreas

The **pancreas** is located behind the stomach and is connected to the small intestine by the pancreatic duct, as shown in **Figure 9.19**. Much of the pancreatic tissue secretes digestive enzymes into the small intestine. However, the pancreas also functions as an endocrine gland, secreting hormones directly into the bloodstream. Scattered throughout the pancreas are more than 2000 clusters of endocrine cells called the **islets of Langerhans**. They are named for Paul Langerhans, the scientist who first described them in 1869.

The islets of Langerhans secrete two hormones, insulin and glucagon, which have opposite effects: they are antagonistic. The **beta cells** of the pancreas secrete insulin, which decreases the level of blood glucose. Glucagon, secreted by the **alpha cells**, increases the level of blood glucose.

Activity 9.1 How Do Vitamins and Amino Acids Affect Homeostasis?

Vitamins and amino acids are essential to good health and maintaining homeostasis. For example, vitamin K is essential for proper blood clotting to occur. A deficiency in vitamin K can result in delayed blood clotting and excessive bleeding. Sources of vitamins and amino acids include food and supplements. Vitamin K is found in spinach, cabbage, and cauliflower. In this activity, you will learn more about how certain vitamins and amino acids affect the endocrine system and homeostasis.

Materials

- reference books
- computer with Internet access

Procedure

1. Use print and Internet resources to research more information about the following vitamins and amino acids:
   - vitamin D
   - vitamin B_1
   - pantothenic acid
   - biotin
   - tyrosine
   - tryptophan
   - valine
2. Construct a table with the following column headings: Vitamin or Amino Acid, Dietary Sources, Functions, Effect on Endocrine System and/or Homeostasis, Effects of Deficiency, Effects of Too Much Intake.
3. Record the results of your research in the table.

Questions

1. Using vitamin D as an example, explain how vitamins can affect homeostasis.
2. How does biotin affect the regulation of blood glucose levels?
3. How could a deficiency in tyrosine affect the body's metabolism?

Both **insulin** and **glucagon** are regulated by negative feedback mechanisms, as shown in **Figure 9.20**. When you eat a meal, your digestive system breaks down the food and releases a substantial amount of glucose into your bloodstream. When the blood glucose levels rise, the pancreatic beta cells secrete appropriate amounts of insulin. Insulin circulates throughout the body and acts on specific receptors to make the target cells more permeable to glucose. It especially affects muscle cells, which use large amounts of glucose in cellular respiration, and liver cells, where glucose is converted into glycogen for temporary storage. Other cells of the body also take in and use glucose for energy. As the glucose levels in the blood return to normal, insulin secretion slows.

Rigorous exercise or fasting (skipped meals) can cause blood glucose levels to drop. Low blood sugar stimulates the alpha cells of the islets of Langerhans to release glucagon. Glucagon stimulates the liver to convert glycogen back into glucose, which is released into the blood. Other hormones, such as hGH, cortisol, and epinephrine, also contribute to increasing the level of blood glucose.

insulin a hormone secreted by the beta cells of the islets of Langerhans in the pancreas to make target cells more permeable to glucose; enables the body to use sugar and other carbohydrates

glucagon hormone produced by the alpha cells of the islets of Langerhans in the pancreas to stimulate the liver to convert glycogen back into glucose, which is released into the blood

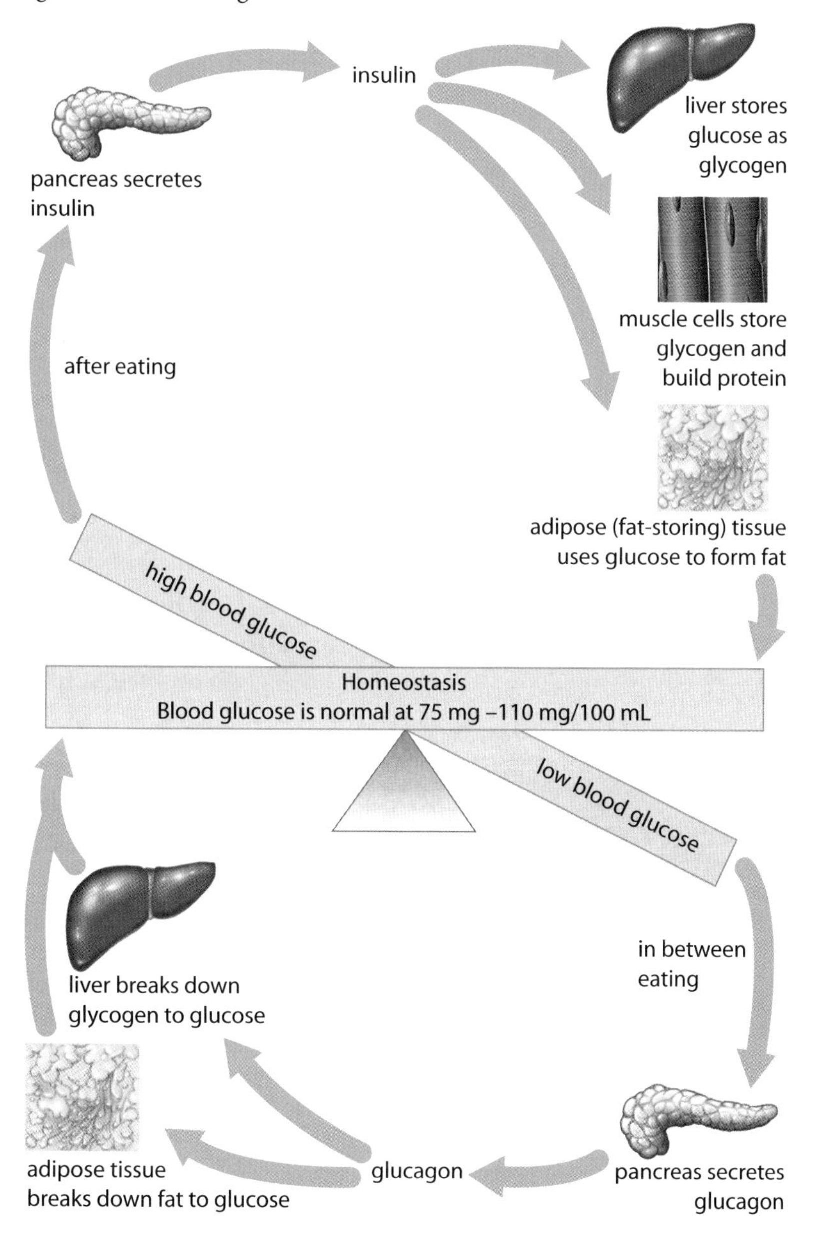
insulin
pancreas secretes insulin
liver stores glucose as glycogen
muscle cells store glycogen and build protein
after eating
adipose (fat-storing) tissue uses glucose to form fat
high blood glucose
Homeostasis
Blood glucose is normal at 75 mg –110 mg/100 mL
low blood glucose
in between eating
liver breaks down glycogen to glucose
adipose tissue breaks down fat to glucose
glucagon
pancreas secretes glucagon

Figure 9.20 Negative feedback regulates blood glucose levels within a very narrow range.
Explain *How are insulin and glucagon able to help maintain homeostasis in the body?*

diabetes mellitus a serious chronic condition that results when the pancreas does not make enough insulin or the body does not respond properly to insulin; levels of blood glucose tend to rise sharply after meals (hyperglycemia) and remain at significantly elevated levels

hyperglycemia condition resulting from high levels of blood glucose; occurs in individuals with diabetes mellitus

The Effects of Glucose Imbalance

Diabetes mellitus is a serious chronic condition with no known cure. It affects over 285 million people worldwide (as of 2009), including over two million Canadians. Diabetes results when the body does not produce enough insulin, or does not respond properly to insulin. As a result, levels of blood glucose tend to rise sharply after meals, and remain at significantly elevated levels. This condition is called hyperglycemia, or high blood sugar, from the Greek word parts *hype*r (too much), *glyco* (sugar), and *emia* (condition of the blood).

Hyperglycemia has various short-term and long-term effects on the body. Without insulin, cells remain relatively impermeable to glucose and cannot obtain enough from the blood. The individual experiences fatigue as the cells become starved for glucose. The body compensates to some degree by switching to protein and fat metabolism for energy. Fats and proteins are less accessible, however, and more difficult than glucose to break down. Fat metabolism also releases ketones, such as acetone, as a toxic by-product, which can be smelled on the breath.

The kidneys are incapable of reabsorbing all of the glucose that is filtered through them from the blood, and so glucose is excreted in the urine. Due to the concentration gradient in the kidneys, large volumes of water follow the glucose into the urine and get excreted. People with untreated diabetes experience low energy and great thirst, and produce large volumes of glucose-rich urine. In the long term, continued high levels of blood glucose can lead to blindness, kidney failure, nerve damage, and gangrene (a severe infection) in the limbs. Also, in many diabetics, the alpha cells that produce glucagon degenerate. Diabetes remains one of the leading causes of death in North America.

Learning Check

25. Identify two cell types of the islets of Langerhans and explain their functions.

26. Describe the roles of insulin and glucagon in maintaining homeostasis.

27. Describe diabetes mellitus.

28. What are the long-term effects of untreated diabetes mellitus on the body?

29. Explain, in terms of cellular respiration, why an individual with diabetes mellitus experiences fatigue.

30. How does hyperglycemia differ from diabetes mellitus?

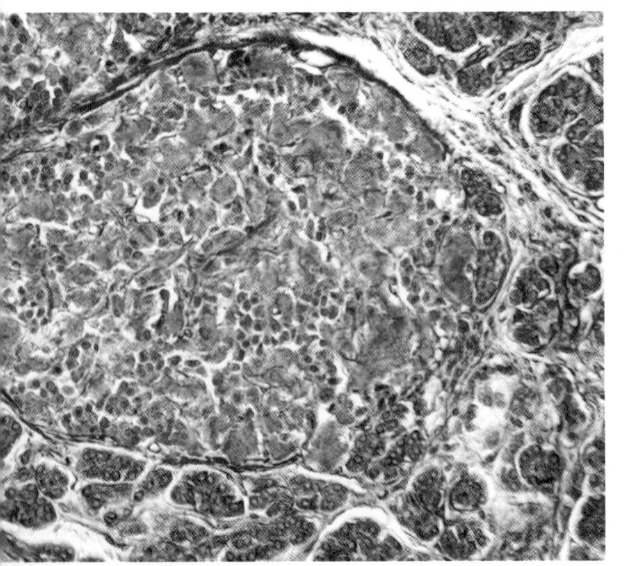

Magnification: 55×

Causes of Diabetes

There are two major types of diabetes mellitus: type 1 diabetes (also called juvenile diabetes and insulin-dependent diabetes) and type 2 diabetes (also called adult-onset diabetes and non-insulin-dependent diabetes). In type 1 diabetes, the immune system produces antibodies that attack and destroy the beta cells of the pancreas. As a result, the beta cells degenerate, like the ones shown in **Figure 9.21**, and are unable to produce insulin. The condition is usually diagnosed in childhood, and people with type 1 diabetes must have daily insulin injections in order to live.

Figure 9.21 A light micrograph of pancreatic beta cells from someone with type 1 diabetes. Many of the beta cells have been destroyed, leaving behind only non-beta cells (stained purple), and so the islet is malformed. ***Identify*** *What causes the destruction of beta cells?*

Type 2 diabetes tends to develop gradually, often because the insulin receptors on the body's cells stop responding to insulin. In other cases, the beta cells of the pancreas produce less and less insulin over time. People who are overweight have a greater chance of developing type 2 diabetes. It is usually diagnosed in adulthood and often can be controlled with diet, exercise, and oral medications. Most people with diabetes—about 90 percent—have type 2. Without proper care, type 2 diabetes can develop into type 1, which is insulin-dependent.

Type 2 diabetes is increasing worldwide at an alarming rate, especially among certain ethno-cultural groups. Among the Aboriginal peoples of Canada, for example, the incidence of type 2 diabetes is rising at three times the national rate. Health scientists describe the increase as an epidemic. One explanation for it could be that people of Aboriginal ancestry have inherited the ability to store food energy very efficiently, since their ancestors traditionally lived through cycles of "feast or famine." In the past, it would have been advantageous to gain weight when there was lots of food, and go through a starvation period when food was scarce. Today, not only is food available year round, but many people are now eating modern Western diets, which are high in refined carbohydrates. Type 2 diabetes is linked closely to unhealthy diet and weight gain, factors that are influencing the rate of diabetes in many populations.

Suggested Investigation

Inquiry Investigation 9-B, Analyzing Endocrine Disorders

Activity 9.2 Blood Glucose Regulation and Homeostasis

How do levels of blood glucose fluctuate throughout the day in someone with diabetes compared to someone without diabetes?

Materials

- graph paper

Procedure

1. Compare the following blood glucose concentration data provided for Maria and Tamika. One of these young women has diabetes. Blood glucose concentrations were monitored over 15 h for both. Both women ate identical meals at the same times, and got equal amounts of exercise at the same times. Neither is currently taking insulin.
2. Plot both sets of data on the same graph and draw a line of best fit for each. Label your graph appropriately.

Maria's and Tamika's Blood Glucose Levels over 15 h

Event/Time	Blood Glucose Concentration (mmol/L)	
	Maria	Tamika
Wake up: 8:00 A.M.	4.0	10.0
1 h after breakfast: 9:00 A.M.	7.0	14.0
Pre-lunch: noon	4.5	10.0
2 h after lunch: 2:00 P.M.	6.0	15.0
Mid-afternoon: 3:00 P.M.	4.5	10.0
1 h after vigorous exercise: 4:00 P.M.	4.0	4.0
Pre-supper: 6:00 P.M.	4.5	9.0
1 h after supper: 7:00 P.M.	6.5	18.0
Bedtime: 11:00 P.M.	4.5	12.0

Questions

1. A healthy range for blood glucose is between 4.5 and 5.0 mmol/L. In general, a person with moderate diabetes would take an insulin shot if the blood glucose level went above 13–15 mmol/L. On your graph, indicate which woman is diabetic and which is not. Write a paragraph to explain your answer.
2. Indicate the times and activities during which the pancreas of the healthy person would release insulin. How did insulin affect her body at these times?
3. Indicate the times and activities during which the pancreas of the healthy person would release glucagon. How did glucagon affect her body at these times?
4. Suggest a medication that the woman with diabetes could take to help her blood glucose levels return to healthy levels after a meal. Explain how this treatment would work.
5. During exercise, Tamika's blood glucose drops dramatically. What could she do to help raise her blood glucose to a healthy range?

Toward a Cure for Diabetes

It was not until 1889, when physician Oscar Minkowski removed the pancreas from a healthy dog and it developed the symptoms of diabetes, that the relationship between the pancreas and diabetes was established. For the next two decades, researchers attempted to isolate a substance from the pancreas that could be used to treat diabetes, but were unsuccessful.

In 1921, a research team from the University of Toronto led by Frederick Banting and his assistant Charles Best made a breakthrough. By tying off a dog's pancreatic duct with some string, they were able to remove some islets of Langerhans from the dog's pancreas, and then isolate insulin from the islets.

Banting and his research team soon found a way to isolate insulin from the pancreases of embryonic calves that were a by-product of the beef industry. Working with a biochemist from the University of Alberta, J.B. Collip, they further purified the extracted insulin and used it to successfully treat a boy with diabetes.

Today, synthetic insulin is produced by genetically engineered bacteria and other organisms. Furthermore, the Edmonton Protocol, led by James Shapiro at the University of Alberta, has pioneered the first successful islet cell transplants to restore functioning beta cells to the pancreas.

The technology of blood glucose monitoring devices is also improving. Many people with diabetes use digital blood glucose monitors. Advances in insulin injection technology have led to the development of the insulin pump, shown in **Figure 9.22**, which mimics the pattern of release of insulin from a healthy pancreas.

Although we all have in common the same types of body systems and the requirement for homeostasis, our particular perceptions, and conscious and autonomic responses, are unique. It is likely that, in the future, pain medication, medications to correct hormonal imbalances, and other pharmaceuticals will be tailor-made for individuals, taking into account our genes. As imaging techniques continue to improve, scientists will have more tools to solve medical problems, and to piece together the many facets of homeostasis.

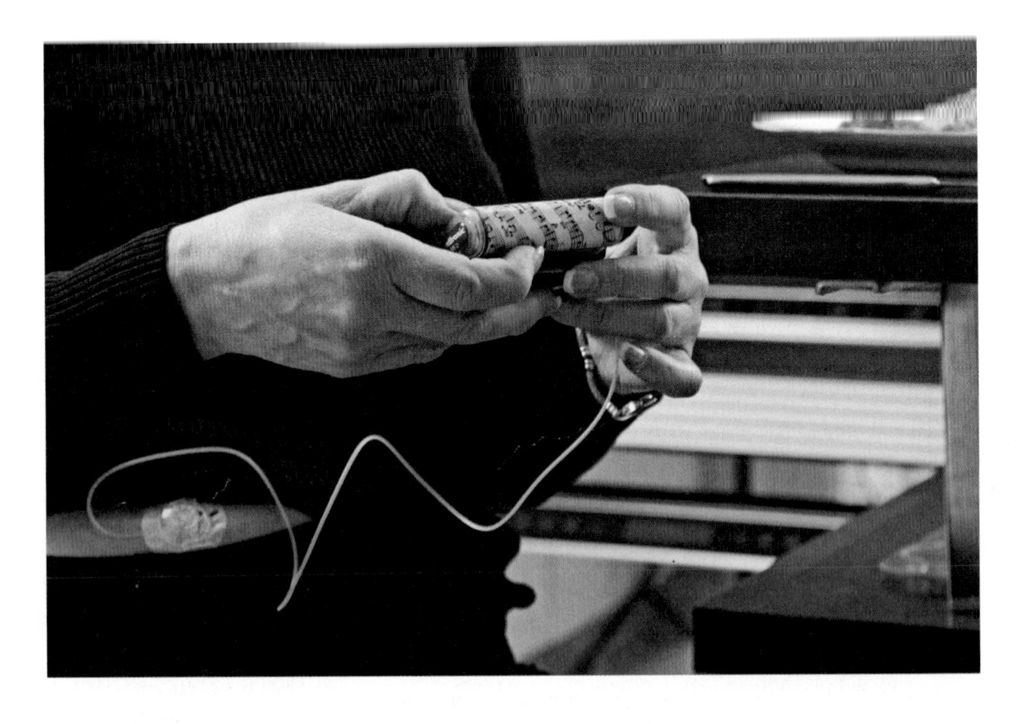

Figure 9.22 A continuous blood glucose monitor and insulin pump. The pump releases small amounts of insulin throughout the day, which minimizes the need for insulin injections.

Section 9.3 Review

Section Summary

- In the short-term stress response, also called the fight-or-flight response, the adrenal medulla is stimulated to release epinephrine and norepinephrine.
- When the body is under stress, the hypothalamus secretes a releasing hormone. The releasing hormone stimulates the anterior pituitary to release adrenocorticotropic hormone (ACTH). ACTH in turn stimulates the adrenal cortex to release cortisol, a steroid hormone.
- The islets of Langerhans contain beta cells, which secrete insulin in response to high levels of blood glucose.
- The islets of Langerhans contain alpha cells, which secrete glucagon in response to low levels of blood glucose.
- Glucagon stimulates the liver cells to break down glycogen, which releases glucose.
- If the beta cells are destroyed, type 1 diabetes results. Type 2 diabetes develops when the insulin receptors on the cells do not respond properly to insulin.

Review Questions

1. **C** Create a T-chart in your notebook. Label one side "short-term stress response" and the other side "long-term stress response."
   a. Indicate which part of the adrenal gland is involved in each response.
   b. Note which system (nervous or endocrine) stimulates the adrenal glands in each response. What hormones are involved in either pathway?
   c. List the substances secreted by the adrenal gland. Briefly compare their effects on the body.
2. **A** If you found yourself in each of the following situations, would the fight-or-flight response be useful or unhelpful? Explain your reasoning.
   a. while playing soccer
   b. during a final exam
   c. when late for your bus
   d. just before heading on stage to act in the school play
3. **A** Name or briefly describe a common but stressful situation that could occur over many weeks or months. Why is it that the body's response in this situation could result in ill health?
4. **A** How can synthetic cortisol be used to help athletes suffering from joint injuries? How could its overuse make an injury worse?
5. **K/U** Is norepinephrine a neurotransmitter or a hormone? Explain your answer.
6. **K/U** Some skiers and snowboarders report feeling an adrenaline rush when they perform their sport. What is an "adrenaline rush" and how does it affect the body?
7. **C** Draw a negative feedback mechanism to show how a healthy pancreas regulates blood glucose levels after a meal high in carbohydrates.
8. **K/U** Suppose you wake up late and skip breakfast in order to get to school on time. How would your pancreas enable you to have enough energy to get through the morning until you can eat lunch?
9. **A** The graph below shows a person's insulin and glucagon levels during a four-hour hike with no break for food.
   a. When does the level of insulin drop? What is the effect on the body?
   b. When does the level of glucagon rise? What is the effect on the body?
   c. How would having a large meal at the 4 h time point affect the person's levels of insulin and glucagon?
   d. Hypothesize what this graph would look like if this person had untreated type 1 diabetes mellitus.

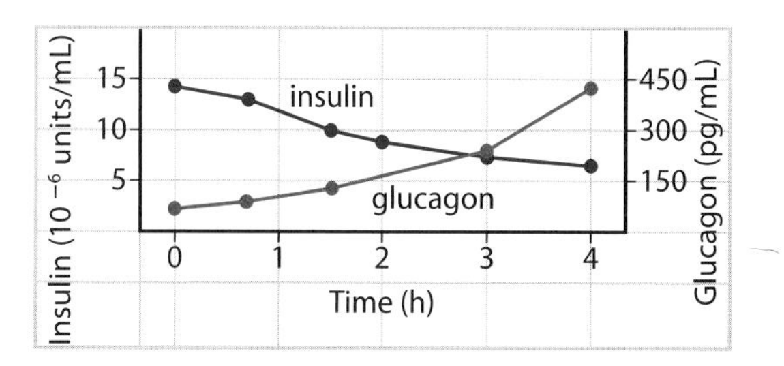

10. **T/I** Researchers believe they are just a few years away from developing an artificial pancreas that would allow people with type 1 diabetes to live a nearly carefree lifestyle. Predict the technologies these researchers would likely have to combine to make an artificial pancreas and explain your prediction.
11. **T/I** Stem cells have the ability to grow into different kinds of cells, including insulin-producing cells. There are documented cases where patients have gone as long as four years (and counting) without needing insulin injections. Predict why the use of stem cells to treat diabetes would benefit the patient.

SECTION 9.4

Hormonal Regulation of the Reproductive System

Key Terms

gonad
sex hormone
gonadotropin-releasing hormone (GnRH)
follicle-stimulating hormone (FSH)
luteinizing hormone (LH)
inhibin
testosterone
andropause
estrogen
progesterone
menstrual cycle
menopause
hormone replacement therapy (HRT)

gonad organ that produces reproductive cells (gametes); the ovary produces eggs (ova), and the testes produce sperm

sex hormone one of several chemical compounds that control the development and function of the reproductive system or secondary sex characteristics

The human reproductive system is adapted to unite a single reproductive cell from a female parent with a single reproductive cell from a male parent. To achieve this outcome, the male and female reproductive systems have different structures, functions, and hormones. The two systems also have many features in common.

Both the male and female reproductive systems include a pair of **gonads**. The gonads (testes and ovaries) are the organs that produce reproductive cells: sperm in males and eggs in females. The male and female reproductive cells are also called gametes. The gonads also produce **sex hormones**. Sex hormones are the chemical compounds that control the development and function of the reproductive system.

Structures and Functions of the Male Reproductive System

The male reproductive system includes organs that produce and store large numbers of sperm cells and organs that help to deposit these sperm cells within the female reproductive tract. Some of the male reproductive structures are located outside the body, and others are located inside the body. **Figure 9.23** shows the male reproductive system.

The Testes

The two male gonads are called the testes. The testes are held outside the body in a pouch of skin called the scrotum. The scrotum regulates the temperature of the testes. In humans, sperm production is most successful at temperatures around 35°C, which is a few degrees cooler than normal body temperature. In cold conditions, the scrotum draws close to the body, so the testicles stay warm. In hot conditions, the scrotum holds the testicles more loosely, allowing them to remain cooler than the body.

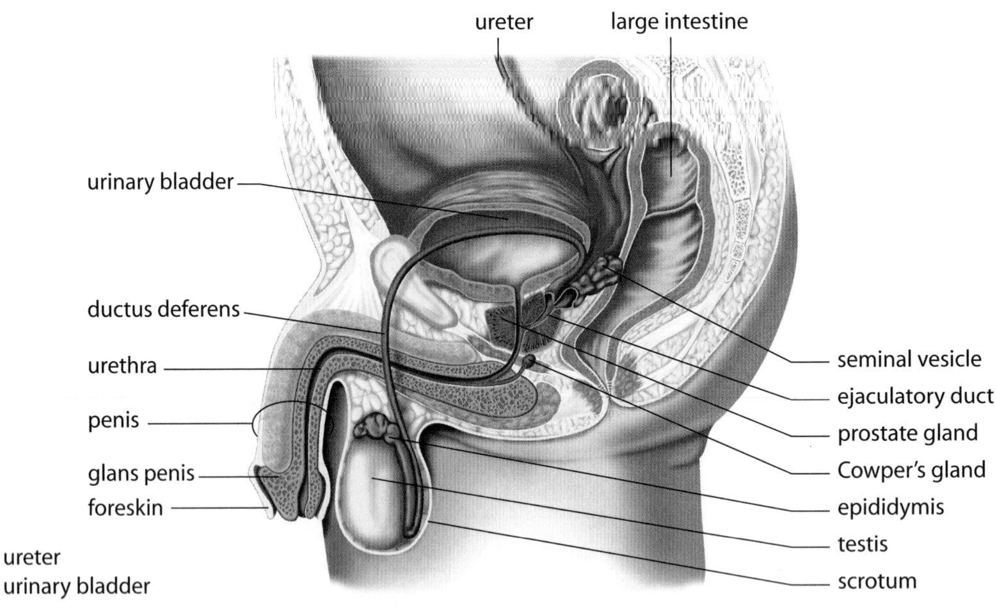

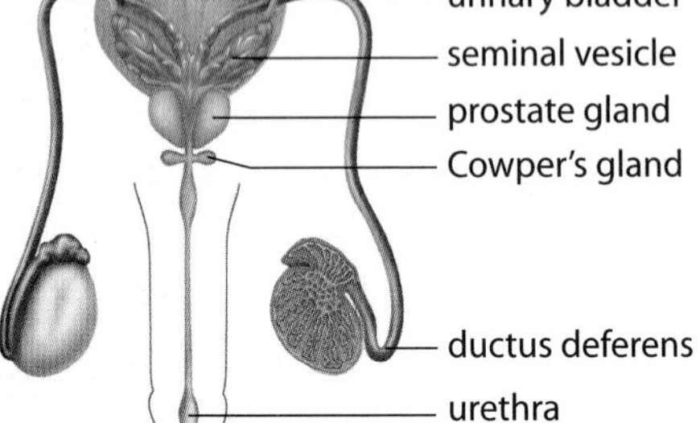

Figure 9.23 The illustration shows the structures of the male reproductive system. Sperm originate in the testes and leave the male body through the penis. The testes and the penis are located outside the body, while most of the other reproductive structures are located inside the body.

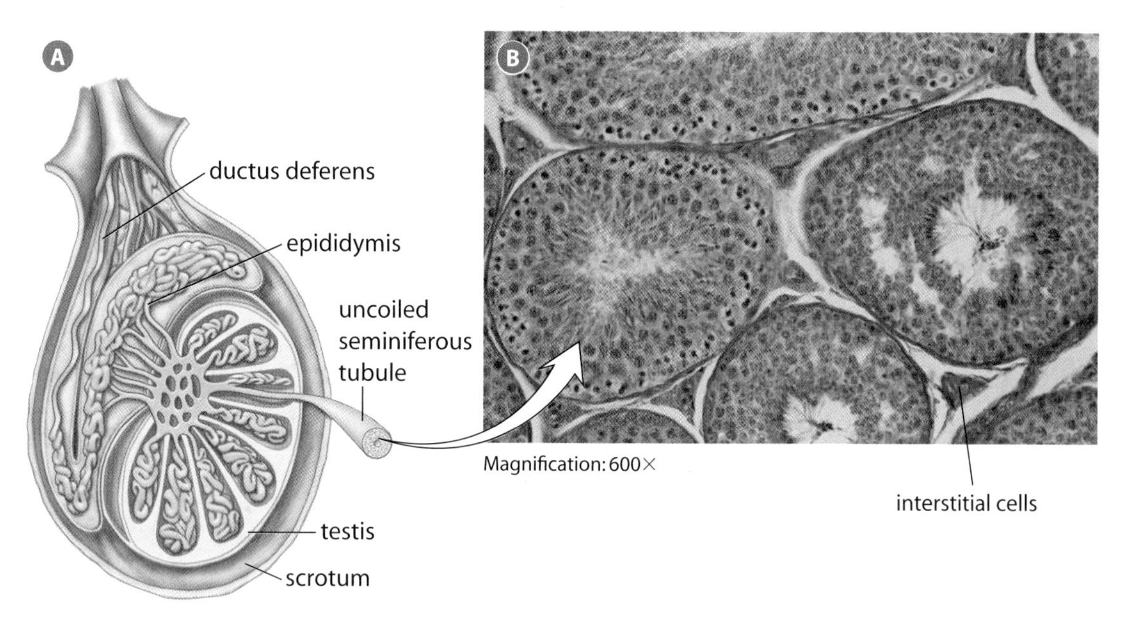

Figure 9.24 A Each testis contains several compartments, packed with seminiferous tubules. **B** This light micrograph shows a cross section of a seminiferous tubule. Some of the cells are engaged in sperm formation (spermatogenesis).

As shown in **Figure 9.24**, the testes are composed of long, coiled tubes, called seminiferous tubules, as well as hormone-secreting cells, called interstitial cells, that lie between the seminiferous tubules. The interstitial cells secrete the male hormone testosterone. The seminiferous tubules are where sperm are produced. Each testis contains more than 250 m of seminiferous tubules and can produce more than 100 million sperm each day.

From each testis, sperm are transported to a nearby duct called the epididymis. Within each epididymis, the sperm mature and become motile. The epididymis is connected to a storage duct called the *ductus deferens* (plural: *ductus deferentia*), which leads to the penis via the ejaculatory duct. (The ductus deferens is also known by an older term, *vas deferens.*)

The Penis

The penis is the male organ for sexual intercourse. Its primary reproductive function is to transfer sperm from the male to the female reproductive tract. The penis has a variable-length shaft with an enlarged tip called the *glans penis.* A sheath of skin called the foreskin surrounds and protects the glans penis. The foreskin does not have any reproductive function. Circumcision, the surgical removal of the foreskin, is a common practice in some cultures and families.

During sexual arousal, the flow of blood increases to specialized erectile tissues in the penis. This causes the erectile tissues to expand. At the same time, the veins that carry blood away from the penis become compressed. As a result, the penis engorges with blood and becomes erect. Sperm cells move out of each epididymis though the ductus deferens.

Seminal Fluid

As the sperm cells pass through the ductus deferentia, they are mixed with fluids from a series of glands (the seminal vesicles, the prostate gland, and Cowper's gland). The combination of sperm cells and fluids is called semen. If sexual arousal continues, semen enters the urethra from the ductus deferentia. The urethra is a duct that carries fluid through the penis. The movement of semen is the result of a series of interactions between the sympathetic, parasympathetic, and somatic nervous systems. Sensory stimulation, arousal, and coordinated muscular contractions combine to trigger the release, or ejaculation, of semen from the penis. The semen is deposited inside the vagina.

Learning Check

31. What is the function of the reproductive system?
32. What is a sex hormone?
33. Use a graphic organizer to show the relationships among the following: ductus deferens, epididymis, scrotum, seminiferous tubules, and testes.
34. What are the interstitial cells in the testes and what is their function?
35. Describe how the nervous system and male reproductive system work together.
36. The most common identifiable cause of infertility in men is varicocele. This is a condition of enlarged veins in the scrotum that causes abnormalities in the temperature regulation of the testes. Infer why this condition leads to infertility in men.

gonadotropin-releasing hormone (GnRH) hormone that acts on the anterior pituitary gland to cause it to release two different sex hormones: luteinizing hormone (LH) and follicle-stimulating hormone (FSH)

follicle-stimulating hormone (FSH) reproductive hormone that stimulates the development of the sex organs and gamete production in males and females

luteinizing hormone (LH) reproductive hormone that triggers ovulation, stimulates the formation of the corpus luteum, and (with follicle-stimulating hormone) stimulates estrogen production in the ovaries and stimulates the release of testosterone in the testes

inhibin hormone that acts on the anterior pituitary to inhibit the production of follicle-stimulating hormone (FSH); produces a negative feedback loop that controls the rate of sperm formation

Sex Hormones and the Male Reproductive System

The development of the male sex organs begins before birth. In embryos that are genetically male, the Y chromosome carries a gene called the testis-determining factor (TDF) gene. The action of this gene triggers the production of the male sex hormones. The male sex hormones are also known as androgens. The prefix *andro-* comes from a Greek word that means "man" or "male." The presence of androgens initiates the development of male sex organs and ducts in the fetus.

As the reproductive structures develop, they migrate within the body to their final locations. For example, the testes first develop in the abdominal cavity. During the third month of fetal development, the testes begin to descend toward the scrotum. This process is not complete until shortly before birth.

Maturation of the Male Reproductive System

A boy's genitalia are visible at birth, but his reproductive system will not be mature until puberty. Puberty is the period in which the reproductive system completes its development and becomes fully functional. Most boys enter puberty between 10 and 13 years of age, although the age of onset varies greatly. At puberty, a series of hormonal events lead to gradual physical changes in the body. These changes include the final development of the sex organs, as well as the development of the secondary sex characteristics.

Puberty begins when the hypothalamus increases its production of **gonadotropin-releasing hormone (GnRH)**. GnRH acts on the anterior pituitary gland, causing it to release two different sex hormones: **follicle-stimulating hormone (FSH)** and **luteinizing hormone (LH)**. In males, these hormones cause the testes to begin producing sperm and to release testosterone. Testosterone acts on various tissues to complete the development of the sex organs and sexual characteristics.

Hormonal Regulation of the Male Reproductive System

From the end of puberty, the male reproductive system is usually capable of producing millions of sperm every hour of the day, seven days a week until death. The same hormones that trigger the events of puberty also regulate the mature male reproductive system over a person's lifetime. Hormone feedback mechanisms control the process of sperm production, and they maintain the secondary sexual characteristics. Refer to **Figure 9.25**, on the next page, as you read the following paragraphs.

As shown, the release of GnRH from the hypothalamus triggers the release of FSH and LH from the anterior pituitary. FSH causes the seminiferous tubules in the testes to produce sperm. At the same time, FSH causes cells in the seminiferous tubules to release a hormone called **inhibin**. Inhibin acts on the anterior pituitary to inhibit the production of FSH. The result is a negative feedback loop. As the level of FSH drops, the testes release less inhibin. A decrease in the level of inhibin causes the anterior pituitary to release more FSH. This feedback loop keeps the level of sperm production relatively constant over time.

A similar feedback loop maintains the secondary sex characteristics. LH causes the interstitial cells in the testes to release **testosterone**, which promotes changes such as muscle development and the formation of facial hair. As well, testosterone acts on the anterior pituitary to inhibit the release of LH. This feedback loop keeps the testosterone level relatively constant in the body.

Reproductive function and secondary sex characteristics both depend on the continued presence of male sex hormones. Substances that interfere with the hormonal feedback system can cause changes in the reproductive system. For example, anabolic steroids mimic the action of testosterone in promoting muscle development. For this reason, some athletes illegally use steroids to increase their speed or strength. Steroids, however, also disrupt the reproductive hormone feedback systems. The side effects of steroid use in men may include shrinking testicles, low sperm count, and the development of breasts.

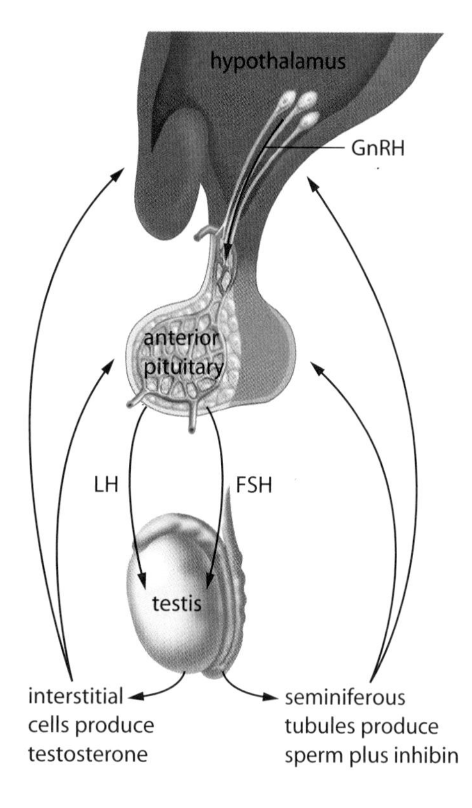

Figure 9.25 The release of GnRH triggers a series of events in the male reproductive system.

Identify *Which type of feedback system controls the release of hormones in the male reproductive system?*

testosterone reproductive hormone that stimulates the development of the male reproductive tract and secondary sex characteristics; only minor effects in females

Activity 9.3 Testosterone and Male Development

Blood tests of testosterone levels are sometimes used to help diagnose disorders of the male reproductive system, as well as other conditions that affect hormone balance in the body. In this activity, you will plot and analyze blood testosterone data for male children and young adult males.

Materials

- graph paper
- reference books
- computer with Internet access

Procedure

1. Examine the data in the table. Graph these data using the type of graph you think is most appropriate.
2. Based on the data, at what age does puberty begin? Mark this on your graph.
3. Use print or Internet resources to research the physiological changes that take place during male puberty. Write a brief description of the changes that are associated with changing levels of testosterone. Identify which of these changes are directly related to reproductive function.

Normal Blood Testosterone Levels in Males

Age (years)	Blood Testosterone level (ng/dl)
1 to 7.9	40
8 to 10.9	42
11 to 11.9	260
12 to 13.9	420
14 to 17.9	1000
18 to 29	1100

Questions

1. Is it possible to use blood hormone data to identify the end of puberty? Explain your answer.
2. In young men, the growth of facial hair begins at the same time as blood testosterone levels start to increase.
   a. From this evidence, can you conclude that testosterone causes facial hair growth? Justify your answer.
   b. Design an investigation to test the hypothesis that testosterone causes facial hair growth in men.

andropause in men, a gradual decline in their testosterone level beginning around age 40; symptoms include fatigue, depression, loss of muscle and bone mass, and a drop in sperm production

Aging and the Male Reproductive System

A man in good health can remain fertile for his entire life. Even so, most men experience a gradual decline in their testosterone level beginning around age 40. This condition is called **andropause**. In some men, the hormonal change may be linked to symptoms such as fatigue, depression, loss of muscle and bone mass, and a drop in sperm production. However, some studies suggest that low doses of testosterone can help to counter the symptoms of andropause. Because not all men experience symptoms of andropause, and because the symptoms can vary widely, this condition is difficult to diagnose accurately.

Other hormonal changes associated with aging can also affect the male reproductive system. For example, the prostate gland often begins gradually to grow in men over age 40. This can lead to discomfort and urinary difficulties, because the prostate squeezes on the urethra as it grows. Older men have an increased risk of cancer of the prostate gland, as well. Surgery may be used to provide relief and to reduce the cancer risk.

Structures and Functions of the Female Reproductive System

In contrast to the male reproductive system, the female reproductive system does not mass-produce large numbers of gametes. The two female gonads, or ovaries, produce only a limited number of gametes. The female gametes are called eggs, or ova (singular: ovum). The other female sexual organs are adapted to provide a safe environment for fertilization, for supporting and nourishing a developing fetus, and for allowing the birth of a baby. Most of the structures of the female reproductive system are located inside the body. **Figure 9.26** shows the main structures of the female reproductive system. Refer to this figure as you read the following paragraphs.

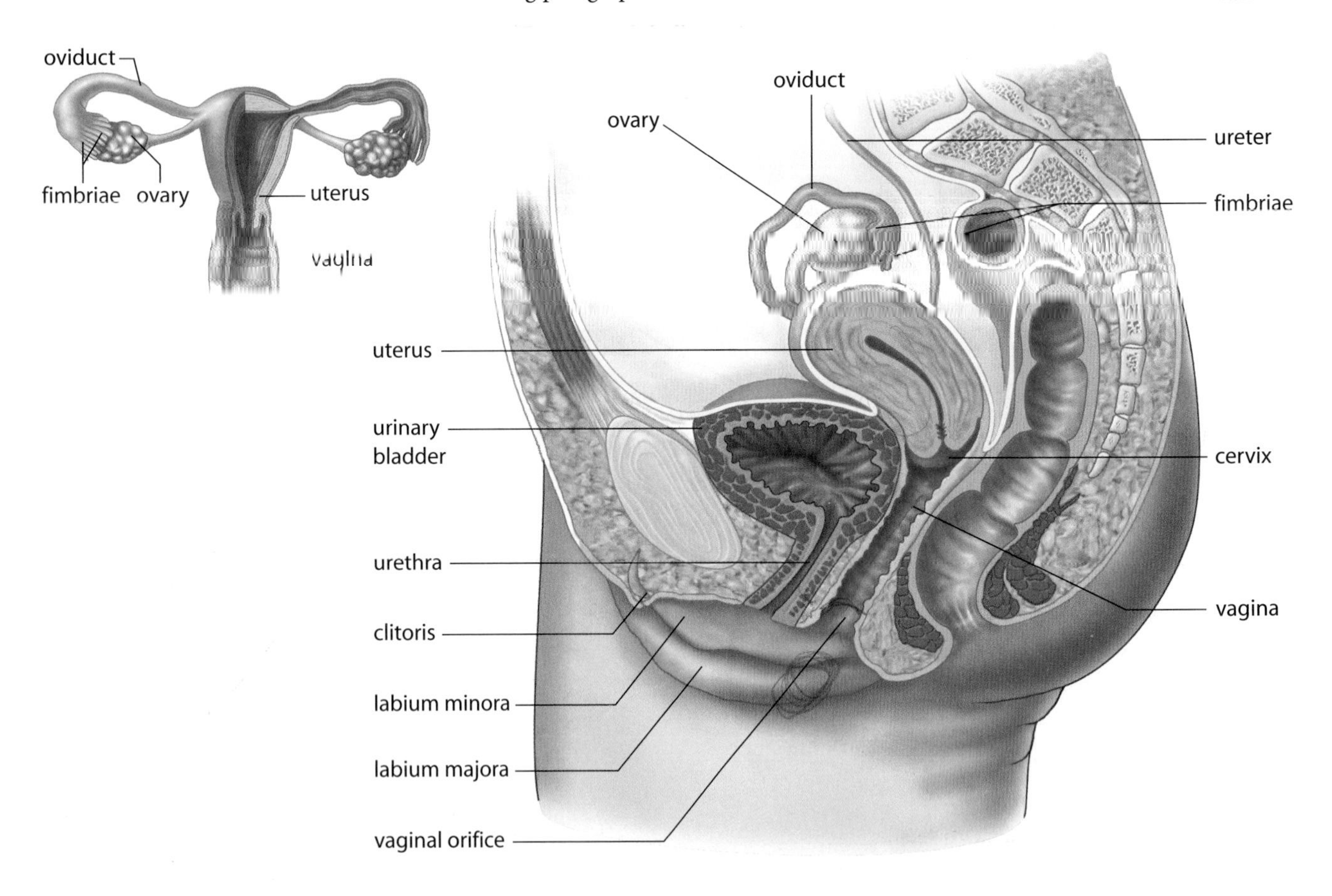

Figure 9.26 In the female reproductive system, the reproductive organs and ducts are inside the body. Gametes are produced in the ovaries, fertilization occurs in the oviduct, and fetal development takes place in the uterus. The vagina is both the organ of sexual intercourse and the birth canal.

The Ovaries

The two ovaries are suspended by ligaments within the abdominal cavity. The ovaries are the site of oogenesis—the production of an ovum. Oogenesis comes from two Greek words that mean "egg-creation." Ova are also called oocytes. In contrast to the male reproductive system, in which both testes function at the same time, the ovaries usually alternate so that only one produces an egg each month.

The ovary contains specialized cell structures called follicles. A single ovum develops within each follicle. Each month, a single follicle matures and then ruptures, releasing the ovum into the oviduct. This event is called ovulation. Thread-like projections called fimbriae continually sweep over the ovary. When an ovum is released, it is swept by the fimbriae into a cilia-lined tube about 10 cm long called an oviduct. The oviduct carries the ovum from the ovary to the uterus. Within the oviduct, the beating cilia create a current that moves the ovum toward the uterus.

A mature ovum is a non-motile, sphere-shaped cell approximately 0.1 mm in diameter (that is, over 20 times larger than the head of a sperm cell). The ovum contains a large quantity of cytoplasm, which contains nutrients for the first days of development after fertilization. The ovum is encased in a thick membrane that must be penetrated by a sperm cell before fertilization can take place.

The Uterus and Vagina

The uterus is a muscular organ that holds and nourishes a developing fetus. The uterus is normally about the size and shape of a pear, but it expands to many times its size as the fetus develops. The lining of the uterus, called the endometrium, is richly supplied with blood vessels to provide nutrients for the fetus. At its upper end, the uterus connects to the oviducts. At its base, the uterus forms a narrow opening called the cervix. The cervix, in turn, connects to the vagina. The vagina serves as an entrance for an erect penis to deposit sperm during sexual intercourse. The vagina also serves as an exit for the fetus during childbirth.

The ovum survives in the oviduct for up to 24 hours after ovulation. If a living egg encounters sperm in the oviduct, fertilization may take place. The fertilized egg, now called a zygote, continues to move through the oviduct for several days before reaching the uterus. During this time, the endometrium thickens as it prepares to receive the zygote. The zygote implants itself in the endometrium, and development of the embryo begins. If the egg is not fertilized, it does not implant in the endometrium. The endometrium disintegrates, and its tissues and blood flow out the vagina in a process known as menstruation.

The vagina opens into the female external genital organs, known together as the vulva. The vulva includes the labia majora and labia minora, which are two pairs of skin folds that protect the vaginal opening. The vulva also includes the glans clitoris.

Learning Check

37. Genetically male individuals have an X and a Y chromosome combination (XY). Explain how the XY chromosome combination is responsible for the development of the male sex organs.

38. Summarize the hormonal changes that occur as a male begins puberty.

39. Describe the role the hormone inhibin plays in maintaining a relatively constant level of sperm development.

40. What is andropause and what are the symptoms of this condition?

41. In a table, identify the structures of the female reproductive system and list the function of each structure.

42. How does the path of an unfertilized ovum differ from the path of a fertilized ovum?

estrogen female sex hormone produced in the ovary; helps maintain sexual organs and secondary sexual characteristics

progesterone female sex hormone produced first by the corpus luteum of the ovary to prepare the uterus for the fertilized egg (ovum), and later by the placenta to maintain pregnancy

menstrual cycle in a human female, period of 20–45 days during which hormones stimulate the development of the uterine lining, and an egg (ovum) is developed and released from an ovary; if the egg is not fertilized, the uterine lining is shed as the cycle begins again; can be divided into the ovarian cycle and the uterine cycle

Sex Hormones and the Female Reproductive System

Our understanding of the specific factors that trigger the development of female sex organs in a genetically female embryo is incomplete. Until recently, scientists assumed that the development of female sex organs was a "default" pattern—that is, if there is no Y chromosome, then female organs will develop. Researchers now suspect that the processes of female sex development are more complex and that specific hormonal triggers cause female sex organs to develop.

Like a baby boy, a baby girl has a complete but immature set of reproductive organs at birth. North American girls usually begin puberty between 9 and 13 years of age. The basic hormones and hormonal processes of female puberty are similar to those of male puberty. A girl begins puberty when the hypothalamus increases its production of GnRH. This hormone acts on the anterior pituitary to trigger the release of LH and FSH.

In girls, FSH and LH act on the ovaries to produce the female sex hormones **estrogen** and **progesterone**. These hormones stimulate the development of the female secondary sex characteristics and launch a reproductive cycle that will continue until about middle age.

Hormonal Regulation of the Female Reproductive System

In humans, female reproductive function follows a cyclical pattern known as the **menstrual cycle**. The menstrual cycle ensures that an ovum is released at the same time as the uterus is most receptive to a fertilized egg.

The menstrual cycle is usually about 28 days long, although it may vary considerably from one woman to the next, and even from one cycle to the next in the same woman. By convention, the cycle is said to begin with menstruation and end with the start of the next menstrual period. The menstrual cycle is actually two separate but interconnected cycles of events. One cycle takes place in the ovaries and is known as the ovarian cycle, shown in **Figure 9.27**. The other cycle takes place in the uterus and is known as the uterine cycle. Both cycles are controlled by the female sex hormones estrogen and progesterone, which are produced by the ovaries.

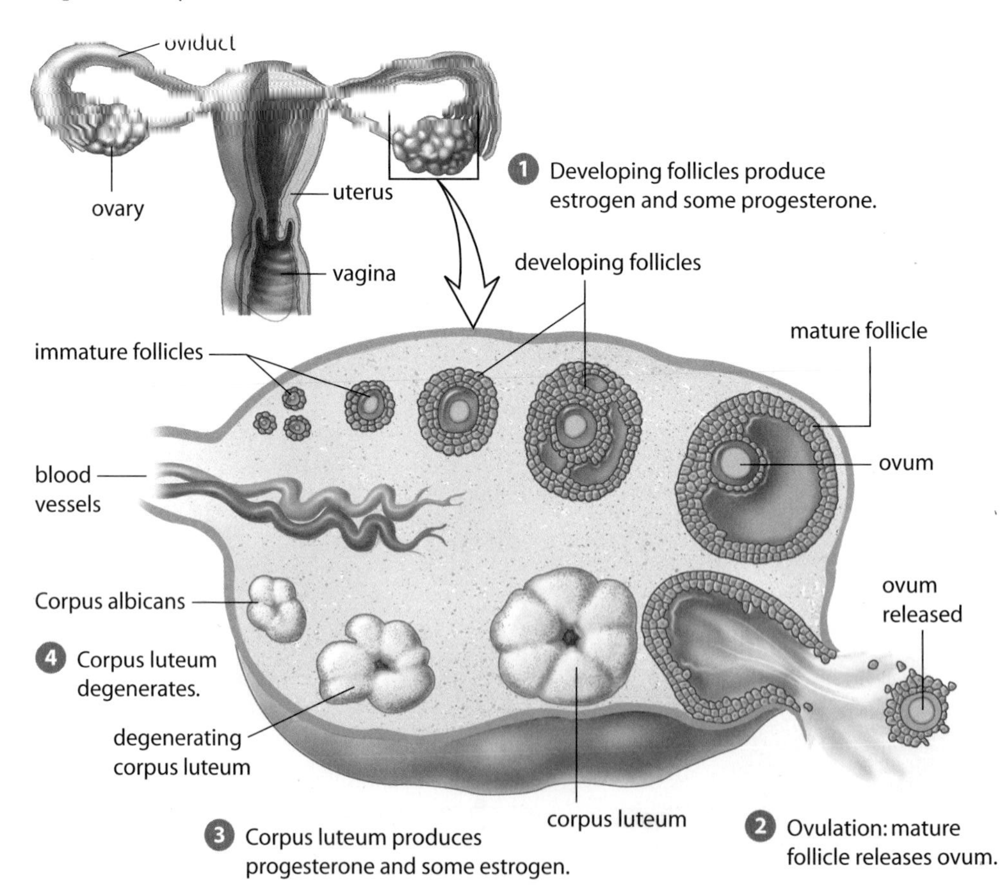

Figure 9.27 A follicle matures by growing layers of follicular cells and a central fluid-filled vesicle. The vesicle contains the maturing ovum. At ovulation, the follicle ruptures and the ovum is released into the oviduct. The follicle develops into a corpus luteum. If pregnancy does not occur, the corpus luteum starts to degenerate after about 10 days. Note that the follicle does not migrate around the ovary, as shown here for clarity, but goes through all the stages in one place.

The Ovarian Cycle

The ovary contains cellular structures called follicles. Each follicle contains a single immature ovum. At birth, a baby girl has more than 2 million follicles. Many degenerate, leaving up to about 400 000 by puberty. During her lifetime, only approximately 400 of these follicles will mature to release an ovum. In a single ovarian cycle, one follicle matures, releases an ovum, and then develops into a yellowish, gland-like structure known as a corpus luteum. The corpus luteum then degenerates. **Figure 9.27** illustrates the ovarian cycle, and **Figure 9.28** illustrates the hormone systems that control this cycle.

The ovarian cycle can be roughly divided into two stages. The first stage is known as the follicular stage. It begins with an increase in the level of FSH released by the anterior pituitary gland. FSH stimulates one follicle to mature. As the follicle matures, it releases estrogen and some progesterone. The rising level of estrogen in the blood acts on the anterior pituitary to inhibit the release of FSH. At the same time, the estrogen triggers a sudden release of GnRH from the hypothalamus. This leads to a sharp increase in LH production by the anterior pituitary triggering ovulation—the follicle bursts, releasing its ovum.

Ovulation marks the end of the follicular stage and the beginning of the second stage. The second stage is called the luteal stage. Once the ovum has been released, LH causes the follicle to develop into a corpus luteum. The corpus luteum secretes progesterone and some estrogen. As the levels of these hormones rise in the blood, they act on the anterior pituitary to inhibit FSH and LH production. The corpus luteum degenerates, leading to a decrease in the levels of estrogen and progesterone. The low levels of these sex hormones in the blood cause the anterior pituitary to increase its secretion of FSH, and the cycle begins again.

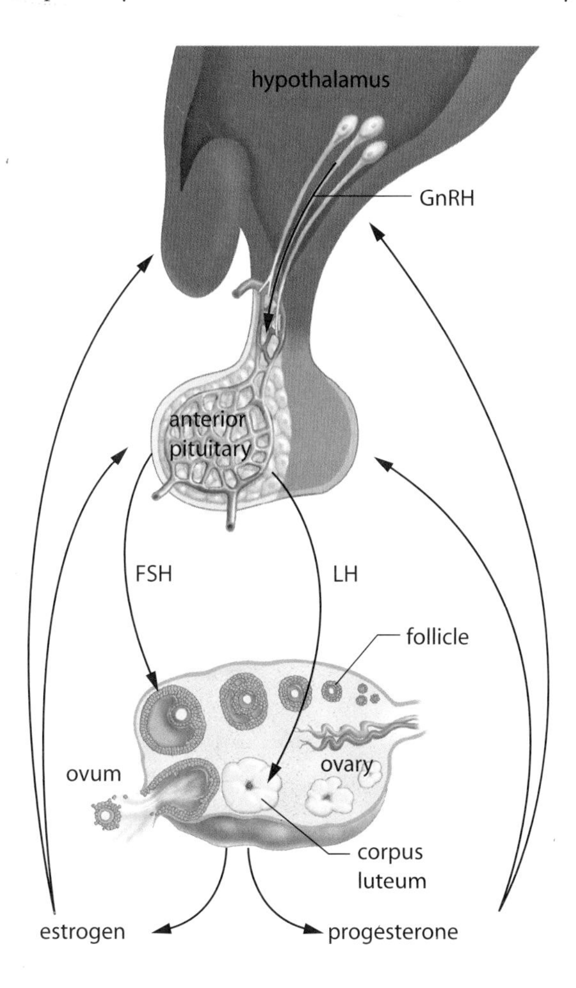

Figure 9.28 The hypothalamus produces GnRH, which stimulates the anterior pituitary to produce FSH and LH. FSH stimulates the follicle to produce estrogen. LH stimulates the corpus luteum to produce progesterone.

If the ovum is fertilized and implants in the endometrium, blood hormone levels of progesterone and estrogen remain high under stimulus of hormones released by embryo-supporting membranes. The continued presence of progesterone maintains the endometrium to support the developing fetus. The continued presence of estrogen stops the ovarian cycle so no additional follicles mature.

The Uterine Cycle

The uterine cycle is closely linked to the ovarian cycle. As you have seen, ovulation takes place about halfway through the ovarian cycle, around day 14. The ovum survives for up to 24 h after ovulation. If fertilization occurs, the fertilized egg completes the passage through the oviduct and arrives at the uterus a few days later. The timing of the uterine cycle ensures that the uterus is prepared to receive and nurture a new life. The events of the uterine cycle cause a build-up of blood vessels and tissues in the endometrium. If fertilization does not occur, the endometrium disintegrates and menstruation begins.

The uterine cycle begins on the first day of menstruation (which is also the first day of the ovarian cycle). On this day, the corpus luteum has degenerated and the levels of the sex hormones in the blood are low, as shown in **Figure 9.29**. Menstruation lasts for the first 5 days of the uterine cycle and by the end, the endometrium is very thin, also shown in **Figure 9.29**. As a new follicle begins to mature and release estrogen, the level of estrogen in the blood gradually increases.

Suggested Investigation

ThoughtLab Investigation 9-C, The Menstrual Cycle

Beginning around the sixth day of the uterine cycle, the estrogen level is high enough to cause the endometrium to begin thickening, also shown in **Figure 9.29**. After ovulation, the release of progesterone by the corpus luteum causes a more rapid thickening of the endometrium. Between days 15 and 23 of the cycle, the thickness of the endometrium may double [illegible] the corpus luteum degenerates. The levels of the sex hormones drop, the endometrium breaks down, and menstruation begins again. You have seen that the menstrual cycle involves a number of different hormones, each of which triggers different events in the body. **Figure 9.29** summarizes the hormonal and physical changes that occur in the body throughout the menstrual cycle.

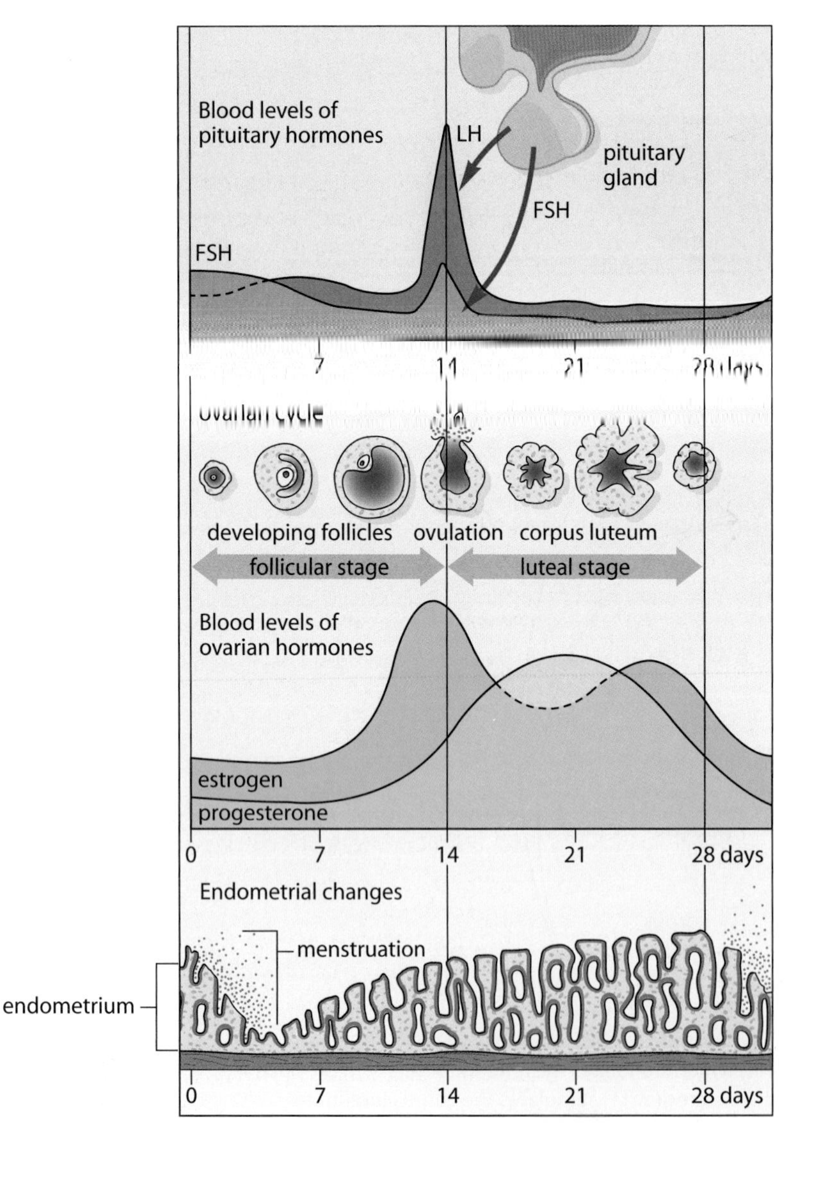

Figure 9.29 Changes in the levels of FSH, LH, estrogen, and progesterone are shown along with changes in the follicle and the thickness of the endometrium throughout the menstrual cycle.

Aging and the Menstrual Cycle

After puberty, the male reproductive system can continue to produce viable sperm for a lifetime. In contrast, the number of functioning follicles in the female reproductive system decreases with age. This, in turn, leads to a gradual overall decline in the amount of estrogen and progesterone in the blood. As hormone levels drop, a woman's menstrual cycle becomes irregular. Within a few years, it stops altogether. The end of the menstrual cycle is known as menopause. Among North American women, the average age of **menopause** is approximately 50, but menopause can begin earlier or later.

menopause period in a woman's life during which a decrease in estrogen and progesterone results in an end of menstrual cycles, usually occurring around age 50

A woman who has completed menopause no longer produces ova, so she is no longer fertile. As well, the decrease in the sex hormones disrupts the homeostasis of a number of hormone systems. This has a range of effects on the body. During menopause, blood vessels alternately constrict and dilate, resulting in uncomfortable sensations for some women known as "hot flashes." Some women also experience variable changes in mood. Over the longer term, menopause is associated with rising cholesterol levels, diminishing bone mass, and increased risk of uterine cancer, breast cancer, and heart disease. For these reasons, many women consider hormone replacement therapy during or following menopause.

Learning Check

43. Summarize the events leading to puberty in a girl.

44. Compare and contrast the functions of estrogen and progesterone.

45. Using a diagram, summarize the events of the ovarian cycle. Write a caption for your diagram.

46. Summarize, in point form, the events that take place during the uterine cycle.

47. A couple is trying to conceive a baby. They purchase a Home Ovulation Test Kit from their local pharmacy. It is a urine-based test kit. Predict the hormone that this urine test kit would be trying to identify. Explain your prediction.

48. What is the meaning of the term "menopause", when does it typically occur, and what is the principal factor responsible for its occurrence?

Activity 9.4 Therapy Options for Menopause

For many years, doctors in North America routinely prescribed estrogen, or a combination of estrogen and progesterone, to help alleviate the symptoms of menopause. In 2000, the U.S.-based National Institutes of Health (NIH) began a detailed investigation of the effects of hormone replacement therapy (HRT). Researchers stopped the study several years early, however, when they found that HRT was associated with a significant increase in the risk of strokes, heart disease, and breast cancer among their test subjects. The findings led scientists to look for a safer alternative to hormone replacement therapy.

Materials

- reference books
- computer with Internet access

Procedure

1. Using print or Internet resources, research two or three different kinds of therapies available to women who want to alleviate the symptoms of menopause.
2. For each of the therapies you are studying, gather information about
   - how the treatment affects cells and tissues in the body
   - how the treatment affects hormone feedback systems
   - any known health risks and benefits
3. Organize the information you have gathered. Create a short report or presentation to summarize and communicate your findings.

Questions

1. Based on your research, is it possible to claim that one of these therapies is better than the others?
2. Many people argue that women should never have been prescribed hormone replacement therapy in the years before the NIH study was completed. Do you agree? What, if anything, should health practitioners and pharmaceutical companies have done differently?

hormone replacement therapy (HRT) administration of low levels of estrogen and/or progesterone to alleviate symptoms of menopause in females

Hormone Replacement Theory

Hormone replacement therapy (HRT) is a prescription of low levels of estrogen with or without progesterone. However, while this therapy can ease some symptoms of menopause, the treatment also carries a number of health risks. In recent studies, hormone replacement therapy has been linked to

- an increased risk of coronary heart disease, strokes, and blood clots
- an increased risk of breast cancer and colorectal cancer
- an increased risk of dementia

For this reason, Health Canada advises that a woman should not begin hormone replacement therapy without a thorough medical evaluation and a careful assessment of her own particular needs, health, and medical history. In some cases, the benefits of the therapy may outweigh the risks. In other cases, the reverse is true. Scientists continue to search for other ways to alleviate the symptoms and long-term health effects of menopause.

Summarizing Reproductive Hormones

Table 9.2 summarizes the reproductive hormones and their functions in the male and female reproductive systems.

Table 9.2 Summary of Key Reproductive Hormones and Their Functions

Hormone	Production Site	Target Organ(s)	Function in Male Reproductive System	Function in Female Reproductive System
Gonadotropin-releasing hormone (GnRH)	Hypothalamus	Anterior pituitary gland	Stimulates the release of FSH and LH from the anterior pituitary	Stimulates the release of FSH and LH from the anterior pituitary
Follicle-stimulating hormone (FSH)	Anterior pituitary	Ovaries and testes	Stimulates the development of the sex organs and gamete production	Stimulates the development of the sex organs and gamete production
Luteinizing hormone (LH)	Anterior pituitary	Ovaries and testes	Stimulates the production of testosterone	Triggers ovulation and (with FSH) stimulates estrogen production
Estrogen	Ovary (follicle)	Entire body	Minor	Stimulates the development of the female reproductive tract and secondary sex characteristics
Progesterone	Ovary (corpus luteum)	Uterus	Minor	Causes uterine thickening
Testosterone	Testes (interstitial cells)	Entire body	Stimulates the development of the male reproductive tract and secondary sex characteristics	Minor
Inhibin	Testes (seminiferous tubules)	Anterior pituitary and hypothalamus	Inhibits FSH production	Inhibits FSH production

Section 9.4 Review

Section Summary

- Sex hormones work to stimulate the development of male and female reproductive systems and regulate the function of the mature reproductive system.
- In males, the main sex hormone is testosterone. In females, the main sex hormones are estrogen and progesterone.
- In the male reproductive system, a negative feedback hormone system maintains a relatively constant level of sperm production and testosterone.
- In females, hormone systems interact to regulate a monthly menstrual cycle.

Review Questions

1. **K/U** What are the two main purposes of the gonads in both males and females?
2. **K/U** For each of the following structures, write a short description to indicate whether it is found in the male or female reproductive system and to summarize its function:
   a. fimbriae
   b. ductus deferens
   c. endometrium
   d. epididymis
   e. testis
   f. uterus
3. **C** Draw a labelled diagram to describe the pathway of a sperm cell through the male reproductive system, beginning from the testes.
4. **C** Use a diagram to summarize the hormone interactions that regulate the male reproductive system.
5. **A** Assuming that all other body systems remain unaffected, what physiological effects would you expect to find in an adolescent male whose anterior pituitary produced FSH but not LH? Use a flowchart or labelled diagram to explain your reasoning.
6. **T/I** Predict what would happen to testosterone production in the testes in response to an injection of a large amount of testosterone in an adult male.
7. **K/U** Briefly describe the main hormonal and physiological events of
   a. male puberty
   b. female puberty
   c. the ovarian cycle
   d. the uterine cycle
   e. sperm production
   f. andropause
8. **K/U** Which hormone stimulates the release of FSH and LH in both males and females?
9. **A** Examine the graphs and answer the following questions:
   a. Identify the hormones represented by the letters A, B, C, and D.
   b. Describe the events that are occurring in the region of the diagram labelled E.
   c. Identify what is happening at the region of the diagram labelled F.
   d. Describe the events that are occurring in the region of the diagram labelled G.
   e. Write suitable labels for the regions of the diagram labelled H and I.
   f. Write a caption that briefly and accurately summarizes what is being depicted in this whole diagram.

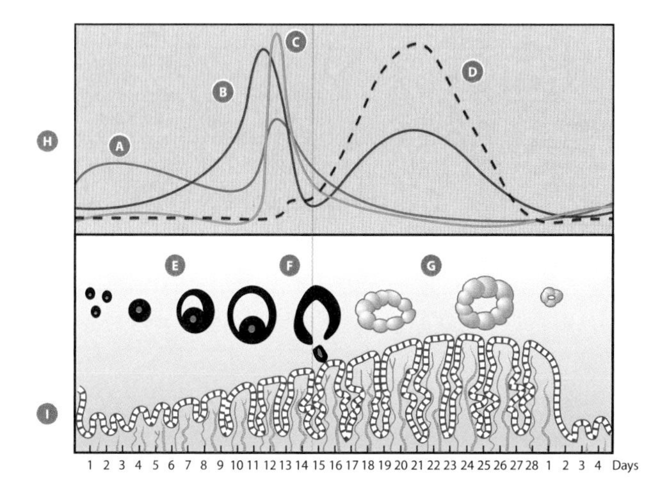

10. **K/U** Why is the timing of the uterine cycle closely linked to the timing of the ovarian cycle?
11. **T/I** Infer why a home pregnancy test kit that identifies hormones in urine will only work a few days after the fertilized ovum has implanted itself in the endometrium.
12. **K/U** Compare and contrast the effects of testosterone on the male body with the effects of estrogen on the female body. In what ways are these effects similar? In what ways do they differ?

Case Study

Anabolic Steroids

Assessing the use of performance-enhancing steroids in sports

In June 2010, the University of Waterloo announced that it was cancelling its 2010 football program after nine players tested positive for anabolic steroids and other banned substances. In 1988, sprinter Ben Johnson tested positive for anabolic steroids after winning an Olympic gold medal and setting a world record in the 100-metre race. He was subsequently stripped of both the medal and the record.

Anabolic steroids are synthetic versions of the naturally occurring male hormone testosterone and they are currently banned from major sports including the Olympics, International Federation of Association Football (FIFA), the National Hockey League (NHL), and other professional sports leagues. Athletes caught using anabolic steroids face penalties that include suspensions, bans, and records being overturned, as well as possible legal consequences.

Although anabolic steroids have legitimate medical uses, including treatment for delayed onset of puberty or the loss of lean muscle mass from diseases such as cancer and AIDS, they can also be illegally abused by athletes and body builders. Steroids may be used by athletes to help build up lean muscle to be stronger, faster, or, in the case of some body builders, to have muscles appear more prominent. Anabolic steroids are classified as a controlled substance in Canada and the United States, but they are legal in Mexico and parts of Europe and Asia.

The ban on the use of performance-enhancing drugs in sports is not without controversy. Read the articles that follow to understand more about different aspects of the issue, then use your research and critical thinking skills to form your opinion on the matter.

Page 6 Daily News   Friday, March 12

Why Not Let Athletes Use All Available Resources?

■ By Mike Johnson, Sportswriter

I think that we should remove the ban on anabolic steroids in sports and allow athletes to use all of the resources available to them. Athletes worldwide have already benefited from products that enhance performance, such as the specially-designed racing suits swimmers wore in the 2008 Olympics. Both the material and the design of the suits reduced drag, allowing swimmers to achieve some of the fastest times in history.* Bicycle helmets designed to reduce drag and bicycles made from carbon fibre, a material that is both lightweight and strong, help riders in the Tour de France stay on the cutting edge of technology and post record race times. Competitive downhill skiers make use of wind tunnels to learn the best way to move to minimize drag. If science and technology is already providing elite athletes with an edge, why should the use of performance enhancing substances, such as steroids, be viewed any differently?

One advantage to legalizing the use of steroids in sports is that it would help protect an athlete's health. Currently, many anabolic steroids are purchased from an unregulated market in which there is no guarantee that a substance is untainted and actually is what it says it is. There is no oversight from medical professionals in terms of safe dosage. In some cases, people taking steroids illegally may be taking between 10 to 100 times the dosage that would be prescribed by a doctor. If we legalize the use of performance-enhancing drugs in sports, we may be able to better protect athletes from endangering their health in order to win.

* Editor's Note: In 2010, these high-tech suits and others like them were banned by FINA, the international governing body of swimming, from pool competitions.

Athletes, body builders, and professional wrestlers may be motivated to take anabolic steroids to increase their muscle mass, gain weight, become faster, or become stronger.

LETTERS

Winning at the Cost of Personal Health?

By Sheila Simon, M.D.

Having been a practising physician in the field of sports medicine for twenty years, I feel strongly that the health risks involved in taking anabolic steroids in order to win are not worth it. The short-term health effects of steroid use in males include reduced sperm count, difficultly or pain while urinating, and damage to the heart muscle. In females, short-term health risks include excessive hair growth and reduced breast size. In both males and females, other short-term effects include increased blood pressure, insomnia, headaches, and acne.

Long-term use in males can result in a shrinking of the testicles, the development of breasts, and permanent damage to the liver. In females, long-term use can lead to a deepened voice and abnormal menstrual cycle. In both males and females, long-term use can disrupt homeostasis, resulting in blood clotting disorders; an enlarged heart; heart attack; damage to the liver, kidneys, and reproductive organs; and an increased risk of injury to muscles, ligaments, and tendons. Anabolic steroid use is also associated with severe mood swings, during which a user may become extremely aggressive or violent.

A 2010 study published in the scientific journal *Circulation: Heart Failure* found that long-term use of steroids resulted in a weakened ability of the heart to pump blood. The type of weakening described in the study had been previously linked to an increased risk of heart failure and sudden death due to heart attack. A 2009 study, conducted by researchers at Columbia University Medical Center, showed that long-term steroid use resulted in severe reduction in kidney function due to scarring of kidney tissue. Researchers hypothesize that the scarring is due to the increased workload put on the kidneys when the steroid users gained extreme amounts of extra muscle mass. They also believe that the steroids acted as a toxin in the kidneys, causing further damage to the delicate tissue.

Although winning has many rewards, both emotionally and financially, the risks to a person's heart, kidneys, and mental health are far greater.

Research and Analyze

1. Research more information about whether or not steroids are addictive. How do they affect neurotransmitters in the brain? How do they disrupt homeostasis within the nervous system? Based on your research, state your opinion on whether anabolic steroids are addictive. Support your opinion with information from your research.
2. Conduct research on the legal penalties, as well as penalties imposed by various athletic organizations, if athletes are caught using steroids. How much time and money is spent on policing steroid abuse? What resources are used to track and deter the smuggling of illegal steroids across international borders? Are the existing penalties effective deterrents? Identify changes to current policies that would need to occur to effectively address the use of steroids in sports.
3. Using information presented in the case study and from your research in questions 1 and 2, perform a risk/benefit analysis of an athlete using anabolic steroids. State your conclusions in a short report.

Take Action

4. **PLAN** Find out your school's policy on the use of anabolic steroids or other performance-enhancing drugs by student athletes. Does testing occur? Is there a need for testing? Create a five-question survey to gather the opinion of the student population at your school. Craft the survey to determine how people feel about steroid use in sports in general as well as what they think the appropriate policy at your school should be.

5 **ACT** Based on the information you gathered in question 4, make improvements to your school's existing anti-doping policy. If no policy exists, write a new policy for your school. Present your proposed amendments or new policy to your principal. If you feel that no improvements are needed to your school's policy, write a summary of the policy, highlighting any important points, that could be included in your school's student agenda.

STSE

BIOLOGY Connections

Endocrine Disruptors in the Environment

Hormones influence nearly all aspects of the body processes in animals, notably cellular development, growth, and reproduction. Thus, when people noticed that animals in areas that are contaminated by pollutants began to exhibit certain types of abnormalities, scientists began to wonder if there might be a link between certain pollutants and endocrine effects on body systems.

EXAMINING THE EVIDENCE Substances that interfere with the normal functions of hormones are called *endocrine disruptors*. These chemicals upset the growth, development, and reproduction of organisms by mimicking natural hormones or by blocking their effects. For example, when fish populations are exposed to reproductive endocrine disruptors, the sexual organs of young males fail to fully develop. In extreme cases, the males produce eggs! Both effects are examples of feminization. Endocrine disruptors that mimic estrogen can cause feminization. Substances that block the action of male reproductive hormones can also have feminizing effects.

While the effects of known endocrine disruptors can be demonstrated in a laboratory, it is not always easy to assess their impact in the environment. One problem is that endocrine disruptors tend to be diluted in lakes and rivers. Even so, scientists have observed effects such as the [illegible] have also observed impaired reproduction and development of fish near pulp and paper mills. As shown in the table below, sewage and mill wastes both contain endocrine disruptors. Even treated sewage contains obvious endocrine disruptors such as synthetic estrogen from birth control pills.

Selected Products Containing Endocrine Disruptors

Source/Product	Endocrine Disruptors
flame retardants	polybrominated diphenyl ethers (PBDEs)
paint (for ships' hulls)	tributyltin
pesticides	DDT, lindane, permethrin
soft plastics	phthalates
pulp and paper mill effluent	phytoestrogens
perfumes and soaps	polycyclic musks
shampoo and other cosmetics	phthalates

Many cosmetics—including some, but not all, brands of nail polish—contain endocrine-disrupting chemicals in amounts that are comparable to, and sometimes greater than, hormone levels in the body. Should this be cause for concern?

ASKING QUESTIONS Are estrogen disruptors to blame for the increase in abnormalities in frogs in some regions and the decrease in amphibians worldwide? Why do some female black bears and polar bears develop male sexual traits? It could be that increased exposure to UV light is harming amphibians, and that some bears simply inherit the trait of showing both male and female sexual characteristics.

ASSESSING THE RISKS If endocrine disruptors in the environment are affecting wild animals, are they also affecting people? Scientists and representatives of industrial [illegible] pose a health risk to humans. There is evidence that endocrine disruptors are leading to lower sperm counts, reduced fertility in both men and women, and increased rates of certain types of cancers. Some studies have suggested a link between endocrine disruptors and learning and behaviour problems in children. To date, however, direct (causal) links between environmental exposure to endocrine disruptors and human health effects have not been established. Continued research that includes the cooperation of industry and financial support of governments is necessary to identify and mediate the risks that these chemicals pose to the environment and organisms.

Connect to the Environment

1. Should a potential endocrine disruptor be considered "guilty until proven innocent" or "innocent until proven guilty" before being put on the market? Justify your response.
2. What would you need to know to link a particular substance to specific endocrine disrupting effects in a population, such as unusually high cancer rates?

ThoughtLab INVESTIGATION

9-A

Skill Check

✓ Initiating and Planning

✓ Performing and Recording

✓ Analyzing and Interpreting

✓ Communicating

Suggested Materials

- reference materials
- computer with Internet access

Regulation of Melatonin

The pineal gland is a small, pinecone-shaped gland found deep in the brain. It is critical in the production of the hormone melatonin. Melatonin influences the body's daily sleep/wake cycle and is thought to promote sleepiness.

Pre-Lab Questions

1. What is the primary function of melatonin in the human body?
2. Which endocrine gland secretes melatonin in humans?

Question

How can you construct a model to illustrate the feedback mechanism that controls the hormone melatonin?

Organize the Data

1. Use reference books and the Internet to research the pineal gland and the release of melatonin. Some questions you may use to guide your research include: What is the stimulus that causes sensors to send messages to the control centre? What is the role of the suprachiasmatic nucleus in the secretion of melatonin? What stimulates and inhibits the secretion of melatonin?
2. Create a flowchart to represent the different steps in the feedback mechanism that controls the release and inhibition of melatonin by the pineal gland.
3. Research more information about Seasonal Affective Disorder (SAD). Represent the development of SAD on your flowchart.

Analyze and Interpret

1. What is the ultimate cause of melatonin secretion?
2. Describe how the nervous system and endocrine system interact to maintain homeostasis regarding the release of melatonin and the control of the sleep/wake cycle.
3. Describe the type of feedback system that controls the secretion and inhibition of melatonin.

Conclude and Communicate

4. Based on the results of your research and your flowchart, explain if SAD is more likely to occur in the winter or summer.

Extend Further

5. **INQUIRY** Devices that emit blue light exist to combat jet lag, which occurs when a person travels across several time zones. Explain the physiology behind why these devices might work. Design a study that would test the effectiveness of one of these devices.
6. **RESEARCH** Find out more information about melatonin supplements. How do they work? Why might a person take melatonin supplements?

Inquiry INVESTIGATION

9-B

Skill Check

Initiating and Planning

✓ Performing and Recording

✓ Analyzing and Interpreting

✓ Communicating

Safety Precautions

- Do not drink any of the solutions used in the laboratory.
- Clean up any spills and wash your hands after each trial.
- Benedict's solution is toxic and an irritant. If you get it on your skin or in your eyes, immediately inform your teacher and flush your skin or eyes with clean water.
- Use caution when working around the hot plate and boiling water.
- Be extremely careful around open flames.

Materials

- simulated samples of blood (5)
- simulated samples of urine (5)
- digital blood glucose monitor (if available)
- blood and urine test strips (if using a monitor)
- 10 mL test tubes (10)
- test tube rack
- 10 mL graduated cylinder
- Benedict's solution (if not using a monitor)
- medicine dropper
- 400 mL beaker
- water
- hot plate
- test-tube clamp
- beaker tongs
- Bunsen burner
- cotton swab

Analyzing Endocrine Disorders

In this activity, you will determine which of five hypothetical patients (referred to as patients A, B, C, D, and E) has:

- pituitary gland disorder (limited hGH, epinephrine, and cortisol)
- no hormonal imbalance (is healthy)
- diabetes insipidus
- diabetes mellitus
- Addison's disease

Note: Diabetes insipidus results from insufficient activity of antidiuretic hormone (ADH) and is unrelated to diabetes mellitus.

Pre-Lab Questions

1. What are some of the symptoms of diabetes insipidus?
2. How do the blood glucose levels of a person with diabetes mellitus compare to those of a healthy person?
3. Why is it important to clean up any spills in the laboratory immediately?

Question

Using the information table provided and some simulated blood and urine samples, how can you diagnose hormonal imbalances?

Symptoms of Various Endocrine Imbalances

Patient's Condition	Substances Identified	Blood Levels (mmol/L)	Present or Absent in Urine	Additional Information
Healthy	Glucose	5.0	Absent	No additional symptoms
	Sodium	140	Absent	
Diabetes mellitus	Glucose	25	Present	Person is thirsty and must urinate frequently
	Sodium	138	Absent	
Diabetes insipidus	Glucose	4.5	Absent	Producing large volumes of dilute, pale urine
	Sodium	150	Absent	
Addison's disease	Glucose	4.0	Absent	Person is under stress; urine output is high; there is sodium in the urine
	Sodium	130	Present	
Pituitary gland disorder	Glucose	3.5	Absent	Older person whose glucagon-producing cells have deteriorated
	Sodium	142	Absent	

Source: Data provided by Dr. Edmond A. Ryan, Professor of Medicine, University of Alberta, Medical Director of the Clinical Islet Cell Transplant Program

Procedure

Make a table in your notebook like the one below to record your data.

Patient (A, B, C, D, or E)	Blood Glucose Concentration	Glucose Present or Absent in the Urine	Sodium Present or Absent in the Urine	Name of the Disorder

Part A: Testing for Glucose Concentrations in the Blood and Urine

1. If your school has a glucose monitor, place a drop of the first sample of simulated blood or urine on a clean test strip. Plug the strip into the monitor and take a glucose reading. Record the value that you obtain in your data table. Repeat the procedure for the other samples.
2. If a glucose monitor is not available, you can use the Benedict's test to determine the concentration of glucose in each of the blood samples, and to detect the presence or absence of glucose in the urine. Benedict's solution identifies simple sugars, such as glucose, by causing a colour change. As the concentration of glucose changes, so will the colour of the sample mixed with Benedict's solution, according to the following table.

Benedict's Test Colour Equivalence Table

Colour of Solution	Glucose Concentration (%)	Glucose Concentration (mmol/L)
Blue	0.0	0
Light green	0.1–0.5	5.56–27.8
Olive green	0.5–1.0	27.8–55.6
Yellow	1.0–1.5	55.6–83.3
Orange	1.5–2.0	83.3–111
Red-brown	2.0+	111+

   a. Test the 5 blood samples first. Label 5 test tubes A through E. Use the 10 mL graduated cylinder to measure 5 mL of Benedict's solution into each test tube.

   b. With a medicine dropper, add 5 drops of simulated blood from each of the patient's samples to the appropriately labelled test tube. Rinse out the medicine dropper with clean water between samples.

   c. Fill a 400 mL beaker about two thirds full with water. Place the beaker on a hot plate and turn it on. Allow the water in the beaker to boil. Use test-tube clamps to place the test tubes with the samples and Benedict's solution into the beaker. Leave the test tubes in the beaker until there is a colour change, or a maximum of 5 min.

   d. Use the test-tube clamps to remove the test tubes from the water bath. Record your results.

   e. Next, test the 5 urine samples. Use the procedural steps (a) through (d).

Part B: Testing for Sodium in the Urine (Teacher Demonstration)

3. Have your teacher ignite a Bunsen burner. Your teacher will dip a cotton swab in one of the urine samples, then immediately place the wet end of the swab in the flame. If sodium is present in the urine, the flame should flare bright orange. If not, the flame should stay blue. Record your observations. Your teacher will repeat this step for the remaining urine samples, using a new cotton swab for each sample.

Analyze and Interpret

1. Which patient in this investigation acted as a control?
2. Why were simulated blood and urine samples used in this investigation instead of real samples?
3. Use the table listing the symptoms of different endocrine imbalances to diagnose the condition of each of the patients (A through E).
4. List the hormones that are imbalanced in each of the patients. For each hormone, describe its effect on blood glucose regulation.

Conclude and Communicate

5. For the patient with the pituitary disorder, how would you account for the lack of hGH, epinephrine, and cortisol in the patient's blood? Could another hormone have compensated for these three? If so, how does this other hormone affect the body?

Extend Further

6. **INQUIRY** Identify possible sources of error in your investigation. How could you reduce error if you were to repeat the investigation?
7. **RESEARCH** For each of the hormonal imbalances identified in the investigation, research and suggest a possible treatment.

Thoughtlab INVESTIGATION

9-C

Skill Check

Initiating and Planning

✓ Performing and Recording

✓ Analyzing and Interpreting

✓ Communicating

The Menstrual Cycle

LH and FSH are *pituitary hormones* because they are produced by the pituitary gland. Similarly, progesterone and estrogen are *ovarian hormones* because they are produced in the ovaries. In this investigation, you will see how pituitary and ovarian hormones affect, and are affected by, ovarian and uterine events during the menstrual cycle.

Pre-Lab Questions

1. What are two pituitary hormones that affect the menstrual cycle?
2. What is the effect of estrogen on the menstrual cycle?
3. How does the endometrium change throughout the menstrual cycle?

Question

How do pituitary and ovarian hormones interact with ovarian and uterine events during the menstrual cycle?

Procedure

1. Study the graphs on the next page, and observe how the levels of hormones affect each other as well as the follicle and endometrium.
2. Answer the questions below to help you analyze and interpret the graphs.

Analyze and Interpret

1. During which days of the menstrual cycle does the level of FSH increase? What happens to the follicle during this time?
2. On which day is the level of LH in the bloodstream at its highest? What event occurs immediately after this peak?
3. What event is associated with the decline of LH in the blood?
4. During which days of the cycle does the level of estrogen in the blood increase most rapidly? What happens in the uterus during this time?
5. During which days of the cycle does the level of progesterone in the blood increase most rapidly? What happens in the uterus during this time?
6. During which days of the cycle are the levels of estrogen and progesterone at their lowest? What happens in the uterus during this time?

Conclude and Communicate

7. How do increased levels of estrogen and progesterone appear to affect the level of FSH in the blood?
8. Do the names of the hormones FSH and LH correspond to their functions? Explain your answer.
9. At which time in the menstrual cycle is a woman most fertile? Explain your answer.

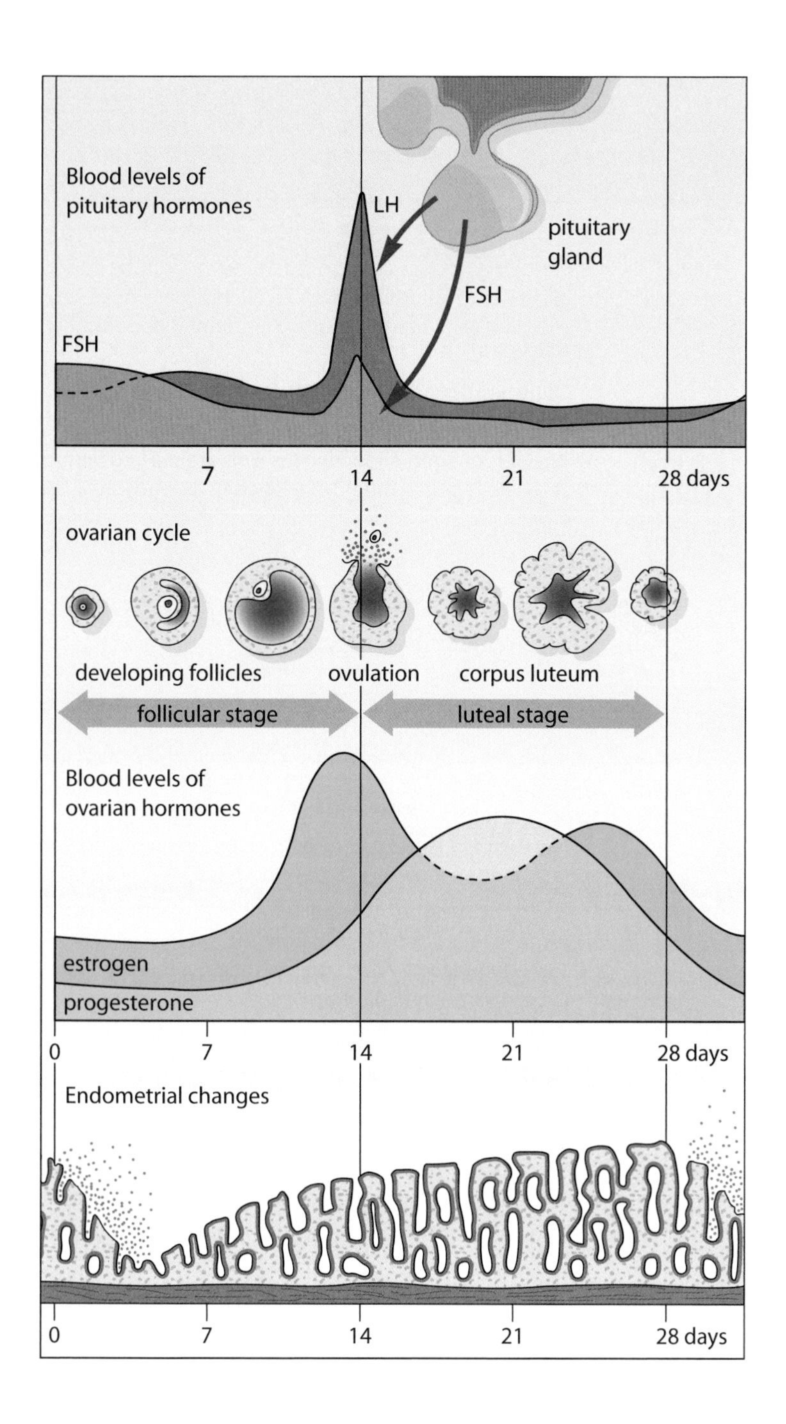

Extend Further

10. INQUIRY Use a graphic organizer to compare and contrast the functions of estrogen and progesterone in the menstrual cycle.

11. RESEARCH Polycystic ovary syndrome (PCOS) is a condition in which the pituitary gland may secrete high levels of LH and the ovaries may overproduce androgens. Research more information about PCOS. What factors may play a role in the development of PCOS? What are the symptoms? How is homeostasis disturbed by this condition?

Chapter 9 SUMMARY

Section 9.1 The Glands and Hormones of the Endocrine System

Hormones and glands comprise the endocrine system.

Key Terms

endocrine gland
hormone
endocrine system
tropic hormone

Key Concepts

- The nervous system rapidly affects specific tissues, to which it is directly connected by neurons. The endocrine system relies on chemical messengers called hormones, which circulate in the blood and have broad, long-lasting effects.
- Steroid hormones can easily diffuse through the lipid bilayer of cell membranes. Water-soluble hormones, such as amino acid–based hormones, cannot diffuse across the cell membrane.
- Many hormones are regulated by negative feedback mechanisms.
- The nervous and endocrine systems are self-regulating and help regulate other body systems, thereby maintaining homeostasis.

Section 9.2 Hormonal Regulation of Growth, Development, and Metabolism

The pituitary gland and the thyroid gland secrete hormones that regulate growth (such as hGH) and cell metabolism (such as T_4).

Key Terms

posterior pituitary
anterior pituitary
human growth hormone (hGH)
thyroid gland
thyroxine (T_4)
hypothyroidism
hyperthyroidism
thyroid-stimulating hormone (TSH)
goitre

Key Concepts

- The hypothalamus controls the pituitary gland. The pituitary gland has two lobes that store and release tropic hormones
- The anterior pituitary gland releases human growth hormone (hGH), which stimulates fat metabolism and targets the liver to release hormones that stimulate protein synthesis and muscle and bone growth.
- The thyroid gland secretes hormones that regulate cell metabolism, growth, and development.
- Thyroxine secretion is regulated by the release of thyroid-stimulating hormone (TSH) from the anterior pituitary. TSH is regulated by negative feedback by thyroxine on the hypothalamus and pituitary.
- The parathyroid glands secrete parathyroid hormone (PTH), which raises blood calcium levels.

Section 9.3 Hormonal Regulation of the Stress Response and Blood Sugar

The hormones released by the adrenal glands are involved in the stress response, and the hormones released by the pancreas regulate blood glucose levels.

Key Terms

adrenal gland
adrenal medulla
epinephrine
short-term stress response
adrenal cortex
long-term stress response
cortisol
adrenocorticotropic hormone (ACTH)
aldosterone
pancreas
islets of Langerhans
beta cell
alpha cell
insulin
glucagon
diabetes mellitus
hyperglycemia

Key Concepts

- In the short-term stress response, also called the fight-or-flight response, the adrenal medulla is stimulated to release epinephrine and norepinephrine.
- When the body is under stress, the hypothalamus secretes a releasing hormone. The releasing hormone stimulates the anterior pituitary to release adrenocorticotropic hormone (ACTH). ACTH in turn stimulates the adrenal cortex to release cortisol, a steroid hormone.
- The islets of Langerhans contain beta cells, which secrete insulin in response to high levels of blood glucose.
- The islets of Langerhans contain alpha cells, which secrete glucagon in response to low levels of blood glucose.
- Glucagon stimulates the liver cells to break down glycogen, which releases glucose.
- If the beta cells are destroyed, type 1 diabetes results. Type 2 diabetes develops when the insulin receptors on the cells do not respond properly to insulin.

Chapter 9 SUMMARY & REVIEW

Section 9.4 Hormonal Regulation of the Reproductive System

Both the male and female reproductive systems are regulated by hormones, including FSH, LH, testosterone, estrogen, and progesterone.

Key Terms

gonad
sex hormone
gonadotropin-releasing hormone (GnRH)
follicle-stimulating hormone (FSH)
luteinizing hormone (LH)
inhibin
testosterone
andropause
estrogen
progesterone
menstrual cycle
menopause
hormone replacement therapy (HRT)

Key Concepts

- Sex hormones work to stimulate the development of male and female reproductive systems and regulate the function of the mature reproductive system.
- In males, the main sex hormone is testosterone. In females, the main sex hormones are estrogen and progesterone.
- In the male reproductive system, a negative feedback hormone system maintains a relatively constant level of sperm production and testosterone.
- In females, hormone systems interact to regulate a monthly menstrual cycle.

Knowledge and Understanding

Select the letter of the best answer below.

1. Which type of hormone stimulates other endocrine glands?
   a. neurotransmitter hormone
   b. negative feedback hormone
   c. positive feedback hormone
   d. trophic hormone
   e. tropic hormone

2. Which of these is a pair of antagonistic hormones?
   a. insulin and glucagon
   b. thyroid-stimulating hormone (TSH) and thyroxine
   c. cortisol and human growth hormone (hGH)
   d. epinephrine and norepinephrine
   e. estrogen and progesterone

3. Which hormone(s) is controlled through negative feedback by high levels of cortisol in the blood?
   a. thyroid-stimulating hormone (TSH) from the anterior pituitary
   b. thyroid-stimulating hormone (TSH) from the hypothalamus
   c. adrenocorticotropic hormone (ACTH) from the anterior pituitary
   d. adrenocorticotropic hormone (ACTH) from the hypothalamus
   e. both thyroid-stimulating hormone (TSH) and adrenocorticotropic hormone (ACTH) from the hypothalamus

4. Which statement is *true*?
   a. Exocrine glands secrete chemical messengers called hormones directly into the bloodstream, which transports the hormones throughout the body.
   b. Compared to the actions of the nervous system, the hormones of the endocrine system have faster and longer-acting effects on a broader range of cell types.
   c. The concentration of hormones in the blood remains constant to maintain homeostasis.
   d. When a hormone binds to a specific receptor in the target cell, it triggers a series of reactions inside the cell.
   e. Steroid hormones bind to a receptor in the plasma membrane while water-soluble hormones pass directly through the target cell's membrane.

5. Which endocrine gland is not involved in secreting a hormone that directly affects blood glucose levels?
   a. pancreas
   b. adrenal medulla
   c. adrenal cortex
   d. parathyroid gland
   e. thyroid gland

6. Which *best* explains acromegaly in adults?
   a. increased production of TSH
   b. increased production of ACTH
   c. increased production of ADH
   d. increased production of hGH
   e. increased production of FSH

Use the diagram to answer questions 7 and 8.

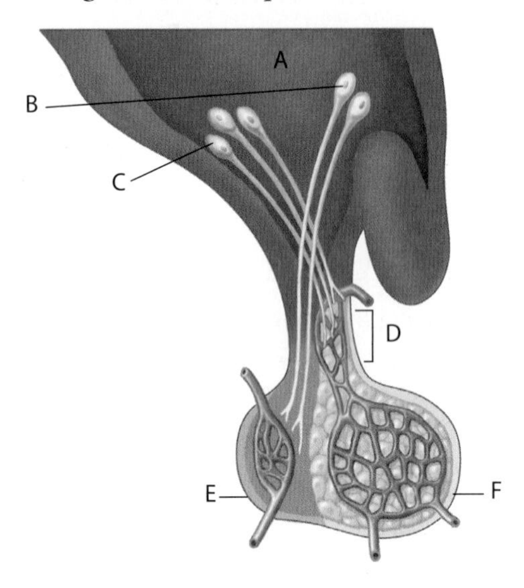

7. Identify structure A and structure F, respectively.
a. hypothalamus and posterior pituitary
b. posterior pituitary and hypothalamus
c. hypothalamus and anterior pituitary
d. anterior pituitary and hypothalamus
e. posterior pituitary and anterior pituitary

8. Which is the function of the structure labelled B?
a. to produce releasing and release-inhibiting hormones
b. to produce antidiuretic hormone and oxytocin
c. to produce thyroid-stimulating hormone (TSH) and adrenocorticotropic hormone (ACTH)
d. to produce prolactin and human growth hormone (hGH)
e. to produce follicle-stimulating hormone (FSH) and luteinizing hormone (LH)

9. Which structure(s) secrete the male hormone testosterone?
a. anterior pituitary
b. interstitial cells of the testes
c. hypothalamus
d. follicle-stimulating cells of the testes
e. luteinizing cells of the testes

10. Which hormone is directly responsible for ovulation in a human female?
a. luteinizing hormone (LH)
b. follicle-stimulating hormone (FSH)
c. estrogen
d. progesterone
e. gonadotropin-releasing hormone (GnRH)

11. Which is an example of an endocrine gland that is not a permanent structure in the human body?
a. hypothalamus
b. corpus luteum
c. testes
d. ovary
e. anterior pituitary gland

Answer the questions below.

12. Compare the general roles of the nervous system and the endocrine system in maintaining homeostasis.

13. Explain why the pituitary gland is sometimes called the "master gland."

14. Using ACTH as an example, explain what a tropic hormone is.

15. Suppose you are tutoring a younger student in biology. The student tells you that the beta cells of the pancreas secrete glucagon, and the alpha cells secrete a hormone that lowers blood glucose levels. Is this statement correct? If not, rewrite it to make it correct.

16. Compare how cortisol, epinephrine, insulin, and glucagon affect blood glucose levels.

17. Assuming there is adequate iodine in the diet, how would secretion of thyroid-stimulating hormone (TSH) affect the production of thyroxine by the thyroid gland? How would increased levels of thyroxine affect the production of TSH?

18. Distinguish between the structure and function of the adrenal medulla and adrenal cortex.

19. How do levels of hGH change as people age? How do the changing levels of hGH affect the body?

20. Explain why glucagon and insulin are considered antagonistic hormones.

21. Compare and contrast the role of norepinephrine in the nervous system with its role in the endocrine system.

22. How does aldosterone help the body cope with an ongoing stressful situation? How is this response different from the fight-or-flight response?

23. In which part of the body are each of the following produced?
a. testosterone
b. progesterone
c. sperm
d. luteinizing hormone
e. follicle-stimulating hormone
f. ovum

24. Compare the levels of sex hormones in the blood of a female before and after the onset of puberty, and describe how these changes affect the reproductive system.

Use the graph below, which shows the average blood concentration of four circulating hormones collected from 50 healthy adult women who were not pregnant, to answer questions 25–27.

Concentration of Hormones

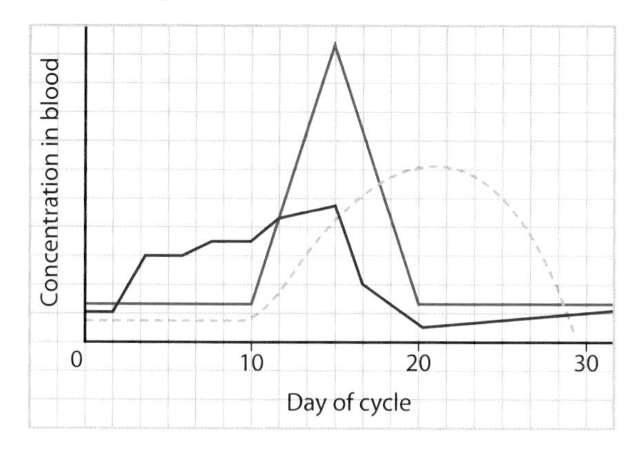

25. Which line represents luteinizing hormone?

26. Which line represents progesterone?

27. Which hormone increases during the last half of the menstrual cycle?

Thinking and Investigation

28. Suppose an early researcher was trying to determine the function of the pituitary. The researcher removed the pituitary from a laboratory animal and observed the effects on the animal. Suggest two or more reasons why this approach probably would have given the researcher confusing results.

29. Studies with rats suggest that overcrowding causes behaviour changes due to increased stress levels.

- **a.** Hypothesize how overcrowding might affect hormone levels in humans.
- **b.** How could you investigate whether people living in cities experience more or less stress than people living in less crowded environments? Identify what the independent and dependent variables would be in such an investigation.
- **c.** Supposing city dwellers were found to be more prone to stress, could you conclusively link this observation to overcrowding? Explain why or why not.

30. BIG IDEAS Environmental factors can affect homeostasis. There is a strong correlation between people with type 2 diabetes and obesity. Identify some societal factors that might be contributing to the rise of type 2 diabetes in countries such as Canada. Suggest how one of these factors could be addressed.

Use the information below to answer questions 31 and 32.

A person can conduct a simple investigation to see if his body has enough iodine to ensure thyroid hormone production. Lugol's iodine solution can be purchased at a pharmacy. A medicine dropper or brush is used to paint an 8 cm stain on the belly or upper thigh. The stain is checked frequently during a 24-hour period. If the stain disappears in less than 24 hours, it means the body absorbed the iodine. The absorption can indicate that the person is iodine deficient.

31. Critique the experimental method used in this investigation.

32. Design an investigation that would produce valid results.

Use the information below to answer questions 33–35.

A rapid breakdown of fat results in elevated levels of ketone bodies in the blood. Ketosis is a condition that is the result of high levels of ketone bodies in the blood. If there are sufficient amounts of ketone bodies in the blood to lower the blood pH, the condition is called ketoacidosis which, when severe enough, can lead to coma and death.

33. Explain why a person with diabetes mellitus would be susceptible to ketosis.

34. Predict the type of test a health-care professional may order to diagnose ketoacidosis.

35. Once ketoacidosis has been diagnosed, infer what other tests may be performed to determine the underlying cause of the disorder.

36. Certain high doses of chemotherapy can damage the ovaries. The ovaries may not function properly due to the damage.

- **a.** Infer how damage to the ovaries would affect hormone release and regulation in a female's body.
- **b.** Predict some possible health effects if the hormones released by the ovaries were no longer secreted.

37. A doctor has a patient with very low levels of thyroxine in the blood but high levels of TSH.

- **a.** Is the person's problem in the thyroid gland or the pituitary gland? Explain your answer.
- **b.** What condition might this imbalance cause?

Use the graph below, which shows the blood glucose levels of two individuals, to answer questions 38–44.

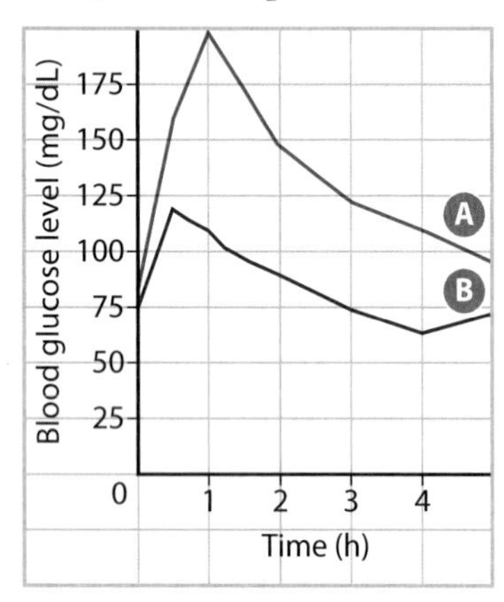

38. Identify what event most likely occurred at time 0.

39. Identify which of these individuals has diabetes and which does not. Explain your reasoning.

40. Explain which hormone the person with diabetes took to cause the drop in glucose from 1 to 5 h.

41. What caused the drop in glucose from 1 to 5 h in the person without diabetes?

42. If both people exercised heavily at 5 h, what would you predict would happen to their blood glucose levels?

43. Identify the hormone that would be released to regulate the blood glucose levels at 5 h.

44. Identify the substance the person with diabetes could take following [illegible] the blood glucose to a healthy level.

Communication

45. In 1996, scientists at the Mahidol University of Bangkok, in Thailand, conducted a study on the effects of aloe vera on high blood sugar levels. Research information about this study and any others that have examined the potential effectiveness of aloe vera as a treatment for diabetes. Summarize the results of the investigations in a short essay. Include your opinion on whether aloe vera is an effective treatment for diabetes.

46. Some parents are asking their doctors to prescribe synthetic human growth hormone (hGH) treatments to their children, to improve their children's chance of winning sports scholarships for university. Write a short essay that includes the following points.

a. Infer the reasoning behind this request.

b. State whether you think this practice should be allowed in school-sponsored sports.

c. Provide justification for your response.

47. Using a graphic organizer, compare and contrast type 1 diabetes and type 2 diabetes.

48. Using the headings in the example table below, make a similar table in your notebook. In your table, write the name of each endocrine gland indicated on the following diagram by a letter. Complete the table, and include the hormonal imbalances associated with each gland, including diabetes mellitus, diabetes insipidus, acromegaly, hyperthyroidism, hypothyroidism, Addison's disease, and goitre.

Diagram Letter	Name of Hormonal Imbalance	Endocrine Gland or Glands Involved	Hormones Involved	Symptoms of the Condition

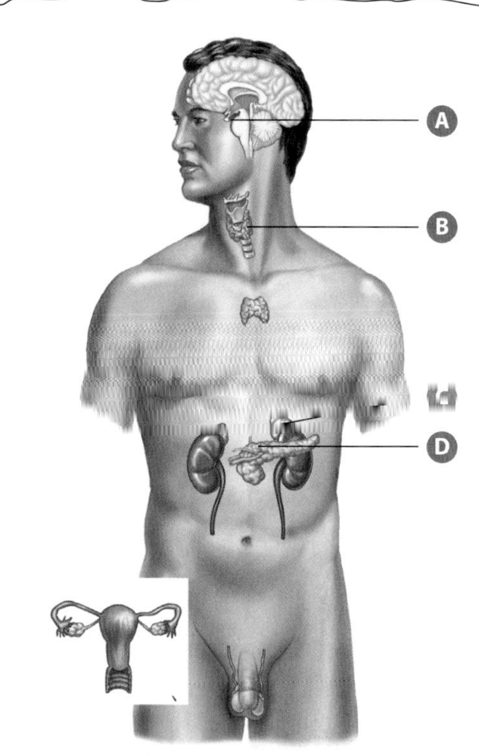

49. High cholesterol can disrupt homeostasis.

a. Research the effects of high cholesterol on homeostasis.

b. Research a class of medications called statins that are used to treat high cholesterol. What are the possible side effects of statin medications?

c. Complete risk-benefit analysis for taking a statin for high cholesterol.

d. Create a 1-page information sheet about high cholesterol, its health effects and treatments. Include the results of your risk analysis about statin.

50. BIG IDEAS Systems that maintain homeostasis rely on feedback mechanisms. Copy the following flowchart into your notebook. Label the flowchart "Regulation of ACTH." On the flowchart, identify the following.

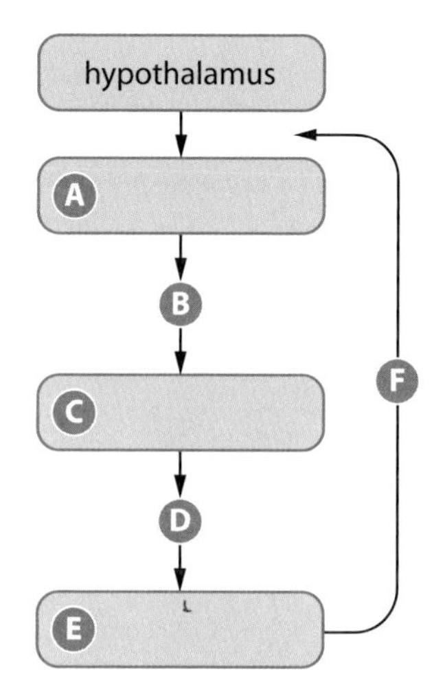

a. lobe of the pituitary affected
b. hormone released from the pituitary gland
c. endocrine gland affected
d. hormone released from this endocrine gland
e. effects on the body systems and tissues
f. what regulates the hormone

51. Summarize your learning in this chapter using a graphic organizer. To help you, the Chapter 9 Summary lists the Key Terms and Key Concepts. Refer to Using Graphic Organizers in Appendix A to help you decide which graphic organizer to use.

Application

52. Suppose you are lost in concentration while studying biology. Suddenly, the phone rings. Outline the physiological changes that occur in your body due to release of the stress hormones cortisol and epinephrine. What triggers the release of each of these hormones?

53. An individual is playing hockey without a helmet. She receives a severe blow to the head that causes severe damage to the anterior pituitary gland. List all the hormones that might be affected, and how this might affect the body in each case.

54. A tumour (overgrowth of cells) in an endocrine gland can sometimes cause the gland to become overactive. What hormonal effects might occur in someone with a tumour in the adrenal gland?

Use the information below to answer questions 55 and 56.

A patient is sluggish, depressed, and intolerant to the cold. After eliminating several other possible causes, he is diagnosed with a hormonal imbalance.

55. What condition would produce these symptoms?

56. Infer the endocrine gland that is probably involved.

Use the information below to answer questions 57 and 58.

A person reports having extreme thirst and fatigue, drinks water almost constantly, and urinates a great deal.

57. Name two hormonal imbalances that could produce these symptoms.

58. How could a doctor determine what the disorder is?

59. Would a person with diabetes mellitus be likely to require more insulin or more sugar following strenuous exercise? Explain your answer.

60. Young female athletes often experience delayed puberty. What would you expect to find if you compare the blood hormone levels of a 14-year-old gymnast with those of less athletic girls of the same age?

61. Some male athletes take anabolic steroids to enhance their performance. These substances mimic the action of testosterone. The side effects of steroid use can include changes associated with an increase in sex hormones, such as increased muscular development and aggressiveness. However, side effects can also include changes associated with a decrease in sex hormones, such as the shrinking of testicles and the loss of facial hair. Using hormone feedback systems, explain how a drug that mimics the effect of testosterone could have these contradictory effects on the male body.

62. The speed skaters in **Figure 9.1** maintain their balance while flying over the ice at speeds around 50 km/h. If one of these skaters falls while rounding the corner, he may cause the others to fall as well. Make an analogy between this situation and the feedback systems that regulate hormones in the human body.

63. BIG IDEAS Organisms have strict limits on the internal conditions that they can tolerate. Cushing's syndrome occurs when the body's tissues are exposed to high levels of cortisol for too long. Many people develop Cushing's syndrome because they take glucocorticoids—steroid hormones that are chemically similar to naturally produced cortisol—such as prednisone for asthma, lupus, and other inflammatory diseases. List five symptoms a person with Cushing's syndrome might display.

Select the letter of the best answer below.

1. K/U Which is a ductless gland that secretes hormones directly into the bloodstream?
   a. an exocrine gland
   b. an endocrine gland
   c. an autonomic gland
   d. a salivary gland
   e. a neurosecretory gland
2. K/U Insufficient production of which hormone during childhood results in pituitary dwarfism?
   a. insulin
   b. glucagon
   c. thyroid-stimulating hormone (TSH)
   d. human growth hormone (hGH)
   e. adrenocorticotropic hormone (ACTH)
3. K/U Hypothyroidism that developed in childhood would be characterized by which of the following?
   a. increased blood pressure and increased heart rate
   b. impaired physical and mental development
   c. thirst and consumption of large volumes of water
   d. suppression of the immune system and the development of a goitre
   e. the development of a person of extremely small stature as an adult
4. K/U Which is NOT involved in metabolic processes?
   a. insulin
   b. glucagon
   c. thyroxine
   d. human growth hormone
   e. parathyroid hormone (PTH)
5. K/U Which correctly matches the endocrine gland to the hormone it secretes?
   a. posterior pituitary – ADH
   b. adrenal cortex – epinephrine
   c. pancreas – insulin
   d. anterior pituitary – oxytocin
   e. thyroid gland – TSH
6. K/U Which is caused by severe hypothyroidism?
   a. Addison's disease
   b. cretinism
   c. hyperglycemia
   d. type 1 diabetes
   e. type 2 diabetes
7. K/U Which of the following hormones is NOT produced by the anterior pituitary gland?
   a. human growth hormone (hGH)
   b. thyroid-stimulating hormone (TSH)
   c. follicle-stimulating hormone (FSH)
   d. luteinizing hormone (LH)
   e. antidiuretic hormone (ADH)
8. K/U Which hormone is primarily responsible for the production of sperm in the testes?
   a. inhibin
   b. follicle-stimulating hormone (FSH)
   c. testosterone
   d. luteinizing hormone (LH)
   e. adrenocorticotropic hormone (ACTH)

Use the diagram below to answer questions 9 and 10.

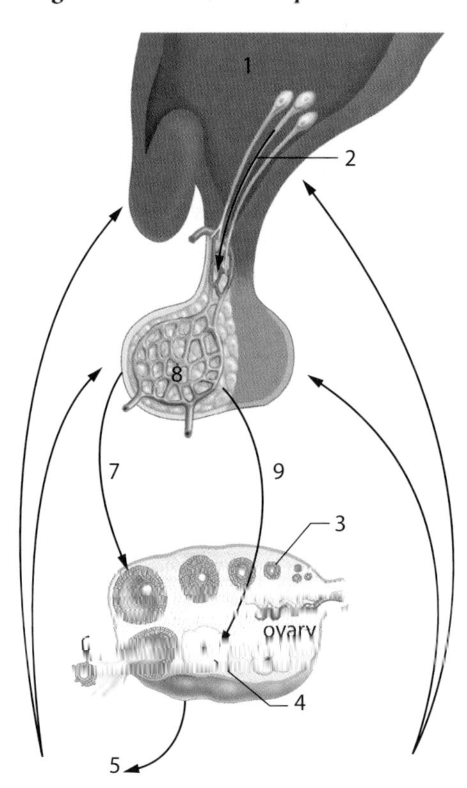

9. K/U Follicle-stimulating hormone (FSH) and luteinizing hormone (LH) are labelled __ and __, respectively.
   a. 7 and 6
   b. 1 and 2
   c. 9 and 3
   d. 4 and 5
   e. 7 and 9
10. K/U The hormones GnRH and estrogen are labelled __ and __, respectively.
    a. 7 and 6
    b. 1 and 2
    c. 9 and 3
    d. 2 and 5
    e. 8 and 4

Use sentences and diagrams as appropriate to answer the questions below.

11. K/U How are the actions of the endocrine system different from the actions of the nervous system?

12. C Using a graphic organizer, summarize the action of a water-soluble hormone such as epinephrine.

13. C Write a paragraph explaining how the hypothalamus controls the production and secretion of hormones from the anterior pituitary gland.

14. T/I Higher than normal thyroid hormone levels can occur from taking too much thyroid hormone medication, a condition referred to as factitious hyperthyroidism. For example, taking thyroid medication to lose weight is extraordinarily dangerous and ineffective in the long term. Infer why a heart attack could result if factitious hyperthyroidism remains untreated.

15. C Using a diagram, describe the adrenal glands. Write a caption for your diagram.

16. K/U Explain how the hypothalamus is involved in the release of epinephrine and norepinephrine in the stress response.

17. C Use a table to compare and contrast the major physiological changes that occur in the short-term and long-term stress responses.

18. A Addison's disease is a disorder characterized by insufficient production of glucocorticoids. The symptoms of adrenal insufficiency usually begin gradually. One of the most common symptoms is chronic, worsening fatigue. Explain why a person with Addison's disease would suffer from chronic fatigue.

19. T/I Certain endocrine disorders, such as Cushing's syndrome, can be caused by excessive secretion of a hormone, in this case ACTH, by the pituitary gland, or by a problem with the adrenal gland itself. If you were able to measure the ACTH levels of a Cushing's patient, how could you tell the difference between a pituitary problem and an adrenal gland problem?

20. A Because some of their functions overlap, why it is necessary to have both a nervous system and an endocrine system?

21. A A fasting blood sugar level of 7 mmol/L or higher usually indicates an individual has diabetes mellitus. To confirm this when the blood sugar level is borderline or slightly elevated, a fasting glucose tolerance test (GTT) is performed. The individual is asked to fast overnight. In the morning, a fasting blood glucose level is recorded. The individual then drinks a very concentrated glucose solution, and blood glucose levels are sampled at several intervals for the next 3 hours. A healthy range for blood glucose levels is between 4.5–5.0 mmol/L.

Blood Glucose Concentration (mmol/L)					
	0 min	30 min	1 hr	2 hr	3 hr
Patient 1	4.7	7.8	8.4	5.6	4.5
Patient 2	8.4	12.3	13.7	14.3	14.5

a. Plot both sets of data on the same graph and label your graph appropriately.

b. How does the fasting blood glucose level differ in these two patients?

c. Predict which of these patients is diabetic and which one is not diabetic.

d. When blood glucose levels drop back to baseline levels, where has the glucose gone?

e. Suggest a medication that the patient with diabetes should take to help her blood glucose levels remain healthy. Explain how this treatment would work.

22. C Using a table, identify the five male sex hormones and briefly describe their significance.

23. K/U List the steps involved in the follicular stage of the ovarian cycle.

24. T/I Hypoparathyroidism occurs when the parathyroid glands do not produce enough parathyroid hormone. The decrease in blood calcium that results causes neurons and muscle cells to depolarize and propagate action potentials spontaneously. Infer a symptom a patient with hypoparathyroidism might report.

25. A Cancer of the pituitary gland can affect any of its normal functions. Why might these tumours be difficult to diagnose?

Self-Check

If you missed question...	1	2	3	4	5	6	7	8	9	10	11	12	13	14	15	16	17	18	19	20	21	22	23	24	25
Review section(s)...	9.1	9.2	9.2	9.1	9.1	9.2	9.3	9.4	9.4	9.4	9.1	9.1	9.1	9.2	9.3	9.3	9.3	9.3	9.3	9.4	9.4	9.2	9.2	9.3	9.1

CHAPTER

10 Excretion and the Interaction of Systems

Specific Expectations

In this chapter, you will learn how to...

- E1.1 **assess**, on the basis of findings from a case study, the effects on the human body of taking chemical substances to enhance performance or improve health (10.3)
- E1.2 **evaluate**, on the basis of research, some of the human health issues that arise from the impact of human activities on the environment (10.3)
- E2.1 **use** appropriate terminology related to homeostasis (10.1, 10.2, 10.3)
- E2.2 **plan** and **construct** a model to illustrate the essential components of the homeostatic process (10.2)
- E2.3 **plan** and **conduct** an investigation to study a feedback system (10.2, 10.3)
- E3.1 **describe** the anatomy and physiology of the endocrine, excretory, and nervous systems, and explain how these systems interact to maintain homeostasis (10.1, 10.2, 10.3)
- E3.3 **describe** the homeostatic processes involved in maintaining water, ionic, thermal, and acid–base equilibrium, and **explain** how these processes help body systems respond to both a change in environment and the effects of medical treatments (10.2, 10.3)

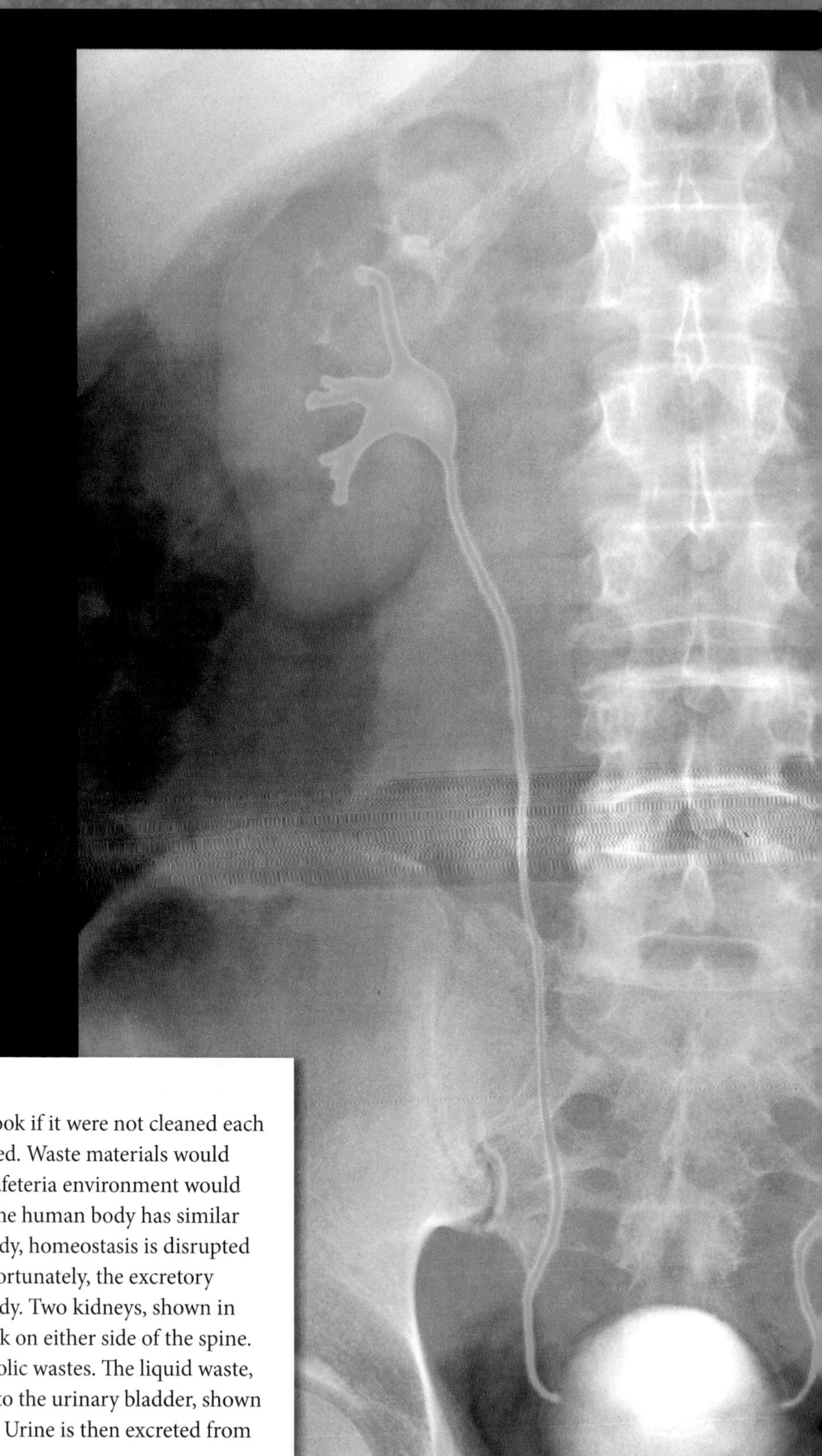

Imagine how your school's cafeteria would look if it were not cleaned each day and if the garbage cans were left unemptied. Waste materials would begin to build up over time. Eventually, the cafeteria environment would become unhealthy. The buildup of wastes in the human body has similar effects. If wastes are not removed from the body, homeostasis is disrupted and serious illness or even death can result. Fortunately, the excretory system performs this vital function for the body. Two kidneys, shown in blue in the photo, are located in the lower back on either side of the spine. The kidneys filter the blood, removing metabolic wastes. The liquid waste, called urine, is transported from the kidneys to the urinary bladder, shown in pink, through two ureters, shown in green. Urine is then excreted from the body.

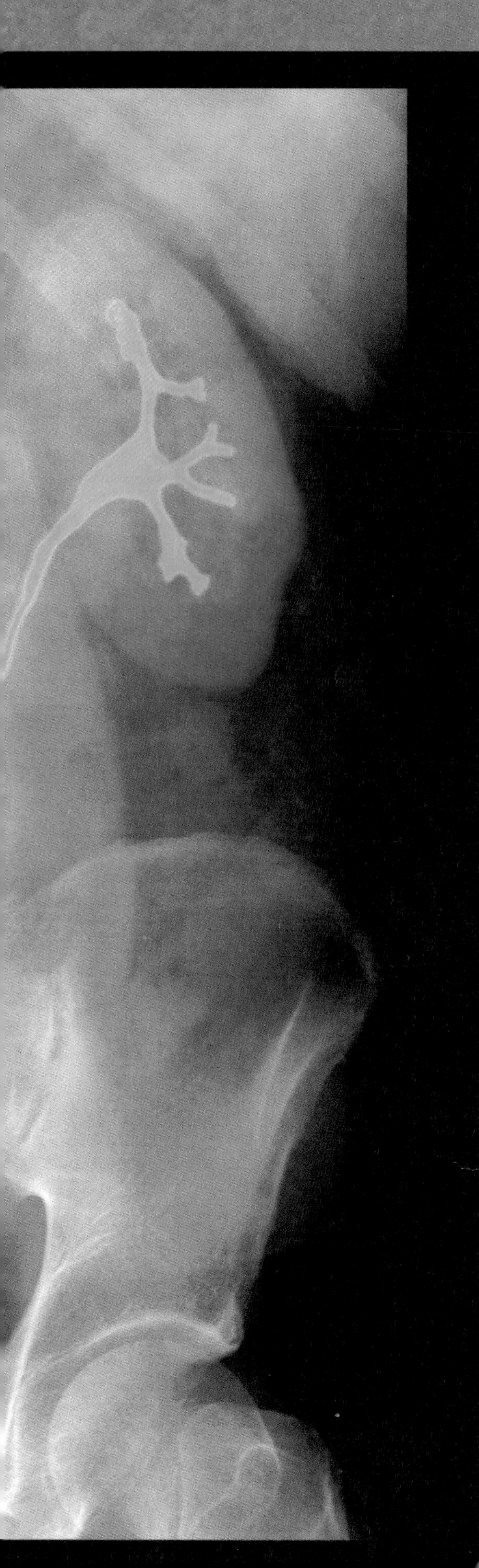

Launch Activity

Dehydration and Urine Colour

Athletes whose sweat loss exceeds fluid intake may become dehydrated during activity. Even slight dehydration (a 1–2 percent loss in body weight) has a negative effect. A doctor of sports medicine, L.E. Armstrong, devised a standardized reference chart for urine colour. Athletes can use this chart to determine their hydration. People who are well-hydrated produce urine that is "very pale yellow," "pale yellow," or "straw coloured."

Safety Precautions

- Wash your hands upon completion of the lab activity.

Materials

- 3 test tubes of simulated urine sample
- urine colour rating chart
- unlined white paper

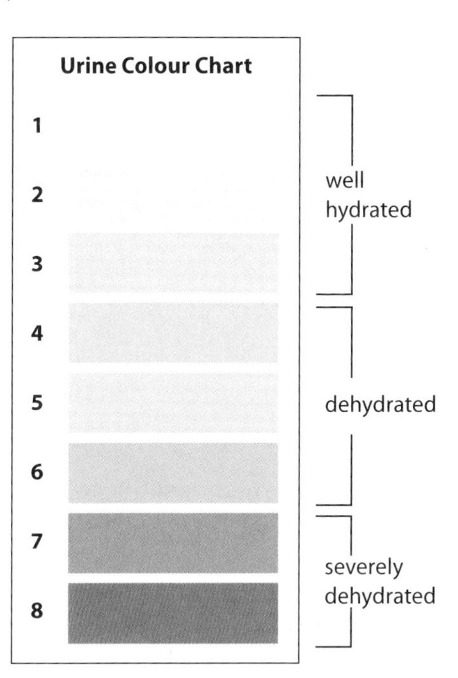

Procedure

1. Draw a data table in your notebook similar to the one shown below.

Urine Colour Results

Simulated Urine Sample	Colour Number	Interpretation
1		
2		
3		

2. Gather the materials listed.
3. Hold each sample in front of the unlined white paper.
4. Match the sample to the Urine Colour Chart.

Questions

1. Based on your observations, infer which sample(s) would indicate a person who is well-hydrated. Which sample indicates a person who is poorly hydrated?
2. Why would it be important for athletes to be able to assess their hydration before, during, and after a game or practice?

SECTION 10.1

Overview of the Excretory System

Key Terms

excretion
excretory system
urine
kidney
ureter
urinary bladder
urethra
renal artery
renal vein
nephron
Bowman's capsule
glomerulus
filtrate
collecting duct

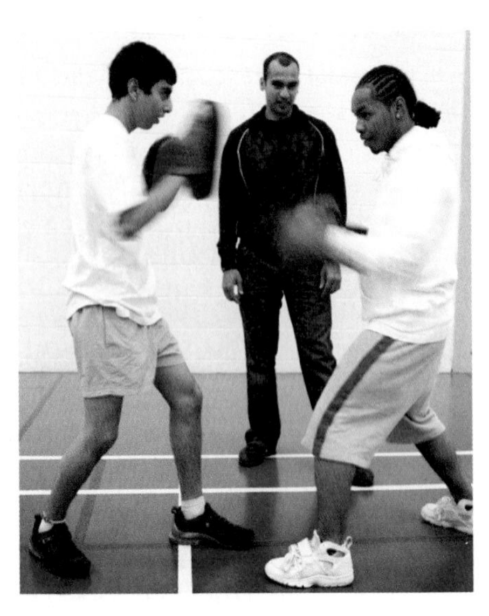

Figure 10.1 Everyday activities—including eating, exercising, and taking a test—all affect homeostasis. The foods you consume may lead to changes in the pH of your blood. Perspiring as you exercise can change the osmotic pressure of your blood. The stress of taking a test can affect hormones whose target cells are in the kidneys. The excretory system helps return these variables and others to within a normal range. Like other body systems, the excretory system plays an important role in maintaining homeostasis.

excretion process of separating wastes from body fluids and eliminating them from the body, performed by several body systems, including the respiratory, integumentary, digestive, and excretory systems

excretory system in animals, the system that regulates the volume and composition of body fluids by excreting metabolic wastes and recycling some substances for reuse; main organs include the kidneys, ureters, bladder, and urethra; also known as the urinary system

urine in the kidneys, filtrate of the nephron upon leaving the collecting duct; exits the body through the urethra

As you go about your daily activities, such as those shown in **Figure 10.1**, your cells are provided with nutrients and oxygen by the activities of the digestive and respiratory systems. The circulatory system ensures that each of the trillions of cells in the body receives the substances it needs for metabolic activities. These activities—energy release, maintenance, and repair—result in waste products that change the balance of volume of water and the concentration and composition of dissolved substances in the body's fluids.

Excretion is the process of separating wastes from body fluids, then eliminating the wastes from the body. Several body systems perform this function. The respiratory system excretes carbon dioxide and small amounts of other gases, including water vapour. The skin excretes water, salts, and some urea in perspiration. The digestive system excretes water, salts, lipids, and a variety of cellular chemicals. Note that the elimination of food residue—feces—is *not* considered to be a process of excretion. Most metabolic wastes are dissolved or suspended in solution and are excreted by the **excretory system** (also called the urinary system).

Functions of the Excretory System

The excretory system produces **urine** and conducts it to outside the body. As the kidneys produce urine, they carry out the following four functions that contribute to homeostasis:

- **Excretion of Metabolic Wastes** The kidneys excrete metabolic wastes, notably nitrogenous (nitrogen-containing) wastes. Nitrogenous wastes include ammonia, urea, and uric acid. Ammonia is highly toxic but is converted in the liver to the less toxic compound urea. Urea makes up the majority of nitrogenous waste in the body, and about half of it is eliminated in urine. Uric acid is present in much lower concentrations, and is contained in urine.
- **Maintenance of Water–Salt Balance** Another important function of the kidneys is to maintain the appropriate balance of water and salt in the blood. Blood volume is closely tied to the salt balance of the body. By regulating salts in the blood, the kidneys are also involved in regulating blood pressure. The kidneys also help maintain the appropriate level of potassium (K^+), bicarbonate (HCO_3^-), and calcium (Ca^{2+}) in the blood.

- **Maintenance of Acid–Base Balance** The kidneys regulate the acid–base balance of the blood. The kidneys monitor and help keep the blood pH at about 7.4, mainly by excreting hydrogen ions (H^+) and reabsorbing the bicarbonate ions (HCO_3^-) as needed. Human urine usually has a pH of 6 or lower because our diet often contains acidic foods.
- **Secretion of Hormones** The kidneys assist the endocrine system in hormone secretion. The kidneys secrete two hormones: calcitriol and erythropoietin. Calcitriol is the active form of vitamin D. Vitamin D promotes calcium (Ca^{2+}) absorption from the digestive tract. Erythropoietin, which stimulates the production of red blood cells, is released in response to increased oxygen demand or reduced oxygen-carrying capacity of the blood. The kidneys also secrete renin, a substance that leads to the secretion of the hormone aldosterone from the adrenal cortex.

The Organs of the Excretory System

As shown in **Figure 10.2**, the human excretory system consists of the kidneys and ureters, the urinary bladder, and the urethra. Two fist-sized **kidneys** are located in the area of the lower back on each side of the spine. If you stand up and put your hands on your hips with your thumbs meeting over your spine, your kidneys lie just above your thumbs. A large cushion of fat usually surrounds the kidneys. This fat layer, along with the lower portion of the ribcage, offers some protection for these vital organs. Although most people have two kidneys, humans are capable of functioning with only one. If one kidney ceases to work or if a single kidney is removed due to disease or because it is being donated to someone in need of a kidney, the single kidney increases in size to handle the increased workload.

The kidneys release urine into two muscular, 28-cm-long tubes called **ureters**. From the ureters, urine is moved by the peristaltic actions of smooth muscle tissue to the muscular **urinary bladder** where it is temporarily stored. Drainage from the bladder is controlled by two rings of muscles called sphincters. Both sphincters must relax before urine can drain from the bladder. The innermost sphincter is involuntarily controlled by the brain. During childhood we learn to voluntarily control relaxation of the other sphincter.

Urine exits the bladder and the body through a tube called the **urethra**. In males, the urethra is approximately 20 cm long and merges with the ductus deferens of the reproductive tract to form a single passageway to the external environment. In females, the urethra is about 4 cm long and the reproductive and urinary tracts have separate openings.

kidney in vertebrates, one of a pair of organs that filters waste from the blood (which is excreted in urine) and adjusts the concentrations of salts in the blood

ureter in mammals, a pair of muscular tubes that carry urine from the kidneys to the bladder

urinary bladder organ where urine is stored before being discharged by way of the urethra

urethra the tube through which urine exits the bladder and the body

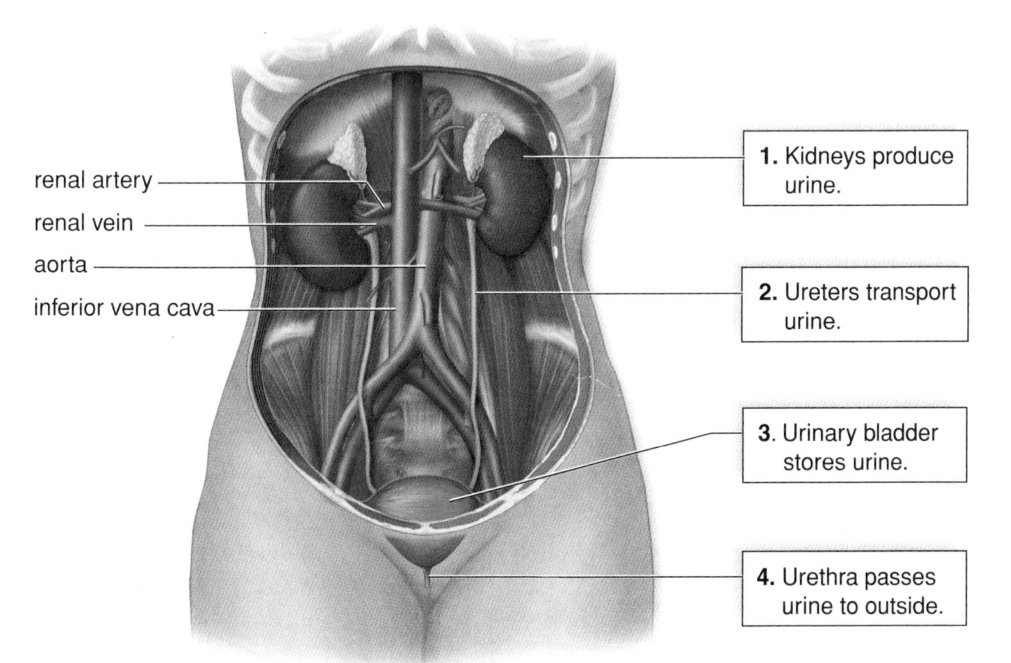

Figure 10.2 The organs of the excretory system are two kidneys, two ureters, the urinary bladder, and the urethra. Some vessels of the circulatory system also are shown in this illustration.

Infer *Although the vessels of the circulatory system are not part of the excretory system, they are intimately connected with it. Why might that be?*

The Kidneys

renal artery blood vessel that originates from the aorta and delivers blood to the kidneys; splits into a fine network of capillaries (the glomerulus) within the Bowman's capsule of the nephron

renal vein blood vessel that drains from the kidney; returns to the body the solutes and water reabsorbed by the kidney

nephron microscopic tube-like filtration unit found in the kidneys that filters and reabsorbs various substances from the blood; produces urine

As illustrated in **Figure 10.3**, the kidneys are bean shaped and reddish-brown in colour. The concave side of each kidney has a depression where a **renal artery** enters and a **renal vein** and a ureter exit the kidney. A lengthwise section of a kidney shows that many branches of the renal artery and renal vein reach inside a kidney.

A kidney has three regions. The renal cortex is an outer layer that dips down into an inner layer called the renal medulla. As shown in **Figure 10.3**, the renal medulla contains cone-shaped tissue masses. The renal pelvis is a central space, or cavity, that is continuous with the ureter.

Embedded within the renal cortex and extending into the renal medulla are more than one million microscopic structures called **nephrons**. Closely associated with these nephrons is a network of blood vessels. The nephrons are responsible for filtering various substances from blood, transforming it into urine. To perform this function, each nephron is organized into three main regions: a filter, a tubule, and a collecting duct. These regions are highlighted in **Figure 10.4** and discussed in further detail below the figure.

Figure 10.3 The view in **A** includes some blood vessels to reinforce the connection between the circulatory and excretory systems. The views in **B** and **C** do not include blood vessels, and identify the three regions of the kidney: renal cortex, renal medulla, and renal pelvis. The view in **C** introduces the functional unit of the kidney, the nephron, which you will examine in greater detail in **Figure 10.4**.

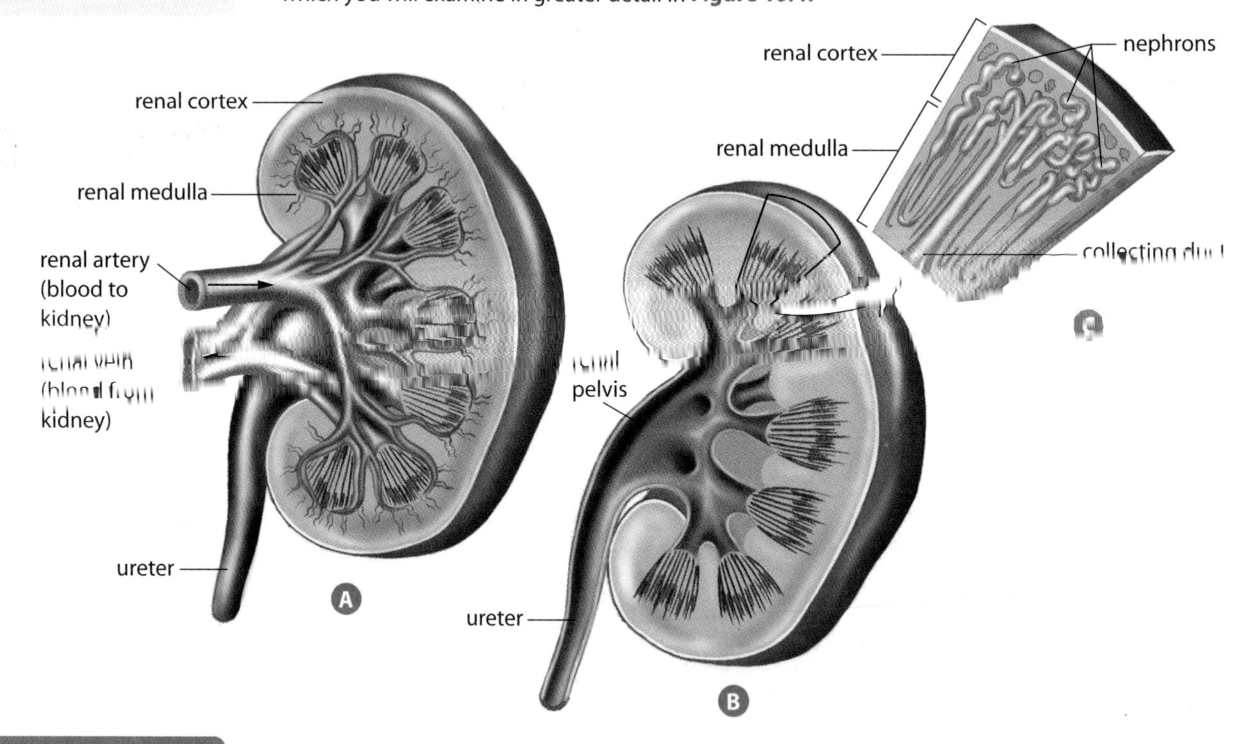

Learning Check

1. Identify four systems in the human body that are involved in the excretion of wastes and briefly explain the role of each system.
2. Explain why is it necessary for wastes to be removed from the body.
3. Explain the relationship between the kidneys and the endocrine system.
4. Use a flowchart to trace the flow of urine from the kidney to the external environment.
5. Infer why humans cannot control the outermost sphincter in the urinary bladder for the first few years of life.
6. Write a paragraph explaining how the circulatory and excretory systems interact to maintain homeostasis.

An Overview of the Nephron and Its Three Functional Regions

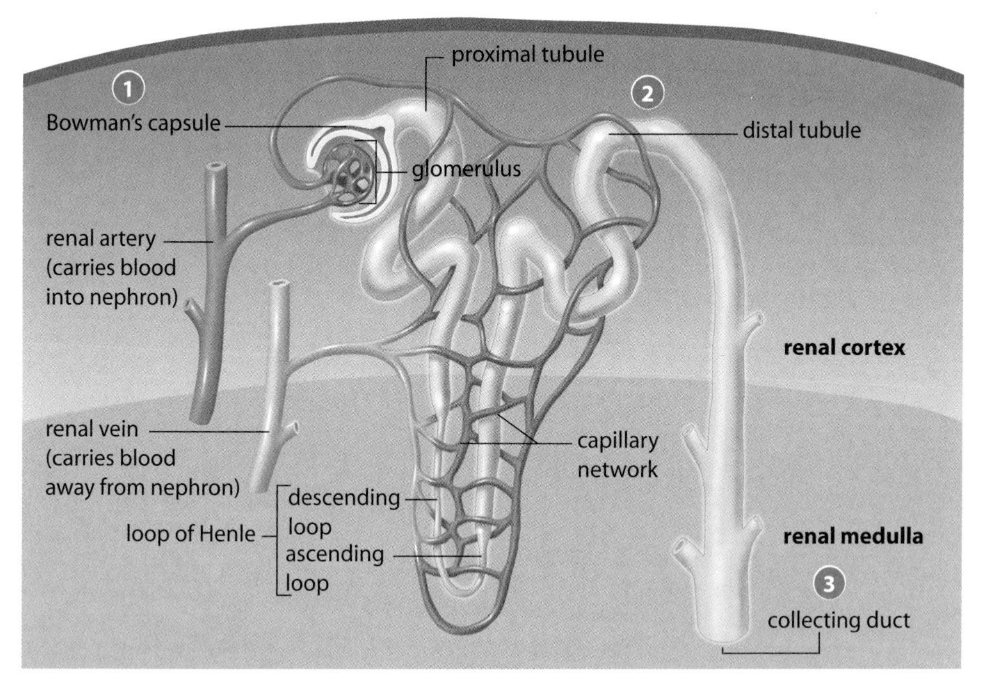

Figure 10.4 The structures of the nephron are labelled here to help outline the processes by which blood becomes urine in the nephron. The word "proximal" (in proximal tubule) means *nearby* and refers to the fact that this part of the tubule is located *near* the Bowman's capsule. The word "distal" (in distal tubule) means *distant* and refers to the fact that this part of the tubule is located more *distantly* from the Bowman's capsule.

***Identify** the sequence of structures through which the filtrate passes, beginning with Bowman's capsule.*

1. A Filter: The filtration structure at the top of each nephron is a cap-like formation called the **Bowman's capsule**. Within each capsule, the renal artery enters and splits into a fine network of capillaries called a glomerulus [pronounced glow-MEER-you-lus] (the term means "little ball" in Latin). The walls of the **glomerulus** act as a filtration device. They are impermeable to proteins, other large molecules, and red blood cells, so these remain within the blood. Water, small molecules, ions, and urea—the main waste products of metabolism—pass through the walls and proceed further into the nephron. The filtered fluid that proceeds from the glomerulus into the Bowman's capsule of the nephron is referred to as **filtrate**.
2. A Tubule: The Bowman's capsule is connected to a small, long, narrow tubule that is twisted back on itself to form a loop. This long hairpin loop is a reabsorption device. The tubule has three sections: the proximal tubule, the loop of Henle, and the distal tubule. Like the small intestine, this tubule absorbs substances that are useful to the body, such as glucose and a variety of ions, from the filtrate passing through it. Unlike the small intestine, this tubule also secretes substances into the tissues surrounding it. You will find out more about these twin processes of reabsorption and secretion in Section 10.2.
3. A Duct: The tubule empties into a larger pipe-like channel called a **collecting duct**. The collecting duct functions as a water-conservation device, reclaiming water from the filtrate passing through it so that very little water is lost from the body. The filtrate that remains in the collecting duct is a suspension of water and various solutes and particles. It is now called urine. Its composition is distinctly different from the fluid that entered the Bowman's capsule. The solutes and water reclaimed during reabsorption are returned to the body via the renal vein.

A more detailed account of the nephron's function in forming urine follows in Section 10.2.

Bowman's capsule in the kidney, the cap-like formation at the top of each nephron that serves as a filtration structure; surrounds the glomerulus

glomerulus in the kidney, a fine network of capillaries within the Bowman's capsule of the nephron; arising from the renal artery, the walls of the glomerulus act as a filtration device

filtrate in the kidney, filtered fluid that proceeds from the glomerulus into the Bowman's capsule of the nephron

collecting duct in the kidney, the large pipe-like channel arising from the tubule connected to the Bowman's capsule in the nephron; functions as a water-conservation device, reabsorbing water from the filtrate in the nephron

Section 10.1 Review

Section Summary

- The human excretory system is responsible for removing liquid waste from the body. The excretory system also regulates the acid–base balance and water–salt balance of the blood and secretes some hormones.
- The kidneys are the primary excretory organs. Kidney tissue is organized into three regions: the renal cortex, the renal medulla, and the renal pelvis.
- The kidneys are composed of millions of functional units called nephrons that filter the waste from the blood and produce urine.
- Urine leaves the kidney through tubes called ureters. It passes into the bladder and exits the body through the urethra.

Review Questions

1. K/U Describe the functions of the excretory system and explain how each function contributes to homeostasis.
2. K/U Identify four examples of metabolic wastes produced in the human body.
3. C In a paragraph, summarize the relationship among the human digestive, respiratory, circulatory, and excretory systems.
4. K/U Which is the filtrate of the nephron that exits the body through the urethra?
   a. salt
   b. sweat
   c. urea
   d. urine
   e. water
5. T/I Blood doping is the practice of boosting the number of red blood cells (RBC's) in the bloodstream in order to enhance athletic performance. Some unethical athletes have been caught taking hormone injections to increase the number of red blood cells. Predict the role the kidneys would play in this scenario.
6. K/U The diagram below shows a cross section of a kidney.

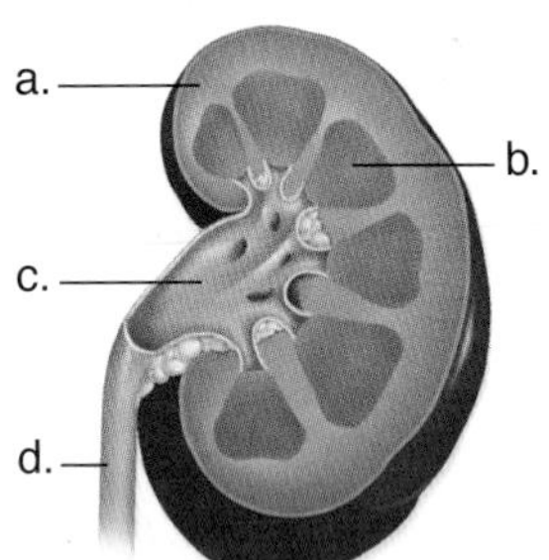

   a. Identify the regions of the kidney shown in the diagram.
   b. Identify the region(s) of the kidney where you would find the following structures:
      i. glomerulus
      ii. proximal tubule
      iii. loop of Henle
      iv. distal tubule
      v. collecting duct
7. C Sketch the basic plan of the nephron. On your sketch indicate the part of the nephron where:
   a. blood enters
   b. filtrate is formed
   c. urine is excreted
8. K/U Distinguish the proximal tubule from the distal tubule of the nephron.
9. A The following diagram summarizes how the parasympathetic nervous system and the somatic nervous system regulate urination. Based on this information, predict the impact of a spinal cord injury (SCI) on an individual's ability to control urination.

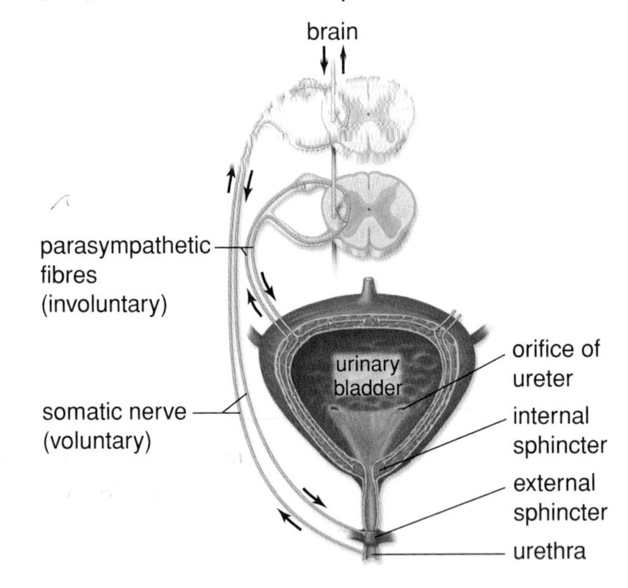

10. T/I Normally, as the urinary bladder fills with urine, nerve signals are sent to the brain, resulting in the feeling that you have to urinate. During urination, the muscles of the bladder contract, forcing urine out of the body. Overactive bladder is a condition in which the muscles of the bladder involuntarily contract, making a person feel as if he or she has the need to urinate, even when the urinary bladder is not full. Predict how medications used to treat overactive bladder would work.

SECTION 10.2

Urine Formation in the Nephron

Refer again to **Figure 10.4**. Note that the upper portions of each nephron are located in the renal cortex of the kidney, while the lower portions are located in the renal medulla of the kidney. Note also the presence of vessels of the circulatory system in association with the nephrons. These details indicate that nephrons are surrounded by the tissues of the renal cortex and the renal medulla. Nephrons are also closely associated with a network of blood vessels that spreads throughout this surrounding tissue. Thus, any substances that are secreted from the nephrons enter the surrounding tissues of the kidney. Most of these substances return to the bloodstream through the network of blood vessels. The remainder leave the body in the form of urine.

Key Terms

glomerular filtration
tubular reabsorption
tubular secretion
water reabsorption
proximal tubule
loop of Henle
distal tubule

How Urine Forms

Four processes are crucial to the formation of urine. These processes are outlined below.

- **Glomerular filtration** moves water and solutes, except proteins, from blood plasma into the nephron. Recall that this filtered fluid is called filtrate.
- **Tubular reabsorption** removes useful substances such as sodium from the filtrate and returns them into the blood for reuse by body systems.
- **Tubular secretion** moves additional wastes and excess substances from the blood into the filtrate.
- **Water reabsorption** removes water from the filtrate and returns it to the blood for reuse by body systems.

glomerular filtration in the kidney, process that results in the movement of water and solutes, except proteins, from the blood plasma into the nephron down a pressure gradient

tubular reabsorption in the kidney, process in which water and useful solutes are reabsorbed from the filtrate in the nephron and transported into capillaries for reuse by the body

tubular secretion in the kidney, process that moves additional wastes and excess substances from the blood into the filtrate in the nephron; uses mainly active transport

water reabsorption in the kidney, process that removes water from the filtrate in the nephron and returns it to the blood for reuse by body systems

Glomerular Filtration Filters Blood

The formation of urine starts with glomerular filtration. This process forces some of the water and dissolved substances in blood plasma from the glomerulus, shown in **Figure 10.5**, into the Bowman's capsule. Keep in mind that this process is occurring in *millions* of nephrons all at the same time. Here, you are focussing your attention on only a single nephron.

Two factors contribute to this filtration. One factor is the permeability of the capillaries of the glomerulus. Unlike capillaries in other parts of the body, capillaries of the glomerulus have many pores in their tissue walls. These pores are large enough to allow water and most dissolved substances in the blood plasma to pass easily through the capillaries and into the Bowman's capsule. On the other hand, the pores are small enough to prevent proteins and blood cells from entering. The other factor is blood pressure. Blood pressure within the glomerulus is about four times greater than it is in capillaries elsewhere in the body. The great rush of blood through the glomerulus provides the force for filtration. The process of glomerular filtration, as well as the remaining processes that form urine, are summarized in **Figure 10.6**, on the next page.

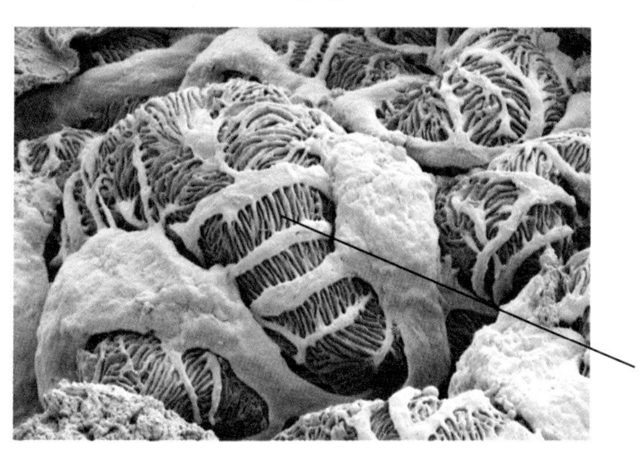

Figure 10.5 The pores in the capillaries of the glomerulus and the filtration slits shown here allow water, glucose, vitamins, amino acids, ammonia, urea, and ions from the blood to pass through into Bowman's capsule.

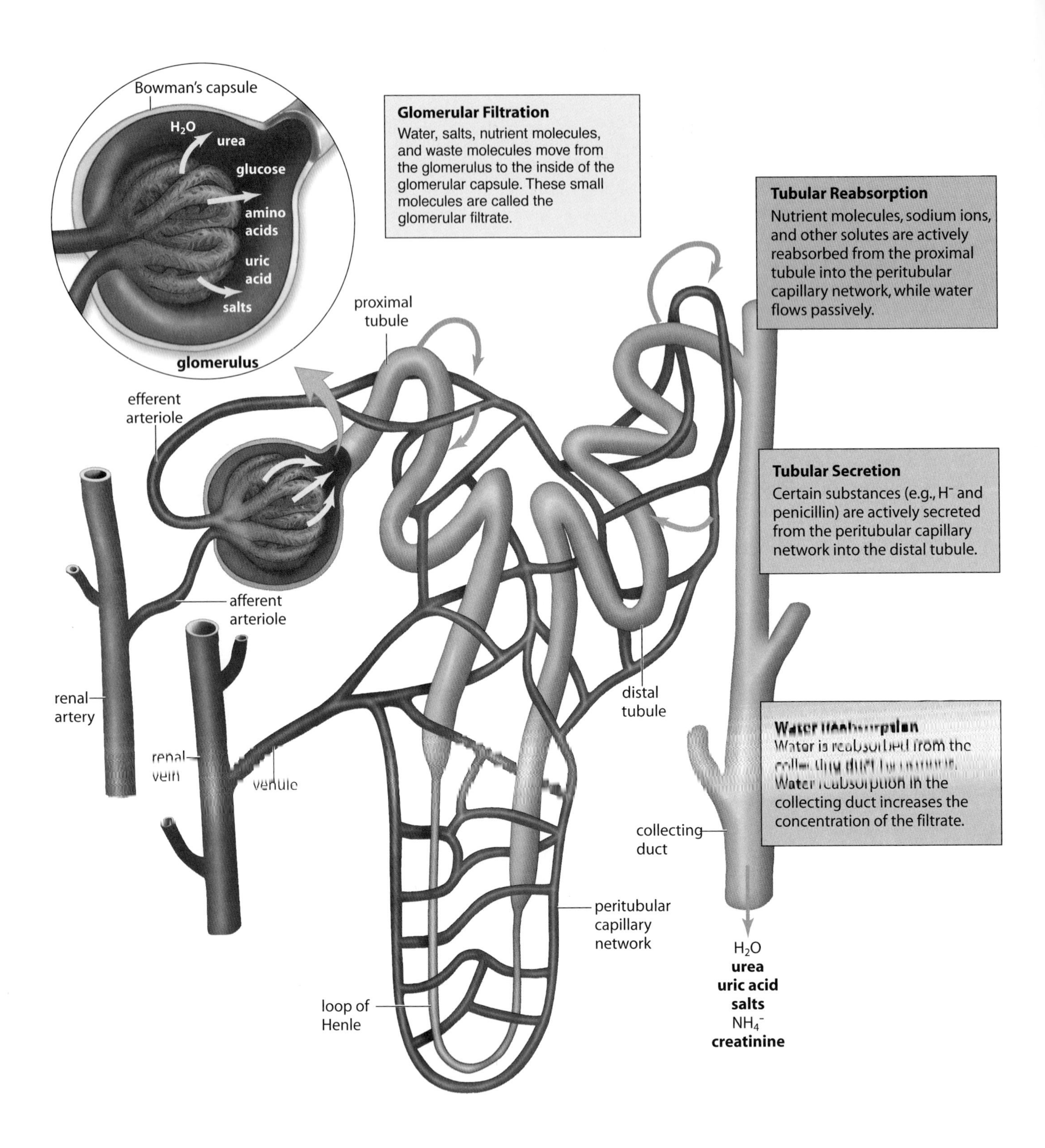

Bowman's capsule
H_2O
urea
glucose
amino acids
uric acid
salts
glomerulus
Glomerular Filtration
Water, salts, nutrient molecules, and waste molecules move from the glomerulus to the inside of the glomerular capsule. These small molecules are called the glomerular filtrate.
Tubular Reabsorption
Nutrient molecules, sodium ions, and other solutes are actively reabsorbed from the proximal tubule into the peritubular capillary network, while water flows passively.
proximal tubule
efferent arteriole
Tubular Secretion
Certain substances (e.g., H^- and penicillin) are actively secreted from the peritubular capillary network into the distal tubule.
afferent arteriole
renal artery
distal tubule
Water Reabsorption
Water is reabsorbed from the
Water reabsorption in the collecting duct increases the concentration of the filtrate.
renal vein
venule
collecting duct
peritubular capillary network
loop of Henle
H_2O
urea
uric acid
salts
NH_4^-
creatinine

Figure 10.6 The four main processes of urine formation are described in boxes and colour-coded to arrows that show the movement of molecules into or out of the nephron at specific locations. In the end, urine is composed of the substances within the collecting duct (see blue arrow).

Describe *what occurs in the proximal and distal tubules.*

Each day, 1600 L to 2000 L of blood pass through your kidneys, producing about 180 L of glomerular filtrate. This filtrate is chemically very similar to blood plasma, as you can see in **Table 10.1**. Essentially, the filtrate is identical to blood plasma, minus proteins and blood cells. If the composition of urine were the same as that of the glomerular filtrate, the body would continually lose water, salts, and nutrients. Therefore, the composition of the filtrate must change as this fluid passes through the remainder of the tubule.

Table 10.1 Concentration of Selected Chemicals in Plasma and Glomerular Filtrate

Chemicals	Blood Plasma (g/L)	Glomerular Filtrate (g/L)
Protein	44.4	0.0
Sodium (Na^+)	3.0	3.0
Chloride (Cl^-)	3.5	3.5
Glucose	1.0	1.0
Urea	0.3	0.3

Tubular Reabsorption: Recovery of Substances in the Proximal Tubule

About 65 percent of the filtrate that passes through the entire length of the **proximal tubule** (including the loop of Henle) is reabsorbed and returned to the body. **Figure 10.7** shows that this process of reabsorption involves both active and passive transport mechanisms. The cells of the proximal tubule contain many mitochondria, which use the energy-releasing power of ATP to drive the active transport of sodium ions (Na^+), glucose, and other solutes back into the blood. Negatively charged ions tag along passively, attracted by the electrical charge on the transported substances. Water follows the ions by osmosis, so it, too, is reabsorbed into the blood flowing through the capillaries.

proximal tubule in the kidney, tubular portion of the nephron that lies between the Bowman's capsule and the loop of Henle; main function is reabsorption of water and solutes, as well as secretion of hydrogen ions

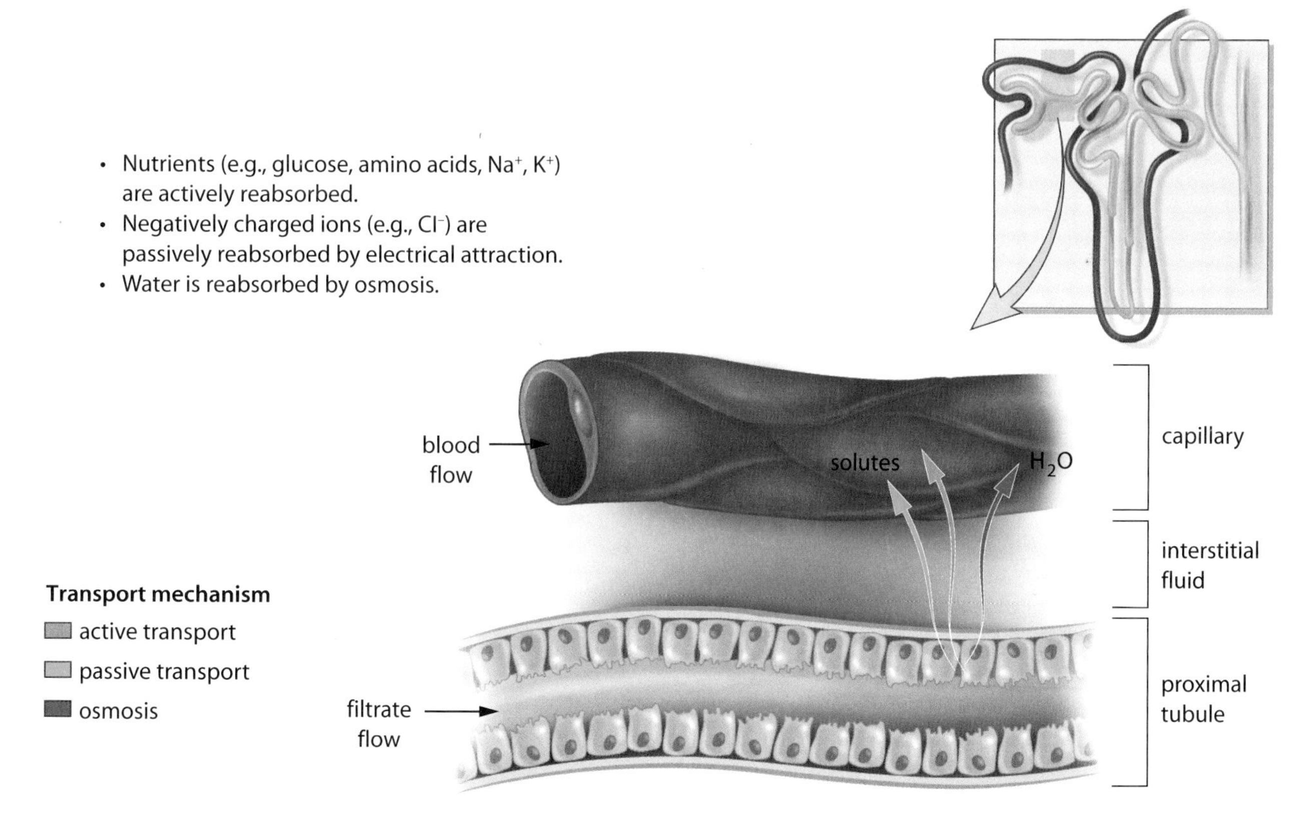

Figure 10.7 Reabsorption of nutrients, including glucose, Na^+, and K^+, occurs in the proximal tubule.

loop of Henle in the kidney, tubular portion of the nephron that lies between the proximal tubule and the distal tubule; main function is reabsorption of water and ions

Focussing on the Loop of Henle in the Proximal Tubule

The function of the **loop of Henle** is to reabsorb water and ions from the glomerular filtrate. As the descending limb of the loop of Henle plunges deeper into the medulla region, it encounters an increasingly salty environment. The cells of the descending limb are permeable to water and only slightly permeable to ions. As a result of the salty environment of the medulla and permeability of the descending limb, water diffuses from the filtrate to the capillaries by osmosis, as shown in **Figure 10.8A**. As water moving through the descending limb leaves the filtrate, the concentration of sodium ions (Na^+) inside the tubule increases, reaching its maximum concentration at the bottom of the loop.

As the filtrate continues around the bend of the loop of Henle and into the ascending limb, the permeability of the nephron tubule changes. Near the bend, the thin portion of the ascending tubule is now impermeable to water and slightly permeable to solutes. Sodium ions diffuse from the filtrate along their concentration gradient and pass into nearby blood vessels, as shown in **Figure 10.8B**.

At the thick-walled portion of the ascending limb of the loop of Henle, sodium ions are moved out of the filtrate by active transport, as illustrated in **Figure 10.8C**. This transport of Na^+ out of the filtrate has two consequences:

- First, it helps replenish the salty environment of the medulla, which aids in the absorption of water from filtrate in the descending limb.
- Second, the removal of sodium ions from the filtrate in the thick-walled portion of the tubule makes the filtrate less concentrated than the tissues and blood in the surrounding cortex tissue.

By now, about two thirds of the Na^+ and water from the filtrate has been reabsorbed.

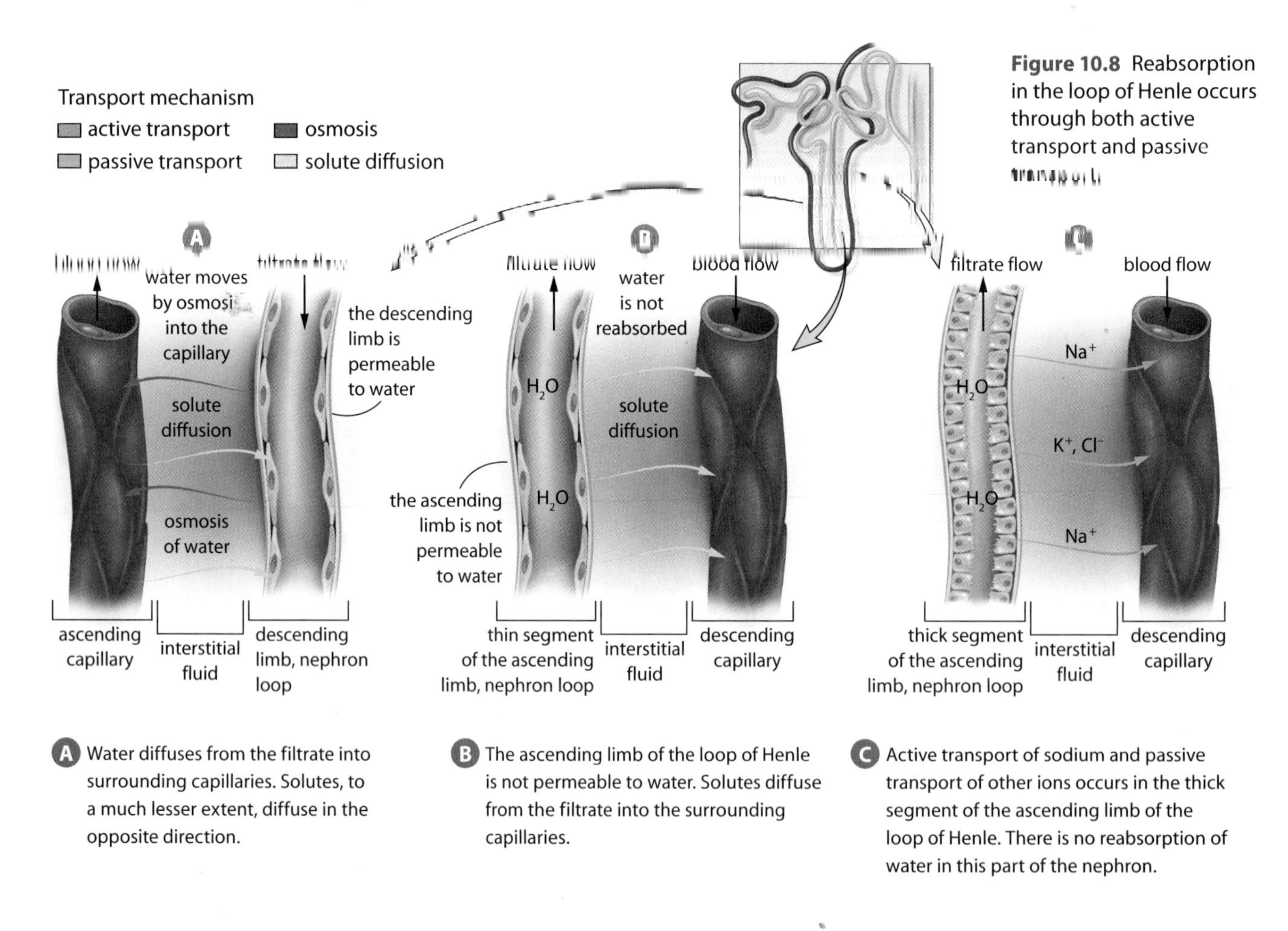

Figure 10.8 Reabsorption in the loop of Henle occurs through both active transport and passive

A Water diffuses from the filtrate into surrounding capillaries. Solutes, to a much lesser extent, diffuse in the opposite direction.

B The ascending limb of the loop of Henle is not permeable to water. Solutes diffuse from the filtrate into the surrounding capillaries.

C Active transport of sodium and passive transport of other ions occurs in the thick segment of the ascending limb of the loop of Henle. There is no reabsorption of water in this part of the nephron.

Learning Check

7. Summarize the four processes that are crucial to the formation of urine.

8. Describe the two factors that contribute to glomerular filtration.

9. Compare the composition of the filtrate once it leaves the glomerulus with that of blood plasma.

10. What role does ATP play in the processes occurring in the proximal tubule?

11. Using a diagram, summarize what happens to the filtrate as it moves through the proximal tubule.

12. In point form, describe what happens to the filtrate as it moves through the loop of Henle.

Tubular Reabsorption and Secretion in the Distal Tubule

The active reabsorption of sodium ions from the filtrate into the capillaries depends on the needs of the body. Passive reabsorption of negative ions such as chloride occurs by electrical attraction. The reabsorption of ions decreases the concentration of the filtrate, which causes water to be reabsorbed by osmosis, as shown in **Figure 10.9**.

Potassium ions (K^+) are actively secreted into the **distal tubule** from the bloodstream in the capillaries. Hydrogen ions (H^+) are also actively secreted from the blood into the distal tubule as necessary in order to maintain the pH of the blood. Other substances that are not normally part of the body, such as penicillin and other medications, are secreted from the blood into the distal tubule. Reabsorption and secretion in the distal tubule are under the control of hormones, as you will see in Section 10.3.

distal tubule in the kidney, tubular portion of the nephron that lies between the loop of Henle and the collecting duct; main function is reabsorption of water and solutes, and secretion of various substances

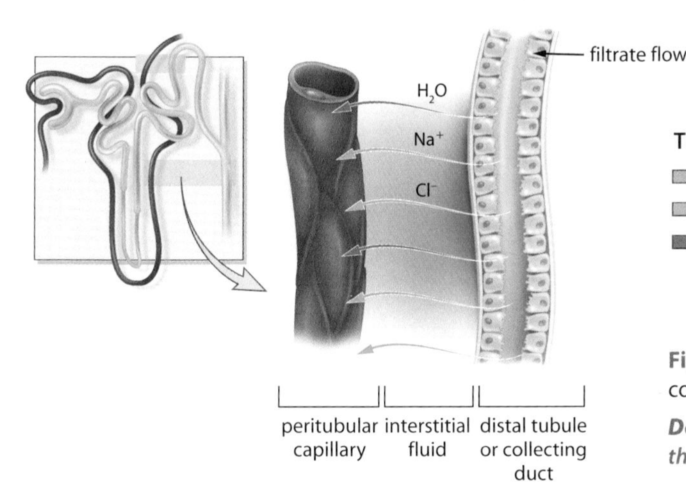

Transport mechanism
- active transport
- passive transport
- osmosis

Figure 10.9 Reabsorption of water and ions continues in the distal tubule and collecting duct.
***Describe** the movement of water in the region of the collecting duct.*

Activity 10.1 Water Loss

Your body loses water through the urinary system when you urinate, through the skin when you perspire, and through the respiratory system when you exhale.

Procedure

1. Study **Table 10.2**, which shows the average daily water loss in humans under different conditions.
2. Determine the total water loss for a person in each case.
3. Calculate the percentage of water lost from each source for each environmental condition.

Questions

1. Summarize the differences in water loss among the three conditions.
2. Infer why more water is lost in perspiration during rigorous exercise and high temperatures than in urine under those same conditions.

Table 10.2 Average Daily Water Loss in Humans (mL)

Source	Normal Temperature	High Temperature	Intense Exercise
Kidneys	1500	1400	750
Skin	450	1800	5000
Lungs	450	350	650

Source: Beers, M. 2003. *The Merck Manual of Medical Information, Second Edition* West Point, PA: Merck & Co. Inc.

Reabsorption from the Collecting Duct

The filtrate entering the collecting duct still contains a lot of water. Because the collecting duct extends deep into the medulla, the concentration of ions along its length increases. This concentration of ions is the result of the active transport of ions from the ascending limb of the loop of Henle. This causes the passive reabsorption of water from the filtrate in the collecting duct by osmosis. If blood plasma is too concentrated (for example, if a person is dehydrated), the permeability to water in the distal tubule and the collecting duct is increased. This causes more water to be reabsorbed into the surrounding capillaries in order to conserve water in the body. In the collecting duct, as in the distal tubule, hormones control reabsorption and secretion.

Suggested Investigation

Inquiry Investigation 10-A, Identifying Structures of the Excretory System

The reabsorption of water in the collecting duct causes the filtrate to become about four times as concentrated by the time it exits the duct. This filtrate—which is approximately 1 percent of the original filtrate volume—is now called urine. **Table 10.3** summarizes the main functions of the nephron and where they occur.

Table 10.3 A Summary of Nephron Functions

Part of the Nephron	Function
Glomerulus	**Filtration** • Glomerular blood pressure forces some of the water and dissolved substances from the blood plasma through the pores of the glomerular walls
Bowman's capsule	Receives filtrate from glomerulus
Proximal tubule	**Reabsorption** • Active reabsorption of all nutrients, including glucose and amino acids • Active reabsorption of positively charged ions such as sodium, potassium, calcium • Passive reabsorption of water by osmosis • Passive reabsorption of negatively charged ions such as chloride and bicarbonate by electrical attraction to positively charged ions **Secretion** • Active secretion of hydrogen ions
Descending loop of Henle	**Reabsorption** • Passive reabsorption of water by osmosis
Ascending loop of Henle	**Reabsorption** • Active reabsorption of sodium ions • Passive reabsorption of chloride and potassium ions
Distal tubule	**Reabsorption** • Active reabsorption of sodium ions • Passive reabsorption of water by osmosis • Passive reabsorption of negatively charged ions such as chloride and bicarbonate **Secretion** • Active secretion of hydrogen ions • Passive secretion of potassium ions by electrical attraction to chloride ions
Collecting duct	**Reabsorption** • Passive reabsorption of water by osmosis

Section 10.2 Review

Section Summary

- In the glomerulus, filtration moves water and solutes (except for protein) from blood plasma into the nephron.
- Solutes are actively transported from the filtrate in the proximal tubule back into the blood.
- Approximately 65 percent of the filtrate that passes through the entire length of the proximal tubule is reabsorbed and returned to the body, while the urine becomes concentrated.
- Filtrate moves through the nephron, and the processes of glomerular filtration, tubular reabsorption, tubular secretion, and water reabsorption modify the filtrate so that cleansed plasma and substances such as glucose, amino acids, and Na^+ return to the blood.

Review Questions

1. **C** Create a table that identifies the major parts of the nephron and summarizes the function of each part.
2. **K/U** Where does the formation of urine in the kidney start?
3. **K/U** Explain the difference between reabsorption and secretion in the nephron.
4. **C** Sketch a simplified nephron. Include a series of captions on your diagram that identify the part(s) of the nephron responsible for each of the following:
   a. movement of sodium ions from the nephron to the surrounding capillaries
   b. movement of water from the nephron to the surrounding capillaries
   c. movement of glucose out of the nephron
   d. movement of penicillin and potassium ions into the nephron
5. **K/U** Why is it important that sodium ions be removed from the filtrate as it moves through the ascending limb of the loop of Henle?
6. **C** Using a diagram, summarize what happens to the filtrate as it moves through the distal tubule and the collecting duct. Include a caption with your diagram.
7. **T/I** In relation to kidney function, infer why a doctor would ask you to take an antibiotic such as penicillin for several days rather than in one single dose.
8. **C** Use a graphic organizer to compare and contrast the composition of blood entering the kidney to urine leaving the kidney in terms of the substances and ions each contain.
9. **K/U** Explain why you agree or disagree with the following statement: Urine contains substances that have undergone glomerular filtration but have not been reabsorbed, as well as substances that have undergone tubular secretion.
10. **A** In diabetes mellitus, the liver and muscles fail to store glucose as glycogen. As a result, the kidneys cannot reabsorb all of the glucose in the filtrate. The data shown in the table below were collected from a healthy individual. What changes would you expect to see in the data from a person who has untreated diabetes mellitus? Explain your reasoning.

Substance	Amount Filtered (per day)	Amount Excreted (per day)	Reabsorption (%)
Water (L)	180	1.8	99.0
Sodium (g)	630	3.2	99.5
Glucose (g)	180	0.0	100.0
Urea (g)	54	30.0	44.0

11. **T/I** Based on what you have learned so far in this chapter, predict how perspiring heavily on a hot day would affect the composition and production of a person's urine. (Assume that the person has not yet consumed water to replenish body fluids.)
12. **A** The following diagram is a diagrammatic representation of the cells found in the proximal tubule in a nephron. Explain why these cells have a high number of mitochondria.

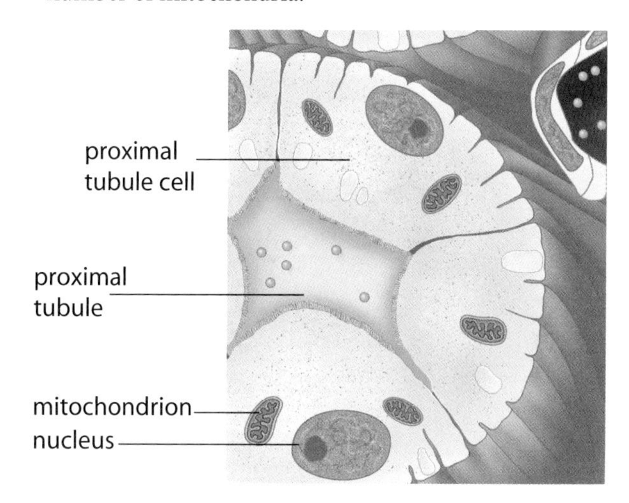

SECTION 10.3

Other Functions and Disorders of the Excretory System

Key Terms

antidiuretic hormone
renal insufficiency
dialysis
hemodialysis
peritoneal dialysis

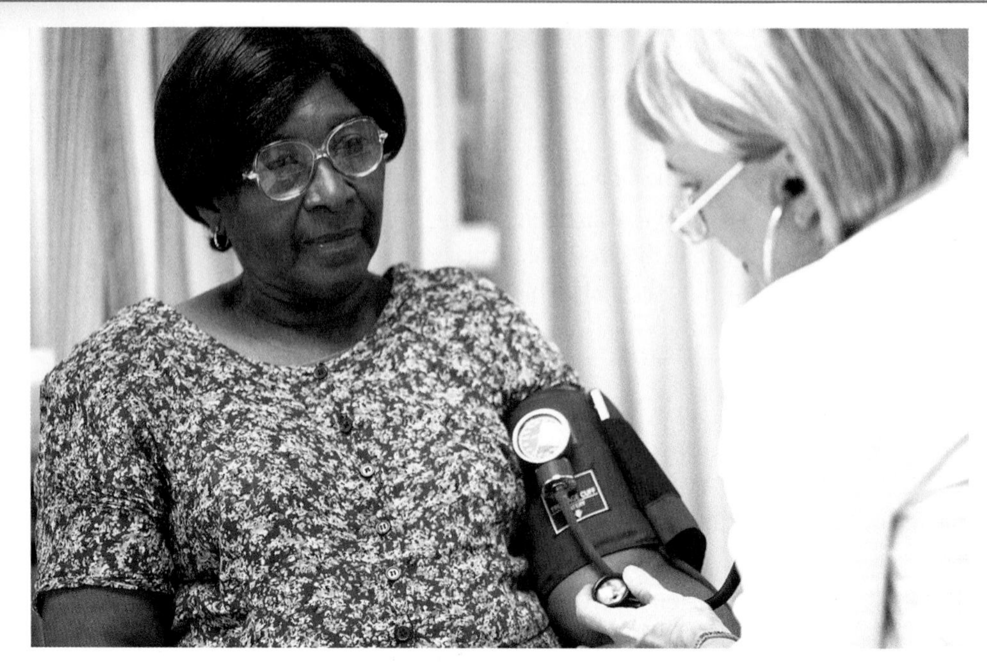

Figure 10.10 By controlling the reabsorption and secretion of water and salts in the blood, the kidneys help regulate blood volume and blood pressure. High blood pressure can be a warning sign that the kidneys are not functioning properly.

Recall from Section 10.1 that the kidneys not only filter wastes from the blood, but also carry out several other important homeostatic functions including maintaining the water–salt balance of the blood, regulating blood pH, and secreting some hormones. Although the connection between the kidneys and cardiovascular health may not seem obvious at first, the kidneys also play an important role in maintaining blood pressure and can be damaged if blood pressure gets too high. The person in **Figure 10.10** is getting her blood pressure checked using a blood pressure cuff. Along with blood tests and a urinalysis, a blood pressure measurement is used to help determine whether the kidneys are functioning properly.

Regulating Water–Salt Balance

When functioning properly, the kidneys keep the water–salt balance of the blood within normal limits. In this way, they also maintain the blood volume and blood pressure. Most of the water and salt (NaCl) present in the filtrate is reabsorbed across the wall of the proximal convoluted tubule.

Reabsorption of Water

The force generated as water moves by osmosis is called osmotic pressure. Osmotic pressure affects many cellular activities, especially the exchange of materials between cells and blood. Osmoreceptors are cells that are sensitive to osmotic pressure. Most osmoreceptors are located in the hypothalamus. Recall from Chapter 8 that the hypothalamus regulates mechanisms that enable the body to maintain homeostasis, such as hunger, thirst, blood pressure, body temperature, fluid balance, and salt balance.

When blood plasma becomes too concentrated (for example, if you are dehydrated), osmotic pressure increases. In response, osmoreceptors in the hypothalamus send impulses to the adjacent pituitary gland in the brain that causes the release of **antidiuretic hormone** (ADH). (*Diuresis* means "increased excretion of urine." Since the "anti-" prefix means "against" or "opposed to," *antidiuresis* means "decreased excretion of urine.") As shown in **Figure 10.11**, ADH travels through the blood to the kidneys, where it increases the permeability of the distal tubule and the collecting duct, allowing more water to be reabsorbed into the blood. This dilutes the blood and lowers osmotic pressure to normal.

antidiuretic hormone (ADH) hormone regulated by the hyphothalamus and released by the pituitary gland that increases the permeability of the distal tubule and the collecting duct in the nephrons of the kidneys, allowing more water to be reabsorbed into the blood from the filtrate

Body fluids too dilute (osmotic pressure too low).

Osmotic pressure of body fluids.

Body fluids too concentrated (osmotic pressure too high).

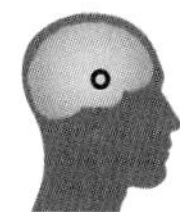

Osmoreceptors in hypothalamus send signal to decrease release of ADH.

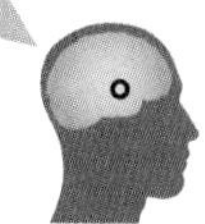

Osmoreceptors in hypothalamus sense increased osmotic pressure, and send signals to the pituitary gland to release ADH into bloodstream.

Osmotic pressure of body fluids increases.

Osmotic pressure of body fluids decreases.

Decreased reabsorption of water in kidney tubules and collecting ducts; increased water in urine.

Increased reabsorption of water in kidney tubules and collecting ducts; decreased water in urine.

Figure 10.11 The release of ADH controls the amount of water reabsorbed or excreted in urine.

Identify *Which type of feedback mechanism controls the release of ADH? Explain your reasoning.*

Conversely, if blood plasma is too dilute (that is, if the osmotic pressure is too low) osmoreceptors in the hypothalamus stop or prevent the release of ADH. As a result, the distal tubule and the collecting duct become less permeable to water. This allows more water to be excreted in the urine, concentrating the solutes in the blood. The osmotic pressure of the plasma and tissue fluids rises to normal.

In a condition called *diabetes insipidus*, ADH activity is insufficient, so a person urinates excessively—perhaps as much as 4 L to 8 L per day. Thirst is intense, but water is excreted more quickly than it is consumed, leading to severe dehydration and ion imbalances. People who have diabetes insipidus may take synthetic ADH to restore the balance of water reabsorption.

The ethanol in alcoholic beverages is a diuretic, so it increases the volume of urine. Alcohol stimulates urine production partly by inhibiting the release of ADH, which decreases of the permeability of the tubules and collecting ducts. Because it increases water loss to urine, drinking an alcoholic beverage actually *intensifies* thirst and leads to dehydration. Caffeine, a substance in coffee and many carbonated drinks, is also a diuretic.

Activity 10.2 How Can Diuretics Disrupt Homeostasis?

Some medications, such as those used to control high blood pressure, edema, and heart failure, are diuretics. Besides their medical uses, diuretics can also be abused as a method to lose weight or to disguise the use of illegal performance-enhancing drugs in sports.

Materials

- reference books
- computer with Internet access

Procedure

1. Use print and Internet resources to research more about the medical uses of diuretics. Determine how diuretics are abused as well.
2. Construct two data tables to organize the results of your research. The first table should include the following column titles: Medical Uses of Diuretics, Description of the Medication, How the Medication Works, and Possible Side Effects.

   The second table should include these column titles: Nonmedical Abuses of Diuretics, Desired Effects, and Possible Side Effects and Health Risks.

Questions

1. What are some of the benefits and risks of using diuretics for a medical reason (while under a doctor's care)?
2. What are some of the risks of using diuretics to lose weight? Do diuretics have long-term effectiveness for weight loss? Explain why or why not.
3. Why is the use of diuretics banned in many sports? Write a paragraph explaining why you agree or disagree with this policy. Consider the potential risks and benefits of diuretics as part of your answer.

Reabsorption of Salts

The kidneys regulate salt balance in the blood by controlling the excretion and reabsorption of various ions. The sodium ion (Na^+) is the most abundant ion in blood plasma, but its concentration can fluctuate dramatically depending on diet and the consumption of beverages with diuretic effects.

Hormones regulate the reabsorption of sodium at the distal tubule. Recall from Chapter 9 that aldosterone is a hormone secreted by the adrenal cortex. Aldosterone stimulates the excretion of potassium ions (K^+) and the reabsorption of sodium ions (Na^+). The release of aldosterone is set in motion by the kidneys themselves. When blood volume, and therefore blood pressure, is too low to promote glomerular filtration, the kidneys secrete renin. The presence of renin, an enzyme, starts a reaction that eventually triggers the release of aldosterone from the adrenal cortex. Aldosterone stimulates the distal tubules and collecting ducts to reabsorb Na^+. Because the reabsorption of Na^+ is followed passively by chloride ions and water, aldosterone has the net effect of retaining both salt and water. As a result, blood volume and blood pressure increase.

Maintaining Blood pH

The normal pH of body fluids is about 7.4. This is the pH at which the enzymes in our bodies function optimally. If homeostasis is not maintained and blood pH goes above or below 7.4, serious medical conditions can result. Recall from **Figure 10.1** that the foods and liquids we consume can result in changes to blood pH. Other metabolic processes, including cellular respiration, can also alter blood pH. However, due to three main homeostatic mechanisms, blood pH remains at about 7.4. The three mechanisms are the acid–base buffer system, respiration, and the function of the kidneys.

Acid–Base Buffer System

The acid–base buffer system buffers the blood, preventing changes in pH by taking up excess hydrogen ions (H^+) or excess hydroxide ions (OH^-) that enter the blood. One of the key buffering reactions in the blood involves carbonic acid (H_2CO_3) and bicarbonate ions (HCO_3^-). When hydrogen ions (H^+) are added to the blood, the following reaction occurs:

$$H^+ + HCO_3^- \rightarrow H_2CO_3$$

When hydroxide ions (OH^-) are added to the blood, this reaction occurs:

$$OH^- + H_2CO_3 \rightarrow HCO_3^- + H_2O$$

These reactions temporarily prevent any significant change in blood pH. A blood buffer, however, can be overwhelmed unless some more permanent adjustment is made. The next adjustment to keep the pH of the blood constant occurs in the lungs.

Activity 10.3 The Renin-Angiotensin-Aldosterone System

The release of aldosterone is stimulated directly by a rise or fall in Na^+ and K^+ concentrations in the blood. Aldosterone is stimulated indirectly through a drop in blood pressure.

Materials

- reference books
- computer with Internet access

Procedure

1. Use print or Internet resources to research the main steps involved in the renin-angiotensin-aldosterone system of hormone control.
2. Sketch a flowchart to show how this system responds to a rise or fall in blood pressure and blood volume.

Questions

1. What types of conditions in the body can result in a decrease in blood volume?
2. Addison's disease results in a progressive destruction of the adrenal cortex and the loss of aldosterone secretion. Predict some of the symptoms of Addison's disease.
3. Atrial natriuretic hormone (ANH) is secreted by the atria of the heart when cardiac cells are stretched due to increased blood volume. ANH inhibits the secretion of renin and aldosterone. Add this step to your flowchart.
4. What type of feedback system controls the renin-angiotensin-aldosterone system? Explain your reasoning.

Respiratory Centre

If the hydrogen ion concentration of the blood rises, the respiratory centre in the medulla oblongata increases the breathing rate. Increasing the breathing rate rids the body of hydrogen ions because the following reaction takes place in capillaries within the lungs:

$$H^+ + HCO_3^- \rightleftarrows H_2CO_3 \rightleftarrows H_2O + CO_2$$

blood pH decreases $\leftarrow$ $\rightarrow$ blood pH increases

An increased breathing (respiration) rate pulls the reaction to the right to generate CO_2 more quickly. When carbon dioxide is exhaled, the number of hydrogen ions is reduced.

It is important to have the correct proportion of carbonic acid and bicarbonate ions in the blood. Breathing readjusts this proportion so that this particular acid–base buffer system can continue to absorb both H^+ and OH^- as needed.

The Kidneys

The acid–base buffer system and respiration are aided by the more powerful actions of the kidneys to control the acid–base balance in the blood. Only the kidneys can rid the body of a wide range of acidic and basic substances and otherwise adjust the pH. The kidneys are slower acting than the other two mechanisms, but they have a more powerful effect on pH.

Think of the kidneys as excreting H^+ and reabsorbing HCO_3^- as needed to maintain normal blood pH, as shown in **Figure 10.12**. If the blood is too acidic, H^+ is excreted and HCO_3^- is reabsorbed. If the blood is too basic, H^+ is not excreted and HCO_3^- is not reabsorbed. Since urine is usually acidic, it follows that H^+ is usually excreted.

Ammonia (NH_3) provides another means of buffering and removing the hydrogen ions in urine ($NH_3 + H^+ \rightarrow NH_4^+$). Ammonia, the presence of which is quite obvious in the diaper pail or kitty litter box, is produced in tubule cells by the breakdown of amino acids. The ability of the kidneys to control blood pH is crucial to maintaining an internal environment in which cell enzymes continue to function properly.

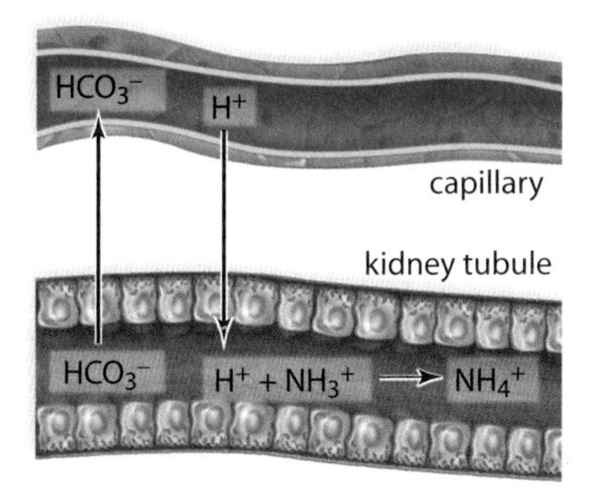

Figure 10.12 In the kidneys, bicarbonate ions (HCO_3^-) are reabsorbed and hydrogen ions (H^+) are excreted as needed to maintain the pH of the blood. Excess hydrogen ions are buffered, for example, by ammonia (NH_3), which is produced in tubule cells by the breakdown of amino acids.

Learning Check

13. Describe the role of osmoreceptors in regulating water reabsorption.

14. Use a graphic organizer to summarize what happens when the osmotic pressure of body fluids becomes too low.

15. Write the chemical reaction that occurs in the blood when excess hydroxide ions are added to the blood and explain this reaction briefly.

16. Explain how the kidneys help to maintain the pH levels of the blood.

17. Explain why an individual who drinks several cups of regular coffee would likely experience greater urine output than a person who drinks the same amount of decaffeinated coffee.

18. Explain whether you agree or disagree with the following statement: "The solute concentration of the blood remains constant despite variations in the amount of water people consume in foods and liquids."

Releasing Hormones

You have already read that the kidneys release renin, a substance that leads to the secretion of the hormone aldosterone. In addition to being part of the feedback loop of aldosterone, the kidneys release two hormones of their own: erythropoietin [pronounced e-RITH-ro POI-e-tin] and calcitriol.

If the oxygen-carrying capacity of the blood is reduced, or oxygen demand increases, sensors in the kidneys stimulate kidney cells to release erythropoietin. Erythropoietin stimulates the production of red blood cells in bone marrow. As the number of red blood cells circulating in the blood increases, the oxygen-carrying capacity of the blood increases. When oxygen delivery to the kidneys returns to normal, the kidneys stop releasing erythropoietin.

The kidneys also play a role in calcium regulation in the blood. If the level of calcium in the blood falls below normal, parathyroid hormone (PTH) is released by the parathyroid gland. PTH stimulates the release of calcitriol by the kidneys. Calcitriol (the active form of vitamin D) promotes calcium (Ca^{2+}) absorption from the digestive tract.

Disorders of the Excretory System

The excretory system is vital to maintaining homeostasis, so when it is affected by a disorder the proper functioning of other body systems may be jeopardized.

Urinary Tract Infection

One of the most common disorders of the excretory system is a urinary tract infection. If the bladder has a bacterial or viral infection, the disorder is called cystitis; if only the urethra is involved, the condition is called urethritis. Urinary tract infections are more common in females than in males, primarily because of the differences in anatomy. In females, the urethral and anal openings are closer together, making it easier for bacteria from the bowels to enter the urinary tract and start an infection.

Symptoms of a urinary tract infection include a painful burning sensation during urination, a need to urinate frequently even when no urine is present, and bloody or brown urine. The upper abdomen or lower back may be tender, and chills, fever, nausea, and vomiting may be present. Urinary tract infections have the potential to become serious, and they can result in permanent damage to the kidneys and possible kidney failure. Treatment usually is with an antibiotic, but in serious cases of kidney infection surgery may be needed. Preventive measures include maintaining hygienic personal behaviours, such as proper wiping from front to back after a bowel movement, and drinking lots of water.

Kidney Stones

Another fairly common disorder of the excretory system involves the development of crystalline formations called kidney stones, shown in **Figure 10.13**. Most kidney stones form due to excess calcium in the urine. In fact, about 85 percent of kidney stones are made up of calcium compounds. Recurrent urinary tract infections, insufficient water consumption, and low activity levels contribute to kidney stone formation. Treatment varies depending on the size of the stones. Many stones pass through the urinary tract on their own. Depending on the cause of the stone formation, medications may help to break down the crystals. If the stones are less than 20 mm in diameter, ultrasound shock waves can be used to disintegrate the crystalline structure of the stones so that they can be passed naturally in the urine. For larger stones, surgery may be needed to remove them.

renal insufficiency the state in which the kidneys cannot maintain homeostasis due to nephron damage

Renal Insufficiency

Renal insufficiency is a general term used to describe the state in which the kidneys cannot maintain homeostasis due to damage to their nephrons. Some causes of nephron damage include:

- kidney infection
- high blood pressure

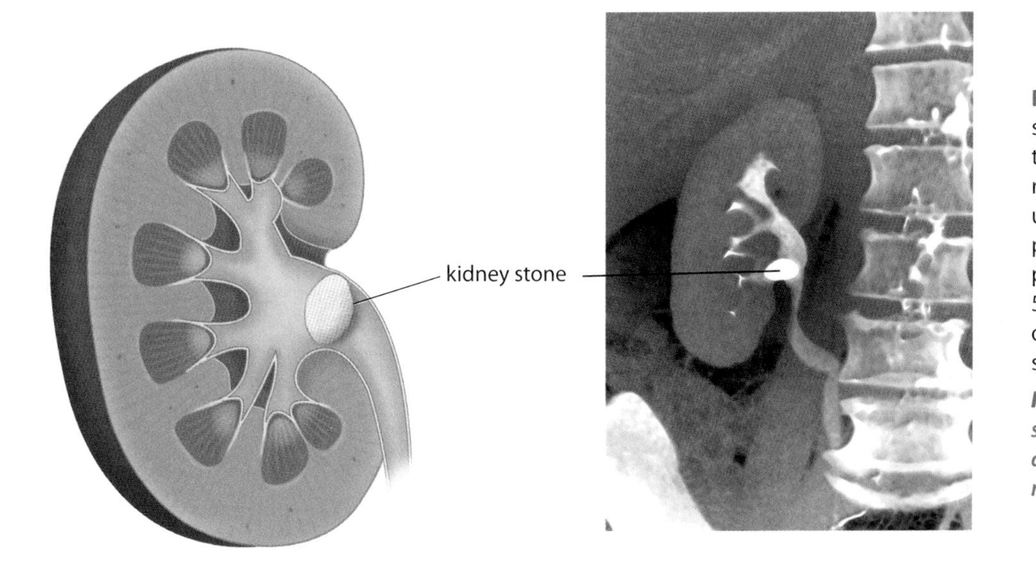

Figure 10.13 Kidney stones are hard granules that can form in the renal pelvis. They are usually made of calcium, phosphate, uric acid, and protein. It is estimated that 5–15 percent of people will develop kidney stones at some time in their lives.

***Infer** why larger kidney stones need to be broken apart or surgically removed.*

- diabetes mellitus
- polycystic kidney disease (PKD) (a genetic disorder in which cysts grow in the kidneys, impairing proper functioning)
- trauma from a blow to the lower back or constant vibration from machinery
- poisoning (either from skin contact, inhalation of fumes, or ingestion of contaminated food) by heavy metals such as mercury and lead or solvents such as paint thinners
- atherosclerosis (which reduces blood flow to the kidneys)
- blockage of the tubules

Nephrons can regenerate and restore kidney function after short-term injuries. Even when some of the nephrons are irreversibly damaged, others can compensate for their lost function. In fact, a person can survive on as little as one third of one kidney. If 75 percent or more of the nephrons are destroyed, however, urine output is inadequate to maintain homeostasis. Under these circumstances, a person requires a means for replacing kidney function. This is achieved either with a kidney transplanted from a donor, if one is available, or with an artificial kidney that performs a blood-cleansing process called dialysis.

dialysis procedure that removes wastes and excess fluid from the blood when kidney function is lost due to renal failure

hemodialysis type of renal (kidney) dialysis that utilizes an artificial membrane in an external device and is connected to an artery and a vein in a person's arm to remove waste and excess fluid from the blood when kidney function is lost due to renal failure

peritoneal dialysis type of renal (kidney) dialysis that utilizes the lining of the intestines, called the peritoneum, as the dialysis membrane to remove waste and excess fluid from the blood when kidney function is lost due to renal failure

Hemodialysis and Peritoneal Dialysis

The diffusion of dissolved substances through a semipermeable membrane is referred to as **dialysis**. These substances move across a membrane from the area of greater concentration to one of lower concentration. Substances more concentrated in blood diffuse into the dialysis solution, called the dialysate. Substances more concentrated in the dialysate diffuse into the blood. Other substances can be added to the blood following this same principle. For example, if the acid–base balance of the blood is off and the blood is too acidic, bicarbonate ions can be added to the dialysate where they will diffuse into the blood and reduce its acidity.

There are two main types of renal (kidney) dialysis: hemodialysis and peritoneal dialysis. **Hemodialysis** utilizes an artificial membrane in an external device—in essence, an artificial kidney—that is connected to an artery and a vein in a person's arm. **Peritoneal dialysis** utilizes the lining of the intestines, called the peritoneum, as the dialysis membrane. Dialysate is introduced to the abdominal cavity, where the large surface area and rich supply of capillaries of the peritoneum slowly filter the blood. **Figure 10.14**, on the next page, illustrates these two methods of dialysis.

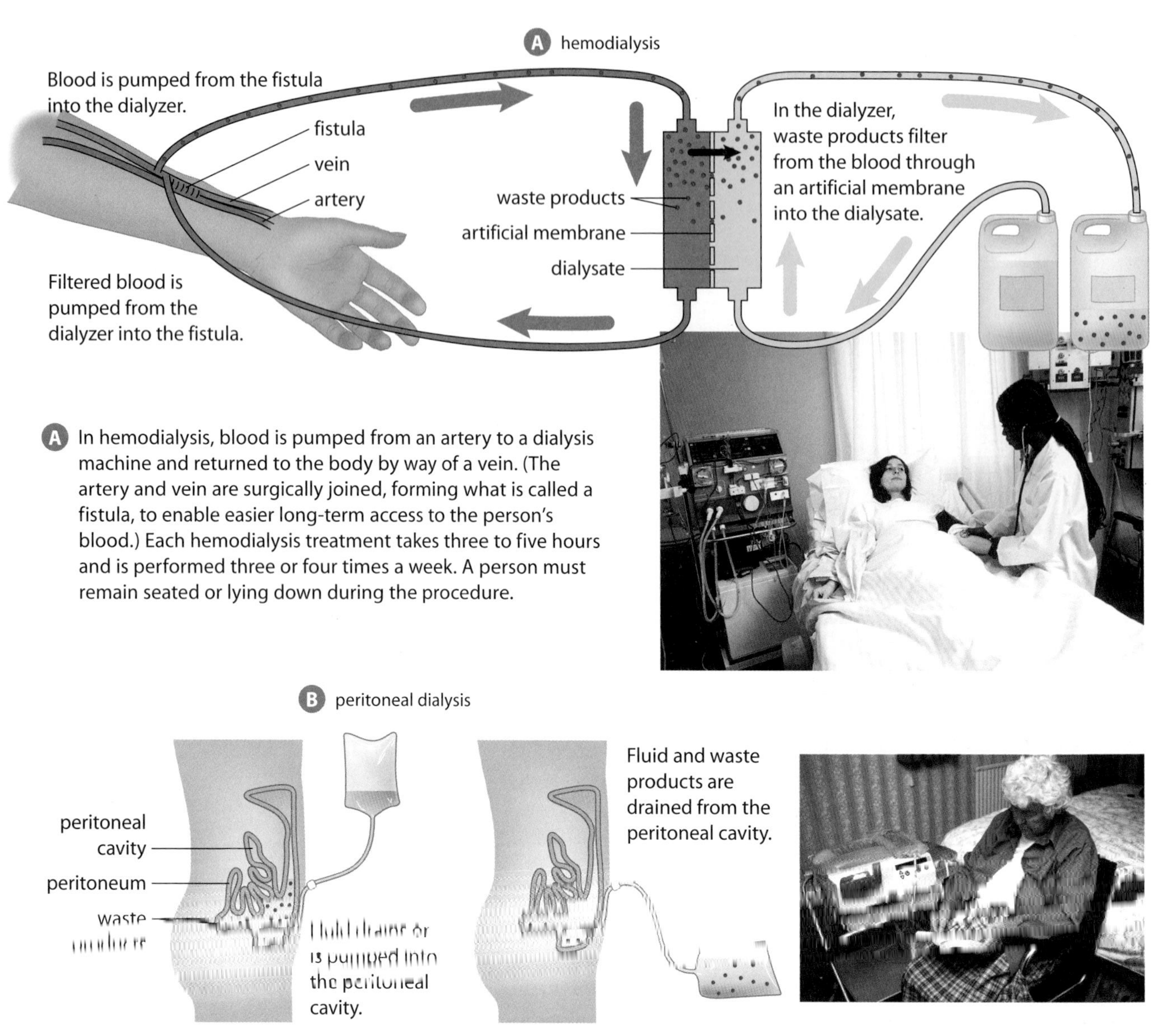

A hemodialysis
Blood is pumped from the fistula into the dialyzer.
fistula
vein
artery
waste products
artificial membrane
dialysate
In the dialyzer, waste products filter from the blood through an artificial membrane into the dialysate.
Filtered blood is pumped from the dialyzer into the fistula.
B peritoneal dialysis
peritoneal cavity
peritoneum
waste
cavity.
Fluid and waste products are drained from the peritoneal cavity.

A In hemodialysis, blood is pumped from an artery to a dialysis machine and returned to the body by way of a vein. (The artery and vein are surgically joined, forming what is called a fistula, to enable easier long-term access to the person's blood.) Each hemodialysis treatment takes three to five hours and is performed three or four times a week. A person must remain seated or lying down during the procedure.

B In peritoneal dialysis, a catheter (flexible tube) is surgically inserted into the abdominal cavity and dialysate may be delivered, removed, and replaced. Because dialysate is always present, the blood is continuously filtered. The full name for this type of dialysis is continual peritoneal dialysis, or CPD. There are several types of CPD. In continuous ambulatory peritoneal dialysis (CAPD), the procedure can be done at home, work, or school—any place that is clean and convenient. Usually, three to five exchanges of fresh dialysate for used dialysate are needed each day. In automated peritoneal dialysis (APD), a machine performs the exchange, which often is done at night for a period of up to 12 hours.

Figure 10.14 **A** Hemodialysis and **B** peritoneal dialysis.

Kidney Transplants

Dialysis enables people with kidney disease to continue many of their daily activities, such as going to a job or attending school. However, dialysis is not a cure, and it is not intended to be a long-term solution to the problem of kidney disease. Individuals with 10 percent or less kidney function will eventually have to replace their kidneys. The need for kidneys is much greater than the available supply. In Canada, the overall rate of organ donation is low compared to other developed countries, with about 14 donors per million people. There are thousands of people in Canada waiting for organs, and the vast majority of them—more than 75 percent—are waiting for kidneys.

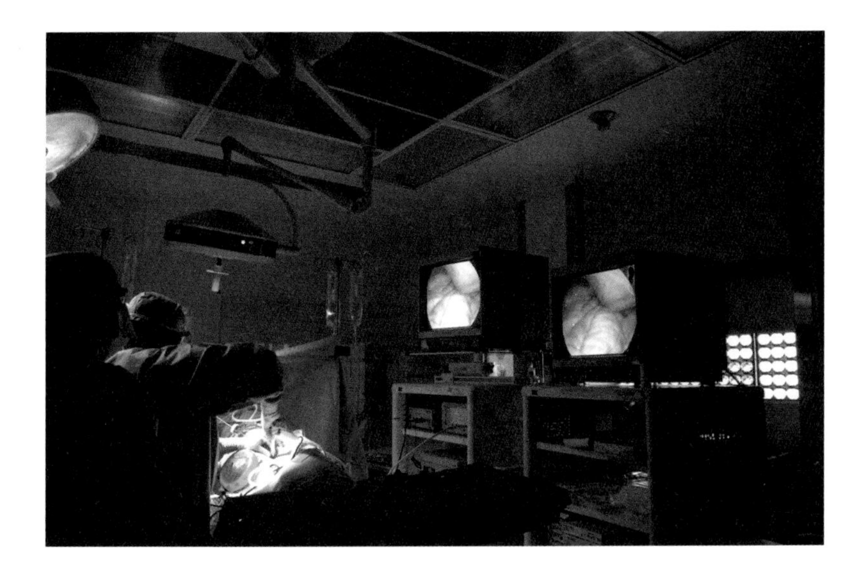

Figure 10.15 Laparoscopic surgery allows a kidney to be removed from a living donor. The location of the kidney behind the abdominal organs makes it difficult to access using traditional laparoscopic instruments. Specially designed hand-assisted laparoscopes are being used with great success.

The success rate of organ transplantation, particularly of kidneys, is fairly high. The success rate for living-donor transplants is about 98 percent, compared with up to 95 percent success with transplants of kidneys from organ donors that have suddenly died. In the future, these results likely will improve. Surgical techniques, such as the laparoscopic surgery shown in **Figure 10.15**, are constantly being improved, and new medicines to prevent rejection of the new organ are constantly being developed and improved.

The Kidney–Coronary Connection

High blood pressure is one of the main reasons that kidneys begin to fail. When blood pressure is high for a prolonged period of time, the heart must pump a greater volume of blood, and blood vessels can be damaged. The blood vessels in the kidneys are very sensitive to changes in blood pressure, and if they become damaged by high blood pressure the amount of waste and extra fluid that can be filtered from blood will be reduced. As the extra fluid accumulates in the body, it will increase the blood volume even more and cause the blood pressure to rise further. This cycle can continue until the kidney function is so reduced that symptoms become obvious. Unfortunately, both high blood pressure and kidney impairment do not have obvious symptoms until the damage is well underway.

Maintaining a healthy lifestyle supports the overall health of all of your body's systems. Remember that none of these systems functions in isolation, so any activity that affects one of your systems will affect other systems as well. Ensuring that you have adequate physical exercise, for example, can help to make your heart and circulatory system stronger and healthier. A stronger heart can pump more blood throughout the body with less effort. This reduced effort translates to less force on the arteries, keeping blood pressure low. High blood pressure has such an impact on kidney function that exercise to reduce blood pressure ultimately reduces the likelihood of kidney damage.

Learning Check

19. Use a graphic organizer to show how the kidneys are involved in maintaining the oxygen-carrying capacity of the blood.

20. Explain how the kidneys help maintain blood calcium levels.

21. Summarize the various methods that can be used to treat kidney stones.

22. Use a Venn diagram to compare hemodialysis to peritoneal dialysis.

23. Relate high blood pressure to kidney disease.

24. Infer how untreated renal insufficiency can lead to uremia, a condition in which urea and other waste products accumulate in the blood.

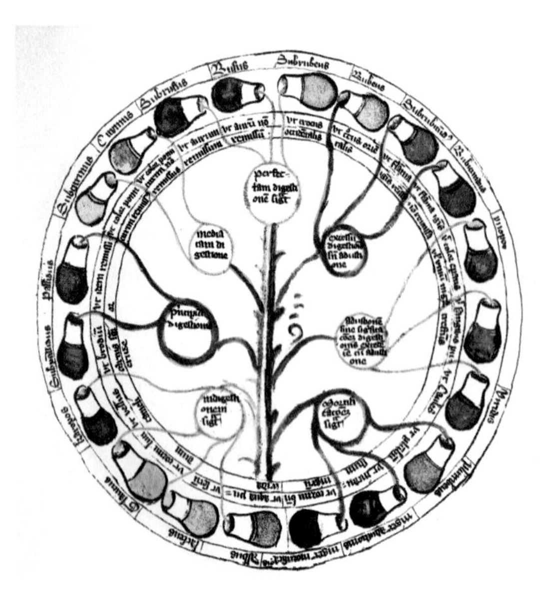

Figure 10.16 Urine wheels were used by physicians several hundred years ago. A typical urine wheel contained information about the colour, smell, and taste of urine.

Evaluating Kidney Function

The composition of urine reflects the amounts of water and solutes that the kidneys must remove from or retain in the body to maintain homeostasis. Analyzing the physical and chemical composition of urine, therefore, enables physicians to make reasoned inferences and hypotheses about a person's health and kidney function. In fact, physicians have been using urinalysis as a tool for thousands of years. Examining the characteristics of urine, including colour, odour, and taste, gave clues as to the internal conditions of a person's body. The wheel shown in **Figure 10.16** was used by physicians in the 1500s as a guide to diagnose disease based on urine characteristics.

Urinalysis

Table 10.4 provides values for selected tests that are performed in a modern-day urinalysis. These values are consistent with those of urine from a healthy adult. Note, however, that urine composition varies greatly over the course of a day due to factors such as dietary intake, physical activity, emotional stress, and fatigue. In addition, unhealthy constituents of urine may not necessarily indicate illness or disease. For example, the presence of glucose in urine may result from a sugary meal. Proteins may appear in urine following vigorous exercise. Ketones—acids that result from the digestion of fats when the body lacks sufficient stores of carbohydrates—may result from a short-term fast or a specially designed low-carbohydrate diet. Because so many factors can influence the presence and amounts of substances in urine, trained professionals must consider a wide variety of variables when evaluating a sample of urine. You will have a chance to examine a much narrower range of variables and substances in Inquiry Investigation 10-B.

Table 10.4 Normal Values from Selected Common Urine Tests

Urine Test	Accepted Healthy Value*
Acetone and ketones	0
Albumin (protein)	0–trace
Bilirubin (a breakdown product of hemoglobin)	0
Calcium	< 150 mg/day
Colour and clarity	Pale yellow to light amber; transparent
Glucose	0
pH	4.5–8.0
Urea	25–35 g/day
Uric acid	0.5–1.0 g/day

** These values may vary with the type of equipment used for analysis.*

Blood Tests

Blood tests can also reveal information about kidney function. For example, health-care professionals may screen the blood to measure the amount of urea nitrogen it contains. If levels are higher than normal, it can indicate that the kidneys are not working properly. As you have learned, when the kidneys function normally urea is filtered from the blood. Another measurement that may be monitored is the amount of creatinine in the blood. Creatinine is waste produced by muscles during metabolic processes. A high level of creatinine in the blood indicates that the kidneys are not filtering it properly.

Suggested Investigation

Inquiry Investigation 10-B, Urinalysis

Section 10.3 Review

Section Summary

- The kidneys maintain the water–salt balance in the body.
- ADH regulates the amount of water in the body by reabsorbing water in the distal convoluted tubule and the collecting duct.
- Aldosterone regulates the amount of salt in the body by reabsorbing sodium ions in the distal convoluted tubule.
- The kidneys maintain blood pH within narrow limits by excreting excess hydrogen ions and reabsorbing bicarbonate ions.
- The kidneys release two hormones: erythropoietin and calcitriol.
- Disorders of the excretory system include urinary tract infections, kidney stones, and renal insufficiency.
- Treatments for kidney failure include peritoneal dialysis and hemodialysis. Kidney transplants are the only way to cure chronic kidney failure.
- The health of the excretory system affects the health of other body systems. Maintaining healthy blood pressure is a key way to protect the excretory system.

Review Questions

1. K/U Identify how the amount of water reabsorbed from the filtrate influences two characteristics of blood.
2. K/U The diagram below illustrates how the release of antidiuretic hormone (ADH) controls the amount of water reabsorbed or excreted in the urine.
   a. Complete captions for the areas on this diagram labelled A to H.
   b. Use the diagram to explain why drinking alcoholic beverages stimulates urine production.

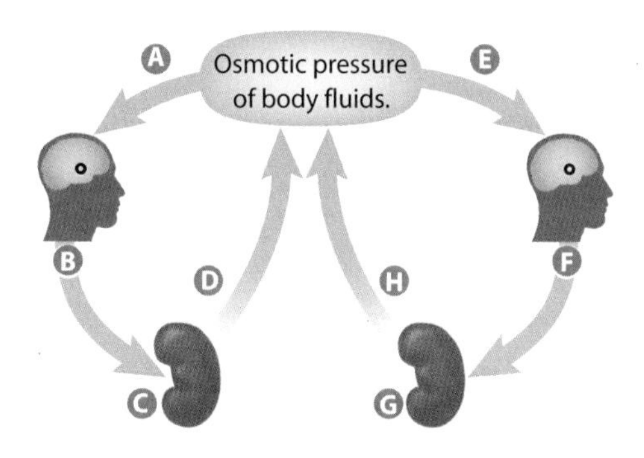

3. K/U Explain why an individual with diabetes insipidus urinates as much as 4 L to 8 L per day. Why would this individual have "intense thirst"?
4. A Explain what would happen if the body were unable to release the hormone aldosterone.
5. T/I Based on the data in the table below, identify the substances that are secreted and those that are reabsorbed. (You do not need to know what these substances are to answer this question.) Justify your answer.

Substance	Plasma (g/L)	Filtrate (g/L)	Urine (g/L)
Creatine	0.01	0.01	1.9
Uric acid	0.05	0.05	1.0
Bicarbonate ion	1.7	1.7	0.4

6. K/U Describe briefly how the acid–base balance of the blood is maintained.
7. T/I Predict the effect of very low blood pressure on kidney function.
8. A Imagine that you are adrift at sea and your supply of fresh water has run out. You are surrounded by water. Explain why you will become dehydrated if you drink the seawater.
9. C Design a table that summarizes the causes, symptoms, and treatments for the following disorders of the excretory system:
   a. urinary tract infections
   b. kidney stones
   c. renal insufficiency

Use the following information to answer question 10.

The colour and appearance of the urine specimen is recorded. Usual colours are colourless, straw, yellow, amber; less commonly pink, red, brown. Usual appearances (opacity) are clear or hazy; less commonly turbid, cloudy, and opaque, unless the specimen has remained at room or refrigerated temperatures.

The common chemical testing of urine utilizes commercial disposable test strips. Strips test for: glucose, ketone, blood, pH, protein, nitrite, and white blood cells.

10. A The urine of different individuals is analyzed.
    a. Describe the results you would expect in a urinalysis of a person who has no excretory system disorders.
    b. What might the presence of white blood cells in urine indicate?

BIOLOGY Connections

What's in Your Drinking Water?

Pharmaceutical medications are widely used by Canadians: each person in Canada will fill an average of 10 prescriptions annually. We rely on these drugs to act on our body as needed, and in turn we depend on our body systems to eliminate these drugs once this work is done.

In some areas of North America and Europe, residue from pharmaceutical medications has been found in trace amounts in drinking water.

Many medications that we consume, such as anti-depressants, birth control pills, and antibiotics, are excreted by our kidneys in urine. Medications excreted in urine are generally those that are water-soluble, or that have been metabolized by the body to become more water-soluble. This helps to prevent them from being reabsorbed from the kidneys back into the bloodstream. Some medications may be eliminated through other less common routes, such as in bile; these drugs may then exit the body in feces.

Pharmaceuticals in the urine and feces are literally flushed away, entering into the sewage system as wastewater. But the passage of pharmaceutical wastes through sewage pipes is only the beginning of the journey. Government agencies, researchers, and environmental groups in Canada are taking a look at the output from wastewater treatment plants (WWTPs). Traces of pharmaceutical medications are being found in the discharges that flow out into the waterways and may ultimately end up in our drinking water.

COMING THROUGH THE PIPES As the pharmaceuticals in urine and feces enter WWTPs, the liquid portion is separated from the solid waste. After several treatment steps, the liquid waste is discharged from the WWTP as treated water, or effluent. Release of effluent into waterways is the most direct route for pharmaceuticals to enter from WWTPs into the environment. Pharmaceuticals may also enter waterways indirectly through solid waste. During WWTP processing, the solid wastes form a semi-solid sludge with some liquid still remaining. The sludge, which may have some pharmaceuticals trapped or absorbed within, is then processed in the WWTP to form biosolids. Biosolids can be applied to agricultural land to improve soil quality, placed in landfills, or incinerated. When biosolids are land-applied, pharmaceuticals may travel through runoff into waterways. Waterways that capture discharges from WWTPs are used as source waters for drinking-water plants. Further treatment is needed before the water is to be used for drinking, but depending on the type of processing that occurs pharmaceuticals may remain in the finished water.

PHARMACEUTICALS IN ONTARIO DRINKING WATER In 2010, the Ontario Ministry of the Environment conducted a survey looking at selected pharmaceutical products in source and drinking water from WWTPs across the province. Most of the 46 types of compounds surveyed were rarely detected in source or drinking water. Pharmaceuticals most commonly found in the drinking water samples included anti-epileptic medication, cholesterol-regulating agents, and non-steroidal anti-inflammatory medications. Levels of these drugs were well below the maximum acceptable daily intake for drinking water. However, risks to human health and ecology from long-term consumption of traces of these other drugs found in tap water are largely unknown.

Connect to the Environment

1. Research studies are underway to identify and quantify pharmaceutical wastes in waterways, drinking water, and biosolids. What types of pharmaceutical wastes are causing the most concern? What are the main concerns to ecology and human safety, including the potential effects of traces of pharmaceuticals in drinking water on human homeostasis? What guidelines/regulatory standards are currently in place in Canada? What are some of the strategies being considered and implemented to manage risk associated with pharmaceutical wastes?
2. Opponents of land application of biosolids are concerned about the potential health impact of drugs that may persist in biosolids. Conduct research into the treatment of biosolids in your region. Develop a table listing the advantages and disadvantages of biosolids disposal through land application, incineration, and landfill. How would you respond if you were surveyed for the "Biosolids Master Plan" drafted by the City of Toronto in 2009: "Using a scale from 1–10 where 1 means you strongly oppose it and 10 means you strongly support it, please tell me how much you support/oppose application of biosolids on land? Incineration of biosolids? Landfilling biosolids?" Explain your rationale.

Inquiry
INVESTIGATION

Skill Check

Initiating and Planning

✓ Performing and Recording

✓ Analyzing and Interpreting

✓ Communicating

10-A

Identifying Structures of the Excretory System

In this investigation, you will perform a dissection of a sheep's kidney in order to identify the major parts of the organ.

Pre-Lab Questions

1. Describe the appearance of the renal capsule.
2. Where are the renal medulla and renal cortex in relation to each other?

Question

Which features of a mammalian kidney can you identify?

Safety Precautions

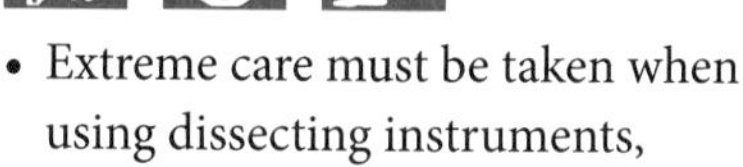

- Extreme care must be taken when using dissecting instruments, particularly scalpels.
- Dispose of all materials as instructed by your teacher, and clean your work area.

Materials

- dissecting tray
- newspapers and/or paper towels
- large tongs
- preserved sheep kidney
- dissecting instruments
- plastic bag and tie (to store your specimen if necessary)

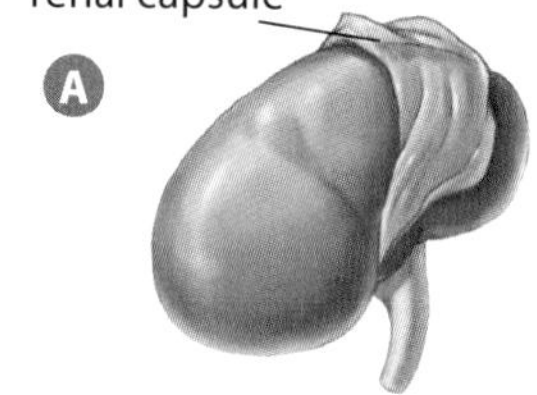

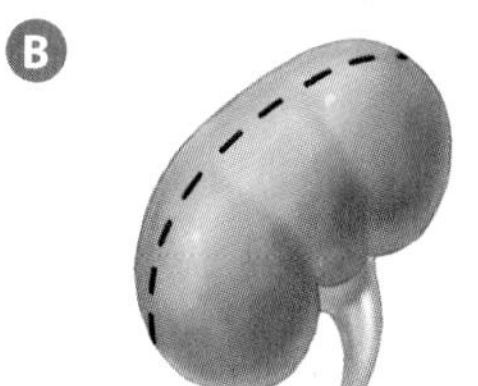

Cut away from you as you open the kidney.

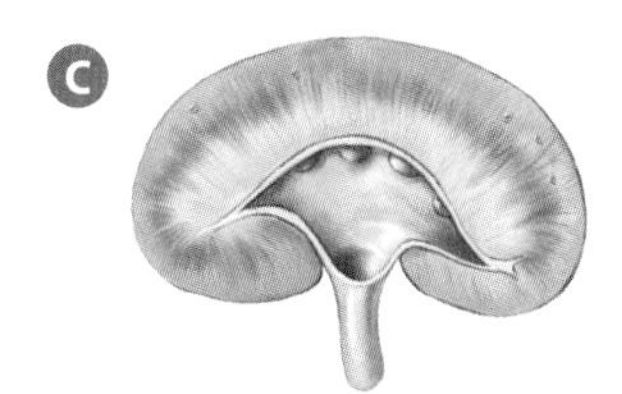

Internal features of the kidney

Procedure

1. Your teacher will give you a kidney. Observe its external features. The renal capsule is a smooth, semi-transparent membrane that is tightly bound to the outer surface of the kidney. Identify and remove the renal capsule.
2. Under the renal capsule is the surface of the renal cortex. Locate the area where the renal blood vessels and the ureter are attached to the kidney.
3. Cut through the kidney lengthwise as shown in the illustration on the left. Identify the renal cortex.
4. Locate the renal medulla. The renal medulla contains the collecting ducts. They are visible as a striped pattern throughout the medulla.
5. Locate the renal pelvis, which is continuous with the ureter.

Analyze and Interpret

1. Based on your specimen, draw a labelled sketch of the kidney that includes the following structures: renal capsule, renal cortex, renal pelvis, renal medulla, renal vein, and renal artery.

Conclude and Communicate

2. Refer to **Figure 10.3** and **Figure 10.4** and your sketch from question 1. Draw a new sketch of your specimen that shows the regions of the kidney in which you would expect to observe the following structures: glomerulus, proximal tubule, loop of Henle, distal tubule, and collecting duct.

Extend Further

3. **INQUIRY** Using available materials, make a model of a nephron. The basic structures of a nephron should be represented in the model. Include a card with the model that briefly details the passage of filtrate through a nephron.
4. **RESEARCH** Polycystic kidney disease (PKD) is a genetic disorder in which cysts grow in the kidneys. Research more information about PKD, including how it interferes with kidney function, its symptoms, how it is diagnosed, and how it is treated. Prepare a pamphlet about the disorder, similar to one you might see in the waiting room of a doctor who specializes in kidney disorders.

Inquiry INVESTIGATION 10-B

Skill Check

- Initiating and Planning
- ✓ Performing and Recording
- ✓ Analyzing and Interpreting
- ✓ Communicating

Safety Precautions

- Do not taste the simulated urine.
- If observing odour, follow safe and proper methods.
- Be careful when handling the simulated urine. Clean up spills immediately, and notify your teacher if a spill occurs.

Materials

- simulated urine samples
- 50 mL graduated cylinder
- 100 mL beaker
- 1 test tube rack
- 5 test tubes
- hot water bath
- thermometer
- 1 medicine dropper
- universal indicator paper with colour charts
- glucose test strips with colour charts

Urinalysis

Urinalysis is the physical, chemical, and sometimes microscopic examination of urine. Many diseases with no obvious symptoms can be revealed during urinalysis. Long before modern techniques such as chemical dipsticks and microscopic analysis were developed, medical practitioners used the appearance, odour, and even taste of urine to help them make inferences about a person's health. In this investigation, you will test samples of simulated urine to identify an imaginary criminal. In so doing, you will perform several of the tests that are performed when health professionals do a urinalysis.

Pre-Lab Questions

1. What information can be learned from the odour of urine?
2. How will you test the pH of the urine samples?
3. What is the proper procedure to carry out if a spill occurs?

Question

What physical and chemical tests can you use, and what data do they provide, in the analysis of urine?

Procedure

Consider this scenario. A theft was committed in the washroom of a community building. Forensic specialists collected a urine sample at the scene of the crime. The police have four suspects in custody. Your task is to determine who committed the crime.

1. Copy the data table at the top of the next page into your notebook.
2. You will start by doing a trial run of four tests to determine what information they provide. This will serve as the control for this investigation.
3. Perform the following tests on your simulated urine sample.

Test 1—Colour, Odour, Clarity: Normal urine is a clear, straw-coloured liquid. Urine may be cloudy because it contains red or white blood cells, bacteria, or pus from a bladder or kidney infection. Normal urine has a slight odour. Foul-smelling urine is a common symptom of urinary tract infection. A fruity odour is associated with diabetes mellitus. Determine the colour, odour, and clarity of your simulated urine.

a. Use the graduated cylinder to obtain 20 mL of the control urine sample.
b. Place the control sample into the beaker.
c. Examine the urine carefully. Record the colour and the clarity (clear or cloudy) in your data table.
d. Using the proper technique, as instructed by your teacher, determine the odour of the urine. Record your observations in your data table.

Test	Control Tests	Crime Scene	Suspect 1	Suspect 2	Suspect 3	Suspect 4
1. Colour/odour/clarity						
2. Protein						
3. pH						
4. Glucose						

Test 2—Protein: One sign of kidney damage is the presence of protein in urine. Determine whether the sample contains protein by doing the following.

a. Use the graduated cylinder to divide the sample equally between two test tubes (10 mL into each test tube).
b. Put one test tube into the hot water bath with a temperature of 70°C, and leave the other at room temperature.
c. After a few minutes, remove the test tube from the hot water bath and compare the heated and unheated urine.
d. If the heated sample is cloudier, it contains protein. Record your observations.
e. Dispose of the heated sample as directed by your teacher. Use the unheated sample for the next test.

Test 3—pH: The wide range of pH values (pH 4.7 to 8.5) makes this the least useful parameter for diagnosis of kidney disorders. Kidney stones are less likely to form and some antibiotics are more effective in alkaline urine. There may be times when acidic urine may help prevent some kinds of kidney stones. Bacterial infections also increase alkalinity, producing a urine pH in the higher 7–8 range.

a. Use a clean medicine dropper to place a drop of the urine on a small piece of universal indicator (pH) paper.
b. Leave the paper for about 30 seconds.
c. Determine the pH by comparing the new colour with the colour chart provided.
d. Record the pH of your urine sample in your data table.

Test 4—Glucose: One sign that a person has diabetes mellitus is the presence of glucose in urine. Determine whether the sample contains glucose by doing the following.

a. Dip a glucose test strip into the test tube of unheated urine sample and immediately take it out.
b. Count to 10, then check the colour with the glucose colour chart.
c. Record whether the results are negative, light, medium, or dark. (The darker the colour, the greater the amount of glucose.)

4. Your group will be assigned to test one of the remaining samples of simulated urine. One sample was collected at the crime scene. The others have been provided by the four suspects in police custody. Run the four urinalysis tests and record your observations.
5. Your teacher will provide instructions on how to compile each group's data for analysis.

Analyze and Interpret

1. Which suspect do you think committed the crime?
2. Explain how you arrived at this conclusion.

Conclude and Communicate

3. Based on your urinalysis, identify the disease that Suspect 4 might have. Explain.
4. List at least three other characteristics of urine that you would expect to observe (or not) in a healthy urine sample.
5. In what ways were the data that you collected in this urinalysis limited? What additional data would provide a more comprehensive picture of a urine sample?

Extend Further

6. **INQUIRY** The urine of athletes is routinely tested for evidence that they may have taken performance-enhancing drugs. Based on your understanding of urine formation, hypothesize how molecules of a drug could appear in a person's urine.
7. **RESEARCH** Gather information about glomerulonephritis, inflammation of the glomeruli. What abnormalities would you expect to find in the urine of a patient with this condition? Why? What are some possible causes of glomerulonephritis? How can this condition affect blood pressure?

Chapter 10 SUMMARY

Section 10.1 Overview of the Excretory System

The excretory system removes liquid waste from the body and consists of the kidneys, ureters, the urinary bladder, and the urethra.

Key Terms

excretion
excretory system
urine
kidney
ureter
urinary bladder
urethra
renal artery
renal vein
nephron
Bowman's capsule
glomerulus
filtrate
collecting duct

Key Concepts

- The human excretory system is responsible for removing liquid waste from the body. The excretory system also regulates the acid–base balance and water–salt balance of the blood and secretes some hormones.
- The kidneys are the primary excretory organs. Kidney tissue is organized into three regions: the renal cortex, the renal medulla, and the renal pelvis.
- The kidneys are composed of millions of functional units called nephrons that filter the waste from the blood and produce urine.
- Urine leaves the kidney through tubes called ureters. It passes into the bladder and exits the body through the urethra.

Section 10.2 Urine Formation in the Nephron

Urine forms after the filtrate passes through the nephrons.

Key Terms

glomerular filtration
tubular reabsorption
tubular secretion
water reabsorption
proximal tubule
loop of Henle
distal tubule

Key Concepts

- In the glomerulus, filtration moves water and solutes (except for protein) from blood plasma into the nephron.
- Solutes are actively transported from the filtrate in the proximal tubule back into the blood.
- Approximately 65 percent of the filtrate that passes through the entire length of the proximal tubule is reabsorbed and returned to the body, while the urine becomes concentrated.
- Filtrate moves through the nephron, and the processes of glomerular filtration, tubular reabsorption, tubular secretion, and water reabsorption modify the filtrate so that cleansed plasma and substances such as glucose, amino acids, and Na^+ return to the blood.

Section 10.3 Other Functions and Disorders of the Excretory System

The kidneys play an important role in the water–salt balance of the body and the pH of the blood and can affect the health of other body systems.

Key Terms

antidiuretic hormone
renal insufficiency
dialysis
hemodialysis
peritoneal dialysis

Key Concepts

- The kidneys maintain the water–salt balance in the body.
- ADH regulates the amount of water in the body by reabsorbing water in the distal convoluted tubule and the collecting duct.
- Aldosterone regulates the amount of salt in the body by reabsorbing sodium ions in the distal convoluted tubule.
- The kidneys maintain blood pH within narrow limits by excreting excess hydrogen ions and reabsorbing bicarbonate ions.
- The kidneys release two hormones: erythropoietin and calcitriol.
- Disorders of the excretory system include urinary tract infections, kidney stones, and renal insufficiency.
- Treatments for kidney failure include peritoneal dialysis and hemodialysis. Kidney transplants are the only way to cure chronic kidney failure.
- The health of the excretory system affects the health of other body systems. Maintaining healthy blood pressure is a key way to protect the excretory system.

Chapter 10 REVIEW

Knowledge and Understanding

Select the letter of the best answer below.

1. All of the following are examples of excretion *except*
   - **a.** the lungs exhaling carbon dioxide.
   - **b.** the elimination of solid wastes (feces) from the digestive system.
   - **c.** the removal of nitrogenous wastes in urine.
   - **d.** the excretion of urea in perspiration.
   - **e.** the excretion of lipids, water, and salts from the digestive system.

Use the diagram below to answer questions 2 and 3.

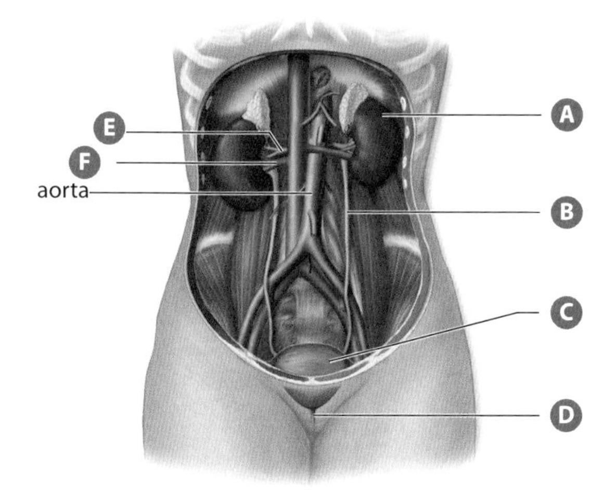

2. Which choice below includes a structure that is correctly matched to its function?
   - **a.** Structure A—transport of urine to urinary bladder
   - **b.** Structure B—delivery of blood to kidney
   - **c.** Structure C—reabsorption of water
   - **d.** Structure D—passage of urine to the environment
   - **e.** Structure E—production of urine
3. In which structure are nephrons located?
   - **a.** Structure A
   - **b.** Structure B
   - **c.** Structure C
   - **d.** Structure D
   - **e.** Structure E
4. Which statement is *false*?
   - **a.** Water is found in the glomerular filtrate.
   - **b.** Water is reabsorbed from the collecting duct by active transport.
   - **c.** A person with diabetes insipidus loses as much as 4 L to 8 L of water per day.
   - **d.** The amount of water lost in urine is controlled by osmoreceptors in the hypothalamus.
   - **e.** Renal insufficiency is a condition in which the kidneys cannot maintain homeostasis.
5. How do the kidneys respond if the blood is too acidic?
   - **a.** excrete hydrogen ions (H^+) and reabsorb bicarbonate ions (HCO_3^-)
   - **b.** excrete bicarbonate ions (HCO_3^-) and reabsorb hydrogen ions (H^+)
   - **c.** absorb both hydrogen ions (H^+) and bicarbonate ions (HCO_3^-)
   - **d.** increase respiration rate to rid the body of CO_2
   - **e.** decrease respiration rate to slow the loss of CO_2
6. By which transport process are most molecules secreted from the blood into the distal tubule?
   - **a.** osmosis
   - **b.** diffusion
   - **c.** facilitated diffusion
   - **d.** active transport
   - **e.** passive transport
7. All of the following are symptoms of a urinary tract infection *except*
   - **a.** the development of crystalline formations in the kidney.
   - **b.** painful burning sensation during urination.
   - **c.** a need to urinate frequently even when no urine is present.
   - **d.** bloody or brown urine is produced.
   - **e.** the upper abdomen or lower back may be tender.
8. Which statement about kidney transplants is *false*?
   - **a.** A minimally invasive surgical technique used to remove the kidney is called laparoscopy.
   - **b.** The success rate for kidney transplants from living donors is generally higher than the success rate for transplants from organ donors that have suddenly died.
   - **c.** New medicines to prevent rejection of the new kidney are constantly being developed and improved.
   - **d.** In Canada, there is an oversupply of kidneys that are available for kidney transplants.
   - **e.** Individuals with 10 percent or less kidney function will eventually have to replace their kidneys.

Answer the questions below.

9. Explain how the process of active transport within the kidney helps control the volume of urine produced during the day.
10. Explain how increased secretion of ADH affects urine concentration and urine volume. Identify a possible situation in which the hypothalamus may be stimulated to release ADH.

Chapter 10 REVIEW

11. Explain why proteins and blood are not normally found in urine.

12. Identify the parts of the nephron that are most involved in regulating the pH level of the blood.

13. Study the diagram of the excretory system below.
   a. Identify the structures labelled A to F in the diagram.
   b. Describe their function in terms of the excretory system.

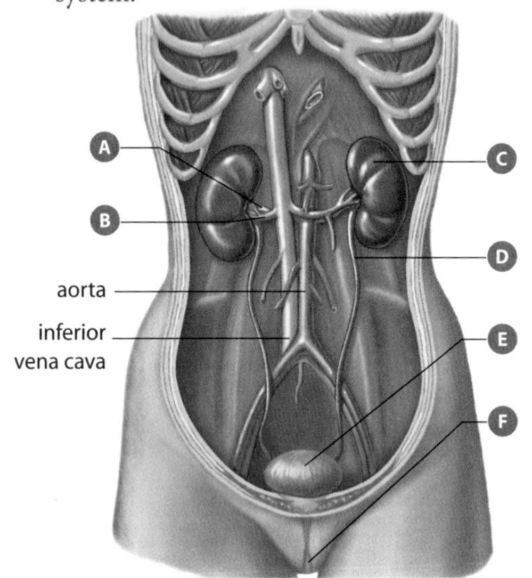

14. Identify four differences between the blood entering the kidney in the renal artery and the blood leaving the kidney in the renal vein.

15. Identify four types of dissolved substances found in the filtrate that forms in the Bowman's capsule of the nephron.

16. A kidney stone is a hard granule that can form in the renal pelvis.
   a. Describe possible factors involved in the development of a kidney stone.
   b. Identify treatments for this condition.

17. Draw a flowchart or other graphic organizer to show the response of the body after drinking several cups of water on a cool evening.

18. Distinguish among the processes of filtration, tubular secretion, reabsorption, and osmosis as they relate to the process of urine formation.

19. How would each of the following processes in the kidney change in response to serious dehydration in the body caused by excessive sweating on a hot day?
   a. glomerular filtration
   b. tubular reabsorption
   c. ADH secretion

20. The diagram below shows a nephron in detail. Identify each of the structures labelled A to K.

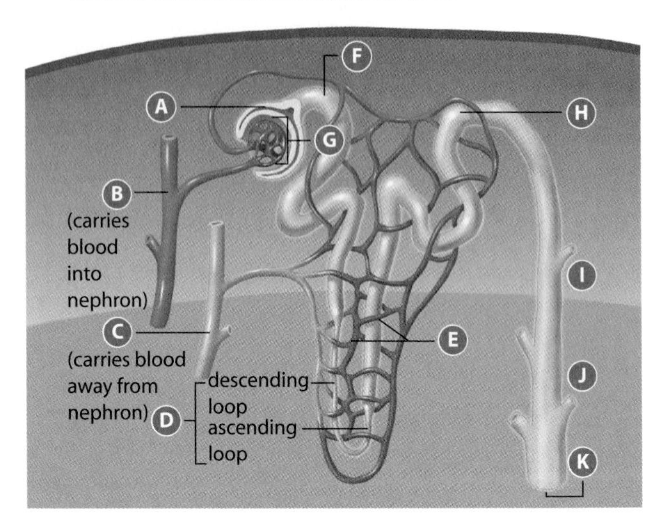

21. Describe the functions of the following structures.
   a. glomerulus
   b. Bowman's capsule
   c. proximal tubule
   d. descending loop of Henle
   e. ascending loop of Henle
   f. distal tubule
   g. collecting duct

Thinking and Investigation

Use the information below to answer questions 22–25.

A diminished number of nephrons has been proposed as one of the factors contributing to the development of high blood pressure. To test this hypothesis, a group of scientists compared the number and volume of glomeruli in 10 patients with a history of high blood pressure to the number and volume of glomeruli in 10 subjects without high blood pressure. Subjects in both groups were matched for ethnicity, gender, age, height, and weight. All 20 subjects had died in accidents. The results showed that patients with hypertension had significantly fewer glomeruli per kidney than matched controls. As well, patients with hypertension also had a significantly greater glomerular volume than did the controls.

22. Identify the hypothesis tested in this investigation.

23. List five variables that were controlled in this investigation.

24. Did the results support or reject the hypothesis of the investigators? Explain.

25. Infer why the patients with hypertension had significantly greater glomerular volume (larger glomeruli) than the control subjects.

Use the information below to answer questions 26–30.

Two subjects participated in a urinalysis investigation. Subject A was given 1 L of distilled water to drink while Subject B was given nothing to drink during the investigation. The subjects each ate a light meal the evening prior to the investigation, after which they ate nothing eight hours prior to the start of the investigation. The types of food and fluid taken in prior to the investigation were noted. Wide mouth collecting bottles with 1 L capacity were used for urine collection. Graduated cylinders were used for measurement of the urine volume. Information about how the samples were collected and the volume of each sample is listed below.

- Sample #1: The subjects collected urine at the start of the laboratory procedure. Urine from the first morning urination was not used in the investigation. The time of collection was noted.
- Sample #2: After 30 min, the subjects collected the second sample. As soon as the second sample was collected, one of the subjects drank 1 L of distilled water.
- Sample #3: The third sample was collected 30 min after drinking the water.
- Sample #4: The fourth sample was collected approximately 90 min after the one subject drank the distilled water.

Effect of Drinking Distilled Water on Urine Output

Sample	Subject A Volume of Urine (mL)	Subject B Volume of Urine (mL)
Sample #1	133	17
Sample #2	12	7
Sample #3	44	8
Sample #4	504	13

26. Graph the results of this investigation.

27. Identify the independent and dependent variables for this investigation.

28. Which subject drank the 1 L of distilled water? Explain your answer in terms of antidiuretic hormone production and water reabsorption.

29. Which subject did not drink the distilled water? Explain your answer in terms of antidiuretic hormone production and water reabsorption.

30. Identify two variables that were controlled in this investigation.

Use the following information to answer questions 31 and 32.

In diseases and conditions classified as *glomerulonephritis* (also called *nephritis* or *nephritic syndrome*), the glomerulus becomes inflamed and scarred. Most often it is caused by an autoimmune disease, but it can also result from infection.

31. Predict how nephritis might affect urinary output and the removal of nitrogenous wastes from the blood.

32. Provide a possible explanation why the urine of a person with nephritis is usually brown instead of yellow in colour.

Communication

33. The process of dialysis filters blood in people with reduced kidney function or kidney failure. Use a graphic organizer to compare the process of dialysis to that of blood filtration in the human kidney.

34. Create a flowchart that illustrates how the filtrate is modified by the processes of active transport and passive transport as it flows through a single nephron in the human kidney.

35. Using a labelled diagram, illustrate the regions of the kidney and the interaction between the circulatory system and the excretory system. Write a caption for your diagram.

Use the information below to answer questions 36 and 37.

Consider the following cost comparison of dialysis versus kidney transplants. Dialysis cost approximately $50 000 per year in 2006. A kidney transplant operation at this time cost approximately $20 000 and required $6000 per year in follow-up treatments. Kidney transplants have a 98 percent success rate using kidneys from a living donor and a 95 percent success rate using kidneys from a deceased donor.

36. In your opinion, to what extent should cost be considered as an important variable when people make decisions about whether to choose a transplant or dialysis?

37. What additional information would you like to have to help you write an informed, reasoned answer to this question?

38. Use the Internet or other references to research polycystic kidney disease (PKD). Write a paragraph summarizing the causes, symptoms, and treatments for this disorder.

39. Explain the following statement in your own words.

"Generally, the greater the concentration of atypical substances such as increased amounts of glucose, protein, or red blood cells in the urine, the more likely it is that there is a problem that needs to be addressed."

40. BIG IDEAS Systems that maintain homeostasis rely on feedback mechanisms. Using a flowchart or other graphic organizer, summarize what happens when the osmotic pressure of body fluids becomes too high.

Use the information below to answer questions 41–43.

Chronic kidney disease (CKD) is the slow loss of kidney function over time. CKD slowly gets worse over time. In the early stages, there may be no symptoms. The loss of function usually takes months or years to occur. It may be so slow that symptoms do not occur until kidney function is less than one tenth of normal. Chronic kidney disease leads to a buildup of fluid and waste products in the body. This condition affects most body systems and functions, including red blood cell production, blood pressure control, and vitamin D and bone health.

41. Explain how CKD would impact red blood cell production.

42. Consider the role the kidneys play in the production of vitamin D and bone health. Summarize this information in a form that a person without a science or medical background would understand. Include an explanation of how CKD would affect kidney function in this capacity.

43. Explain the following statement: "Hypertension (high blood pressure) causes CKD and CKD causes hypertension."

44. Many school management systems already do, or have considered implementing, random drug testing (collecting a urine sample, then testing it for the presence of certain illegal and/or performance-enhancing drugs).

a. What ethical issues must be considered before implementing such a policy?

b. What is your opinion on randomly testing high school students for the presence of drugs in their urine? Support your position.

45. Summarize your learning in this chapter using a graphic organizer. To help you, the Chapter 10 Summary lists the Key Terms and Key Concepts. Refer to Using Graphic Organizers in Appendix A to help you decide which graphic organizer to use.

Application

Use the information below to answer questions 46–49.

Research shows that dehydration can lead to reduced performance levels in athletes and slow reaction time by up to 30 percent. Studies on dehydration and mental performance show that mental acuity and coordination are already decreasing at 1 percent dehydration, and are consistently and significantly decreased at 2 percent dehydration. The brain is 75 percent water and, when it needs to replace lost fluid, it can manifest certain symptoms including headaches, lightheadedness, and fatigue. Dehydration can also contribute to "fuzzy thinking," poor decision making, dizziness, and muscle fatigue.

46. BIG IDEAS Environmental factors can affect homeostasis. Identify some causes of dehydration. Use reference materials as needed to answer this question.

47. A runner drinks three cups of caffeinated coffee to prevent becoming dehydrated while completing a marathon. Do you think this practice will help prevent dehydration or promote dehydration? Explain your answer.

48. BIG IDEAS Organisms have strict limits on the internal conditions they can tolerate. Explain how the kidneys respond directly to dehydration and the resulting reduction in blood volume and blood pressure.

49. Explain the following statement: "Because thirst begins after we are mildly dehydrated, a person should not wait until he or she is thirsty to begin hydrating."

50. Use the Internet or other references to research hemolytic uremic syndrome (HUS), commonly referred to as "hamburger disease."

a. Write a paragraph summarizing the causes, symptoms, and treatments for this disorder.

b. What changes to food safety practices, including inspections or regulations, might help reduce the incidence of HUS?

c. Outline a plan to notify the public and contain an outbreak of HUS as quickly as possible.

51. If you were to perform a urinalysis on the filtrate removed from a peritoneal dialysis machine, what would you expect to find in a chemical analysis? Assume the person is healthy other than suffering kidney failure.

Use the information below to answer questions 52–57.

Glomerular filtration rate (GFR) is the volume of fluid filtered from the renal (kidney) glomerular capillaries into the Bowman's capsule per unit time. Creatinine clearance rate (C_{Cr}) is the volume of blood plasma that is cleared of creatinine per unit time and is a useful measure for approximating the GFR. Creatinine is derived from the metabolism of creatine in skeletal muscle and from dietary meat intake; it is released into the circulation at a relatively constant rate and has a stable plasma concentration. Creatinine is freely filtered across the glomerulus and is neither reabsorbed nor metabolized by the kidney. However, approximately 15 percent of urinary creatinine is derived from tubular secretion in the proximal tubule. If the effect of secretion is ignored, then the C_{Cr} can be calculated from the creatinine concentration in the collected urine sample (U_{Cr}), urine flow rate (V), and the plasma concentration (P_{Cr}). The rate is expressed in mL/min. Normal values for adult males are about 120 mL/min (+/–25 mL/min), while normal values for adult females are about 110 mL/min (+/–20 mL/min).

The formula to calculate the C_{Cr} is below:

$$C_{Cr} = \frac{U_{Cr} \times V}{P_{Cr}}$$

52. A male patient has a plasma creatinine concentration of 0.01 mg/mL and produces urine at a rate of 1 mL/min. The urine produced has a creatinine concentration of 1.25 mg/mL. Calculate the C_{Cr} for this individual.

53. Are the kidneys of the patient in question 52 working properly? Explain how you reached your conclusion.

54. A female patient comes into the hospital after losing a lot of blood in a motorcycle accident. After her condition is stabilized she is given a creatinine clearance test to estimate her glomerular filtration rate. The results of her C_{Cr} are shown below. Use the formula to calculate her creatinine clearance rate in (mL blood/min).

- Plasma creatinine levels = 0.1 mg creatinine/mL blood
- Urine creatinine levels = 16 mg creatinine/mL urine
- Urine production = 0.5 mL urine/min

55. Are the kidneys of the patient in question 54 working properly? Explain your answer.

56. Infer what might happen in terms of reabsorption as the filtrate passes through the renal tubules if GFR is too high.

57. Infer what might happen in terms of reabsorption as the filtrate passes through the renal tubules if GFR is too low.

Use the information below to answer questions 58–60.

Urine testing is still is one of the more important tests to diagnose diseases and illnesses in some individuals. For example, maple syrup urine disease (MSUD) is an inherited disorder in which the body is unable to process the amino acids leucine, isoleucine, and valine properly. The more common form of the disease produces symptoms in infants between the ages of 4 and 7 days. The by-product of isoleucine has a characteristic sweet smell. The condition gets its name from the distinctive maple syrup aroma of an affected infant's urine. If left untreated, this leads to brain damage and progressive nervous system degeneration.

58. Describe what parents should do if they notice their newborn baby has abnormal-smelling urine.

59. Infer how this condition could be treated through a special diet that includes dietary restrictions.

60. Use the Internet to research "electronic noses for diagnosing diseases." Summarize these technologies and explain how they could be used to diagnose MSUD.

61. Explain how alcohol, which is a diuretic, affects the concentration of urine produced.

62. Use the information in the table below to explain what is happening to each substance as it passes through the different regions of the nephron.

a. water

b. sodium (salts)

c. glucose

d. urea

Contents of Filtrate and Urine

Substance	Glomerular Filtrate (per day)	Urine (per day)	Reabsorption (percent)
Water (L)	180	1.8	99.0
Sodium (g)	630	3.2	99.5
Glucose (g)	180	0.0	100.0
Urea (g)	54	30.0	44.0

63. People with hyperparathyroidism have an increased risk of developing kidney stones. Explain the connection between these two disorders.

Chapter 10 SELF-ASSESSMENT

Select the letter of the best answer below.

1. **K/U** Urine is produced from blood that enters the kidney through which structure?
   a. nephron
   b. renal artery
   c. renal vein
   d. ureter
   e. urethra

2. **K/U** Which is the structure into which filtrate first passes?
   a. renal artery
   b. glomerulus
   c. Bowman's capsule
   d. distal tubule
   e. proximal tubule

3. **K/U** Which structure carries urine into the renal pelvis?
   a. renal artery
   b. glomerulus
   c. Bowman's capsule
   d. renal vein
   e. collecting duct

4. **K/U** The excretion of concentrated (hypertonic) urine in humans is associated primarily with the action of antidiuretic hormone (ADH) in the
   a. glomerular capsule and proximal tubule.
   b. proximal tubule only.
   c. descending loop of Henle and the collecting duct.
   d. ascending loop of Henle and the distal tubule.
   e. distal tubule and the collecting duct.

5. **[illegible]** Identify the statement that explains the homeostatic response of an adult body that is well hydrated.
   a. Osmoreceptors stimulate the release of ADH and the collecting duct becomes more permeable to water.
   b. Osmoreceptors stimulate the release of ADH and the collecting duct becomes less permeable to water.
   c. Osmoreceptors inhibit the release of ADH and the collecting duct becomes more permeable to water.
   d. Osmoreceptors inhibit the release of ADH and the collecting duct becomes less permeable to water.
   e. Because the individual is well hydrated, no actions are necessary to maintain homeostasis.

6. **K/U** The high level of Na^+ ions in the tissues of the renal medulla is primarily the result of active transport of Na^+ ions out of which of the following?
   a. descending loop of the Henle
   b. ascending loop of the Henle
   c. collecting duct
   d. proximal tubule
   e. Bowman's capsule

7. **K/U** Tubular secretion of glucose from the glomerular filtrate into the capillary bed surrounding the nephron occurs in which of the following?
   a. distal tubule
   b. glomerular capsule
   c. proximal tubule
   d. loop of Henle
   e. Bowman's capsule

Use the information in the table below to answer questions 8 and 9.

Composition of Plasma, Glomerular Filtrate, and Urine of a Healthy Adult (g/100 mL fluid)

Component	Plasma	Glomerular Filtrate	Urine
Urea	0.03	0.03	2.00
Uric acid	0.004	0.004	0.05
Glucose	0.10	0.10	0.00
Amino acids	0.05	0.05	0.00
Salts	0.72	0.72	1.50
Proteins	8.00	0.00	0.00

8. **K/U** Which two components are completely reabsorbed from the glomerular filtrate before it becomes urine?
   a. urea and uric acid
   b. proteins and salts
   c. glucose and amino acids
   d. urea and glucose
   e. amino acids and salts

9. **K/U** Which statement is *true*?
   a. Glucose is the only component of the filtrate to be completely reabsorbed and returned to the cardiovascular system.
   b. Uric acid is the most abundant component found in the urine.
   c. Glomerular filtrate can be described as plasma without the proteins.
   d. Most of the urea found in the filtrate is reabsorbed and returned to the cardiovascular system.
   e. Salts in the filtrate are completely reabsorbed and returned to the plasma.

10. **K/U** A treatment for renal insufficiency is to have a flexible tube surgically inserted into the abdominal cavity. Dialysate is always present so the blood is continuously filtered. Which treatment is described?
    a. hemodialysis
    b. kidney transplant
    c. peritoneal dialysis
    d. ultrasound technology
    e. urinalysis

Use sentences and diagrams as appropriate to answer the questions below.

11. K/U What is the basic function of the excretory system?

12. C Nephrons are the filtration units of the kidneys.

a. Illustrate the locations of the nephron that are involved in water absorption.

b. Explain how the excretory system would respond if you drank very little water on a hot day. Modify your answer in part (a) to include this information.

13. K/U Describe the composition of the filtrate as it moves up the ascending limb of the loop of Henle toward the distal tubule.

14. C Acidosis and alkalosis are terms used to describe the abnormal conditions that result from an excess of acid or alkali (base) within the blood. Use the Internet or other reference materials to research these two disorders and then write a paragraph briefly explaining each of them.

15. T/I Design a simple investigation that would support the hypothesis that if a healthy individual drinks one litre of a 5 percent glucose solution, then their urine will have no glucose in it.

16. K/U Identify the two characteristics of blood that influence how much water is reabsorbed from the filtrate.

17. T/I Explain the results of an investigation where researchers collected four urine samples. The first two samples were taken prior to an individual drinking one litre of a 0.9 percent saline solution (NaCl). Samples 3 and 4 were taken after the same individual drank the saline solution. The results are shown in the table below.

Urine Volume Results

Sample	Volume of Urine Produced (mL)
Sample #1	28
Sample #2	9
Sample #3	no urine produced
Sample #4	9

18. T/I Albumin, a protein produced by the liver, makes up about 60 percent of the protein in the blood. Creatinine is a breakdown product of creatine phosphate in muscle. Creatinine is mostly filtered out of the blood by the kidneys. Protein is not normally found in the urine, and a urine protein test detects and/or measures the amount of protein being excreted in the urine. The test result is reported as a ratio of protein to creatinine. The normal ratio of albumin to creatinine is less than 30 mg albumin per gram of creatinine. A doctor orders a routine urine test for an individual. The albumin to creatinine ratio comes back at 250 mg/g. Predict what might be occurring in this individual's kidneys and explain why.

19. C Using a series of diagrams, summarize what happens to the filtrate as it moves through the various regions in the loop of Henle. Include captions with your diagrams.

20. A Compare the reabsorbed filtrate components to the non-reabsorbed filtrate components.

21. A Atherosclerosis is a condition in which patchy deposits of fatty material develop in the walls of medium-sized and large arteries, leading to reduced or blocked blood flow. Atherosclerosis is a vascular disease frequently associated with acute kidney failure. Explain how atherosclerosis developing in the renal artery is related to acute kidney failure.

22. K/U How do the excretory and respiratory systems work together to regulate blood pH?

23. A As with other organs, kidney function may be slightly reduced with aging. As the kidneys age, the number of nephrons decreases, the overall amount of kidney tissues also decreases, and the blood vessels supplying the kidney can harden. Explain how these anatomical changes could affect renal functions.

24. C Write a short presentation about dialysis that you could present to a Grade 6 class. Include information about why an individual would need dialysis and the two types of dialysis.

25. C Write three original multiple-choice questions about disorders of the excretory system. Each question should have five answer choices.

Self-Check

If you missed question...	1	2	3	4	5	6	7	8	9	10	11	12	13	14	15	16	17	18	19	20	21	22	23	24	25
Review section(s)...	10.1	10.1	10.1	10.3	10.3	10.2	10.2	10.2	10.2	10.3	10.1	10.2	10.2	10.3	10.2	10.3	10.3	10.3	10.2	10.2	10.3	10.3	10.3	10.3	10.3

Unit 4 Project

Conduct an Inquiry

Testing the Effects of Caffeine on Human Homeostasis

You may have already made some observations about the effects of caffeine on humans. For example, if you drink caffeinated beverages, such as soda or coffee, you may have noticed that you feel more alert or energized after consuming them. Also, recall that caffeine is a diuretic. In this inquiry, you will design an investigation to test the effects of caffeine on human homeostasis. As part of the requirements, your experimental design will include a plan for conducting trials using adult human participants, not chemicals in a test tube or invertebrates, as you have often done in previous investigations. Because of this, the design of your study must ensure clear, unbiased, observable results, as well as minimize the risks to the health and welfare of the test subjects. Note that you will design this experiment, but will not carry it out unless specifically instructed to do so by your teacher.

How can you design an investigation to observe the effects of caffeine on homeostasis?

Initiate and Plan

1. To be statistically reliable, experimental designs involving humans require large numbers of test volunteers. One way to increase the number of participants is to have each student include in his or her design a plan to recruit a small number of adult volunteers. However, to analyze group data you need to ensure that individual investigations are standardized. As you design your investigation, work through the following questions to ensure your results would be compatible with those obtained by other students.

Perform and Record

1. To help avoid bias, research what a double blind investigation is and when it is used. As you continue with the design process, consider whether a double blind investigation is practical in this case.
2. Research more about the ethical considerations and requirements of using humans in an investigation. How will this affect your experimental design? What type of consent form will you create?
3. Research the known effects of short-term and long-term use of caffeine on humans, including the dosages considered safe and the LD50. As a class, discuss what safety considerations might be necessary when using a drug, even a reportedly safe one like caffeine, as part of an investigation.
4. Begin your experimental design by considering what variables you would measure to observe the effects of caffeine on body systems. Some questions that may help guide you include:
   - Which body systems would you observe? Caffeine can affect the endocrine system, the nervous system, the excretory system, and possibly more.
   - How would you collect quantitative data, such as blood pressure, pulse rate, or urine output, to observe the effects of caffeine on body systems? What would you do to keep the observations consistent?

Nearly 90 percent of adults are thought to consume caffeine every day, often in the form of a caffeinated beverage such as coffee, tea, soda, or energy drinks. The effects often felt when someone skips their morning "dose" of this drug would suggest that there is at least a psychological, if not physiological, dependence that comes with long-term use.

- How would you collect qualitative data, such as increased alertness or nervousness after consumption of caffeine? Consider having volunteers complete a self-assessment checklist. To avoid response bias, your checklist should have some symptoms that you would expect to see, some that you would not, and some that are contradictory, such as increased calmness and increased nervousness.
- Whom would your test group include? How would you control for variables such as gender, age, ethnicity, the presence of existing health conditions, and the quantity of caffeine a subject normally consumes?
- What would be the duration of the data collection? Would you conduct repeated trials?

5. Decide as a class whether you would standardize the volume of liquid that is ingested and/or the amount of caffeine that is consumed. Points to consider:
   - What would you use for a control? Would it be something known to be harmless, like an equivalent volume of distilled water, or something as similar to the caffeinated drinks as possible, such as the same drink, only decaffeinated? List the advantages and disadvantages of each and decide on your control.
6. Create a data table in which you would record your data. It should include any independent and dependent variables. It should contain a unique identifying number for each patient.
7. Consider if or how you would pool your data with the rest of the class. The greater the amount of data, the more reliable the results of an investigation are considered to be.
8. Create a consent form for your participants. It should inform them of the nature of the investigation, the times to drink their fluids, the variables they would have to observe, how their information should be measured and recorded, as well as information about how their privacy will be protected. Finally, you may have to make a provision to share some of the final results with those people who wish to know what you have learned.
9. Make a list of any materials and equipment you would need to conduct the investigation and collect data. Would it be necessary to train your subjects on how to use any of the equipment?
10. Write a comprehensive description of your design.

Analyze and Interpret

1. Determine how you would identify whether the independent variable had a significant effect on the dependent variable. For example, if 90 percent of test subjects experienced a certain reaction to the caffeine, would you consider this to be significant?
2. Review your design and identify sources of error in your investigation. What might cause bias in your data?

Communicate Your Findings

3. Write a one-page summary of your design that would be shared with potential participants of the study.
4. Some schools have banned drinks containing caffeine. Based on your research, justify whether you do or do not support this decision.
5. Imagine that caffeine is a new discovery. Based on your research, create an advertisement for your new product. What property will you emphasize? How will you inform people about the side effects and/or addictive qualities?
6. Write a paragraph describing what you learned about investigations involving human test subjects.

Assessment Criteria

After you complete your project, ask yourself these questions. Did you…

- ✓ **K/U** gather information about the effects of caffeine from a variety of respected sources?
- ✓ **T/I** plan and conduct your investigation carefully, keeping appropriate safety and ethical requirements in mind?
- ✓ **C** design an appropriate and accurate recording system?
- ✓ **C** create a consent form that includes clear instructions for your test subjects?
- ✓ **T/I** include in your design a way to protect the privacy of the test subjects, while allowing their information to be gathered and shared?
- ✓ **C** write a comprehensive description of your experimental design?
- ✓ **T/I** analyze your design for sources of error?
- ✓ **A** recognize the differences between investigations using humans as subjects versus those using other organisms?

UNIT 4 SUMMARY

BIG IDEAS

- Organisms have strict limits on the internal conditions that they can tolerate.
- Systems that maintain homeostasis rely on feedback mechanisms.
- Environmental factors can affect homeostasis.

Overall Expectations

In this unit you learned how to…

- **evaluate** the impact on the human body of selected chemical substances and of environmental factors related to human activity
- **investigate** the feedback mechanisms that maintain homeostasis in living organisms
- **demonstrate** an understanding of the anatomy and physiology of human body systems, and **explain** the mechanisms that enable the body to maintain homeostasis

Chapter 8 The Nervous System and Homeostasis

Key Ideas

- Homeostasis is the tendency of the body to maintain a relatively constant internal environment and is critical for survival.
- Homeostasis is maintained through feedback systems that continually monitor, assess, and adjust variables in the body's internal environment.
- The human nervous system is a complex system composed of many subsystems that all work together to maintain homeostasis in the body.
- The neuron is the functional unit of the nervous system. Neurons can be classified based on structure (multipolar, bipolar, or unipolar) or function (sensory neurons, interneurons, and motor neurons).

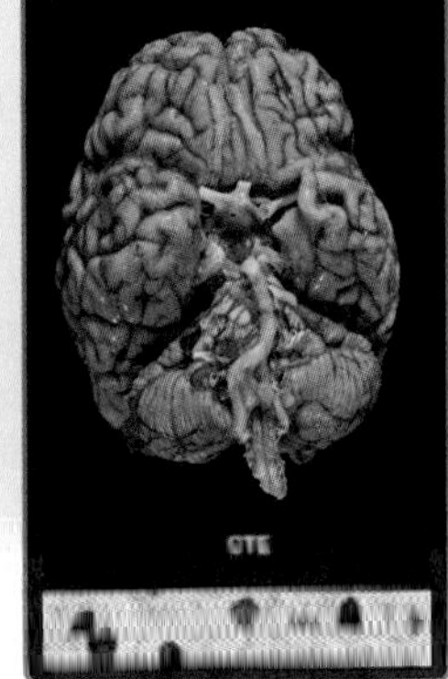

- The central nervous system, which consists of the brain and spinal cord, is the control centre of the nervous system.
- The brain can be subdivided into three general regions: the hindbrain, the midbrain, and the forebrain.
- The peripheral nervous system consists of the somatic and autonomic nervous systems.
- Homeostasis is maintained in the body by the often-antagonistic actions of the sympathetic and parasympathetic nervous systems.

Chapter 9 The Endocrine System

Key Ideas

- The nervous system rapidly affects specific tissues, to which it is directly connected by neurons. The endocrine system relies on chemical messengers called hormones, which circulate in the blood and have broad, long-lasting effects.
- Many hormones are regulated by negative feedback mechanisms.
- The hypothalamus controls the pituitary gland. The anterior pituitary gland releases human growth hormone (hGH), which stimulates fat metabolism and targets the liver to release hormones that stimulate protein synthesis and muscle and bone growth.
- The thyroid gland secretes hormones that regulate cell metabolism, growth, and development.
- In the short-term stress response, also called the fight-or-flight response, the adrenal medulla is stimulated to release epinephrine and norepinephrine.

- In the long-term stress response, ACTH stimulates the adrenal cortex to release cortisol, a steroid hormone.
- The islets of Langerhans in the pancreas contain beta cells, which secrete insulin in response to high levels of blood glucose. The islets of Langerhans also contain alpha cells, which secrete glucagon in response to low levels of blood glucose.
- In the male reproductive system, a negative feedback hormone system maintains a relatively constant level of sperm production and testosterone.
- In females, hormone systems interact to regulate a monthly menstrual cycle.

UNIT 4 SUMMARY & REVIEW

Chapter 10 Excretion and the Interaction of Systems

Key Ideas

- The human excretory system is responsible for removing liquid waste from the body. The excretory system also regulates the acid–base balance and water–salt balance of the blood and secretes some hormones.
- The kidneys are the primary excretory organs. The kidneys are composed of millions of functional units called nephrons that filter the waste from the blood and produce urine.
- Filtrate moves through the nephron, and the processes of glomerular filtration, tubular reabsorption, tubular secretion, and water reabsorption modify the filtrate so that cleansed plasma and substances such as glucose, amino acids, and Na^+ return to the blood.
- The kidneys maintain the water–salt balance in the body, maintain blood pH within narrow limits, and release two hormones: erythropoietin and calcitriol.
- Disorders of the excretory system include urinary tract infections, kidney stones, and renal insufficiency.
- The health of the excretory system affects the health of other body systems. Maintaining healthy blood pressure is a key way to protect the excretory system.

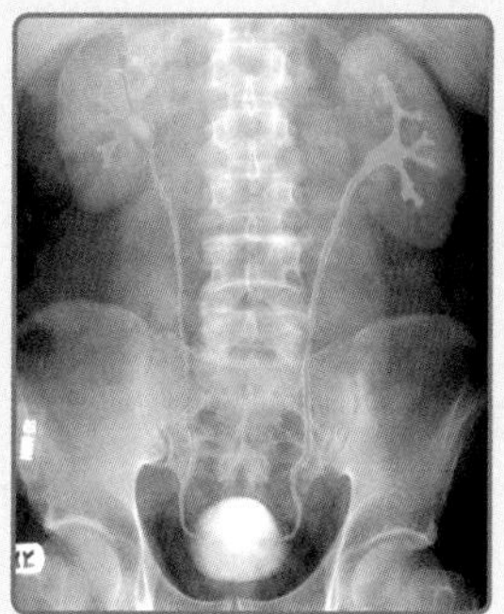

Knowledge and Understanding

Select the letter of the best answer below.

1. Which list represents the correct order for a negative-feedback homeostatic response?
   a. sensory detection, control centre, effector brings about change that reverses the change and brings condition back to normal again
   b. control centre, sensory detection, effector brings about change in internal environment
   c. sensory detection, control centre, effector causes no change in internal environment
   d. effector detection, sensory identification, control centre brings about change in internal environment
   e. sensory detection, control centre, effector brings about change that increases the change
2. The organization of neurons that enable you to react rapidly and involuntarily in times of danger is called
   a. a sensory pathway.
   b. a positive feedback loop.
   c. the central nervous system.
   d. a reflex arc.
   e. the autonomic nervous system.
3. Which lobe of the cerebral cortex receives and processes sensory information from the skin?
   a. frontal lobe
   b. temporal lobe
   c. parietal lobe
   d. occipital lobe
   e. medulla oblongata
4. An individual with diabetes mellitus has levels of blood glucose that tend to rise sharply after a meal and remain at significantly elevated levels. The condition of elevated blood glucose levels is called
   a. diabetes insipidus.
   b. hypoglycemia.
   c. hyperglycemia.
   d. pancreatitis.
   e. acromegaly.
5. Which describes the potential difference across the membrane in a neuron that is not active?
   a. resting membrane potential
   b. action potential
   c. saltatory conduction model
   d. sodium-potassium pump
   e. refractory period
6. Suppose a person is relaxing in front of the television after eating a large meal. Which component of the autonomic nervous system is activated when the body is calm and at rest?
   a. central nervous system
   b. peripheral nervous system
   c. somatic nervous system
   d. sympathetic nervous system
   e. parasympathetic nervous system

7. Thyroid-stimulating hormone (TSH) is an example of which category of hormones?

a. hypothalamic releasing hormones
b. tropic hormones
c. stress hormones
d. growth hormones
e. sex hormones

8. Which hormone stimulates the distal tubule and the collecting ducts in the kidney to increase the absorption of sodium into the bloodstream?

a. adrenocorticotropic hormone (ACTH)
b. cortisol
c. glucagon
d. aldosterone
e. calcitriol

9. In the short-term stress response, the adrenal medulla releases which two hormones?

a. ephinephrine and norepinephrine
b. adrenocorticotropic hormone (ACTH) and cortisol
c. insulin and glucagon
d. acetylcholine and acetylcholine esterase
e. aldosterone and testosterone

10. In females, follicle-stimulating hormone (FSH) and luteinizing hormone (LH) act on the ovaries to produce which two hormones?

a. testosterone and inhibin
b. insulin and glucagon
c. gonadotropin-releasing hormone (GnRH) and adrenocorticotropic hormone (ACTH)
d. erythropoietin and calcitriol
e. estrogen and progesterone

11. Ovulation marks the end of the follicular stage of the menstrual cycle. What is the name of the second stage of the menstrual cycle?

a. ovarian cycle
b. luteal stage
c. uterine cycle
d. menopausal stage
e. endometrial stage

12. Which substance is not normally found in the filtrate inside Bowman's capsule?

a. protein
b. amino acids
c. glucose
d. sodium ions
e. water

13. Which list represents the correct order of structures that urine flows through to reach the external environment?

a. renal medulla, renal artery, renal cortex
b. renal cortex, renal vein, renal medulla
c. kidneys, ureters, urinary bladder, urethra
d. urethra, urinary bladder, ureters, kidneys
e. renal vein, glomerulus, Bowman's capsule, collecting duct

14. Which structure is correctly matched to its function?

a. Bowman's capsule – tubular reabsorption
b. glomerulus – tubular secretion
c. collecting duct – tubular secretion
d. glomerulus – filtration
e. Bowman's capsule – water reabsorption

Answer the questions below.

Use the diagram below, which shows the nerve pathway that would be involved if you accidentally caught your finger in a door as it closed, to answer questions 15–20.

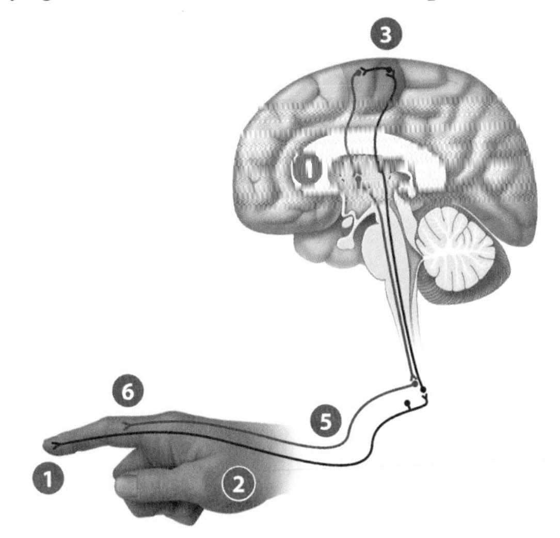

15. Name the types of neurons indicated by numbers 2, 3, and 5 on the diagram, and describe their functions.

16. Describe what is occurring at each number from 1 through 6. Identify the stimulus and effector.

17. Identify the two brain structures at number 3, and describe their functions.

18. Name the brain structure at number 4, and describe its function.

19. Explain how this nerve pathway differs from a reflex arc.

20. What medication might be prescribed to reduce inflammation if one of the finger joints is injured?

21. Which type of glial cell is responsible for increasing the speed of nerve impulses? Identify the structure that this type of glial cell forms, and describe how this structure speeds impulse transmission.

22. Describe the hormonal response that is involved in a short-term response to a stressful situation, such as being surprised by a fire alarm. What is this response called?

23. In general terms, how is a hormone able to "recognize" and stimulate its target cells?

24. People with diabetes mellitus tend to avoid foods that are high in sugar. Explain why someone with type 1 diabetes would keep juice or a chocolate bar handy.

25. List three hormones that are produced by the adrenal cortex. Compare and contrast the functions of these hormones and their effects on the body.

26. Which of these hormones—estrogen, GnRH, progesterone, FSH, LH—is the correct match for each of the following descriptions?
    a. secreted by the follicular cells
    b. stimulates maturing of female sex organs
    c. secreted by the corpus luteum
    d. stimulates the development and function of the corpus luteum
    e. promotes thickening of the endometrium
    f. stimulates development of ovarian follicles
    g. inhibits release of GnRH secretions in high concentrations
    h. maintains the uterine lining during pregnancy

27. Explain the relationship between internal feedback mechanisms and the regulation of male reproductive hormones.

28. Clearly distinguish among the following three processes involved in the formation of urine: filtration, reabsorption, and secretion.

29. Explain what prevents the interstitial fluid from becoming diluted when water from the collecting duct is reabsorbed.

Thinking and Investigation

30. Substance X is found to be present in a urine sample. Is it reasonable to infer, therefore, that Substance X should have been (but was not) filtered at the glomerulus? Justify your answer.

Use the following graph, which illustrates the changes in a person's blood glucose concentrations before and after a meal, to answer questions 31–35.

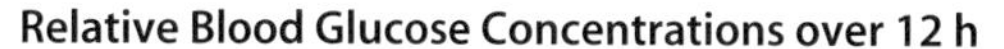

Relative Blood Glucose Concentrations over 12 h

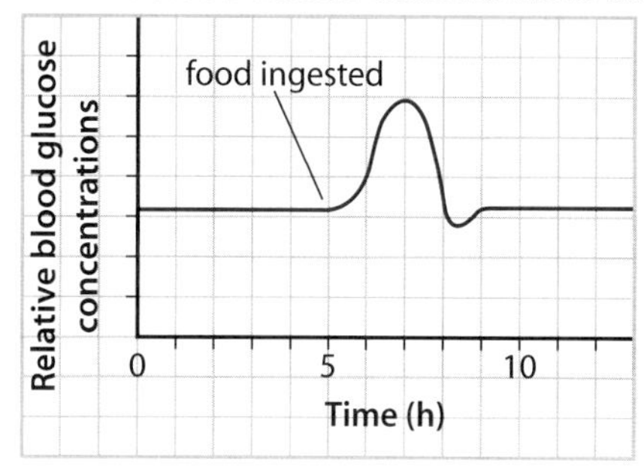

31. Describe what happened to the person's blood glucose concentration at 5 h.

32. Identify the hormone that was released at 7 h, the gland that released it, and the stimulus.

33. Identify the hormone that was released at 8 h, the gland that released it, and the stimulus.

34. Predict how the graph would look if the person engaged in strenuous exercise at 12 h. What hormone would be released during exercise?

35. Suppose that the person has type 1 diabetes. When would he have taken insulin?

Use the information below to answer questions 36–38.

A person's blood potassium concentration is determined to be 4 mmol/L (millimoles per litre) when she is eating food containing 150 mmol of potassium per day. One day she doubles her potassium intake and continues to eat that amount every day afterward. Potassium is regulated homeostatically in the body.

36. If her blood is analyzed, predict whether you think her blood potassium concentration is likely to be 8 mmol/L or about 4 mmol/L. Explain. (Note: The answer to this question does not require any knowledge about potassium or familiarity with the unit "mmol/L".)

37. Think about the role of potassium in the propagation of an action potential. Predict how a severe deficit of potassium in the body could affect muscle function.

38. Suppose the patient's urine is analyzed. Predict the results of the urinalysis with regard to potassium levels. Explain your reasoning.

39. Researchers used a giant squid axon to study changes in the neuron's membrane during nerve impulse transmission. They manipulated the ion concentrations around the neuron, but kept ion concentrations within the neuron constant.

a. Predict the results they obtained when they added sodium ions to the fluid surrounding the axon.

b. Predict the results they obtained when they added potassium ions to the fluid surrounding the axon.

c. Suggest a third change that the researchers could make to ion concentrations around the neuron and the effect this change would have.

Use the table below to answer questions 40–43.

The following data were collected from a female subject over 40 years as part of an experimental study. Results were always collected 3 h after a main meal.

Daily Blood Glucose Concentrations over 40 Years

Age	Average daily blood glucose concentration (mmol/L)
10	4.5
20	5.0
30	6.5
40	8.0
50	[illegible]

40. Why were the readings always taken 3 h after the main meal of the day?

41. The concentrations given are averages of 10 readings taken over one month. Why are averages given instead of raw data?

42. What has happened to the person's blood glucose concentrations over the course of the 40-year study?

43. What condition is associated with the woman's symptoms? What causes this condition?

44. Amenorrhea is the absence of menstruation. It is a symptom that reflects a condition, rather than a disorder itself. Primary amenorrhea is the absence of menstruation by age 16. Secondary amenorrhea is the absence of three or more menstrual periods in a row. For each possible cause below, infer how the problem could lead to amenorrhea.

a. low levels of FSH

b. a tumour on the pituitary gland

c. a disorder of the thyroid gland

d. a deficiency of GnRH

Communication

45. In your notebook, draw a diagram of a neuron.

a. Label the dendrites, cell body, axon, and synaptic terminals.

b. Identify the sodium-potassium pump.

c. Write a caption for your diagram.

46. Draw a graph to show the change in voltage across a neuron's membrane during impulse transmission.

a. Label the x-axis "Voltage."

b. Label the y-axis "Time."

c. Label polarization, depolarization, repolarization, and then return to polarization on your graph.

d. Indicate the voltage during each of the events.

e. Give your graph a title.

47. In a table, list the ways in which the sympathetic and parasympathetic nervous systems act antagonistically to maintain homeostasis in the body.

48. Draw and annotate a feedback loop to explain how a lack of dietary iodine can result in a goitre.

49. BIG IDEAS Systems that maintain homeostasis rely on feedback mechanisms. Use a flowchart to show why thyroid-stimulating hormone (TSH) is considered to be a tropic hormone. What regulates the release of TSH?

50. Construct a table with the following headings. Then fill in your table appropriately with the terms in the box below.

- *Controlling structure or tropic hormone*
- *Endocrine gland*
- *Hormone secreted*
- *Target tissues/organs*

hypothalamus, anterior pituitary, posterior pituitary, thyroid, adrenal medulla, adrenal cortex, kidneys, pancreas, hGH, ADH, TSH, thyroxine, ACTH, cortisol, insulin, glucagon

51. Suppose you stub your foot on a chair leg. Your foot recoils before you feel any pain.

a. Draw a flowchart of the nerve pathway that is involved in this reaction.

b. What is this reaction called?

c. How does this reaction protect you?

52. Use a labelled diagram that shows a presynaptic neuron, a neuromuscular junction, and the associated muscle. Illustrate the transmission of a nerve impulse at this neuromuscular junction. Label the synaptic vesicles, acetylcholine, cholinesterase, and receptor proteins. Write a series of captions for your diagram.

53. A neurosurgeon is probing a person's brain in order to map the brain's functions. The person is awake and feels no pain during this procedure. As each area of the brain is probed, the person perceives a different sensation, as described in the following table. Use a table to list the structures in the brain that are being stimulated by the probes, and the function of each structure.

Patient's Response during Brain Probe

Area Probed	Person's Response
A	"I can hear a radio playing."
B	"I see a flash of bright light."
C	"I can smell the flowers in my garden."
D	"I remember a happy moment from my childhood."
E	"I can feel pain in my foot."
F	"My finger just twitched."

54. Use a labelled diagram to distinguish the following structures: scrotum, testes, seminiferous tubules, and interstitial cells.

55. Use a table to compare the function of GnRH, FSH, and LH in males and females.

56. Copy the following table into your notebook and complete it.

Hormone	Produced By	Target Organ(s)	Effect(s)
Testosterone			
	Hypothalamus		• Stimulates release of FSH and LH
		Ovaries, uterus	• Inhibits ovulation • Stimulates thickening of the endometrium
			• Stimulates development of sex organs • Stimulates gamete production

57. Use a labelled diagram to summarize the feedback mechanism involved in the control of the ovaries.

58. Draw a flowchart that outlines the path of urine through the organs of the excretory system to the external environment. (Limit your answer to an organ-level response; do not trace the flow of blood or urine through the nephrons.)

59. Use a flowchart to identify the structures through which a molecule of water would travel as it moves from the renal artery to the renal pelvis.

60. The Student Council at your school has raised $5000 for medical research. Students are being polled to determine what research they would like to support: research on multiple sclerosis, Alzheimer's disease, or Creutzfeldt-Jakob disease.
   a. Research how each of these diseases affects the nervous system.
   b. Argue why each disease is worthy of money for research.

61. Researchers continue to study medications that could be used to help treat an addiction to the drug cocaine.
   a. Suggest how a medication that binds and interferes with dopamine could be used to help someone recover from a cocaine addiction.
   b. Use a diagram to show how the medication might work.
   c. Do you think that someone with a cocaine addiction should be forced to take the medication? Justify your response.

Application

62. Many people consider the Ironman Triathlon to be the most gruelling endurance event in the world. Sometimes contenders push their body systems well past their homeostatic limits. In one case, a competitor collapsed just metres before the finish line and had to be rushed to hospital. Doctors found that many of his body's systems had been damaged from severe dehydration and heat.
   a. Name five physiological conditions that were probably not operating at homeostasis in this competitor's body by the end of the race.
   b. Which responses of the nervous and endocrine systems would ordinarily have regulated these physiological conditions in order to restore homeostasis in his body?
   c. Infer which body systems may have been permanently damaged by the severe dehydration. Explain your reasoning.

UNIT 4 REVIEW

Use the information below to answer questions 63–65.

Kidney failure (or end-stage renal disease, ESRD) is rare, occurring in about one of every 2000 people. Some of the problems associated with kidney failure include reduced production of red blood cells, as well as the inability to remove nitrogenous and other wastes, regulate the volume of water, balance chemicals, control blood pressure, and produce the active form of vitamin D. There are two types of renal dialysis commonly used today: hemodialysis and peritoneal dialysis. Both are equally effective at maintaining homeostasis.

Hemodialysis is the most common type of renal dialysis used to treat people suffering from ESRD. The patient's blood is circulated through the dialysis machine, which contains a *dialyzer* (also called an artificial kidney). The patient's bloodstream is connected to the dialysis machine using a surgically constructed path called a vascular access. The vascular access creates a way for blood to be removed from the body, circulate through the dialyzer, and then be returned to the body.

The dialyzer has two spaces, separated by a thin, semipermeable cellulose membrane. During hemodialysis, a small quantity of blood passes on one side of the membrane, while a dialysis solution is on the other side of the membrane. The dialysis solution consists only of glucose, amino acids, and mineral ions. The concentrations of these substances are either similar to those of normal plasma or slightly higher.

The difference in osmotic pressure between the blood and the dialysis solution allows water from the blood to diffuse across the membrane (by osmosis) to dilute the dialysis solution. The semipermeable membrane allows small molecules to pass out of the blood into the dialysis solution as well. However, large molecules, such as proteins, stay in the blood. The dialysis fluid is then discarded, along with wastes and excess water. The cleansed blood is returned to the cardiovascular system.

63. Describe how nitrogenous wastes are removed from the blood of a person on dialysis.

64. In a healthy kidney, nutrients are actively removed from the filtrate and returned to the cardiovascular system. Identify where this process occurs in the healthy kidney, and explain how these nutrients are replaced during dialysis.

65. **BIG IDEAS** Organisms have strict limits on the internal conditions that they can tolerate. Explain how ESRD can affect the endocrine system. How do these affects influence other body systems?

66. Suppose that Alan, a six-year-old boy, is growing so quickly that he is 80 percent taller than his Grade 1 peers. An MRI scan reveals that he has a tumour in the anterior pituitary gland.

a. What is most likely causing his rapid growth?

b. What condition could Alan develop if corrective measures are not taken to deal with his hormonal imbalance?

67. When children are between six months and 10 years of age, the frontal lobe of their brain consumes twice as much energy as the frontal lobe of an adult brain.

a. Name the major functions of the frontal lobe of an adult brain.

b. Suggest why the frontal lobe of a child's brain needs more energy than the frontal lobe of an adult's brain.

68. On a dare, a high-school student drinks 4 L of water in 20 min.

a. How would this affect the concentrations of sodium in the student's blood?

b. Describe how the endocrine system would help to return the body systems to homeostasis.

c. Name the hormones and gland(s) that would be involved.

69. Suppose that you belong to a health promotion committee. The committee is preparing a campaign to teach people about treatments for cancer, high blood pressure, and heart disease. Explain to the committee how the endocrine system reacts to long-term stress and how this reaction affects the body. Explain why the campaign should also teach people ways to reduce stress.

70. Six months ago, your friend started a sodium-free diet. She has eliminated table salt, seafood, and dairy products from her diet. She tells you that recently her throat has been swollen, she has been feeling more tired and cold than usual, and she has gained weight.

a. Based on her symptoms, what do you think has occurred in her endocrine system?

b. Why should she see a doctor?

71. If a woman has one ovary surgically removed, can she still become pregnant? Why or why not?

72. Describe the function of the corpus luteum

a. if an egg is fertilized and implants in the endometrium.

b. if an egg is unfertilized.

Use the table below to answer questions 73 and 74.

Hormone Concentrations during a Menstrual Cycle

Day of Menstrual Cycle	Relative Hormone Concentration		
	A	B	C
1	12	5	10
5	14	5	14
9	14	5	13
13	70	10	20
17	12	60	9
21	12	150	8
25	8	100	8
1	12	5	10

73. Two main events of the menstrual cycle are the release of an egg from an ovary and the build-up of the uterine lining. Use the data table to predict which of the hormones A, B, or C would be associated with each of these two events. Assume the cycle length is 28 days.

74. A woman is having difficulty becoming pregnant. For the last month, her hormone levels have been measured on each of the eight days listed in the data table. Hormone A remained steady at 12. Hormone B remained steady at 5. Hormone C had the levels shown in the data table. How could these hormone levels explain her fertility difficulties?

75. Because one of the functions of hGH in the body is to build lean muscle mass, its use has become widespread among various athletes. Despite its expense, many athletes from baseball players to weightlifters are acquiring synthetic hGH because it is difficult for drug testers to detect.

a. Research more information about the health side effects of hGH use for non-medical purposes.

b. What are some of the penalties athletes face if caught using hGH?

c. Should competitive athletes be allowed legal access to synthetic hGH? Explain your answer.

76. If a male's testes were removed, would he still be able to produce male sex hormones? Explain.

77. Which two anterior pituitary gland hormones are described by the following statement? "These hormones are released from the anterior pituitary, travel to the ovaries, and stimulate oogenesis, ovulation, and the production of hormones responsible for the development of secondary sexual characteristics in humans."

78. Through the use of technology, many body-system disorders and their symptoms may be modified and sometimes cured. Select two body systems, and identify one disorder that affects each. Explain how a particular technology is used to address the disorder, noting how successful the technology is in solving the problem it is intended to solve.

79. The table below shows the sodium ion and potassium ion concentrations inside and outside a neuron.

Sodium Ion and Potassium Ion Concentrations Inside and Outside a Neuron

Ion	Concentration Inside Neuron (mmol/L)	Concentration in Body Fluid Outside Neuron (mmol/L)
Sodium	12	145
Potassium	140	4

a. What do the ion concentrations in the table suggest about the state of polarization of the neuron?

b. Indicate the voltage across the membrane with respect to the inside of the neuron.

c. How does a neuron establish this charge difference across the membrane?

d. How might a strong stimulus affect the state of polarization of the neuron?

e. What would the ion concentrations and voltage across the membrane become?

f. What would the generated response be called?

g. Qualitatively describe how the ion concentrations would change during repolarization.

h. How would this affect the voltage across the membrane?

80. Acupuncture is a traditional Chinese medical practice in which needles are inserted at specific points on the body. Based on Western scientific thinking, it is hypothesized that the resulting stimulation of these points triggers the brain to release endorphins, neurotransmitters that ease the sensation of pain. Research more information about endorphins and how they work.

81. BIG IDEAS Environmental factors can affect homeostasis. After playing a game of baseball on a hot, sunny day, you are perspiring and very thirsty.

a. List the glands and hormones that help your body maintain homeostasis in this situation.

b. Describe the effects of these glands and hormones.

Select the letter of the best answer below

1. **K/U** Which system in the human body plays the biggest role in fluid balance?
   a. cardiovascular
   b. excretory
   c. digestive
   d. integumentary
   e. nervous

Use the diagram below to answer question 2.

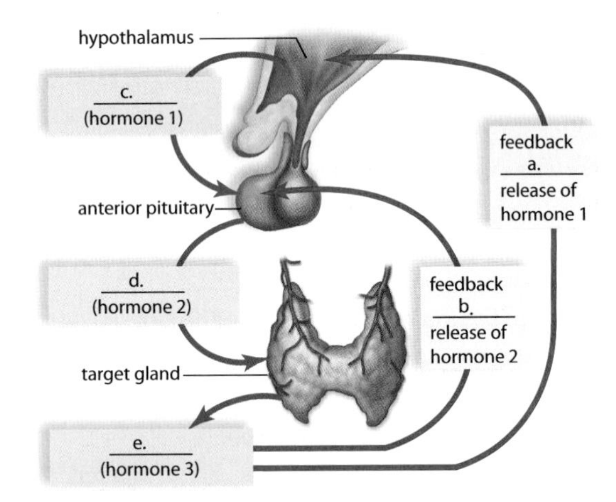

2. **K/U** In the diagram, hormones "c" and "d" are, respectively,
   a. thyroid-stimulating hormone (TSH) and thyroxine
   b. parathyroid-stimulating hormone (PSH) and parathyroid hormone (PTH).
   c. neurosecretory-releasing hormone and adrenocorticotropic hormone (ACTH).
   d. thyroid-releasing hormone (TRH) and thyroid-stimulating hormone (TSH).
   e. gonadotropin-releasing hormone (GnRH) and follicle-stimulating hormone (FSH).

3. **K/U** Which allows rapid change in one direction and does not achieve stability?
   a. homeostasis
   b. positive feedback loop
   c. negative feedback loop
   d. sympathetic nervous stimulation
   e. parasympathetic nervous stimulation.

4. **K/U** Which is considered a tropic hormone?
   a. calcitonin
   b. oxytocin
   c. glucagon
   d. melatonin
   e. follicle-stimulating hormone

5. **K/U** Which correctly describes the distribution of ions on either side of an axon when it is not conducting a nerve impulse?
   a. more Na^+ outside and more K^+ inside
   b. more K^+ outside and less Na^+ inside
   c. charged protein outside and Na^+ and K^+ inside
   d. Na^+ and K^+ outside and water only inside
   e. chloride ions (Cl^-) outside and K^+ and Na^+ inside

6. **K/U** Which of the following statements about the cerebellum are *true*?
   a. It coordinates skeletal muscle movements.
   b. It receives sensory input from the joints and muscles.
   c. It receives motor input from the frontal lobe of the cerebral cortex.
   d. It is involved in the unconscious coordination of posture, reflexes, and body movements.
   e. All of the above are true.

7. **K/U** Which is not regulated by secretions of the anterior pituitary gland?
   a. adrenal medulla
   b. testes
   c. ovaries
   d. adrenal cortex
   e. thyroid gland

8. **K/U** Reabsorption of which substance from the nephrons is, ordinarily, 100 percent?
   a. sodium
   b. glucose
   c. water
   d. urea
   e. protein

9. **K/U** The function of erythropoietin is to
   a. reabsorb sodium ions.
   b. excrete potassium ions.
   c. reabsorb water.
   d. stimulate red blood cell production.
   e. ncrease blood pressure.

10. **K/U** Production of testosterone in the interstitial cells in the testes is controlled by
   a. luteinizing hormone (LH).
   b. follicle-stimulating hormone (FSH).
   c. gonadotropin-releasing hormone (GnRH).
   d. inhibin.
   e. estrogen.

Use sentences and diagrams as appropriate to answer the questions below.

11. K/U Using a specific example, explain how hormones of the adrenal medulla complement the actions of the sympathetic branch of the autonomic nervous system.

12. T/I Various endocrine glands and their hormones regulate the concentration of glucose in the blood.
  a. List these hormones, and explain their effects on blood glucose levels.
  b. Form a hypothesis to explain why several hormones in the body, rather than just one, control blood glucose levels.

13. C Suppose you are driving down a highway at night when a deer jumps in front of your vehicle. You slam on the brakes and avoid a collision. Draw a diagram of the impulse pathway that would occur in your body during this incident.

14. A Suppose a person eats some chocolate. Sensory neurons relay the satisfying taste information to his brain. His brain releases endorphins, neurotransmitters that ease the sensation of pain and produce a feeling of well-being. This person now associates these positive feelings with eating chocolate. The next time he eats chocolate, the same thing happens and he eats more chocolate. Now he feels even better. Which feedback system is described in this scenario?

15. C Write a paragraph explaining why it is important for an organism to maintain homeostasis.

16. K/U What is the function of the seminiferous tubules?

17. K/U Explain how the nephron maintains normal body fluid with respect to each of the following: water, ions, and pH.

18. C Choose one example of homeostasis being maintained by a negative feedback system. Make a diagram that explains the regulation of the variable you choose.

19. A What types of tests might a doctor perform to determine if a person is diabetic? What variables would she examine in each test?

20. A An excess of cortisol in the body can lead to glucose intolerance by cells and diabetes. Explain the physiology that would cause this to happen.

Use the information below to answer questions 21–22.

The neurotransmitter serotonin is associated with depression. Certain antidepressant medications, referred to as selective serotonin reuptake inhibitors (SSRIs), seem to relieve symptoms of depression by blocking the reabsorption (reuptake) of serotonin by certain nerve cells in the brain. This leaves more serotonin available in the brain and improves mood. SSRIs are called "selective" because they seem to affect only serotonin, not other neurotransmitters.

21. C Draw a diagram that includes a comparison of the processes occurring at a synapse in the brain of a person who has depression and is not being treated to that of a synapse in the brain of a person who has depression but is taking an SSRI.

22. T/I Form a hypothesis that explains how SSRIs work.

Use the information below to answer questions 23–24.

Parkinson's disease is characterized by a gradual loss of motor control, beginning between the ages of 50 and 60. In people who have Parkinson's disease, areas of grey matter in the cerebrum are overactive because of a degeneration of the dopamine-releasing neurons in the brain.

23. T/I Infer how the decrease in dopamine could result in the symptoms of Parkinson's disease. (Hint: Dopamine is an inhibitory neurotransmitter.)

24. T/I It is not possible to give Parkinson's disease patients dopamine because of the impermeability of the capillaries serving the brain. Think about the structure and function of the blood-brain barrier and suggest a way to get dopamine past it.

25. A Patients who receive a kidney transplant must take steroid medications, similar to glucocorticoids, to prevent their immune system from attacking the new organ. Suggest some possible side effects patients may experience from taking steroid medications long-term.

Self-Check

If you missed question...	1	2	3	4	5	6	7	8	9	10	11	12	13	14	15	16	17	18	19	20	21	22	23	24	25
Review section(s)...	10.1	9.2	8.1	9.1	8.2	8.3	9.1	10.2	10.3	9.4	9.3	9.3	8.2	8.1	8.1	9.4	10.2	8.1	10.3	9.3	8.2	8.2	8.2	8.3	9.3

UNIT 5 Population Dynamics

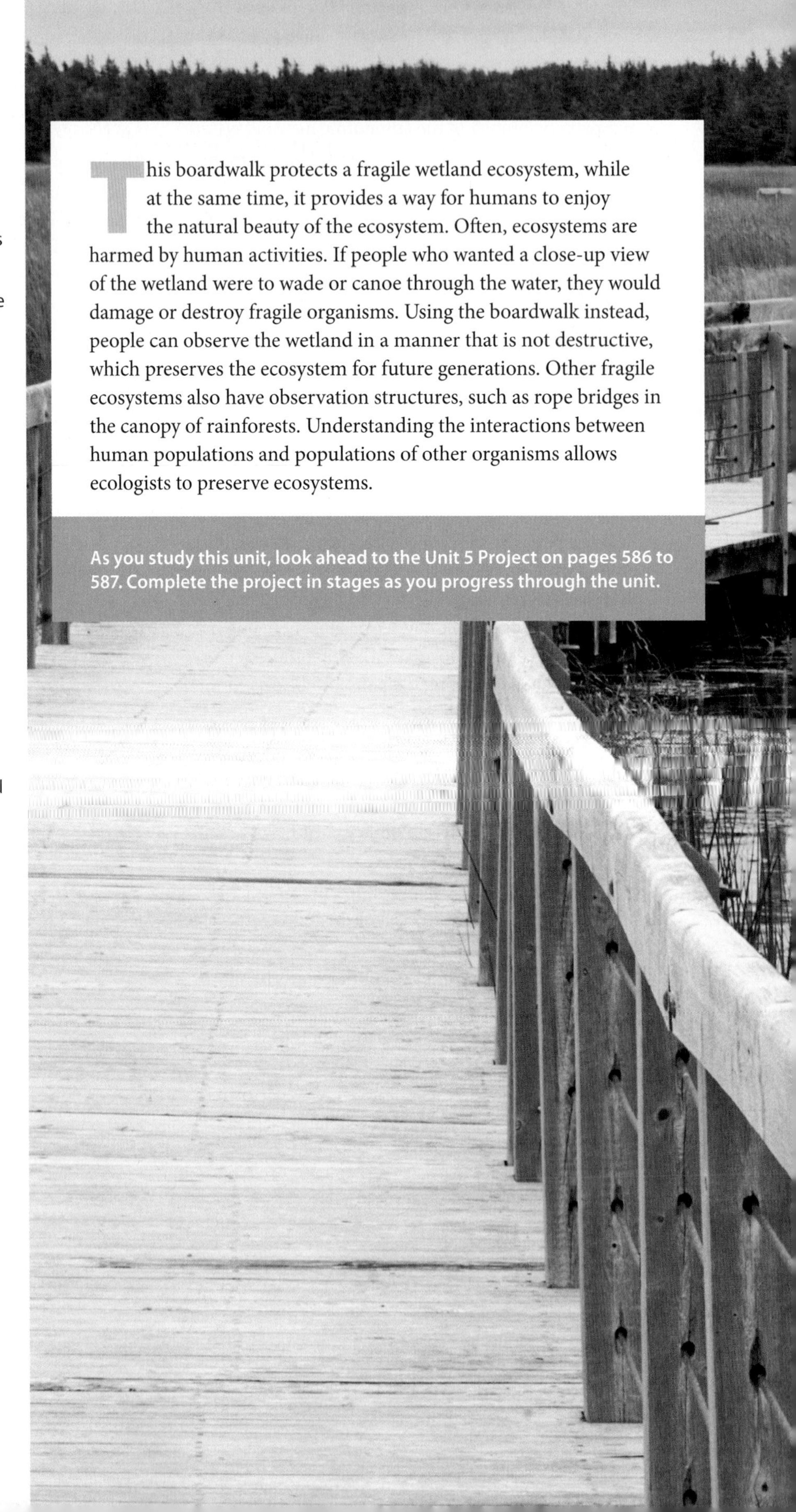

This boardwalk protects a fragile wetland ecosystem, while at the same time, it provides a way for humans to enjoy the natural beauty of the ecosystem. Often, ecosystems are harmed by human activities. If people who wanted a close-up view of the wetland were to wade or canoe through the water, they would damage or destroy fragile organisms. Using the boardwalk instead, people can observe the wetland in a manner that is not destructive, which preserves the ecosystem for future generations. Other fragile ecosystems also have observation structures, such as rope bridges in the canopy of rainforests. Understanding the interactions between human populations and populations of other organisms allows ecologists to preserve ecosystems.

As you study this unit, look ahead to the Unit 5 Project on pages 586 to 587. Complete the project in stages as you progress through the unit.

BIG IDEAS

- Population growth follows predictable patterns.
- The increased consumption of resources and production of waste associated with population growth result in specific stresses that affect Earth's sustainability.
- Technological developments can contribute to or help offset the ecological footprint associated with population growth and the consumption of resources.

Overall Expectations

In this unit, you will...

- **analyze** the relationships between population growth, personal consumption, technological development, and our ecological footprint, and **assess** the effectiveness of some Canadian initiatives intended to assist expanding populations
- **investigate** the characteristics of population growth, and use models to calculate the growth of populations within an ecosystem
- **demonstrate** an understanding of concepts related to population growth, and **explain** the factors that affect the growth of various populations of species

Unit Contents

Focussing Questions

1. *In what ways do populations grow and change?*
2. *How are human population growth, personal consumption, technological development, and ecological footprint related?*

Go to **scienceontario** to find out more about population dynamics.

UNIT 5 Preparation

Ecosystems and Their Components

- A population is a group of organisms of one species that lives in the same place, at the same time, and can successfully reproduce. All populations that interact in a given area form a community.
- Ecosystems are made up of all the interacting parts of a community and its environment. Thus, ecosystems are composed of both biotic (living) and abiotic (non-living) components.
- An area with a particular set of biotic and abiotic components, in which an organism is able to survive and reproduce, is that organism's habitat. The role an organism has within its habitat, including its biotic relationships and the abiotic resources it uses, makes up its ecological niche.

1. How does a population differ from a community?

2. An example of a biotic component of a mountainous ecosystem is
   a. light
   b. altitude
   c. oxygen
   d. soil nutrients
   e. soil microorganisms

3. The diagram below shows the migration pattern and life cycle of the American eel (*Anguilla rostrata*). American eels hatch in the Sargasso Sea in the Atlantic Ocean and migrate towards the St. Lawrence River and the Great Lakes. After 10 to 15 years, the eels return to the Atlantic to reproduce and die. Compare and contrast the different ecosystems in which American eels spend their lives.

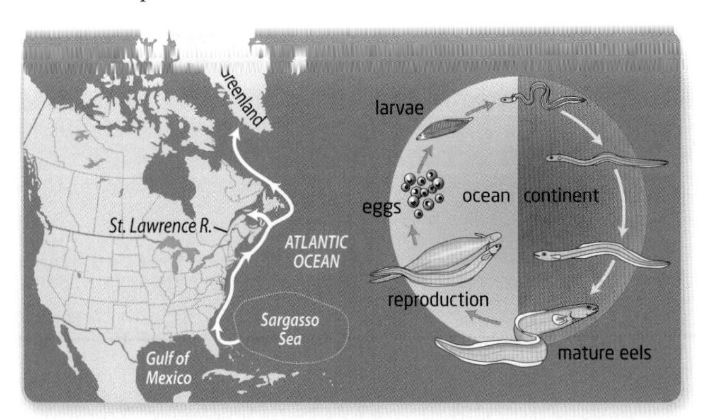

4. Correct any of the statements below that are false.
   a. The fact that most organisms are limited to particular ecological niches partly explains why different species are found in particular types of ecosystems in particular regions of the world.
   b. Because carnivorous plants are well-adapted to life in a bog, they would do well in another habitat with different conditions.
   c. Two or more species can occupy the same ecological niche.

5. Little brown bats (*Myotis lucifugus*), like the one shown in the photograph below, eat certain insects. This behaviour is a component of the bats'
   a. environmental niche
   b. biotic potential
   c. ecosystem
   d. ecological niche
   e. habitat

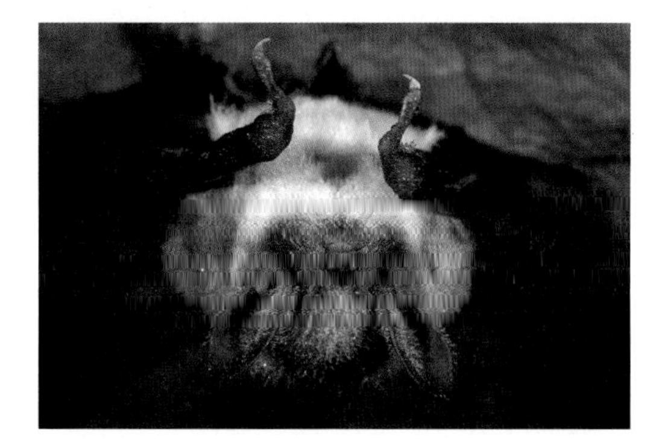

6. Yellow wild indigo (*Baptisia tinctoria*) is a plant that is native to Ontario. It is also called horsefly weed. This plant grows to about a metre in height and has yellow flowers from late spring to late summer. It thrives in dry soil and full Sun.
   a. What are three abiotic factors that could affect this plant?
   b. What are three biotic factors that could affect this plant?

7. Lake Superior is a fresh water lake on the southern boundary of Ontario. Lake trout (*Salvelinus namaycush*), lake whitefish (*Coregonus clupeaformis*), sea lamprey (*Petromyzon marinus*), and rainbow smelt (*Osmerus mardax*) are native and non-native species found in the lake. Using these organisms, explain what is considered a population and what is considered a community in Lake Superior.

Interactions between Organisms

- Interactions that occur between organisms in ecosystems are called biotic interactions. Biotic interactions include predation, competition, and symbiosis. Symbiosis includes mutualism and parasitism.
- Organisms at different trophic levels interact in a feeding relationship called a food chain. Numerous interacting food chains make up a food web.
- Energy is transferred from one trophic level to another within food chains and food webs. Only about 10 percent of the energy available at one trophic level is transferred to the next trophic level. This is because organisms use much of the energy they consume for life functions. Additionally, much energy is lost as heat, while some is lost in waste.

8. The Red-billed oxpecker (*Buphagus erythrorhynchuys*) removes parasites from the impala (*Aepyceros melampus*). The oxbill gets a meal and the impala gets parasites removed.

This relationship is an example of

a. predation
b. competition
c. mutualism
d. parasitism
e. disease

9. Common dandelions (*Taraxacum officinale*) and grass both take up the same water and nutrients from a lawn. This relationship is an example of

a. mutualism
b. competition
c. trophic efficiency
d. predation
e. parasitism

10. Write a caption that describes the interactions shown in the photograph below.

11. A *keystone species* is one that can greatly affect the health of an ecosystem. Explain how such a species can affect other populations within an ecosystem.

12. Study the food chain below.

a. Name the relationship that occurs between the frog and the hawk in this food chain.
b. Which of the organisms is a producer?
c. Which of the organisms is a secondary consumer?
d. Which of the organisms is a herbivore?
e. What is being transferred in this food chain?
f. How many trophic levels are in this food chain?

13. Where would decomposers fit in the food chain shown in question 12?

14. A student is studying a specific food chain in a forest ecosystem in northern Ontario. Through her research, she determines that about 50 percent of the energy available in the first trophic level is transferred to the second trophic level.

a. Do you think the student's results are accurate? Explain your reasoning.
b. What happens to energy that is not transferred to the second trophic level?

15. Use the concept of energy transfer between trophic levels to explain why there are many more deer in a forest ecosystem than there are bears.

16. Draw a simple food web for each of the following ecosystems.

a. freshwater lake
b. forest
c. prairie

Biodiversity

- Biodiversity refers to the number and variety of organisms found within a specific region.
- Threats to biodiversity include habitat loss, alien or invasive species, overexploitation, and breaking the connectivity among ecosystems.
- Extinction is the state in which a species that has been present in the past no longer exists on Earth.
- Current extinction rates appear to be accelerated as a consequence of human activities.

17. What is the difference between *background extinction* and *mass extinction*?

18. About 24 percent of the world's wetlands are in Canada.

a. Explain why these wetlands are important to protect in terms of biodiversity.

b. Suggest two ways that human activities may reduce biodiversity in these wetlands.

19. Compare and contrast *alien species* and *invasive species*.

20. Define *overexploitation* in your own words and explain how it threatens biodiversity.

21. Which factors can result in species extinction?

a. air pollution
b. water pollution
c. habitat destruction
d. climate change
e. all of the above

22. Which human activity threatens biodiversity?

a. removal of rainforests to plant crops
b. releasing a pet python into wetlands
c. overharvesting a specific fish species
d. draining wetlands to build housing
e. all of the above

Evolution

- Organisms face environmental challenges that affect their ability to survive and reproduce. Adaptations that help organisms survive these challenges include mimicry and variations within species. Species that do not overcome these challenges become extinct.
- Natural selection is a process by which characteristics of a population change over generations. Natural selection occurs when individuals that have inherited a trait that helps them survive environmental challenges are able to pass the alleles responsible for this trait to their offspring. The genetic change that results over time is called evolution.

23. When species that were once similar become more distinct, the result is a type of evolution known as

a. divergence or divergent evolution
b. convergence or convergent evolution
c. gradualism
d. punctuated equilibrium
e. distinction

24. A stick insect looks like a twig on a branch.

a. What type of adaptation does this insect display?

b. How is this adaptation advantageous to the insect?

25. When discussing evolution, explain why it is necessary to keep in mind that populations evolve; individuals do not evolve.

26. The palatable viceroy butterfly looks very similar to the unpalatable monarch butterfly. This adaptation is an example of

a. modification
b. artificial selection
c. variation
d. mimicry
e. convergence

27. A biologist studied a population of moles for 10 years. During that time, the population was never fewer than 20 moles and never more than 50. Her data showed that over half the moles born did not survive to reproduce because of selection factors, such as competition for food or predators. Then, in a single generation, 90 percent of the moles that were born lived to reproduce. The population doubled in size. Why do you think this happened?

Population Change

- Populations that exhibit exponential growth have accelerated growth that produces a J-shaped curve when population size is graphed against time.
- Limiting factors limit population size, as well as the growth or distribution of a population within an ecosystem.
- Carrying capacity refers to the size of a population that can be supported indefinitely by the resources and services available in an ecosystem.
- Human activity, biotic interactions, and environmental factors can affect populations of terrestrial and aquatic ecosystems.

28. Populations tend to increase in size when individuals reproduce at rates that are

a. less than the rate required to replace individuals who have died or left the population

b. greater than the rate required to replace individuals who have died or left the population

c. equal to the rate required to replace individuals who have died or left the population

d. both a and c

e. both b and c

29. Exponential population growth

a. occurs when a population enters a new habitat with a large number of resources

b. typically occurs for a long time in nature

c. occurs when there are pressures such as predation, on populations

d. results in an L-shaped curve

e. all of the above

30. Examine the graph below. Growth of this elephant population became exponential when it was protected in a national park after years of being hunted.

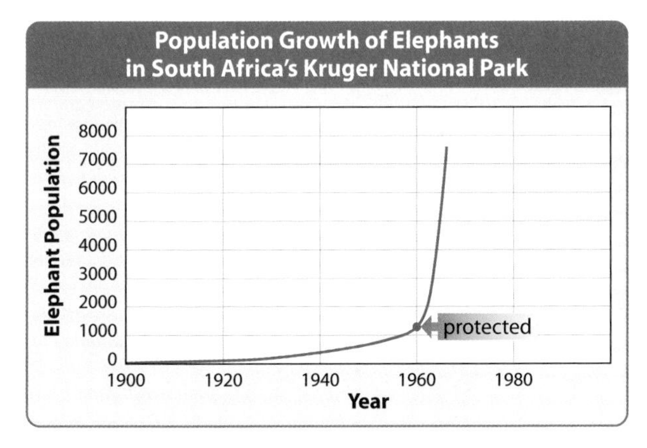

a. Identify the time period on the graph in which exponential growth took place.

b. Will exponential growth continue indefinitely in this population? Explain your answer.

31. Limiting factors for a population of aquatic plants in a pond ecosystem include

a. temperature

b. light levels

c. water pH

d. all of the above

e. none of the above

32. The graph below shows how the size of a population of northern fur seals (*Callorhinus ursinus*) changed between 1910 and 1950. In 1911, an international treaty to conserve wildlife reduced seal hunting for the fur trade.

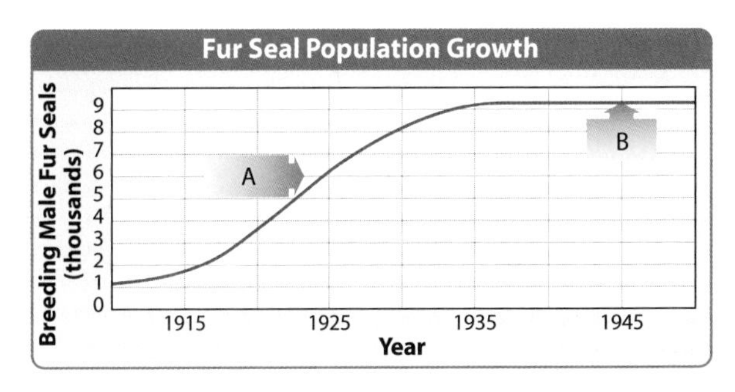

a. Identify A and B on the graph.

b. How and why did population size change between 1911 and 1935?

c. How did population size change after 1935?

d. Suggest two limiting factors that might have caused the change you observed in c.

33. When a population is maintained at its carrying capacity, the size of the population is

a. steadily increasing

b. steadily decreasing

c. maintained at a steady equilibrium

d. experiencing exponential growth

e. zero

34. Explain the relationship among population growth, limiting factors, and carrying capacity.

CHAPTER

11 Describing Populations and Their Growth

Specific Expectations

In this chapter, you will learn how to . . .

- F2.1 **use** appropriate terminology related to population dynamics (11.1, 11.2, 11.3)
- F2.2 **use** conceptual and mathematical population growth models to calculate the growth of populations of various species in an ecosystem (11.2)
- F2.3 **determine**, through laboratory inquiry or using computer simulations, the characteristics of population growth of two different populations (11.3)
- F3.1 **explain** the concepts of interaction between different species (11.3)
- F3.2 **describe** the characteristics of a given population, such as its growth, density, distribution, and minimum viable size (11.1,11.2)
- F3.3 **explain** factors such as carrying capacity, fecundity, density, and predation that cause fluctuation in populations, and analyse the fluctuation in the population of a species of plant, wild animal, or microorganism (11.1, 11.2, 11.3)
- F3.5 explain how a change in one population in an aquatic or terrestrial ecosystem can affect the entire hierarchy of living things in that system (11.3)

Does the mosquito shown above make you itch just thinking about it? Few people have escaped the itchy bite of a mosquito. In many places, these creatures are more than an inconvenience. They transmit diseases, such as malaria, yellow fever, filariasis, dengue fever, West Nile virus, and encephalitis. Almost one million people die annually from malaria, most of them young children in Africa. Understanding and controlling mosquito populations is an important factor in saving lives in regions where mosquitos live and breed. This is just one example of the importance of understanding populations and population growth.

Launch Activity

Reproductive Strategies and Population Growth

The female *Aedes* sp. mosquito uses blood sucked from vertebrate animals to nourish her developing eggs. In her lifetime, she may lay up to 600 eggs. These eggs will remain dormant until conditions are favourable for their growth. Once an *Aedes* sp. egg hatches, the larva takes about a week to develop into an adult mosquito, which will live about another 14 days. The wolf, *Canis lupus*, on the other hand, lives an average of about 5 years in the wild. A female wolf reaches sexual maturity about 2 to 3 years of age and she produces a litter averaging 7 cubs each spring.

Procedure

Use the data in the tables to create two graphs (one for each population) showing population size over time. Then answer the Questions below.

Size of a Hypothetical *Aedes* sp. Mosquito Population over One Growing Season

Day	Number of adult mosquitoes
May 1	20
May 8	40
May 15	80
May 22	160
May 29	320
June 5	640
June 12	1280
June 19	2560
June 26	5120
July 3	10 240

Number of Individuals in the Wolf (*Canis lupus*, subspecies *lycaon*) Population of Algonquin Provincial Park, Ontario

Year	Estimated number of wolves
1988–89	86
1989–90	62
1990–91	65
1991–92	93
1992–93	63
1993–94	45
1994–95	71
1995–96	64
1996–97	54
1997–98	70
1998–99	38

Questions

1. Compare the shapes of your two graphs. Describe the growth of both populations during the given time intervals.
2. Make and record a hypothesis to account for the shape of your graph for the mosquito population and the wolf population.
3. Predict the number of mosquitos and wolves on the following dates: a. July 10, b. December 31, and c. May 22 the following year. Explain your reasoning.
4. Compare your graph with the graph of one of your classmates. If there is a difference in your graphs, explain why.

SECTION 11.1

Characteristics of Populations

Key Terms

population size (N)
population density (D_p)
transect
quadrat
mark-recapture
distribution pattern
life history
fecundity
survivorship

Ecologists use quantitative measurements to study, predict, and describe populations. Quantitative measurements are those that can be measured or presented numerically, and include population size, density, distribution, and life history. These measurements are like snapshots, or static images. Put together like images in a movie, the measurements reveal changes over time in populations and communities.

Determining Population Size and Density

Imagine trying to count the number of individual birds in a swirling flock of crows, or being asked to calculate the population size of moose in northern Ontario. Determining **population size (N)**—the number of individuals of the same species living within a specific geographical area—can be a challenge. (In comparison, counting human populations is relatively easy. Demographers can use censuses, voter lists, and other tools to count human population with more precision.) **Population density (D_p)** is the number of individuals per unit of volume or area. For example, the number of squirrels per square kilometre on the Bruce Peninsula is a population density.

Other than in very small populations, counting each individual within a population is impractical. Such an endeavour is prohibitive in terms of time, money, and human resources. This is why ecologists use other methods to estimate population size. They count or estimate the number of individuals in a number of smaller samples, or subsets, then calculate the average. The results are then applied, or extrapolated, to the entire area occupied by the population to give the population density. Instead of trying to count all the trees in the forest in **Figure 11.1**, for example, researchers can estimate population size and density by sampling several smaller areas.

Techniques used to sample subsets of larger populations include transects, quadrats, and mark-recapture.

population size (N) the number of individuals of the same species living within a specific geographical area

population density (D_p) the number of individuals per unit of volume or area

transect a long, relatively narrow rectangular area or line used for sampling a population

Figure 11.1 Scientists do not count each individual tree in a forest. They use alternate methods to estimate population sizes in large areas such as forests.

Explain *why an alternate method of estimating population size is appropriate for this forest.*

Transects

Members of a population may be sampled along a long rectangular area or line known as a **transect**. Researchers choose a line of a specific length, such as 100 m, then randomly determine a starting point and the direction they will travel. A random starting point and travel direction ensures a result that is more statistically accurate.

To sample a transect, the researcher walks its length, counting the species being monitored. The researcher records individuals within a certain distance of the transect line. This distance can vary. For instance, for sessile (stationary) organisms such as plants, only those within 1 m of the line might be sampled. For mobile organisms such as birds and mammals, the distance will be greater—perhaps up to 50 m. A sample transect is shown in **Figure 11.2** at the top of the next page. Transects are useful when the density of a species is low, or when individual organisms are very large (such as trees).

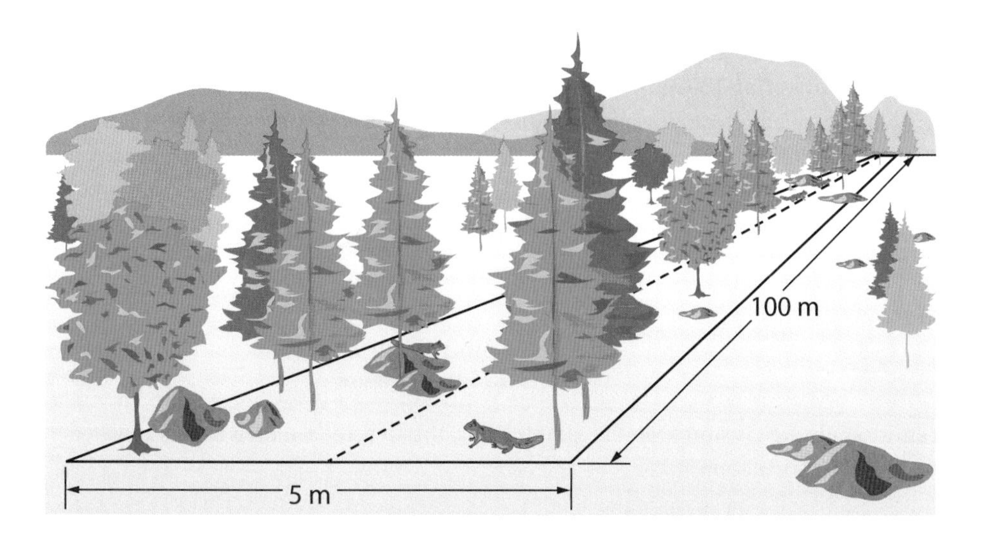

Figure 11.2 In this transect, researchers count individuals of one species within 5 m of a 100 m long transect.

Quadrats

For organisms that are sessile or move very little, ecologists use **quadrats** to sample their populations. They are particularly useful for sampling plant populations and those of sessile animals. Several sample sites are randomly chosen and quadrats of a known size, such as 1 m^2, are marked. The students in **Figure 11.3** are using a quadrat to estimate the population sizes on a rocky shore. The number of individuals within the boundaries is counted. To determine population density, calculate the sum of individuals in the quadrants (N), and then divide by the total area of the quadrats (A).

quadrat an area of specific size used for sampling a population; often used to sample immobile organisms or those that move very little

$$D_{\mathrm{p}} = \frac{N}{A}$$

The size of the population can then be estimated by extrapolating the density to the entire study area. When densities are very high, such as on a grassy field, it is impractical for the students to count each blade of grass, but they can see that grass covers 100 percent of the area. An example using this equation is on the next page.

Figure 11.3 Quadrats can be used to count the number of individuals of a species within a particular area.

Estimating Population Size

Suppose that ecologists are sampling the distribution of flowering white trillium plants in a section of forest covering 100 m by 100 m (10 000 m^2). They place four 1.00 m by 1.00 m (1.00 m^2) quadrats randomly within the study area. The number of white trillium plants in each of the sample quadrants is 5, 3, 2, and 1. You know the population density is found by the following equation:

$$D_p = \frac{N}{A}$$

$$\text{Estimated population density} = \frac{5 + 3 + 2 + 1}{4 \times (1.00 \text{ m}^2)} = 2.75 \text{ trilliums/m}^2$$

If the entire study area is similar to the sample areas, this density can then be extrapolated to determine the population size.

$$\begin{aligned}\text{Estimated population size} &= D_p \times \text{total study area}\\ &= 2.75 \text{ trilliums/m}^2 \times 10\,000 \text{ m}^2\\ &= 27\,500 \text{ trilliums}\end{aligned}$$

Sampling Using Mark-Recapture

mark-recapture a method in which animals are captured, marked with a tag, collar, or band, released, then recaptured at a later time to determine an estimate of population size

A quantitative determination for the size of a population-the number of individuals per unit area-is referred to as absolute density. Since it may not be possible to observe and count individuals directly, scientists also use indirect indicators of individuals such as tracks, nests, burrows, and scats (droppings, or feces). In cases where the actual size of a population is not needed for a particular study being conducted, scientists may instead refer to the relative density, which compares the number of individuals in one area or unit of study to another.

Some wildlife populations are sampled using **mark-recapture**, a method in which animals are trapped temporarily, marked using a tag or transmitter, then released, as shown in **Figure 11.4**. At a later date, perhaps a few days or a few weeks later, the same traps or nets are set again. The scientists then compare the proportion of marked to unmarked animals captured during the second trapping to give an estimate of the population size.

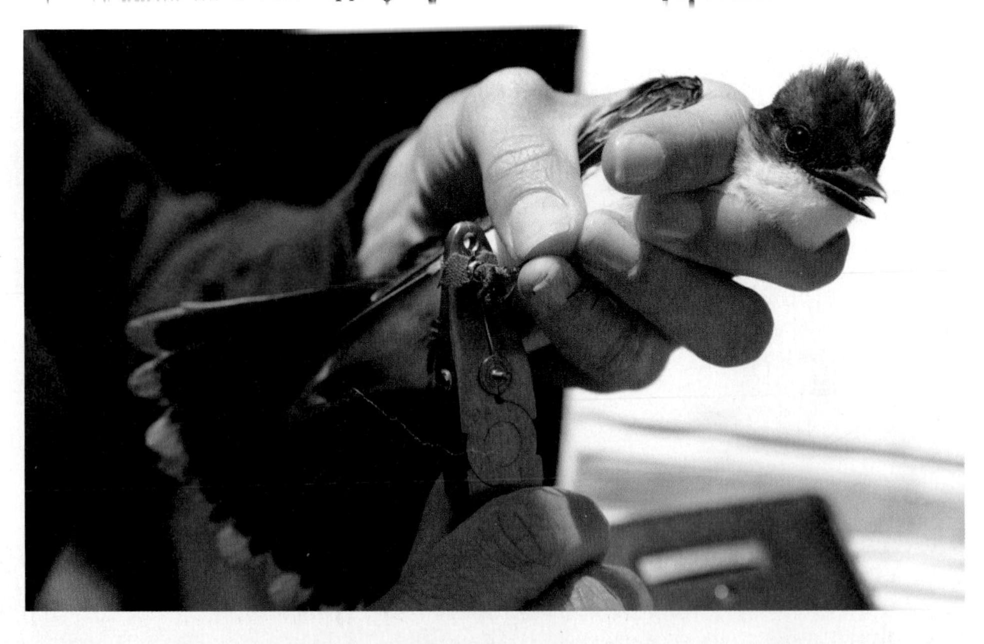

Figure 11.4 The bird was trapped in a fine nylon net, or a mist net, and banded in a mark-recapture program.

For example, suppose that in one study 20 warblers are captured in mist nets. The birds are then marked with leg bands and released. One week later, the nets are reset and 50 warblers are captured.

Of these 50 birds, 10 were *recaptured*—they were banded the week before. The equation below shows how to estimate the population size.

$$\text{Population size } (N) = \frac{(\text{number originally marked}) \times (\text{total individuals in recapture})}{(\text{marked individuals in recapture})}$$

$$= \frac{20 \times 50}{10} = 100$$

The estimated population is 100 warblers. Mark-recapture is particularly useful for highly mobile populations, such as fish or birds.

Population Distribution

Populations are rarely distributed evenly throughout their habitat. A population of little brown bats (*Myotis lucifugus*), for instance, will spread out during the evening to feed. During the day, however, they cluster together in their roosts. Even though these bats are known to live in Wessex County, they are not found everywhere within this region. Ecologists recognize three **distribution patterns** for populations: uniform, random, and clumped. Examples of these patterns, which refer to the ecological density of a population, are shown in **Figure 11.5**.

distribution pattern the pattern in which a population is distributed or spread in an area; three types are uniform, random, and clumped

Figure 11.5 Patterns of population distribution. (**A**) *Uniform Distribution*: Bald eagles are territorial. Pairs are distributed uniformly. (**B**) *Random Distribution*: In summer, female moose with calves tend to be distributed randomly. (**C**) *Clumped Distribution*: Aspens can reproduce sexually by producing seeds or asexually by sprouting new plants via suckers—shoots that grow off their extensive root system. This results in a clumped pattern of distribution.

***Explain** how each of these species benefits from its population distribution pattern.*

Distribution Patterns

Distribution patterns are influenced by two main factors: the distribution of resources such as food and water within a habitat, and the interactions among members of a population or community.

Clumped Distribution

Since resources are typically unevenly distributed, populations tend to gather near them. This results in *clumped distribution*. Animals may gather near a water source, for instance, and plants tend to cluster in locations where moisture, temperature, and soil conditions are optimal for growth.

Clumped distribution is also common among species in which individuals gather into groups for positive interactions, such as protection from predators or to increase hunting efficiency. Shorebirds find "safety in numbers." They fly in large flocks as a way of minimizing the chances of being caught by aerial predators such as falcons. And, at times, humpback whales exhibit co-operative feeding behaviour in which they work in groups to catch prey by blowing bubble nets. The meerkats shown in **Figure 11.6** also use clumped distribution as a way to benefit the population. These highly socialized members of the mongoose family live in groups in which all adults collectively raise the young, enabling parents to carry out other jobs that benefit the population.

Figure 11.6 Meerkats (*Suricata suricatta*) are social animals that live in groups. They stand on their hind legs to watch for predators. If a predator is spotted, the meerkat warns the other members of the group.

Uniform Distribution

In situations where resources are evenly distributed but scarce, populations exhibit *uniform distribution*. This form of distribution is often a consequence of competition between individuals. For example, if water and nutrients are evenly spread throughout a forest but are still in short supply, plants must compete for the resources. They become spread out uniformly as the stronger plants out-compete less robust individuals.

Uniform distribution is also seen in birds of prey and other organisms that behave territorially to defend the food and shelter they need for survival, mating, or raising young. By defending their territory, they keep other individuals out of the area. This results in a more uniform distribution. A wolverine, such as the one in **Figure 11.7** for instance, aggressively defends its territory, which can include a home range as large as 500 km^2.

Figure 11.7 Wolverines are predators and scavengers. They keep others out of their territory to ensure adequate resources for their own and their young's survival.

Plants can also "defend" their territories by using chemicals that discourage the growth of other plants nearby. The black walnut tree, for instance, secretes a chemical called *juglone* into the soil to deter other plants. This ensures that the walnut tree has its own territory, with the water and soil nutrients it requires for optimum growth.

Regardless of whether a species is using chemical warfare or aggression to ensure its share of resources, uniform distribution is usually a result of *negative* interactions among population members. This is in contrast to clumped distribution, which is usually the result of *positive* interactions such as co-operative feeding.

Random Distribution

If resources are plentiful and uniformly distributed across an area, populations exhibit *random distribution.* Since resources are abundant and well distributed, there is no need for individuals to defend their share. Random distribution also requires that interactions between individuals are neutral—neither positive nor negative—and that any young disperse more or less equally throughout the area in question. These conditions are rarely met in nature. Distribution patterns are summarized in **Table 11.1** Use Activity 11.1 on the next page to learn more about distribution patterns.

Table 11.1 Distribution Patterns and Population Dynamics

Distribution Pattern	Resource Distribution	Resource Abundance	Interactions Between Members of a Population
Clumped	Clumped	Varies	Positive
Uniform	Uniform	Scarce	Negative
Random	Uniform	Abundant	Neutral

Activity 11.1 Birds of Prey in Ontario

The birds at the top of the food chain in Ontario have historically been the bald eagle (*Haliaeetus leucocephalus*) and peregrine falcon (*Falco peregrinus anatum/tundrius*). These birds of prey experienced a decline in their populations due in large part to the widespread use of chlorinated hydrocarbon pesticides, such as DDT. As a result, peregrine falcons were locally extinct in Ontario and bald eagles were reduced to a limited range at the northern edge of human settlement in the province. Both species have made impressive comebacks, and have been removed from the endangered list in parts of Ontario within the last decade.

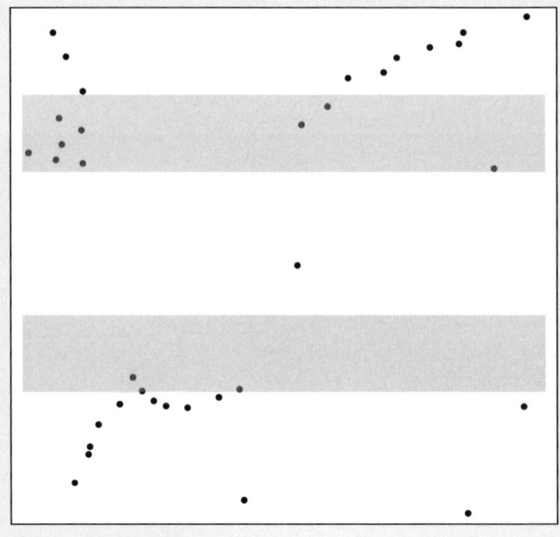

Distribution Pattern 1

Procedure

1. Examine the three diagrams. Each dot represents 1 pair of eagles. Which method would be most useful for estimating populations of eagles, and why?
2. Examine the three diagrams of eagles that were found. Classify them as clumped, uniform, or random.
3. What might cause each of these distributions for a population of bald eagles?
4. The shaded lines represent the transects that were used to sample the population. Assume a scale of 1 cm = 1 km and calculate the area of each transect.
5. Calculate the hypothetical population density for each transect by dividing the number of eagles in the transect by the area of the transect.
6. Calculate the total area inhabited by each population of eagles (the area of the box). Using ratios, predict the number of eagles that would be found in each area.
7. Count the number of eagles actually present.

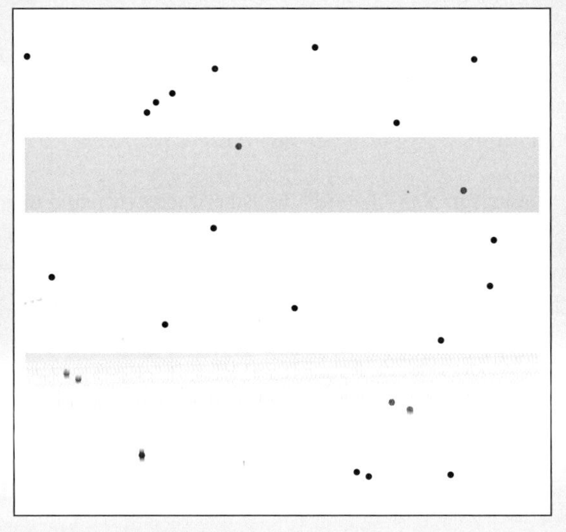

Distribution Pattern 2

Questions

1. How close were your predictions to the actual numbers present?
2. Explain the reason for the differences between your estimates and the numbers for each population.
3. As these birds are recolonizing their old territories, people are also reintroducing them. There is concern that an introduced population of eagles may deplete the resources in its home range. Which distribution would give the most accurate assessment of a sustainable maximum population?

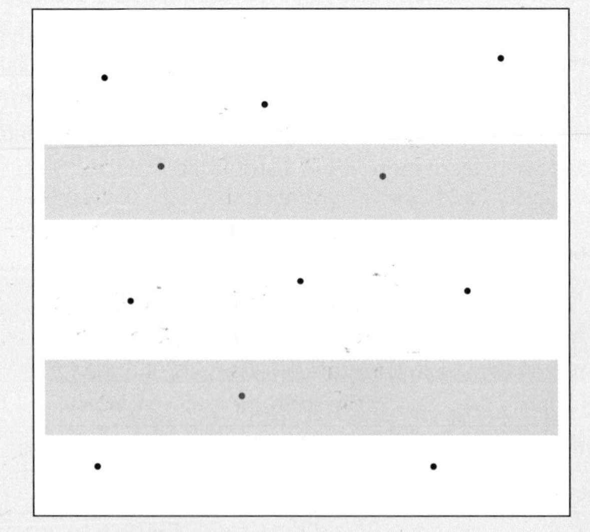

Distribution Pattern 3

Distribution Patterns Are Fluid

When the wind sends dandelion "parachutes" flying with tiny seeds attached, as shown in **Figure 11.8**, the seeds are initially distributed in a random pattern. Soon, however, growing conditions such as available moisture in a field may result in a more clumped distribution. The distribution patterns described above are *models* that help ecologists describe populations. In reality, distribution patterns displayed by a population are fluid and can change with the passing seasons or even over the course of a day.

For each population, the distribution pattern is a product of complex interactions between behaviours and other characteristics that increase each individual's chances of reproduction and survival. Moose, for example, may cluster in small groups near food and shelter during the winter but disperse more randomly in the summer when resources are more plentiful. Distribution patterns often are also dictated by an organism's life stage. Insects such as mayflies often swarm during the mating season. This behaviour increases an adult's reproductive success, but lasts for only a few days before they disperse.

Figure 11.8 Population distribution patterns are not constant. Many factors and interactions contribute to the fluid nature of the patterns.

Learning Check

1. Why do ecologists estimate populations rather than count individual members of a population?
2. What techniques do biologists use to help them measure population size?
3. Describe how the mark-recapture method is used to determine population size.
4. Refer to **Figure 11.5**. Explain why aspens have a clumped distribution.
5. Compare the resources among the three main distribution patterns.
6. Why are interactions in clumped distribution patterns considered to be positive?

Life Histories and Populations

Ecologists also use life histories to add to their understanding of populations. But these are not like a life history or story you might tell about your own life. Rather, these are quantitative measures of the vital statistics that determine a population size. By definition, the survivorship and reproductive patterns shown by an individual in a population is its **life history**.

life history the survivorship and reproductive patterns shown by individuals in a population

In population ecology, life histories include the age at which an organism is sexually mature (when it can reproduce), how often it reproduces, and how many offspring it has at a time. Life span of an organism is also part of the analysis.

Life histories vary between species, but natural selection favours features that maximize evolutionary fitness. This is the number of offspring that survive to reproduce. It makes sense—an organism needs to survive long enough to reproduce and ensure the existence of the next generation.

Ecologists use two main measures to describe life history: fecundity and survivorship.

Fecundity

fecundity the average number of offspring produced by a female member of a population over her lifetime

The average number of offspring produced by a female over her lifetime is called the **fecundity** of that population. This varies widely between species. A female salmon, for instance, produces her lifetime quota of eggs all at once and then dies. Annual plants—those that only grow for one season—have a similar pattern. They live for one season, reproduce, and then die. In other populations, such as songbirds and elephants, females typically survive their first reproductive event (whether laying eggs or giving birth to live young) and go on to reproduce several more times. Perennial plants survive for many years and reproduce each year.

The actual number of offspring produced can also vary widely between species. Organisms that reproduce once then die often have the potential to bear huge numbers of offspring. The salmon in **Figure 11.9**, for instance, has a high fecundity. She may lay over 15 000 eggs before she dies. This is in contrast to a mammal such as a black bear, which has a much lower fecundity. A black bear will give birth to between one and three cubs at a time, which it then must raise for several years.

Figure 11.9 Salmon live for only three or four years. They reproduce once, releasing thousands of eggs before dying.

Measures of fecundity are also affected by the age at which an organism becomes sexually mature. Rabbits can reproduce when they are as young as three months old, while a female elephant is usually not sexually mature until about 12 years of age.

Reproduction always has a cost and the cost is paid in energy. Consider the examples given above. Each species has a slightly different reproductive strategy, each with a different output of energy. Studies of fecundity have shown that the number of offspring tends to be inversely related to the amount of care parents provide. A mother that produces hundreds of young cannot feed or protect each one for long. And many mothers, such as the salmon, provide no protection at all. But other animals with longer life spans produce only one offspring at a time, and devote huge amounts of time and energy to raising their young.

Survivorship

survivorship the number or percentage of organisms that typically live to a given age in a given population

Despite the 15 000 eggs that could potentially be produced by the salmon in **Figure 11.8**, there is no guarantee they all will survive. In fact, it is likely that less than 1 percent will actually survive and grow to sexual maturity. **Survivorship** is the proportion, or percentage, of individuals in a population that survive to a given age.

To study survivorship, ecologists study a large group of individuals all born at the same time. This is called a *cohort*. Ecologists monitor the group over its lifetime and record the age of death for each organism. By studying cohorts, they have determined three general patterns of survivorship. These patterns are visually represented as the survivorship curves shown in **Figure 11.10**.

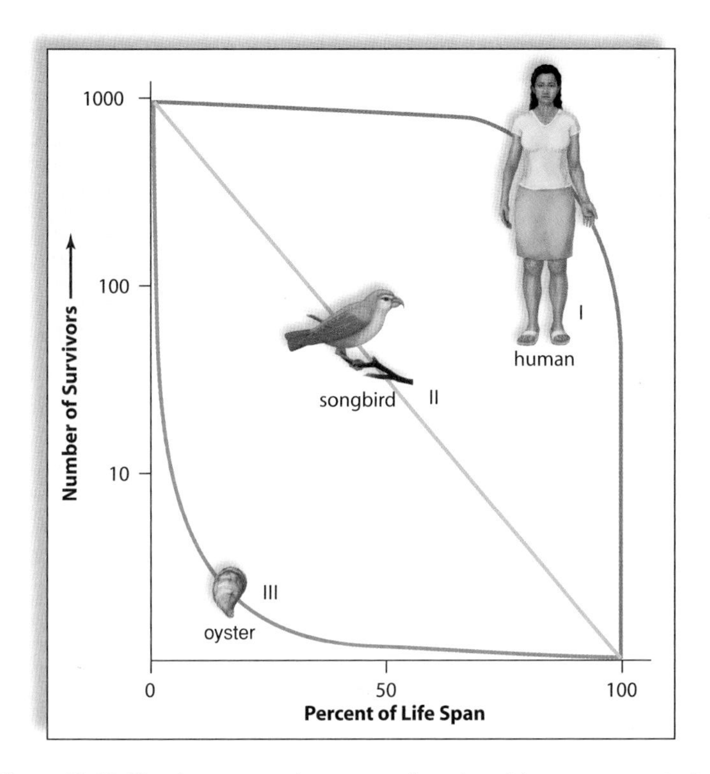

Figure 11.10 The three general patterns of survivorship are represented as Types I, II, and III. In this example, the number of individuals in a population is standardized to 1000, regardless of the true size of the cohort.

In populations that have a Type III survivorship pattern, most individuals die as juveniles, perhaps even before they sprout, hatch, or are born. Only a few members of such a population actually live long enough to produce offspring, and fewer live to old age. However, these individuals can produce large numbers of offspring. Oysters fit this survivorship pattern. An adult female oyster will send thousands of eggs into the water. Many will never even be fertilized, while countless others will be consumed by predators either as eggs or as juveniles. However, the few that do survive until adulthood often have a high survivorship and few predators. These survivors continue the cycle, producing thousands of potential offspring (eggs or seeds) in a sort of "mass mailing" strategy. Many insects, plants, and invertebrate organisms fit this pattern.

The Type I survivorship curve is the mirror opposite of the Type III curve. This pattern shows a high rate of juvenile survival and individuals that live until sexual maturity and beyond. As you can see, the example in the graph shows a human. Humans provide a high level of parental care for their young. Most mammals fit into this pattern of survivorship. While they produce fewer young, the juveniles of these species are well cared for and have a high survivorship.

Type II populations lie between the other two. The risk of mortality is constant throughout an individual's lifetime.

As with distribution patterns, survivorship curves are models. They are tools ecologists can use to simplify patterns to better understand trends. Not all populations precisely fit one of the three patterns. More often, population curves fall somewhere between Types II and III.

Section 11.1 Review

Section Summary

- Ecologists use quantitative measurements to study and describe populations, such as population size and population density.
- Ecologist use techniques to measure population sizes that include using transects, quadrats, and mark-recapture.
- Populations usually are not evenly distributed throughout their habitat. Ecologists use three distribution patterns to describe populations: uniform, random, and clumped.
- Ecologists use life histories, which are quantitative measures of vital statistics that determine population size to aid in understanding populations.
- Two primary measures are used to describe life history: fecundity and survivorship.

Review Questions

1. **K/U** What quantitative measures might an ecologist use to represent the changes in a population over time?
2. **K/U** How do ecologists generally determine population size?
3. **T/I** In order to estimate the number of flowering dogwood in a 50 000 m^2 nature park, a researcher randomly placed six 10-metre by 10-metre quadrats in the park. The researcher counted the number of flowering dogwood in each quadrat as follows: 12, 8, 16, 14, 15, and 19. What are the best estimates for the density and the population of flowering dogwood in that nature park?
4. **K/U** A zoologist recorded the number of deer tracks and the number of coyote tracks along transects in a certain area. What can the zoologist determine from this data?
5. **[illegible]** A biologist wanted to estimate the number of endangered Blanding turtles living in a marsh. The biologist caught, tagged, and released a sample of six Blanding turtles in the marsh. After one month, the biologist captured eight Blanding turtles in the same area and found that three of them had tags, showing they were in the original sample. Based on this information, estimate the population of Blanding turtles in the marsh. Explain whether another sampling procedure would have been better for the biologist to use.
6. **K/U** What factors contribute to the estimated population density of a population?
7. **C** Use a table to summarize the life history that is typical of the three survivorship patterns.
8. **C** Ecologists often use the terms ecological density, which includes only the portion of the organism's range that can be colonized by it, and crude density, which includes all the area within an organism's range. Describe a scenario in which using these terms to describe a population would be useful.
9. **C** Use a Venn diagram to compare and contrast the distribution and abundance of resources, and the interaction between individuals in the different distribution patterns.
10. **C** Consider a herd of cattle in a large fenced pasture. Draw and explain a model for the distribution of the cattle during the rainy season and during the dry season.
11. **A** Wolves travel in packs and are territorial. Make a hypothesis concerning the distribution patterns you would expect to find in wolf populations and the reasons these types of distributions would occur.
12. **A** A part of the human life cycle includes childhood, adolescence, adulthood, and old age. At which of these stages would the highest fecundity rate occur in humans? Explain your answer.
13. **K/U** How does the number of offspring an organism produces generally relate to the amount of energy spent taking care of those offspring? Explain why.
14. **A** The life cycle of the frog shown below represents a Type III survivorship pattern. Based on this information, predict the fecundity of this population. Explain your answer.

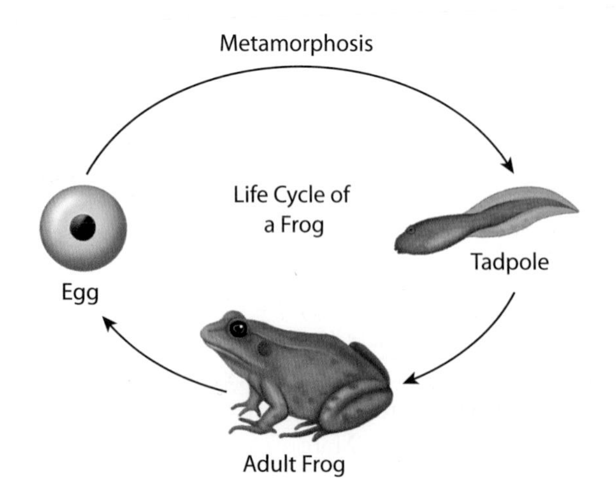

SECTION 11.2

Changes in Population Size

The pond shown in **Figure 11.11** provides the perfect habitat for the western painted turtles (*Chrysemys picta bellii*). There is enough food, space, and access to mates for the population that lives here. But that can change. Environmental changes, such as a severe winter or a disease that affects a food source, can lead to changes in the population. Other factors can affect the population size. For example, a particularly cold winter might cause the population to decrease through mortality or a few turtles might migrate from a nearby pond to increase the population. Populations are always in flux—they are dynamic and changing.

Key Terms

immigration
emigration
biotic potential
exponential growth
carrying capacity
logistic growth
r-selected strategy
K-selected strategy

Measuring Population Change

In 1911, the population of Canada was about 7.2 million. In 2011, it was 34 million. Several factors have influenced this dramatic change. In general, people are living longer. Fewer women die in childbirth. In addition, child mortality and catastrophic diseases such as influenza have been absent in the past decades. Disease can kill many people in a short period of time. For example, the Spanish flu pandemic swept across the world in 1918–1919. It killed an estimated 60 to 80 million people worldwide, including 50 000 Canadians.

People have also moved to Canada, or immigrated. Immigrants from dozens of countries over the past century have increased the country's population. And, of course, some people have also moved away, or emigrated. Birth, death, immigration, and emigration are the four basic processes that cause change in the size of all populations. **Immigration** is the movement of individuals into a population, while **emigration** is the movement of individuals out of a population. Birth (or *natality*), death (or *mortality*), immigration, and emigration can be used when measuring changes in population size.

immigration the movement of individuals into a population

emigration the movement of individuals out of a population

Figure 11.11 Western painted turtles are common in northwestern Ontario. They live in ponds, marshes, and slow-moving rivers. Even though the turtle is common, an ecosystem can sustain only a limited number of them.

Calculating Change in Population Size

Humans are unusual when considering population dynamics because they tend to move more than is typical for most species. In most populations, immigration and emigration are roughly equal. For this reason, and for simplicity's sake, ecologists tend to focus only on birth and death when considering how population size changes.

Expressed as a word equation, change in population size could be written as

Change in population size =
number of births − number of deaths

Expressed mathematically, the equation is

$$\Delta N = B - D$$

Δ = change
N = population size
B = births
D = deaths

Suppose an ecologist is studying a population of painted turtles, like the ones in **Figure 11.11**. If 78 new turtles were born and 12 turtles died, the change in population size is calculated as

$$\begin{aligned}\Delta N &= B - D \\ &= 78 - 12 \\ &= 66 \text{ turtles}\end{aligned}$$

Population Change with Immigration and Emigration

As mentioned above, for most species, immigration and emigration do not affect the sizes of populations. The calculations tend to cancel each other out. There are exceptions, such as some populations of butterflies with populations that merge at times. Human populations, however, are greatly affected by immigration and emigration. When these factors must be considered in a calculation, population change can be determined using the following equation:

$$\Delta N = [B + I] - [D + E]$$

I = immigration
E = emigration

Rate of Population Growth

Ecologists studying populations of zebra mussels (*Dreissena polymorpha*) in the Great Lakes monitor the population size, but also the rate at which the population is growing. Understanding the rate of change, whether increasing or decreasing, can help ecologists make management decisions. In the case of the zebra mussels, shown in **Figure 11.12**, ecologists studying the rate of change could show the severity of the rapid population growth of this invasive species.

Figure 11.12 Zebra mussels were introduced to Ontario lakes in the ballast of ships.

Zebra mussels are an invasive species that spread throughout the Great Lakes at an astonishing rate, often out-competing native species for available habitat. A population undergoes a population explosion when it increases so rapidly that it spreads before it can be contained. The opposite of population explosion is a *population crash.*

The change in the number of individuals in a population (ΔN) over a specific time period (Δt) is the population's growth rate (gr), which is represented by the equation below:

$$gr = \frac{\Delta N}{\Delta t}$$

gr = growth rate
ΔN = change in population size
Δt = change in time

Measuring growth rates is of particular use for populations that are expanding quickly, but also for populations that might be endangered. The rate provides a measurement that shows increases or decreases over time. Understanding the growth rate helped biologists in Banff National Park make management decisions to protect the Banff Springs snail, (*Physella johnsoni*) shown in **Figure 11.13**. The population of the snail in the Lower Cave and Basin springs was estimated to be 3800 in 1997. Two years later, the population was about 1800.

The first step in determining the growth rate was to find the change in population size. You have already learned that change in population size can be written as $\Delta N = B - D$. However, a change in population size can also be determined by comparing the size of the population at different points over time (N_2 and N_1).

$$\begin{aligned} \text{Thus, } \Delta N &= N_2 - N_1 \\ &= 1800 - 3800 \\ &= -2000 \text{ snails} \end{aligned}$$

Since the change occurred over two years, the growth rate (gr) was

$$\frac{-2000}{2} = \frac{-1000 \text{ snails}}{\text{year}}$$

The population decreased at a rate of 1000 snails per year.

Figure 11.13 Understanding the change in the growth rate of the Banff Springs snail helped ecologists protect its habitat.

Predict *what might happen to the Banff Spring snail if its population continues to decline.*

Measuring Per Capita Growth Rate

Note that the growth rate of a population does not take into consideration how the initial size of the population may affect population growth. As long as nothing limits it, the growth of a larger population will always be greater than that of a smaller population, since this population typically has more individuals that can reproduce. For example, **Figure 11.14** shows the growth rate of two populations. The population in Bowl A was 10 bacterial cells. The population in Bowl B was 2 cells. If the bacterial population doubled every 30 minutes, then after 6 hours, the population in Bowl A would have almost 33 000 more bacterial cells than Bowl B.

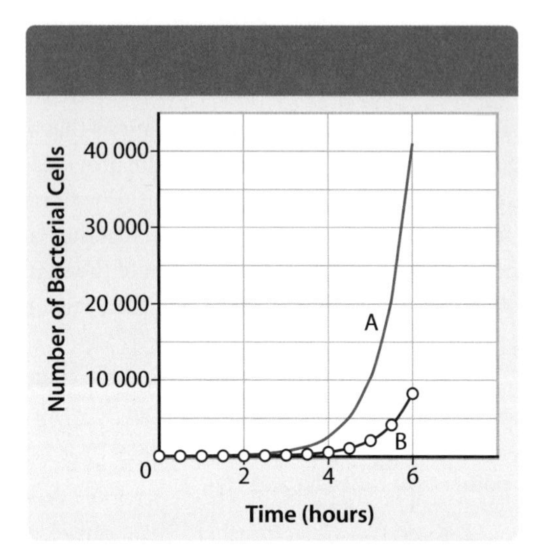

Figure 11.14 This graph shows the growth of two bacterial populations over 6 h. The larger population has the larger population growth.

Explain *why the population size of A was so much greater than the population size of B.*

To make comparisons with more meaning, it is important to express the change in population size as the rate of change per individual or *per capita*. Per capita growth rate (*cgr*) is determined by calculating the change in the number of individuals (ΔN) relative to the original number of individuals (N) as shown in the equation below:

$$cgr = \frac{\Delta N}{N}$$

As an example, a town has a population of 1000 people at the beginning of the year. At the end of the year, it has a population of 1020. The per capita growth rate (cgr) for the population during this time interval is

$$cgr = \frac{\Delta N}{N} = \frac{1020 - 1000}{1000} = 0.02$$

So, the per capita growth rate is 0.02, or 2 percent. In this situation, the population is growing. In other situations, this figure could be negative, which would mean the population is declining over time.

Learning Check

7. What are the four main processes that can cause changes in population size? Write an equation for the relationship among them.
8. Why is immigration and emigration often not considered when determining the change in population size?
9. Why is it important to understand how populations are changing?
10. An ecologist estimated 1500 individuals in a population of geese. Three years later, the ecologist estimated 3600 individuals in the same population. Determine the growth rate of the population during this time interval.
11. Refer to **Figure 11.14**. Population A and B grow at the same rate and there are no factors limiting their growth. Why do you think population A shows a greater increase than population B? How could you make a more meaningful comparison of the changes in population size for the two populations?
12. The per capita growth rate for a certain population is –4.5 percent. Explain what this means, and what processes are the likely causes of these changes.

Population Growth Models

Compare the growing environment of the two populations in **Figure 11.15**. The growing conditions for the bacterial culture are highly controlled in the laboratory environment. A scientist can attain ideal growing conditions for the culture as a result of unlimited access to the resources it requires. But in natural environments, such as those of the pitcher plants, populations are subject to the unpredictability of nature. At times, conditions may be ideal and growth will be optimized but, in reality, that situation is rarely achieved. When ecologists study population growth, it is important to consider the conditions under which the population is growing.

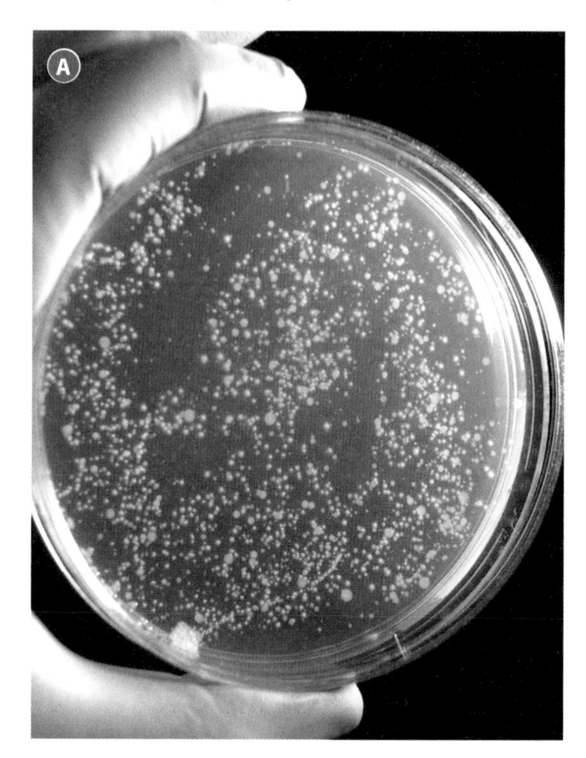

Figure 11.15 This bacterial culture **(A)** grows under highly controlled conditions in a laboratory, whereas these pitcher plants **(B)** are subject to whatever conditions nature delivers.

Population Growth in Unlimited Environments

biotic potential the highest possible per capita growth rate for a population

exponential growth the growth pattern exhibited by a population growing at its biotic potential

The calculations described previously provide information for populations under ideal conditions, with no predators and where resources—including food, water, shelter, and space—are available in unlimited quantities. Under these living conditions, each species has its highest possible per capita growth rate. This is called its **biotic potential**. The factors that determine a species' biotic potential are all related to its fecundity, and include

- the number of offspring per reproductive cycle
- the number of offspring that survive long enough to reproduce
- the age of reproductive maturity
- the number of times the individuals reproduce in a life span
- the life span of the individuals

Figure 11.16 shows a population growing at its maximum biotic potential. The population has no limits; there is no predation and resources are unlimited. A population growing at its biotic potential grows exponentially. There is a brief lag phase, followed by a steep increase in the growth curve, which is called an **exponential growth** pattern. This growth pattern is represented graphically by a J-shaped growth curve.

Populations growing at their biotic potential are often found in laboratory conditions. Examples include bacteria and other micro-organisms, small invertebrates, and plants. In contrast to a population of wild swallows, for instance, the micro-environment of a bacterial culture growing in a Petri dish is much easier to control and to make "ideal." For this reason, scientists can only estimate the biotic potential of species living in natural environments.

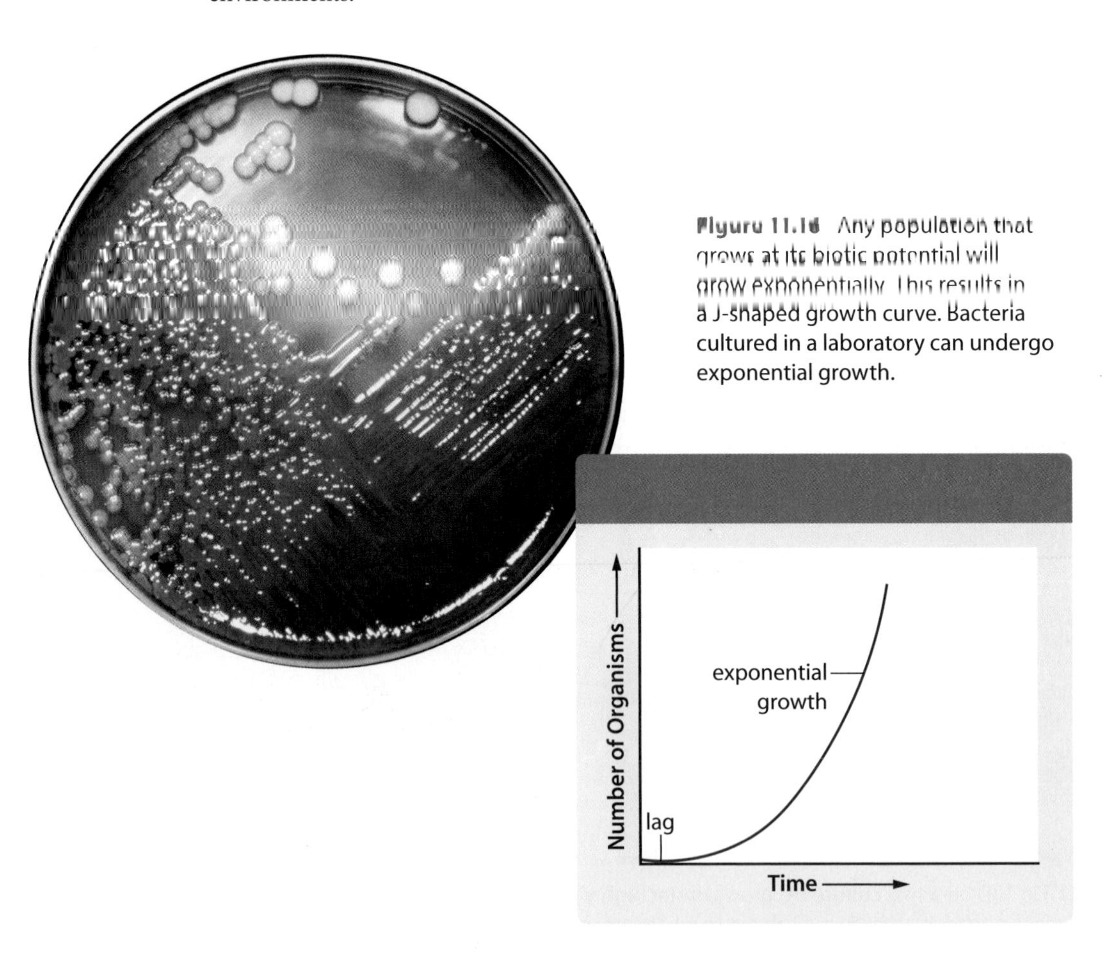

Figure 11.16 Any population that grows at its biotic potential will grow exponentially. This results in a J-shaped growth curve. Bacteria cultured in a laboratory can undergo exponential growth.

Populations Growing in Limited Environments

If one bacterium divided every 30 minutes, in 20 hours there would be 1.1×10^{12} bacteria. In four days, the mass of the colony would be larger than our entire planet. It sounds like a horror film and, in reality, this scenario cannot occur. A population cannot grow at its biotic potential because resources will quickly become limited. Eventually, members of the population will compete for resources and the growth rate will slow. Lack of food, for example, will limit the energy available for survival and also the energy individuals can put toward reproduction.

In conditions where resources are limited, a population initially experiences a *lag phase* of slow growth as shown in **Figure 11.17**. Since there are only a few individuals in the population able to reproduce, the population grows slowly. This is followed by a period of rapid growth. Eventually, however, resources become limited. As individuals begin to expend energy competing for the resources, the rate of growth slows. Eventually the population reaches a state in which the birth rate and death rate are the same. At this point, the habitat is at its **carrying capacity**, which is the maximum population size the available resources can sustain. This pattern of population growth is represented in a *sigmoidal*, or S-shaped curve, as shown in **Figure 11.17**. The graph shows a **logistic growth** pattern.

carrying capacity maximum population size that a habitat can sustain over an extended period of time

logistic growth the growth pattern exhibited by a population for which growth is limited by carrying capacity, or limited availability of resources

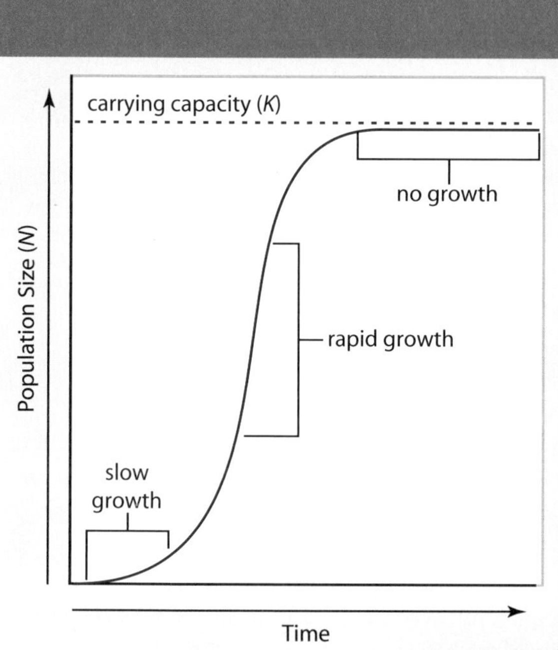

Figure 11.17 A population that lives in an environment where food or other resources are limited grows in an S-shaped curve. This population of snow geese, *Chen caerulescens*, will increase until the population reaches the carrying capacity of their habitat.

Effect of Environmental Resistance on Population Growth

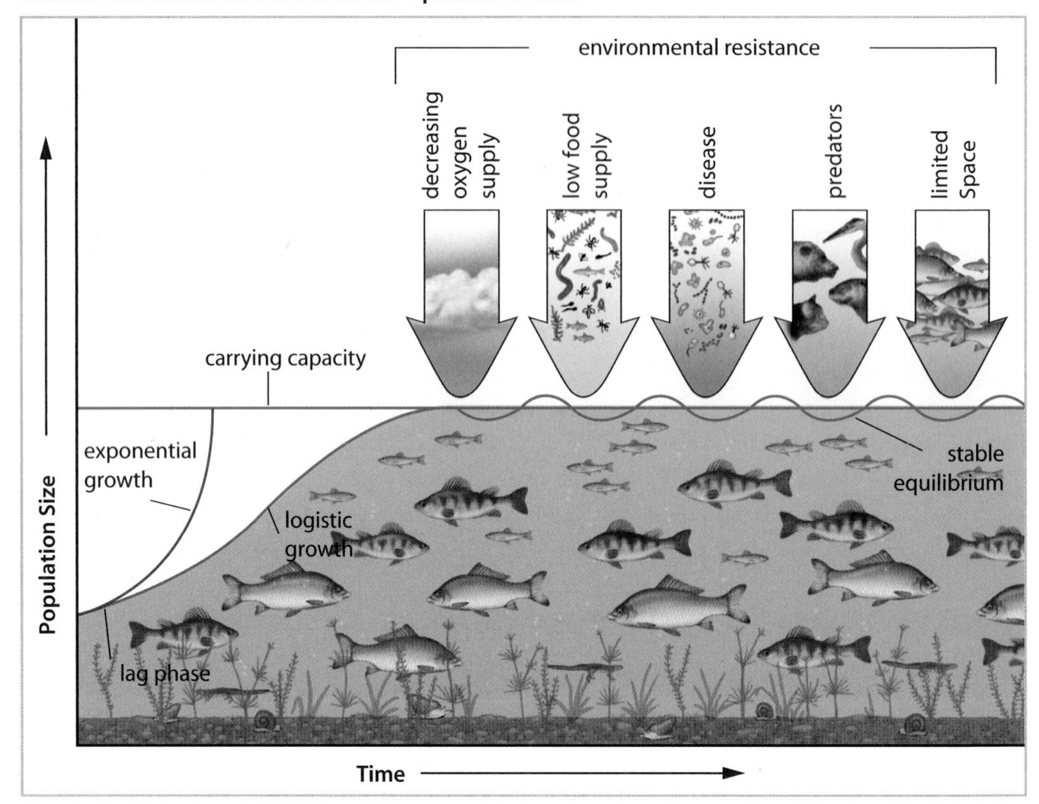

Figure 11.18 Limiting factors keep a population from growing exponentially. Although the number of individuals that an environment will support changes from time to time, such as from season to season, there is always a limit to what the environment can support.

As a population increases in size, limiting factors such as disease, predation, and competition for limited resources reduce the amount of energy that is available for reproduction. This causes the growth rate of the population to decrease. **Figure 11.18** illustrates the logistic growth of a population. The green line running through the S-curve represents the carrying capacity of the habitat and shows the number of individuals it can sustain given its limited resources. Carrying capacity is not a static condition, however. From season to season, the carrying capacity changes as the population responds to changing conditions, such as a decreasing oxygen supply in the pond, low food supply during the winter, disease, predation, and limited space. Over time, the population size fluctuates around the carrying capacity of the habitat in a stable equilibrium.

Population Growth Models and Life History

In North America, most organisms, including the loons in **Figure 11.19** at the top of the next page, reproduce in the spring. The harsh winter is over, days are longer and warmer, plants are growing, and ponds and oceans are flush with life as photosynthesis drives the production of phytoplankton. All of these conditions increase the potential for survival. Life history traits are genetically controlled, but organisms use strategies to maximize the number of offspring that survive to a reproductive age. In theory, the best strategy would be for an organism to reach sexual maturity at a young age, have a long life span, produce large numbers of offspring, and be able to provide these offspring with a high level of care until they reach an age at which they can reproduce.

Figure 11.19 Ontario's provincial bird, the common loon *(Gavia immer)*, lays eggs in the spring like most North American birds.
***Explain** why this strategy is beneficial to the survival of this organism.*

Of course, this is not realistic. As you have learned, resources are limited and how an organism uses its time and energy will vary. In reality, organisms have to make trade-offs to maximize the number of offspring that survive. They use different *life strategies*. Life strategies employed by organisms can be correlated to the type of environment in which the organism lives. An infectious bacterium in a human body will multiply rapidly, producing a huge number of offspring in a very short period of time. To the bacterium, the human body is an unstable environment. In unstable environments, conditions are variable and often unpredictable. The most advantageous way for the bacterium to ensure the survival of its offspring is to reproduce close to its biotic potential—produce as many offspring as possible in as short a period of time as possible.

In contrast to a bacterium are organisms such as wood bison or northern fur seals that live close to their carrying capacity. In comparison to the bacterium, they live in more stable environments. Their strategy is to have fewer offspring and to provide their young with more care, thus investing their energy into ensuring their offspring reaches reproductive age.

r- and *K*-selected Strategies

Ecologists explain life strategies as either being a ***r***-selected strategy or a *K*-selected strategy. Species that have an ***r*-selected strategy** live close to their biotic potential (*r*). In general, these organisms

- have a short life span
- become sexually mature at a young age
- produce large broods of offspring
- provide little or no parental care to their offspring

Insects, annual plants, and algae are examples of organisms that use r-selected strategies. They take advantage of favourable environmental conditions, such as the availability of food, sunlight, and warm temperatures, to reproduce quickly. They experience exponential growth during the summer, but die in large numbers at the end of the season.

***r*-selected strategies** life strategies used by populations that live close to their biotic potential

K-selected strategies life strategies used by populations that live close to the carrying capacity of their environment

Organisms with a **K-selected strategy** live close to the carrying capacity (*K*) of their habitats. In general, these organisms

- have a relatively long life span
- become sexually mature later in life
- produce few offspring per reproductive cycle
- provide a high level of parental care

Mammals and birds are organisms that tend to use this strategy. They have few offspring, but invest large amounts of time and energy to ensure those offspring reach an age at which they can reproduce. Although organisms tend toward either an *r*-selected or a *K*-selected strategy, in reality most populations are somewhere between the two groups. The balsam fir (*Abies balsamea*) in **Figure 11.20**, for instance, is large and can live for many years, yet it produces hundreds of gamete-bearing seeds in cones.

Figure 11.20 The balsam fir shows both *r*-selected and *K*-selected strategies.

As well, properly describing whether a population uses an *r*-selected or a *K*-selected life strategy requires that it be compared to another population. For example, a rabbit population could be described as *K*-selected when compared to a population of mosquitoes, even though rabbits are well known for their fecundity. Rabbits become sexually mature when they are as young as three months old, and can have several litters of up to 12 kits a year. However, when a population of rabbits is compared to a population of black bears, rabbits are better described as being *r*-selected.

Ecologists use an understanding of life strategies to predict the success of a population in a particular habitat. For example, two populations of Antarctic fur seals, like those in **Figure 11.21**, living on the same island nurse their young for different lengths of time. One population nurses their young for 10 months, while another population nurses for four months. The researchers are trying to determine how the time and energy invested by the population that nurses their young for 10 months pays off in terms of survivorship when compared to a population that nurses for only four months.

Suggested Investigation

ThoughtLab Investigation 11-A, Estimating Population Sizes

Figure 11.21 Scientists studying populations of Antarctic fur seals *(Arctocephalus australis)* have found that different populations use slightly different life strategies. Scientists are trying to determine how these different strategies affect survivorship of the pups.

Section 11.2 Review

Section Summary

- The factors that change population size are births, deaths, immigration, and emigration.
- Population growth rates are calculated on a per capita basis to standardize the measurements.
- Two population growth models are exponential growth and logistic growth.
- All habitats have a carrying capacity, which is the maximum population size that it can support.
- Organisms usually exhibit two types of life strategies: *r*-selected strategy or *K*-selected strategy.

Review Questions

1. **K/U** What value can be assigned to the growth rate in a stable population? What has to be true of the birth and death rates?
2. **K/U** What effect does birth rate have on per capita growth rate? Explain your answer.
3. **T/I** In a population of 5000 individuals, 600 newborns are produced annually and 800 individuals die annually. Determine the per capita growth rate of this population.
4. **A** Human populations live in different parts of the world and face different conditions. These populations grow at different rates. How might you measure and compare the growth rates of different human populations?
5. **K/U** Does exponential growth last long in nature? Why or why not?
6. **K/U** What is the growth rate of a population in a habitat that has reached carrying capacity? Explain your answer.
7. **A** A researcher observing a certain population found that it could not grow to its biotic potential. Explain why, and predict the resulting curve that represents the researcher's data. What do you think would happen if the researcher tries to control the environment to make it "ideal"?
8. **C** A farmer noticed that the weeds in a field were becoming numerous. The farmer used a herbicide to control the weeds. Draw a graph that represents the population of weeds in this situation. Explain your reasoning, and include the terms *population explosion* and *population crash*.
9. **A** You are asked to catalogue several species by their life strategies. What criteria would you use to determine the predominant life strategy of each species?
10. **A** Weeds, such as dandelions, have a *r*-selected life strategy. Use the characteristics of this type of strategy to explain why the term opportunistic population is suitable for these populations.
11. **A** One of the curves on the graph below represents a population of a certain insect, capable of infesting an entire area, when growing at its biotic potential. The other curve represents a population of a certain mammal in a limited environment. Identify the curve that represents the insect population and the one that represents the mammal population. For each curve, explain what is occurring in each phase and why this is occurring.

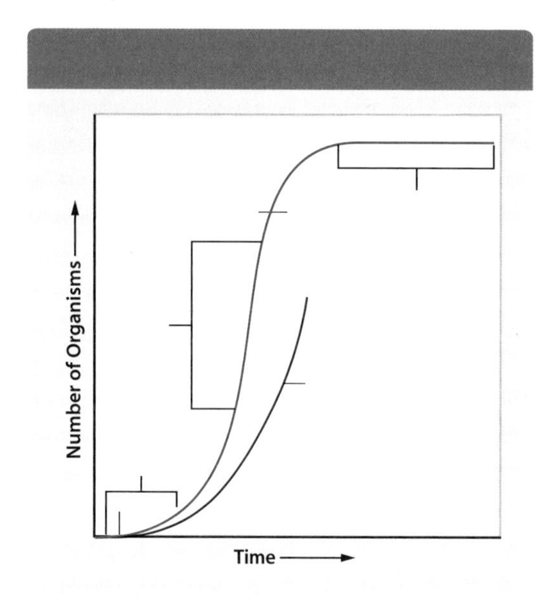

12. **C** The carrying capacity of a habitat is the maximum population size the habitat can sustain over an extended period of time. Use a diagram to show the pattern of population growth that reaches carrying capacity. Include on the diagram the carrying capacity and the limiting factors.
13. **A** Many of the bird populations in Ontario are threatened, endangered, or otherwise at risk. These populations tend to have a *K*-selected life strategy. Explain why populations with this type of life strategy are often at risk.
14. **K/U** What factors determine the biotic potential of a population?

SECTION 11.3

Factors That Regulate Natural Populations

Key Terms

density-independent factor
density-dependent factor
intraspecific competition
interspecific competition
population cycle
sinusoidal growth
protective colouration
symbiosis
parasitism
mutualism
commensalism

The ice storm shown in **Figure 11.22** swept through Ontario and Québec in 1998. It was a serious weather event that affected thousands of people, leaving many without electricity and phones for days. But the storm was devastating for natural systems, too. It killed populations of maple, birch, cedar, and other tree species. When populations suddenly stop growing, or crash, it is often the result of an *abiotic* factor such as severe weather.

Figure 11.22 In 1998, a severe ice storm hit parts of Ontario and Québec. It made roads treacherous and impassable and knocked out the power in some locations for days.
Explain *how this storm might have affected populations of living things.*

In another example, bark beetles usually grow exponentially until very cold weather in winter kills many of the individuals that would produce the next generation. Because temperature affects a bark beetle population *regardless of its density*, it is called a **density-independent factor**. It is interesting to note that, in the last decade, climate change producing warmer-than-usual winters has prevented the die-off of bark beetles. As a result, forests in much of British Columbia have been killed by mountain pine beetle (*Dendroctonus ponderosae*) damage. For now, it seems, cold is becoming less and less of a limiting factor for this population.

density-independent factor an abiotic event that affects population growth in the same way, regardless of population density

Besides weather, density-independent limiting factors include other abiotic events such as floods or droughts and, for populations within a small geographic area, forest fires, hurricanes, or tornadoes.

If the events are severe enough, density-independent factors result in a population crash, a sharp drop in population growth as shown in **Figure 11.23** at the top of the next page. Note that the temperature decrease shown in the figure would have caused this population to crash regardless of its density. Instead, it is the severity of the abiotic event that determines whether a crash occurs. Density-independent factors are not always so severe that a total collapse of population numbers results; however, the influence such a factor has on population growth is always independent of population size or density.

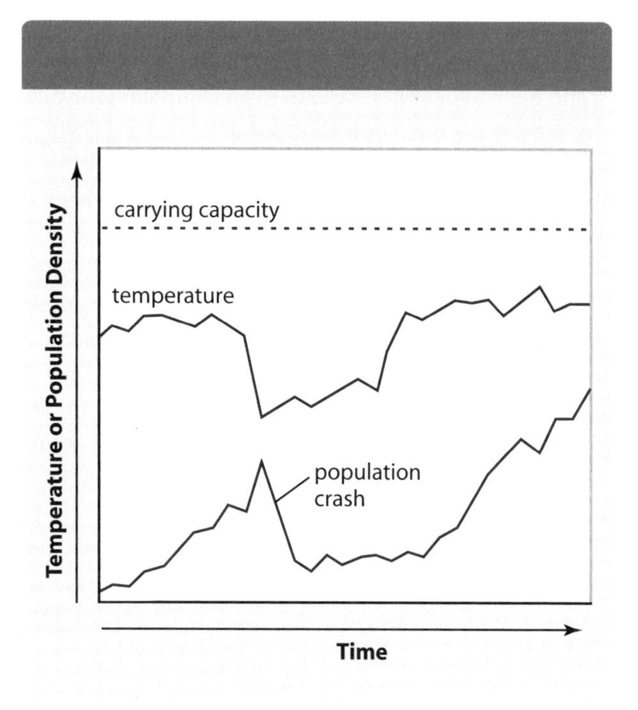

Figure 11.23 This graph shows changes in temperature—a density-independent factor—that result in a population crash, a steep decline in population growth. The decline begins well before the population is halfway to its carrying capacity and well before density would inhibit its growth. This decline is thus density-*independent*.

Density-dependent Factors

As you read in Section 11.2, most populations follow a logistic growth pattern rather than an exponential growth pattern. These populations are limited by **density-dependent factors**. Density-dependent factors are *biotic*, such as competition or predation. The strength at which they slow a population's growth depends directly on the density of the population. If a population is far below its carrying capacity, density-dependent factors will have no effect. However, once a population reaches a particular density, factors such as competition for food start to have an effect. At that point, the population growth slows and eventually stops when the carrying capacity is reached.

density-dependent factor a biotic interaction that varies in its effect on population growth, depending on the density of the populations involved

As an example, consider the spread of parasites or disease in a population. The tapeworm (*Taenia sp.*) in **Figure 11.24** is a parasite that can lay up to 10 000 eggs a day. The eggs are shed in the feces of infected individuals and enter the environment where they can be picked up by another host. Parasites need a host to survive. In a population with a low density, it is more difficult for the parasite to spread. However, when populations have a high density, transfer of the parasite from host to host is more probable.

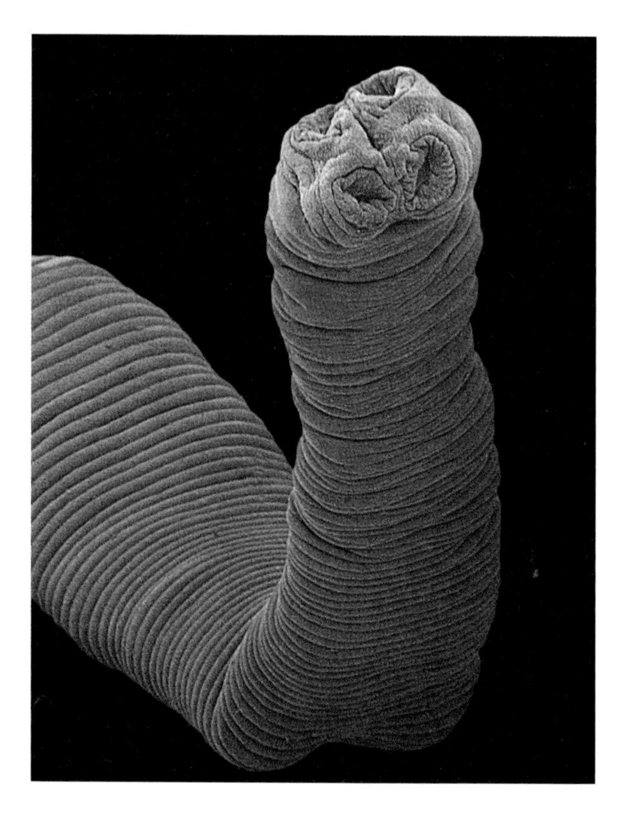

Figure 11.24 Tapeworms are a type of parasite. They are transferred from livestock to humans when people eat infected and undercooked beef or pork.

Explain *why a high-density population allows parasites to spread more easily.*

Figure 11.25 Each tree in this forest will release hundreds, perhaps thousands, of seeds. Of these seeds, only a small percentage germinate into young trees.

Competition

The young trees in **Figure 11.25** all require resources such as water, soil nutrients, and light. When these resources are abundant, the population grows quickly. They are in the rapid growth phase of a logistic growth, or S-shaped curve. However, once resources become limited and the habitat approaches its carrying capacity, the members of the population must begin to compete. The result is the birth rate decreases or the death rate increases, or both, and the population growth slows. In this example, the competition is *intraspecific*. **Intraspecific competition** is among members of the same population and could include competition for resources such as water, nutrients, breeding sites, light, shelter, and reproductive opportunities.

The Mackenzie Bison Sanctuary covers 10 000 km^2 on the shore of Great Slave Lake in the Northwest Territories. It is the habitat for a population of over 2000 wood bison (*Bison bison athabascae*), as shown in **Figure 11.26**. The population's growth is regulated primarily by the availability of food. In a lean year, the bison begin to compete with each other for food and the competition intensifies as population density increases. If food is scarce enough, population growth stops and may even become negative until food once again becomes more abundant.

Figure 11.26 The growth of the wood bison population in the Mackenzie Bison Sanctuary is regulated mainly by the availability of food, which in turn is influenced by abiotic factors.

Figure 11.27 Garlic mustard is an invasive species that out competes native species for resources.

In some situations, there is **interspecific competition** in which two or more populations compete for limited resources. Often, one species, such as the garlic mustard (*Alliaria petiolata*) in **Figure 11.27**, will out-compete another. This plant was introduced from Europe and competes with native plants in woodland habitats. In this example, two plants have the same habitat requirements and similar ecological niches and one plant (the garlic mustard) "wins" and excludes the native plant or plants from the habitat. Ecologists describe the effects of interspecific competition in this example as the *competitive exclusion principle*. This principle states that two species with overlapping niches cannot coexist. However, when competing species have niches that are quite different, they can both live in the same area. The density of one species' population, however, may be lower than that of the other.

intraspecific competition a situation in which members of the same population compete for resources

interspecific competition a situation in which two or more populations compete for limited resources

Interspecific competition is one of the driving forces of evolutionary change. In competing species, individuals that are most different from their competitors will be best able to avoid competitive interactions and will therefore obtain the most resources. Ultimately, this means the individual will be more likely to survive and reproduce, thus their alleles will increase in frequency in subsequent generations.

Natural selection and *resource partitioning*—the utilization of different resources within a habitat—can also produce increased divergence between competing species. Consider the five species of warbler in **Figure 11.28**. These warblers can successfully forage in the same spruce tree because each species of warbler tends to feed on insects that are found in a different part of the tree.

Figure 11.28 Each of these warblers occupies a slightly different ecological niche, foraging for insects in different parts of a spruce tree. As a result, interspecific competition is reduced. ***Explain*** *how the differences in ecological niches aid in the survival of each species.*

Producer-Consumer and Predator-Prey Interactions

Not all interspecific interactions in a community are competitive and thus negative. Primary producers have a direct relationship with the primary consumers that eat them. In turn, primary consumers have a direct relationship with their predators—the secondary consumers.

Producers and prey use defensive strategies and "weapons" against their consumers and predators. These producer-consumer and predator-prey relationships put selective pressure on both parties—the more successful predators and consumers drive the natural selection of the producers and prey. For example, producers and prey that are more difficult to catch or less desirable to consume, such as toxic plants or butterflies, are more likely to survive.

The scarcity of a producer or prey species will limit the growth of a consumer or predator species' population. The corollary can also be true: a large population of consumers or predators may control the growth of producer and prey populations. For example, some grey wolves (*Canis lupus*) prey mainly on elk (*Cervus elaphus*). Following the extirpation (local extinction) of grey wolves from Banff National Park in the 1970s, the elk population of the park soared. Ecologists found that the artificially high elk population was overgrazing and damaging willow and aspen trees. In turn, they linked the loss of these trees to declines in local populations of songbirds and beavers.

population cycle alternating periods of large and small population sizes

sinusoidal growth a wavelike oscillating growth pattern that is typical of predator-prey interactions

The absence of the wolves, a top predator, trickled down to affect many parts of the ecosystem. As grey wolves were encouraged back into the area, the composition of species shifted again. By 2003, for instance, the elk population had declined significantly, and overgrazing of trees and bushes was reduced. Although the wolves played a key role in this scenario, they were just one part of a complex and interconnected system.

Population Cycles

Predator-prey interactions are one of many factors that can result in regular **population cycles**—alternating periods of large and small population sizes. **Figure 11.29** presents a simplified model of such a cycle. The resulting growth pattern is referred to as **sinusoidal growth**. One explanation for the wavelike pattern reflected in sinusoidal growth is the density-dependent effect each population has on the other. Hypothetically, the larger a prey population is, the more food that is available to its predators. Thus, the predators have a higher survival rate and more energy to reproduce and care for their young. This allows the predator population to increase. With a greater number of predators, the prey population will decline. This results in more intense competition among the predators for food, which limits population growth. The predator population therefore declines and, with fewer predators, the prey population increases.

Figure 11.29 A simplified graph of predator-prey population cycles, such as elks and wolves, is shown here. An increase in prey increases the resources that are available to predators (**A**), so the predator population increases (**B**). This leads to a reduction in the prey population (**C**), followed by a reduction in the predator population (**D**). The cycle repeats itself over time.

Some predator and prey populations exhibit regular density fluctuations that roughly mirror a sinusoidal growth pattern. For instance, the stoats (*Mustela ermine*) and lemmings (*Dicrostonyx groenlandicus*) in **Figure 11.30c** on the next page share a predator-prey relationship in arctic ecosystems that experiences cyclical changes in population size. The cycles, shown in **Figure 11.30**, have an oscillating pattern that strongly resembles the model sinusoidal growth curve. While the pattern is somewhat less regular than the model, the same cycling of prey and predator is evident. An increase in the lemming population is followed by an increase in the stoat population. This results in a decrease in the lemming population that the stoat population soon mirrors.

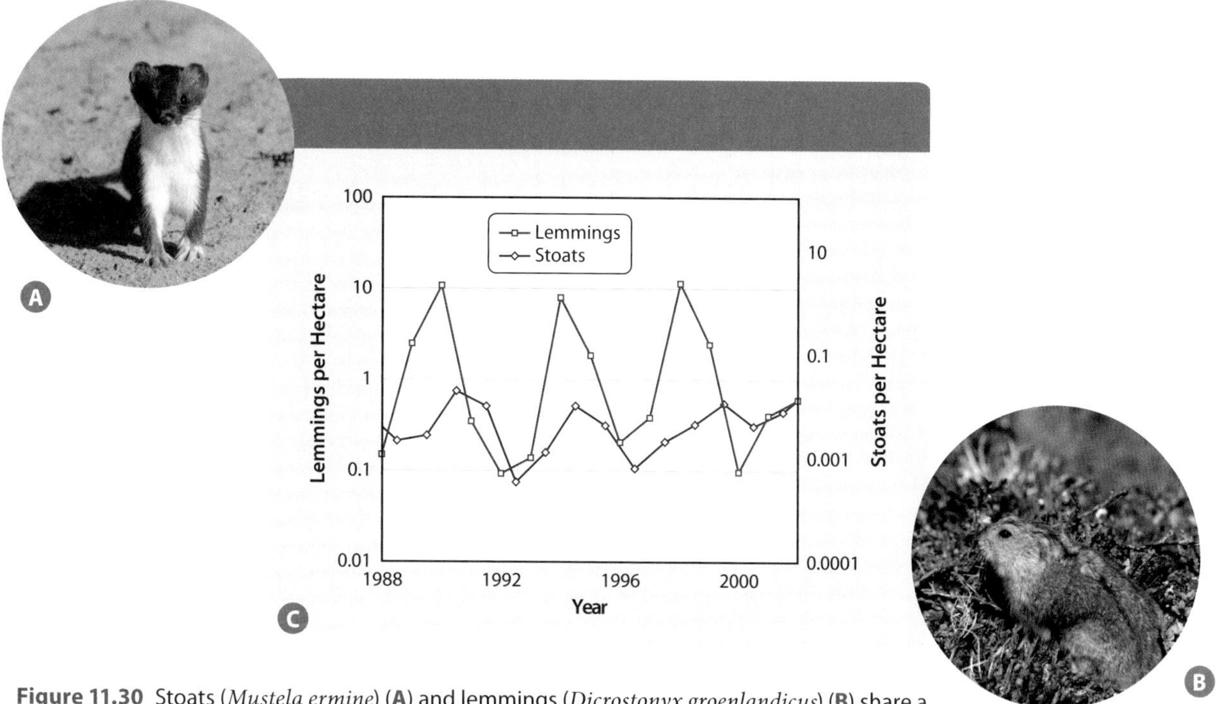

Figure 11.30 Stoats (*Mustela ermine*) (**A**) and lemmings (*Dicrostonyx groenlandicus*) (**B**) share a predator-prey relationship in arctic ecosystems. The predator-prey relationship between stoats and lemmings follows an oscillating pattern (**C**) that resembles a sinusoidal growth curve.

Are sinusoidal population growth cycles completely explained by predator-prey interactions? Some extensive, long-term ecosystem studies suggest they are not. One of the longest-studied sets of sinusoidal population cycles is the oscillating 10-year cycle set seen in populations of snowshoe hare (*Lepus americanus*) and Canada lynx (*Lynx canadensis*). **Figure 11.31** shows the numbers of Canada lynx and snowshoe hare pelts that were traded to the Hudson Bay Company over 100 years. The number of pelts would have been affected by the demand for pelts, the number of trappers, and the locations of the traps. Nevertheless, ecologist Charles Elton studied the hare and lynx data and found that the cycles were too regular to simply be due to abiotic factors. Furthermore, the increases and decreases in the lynx population closely followed increases and decreases in the hare population. This observation led scientists to hypothesize that predator-prey interactions were causing, or at least affecting, the population cycles.

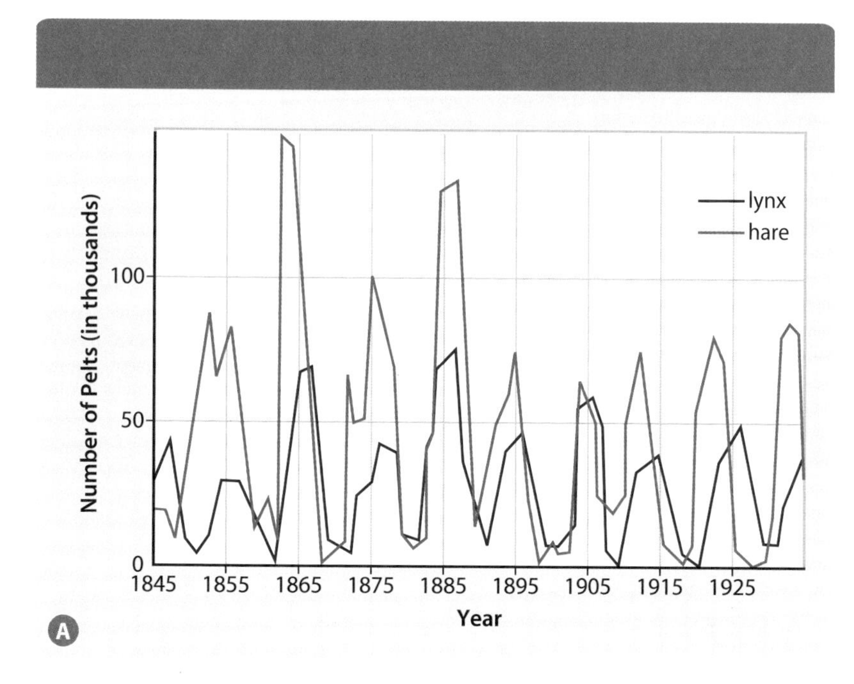

Figure 11.31 Canada lynx and snowshoe hare populations exhibit staggered sinusoidal population cycles (**A**) that oscillate every 10 years. This lynx is about to catch its prey (**B**).

Suggested **Investigation**

ThoughtLab Investigation 11-B, Sampling Hare Populations

Another hypothesis to explain the cycling of the snowshoe hare population was that, at higher population numbers, the hares were depleting their food supply. The reduction in the quantity and quality of the vegetation they browsed would cause the snowshoe hare population to crash.

Charles Krebs and a team from the University of British Columbia completed an eight-year experiment to test the two hypotheses. Their results are shown in **Figure 11.32**. They found that the hare population increased when ground-dwelling predators were excluded from the study site. The hare population also increased when food was abundant (at levels higher than "normal"). When the population was protected from ground-dwelling predators, *and* had an increased food supply, the hares reached their highest relative density. Based on these results, the Canadian scientists concluded that the periodic dips in snowshoe hare populations are probably due to a combination of food availability and increased predation.

Studies such as this one illustrate the intricate connections and interrelationships among populations in food webs. Note also that, although the Krebs study focussed on biotic interactions, abiotic factors such as weather, temperature, and shelter, likely play a role in fluctuations in natural populations.

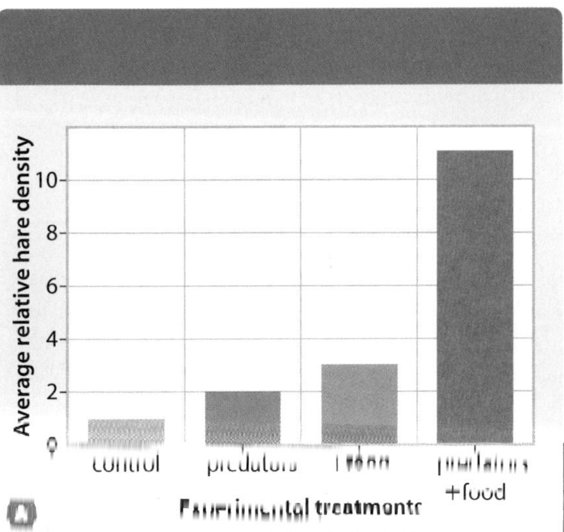

Figure 11.32 The results of the snowshoe hare-Canada lynx field study **(A)** by Krebs et al. are shown here in graph form. Plots in which the hares either had unlimited food **(B)** or were safe from ground-dwelling predators, or both, were compared to control plots in undisturbed sections of boreal forest.

***Describe** other factors that might play a role in this relationship.*

Activity 11.2 Predator-Prey Relations on Isle Royale

The most common way to learn about something is to study a simplified system. That can be difficult in population biology, as most organisms are part of a complex food web, eating a variety of species and being eaten by a variety of others. Isle Royale is a large island in Lake Superior. In the early 1900's a small herd of moose (*Alces alces*) crossed the frozen lake to reach this island, where their population exploded due to an abundance of food and a lack of predators. Their population crashed after about 30 years, because the food source was exhausted. In 1949, a pair of wolves (*Canis lupus*) crossed to the island in a cold winter. The populations of both species have been tracked continuously since 1958, making it one of the most studied of predator-prey interactions. Recently, however, scientists learned some surprising news about the wolf population on Isle Royale. Analysis of DNA from wolf scat (feces) revealed the presence of alleles that were not present prior to 1997. Field observations linked these alleles to a newcomer to Isle Royale, a male wolf that apparently crossed an ice bridge to the island that same year. The wolf, known as "The Old Gray Guy" due to his light-coloured fur, sired 34 offspring before his death in 2010. These offspring have reproduced as well, adding further genetic variation to the Isle Royale wolf population.

Procedure

1. Study the graph below. Assume this data is typical. Average the numbers visually and estimate the population of moose and wolves you expect there to be once the numbers stabilized.
2. Provide an approximate value for the natality, immigration, mortality, and emigration numbers over the years given in the chart.

Questions

1. Look at the relationship between the wolf and moose populations between 1965 and 1980. Given that there is no other moose predator on the island, and that moose make up most of the wolves' diet, what seems to be happening?
2. What might have caused the dramatic decline of wolves around 1983?
3. If you were to chart a graph of the population of balsam fir (*Abies balsamea*), which makes about 60 percent of the moose's diet, what would it look like?
4. Explain at least one possible reason why biologists think that this relationship between wolves and moose is not going to be stable over long time periods.
5. Moose have *K*-selected reproductive strategies. What evidence do you have for this, based on their populations?
6. Are the wolves on the island displaying a *K*- or *r*-selected reproductive strategy? Explain.
7. Why is interspecific competition for moose not a factor on this island?
8. In the absence of wolves, what limited the moose population?
9. Speculate how the introduction of a breeding male wolf in 1997 could have affected the wolf and moose populations.
10. Due to its size and isolation, the wolf population has generally been an inbreeding population. Inbreeding is the ongoing breeding of closely related individuals. In conservation biology, the introduction of unrelated individuals into an inbreeding population is called genetic rescue. Complete research to determine the following
    a. the impact of inbreeding on a population
    b. the effectiveness of genetic rescue, such as that which occurred with the island's wolf population

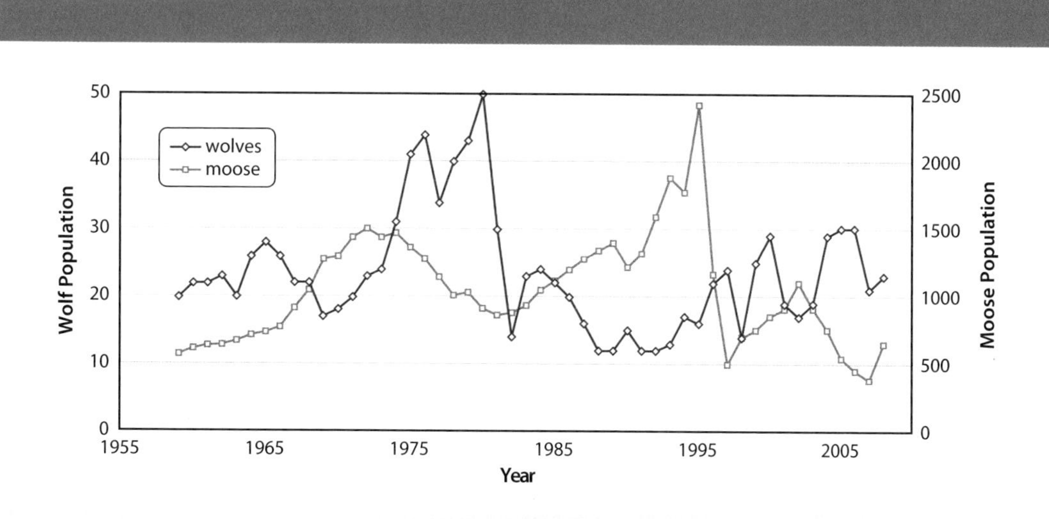

protective colouration adaptations that help individuals avoid predation; includes camouflage, mimicry, and body colouration used as a warning signal

Defence Mechanisms

Most organisms have adaptations that help protect them against their predators—cacti have thorns, porcupines have quills, and monarch butterflies have toxins and **protective colouration**. The interactions between producers and consumers typically result in *co-evolution*. The milkweed plant, for example, produces bitter-tasting chemicals that discourage many herbivores. Some of these herbivores, however, have adapted to tolerate the toxin.

Many organisms use protective colouration as a natural defence mechanism. This can include camouflage, mimicry, and warning colouration. The dead leaf butterfly (*Kallima paralekta*) in **Figure 11.33**, for example, is nearly invisible to predators. When it remains motionless, its brown colouration and the veined pattern on its wings camouflage it from predators.

Figure 11.33 The dead leaf butterfly of Southeast Asia fools predators through protective colouration that makes it resemble a dead leaf.

Some species use warning colours, with such as red, yellow, and black. For instance, the highly venomous eastern coral snake (*Micrurus fulvius*) has red, yellow, and black stripes and yellow jacket wasps are black and yellow. Other species that are not toxic or poisonous, use these colours to their advantage. The syrphid fly, for instance, has a similar yellow-and-black colouration to a yellow jacket wasp but is not dangerous. It uses mimicry to deter predators and enhance its survival. The non-venomous scarlet king snake (*Lampropeltis triangulum*) also uses this strategy. As you can see in Figure 11.34, it has a similar colouration to the venomous coral snake. This type of mimicry, where a species looks like another species that has an effective defence strategy, is called *Batesian mimicry*.

Figure 11.34 The eastern coral snake **(A)** has black, red, and yellow colouration that warns other organisms that it is venomous. The non-venomous Scarlet king snake, **(B)** mimics this colouration and pattern as a means of protection.

Even two species that are poisonous, harmful, or unpalatable may benefit from mimicking each other. For example, in **Figure 11.35**, the Zimmerman's poison frog or poison dart frog (*Dendrobates variabilis*), closely resembles the mimic poison arrow frog (*Dendrobates imitator*). An animal that becomes sick preying on the Zimmerman's poison frog will avoid all frogs with that colouration, including the mimic frog. Scientists hypothesize that the converse is also true: predators finding the mimic frog distasteful, will also avoid Zimmerman's poison frogs. This co-evolved defence mechanism is called *Müllerian mimicry*.

Figure 11.35 The Zimmerman's poison frog (**A**) and the mimic poison frog (**B**) have similar warning colouration that deters predators. Even though both frog species are poisonous, this similar colouration provides additional protection against predators.

Learning Check

13. How do abiotic factors differ from biotic factors in limiting population growth?
14. What is the difference between intraspecific and interspecific competition?
15. How do predator-prey relationships put selective pressure on both partners?
16. Why does the size of some predator-prey populations follow a cycle?
17. What generally happens to other populations in an environment when there is a change in one population?
18. Use **Figure 11.32** to explain how the dead leaf butterfly protects itself from predators.

Symbiotic Relationships

Close interactions between two species living in direct contact often result in an ecological relationship called **symbiosis**. Symbiosis means *living together*. Symbiotic relationships have one organism, the *symbiont*, which lives or feeds in or on another organism—the *host*. There are three forms of symbiosis: parasitism, mutualism, and commensalism.

symbiosis an ecological relationship between two species living in direct contact; includes parasitism, mutualism, and commensalism

Figure 11.36 Mistletoe does not photosynthesize. Instead, its roots grow directly into a host tree where it obtains nutrients from the host's sap.

Parasitism

In **parasitism**, a symbiont (the parasite) benefits from the relationship but the host is harmed by it. The mistletoe in **Figure 11.36** is a parasite. It obtains food by growing roots directly into the host tree and gaining nutrients from its sap. The interaction weakens the tree and predisposes it to disease.

Parasites include viruses, unicellular organisms, insects, and various types of worms. Some parasites such as the mistletoe are ectoparasites—they live outside of their hosts. Others are endoparasites that live inside their hosts. These internal parasites are often unicellular organisms such as the protist *Plasmodium falciparum*, which causes malaria. Internal parasites usually depend on their interactions with their hosts to survive and therefore cannot exist outside the host.

The parasite-host cycles are similar to predator-prey cycles and show a direct relationship to population density. An increase in the host population results in an increase in the parasite population. In turn, the increase in parasites eventually reduces the host population growth, either through decreasing the hosts' abilities to reproduce or by reducing survivorship. The cycle continues as the survivors in a now-reduced population of hosts do not have to compete with as many individuals for resources. **Figure 11.37** illustrates the population cycles of one host-parasite relationship—the adzuki bean weevil and its wasp parasitoid. The adult female wasp lays her eggs into or on the host bean weevil. The larva hatches and eats the tissue of its host. This type of wasp parasite that kills their host is called a parasitoid.

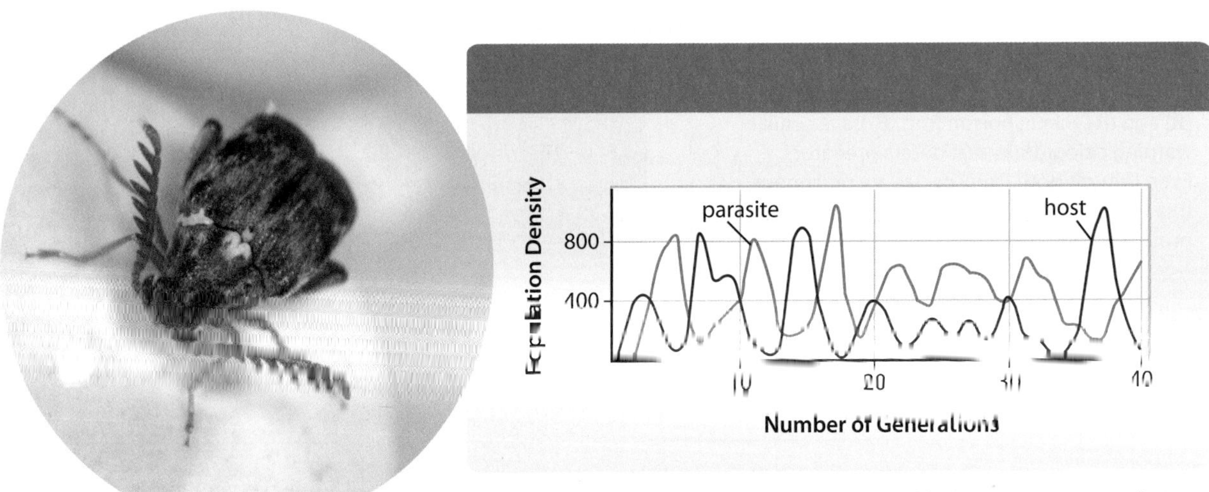

Figure 11.37 The population cycles exhibited by host-parasite populations are similar to those seen in predator-prey relationships. This graph shows the cycling of the host adzuki bean weevil (*Callosobruchus chinensis*) and its wasp parasitoid (*Dinarmus basalis*).

parasitism a symbiotic relationship in which a symbiont lives off and harms the host

mutualism a type of symbiotic relationship in which both species benefit from the relationship

Mutualism

When both partners in a symbiotic relationship benefit from the relationship, or depend on it to survive, their relationship is called **mutualism**. Such relationships are common in nature. A lichen, for example, is actually a combination of an alga and a fungus. Their mutualistic relationship allows them to grow on exposed, bare rock, where neither one would survive on its own. While the algal partner in the relationship carries out photosynthesis to feed both organisms, the fungus protects the alga from drying out or blowing away. It also produces an acid that dissolves rock, releasing minerals the alga requires.

Mutualism is common in aquatic ecosystems, as well. The hermit crab (*Clibanaarius erythropus*) and sea anemone (*Actinia equina*) in **Figure 11.38** have a mutualistic relationship. The sea anemone's stinging tentacles protect the crab from predators. In turn, the crab provides the sea anemone with a ready source of food—the detritus from its meals. In this relationship, the crab also provides a "mobile home."

Animal behaviour is an important part of most mutualistic relationships. In Latin America, for example, bull-horn acacia trees (*Acacia cornigera*) show mutualism with stinging ants. The leaves of the Acacia produce a sugary liquid that the stinging Acacia ants (*Pseudomyrmex ferruginea*) consume. The stinging ants find protection inside the tree's hollow thorns. The ants are beneficial to the tree because they attack any other herbivores that land on it and they cut down the branches of other plants that come in contact with the Acacia. This ensures that the Acacia has adequate light for photosynthesis.

How do mutualistic relationships affect the growth of the populations involved? Because both partners have co-evolved, growth in one population typically spurs growth in the other population. Similarly, if one population decreases in size, the other population tends to do the same.

Figure 11.38 This relationship between a hermit crab and a sea anemone is an example of mutualism, in which two species benefit from a relationship.

Commensalism

A symbiotic relationship in which one partner benefits and the other partner is unaffected is **commensalism**. For example, the lemon shark (*Negaprion brevirostris*) in **Figure 11.39** does not appear to benefit or suffer from its relationship with the remora (*Echeneis naucrates*), who use a modified, sucker-like dorsal fin to hold fast to the shark's body. The remora, however, receive protection and bits of food from the shark and also gain a source of transportation. As the shark seems unaffected, the relationship appears to be one of commensalism.

commensalism a symbiotic relationship in which one partner benefits and the other partner neither benefits nor is harmed

Figure 11.39 Remora attaches to the surface of a shark's body, gaining bits of food and a free ride.

Figure 11.40 Cattle egrets engage in commensalism with various large mammals including this rhinoceros, cattle, and kangaroos.

Another example of commensalism is seen in the relationship between the cattle egret (*Bubulcus ibis*) and cattle. The birds follow the cattle around, feeding on insects roused by the cattle's movement. The cattle seem unaffected by the ever-present birds, neither profiting nor being harmed by the relationship. Cattle egrets do not just limit their relationships to cattle, but also seem to engage in commensalism with other large animals, such as rhinoceroses and even kangaroos. As **Figure 11.40** shows, these creatures appear to take little notice of the birds.

In cases of commensalism, it is often difficult to determine how each species is affected. Some ecologists argue that there are few true cases of commensalism. They believe both partners in symbiotic relationships are usually affected in some way, making the relationship a mutualistic one, although how both organisms are affected is not always clear. If true commensalism does exist, however, growth of the host population would affect growth of the symbiont population in a positive way. However, growth of the symbiont population would have no effect on the host population whatsoever.

Table 11.2 summarizes the effects interspecific interactions—competition, predator-prey relationships, and symbiotic relationships—can have on population growth. In this table, a plus sign (+) indicates an increase in population density and a minus sign (−) indicates a decrease.

Table 11.2 Interspecific Interactions

Nature of Relationship between Populations	Effect of Growth in One Population on the Other Population
competitive	–/– (both are negatively affected)
predator-prey or herbivore-plant	+/– (one population gains at the expense of the other)
host-parasite	–/+ (one population gains at the expense of the other)
mutualistic	+/+ (both are positively affected)

Section 11.3 Review

Section Summary

- Population numbers are usually controlled by abiotic factors such as temperature, floods, drought, forest fires, hurricanes, and tornadoes.
- Density-independent factors are abiotic events that control population growth.
- Density-dependent factors that control population growth include biotic factors such as competition and predation.
- Some organisms live in symbiotic relationships with other organisms.
- There are three types of symbiotic relationships: parasitism, mutualism, and commensalism.

Review Questions

1. K/U Compare and contrast density-independent and density-dependent factors.
2. A What factors could limit the population of gypsy moths growing exponentially in northern Ontario during the summer months?
3. A Provide a real-world example of a situation in which a density-independent factor limits a population.
4. K/U Songbirds live in a density-dependent population. Why does the population of the songbirds decline when population density gets to a certain point?
5. A Analyze the picture of gannets nesting on a rocky island. Explain the type of competition that you would expect, and the effects of this competition.

6. A Poison hemlock, commonly known as snakeweed, is a poisonous plant that is not native to North America. However, it can now be found in many locations on this continent. Explain how this plant has come to occupy so many habitats in North America and what its effect is on native plants.
7. K/U Explain why the relationship between consumer and producer populations is similar to the relationship between predator and prey populations.
8. C Draw a model of predator-prey populations to teach a Grade 6 class about sinusoidal growth patterns. Explain the model.
9. T/I Refer to the **Figure 11.31**. Predict what would happen if biologists kept introducing hares into the environment to keep the hare population at a high and constant level. Predict what would happen if they kept only the lynx population at a high and constant level.
10. T/I The snowshoe hare uses protective colouration as a defence mechanism against predators. If you were trying to observe snowshoe hares in their natural habitat, explain what you would expect to see in their appearance during the spring and during the winter.
11. K/U How does mimicry relate to warning colouration?
12. K/U Compare and contrast Batesian mimicry and Müllerian mimicry.
13. K/U Describe how the parasite-host population cycle is similar to predator-prey population cycles.
14. C Give an example of mutualism, and explain how each partner in your example benefits from the relationship.
15. A How would you expect the growth of one population to affect the other in a mutualistic relationship? Explain your answer.
16. A The bacterium *Helicobacter pylori* lives in the stomach of humans. This bacterium can prevent the development of esophageal cancer and acid reflux. However, it can also cause stomach ulcers. Identify the symbiotic relationships between this bacterium and humans. Explain your reasoning.
17. A Give examples of intraspecific competition and interspecific competition of populations not mentioned in the text that occur in your area.

ThoughtLab INVESTIGATION **11-A**

Skill Check

- Initiating and Planning
- Performing and Recording
- ✓ Analyzing and Interpreting
- ✓ Communicating

Estimating Population Sizes

The spotted turtle (*Clemmys guttata*) was once common in southern Ontario, but sightings have become less frequent because of urbanization of the province. Deciding whether it is in danger of becoming extinct or whether it is just not seen as much is a job for biologists.

Pre-Lab Questions

1. Spotted turtles live in shallow marshes, bogs, and beaver ponds. What distribution pattern would you expect them to have, and why?
2. These turtles can live very long lives (110 years for females, 65 years for males). Explain whether they would be expected to have a *r*-selected or *K*-selected reproductive strategy.
3. This data was measured using the mark-recapture method. What is the mark-recapture method for studying populations?
4. Explain whether the researchers would have likely used transects or quadrats to obtain this data.

Question

What demographic information does a study of the spotted turtle in Ontario provide?

Organize the Data

The following data is from a field study done by Dr. Dan Reeves and Dr. Jacqueline Litzgus of Laurentian University in Sudbury. They studied populations of spotted turtles on a small island in Georgian Bay over three years.

Student Data

Study Date	Females Captured	Females Recaptured	Males Captured	Males Recaptured	Juveniles Captured	Juveniles Recaptured	Total Captured	Total Recaptured
10 June 2005	4	0	2	0	4	0	10	0
15 May 2006	3	0	0	0	2	1	5	1
20 June 2006	14	6	2	0	6	3	22	9
21 June 2006	7	3	1	0	3	0	11	3
6 July 2006	4	2	0	0	0	0	4	2
22 July 2006	9	8	2	2	5	3	16	13
10 September 2007	2	1	1	0	1	0	4	1

Source: Reeves, Dan J. and Jacqueline D. Litzgus. 2008. Demography of an Island Population of Spotted Turtles (Clemmys guttata) at the Species' Northern Range Limit. Northeastern Naturalist 15(3): 417–430.

1. Calculate the populations of males, females, and all turtles found in the study. Include information as to how many were recaptured.
2. The island was observed to be 23.2 hectares (0.232 km^2). If all the turtles on the island were found in the study, what would the population density be?
3. A student found the population density to be 72 turtles/0.232 km^2 = 310 turtles/km2. What mistake did they make?
4. The sampling of July 22, 2006 found 9 turtles in a 0.42 hectare wetland, which is considered prime habitat for these turtles. What was the local population density?
5. Based on that answer, what distribution pattern do turtles follow?
6. Explain whether you can use this data to reliably calculate the population growth rate, and do so if possible.
7. What is the ratio of male to female turtles in this study?

Analyze and Interpret

1. Why were there no recaptures in the June 2005 measurement?
2. Calculate the expected number of turtles on this island as of July 22, 2006. What would the real population density be using this number?
3. These turtles tend to swim some distance and hibernate over the winter, and do not necessarily return to the same place in the next year. However, the populations of spotted turtles in Ontario tend to be isolated from each other. Use the data from September 2007 to estimate the local turtle population.
4. What does this isolatation mean in terms of the population change equation?
5. Estimate the carrying capacity of an ideal environment for spotted turtles.
6. On average, a female spotted turtle will lay 10.5 eggs per clutch, and lays them for approximately 66 years. What is her fecundity?
7. A spotted turtle begins laying eggs at about age 5. What can you conclude about the typical survivorship to that age?

Conclude and Communicate

8. What sort of survivorship curve is implied by this data?
9. Why would the researchers not have made a visit to the island in August?

Propose a Course of Action

The spotted turtle is listed as endangered in Ontario but not in the states bordering Ontario. Describe the sort of study that would be needed to see whether the turtle is really endangered in Ontario but not on the other side of the Great Lakes.

Extend Further

10. **INQUIRY** What other information would be useful if you wanted to study the population of these turtles?
11. **RESEARCH** These turtles have a rare reproductive strategy that is neither *r*-selected nor *K*-selected, called a "bet hedging strategy". Research this strategy and explain what makes this turtle well suited to this approach.

ThoughtLab INVESTIGATION 11-B

Skill Check

Initiating and Planning

Performing and Recording

✓ Analyzing and Interpreting

✓ Communicating

Materials

- notebook paper
- graphing paper
- calculator (optional)
- spreadsheet or graphing software (optional)

Sampling Hare Populations

The snowshoe hare (*Lepus americanus*) is a North American mammal, residing predominantly in the boreal forests of Canada but also extending into the tundra. It eats various plants, depending on what grows where it lives, and in the winter survives on a diet of buds, twigs, bark and even meat under severe conditions. Predators include the lynx (*Lynx canadensis*), red fox (*Vulpes vulpes*), arctic fox (*Vulpes lagopus*), great horned owls (*Bubo virginianus*), and various hawks, ravens, and crows. One of the longest running studies of hare populations began in 1976 in a boreal forest habitat in the Kluane Region of the Yukon Territory. The goal of the Kluane Ecological Monitoring Project was to measure long-term changes in the communities in this region. As shown by the food web diagram below, the researchers thus monitored not only hares, but several other species, all of which affected or were affected by the hares, either directly or indirectly.

Pre-Lab Questions

1. How can a change in the dominant species of primary producers affect the rest of the food web?
2. If a population of about 200 000 herbivores supports about 200 predators of those herbivores for long periods of time, what would the effect be of a flash flood causing 1000 herbivores to die?
3. If the same flood killed off 100 of the predators, what might happen?

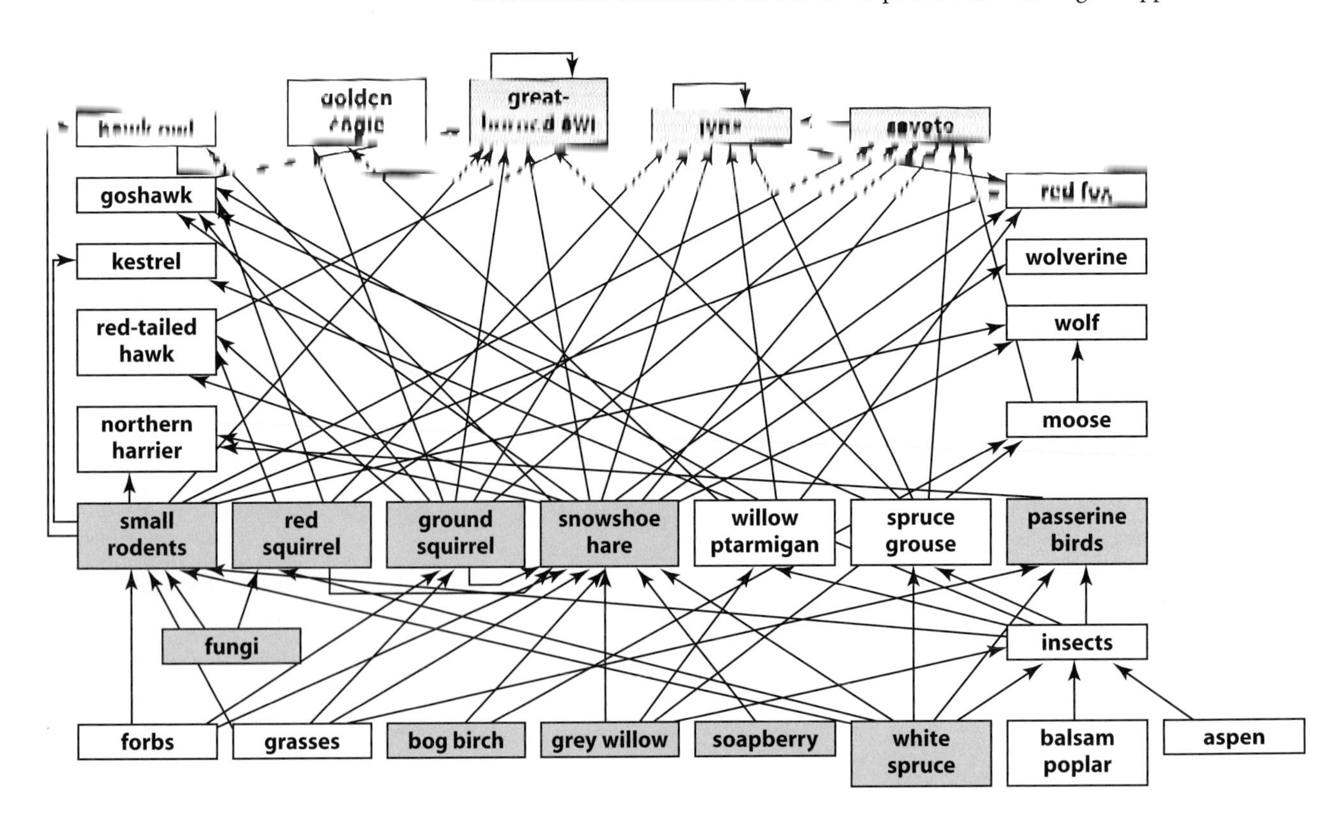

A simplified food web (only the major feeding links for boreal forest of the Kluane Region, Yukon Territory). The shaded boxes indicate which species were studied as part of the Kluane Ecological Monitoring Project.

Question

How can data be found and used to estimate population numbers and densities in a given year, and how can the fluctuation pattern of the population be determined?

Organize the Data

1. Study the data in the table. Although these data have been simulated, they are representative of the densities observed in each year of the Kluane study. In this hypothetical study, traps were laid out at regular intervals in a grid pattern in four study areas each 60 hectares in size, as shown in the figure on the next page. In each area, trapping was done for the same number of days in both spring and fall, and the same number of traps was used in each area. Traps were opened in the evening, and were checked again in the early morning. When a hare was caught for the first time it was tagged and various variables were recorded (such as sex, weight, grid location where caught) before it was released. If a hare was recaptured its tag number was recorded and the same measurements were taken again. On the basis of these captures and recaptures an estimate of the number of animals in each study area was calculated. These estimates from the fall of 1976 to the spring of 2001 are shown in the table.

Data Table

Year of Study (fall/spring)	Estimated Number of Hares in Each Study Area in Fall				Estimated Number of Hares in Each Study Area the Next Spring			
	Area A	Area B	Area C	Area D	Area A	Area B	Area C	Area D
1976/77	1.2	3.4	2	3	3	2	5	4
1977/78	11	16.6	15.2	19.5	18	17	19	18
1978/79	68	59.75	65	78	50	42	39	44
1979/80	191.33	202.5	226.25	245.55	117	112	125	117
1980/81	202	190	214	204	172	182	162	172
1981/82	252	242	288	280	40	50	45	45
1982/83	22	20.33	9	20	8	9	8	9
1983/84	15.33	21.25	11.5	18.66	11	10	9	9
1984/85	11	16.6	15.2	19.5	9	9	10	11
1985/86	14	17.2	17.66	19.5	7	8	6	7
1986/87	1.2	3.33	4.66	3	7	8	9	10
1987/88	52	52	62	70	60	50	70	50
1988/89	130	137.5	122.66	157.33	45	55	35	45
1989/90	142	121	152	144	88	98	68	88
1990/91	87	110	86	103	50	40	60	50
1991/92	58	59.75	75.5	68	17	12	20	19
1992/93	10	5.5	7.25	8.75	5	4	6	5
1993/94	11	14	15.2	9	4.5	5.5	3	5
1994/95	32	36	35	28	11	6	17	10
1995/96	70	52	64	67	30	40	20	30
1996/97	100	99	86	111	45	65	35	45
1997/98	122	132	116	132	120	130	110	120
1998/99	149	166	184	157.33	80	90	70	80
1999/2000	66	63	72	76	14	16	6	9
2000/01	3.75	5.5	8.75	3				

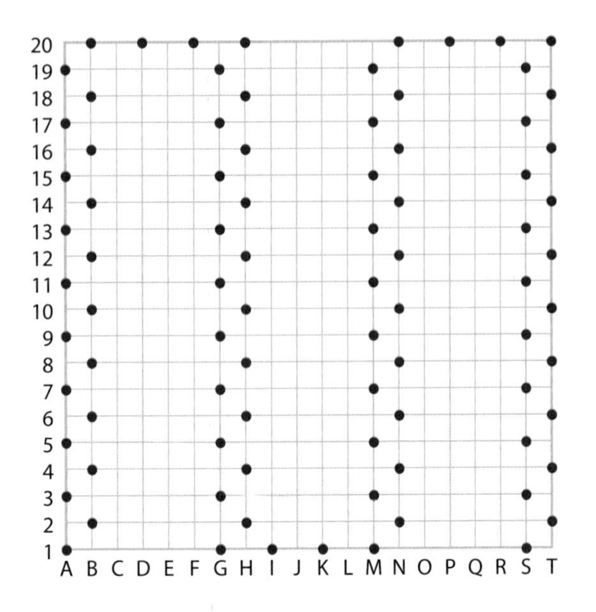

Each dot on this grid represents a trap. There are approximately 30 m between each point on the grid. The study area considered to be "covered" by these traps is larger than the grid itself, being about 60 ha in size.

2. Calculate the average number of hares captured per 60-hectare area in the spring of each year, then do the same calculation for fall. You might decide to work in groups, with some members of a group doing the spring data and others doing calculations for fall. Using this average figure, determine the density of hares per hectare in each season.

3. Make a graph of your data—for spring or fall densities or both—showing how the population density changes over time. Year (this is the manipulated or independent variable) should be plotted on the x-axis and hare density (this is the responding or dependent variable) on the *y*-axis.

Analyze and Interpret

4. For step 1, what is the advantage of sampling the number of hares in relatively small areas rather than trapping over the entire area? Why was the trapping done the same way in each area, or why was each area the same size?

5. For step 2, why were four areas studied instead of just one? Why is the density you calculated really an estimate of the density in this area, rather than a measure of the actual density?

6. Examine your graph from step 3. Describe the period of the hare population cycle in the Kluane Region. That is, how many years are there between successive density peaks or lows? Was there a difference between the periods of the spring and fall cycles? Suggest a reason why or why not.

7. In the Kluane Region, the snowshoe hare is what many ecologists refer to as a keystone species: one whose disappearance would have catastrophic effects on many other species in the community in which it lives. Consider the food web shown on the previous page. How do you think removal of all hares from this area would affect the structure of the community?

Conclude and Communicate

8. What might cause the difference between the spring and fall cycles?

9. What might cause the multiyear cycle?

10. How would you expect the population of predators to vary over the same time frames? Explain a possible means for the predator population to undergo a cyclic change similar to the hares. Which predators might do this?

11. Explain a possible means for the predator population to remain constant throughout the year. Which predators might do this?

12. Would the predator population show the same population curve on the multi-year cycle as the hares do? Explain your answer.

13. In this lab, the method used for gathering data was "mark and release". If you were to modify this study to gather data about lynx in the same region, what would the best method be to estimate their population, and why?

Propose a Course of Action

Research the rationale for introducing rabbits to Australia, their effects on the marsupial population, and the effects of introducing other species to try to control rabbit populations there. What suggestions do you have to deal with this situation?

Extend Further

14. INQUIRY This is a "textbook" example of population curves, meaning that it seems simple and the real world data match what the model would predict. Research a "chaotic population bifurcation" and write a paragraph to explain what you found.

15. RESEARCH The snowshoe hare was introduced to Newfoundland in the mid 19th century. What effects did that introduction have on Newfoundland's indigenous arctic hare population? Why were the effects so much less than the introduction of rabbits to Australia?

STSE

BIOLOGY Connections

Keystone Species

A keystone species is one that significantly shapes its ecosystem, helping to keep all parts of the ecosystem in balance. Just as in architecture, where builders of a stone arch use a central wedge-shaped keystone to lock together the other stones, keystone species are crucial to a stable ecosystem.

A crayfish is a good example of a keystone species. It is essential to ecosystem stability because it eats plants in the lower trophic levels of the food web and is, in turn, eaten by animals in the higher trophic levels.

While many species eat other organisms and are in turn eaten by different organisms, keystone species can actually affect the population levels of other species. When a keystone species disappears from an ecosystem, species in the lower trophic levels may become overly abundant because their main predator is no longer keeping their numbers in check. Alternatively, species in the higher trophic levels may experience drastic population declines as their favoured prey disappears.

Beavers are another good example of a keystone species because they are ecosystem engineers. Keystone species such as beavers are critical to a healthy ecosystem because the species shapes the ecosystem in ways that make it suitable for a great number of other plants and animals.

THE LINK BETWEEN CRAYFISH AND SMALLMOUTH BASS Crayfish are the largest freshwater crustaceans in North America. They are omnivores and, in fact, will eat just about anything, including worms, fish eggs, insects and plants, as well as decomposing animals and vegetation. In turn, crayfish are eaten by fish, birds, water snakes, salamanders, frogs, raccoons, muskrats, river otters, people, and even other crayfish. This makes them an important part of the aquatic food web, transferring energy and nutrients from lower trophic levels to higher ones.

But in many lakes in the northern part of North America, an aggressive species called the rusty crayfish (*Orconectes rusticus*) is extirpating smaller species of crayfish. Rusty crayfish are native to Ohio, Kentucky, and Tennessee, but fishers using them as bait have introduced them to lakes and streams in Ontario and in the northern United States. .

Not only do rusty crayfish out-compete smaller crayfish in finding food, but they also eat the eggs of the smaller species. Because rusty crayfish grow quickly, they are too big for fish, such as smallmouth bass (*Micropterus dolomieu*), to eat. Their harder shell also makes them less palatable to fish than the smaller crayfish.

While rusty crayfish were first spotted in Ontario's Kawartha Lakes in the 1960s, they have since found their way into several waterways in Ontario.

THE ROLE OF BEAVERS IN AN ECOSYSTEM The beaver (*Castor canadensis*) also plays a vital role in its ecosystem. Beavers build dams in bodies of water because they are clumsy on land, making them vulnerable to predators. They are good swimmers, so they tend to prefer aquatic environments. Flooding the land creates ponds where other food sources such as water lilies grow and where beavers can more easily evade predators. It also allows them to more safely reach trees that were once too far from the water's edge for them to cut.

In the process, the beaver creates an ideal growing environment for succulent vegetation that attracts deer and moose. Thus, when beavers disappear from an ecosystem, populations of various plant species, deer, moose, and wolves ultimately decline as well.

Connect to the Environment

1. The resources available for maintaining healthy ecosystems are limited. How important is it to focus conservation efforts on keystone species? Does the importance of protecting keystone species change when other species are also at risk of extirpation or extinction? How should resources be allocated under these circumstances?
2. Some people see beavers as destructive pests because they can flood farmlands and dam water sources necessary to run hydroelectric turbines. As a result, some provinces have enacted bounties to control beaver populations. How do you weigh the ecosystem benefits of this keystone species with the needs of farmers and with society's need for hydroelectricity? Defend your point of view.

Chapter 11 SUMMARY

Section 11.1 Characteristics of Populations

Ecologists use quantitative measurements to study and describe populations.

Key Terms

population size (N)
population density (D_p)
transect
quadrat
mark-recapture
distribution pattern
life history
fecundity
survivorship

Key Concepts

- Ecologists use quantitative measurements to study and describe populations, such as population size and population density.
- Ecologist use techniques to measure population sizes that include using transects, quadrats, and mark-recapture.
- Populations usually are not evenly distributed throughout their habitat. Ecologists use three distribution patterns to describe populations: uniform, random, and clumped.
- Ecologists use life histories, which are quantitative measures of vital statistics that determine population size to aid in understanding populations.
- Two primary measures are used to describe life history: fecundity and survivorship.

Section 11.2 Changes in Population Size

Factors, such as births, deaths, immigration, and emigration, change population sizes.

Key Terms

immigration
emigration
biotic potential
exponential growth
carrying capacity
logistic growth
r-selected strategy
K-selected strategy

Key Concepts

- The factors that change population sizes are: births, deaths, immigration, and emigration.
- Population growth rates are calculated on a per capita basis to standardize the measurements.
- Two population growth models are exponential growth and logistic growth.
- All habitats have a carrying capacity, which is the maximum population size that it can support.
- Organisms usually exhibit two types of life strategies: *r*-selected strategy or *K*-selected strategy.

Section 11.3 Factors that Regulate Natural Populations

Population sizes are always changing due to biotic and abiotic factors.

Key Terms

density-independent factor
density-dependent factor
intraspecific competition
interspecific competition
population cycle
sinusoidal growth
protective colouration
symbiosis
parasitism
mutualism
commensalism

Key Concepts

- Population numbers are usually controlled by abiotic factors such as temperature, floods, drought, forest fires, hurricanes, and tornadoes.
- Density-independent factors are abiotic events that control population growth.
- Density-dependent factors that control population growth include biotic factors such as competition and predation.
- Some organisms live in symbiotic relationships with other organisms.
- There are three types of symbiotic relationships: parasitism, mutualism, and commensalism.

Chapter 11 | REVIEW

Knowledge and Understanding

Select the letter of the best answer below.

1. An ecologist measured the population density of a certain species. Which of these does this measurement reveal?
   a. the life history of the population
   b. the number of individuals per unit area
   c. the changes in the population over time
   d. the total number of individuals in the area
   e. the distribution of the population in the area
2. A biologist is sampling a plant population to estimate population size. The biologist counts individuals within 10 m of a 100 m long randomly selected line. Which technique is the biologist using?
   a. mark-recapture
   b. indirect indicators
   c. quadrats
   d. transects
   e. tracks
3. Which is true of animals living in a clumped distribution pattern?
   a. They may use chemical warfare.
   b. They aggressively defend their territory.
   c. Their food is scarce, but evenly distributed.
   d. They have the potential for social interactions.
   e. Their young are distributed fairly evenly throughout the area.
4. Zebrafish (a common name for several species) have a high fecundity. Based on this statement, identify one reproductive characteristic of the zebrafish.
   a. lays many eggs at once
   b. reproduces about once each year
   c. feeds and protects its young for a long time
   d. becomes sexually mature late in its life span
   e. invests a large amount of energy in raising its young
5. Which species is most likely modeled by a Type I survivorship curve?
   a. a fish that reproduces once in its lifetime
   b. a plant in which most seedlings do not germinate
   c. an animal that provides a high level of parental care
   d. an insect in which juveniles have a high probability of dying
   e. an invertebrate in which the probability of dying is always constant
6. A population of bacteria is growing exponentially. Which of these must be true?
   a. The resources are limited.
   b. The habitat is at its carrying capacity.
   c. The birth rate is the same as the death rate.
   d. The population is growing at its biotic potential.
   e. The growth pattern can be represented by an S-shaped curve.
7. Which is characteristic of species with a *r*-selected life strategy?
   a. long life span
   b. few offspring
   c. late maturation
   d. high level of prenatal care
   e. live close to biotic potential
8. Which of these is a density-independent limiting factor?
   a. availability of resources
   b. competition
   c. population size
   d. predation
   e. weather
9. The seedlings shown below are growing on the floor of an oak forest. They require water, nutrients, sunlight, and space to grow. Only a few seedlings will compete well enough to survive. What type of competition is occurring among the seedlings?

   a. biotic
   b. competitive exclusion
   c. intraspecific
   d. interspecific
   e. resource partitioning

10. On a graph that shows predator-prey population cycles, an increase in predator population generally shows which of these?
    - **a.** a decline in prey population followed by a decline in predator population
    - **b.** a decline in prey population followed by an increase in predator population
    - **c.** an increase in prey population followed by an increase in predator population
    - **d.** an increase in prey population followed by a decline in predator population
    - **e.** no changes in the populations

11. A tasty nonpoisonous insect that mimics the colour of a poisonous insect is using which defense mechanism?
    - **a.** Batesian mimicry
    - **b.** body colouration
    - **c.** camouflage
    - **d.** Müllerian mimicry
    - **e.** warning colouration

12. In which of these relationships do both partners benefit?
    - **a.** commensalism
    - **b.** interspecific
    - **c.** mutualism
    - **d.** parasitism
    - **e.** predation

13. Which type of relationship do the bird and tick below share?

   - **a.** commensalism
   - **b.** intraspecific competition
   - **c.** interspecific cooperation
   - **d.** mutualism
   - **e.** parasitism

14. Which is *not* a good indirect indicator for estimating the relative density of bears in an area?
    - **a.** tracks
    - **b.** clumps of hair
    - **c.** claw marks
    - **d.** carcasses
    - **e.** scat

Answer the questions below.

15. Why is population density not always a reliable method for estimating the number of individuals in a population?
16. Describe how ecologists determine population size.
17. When are quadrats useful to determine population size?
18. Describe how the distribution and abundance of resources might influence how populations are distributed in a certain habitat.
19. Do humans have a high fecundity or a low fecundity? Explain your answer.
20. Describe the risk of mortality in the three survivorship patterns.
21. Describe three types of survivorship curves and give an example of a species with each type.
22. What does biotic potential have to do with fecundity? List three factors that determine the biotic potential of a species.
23. Why is it advantageous for species in unstable environments to use a *r*-selected life strategy?
24. What are density-independent factors? How do these factors regulate populations?
25. How does interspecific competition affect population growth?
26. Compare parasitism and predation with respect to their effect on the growth of the partner populations.
27. Both interspecific competition and intraspecific competition are density-dependent limiting factors. Explain why extinction is a possible outcome of interspecific competition, but not of intraspecific competition.

Thinking and Investigation

28. A researcher captures 24 geese and places a special leg band on each one before releasing them. A week later, the researcher captures 30 geese in the same area and finds the special leg band on 9 of them. Which is the best estimate for the population of geese in that area?
29. A population of mice contains 200 animals. If the birthrate is 21 animals per year and the death rate is 15 animals per year, what is the per capita growth rate?
30. The population of beetles in a certain area was estimated to be 1600. Two years later, the population was estimated to be 2200. What is the growth rate of the population?

31. To determine the population of butterflies in two different meadows, a researcher sampled five different sites in each meadow. The researcher's data is shown in the table below.

Meadow Data

Site	Population Density in Meadow A (butterflies per plant)	Population Density in Meadow B (butterflies per plant)
I	6	9
II	10	5
III	8	10
IV	5	8
V	11	3

If there are about 10 000 plants in each meadow, determine the population of butterflies in each one.

32. During the summer months, individuals in a moose population are randomly distributed throughout their habitat. Design a sampling procedure to determine the population of moose within a 5 km by 5 km area.

33. Ecologists are trying to determine the population of a certain species of wildflower in a 1 km^2 conservation area. They randomly sample eight 25 m^2 quadrats and count the following number of this species of wildflower: 8, 6, 9, 12, 4, 10, 8, and 7. Estimate the population of this species of wildflower in the conservation area.

34. A biology student caught and tagged 50 trout in a small lake. After two weeks, the student caught another 40 trout and found that 4 of them were tagged—these 4 had been in the original sample and were caught a second time. What is the best estimate of the population of trout in the lake? What might have contributed to the low number of tagged trout caught the second time? How could the student reduce this factor?

35. In a pond with a population of about 300 turtles, 10 of them leave the population and 5 migrate into the population from a nearby pond. In the same year, there are also 90 births and 6 deaths. What is the per capita growth rate for this population?

36. The per capita growth rate of a particular species of fish in a river is 27. The change in the size of the population of these fish over a period of seven years was 54. Determine the growth rate of this population of fish.

37. Predators tend to cause a decline in the densities of prey populations. What prevents predators from reducing prey populations to such low levels that the predators become extinct?

38. The non-venomous Pueblan milk snake (*Lampropeltis triangulum campbelli*) developed specific colouration to look like the venomous coral snake. This protects the milk snake from predators. Predict what would happen to the population of milk snakes if the coral snakes were eliminated from the area? Justify your prediction.

39. You capture two species of snakes that have bright colouration and similar colour patterns. You know that this is an example of mimicry. Design an experiment to determine whether this is Batesian mimicry or Müllerian mimicry.

40. Suppose you placed several mating pairs of hares on a small island having abundant food and resources, with no predators. Describe the changes you would expect to see in the hare population over time. Include a discussion of any limiting factors in your description.

41. The habitat for a population of hares living on a small island has reached its carrying capacity. A scientist then introduces several mating pairs of foxes to the island. What effect will this have on the hare population on the island? Describe the changes in hare and fox populations you would expect to take place over time.

42. You are part of a team that has been selected to study deer populations. Choose an abiotic factor that regulates deer population. Explain how this factor limits the growth of the population.

Communication

43. The golden eagle (*Aquila chrysaetos*) is endangered in Ontario. Do research to find out about the golden eagle. Your goal is to prepare a recovery plan for this species in Ontario. Write a brief report detailing your recovery plan.

44. Select two species of plants (or animals) that live in Ontario. Compare the two species with respect to their life history and their success in their habitat. Explain what contributes to their successes or failures in their habitat. Summarize these factors in a table.

Chapter 11 REVIEW

45. Choose two mammals in Ontario and do research to compare their life histories. Write a report explaining which one has a more *K*-selected life strategy. Explain whether either is at risk, and how their life strategy contributes to the mammal being at risk.

46. The moose (*Alces* sp.) range includes most of Canada and the northern United States. However, moose are not native to some areas, such as Newfoundland. In 1904, two mating pairs were introduced into this province. The moose population in Newfoundland is now somewhere near 150 000. Likewise, in 1978, a few breeding pairs were introduced into western Colorado—the population is now more than 1000 (and growing). Conduct research to investigate moose populations. Write a report detailing why the moose is so successful.

47. Summarize your learning in this chapter using a graphic organizer. To help you, the Chapter 11 Summary lists the Key Terms and Key Concepts. Refer to Using Graphic Organizers in Appendix A to help decide which graphic organizer to use.

48. You are preparing a lesson to teach a Grade 6 class about symbiotic relationships. Write the main points you would include in your lesson.

49. During an experiment, you grow a bacterial culture under ideal conditions. Draw a graph to show the trend you would expect in your data. Explain why.

50. Use a labelled diagram to compare and contrast exponential and logistic growth patterns.

51. **BIG IDEAS** Population growth follows predictable patterns. It is likely that the spiny water flea (*Bythotrephes longimanus*), shown below, like the zebra mussel, was introduced into the Great Lakes from the discharge of ship ballast water. It was first found in Lake Ontario in 1982, and by 1987 it was present in all of the Great Lakes. Now, the spiny water flea can be found in many inland lakes and waterways throughout Ontario. Create a pamphlet to educate the public about the spiny water flea, making reference to their life history, life strategies, and the reasons they flourish in their new homes.

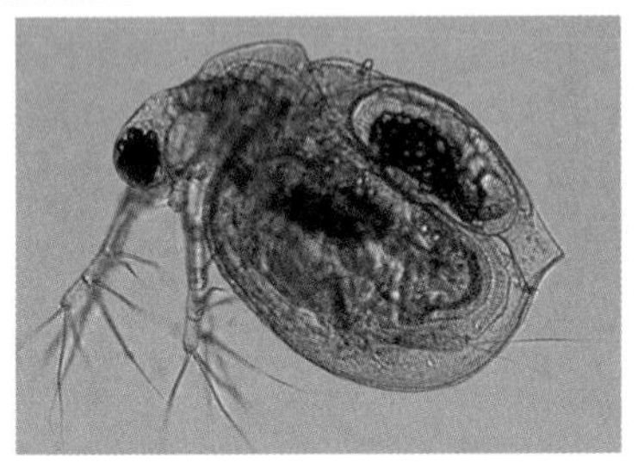

52. Use a concept map to illustrate the relationships among the four processes that cause changes in the population of a species and in the predator population that feeds on the species.

53. Draw a comic strip or a cartoon to illustrate the population cycling in predator-prey interactions or in parasite-host relationships.

54. Use a table to compare and contrast the methods used to sample subsets of a larger population. Include the description of the sampling method, the type of organisms that this method is useful for sampling, and how the population size is determined using this method.

55. One of your peers asks you, "Why didn't organisms simply evolve to reproduce shortly after birth, produce large families, care for the young for a very long time, and outcompete others? Write a paragraph stating what you would say to your peer.

56. Create a poster that identifies and illustrates the distribution patterns in
    a. plants whose seeds are dispersed by wind
    b. plants that reproduce asexually by sending out runners from the parent
    c. plants that use chemicals to discourage the growth of other plants

57. Use a table to compare birth rates and death rates for a population during the slow growth, rapid growth, and no growth phases of a logistic growth pattern.

58. Imagine that populations in each of the three distribution patterns had to justify why the population exhibits this form of distribution pattern. Write a short summary for the key points they should use.

59. Working with a partner, debate the statement, "Competition does not benefit any individual." Then write a short report identifying what you think were the two best points made by each side in the debate.

Application

60. You are conducting an investigation to examine the effects of intraspecific competition on a certain species of plant seedlings. Make a hypothesis about how increasing intraspecific competition will affect the growth of individuals in this population.

61. Though the adults of some animals, such as sponges, are sessile, their larvae are motile. Why would motility of the larvae be important to such animal populations?

62. You are hired to eliminate insect pests on local farms. You know that using the same pesticide year after year can cause resistance in the insect population. From what you have learned about population growth and the factors that affect population growth, what are some strategies you might use to control the insect population?

63. Grasshoppers are found across North America. There are about 40 species in the canadian prairies. One species belonging to the subfamily *Melanoplinae* is found in huge numbers across North America. List three factors that might limit the exponential growth rate of the population of this species, and explain how these factors limit growth.

64. The graphs show the population growth for *Paramecium aurelia* and *Paramecium caudatum* when grown alone and when grown together. Explain the curves in the three graphs.

Competition between Two Laboratory Populations of Paramecia

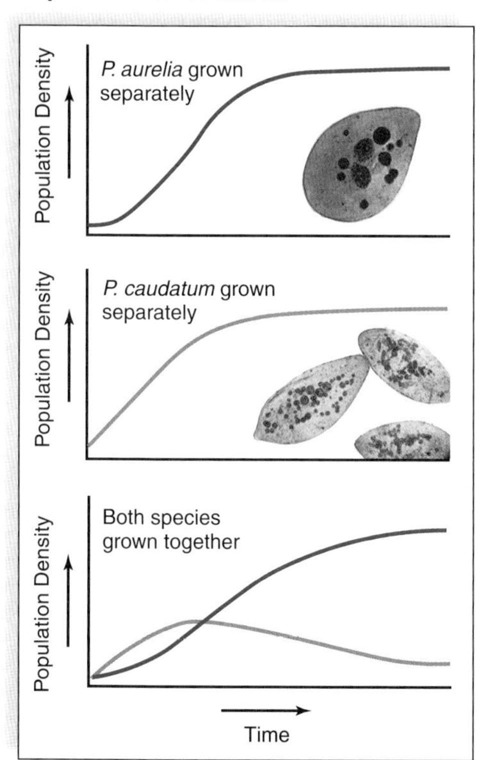

65. Many species that present a problem for homeowners, like pests and weeds, have a *r*-selected life strategy. Give two reasons why such species become a nuisance, explaining your reasoning.

66. Evergreen trees, like lodgepole pine (*Pinus contorta*) and white spruce (*Picea glauca*) tend to grow well in provinces like Alberta. However, white spruce may compete with lodgepole pine for light and living space. Eventually, the white spruce may take over. Do research on both of these species and provide reasons why the white spruce may out-compete the lodgepole pine. What kind of competition is taking place?

67. Monarch caterpillars (*Danaus plexippus*) feed on the bitter leaves of milkweed plants (*Asclepias ovalifolia*), storing the bitter chemicals from the plant. This makes monarch butterflies tolerant to milkweed and causes the butterflies to be unpalatable to predators. The viceroy butterfly (*Limenitis archippus*) has the same colouration as the monarch butterfly, but is quite tasty to its predators. What kind of mimicry is this, and how does it help to keep the viceroy butterfly safe from predators? How would the type of mimicry change if the viceroy was also unpalatable?

68. In South Africa, some plants have co-evolved with ants that disperse their seeds. The seeds are produced with a little food package attached. The ants bring the seeds back to their nest, eat the food package and leave the seeds in the underground nest, where the seeds are protected from fire and seed-eating organisms. *Linepithema humile*, an invasive species of ant native to Argentina, is competing with the native South African ants. They eat the food packages from the larger seeds only and drop the seeds on the ground, where they are exposed to fire and seed-eating organisms.
    a) Identify the symbiotic relationship between the plant and the South African ants. Explain.
    b) Identify the type of relationship that exists between:
    - the Argentine ant and the plant
    - the Argentine ant and the South African ant

69. Over the last eight years, the population of walleye (*Sander vitreus*) in a small lake has remained roughly around 1000. A visitor to the lake suggested that the lake be stocked with 1000 more individuals of the same species. Explain to the visitor what you think will happen.

70. The life history of a bear can be described as *equilibrial*. Research the meaning of this term and explain why this is a suitable term for the characteristics of the life history of the bear.

Chapter 11 SELF-ASSESSMENT

Select the letter of the best answer below.

1. **T/I** A field ecologist studying a population of plants in their native habitat drew the graph below to represent the pattern of growth.

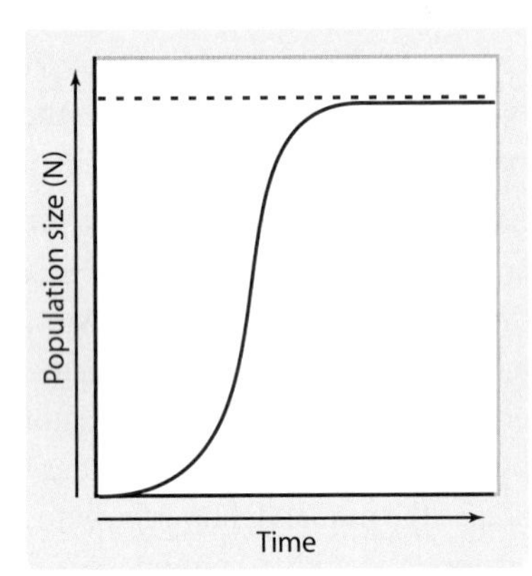

Which of the following can the ecologist conclude will occur at the point where the horizontal line appears at the top of the graph?

a. Unlimited resources will exist in the habitat.
b. The population will greatly increase past this point.
c. The soil will lack sufficient nutrients to sustain the population.
d. The resources are just enough to maintain the maximum population size.
e. A forest fire most likely occurred when the population reached this point.

2. **K/U** Population ecologists use various parameters to describe a population quantitatively. Which of the following is used to determine the population density of an area?

a. $\text{Population density} = \dfrac{\text{space}}{\text{number of individuals}}$

b. $\text{Population density} = \dfrac{\text{time}}{\text{number of individuals}}$

c. $\text{Population density} = \dfrac{\text{number of individuals}}{\text{space}}$

d. $\text{Population density} = \dfrac{\text{number of individuals}}{\text{time}}$

e. $\text{Population denisty} = \dfrac{\text{time}}{\text{space}}$

3. **K/U** A population of fruit flies is growing under ideal conditions. Which of these is the term used for the maximum growth rate achieved by the population?

a. biotic potential
b. carrying capacity
c. logistic growth
d. population density
e. population distribution

4. **T/I** After a flood, the population of grasses in a habitat declined. Which best describes this event?

a. biotic
b. density-independent
c. explosion
d. intraspecific
e. resource partitioning

5. **A** Several species of finches occupy the same area and feed upon insects. Each species has had a stable population size for many years. Which is most likely true?

a. Each species feeds on the same kind of insects.
b. Each species feeds on different kinds of insects.
c. Each species competes for the exact same resources.
d. One species will eventually become extinct.
e. Only one species will eventually remain.

6. **K/U** Which is likely to be the result of an increase in the size of a predator population?

a. an increase in the prey population
b. a decrease in the prey population
c. no change in the prey population
d. extinction of the prey population
e. an increase in natality in the prey population

7. **A** A researcher studying a population of warblers notices that the change in population size is negative after one year. Which statement best explains this change?

a. The birth rate is higher than the mortality rate.
b. The rate of migration is higher than the rate of emigration.
c. The sum of the births and immigration is higher than the sum of the deaths and emigration.
d. The sum of the births and immigration is lower than the sum of the deaths and emigration.
e. The sum of the births and emigration is lower than the sum of the deaths and immigration.

8. **A** Barnacles often anchor themselves to the shells of scallops to feed. The scallops are not affected. What type of relationship is this?

a. commensalism
b. competition
c. mutualism
d. parasitism
e. predation

9. **T/I** A population of snakes contains 500 individuals. In one year, there are 52 births, 20 deaths, 12 immigrations, and 4 emigrations. Calculate the per capita growth rate.

a. 3.2 percent
b. 8.0 percent
c. 10.4 percent
d. 11.2 percent
e. 17.6 percent

10. **K/U** Both intraspecific and interspecific competition affect population growth. Which of the following statements about intraspecific competition is *not* correct?
    a. The death rate of members may increase.
    b. Competition limits the growth of the species.
    c. Species members compete for food and space.
    d. Competition exists between members of the same species.
    e. Species members do not compete for the opportunity to breed.

Use sentences and diagrams as appropriate to answer the questions below.

11. **K/U** How can parasitism regulate population size?
12. **T/I** You want to determine the number of largemouth bass living in a pond. Design a method to determine the population.
13. **A** Between 1976 and 1990, the ring-billed gulls on the Canadian side of the lower Great Lakes increased from about 56 000 to 283 000. Explain why you believe this occurred, and justify your reasoning.
14. **C** Draw a diagram to illustrate the differences among parasitism, mutualism, and commensalism. Provide a caption for your illustration that identifies each example.
15. **C** The graph below shows the growth of a particular population.

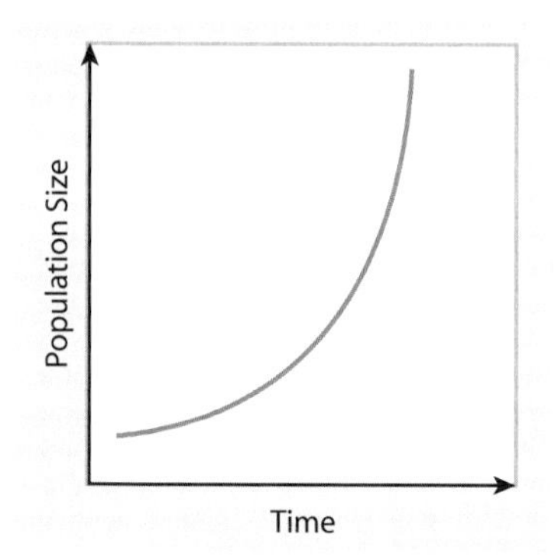

    a. Write a suitable caption for this graph.
    b. Infer the conditions in which the population is growing. Explain your reasoning.

16. **T/I** A researcher counted the number of individuals of a certain species of trees in five 100 m by 5 m transects. The data is as follows: 28, 32, 50, 45, and 41. Determine the density of the population. Is this a good method to use to sample this population? Explain your answer.
17. **A** A population of houseflies is growing at its biotic potential. What factors can limit the growth of the houseflies?
18. **C** Draw and label a graph that represents the three types of survivorship patterns. Write a caption for the graph that summarizes the main characteristics of each pattern.
19. **A** A population of crickets is living in an unstable environment. The crickets lay many eggs so that some will survive the unpredictable forces of nature. Based on this information, identify the life strategy the crickets use, and the type of survivorship pattern. Explain your answer.
20. **C** Most of the seedlings from a certain plant species will never germinate. Draw the survivorship curve that you would expect for this species.
21. **C** Make a table to compare and contrast intraspecific and interspecific competition. Include competing population(s), what they are competing for, and the result of competition.
22. **K/U** Describe the differences among uniform distribution, random distribution, and clumped distribution.
23. **C** Write a hypothesis about how decreasing the population of aquatic plants will affect a pond.
24. **C** You and a friend are studying for a test that will cover defence mechanisms organisms use against predators. Write the information you would put on your study note cards, including examples.
25. **A** Give two real-world examples of mutualism. State how each partner benefits.

Self-Check

If you missed question...	1	2	3	4	5	6	7	8	9	10	11	12	13	14	15	16	17	18	19	20	21	22	23	24	25
Review section(s)...	11.2	11.1	11.2	11.3	11.3	11.3	11.2	11.3	11.2	11.2	11.3	11.1	11.2 11.3	11.3	11.2	11.1	11.2 11.3	11.1	11.1 11.2	11.1	11.3	11.1	11.2	11.3	11.3

CHAPTER 12 Human Populations

Specific Expectations

In this chapter, you will learn how to . . .

- F1.1 **analyse** the effects of human population growth, personal consumption, and technological development on our ecological footprint (12.1, 12.2)
- F1.2 **assess**, on the basis of research, the effectiveness of some Canadian technologies and projects intended to nourish expanding populations (12.2)
- F2.1 **use** appropriate terminology related to population dynamics, including, but not limited to: *carrying capacity, population growth, population cycle, fecundity,* and *mortality* (12.1, 12.2)
- F2.2 **use** conceptual and mathematical population growth models to calculate the growth of populations of various species in an ecosystem (12.1, 12.2)
- F2.3 **determine**, through laboratory inquiry or using computer simulations, the characteristics of population growth of two different populations (12.1, 12.2)
- F3.2 **describe** the characteristics of a given population, such as its growth, density (e.g., fecundity, mortality), [illegible] (12.1, 12.2)
- F3.4 **explain** the concept of energy transfer in a human population in terms of the flow of food energy in the production, distribution, and use of food resources (12.2)

A study of population dynamics is not complete without a study of the human population. Unlike other populations, humans have increased the carrying capacity of their environment and have grown exponentially. Unlike other species, humans are altering Earth in ways that might be detrimental to themselves and to other populations. For example, most humans around the world live in large urban centres such as the one shown in the photograph. What factors might reduce the carrying capacities of large cities as human populations increase? You will consider such questions in this chapter.

What Factors Affect the Growth of a Human Population?

Technological advances have resulted in a rapid growth in human population. However, human population growth is not equal in all countries.

Procedure

As a small group, study the graph and answer the questions below.

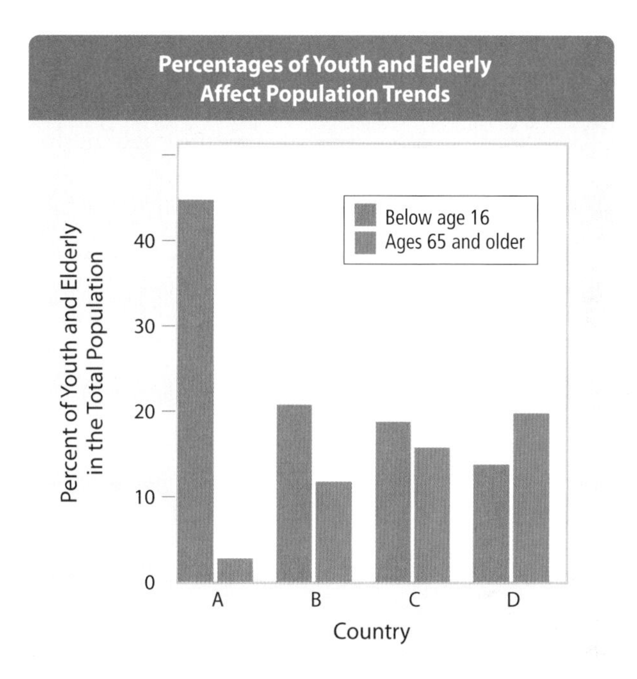

Questions

1. This graph shows one factor affecting human population growth. What is that factor?
2. Use the data to predict how this factor will affect the population in each country between now and the year 2050.
3. Brainstorm a list of factors, events, or conditions that might affect the growth of human populations in these countries. Predict the effect of each factor on the population growth rate.
4. As a group, make a list of factors or groups of factors that have the greatest impact on population growth. Justify your answers.
5. Consider the population of Ontario in 1867 (1.5 million) and the population today (13 million). What factors do you think have contributed to a longer lifespan for individuals within this population? What factors have contributed to changes in family sizes? Do you think the lifespan of the average adult is expected to increase for the next few generations? Explain your answers.

SECTION
12.1

Human Population Growth

Key Terms

demography
doubling time
population pyramid
ecological footprint
available biocapacity

You have read in Chapter 11 about how ecologists study populations of plants, birds, insects, bacteria, and other organisms. Ecologists also study human populations. Human populations exhibit some of the same characteristics as other populations. However, unlike other populations, humans impact their environment to a much greater extent. You might be asking yourself, how have humans grown as a population? What factors have made humans so successful? In this chapter, you will find the answers to those questions and answers to many more.

demography the study of statistics related to human populations, such as population size, density, distribution, movement, births, and deaths

Trends in Human Population Growth

Humans have many of the same characteristics as other *K*-selected species. Humans have a small number of offspring, begin reproduction later in life, and provide a great deal of parental care to their offspring. This strategy has been very successful for humans as a population, as shown in **Figure 12.1**. This graph shows demographers' estimates of human population size over time. The study of human populations, including population size, density, distribution, movement, and birth and death rates is called **demography**.

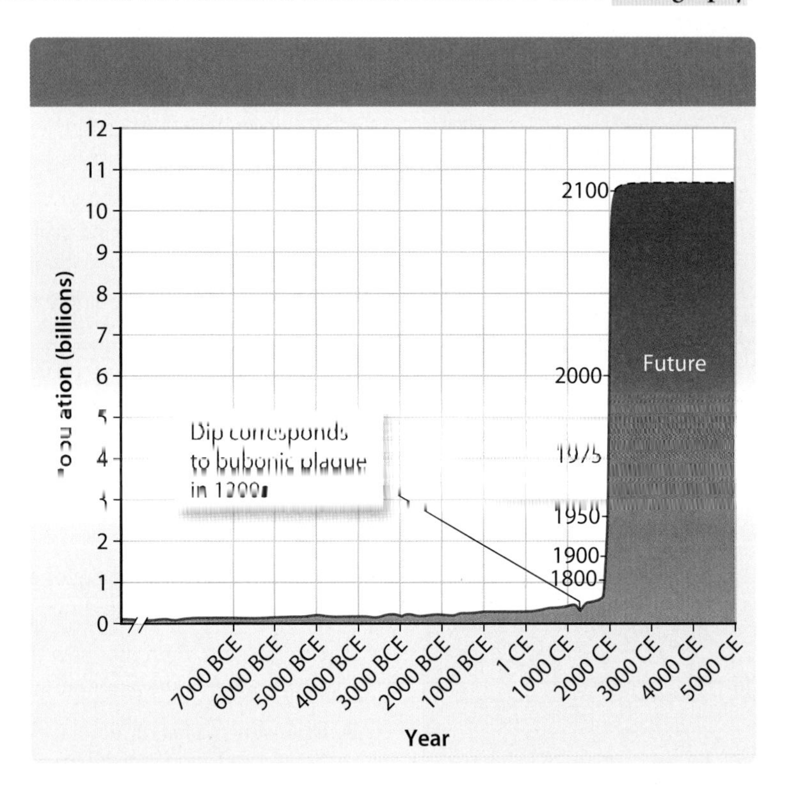

Figure 12.1 The human population was relatively stable until recent times. Then, starting around the 1700s, the population began to grow at an exponential rate.

The graph illustrates some important points about human population growth over time. Throughout most of history, the human population size was very stable. Population size and growth were regulated by factors that affect other populations, such as food availability, disease, and predators.

The slight dip in the population from 1347 to 1350 shows the decrease in the population due to bubonic plague. The plague killed an estimated one third of the population of Europe. A short time after the plague, the population started to grow exponentially. This explosive growth was due to a variety of technological factors.

Factors That Affect Growth

Many advances in technology occurred in a relatively short period of time and dramatically affected human population growth. Starting in the early 1700s in Europe and a little later in North America, humans were able to increase their food supply by improved agricultural methods and the domestication of animals. Breakthroughs in medicine in the late 1800s and early 1900s enabled people to be successfully treated for once-fatal illnesses. Better shelter protected people from the weather, and improvements in the storage capacity of food helped humans survive times when food was less plentiful. All of these factors allowed humans to increase the carrying capacity of their environment and change from a logistic growth pattern to an exponential growth pattern.

Exponential Growth

Because human population growth was no longer as constrained by environmental factors, the growth rate has dramatically increased over the last 300 years. In that period of time, the birth rate has remained about the same—about 30 births per 1000 people per year. The death rate has decreased from about 20 deaths per 1000 people per year to about 13 per 1000 per year. The differences between the birth rate and the death rate resulted in a population growth of about 2 percent per year. In recent years, the growth rate has declined to about 1.2 percent per year. This might sound like a small number but, when applied to populations consisting of millions of people, this change can be significant.

Another way to look at population growth is to consider **doubling time**—the time it takes for a population to double in number. In 1650, approximately 500 million people lived on Earth. In the next 200 years, the population doubled to 1 billion people. Between 1850 and 1930, the population doubled again to 2 billion. Between 1930 and 1975, the world's population doubled again to 4 billion people. Today, the world's population is nearly 7 billion people. At a growth rate of 1.2 percent per year, the world's population is expected to double in about 58 years. To maintain the present standard of living, twice the number of jobs, water, food, energy, homes, and other human requirements must be doubled to accommodate the increase in the number of people.

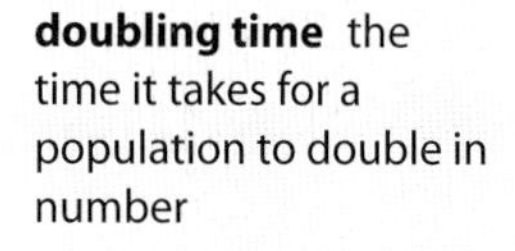

doubling time the time it takes for a population to double in number

Although the worldwide human population is growing, it is not growing uniformly across the globe. **Figure 12.2** shows how the population growth varies in different types of economies. Developed countries, such as Canada, the United States, and countries in Western Europe, are growing more slowly than developing countries, such as Mexico, the Republic of Congo, and the Republic of Honduras.

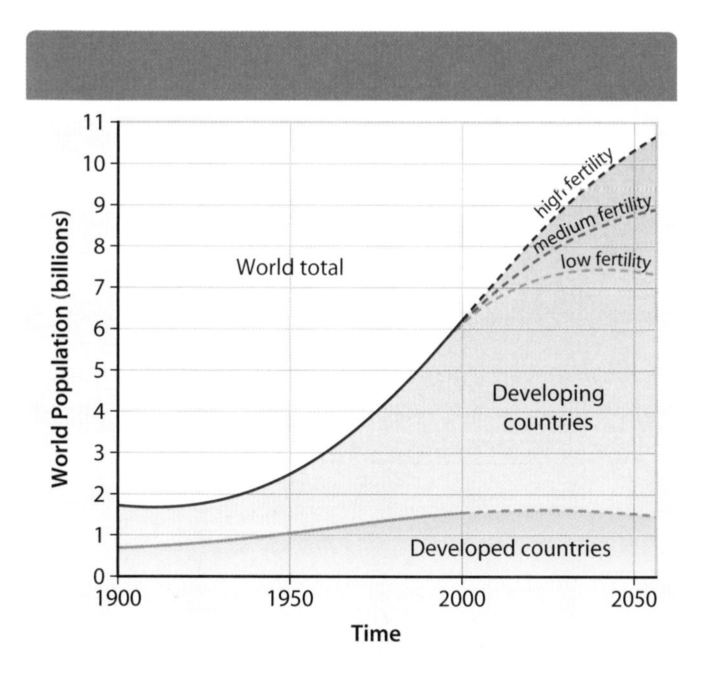

Figure 12.2 In developing countries, such as many of the countries in Asia, Africa, and Latin America, populations are growing much faster than those in developed countries, such as Canada, the United States, England, France, and Sweden.

population pyramid a type of bar graph that shows the age distribution in a population, which demographers use to study a population

Population Age Structure

Just as population growth varies from country to country, the age structure from country to country varies. **Population pyramids** are bar graphs that demographers use to help assess a population's potential for growth. For example, the population pyramids shown in **Figure 12.3** show the profiles for two different countries with two different types of growth. A population pyramid shows the percentage of males (usually shown on the left) and the percentage of females (usually shown on the right) in different age categories (usually five-year intervals). In the graphs below, the male population is shown in blue and the female population is shown in orange.

The shape of the population pyramid is used to predict demographic trends in the population. For example, the population pyramid for Kenya is a triangular shape. A triangular shape predicts a future of explosive growth because a large portion of the population will enter their reproductive years at the same time. The pyramid shape also indicates a decreased average life span. By contrast, the rectangular shape of Sweden's graph indicates that the population is not expanding and it is stable. A further analysis of Sweden's age pyramid shows that only 16 percent of Sweden's population is less than 15 years old. In comparison, nearly half of all Kenyans are less than 15 years old. The fertility rate of the two countries also yields information about the two populations. The average number of offspring per female in Sweden is 1.7, while the average number of offspring in Kenya is 4.7. Because of the high fertility rate and the large number of women entering their reproductive years at about the same time, Kenya could double its population in less than 35 years. Compared to the estimated doubling time of the world's population—58 years—the doubling time for a country such as Kenya demonstrates the potential for explosive growth in a short period of time.

Another possible shape for a population pyramid is an inverted triangle. An inverted triangle indicates a population that is shrinking. In this type of population, there are a large number of individuals that are past their reproductive years and few individuals in or about to enter their reproductive years. As a result, the population shrinks because there are more deaths than births in the population. Use the activity on the next page to explore populations in various countries.

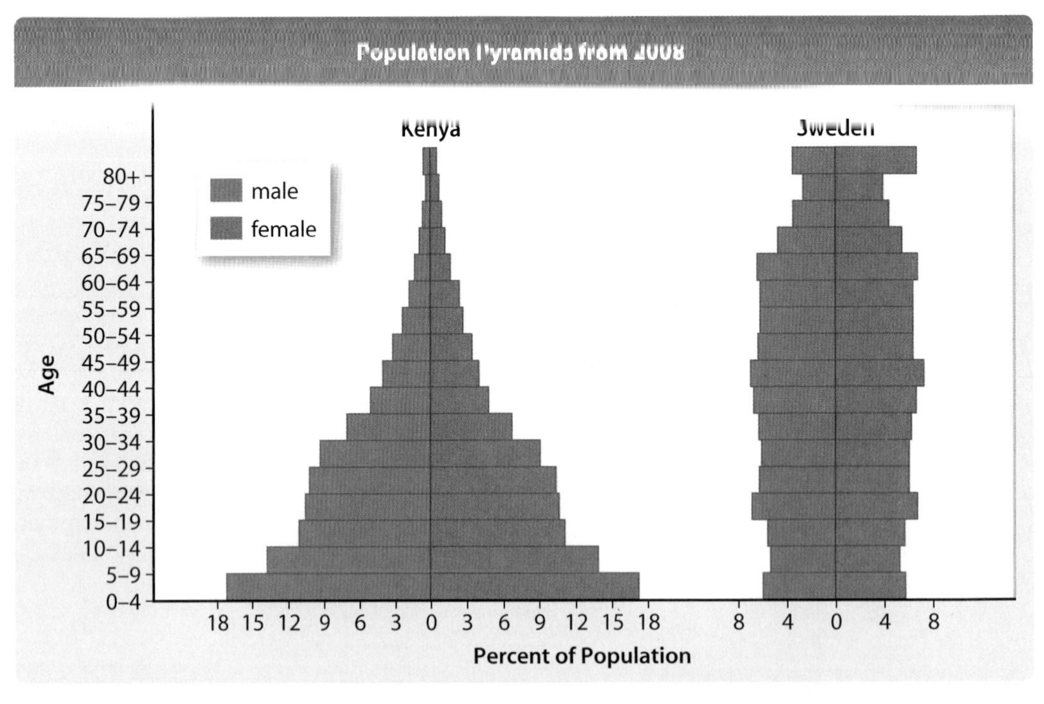

Figure 12.3 Population pyramids are bar graphs that show the age distribution in a population. The age distribution can be used to predict the growth of the population.

List *some factors that cause these two types of population profiles.*

Activity 12.1 Population Growth Rates in Different Countries

Because human populations in different parts of the world live in different environments, where they have variable access to food, clean water, technologies, and medical aid, their per capita birth rates and death rates—and thus their growth rates—also vary. In this activity, you will compare the growth rates of several different human populations.

Procedure

Use the table below to answer the following questions.

Human Demographic Information for Selected Countries in 2001

Country	Population Size (millions) (*N*)	Number of Births per 1000 Individuals (*b*)	Number of Deaths per 1000 Individuals (*d*)
Canada	32.2	10	7
Ethiopia	77.4	41	16
Finland	5.2	11	9
Germany	82.5	9	10
Greece	11.1	9	10
India	1103.6	25	8
Nigeria	131.5	43	19

Source: 2005 World Population Data Sheet of the Population Reference Bureau.

Questions

1. Create a data table or computer spreadsheet with the following title and column headings:

Predicted Population Growth from 2001 to 2011 in Selected Countries

Country	Annual per Capita Growth Rate (*cgr*)	Population Size (*N*) at One-Year Intervals										
		2001	2002	2003	2004	2005	2006	2007	2008	2009	2010	2011

2. The table of demographic information on this page shows the total population size and the number of births and deaths that occur annually per 1000 people in different populations. In other words, the table shows birth and death rates for each population. Subtract the deaths per 1000 individuals from the births per 1000 individuals each year to calculate the annual per capita growth rate (*cgr*) for each population:

$$cgr = \frac{b}{1000} - \frac{d}{1000}$$

Note that this estimate of *cgr* does not take into account emigration or immigration.

3. Use Canada's 2001 population size and annual *cgr* to calculate the predicted population size for 2002:

$$N_{(\text{Canada in 2002})} = N_{(\text{Canada in 2001})} + (cgr)_{(\text{Canada in 2001})}$$
$$= (1 + cgr)(N_{(\text{Canada in 2001})})$$
$$= (1 + cgr)(32.2 \times 10^6)$$

Then use Canada's 2002 population size and the 2001 annual *cgr* to calculate the predicted population size for 2003. Repeat this step for the rest of the years listed in your data table.

4. Repeat the calculations in step 3 for each country listed in your data table.
5. Using a full sheet of graph paper (or a computer graphing program), graph the size of Canada's population from 2001 through 2011. This graph is a hypothetical population growth curve for Canada for 2001 through 2011. Remember to label each axis and include a title for your graph.
6. On the same graph, plot population growth curves for the six other countries listed in your data table. Use a different symbol or colour for each growth curve and provide a legend.
7. Compare the slope, or steepness, of the different growth curves. Describe how annual *cgr* affects the slope of a growth curve.
8. What assumption was made in the calculations of population size for each year? How will this affect the population growth curve?
9. Describe the effect of a population's initial size on the slope of its growth curve.
10. Why is the annual *cgr* negative for some populations? Describe the growth curve for a population with a negative *cgr*.
11. Based on your graph, which populations are currently undergoing exponential growth?
12. Classify the countries in your data table based on whether you would consider them to be highly industrialized or less industrialized. Compare the growth curves that are typical of each group. Explain the differences between the two types of growth curves.

Learning Check

1. Refer to **Figure 12.1** and describe the trend in human population growth through time.
2. What is "doubling time"?
3. How have humans increased the carrying capacity of their environment?
4. Which type of population pyramid do you think Canada has? Explain your answer.
5. What type of problems would a population with an inverted population pyramid have? Explain.
6. There is a dip in the human population graph when bubonic plague killed an estimated 25 million people in Europe. During World War I (1914–1918) and the Spanish flu pandemic (1918), 60-80 million people were killed. Why does this not show up on the graph?

ecological footprint the amount of productive land that is required for each person in a defined area, such as a country, for food, water, transportation, housing, waste management, and other requirements

Earth's Carrying Capacity

Calculating and predicting human population growth is not a philosophical exercise. You have read in Chapter 11 that populations can grow rapidly until resources become limited. Populations reach the maximum amount that the available resources can support. Recall that this is called the carrying capacity of the environment. So far, humans have been able to increase the carrying capacity of Earth, and the population continues to grow exponentially. However, all environments have a limit and there is no evidence to suggest that Earth is an exception.

Today, there is a great disparity among countries in the amount of resources that are used per person. People living in the industrialized world use far more resources than people living in developing countries. In fact, the wealthiest 20 percent of the population consumes 86 percent of the world's resources and produces 53 percent of the world's carbon dioxide emissions. People living in the poorest countries use about 1.3 percent of the world's resources and produce about 3 percent of the world's carbon dioxide emissions.

Ecological Footprint

Consider these statistics another way, using the concept of ecological footprint. An **ecological footprint** is the total amount of land needed to support one person and it includes six major categories of demand: cropland, grazing land, fishing grounds, forest land, carbon absorption land, and building area. The estimated average ecological footprint per person globally is about 2 hectares of land. One hectare equals 10 000 m². However, this ecological footprint varies widely around the world as shown in **Figure 12.4**.

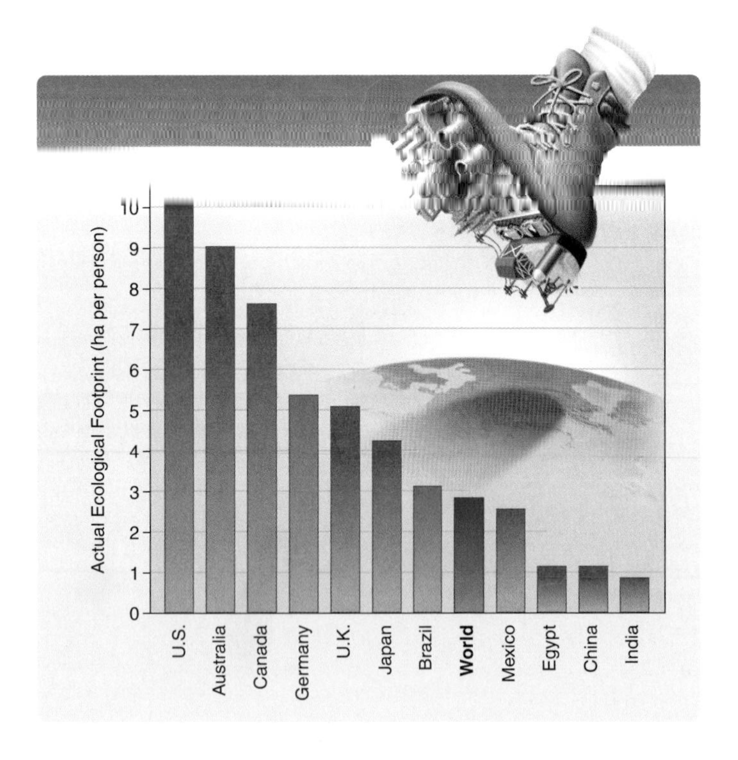

Figure 12.4 People around the world have different ecological footprints. The industrialized countries have a greater footprint than developing countries.

The average Canadian requires about 7.5 hectares of land, while the average American requires about 10 hectares. In most developed countries, such as Canada and the United States, the largest component of the footprint is land used for energy production, followed by land for food production, and then forestry requirements. Much of the forestry requirement results from the need to absorb the carbon dioxide emitted during to the combustion of fossil fuels.

Available Biocapacity

While it is difficult to estimate Earth's carrying capacity, ecologists have attempted to do so. Earth's carrying capacity is called **available biocapacity** and it includes the following factors: cropland, grazing land, fishing grounds, forest land, carbon-absorption land, and building area. Low-productivity areas, such as arid regions and open oceans, are not considered biologically productive areas in this calculation. It is estimated that about one-quarter of Earth's surface, or about 11 billion hectares, constitutes Earth's biocapacity.

available biocapacity Earth's carrying capacity for the human population

Comparing the ecological footprint of the population on Earth in 2002 to the biocapacity reveals that the ecological demand exceeds the ecological supply by about 23 percent. In 1961, the global demand was about 50 percent of the biocapacity. Sometime in the mid-1980s, the human population exceeded the available biocapacity. Since that time, the ecological demand has exceeded biocapacity. The ecological footprint and biocapacity are not uniform throughout the globe as shown in **Figure 12.5**. As you can see, the ecological footprint far exceeds the biocapacity in North America and Western Europe.

Compare the ecological footprint and biocapacity for North America and Latin America. The graphs show that North America exceeds its biocapacity while Latin America does not. Consider the standard of living in the two places. Generally speaking, the standard of living is much higher in North America than in Latin America. The strength of a local economy often is an indicator of the ecological footprint of that location. If the population is subsisting on very little, their ecological footprint often is below the biocapacity of their environment. In Section 12.2, you will think about and investigate how the human population has affected the biosphere and what is being done to sustain Earth's resources.

Suggested Investigations

Investigation 12-A, Do You Tread Lightly on the Earth?

Investigation 12-B, First Impressions Count

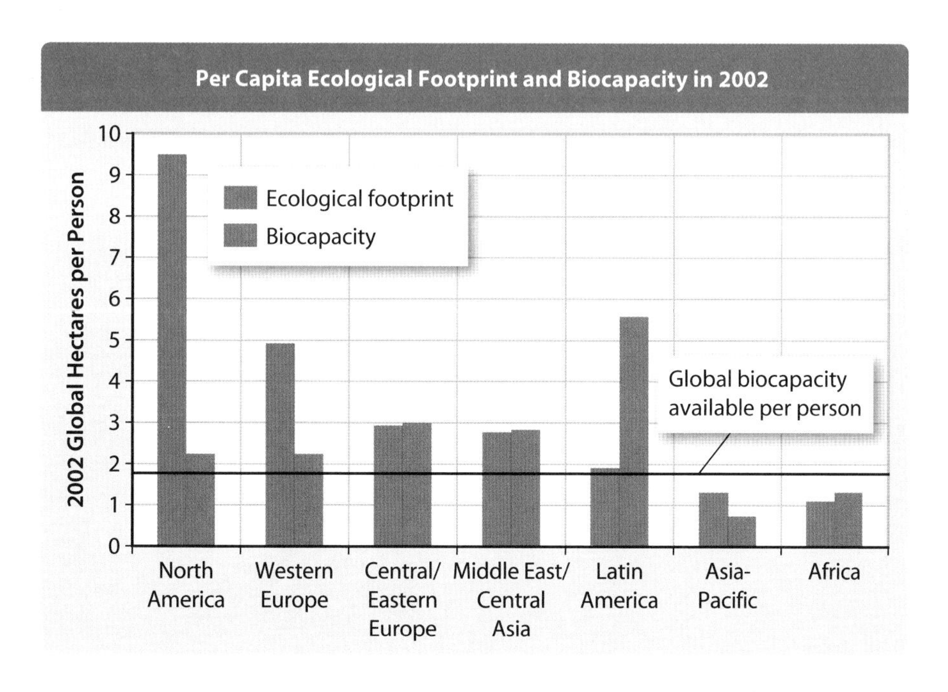

Figure 12.5 Human populations in many countries exceed the biocapacity of their environment. In 2002, the entire human population on Earth exceeded Earth's biocapacity by about 23 percent.

***Infer** what effect humans might have on resources over time if they continue to exceed the biocapacity of Earth.*

Section 12.1 Review

Section Summary

- Demography is the study of statistics on human populations, such as population size, density, distribution, movement, births, and deaths.
- Prior to modern technology, the human population size was stable; however, it is now growing exponentially.
- The age structure of a population varies from country to country.
- The ecological footprint of industrialized countries is much greater than the ecological footprint of developing countries.
- Humans exceed the biocapacity of the world and are depleting many resources.

Review Questions

1. **K/U** What characteristics do humans have that classify them as *K*-selected?
2. **A** When the glaciers of the last ice age receded, the human population was estimated at about 5 million. Their survival depended on the available plants in their environment and their ability to hunt. Why do you think the human population did not grow rapidly at that time?
3. **K/U** Use **Figure 12.1** to describe the pattern of human population growth throughout history. Explain why this pattern occurred. Use the terms *logistic* and *exponential* in your explanation.
4. **C** In a table, compare developed countries to developing countries with respect to growth rate, general age structure, fertility rate, shape of population pyramid, doubling time, and use of resources.
5. **K/U** How does age structure influence population growth?
6. **K/U** Why should people be concerned that the doubling time for the world's population is only about 58 years?
7. **T/I** North America and Europe together cover about 35 million km^2 while Africa covers about 30 million km^2. Use your knowledge of population growth rates in different countries to predict whether North America and Europe together will contribute more to the world's future human population growth than Africa.
8. **C** Draw and describe the population pyramid that you would expect for an industrialized country, and one for a less-developed country. Explain the population growth that will most likely occur in each country.
9. **A** Building a sustainable world is an important task for the survival of the human population. On what factors will the quality of life for future generations depend?
10. **A** If the worldwide growth rate has been decreasing, why is the human population expected to continue to increase?
11. **A** Compared to an increase in the growth of people in less-developed countries, an increase in the growth rate of people in industrialized countries leads to a more uncertain future for humanity. Explain why.
12. **T/I** How much more land does it take to support the average Canadian than it takes to support the average person on Earth?
13. **K/U** Considering the three largest components of the ecological footprint in most developed countries, what steps can each person take to reduce their ecological footprint?
14. **C** Compare and contrast the ecological footprint and available biocapacity in 2002 for North America, Middle East/Central Asia, and Africa.
15. **A** Do you think the carrying capacity of Earth is static? Explain your reasoning.
16. **C** Draw two different graphs showing your prediction of the human population over the next 1000 years. Explain your reasoning.
17. **T/I** What are the advantages and disadvantages of a population that has the type of age structure shown below?

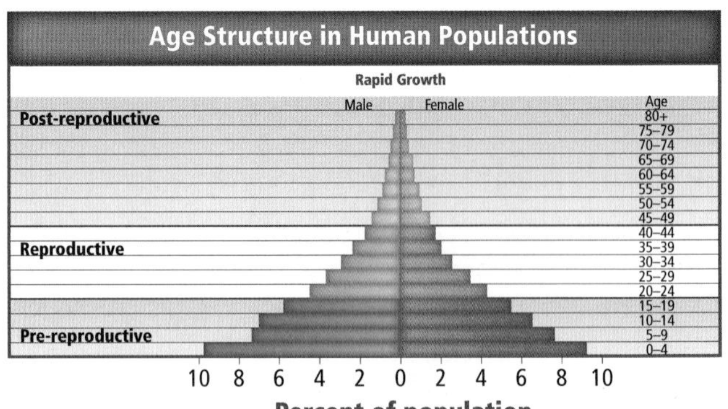

Managing Growth and Resources

The term *ecological footprint* might sound abstract and it might be difficult to understand what that means. Think about your needs. You need a home or shelter. You need energy to heat and cool that home and to cook your meals. You also need energy for transportation and electricity for your electronics. You need food and you need space to dispose of your waste. You need resources to construct your home, your clothing, and all your possessions. Now, multiply those needs by 7 billion.

Like all populations, humans obtain the necessities for survival from their environment. Unlike other species, for many humans their environment is the entire globe. The explosive growth of the human population has created some problems. A few of those problems are discussed in this section. As you read about these global issues, keep in mind that these environmental issues have solutions. Some of the people of your generation will develop those solutions and help provide for human population growth.

Key Terms

biomagnification
deforestation
sustainable
bycatch
biodiversity
overexploitation
minimum viable population size

Energy Requirements

One cost of a growing population is a demand for more energy. Globally, energy generation is predicted to increase an average of 2.3 percent each year until 2035. Despite efforts to the contrary, coal burning continues to be the main source of electricity production around the world. Its use will increase as the human population grows, as shown in **Figure 12.6**, unless viable and affordable alternatives are found.

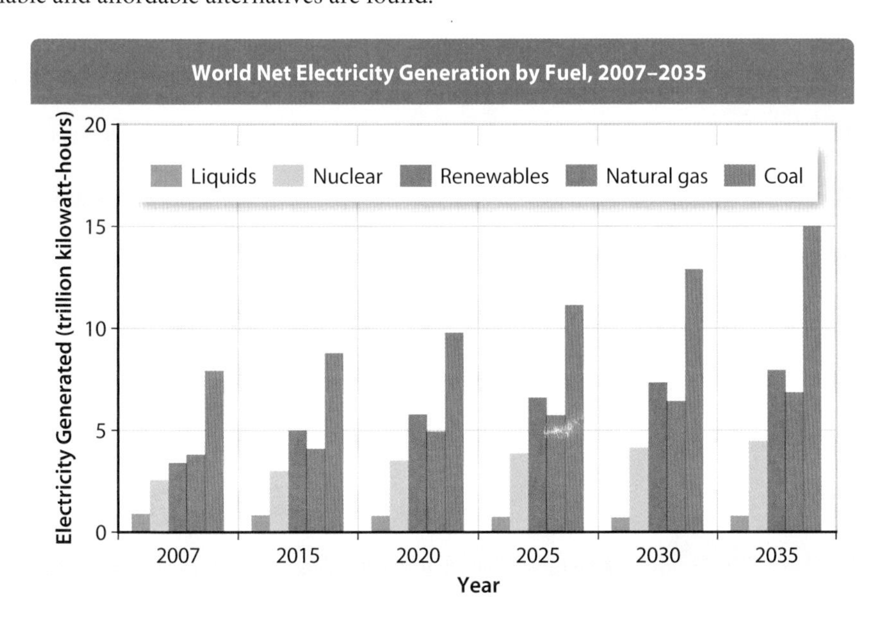

Figure 12.6 Note that net global coal-fired generation is expected to almost double from 2007 to 2035. One reason for this may be that coal is relatively inexpensive in comparison to oil, natural gas, and alternative forms of energy, such as wind and solar power.

Mercury Contamination

The environmental cost of burning coal is significant. Most of the public is aware that coal combustion produces harmful air pollutants such as carbon dioxide (a chief contributor to global warming), sulfur dioxide, and nitrogen oxides, which result in acid rain and photochemical smog. Few people know that burning fuels, such as coal, releases significant amounts of toxic metals into the atmosphere. One of the most dangerous of these is mercury. In fact, coal-burning power plants are the greatest human-produced source of mercury emissions.

Mercury is a naturally occurring element found in many of the rocks that contain coal. Each year, the average coal-burning plant produces over 70 kg of mercury. To put this in perspective, adding 2.7 g of mercury to a lake that covers about 10 hectares makes fish unsafe to consume.

Toxicity and Biomagnification

Although mercury is toxic in its elemental form, a more toxic transformation occurs when this chemical enters ecosystems. When mercury enters the atmosphere, it quickly enters waterways, either settling directly into lakes and streams, or washing into them after being deposited on land. Once in the water, certain micro-organisms convert the mercury into methylmercury. Living organisms easily absorb this highly toxic form of the metal. Methylmercury quickly biomagnifies up the food chain to the higher-level consumers, including humans who eat fish and shellfish.

biomagnification the increase in concentration of a substance, such as methylmercury or DDT, that occurs in a food chain and is not broken down by environmental processes

Biomagnification is the increasing concentration of toxic substances that enters the food chain or food web at low levels. The concentration increases as organisms higher in the food chain consume lower-level organisms, as shown in **Figure 12.7**. Each section of the pyramid represents a trophic level or a step in the food chain. Energy flows from the lower trophic levels at the bottom of the pyramid through each level until it reaches the top trophic level of consumers. The consumers in the bottom trophic levels absorb mercury and transform it into methylmercury. Higher-level consumers eat these lower-level organisms and the concentration of methylmercury builds up in their tissues because it is not removed from their bodies. The biomagnification of methylmercury in organisms is very similar to the biomagnification that occurs with the pesticide DDT. This pesticide biomagnification often is used to explain this process and you have probably read about it in previous courses. The biomagnification of DDT has led to a serious decline in predatory birds such as the bald eagle. DDT and methylmercury biomagnify in organisms in a similar way.

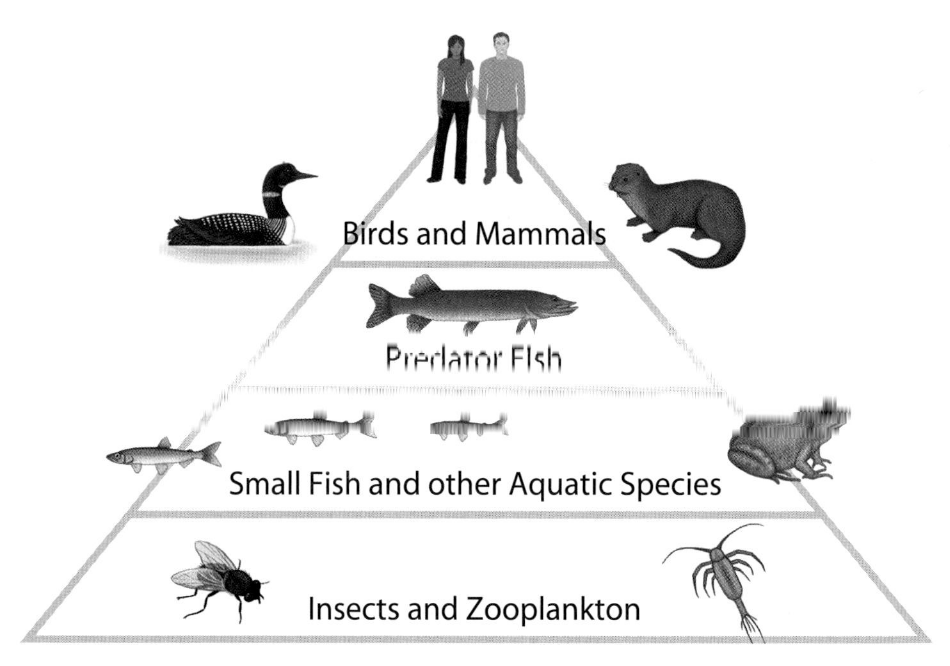

Figure 12.7 Methylmercury increases in concentration in organisms as the trophic level increases.

When methylmercury reaches high enough levels in these animals, the chemical typically interferes with growth, development, and reproduction. Harm to the nervous system may also result in abnormal behaviour and eventual death. Chronic exposure to methylmercury in humans damages the kidneys, liver, and lungs, as well as the immune and nervous systems. In pregnant women, it interferes with the brain development of the fetus. In young children, high levels of methylmercury may also interfere with brain development and the ability to learn.

Pollution-Free Energy

One solution to reducing the effects of burning fossil fuels, such as coal, is to increase the use of other forms of energy. Renewable forms of energy, such as wind, solar, geothermal, and hydroelectric energies, are less harmful to the environment. As **Figure 12.6** shows, renewable energy is used, but there is room for growth. Replacing fossil fuels, such as coal and natural gas, with renewable forms of energy will reduce the amount of pollutants released into the atmosphere. These pollutants include carbon dioxide, sulfur dioxide, nitrogen oxides, and toxic metals such as mercury.

Food Requirements

Another consequence of the expanding size of the human population is a greater need for food. Generally, this need is the greatest where population growth and poverty are also greatest. As you read in Section 12.1, population growth is greatest in developing countries. Much of the tropical rainforest land in developing countries is being cleared and converted to farmland and rangeland for food, as shown in **Figure 12.8**. This process is called **deforestation**.

deforestation the cutting, clearing, or removal of trees so land can be used as pastureland or cropland

Figure 12.8 These mountains were covered with the lush landscape of a rainforest. The removal of the trees, mostly for the production of charcoal, has resulted in heavy erosion of the hillsides.

Deforestation

Tropical rainforests are among the most productive ecosystems in the world. As a result, one might imagine that the 400 km^2 cut down per day would produce a bounty of crops. Surprisingly, this is not the case, as the soil in this ecosystem is extremely poor. In fact, its estimated that 2.8 million hectares of rainforests have been cut down for farming and this has provided less than 1.2 million hectares of arable land on which crops can be grown. Typically, this land is only fertile for 10 to 20 years or less.

The infertility of the soil in this lush ecosystem is difficult to understand. How does such poor soil support so much life? The answer is that it doesn't. Instead, the nutrients and water needed by organisms in the rainforest are recycled within these biotic components. Few nutrients ever reach the soil. Bacteria, which play a large role in this process, make up a greater mass of living organisms in this ecosystem than all other life forms. In addition, abundant rainfalls in these ecosystems results in erosion in areas that are cleared of trees, especially on slopes. Subsequently, mudslides can carry both soil and vegetation into waterways. For instance, in both Haiti and Madagascar, the consequences of soil erosion have been catastrophic, because thousands of tonnes of soil are washed away every year.

Poverty in these nations often means people have no choice but to clear land. The loss of rainforest has other consequences, as the trees and other lush vegetation are estimated to remove about one-third of the atmosphere's carbon dioxide, much of which is a result of human activities such as burning fossil fuels.

sustainable the use of resources, such as food, energy, timber, and other items acquired from the environment, at a level that does not exhaust the supply or cause ecological damage

The rainforests are not the only forests at risk of deforestation. The Canadian Boreal Forest is a 567 million hectare tract and is one of the largest intact forests in the world. While about 10 percent of the forest is protected, the remaining 90 percent is at risk due to unsustainable industrial and forestry activity. **Sustainable** harvesting of resources is a method of obtaining resources that maintains ecological balance and conserves the resources by avoiding their depletion. The Canadian Boreal Forest contains about 22 percent of the carbon stored on Earth's surface. This carbon is stored in the form of carbon compounds that make up the trees, plants, and other living organisms in the forest ecosystem and in the form of peat. The Canadian Boreal Forest also removes carbon dioxide from the atmosphere as the plants undergo photosynthesis, in the same way as the plants in the rainforests do.

Stopping Deforestation

Stopping or reducing deforestation in impoverished countries poses significant challenges. For example, in Haiti, where much of the population has lived in poverty for many decades, large tracts of forests have been cut down to clear space for growing subsistence crops to feed rapidly growing rural populations and, more substantially, to make charcoal to sell for fuel to people in urban centres. Efforts are underway, however, to combat deforestation through the planting of trees, as shown in **Figure 12.9**, as well as through the terracing of hills to stop erosion and prevent mudslides and landslides during the rainy season.

Figure 12.9 Some of the people in Haiti are working to replant trees on the hillsides in an effort to stop the damage caused by deforestation.

Overharvesting Fish

As the size of the human population increased, so did demands on global fisheries. By 1950, humans had managed to remove a large percentage of common fish stocks from the world's oceans. This occurred mainly through the adoption of industrialized fishing technologies.

The decline of fish stocks has not gone unnoticed by scientists, including the research team of Canadian biologists Boris Worm and Ransom Myers. In 2003, they published the results of their ten-year study of five decades of ocean data. Their findings confirmed that the mass of large predatory fish, including popular food fish such as blue fin tuna and cod, is now only about 10 percent of what it was before 1950.

New technologies often solve problems, but they frequently introduce new problems. Worm and Myers found that within 15 short years, industrialized fishing technology decreased the mass of other living organisms in marine communities by approximately 80 percent. How did that occur? Advances in technology have brought with them destructive fishing techniques. These techniques, such as dredging and trawling the sea floor, remove nearly all animal life from the sea floor. Millions of tonnes of other ocean species are extracted from global oceans unintentionally in this way. Unwanted species that are caught in nets during this industrial fishing practice and discarded as waste are called **bycatch** (**Figure 12.10**). Species are not limited to fish, but include such varied organisms as sea birds, octopuses, sea turtles, dolphins, sea grass, and starfish.

bycatch aquatic organisms that are caught unintentionally by fishing gear or nets and often are discarded as waste

One example of trawling occurs in the tropics. Shrimp trawlers trap an estimated 5–20 kg of bycatch for every 1 kg of shrimp. Trawling also occurs in the North Atlantic. Atlantic cod was once plentiful, but overharvesting has caused a serious decline in the population. Recovery of cod is difficult because fishers that are targeting other fish, such as haddock, catch large numbers of juvenile cod as bycatch.

Figure 12.10 Trawling is an unsustainable practice of harvesting fish and shellfish. Unwanted organisms often far outnumber the desired ones and are discarded as waste.

Providing Food Using Sustainable Methods

In order to preserve fish and shellfish, as well as other marine life, they must be harvested using sustainable methods. Public awareness is a key factor in bringing about change to overharvesting and trawling. Reducing illegal and unregulated fishing will help marine populations recover and improve the health of the world's oceans.

Other foods must be grown, harvested, and transported using sustainable methods. The food demands, or energy demands, of a growing human population must be met in ways that do not harm Earth and will provide for future populations. A recent study suggests there is some recovery happening.

Activity 12.2 Living and Eating

In this activity, you will consider the impact on food supply of some of the choices people make. The units used in the activity, kcal, are not SI units; however, the kilocalorie is still commonly used to report the energy content of food due to its widespread usage in the United States, especially. Note that the kcal has the symbol C (capital C). Thus, 1 kcal (or 1 C) is equal to 4.186 kJ.

Procedure

1. Statistics Canada estimates that the average Canadian eats about 2400 kcal/day. How much would they eat in one year?
2. What area of agricultural land would be needed to feed an average Canadian for one year? Assume that humans can use 15 percent of the net primary productivity of farmland for food (the rest are roots, stalks, leaves, or other inedible parts of the plants).

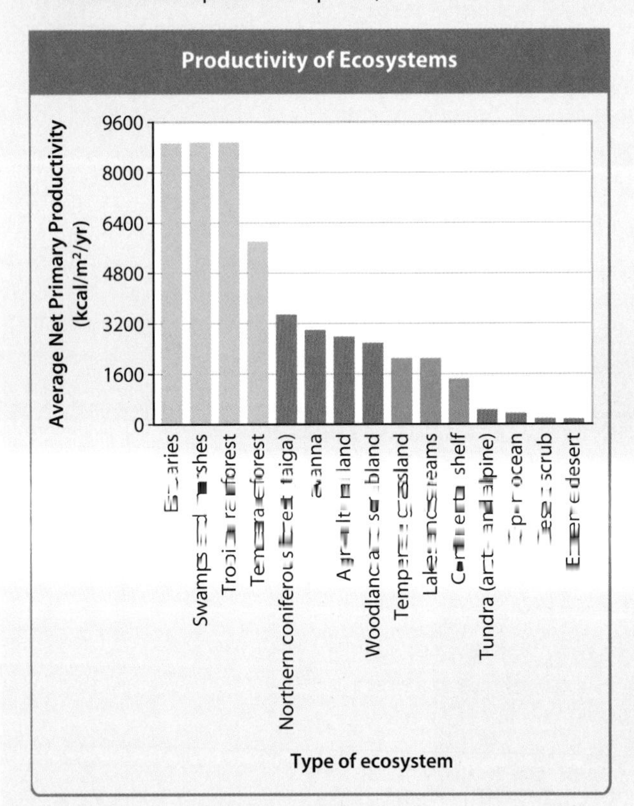

3. There are about 5 billion hectares of agricultural land on Earth (5×10^{13} m^2). If all of this was used to feed humans a strictly vegetarian diet, how many people could it support?
4. Compared to primary producers, the average omnivore requires about 5.5 times as many calories to support itself than a herbivore, because of the extra energy needed to support food species higher on the food chain. If the world were to adopt a typical Canadian diet, which is omnivorous, how many people could be fed?
5. Given that the current world population is around 7 billion people, what conclusion can you draw?

Questions

1. Why do you suppose the average net productivity for open oceans, shown in the graph below, is so low? Why do some people still consider fish a part of an ecologically friendly diet?
2. A vegan diet relies exclusively on plant products. Explain the rationale for a vegan diet being good for the planet.
3. Sometimes the answers for questions 3 and 4 from the procedure are given as a rationale for everyone adopting a vegetarian diet. Explain why this argument is not entirely sound by pointing out some of the assumptions not taken into account in the simple calculation you have done.
4. A study in Iowa found that the average food produced locally was transported about 72 km from farm to home, while the average food in a grocery story was transported about 2500 km. A *localvore* is someone who tries to eat as much of his or her food as possible within 100 km of where the food is produced. Explain the rationale for a localvore diet being good for the planet.
5. If the world could produce food for as many people as suggested in step 3 of the procedure, why do we say that the world is at its food-production limit? There are multiple reasons.
6. Sugarcane produces more available food energy for human consumption than almost any other crop. Could we feed the world sugar as a way of producing enough food on limited cropland? Why or why not?
7. It takes an average of 8 kcal of energy to produce, fertilize, transport, store, and prepare 1 kcal of food for human consumption. We get most of this energy from fossil fuels. Explain why it took about 80 percent of the human population to produce our food about a century ago whereas it now takes barely 1 percent to do so. Why are some people concerned that this might not be sustainable?
8. What can you personally do to help rectify this situation?

Waste Disposal

As the human population grows, so does the need for adequate waste disposal. Unfortunately, the oceans often have served as waste-disposal sites. The earlier examples about fishing and bycatch explained how humans have removed millions of tonnes of ocean species from Earth's seas. In exchange for these life forms, we have dumped garbage into the sea. One problematic type of garbage, or marine debris, is plastics, as shown in **Figure 12.11**. Plastics are a human-made product that takes years to decay. While indestructible, long-lasting plastic products are desirable to the consumer, they are not desirable in the world's oceans.

The exact amount of garbage that has been dumped is unknown and is extremely difficult to calculate. Estimates have been published and then disputed by other sources. According to the National Oceanic and Atmospheric Administration, NOAA, one problem in making an accurate estimate is that there is no standard method for monitoring at-sea plastic marine debris.

You might have heard or read about large garbage patches in the oceans. The one in the North Pacific Ocean, the North Pacific Gyre, has gotten considerable media attention. These garbage patches accumulate where currents move in circular patterns. The garbage becomes trapped in these areas and accumulates over time. The exact sizes of the garbage patches are unknown and the exact density of debris, especially plastic, is disputed. Much of the debris in these garbage patches is small pieces of plastic and abandoned fishing nets.

Figure 12.11 All types of garbage are found in the oceans, but some of the most harmful are the plastic items. These items do not decompose easily and they remain in the environment for a very long time.

Sources of Marine Garbage

The sources of this marine debris include cargo and passenger ships, oil platforms, runoff from rivers, and other means. Of all the forms of waste found in the sea, plastic is undoubtedly the most dangerous. This includes plastic such as water bottles and shopping bags that are used briefly before being tossed away, as well as fishing line and abandoned fishing nets. This plastic ultimately finds its way into and onto marine organisms. In 2007, a whale that washed up on the coast of California was found with over 200 kg of plastic in it. The fishing lines and nets also entangle marine organisms such as whales, seals, sea lions, and otters. In 2005, a Russian submarine had to be rescued in the Bering Sea after finding itself in a similar predicament.

Figure 12.12 Plastic pellets called *nurdles* are consumed by marine life but their bodies do not break them down. As a result, the plastics stay in their bodies and toxins accumulate.

Pre-production plastics, small, lightweight spherical nodules called *nurdles*, shown in **Figure 12.12**, also present a growing problem. Each year, 100 billion kg of these small bits of plastic are transported around the world, often by freighter. They are sent to manufacturing facilities where they are used to create plastic lids, toys, and virtually anything else made from plastic. Nurdles are so small and lightweight that they escape containers easily and are the perfect size to be swallowed by fish. Researchers have found that all of the fish they sampled from the North Pacific Gyre had ingested small bits of plastic. Ingesting plastic is dangerous to marine animals, but the consequences of carrying a load of plastic for extended periods of time are often more harmful to entire ecosystems. Plastics contain [illegible] An overabundance of this hormone has been shown to disrupt the endocrine system of organisms, to lead to the feminization of males, and to cause reproductive problems. Plastics also attract and concentrate toxins such as mercury and pesticides. Studies have shown that nurdles in oceans frequently have concentrations of these chemicals that are a million times greater than the seawater they float in. Once an organism ingests the plastic, it starts to absorb these chemicals that leach from the plastics into the rest of its body. When predators consume these organisms, the concentrated toxins are transferred to the predator. In this way, the toxins biomagnify with each transfer along the marine food chain, harming top predators the most.

A Waste Solution

Understanding the extent of the problem is easier than fixing it. The solution likely lies in a combination of increased awareness, behaviour change, and technology. Not everyone is aware of the growing issue of marine garbage, but attempts are being made to inform the world. In 2009, the *Plastiki*, an 18 m catamaran, made almost entirely of reclaimed plastic bottles, sailed from San Francisco to Sydney, Australia, through the North Pacific Gyre, to help raise funds and awareness about marine garbage and other environmental issues. As knowledge of this issue increases, personal choices, as well as new laws can help slow the accumulation of garbage in our oceans, while cleanup efforts may help reverse the problem. But change takes time and laws are difficult to enforce, especially on the high seas. Until our behaviours change, technology may be the only hope for saving the seas.

Other Waste Problems

Garbage in the oceans is only one waste issue. As the human population grows, so does the need to dispose of garbage on land. Landfills are filling up and new solutions to garbage disposal must be found. One way to prolong the life of the current landfills is to reduce the amount of garbage that goes into them. Recycling materials is one way to do this. Another part of the solution is to manufacture and purchase items that have a long life. Electronic devices should be designed to be easily upgraded and updated, so the majority of the components are reused. Appliances and other commonly used items should be manufactured to last for years and, once their useful life is over, the components should be easily recycled.

Education and public awareness of the problems related to garbage disposal is one solution to the problem. As people become more aware of the issue, additional solutions are sure to follow.

Learning Check

7. Define biomagnification in your own words.
8. Describe the connection between burning coal and methylmercury.
9. Explain why deforestation is difficult to stop in impoverished countries.
10. Infer a method in which rainforests could be used in a sustainable manner.
11. Refer to **Figure 12.6** and infer why the burning of coal continues even though it is harming the environment.
12. Explain why plastic items are harmful to marine life and why plastics will remain an environmental issue in the oceans for years to come.

Preserving Biodiversity

Another problem that exponential human population growth has caused is a decline in biodiversity on Earth. **Biodiversity** encompasses species diversity (the variety of plants, animals, and other organisms on Earth), the genetic diversity that exists within each of these species, and the diversity of ecosystems (ecosystem diversity) to which these species belong.

biodiversity encompasses species diversity (the variety of plants, animals, and other organisms on Earth), the genetic diversity that exists within each of these species, and the diversity of ecosystems (ecosystem diversity) to which these species belong

Why should we measure and be concerned about biodiversity? To answer this question, it is important to understand that biodiversity stabilizes ecosystems, making them more resilient to change and degradation. For instance, an ecosystem is much more likely to bounce back from change if it has a greater variety of species of which some are able to withstand this change. Such ecosystem change may refer to extreme natural events, such as drought and flood, as well as harmful human-caused events, such as oil spills and acid rain. The greater the diversity of species in an ecosystem, the more resistant an ecosystem is in the face of such an event.

Ecosystem stability is essential to the survival of our species, especially as it reaches record population numbers, since our livelihood depends on services that ecosystems provide. For instance, wetland ecosystems filter and purify the water we drink. Marine and forest ecosystems take up atmospheric carbon dioxide, which helps regulate our climate. Other ecosystems also provide us with materials to build shelters, food resources, and even medicine, as shown in **Figure 12.13**. If an ecosystem cannot weather change, the consequences can be devastating for humans that rely on its services.

Figure 12.13 The bark from the quinine tree shown here is used to treat malaria. Other medicines developed from plants are used to treat other diseases.

Spiritual Benefits

Biodiversity and the resulting ecosystem stability are also important to us for other reasons: they provide us with aesthetic, spiritual, and psychological benefits. Not only would the Earth be a much poorer place without the beauty brought by the great variety of plants and animals that inhabit the world today, but we would be much poorer as well. Humans have an emotional link to nature that brings us both happiness and peace. For example, consider your reaction to a butterfly landing on a flower or a hummingbird taking nectar from a flower. These visual experiences usually are calming and peaceful for humans. Consider the practice of taking flowers to a friend or family member in the hospital. The flowers are in stark contrast to the sterile environment of the hospital room. The presence of nature in the hospital room usually provides peaceful and calming feelings for the patient.

Why do large cities, such as New York City or London, maintain large city parks and green spaces? People flock to these spaces seeking interactions with nature that is lacking in concrete cities. The parks are peaceful settings where people go to jog, walk their pets, read a book, or leisurely stroll through the landscape. The presence and biodiversity of nature provide an indirect value that is difficult to quantify, but we know that it is important to the well being of humans.

The Cost of Limiting Biodiversity

Some humans try to control the biodiversity in their environment. For example, some modern agricultural techniques are designed to limit biodiversity in a specific area. Crops grown using monoculture result in a higher crop yield per area, but are particularly vulnerable to change. Monoculture is a technique of growing a single type or cultivar of crop, as shown in **Figure 12.14**. Recall that the less diverse an ecosystem is, the less resilient it is to change. Two examples of dramatic change took place in the 1800s. In Ireland, a fungal disease called *Phytophthora infestans* attacked potato crops. Unfortunately, these potato crops were grown as a monoculture crop, with only a single type of crop being grown. Because there was little biodiversity in the ecosystem, the crop was devastated by the fungus. Over the next few years, about 25 percent of the human population starved or were forced to leave Ireland in search of new homes. Further devastation occurred in the European wine industry when the majority of vineyards in Europe, grown in monoculture, were killed by the sap-sucking insect *Phylloxera* and secondary fungal infections. These extreme examples illustrate how the dependence on one crop can result in devastating consequences. Although monoculture techniques might increase yield, the resilience of the human-made ecosystem is limited.

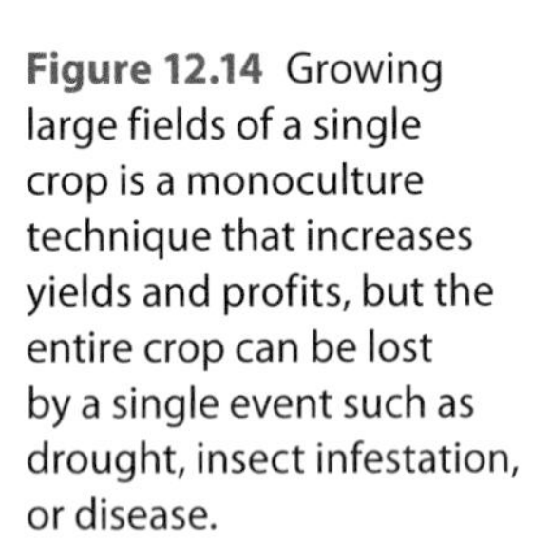

Figure 12.14 Growing large fields of a single crop is a monoculture technique that increases yields and profits, but the entire crop can be lost by a single event such as drought, insect infestation, or disease.

Threats to Biodiversity and Habitat Loss

As the human population has grown, unfortunately so have the threats to biodiversity. One of the primary threats to biodiversity is habitat loss due to destruction, fragmentation, or degradation. One form of destruction you have already read about is deforestation. When the forests were cut down, all of the organisms that lived in the forest lost their home. Some of the organisms possibly migrated to nearby forests or other habitats. Those that could not move or did not make it to a new habitat died. Any organism that was reliant on the dead organisms also died. Competition in the remaining habitat and in any new habitat would increase significantly and more organism deaths would occur. Other forms of habitat loss include the filling of wetlands, dredging of rivers, and mowing of fields.

Another form of habitat loss is *habitat fragmentation*. When roads are constructed or dams and water diversions are built, the habitat is cut into pieces or fragments. These fragments limit interactions among populations, restrict the movements of large animals that need large areas to find food and mates, and make it difficult or impossible for migratory species to complete their migrations (**Figure 12.15**). One very disruptive form of habitat loss is *habitat degradation*. Pollution can completely destroy habitats, and invasive species can out-compete native species to the point of eliminating them from the habitat.

Figure 12.15 Fish ladders are built so that salmon can overcome fragmentation caused by dams in the river and swim upstream to spawn.

Overexploitation

People have always depended on wildlife for survival to meet their needs for food, clothing, medicine, and shelter. However, disregard for the health of ecosystems and greed have led to the overexploitation of many species. **Overexploitation** is the excessive use or removal of a species from their natural environment, until the species no longer exists or has a very small population.

overexploitation the excessive harvesting or killing of a species until it no longer exists or is reduced to a very small population

Species other than fish have also been exploited. The destruction of the North American buffalo herds in the late 1800s is a good example. At one time, there were an estimated 50 million buffalo roaming North America. By 1889, there were an estimated 1000 buffalo left. These animals were slaughtered for their meat and hides. Many were slaughtered and their flesh rotted where they fell. Fortunately, conservation projects have protected the buffalo and some herds still roam parts of North America. The Carolina parakeet was not as lucky. The Carolina parakeet (*Conuropsis carolinensis*) was a North American parrot. Humans hunted it to extinction. The last captive bird died in a zoo in 1914.

Two risks of overexploitation of a species are the *extinction* of a species, which is the disappearance of all members of a species from Earth, and the reduction of the population below the **minimum viable population size**, which is the lowest number of individuals that can persist in an environment for a long period of time without the species going extinct. If the population goes below the minimum viable population size, the individuals might have trouble finding mates. The population's genetic diversity will be low and inbreeding will occur, producing weakened or abnormal offspring. In this situation, genetic diversity might be too low for individuals to adapt to changing conditions.

minimum viable population size the lowest number of individuals that can sustain a population in the wild for a long period of time

Figure 12.16 Giant Hogweed (*Heracleum mantegazzianum*) is an invasive plant that was introduced to North America in the early 1900s as an ornamental plant. It outcompetes native plants for space, sunlight, and nutrients, which reduces biodiversity in the ecosystem.

Invasive Species

Biodiversity is also threatened by invasive species. Invasive species are non-native species that relocate to an area and outcompete the native species for resources. Because invasive species usually do not have predators in their new environments, they reproduce in large numbers. For example, zebra mussels (*Dreissena polymorpha*) in the Great Lakes are a thriving invasive species. These mussels out-compete the native species for food and habitat. They also clog water intake and discharge pipes, resulting in expensive repair costs.

Invasive species are usually transported by human activities. For example, zebra mussels are transported in the ballast water of cargo ships, and fire ants are carried in wood products such as pallets and crates. Some plant species such as kudzu are transported as ornamental species for gardens and greenhouses. Once an invasive species establishes itself in its new environment, it is difficult to remove it. Invasive species can cause damage to the economy, as is the case with zebra mussels in the Great Lakes. In addition, biodiversity within an ecosystem decreases when invasive species threaten the existence of native species , as shown in **Figure 12.16**.

Pollutants

You have read about plastic pollutants in marine ecosystems and methylmercury in aquatic and marine ecosystems. Many other pollutants threaten biodiversity by killing large numbers of organisms. You are probably aware of many pollutants, such as DDT, a pesticide that threatened top-of-the-food-chain predators like the bald eagle. Other organic pollutants, such as PCBs and dioxins, also threaten biodiversity just like DDT and methylmercury. They enter the food chain and biomagnify. They accumulate in organisms' tissues because they do not naturally decompose. Pollutants threaten biodiversity by reducing the number of individuals in a population, which can lead to a species' extinction.

Disease

When ecosystems are weakened by decreased biodiversity, either genetic biodiversity or diversity of species, disease can be devastating. The Irish potato famine is an example of what can happen when disease strikes an ecosystem that no longer has diversity. Another example is *chronic wasting disease*, which is a contagious neurological disease primarily found in elk and deer. This disease appears to have been introduced to Canadian stock from infected elk imported from the United States in the late 1980s and early 1990s. Management of this disease is difficult in wild populations.

Solutions to Sustainability

There is not one solution for providing resources for a growing human population. Many solutions are needed that work together to provide the needs of the human population and, at the same time, preserve Earth's resources. The first step to finding viable solutions is recognizing the problems humans are creating as they use the resources and deposit wastes haphazardly. The problems are as varied as the populations themselves. Industrialized countries must curb their use of all resources and dispose of wastes in a responsible manner. Developing countries must grow their economies so people have the resources they need. At the same time, they must learn from mistakes and protect their environment so it will continue to provide and sustain all populations. Sustainable use must be practised by all populations in the years to come.

Section 12.2 Review

Section Summary

- Like all populations, the environment provides resources needed by humans.
- Burning coal to provide energy creates toxins.
- Human activities cause environmental problems and many human activities are unsustainable.
- Preserving biodiversity is important for the overall health of all populations on Earth.
- Human activities have decreased biodiversity.
- Many solutions are needed to provide the needs of a growing human population and, at the same time, preserve the environment.

Review Questions

1. K/U As the human population grows, so does the demand for energy. Describe the general trend in the resources used to generate electricity. Why should we be concerned about this trend?
2. A China and the United States together produce just under half of all the carbon dioxide emissions.
   a. Based on this knowledge, do you think that all countries should be required to reduce their carbon dioxide emissions? Explain your reasoning.
   b. Devise a fair method for calculating the maximum amount of carbon dioxide that each country should be allowed to emit.
3. C Draw an example of an aquatic food chain or a food web that includes sea birds. Describe how a toxic substance dumped into the water ends up in the tissues of the sea birds.
4. T/I Today less than 3 percent of Haiti's original rainforests remain and the per capita income is about $500 per year. Most Haitians live in abject poverty. Collecting wood from the hillsides and making it into charcoal is a source of income for many families, which has resulted in deforestation. You have been hired to head a reforestation project in Haiti. Summarize how you would do this.
5. K/U Trawling can be described as clearcutting the ocean floor. Explain why.
6. T/I In some places, trawlers are required to use sea turtle excluder devices to reduce their catch of sea turtles. Infer why this is necessary.
7. K/U Every year, more and more garbage patches similar to the Great Pacific Garbage Patch are found. How are these patches created? What problems do such garbage patches cause?
8. K/U Describe two ways in which nurdles affect marine life.
9. C Conserving biodiversity may increase costs and threaten people's source of income. Write an argument for conserving biodiversity.
10. A Canada has taken steps to reduce the number of plastic bags people use.
    a. Discuss some of the steps that you have seen to reduce plastic bags. Why is this effort necessary?
    b. Describe methods that you think could help to reduce the amount of other types of plastics we use.
11. A A farmer plants hectares of corn year after year using a monoculture technique, without rotating with different crops. Corn rootworm, an insect pest, is introduced to the cornfield. Write a hypothesis about what you would expect to occur.
12. K/U Describe three threats to biodiversity as the human population grows.
13. C When invasive species such as zebra mussels enter a habitat, they often out-compete native species for resources. Draw an illustration that demonstrates an accidental way and a purposeful way that invasive species enter a habitat. Include the effect on both species.
14. A What should the human population do to make sure species do not become threatened or extinct due to human activity?
15. A If you had to design your own city, how would you make sure its residents have the resources to maintain a good quality of life while reducing negative impacts on the environment?
16. A Explain how using the item shown in the photograph plays a part in managing Earth's resources and providing for human population growth.

ThoughtLab
INVESTIGATION

12-A

Skill Check

Initiating and Planning

Performing and Recording

✓ Analyzing and Interpreting

✓ Communicating

Materials

- computer with internet access
- paper and pen

Do You Tread Lightly on the Earth?

An ecological footprint is a method for calculating your impact on the natural environment by assessing how much land is required to produce the natural resources that you use. We use these resources for food, clothes, housing, heat, transportation, education, and recreation. Your teacher will direct you to a number of websites that calculate ecological footprints. They tend to give answers in terms of the area needed to sustain an individual, and the number of Earth-like planets needed to support humanity if everyone had the same lifestyle as the respondent.

Pre-Lab Questions

1. What is the definition of carrying capacity?
2. What is the surface area of the Earth? If about 13.4 billion hectares are ecologically productive, what percentage of the Earth's surface is biologically productive to a significant extent?
3. What is the current population of the Earth?
4. What is the theoretical maximum for the average ecological footprint if the Earth is to be used sustainably (i.e., inhabited by humans into the indefinite future)?
5. Survey five people not in this class as to the three most important issues affecting them. Pool your data with the rest of the class. It will be used in a later question.

Question

Can people live with lifestyles like ours indefinitely? What would a sustainable lifestyle for human beings look like?

Hypothesis

Pose the question above in the form of a hypothesis before you begin the lab.

Organize the Data

1. Your teacher will direct you to a series of ecological footprint calculator sites on the internet. Copy the data table shown on the next page into your notebook and fill in the answers to the questions as you go to each ecological footprint site.
2. What is the size of the average Canadian's footprint?
3. What percentage of the average Canadian ecological footprint do you leave?

Analyze and Interpret

1. How many planets would we need if everyone used the same amount of resources that you use? How does this make you feel? Explain why.
2. What is your intellectual reaction to this number?

Sample Data Table

Questions	Footprint Calculator 1	Footprint Calculator 2	Footprint Calculator 3
URL of ecological calculator site			
Name of the organization who sponsors the site			
What is the size of your footprint (in hectares) using this calculator?			
To which questions did you not have the answers? Find those answers and redo the survey. How accurate were your initial guesses, and how did the correct data affect the final answer?			
Try the survey again, adjusting your answers to make it so everyone on Earth can have a footprint like yours. Were you able to do this? What did you have to change to do so?			
Redo the survey as someone with your lifestyle in a tropical developing country (i.e., don't change your answers to the questions but pretend that you live in a tropical developing country). What happens to your footprint? Why does this happen?			
What was the ecological footprint for a person using the same resources that you use while living in the developing country that you chose?			
What questions did they include that were not found in the others?			
Change some of the parameters by changing the answers to some of the questions. What seems to make the greatest difference in the final ecological footprint answer?			
How many planets would we need if everyone used the same amount of resources that you use?			

3. Scientific studies show that we currently use 40 percent of the net product of terrestrial photosynthesis and 25-35 percent of coastal shelf primary production, and that these might represent unsustainable use of ecosystems—for example, global fisheries yields have fallen for over 20 years. At the same time some global waste sinks seem full to overflowing. Half of the global population (3.5 billion people) lives on the equivalent of less than $2/day. About 1 billion people live with a lifestyle comparable to Canada. Explain how these facts make it impossible to have all the people on Earth live with a lifestyle like ours.

Conclude and Communicate

4. Why has every value for carrying capacity so far been an underestimate?

5. What is the likely consequence of not resolving the problem of people exceeding their carrying capacity in the next 20 years?

6. Decide on the relative importance of the results you have found here compared to the issues that your survey revealed people to be concerned about prior to doing this lab. Do you agree with the survey as to the most important issues?

7. If the issue of people overextending the carrying capacity of the Earth is not resolved, which of the issues that your survey found will still be important?

8. Why is the study of an issue as important as this relegated to a course that only a small proportion of high school students take?

Propose a Course of Action

Propose a course of action to increase the awareness and urgency of this problem to at least one person who might be influential in assisting to solve the problem (i.e., a call to a politician, talk to a business owner, or run a public information campaign). If you choose not to follow out your plan of action, explain why you made that decision.

Extend Further

9. INQUIRY What sort of question would you like to have more detail on in the ecological footprint calculator? Why?

10. RESEARCH Find out about "Earth Overshoot Day" and make a chart of what day of the year it falls on. Draw a conclusion based on this chart.

ThoughtLab INVESTIGATION

12-B

Skill Check

Initiating and Planning

Performing and Recording

✓ Analyzing and Interpreting

✓ Communicating

Suggested Materials

- calculator, or preferably a computer with a spreadsheet program
- data provided by teacher

First Impressions Count

Human dynamics, population growth, and environmental/ecological concerns tend to elicit an emotional response that is quite strong. Many people approach these issues with their mind mostly made up before they know any facts. For this reason, it is important to keep a critical eye on the facts presented, because even if the information is true it can be misleading. In this investigation, you will be taking real-world data that is presented in different ways and drawing different, sometimes contradictory, inferences from it.

Pre-Lab Questions

1. List two rules to follow when drawing axes on graphs.
2. Research and find out what the "Rule of 72" means. Explain how the "Rule of 72" relates to doubling times.
3. What does "per capita" mean?

Question

How accurate is your initial perception based on numerical and graphical data?

Organize the Data

Part 1: Observing and Interpreting Line Graphs

1. Use the data provided by your teacher to graph the Year vs. Population for each country listed excluding China and Qatar.
2. Calculate the annual change in population for each country and the annual percentage of change in population for each country.
3. Prepare a data table, like the one shown on the next page, in your notebook. The "Initial Impression" side of the chart is for you to answer after no more than 5 seconds of looking at the graph. It represents both an uncritical look at the data, and the sort of result that people would remember without critical examination. There is no wrong answer here, and some things have been chosen because they are misleading, though all data is accurate. The "Analysis" side is for you to fill in after considering the data for as long as needed to get an answer in which you are confident. It is not expected that both columns will be the same.
4. Observe your graph and write your responses to the graph in your data table.
5. Study the data for China and Qatar. Why is the data for these countries difficult to add to the graph?

Go to ***Constructing Graphs*** in ***Appendix A*** *for help with drawing graphs.*

Part 2: Observing and Interpreting Maps

6. Look at the four maps dealing with carbon dioxide, CO_2 emissions on the next several pages. All deal with the same data, but have been adapted to show different things. Record your first impressions of each map in your science notebook.

Data Table for Population Data Responses

Question	Initial Impression		After Accurate Analysis
	Graph	Data	
Which country has the highest percentage growth rate?			
Which countries are showing a decrease in population?			
Which nations are experiencing a growth rate in the last 5 years that is faster than their average growth rate for the last 50 years?			
Which nation has the fastest doubling time?			
Which nation(s) appear to have a carrying capacity of about 10 million people?			
Which nations are in the logarithmic phase of their growth?			

Change of Carbon Dioxide Emissions from 1990 to 2007

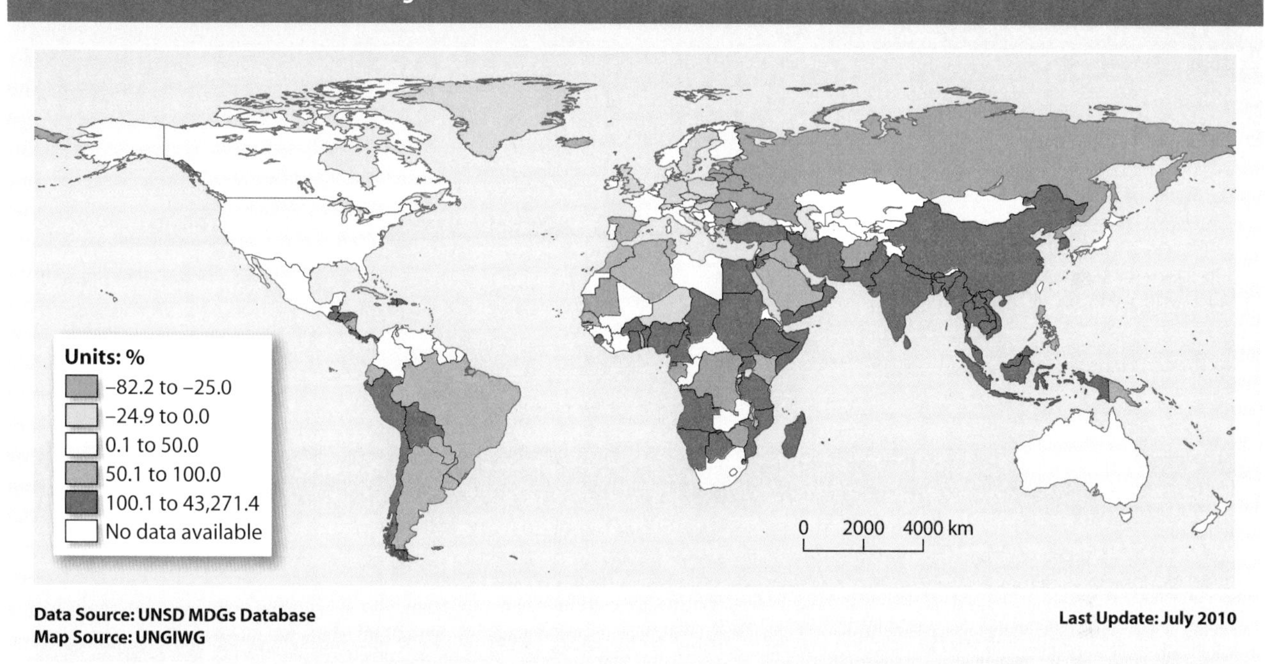

Carbon Dioxide Emissions in 2007

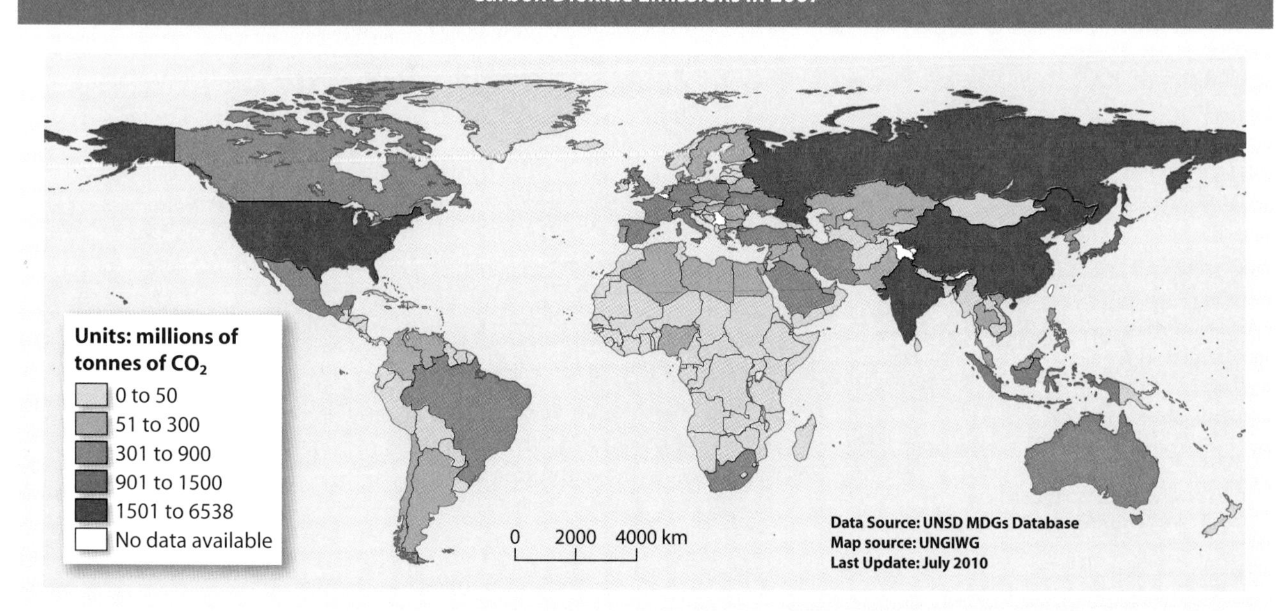

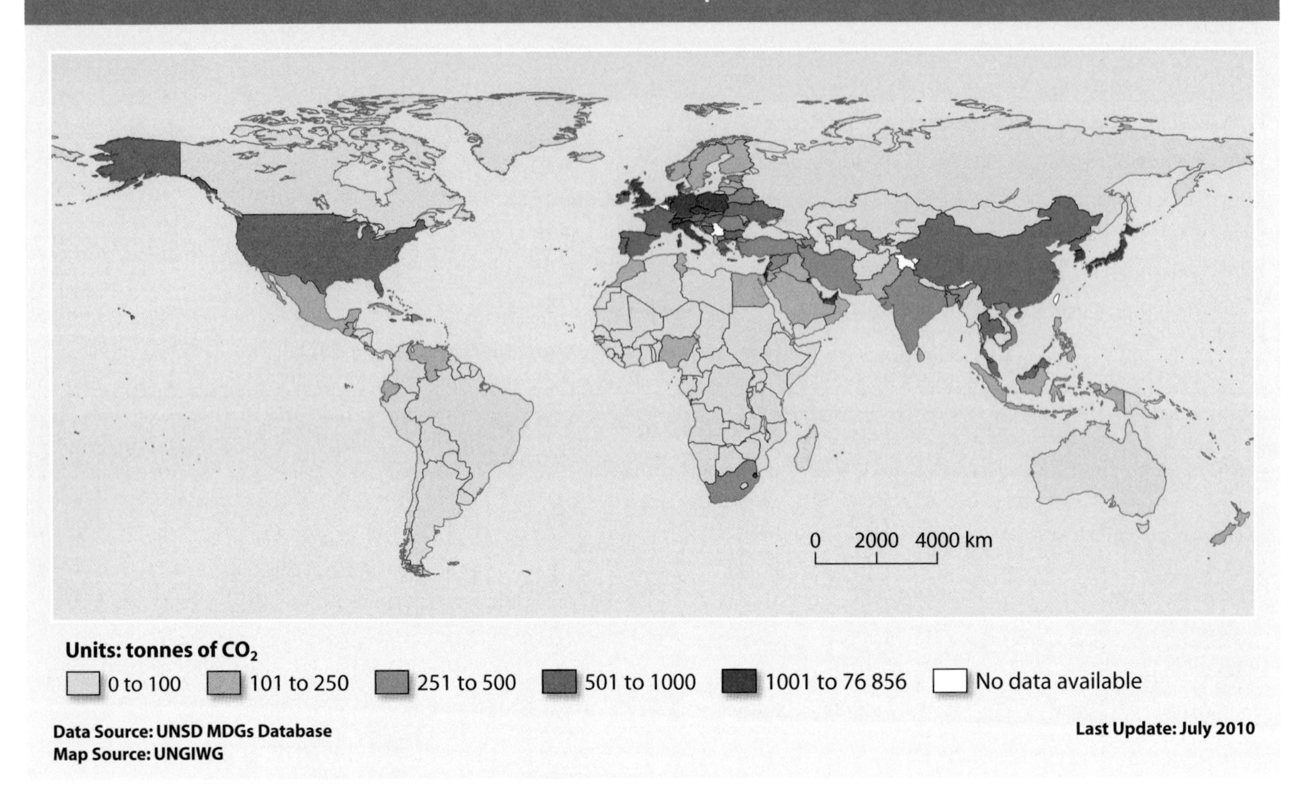

Carbon Dioxide Emissions per km² in 2007
0 2000 4000 km
Units: tonnes of CO₂
0 to 100
101 to 250
251 to 500
501 to 1000
1001 to 76 856
No data available
Data Source: UNSD MDGs Database
Map Source: UNGIWG
Last Update: July 2010

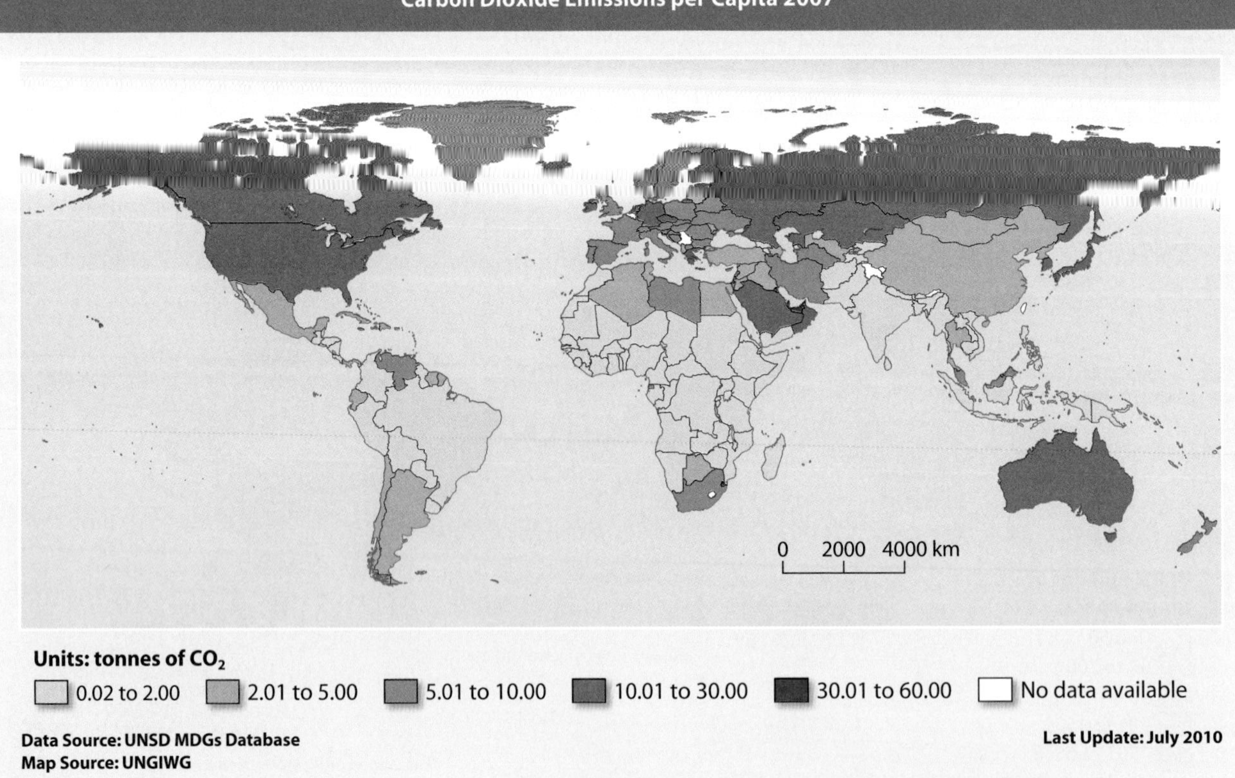

Carbon Dioxide Emissions per Capita 2007
0 2000 4000 km
Units: tonnes of CO₂
0.02 to 2.00
2.01 to 5.00
5.01 to 10.00
10.01 to 30.00
30.01 to 60.00
No data available
Data Source: UNSD MDGs Database
Map Source: UNGIWG
Last Update: July 2010

Analyze and Interpret

1. In 1995, the Canadian death rate (15.4%) exceeded our birth rate (15.3%). However, Canada's population grew that year. How can this be?
2. What sort of questions did you interpret the graphs correctly for on your initial impression?
3. Which questions about populations did you have the hardest time figuring out?
4. Compare your results with at least three classmates. What do you notice?
5. Carbon dioxide emissions are often used as an indicator for environmental impact. Give two reasons why this data is used.
6. One of the reasons that the original Kyoto Protocol on limiting carbon dioxide emissions failed was disagreement about the responsibilities of different countries for adding this gas to the atmosphere. Assume that your job is to "spin" the data by presenting each nation shown in the table in as good or bad a light as possible. Copy the data table below into your notebook and explain your reason to use each map for each purpose. Justify you answers.
7. Explain what might account for the decrease in emissions in Afghanistan, Russia, and Sweden.

Conclude and Communicate

8. Was it faster to comprehend the information presented as numbers or as visuals (maps, graphs)?
9. Were the numbers or graphs less ambiguous and why?
10. What was the most surprising result from the population graph analysis? Why?

Extend Further

11. **INQUIRY** It was mentioned that carbon dioxide emissions are used as an indicator for pollution data. Find a measurement that is used to estimate another hard to quantify quality, such as the Living Planet Index (LPI), and explain how it was formed, what it measures, and what the justification is for this quality.
12. **RESEARCH** Find an actual graph or statistic other than those presented here to justify action (or inaction) on a topic related to the sustainable use of resources. Assess whether the data is presented fairly and where the group has manipulated the data to justify their desired outcomes.

Data Table for Carbon Dioxide Responses

Nation	To Minimize Their Responsibility for Carbon Dioxide Emissions:	To Maximize Their Responsibility for Carbon Dioxide Emissions:
Canada		
China		
India		
Russia		
Saudi Arabia		
USA		

Case Study

Micronutrient Deficiencies in Expanding Populations

Assessing the Scope of the Problem

Scenario

Micronutrients are vitamins and minerals that are essential to good health. These substances, including vitamin A, iodine, iron, and folic acid, are only needed in small amounts, but they are critical to life functions, such as producing proteins and hormones and the proper development of nervous system tissue. A deficiency in a micronutrient can have severe health consequences, including blindness, mental impairment, and reduced immune system function.

Often referred to as "hidden hunger," micronutrient deficiencies affect over 2 billion people in expanding populations around the world. Pregnant women, infants, and young children are particularly vulnerable. For example, worldwide, it is estimated that over 2 million children die each year as a result of vitamin A, iron, and zinc deficiencies.

Micronutrient deficiencies can affect not only the health of a nation, but its productivity and economy as well. Statistics show that reducing iron deficiency alone can increase a country's health and productivity levels by 20 percent.

The Canada-based Micronutrient Initiative (MI), the World Health Organization (WHO), and the United Nations Children's Fund (UNICEF) are three agencies that have programs focused on increasing the availability of micronutrients to target populations. Read the following fact sheets about four of the main micronutrients missing from people's diets. Use the information in this case study, as well as your own research, to assess and recommend sustainable solutions to this problem.

Vitamin A

Vitamin A is critical to the production of proteins in the retina of the eye that absorb light and initiate the formation of images. Natural sources of vitamin A include eggs, milk, and liver. A deficiency of vitamin A leads to blindness, as well as reduced immune system function.

Impact and Scope of the Problem

- Between 250 000 and 500 000 children become blind each year as a result of vitamin A deficiency.
- About 40 percent of children under age 5 in developing countries have compromised immune systems due to vitamin A deficiency. This places them at an increased risk of infection from malaria and other diseases.
- Vitamin A deficiencies result in the death of about 1 million children annually.

Solutions So Far

In 1998, several agencies, including MI, UNICEF, and WHO, began working with governments around the world to supply a preventative vitamin A supplement for children every six months. Within six years, almost 60 percent of populations worldwide had been reached.

Obstacles

- Geographical barriers make some areas hard to reach.
- More outreach and monitoring programs are needed.

Iodine

Iodine is needed for the normal metabolic functions of all body cells. It is also critical to thyroid function and the production of thyroxine (T_4). Iodine is found naturally in seafood, fish, kelp, and dairy products. A deficiency of iodine in developing fetuses and young children can lead to brain damage and developmental delays.

Impact and Scope of the Problem

- About 18 million babies worldwide are born with mental impairments due to iodine deficiency during fetal development.
- People in over 50 countries have iodine deficiencies.

Solutions So Far

In the last 20 years, organizations including WHO, MI, and the International Council for the Control of Iodine Deficiency Disorders (ICCIDD) have worked together to establish universal salt iodization (USI) programs. Because iodine is added to salt, iodine now reaches about 66 percent of households worldwide.

Obstacles

- There are limitations on production and supply of iodized salt.
- The enforcement of regulations is weak in some areas.
- Greater consumer awareness is needed.

Iron

Iron is an essential component of hemoglobin, the oxygen-carrying protein found in red blood cells. Dietary sources of iron include beans, red meat, eggs, and whole grains. Iron deficiency results in a condition called anemia, the symptoms of which include fatigue and an irregular heartbeat. Iron deficiency in children can lead to mental impairment.

Impact and Scope of the Problem

- Iron deficiency impedes mental development in over 40 percent of infants and young children in developing countries.
- Approximately 500 million women suffer from iron deficiency worldwide. This affects their energy and productivity levels and leads to poor health and lost earnings.

Solutions So Far

MI and WHO promote the distribution of micronutrient packets that include iron supplements, as well as the fortification of wheat and corn flour with iron and folic acid. MI has also begun an initiative to produce and distribute double fortified salt (DFS)—salt that is fortified with both iodine and iron.

Obstacles

- There is a need for increased nutrition education in certain regions.
- Sometimes it is difficult to balance supply and demand with food prices

Folic Acid

Folic acid is essential for the production of both red and white blood cells. It is also involved in building DNA. Folic acid is found in green leafy vegetables and dried beans. Children may be born with severe brain and spinal cord defects, referred to as neural tube defects (NTD), if a woman has folic acid deficiency during pregnancy.

Impact and Scope of the Problem

- Approximately 150 000 children per year are born with severe neural tube defects due to folic acid deficiency in the pregnant mother.

Solutions So Far

The Flour Fortification Initiative (FFI) is an international network of organizations working to make it a standard practice to fortify flour with micronutrients, including folic acid, iron, and zinc. Since 2004, work done by the FFI has led to a 12 percent increase in the amount of fortified flour produced, and increased the number of countries with national regulations for mandatory wheat flour fortification from 33 to 57. A target date of 2015 has been set to increase to 80 percent the amount of milled wheat flour fortified with folic acid.

Obstacles

- More countries need a national requirement for flour fortification.
- Prices must be kept low enough for people to afford the fortified flour.

Research and Analyze

1. Research more information about the Micronutrient Initiative. How does this agency develop, implement, and monitor programs aimed at eliminating vitamin and mineral deficiencies in expanding populations? Choose a single program of the MI and assess its success at creating a sustainable solution to the problem. What are the strengths and weaknesses of the program? What resources or actions are needed to improve the program? Could the program be used as a template in other areas of the world that are experiencing a similar problem? Why or why not? Summarize the results of your assessment in a brief report.
2. Research more information about the Flour Fortification Initiative. What are the goals of the initiative? What is Canada's role in the initiative? How effectively does this initiative meet its goal of nourishing expanding populations? Suppose you are a faculty member in the food sciences department at a university. The university is considering becoming a partner in the FFI. Prepare a brief report to either persuade or dissuade the university from proceeding with the partnership.
3. Golden rice, named for its colour, is rice that has been genetically modified to contain beta-carotene—a pigment that the body converts to vitamin A. Although golden rice is considered by many people to be part of the solution to vitamin A deficiency, regulatory requirements have slowed the product from reaching the market. Research more information about the current status of golden rice. What are the potential benefits and risks associated with its use? Do you believe golden rice should be used to help combat vitamin A deficiency? Why or why not? Justify your opinion.

Take Action

1. **PLAN** Choose one micronutrient and one obstacle associated with getting the micronutrient to a certain population. Recommend a sustainable solution to the obstacle. Be sure to consider appropriate social, economic, and environmental factors as you plan your solution.
2. **ACT** Suppose you are asked to present your plan to the board of directors at the Micronutrient Initiative. Create a presentation that explains your plan in a media format of your choice.

CANADIAN RESEARCH IN ACTION

Farming and Food Security in Honduras

Dr. Sally Humphries with a group of local women in Honduras

Poverty and famine affect many rural areas in Honduras. Per capita, there are more rural poor in Honduras than in almost any other Central American country. More than half of the population works the land to support their families. However, the best land in the valleys is devoted to industrial-scale farming, so subsistence farmers must often work the poor soil found on the steeply sloped land high above sea level. Honduras also faces severe weather events that frequently damage farmers' crops, such as hurricanes that deliver destructive rains and winds that rage at over 200 km/h. These challenges leave many farmers struggling to feed their families. Local farmers with the expertise to combat these and other challenges are relatively few, but for many years they have been assisted by a Canadian, Sally Humphries, whose commitment to ecology as well as to issues involving social justice, have helped to establish a new rural reality characterized by self-reliance and food security.

Dr. Sally Humphries is a professor with the Department of Sociology and Anthropology at the University of Guelph in Ontario. Humphries says, "My interdisciplinary background in social science and rural and agricultural development led to a Rockefeller Social Science Fellowship in Agriculture at the International Centre for Tropical Agriculture (CIAT). At CIAT I set up a program in Honduras in the early 1990s which subsequently evolved into La Fundacion para la Investigacion Participativa con Agricultores de Honduras or FIPAH." Through FIPAH, Honduran farmers work together in agricultural research teams to develop new varieties of crops in a process called participatory plant breeding. According to Humphries, "through farmer selection of plants with different traits and crossing, new varieties are developed which improve on local or *landrace* varieties." For instance, new varieties of corn were developed with improved adaptability to a variety of altitudes, shorter stature to reduce wind damage, and greater resistance to disease. The overall result was improved yield in a mountainous habitat.

Today FIPAH, includes over 80 research teams and 800 farmers. As a result of their involvement with FIPAH, many Honduran farmers have improved local crops varieties to such an extent that hunger has been significantly reduced.

Related Career

Crop consultants are professionals who provide agricultural advice based on their areas of expertise. Crop consultants may advise farmers on such diverse topics as plant research, farm waste management, and pest control. They use their scientific and technological knowledge to improve farm output and profitability, while also enhancing sustainable farming practices. Crop consultants often have varied educational backgrounds, including advanced science degrees in agriculture. They are generally certified by a professional organization, such as the Canadian Certified Crop Advisor Association in Elmira, Ontario.

QUESTIONS

1. Explain the nature of participatory plant breeding in your own words.
2. Farmers on small farms—in both developed and developing countries—tend to face similar challenges. Many small farmers are now working more cooperatively, sharing knowledge and methods, independent of government assistance or, in some cases, interference. What is the value of this approach?
3. Careers in agriculture focus on feeding people in Canada and around the world. Research one agricultural career that interests you. Write a brief paragraph explaining why you find this career interesting and appealing.

Chapter 12 SUMMARY

Section 12.1 Human Population Growth

Human population growth was very stable for most of history; however, the population grew exponentially after several technological advances.

Key Terms

demography
doubling time
population pyramid
ecological footprint
available biocapacity

Key Concepts

- Demography is the study of statistics of human populations, such as population size, density, distribution, movement, births, and deaths.
- Prior to modern technology, the human population was stable; however, it is growing exponentially now.
- The age structure of a population varies from country to country.
- The ecological footprint of industrialized countries is much greater than the ecological footprint of developing countries.
- Humans exceed the biocapacity of the world and are depleting many resources.

Section 12.2 Managing Growth and Resources

As the human population grows, more resources are required to sustain the population, which is taking a toll on Earth.

Key Terms

biomagnification
deforestation
sustainable
bycatch
biodiversity
overexploitation
minimum viable population size

Key Concepts

- Like all populations, the environment provides resources needed by the humans.
- Burning coal to provide energy creates toxins.
- Human activities cause environmental problems and many activities are unsustainable.
- Preserving biodiversity is important for the overall health of all populations on Earth.
- Human activities have decreased biodiversity.
- Many solutions are needed to provide the needs of a growing human population and at the same time, preserve the environment.

Chapter 12 REVIEW

Knowledge and Understanding

Select the letter of the best answer below.

1. A graph of the human population growth is similar to which of these?
   a. a stable population
   b. a declining population
   c. a population exhibiting logistic growth
   d. a population that has reached its carrying capacity
   e. a population growing without limiting factors

2. Which of the following contributes *most* to the current human population growth?
   a. decrease in birth rate
   b. increase in birth rate
   c. decrease in death rate
   d. increase in death rate
   e. decrease in death rate and increase in birth rate

3. A population pyramid is a graph, that shows the age structure of a human population. Which is *true* of a population pyramid in the shape of an inverted triangle?
   a. The population is stable.
   b. The birth rate is increasing.
   c. The population is growing slowly.
   d. The population has a short life expectancy.
   e. A large number of the population is past the reproductive years.

4. Which is *true* of human populations in different parts of the world?
   a. Their growth rates are similar.
   b. They may have different population profiles.
   c. Their per capita birth and death rates are the same.
   d. The graph of each population pyramid is rectangular.
   e. They are expected to contribute to the human population to the same degree.

5. Ecological footprint is
   a. the amount of productive land needed to support one person.
   b. made up primarily of housing requirements.
   c. dependent on the country's population.
   d. the same for everyone in the world.
   e. always equal to the biocapacity.

6. Scientists have attempted to calculate Earth's carrying capacity. Which of these did scientists find?
   a. ecological demand equals ecological supply
   b ecological demand is less than ecological supply
   c. ecological demand is greater than ecological supply
   d. ecological demand is proportional to ecological supply
   e. ecological demand and supply cannot be measured

7. Which type of energy reduces the amount of pollutants we contribute to our atmosphere?
   a. coal
   b. fossil fuel
   c. renewable
   d. metal
   e. oil

8. Which of these often occur when trees are removed from rainforests?
   a. The climate is regulated.
   b. Mudslides carry away the soil.
   c. The wealth of the country increases.
   d. The ecosystem becomes more stable.
   e. Most of the land is rich and used for farming.

9. The agricultural technique shown below

   a. creates a strong ecosystem.
   b. reduces yield but increases profit.
   c. is a system that rotates crops each year.
   d. promotes biological resistance among crops.
   e. makes crops susceptible to disease and pests.

10. Which country has the smallest ecological footprint?
   a. Australia
   b. Canada
   c. Germany
   d. India
   e. United States

11. Over the years, the destruction of forests has been associated with the rise of carbon dioxide in the atmosphere. How are the two related?
    a. The trees release carbon dioxide when they are cut.
    b. The process of cutting wood from trees adds carbon dioxide to the air.
    c. Fewer trees means less photosynthesis and therefore more carbon dioxide is left in the air.
    d. Respiration by the trees that remain adds more carbon dioxide than photosynthesis can remove.
    e. The soil that remains contains a great deal of carbon dioxide, which is released into the atmosphere.
12. Which is *not* a cause of loss of biodiversity?
    a. deforestation
    b. invasive species
    c. overharvesting
    d. pollution
    e. all are causes
13. Invasive species are introduced into new environments by which method?
    a. wind
    b. pollination
    c. human activity
    d. wind and human activity
    e. pollination, wind, and human activity
14. Which of the following can result from human activity?
    a. a decrease in the carrying capacity of Earth for humans
    b. an increase in the carrying capacity of Earth for humans
    c. an effect on the carrying capacity of Earth for other organisms
    d. both a and b.
    e. a, b, and c are true
15. Which population is *least* vulnerable to disease?
    a. A small population of trees.
    b. A population of ground squirrels in a field.
    c. A population of crops grown in monoculture.
    d. A population of deer with a road being built through its habitat.
    e. A population of ducks in which less resilient ducks die, reducing genetic diversity.

Answer the questions below.

16. What factors contributed to the human population growth over the last 300 years?
17. The birth rate of the human population has remained the same over the last few hundred years. Why then does the population continue to increase rapidly?
18. What has happened to the doubling time of the population over the years? Why does this present a problem?
19. How can differences in population pyramids explain differences in population growth rates between more developed countries and less developed countries?
20. Comparing the ecological footprint to Earth's biocapacity shows that ecological demand has exceeded ecological supply. However, when populations reach the maximum their environment can support, the population eventually stops growing. If this is true, explain why the human population continues to grow?
21. How do sulfur dioxide and nitrogen oxides get into water systems?
22. The kelp shown below is a seaweed that humans consume. Why should people be more concerned about pollutants in fish than in kelp?

23. How are the air and soil affected by clear cutting forests?
24. How has technology harmed the fishing industry?
25. What is one of the most dangerous waste products found in the sea? Why is it so dangerous?
26. Which parts of Canada do you think produce the most pollution? Explain why.
27. Some modern agricultural techniques increase crops dramatically. What problems can they cause?

28. What are the negative effects of the loss of biodiversity?

29. In what ways have humans altered the world in which we live?

Thinking and Investigation

30. Use the populations in the table below to draw your own graph of the world population and the population of Canada from 1750 to 2010. Compare the trends in the two graphs. Use the graphs to predict the world population and the Canadian population in 100 years. Explain your predictions.

Year	World Population (billions)	Canadian Population (millions)
1750	0.8	1.9
1850	1.2	2.4
1900	1.6	5.3
1950	2.5	13.7
1960	3.0	17.8
1970	3.6	21.3
1975	4.0	22.7
1980	4.4	24.5
1990	5.1	27.5
2000	6.0	30.7
2005	6.4	32.4
2010	6.8	34.2

31. In many developed countries, the average number of births per couple is 1.5 to 2.5. Assume that the average number of births per couple for the entire world falls to about 2 (to replace the parents). What do you think will happen to the world population?

32. Describe the shape of the population pyramid for a country with a relatively short life expectancy. Explain your reasoning. Predict what will happen to the size of the population over time.

33. Use the data showing the age categories in Canada in 2010 to construct a population pyramid. Your teacher will give you the website to use. Based on the pyramid you have constructed, predict the future of the Canadian population.

34. Assume that there are 13.5 births per 1000 individuals and 8.0 deaths per 1000 individuals in the United States. If the current population is about 310 million, calculate the population of the United States over the next five years. About how many people will be added to the population over that time?

35. The world population growth rate has decreased in recent years. Infer the reasons the rate has decreased.

36. If the average size footprint for each member of the human population is 2 hectares, calculate the number of people Earth should be able to support. How does this compare to the current population?

37. It is difficult for ecologists to determine Earth's true carrying capacity. What variables need to be considered when determining Earth's carrying capacity?

38. Burning coal produces many negative effects on our environment. Infer why burning coal continues even though it harms our environment.

39. Design a study that would allow you to determine the amount of carbon dioxide contributed by different types of cars.

40. What are two ways in which deforestation might be stopped.

41. Conduct research to find the amount of waste deposited by your city, town, or region to landfills each year. Suggest ways to reduce the amount of material.

42. Plastic is one of the most widely used materials in the world. Infer the reason why it is so widely used and why it so harmful at the same time.

43. Write a hypothesis concerning human impact on Earth's carrying capacity. Justify your hypothesis.

44. Create a model of a sustainable ecosystem. Explain what makes the ecosystem work. What would make your ecosystem unsustainable?

45. If the animals below were being transported to populate a newly discovered Earth-like planet, what are some of the factors that should be considered when determining the number of animals to be transported?

Communication

46. "The wealthiest 20 percent of the population consumes 86 percent of the world's resources and produces 53 percent of the world's carbon dioxide." With that statement in mind, write two sides to debate the following: Earth is overpopulated, which is causing stress on the planet. Write down your main points.

47. Use the graph in the Launch Activity to predict the structure of the population pyramid for each population. Justify your reasoning.

48. **BIG IDEAS** Technological developments can contribute to or help offset the ecological footprint associated with population growth and the consumption of resources. The burning of fossil fuels results in ozone that irritates our eyes and lungs, and lets harmful radiation reach the surface of Earth. Explain why coal-fired electricity generation should be stopped and suggest other energy sources as replacements.

49. You are asked to teach a class of fifth graders about pollution-free energy. Write the main points you want to cover.

50. List of all the ways humans use energy to obtain food.

51. About 80 percent of the world's population eats mostly grain-based food.
    a. Draw three different food chains for humans.
    b. Why do you think so many of the population eat mostly grain-based food?

52. When habitats are destroyed, many species become extinct before we have even discovered them. Use your knowledge of the advantages of high levels of species diversity to write a convincing argument to a country with high levels of biodiversity being lost, and convince them to preserve habitats and species.

53. Crops grown for the purpose of harvesting their seeds were introduced to Canada by early settlers. They are environmentally friendly and provide food for humans and animals. Garlic mustard is an invasive species brought by settlers as a medicinal plant. However, Garlic mustard suppresses native seedlings, like maple and ash.
    a. Compile a list of introduced species, and include how they were introduced and whether they are beneficial, invasive, or neither.
    b. What impact does each one have on native species? What methods are used to control them, if any? Why would they need to be controlled?

54. As long as there is a market for tropical timber, rainforests will continue to be destroyed. Write a campaign to educate consumers on the destruction of rainforests. Include alternatives.

55. **BIG IDEAS** The increased consumption of resources and production of waste associated with population growth result in specific stresses that affect Earth's sustainability. Research deforestation in Canada, or overharvesting fish in or near Canadian waters. Write a report on the problem. Do you think Canada should spend money to solve the problem? Explain your answer.

Application

56. Describe the population age structure in which HIV would spread most rapidly. Explain your reasoning.

57. In 1980, the population of China passed 1 billion. To limit the growth of the population, the Chinese government implemented the one-child policy that limited a large number of families to one child.
    a. What are the advantages of this policy?
    b. What are the disadvantages of this policy?

58. Over the last decade, there has been an increase in deforestation in the upper part of the Amazon. The height of the flood crest has also increased in this area. What suggestions would you make to the people in this area? Why?

59. Fish and other marine species are being removed from habitats faster than they can reproduce. Devise some strategies to help create sustainable fisheries.

60. The objects below are hard to break down, fill our landfills, and harm marine life.
    a. What makes these items popular with consumers?
    b. Describe some ways in which these items might harm terrestrial life.

61. What type of environmental problems are associated with cars? How can we reduce these?

Chapter 12 SELF-ASSESSMENT

Select the letter of the best answer below.

1. **K/U** A demographic study may include which of these?
   - **a.** population density
   - **b.** population distribution
   - **c.** population growth rates
   - **d.** population size
   - **e.** all of these
2. **K/U** Food availability, diseases, and predators
   - **a.** limit only animal populations.
   - **b.** limit only the human population.
   - **c.** limit all populations, except humans.
   - **d.** limit all populations.
   - **e.** do not limit any population.
3. **K/U** Which is the result of burning fossil fuels?
   - **a.** carbon dioxide in the air
   - **b.** toxins in our oceans
   - **c.** acid rain and smog
   - **d.** mercury in fish
   - **e.** all of these
4. **K/U** The world-wide human population most closely resembles which growth pattern?
   - **a.** exponential
   - **b.** infinite
   - **c.** linear
   - **d.** logistic
   - **e.** stable
5. **K/U** Which process does the diagram below illustrate?

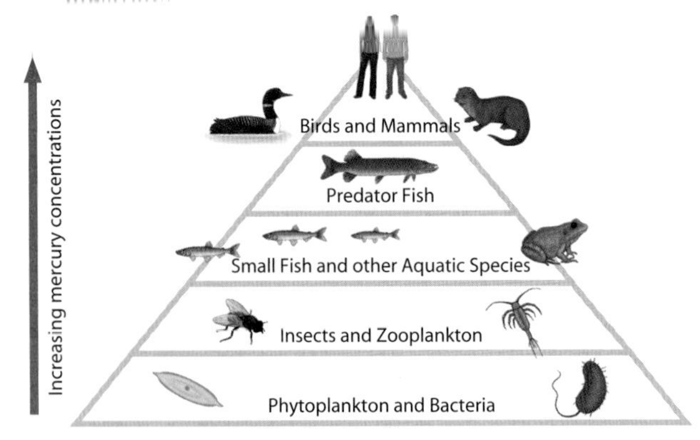

   - **a.** biomagnification
   - **b.** monoculture
   - **c.** overexploitation
   - **d.** pollution
   - **e.** toxicity
6. **A** Which would you expect of the ecological footprint of the average Canadian?
   - **a.** It is less than the ecological footprint of the average Ethiopian.
   - **b.** It is greater than the ecological footprint of the average Kenyan.
   - **c.** It is greater than the ecological footprint of the average American.
   - **d.** It is less than the ecological footprint of the average person in the world.
   - **e.** It is the same as the ecological footprint of the average person in the world.
7. **K/U** Which is most likely the shape of the population pyramid of a developed country in which the population is growing?
   - **a.** a rectangle
   - **b.** a triangle
   - **c.** an inverted triangle
   - **d.** a broad base and a narrow centre
   - **e.** a narrow base and a broad centre
8. **K/U** In most cases, what is the source of the matter and energy that humans need to survive?
   - **a.** their home
   - **b.** the entire world
   - **c.** their own country
   - **d.** abiotic factors
   - **e.** the continent on which they live
9. **K/U** Which is one way to improve the health of the world's oceans?
   - **a.** reduce bycatch
   - **b.** increase trawling
   - **c.** remove predator fish
   - **d.** create more fish farms
   - **e.** catch only a few species
10. **K/U** Which is most likely to occur when organisms ingest plastics?
    - **a.** The female organisms are defeminized.
    - **b.** The plastics decompose and are eliminated.
    - **c.** The organisms have reproductive problems.
    - **d.** The plastics help to remove other toxins.
    - **e.** All are likely to occur.

Use sentences and diagrams as appropriate to answer the questions below.

11. C Write a short letter to Fisheries and Oceans Canada to explain why it is important to assist with the cleanup of garbage dumps in the world's oceans.

12. A There is a quote that says, "History repeats itself." If this also pertains to the human population, then the human population should stabilize at some point in the future and then experience rapid growth. With the quote in mind, explain the factors that would lead to this rapid growth.

13. T/I Draw a graph that represents the human population over history. Explain the graph.

14. T/I The table below shows the demographic information for two countries. Compare the population size at the start and after two years. Show your work.

Country	Population Size (millions)	Annual Number of Births per 1000 Individuals	Annual Number of Deaths per 1000 Individuals
Country A	21	8	12
Country B	19	16	8

15. T/I Explain the term ecological footprint. Infer the reasons for the difference in ecological footprint for different countries.

16. C Make a list of some of the features of a sustainable world.

17. T/I Draw a bar graph showing the age populations of a fictitious country that is growing rapidly. Explain how you can use your graph to tell that the population is growing rapidly.

18. T/I How do you think scientists determine whether the human population has exceeded the available biocapacity? What assumptions must scientists make when making this determination?

19. C Make a diagram to show how burning fossil fuels result in mercury toxins in humans.

20. A Tropical forests, like the one shown below, are disappearing quickly as humans clear them to make room for farms, pastures, roads, and to harvest timber. Large expanses of forests are often chopped into smaller blocks. Describe the risks to the plants and animals in the fragments of forest that remain.

21. A Rebuilding of fish stocks in oceans requires the understanding of the influences that alter marine communities. Propose different ways in which fish stocks can be rebuilt.

22. A Why is there such a strain on Earth's resources? How can we solve this problem?

23. T/I In what way do humans try to control biodiversity when farming? What techniques do you think farmers could use to increase biodiversity when growing crops?

24. C Complete the table below to compare the population age structure for the three countries.

Country	General Age Structure	Pyramid Shape	Future Growth
Kenya			
Sweden			
Germany			

25. T/I If you tested animals with high levels of methylmercury, what biological changes would you expect to see?

Self-Check

If you missed question...	1	2	3	4	5	6	7	8	9	10	11	12	13	14	15	16	17	18	19	20	21	22	23	24	25
Review section(s)...	12.1	12.1, 12.2	12.2	12.1	12.2	12.2	12.1	12.1	12.2	12.2	12.2	12.1, 12.2	12.1	12.1	12.1, 12.2	12.1	12.1	12.1	12.2	12.2	12.2	12.1, 12.3	12.2	12.1	12.2

Unit 5 Project

An Issue to Analyze

Minimizing Negative Effects of Human Population Growth on Other Species

Demand for resources to meet the needs of humans and inattentiveness to the stability of ecosystems have led to overexploitation of many species. Increased consumption of resources and increased waste production by humans are activities that affect Earth's sustainability. Is human population growth pushing the planet past its sustainable limits? How can the needs of a growing human population be met while at the same time preserving biodiversity and preventing the decline or possible extinction of other species? In this project, you will research and analyze the effects of human population growth on three different populations of organisms: one local species that is endangered, one endangered species whose habitat is in another country far from where you live, and one species whose population decline has been halted through human intervention. You will then determine what personal actions you and others in your community might take to help protect the endangered species you have investigated.

What are some issues related to the effects of human population growth on populations of other species, and how can human lifestyle changes minimize the negative effects of this growth on Earth's ecosystems and biodiversity?

Initiate and Plan

1. Choose one local species that is either endangered or that is undergoing a rapid decline in population. A partial list of species for Ontario includes the honeybee (*Apis mellifera*), monarch butterfly (*Danaus plexippus*), polar bear (*Ursus maritimus*), eastern cougar (*Puma concolar couguar*), sturgeon (*Acipenser sturio*), spiny softshell turtle (*Apalone spinifera*), Massasauga rattlesnake (*Sistrurus catenatus*), barn owl (*Tyto alba affinis*), the small white lady's slipper (*Cypripedium candidum*), swamp rosemallow (*Hibiscus moscheutos*), spotted wintergreen (*Chimaphila maculata*), and American chestnut tree (*Castanea sativa*).
2. Choose one species that is endangered or that is undergoing a rapid population decline elsewhere in the world—for example, coral (*Class Anthozoa*), any subgroup of Asian elephant (*Elaphus maximus maximus*), various fish species (especially those used for food), or the Sumatran tiger (*Panthera tigris sumatrae*).
3. Choose one endangered species whose populations have recovered due to human efforts—for example, the peregrine falcon (*Falco peregrinus*), the California condor (*Gymnogyps californianus*), the whooping crane (*Grus americana*), the plains bison (*Bison bison bison*), or the white rhinoceros (*Ceratotherium simum*).
4. For each species, plan how you will find reliable and unbiased information that will help you examine the population dynamics that led to its decline.

The small white lady's slipper is an endangered species in Ontario.

Human efforts have saved the peregrine falcon from extinction.

Perform and Record

1. Research and describe the characteristics of each of the three populations you have chosen, including growth, density, fecundity, mortality, distribution, and minimum viable size.
2. Examine the fluctuations of the population of each species and describe factors, such as carrying capacity and predation, that may have caused the fluctuations.
3. Investigate the probable causes for the species' population decline. What role, if any, has human population growth played in this decline?
4. Conduct research to learn about conservation groups that are taking action to mediate the needs of expanding human populations and the needs of other species. Three of these organizations are Conservación y Desarrollo in Ecuador, the Rainforest Alliance, and the World Wildlife Fund. What efforts are underway to reach a compromise between the needs of humans and those of other species?

Analyze and Interpret

1. Determine the three major pressures on the population of each endangered species and summarize details related to those pressures in a graphic organizer of your choice.
2. Assess what types of human activities are placing each species at increased risk of extinction.
3. Based on your research findings, evaluate the success of various actions on the part of conservation groups to increase the survival rate of an endangered local species or an endangered species in another part of the world.
4. Decide on personal actions that you can undertake to help an endangered local species survive and co-exist with the human population in your area.
5. Decide on personal actions that you can undertake to help your chosen endangered species in a different part of the world to survive.

Communicate Your Findings

6. Create a short presentation (for example, a speech, a web page, or a podcast) explaining practical and effective actions that students can undertake at home or at school to help preserve biodiversity for at least one endangered species. For example, create a list of local plants that are attractive to birds, bees, and butterflies for inclusion in a community garden or school flowerbed, or prepare a list of invasive plants such as purple loosestrife that need to be monitored and controlled.
7. Prepare and present a short report relating how the actions that you undertake could help protect an endangered species elsewhere in the world.
8. Share your action plans with at least three people in your community who are not classmates. How do their perceptions of these issues compare with those of your classmates? Conclude your report with a statement about whether your point of view regarding the effects of human population growth on other species may have changed in the course of doing this project, and if so, how.

Assessment Criteria

Once you complete your project, ask yourself these questions. Did you…

- ✓ **K/U** research information about the pressures on both a local species and an endangered species in another part of the world?
- ✓ **T/I** evaluate your information sources for bias?
- ✓ **K/U** determine the relationship between human population growth and the population growth or decline of your chosen species?
- ✓ **T/I** assess what types of human activities are placing each species at risk of extinction?
- ✓ **T/I** evaluate the effectiveness of various actions on the part of conservation groups to protect endangered species?
- ✓ **A** make recommendations for practical and effective personal actions to protect endangered species?
- ✓ **C** present viable action plans to your class and to at least three people other than those in your class?
- ✓ **C** use scientific terminology, appropriate to both your purpose and your audience, in your presentations?

UNIT 5 SUMMARY

BIG IDEAS

- Population growth follows predictable patterns.
- The increased consumption of resources and production of waste associated with population growth result in specific stresses that affect Earth's sustainability.
- Technological developments can contribute to or help offset the ecological footprint associated with population growth and the consumption of resources.

Overall Expectations

In this unit you learned how to…

- **analyze** the relationships between population growth, personal consumption, technological development, and our ecological footprint, and **assess** the effectiveness of some Canadian initiatives intended to assist expanding populations
- **investigate** the characteristics of population growth, and **use models** to calculate the growth of populations within an ecosystem
- **demonstrate** an understanding of concepts related to population growth, and **explain** the factors that affect the growth of various populations of species

Chapter 11 Describing Populations and Their Growth

Key Ideas

- Ecologists use quantitative measurements to study and describe populations, such as population size and population density.
- Ecologists use three distribution patterns to describe populations: uniform, random, and clumped.
- Ecologists use life histories, which are quantitative measures of vital statistics that determine population size to aid in understanding populations.
- The factors that change population size are: births, deaths, immigration, and emigration.
- Two population growth models are exponential growth and logistic growth.
- All habitats have a carrying capacity, which is the maximum population size that it can support.
- Organisms usually exhibit two types of life strategies: *r*-selected strategy or *K*-selected strategy.
- There are three types of symbiotic relationships: parasitism, mutualism, and commensalism.

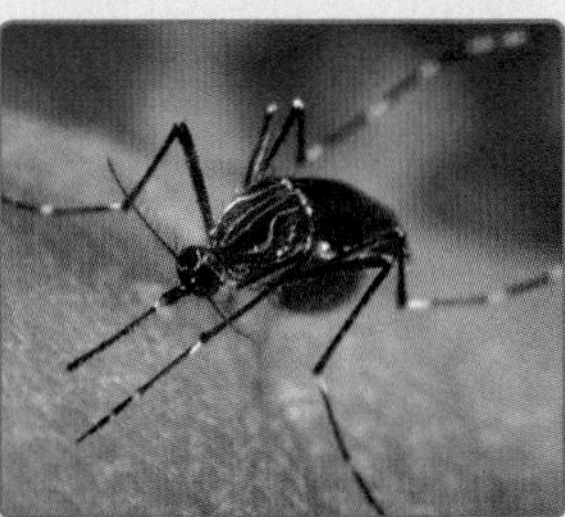

Chapter 12 Human Populations

Key Ideas

- Demography is the study of statistics on human populations, such as population size, density, distribution, movement, births, and deaths.
- The age structure of a population varies from country to country.
- The ecological footprint of industrialized countries is much greater than the ecological footprint of developing countries.
- Humans exceed the biocapacity of the world and are depleting many resources.
- Like all populations, the environment provides resources needed by humans.
- Human activities cause environmental problems and many activities are unsustainable.
- Human activities have decreased biodiversity.
- Many solutions are needed to provide the needs of a growing human population and at the same time, preserve the environment.

Knowledge and Understanding

Select the letter of the best answer below.

1. A biology student wants to determine the population of sea gulls on a small island in the middle of a lake. Which is the best method for the student to use?
a. mark-recapture
b. nest monitoring
c. quadrats
d. scats
e. transects

2. In a population that exhibits a uniform distribution of individuals
a. there is a high level of cooperation among individuals.
b. a high level of immigration from other populations is likely.
c. the population is most likely well below its carrying capacity.
d. the interactions among individuals are neither positive nor negative.
e. features that maximize the ability to compete for resources are likely to be naturally selected.

3. When individuals of a population move from one region to another and settle permanently in the new region, they are considered
a. emigrants of the new region
b. immigrants of the new region
c. migrants of the new region
d. both immigrants and emigrants of the new region
e. both migrants and immigrants of the new region

4. Zero population growth can be reached only if which of these occur?
a. the death rate is increased
b. the birth rate is reduced to zero
c. the population becomes subject to density-dependent factors
d. the birth rate and death rate are equal for a long period of time
e. the population becomes subject to density-independent factors

5. The difference between a population that is represented by a logistic growth curve and one that is represented by an exponential growth curve results primarily from which of these?
a. abiotic factors
b. biotic factors
c. environmental factors
d. life history
e. symbiotic relationships

6. Biotic potential is the highest possible per capita growth rate for a population. Which of the following would limit the biotic potential of an organism?
a. limited resources
b. numerous reproductive cycles in a life span
c. numerous offspring per reproductive cycle
d. a high percentage of offspring surviving long enough to reproduce
e. All of these would limit the biotic potential.

7. On the graph below, what does the dashed line at the top indicate?

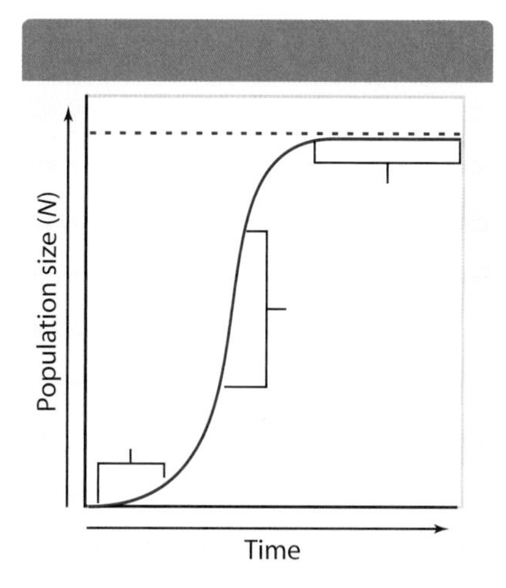

a. carrying capacity
b. growth rate
c. lag phase
d. linear growth
e. rapid growth

8. Life strategies of various populations can be represented as *r*-selected or *K*-selected. Which of the following is *not* a characteristic of a *r*-selected population?
a. The growth rate is unpredictable.
b. The organisms have a short life span.
c. The death rate of the population is very high.
d. The population size rarely exceeds the carrying capacity of its environment.
e. All of the above are characteristic of a *r*-selected population.

9. The growth of a population of paramecia is limited by heating the culture. Which of these is true?
a. Heat is a density-dependent limiting factor.
b. Heat is a density-independent limiting factor.
c. Paramecia grow at its biotic potential when heated.
d. The temperature was optimal for the growth of the paramecium.
e. None of these are true.

10. Which of these can be inferred from the two scenarios below?
Scenario 1: Two species of barnacles occupying the same habitat resulted in the survival of both species, although the density of one was greater than the other.
Scenario 2: Two species of trout occupying the same habitat resulted in the local extinction of one of the species.
   a. Scenario 1: The ecological niches of the two species were similar.
   Scenario 2: The ecological niches of the two species were very dissimilar.
   b. Scenario 1: The ecological niches of the two species were very dissimilar.
   Scenario 2: The ecological niches of the two species were similar.
   c. Scenario 1: The ecological niches of the two species were similar.
   Scenario 2: The ecological niches of the two species were overlapping, but not completely similar.
   d. Scenario 1: The ecological niches of the two species were overlapping, but not completely similar.
   Scenario 2: The ecological niches of the two species were very similar.
   e. Nothing can be inferred about the ecological niches.

11. *Paradoxophyla palmata* is a frog that is native to Madagascar. The colour of its body matches mud and tree trunks, which camouflages the frog from its predators. Camouflage is a type of protective mechanism known as
   a. co-evolution
   b. competition
   c. deceptive behaviour
   d. mimicry
   e. protective colouration

12. Which is true regarding the pattern in the world's population growth?
   a. There is exponential growth.
   b. There is a change from logistic growth.
   c. There is an explosion due to technology.
   d. The pattern has changed over the last 400 years.
   e. All of the above are true.

13. Which term is *not* related to the other terms?
   a. commensalism
   b. symbiosis
   c. mimicry
   d. mutualism
   e. parasitism

14. The population whose age pyramid is shown below does not have which of these?

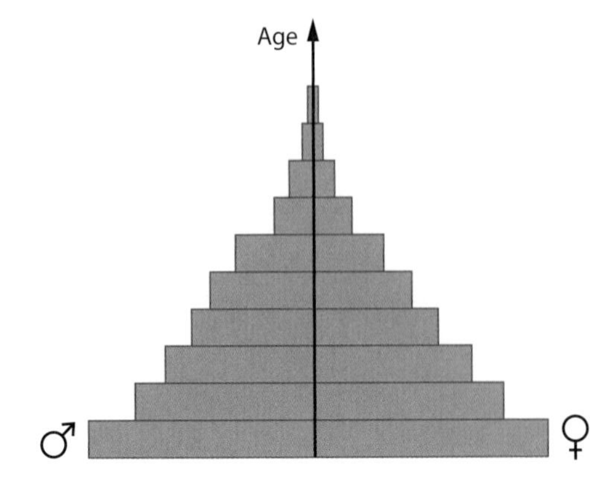

   a. a high birth rate
   b. a high death rate
   c. a long life expectancy
   d. a large number of children
   e. a high percentage of young people

15. Which of these is not a threat to biodiversity?
   a. bycatch
   b. deforestation
   c. genetic diversity
   d. invasive species
   e. pollution

16. Which of these practices can help to build a sustainable society?
   a. use resources efficiently
   b. [illegible]
   c. take steps to return ecosystems to their states prior to degradation
   d. none of the above
   e. a and c only

Answer the questions below.

17. Describe the factors that give rise to each of the three distribution patterns.

18. Garden phlox is a flowering plant, which can be represented with a Type III survivorship curve. What does the survivorship curve tell you about the life history of this species?

19. What does the growth rate of a population measure?

20. What is the typical shape of the growth curve for most populations? How is the biotic potential of the population represented on a graph?

21. How is interspecific competition different from intraspecific competition?

22. A parasitic mistletoe plant attaches its roots to an oak tree and absorbs nutrients from the tree. What effect will the mistletoe have on the oak?

23. Why is it possible to predict whether a population pyramid belongs to a less-developed country or a wealthy industrialized country?

24. The human population growth rate has declined in recent years. Give reasons for this.

25. What areas of Earth are used in determining its available biocapacity?

26. If renewable forms of energy are not utilized on a much greater scale in the future, what impact would this have on the health of the human population?

27. List three consequences of the expanding size of the human population.

28. Does a monoculture technique contribute greatly to biodiversity in an ecosystem? Why or why not?

29. Why are ecosystems with great biodiversity more stable than ones with less biodiversity?

30. What is the usual end result of overexploitation?

31. How do toxins, such as mercury and pesticides, affect the environment?

Thinking and Investigation

32. If you wanted to give the description of the population of mice in a grassy field, what measurements would you use?

33. Environment Canada is concerned about the declining population of burrowing owls (*Athene cunicularia*), shown below, in the prairie provinces. You are hired to estimate the population of burrowing owls in a 200 square kilometre grassland region of Saskatchewan.

   a. Describe the method you would use to estimate the population size.
   b. Show a table containing sample data.
   c. Use the sample data in your table to estimate the population size. Include any calculations you made.

34. How is population density defined? If a population of 1500 mice occupies a 100 m by 200 m field, what is the density of the population?

35. You want to find the population of cattails in a 10 000 m^2 marsh. You use ten quadrats that are each 1 m by 1 m and count 150 cattails altogether. What is the population of cattails in this marsh?

36. An ecologist is trying to determine the population of white birch in a forest in northern Saskatchewan. What technique should the ecologist use? Explain why you chose this technique. Write a procedure for the biologist to use to determine the population of white birch in this area.

37. The number of elephants in a population changed from 15 916 in 2003 to 15 872 in 2007. What was the growth rate of the elephant population from 2003 to 2007?

38. A population of swift foxes was under observation for two years. The per capita growth rate of this population was 3.1 percent and the change in the size of the population at the end of the study was 14. What is the final number of foxes in the population?

39. A population of 500 fish faced a problem of biological magnification resulting in a large number of deaths. But 200 births also took place and 27 fish immigrated into the population while 15 fish migrated out. How many deaths took place in the population if the final population number was 350?

40. A scientist is studying the effects of water, light, and temperature on the growth of certain populations in an ecosystem. What do you think the scientist will find? Explain your reasoning.

41. You are monitoring the growth of *E. coli* bacteria growing under ideal conditions in a Petri dish. You are also monitoring *E. coli* growing under conditions that simulate an animal's intestine. Draw a graph to show the results you expect. Explain your graphs.

42. A biologist is conducting fertility experiments on a population of rats in which a proportionately large number of rats are not yet mature. Each pair in the population will reproduce only once in its lifespan. Predict what will happen to this population of rats. Explain your prediction.

43. Use the Internet to find data for the ages of the world population. Construct a population pyramid. Make a prediction for the future of the world population based on your pyramid. Justify your prediction.

44. If you were creating a population pyramid for a population of wolves, what type of measurements would you take? Describe both quantitative and qualitative measures. Why would you take these measures?

45. The population of a city is growing at a rate of 2.5 percent yearly. The population today is 85 000. What will be the population in two years?

46. Estimate the doubling time of the Canadian population, if the population is now 34 million and has a growth rate of 1 percent.

47. Your teacher asks you to estimate the average ecological footprint of Canadians. Write a detailed procedure to estimate the ecological footprint. Why might your estimate differ from a student's in the Yukon Territory?

48. Describe a hypothesis to explain the population cycles of the snowshoe hare and the Canada lynx.

49. Scientists are researching ways to reduce or control an invasive algal species, *Caulerpa taxifolia*, spreading along the floor of the Mediterranean. The area supports a native species of mollusk in low numbers, which feed slowly on the alga. What steps do you suggest that the scientists take to resolve this problem? Can you predict any possible dangers that may arise from your solution?

50. As the population grows, especially in dense urban areas, there is an increased risk of transmission of conta[illegible] the pandemic avian flu have raised international concerns in the past. Devise a study that could determine the rate of transmission for a contagious disease.

51. Your teacher asks you to conduct an investigation to determine how overcrowding affects the growth and development of bean plants.
   a. What do you think you would need to do in this investigation?
   b. How should overcrowding affect the plants?

Communication

52. Into what categories can you classify waste? Explain your answer.

53. Humans play a large role in introducing organisms to new habitats, where they did not exist before. List a few organisms that were introduced to Ontario by humans, describing how they were introduced. What is the consequence to the environment?

54. **BIG IDEAS** Population growth follows predictable patterns. The Vancouver Island marmot (*Marmota vancouverensis*), shown below, is an important part of Canada's biodiversity and our most endangered animal. Conduct research on this species. Put together snapshots that will tell a story about the marmots. You may want to include population size, distribution patterns, life history, and so on.

55. Use a Venn diagram to compare and contrast techniques you can use to estimate population size.

56. Draw a poster to show the distribution patterns of three different species. Make sure to represent each distribution pattern.

57. A species of rabbits shares the same physical environment as a species of tortoises. The tortoises take time to grow a hard shell as a shield from predators and develop a slow metabolism to survive for periods without food. The rabbits are easily killed by predators and by temporary periods without food. Use a table to compare the reproductive strategies of the rabbits and tortoises living in this environment.

58. Use a graph to illustrate the variations in numbers you would expect to see in a population of rabbits and foxes living in the same habitat. Explain your graph.

59. Draw a three-panel illustration that will help a Grade 5 student understand Batesian mimicry.

60. Write your own definition of symbiosis. Show illustrations of the three forms of symbiotic relationships.

61. Use diagrams to illustrate the general structure of a rapidly growing population, a stable population, and a declining population.

62. A Grade 5 student asks you, "What is biodiversity?" How would you explain biodiversity in your own words to him or her?

63. Make a list of the main causes of the loss of biodiversity on Earth today.

64. If you were hired to help world leaders create a plan that will help to slow the human population growth, what would you tell them? Write a letter that outlines your main points.

65. **BIG IDEAS** Technological developments can contribute to or help offset the ecological footprint associated with population growth and the consumption of resources. You are a conservationist who has the task of convincing people on Earth to preserve biodiversity and use sustainable practices. You decide to help people see what it would be like on Earth in the year 2100, if they continue to use up resources. Describe Earth in the year 2100 due to unsustainable practices.

66. A kindergarten teacher has asked you to help to teach his class about problems caused by the vast amounts of waste humans produce. You decide to put on a puppet show to teach students about this topic. Write a script for a short puppet show. Be sure to include solutions to the problems.

Application

67. Random distribution patterns are rarely found in nature. Explain why.

68. An ecological community generally changes over time as different plants get established. The first species to invade an area and begin this process usually use *r*-selected strategies. What are the reproductive strategies of *r*-selected species that make them ideal for this role? Explain your reasoning.

69. What is meant by a "development-reproduction trade-off"? Conduct research to find and assess one example of such a trade-off.

70. Why does the flu spread more rapidly among office workers than among people who work from home?

71. Grizzly bears (*Ursus arctos*) are sexually mature when they reach five years of age. When resources are abundant, females average two cubs per litter every other year. They reproduce less when resources are scarce. How is the reproductive strategy of grizzlies a challenge for people working to conserve the grizzly bear population?

72. Mercury is a well-known environmental pollutant.
   a. Name some sources of mercury pollution.
   b. How does mercury harm the environment?

73. Aphids feed on plant sap by piercing the plants with their mouthparts. This causes damage to the plant. The plant sap is rich in carbohydrates but deficient in amino acids. Since aphids cannot synthesize amino acids, bacteria inside the aphid's cells synthesize amino acids as the aphid supplies the bacteria with energy. Explain these symbiotic relationships.

74. Waste is one of the major environmental concerns in our society. Explain why.

75. How could fertilizers that run off into a pond affect the pond's ecosystem?

76. Identify three limiting factors that you think might slow down the growth of the human population. Explain how these factors would slow the population growth.

77. How might the leaders of a country use population pyramids to predict their economic future?

78. How can countries with very diverse ecosystems, such as a tropical rainforest, benefit economically without destroying the ecosystems?

79. The Amazon Rainforest has great biodiversity, but may become a desert. Explain why.

80. Conduct research on organic pollutants, such as PCBs and dioxins. Describe what they are and explain their hazards.

81. **BIG IDEAS** The increased consumption of resources and production of waste associated with population growth result in specific stresses that affect Earth's sustainability. Explain how further urbanization in Ontario can affect environmental conditions in the province.

UNIT 5 SELF-ASSESSMENT

Select the letter of the best answer below.

1. K/U Which of these is not a limiting factor for populations?
   a. biotic potential
   b. density-dependent factors
   c. density-independent factors
   d. interspecific competition
   e. intraspecific competition

2. K/U Species with *r*-selected life histories
   a. have relatively long life spans.
   b. live close to the carrying capacity.
   c. have a fairly stable population size.
   d. have a high population growth rate.
   e. produce few offspring at each reproductive event.

3. K/U Age pyramids are histograms that show the age structure of a human population. Which of the following statements is true regarding a population that is represented by an inverted triangular pyramid?
   a. The population has low natality and high mortality.
   b. The population has high natality and high mortality.
   c. The population is growing slowly.
   d. The population is shrinking.
   e. The population is stable.

4. K/U Biodiversity provides the human population with
   a. psychological benefits **d.** b and c only
   b. medicinal benefits **e.** a, b, and c
   c. stable ecosystems

5. K/U If a certain population has a great deal of predators
   a. the individuals should stop breeding altogether.
   b. the individuals should invest a lot of time in raising their offspring.
   c. the individuals should produce few offspring and invest little in any of them.
   d. the individuals should invest little energy in reproduction to maximize their survival.
   e. the individuals should invest greatly in reproduction since they may not survive to another breeding season.

6. K/U The number of organisms of a certain species that a habitat can support is the habitat's
   a. biotic potential **d.** exponential growth
   b. carrying capacity **e.** density
   c. logistic growth

7. K/U The bluestreak cleaner wrasse (*Labroides dimidiatus*) gets energy by eating parasites from the scales and gills of large fish, such as the blue-spotted rock cod (*Cephalopholis miniata*). Although the wrasse is suitable prey for the larger fish, it is not eaten. This is an example of

   a. commensalism **d.** a and b
   b. mutualism **e.** b and c
   c. symbiosis

8. K/U An original count of a deer population in an Alberta forest determined that 238 deer were present. In a later study, this number increased to 431. Using this data, one could determine the
   a. births in the population.
   b. deaths in the population.
   c. density of the population.
   d. growth rate of the population.
   e. per capita growth rate of the population.

9. K/U Which of these has affected the human population growth most?
   a. diseases **d.** natural resources
   b. technology **e.** lack of predators
   c. high birth rates

10. K/U Dolphins are social animals. In which distribution pattern would you expect to find dolphins?
    a. clumped **d.** clumped and random
    b. random **e.** random and uniform
    c. uniform

Use sentences and diagrams as appropriate to answer the questions below.

11. C Use a diagram to illustrate resource partitioning. Explain how interspecific competition leads to resource partitioning.

12. T/I Gardeners must consider the effects of both interspecific and intraspecific competition when planning out their gardens. Design an experiment that will demonstrate to gardeners the effects of interspecific and intraspecific competition.

13. A Namibia is a southern African country with a border on the Atlantic Ocean. Why does the ecological footprint of the average person in Namibia differ from the ecological footprint of the average person in Canada?

14. T/I An unusual cold spell over a four-year period in an area supporting a population of weasels was found to be responsible for the drop in the size of the population from 27 to eight individuals over this time period. What is the per capita growth rate of the population of weasels?

15. T/I The density of bees in a garden is $13/m^2$. The area of the garden is $175\ m^2$. What is the population of bees in the garden?

16. A Draw three population pyramids. The first one should show rapid population growth. The second one should show slow population growth and the third should show negative population growth. Label each pyramid.

a. How would a decrease in birth rate affect each population? Explain your answer.

b. Suggest how the age structure of each population could affect the economic well-being of the country.

17. C Use a bar graph to illustrate the proportion of coal, natural gas, and renewable resources we use to generate energy in today's society. Use an illustration to show the proportion that you would consider "ideal." Explain why you consider this ideal.

18. T/I A researcher wanted to determine the population of shorebirds on a small island. The researcher caught 120 shorebirds, and then tagged and released them. Six months later, the researcher caught 65 shorebirds and noted that 8 of them were tagged. The researcher used the data to determine that the population of shorebirds is 975. Explain any flaw in the researcher's procedure.

19. T/I An ecologist is studying a population of 100 plants growing in its natural environment. The ecologist determined that the growth rate of the population is 8 percent. Determine the population of plants after five years.

20. C Moose (*Alces alces*) are the largest species in the deer family. They are solitary animals with few natural predators. Their resources are plentiful during the summer, but scarce during the winter. Draw the distribution patterns for which you would expect to find moose during the summer months and during the winter months.

21. A DDT is a pesticide that was widely used in agriculture. However, agricultural use has been banned in most developed countries. What impact does DDT have on the environment when it is used?

22. A We should always expect some kind of consequence when we alter nature. Explain this statement using examples.

23. T/I Design an experiment to tell whether two species that are very similar in appearance are examples of Batesian mimicry or Müllerian mimicry.

24. C Illustrate the growth of a population of insects during the summer months. Explain why the population has this growth pattern and what happens to the population after the summer months.

25. T/I Suppose an ecologist introduced a population of deer to a small island where resources were plentiful and the deer had no predators. The graph below is created to show the deer population after 40 years. Interpret the graph. What conclusions can you draw from the graph?

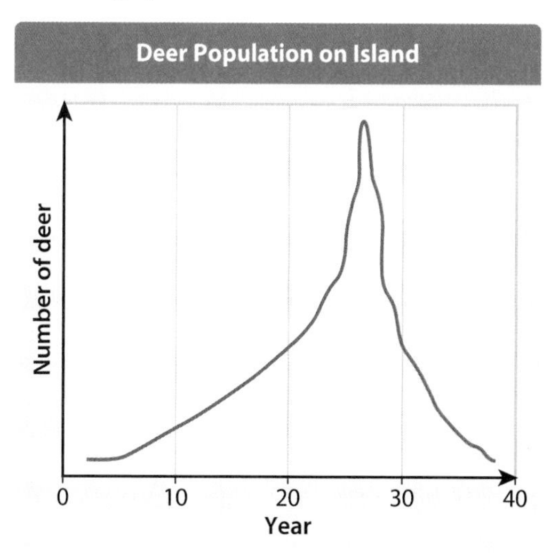

Self-Check

If you missed question...	1	2	3	4	5	6	7	8	9	10	11	12	13	14	15	16	17	18	19	20	21	22	23	24	25
Review section(s)...	11.3	11.2	12.1	12.2	11.3	11.2	11.3	11.2	12.1	11.1	11.3	11.3	12.1 12.2	11.2	11.2	12.1	12.2	11.1	11.2 12.1	11.1	12.2	12.2	11.3	11.2 11.3	11.2

Guide to the Appendices

Guide to the Appendices

Analyzing STSE Issues

STSE is an abbreviation for **s**cience, **t**echnology, **s**ociety, and the **e**nvironment. In *Biology 12,* you are frequently asked to make connections between scientific, technological, social, and environmental issues. Making such connections often involves, for example, assessing the impact of science on technology, people and other living things, and the environment. Analyzing STSE issues involves researching background information about a problem related to science, technology, society, and the environment; evaluating differing points of view concerning the problem; deciding on the best response to the problem; and proposing a course of action to deal with the problem.

The following flowchart outlines one process that can help you to focus your thinking and organize your approach to analyzing STSE issues. The most effective analyses result in decision making and, ultimately, an action plan. Group discussion and collaborative analysis can also play a role in analyzing an STSE issue.

A Process for Analyzing Issues

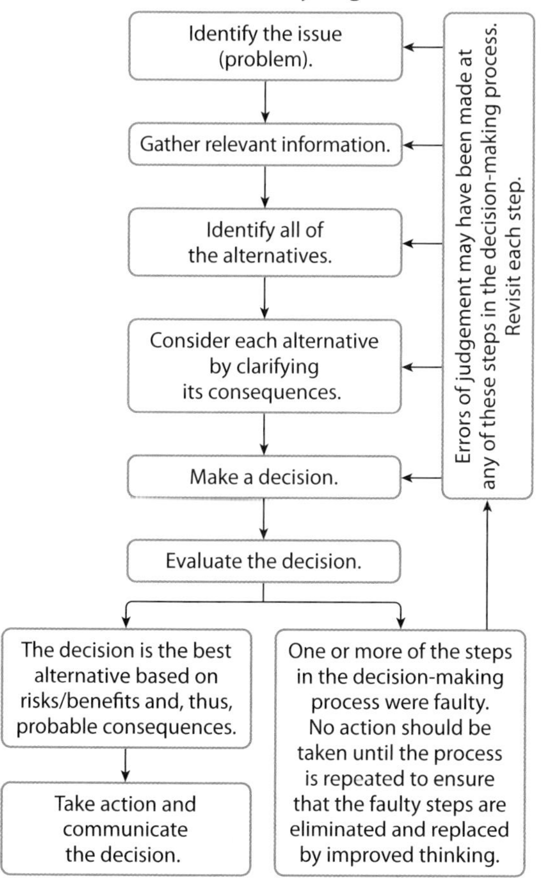

Identify the Issue (Problem)

An STSE issue is a topic that is debatable—it can be viewed from more than one perspective. When you encounter an issue related to STSE, such as a medical breakthrough, a health-care policy, or an environmental regulation, you need to try to understand it from multiple points of view.

Suppose you have learned that certain genetically modified crops have higher resistance to pests and diseases, leading to improved crop yields.

Assess whether there is any controversy associated with this situation. Could there be different viewpoints concerning the cause of the situation and how to respond to it?

You read a blog discussing health and environmental concerns related to the production and use of genetically modified crops. You decide that this situation does represent an STSE issue, because there are both benefits and risks that need to be considered.

Try to sum up the issue in a specific question.

For example, can or do government regulations ensure that the benefits of genetically modified organisms (GMOs) outweigh the risks?

Gather Relevant Information

You will need to do some research to gain a better understanding of the issue. Go to Developing Research Skills in Appendix A for help with finding information.

For example, what current Canadian regulations are in place to ensure that GMOs are safe to consume? What are the most common GMOs in Canadian supermarkets? What scientific information is present to back up people who consider GMOs safe, and what scientific information is present to back up those who consider GMOs unsafe?

Identify Possible Solutions to the Problem

In order to make an informed decision about how to respond to the issue, you will need to assess the possible solutions to the problem. Your research should reveal some alternative solutions.

For example, you see a news report in which Health Canada announces new research into the effects on the human body of eating genetically modified crops. You read a blog that states that people should be given a choice as to whether they eat GMOs or not. One article you read discusses the difficulty Canadians find in discovering whether or not food items contain GMO components. Perhaps the government should require food companies to clearly label foods containing GMOs.

Clarify the Consequences of Each Possible Solution

You may need to do additional research to identify potential consequences of each alternative solution and the reactions of the various stakeholders (that is, the individuals or groups affected by the issue).

For example:

- How many years might it take to see any negative effects from eating GMOs? Do studies take this time into account?
- If the public were better informed, would they fear the effects of eating GMOs less?
- How might labelling GMO foods affect the agricultural industry?

You can sort the potential consequences of an alternative into benefits (positive outcomes) and risks (negative outcomes).Use a risk-benefit analysis table like the one below to help you analyze the alternative solutions. For each possible solution, assess the impact on various stakeholders. The potential consequences of each solution could be different for each stakeholder. For some issues, you might choose to assess differing perspectives rather than differing effects on stakeholders. For example, you could assess benefits and risks from economic, environmental, social, scientific, and ethical perspectives. Each perspective could reveal different consequences.

Risk-Benefit Analysis

Issue: Are foods containing genetically modified organisms safe for consumers to eat?

Possible Solutions	Stakeholders	Potential Benefits (positive outcomes)	Potential Risks (negative outcomes)
1. The Canadian government should require manufacturers to label foods containing GMOs.	Government	• Reduced risk of costly medical bills and even lawsuits in years to come if health risks become evident	• Would affect the Canadian economy, perhaps negatively
	Farmers	• Increased sales for organic farmers	• Decreased sales for farmers producing GMO crops
	Food manufacturers	• Good public relations	• Loss of sales on GMO products
	Citizens	• People will feel safer, being able to make their own choices	• Many people may not choose to eat GMO foods
2. Health Canada and other centres should continue to research possible health risks of eating GMOs, and make their findings easily accessible to the general public.			

Make a Decision

Once you have identified potential outcomes for each possible solution, you are faced with the task of making a decision. Which alternative promises the greatest benefits and the least risks or lowest costs? Your personal values will influence your assessment. You will need to decide whether the benefits of a particular alternative are major or minor. You will also need to decide what an acceptable level of risk is. You might find it helpful to write down a list of questions to help you evaluate the alternative solutions. Some factors to consider are listed here:

- How likely is it that a potential outcome will occur?
- Is there evidence to support the likelihood of a predicted outcome?
- How many people (or other organisms) will the proposed course of action affect?
- Is there an estimated sum of money associated with the benefits or costs of each solution?
- Is the outcome of a proposed solution short-term (a one-time benefit/risk) or long-term (ongoing)?
- According to your analysis, how important are the risks of a possible solution compared to its potential benefits?
- How do the benefits and risks of one possible solution compare with the risks and benefits of other possible solutions?

After considering all the alternatives, you might decide that requiring food manufacturers to label foods containing GMOs is reasonable, based on the rights of Canadian citizens and the uncertainties still being researched. Your research suggests that this labelling might harm the agricultural and food industries, so government support for these industries over an adjustment period of a few years would need to be part of this solution.

Evaluate the Decision

Once you have made a decision, evaluate whether you can justify it with logic and verifiable information. If you discover that some of the information you used to make the decision was incorrect, you should reconsider the alternatives. If new information becomes available, that could also affect your decision.

Suppose a new study reveals that the consumption of GMOs helps to strengthen the human immune system, resulting in significant new benefits. How might this new information affect your decision?

Also assess whether you have taken all perspectives into account in your analysis. Is there another stakeholder that is strongly affected by a particular alternative? If you decide that you are not confident in the decision you have made, you will need to revisit each step in your analysis.

Act on Your Decision

If you are confident in your decision, the next step is to propose and implement a course of action.

For example, you could start a community letter-writing campaign urging your Member of Parliament to propose that all foods containing GMOs be clearly labelled.

Instant Practice

1. Consider the second possible solution listed in the risk-benefit analysis table on the previous page. Create a table in your notes to analyze the benefits and risks of this possible solution. Fill in the Stakeholders, Potential Benefits, and Potential Risks columns.
2. Look for a science-related STSE issue in the news. Apply the analysis method outlined in this appendix to determine your response to the issue. Write a brief paragraph to explain your viewpoint and a proposed course of action.

Scientific Inquiry

Scientific inquiry is a process that involves making observations, asking questions, performing investigations, and drawing conclusions.

A Process for Scientific Inquiry

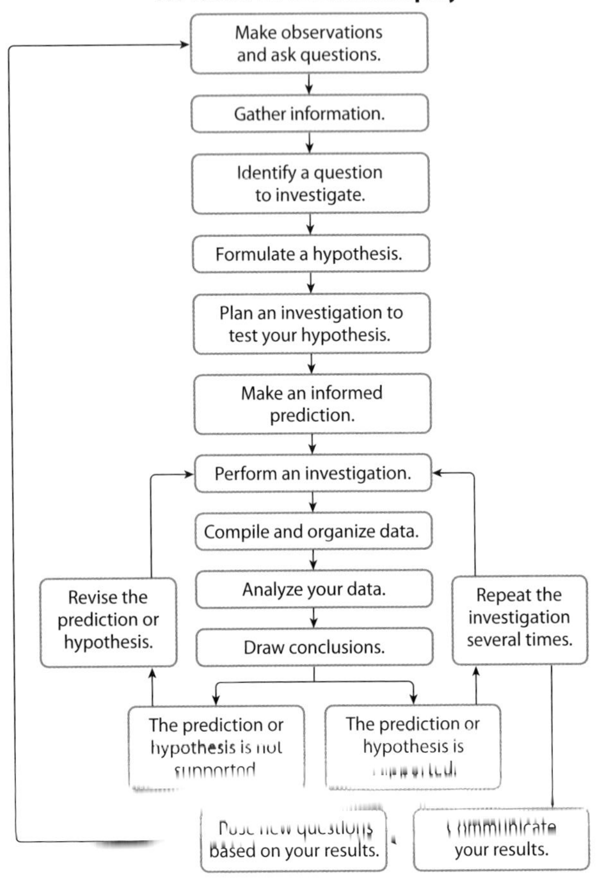

Make Observations and Ask Questions

Scientific inquiry usually starts with observations. You notice something that sparks your curiosity and prompts you to ask questions. You try to make sense of your observations by connecting them to your existing knowledge. When your existing knowledge cannot explain your observations, you ask more questions.

For example, suppose a train derailment has resulted in an acid spill near the shore of a local lake. You wonder what effect the spill will have on the fish and other organisms that live there. Has the spill killed all of the organisms in the lake? How have other organisms, such as aquatic plants, been affected by the spill? How can you find answers to your questions without endangering your safety?

Gather Information

Background research may help you to understand your observations and answer some of your questions. Go to Developing Research Skills in Appendix A for guidance on conducting research. You may also be able to gather information by making additional observations.

For example, you read a news report about an environmental assessment of the spill site. You discover that the pH of the lake water before the accident was 6.7. Measurements taken after the accident indicate that the pH dropped to 4.1. You do additional research to find out what kinds of organisms inhabit the lake and the optimal pH for their survival and growth.

Identify a Question to Investigate

You need to have a clear purpose and decide on a specific question that you are able to investigate with the resources available. If a question is provided for you, make sure you understand the science behind the question.

You decide to investigate the effect of acidity on living organisms. You do not wish to risk harming fish or other animals, so you decide to use aquatic plants as your test organism. You pose the scientific question, "What effect will increasing acidity have on aquatic plants grown in an aquarium?"

Formulate a Hypothesis

A hypothesis attempts to answer the question being investigated. It often proposes a relationship that is based on background information or an observed pattern of events.

You hypothesize that because plants can remove some impurities from polluted water, aquatic plants will be able to reduce the effect of small amounts of acid. Because highly acidic water will damage or kill most organisms, however, you hypothesize that the aquatic plants will not be able to counteract the addition of large quantities of acid.

Plan an Investigation

Some investigations lay out steps for you to follow in order to answer a question, analyze a set of data, explore an issue, or solve a problem. In planning your own investigation, however, *you* must decide how to approach a scientific question. Taking time to plan your approach thoroughly will ensure that you address the question appropriately.

Design a Procedure Write out step-by-step instructions for performing the investigation. Include instructions for repeat trials, if appropriate. Ensure that the procedure is written in a logical sequence, and that it is complete and clear enough that someone else could carry it out. Create diagrams, if necessary. Ask someone else to read through the procedure and explain it back to you, to ensure you have not omitted any important details.

> You decide to investigate the change in the pH of water when you add acid to a large glass bottle containing water and aquatic plants. You will measure the pH of the water and observe the physical appearance of the plants twice a day for three days.

Identify Variables Many investigations study relationships between variables (quantities or factors that can change). An *independent variable* is changed by the person conducting the investigation. A *dependent variable* is affected by changes in the independent variable. *Controlled variables* are kept the same throughout an experiment.

A simple controlled experiment shows relationships especially clearly because it has a single independent variable and a single dependent variable. All other variables are controlled. Changes in the dependent variable occur only in response to changes in the independent variable. When you are planning your investigation, you will need to identify the variables and decide which ones to control.

If possible, investigations include a *control:* a situation identical to the one being tested, except that the independent variable is not changed in any way. There is no reason, therefore, for the dependent variable to change. If it does, the reasoning behind your hypothesis, prediction, and variable analysis may be faulty. Look at the illustration at the top of the next column to see some examples of independent and dependent variables, as well as two examples of a control (no independent variable).

a. A test to find the best filter for muddy water

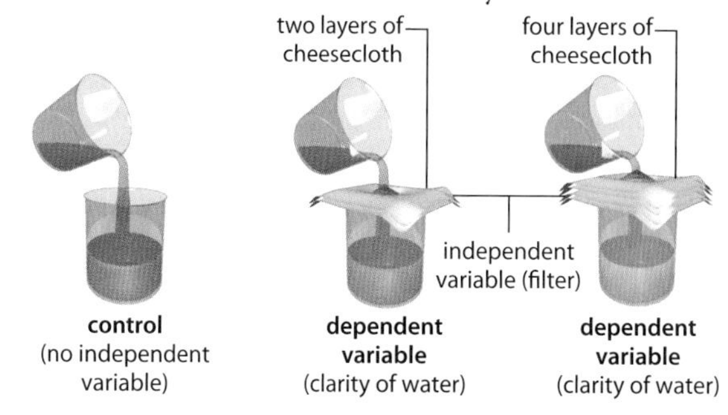

b. A test to find the best plant food for plant growth

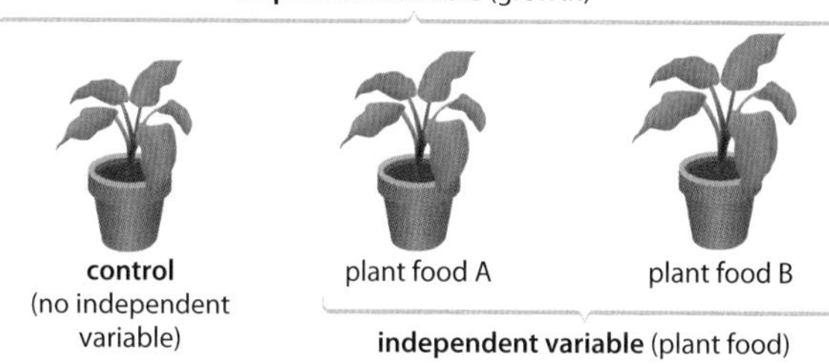

In planning your investigation, you decide to manipulate (change) the quantity of acid added to the aquarium you have made. Therefore, the quantity of acid is the independent variable. The pH of the water will be the dependent variable. Water temperature, lighting conditions, the particular species of aquatic plant used, and nutrients will be the controlled variables. In addition, you decide to set up an identical aquarium as a control. No acid will be added to the water in this aquarium, so you expect the pH of the water it contains to stay constant.

You decide to set up three different experimental aquariums. You plan to add a different amount of acid to the water in each of these aquariums. You will add no acid to the fourth aquarium—the control. Your teacher suggests using 1.0 mol/L hydrochloric acid (HCl). You will add 5 drops of hydrochloric acid to Aquarium 1, 10 drops to Aquarium 2, and 15 drops to Aquarium 3. You will measure the pH of the water in each aquarium at 9:00 A.M. on the first day, immediately before adding the acid, and then at 2:00 P.M. You will measure the pH again at 9:00 A.M. and 2:00 P.M. on Day 2 and Day 3.

List Materials and Safety Precautions Develop a list of materials and apparatus you will need. Include measuring and recording instruments. Examine your procedure for safety hazards and plan any necessary precautions. (Go to Safety in Your Biology Classroom at the front of this book for information about safety hazards and precautions.) **Note:** Before doing any experimental work, ask your teacher to examine and approve your plan.

Your materials list will include safety glasses, a lab apron, protective gloves, four glass jars, aquatic plants, hydrochloric acid, water, and a pH meter. Safety precautions include handling glassware carefully to avoid breakage; wearing safety glasses, gloves, and protective clothing to protect yourself from any acid spillage; storing the acid safely after use; and disposing of the aquarium water at the end of the investigation according to your teacher's instructions.

Make an Informed Prediction

A clear hypothesis often leads to a specific, testable prediction about what the investigation will reveal. You need to determine how to test your question before you can predict what will happen.

You predict that an aquarium full of a certain species of aquatic plants will maintain a stable pH of about 7 when a small quantity of acid is added. When greater [illegible] added, however, [illegible] [illegible] will [illegible] decrease rapidly and the plants will eventually die.

Perform an Investigation

Be responsible whenever you conduct an investigation. Think before acting, and follow all safety precautions. Carry out your procedure carefully. Ask for assistance if you are unsure how to proceed or if you encounter an unexpected difficulty. Report any accidents to your teacher immediately. Keep your workspace neat and clean it up when you have finished your investigation.

Compile and Organize Data

Record your results carefully and organize them in a logical way. Go to Organizing Data in a Table in Appendix A for help with recording and organizing the results of an investigation. As part of your observations, keep careful notes of any unexpected occurrences, problems with equipment, or unusual circumstances that might affect your results. If you are working with a partner, ensure that both of you have a copy of all observations and results.

Your results may include either qualitative or quantitative observations, or both. *Quantitative observations* are measurable and involve numbers. *Qualitative observations* involve descriptions rather than numbers or measurements. When making qualitative observations, try to record specific characteristics so that you can make comparisons between different trials.

In your investigation, you will record both qualitative and quantitative results. The pH values that you record are quantitative observations. Your descriptions of the physical appearance of the aquatic plants are qualitative observations. Looking at specific plant characteristics such as colour (green or brown) and vigour (robust or spindly) will help you to compare the physical appearance of the plants in each aquarium.

You might use a table like the one below to record and organize the data from your investigation.

Effect of Increasing Acidity on the Physical Appearance of Aquatic Plants in an Aquarium

	Physical Appearance of Plants (colour and vigour)			
	Control	+ 5 drops of acid	+ 10 drops of acid	+ 15 drops of acid
Day 1, [illegible] (before [illegible])	green, [illegible]	green, robust	green, [illegible]	green, robust
Day 1, 2:00 P.M.	green, robust	green, robust	green, robust	brownish spots, spindly
Day 2, 9:00 A.M.	green, robust	green, robust	brownish spots, less robust	brown, spindly (looks dead)
Day 2, 2:00 P.M.	green, robust	green, robust	mostly brown, less robust	brown, spindly (looks dead)
Day 3, 9:00 A.M.	green, robust	green, robust	brown, spindly (looks dead)	brown, spindly (looks dead)
Day 3, 2:00 P.M.	green, robust	green, robust	brown, spindly (looks dead)	brown, spindly (looks dead)

Analyze Your Data

Perform any necessary graph work or calculations. Go to Constructing Graphs in Appendix A for help with graphing. Then consider and interpret your results. Do your data and observations support or refute your hypothesis or prediction? Are additional data needed before you can draw definite conclusions? Identify any possible sources of error or bias in your investigation. Does the procedure or apparatus need to be modified to obtain better data?

Using the pH data from your investigation, you construct the graph shown below.

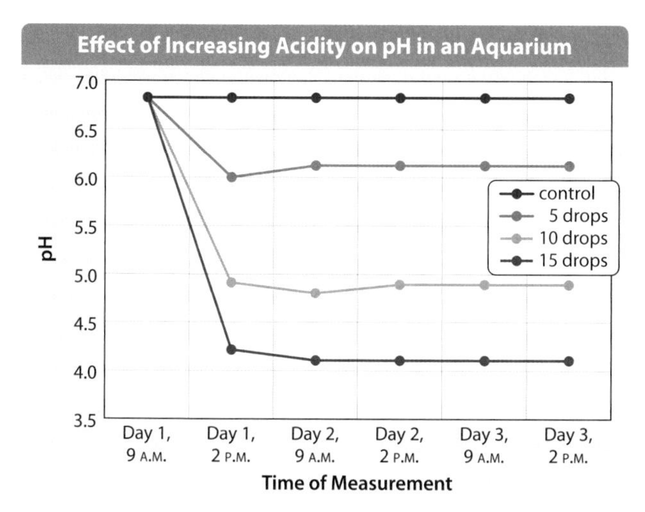

Draw Conclusions

Conclusions usually answer several questions:

- What has the investigation revealed about the answer to the question?
- How well does your prediction agree with the data?
- How well is your hypothesis supported by the data? Are the observations explained by the hypothesis?
- How precise were the measuring instruments and resulting observations?
- What improvements could be made to the investigation?

Relate your conclusions to your background knowledge of the scientific principles involved.

You conclude that your hypothesis and prediction were supported by the data. The aquatic plants were able to survive the addition of small amounts of hydrochloric acid (5 drops), but when more acid was added to the aquarium, the pH decreased rapidly and the plants soon looked brown and spindly.

You are unsure whether the plants in the highly acidic water were actually dead, however. You think perhaps you could place them in fresh water to see whether they could recover.

Now you can relate your results to the original problem of the acid spill in the lake. Are the aquatic plants in the lake likely to survive? Do your results reflect the conditions in the lake? What other factors might you need to consider?

Communicate Your Results

Communicate the results of your investigation. Always include a summary of your findings and an evaluation of the investigation. Be sure to round answers to the proper number of significant digits. Go to Significant Digits and Rounding in Appendix A for help with significant digits. Demonstrate your results clearly using graphs, tables, or diagrams, as appropriate. Go to Constructing Graphs or Organizing Data in a Table in Appendix A for help with communicating your results. Be sure to include units when expressing measurements. Go to Measurement in Appendix A for information on units and measurements.

Pose New Questions Based on Your Results

The conclusion of an investigation is not the end of scientific inquiry. Scientific inquiry is a continuous process in which results and conclusions lead to new questions. What new research questions might arise from your investigation? How might you find an answer to one of these questions?

After performing this investigation, you wonder how much the aquatic plants themselves affected the pH of the water. Would the decrease in pH have been more noticeable if there were no plants present? How would you test this?

Instant Practice

You are asked to plan an investigation on the response mechanism of an invertebrate to external stimuli such as light or sound. Think about how you could test a hypothesis related to this question.

1. Will your results include qualitative or quantitative observations, or both?
2. State a hypothesis for this investigation.
3. What will your independent variable be? What will your dependent variable be? What control will you set up?

Developing Research Skills

In this course, you will need to conduct research to answer specific questions and to explore broad research topics. The following skills will take you through the research process from start to finish:

- focussing your research
- searching for resources that contain information related to your topic
- evaluating the reliability of your information sources
- gathering, recording, and organizing information in an appropriate format
- presenting your work

Focussing Your Research

- Start by carefully reading your assignment. Pick out key words and phrases, such as *apply*, *analyze*, *argue*, *compare and contrast*, *describe*, *discuss*, *evaluate*, *explain*, *identify*, *infer*, *interpret*, and *predict*. These key words and phrases will guide you on what kind of information you need to collect, and what you need to do with the information.
- Jot down ideas on your own, and then get additional input from others, including your teacher.
- Once you have done some general research, narrow down your topic until you can express it in one specific question. This will help you focus your research.
- Ensure that the question you are researching fulfills the guidelines of the assignment provided by your teacher.

Searching for Resources

- It is important to find reliable resources to help you answer your question. Potential sources of information include print and on-line resources such as encyclopedias, textbooks, non-fiction books, journals, websites, and newsgroups.
- The library and the Internet can both provide information for your search. Whether you are looking at print or digital resources, you need to evaluate the accuracy and objectivity of the information.

Evaluating the Reliability of Your Information Sources

Assess the reliability of your information sources to help you decide whether the information you find is likely to be accurate. To determine the validity of a source, check that the author is identified, a recent publication date is given, and the source of facts or quotations is identified. An author's credentials are important. Look for an indication of educational background, work experience, or professional affiliation. If the information is published by a group, try to find out what interests the group represents. The following guidelines may be helpful in assessing your information sources:

- On-line and print scientific journals provide data that have been reviewed by experts in a field of study (peer-reviewed), so they are usually a reliable source. Be aware, however, that the conclusions in journal articles may contain opinions as well as facts.
- Data on the websites of government statistical departments tend to be reliable. Be sure to read carefully, however, to interpret the data correctly.
- University resources, such as websites ending in ".edu" are generally reliable.
- Reliable experts in a field of study often have a PhD or MSc degree, and their work is regularly cited in other publications.
- Consumer and corporate sources may present a biased view. That is, they may present only data that support their side of an issue. Look for sources that treat all sides of an issue equally and fairly, or that clearly specify which perspective(s) they are presenting.
- Some sources, such as blogs and editorials, provide information that represents an individual's point of view or opinion. Therefore, the information is not objective. However, opinion pieces can alert you to controversy about an issue and help you consider various perspectives. The opinion of an expert in a field of study should carry more weight than that of an unidentified source.
- On-line videos and podcasts can be dynamic and valuable sources of information. However, their accuracy and objectivity must be evaluated just as thoroughly as all other sources.

- A piece of information is generally reliable if you can find it in two other sources. However, be aware that several on-line resources might use the same incorrect source of information. If you see identical wording on multiple sites, try to find a different source to verify the information.

Gathering, Recording, and Organizing Information

- As you locate information, you may find it useful to jot it down on large sticky notes or colour-coded entries in a digital file so you can group similar ideas together. Remember to document the source of your information for each note or data entry.

Avoid Plagiarism Copying information word-for-word and then presenting it as your own work is called *plagiarism*. Instead, you must cite every source you use for a research assignment. This includes all ideas, information, data, and opinions that appear in your work. If you include a quotation, be sure to indicate it as such, and supply all source information. Avoid direct quotations whenever possible—put information in your own words. Remember, though, that even when you paraphrase, you need to cite your sources.

Record Source Information A research paper should always include a bibliography—a list of relevant information sources you have consulted while writing it. Bibliographic entries include information such as the author, title, publication year, name of the publisher, and city in which the publisher is located. For magazine or journal articles, the name of the magazine or journal, the name of the article, the issue number, and the page numbers should be recorded. For on-line resources, you should record the site URL, the name of the site, the author or publishing organization, and the date on which you retrieved the information. Remember to record source information while you are taking notes to avoid having to search it out again later! Ask your teacher about the preferred style for your references.

- You might find it helpful to create a chart to keep track of detailed source information. For on-line searches, a tracking chart is useful to record the key words you searched, the information you found, and the URL of the website where you found the information.
- Write down any additional questions that you think of as you are researching. You may need to refine your topic if it is too broad, or take a different approach if there is not enough information available to answer your research question.

Presenting Your Work

- Once you have organized all of your information, you should be able to summarize your research so that it provides a concise answer to your original research question. If you cannot answer this question, you may need to refine the question or do a bit more research.
- Check the assignment guidelines for instructions on how to format your work.
- Be sure that you fulfill all of the criteria of the assignment when you communicate your findings.

Instant Practice

1. Your assignment asks you to research the effectiveness of some Canadian technologies in providing food for growing populations.
   a. What search terms might you use for your initial research on the Internet or at the library?
   b. How might you narrow down this assignment into a research question?
2. How could you verify the information in an article about genetically modified organisms that you found on a wiki site?
3. Suggest two or more clues that could indicate that the information in an on-line video might not be reliable.

Appendix A

Writing a Lab Report

A lab report is one format for communicating the results of an investigation clearly. Use the following headings and guidelines to create a neat and legible lab report.

Title

- Choose a title that clearly states the independent variable and the dependent variable, but not the outcome of the investigation. For example, "The Effect of Temperature on Stem Growth in Bean Seedlings."
- Under the title, write the names of all participants, the name of your teacher, and the date(s) of the investigation.

Introduction

- Summarize the background of the problem.
- Cite any relevant scientific principles or literature related to the question being investigated.

Question/Problem

- Clearly state the question being investigated or the problem for which you are seeking a solution. For example, "Does temperature have an effect on stem growth in bean seedlings?"

Hypothesis

- State, in general terms, the relationship that you believe exists between the independent variable and the dependent variable. For example, "Temperature has a positive effect on stem growth in bean seedlings."

Prediction

- State, in detailed terms, the specific results you expect to observe. For example, "Seedlings grown for three days at 20°C will be taller than seedlings grown for the same length of time, under the same conditions, at 10°C."

Materials

- List all of the materials and equipment you used, or refer to the appropriate page number in your textbook, and note any additions, deletions, or substitutions you have made.

Procedure

- Write your procedure in the form of precise, numbered steps, or refer to the appropriate page number in your textbook, and note any changes you have made to the procedure. Remember to include any safety precautions.

Results

- Set out the observations and/or data in a clearly organized table(s). Give your table a title.
- If appropriate, construct a graph that shows the data accurately. Label the x-axis and the y-axis of the graph clearly and accurately, and use the correct scale and units. Give your graph a title.

Data Analysis

- Analyze all the results you have gathered and recorded, and ensure that you can defend your analysis. For example, "As shown in Figure 1, the average growth at 20°C was 2.7 cm greater than the average growth at 10°C..."
- Show sample calculations for any mathematical data analysis.

Conclusion

- State a conclusion based on your data analysis. Relate your conclusion to your hypothesis. For example, "Based on the results of this investigation, temperature has a positive effect on stem growth in bean seedlings."
- Compare the results you obtained with those you expected, or those obtained by other researchers.
- Examine and comment on experimental error.
- Assess the effectiveness of the experimental design.
- Indicate how the data support your conclusion.
- Make recommendations for how your conclusion could be applied, or for further study of the question you investigated.

References

- Cite your information sources according to the reference style your teacher suggests. The American Psychological Association (APA) style is frequently used for documenting sources.
- Sources that need to be cited include background information for your introduction, a materials list or procedure from a textbook, any specialized methods of data analysis, results from other studies that you used for comparison with your own results, and any other sources used in your conclusion.

Organizing Data in a Table

Scientific investigation is about collecting information to help you answer a question. In many cases, you will develop a hypothesis and collect data to see if your hypothesis is supported. An important part of any successful investigation is recording and organizing your data. Often, scientists create tables in which to record data.

Planning to Record Your Data Suppose you are doing an investigation on the water quality of a stream that runs near your school. You will take samples of the numbers and types of organisms at three different locations along the stream. You need to decide how to record and organize your data. Begin by making a list of what you need to record. For this experiment, you will need to record the sample site, the pH of the water at each sample site, the types of organisms found at each sample site, and how many of each type of organism you collected.

Creating Your Data Table Your data table must allow you to record your data neatly. To do this you need to create

- headings to show what you are recording
- columns and rows that you will fill with data
- enough cells to record all the data
- a title for the table

In this investigation, you will find multiple organisms at each site, so you must make space for multiple recordings at each site. This means every row representing a sample site will have at least three rows associated with it for the different organisms.

If you think you might need extra space, create a special section. In this investigation, leave space at the bottom of your table, in case you find more than three organisms at a sample site. Remember, if you use the extra rows, make sure you identify which sample site the extra data are from. Finally, give your table an appropriate title. Your data table might look like the one at the top of this page.

Reading a Table A table can be used to organize observations and measurements so that data are represented neatly and clearly. However, a table can also show relationships among the data presented. When you are reading a table, be sure to start by reading the column and row headings carefully. If the table contains measurements, look for the units in which they are reported. Follow vertically down a column or horizontally across a row to look for trends in the data. If the table contains numbers, do the numbers increase or decrease as you look down the column or across the row?

Observations Made at Three Sample Stream Sites

headings show what is being recorded

columns and rows contain data

Sample Site	pH	Type of Organism	Number of Organisms
1		beetle	3
		snail	1
		dragonfly larva	8
2		beetle	6
		dragonfly larva	7
3		snail	5
		leech	1
		dragonfly larva	2

extra rows to collect data in case you need to add observations

Also look for relationships between columns or rows. Do the numbers in one column increase as the numbers in another column decrease? Is there one piece of data that does not fit the pattern in the rest of the table? Think about why this might be the case.

Instant Practice

1. You want to investigate the population cycles of two different bird populations that compete for food. Construct a table to record the average population of each bird species on the following bimonthly dates:
   - February 1
   - April 1
   - June 1
   - August 1
2. Now you wish to refine your investigation to record the number of males and females of each species. Draw a new table to record these data.
3. Examine the table at the top of the page. What does it tell you about dragonfly larva in these regions?

Appendix A

Constructing Graphs

A graph is a diagram that shows relationships among variables. Graphs help you to interpret and analyze data. The three basic types of graphs used in science are the line graph, the bar graph, and the circle graph.

The instructions given here describe how to construct graphs using paper and pencil. You can also use computer software to generate graphs. Whichever method you use, the graphs you construct should have the features described in the following pages.

Line Graphs

A line graph is used to show the relationship between two variables. The independent variable is plotted on the horizontal axis, called the x-axis. The dependent variable is plotted on the vertical axis, called the y-axis. The dependent variable (y) changes as a result of a change in the independent variable (x).

Suppose a school started a bird-watching group to observe the number of birds in the school courtyard. The number of birds in the courtyard was recorded each day for four months. The average number of birds per month was calculated. A table of the birds' visitations is shown below.

Average Number of Birds Viewed

Time (days)	Average Number of Birds per Day
30	24
60	27
[illegible]	30
120	32

To make a graph of the average number of birds over a period of time, start by determining the dependent and independent variables. The average number of birds after each period of time is the dependent variable and is plotted on the y-axis. The independent variable, or the number of days, is plotted on the x-axis.

Give your graph a title and label each axis, indicating the units if appropriate. In this example, label the number of days on the x-axis. Your x-axis will need to be numbered to at least 120 days. Because the lowest average of birds viewed was 24 and the highest was 32, you know that you will have to start numbers on the y-axis from at least 24 and number to at least 32. You could decide to number 20–40 by intervals of 4 spaced at equal distances. Look at the example at the top of the page to see how you could label your axes.

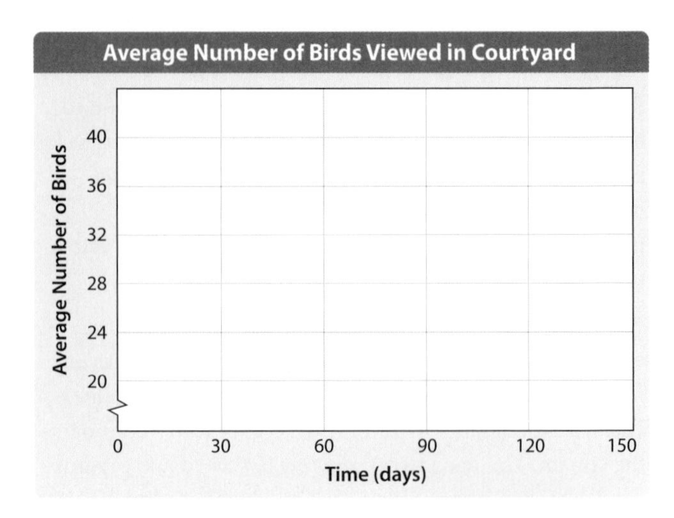

Begin plotting points by locating 30 days on the x-axis and 24 on the y-axis. Where an imaginary vertical line from the x-axis and an imaginary horizontal line from the y-axis meet, place the first data point. Place other data points using the same process. After all the points are plotted, draw a "best fit" straight line through the points.

A best fit line should be drawn to represent the general trend of the data. Try to draw the line so that there are as many points above it as there are below. Do not change the position or slope of the line dramatically just to include an outlier—a single data point that does not seem to be in line with all the others.

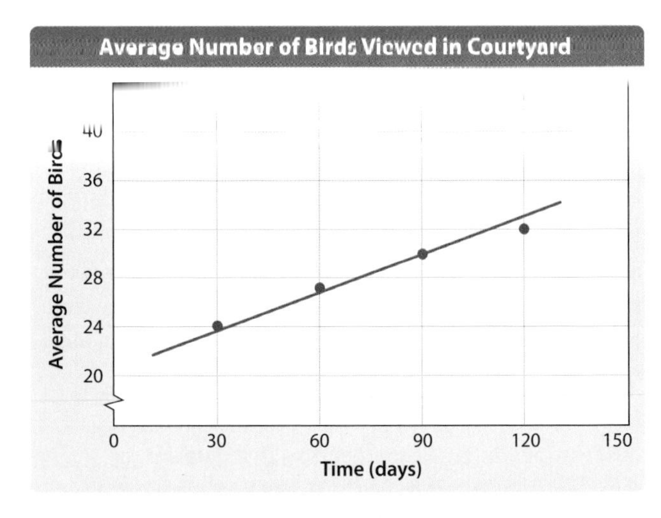

The bird-watching group also recorded the average number of brown-feathered birds they observed in the school courtyard. Their observations are shown in the table at the top of the next page.

Average Number of Brown-Feathered Birds Viewed

Time (days)	Average Number of Birds per Day
30	21
60	24
90	28
120	30

What if you want to compare the average number of birds viewed with the average number of brown-feathered birds? The average brown-feathered bird data can be plotted on the same graph. Include a key with different lines indicating different sets of data.

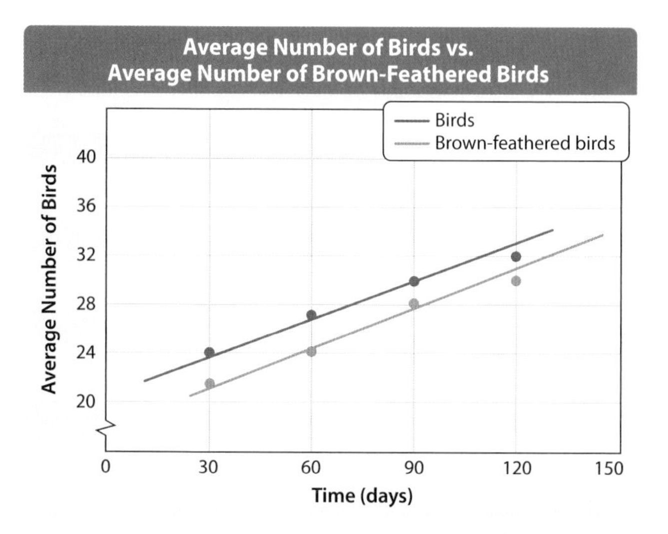

Instant Practice

1. What is the difference between the number of brown-feathered birds and the total number of birds on the 90th day?
2. Compare the increase in the average number of brown-feathered birds with the increase in the average number of birds over the first 60 days.
3. Construct a line graph for the following data:

The Effect of pH on Enzyme Function

pH of Reaction Mixture	Rate of Reaction (mol/s)
5.5	0.45
6.0	0.48
6.5	0.52
7.0	0.55
7.5	0.59

Slope of a Linear Graph The slope of a line is a number determined by any two points on the line. This number describes how steep the line is. The greater the absolute value of the slope, the steeper the line. Slope is the ratio of the change in the y-coordinates (rise) to the change in the x-coordinates (run) as you move from one point to the other.

The graph below shows a line that passes through points (5, 4) and (9, 6).

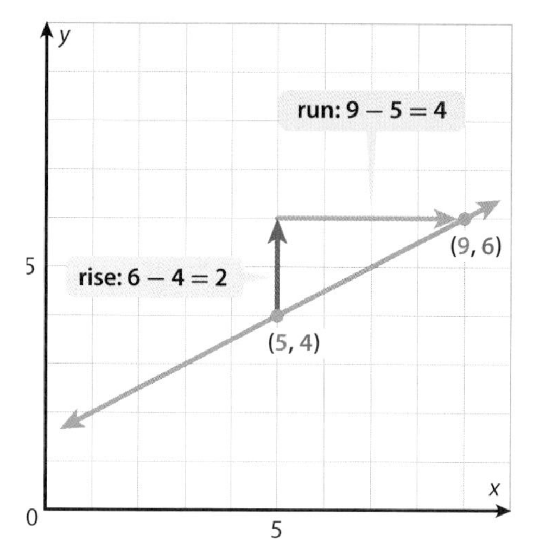

$$\text{Slope} = \frac{\text{rise}}{\text{run}}$$
$$= \frac{\text{change in } y\text{-coordinates}}{\text{change in } x\text{-coordinates}}$$
$$= \frac{6-4}{9-5}$$
$$= \frac{2}{4} \text{ or } \frac{1}{2}$$

So, the slope of the line is $\frac{1}{2}$.

A positive slope indicates that the line climbs from left to right. A negative slope indicates that the line descends from left to right. A slope of zero indicates that there is no change in the dependent variable as the independent variable increases. A horizontal line has a slope of zero.

Linear and Exponential Trends Two types of trends you are likely to see when you graph data in biology are linear trends and exponential trends. A linear trend has a constant increase or decrease in data values. In an exponential trend the values are increasing or decreasing more and more rapidly. The graphs shown on the next page are examples of these two common trends.

In the graph below, there are two lines describing two frog species. Both lines show an increasing linear trend. As the temperature increases, so do the call pulse rates of the frogs. The rate of increase is constant.

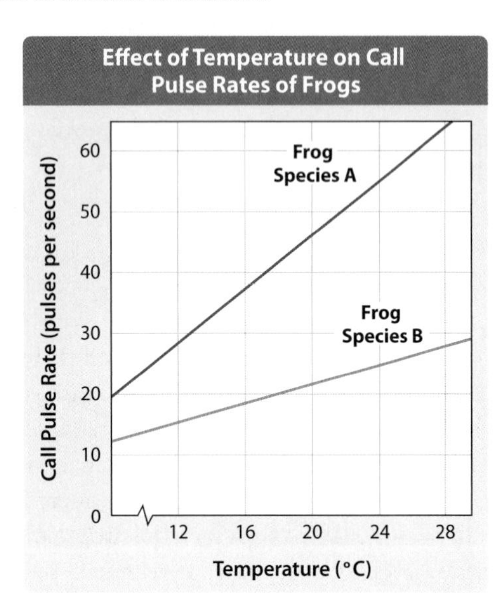

The example below shows how a mouse population grows if the mice are allowed to reproduce unhindered. At first the population grows slowly. The population growth rate soon accelerates because the total number of mice that are able to reproduce has increased. Notice that the portion of the graph where the population is increasing more and more rapidly is J-shaped. A J-shaped curve generally indicates exponential growth.

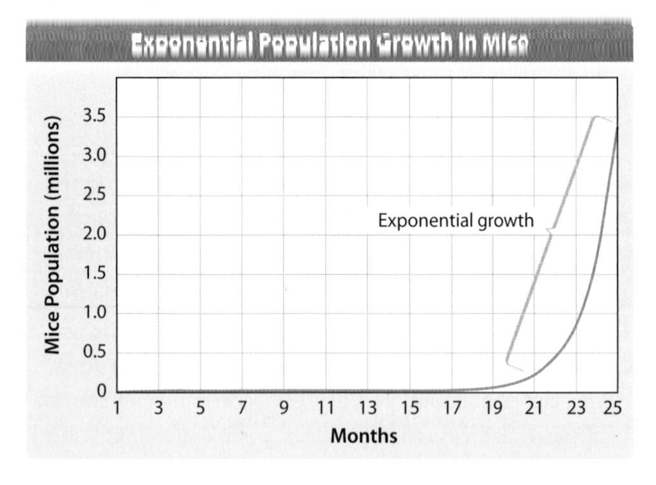

When you are drawing a curve to represent an exponential trend, you should not connect the data points. Instead, draw a best fit smooth curve that shows the general trend of the data. Try to draw the curve so there are as many points above it as there are below. The curve should change smoothly. It should not have a dramatic change in direction just to include a single data point that does not fit with the others.

Bar Graphs

A bar graph displays a comparison of different categories of data by representing each category with a bar. The length of the bar is related to the category's frequency. To make a bar graph, set up the *x*-axis and *y*-axis as you did for the line graph. Plot the data by drawing thick bars from the *x*-axis up to an imaginary line representing the *y*-axis point.

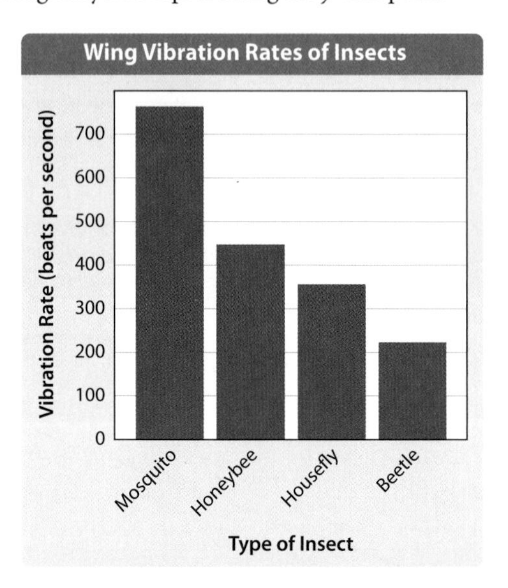

Look at the graph above. The independent variable is the type of insect. The dependent variable is the number of wing vibrations per second.

Bar graphs can also be used to display multiple sets of data in different categories at the same time. A bar graph that displays two sets of data is called a double bar graph. Double bar graphs like the one below have a legend to denote which bars represent each set of data.

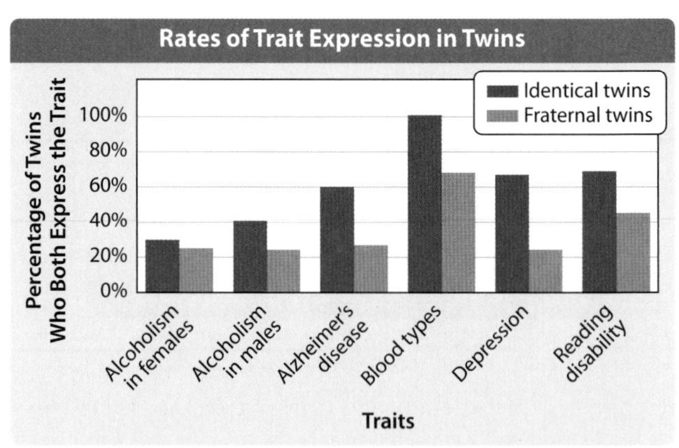

Instant Practice

In the graph above, which trait is most likely to be expressed in both of a set of fraternal twins?

Circle Graphs

A circle graph is a circle divided into sections that represent parts of a whole. When all the sections are placed together, they equal 100 percent of the whole.

Suppose you want to make a circle graph to show the number of seeds that germinate in a package. You would first determine the total number of seeds in the package. Then determine the number of seeds that germinate out of the total. You plant 143 seeds. Therefore, the whole circle represents this amount. You find that 129 seeds germinate. The seeds that germinate make up one section of the circle graph, and the seeds that do not germinate make up another section.

To find out how much of the circle each section should cover, divide the number of seeds that germinate by the total number of seeds. Then multiply the answer by 360, the number of degrees in a circle. Round your answer to the nearest whole number. The sum of all the segments of the circle graph should add up to 360°.

Segment of circle for seeds that germinated $= \frac{\text{seeds that germinated}}{\text{total number of seeds}}$

Divide $= \frac{129}{143}$

Multiply by number of degrees in a circle $= 0.902 \times 360°$

$= 324.72°$

Round to nearest whole number $= 325°$

Segment of circle for seeds that did not germinate $= 360° - 325°$

$= 35°$

To draw your circle graph, you will need a compass and a protractor. First, use the compass to draw a circle.

Then, draw a straight line from the centre to the edge of the circle. Place your protractor on this line, and mark the point on the circle where an angle of 35° will intersect the circle. Draw a straight line from the centre of the circle to the intersection point. This is the section for the seeds that did not germinate. The other section represents the group of seeds that did germinate.

Next, determine the percentages for each part of the whole. Calculate percentages by dividing the part by the total and multiplying by 100. Repeat this calculation for each part.

Percentage of seeds that germinate $= \frac{\text{seeds that germinated}}{\text{total number of seeds}}$

$= \frac{129}{143}$

Multiply by 100% $= 0.902 \times 100\%$

$= 90.2\%$

Percentage of seeds that did not germinate $= 100\% - 90.2\%$

$= 9.8\%$

Complete the graph by labelling the sections of the graph with percentages and giving the graph a title. Your completed graph should look similar to the one below.

If your circle graph has more than two sections, you will need to construct a segment for each entry. Place your protractor on the last line segment that you have drawn and mark off the appropriate angle. Draw a line segment from the centre of the circle to the new mark on the circle. Continue this process until all of the segments have been drawn.

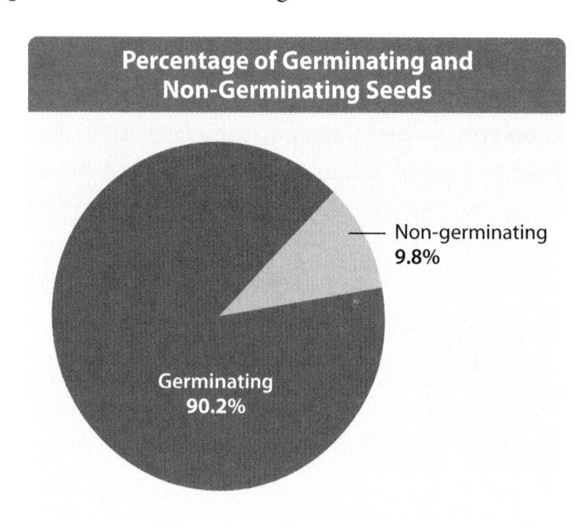

Instant Practice

The main components of the human body are water, proteins, fats, minerals, and carbohydrates. Construct a circle graph showing the percentage of each type of molecule in the body. A person with a mass of 70.0 kg contains about 44.8 kg water, 14.0 kg protein, 7.0 kg fat, 0.7 kg carbohydrate, and 3.5 kg minerals.

Graphing Independent and Dependent Variables

Line graphs and bar graphs are often used to show the relationship between two variables. The data from an experiment usually contain an independent variable and a dependent variable. The independent variable is the one that the experimenter changes. This variable is usually plotted on the x-axis (horizontal axis). The dependent variable responds to the change in the independent variable. The dependent variable is usually plotted on the y-axis (vertical axis), as shown below.

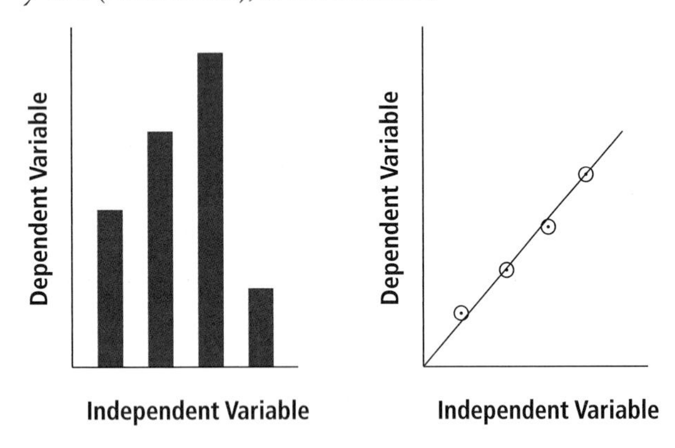

The first step in drawing a graph of experimental data is to analyze the procedure and data to determine the dependent and independent variables. For example, you may have noticed when driving that after you step on the brakes, the distance your car travels before it stops depends on how fast you were travelling. Suppose we set up an experiment in which a car is driven at different speeds and the distance it travels, after the brakes are applied, is recorded. The data for this experiment are shown below.

Relation of Braking Distance to Speed

Speed (km/h)	Braking Distance (m)
30	6.0
40	10.5
50	16.5
60	23.6
70	32.0
80	42.0

In this experiment, the procedure was to change the speed and measure the braking distance. The variable that the experimenter changed was the speed. Therefore, speed is the independent variable and should be plotted on the x-axis. The braking distance varied as a result of the changes in speed and therefore is the dependent variable. The braking distance is plotted on the y-axis.

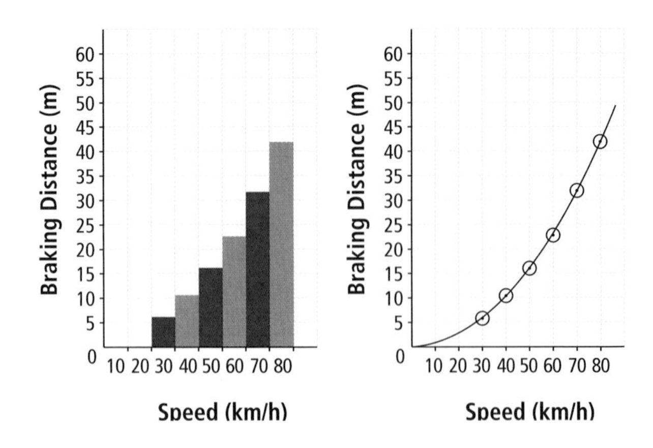

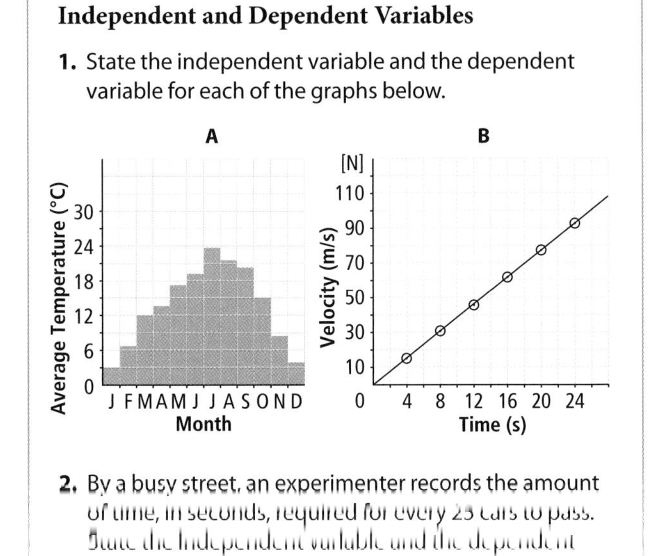

Instant Practice

Independent and Dependent Variables

1. State the independent variable and the dependent variable for each of the graphs below.

2. By a busy street, an experimenter records the amount of time, in seconds, required for every 25 cars to pass. State the independent variable and the dependent variable in this description.

Scatter Plots A scatter plot is a graphical display of how much one variable is affected by another variable. The relationship between two variables is called their correlation. Scatter plots, like line graphs, use horizontal and vertical axes to plot data points. If the data points cluster together closely, you may be able to see a clear trend in the data.

Data points that cluster in a band running from lower left to upper right indicate a positive correlation. In other words, if x increases, y increases.

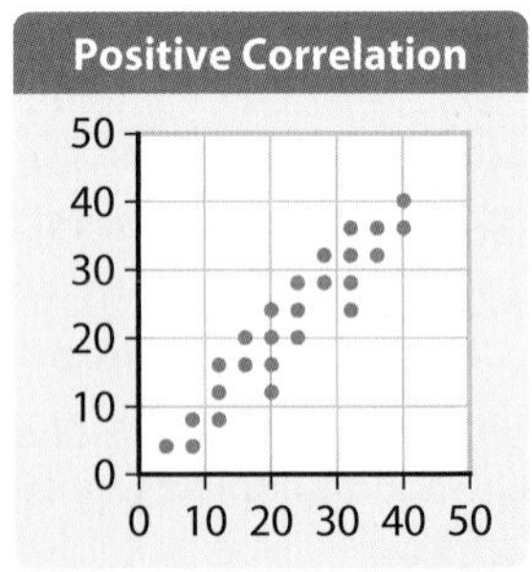

Data points that cluster in a band running from upper left to lower right indicate a negative correlation. In other words, if x increases, y decreases.

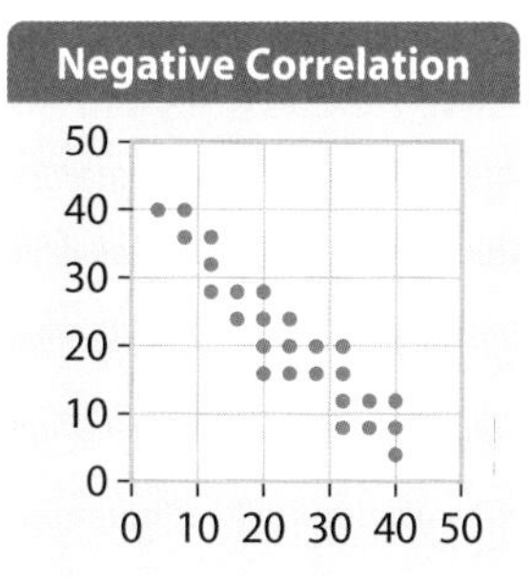

Data points that show no significant clustering usually indicate no correlation.

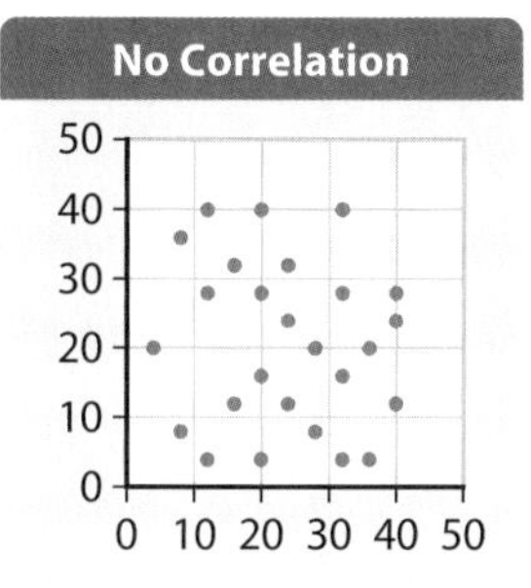

When there is a high correlation between two variables, the plotted data points will be closer to making a straight line, than when there is low correlation. Imagine drawing a straight line or curve through the data so that it "fits" as well as possible. The more the points cluster closely around the imaginary line of best fit, the stronger the relationship that exists between the two variables (see below).

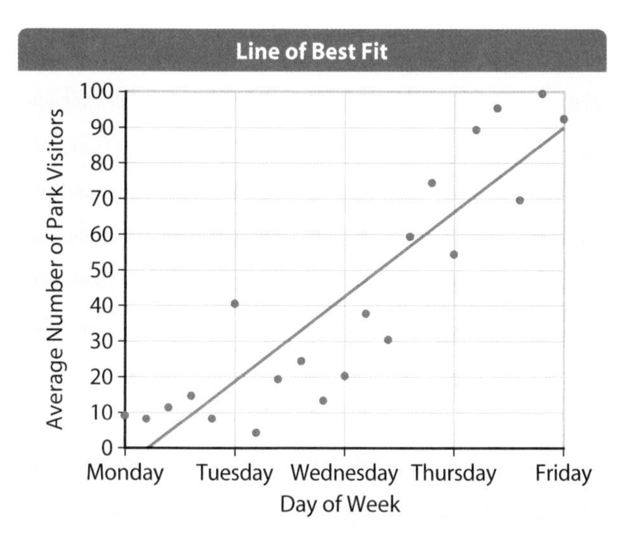

It is not always possible to draw a line that connects all of the points on your graph. Scientific investigations often involve quantities that do not change smoothly. On a graph, this means that you should draw a smooth curve (or straight line) that most closely fits the general shape outlined by the points. This is called a line of best fit. A best-fit line often passes through many of the points, but sometimes it goes between points. Think of the dots on your graph as clues about where the perfect smooth curve (or straight line) should go. A line of best fit shows the trend of the data. It can be extended beyond the first and last points to indicate what might happen.

Note that although a scatter plot shows the relationship between two variables, it does not necessarily prove that one variable causes the other. The relationship may be due to chance or it may be due to a third variable that is not shown in the data.

Instant Practice

In the Line of Best Fit graph above, what relationship would you suggest exists between the days of the week and number of park visitors?

Graphing on a Computer

Computers are a useful tool for graph preparation for the following reasons.

1. Data need be entered only once. As many graphs as you need can then be prepared without any more data entry.
2. Once the data are entered, you can use the computer to manipulate them. You can change the scale, zoom in on important parts of the graph, graph different parts of the data in different ways, and so on—all without doing any calculations.
3. Computers prepare graphs far more quickly than people working carefully.
4. Computers can be hooked up to sensors (thermometers, timers, etc.), so you do not need to read instruments and enter data by hand, with all the resulting possibilities for error. The computer can display the readings on a graph as data are collected (in "real" time), so you can quickly get a picture of how your experiment is going.
5. Errors can be corrected much more easily when working with a computer. Just correct the error and print again.
6. Computer graphs can be easily inserted into written lab reports, magazine articles, or Internet pages. It is possible to scan hand-drawn graphs into a computer, but it is not easy to do it well, and the resulting files are very large.
7. Once data have been entered into a computer, the computer can determine a line of best fit and a mathematical equation that describes the line. This line can help you to discover patterns in your data and make predictions to test your inferences.

Appendix A

Using Statistics to Test Hypotheses

You may be interested in answering many types of science questions. Is Earth's surface getting warmer? Which air bag most effectively reduces the severity of injuries in a car accident? Does the public prefer a certain colour in a new line of fashion? To answer those questions you need to conduct a study or experiment.

Statistics is the science of conducting studies to collect, organize, summarize, analyze, and draw conclusions from data. Data are the values (measurements or observations) that a variable can assume. A collection of data values is called a data set.

Suppose you wanted to know: Does a new medication affect people's pulse rates? You cannot give the drug to everyone in the population and measure their pulse rate. You need to take a sample. A sample is a group of subjects selected from a population.

If you could sample the entire population, you would know exactly what the effect of the medication is on the pulse rate of the population. Since this is not possible, you use the observations from the sample to approximate the effect on pulse rate of the population.

Outliers Once you have conducted your study and collected the data, you need to check the data for extremely high or low values, called *outliers*. For example, suppose that in your pulse rate study the data you collected included

[illegible]

The value 20 is an outlier and you might be suspicious of whether it is valid.

- An outlier may have resulted from a measurement or observational error.
- An outlier may have resulted from a recording error.
- An outlier may have been obtained from a subject that is not in the defined population.
- An outlier might be a legitimate value that occurred by chance (although the probability is extremely small).

There are no hard and fast rules on what to do with outliers, nor is there complete agreement among statisticians on ways to identify them. If outliers occurred as a result of an error, you should attempt to correct the error or omit the data value. When outliers occur by chance, you need to make a decision about whether or not to include them in the data set.

Describing the Data

The data you collect in your study is called raw data. To draw conclusions from the data, you need to look at how it is distributed. The two most important statistics for describing the data are the mean and the standard deviation.

- The *mean* describes the average. You find the mean by adding all values of the data and dividing by the total number of values.
- *Standard deviation* describes the spread of the data.

The distribution that is defined by these two parameters from the sample describes the likely effect on the pulse rate in the population. The most likely average pulse rate of the population with the drug is the mean of the sample. However, since you did not measure the pulse rate of the entire population, there is some uncertainty in your measurements. As you get further away from the mean, the likelihood that the true population average has that value diminishes. The shape of the likelihood curve is described by [illegible] ve, sh[illegible]

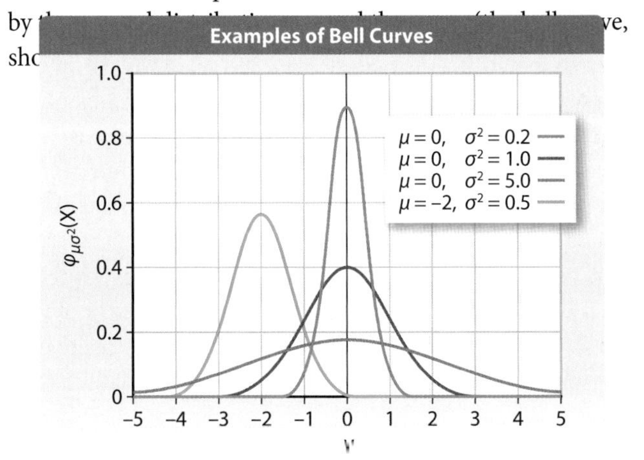

Examples of normal distribution of data

Testing the Research Question

The basic question you are testing in describing your data is: How likely is the chance that the mean pulse rate of the sample with the drug could actually be the same as the population average pulse rate without the drug? The population average pulse rate would have been obtained from a previous study, which also studied a sample and not the entire population.

The statistical method for examining whether there is a difference in pulse rates between the two samples is called hypothesis testing. A hypothesis is a proposed effect of a phenomenon or treatment. Hypothesis testing always include two hypotheses. Note that you only investigate the null hypothesis. The alternative hypothesis is just the complement.

- The *null hypothesis*, symbolized by H_0, states that there is no effect (the drug does not affect the pulse rate).

- The *alternative hypothesis*, symbolized by H_1, states that there is an effect (the drug does affect the pulse rate).

One of these hypotheses needs to be true. The result of a hypothesis test is whether we accept or reject the null hypothesis.

- By accepting the null hypothesis, we say that there is no statistical difference between the mean pulse rate of the sample with the drug and the average pulse rate of the population.
- By rejecting the null hypothesis, we say that there is a statistical difference between the mean pulse rate of the sample with the drug and the average pulse rate of the population.

The decision rule for accepting or rejecting the null hypothesis is defined by the significance level. Since you only measured the pulse rate of the sample and not the entire population, you can never be certain that your measurements actually truly represent the population. The significance level is the probability of a type I error you are willing to accept for your decision. A type I error is rejecting the null hypothesis when it was actually true (i.e., you say there is a difference when there is actually none). Typically, researchers only reject the null hypothesis when the probability of a type I error is smaller than 5%. In other words, they reject when there is only a 1 in 20 chance that the drug does not affect pulse rate.

Representing the Hypotheses Mathematically

as shown in the table and examples below.

Common Phrases for Hypothesis-Testing	
$>$ Is greater than Is above Is higher than Is longer than Is bigger than Is increase	$<$ Is less than Is below Is lower than Is shorter than Is smaller than Is decreased or reduced from
$\geq$ Is greater than or equal to Is at least Is not less than	$\leq$ Is less than or equal to Is at most Is not more than
$=$ Is equal to Is exactly the same as Has not changed from Is the same as	$\neq$ Is not equal to Is different from Has changed from Is not the same as

Example: Will it change?

Suppose the mean pulse rate for the population under study is 82 beats per minute. The null hypothesis for your pulse rate study states that the medication has no effect on the pulse rate.

$$H_0\ \mu = 82$$

(The Greek letter μ (mu) is used to represent the mean of a population.)

The alternative hypothesis states that the medication has an effect on the pulse rate.

$$H_1\ \mu \neq 82$$

Example: Will it increase?

A chemist invents an additive to increase the life of a car battery. The mean lifetime of the battery is 36 months. In this situation, the chemist in only interested in increasing the lifetime of the battery.

The null hypothesis states that the additive has no effect on the battery life or even shortens it.

$$H_0\ \mu \leq 36$$

The alternative hypothesis is that the additive increases the battery life.

$$H_1\ \mu > 36$$

Example: Will it decrease?

A contractor wishes to lower heating bills by using a special type of insulation in homes. The average monthly heating bill is $78.00.

The null hypothesis states that the insulation has no effect on the monthly heating bill or even increases it.

$$H_0\ \mu \geq \$78$$

The alternative hypothesis is that the insulation decreases the monthly heating bill.

Instant Practice

1. State the null hypothesis and alternative hypothesis for each of the following situations. Give your answer in words first and then in mathematical symbols.
   a. A psychologist is curious whether the playing of soft music during a test will change the results of the test. The psychologist is not sure whether the grades will be higher or lower. In the past, the mean of the scores without soft music was 73.
   b. An engineer hypothesizes that the mean number of defects can be decreased during the manufacture of compact discs by using robots instead of humans for certain tasks. The mean number of defective discs in human production per 1000 is 18.
   c. A researcher thinks that if expectant mothers use vitamin pills, the birth weight of the babies will increase. The average birth weight of babies from mothers who are not using vitamin pills is 3.9 kg.

Appendix A

Tips for Answering Written Response Type Exam Questions

Answers to written response type questions on an exam are assessed on the basis of how well you communicate both your understanding of the information presented and your understanding of the applicable science.

Key Exam Skills

Evaluators will be looking for examples of your understanding of scientific principles and techniques. In order to successfully answer the questions, you must be able to do the following.

1. Read critically and identify
- key words, phrases, and data that deliver useful information
- distractor information and data that can be ignored because they do not have any bearing on the answer to the question
- if the question is an open-response style that requires a unified response, or if it is a closed-response style that requires a more analytical approach
- precisely what the question is asking
  - pay close attention to the process words (see below)
  - pay close attention to the directing words (see list following) to determine how you should answer the question. The directing words are sometimes highlighted in boldface type.
- the scientific concept(s) that you should include in your answer
- any formulas that you should include in your answer

2. Interpret and Analyze
- process words and directing words
- information including the key words, phrases, and data presented in the information box
- information that is presented in charts, tables, and graphs

3. Communicate
- conclusions by making a formal statement
- results in the form of charts, graphs, or diagrams
- ideas or answers to questions in the form of complete sentences, paragraphs, or short essays

4. If you are asked to perform an experiment, write the experimental design as follows.
- State the problem or questions to be answered.
- Formulate the hypothesis or make a prediction.
- Identify the independent and dependent variables if required.
- Provide a method for controlling variables.
- Identify the required materials clearly.
- Describe any applicable safety procedures.
- Provide a sketch of the apparatus to help make the set-up clear.
- Provide a method for collecting and recording pertinent data—include the units for the data being collected.

Process Words

You may find this list of words helpful in both understanding what the questions are asking and what is expected in your answer.

Hypothesis: A single proposition intended as a possible explanation for an observed phenomenon, e.g., a possible cause for a specific effect

Conclusion: A proposition that summarizes the extent to which a hypothesis and/or theory has been supported or contradicted by evidence

Experiment: A set of manipulations and/or specific observations of nature that allow the testing of hypotheses and/or generalizations

Variables: Conditions that can change in an experiment. Variables in experiments are categorized as

> **Independent variable (manipulated variable):** Condition that was deliberately changed by the experimenter
>
> **Controlled variables:** Conditions that could have changed but did not, because of the intervention of the experimenter
>
> **Dependent variable (responding variable):** Condition that changed in response to the change in the independent variable

Directing Words

Algebraically Using mathematical procedures that involve letters or symbols to represent numbers

Analyze To make a mathematical, chemical, or methodical examination of parts to determine the nature, proportion, function, interrelationship, etc. of the whole

Compare Examine the character or qualities of two things by providing characteristics of both that point out their similarities and differences

Conclude State a logical end based on reasoning and/or evidence

Contrast/ Distinguish Point out the differences between two things that have similar or comparable natures

Criticize Point out the demerits of an item or issue

Define Provide the essential qualities or meaning of a word or concept; make distinct and clear by marking out the limits

Describe Give a written account or represent the characteristics of something by a figure, model, or picture

Design/Plan Construct a plan, i.e., a detailed sequence of actions for a specific purpose

Determine Find a solution to a specified degree of accuracy, to a problem by showing appropriate formulas, procedures, and calculations

Enumerate Specify one-by-one or list in concise form and according to some order

Evaluate Give the significance or worth of something by identifying the good and bad points or the advantages and disadvantages

Explain Make clear what is not immediately obvious or entirely known; give the cause of or reason for; make known in detail

Graphically Use a drawing that is produced electronically or by hand and that shows a relation between certain sets of numbers

How Show in what manner or way, with what meaning

Hypothesize Form a tentative proposition intended as a possible explanation for an observed phenomenon i.e., a possible cause for a specific effect. The proposition should be testable logically and/or empirically.

Identity Recognize and select as having the characteristics of something

Illustrate Make clear by giving an example. The form of the example must be specified in the question i.e., word description, sketch, or diagram

Infer Form a generalization from sample data; arrive at a conclusion by reasoning from evidence

Interpret Tell the meaning of something; present information in a new form that adds meaning to the original data

Justify/ Show How Show reason for or give facts that support a position

Model Find a model that does a good job of representing a situation

Outline Give, in an organized fashion, the essential parts of something. The form of the outline may be specified in the question, e.g., list, flow chart, or concept map.

Predict Tell in advance on the basis of empirical evidence and/or logic

Prove Establish the truth or validity of a statement for the general case by giving factual evidence or logical argument

Relate Show logical or causal connection between things

Sketch Provide a drawing that represents the key features of an object or graph

Solve Give a solution of a problem, i.e., explanation in words and/or numbers

Summarize Give a brief account of the main points

Trace Give a step-by-step description of the development

Verify Establish, by substitution for a particular case or by geometric comparison, the truth of a statement

Why Show the cause, reason, or purpose

Appendix A

Measurement

Scientists have developed globally agreed-upon standards for measurement, and for recording and calculating data. These are the standards that you will use throughout this science program.

Units of Measurement

When you take measurements for scientific purposes, you use the International System of Measurement (commonly know as SI, from the French *Système international d'unités*). SI includes the metric system and other standard units, symbols, and prefixes, which are reviewed in the tables on this page.

In SI, the base units include the metre, the kilogram, and the second. The size of any particular unit can be determined by the prefix used with the base unit. Larger and smaller units of measurement can be obtained by either multiplying or dividing the base unit by a multiple of 10.

For example, the prefix *kilo-* means multiplied by 1000. So, one kilogram is equivalent to 1000 grams:

$$1 \text{ kg} = 1000 \text{ g}$$

The prefix *milli-* means divided by 1000. So, one milligram is equivalent to one thousandth of a gram:

$$1 \text{ mg} = \frac{1}{1000 \text{ g}}$$

The following tables show the most commonly used metric prefixes, as well as some common metric quantities, units, and symbols.

Commonly Used Metric Prefixes

Prefix	Symbol	Relationship to the Base Unit
tera-	T	10^{12} = 1 000 000 000 000
giga-	G	10^{9} = 1 000 000 000
mega-	M	10^{6} = 1 000 000
kilo-	k	10^{3} = 1 000
hecto-	h	10^{2} = 100
deca-	da	10^{1} = 10
—	—	10^{0} = 1
deci-	d	10^{-1} = 0.1
centi-	c	10^{-2} = 0.01
milli-	m	10^{-3} = 0.001
micro-	μ	10^{-6} = 0.000 001
nano-	n	10^{-9} = 0.000 000 001
pico-	p	10^{-12} = 0.000 000 000 001

Commonly Used Metric Quantities, Units, and Symbols

Quantity	Unit	Symbol
Length	nanometre micrometre millimetre centimetre metre kilometre	nm μm mm cm m km
Mass	gram kilogram tonne	g kg t
Area	square metre square centimetre hectare	m^2 cm^2 ha (10 000 m^2)
Volume	cubic centimetre cubic metre millilitre litre	cm^3 m^3 mL L
Time	second	s
Temperature	degree Celsius	°C
Force	N	newton
Energy	joule kilojoule*	J kJ
Pressure	pascal kilopascal**	Pa kPa
Electric current	ampere	A
Quantity of electric charge	coulomb	C
Frequency	hertz	Hz
Power	watt	W

* Many dieticians in North America continue to measure nutritional energy in Calories, also known as kilocalories or dietetic Calories. In SI units, 1 Calorie = 4.186 kJ.

** In current North American medical practice, blood pressure is measured in millimetres of mercury, symbolized as mmHg. In SI units, 1 mmHg = 0.133 kPa.

Accuracy and Precision

In science, the terms accuracy and precision have specific definitions that differ from their everyday meanings.

Scientific *accuracy* refers to how close a given quantity is to an accepted or expected value. For example, under standard (defined) conditions of temperature and pressure, 5 mL of water has a mass of 5 g. When you measure the mass of 5 mL of water under the same conditions, you should, if you are accurate, find the mass is 5 g.

Scientific *precision* refers to the exactness of your measurements. The precision of your measurements is directly related to the instruments you use to make the measurements. While faulty instruments (for example, a balance that is not working properly) will likely affect both the accuracy and the precision of your measurements, the calibration of the instruments you use is the factor that most affects precision. For example, a ruler calibrated in millimetres will allow you to make more precise measurements than one that shows only centimetres.

Precision also describes the repeatability of measurements. The closeness of a series of data points on a graph is an indicator of repeatability. Data that are close to one another, as in graph A, below, are said to be precise.

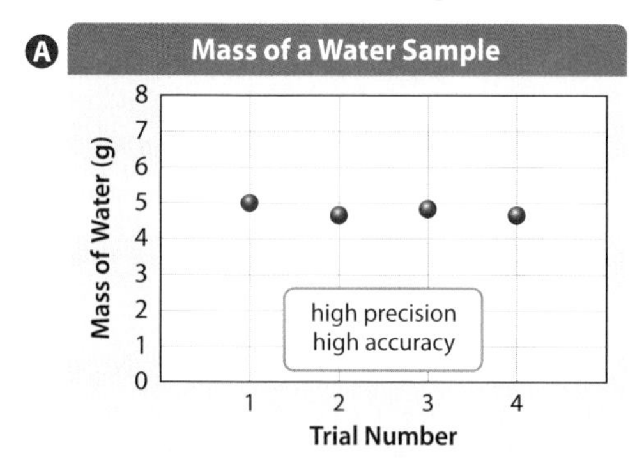

Graph A shows a group of data with high accuracy, since the data points are all grouped around 5 g.

There is no guarantee, however, that the data are accurate until a comparison with an accepted value is made. For example, graph B shows a group of measurements that are precise, but not accurate, since they report the mass of a 5 g sample of water as approximately 7 g.

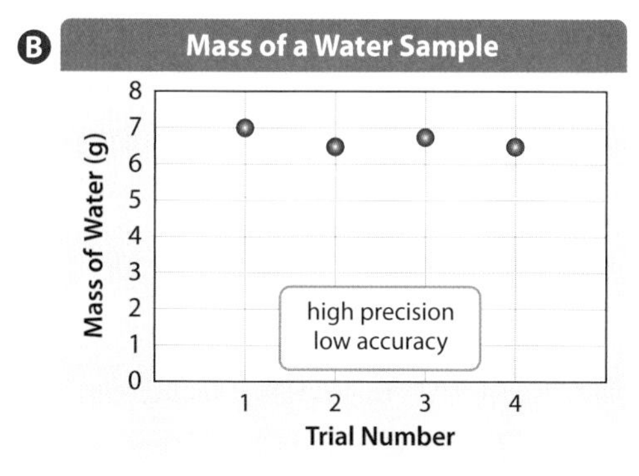

Graph B shows data with low accuracy, since the data points are grouped around 7 g.

In graph C, the data points give an accurate value for average mass, but they are not precise.

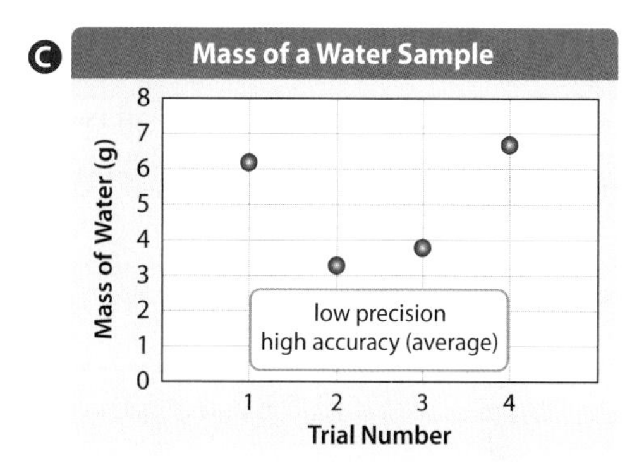

In graph C, the individual data points are not very accurate, since they are all more than 1 g away from the expected value of 5 g. However, taken as a group, the data set in graph C has high accuracy, since the average mass from the four trials is 5 g.

Error

Error exists in every measured or experimentally obtained value. Even the most careful scientist cannot avoid having error in a measurement. *Random error* results from uncontrollable variation in how we obtain a measurement. For example, human reflexes vary, so it is not possible to push the stem of a stopwatch exactly the same way every time. No measurement is perfect. Repeating trials will reduce but never eliminate the effects of random error. Random error affects precision and, usually, accuracy.

Systematic error results from consistent bias in observation. For example, a scale might consistently give a reading that is 0.5 g heavier than the actual mass of a sample, or a person might consistently read the scale of a measuring instrument incorrectly. Repeating trials will not reduce systematic error. Systematic error affects accuracy.

Percentage Error

The amount of error associated with a measurement can be expressed as a percentage, which can help you to evaluate the accuracy of your measurement. The higher the *percentage error* is, the less accurate the measurement. Percentage error is calculated using the following equation:

$$\text{Percentage error} = \left|\frac{\text{measured} - \text{expected value}}{\text{expected value}}\right| \times 100\%$$

(Note that the vertical lines surrounding the fraction mean *the absolute value of* the expression within the lines. That is, the expression's numerical value should be reported without a positive or negative sign.) As an example, a student measures a 5 mL sample of water and finds the mass to be 4.6 mL.

$$\begin{aligned}\text{percentage error} &= \left|\frac{4.6\text{ mL} - 5\text{ mL}}{5\text{ mL}}\right| \times 100\% \\ &= \left|\frac{-0.4\text{ mL}}{5\text{ mL}}\right| \times 100\% \\ &= 8\%\end{aligned}$$

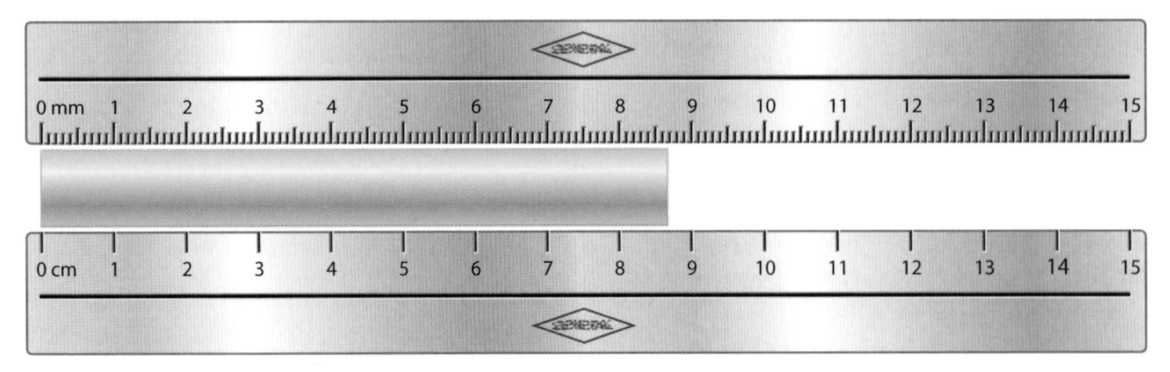

Estimated uncertainty is half of the smallest visible division. In this case, the estimated uncertainty is ±0.5 mm for the top ruler and ±0.5 cm for the bottom ruler.

Uncertainty

Estimated uncertainty describes the limitations of a measuring device. It is defined as half of the smallest division of the measuring device. For example, a metre stick with only centimetres marked on it would have an error of ±0.5 cm. A ruler that includes millimetre divisions would have a smaller error of ±0.5 mm (0.05 cm, or a 10-fold decrease in error). A measurement can be recorded with its estimated uncertainty. In the diagram at the top of the page, for example, the top ruler gives a measurement of 8.69 ±0.05 cm, while the bottom ruler gives a measurement of 8.7 ±0.5 cm.

You can convert the estimated uncertainty into a percentage of the actual measured value using the following equation:

$$\text{Relative uncertainty} = \frac{\text{estimated uncertainty}}{\text{actual measurement}} \times 100\%$$

Example

Convert the error represented by 22.0 ±0.5 cm to a percentage.

$$\text{Relative uncertainty} = \frac{0.5 \text{ cm}}{22.0 \text{ cm}} \times 100\%$$
$$= 2\%$$

Estimating

Sometimes it is not practical or possible to make an accurate measurement of a quantity. You must instead make an *estimate*—an informed judgement that approximates a quantity. For example, if you were conducting an experiment to compare the number of weeds in a field treated with herbicide with the number of weeds in an untreated field, counting the weeds would be impractical, if not impossible. Instead, you could count the number of weeds in a typical square metre of each field. You could then estimate the number of weeds in the entire field by multiplying the number of weeds in a typical square metre by the number of square metres in the field. To make a reasonable estimate of the number of weeds in the field, though, you would need to sample many areas, each 1 m^2, and then calculate an average to determine the number of weeds in a typical square metre for each field.

Estimating can be a valuable tool in science. It is important to keep in mind, however, that the number of samples you take can greatly influence the reliability of your estimate. To make a good estimate, include as many samples as is practical.

Instant Practice

1. During an investigation on cellular respiration, you repeat the same experiment three times to obtain three volumes of gases with the measurements 12.03 mL, 12.10 mL, and 9.58 mL. Your calculations indicate that the expected volume is 12.0 mL. Analyze your results in terms of accuracy and precision.
2. Calculate the percentage error for each of the measurements in question 1, and for the average of the three measurements. How does the accuracy of the individual measurements differ from the accuracy of the group of measurements?
3. The estimated uncertainty of the measurements in question 1 is ±0.05 mL. Calculate the relative uncertainty of the average you determined in question 2.

Appendix A

Significant Digits and Rounding

You might think that a measurement is an exact quantity. In fact, all measurements involve uncertainty. The measuring device is one source of uncertainty, and you, as the reader of the device, are another. Every time you take a measurement, you are making an estimate by interpreting the reading. For example, the illustration below shows a ruler measuring the length of a rod. The ruler can give quite an accurate reading, since it is divided into millimetre marks. But the end of the rod falls between two marks. There is still uncertainty in the measurement. You can be certain that if the ruler is accurate, the length of the rod is between 5.2 mm and 5.3 mm. However, you must estimate the distance between the 2 mm and 3 mm marks.

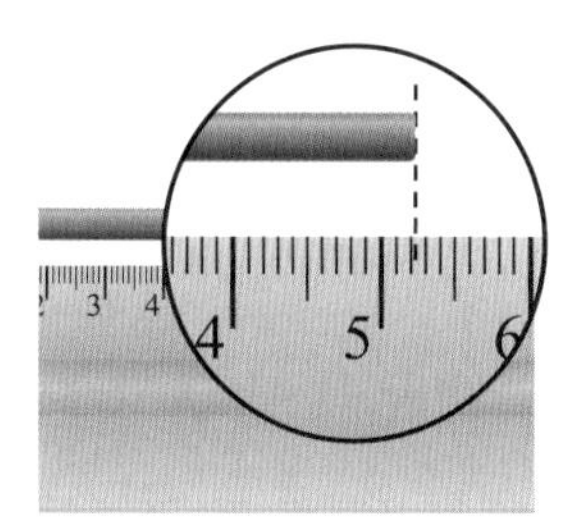

Significant Digits

Significant digits are the digits you record when you take a measurement. The significant digits in a measured quantity include all the certain digits plus the first uncertain digit. In the example above, suppose you estimate the length of the rod to be 5.23 cm. The first two digits (5 and 2) are certain (those marks are visible), but the last digit (0.03) is estimated. The measurement 5.23 cm has three significant digits.

Determining the Number of Significant Digits

The following rules will help you determine the number of significant digits in a given measurement.

1. All non-zero digits (1–9) are significant.
   Examples:
   - 123 m – three significant digits
   - 23.56 km – four significant digits
2. Zeros between non-zero digits are also significant.
   Examples:
   - 1207 m – four significant digits
   - 120.5 km/h – four significant digits
3. Any zero that follows a non-zero digit *and* is to the right of the decimal point is significant.
   Examples:
   - 12.50 m/s^2 – four significant digits
   - 6.0 km – two significant digits
4. Zeros that are to the left of a measurement are not significant.
   Examples:
   - 0.056 – two significant digits
   - 0.007 60 – three significant digits
5. Zeros used to indicate the position of the decimal are not significant. These zeros are sometimes called spacers.
   Examples:
   - 500 km – one significant digit (the decimal point is assumed to be after the final zero)
   - 0.325 m – three significant digits
   - 0.000 34 km – two significant digits
6. In some cases, a zero that appears to be a spacer is actually a significant digit. All counting numbers have an infinite number of significant digits.
   Examples:
   - 6 apples – infinite number of significant digits
   - 125 people – infinite number of significant digits
   - 450 deer – infinite number of significant digits

Instant Practice

Determine the number of significant digits in each measurement.

a. 46 units
b. 2973 L
c. 82.9 cm
d. 9.0034 W
e. 20.380 g
f. 0.0073 mm
g. 0.340 kg
h. 2 s
i. 400 J
j. 439.0001 km

Using Significant Digits in Mathematical Operations

When you use measured values in mathematical operations, the calculated answer cannot be more certain than the measurements on which it is based. Often the answer on your calculator will have to be rounded to the correct number of significant digits.

Rules for Rounding

1. When the first digit to be dropped is less than 5, the preceding digit is not changed.

   Example:

   6.723 m rounded to two significant digits is 6.7 m. The digit after the 7 is less than 5, so the 7 does not change.

2. When the first digit to be dropped is 5 or greater, the preceding digit is increased by one.

   Example:

   7.237 m rounded to three significant digits is 7.24 m. The digit after the 3 is greater than 5, so the 3 is increased by one.

3. When the first digit to be dropped is 5, and there are no following digits, increase the preceding number by 1 if it is odd, but leave the preceding number unchanged if it is even.

   Examples:

   8.345 L rounded to two significant digits is 8.34 L, because the digit before the 5 is even.

   [illegible] L rounded to two significant digits is [illegible] L, because the digit before the 5 is odd.

Adding or Subtracting Measurements

Perform the mathematical operation, and then round off the answer so it has the same number of significant digits as the value that has the fewest decimal places.

Example:

Add the following measured lengths and express the answer to the correct number of significant digits.

$$x = 2.3 \text{ cm} + 6.47 \text{ cm} + 13.689 \text{ cm}$$
$$= 22.459 \text{ cm}$$
$$= 22.5 \text{ cm}$$

Since 2.3 cm has only one decimal place, the answer can have only one decimal place.

Multiplying or Dividing Measurements

Perform the mathematical operation, and then round off the answer so it has the same number of significant digits as the value that has the least number of significant digits.

Example:

Multiply the following measured lengths and express the answer to the correct number of significant digits.

$$x = (2.342 \text{ m})(0.063 \text{ m})(306 \text{ m})$$
$$= 45.149\,076 \text{ m}^3$$
$$= 45 \text{ m}^3$$

Since 0.063 m has only two significant digits, the final answer must also have two significant digits.

Instant Practice

Perform the following calculations, rounding off your answer to the correct number of significant digits.

a. 2.0 cm + 0.52 cm + 3.2 cm

b. 642.0 mg/33.402 mg

c. 23.8 L × 0.00321 L

d. 8.045 kJ − 3.10 kJ

e. 72.1 mm/0.3 mm + 2.09 mm

Appendix A

Scientific Notation and Logarithms

An exponent is the symbol or number denoting the power to which another number or symbol is to be raised. The exponent shows the number of repeated multiplications of the base. In 10^2, the exponent is 2 and the base is 10. The expression 10^2 means 10×10.

Powers of 10

Digits	Standard Form	Exponential Form
Ten thousands	10 000	10^4
Thousands	1 000	10^3
Hundreds	100	10^2
Tens	10	10^1
Ones	1	10^0
Tenths	0.1	10^{-1}
Hundredths	0.01	10^{-2}
Thousandths	0.001	10^{-3}
Ten thousandths	0.0001	10^{-4}

Why use exponents? Consider this: One molecule of water has a mass of 0.000 000 000 000 000 000 000 029 9 g. Using such a number for calculations would be quite awkward. The mistaken addition or omission of a single zero would make the number either 10 times larger or 10 times smaller than it actually is. Scientific notation allows scientists to express very large and very small numbers more easily, to avoid mistakes, and to clarify the number of significant digits.

Scientific Notation

In scientific notation, a number has the form $x \times 10^n$, where x is greater than or equal to 1 but less than 10, and 10^n is a power of 10. To express a number in scientific notation, use the following steps:

1. To determine the value of x, move the decimal point in the number so that only one non-zero digit is to the left of the decimal point.
2. To determine the value of the exponent n, count the number of places the decimal point moves to the left or right. If the decimal point moves to the right, express n as a positive exponent. If the decimal point moves to the left, express n as a negative exponent.
3. Use the values you have determined for x and n to express the number in the form $x \times 10^n$.

Examples

Express 0.000 000 000 000 000 000 000 029 9 g in scientific notation.

1. To determine x, move the decimal point so that only one non-zero number is to the left of the decimal point:

   2.99

2. To determine n, count the number of places the decimal moved:

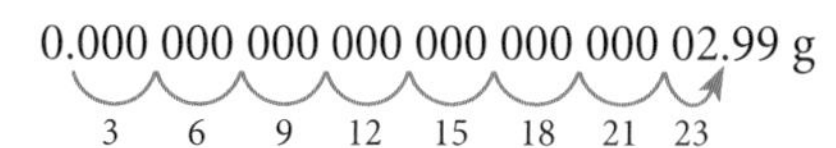

   Since the decimal point moved to the right, the exponent will be negative.

3. Express the number in the form $x \times 10^n$:

   $$2.99 \times 10^{-23}\ \text{g}$$

Express 602 000 000 000 000 000 000 000 in scientific notation.

1. To determine x, move the decimal point so that only one non-zero number is to the left of the decimal point:

   6.02

2. To determine n, count the number of places the decimal moved:

   6.02 000 000 000 000 000 000 000.

   23 21 18 15 12 9 6 3

   Since the decimal point moved to the left, the exponent will be positive.

3. Express the number in the form $x \times 10^n$:

   $$6.02 \times 10^{23}$$

Rules for Scientific Notation

1. To multiply two numbers in scientific notation, add the exponents.

   Example:

   $(7.32 \times 10^{-3}) \times (8.91 \times 10^{-2})$

   $= (7.32 \times 8.91) \times 10^{(-3)+(-2)}$

   $= 65.2212 \times 10^{-5}$

   $= 6.52 \times 10^{-4}$

   (Remember to report the correct number of significant digits in your answer. Also, you will need to move the decimal point to achieve correct scientific notation.)

2. To divide two numbers in scientific notation, subtract the exponents.

   Example:

   $(1.842 \times 10^6) \div (1.0787 \times 10^2)$

   $= (1.842 \div 1.0787) \times 10^{(6-2)}$

   $= 1.707\ 611 \times 10^4$

   $= 1.708 \times 10^4$

3. To add or subtract numbers in scientific notation, first convert the numbers so they have the same exponent. Each number should have the same exponent as the number with the greatest power of 10. Once the numbers are all expressed to the same power of 10, the power of 10 can be ignored (neither added nor subtracted) in the calculation.

   Example:

   $(3.42 \times 10^6) + (8.53 \times 10^3)$

   Express 8.53×10^3 as a number multiplied by 10^6. Then perform the addition.

   $= (3.42 \times 10^6) + (0.008\ 53 \times 10^6)$

   $= 3.428\ 53 \times 10^6$

   $= 3.43 \times 10^6$

Logarithms

Logarithms are a convenient method for communicating large and small numbers. The logarithm, or log, of a number is the value of the exponent to which 10 would need to be raised in order to equal this number. Every positive number has a logarithm. Numbers that are greater than 1 have a positive logarithm. Numbers that are between 0 and 1 have a negative logarithm. The number 1 has a logarithm of 0.

Some Numbers and Their Logarithms

Number	Scientific Notation	As a Power of 10	Logarithm
1 000 000	1×10^6	10^6	6
7 895 900	7.8959×10^6	$10^{6.897\ 40}$	6.897 40
1	1×10^0	10^0	1
0.000 001	1×10^{-5}	10^{-5}	−5
0.004 276	4.276×10^{-3}	$10^{-2.3690}$	−2.3690

For logarithmic values, only the digits to the right of the decimal point count as significant digits. The digit to the left of the decimal point fixes the location of the decimal point of the original value. (Notice in the second row of the table at the bottom of the page that 7 895 900, which has 5 significant digits, has a log of 6.897 40. The log has 5 digits to the right of the decimal point.)

Logarithms are especially useful for expressing values that span a range of powers of 10. The Richter scale for earthquakes, the decibel scale for sound, and the pH scale for acids and bases all use logarithmic scales.

Logarithms and pH

The pH of an acid solution is defined as $-\log[H^+]$. (The square brackets mean *concentration*.) Thus, you can find the pH of a solution by taking the negative log of the concentration of hydrogen ions.

Example:

Find the pH of a solution with a hydrogen ion concentration of 0.004 76 mol/L.

$$-\log[0.004\ 76 \text{ mol/L}] = 2.322$$

The pH scale is a negative log scale. Thus, a decrease from pH 7 to pH 4 is actually an increase of 10^3, or 1000 times, in the acidity of a solution. An increase from pH 3 to pH 6 is a decrease in acidity of 10^3 times.

Instant Practice

1. Convert the following values into scientific notation.
   - **a.** [illegible]
   - **b.** [illegible]
   - **c.** 99 482
   - **d.** 0.000 718
   - **e.** 8 382 441 002
2. Convert the following numbers from scientific notation into ordinary numbers.
   - **a.** 2.3×10^3
   - **b.** 8.003×10^{-2}
   - **c.** 1.3492×10^6
3. Perform the following calculations.
   - **a.** $(5.2 \times 10^5) + (1.32 \times 10^2)$
   - **b.** $(9.231 \times 10^{-3}) - (2.4 \times 10^1)$
   - **c.** $(7.92 \times 10^{-1}) \times (3.05 \times 10^{-3})$
   - **d.** $(8.228 \times 10^4) \times (1.1 \times 10^5)$
4. Determine the pH of a solution, given each hydrogen ion concentration.
   - **a.** 0.00001 mol/L
   - **b.** 1 000 mol/L
   - **c.** 3.28 mol/L

Appendix A

Using a Microscope

Part 1: Care of a Microscope

A compound light microscope is an optical instrument with a series of lenses that greatly magnify objects too small to be seen with the unaided eye. Study the compound light microscope shown on the facing page and review the major parts and their functions.

To keep your microscope in good operating condition, the following points should be observed.

1. To carry a microscope, always use one hand to hold the arm and your other hand to support the base.
2. Do not touch the lens surfaces with your fingers.
3. Use only lens tissue to clean the lens surfaces.
4. Do not adjust any of the focussing knobs until you are ready to use the microscope.
5. Always focus first using the coarse adjustment knob, with the low-power objective lens in position.
6. Do not use the coarse adjustment knob when either the medium-power or high-power objective lens is in position.
7. Cover the microscope when it is not in use.

Part 2: Use of a Microscope

Here, you will use the microscope to view a prepared slide. You will determine the area that can be seen through the eyepiece, called the *field of view*, and calculate the magnification. Finally, you will estimate the actual size of the object you are viewing.

CAUTION: Be sure your hands are dry when handling electrical equipment. Handle microscope slides carefully, since they can break easily and cause cuts.

Materials

- microscope
- prepared microscope slide
- clear plastic ruler

Procedure

1. Place the microscope on a flat surface.
2. The microscope should always be stored with the low-power objective in position. If your microscope has not been stored that way, look from the side to ensure that the objectives do not collide with the stage and rotate the revolving nosepiece until the low-power objective clicks into place.
3. Use the coarse-adjustment knob to lower the low-power objective until the lens is about 1 cm above the stage.
4. Look through the eyepiece and adjust the diaphragm until the view is as bright as possible.

Total Magnification and Field of View

5. To calculate the *total magnification* of an object, multiply the power of the eyepiece by the power of the objective. For example, if the eyepiece magnification is 10×, the low-power objective is 4×, and the high-power objective is 40×, then
   a. The total magnification using the low-power objective is $10 \times 4 = 40\times$.
   b. The total magnification using the high-power objective is $10 \times 40 = 400\times$.
6. To determine the diameter of the field of view, place the clear plastic ruler on the stage.
7. Using the coarse-adjustment knob, focus on the ruler. Position the ruler so that one of the millimetre markings is at the left edge of the field of view, as shown below.

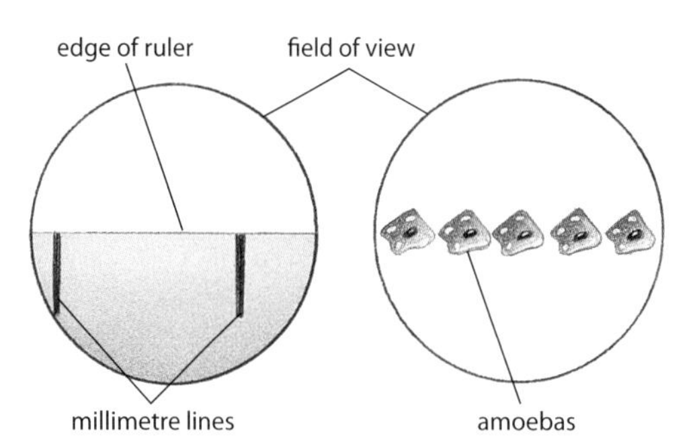

The diameter of the field of view under low power illustrated here is about 1.5 mm.

8. Measure and record the diameter of the field of view in millimetres (mm) for the low-power objective.
9. Use the following formula to calculate the field of view for the medium-power objective:

$$\text{Medium-power field of view} = \text{low-power field of view} \times \frac{\text{magnification of low-power objective}}{\text{magnification of medium-power objective}}$$

For example, if the low-power objective is 4× with a field of view of 2 mm, and the medium-power objective is 10×, then

$$\begin{aligned}\text{Medium-power field of view} &= 2\text{ mm} \times \frac{4}{10}\\ &= 2\text{ mm} \times 0.4\\ &= 0.8\text{ mm}\end{aligned}$$

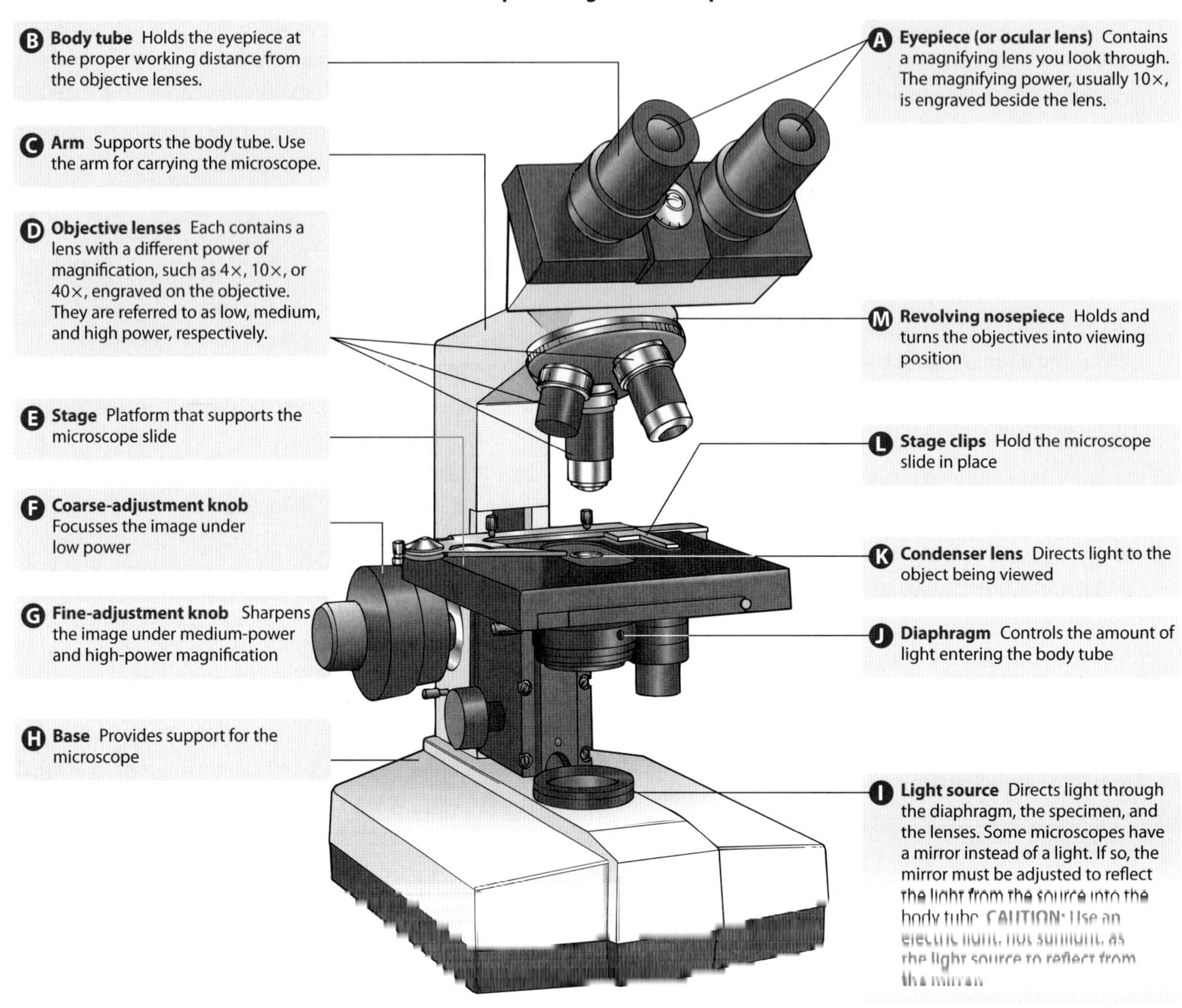

B Body tube Holds the eyepiece at the proper working distance from the objective lenses.
C Arm Supports the body tube. Use the arm for carrying the microscope.
D Objective lenses Each contains a lens with a different power of magnification, such as 4×, 10×, or 40×, engraved on the objective. They are referred to as low, medium, and high power, respectively.
E Stage Platform that supports the microscope slide
F Coarse-adjustment knob Focusses the image under low power
G Fine-adjustment knob Sharpens the image under medium-power and high-power magnification
H Base Provides support for the microscope
A Eyepiece (or ocular lens) Contains a magnifying lens you look through. The magnifying power, usually 10×, is engraved beside the lens.
M Revolving nosepiece Holds and turns the objectives into viewing position
L Stage clips Hold the microscope slide in place
K Condenser lens Directs light to the object being viewed
J Diaphragm Controls the amount of light entering the body tube
I Light source Directs light through the diaphragm, the specimen, and the lenses. Some microscopes have a mirror instead of a light. If so, the mirror must be adjusted to reflect the light from the source into the body tube. CAUTION: Use an
the light source to reflect from

10. Objects in the field of view of a microscope are usually measured in micrometres (μm). One micrometre equals 0.001 mm; or 1000 μm equals 1 millimetre.

    In the example in step 9, the field of view under the medium-power objective would be $0.8 \text{ mm} \times 1000 = 800\ \mu\text{m}$.

Calculating Object Size

11. You can determine the size of a specimen (such as an amoeba) by estimating how many could fit end to end across the field of view. (See the diagram on the previous page.) To do this, divide the field of view by the number of specimens. If the field of view in the illustration is 1500 μm, what is the diameter of each amoeba?

Drawing Magnification

12. Go to Biological Drawing on page 653 for instructions on how to do a scale drawing. If your teacher instructs you to calculate the *drawing magnification*, you can use the following equation:

$$\text{Drawing magnification} = \frac{\text{size of drawing}}{\text{size of object}}$$

You can use the specimen size you calculated in step 11 for the size of the object. For the drawing size, you will need to measure your drawing with a ruler. Be sure that if you measure the width of the object in the field of view, you also measure the width of the drawing at the same point. Also, be sure to express both measurements in the same units.

Viewing a Prepared Slide

13. Place a prepared slide on the stage and secure it in place with the stage clips. The low-power objective should be in position. Make sure the object you intend to view is centred over the opening in the stage.

a. Look through the eyepiece. Slowly turn the coarse-adjustment knob until the object is in focus.

b. Use the fine-adjustment knob to sharpen the focus.

14. Once the object is in focus using low power, carefully rotate the revolving nosepiece to the medium-power objective. Look at the side of the objective as you rotate the nosepiece to be sure the objective lens does not strike the surface of the slide.

a. Adjust the focus using *only* the fine-adjustment knob.

b. Next, view the object using the high-power objective. Carefully rotate the nosepiece until the high-power objective clicks into position. Again, be sure the objective does not strike the surface of the slide as you rotate the nosepiece. Adjust the focus using *only* the fine-adjustment knob.

15. Once you have finished viewing the slide, carefully rotate the nosepiece until the low-power objective is in position. Remove the slide from the stage and return it to its proper container. Unplug the light source and return the microscope to its cabinet.

CAUTION: Never tug on the electrical cord to unplug it.

Part 3: Preparation of a Wet Mount

Now prepare and view slides of a variety of specimens.

CAUTION: Be careful when using sharp objects such as tweezers. Handle microscope slides and cover slips carefully, since they can break easily.

Materials

- small piece of newspaper and other samples
- tap water
- cotton ball
- microscope slides
- cover slips
- tweezers
- medicine dropper
- microscope

Procedure

1. To prepare a wet mount, begin with a clean slide and cover slip. Hold the slide and cover slip by their edges to avoid getting your fingerprints on their surfaces.

2. Tear out a small piece of newspaper containing a single letter. Use an *e*, *f*, *g*, or *h*. Using the tweezers, position the letter in the centre of the slide.

3. Using the medicine dropper, place one drop of water on the sample. Hold a cover slip over the sample at a 45° angle. One edge of the cover slip should touch the surface of the slide near the newspaper letter sample.

4. Slowly lower the opposite edge of the cover slip over the sample. Be sure no bubbles form beneath the cover slip. This type of sample preparation is called a wet mount.

5. With the low-power objective of the microscope in position, place the slide on the stage and secure it with the stage clips. Centre the sample over the opening.

a. Look though the eyepiece. Reposition the slide, if necessary, until you can see the letter. Using the coarse-adjustment knob, focus on the letter. Then, adjust the focus with the fine-adjustment knob.

b. Examine the letter using medium power. Note that it is composed of many small dots.

6. To reveal the structure of small objects, the microscope must do more than magnify—it must also reveal detail. The capacity to distinguish detail is called *resolution*, and the measure of resolution is known as *resolving power*. The resolving power of a microscope is defined as the minimum distance that two objects can be apart and still be seen as separate objects. Prepare another wet mount using several fibres from a cotton ball. Using the low-power objective, locate a part of the slide where two fibres cross each other. Change to the high-power objective. Use the fine-adjustment knob to focus on the fibres. Can both strands of cotton be seen clearly at the same time under high power? How might you explain this result?

Instant Practice

1. Calculate the total magnification of an object viewed with an eyepiece magnification of 10X and a medium-power objective of 20X.

2. The diameter of a field of view under low power measures 1.3 mm.

a. Calculate the medium-power field of view if the magnification of the low-power objective is 10X and of the medium-power objective is 20X.

b. Eight plant cells fit end-to-end across the diameter of the medium-power field of view. What is the width of one plant cell? Express this value in μm.

Biological Drawing

A clear, concise drawing can often replace words in a scientific description. Drawings are especially important when you are trying to explain difficult concepts or describe complex structures. Follow these steps to make a good scientific drawing:

1. Use an unlined (blank) sheet of paper and a sharp lead pencil, ideally 2H, for the drawing, title, and all labels.
2. Make sure your drawing will be large enough to show all the necessary details; a drawing about half a page in size is usually sufficient. Also allow space for the labels, which identify parts of the object you are drawing. Place all labels to the right of your drawing.
3. Make your drawing to scale. Imagine that the object you are viewing or the field of view you are looking at under the microscope is divided into four equal sections, as shown below. Imagine that your drawing is also divided into four sections. The object should occupy the same proportion of space within each quarter of the drawing as it does in each quarter of the field of view. Indicate the total magnification of the object. You can also calculate the actual size of the object if your teacher instructs you to do so. (Go to Using a Microscope in Appendix A.)

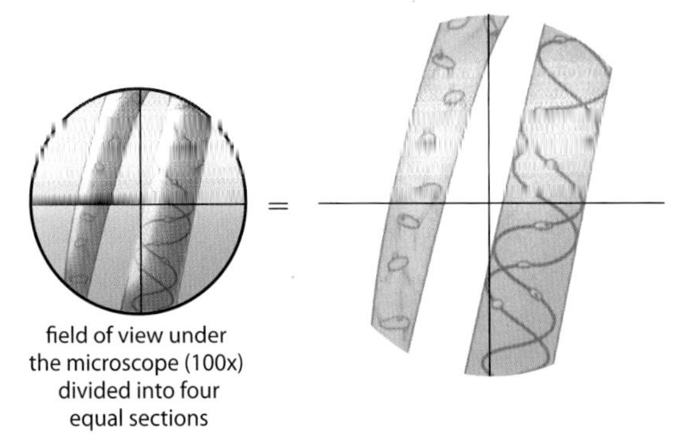

4. Make your drawing as simple as possible, using clean-cut pencil lines. (Do not sketch.) Draw only what you observe. Do not draw parts of the object that are not visible from the angle of view you are observing. If you must show another part of the object, make a second drawing. Indicate the angle of view on each drawing.
5. Most animal and plant tissues are composed of many cells. If you are drawing a representative cell of such tissue, include the boundaries of the other cells surrounding it. This approach will provide context for your drawing.
6. Shading is not usually used in scientific drawings. To indicate darker areas in your drawing, use stippling (a series of dots) as shown in the drawing at the bottom of the page.
7. Label your drawing carefully and completely. All labels should be horizontal, printed in lower case, and placed in a column to the right of your drawing. Imagine for a moment that you know nothing about the object you are drawing. Think about what structures you would like identified if you were seeing the drawing for the first time.
8. Use a ruler to draw a horizontal line from each label to the structure you are identifying. Make sure that none of these label lines cross each other.
9. Give your drawing a title. The title should appear immediately above the drawing. The title should be printed and underlined. Indicate the magnification of the drawing in parentheses. Note: The drawing of onion skin cells shown below is from a student's notebook. The student used stippling to show darker areas, included horizontal labels and label lines for each structure observed, gave the drawing a title, and indicated the magnification—all elements of a complete scientific drawing.

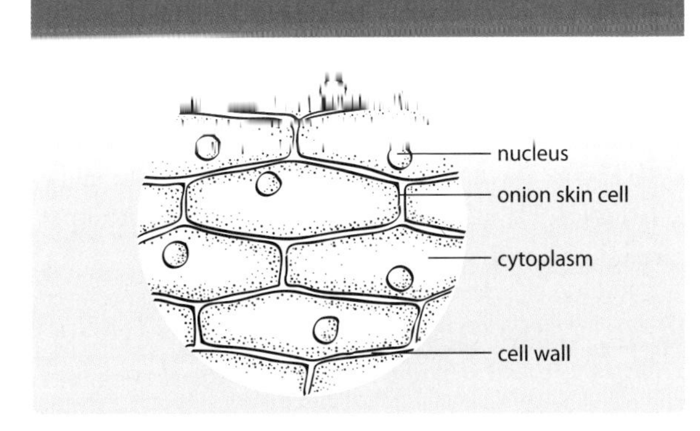

The stippling on this drawing of onion skin cells, as observed under a microscope, shows that some areas are darker in appearance than others.

Instant Practice

Your teacher will provide you with a prepared slide to observe under a microscope. Produce a scale drawing of your observations, complete with title and labels, and indicate the total magnification.

Appendix A

Using Graphic Organizers

When deciding which type of graphic organizer to use, consider your purpose. It may be to brainstorm, to show relationships among ideas, to summarize a section of text, to record research notes, or to review what you have learned before writing a test. Several different graphic organizers are shown here. The descriptions indicate the function or purpose of each organizer.

PMI Chart

PMI stands for Plus, Minus, and Interesting. A PMI chart is a simple three-column table that can be used to state the positive and negative aspects of an issue, or to describe advantages and disadvantages related to the issue. The third column in the chart is used to list interesting information related to the issue. PMI charts help you organize your thinking after reading about a topic that is up for debate or that can have positive or negative effects. They are useful when analyzing an issue.

A diet can help a person to regain a healthy body weight and improve their overall health.

P	M	I
• A healthy body weight reduces strain on the heart and joints. It also reduces the risk of developing some diseases such as diabetes.	• Being too concerned with body weight can lead to low self-esteem and to disorders such as anorexia.	• Healthy body weight varies widely depending on the individual. It is not a case of "one size fits all."
• Calorie-counting can help a person keep track of the energy they consume.	• Fad diet pills and extremely low-calorie diets can lead to destructive metabolic processes and organ failure.	• Individuals have different metabolic rates, and your metabolic rate changes as you age.

Main Idea Web

A main idea web shows a main idea and several supporting details. The main idea is written in the centre of the web, and each detail is written at the end of a line extending from the centre. This organizer is useful for brainstorming or for summarizing text.

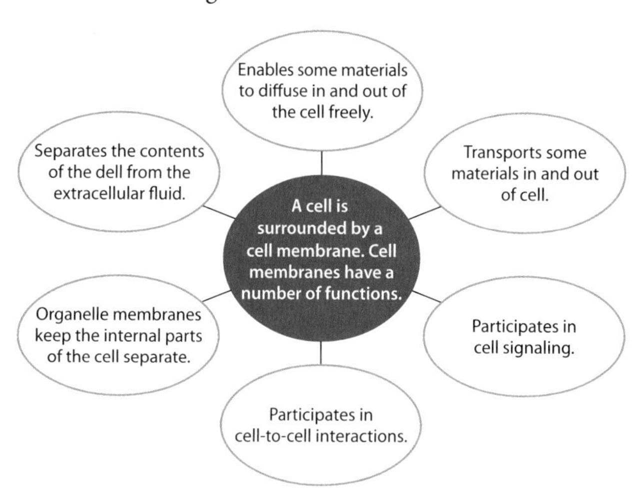

Spider Map

A spider map shows a main idea and several ideas associated with the main idea. It does not show the relationships among the ideas. A spider map is useful when you are brainstorming or taking notes.

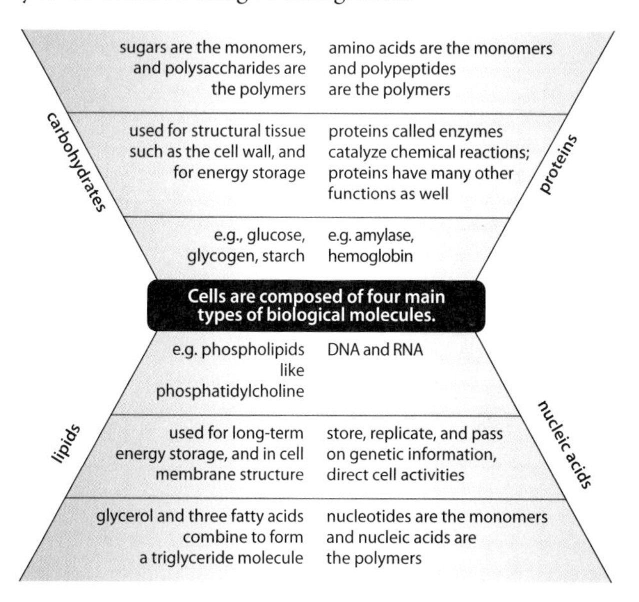

Fishbone Diagram

A fishbone diagram looks similar to a spider map, but it organizes information differently. A main topic, situation, or idea is placed in the middle of the diagram. This is the "backbone" of the "fish." The "bones" (lines) that shoot out from the backbone can be used to list reasons why the situation exists, factors that affect the main idea, or arguments that support the main idea. Finally, supporting details shoot outward from these issues. Fishbone diagrams are useful for planning and organizing a research project. You can clearly see when you do not have enough details to support an issue, which indicates that you need to do additional research.

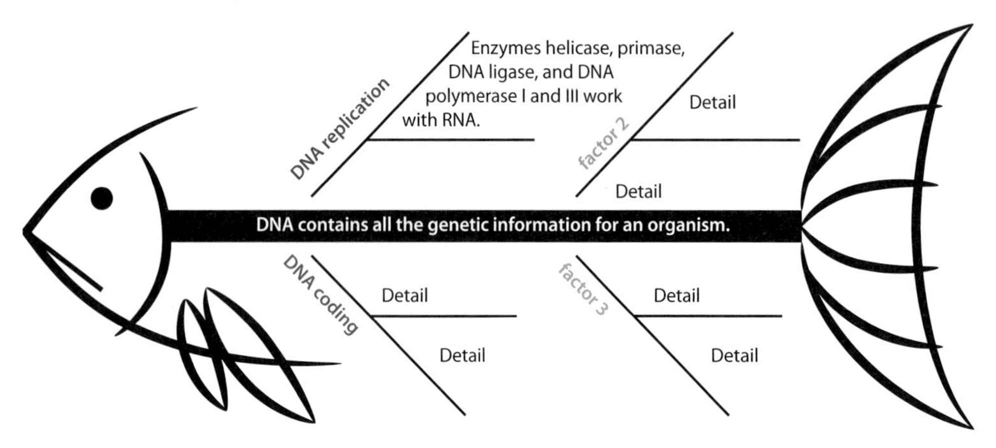

Concept Map

A concept map uses shapes and lines to show how ideas are related. Each idea, or concept, is written inside a circle, a square, a rectangle, or another shape. Words that explain how the concepts are related are written on the lines that connect the shapes.

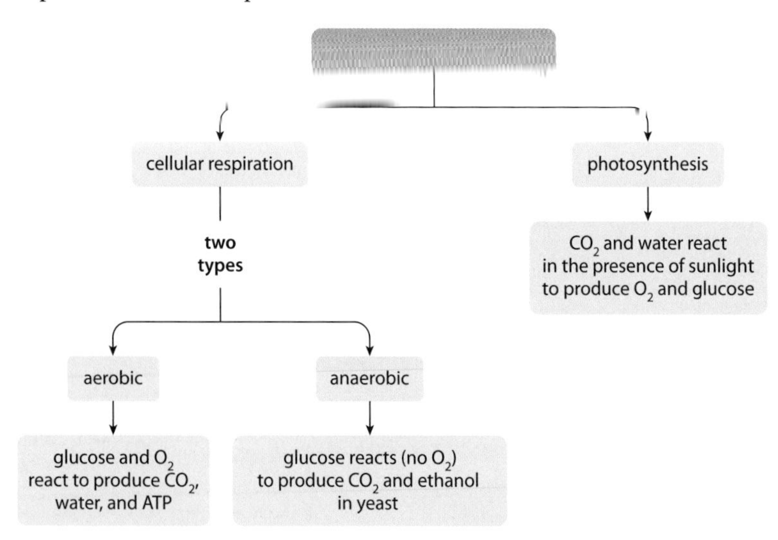

Flowchart

A flowchart shows a sequence of events or the steps in a process. An arrow leads from an initial event or step to the next event or step, and so on, until the final outcome is reached. All the events or steps are shown in the order in which they occur.

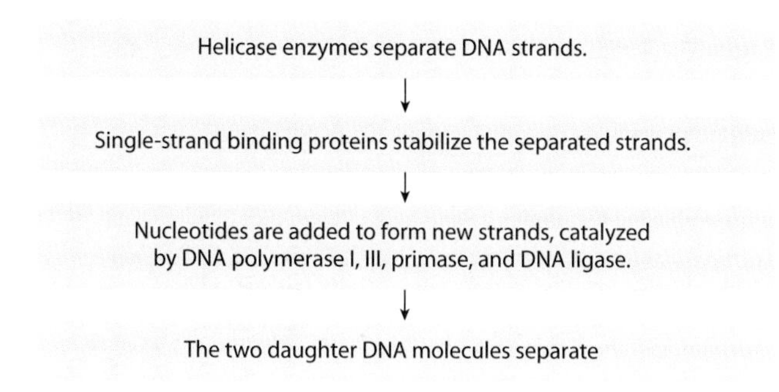

Helicase enzymes separate DNA strands.
Single-strand binding proteins stabilize the separated strands.
Nucleotides are added to form new strands, catalyzed by DNA polymerase I, III, primase, and DNA ligase.
The two daughter DNA molecules separate

Cycle Chart

A cycle chart is a flowchart that has no distinct beginning or end. All the events are shown in the order in which they occur, as indicated by arrows, but there are no first and last events. Instead, the events occur again and again in a continuous cycle. In the cell cycle, shown below, one arrow branches off to show that the cell divides into two. One of the daughter cells starts a new cycle of its own, while the other continues in the original cycle.

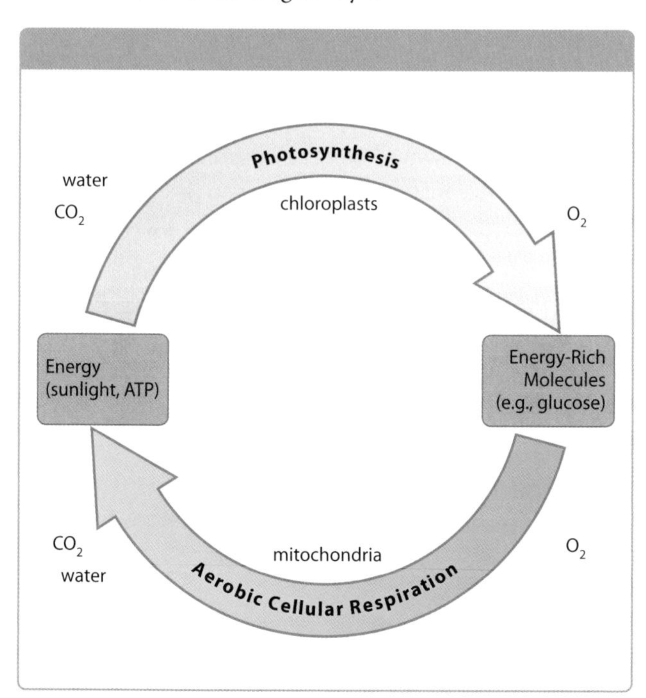

Venn Diagram

A Venn diagram uses overlapping shapes to show similarities and differences among concepts.

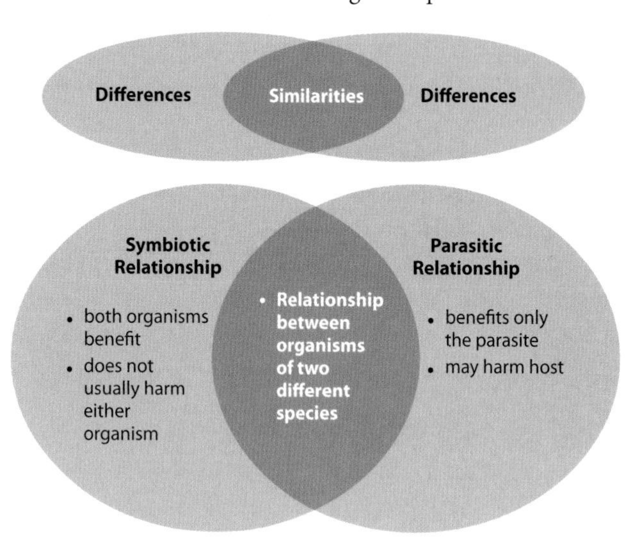

Cause-and-Effect Map

The first cause-and-effect map below shows one cause that results in several effects. The second map shows one effect that has several causes.

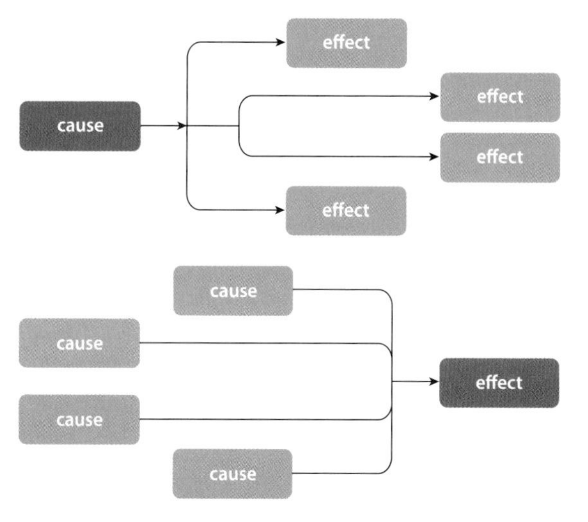

Instant Practice

1. Find and identify three different types of diagrams in this textbook. Discuss with a partner the effectiveness of each diagram in communicating information.
2. Create a flowchart to show the steps involved in the process of aerobic cellular respiration.

Appendix B Useful References

A Brief Chemistry Reference

Matter is anything that takes up space and has mass. All matter can be classified as mixtures or pure substances. Pure substances are either elements or compounds. An element is a form of matter that cannot be broken down further by chemical or physical methods. A compound is made up of two or more elements that have been chemically combined. Mixtures contain two or more pure substances.

Like all matter, the matter in living organisms—whether it is a pure substance or a mixture—is made up of atoms. An ordinary chemical reaction cannot destroy, create, or split an atom. Atoms are made up of subatomic particles—particles that are smaller than an atom.

The Structure of Atoms

To understand and explain the properties of matter and the nature of chemical reactions, you need to know about the subatomic particles called *protons, neutrons,* and *electrons.* Their properties are summarized in the following table.

Protons, Neutrons, and Electrons

Subatomic Particle	Symbol	Type of Charge	Amount of Charge	Mass (u)
Proton	p^+	Positive	+1	1.0
Neutron	n^0	Neutral	0	1.0
Electron	e^-	Negative	–1	0.000 55

In atoms, subatomic particles are arranged in a characteristic structure, as shown in the diagram below. The protons and neutrons are clustered together in the nucleus, which contains over 99 percent of an atom's mass but makes up less than 1 percent of its volume. The electrons surround the nucleus in regions called shells. Electrons make up less than 1 percent of an atom's mass, although the shells they occupy make up over 99 percent of its volume.

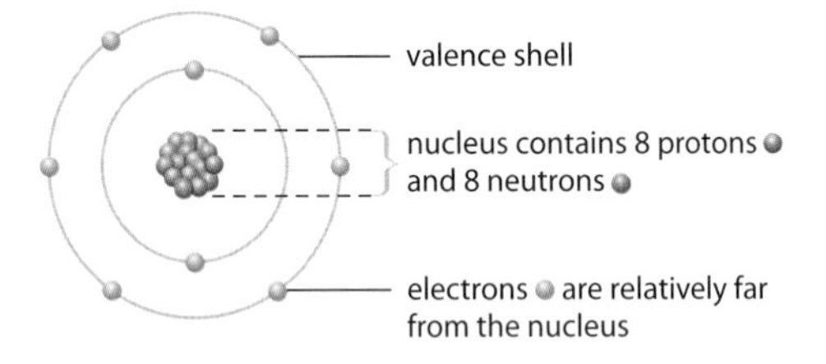

This model of an oxygen atom shows the arrangement of its subatomic particles. Fixed numbers of electrons occupy regions called shells. The outermost shell is called the valence shell.

Different elements, such as hydrogen and oxygen, are distinguished from one another by the number of protons their atoms contain. All atoms of the same element contain the same number of protons. In a neutral atom, the number of its electrons always equals the number of its protons. The periodic table lists and provides information about all the known elements. For a given element, the number of neutrons may vary from one atom to another.

Covalent Bonds

A *covalent bond* forms when the electron shells of two atoms overlap so that the valence electrons of each atom are shared between both atoms. A covalent bond between the two hydrogen atoms in a hydrogen molecule is shown below. Atoms of different elements can also form molecules held together by covalent bonds. Compounds that are made of molecules are called *molecular compounds.*

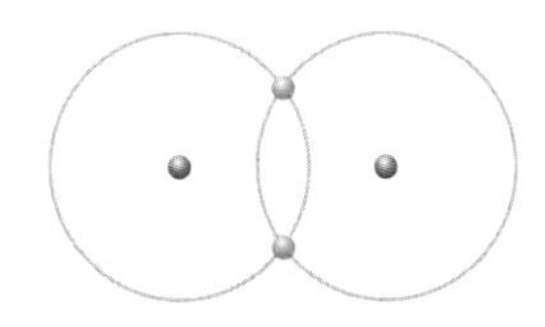

Two hydrogen atoms that share a pair of electrons form a molecule (H_2) with a single covalent bond. The structural formula to show the single covalent bond between the hydrogen atoms is H–H.

The Tendency Toward Stability

The noble gases (Group 18 on the periodic table) are so chemically stable that they are unlikely to take part in chemical reactions. When atoms bond, they share, give up, or gain electrons to achieve the same arrangement of valence electrons as that of the noble gas to which they are closest in the periodic table.

The maximum number of electrons that can occupy the first valence shell outside a nucleus is two (the valence-shell arrangement of the noble gas helium). In a hydrogen molecule, which is made up of two hydrogen atoms, each atom achieves a stable valence-shell arrangement by sharing a valence electron with the other atom. For elements with atomic numbers 3 to 20, the maximum number of electrons that can occupy the valence shell of each atom is eight (the valence-shell arrangement of the other noble gases).

Double and Triple Bonds

The bond between atoms in a hydrogen molecule is a single covalent bond, since it involves a single pair of shared electrons. However, atoms can also share two pairs of electrons or three pairs of electrons in a covalent bond. In a double covalent bond, two atoms share two pairs of electrons. This is shown using two oxygen atoms in the illustration below.

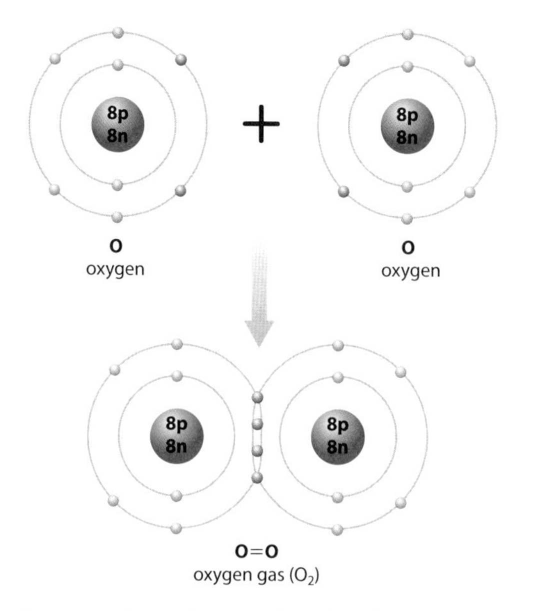

By sharing two pairs of electrons (in a double bond), each of the oxygen atoms has access to eight electrons in its valence shell. This gives each oxygen atom the same stable valence-shell arrangement as the noble gas closest to it in the periodic table, which is neon.

Carbon dioxide (CO_2) is an example of a three-atom molecule held together by double covalent bonds. Examine the CO_2 molecule in the following Lewis structure. Look for evidence that each atom in the molecule has access to a stable valence-shell arrangement.

:Ö: :C: :Ö:

This Lewis structure illustrates a carbon dioxide molecule. The structural formula of the molecule is O=C=O.

In a triple covalent bond, two atoms share three pairs of electrons, as shown in the following Lewis structure for a molecule of nitrogen (N_2).

:N:::N:

This Lewis structure illustrates molecular nitrogen, which consists of two nitrogen atoms joined by a triple bond. The structural formula of the molecule is N–N.

Polar Covalent Bonds and Polar Molecules

Most of the biochemical reactions in a living cell take place in a water solution.

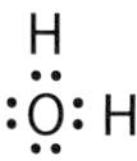

This Lewis structure is a model of a water molecule.

The Lewis structure above shows that a water molecule is held together by covalent bonds. What the diagram does not show is that the oxygen atom attracts electrons with greater force than the hydrogen atoms do, because the oxygen atom has more protons in its nucleus. This type of covalent bond, in which the electrons are unequally shared between atoms, is called a *polar covalent bond*.

There are two polar covalent bonds in water. Because of the bent shape of the water molecule, it has what can be thought of as a "hydrogen end" and an "oxygen end." Due to the polar bonds, the oxygen end of a water molecule has a slight negative charge, while the hydrogen end of the molecule has a slight positive charge. Water is considered a polar molecule. The following diagram provides a representation of the polarity in a water molecule.

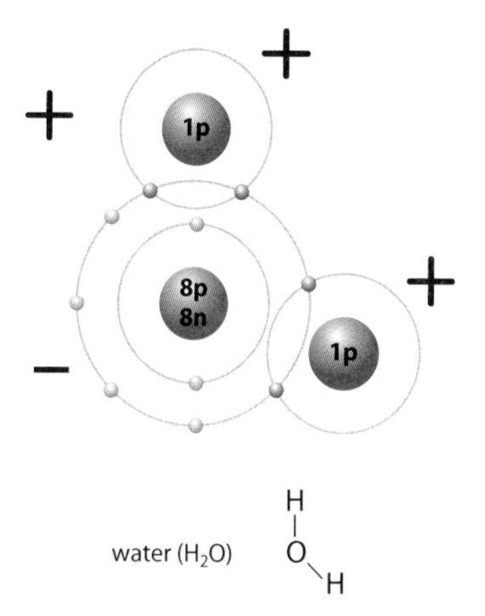

Even though a water molecule is held together with polar covalent bonds and is itself polar, it is electrically neutral overall.

Water and Hydrogen Bonds

When water molecules or other polar molecules are near each other, the slight negative charge on one end of a molecule attracts the slight positive charge on the other end of another molecule.

The attraction that forms between water molecules is called a *hydrogen bond*. Hydrogen bonds are much weaker than covalent bonds, but they are strong compared to other bonds that form between molecules. Hydrogen bonds are responsible for some of the unique properties of water.

For example, hydrogen bonding allows water molecules to "stick together" as they are pulled up the trunk of a tall tree. The illustration below shows the pattern of attractions that forms between liquid water molecules as a result of hydrogen bonding. Hydrogen bonds are found not just in water—they are also in other important biological molecules, such as DNA and proteins.

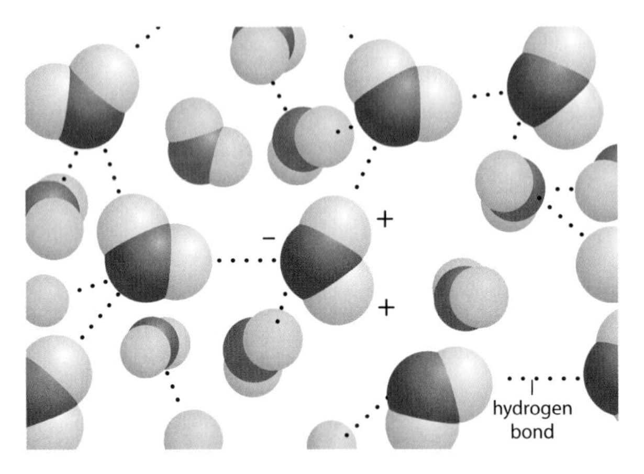

The polarity of water molecules allows attractions called hydrogen bonds to form between the water molecules. The dotted lines represent hydrogen bonds. Dotted lines are used to indicate the weakness of hydrogen bonds relative to covalent and ionic bonds.

Ionic Bonds

Atoms can also form ionic bonds. Ionic bonds form when atoms or groups of atoms transfer electrons rather than share electrons (as in a covalent bond). When an atom or group of atoms gains or loses electrons, it acquires an electric charge and becomes an ion. When the number of electrons is less than the number of protons, the ion is positive (and called a cation). When the number of electrons exceeds the number of protons, the ion is negative (and called an anion). Ions can be composed of only one element, such as the hydrogen ion (H^+), or of several elements, such as the bicarbonate ion (HCO_3^-). The attraction between oppositely charged ions is called an *ionic bond.*

Forming Ionic Compounds

Ionic bonds hold positively and negatively charged ions together within an *ionic compound.* When solid sodium, Na(s), is exposed to chlorine gas, Cl_2(g), there is an explosive reaction that releases both heat and light. In this reaction, electrons are transferred from the sodium atoms to the chlorine atoms. Sodium achieves the stable electron configuration of the noble gas neon, while chlorine achieves the configuration of the noble gas argon.

As shown in the diagram below, two ions are formed simultaneously: Na^+ and Cl^-.

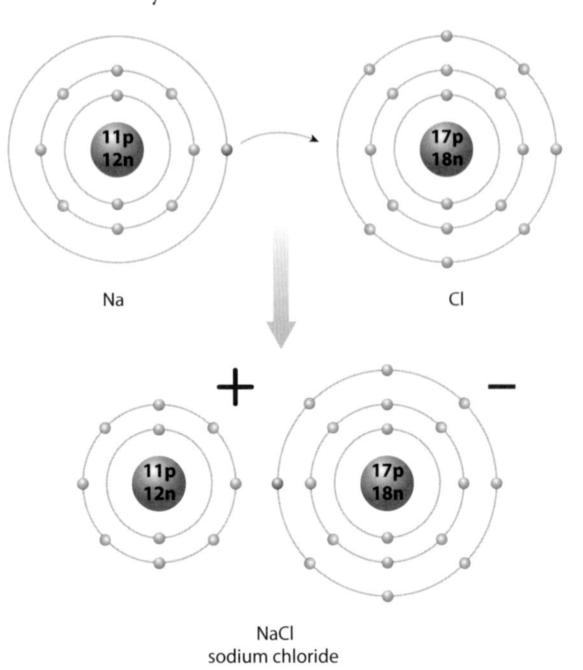

Sodium chloride is created when an ionic bond forms between Na^+ and Cl^-.

The ions align themselves into a regular, repeating pattern based on the size of the individual ions, the amount of charge they carry, and the type of charge they carry (either positive or negative). The ionic compound formed by sodium and chlorine is sodium chloride, NaCl(s), or table salt.

Ionic Compounds in Solution

Table salt and many other ionic compounds dissolve in water. What happens to the ions in sodium chloride when the compound dissolves in water? Attraction by the charged poles of the surrounding water molecules pulls the ions away from the compound and into solution.

Once dissolved in water, the sodium and chloride ions are free to move about and collide with other particles. This makes the ions mobile enough to carry an electric current from one location to another. So, like many ionic compounds, sodium chloride is an electrolyte. An *electrolyte* is a substance that, when dissolved in water, enables the solution to carry an electric current.

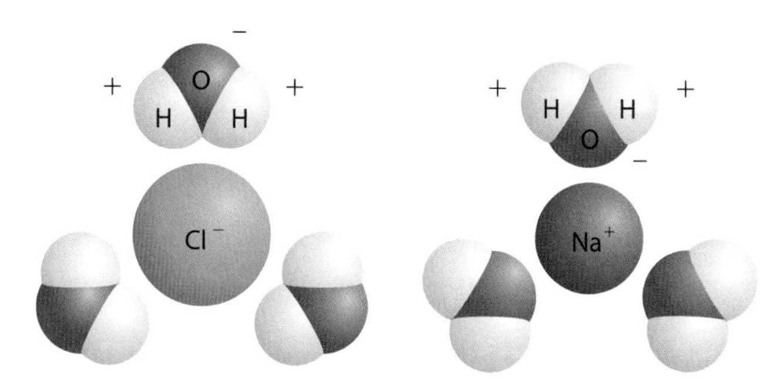

Notice the orientation of the water molecules around the sodium ion and the chloride ion.

The Biological Significance of Ions

Dissolved ions play a vital role in the chemistry of living cells and body systems. The following table identifies the significance of the important ions in your body.

Significant Ions in the Body

Name	Symbol	Special Significanc
Bicarbonate	HCO_3^-	important in acid-base balance
Calcium	Ca^{2+}	found in bones and teeth; important in muscle contraction
Chloride	Cl^-	found in body fluids; important in maintaining fluid balance
Hydrogen	H^+	important in acid-base balance Phosphate
Hydroxide	OH^-	important in acid-base balance
Phosphate	PO_4^{3-}	found in bones, teeth, and the high-energy molecule that cells use for energy
Potassium	K^+	found primarily inside cells; important in muscle contraction and nerve conduction
Sodium	Na^+	found in body fluids; important in muscle contraction and nerve conduction

Understanding pH

Biological processes take place within specific limits of acidity and basicity. If an environment becomes too acidic or too basic for a process to continue at optimum levels, the organism that depends on that process may suffer and die. Freshwater fish, for example, cannot survive in water that is too acidic. Pitcher plants, sundews, and many other plants that grow in acidic soils cannot tolerate basic conditions.

Whether an environment is acidic or basic depends on its pH. The pH scale is used to measure the acidity or basicity of a solution. The scale, shown below, goes from 0 to 14, with 0 representing an extremely acidic solution and 14 representing an extremely basic one. The midpoint of the scale, 7, represents a neutral solution—one that is neither acidic nor basic. Each change in number up or down the scale represents a tenfold increase or decrease in the acidity of the solution.

Measuring pH

For a relative indication of the pH level, a sample can be tested with litmus paper. This simple test will determine whether a solution is acidic or basic. The pH can also be tested by adding an acid-base indicator, such as bromothymol blue. The resulting colour is then compared to a colour chart that indicates relative pH. More precise readings can be determined using a pH meter or probe.

The average pH values for various substances are indicated on this pH scale

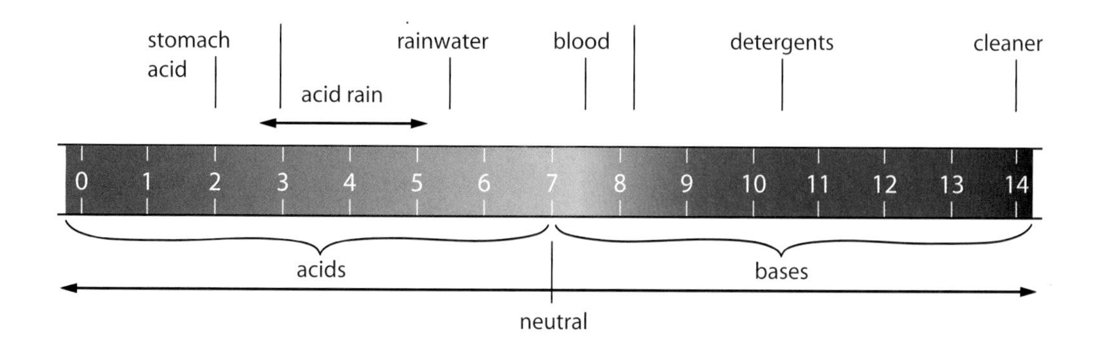

Periodic Table of the Elements

Key: Atomic number (26), Relative atomic mass (u) (55.85), Electronegativity (1.8), Common ion charges (3+,2+), Element Symbol (Fe), Name of element (iron)

Legend

- Alkali metals
- Alkaline earth metals
- Transition metals
- Other metals
- Other non-metals
- Halogens
- Noble gases
- Lanthanoids
- Actinoids
- Metalloids
- Solid
- Liquid
- Gas
- No stable isotopes

Cell format: atomic number Symbol name; atomic mass; electronegativity; common ion charges

main-group elements (groups 1–2, 13–18); transition elements (groups 3–12)

Period	1	2	3	4	5	6	7	8	9	10	11	12	13	14	15	16	17	18
1	1 H hydrogen; 1.01; 2.2; 1+,1−																	2 He helium; 4.00; –; –
2	3 Li lithium; 6.94; 1.0; 1+	4 Be beryllium; 9.01; 1.6; 2+											5 B boron; 10.81; 2.0; –	6 C carbon; 12.01; 2.6; –	7 N nitrogen; 14.01; 3.0; 3−	8 O oxygen; 16.00; 3.4; 2−	9 F fluorine; 19.00; 4.0; 1−	10 Ne neon; 20.18; –; –
3	11 Na sodium; 22.99; 0.9; 1+	12 Mg magnesium; 24.31; 1.3; 2+											13 Al aluminum; 26.98; 1.6; 3+	14 Si silicon; 28.09; 1.9; –	15 P phosphorus; 30.97; 2.2; 3−	16 S sulfur; 32.07; 2.6; 2−	17 Cl chlorine; 35.45; 3.2; 1−	18 Ar argon; 39.95; –; –
4	19 K potassium; 39.10; 0.8; 1+	20 Ca calcium; 40.08; 1.0; 2+	21 Sc scandium; 44.96; 1.4; 3+	22 Ti titanium; 47.87; 1.5; 4+,3+	23 V vanadium; 50.94; 1.6; 5+,4+	24 Cr chromium; 52.00; 1.7; 3+,2+	25 Mn manganese; 5[illegible]; 1.6; 2+,[illegible]	2[illegible] Fe iron; 55.85; 1.[illegible]; 3+,2+	27 Co cobalt; 58.93; 1.9; 2+,3+	28 Ni nickel; 58.69; 1.9; 2+,3+	29 Cu copper; 63.55; 1.9; 2+,1+	30 Zn zinc; 65.38; 1.7; 2+	31 Ga gallium; 69.72; 1.8; 3+	32 Ge germanium; 72.64; 2.0; 4+	33 As arsenic; 74.92; 2.2; 3−	34 Se selenium; 78.96; 2.6; 2−	35 Br bromine; 79.90; 3.0; 1−	36 Kr krypton; 83.80; –; –
5	37 Rb rubidium; 85.47; 0.8; 1+	38 Sr strontium; 87.62; 1.0; 2+	39 Y yttrium; 88.91; 1.2; 3+	40 Zr zirconium; 91.22; 1.3; 4+	41 Nb niobium; 92.91; 1.6; 5+,3+	42 Mo molybdenum; 95.96; 2.2; 6+	43 Tc technetium; [illegible]; 2.1; 7[illegible]	4[illegible] Ru ruthenium; 101.07; 2.[illegible]; 3+	45 Rh rhodium; 102.91; 2.3; 3+	46 Pd palladium; 106.42; 2.2; 2+,3+	47 Ag silver; 107.87; 1.9; 1+	48 Cd cadmium; 112.41; 1.7; 2+	49 In indium; 114.82; 1.8; 3+	50 Sn tin; 118.71; 2.0; 4+,2+	51 Sb antimony; 121.76; 2.1; 3+,5+	52 Te tellurium; 127.60; 2.1; 2−	53 I iodine; 126.90; 2.7; 1−	54 Xe xenon; 131.29; 2.6; –
6	55 Cs cesium; 132.91; 0.8; 1+	56 Ba barium; 137.33; 0.9; 2+	57–71 La–Lu	72 Hf hafnium; 178.49; 1.3; 4+	73 Ta tantalum; 180.95; 1.5; 5+	74 W tungsten; 183.84; 1.7; 6+	75 Re rhenium; 18[illegible]; 1.9; 7[illegible]	7[illegible] Os osmium; 190.23; 2.[illegible]; 4+	77 Ir iridium; 192.22; 2.2; 4+	78 Pt platinum; 195.08; 2.2; 4+,2+	79 Au gold; 196.97; 2.4; 3+,1+	80 Hg mercury; 200.59; 1.9; 2+,1+	81 Tl thallium; 204.38; 1.8; 1+,3+	82 Pb lead; 207.2*; 1.8; 2+,4+	83 Bi bismuth; 208.98; 1.9; 3+,5+	84 Po polonium; (209); 2.0; 2+,4+	85 At astatine; (210); 2.2; 1−	86 Rn radon; (222); –; –
7	87 Fr francium; (223); 0.7; 1+	88 Ra radium; (226); 0.9; 2+	89–103 Ac–Lr	104 Rf rutherfordium; (261); –; –	105 Db dubnium; (262); –; –	106 Sg seaborgium; (266); –; –	107 Bh bohrium; ([illegible]); –; –	108 Hs hassium; (277); –; –	109 Mt meitnerium; (268); –; –	110 Ds darmstadtium; (271); –; –	111 Rg roentgenium; (272); –; –	112 Cn copernicium; (285); –; –	113 Uut* ununtrium; (284); –; –	114 Uuq* ununquadium; (289); –; –	115 Uup* ununpentium; (288); –; –	116 Uuh* ununhexium; (292); –; –		118 Uuo* ununoctium; (294); –; –

*Temporary names

inner transition elements

Period	Series															
6	Lanthanoids	57 La lanthanum; 138.91; 1.1; 3+	58 Ce cerium; 140.12; 1.1; 3+	59 Pr praseodymium; (140.91); 1.1; 3+	60 Nd neodymium; 144.24; 1.1; 3+	61 Pm promethium; (1[illegible]); –; 3+	62 Sm samarium; 150.36; 1.2; 3+,2+	63 Eu europium; 151.96; –; 3+,2+	64 Gd gadolinium; 157.25; 1.2; 3+	65 Tb terbium; 158.93; –; 3+	66 Dy dysprosium; 162.50; 1.2; 3+	67 Ho holmium; 164.93; 1.2; 3+	68 Er erbium; 167.26; 1.2; 3+	69 Tm thulium; 168.93; 1.3; 3+	70 Yb ytterbium; 173.05; –; 3+,2+	71 Lu lutetium; 174.97; 1.0; 3+
7	Actinoids	89 Ac actinium; (227); 1.1; 3+	90 Th thorium; 232.04; 1.3; 4+	91 Pa protactinium; 231.04; 1.5; 5+,4+	92 U uranium; 238.03; 1.7; 6+,4+	93 Np neptunium; (2[illegible]); 1.3; 5+	94 Pu plutonium; (244); 1.3; 4+,6+	95 Am americium; (243); –; 3+,4+	96 Cm curium; (247); –; 3+	97 Bk berkelium; (247); –; 3+,4+	98 Cf californium; (251); –; 3+	99 Es einsteinium; (252); –; 3+	100 Fm fermium; (257); –; 3+	101 Md mendelevium; (258); –; 2+,3+	102 No nobelium; (259); –; 2+,3+	103 Lr lawrencium; (262); –; 3+

Although Group 12 elements are often included in the transition elements, these elements are chemically more similar to the main-group elements.
Any value in parentheses is the mass of the least unstable or best known isotope for elements that do not occur naturally.

Aquaporins

In the early 1990s, American physicist Peter Agre and his colleagues discovered the first of what would eventually be recognized as a new family of proteins, called aquaporins, a discovery that earned Agre the Nobel Prize in Chemistry in 2003. Aquaporins, it is now known, are a subfamily of an even larger family of proteins with membrane transport capabilities, found in all kingdoms of living organisms. The functions of all the members of this family are not yet known, but some aquaporins can transport other small molecules, such as glycerol and urea. There are currently at least 11 known members of the aquaporin family in the human genome.

Aquaporins are proteins with six transmembrane helical domains and two short loops in the membrane (see the figure below). In animals, the two short loops come together to form the three-dimensional core of the water channel. The importance of the loops is reflected in the fact that these portions are the most highly conserved sequences of the channels among different species. Water must pass through a zone of constriction, created by the loops, that reduces the channel opening to about 30 picometres, or just about the width of a water molecule. Scattered along the inner part of the channel are arginine amino acids, which are positively charged. The charged arginines participate in hydrogen bonding with water molecules, facilitating their single-file movement through each channel at rates that have been estimated to be up to billions of water molecules per second.

Within the various extracellular and intracellular domains of aquaporins, are sites that can be modified by enzymes, such as kinases. This suggests that the opening and closing of these channels may be gated by stimuli, like the way ion channels are gated in neurons and other cells. In addition, the promoter region of certain aquaporin genes contains a site that is recognized by transcriptional activator proteins that are responsive to the presence of cAMP, a common intracellular signalling molecule and one that is generated by cells stimulated by ADH. Thus, one mechanism by which ADH promotes osmosis of water out of the renal collecting ducts appears to be by stimulating the transcription of one or more aquaporin genes.

Our understanding of aquaporin functions has allowed us to explain the molecular basis of one form of an inherited human disease in which patients are unable to produce a concentrated urine and who consequently lose large amounts of water. A mutation in one aquaporin genes results in a form of the protein expressing any of several abnormalities: improper folding into its correct shape, impaired ability of the molecule to enter the plasma membrane, or impaired ability of the molecule to form a channel core.

The discovery of aquaporins may have widespread implications for other areas of biology and health research. For example, Canada research chair in genomics David Baillie, at Simon Fraser University, has shown that AQP-8, an aquaporin from the nematode worm *Caenorhabditis elegans*, is expressed in the worm's excretory cell and is involved in water balance. AQP-8 is orthologous to aquaporins that maintain water balance in vertebrates, which suggests that *C. elegans* may well be used as a model system to understand gene-regulatory networks in the developing vertebrate kidney.

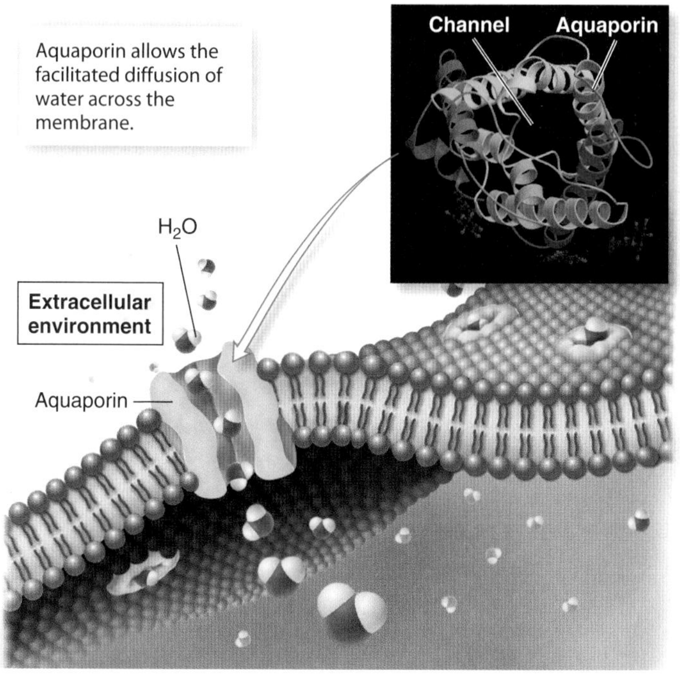

Aquaporin allows the facilitated diffusion of water across the membrane.

Channel

Aquaporin

H_2O

Extracellular environment

Aquaporin

Function and structure of aquaporin. Aquaporin is a transport protein that functions by forming a channel in the membrane of certain cell types and allows the rapid diffusion of water across the membrane. The inset shows the structure of aquaporin, which was determined by X-ray crystallography.

Appendix B

Thermodynamics and the Types of Energy Interconversions

Energy is the capacity to do work, or in other words, the ability to promote change. Physicists often consider energy in two forms: kinetic energy and potential energy. Kinetic energy is energy associated with movement, such as the movement of a baseball bat from one location to another as the batter swings at a pitch. By comparison, potential energy is the energy a substance possess because of its structure or location. The energy contained within covalent bonds in molecules is a type of potential energy called chemical energy. The breakage of those bonds is one way that living cells can harness the energy to perform cellular functions. The table below summarizes chemical and other forms of energy that are important in biological systems.

Types of Energy that Are Important in Biology

Energy type	Description	Biological example
Light	Light is a form of electromagnetic radiation that is visible to the eye. The energy is packaged in photons.	During photosynthesis, light energy is captured by pigments in chloroplasts. Ultimately, this energy is used to reduce carbon, thus producing organic molecules.
Heat	Heat is the transfer of kinetic energy from one object to another or from an energy source to an object. In biology, heat is often viewed as energy that can be transferred because of a difference in temperature between two objects or locations.	Many organisms, such as humans, maintain their bodies at a constant temperature. This is achieved by chemical reactions that generate heat.
Mechanical	Mechanical energy is the energy that is possessed by an object because of its motion or its position relative to other objects.	In animals, mechanical energy is associated with movements caused by muscle contraction, such as walking.
Chemical	Chemical energy is energy stored in the chemical bonds of molecules. When the bonds are broken and rearranged, this can release large amounts of energy.	The covalent bonds in organic molecules, such as glucose and ATP, store large amounts of energy. When these bonds are broken, the chemical energy released can be used to drive cellular processes.
Electrical/ Ion gradient	The movement of charge or the separation of charge can provide energy. Also, a difference in ion concentration across a membrane constitutes a electrochemical gradient, which is a source of potential energy.	High-energy electrons can release energy (that is, drop down to lower energy levels). The energy that is released can be used to drive cellular processes, such as pumping H^+ across membranes.

Energy Interconversions

An important issue in biology is the ability of energy to be converted from one form to another. The study of energy interconversions is called thermodynamics. Physicists have determined that two laws govern energy interconversions:

1. The first law of thermodynamics states that energy cannot be created or destroyed. However, energy can be transferred from one place to another and can be transformed from one type to another.

During photosynthesis, energy in the form of light is transferred from the Sun, some 148 km away, to a pigment molecule in a photosynthetic organism, such as a plant. A complex series of energy transformations occurs in which light energy is transformed into electrochemical energy and then into energy stored within chemical bonds.

2. The second law of thermodynamics states that the transfer of energy or the transformation of energy from one form to another increases the entropy, or degree of disorder in a system. When energy is converted from one form to another, the increase in entropy causes some energy to become unusable by living organisms.

Ecosystem ecology deals with the flow of energy and the cycling of nutrients among organisms in a community and between organisms and the environment. In every energy transformation, free energy is reduced because heat energy is lost from the ecosystem in the process, and the entropy of the universe increases. There is, therefore, a unidirectional flow of energy through an ecosystem, with energy dissipated at every step. An ecosystem needs a recurring input of energy from an external source—in most cases the Sun—to sustain itself. In contrast, chemicals, such as mercury, cycle between abiotic and biotic components of the environment, often becoming more concentrated in organisms in higher trophic levels.

Appendix B

Greek and Latin Prefixes, Suffixes, and Word Roots

The following list of Greek (*G*) and Latin (*L*) prefixes (word beginnings) suffixes (word endings), and word roots will help you to understand some of the scientific terms you encounter in *Biology 12*.

Prefix/Suffix	Meaning	Example
A		
a- (*G*)	not, without	abiotic
ad- (*L*)	to, toward	adapt
aero- (*G*)	air	aerobic
agr/o- (*G*)	field	agriculture
allo- (*G*)	other, different	allosteric
amphi- (G)	on both sides	amphibian
ana- (*G*)	again, up	anabolic
ante- (L)	in front of	anterior
anti- (*G*)	against	antiporter
archae- (*G*)	ancient	archaeon
arthro- (*G*)	jointed	arthropod
auto- (*G*)	self	autotroph
B		
bi- (*L*)	two	bilateral
bio- (*G*)	life	biological
blast- (*G*)	embryo, germ, or bud	blastula
C		
carcin- (G)	cancer	carcinogen
cata- (*G*)	down, under	catabolic
cereb- (L)	pertaining to the brain	cerebrum
chloro- (*G*)	green	chlorophyll
chroma- (*G*)	pigmented	chromosome
-cide (L)	kill	pesticide
circ- (L)	around	circulation
co- (*L*)	with	cotransport
con- (*L*)	together	convergent
-cycle (G)	circle	Krebs cycle
cyto-/-cyte (*G*)	cell	cytoplasm

Prefix/Suffix	Meaning	Example
D		
derm-/-derm (*G*)	skin	dermal
di- (*G*)	two	dichotomous
dia- (*G*)	apart	dialysis
dorm- (*L*)	sleep	dormancy
E		
ecto- (G)	outer, outside	ectoderm
-ectomy (G)	surgical operation or removal	appendectomy
em- (*G*)	inside	embryo
endo- (*G*)	within	endosymbiont
epi- (*G*)	upon	epidermis
erythro- (*G*)	red	erythrocyte
eu- (*G*)	true	eukaryote
exo- (*G*)	outside	exoskeleton
F		
-fer (*L*)	to carry	conifer
G		
gastro- (*G*)	stomach	gastric juice
-genesis (*G*)	to originate	oogenesis
gen-/-gen (*G*)	kind	genotype
gravi- (*L*)	heavy	gravitropism
H		
hapl/o- (*G*)	single	haploid
hemi- (G)	half	hemisphere
hem/o- (*G*)	blood	hemoglobin
heter/o- (*G*)	different	heterotrophic
hist- (G)	tissue	histones
homo-/homeo- (*G*)	same	homeostasis
hydr/o- (*G*)	water	hydrolysis
hyper- (*G*)	above	hypertonic
hypo- (*G*)	below	hypotonic

Prefix/Suffix	Meaning	Example
I		
inter- (*L*)	between	interphase
intra- (*L*)	within	intracellular
iso- (*G*)	equal	isotonic
-ist (G)	one who specializes in something	biologist
L		
leuc/o- (*G*)	white	leukocyte
-logy (*G*)	study of	biology
lymph/o- (*L*)	water	lymphocyte
-lysis (*G*)	break up	glycolysis
M		
macr/o- (*G*)	large	macromolecule
mega- (G)	large	megaspore
meso- (*L*)	in the middle	mesophyll
meta- (*G*)	after	metaphase
micr/o- (*G*)	small	microvilli
mono- (G)	one, single	monosaturated
morph/o-/-morph (*G*)	form	morphology
multi- (*L*)	many	multicellular
N		
nephr- (G)	kidney	nephrons
neur- (G)	nerve	neurons
-nomy (*G*)	system of laws	taxonomy
O		
-oma (G)	tumour or swelling	melanoma
oo- (*G*)	egg	oogonium
orth- (G)	straight, normal, correct	orthodontist
P		
pal(a)e/o- (*G*)	ancient	paleontology
para- (*G*)	beside	parasitic
path- (G)	disease	pathogens
peri- (*G*)	around, about	peristalsis
phago-/-phage (*G*)	eating	phagocytosis
phot/o- (*G*)	light	photosynthesis
pino- (*G*)	drinking	pinocytosis
pneum- (*G*)	lung	pneumonia
pod-/-pod (*G*)	foot	pseudopod
poly- (*G*)	many	polysaccharide
post- (*L*)	after	posterior
pre- (L)	before, in front of	precursor mRNA
pro- (*G*) (*L*)	before	prokaryote
prot/o- (*G*)	first	proton
R		
re- (*L*)	back, again	reproduce
S		
-scope (*G*)	to look	microscope
semi- (L)	one-half, partially	semipermeable
-some (*G*)	body	chromosome
stom-/-stome (*G*)	mouth, opening	stomata
sub- (L)	beneath, under	substrate
super-, supra- (L)	above	supramolecular
sym/syn- (*G*)	together	symbiotic
T		
therm-/-therm (*G*)	heat	endotherm
-tomy (G)	act of cutting	tonsillectomy
trans- (*L*)	across	transpiration
-trophic (*G*)	nourishment	heterotrophic
U		
uni- (*L*)	one	unicellular
ur- (G)	pertaining to urine or the urinary system	urology
V		
visc- (L)	internal	visceral
Z		
-zygous (*G*)	joined together	homozygous

Sensory Receptors

Our senses allow us to navigate and experience the world around us. Our sensory nervous system informs our brain about the condition and position of our body and also provides it with a detailed picture of the surroundings. This information allows our body to maintain homeostasis.

Sensory systems convert chemical or physical stimuli from our body and the external environment into a signal that causes a change in the membrane potential of sensory neurons. *Sensory transduction* is the process by which incoming stimuli are converted into neural signals. *Sensory transduction* involves cellular changes, such as opening of ion channels, which may cause electrical impulses called action potentials in neurons.

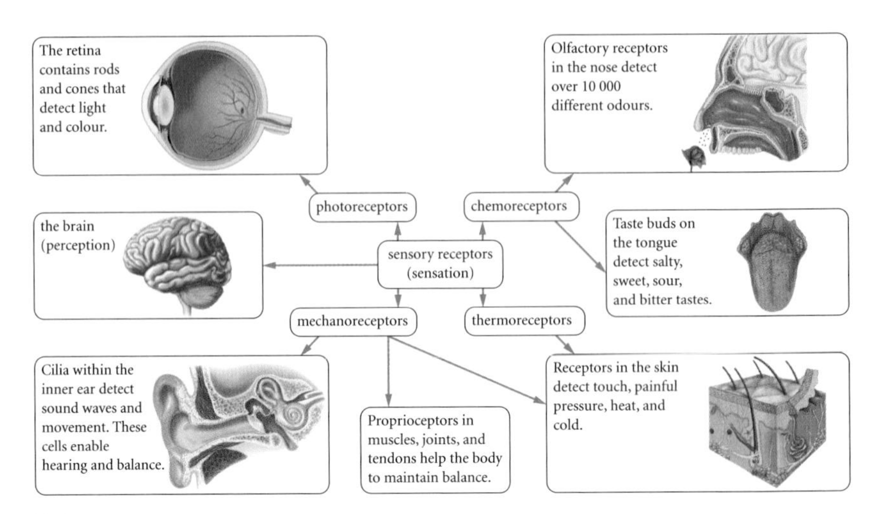

Major sensory receptors

Perception is an awareness of the sensations that you experience. For instance, touching a hot stove generates a thermal sensation, which initiates a neural response, giving you the perception that this stimulus is hot. Perception requires integration of incoming stimuli by the central nervous system. We do not consciously perceive all sensations that we experience. Most of the time, for example, we are not aware of the touch of our clothing. Also, the brain processes some sensory information in areas that do not generate conscious thought. For instance certain neurons constantly monitor blood pressure, but we are not aware of this occurring.

There are specialized cells called *sensory receptors* whose function it is to receive sensory inputs. Sensory receptors are found in higher concentrations in the sense organs than in other parts of the body. A sensory receptor is either a neuron or a specialized epithelial cell that recognizes an internal or environmental stimulus and initiates signal transduction in the same cell or an adjacent cell. When a response is strong enough, sensory receptors can initiate electrical responses to stimuli, such as chemicals, light, touch, heat, and sound, which leads to action potentials that are sent to the central nervous system.

An intense stimulus generates more frequent action potentials. These electrical impulses are transmitted into the central nervous system and carried to the brain for interpretation. The brain interprets a higher frequency of electrical impulses as a more intense stimulus.

Different stimuli produce different sensations because they activate specific neural pathways that are dedicated to processing only that type of stimulus. The central nervous system processes each sense within its own pathway. We know that we are seeing light because the signals generated by visual sensory receptors are transmitted along a neural pathway that sends electrical impulses into areas of the brain that are devoted to processing vision. For this reason, the brain interprets such signals as visual stimuli.

We can classify sensory receptors into general classes, based on the type of stimulus to which they respond: mechanoreceptors, thermoreceptors, photoreceptors, and chemoreceptors. Each type uses a different mechanism to respond to signals and transmits the response to different regions of the central nervous system,

Mechanoreceptors: Hearing, Balance, Touch

Mechanoreceptors respond to physical stimuli, such as sound, movement, pressure, and stretch.

Hearing

Hearing is the ability to detect and interpret sound waves. It is a critical sense for the survival of many types of animals. The human ear has three main compartments called the outer, middle, and inner ear. The outer ear collects sound waves and causes the tympanic membrane to vibrate. The tympanic membrane vibrates the bones of the middle ear, called the ossicles, which in turn vibrate against the oval window. Hair-like mechanoreceptors in the inner ear, called stereocilia, translate these vibrations to action potentials, which travel through the auditory nerve to the auditory processing centres in the brain.

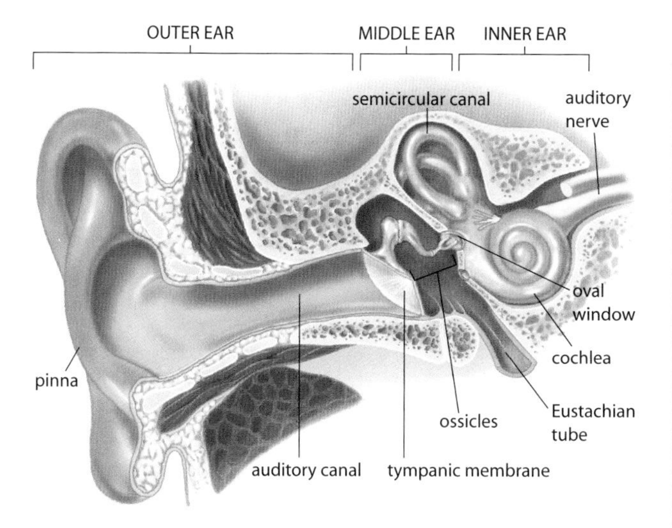

The anatomy of the human ear

When we hear a great number of sound frequencies at once, such as at a music concert, the sound waves move stereocilia all along the basilar membrane in a physical representation of the music. These cells stimulate sensory neurons, which send multiple action potentials to the auditory area of the brain for processing. The most incredible feature of this process, however, is that—unlike a radio, which can only be tuned to one frequency at a time—the human ear and brain can "tune in" to all these frequencies simultaneously.

Sound waves that travel at a high frequency are perceived as having a high pitch. The amplitude of a sound wave is experienced as the intensity or volume of a sound. The louder the noise is, the more pressure that the fluid in the cochlea puts on the hair cells of the basal membrane.

The stereocilia of the hair cells are very delicate. Repeated or sustained exposure to loud noise destroys the stereocilia, and the sound frequencies they interpret can no longer be heard. The damage is permanent. You can reduce this type of hearing loss by limiting your exposure to loud noise, such as machinery and loud music. Noise is measured in decibels (dB), and any noise over 80 dB can damage the hair cells (see the table at right).

While hearing aids to amplify sounds can often help people with conduction deafness, nerve deafness is more difficult to treat. In some cases, a device can be implanted in the ear to pick up sounds and directly relay signals to the auditory nerve. As well, researchers are exploring techniques to regenerate damaged or lost hair cells. One technique is to use a virus to insert a gene into the inner ear cells. The gene causes these cells to "sprout" new hair cells.

Examples of Noises that Affect Hearing

Type of noise	Sound level (dB)	Effect
jet engine or rock concert	Over 125	Noise is beyond the threshold of pain. There is high potential for hearing loss.
boom box, chain saw, or snowmobile	100-125	Regular exposure for short periods of time may cause permanent hearing loss.
farm tractor, lawn mower, or motorcycle	90-100	15 min of exposure may cause hearing loss
food blender or average city traffic	80-90	Continuous daily exposure for longer than 8 h can cause damage.

Source: United States National Institute of Deafness and Other Communication Disorders, 2005

Balance

Three major structures in the inner ear—the semicircular canals, utricle, and saccule—function in our sense of equilibrium. They help us stand upright and move without losing our balance. The semicircular canals contain mechanoreceptors that detect head and body rotation. The semicircular canals are three fluid-filled loops, arranged in three different planes—one for each dimension of space. The base of each semicircular canal ends in a bulge. Inside each bulge, the stereocilia of the hair cells stick into a jelly-like covering. When your head rotates, the fluid inside the semicircular canals moves and bends the stereocilia, causing the hair cells to send rotational information to your brain. On a fast-spinning midway ride, for example, the rapid circular motion causes the fluid within the semicircular canals to rotate and send information confirming this to your brain. When the ride stops, however, the fluid is still moving, which might give you the feeling of dizziness or nausea.

The balance required while moving your head forward and backward is called gravitational equilibrium. Gravitational equilibrium depends on the utricle and the saccule, which together make up the fluid-filled vestibule of the inner ear. Both of these structures contain calcium carbonate granules. When your head dips forward or back, gravity pulls on the calcium carbonate granules. This puts pressure on some of the hair cells, causing them to send a neural impulse to the brain. The brain decodes how far and in what direction the hair cells bend, determines the position of your head, and responds with messages to help your body maintain balance.

Another type of mechanoreceptors involved in coordination are called proprioceptors. Proprioceptors are found in muscles, tendons, and joints throughout the body, and they send information about the body position to the brain.

Touch

The mechanoreceptors associated with touch are located all over the body. The skin contains more than four million mechanoreceptors, with many of them concentrated in the fingers, tongue, lips, and genitals. Different receptors in the skin are sensitive to different stimuli, such as light touch, pressure, pain, and high and low temperatures. These receptors gather information and transmit it back through the sensory neurons to the brain and spinal cord for processing and a possible reaction.

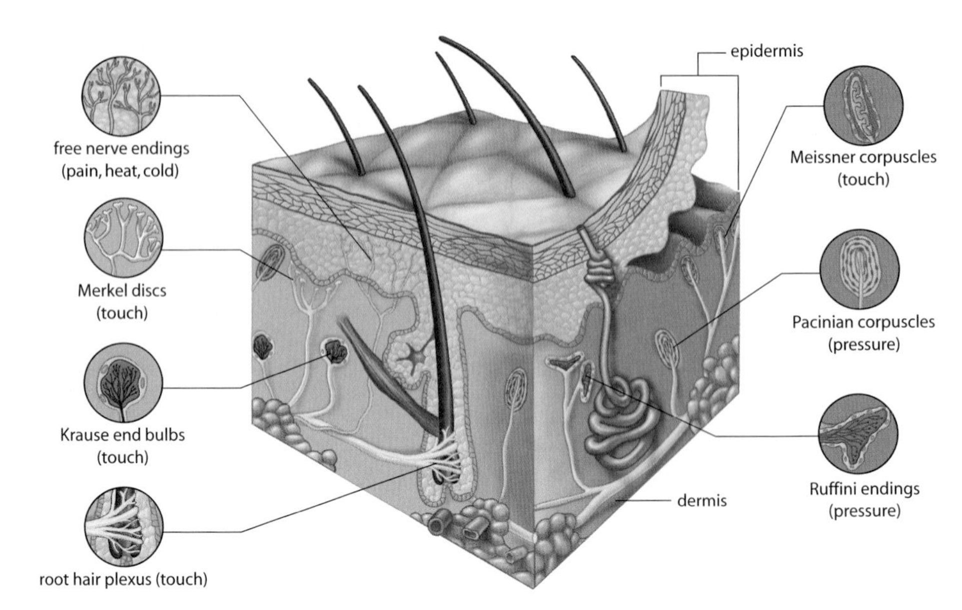

Various types of sensory neurons found in the skin

Nocireceptors, the receptors for pain, are nerve endings in the skin and internal organs. They respond to tissue damage or to stimuli that are about to cause tissue damage. Nocireceptors are unusual because they can respond not only to external stimuli, such as extreme temperatures, but also to internal stimuli, such as molecules released into the extracellular space from injured cells. Damaged cells release a number of substances, such as acids, that cause inflammation and make nociceptors more sensitive to painful stimuli. Painkillers, such as ibuprofen and Aspirin™, block the release of some of these substances, which may help to reduce pain.

Signals arising from nocireceptors travel to the central nervous system and reach the cerebrum, where the type or cause of the pain is interpreted. The signals are also sent to the limbic system, which holds memories and emotions associated with pain, and to the reticular formation, which increases alertness and arousal—an important response to a painful stimulus.

Stretch receptors are a type of mechanoreceptor found in muscles and in the walls of organs, like the stomach and bladder. Stretch receptors are nerve endings that can be distended. When your stomach stretches after a meal, its stretch receptors are deformed, causing them to become depolarized and send action potentials to the brain. The brain interprets the signals as fullness, which reduces appetite. In a similar way, the degree to which certain blood vessels are stretched gives the brain information about your blood pressure.

Thermoreceptors: Detecting Temperature

Sensing the outside temperature is important because body temperature is affected by the external temperature. Animals can survive at body temperature only within certain limits, because the proteins in their cells function optimally within a particular temperature range. Thermoreceptors respond to cold or hot temperatures by activating or inhibiting enzymes within their plasma membranes, which alter membrane channels. Thermoreceptors are often linked with reflexive behaviours, such as when you touch a hot stove and pull your hand away.

In addition to skin receptors that sense the outside temperature, thermoreceptors in the brain detect changes in core body temperature. Activation of skin or brain thermoreceptors triggers physiological and behavioural adjustments that help maintain body temperature.

Photoreceptors: Vision

Photoreceptors are a specialized kind of electromagnetic receptor. Electromagnetic receptors detect radiation within a wide range of the electromagnetic spectrum, including visible, ultraviolet, and infrared light. In some animals, electromagnetic receptors also detect electrical and magnetic fields, an adaptation for long-distance migration or low-light environments. In humans, photoreceptors respond to visible light energy.

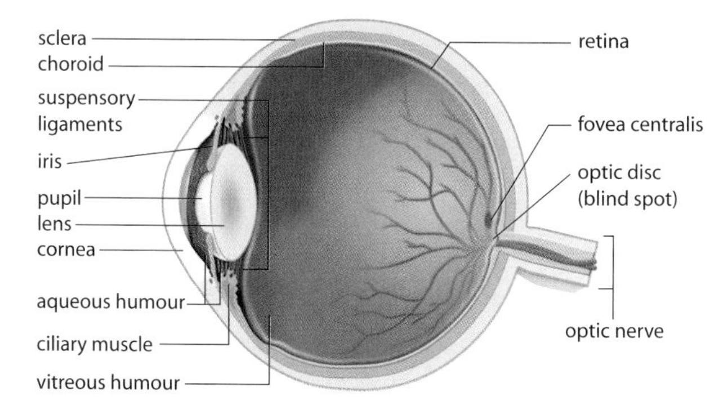

The anatomy of the human eye

Vision requires the eye, the optic nerves, and the visual cortex in the brain. Different patterns of light emitted from images in the animal's field of view are transmitted through a small opening or pupil, through the lens, to a sheet-like layer of photoreceptors called the retina, at the back of the eye. The photoreceptors trigger electrical changes in neurons that pass out of the eye through the optic nerve, which carries the signals to the brain.

Two types of photoreceptors in the retina have names that are derived from their shapes: rods and cones. Rods contain a light-sensitive pigment that allows them to respond to low-intensity light. They can respond to as little as one photon, but they do not discriminate different colours. Rods are utilized mostly at night, and they send signals to the brain that generate a black-and-white visual image. Cones are used in daylight and are less sensitive to low levels of light. Unlike rods, cones can detect colour.

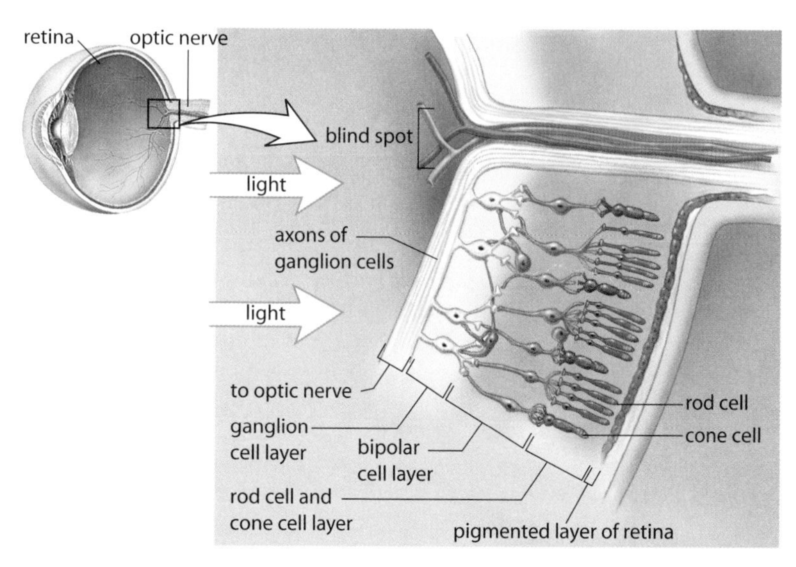

The organization of the principal cells of the retina. Light passes through two layers of cells before it reaches the photoreceptors. Once stimulated, the rods and cones permit a neural impulse to pass through the bipolar cells to the ganglion cells, which form the optic nerve.

Humans have three kinds of cones in the retina that absorb different wavelengths of light and allow a person to perceive all the colours in the visible spectrum. If the eye has faulty or missing cones, then a person may not be able to distinguish certain colours. Because the two types of photoreceptors are specialized for either night or day vision, neither rods nor cones function at peak efficiency at twilight. This accounts for our relatively poor vision at this time.

Chemoreceptors: Taste and Smell

Chemoreception includes the senses of taste and smell (olfaction), which involve detecting chemicals in the air, water, and food. These chemicals bind to chemoreceptors, which initiate electrical responses in other neurons that pass into the brain. Airborne molecules that bind to olfactory receptors must be small enough to be carried in the air and into the nose. Taste molecules can be heavier because they are conveyed in food and liquid. A close relationship exists between taste and smell. About 80% of the perception of taste is actually due to olfactory receptors. This is why food loses its flavour when olfaction is impaired, such as when you have a cold.

Taste

The tongue contains chemoreceptors that allow us to taste substances entering the mouth. Taste buds are a group of chemosensory cells in the bumps on the tongue that detect particular molecules in food. Humans have about 9000 taste buds. Specific taste cells within the taste buds detect molecules from one of the four basic tastes: sweet, sour, salty, and bitter. The combination of taste information sent from different areas of the tongue, as well as from sensory neurons in the nose, allows us to perceive flavours. The salivary glands are connected to the brain stem, which is why they are stimulated whenever we taste, smell, or think about something delicious.

Another type of chemoreception involves hormone-like chemicals called pheromones. Many animals, including humans, release pheromones to aid in the recognition and attraction of a mate, sometimes over long distances. These chemicals are detected in the nose by a structure called the vomeronasal organ. Recently, scientists determined that the human nose also contains a vomeronasal organ, although people cannot consciously smell pheromones.

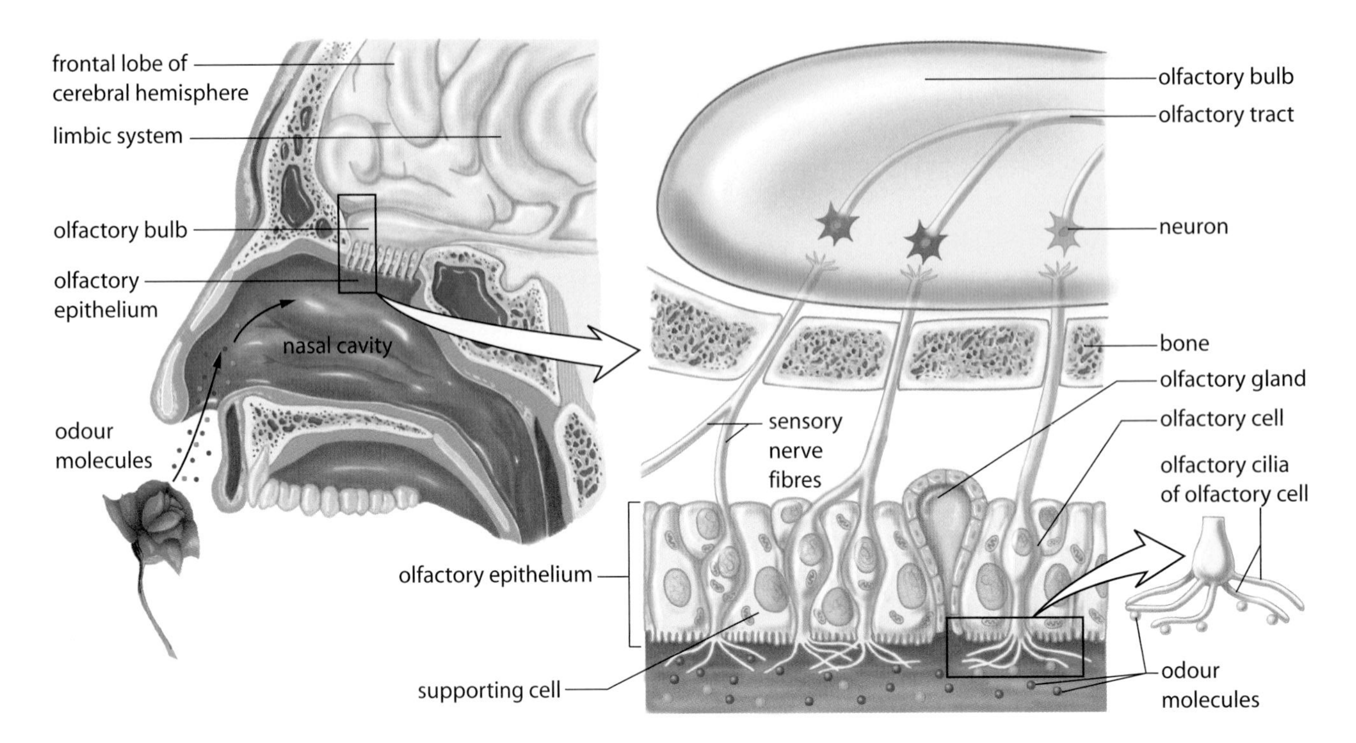

(A) The human olfactory system. (B) The cilia of each olfactory cell can bind to only one type of odour molecule (represented here with colour).

Smell

The human sense of smell can likely distinguish over 10 000 different odours. Amazingly, the binding of a single molecule to a receptor cell can sometimes be perceived [illegible] think that each of these odours is produced from particles that fit, much like a lock and key, into specific chemoreceptors, called olfactory cells. When the particles fit into the olfactory cells, ion channels in the cell membrane open. This generates an action potential in the olfactory cells, which are directly linked to the olfactory bulb of the brain. From there, the impulse is sent to the emotional centres of the brain, called the limbic system, and the frontal lobe, where the perception of odour occurs. Have you notices that particular odours can instantly conjure up scenes and emotions from the past? Perfume experts create fragrances to evoke certain memories and emotions.

Major Sensory Receptors in the Human Body

Category and type of receptor	Examples of receptor	Stimulus
Mechanoreceptors		
touch/pressure/pain	receptors in the skin	mechanical pressure
hearing	hair cells in the inner ear	sound waves
balance	hair cells in the inner ear	fluid movement
body position	proprioceptors in the muscles and tendons and at the joints	muscle contractions, stretching, and movement
Thermoreceptors		
temperature	heat and cold receptors in the skin	change in radiant energy
Photoreceptors		
vision	rods and cones in the eye	visible light
Chemoreceptors		
taste	taste buds on the tongue	food particles in saliva
smell	olfactory receptors in the nose	odour molecules
internal senses	osmoreceptors in the hypothalamus receptors in the carotid artery and aorta	low blood volume blood pH

Sustainability

Sustainability is the practice of using resources, such as food, energy, timber, and other items acquired from the environment, at a level that does not exhaust the supply or cause ecological damage. Ecological damage can result in reduced biodiversity, including declines in species, genetic, and ecosystem diversity. Biodiversity stabilizes ecosystems, making them more resilient to change and degradation. For instance, an ecosystem is more likely to bounce back from change if it has a greater variety of species of which some are able to withstand the change. Ecosystem stability is essential to the survival of our species, especially as we reach record population numbers, since our livelihood depends on services that ecosystems provide (see the tables on this page). If the resources in an ecosystem are not harvested in a sustainable manner, the consequences can be devastating for humans that rely on its services.

Examples of the World's Ecosystem Services

Service	Example
Atmospheric gas supply	Regulation of carbon dioxide, ozone, and oxygen levels
Climate regulation	Regulation of carbon dioxide, nitrogen dioxide, and methane levels
Water supply	Irrigation, water for industry
Pollination	Pollination of crops
Biological control	Pest population regulation
Wilderness and refuges	Habitat for wildlife
Food production	[illegible]
Raw materials	Fossil fuels, timber
Genetic resources	Medicines, genes for plant resistance
Recreation	Ecotourism
Cultural	Aesthetic and educational value
Disturbance regulation	Storm protection, flood control
Waste treatment	Sewage purification
Soil erosion control	Retention of topsoil, reduction of accumulation of sediments in lakes
Nutrient recycling	Nitrogen, phosphorus, carbon, and sulphur cycles

Valuation of the World's Ecosystem Services

Biome	Total global value* ($trillions)	Total value (per ha) ($)	Main ecosystem service
Open ocean	8 381	252	Nutrient cycling
Coastal shelf	4 283	1 610	Nutrient cycling
Estuaries	4 100	22 832	Nutrient cycling
Tropical forest	3 813	2 007	Nutrient cycling/raw materials
Sea grass and algal beds	3 801	19 004	Nutrient cycling
Swamps and other wetlands	3 231	19 580	Water supply/ disturbance regulation
Lakes and rivers	1 700	8 498	Water regulation
Tidal marsh	1 648	9 990	Waste treatment/ disturbance regulation
Grasslands	906	232	Waste treatment/ food production
Temperate forest	894	302	Climate regulation/ waste treatment/ lumber
Coral reefs	375	6 075	Recreational/ disturbance regulation
Cropland	128	92	Food production
Desert	0	0	
Ice and rock	0	0	
Tundra	0	0	
Urban	0	0	
Total	33 260		

*[illegible]

A Concern for Everyone

Biologists Paul Ehrlich and E.O. Wilson have suggested that the loss of biodiversity should be of concern to everyone for at least three reasons. First, they proposed that we have an ethical responsibility to protect what are our only known living companions in the universe. Second, humanity has obtained enormous benefits from foods, medicines, and industrial products derived from plants, animals, and microorganisms, and we have the potential to gain many more. The third reason to preserve biodiversity focuses on preserving the array of essential services provided by ecosystems, such as clean air and water.

How does Canada monitor efforts to increase and improve its sustainable practices? The federal government uses indicators such as local air quality, freshwater quality, and changes in wildlife species disappearance risks to track the long-term trends for issues of key concern to Canadians.

Environmental Performance Index

Many Canadians are surprised to learn that Canada is not among the highest rated countries in the world for sustainable practices. We may think that our fresh air, abundant sources of freshwater, and national parks are indicators that our country has an excellent record in sustainability. In fact, Canada ranked only 46th out of a total of 163 countries on the 25 performance indicators measured on the 2010 Environmental Performance Index (EPI). This index tracks the indicators across both environmental public health and ecosystem vitality to help determine how close countries are to established environmental policy goals.

The top five countries in the 2010 index were Iceland (score 93.5), Switzerland (score 89.1), Costa Rica (score 86.4), Sweden (score 86.0), and Norway (score 81.1). Canada's score was 66.4. Since there is no systematic process for verifying numbers reported by different governments and no system within the EPI for tracking changes over time, and since improvement is needed in data collection and analysis of environmental indicators, a country's score and comparative standing within the index cannot be taken as a definitive value. However, the scores can be used to give a sense of which countries are generally doing the best to develop public policies in support of environmental sustainability.

Environmental Performance Index Indicators

Objectives	Policy categories	Indicators
Environmental Health	Environmental burden of disease	Environmental burden of disease
	Water (effects on humans)	Access to drinking water Access to sanitation
	Air pollution (effects on humans)	Urban particulates Indoor air pollution
Ecosystem vitality	Air pollution (effects on ecosystems)	Sulphur dioxide emissions Nitrogen oxide emissions Volatile organic compound emissions Ozone exceedance
	Water (effects on ecosystems)	Water quality index Water stress Water scarcity index
	Biodiversity and Habitat	Biome protection Critical habitat protection Marine protected areas
	Forestry	Growing stock Forest cover
	Fisheries	Marine trophic index Trawling intensity
	Agriculture	Pesticide regulation Agricultural water intensity Agricultural subsidies
	Climate Change	Greenhouse gas emissions/capita Electricity carbon intensity

Appendix C Interpreting Instructional Visuals

Tips for Reading Graphical Texts

You may have noticed that the way you read a textbook is different from how you read a novel or a magazine. Deciphering a textbook requires more than a casual glance. You need to actively ask yourself questions about what the visuals illustrate, what clues they provide about the concepts they describe, and how that information applies to your previous knowledge of the subject.

The drawings, diagrams, graphs, and other visuals that you encounter in your science textbook and other resources have been created to help communicate information and to make complicated concepts easier to understand. These words and visuals are called *graphical texts*, and they help to show how different pieces of information are related to each other.

Features of Graphical Texts

Graphical texts make use of various features and elements to communicate effectively. You have probably used these same features and designs in work that you have produced, both electronically and by hand. For example, you may have used features, such as different size and style of fonts. Your work has probably included headings, subheadings, labels, and captions. You may have used bullets, colour, and shading, and made use of arrows and lines to show how different parts of a visual are related. You have probably used different patterns and forms, such as organizing information in categories and placing it in a specific order, perhaps in a table or chart form. Each graphical text you encounter uses some combination of elements and features to present its information. An important key to understanding the information is to know how to "read" the elements and features. The following questions and examples can help you improve your understanding.

Before You Read

Before you read a graphical text, take a moment to look over the page. The purpose of this stage is to get a sense of what the page is about. You do not need to decipher the details at this point, you simply take a look over the whole thing. Ask yourself:

- **Why am I reading this text?**

Establish your reason for reading. You might turn a heading into a question. For example, the main heading of the page shown below is, "Triglycerides: Lipids Used for Energy Storage." Your purpose may be to answer the question, "How are triglycerides used for energy storage?"

- **What do I already know about this topic?**

Think about the heading or title. Recall what you already know about the subject and jot down any questions that you might have. For example, you may already know that lipids are large molecules composed of carbon, hydrogen, and oxygen atoms. You may also know that lipids are efficient energy–storage molecules because they have many energy-rich C-H bonds. You might be wondering, "What is the difference between triglycerides and other lipids? Where are triglycerides found?"

- **What elements and features are on this page?**

Read the captions, titles, and labels that accompany the visuals. The title and caption will tell you what the visual is about. Make predictions about what you will learn. Notice whether there are words or phrases on the page in boldface or italics. In the page below you may notice that italics are used for *ester linkage, saturated fatty acid, monosaturated,* and *polyunsaturated*. If there are words or phrases defined in a shaded box, read those. In the example below, you will find the definitions of triglyceride and fatty acid in a shaded box in the margin.

triglyceride a lipid molecule composed of a glycerol molecule and three fatty acids linked by ester bonds

fatty acid a hydrocarbon chain ending in a carboxyl group

Triglycerides: Lipids Used for Energy Storage

As shown in **Figure 1.12**, **triglycerides** are composed of one glycerol molecule and three fatty acid molecules. The bond between the hydroxyl group on a glycerol molecule and the carboxyl group on a fatty acid is called an *ester linkage*, because it results in the formation of an ester functional group. A **fatty acid** is a hydrocarbon chain that ends with an acidic carboxyl group, –COOH. Fatty acids are either saturated or unsaturated. A *saturated fatty acid* has no double bonds between carbon atoms, while an unsaturated fatty acid has one or more double bonds between carbon atoms. If the unsaturated fatty acid has one double bond, it is *monounsaturated*; unsaturated fatty acids with two or more double bonds are *polyunsaturated*. Humans cannot synthesize polyunsaturated fats. Therefore, these essential fats must be consumed in our diet.

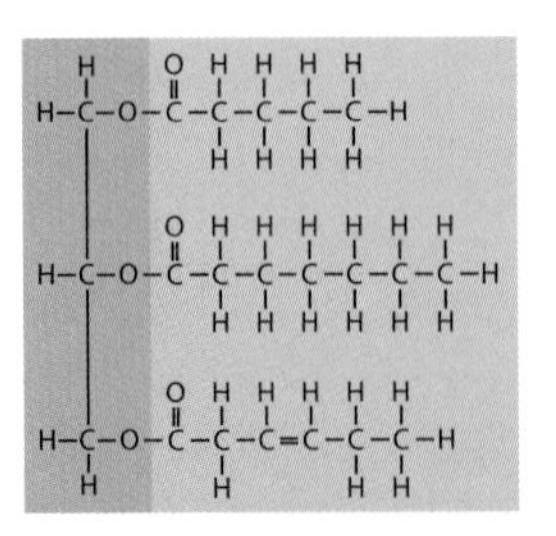

Figure 1.12 A triglyceride is composed of three fatty acids (orange background) and a glycerol molecule (green background). The fatty acids can be saturated (no double bonds) or unsaturated (with one or more double bonds).

An example of a visual text from page 22 of this textbook]

While You Read

Now you are ready to read the section and study the visuals. Read slowly and reread as necessary. Jot down notes and any questions about the content that occur to you. As you read, ask yourself:

- **What does this mean?**

Ask yourself this question often as you read. Pause and ask yourself when you finish reading a paragraph or even after every sentence if there is a lot of information to absorb. When you answer, use the sentence stem "This means..."

- **Which visual connects with this text?**

Find the part of the text that refers to the visual. Usually, a visual follows immediately after the text that describes it, but sometimes the visual is in the margin or on the next page. You can find a clue given in the text about when to refer to the visual. The clue may identify the visual by its number. The text in the example above begins with "As shown in **Figure 1.12**..." In some texts, the clue does not give a figure number but will be worded like "as shown in the following diagram.

- **What is the purpose of this visual?**

Consider why the textbook writer has added a visual. In **Figure 1.12** above its purpose is to help you picture the composition of a triglyceride.

- **How is the information organized?**

Notice in the example above that there are two blocks of colour. That is a clue that there are two different things or processes being described. The caption explains that the orange background is for the fatty acids and the green background is for the glycerol molecule. Sometimes it helps to only look at part of the visual at a time. Cover up the orange section with your hand and examine the glycerol molecule. Then cover up the green section and examine the fatty acids. Uncover the section and look at the whole diagram.

Many visuals, such as **Figure 2.14** on this page, have labels as well as a caption. Read all the labels and examine how they are related to the visual. Some labels are very similar, but they have important differences. The labels on the left hand beaker and tube are 5% and 10%. On the right hand beaker and tube they are >5% and <10%. Notice also the changes in the words "higher," "lower," "more," and "less."

Arrows in science graphics can have different purposes depending on the context. For example, they might to point to the things they name, or show direction, chemical processes, or the consequences of a process. The large arrow in **Figure 2.14** shows you that the contents of the tube on the left undergo a process to become the contents of the tube on the right. This illustration has a special feature that shows an enlarged view of a section of the figure. There are arrows in the enlarged view to indicate the movement of water and the solute across the semi-permeable membrane.

- **How does the visual connect with the written information?**

Many captions provide additional information to what is included in the text. After you finish studying the visual, use the figure number to find the part of the text that refers to it. Reread that part.

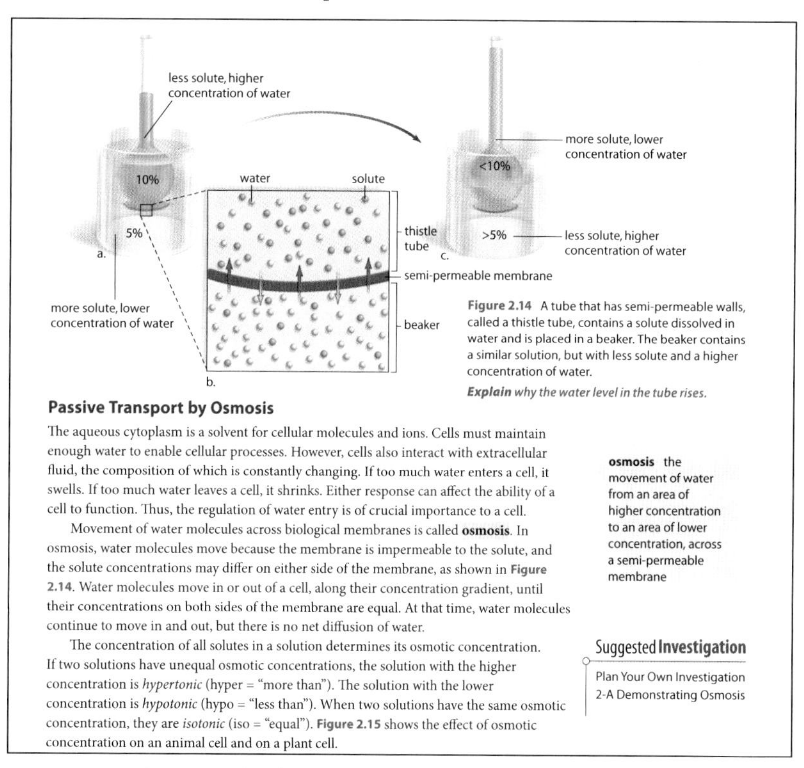

Figure 2.14 A tube that has semi-permeable walls, called a thistle tube, contains a solute dissolved in water and is placed in a beaker. The beaker contains a similar solution, but with less solute and a higher concentration of water.

***Explain** why the water level in the tube rises.*

Passive Transport by Osmosis

The aqueous cytoplasm is a solvent for cellular molecules and ions. Cells must maintain enough water to enable cellular processes. However, cells also interact with extracellular fluid, the composition of which is constantly changing. If too much water enters a cell, it swells. If too much water leaves a cell, it shrinks. Either response can affect the ability of a cell to function. Thus, the regulation of water entry is of crucial importance to a cell.

Movement of water molecules across biological membranes is called **osmosis**. In osmosis, water molecules move because the membrane is impermeable to the solute, and the solute concentrations may differ on either side of the membrane, as shown in **Figure 2.14**. Water molecules move in or out of a cell, along their concentration gradient, until their concentrations on both sides of the membrane are equal. At that time, water molecules continue to move in and out, but there is no net diffusion of water.

The concentration of all solutes in a solution determines its osmotic concentration. If two solutions have unequal osmotic concentrations, the solution with the higher concentration is *hypertonic* (hyper = "more than"). The solution with the lower concentration is *hypotonic* (hypo = "less than"). When two solutions have the same osmotic concentration, they are *isotonic* (iso = "equal"). **Figure 2.15** shows the effect of osmotic concentration on an animal cell and on a plant cell.

osmosis the movement of water from an area of higher concentration to an area of lower concentration, across a semi-permeable membrane

Suggested **Investigation**

Plan Your Own Investigation 2-A Demonstrating Osmosis

An example of a visual text from page 73 of this textbook

After Reading

After you have finished reading you can consolidate and extend your understanding of the content. Ask yourself:

- **Why might this be important?**

Think about how this information relates to what you already know about the topic.

- **How else can I explain this?**

Represent the information in a different way, such as by using a concept map or flow chart.

Practise Your Reading Skills

The following visuals are shown without accompanying text. You can practise using your skills to decipher them.

Reading Graphs

A student conducted a research study about the relationship between the number of cigarettes smoked in a sample population and the incidence of cardiovascular disease over a period of 80 years. She first represented her data in Graph A, and then created Graph B. She chose Graph B to submit along with her report. She titled Graph B, "The Relationship of Cigarette Smoking and Cardiovascular Disease in a Sample Population."

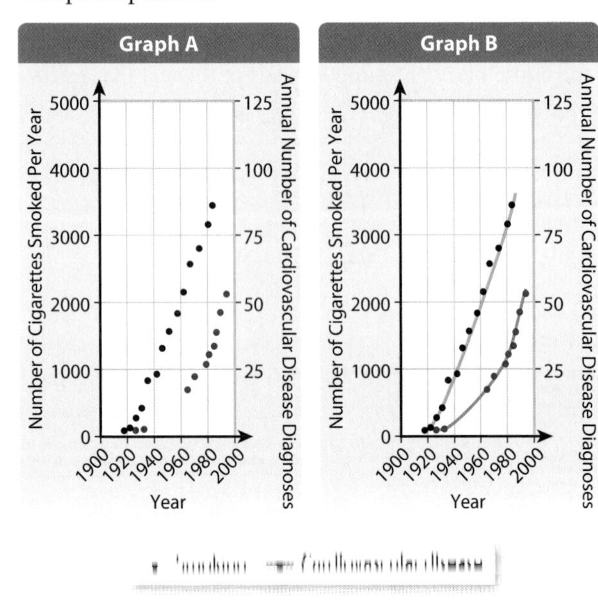

Before

1. Why did the student put two plots on the same grid?
2. What type of graph is Graph B?
3. Why did the student choose this graph to submit?

During

4. **a.** What are the dependent and independent variables?
   b. Where are each of the variables plotted?
5. Why aren't all the data points on the lines in Graph B?
6. The student did not find any data for the year 1960.
   a. Which graph would you use to interpolate that? (Interpolate means to construct new data points within a range of known data.)
   b. Explain the reasons for your choice.
   c. What value do you predict?
   d. Which graph would you use to extrapolate what value might be expected for 2010? (Extrapolate means to infer from something known, assuming that existing trends will continue.)
   e. Explain the reasons for your choice.
   f. What value do you predict?

After

7. What information does Graph B show more clearly than did Graph A?
8. What does this study suggest is the relationship between the number of cigarettes smoked and cardiovascular disease?

Reading Concept Maps

Reading Charts with Several Parts

A concept map uses shapes and lines to show how ideas are related.

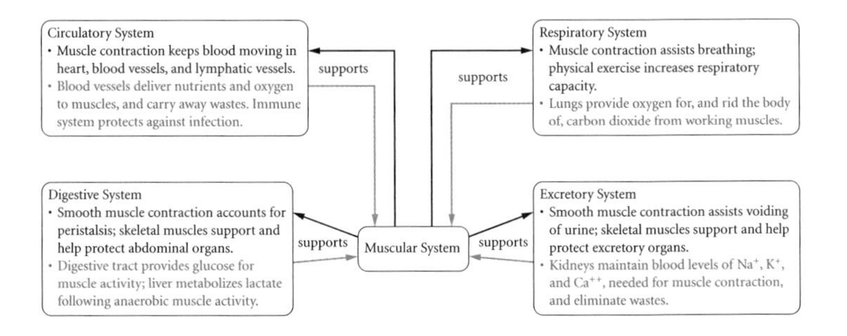

Before You Read

First, take a moment to get a sense of the whole concept map. There is no title, but you will notice that "Muscular System" is right in the middle of the chart and that it is connected to four other boxes.

1. Make a prediction about what this concept map is about.
2. What do you already know about this topic?
3. What elements and features are on this page?

While You Read

4. Why are some words enclosed in boxes while others are written on connecting lines?
5. What is the difference in content between the text printed in green and text printed in black?
6. What is the purpose of this concept map?

After Reading

7. In one or two sentences, summarize the information presented in this concept map.
8. Why might the information in this concept map be important?

Reading Cycle Diagrams

A cycle chart is a type of flowchart that has no distinct beginning or end. All the events are shown in the order in which they occur, as indicated by arrows, but there are no first and last events.

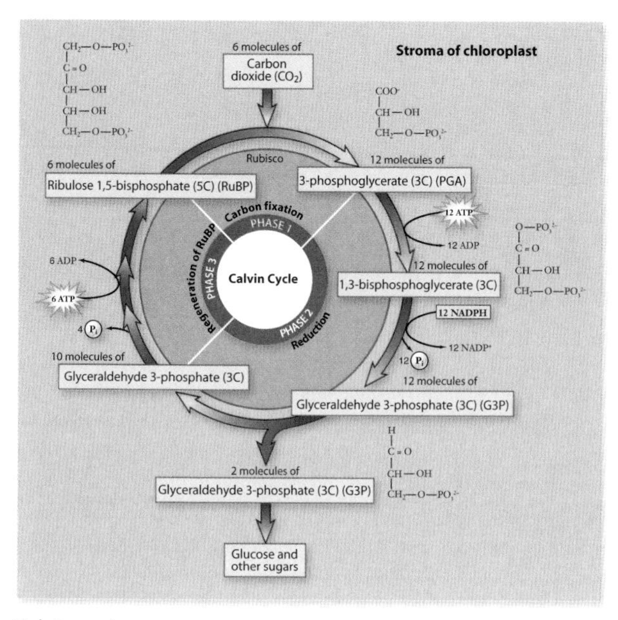

Calvin cycle

Before You Read

1. What cycle does this illustration represent?
2. What do you already know about this topic?
3. What elements and features are on this page?

While You Read

4. What do the arrows around the outside of the circle indicate?
5. What is the name of:
   a. Phase 1
   b. Phase 2
   c. Phase 3
6. Where does this cycle take place?

After Reading

7. Why might understanding this cycle be important?
8. Use an events chain to explain this cycle.

Reading Models

In science, a model can be anything that helps you better understand a concept. For example, a model can be a picture, a mental image, a structure, or even a mathematical formula.

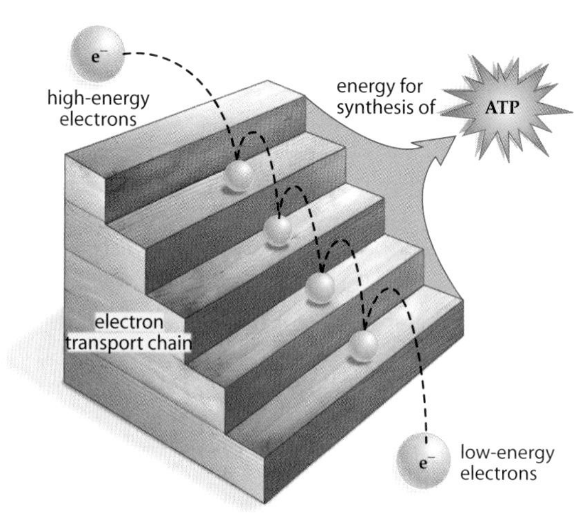

Electron transport chain model

Before You Read

1. The caption of this model is "Electron transport chain model." What does the word "model" tell you about what you will see in the illustration?
2. What is ATP?
3. What do you know about the connection between electrons and energy for synthesis of ATP?

While You Read

4. a. What do the steps in the model represent?
   b. What does the electron represent as it bounces down the steps?
5. What happens to the electron as it goes down the steps?
6. What would be a good title for this model?

After Reading

7. How did this model help you understand the concept?
8. What other model could you use to explain this concept?

Reading Charts with Several Parts

The visual below features a flowchart in the centre that shows a sequence of events. One of the events is represented with a cycle chart. The "Activates" and "Inhibits" labels are part of a cause-and-effect map.

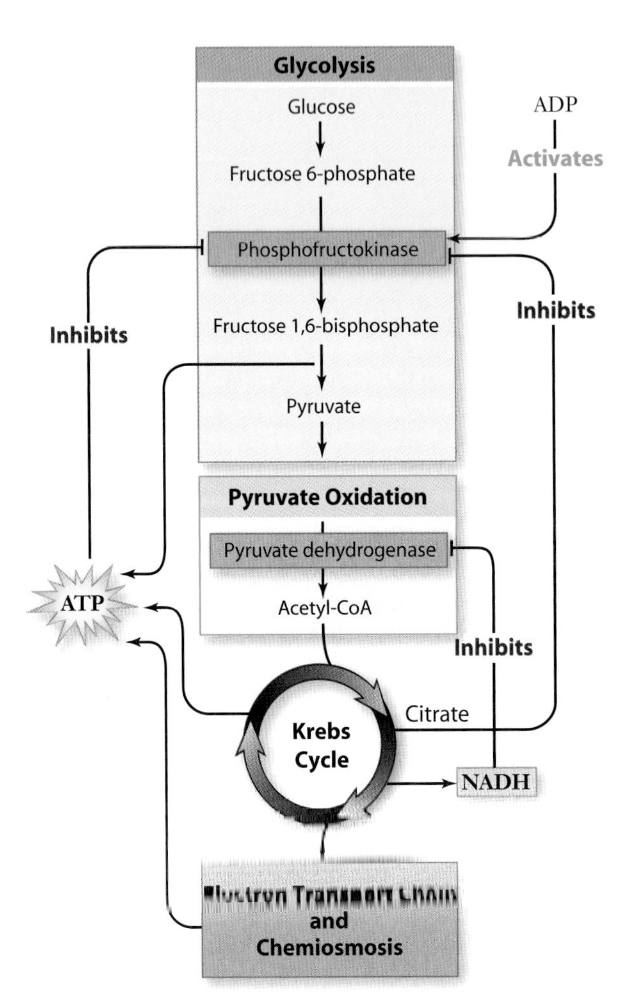

The ATP and NADP used in the Calvin cycle are produced in the light-dependent reactions. For [illegible] of G3P made in the Calvin cycle, two are used [illegible] and other high energy compounds.

Before You Read

1. Read the caption. What do you already know about this topic?
2. What are the four main sections in the middle of the chart?
3. Which cycle is included in this flowchart?

While You Read

4. Which labels are printed in boldface?
5. **a.** Which word is printed in red?
   b. Which word is printed in green?
   c. What is the difference in meaning between the two words?
6. **a.** What do the arrows in the boxes indicate?
   b. What do the arrows on the lines indicate?

After Reading

7. Why do you think the writer of the textbook included this diagram?
8. What would you change about this diagram to make it easier to understand?

Reading Diagrams with "Call-outs"

Complex visuals sometimes include "call-out" boxes that point out parts of a visual that a reader should pay attention to.

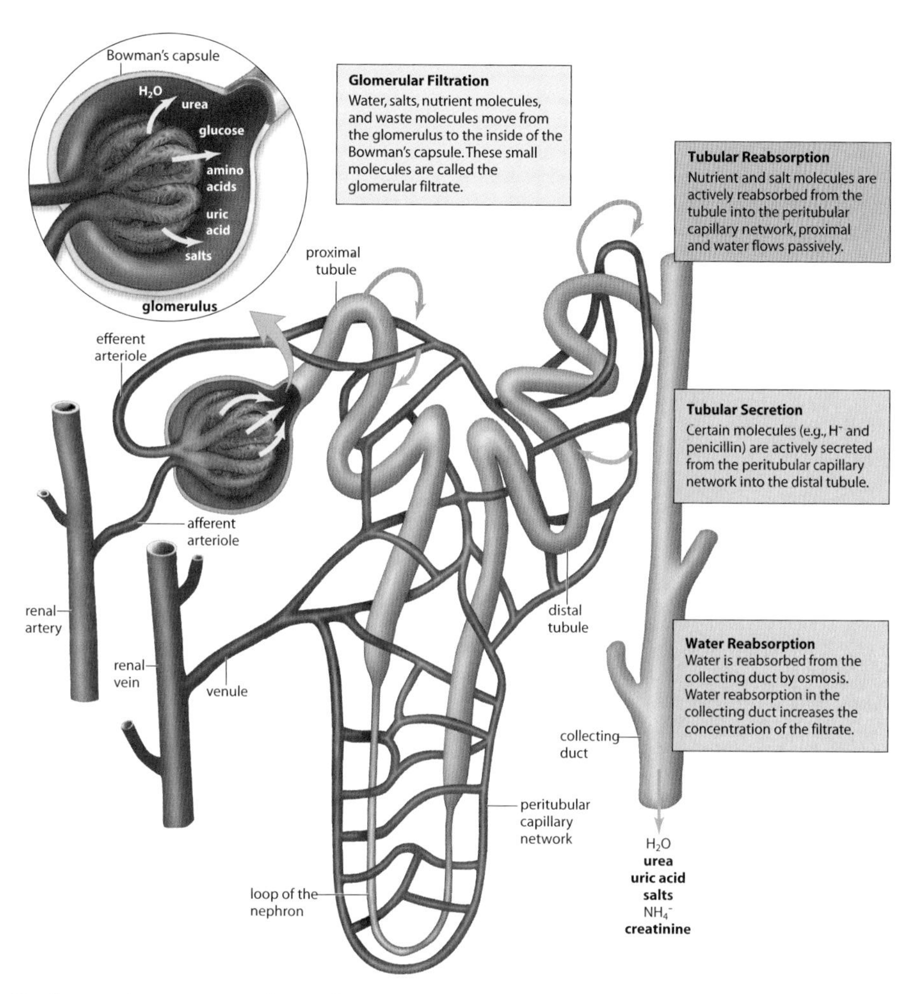

The four main processes of urine formation are described in boxes and colour-coded to arrows that show the movement of molecules into or out of the nephron at specific locations.

Before You Read

1. Which body system does this illustration primarily represent?
2. The caption mentions processes of urine formation. What do you know about this topic?
3. What does each of the different colours of tubes represent?
   a. red
   b. blue
   c. purple
   d. brown

While You Read

4. Which labels are in bold type?
5. Notice the large pink arrow pointing to the glomerulus. Why do you think the glomerulus gets a special, expanded view in this diagram?
6. What information did you get from the four boxes that you would not have known if the boxes were only labels without descriptions?

After Reading

7. Why is it important to understand this information?
8. Use a flowchart or other graphic organizer to describe this topic.

Selected Figures for Interpretation

Now that you have seen Before/During/After questions modelled, you can develop your own skills for reading graphical texts. The following pages present a selection of figures for you to interpret. With a partner or in a small group, choose several of the figures within this section. Work together to create Before/During/After questions about the figures you have chosen. Then answer your questions. Share your questions and figures with another group and then answer the other group's questions. Notice which questions are particularly helpful for interpreting the figures. Discuss why this is so with your group members.

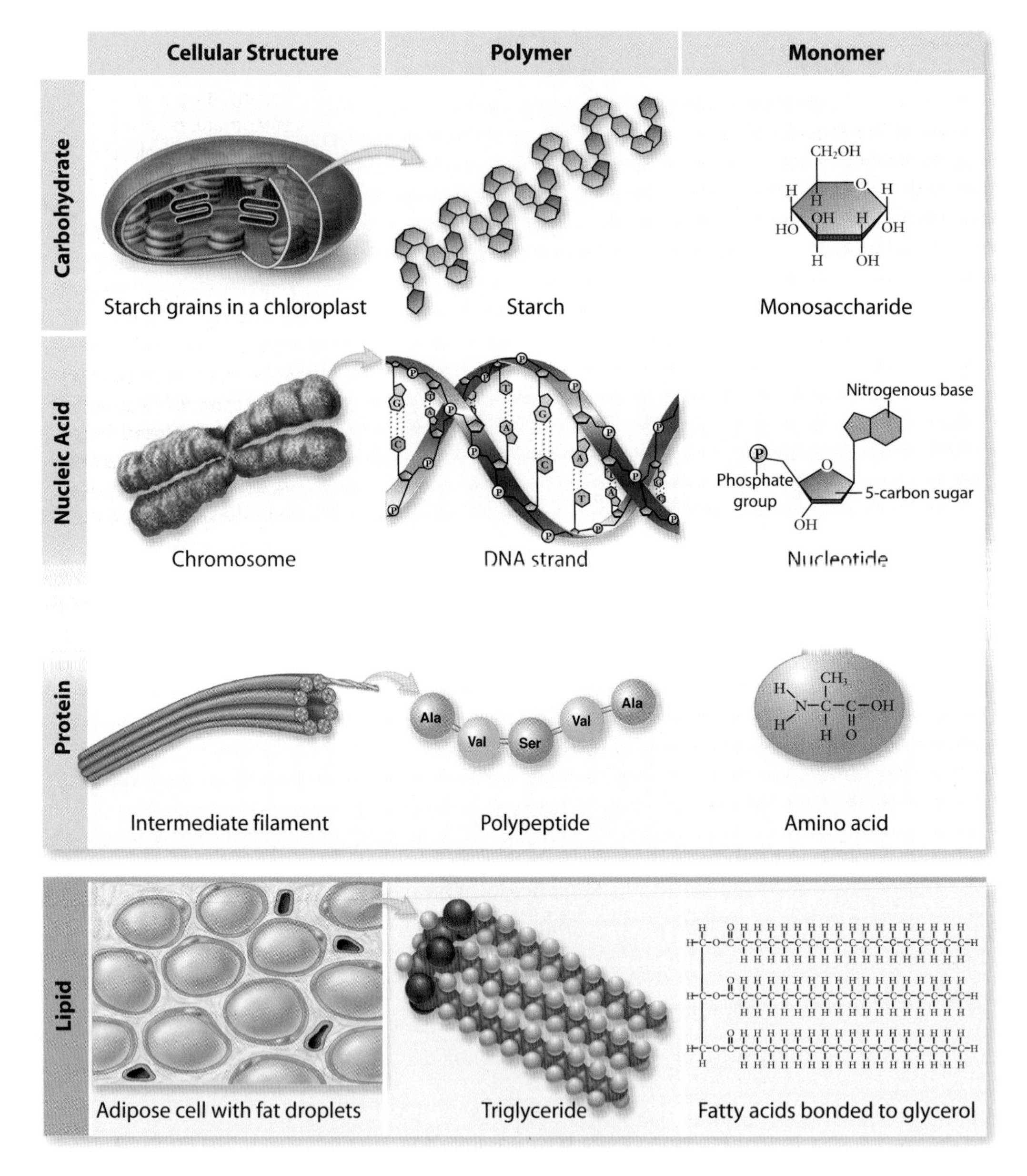

Carbohydrates, nucleic acids, proteins, and lipids (shown here as a triglyceride) are biologically important components of larger structures in the cell.

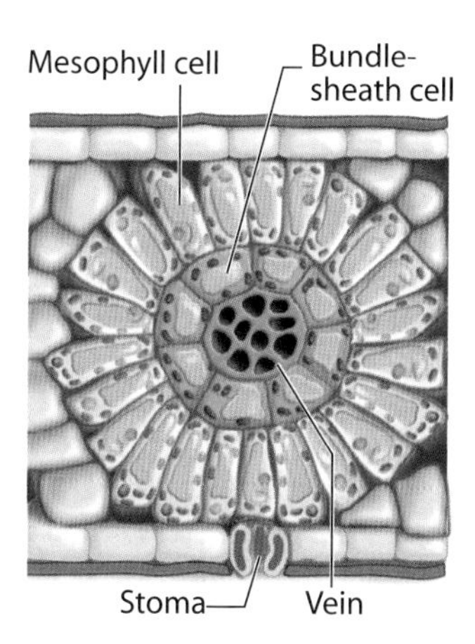

C4 plants use energy to "pump" carbon dioxide into the bundle-sheath cells, where it becomes concentrated. Included among the C4 plants are food crops such as corn, sorghum, sugarcane (shown here), and millet. Also included are grasses such as crabgrass and Bermuda grass.

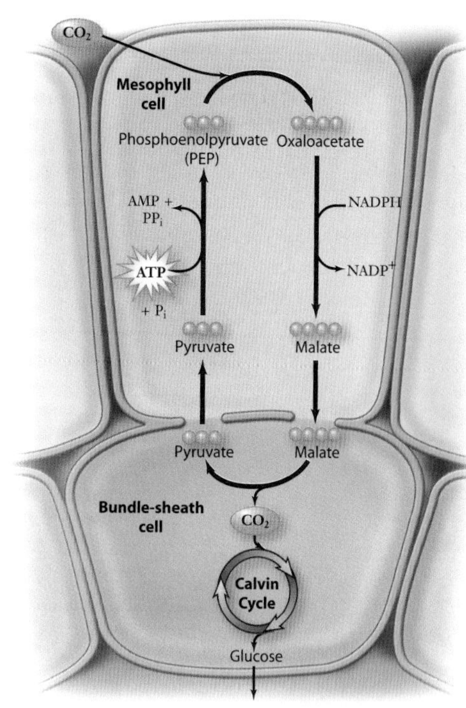

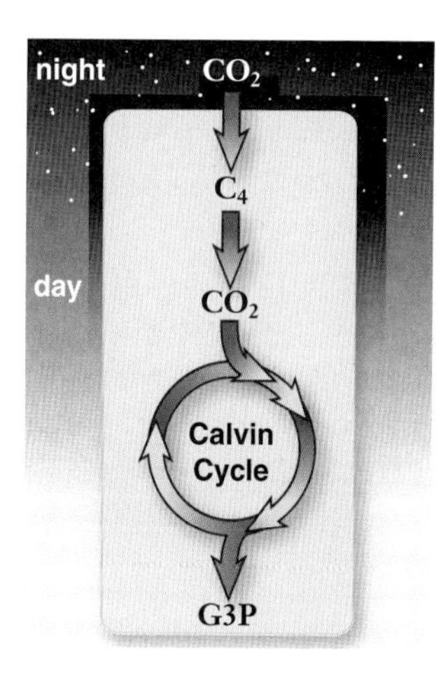

The opening and closing of stomata in CAM plants, such as pineapple and cacti, are opposite from most plants. The stomata are open at night and closed in the daytime. When the carbon dioxide is removed from the four-carbon compound malate in the daytime, it cannot leave the cell because the stomata are closed. In cool climates, CAM plants are very inefficient, because they use energy to drive the reactions that store carbon dioxide.

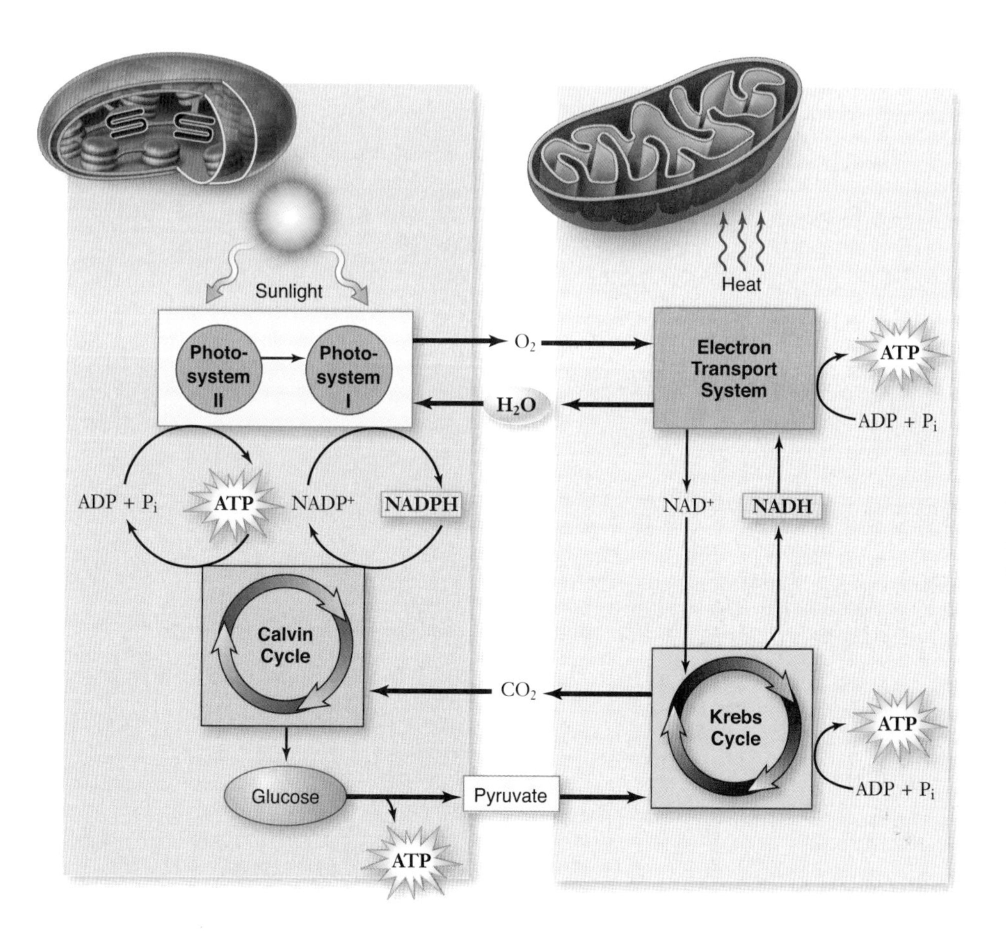

Water, carbon dioxide, glucose, and oxygen cycle between chloroplasts and mitochondria in plant cells.

Triglyceride is composed of three fatty acids (orange background) and a glycerol molecule (green background). The fatty acids can be saturated (no double bonds) or unsaturated (with one or more double bonds).

R group

central carbon atom

amino group

hydrogen

carboxyl group

An amino acid molecule is composed of a central carbon atom bonded to an amino group, a carboxyl group, and a hydrogen atom. Each amino acid also has an *R* group bonded to the central carbon atom, providing the amino acid with its unique identity.

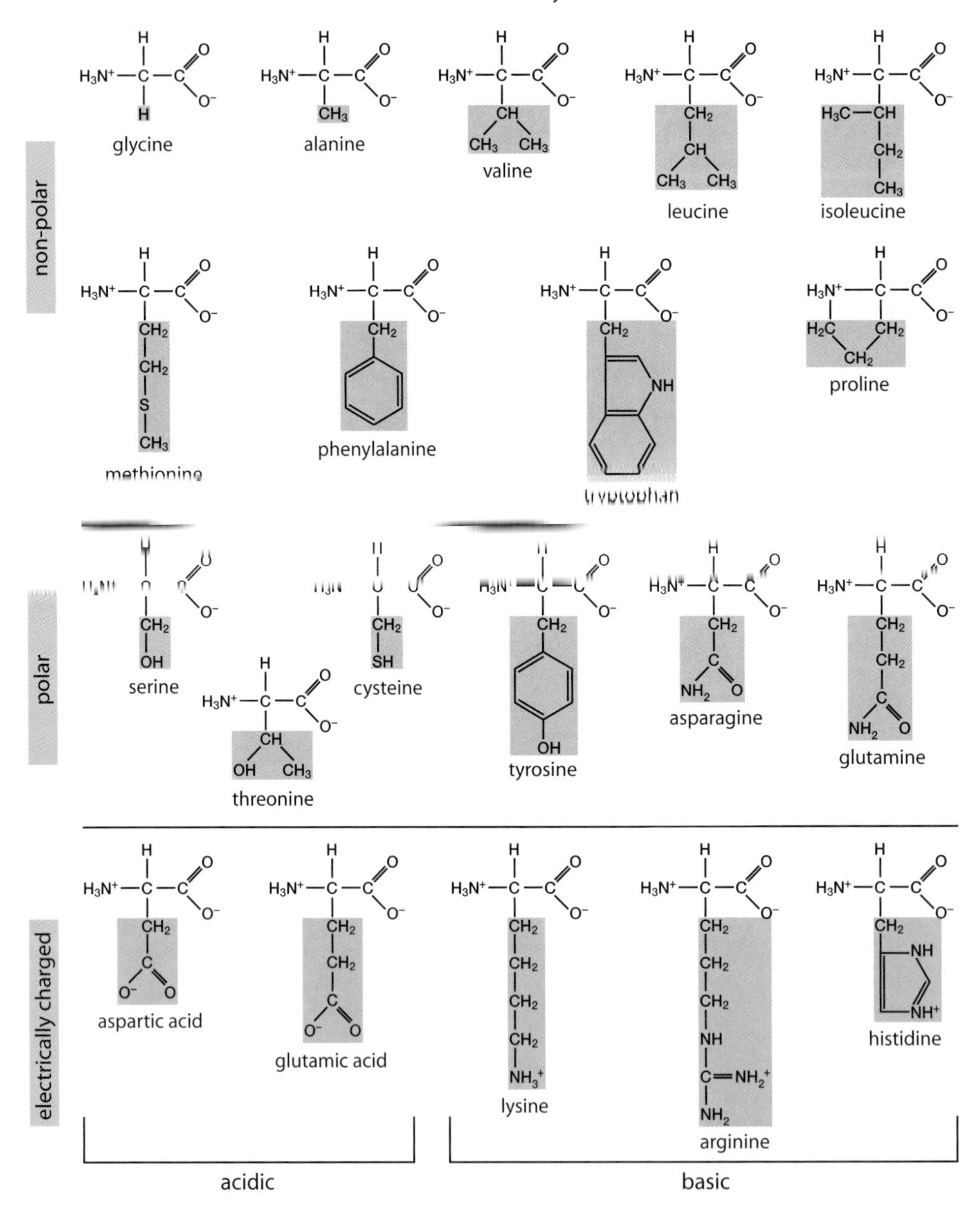

Common amino acids making up most proteins

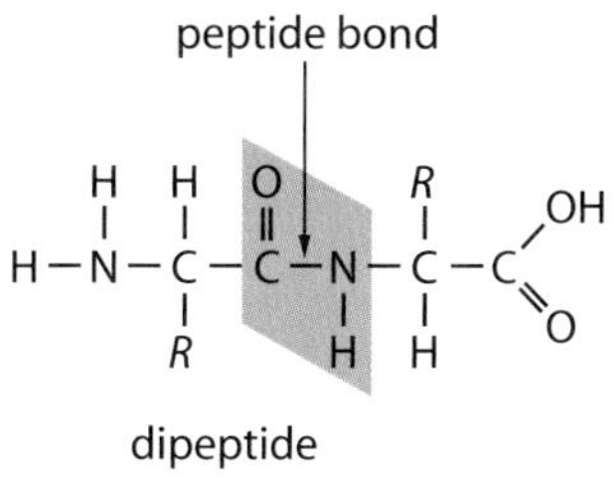

peptide bond

dipeptide

Forming a dipeptide

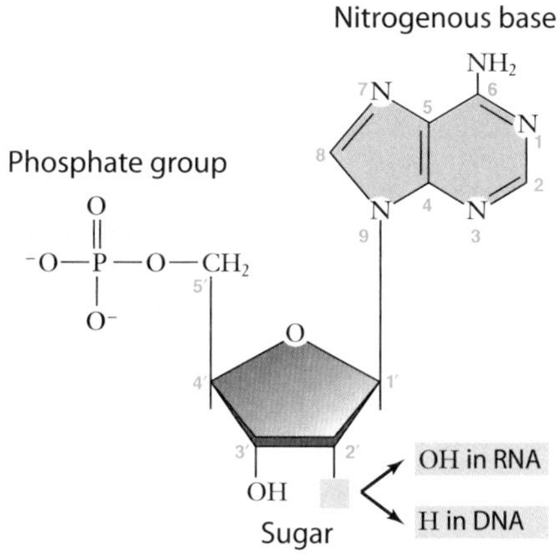

Nitrogenous base

Phosphate group

Sugar

OH in RNA

H in DNA

Nucleotide contents

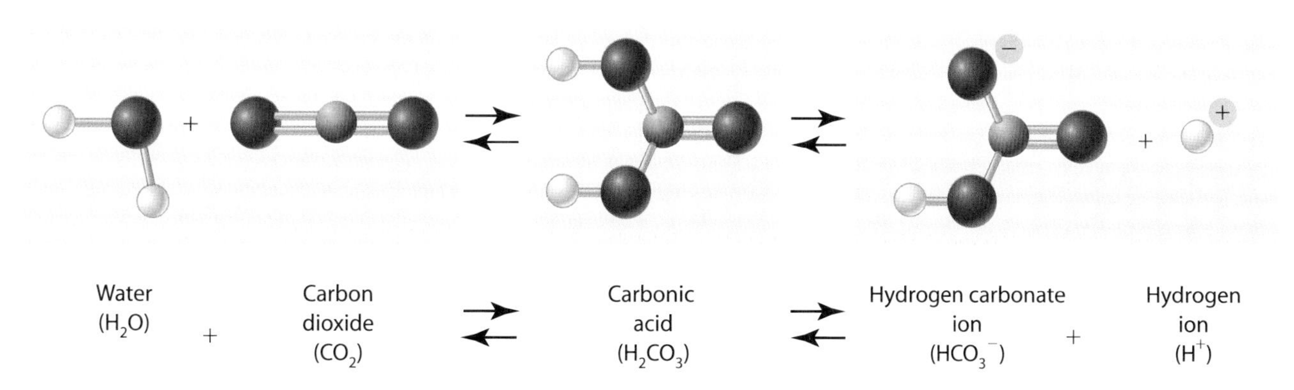

Carbonic acid-hydrogen carbonate ion buffer system

Examples of Biologically Important Molecules

Carbohydrates			
Type	**Structure**	**Examples**	**Some Functions**
Monosaccharide	• Contains a single three- to seven-carbon atom-based structure	Glucose, fructose, galactose	• Glucose is used as a primary energy source
Disaccharide	• Contains two monosaccharides joined by a glycosidic linkage	Sucrose, lactose, maltose	• Sucrose and lactose are dietary sugars that are used for energy
Polysaccharide	• Contains many monosaccharides joined by glycosidic linkages	Starch, glycogen, cellulose	• Glycogen is a form of storing glucose in animals • Cellulose provides structural support in plants
Lipids			
Type	**Structure**	**Examples**	**Some Functions**
Triglyceride	• Contains three fatty acids joined to glycerol by ester linkages	Lard, butter, vegetable oils	• Provides long-term energy storage • Acts to cushion organs and insulate from heat loss
Phospholipid	• Contains two fatty acids and a phosphate group joined to glycerol	Phosphatidylcholine	• Forms the main structure of cell membranes
Steroid	• Contains four carbon-based rings attached to one another	Cholesterol, testosterone, estrogen	• Cholesterol is part of cell membranes • Testosterone and estrogen are sex hormones
Wax	• Contains long carbon-based chains	Earwax, beeswax, spermaceti	• A variety of functions, including protection
Protein			
Type	**Structure**	**Examples**	**Some Functions**
Catalyst	• Contains amino acid monomers joined by peptide bonds • All have primary, secondary, tertiary structure	Amylase, sucrase	• Speeds up chemical reactions
Transport		Hemoglobin, ion channel proteins	• Transports specific substances
Structural		Collagen, keratin	• Provides structure
Movement		Myosin, actin	• Enables movement
Regulatory		Hormones, neurotransmitters	• Carries cellular messages
Defence		Antibodies	• Fights infection
Nucleic Acids			
Type	**Structure**	**Some Functions**	
DNA	• Contains deoxyribonucleotide monomers (A, G, T, C)	Stores genetic information of an organism	
RNA	• Contains ribonucleotide monomers (A, U, G, C)	Participates in protein synthesis	

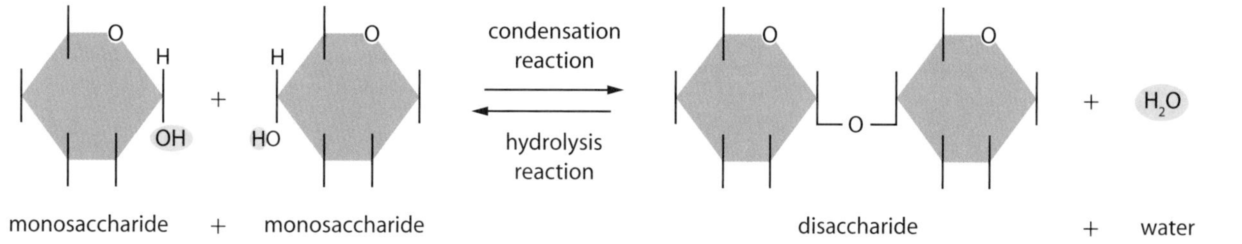

condensation reaction / hydrolysis reaction

monosaccharide + monosaccharide ⇌ disaccharide + water (H_2O)

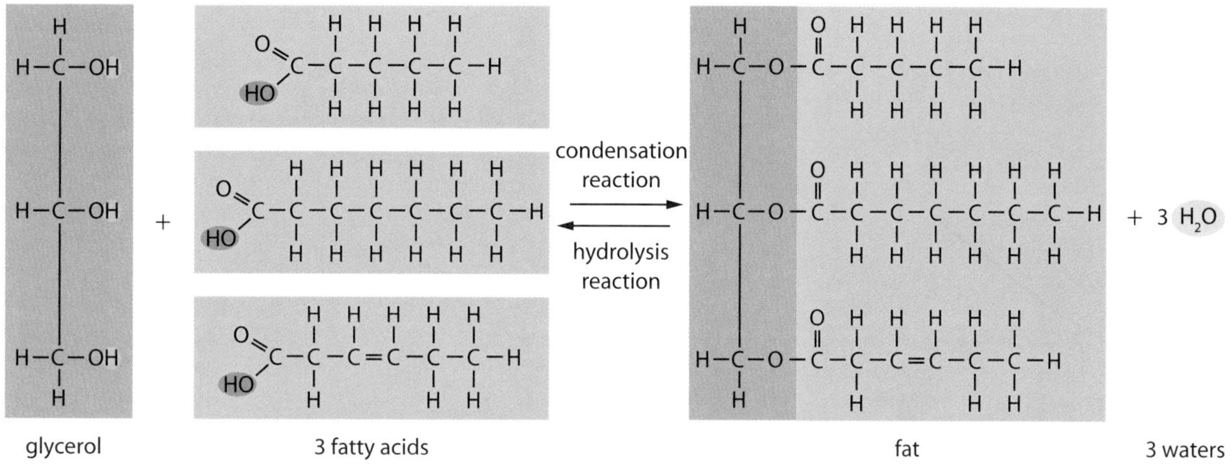

glycerol + 3 fatty acids ⇌ fat + 3 waters ($3\ H_2O$)

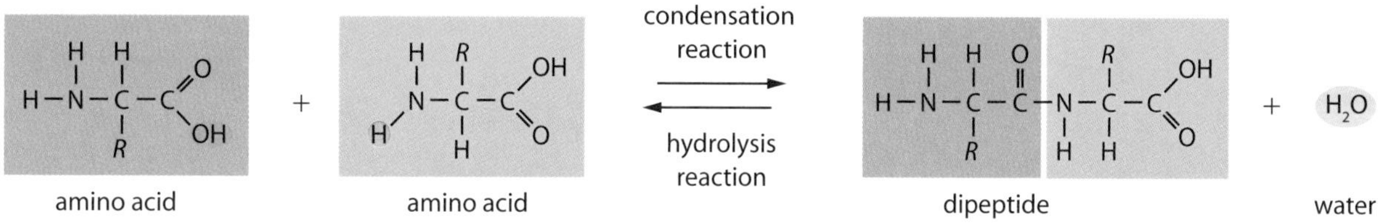

amino acid + amino acid ⇌ dipeptide + water (H_2O)

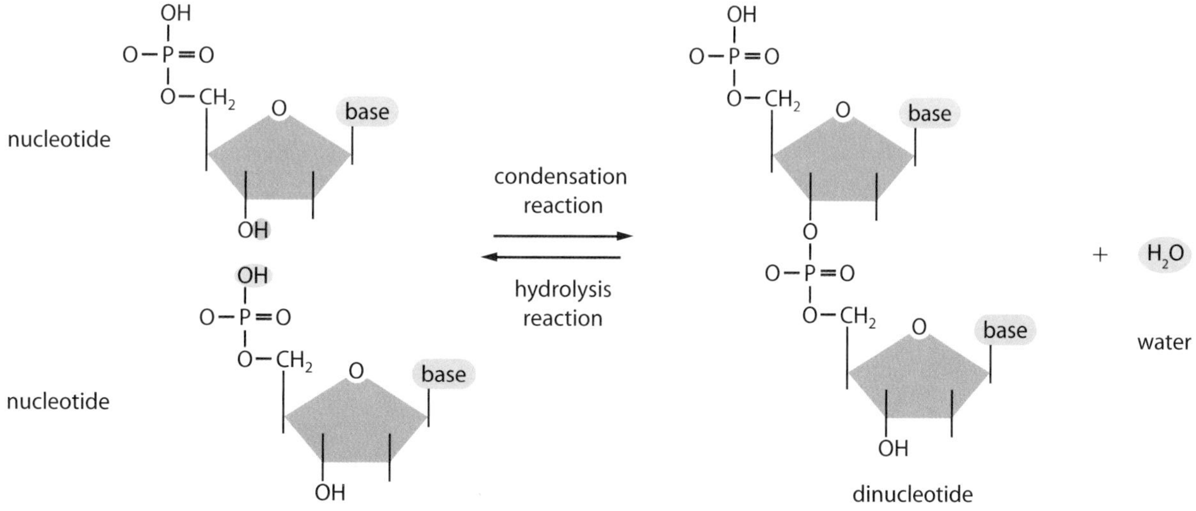

nucleotide + nucleotide ⇌ dinucleotide + water (H_2O)

Carbohydrates, lipids, proteins, and nucleic acids are assembled by condensation reactions among their monomers, and these polymers are broken down by hydrolysis reactions. The double arrow indicates that a chemical reaction can proceed in a "forward" and a "reverse" direction. As written, the forward is a condensation reaction and the reverse is a hydrolysis reaction. Note that the rings in the carbohydrate and nucleotide structures are drawn in this manner so that a particular molecule is not specified.

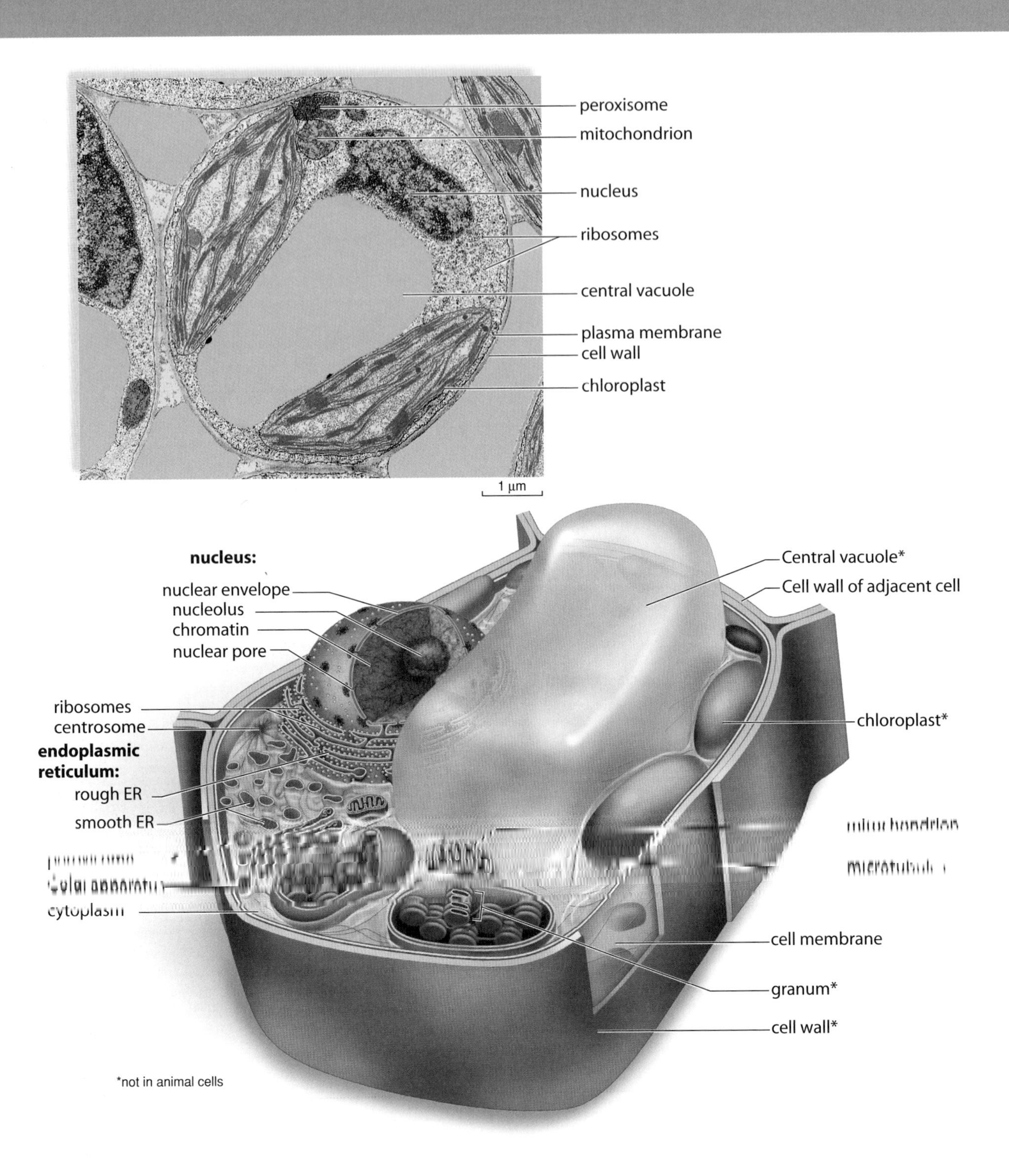

Although most plant cells contain the structures shown here, plant cells also exhibit great diversity in their form, size, and specialized features.

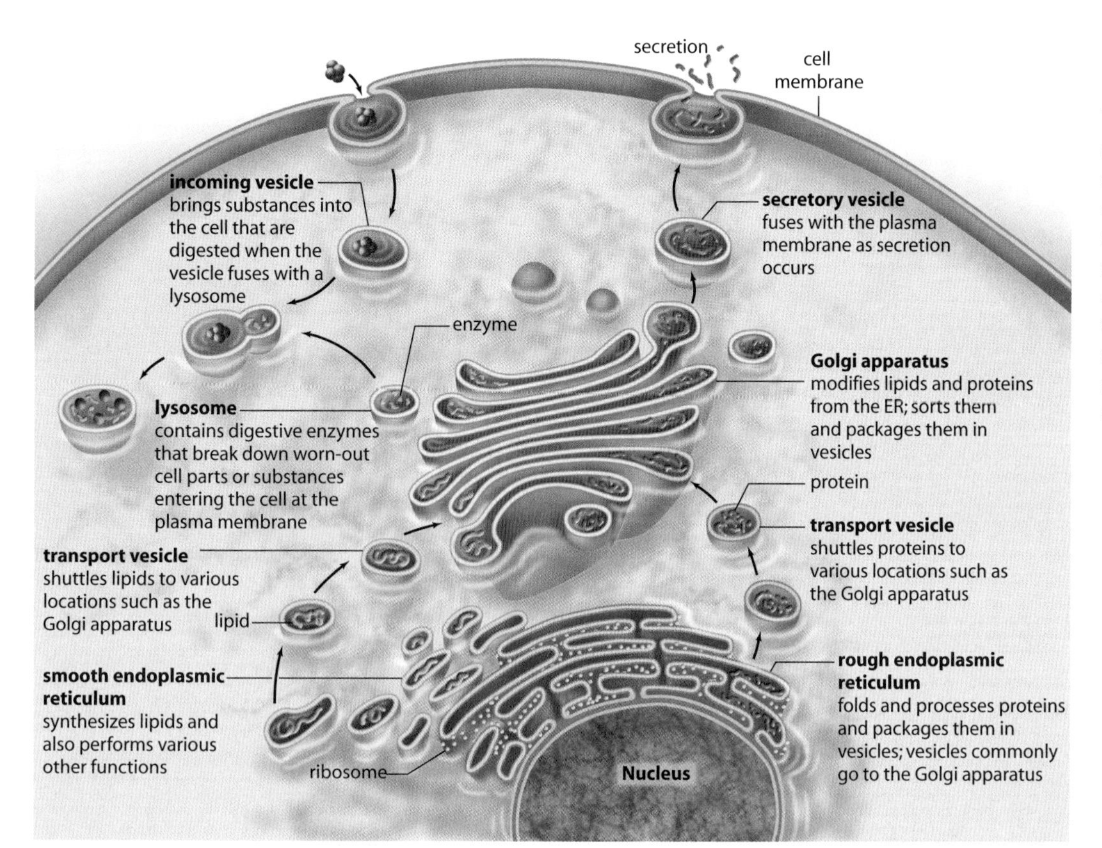

secretion
cell membrane
incoming vesicle
brings substances into the cell that are digested when the vesicle fuses with a lysosome
secretory vesicle
fuses with the plasma membrane as secretion occurs
enzyme
Golgi apparatus
modifies lipids and proteins from the ER; sorts them and packages them in vesicles
lysosome
contains digestive enzymes that break down worn-out cell parts or substances entering the cell at the plasma membrane
protein
transport vesicle
shuttles proteins to various locations such as the Golgi apparatus
transport vesicle
shuttles lipids to various locations such as the Golgi apparatus
lipid
smooth endoplasmic reticulum
synthesizes lipids and also performs various other functions
rough endoplasmic reticulum
folds and processes proteins and packages them in vesicles; vesicles commonly go to the Golgi apparatus
ribosome
Nucleus

The endomembrane system is composed of different organelles that are connected and work together to carry out a number of processes in the cell.

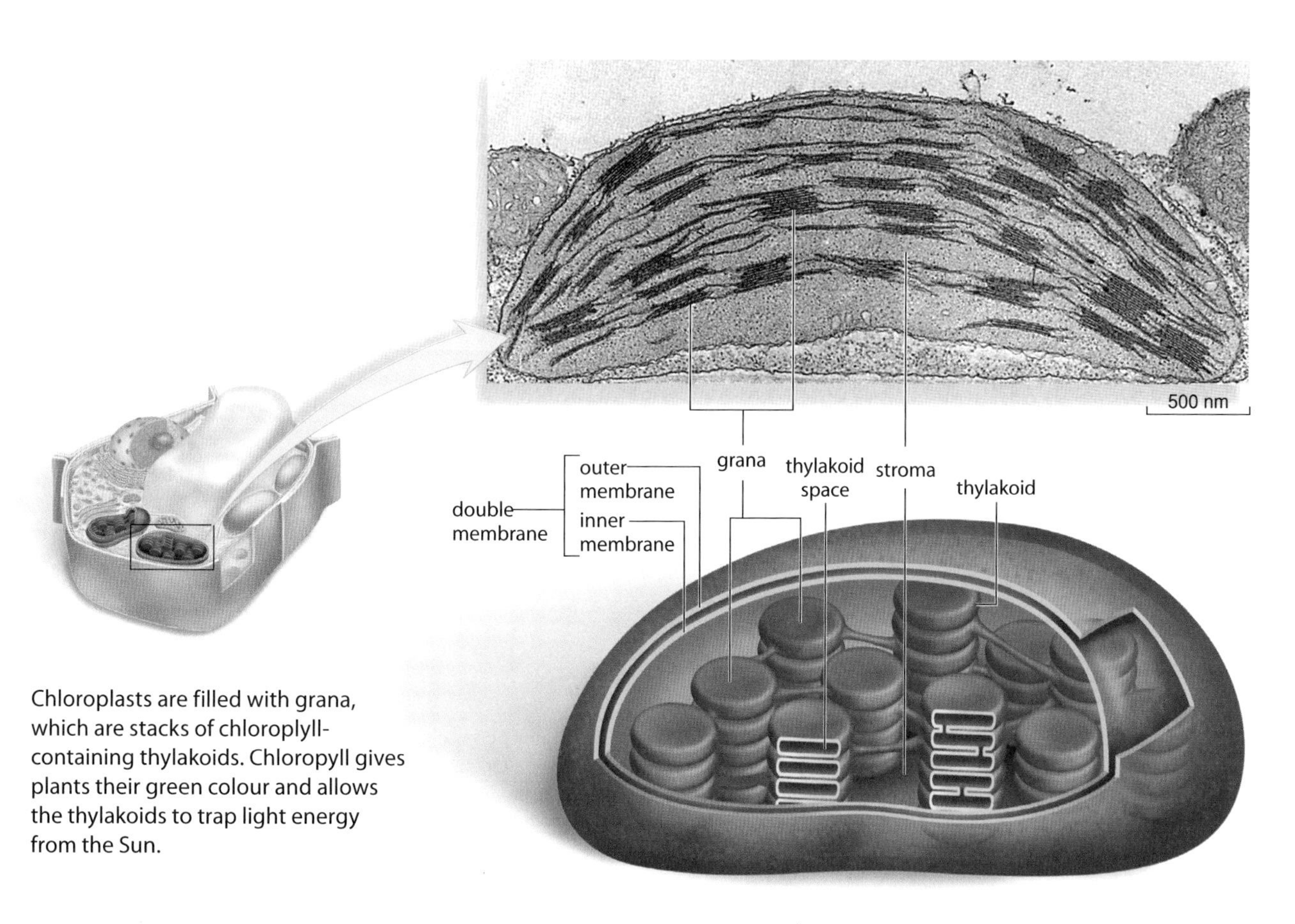

500 nm
grana
thylakoid space
stroma
thylakoid
outer membrane
inner membrane
double membrane

Chloroplasts are filled with grana, which are stacks of chloroplyll-containing thylakoids. Chloropyll gives plants their green colour and allows the thylakoids to trap light energy from the Sun.

Functions of Protein Fibres in the Cytoskeleton

Type of Fibre	Size	Structure	Selected Functions
microtubules	Thickest fibres (average of 25 nm in diameter)	Proteins that form hollow tubes	• Maintain cell shape • Facilitate movement of organelles • Assist in cell division (spindle formation)
intermediate filaments	Intermediate thickness (average of 10 nm in diameter)	Proteins coiled together into cables	• Maintain cell shape • Anchor some organelles • Form the internal scaffolding of the nucleus
microfilaments	Thinnest fibres (average of 8 nm in diameter)	Two strands of actin wound together	• Maintain cell shape • Involved in muscle contraction • Assist in cell division (cleavage furrow)

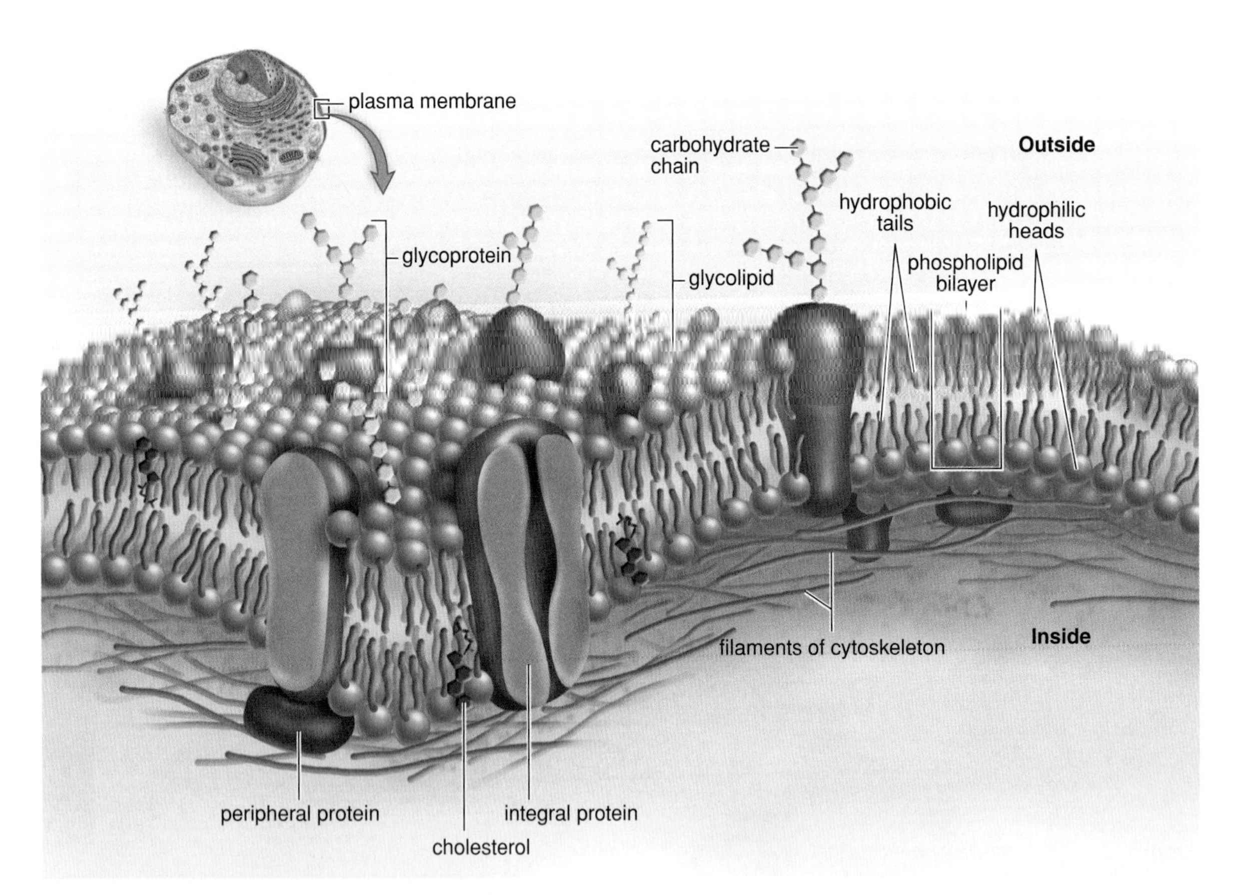

In the modern fluid mosaic model, the basic framework of a cell membrane is a phospholipid bilayer into which proteins are inserted. These proteins may be bound on the surface to other proteins or to lipids, including glycoproteins and glycolipids. Glycoproteins and glycolipids are proteins and lipids covalently bonded to carbohydrates.

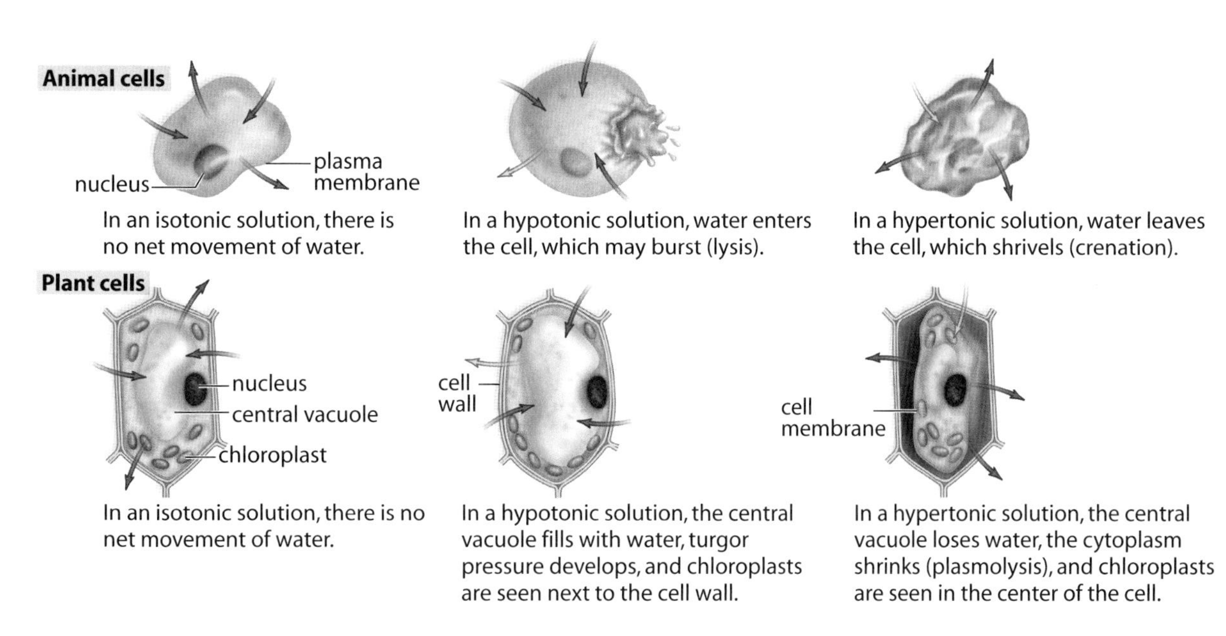

Animal cells
nucleus
plasma membrane
In an isotonic solution, there is no net movement of water.
In a hypotonic solution, water enters the cell, which may burst (lysis).
In a hypertonic solution, water leaves the cell, which shrivels (crenation).
Plant cells
nucleus
central vacuole
chloroplast
cell wall
cell membrane
In an isotonic solution, there is no net movement of water.
In a hypotonic solution, the central vacuole fills with water, turgor pressure develops, and chloroplasts are seen next to the cell wall.
In a hypertonic solution, the central vacuole loses water, the cytoplasm shrinks (plasmolysis), and chloroplasts are seen in the center of the cell.

Isotonic and hypotonic solutions. Arrows indicate the movement of water molecules.

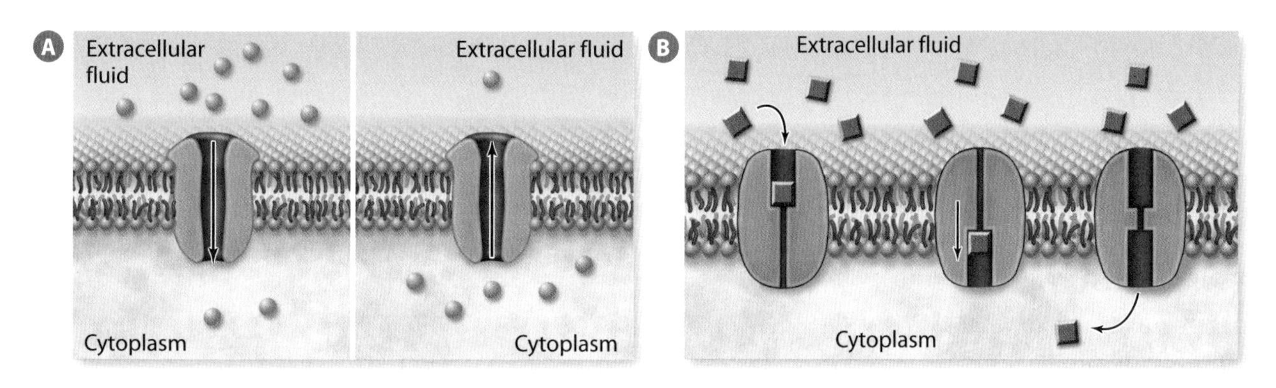

Facilitated diffusion involves membrane proteins. **A** Channel proteins form channels through membranes, which allow passage of specific ions and molecules from areas of higher concentration to areas of lower concentration. **B** Carrier proteins bind to molecules and carry them across a membrane from an area of higher concentration to an area of lower concentration.

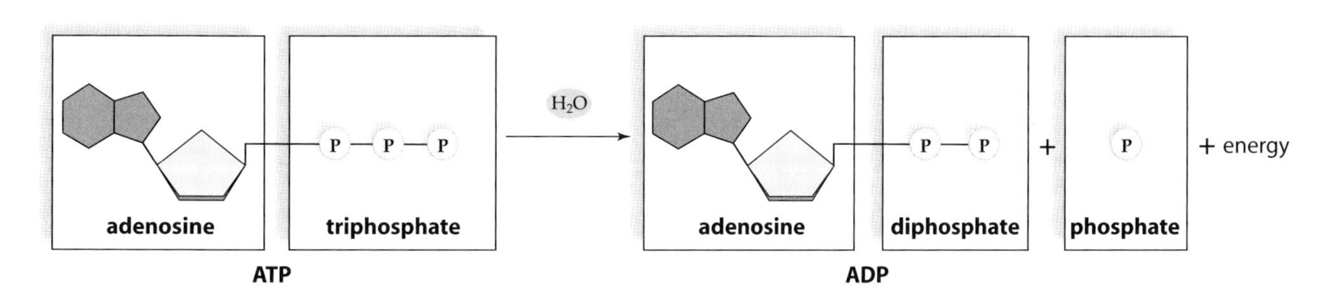

ATP undergoes hydrolysis to form ADP and phosphate, with the release of energy. The cell uses this energy for various functions, including the transport of molecules and ions across the cell membrane against their concentration gradients.

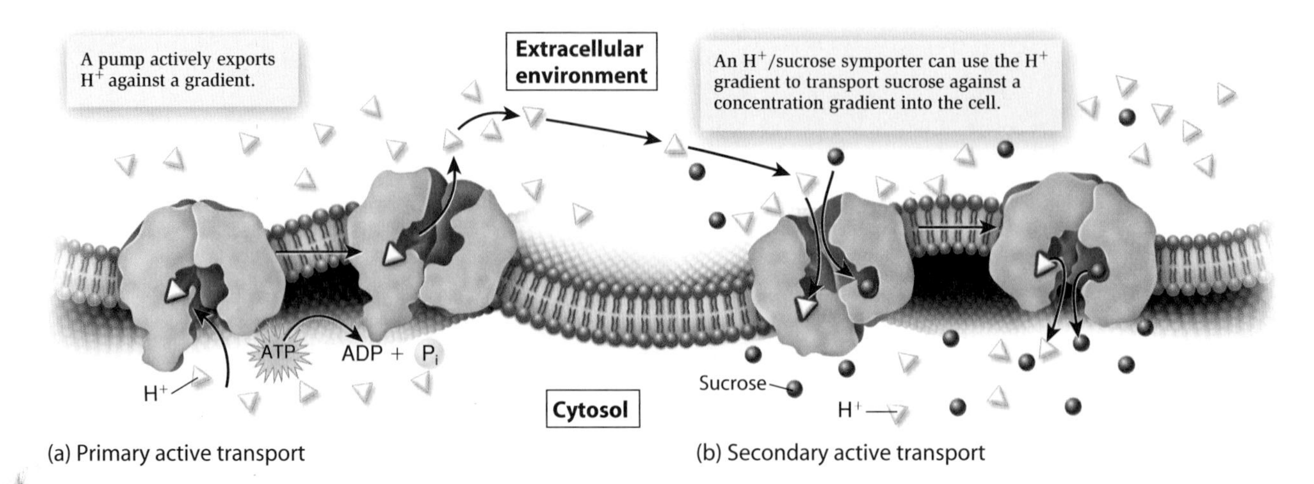

In secondary active transport, the electrochemical gradient created by primary active transport via an ion pump is used by a different protein to transport other molecules across a cell membrane. This kind of transport is common in bacteria and in plant cells.

In phagocytosis, a cell engulfs a large particle along with some of the liquid surrounding it. In pinocytosis, a cell engulfs a liquid and the small particles dissolved or suspended in it. In receptor-mediated endocytosis, receptor proteins in the cell membrane bind to specific molecules outside the cell. The cell membrane folds inward to create a vesicle containing the bound particles. These vesicles are coated with clathrin, a protein that forms a cage around a vesicle.

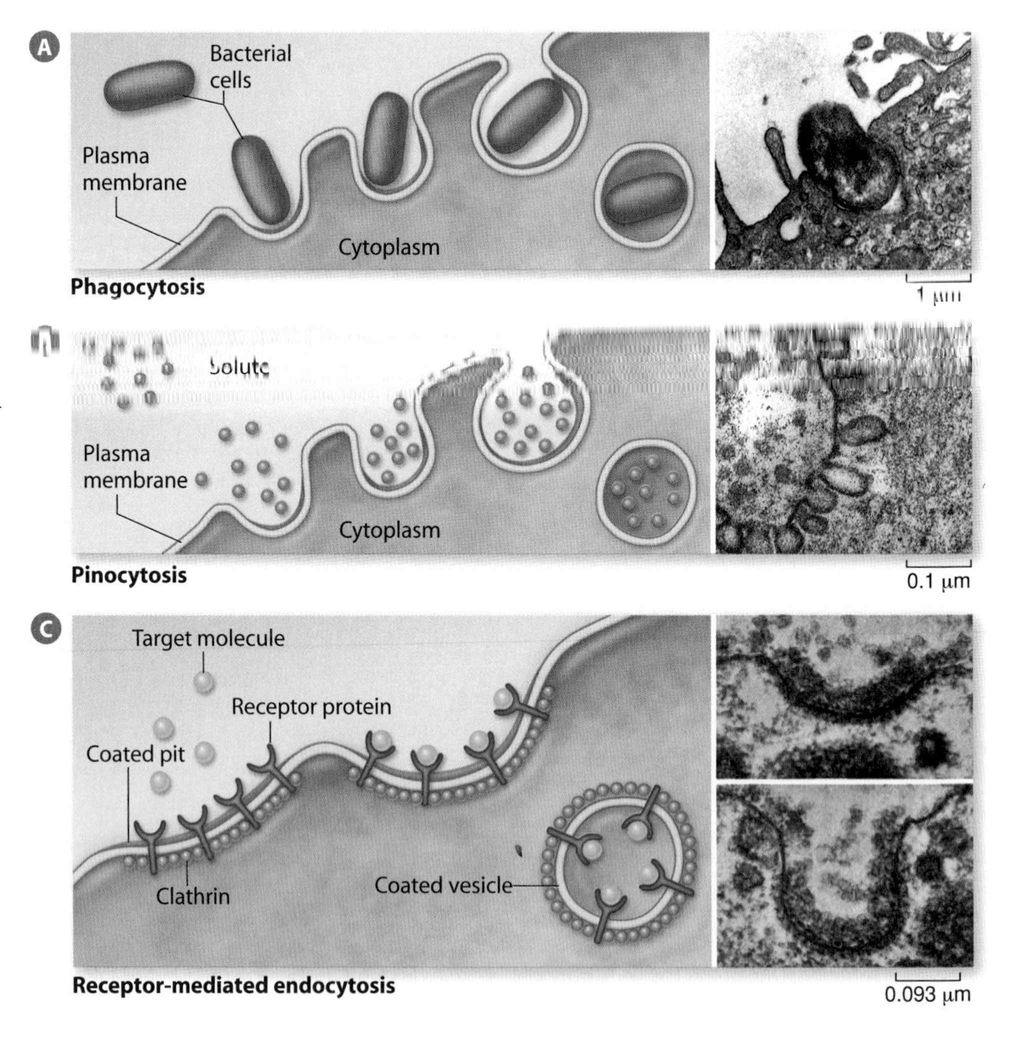

Mechanisms for Transport of Substances Across a Cell Membrane

Is Energy Required for the Mechanism to Function?	Type of Cellular Transport Mechanism	Primary Direction of Movement of Substances	Essential Related Factor(s)	Examples of Transported Substances
No	diffusion	toward lower concentration	concentration gradient	lipid-soluble molecules, water, gases
No	facilitated diffusion	toward lower concentration	channel protein or carrier protein and concentration gradient	some sugars and amino acids
Yes	active transport	toward higher concentration	carrier protein and energy	sugars, amino acids, ions
Yes	endocytosis	toward interior of cell	vesicle formation	macromolecules
Yes	exocytosis	toward exterior of cell	fusion of vesicle with cell membrane	macromolecules

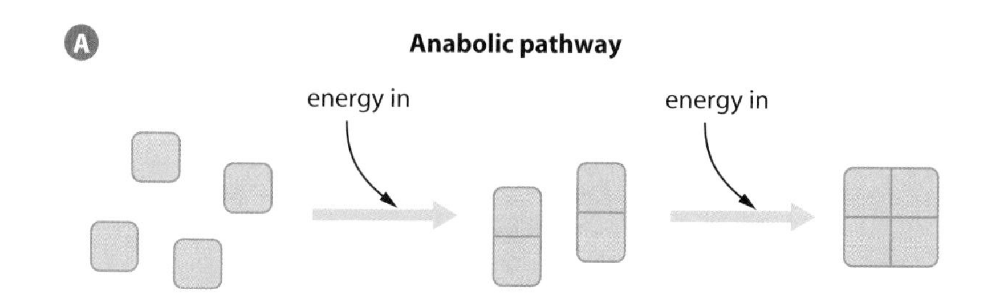

Diagram of a water wave

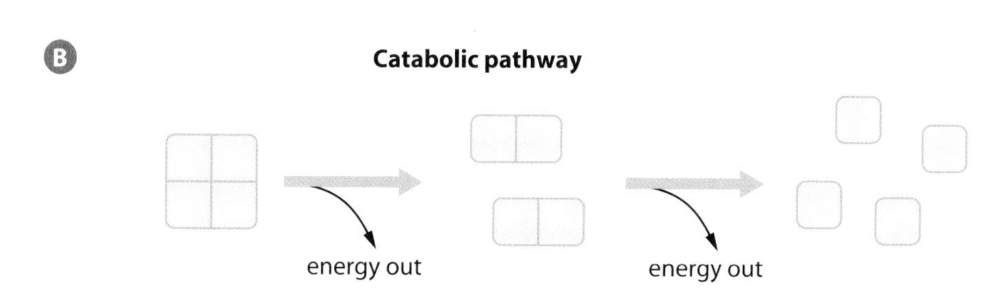

Anabolic reactions (A) build complex molecules, while catabolic reactions (B) reverse that process.

Anabolic reactions

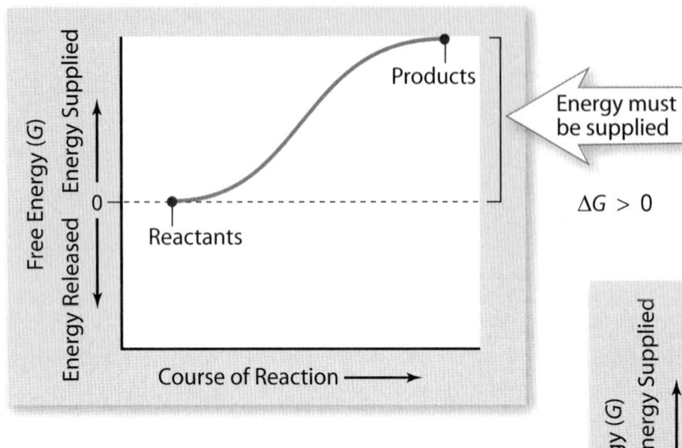

In an endergonic reaction (A), the products of the reaction contain more energy than the reactants, and energy must be supplied for the reaction to proceed. In an exergonic reaction (B), the products contain less energy than the reactants, and excess energy is released.

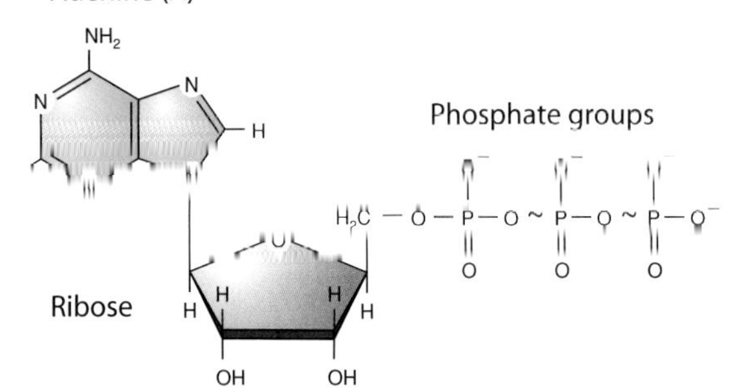

Adenosine triphosphate (ATP)

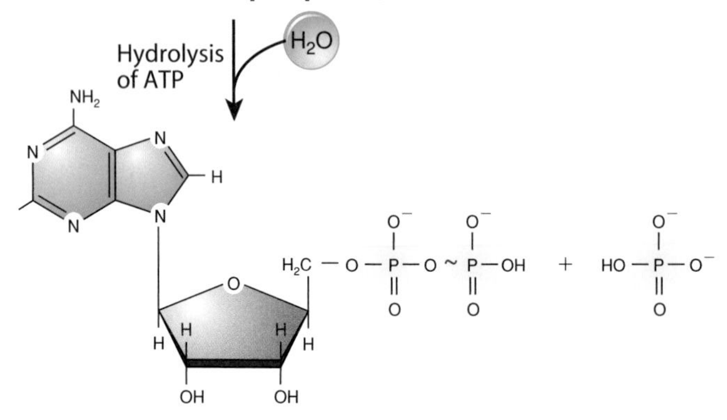

Adenosine diphosphate (ADP) **Phosphate (P_i)**

$\Delta G = -30.6$ kJ/mol

The hydrolysis of ATP to ADP (adenosine diphosphate) and P_i (inorganic phosphate) is a highly exergonic reaction. Repulsion between negative charges on the neighbouring phosphate groups makes the bonds between the first and second and between the second and third phosphate groups unstable. When these bonds are broken, energy is released.

While bound to an enzyme, NAD^+ receives electrons from two hydrogen atoms to become reduced to NADH. One proton (H^+ ion) is released into solution. The product of the energy-rich molecule and the NADH leave the enzyme. The reduced NADH then carries the electrons, or reducing power, to another molecule.

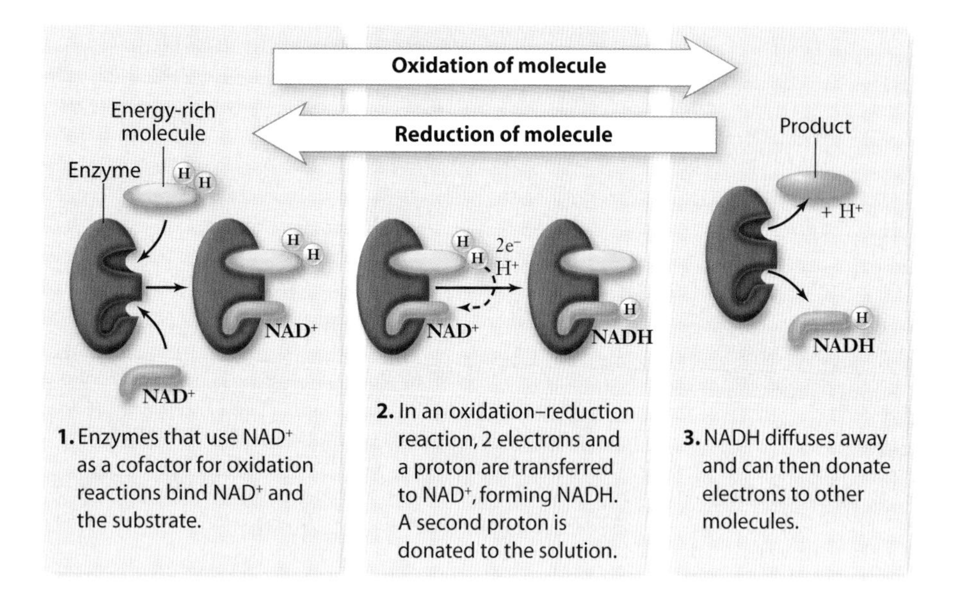

Oxidation of molecule
Reduction of molecule
Energy-rich molecule
Enzyme
H
H
NAD+
2e−
H+
NADH
Product
+ H+
1. Enzymes that use NAD+ as a cofactor for oxidation reactions bind NAD+ and the substrate.
2. In an oxidation–reduction reaction, 2 electrons and a proton are transferred to NAD+, forming NADH. A second proton is donated to the solution.
3. NADH diffuses away and can then donate electrons to other molecules.

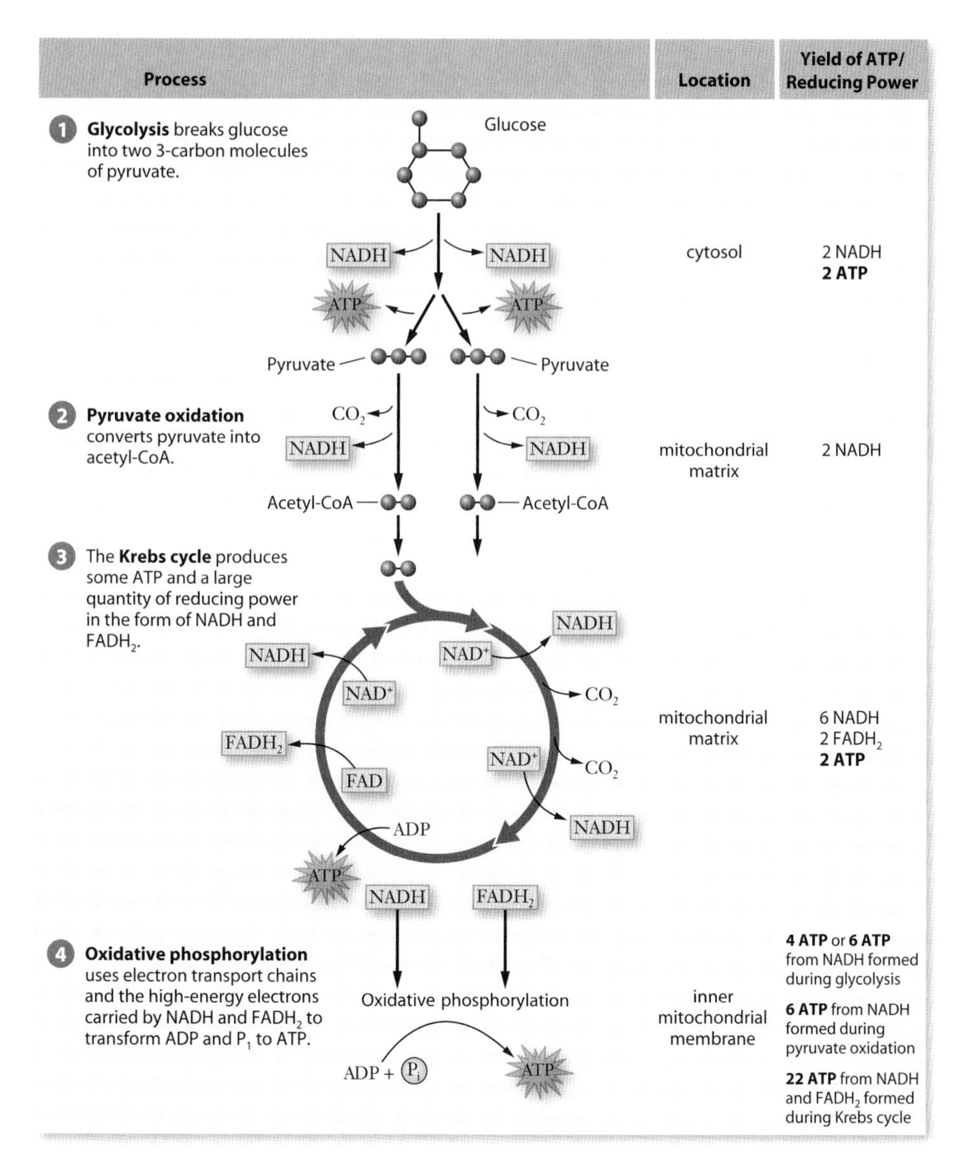

This diagram summarizes the metabolism of glucose by aerobic respiration. The maximum total yield of ATP from cellular respiration in eukaryotes is 36 (2 + 2 + 4 + 6 + 22) and in prokaryotes is 38 (2 + 2 + 6 + 6 + 22). These numbers reflect the fact that the yield of ATP from total glycolytic NADH is 4 ATP in eukaryotes and 6 ATP in prokaryotes.

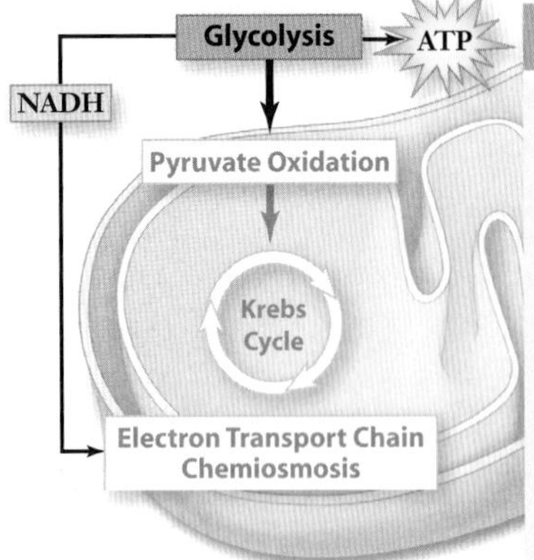

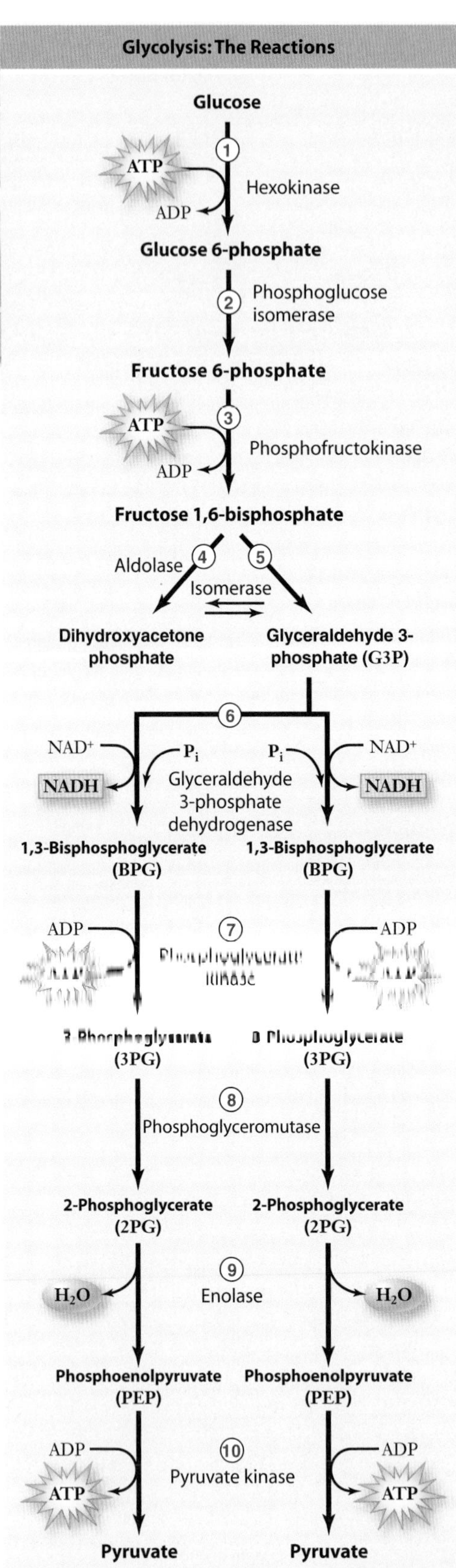

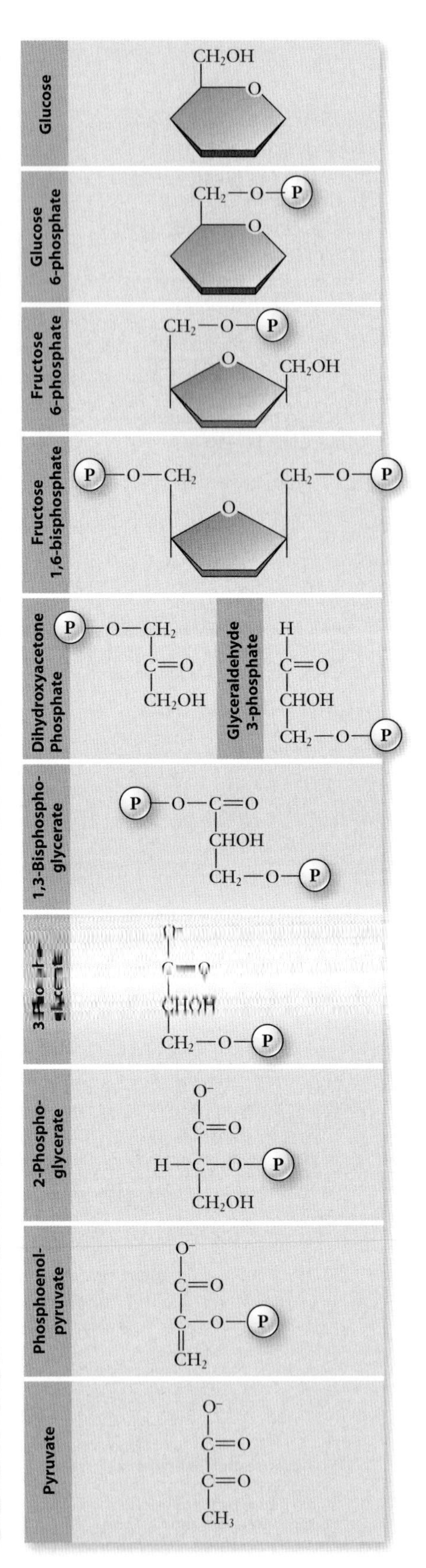

1. Phosphorylation of glucose by ATP.

2–3. Rearrangement, followed by a second ATP phosphorylation.

4–5. The 6-carbon molecule is split into two 3-carbon molecules—one G3P, and another that is converted into G3P in another reaction.

6. Oxidation followed by phosphorylation produces two NADH molecules and two molecules of BPG, each with one high-energy phosphate bond.

7. Removal of high-energy phosphate by two ADP molecules produces two ATP molecules and leaves two 3PG molecules.

8–9. Removal of water yields two PEP molecules, each with a high-energy phosphate bond.

10. Removal of high-energy phosphate by two ADP molecules produces two ATP molecules and two pyruvate molecules.

The 10 reactions of glycolysis. The first five reactions convert a molecule of glucose into two molecules of G3P. The second five reactions convert G3P into pyruvate.

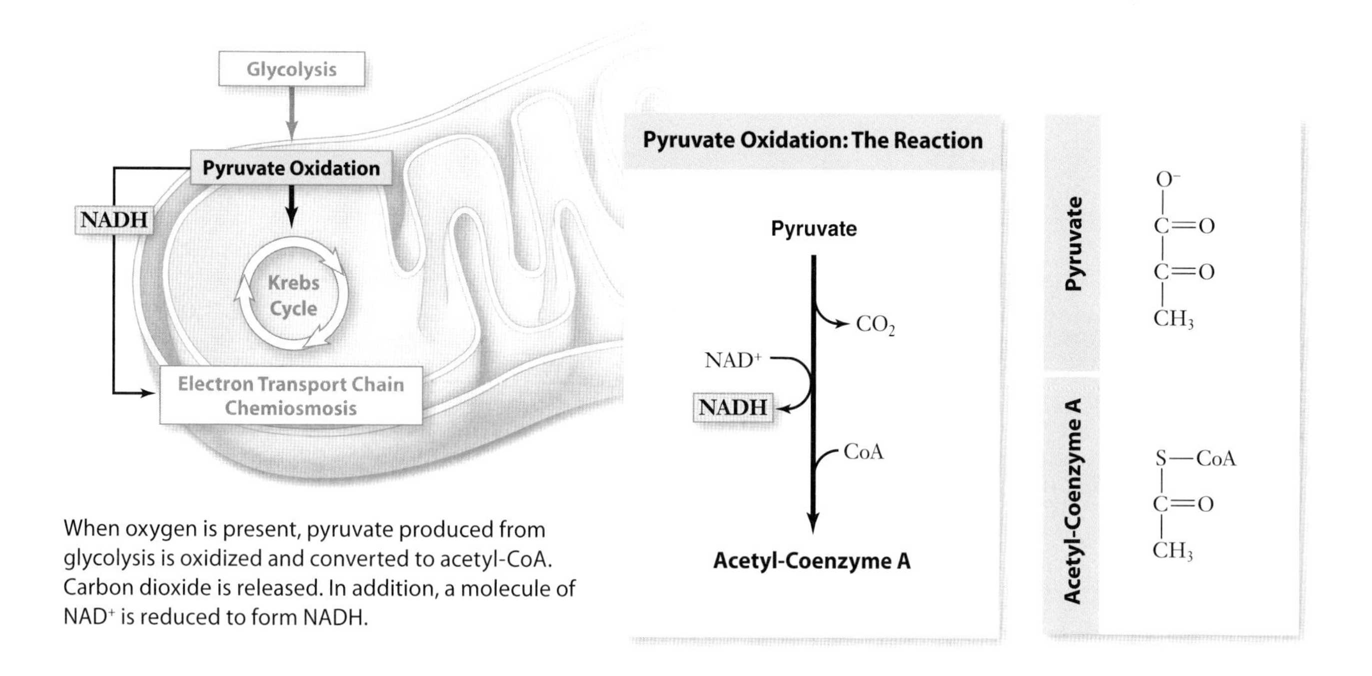

When oxygen is present, pyruvate produced from glycolysis is oxidized and converted to acetyl-CoA. Carbon dioxide is released. In addition, a molecule of NAD^+ is reduced to form NADH.

The electron carriers continuously cycle between their reduced form and their oxidized form while passing electrons from one to the next and finally to oxygen.

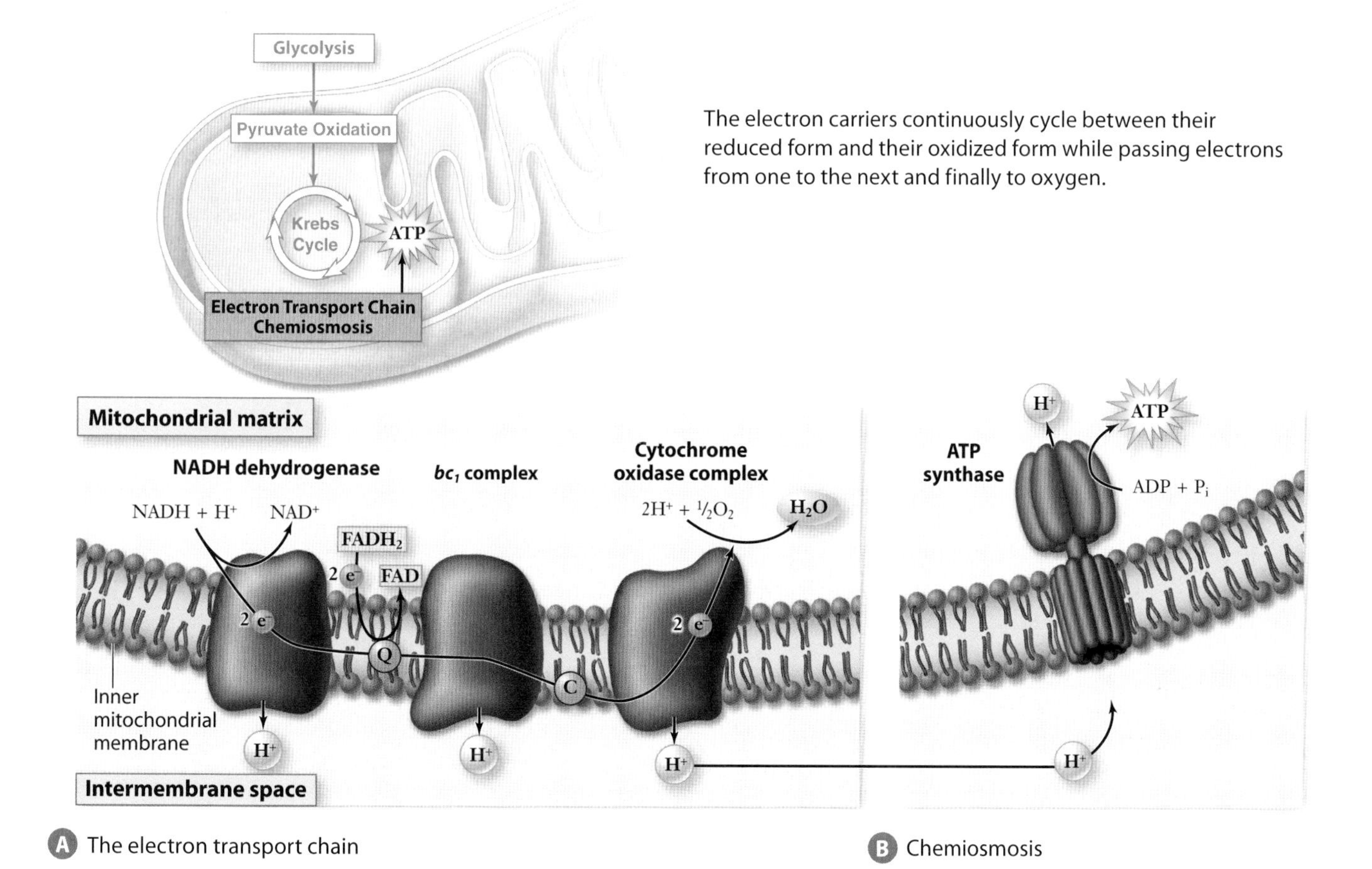

A The electron transport chain

B Chemiosmosis

1. Reaction 1: Condensation
2–3. Reactions 2 and 3: Isomerization
4. Reaction 4: The first oxidation
5. Reaction 5: The second oxidation
6. Reaction 6: Substrate-level phosphorylation
7. Reaction 7: The third oxidation
8–9. Reactions 8 and 9: Regeneration of oxaloacetate and the fourth oxidation

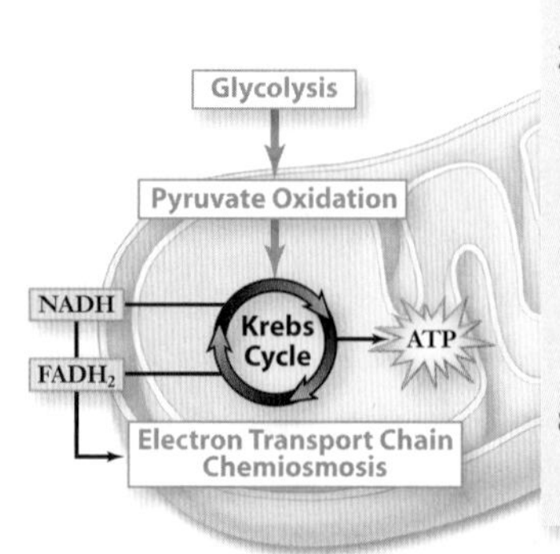

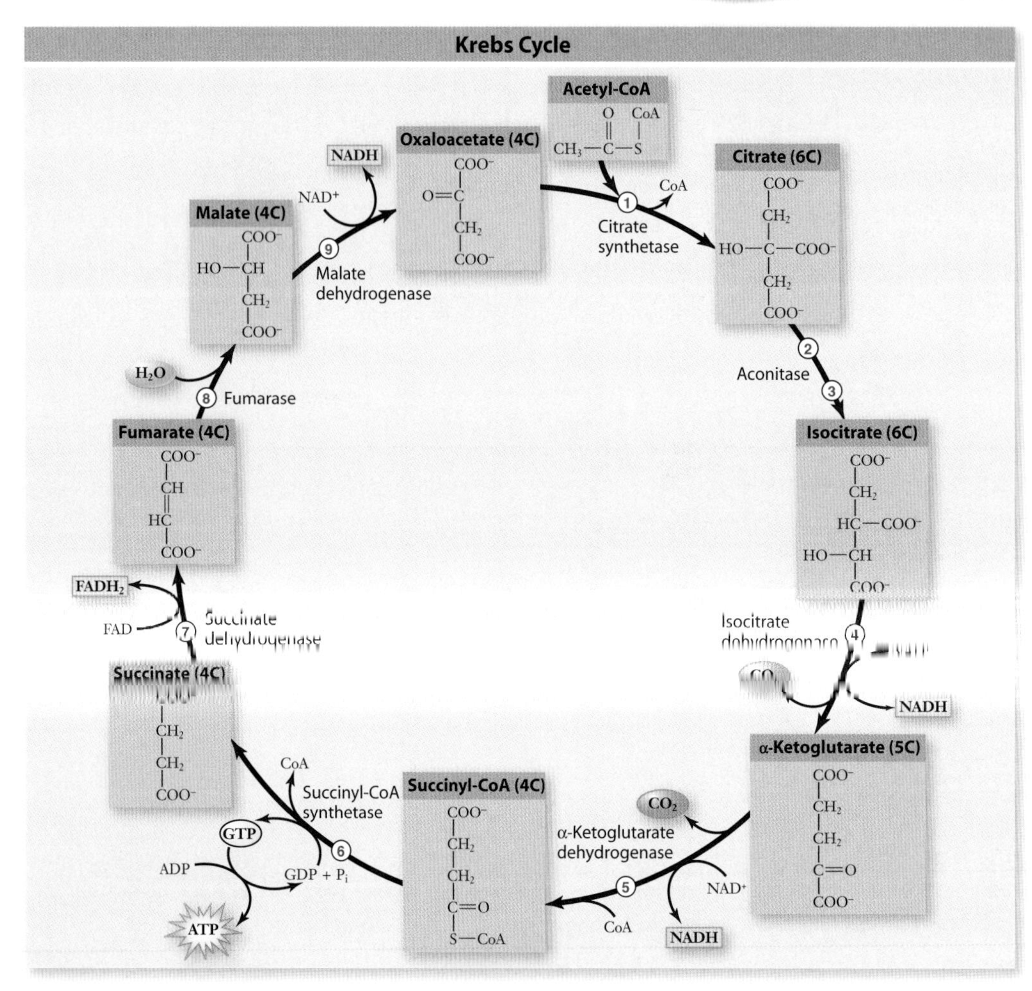

The Krebs cycle. This pathway consists of a number of oxidation reactions and completes the breakdown of glucose to carbon dioxide. Reaction numbers 2 and 3 are written on the same arrow because the reaction takes place in two steps but never leaves the surface of the enzyme. The terms written beside the reaction numbers are the names of the enzymes.

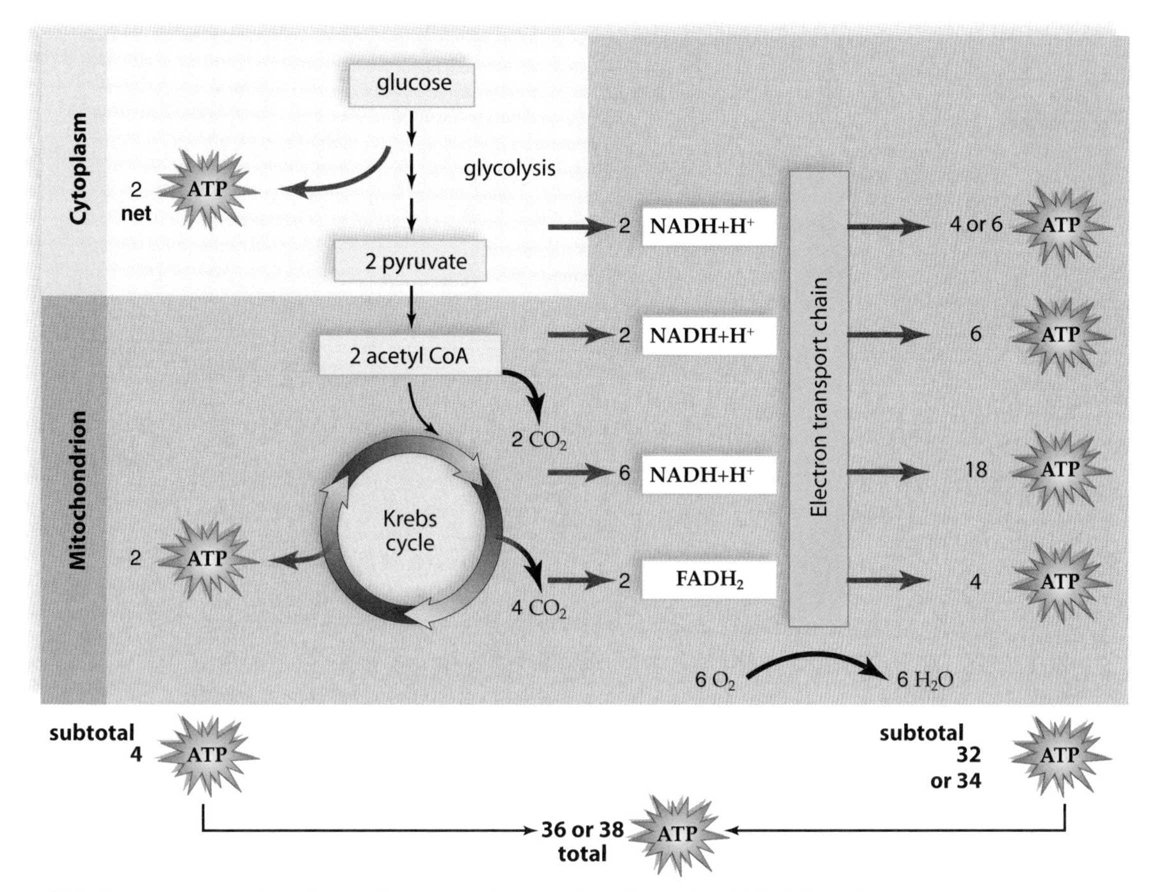

This diagram summarizes the maximum possible number of ways in which ATP can be generated in aerobic respiration.

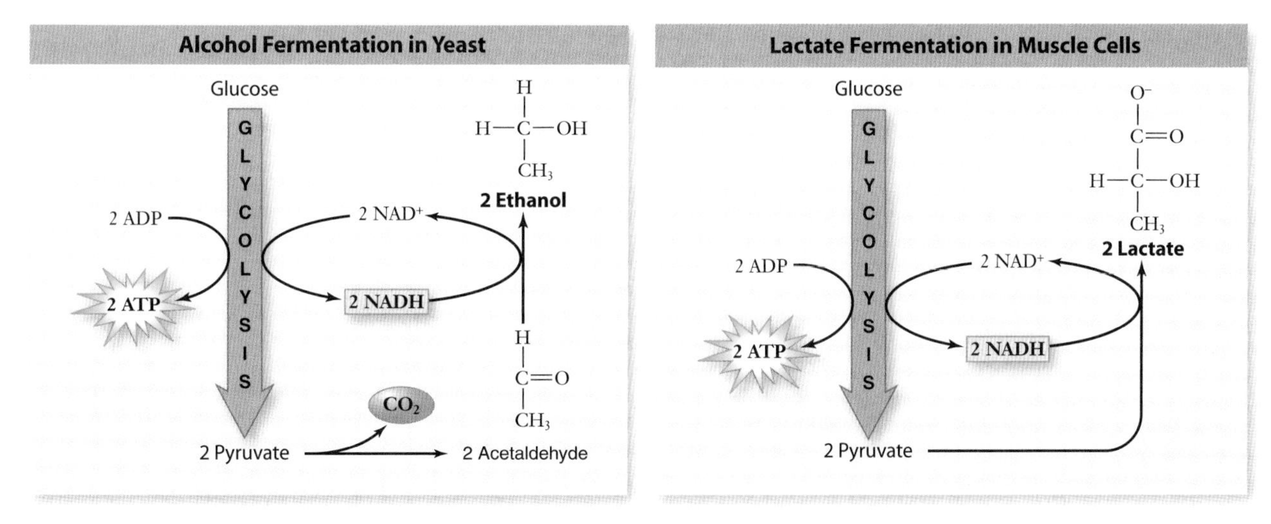

Lactate and ethanol fermentation. Muscle cells convert pyruvate into lactate. Yeasts carry out the conversion of pyruvate to ethanol. In each case, the reduction of a metabolite of glucose has oxidized NADH back to NAD^+ to allow glycolysis to continue under anaerobic conditions.

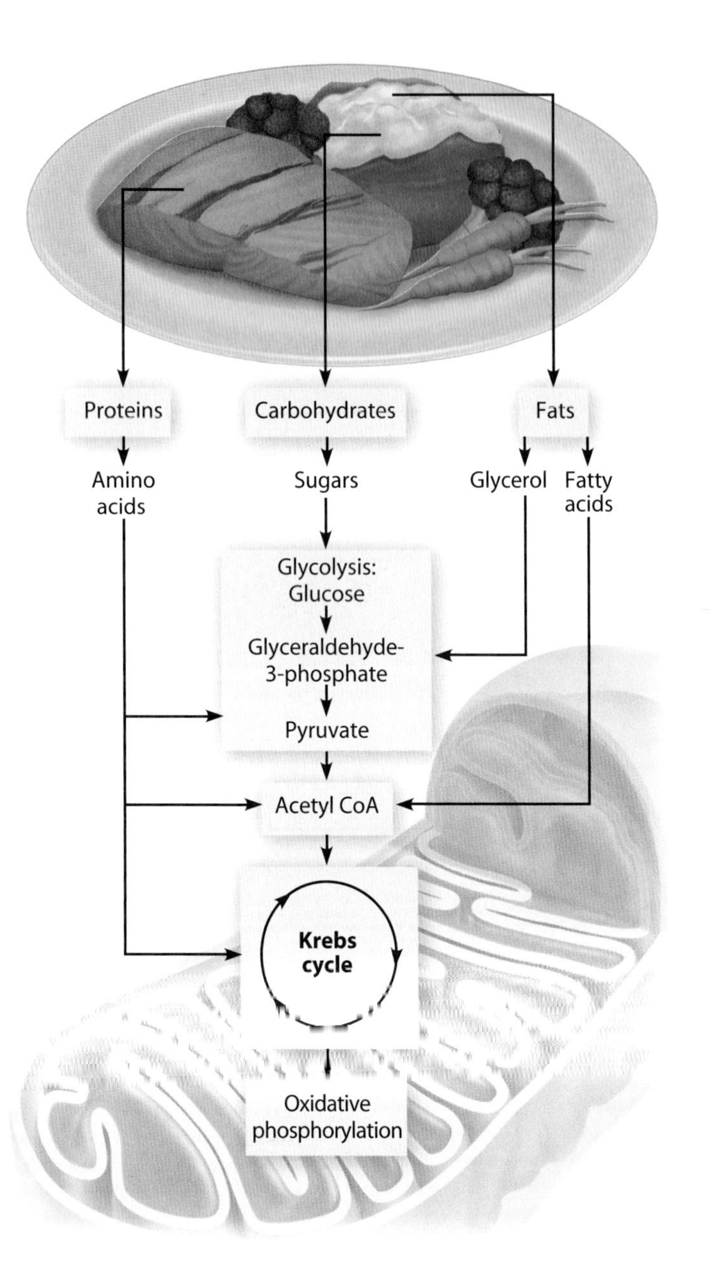

The cellular respiratory pathways are not closed. Compounds from the breakdown of all nutrients can be converted into intermediates in glycolysis and the Krebs cycle and can enter and leave at many different stages of the pathways.

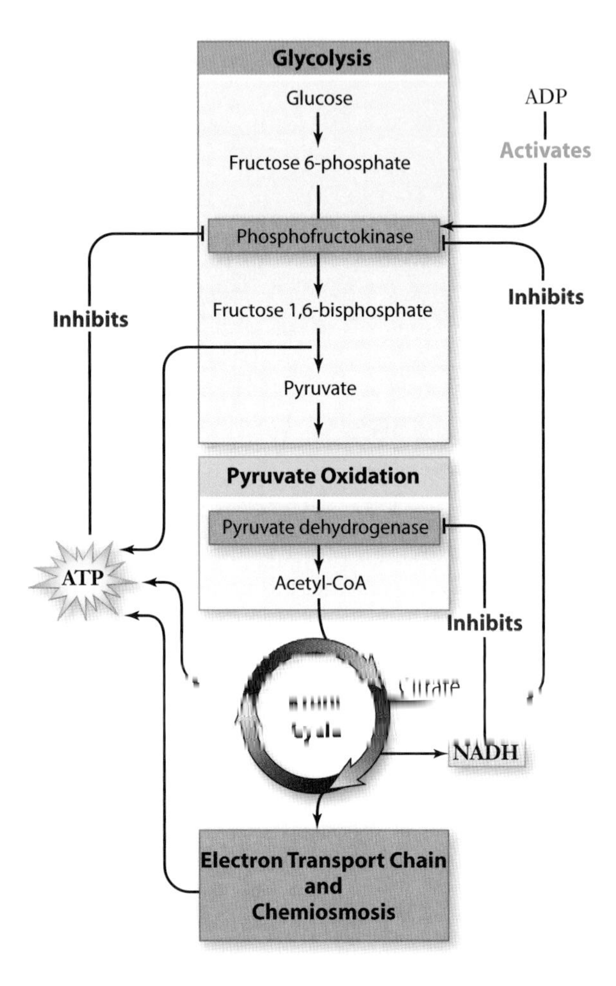

Many of the enzymes in glycolysis and the Krebs cycle are controlled by feedback inhibition of products such as ATP and NADH. They can also be activated by ADP. As a result of these control mechanisms, the ratio of ATP to ADP remains constant in living cells.

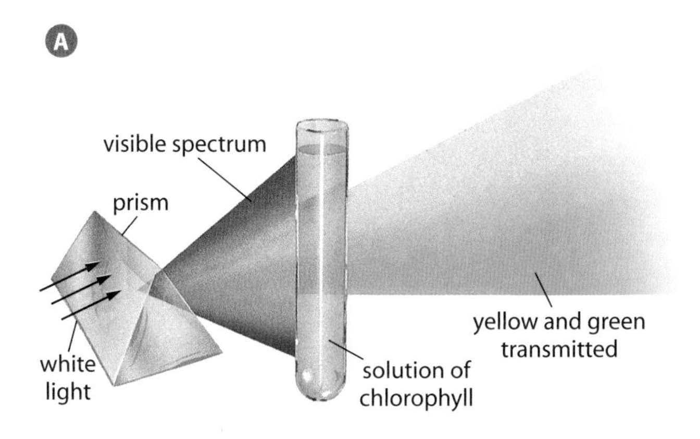

A Leaves appear green, because chlorophyll molecules in leaf cells reflect green and yellow wavelengths of light and absorb other wavelengths (red and blue).
B This absorbance spectrum for three photosynthetic pigments shows that each pigment absorbs a different combination of colours of light.

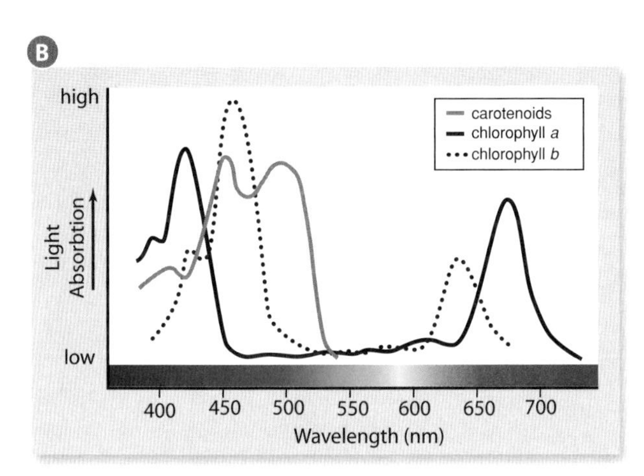

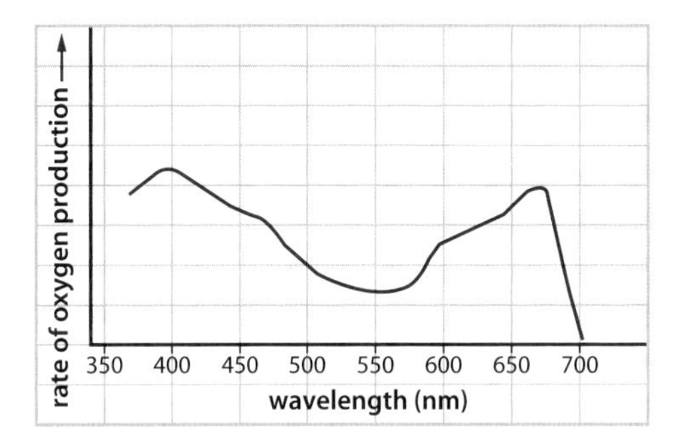

This action spectrum for photosynthesis shows the rate at which oxygen is produced when different wavelengths of light are shown on the leaf.

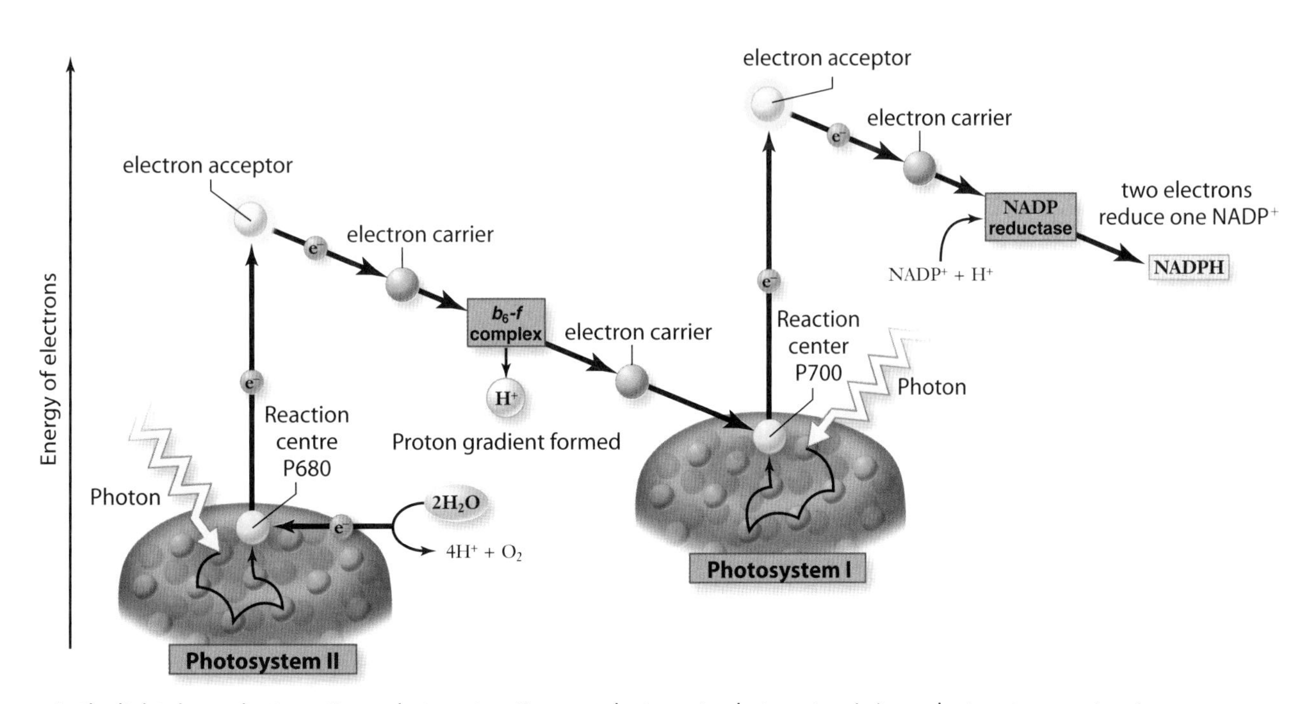

In the light-dependent reactions, photosystem II passes electrons to photosystem I via an electron transport system, which contains the b_6-*f* complex. This complex acts as a proton pump to produce a proton gradient across the thylakoid membrane. The electrons lost from the reaction centre of photosystem II are replenished by the oxidation of water. Photosystem I uses the electrons to reduce $NADP^+$ to NADPH.

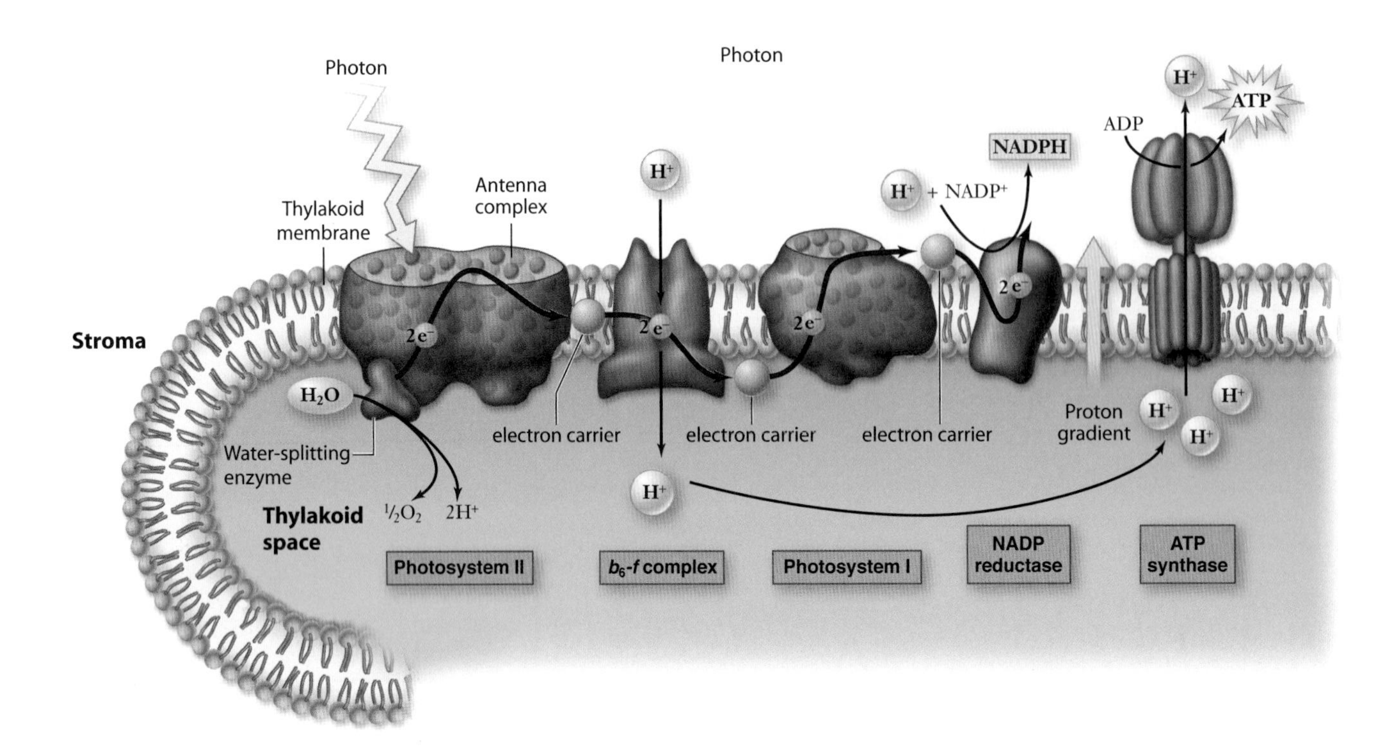

Photosystem I, photosystem II, an electron transport system, and the ATP synthase enzyme are embedded in the thylakoid membrane of chloroplasts. ATP synthesis by chemiosmosis in chloroplasts occurs in a way that is very similar to the way it occurs in mitochondria.

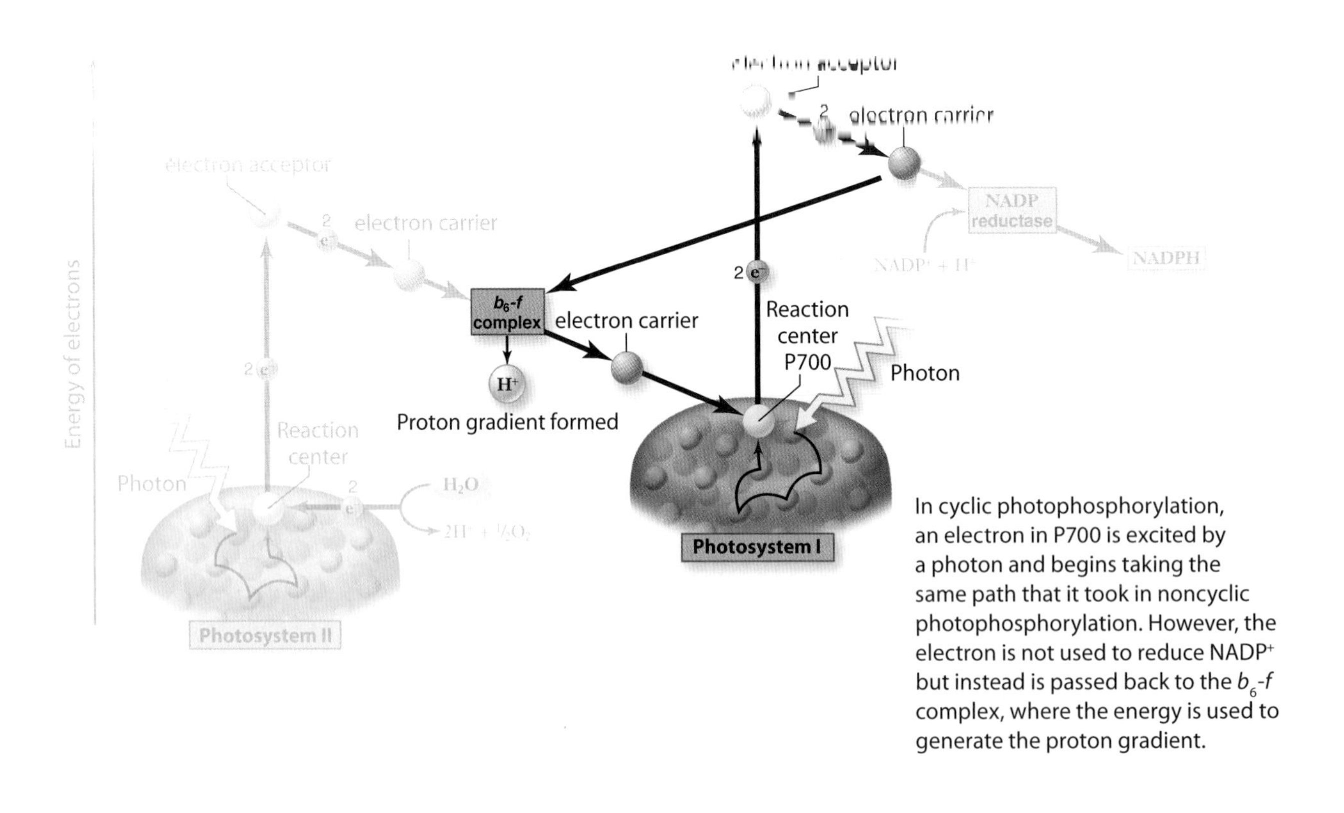

In cyclic photophosphorylation, an electron in P700 is excited by a photon and begins taking the same path that it took in noncyclic photophosphorylation. However, the electron is not used to reduce $NADP^+$ but instead is passed back to the b_6-f complex, where the energy is used to generate the proton gradient.

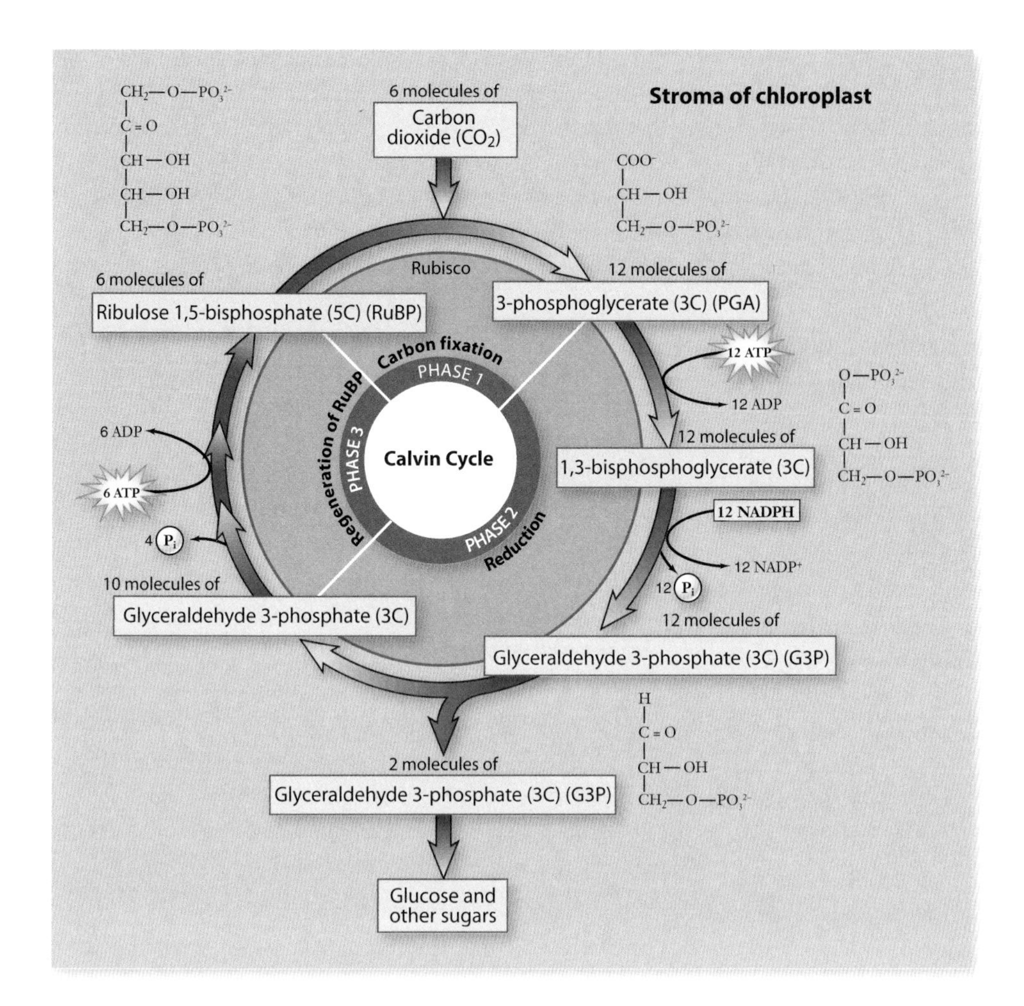

The ATP and NADP used in the Calvin cycle are produced in the light-dependent reactions. For every 12 molecules of G3P made in the Calvin cycle, two are used to make glucose and other high-energy compounds.

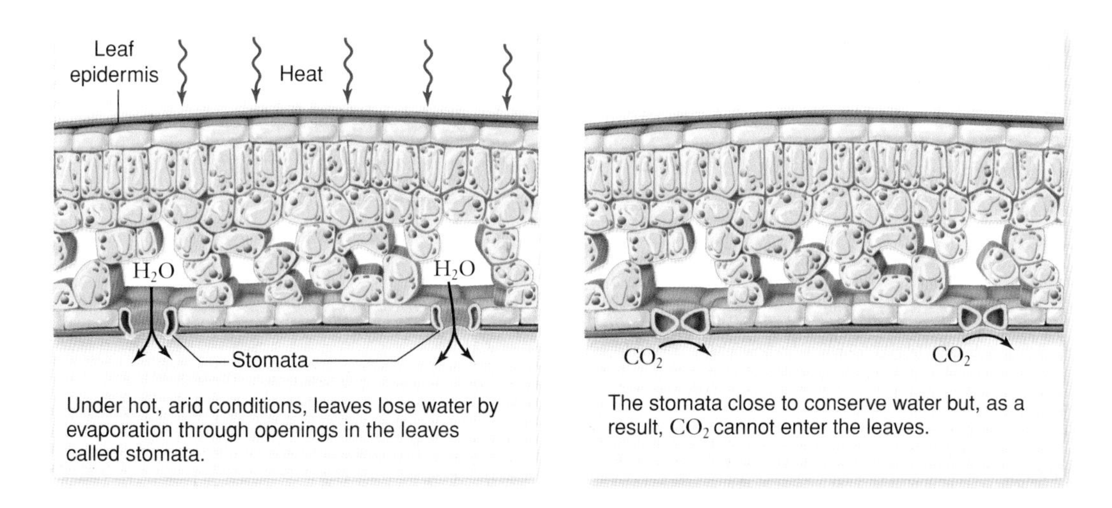

Leaf epidermis
Heat
H_2O
H_2O
Stomata
Under hot, arid conditions, leaves lose water by evaporation through openings in the leaves called stomata.
CO_2
CO_2
The stomata close to conserve water but, as a result, CO_2 cannot enter the leaves.

Leaves prevent water loss in hot, dry conditions by closing their stomata. Water is conserved but carbon dioxide is prevented from entering.

Reading Informational Texts

One of the most important skills you can have is good reading ability. In this section, you will learn basic skills necessary for reading informational texts through the use of six types of questions: main idea, subject matter, supporting ideas, conclusion, clarifying devices, and vocabulary in context.

Main Idea Whenever you read, ask yourself, "What point is the writer trying to make?" If the passage has a title, think about why the writer chose the title. It is usually a clue about the main idea. Sometimes, the first sentence of a paragraph is a topic sentence, and gives the main idea of what the paragraph is about. Other times, the main idea is not stated in a sentence and requires a summary type answer. Try to find the main idea in the following passage.

> *The furry platypus, a native of Australia and Tasmania, looks like a mammal at first glance. Upon studying it more closely, however, one recognizes the birdlike characteristics that have puzzled scientists. For example, like some water birds, the platypus has webbed feet. It also has a leathery bill like a duck. That is how the animal got its name, the "duck-billed platypus." In addition, the semi-aquatic platypus lays eggs like a bird.*

Compare your answer to the following main idea statement: The furry platypus looks like a mammal but has several birdlike characteristics.

Subject Matter It is usually easy to identify the subject matter of what you are reading. A useful tip to help you pay attention to what you read is to ask yourself after the first sentence or two, "What is the subject matter of this passage?" Immediately you start thinking about the passage. If you do not ask yourself this question, your eyes may move across the print, yet your mind may be thinking of other things. Ask yourself, "What is the subject matter of this passage?" as you read the following.

> *The Moon circles Earth on the average every 29 days. Its orbit around Earth is not circular; it is oval. The Moon's distance from Earth can vary a bit. Sometimes, the Moon is 402 000 km from Earth. Other times, it is only 354 000 km away.*

After finishing the first sentence you may have thought, "Ah, a passage about the Moon going around Earth. Maybe I can learn something about this process." By focusing on the subject matter you will be looking for something, and-best of all—you will be understanding, learning, and remembering.

Supporting Details Most of a factual passage is made up of details that support the main idea. You may notice that details come in various forms. They can be examples, explanations, descriptions, definitions, comparisons, contrasts, exceptions, analogies, similes, and metaphors. The main idea is often buried among the details so it may take some work to distinguish between the details and the main idea. If you have trouble finding the main idea, take the passage apart sentence by sentence. Ask yourself, "Does this sentence support something, or is this the idea being supported?" In other words, you must not only separate the main idea from the details, but you must also see how they help one another. The following passage shows how important details are for providing a full picture of what the writer had in mind.

> *The woodpecker pecks at a speed of 2000 km per hour. At this speed, the impact of the bird's beak hitting the wood is almost like the impact of a supersonic jet smashing into a mountain. Each peck takes just a thousandth of a second. The movement is quicker than the human eye can follow. Incredibly, the bird's cherry-sized brain is never injured from all this furious smashing.*

The main idea is in the first sentence. After stating the main idea, the writer gives several examples showing why it is true. The examples are supporting ideas.

Conclusion As you read through a passage, grasping the main idea and the supporting details, it is natural to begin to guess an ending or conclusion. Some passages contain conclusion; others do not. It all depends on the writer's purpose. Often the conclusion is implied. That is, the writer seems to have to a conclusion but has not stated it. It is up to you to draw that conclusion. While reading you have to think, "Where is this writer leading me? What conclusion will I be able to draw?" Like a detective, you must try to guess the conclusion, changing the guess as you get more and more information. In the following passage, the writer implies a conclusion, but does not state it directly.

> *The elephant's great size can sometimes present a heat problem. The larger an object, the more difficultly it has losing heat. Elephants live on the hot plains of Africa, where keeping cool is not an easy task. Elephants' huge ears help them cool their bodies so they can survive in the heat. The large surfaces of the ears have many blood vessels that are very close to the surface of the skin. Blood that is closer to the surface cools more easily.*

From this passage, we can draw the conclusion that, without their large ears, elephants would probably not survive in the African heat.

Clarifying Devices Clarifying devices are words, phrases, and techniques that a writer uses to make main ideas and supporting details clear and interesting. By knowing some of these clarifying and controlling devices, you will be better able to recognize some of them in the passages you read. By recognizing them, you will be able to read with greater comprehension and speed.

Transitional or Signal Words A writer uses signal words such as *first, second, next, last,* and *finally* to keep ideas, steps in a process, or lists in order. Other transitional words include *however, in brief, in conclusion, above all, therefore, since, because,* and *consequently.* When you see transitional words, consider what they mean. A transitional word like *or* tells you that another option or choice is coming. Words like *but* and *however* signal that a contrast, or change in point of view will follow.

Organizational Patterns Writers organize information in different ways, such as using lists of examples, spatial descriptions to tell what things look like, or chronological patterns to describe how events unfold over time.

Textual Devices Textbook writers often use patterns or text styles to make their ideas clear. Bulleted lists, subheads, and boldfaced or italicized words help to highlight important ideas in the text. Concepts shown in charts or diagrams may be easier to understand than concepts explained words alone.

Literal versus Figurative Language Two literary devices that writers use to present ideas in interesting ways are similes and metaphors. Both are used to make comparisons. A simile uses the word *like* or *as*: "She has a mind like a computer." The metaphor makes a direct comparison: "Her mind is a computer."

Vocabulary in Context As you read a passage you may encounter words that are not familiar to you. You can sometimes figure out the meanings of these words from their context—that is, from the words and phrases around them. If that method does not for you, consult a dictionary.

You can practise using these six comprehension skills with the following readings.

DNA Identification

Have you noticed news headlines like these? "Inmates Freed After DNA Tests Prove Innocence." "DNA Tests Confirm Babies Were Swapped."

Where is your DNA? Sneeze into a tissue—your DNA is on the tissue. Lick an envelope—your DNA is on the seal. In fact, DNA is in every cell of your body.

What is DNA? It is a substance found in the chromosomes of cells. A chromosome is a chain of genes. Each gene carries a piece of information for a trait, such as eye colour, hair texture, or nose shape. More than one gene is needed for a trait to be expressed. For example, one gene will contain information for skin colour. But up to six genes that carry skin colour information will produce the colour of your skin. Other genes carry a piece of information for other traits. Traits from your grandparents, great grandparents and so on are passed to you through your parents. All the traits arrange themselves in patterns that are unique. No one, except an identical twin, has the same patterns that you have. DNA acts like a file that stores your unique patterns of traits.

How is DNA used in identification? DNA is obtained from a sample of blood, skin, hair, or saliva. The DNA is treated with a chemical, which breaks the DNA into parts. Each part contains one or more patterns or traits. Next, each part of DNA is copied many hundreds of times. Then, the parts are put on a gel-like substance, and an electrical current is run through them. The current moves the pieces through the gel, leaving a trail of black bars—like bar codes. Scientists call these bars "DNA fingerprints" and just like real fingerprints, these DNA prints are distinct for every person (except an identical twin). Scientists use the DNA identification process to confirm the identity of people both living and dead.

List the numbers from 1-6 in your notebook For question 1, mark the main idea with an "M." Mark the statement that is too broad with a "B." Mark the statement that is too narrow with an "N." For the other questions, choose the best answer from the choices presented.

Main Idea

1. **a.** DNA is found in cells.
   b. No one beside an identical twin has the same DNA as you do.
   c. DNA provides information about traits that is being used to identify people.

Subject Matter

2. This passage is mainly about
   a. where DNA is found
   b. how DNA is used to identify people
   c. DNA and identical twins
   d. DNA stories in the news

Supporting Details

3. DNA stores each person's
   - **a.** patterns of traits
   - **b.** fingerprints
   - **c.** bar codes
   - **d.** body cells

Conclusion

4. From the information in this passage, which of the following traits can you conclude is not a genetic trait?
   - **a.** having blonde hair
   - **b.** having freckles
   - **c.** being late often
   - **d.** having dimples

Clarifying devices

5. The phrase "DNA fingerprint" suggests that DNA
   - **a.** is shaped like a person's thumb
   - **b.** is identified by its swirls
   - **c.** has tin fingers that push it through a cell
   - **d.** is as unique as a person's fingerprints

Vocabulary in context

6. In this passage, <u>confirm</u> means
   - **a.** strengthen a person's beliefs
   - **b.** deny
   - **c.** test
   - **d.** prove the accuracy of

An Essential Scientific Process

All life on Earth depends on green plants. Using sunlight, the plants produce their own food. Then animals feed on the plants. They take in the nutrients the plants have made and stored. But that's not all. Sunlight also helps a plant produce oxygen. Some of the oxygen is used by the plant, but a plant usually produces more oxygen than it uses. The <u>excess</u> oxygen is needed by animals and other organisms.

The process of changing light into food and oxygen is called *photosynthesis*. Besides light energy from the Sun, plants also use water and carbon dioxide. The water gets to the plant through its roots. The carbon dioxide enters the leaves through tiny openings called *stomata*. The carbon dioxide travels to the chloroplasts, special cells in the bodies of green plants. This is where photosynthesis takes place. Chloroplasts contain the chlorophylls that give plants their green colour. The chlorophylls are molecules that trap light energy. The trapped light energy changes water and carbon dioxide to produce oxygen and a simple sugar called *glucose*.

Carbon dioxide and oxygen move into and out of the stomata. Water vapour also moves out of the stomata. More than 90 percent of the water a plant takes in through its roots escapes through the stomata. During the daytime, the stomata of most plants are open. This allows carbon dioxide to enter the leaves for photosynthesis. As night falls, carbon dioxide is not needed. The stomata of most plants close. Water loss stops.

If photosynthesis ceased, there would be little food or other organic matter on Earth. Most organisms would disappear. Earth's atmosphere would no longer contain oxygen. Photosynthesis is essential for life on our planet.

List the numbers from 1-6 in your notebook. For question 1, mark the main idea with an "M." Mark the statement that is too broad with a "B." Mark the statement that is too narrow with an "N." For the other questions, choose the best answer from the choices presented.

Main Idea

1. **a.** Stomata allow carbon dioxide to enter leaves for photosynthesis.
   - **b.** Life on Earth depends on green plants.
   - **c.** The process of changing light into food and oxygen is called photosynthesis.

Subject Matter

2. Another good title for this passage would be
   - **a.** Oxygen and Carbon Dioxide
   - **b.** Plants and Their Roots
   - **c.** How Photosynthesis Works
   - **d.** Why Earth Needs Water

Supporting Details

3. Which of the following does *not* move through a plant's stomata?
   - **a.** carbon dioxide
   - **b.** water vapour
   - **c.** oxygen
   - **d.** food

Conclusion

4. In the title, the term *Essential Scientific Process* refers to
   - **a.** photosynthesis
   - **b.** the formation of glucose
   - **c.** global warming
   - **d.** water getting to the roots of plants

Clarifying Devices

5. This passage is primarily developed by
   - **a.** explaining a process
   - **b.** telling a story
   - **c.** comparing and contrasting
   - **d.** convincing the reader of the importance of plants

Vocabulary in context

6. In this passage, excess means
   - **a.** heavy
   - **b.** extra
   - **c.** green
   - **d.** liquid

Achieving Cardiovascular Fitness

Heart attacks, strokes, and other cardiovascular diseases are the leading cause of death in the developed world. These diseases include disorders of the blood vessels that pump and carry blood throughout the body. In a disorder called atherosclerosis, for instance, the aorta or other major blood vessels become clogged with fatty deposits of cholesterol. As cholesterol builds up, the walls of the vessels harden and thicken. The narrowed passages in the vessels slow the flow of blood cells—and the oxygen they carry—to the brain and to the heart and other muscles. When the heart is deprived of oxygen-rich blood, a heart attack occurs. If oxygen-rich blood does not reach the brain, a stroke results. Such cardiovascular diseases are likely in people who are overweight or smoke cigarettes. Other risk factors include high blood pressure, heart disease, and high cholesterol. Regular exercise can help a person reduce those risk factors.

Aerobic exercise, such as jogging, walking, skating, and swimming, helps people to become cardiovascularly fit. An aerobic workout must be vigorous so that the heart, lungs, blood vessels, and skeletal muscles constantly use energy. During the aerobic activities, the muscle cells undergo aerobic metabolism. In this process, oxygen combines with a fuel source (fats or carbohydrates) to release energy and produce carbon dioxide and water. The muscle cells use the energy to contract, which creates a force that produces movement. The aerobic reaction only occurs if the circulatory and pulmonary systems provide a constant supply of oxygen and fuel to the muscle cells and remove carbon dioxide from them. Most people can achieve cardiovascular fitness by raising their heart rate and breathing rates for 25 to 30 minutes about every other day. Those people will have the energy to do easily all the things the want to do.

List the numbers from 1-6 in your notebook. For question 1, mark the main idea with an "M." Mark the statement that is too broad with a "B." Mark the statement that is too narrow with an "N." For the other questions, choose the best answer from the choices presented.

Main Idea

1. **a.** Cardiovascular diseases cause many deaths.
   - **b.** Cardiovascular diseases are dangerous, but aerobic exercise can prevent them.
   - **c.** Atherosclerosis causes hardening and thickening of major blood vessels.

Subject Matter

2. Another good title for this passage would be
   - **a.** Overweight with High Blood Pressure
   - **b.** Kinds of Cardiovascular Diseases
   - **c.** Fats and Carbohydrates
   - **d.** Avoiding Cardiovascular Disease.

Supporting Details

3. A person will benefit from aerobic exercise if he or she
   - **a.** raises and maintains the heart and breathing rates on a regular basis
   - **b.** monitors the heart rate while exercising
   - **c.** eats fats and carbohydrates for energy
   - **d.** takes a break after every 10 minutes of vigorous exercise

Conclusion

4. We can conclude that the more risk factors a person has for cardiovascular disease, the
   - **a.** fewer the chances of getting the disease
   - **b.** greater the chances of getting the disease
   - **c.** more likely it is that the person smokes
   - **d.** less likely it is that a person is overweight

Clarifying Devices

5. The sentence "If oxygen-rich blood doesn't reach the brain, a stroke results" is an example of
   - **a.** persuasion
   - **b.** narration
   - **c.** definition
   - **d.** cause and effect

Vocabulary in Context

6. In this passage, contract means
   - **a.** an agreement between two people
   - **b.** to form words like *don't* and *can't*
   - **c.** to tighten or make shorter
   - **d.** to get or bring on oneself

Infection Makers

Contraction of a viral disease such as a common cold, the flu, chicken pox, mumps, measles, mononucleosis, polio, or hepatitis is an inescapable part of life. In fact, every living thing—whether it is a plant, fungi, algae, protozoa, animal, or bacteria—can be infected by a type of virus specific to that organism. A virus is an infectious noncellular structure of genetic material, either DNA (deoxyribonucleic acid) or RNA (ribonucleic acid) that is usually surrounded by a protein coat. The sole purpose of a virus is to produce more viruses. A virus cannot, however, multiply or grow independently because a virus is not a cell. A virus must infect, or enter, a living cell (called a host) and use the host's cell structures to reproduce more viruses. Eventually, the infected host cell releases new viruses, and then the host cell usually dies. The released viruses go on to infect other cells.

When a virus infects one or more cells of a body tissue, the infection causes the synthesis and secretion of proteins called intereferons. These proteins strengthen the cell membrane of adjacent healthy cells so the virus cannot penetrate those cells. Sometimes, however, the virus succeeds in spreading to "other" cells. Then the human immune system activates and starts killing the viruses outside the cells as well as any infected cells themselves. Eventually the virus is eliminated, and the organism returns to good health.

Vaccination is often the best protection against viral disease. A vaccine contains weakened or dead viruses that no longer cause the disease. Upon entering the body, a vaccine triggers the immune system to produce antibodies that kill the weakened viruses. Vaccination often results in a lifelong immunity against further infection.

List the numbers from 1-6 in your notebook. For question 1, mark the main idea with an "M." Mark the statement that is too broad with a "B." Mark the statement that is too narrow with an "N." For the other questions, choose the best answer from the choices presented.

Main Idea

1. **a.** A viral infection causes the secretion of proteins from interferons.
   - **b.** Viruses infect people by invading a host cell and reproducing, but then the body's immune system fights back.
   - **c.** Viruses are structures that attack cells.

Subject Matter

2. The purpose of this passage is to
   - **a.** define the role of interferons
   - **b.** explain the penetration of a host cell
   - **c.** describe the impact of a virus on a healthy organism
   - **d.** identify types of viral disease

Supporting Details

3. A virus cannot function independently because it
   - **a.** has too much genetic material
   - **b.** is too small
   - **c.** usually kills the host cell
   - **d.** is not a cell with the structure necessary for growth and reproduction

Conclusion

4. We can conclude from the second paragraph that the "other" cells infected by the virus were
   - **a.** not adjacent to the infected tissue
   - **b.** surrounded by a protein coat
   - **c.** synthesized and secreted proteins
   - **d.** killing the viruses outside the cells

Clarifying devices

5. A device the author uses to help the reader understand the type of genetic material in a virus is
   - **a.** quotation marks
   - **b.** parenthetical notes
   - **c.** a definition
   - **d.** an example

Vocabulary in Context

6. In this passage, the word multiply means
   - **a.** live
   - **b.** reproduce
   - **c.** calculate
   - **d.** magnify

Seeing Double

Twinning, the process that leads to a multiple birth, takes place in the early stages of the reproductive process. The egg cell from the female and the sperm cell from the male are specialized cells called gametes, each of which contains one half of the genetic information needed to form a complete fetus. During fertilization, the female gamete and the male gamete unite to form a zygote, or a fertilized egg, that contains the combined genetic information from both parent cells.

When a zygote undergoes mitosis, or cell division, it splits into two (or more) parts that contain the exact same genetic information. Each part develops into an embryo that is genetically identical to the other; the embryos are of the same sex. Since the separate embryos are formed from a single zygote, the identical twins are called monozygotic (MZ) twins. Usually, MZ twins share a common placenta and amniotic sac.

When two separate zygotes are present to form two unique embryos, dizygotic (DZ) twins, or fraternal twins, form. The presence of two zygotes is a result of multiple ovulations within a single menstrual cycle. As part of the continued birth process, a placenta is formed for each zygote. Since DZ twins come from two separate zygotes, they are no more genetically similar than are other siblings. DZ twins do not have to be of the same sex.

Multiple births have increased since fertility drugs were introduced in the 1960s to help couples who have difficulty conceiving. Fertility drugs often cause the release of more than one egg from the ovary, thus increasing the chance of multiple births of DZ twins.

List the numbers from 1-6 in your notebook. For question 1, mark the main idea with an "M." Mark the statement that is too broad with a "B." Mark the statement that is too narrow with an "N." For the other questions, choose the best answer from the choices presented.

Main Idea

1. **a.** Whether twins are identical or fraternal depends on how many zygotes are present.
   b. Monozygotic twins are genetically identical.
   c. Twin births are fairly common occurrences.

Subject Matter

2. The purpose of this passage is to
   a. define monozygotic and dizygotic
   b. explain the process of twinning
   c. describe fertilization
   d. discuss mitosis

Supporting Details

3. During twinning, the specific process that results in the formation of monozygotic twins is
   a. ovulation
   b. mitosis
   c. reproduction
   d. fertilization

Conclusion

4. If the prefix mono in *monozygotic* means "one," then the prefix *di* in *dizygotic* must mean
   a. multiple
   b. two
   c. split in half
   d. a choice

Clarifying Devices

5. to help the reader understand why identical and fraternal twins are different, the writer
   a. describes the union of two gametes
   b. explains monozygotic twinning and then dizygotic twinning
   c. contrasts fertilization and mitosis
   d. discusses the effects of fertility drugs

Vocabulary in Context

6. The word siblings means
   a. split personalities
   b. female and male gametes
   c. the genetic relationship between twins
   d. brothers and sisters

Using Critical Thinking to Analyze and Assess Research

Critical thinking is a valuable decision-making and problem-solving skill that you can apply to almost every field of life. It is an especially important skill in science, because it helps you determine the validity of a claim and identify the implications and consequences of a conclusion. You can also use critical thinking to help you clearly state a goal, examine data, identify assumptions, and evaluate evidence.

The word "critical" in critical thinking does not mean finding fault or being negative. "Critical" refers to using reason and reflection to analyze and evaluate thinking in order to improve it. A critical thinker displays well-developed intellectual values, such as clarity, accuracy, precision, relevance, logic, significance, depth, breadth, and fairness. Developing skills in critical thinking is a lifelong practice that is self-guided and self-disciplined.

The following points and questions can help you develop your critical thinking skills. You can use the skills to critically evaluate both your own and others' work to determine whether a researcher has conducted fair and objective research. You can also use the questions to evaluate investigations, arguments, advertisements, and science resources. The questions are written as if the research is in a written format, but you can also apply them to spoken formats, such as discussions, or to visual formats, such as [illegible] television shows. If you are evaluating your own work, then you are "the researcher" in each of the categories below. The sentences beginning *Look for* will alert you to which values may be displayed. The sentences beginning *Watch out for* point out some possible shortcomings in the research.

1. What is the researcher trying to accomplish?

Look for clarity and precision. The old expression, "Well begun is half done" applies to stating the purpose. The more accurately and precisely the purpose is stated, the easier it is to develop a plan to achieve it and evaluate whether the investigation has succeeded. The researcher should clearly state the purpose or goal of the inquiry or investigation in a direct statement such as, "The purpose of this study is to describe the characteristics of...." If there is not a subheading called "Purpose" or "Goal" then ask yourself, "What is the problem the researcher is trying to solve? What is the decision the researcher wants to make?"

Watch out for a fuzzy or imprecise purpose, such as "The purpose of this study is to find out something more about..."

You should be able to complete this sentence stem for the inquiry or investigation you are evaluating: **The main purpose of this research is ...**

2. What question does the researcher want to answer?

Look for clarity, precision, and significance. The question in scientific inquiry is directly related to the purpose. If a question is not explicitly stated, you can often identify it from the hypothesis or from a statement about the problem the researcher investigated. The researcher may have described the concerns or issues that led him or her to undertake the investigation. Once you know what the question is, ask yourself, "Is there more than one right answer to this question? What other ways can I think about the question? If I divided the question into sub-questions, what would they be?" If you are evaluating your own research, you may realize that you could improve your question by going more deeply into the complexity of what you are investigating.

Watch out for questions that require judgement rather than just facts to be answered. Is there a bias toward a certain answer implied in the question? For example, the question "Wouldn't it be a good idea to add acidic fertilizer to basic garden soil?" implies that the "correct" answer would be "yes" and that the answer should be based on judgement of what "good" means as well as on facts.

You should be able to complete this sentence stem for the inquiry or investigation you are evaluating: **The question directing this research is ....**

3. What data does the researcher present?

Look for accuracy and relevance. The researcher should have gathered enough relevant data to respond to the research question and reach a reasonable conclusion. The data section may include descriptive statistics, which summarize the raw data, such as means and standard deviations. You may be able to infer conclusions from data called inferential statistics. The researcher may also mention how the data were screened for errors in data entry, outliers and distribution, etc.

As you review and analyze the data, it is important to think about internal validity, which means whether the changes in the dependent variable were the result of something other than, or in addition to, changes in the manipulated variable. Sometimes, the testing procedure itself can cause changes in the dependent variable.

In some investigations that occur over a long period of time, the dependent variable may change due to aging or deterioration. Studies that include data from a comparable control group help to reduce the problems of internal validity.

Watch out for data that is not relevant to the question or investigation.

You should be able to complete this sentence stem for the inquiry or investigation you are evaluating: **The most important data presented in this research are...**

4. What assumptions are the researcher's point of view based on?

Look for logic, breadth, depth, and fairness. Every researcher has a point of view or frame of reference while conducting research. The point of view is based on major assumptions in the research. Try to recognize unstated assumptions and values as well as those that the researcher has clearly acknowledged. As you read the research, ask yourself, "What assumptions is the researcher making? What is the researcher taking for granted? Which of those assumptions would I question? Does the researcher show open mindedness by recognizing and assessing his or her assumptions? What assumption is the conclusion based on?" It is important that a researcher address the strengths and weaknesses of his or her point of view when evaluating data. The researcher also needs to fairly consider objections that arise from alternative relevant points of view or lines of reasoning and to be aware of the strengths and weaknesses of those points of view.

Watch out for a point of view that does not include other possibilities, i.e., "this is the correct point of view and other points of view are not valid." Notice whether the researcher is distorting ideas to try to strength his or her position or to support a point of view.

You should be able to complete this sentence stem for the inquiry or investigation you are evaluating: **The main assumptions underlying the researcher's point of view are...**

5. How well do the conclusions address the research question?

Look for relevance, logic, significance, breadth, and depth. To reach a well-reasoned conclusion, the researcher interprets, appraises, and evaluates data. The researcher considers what the data imply and how those implications address the research question. The conclusions or discussion should always include a clear reference to the original hypothesis or question. The researcher should also attempt to explain data that is contrary to the main conclusion. The conclusions of a research study often include questions that can be the basis of future research.

When you read conclusions in the research you are evaluating, ask yourself, "How did the researcher arrive at these conclusions? What inferences and interpretations did he or she make? What data did he or she use? How well do the data support the conclusions? Do the data suggest other conclusions than what the researcher has provided? In what other way might the data be interpreted?" In order to accept or reject a researcher's conclusions, you need to have a clear idea of how the research was conducted.

All research has implications, which may be either positive or negative. Has the researcher considered and stated the broader implications of his or her findings?

Watch out for unsupported conclusions that go beyond what the data imply. Sometimes, researchers describe insignificant results as though they were significant, or draw conclusions from future studies. For example, a researcher may suggest that even though there was not a significant difference in this particular study, there probably would be in future studies. This is not a valid or supported conclusion.

You should be able to complete these sentence stems for the inquiry or investigation you are evaluating: **The main conclusions in this research are... The implications of these conclusions are...**

Appendix D

Instant Practice

Value of a Human Body

Use your skills in critical thinking to consider the difference in the worth of the human body at the elemental and molecular levels by reading the following and answering the questions.

Biologically, how much are you worth? The answer to this question varies, depending on your perspective. At the elemental level, the specific elements that human bodies are composed of can be purchased for approximately $1,000 (depending on the mass of the individual and elements purchased).

At the molecular level, a human body is worth considerably more. Using highly sophisticated laboratory equipment that incorporates powerful computers, highly trained technicians can manufacture complex biological molecules in the laboratory from the basic elemental building blocks. What is the cost of these biological molecules if purchased in quantities of the human body? Many millions of dollars!

1. Which of life's properties are reflected in the differences in cost between the elemental and biomolecular components of life?
2. Discuss how this difference in cost of life at the elemental and biomolecular levels illustrates life's organization.
3. Manufacturing biological molecules in the laboratory requires the use of relatively pure laboratory grade elements, highly sophisticated laboratory equipment, highly trained laboratory scientists, and high levels of electricity. Draw analogies between these entities and DNA, ATP, and cellular organelles.
4. [illegible] your answers to the first three questions. Assess each answer in terms of whether it displays the intellectual values of clarity, accuracy, precision, relevance, logic, significance, depth, breadth, and/or fairness.

Instant Practice

Medical Research and Ethics

Why is it important that research be critically reviewed and freely shared with peer researchers and the general public? Read the following and answer the questions that follow.

The enormous cost of conducting medical research can be a barrier to the development of effective drugs and medical procedures for the treatment of disease. To bridge this barrier, private drug companies are currently funding much of the research conducted at medical schools. However, many ethicists feel that such partnerships between industry and medical schools may create conflicts of interest that could endanger the lives of patients and shake public confidence in medical research.

A recent study of such collaborations resulted in some disturbing findings. Among the most troubling were instances where the results of research studies were manipulated by individual researchers to make a drug appear more effective than it actually was. In some cases, researchers were discouraged, and sometimes were contractually prohibited, from publishing findings that didn't support the effectiveness of a new drug/treatment.

1. How is the repression of research counter to the purpose of the scientific endeavour?
2. What might cause an individual researcher to manipulate his or her data, changing the conclusions to be drawn from an investigation?
3. What guidelines could be implemented to minimize conflicts of interest, yet still allow collaborations between pharmaceutical companies and medical schools?

Appendix D

Reading a Scientific Research Paper

Scientific researchers publish their findings in peer-reviewed journals in order to share the findings with other members of their profession and the general public. In order to have a study accepted for publication, the researcher's peers (other professionals in the same field) critically examine the study and challenge any inconsistencies, omissions, questionable data, unsupported conclusions, etc. You can find scientific research journals in a library or online. The studies are usually published following a format similar to that outlined below.

Title

The Title usually describes what was done in the study, how it was done, and what subjects were used.

Abstract

The Abstract is a summary of the study's purpose, method, results, and conclusions. By reading the abstract, you can get a sense of what the study is about, how it was conducted, and what the researcher found and recommended. Reading the Abstract can also help you decide whether the rest of the paper applies to what you are researching. Even if the Abstract contains many unfamiliar words or ideas, you still might want to read more of the paper. The Introduction and Discussion sections are sometimes easier to understand. Substantial evidence for the conclusion is not included in the abstract, however, so you if you want to understand how the researcher arrived at those conclusions, you need to read the whole study.

Introduction

The Introduction usually includes a statement of the problem, a literature review, the purpose of the study, and the research question. Read the Introduction carefully if the paper is relevant to your research. It will help you discover why the researcher conducted the research.

The *statement of the problem* describes the questions, concerns, and concepts that led the researcher to undertake the investigation.

The *literature review* identifies the gap between what has been studied and what is yet to be studied in the particular research field and how the researcher's work attempts to close the gap. Sometimes, researchers also use this part of a research paper to critically review other researchers' published works.

The *purpose of the study* is usually stated in a direct, clear statement.

The *research question* is often stated as a hypothesis and reveals the researcher's expectation of results.

Method

The Method section explains how the study was conducted, giving enough details so that another researcher could follow the method to replicate the study. This section often includes a fair bit of specialized terminology that you may not be familiar with. Do not worry about the details here, at least not at first reading. Get a general sense of the method the researcher used. Try to determine the experimental design, which variables were controlled and manipulated, whether there was a control group, and how the sample was selected. After you have read the rest of the paper, you can revisit this section if necessary, to understand it better.

The Method section is often divided into the following parts: subjects, materials, procedure, and data analysis. Sometimes, the Introduction also includes an Assumptions subsection in which the researcher identifies the assumptions the research is based on.

In the *Subjects* subsection, the researcher describes how many subjects were included in the sample and how the subjects were selected.

The *Materials* subsection includes descriptions and sometimes diagrams or photographs of the instruments and apparatus used to conduct the procedure and measure the variables.

The *Procedure* subsection explains exactly how the steps of the study were conducted and how the data were collected.

The *Data Analysis* subsection describes the statistical tests used to analyze the data and how the data were screened for outliers, errors, etc.

Results

The Results section presents the results of the data analysis, often in the form of charts, graphs, and tables. The raw data may be summarized in terms of means and standard deviations. Some of the quantitative data may be difficult to understand at first. Instead, focus on the main points of this section, paying special attention to the researcher's comments about the graphs and charts. This will help you to notice the trends and relationships among the data. The researcher may have provided a summary of the qualitative data, which may be easier to understand.

Discussion

Pay special attention to this section. In the Discussion section, a researcher returns to the research question and hypothesis and explains the meaning of the results and whether the data support the hypothesis. The researcher usually suggests how this study has contributed to the body of research. The Discussion section also provides the researcher with the opportunity to suggest the implications of the study and identify directions and topics for future studies.

Conclusion

The Conclusion section briefly restates the experimental results and implications.

Instant Practice

Read the description of the experiment on the next page which studied proteins used in olfaction, the sense of smell. Answer the questions that follow. You may not be familiar with all the terminology, but by reading slowly or rereading, you should be able to understand the most important points of the experiment.

Buck and Axel Discovered a Family of Olfactory Receptor Proteins that Bind Specific Odour Molecules

How does the olfactory system discriminate between thousands of different odours? American neuroscientists Linda Buck and Richard Axel set out to study this question. When they began, two hypotheses were proposed to explain this phenomenon. One possibility is that many different types of odour molecules might bind to one or just a few types of receptor proteins, with the brain responding differently depending on the number or distribution of the activated receptors. Alternatively, odours might be distinguished at the level of the receptor proteins. The second hypothesis is that olfactory receptor cells can make many different types of receptor proteins, each type binding a particular odour molecule or group of odour molecules.

To begin their study, Buck and Axel assumed that olfactory receptor proteins would be highly expressed in the olfactory receptor cells, but not in other parts of the body. Based on previous work, they also postulated that the receptor proteins would be members of the large family of G-protein-coupled receptors (GPCRs). They isolated olfactory receptor cells from rats and then broke open the cells to release the RNA. The purified RNA was then used to make cDNA via reverse transcriptase. This would generate a large pool of cDNAs, which would represent all of the genes that are expressed in olfactory receptor cells. To determine if any of these cDNAs encoded GPCRs, they used primers that recognized regions within previouslyw known GPCR genes that are highly conserved. A highly conserved region is a DNA sequence that rarely changes among different family members. The primers were used in the technique of PCR to amplify cDNAs that encoded GPCRs. This produced many PCR products that were then subjected to DNA sequencing.

Buck and Axel identified 18 different genes, each encoding a GPCR with a slightly different amino acid sequence. Further research showed that these 18 genes were expressed in nasal epithelia but not in other parts of the rat's body. These results were consistent with the second hypothesis, namely that organisms make a large number of receptor proteins, each type binding a particular odour molecule or group of odour molecules.

Since these studies, researchers have determined that this family of olfactory genes in mammals is surprisingly large. In humans, more than 600 genes that encode olfactory receptor proteins have been identified, though about half of these are pseudogenes that are no longer functional. This value underscores the importance of olfaction even in humans, who have a relatively poor sense of smell compared with certain animals. Each olfactory receptor cell is thought to express only one type of GPCR that recognizes its own specific odour molecule or group of molecules. Most odours are due to multiple chemicals that activate many different types of odour receptors at the same time. We perceive odours based on the combination of receptors that become activated and then send signals to the brain.

The research of Buck and Axel explained, in part, how animals detect odours. In 2004, the received a Nobel Prize for this pioneering work.

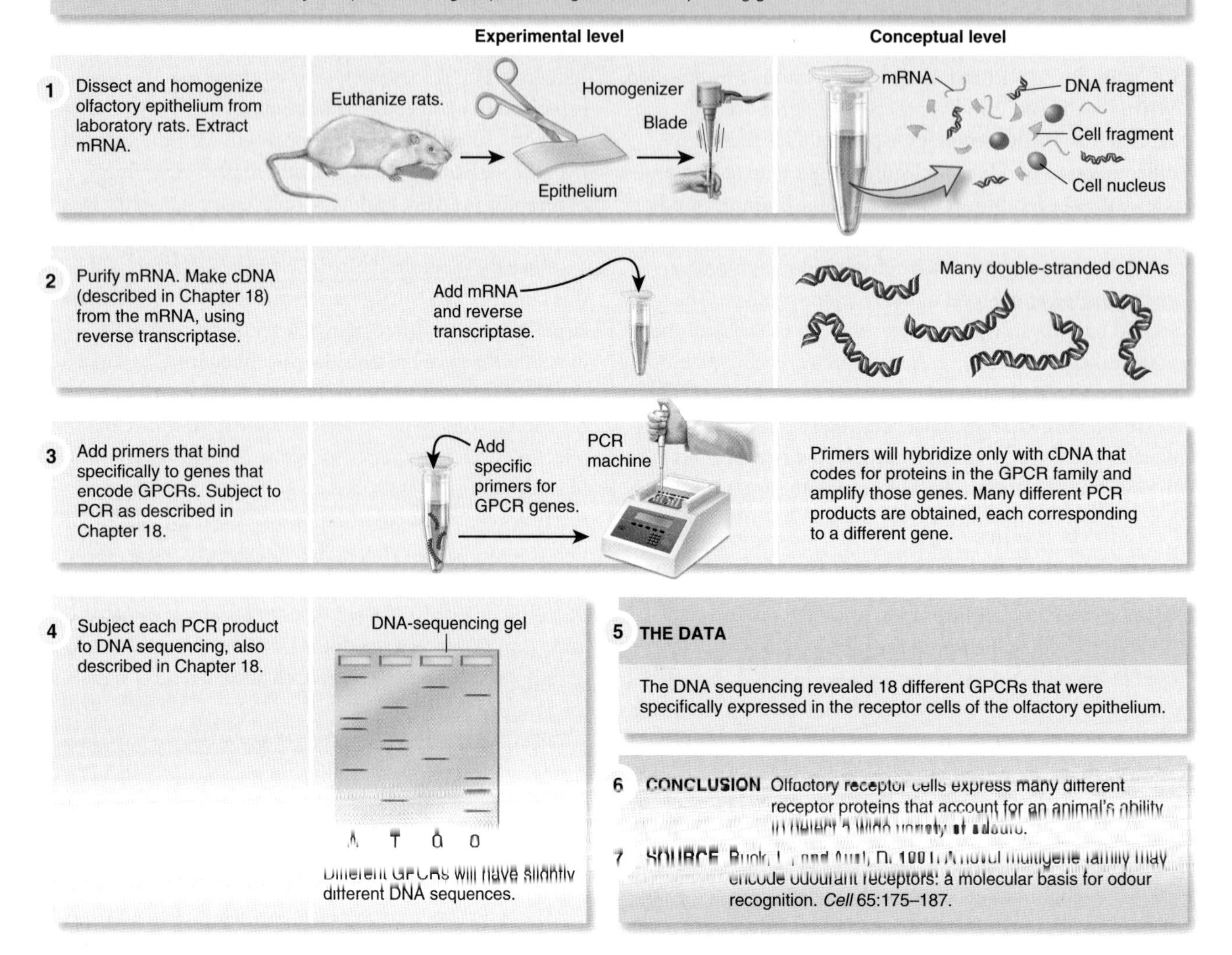

HYPOTHESES 1. Many different types of odour molecules bind to just a few types of receptor proteins. 2. Odour molecules are detected by many specific olfactory receptor proteins belonging to the family of G-protein-coupled receptors (GPCRs).
KEY MATERIALS Laboratory rats (*Rattus norvegicus*), PCR reagents, DNA-sequencing gels
Experimental level
Conceptual level
1 Dissect and homogenize olfactory epithelium from laboratory rats. Extract mRNA.
Euthanize rats.
Homogenizer
Blade
Epithelium
mRNA
DNA fragment
Cell fragment
Cell nucleus
2 Purify mRNA. Make cDNA (described in Chapter 18) from the mRNA, using reverse transcriptase.
Add mRNA and reverse transcriptase.
Many double-stranded cDNAs
3 Add primers that bind specifically to genes that encode GPCRs. Subject to PCR as described in Chapter 18.
Add specific primers for GPCR genes.
PCR machine
Primers will hybridize only with cDNA that codes for proteins in the GPCR family and amplify those genes. Many different PCR products are obtained, each corresponding to a different gene.
4 Subject each PCR product to DNA sequencing, also described in Chapter 18.
DNA-sequencing gel
different DNA sequences.
5 THE DATA
The DNA sequencing revealed 18 different GPCRs that were specifically expressed in the receptor cells of the olfactory epithelium.
6 CONCLUSION Olfactory receptor cells express many different receptor proteins that account for an animal's ability
7 SOURCE
receptors: a molecular basis for odour recognition. *Cell* 65:175–187.

Questions

1. What were the two major hypotheses to explain how animals discriminate between different odours?
2. How did Buck and Axel test the hypothesis of multiple olfactory receptor proteins?
3. What were the results of Buck and Axel's study?
4. Considering the two hypotheses explaining how animals discriminate among different odours, which one was supported by the results of this experiment?
5. With the evidence presented by Buck and Axel, what is the current hypothesis explaining the discrimination of odours in animals?

Assessing Scientific Validity of Advertisements

Manufacturers often make claims of "scientific" or "clinical" evidence for their product's effectiveness. How valid are such claims? Use the following questions to assess the validity of scientific evidence claims.

1. **Control Groups** Control groups provide a reference with which to compare experimental groups to determine if there is an effect caused by the treatment. They provide baseline values for comparison to the measures of the experimental group.
   a. Was a control group used in the investigation?
   b. What is the importance of a control group to a scientific investigation?
2. **Size of Sample** Samples of a population vary with respect to a particular variable—that is, they have a degree of error associated with them. Including many subjects in each study group minimizes the possibility that the differences between groups are due to chance differences between individuals assigned to groups. The greater the sample size, the more variation that is accounted for, and the less error that is associated with them. Studies with larger sample sizes increase the ability to determine a difference between two groups, if one exists.
   a. How many subjects were included in the study?
   b. Why is it important to have as many subjects as possible in an investigation?
3. **Grouping** Subjects must be placed into study groups (control/experimental) in such a way that there is no systematic difference (bias) between them that could affect the results of a study. This makes comparisons between the groups valid.
   a. How were the subjects assigned to the study groups?
   b. Why is the manner in which subjects are assigned to study groups important to the validity of scientific investigations?
4. **Controlling Variables** Controlling for variables reduces the likelihood of alternative explanations and provides for clear interpretation of the results.
   a. What variables were controlled for in the investigation?
   b. Why is it important to the validity of the results that particular variables be controlled for?
5. **Data** Science is a method of inquiry that is based on a critical evaluation of evidence (data). Data must be made available to public scrutiny to be considered valid.
   a. What data are presented?
   b. Why is providing the results of the investigation important?
6. **Replicating Results** Only results that can be reproduced are scientifically valid. If others can reproduce the experiment, they can attempt to verify the data. The methods, as well as the results, of the study are subject to public scrutiny.
   a. Is enough information provided so that the investigation can be replicated?
   b. Why is providing this information vital to a scientific claim?
7. **Analyzing Data** Because most scientific studies use samples, which have an associated degree of error, procedures are needed to determine if the results are significantly different. Statistical procedures can account for the error of samples and determine if significant differences exist between study groups at a given level of confidence.
   a. Is the data statistically analyzed to draw conclusions?
   b. Why is it important that statistical tests be used to analyze experimental results?
8. **Anecdotal Evidence** Anecdotal evidence represents the observations of a limited number of subjects and is of limited scientific value because these subjects are often not representative of the study population.
   a. Is anecdotal evidence (testimonials) presented?
   b. What are the limitations/weaknesses of anecdotal evidence?
9. **Critical Evaluation** A critical evaluation of scientific claims by the peer review process is a hallmark of science as a method of inquiry. A critical evaluation of both the methods and the results of scientific investigations helps to establish the validity of scientific findings. The integrity of science rests on the exposure of new ideas and findings.
   a. What questions would you like to ask the researchers?
   b. What is the role of peer review in the science?

Adapted from: Rutledge, M.L. (2005). Making the nature of science relevant: Effectiveness of an activity that stresses critical thinking skills. The American Biology Teacher, *67(6), 325-329*

Unit 1 Biochemistry

Chapter 1 The Molecules of Life

Answers to Learning Check Questions

(Student textbook page 13)

1. Elements are pure substances that cannot be broken down into simpler components by normal means. Atoms are the smallest units of elements that have all of an element's properties.
2. Carbon-12 has an atomic mass of 12 and has 6 neutrons whereas carbon-14 has 8 neutrons and an atomic mass of 14. Carbon-12 is very stable (and thus very common) compared to carbon-14 which is unstable and radioactive.
3. A polar covalent bond involves the unequal sharing of a pair of electrons between two atoms, where one atom is more electronegative than the other. A polar bond results in one of the atoms having a partial negative charge (the more electronegative atom) and the other atom having a partial positive charge. In an ionic bond, two oppositely charged ions are attracted to one another; one ion has a full negative charge (as a result of its extra electron) and the other has a full positive charge (as a result of having lost an electron). The full charges result from one of the atoms being so much more electronegative than the other that it is able to fully acquire an electron from the more weakly electronegative atom. The transfer of the electron results in ion formation.
4. A water molecule's intramolecular forces make it a polar molecule. As such, its intermolecular interactions are dominated by attraction to other polar molecules (hydrophilic molecules since they are attracted to water) and its inability to attract non-polar molecules (hydrophobic molecules, since they are not attracted to water).
5. The tendency of non-polar molecules or non-polar portions of molecules to stay away from water. The hydrophobic effect is very important in determining the tertiary structure of proteins and in the structure of many other biological molecules.
6. Sample answer: Biotechnology brings together the study of biology and/or biochemistry with technological advancements in many other disciplines, which may include physics, engineering, and chemistry. A good and often overlooked example is the microscope, in which case advances in optics have allowed us to observe cells in great detail, even while they are alive. Molecular genetics is another example of a marriage between disciplines; molecular biology and genetics. Advances in molecular biology, such as the ability to clone genes and sequence DNA, have allowed us to study heritable characteristics at a level of detail never imagined 50 years ago.

(Student textbook page 21)

7. A monomer is the basic building block of a larger polymer; typically a long molecule that may be composed of many hundreds or even thousands of monomeric units joined together. The truly large polymers belong in a class of giant molecules known collectively as macromolecules. Examples include most proteins, larger polysaccharides like starch, glycogen, and cellulose, and many larger nucleic acids such as long double-stranded DNA molecules.
8. Structurally, carbohydrates are organic molecules with many hydroxyl groups, that often contain carbonyl groups, and that generally have the molecular formula $(CH_2O)n$. Functionally, carbohydrates are used as an energy source for cells (e.g., glucose), for energy storage (e.g., starch and glycogen), and to provide structural support (e.g., cellulose).
9. Because they have the same molecular formula ($C_6H_{12}O_6$) but have different three-dimensional structures.
10. Answers should include a Venn diagram or concept map that shows that all three types of molecules are carbohydrates and involved in energy use and/or energy storage, and that disaccharides are composed of two monosaccharides, polysaccharides are composed of more than two monosaccharides, and that monosaccharides are the basic building blocks for the other two.
11. Any two of: energy storage, energy source, and structural support.
12. All three are polysaccharides composed of many monosaccharides linked together, they are composed of many individual glucose monomers linked together covalently, and are macromolecules. They differ in that starch and cellulose are plant products, whereas glycogen is an animal product. Another difference is that the glucose monomers in cellulose are joined by different glycosidic linkages than those found in starch or glycogen. The molecules also differ in function: cellulose is involved in structural support, but starch and glycogen are energy storage molecules.

(Student textbook page 25)

13. Both are organic molecules and have examples that are used for storing energy in cells. Lipids, however, have a greater proportion of carbon and hydrogen atoms and fewer oxygen atoms and, in general, are hydrophobic. Carbohydrates are polar, hydrophilic molecules.
14. Lipids have many energy-rich C-H bonds in their long hydrocarbon chains.
15. Sketches should be similar to **Figure 1.12** or 1.13B on student textbook page 22. The basic structure of a triglyceride is a glycerol molecule with three fatty acid chains linked to it covalently. The fatty acid chains may be all saturated (no carbon-carbon double bonds) or all unsaturated (one or more carbon-carbon double bonds) or a mixture of saturated and unsaturated chains. The unsaturated chain(s) may be monounsaturated or polyunsaturated. The presence of carbon-carbon double bonds in the fatty acid chains makes the chains bent (or kinked) so that they do not pack together as well. Molecules with unsaturated fatty acid chains tend to be liquids at room temperature whereas those with saturated fatty acyl chains (that are quite straight and pack together well) tend to be solids at room temperature.
16. A phospholipid molecule has a dual character in that part of the molecule—its "head" group is polar and thus hydrophilic, but its "tail" portion is non-polar and thus hydrophobic. This structure is essential to its function in that it forces the molecules to orient their head groups

toward water and their tails toward each other when placed in an aqueous environment. This arrangement forms the familiar phospholipid bilayer that is the basic structure of the biological membranes in all cells.

17. Sample answer: Cholesterol is an important component of animal cell membranes and is the precursor used for making many other steroids. Estrogen is an important determinant of sexual function in females and helps regulate the storage of fat.

18. Both are solids at room temperature. In plants such as trees, waxes coat leaves, preventing water loss and offering protection from insects. In animals such as ducks, a waxy layer on feathers prevents them from getting wet, which would add weight and make flight difficult.

(Student textbook page 28)

19. Sample answer: Given the many different ways there are of ordering the 20 common amino acids, proteins exhibit tremendous diversity in structure and function. Examples include the protein fibres that lend structural support to tendons, transport proteins such as hemoglobin that carry oxygen in blood, enzymes that catalyze specific biological reactions, and antibody proteins that fight infections.

20. Sketches should resemble **Figure 1.18** on page 25.

21. The R group of an amino acid may be polar, non-polar, acidic, or basic, which determines many of its properties (e.g., hydrophilic or hydrophobic). When amino acids are part of a protein, the many R groups of those amino acids will largely determine the secondary and tertiary structure of the protein. This structure, in turn, helps determine the protein's function.

22. Because proteins can be built from many different monomers (there are about 20 common amino acids), the many different properties of those amino acids (i.e., polar, non-polar, acidic, or basic, and the reactivity of the specific functional groups found as part of their R groups), and the many different linear sequences of amino acids that are possible (20^n, where n = length of the protein in amino acids).

23.
- Primary: the linear sequence of amino acids.
- Secondary: regions of repetitive structure seen in many different proteins (e.g., alpha helices and beta strands).
- Tertiary: the overall three-dimensional shape of the protein.
- Quaternary: the association of more than one polypeptide to form an intact protein.

24. The overall three-dimensional structure of a protein can be changed by a variety of environmental variables and chemical or physical treatments. Anything that could alter the intramolecular and/or intermolecular interactions that occur in a protein may alter its structure and thus its function as well. For example, a change in temperature (too hot or too cold), in pH, or ionic environment (resulting from changes in salt concentrations). All of these changes could denature a protein either partly or completely. Since a cell depends on the many different functions performed by its proteins, a change in the structure of one or more proteins can be harmful to an organism.

(Student textbook page 36)

25. Products are a salt (an ionic compound) and water. The reaction neutralizes both the acid and the base so that the acid loses its acidic properties and the base loses its basic properties.

26. When a compound is oxidized, its electrons are donated to another molecule (which we say becomes reduced).

27. A redox reaction involves the transfer of electrons from one molecule to another; therefore, both the molecule being oxidized and the molecule being reduced are changed at the same time.

28. A molecule has more energy in its reduced form than when oxidized. As covalent bonds are broken (shared pairs of electrons are separated) energy is released.

29. During hydrolysis, water is consumed (used as a reactant) by the reaction. During a condensation reaction, water is produced (released).

30.
- Condensation reactions—build molecules and release a molecule of water from the H-atom and OH-group that are removed from the combining molecules.
- Hydrolysis reactions—break apart molecules and consume a molecule of water that is used to donate an H-atom and an OH-group to the products.
- Oxidations— remove electrons from molecules, generally by breaking apart covalent bonds and forming a simpler, less complex product or products.
- Reductions—always accompany oxidations. They add electrons to molecules, typically building them up into larger, more complex molecules.

(Student textbook page 39)

31. The activation energy of a reaction is the initial input of energy needed to start the reaction. Its value is significant because reactions with high activation energies occur more slowly than reactions with low activation energies. Anything that lowers the activation energy of a reaction (such as a catalyst) will speed up the rate of the reaction. Many of the chemical reactions that occur in cells have high activation energies and only proceed at the rates needed for life because cells have enzymes that catalyze those reactions.

32. When an enzyme catalyzes a biological reaction, it needs to bind the substrate(s) at its active site to form an enzyme-substrate complex. The enzyme destabilizes the substrate(s) as it binds the active site, lowering the activation energy of the reaction and allowing covalent bonds to be broken and new bonds to form.

33. Even though many important biological reactions can occur naturally without an enzyme, their rates are often too slow to support the needs of the cell or organism. Enzymes are important to biological systems because they help speed these reactions to up the rate required to sustain life.

34. Enzymes are unable to catalyze many different types of reactions because the shape of an active site is usually very substrate-specific. This specificity limits the ability of an enzyme to perform different types of reactions on different molecules.

35. Enzymes can prepare substrates for a reaction by altering the substrate, its environment, or both. Substrate changes can involve the alteration of bond lengths or bond angles (stretching or bending bonds), the addition or removal of electrons (reduction or oxidation), and/or the addition or removal of H-ions (protons) or functional groups. Environmental changes may involve providing an acidic or basic environment at the active site and/or holding substrates close together and in the best orientation relative to one another.

36. A coenzyme is an organic molecule other than the substrate(s) that is required for an enzyme-catalyzed reaction. A cofactor is an inorganic molecule (e.g., metal ion) that is required for an enzyme catalyzed reaction. These molecules assist enzymes in performing the reactions they catalyze. Without them, enzymes could not work. Many enzymes require one or more coenzymes or cofactors for their activity.

Answers to Caption Questions

Figure 1.2 (Student textbook page 11): Based on the partial positive and partial negative charges in the diagram, water molecules would be predicted to interact weakly with each other through the attraction and repulsion of those charges (because like charges repel and unlike charges attract). Since each water molecule has multiple partial charges, each would be predicted to interact with at least three or more other water molecules.

Figure 1.5 (Student textbook page 15): Methane is a non-polar molecule. The two forms of glucose are polar molecules.

Figure 1.7 (Student textbook page 18): Unlike the other molecules, the triglyceride does not have a structure composed of a series of "building blocks" (monomers) linked one after the other to form a long chain (polymer). Although the triglyceride has three fatty acids that are somewhat like monomers in a polymer, they are not linked directly to one another in a chain, but branch off an intermediate molecule (glycerol).

Figure 1.15 (Student textbook page 23): If a phospholipid bilayer contained many lipids with unsaturated fatty acid chains, the bilayer would have a looser arrangement of molecules—since the unsaturated chains would be bent (kinked) and prevent the lipids from packing close together.

Figure 1.25 (Student textbook page 33): If some process in the body began to contribute H^+ (hydrogen ions; protons) to blood, the buffer system would counteract the rise in the concentration of H^+ (the increase in acidity). This would occur because the rise in H^+ concentration would drive the equilibrium of the reaction to the left, toward the formation of water and carbon dioxide.

Figure 1.29 (Student textbook page 37): Even though maltose is also a disaccharide, maltose is not a substrate for this enzyme because it has a different shape than the normal substrate (sucrose) and so would not bind to the enzyme's active site in a manner that would allow the hydrolysis to occur.

Figure 1.30 (Student textbook page 39): It would not be active since the pH is far from optimal for that enzyme.

Answers to Chapter 1 Review Multiple Choice Questions

(Student textbook pages 49-50)

1. b	**2.** c	**3.** b	**4.** b	**5.** d
6. a	**7.** b	**8.** c	**9.** d	**10.** e
11. a	**12.** d	**13.** d	**14.** b	

Answers to Chapter 1 Self-Assessment Multiple Choice Questions

(Student textbook page 54)

1. d	**2.** a	**3.** e	**4.** b	**5.** e
6. d	**7.** b	**8.** d	**9.** c	**10.** b

Chapter 2 The Cell and Its Components

Answers to Learning Check Questions

(Student textbook page 63)

1. Sketches should resemble the centre of **Figure 2.4** on page 60. Sample caption: The double membrane system of the nucleus, the nuclear envelope, surrounds the nucleoplasm (a thick, complex solution of proteins and nucleic acids that contains the chromosomes). Although only visible in dividing cells, the chromosomes are composed of about an equal mass of DNA and protein. A nucleolus is often visible as a dense region containing RNA and other proteins. Nuclear pore complexes stud the nuclear envelope and act as gateways for the passage of materials into and out of the nucleus.

2. These differ from one another structurally by the presence (rough ER) or absence (smooth ER) of ribosomes on the cytoplasmic surface of the membrane. Both are a complex set of highly folded membranes interconnected with one another and with the outer membrane of the nuclear envelope.

   The rough ER is involved in the production of proteins for the endomembrane system. It not only synthesizes proteins for use by the endomembrane system organelles, but also synthesizes transmembrane proteins and proteins for secretion from the cell (e.g., insulin, glucagon, digestive enzymes). The smooth ER is involved in the modification of these newly made proteins and their packaging into vesicles for transport to a Golgi apparatus. Smooth ER also synthesizes phospholipids, steroids, and other lipids.

3. The endomembrane system is a set of eukaryotic membranes and organelles that are either directly connected to one another or engage in the vesicle-mediated transport of substances between one another. The endomembrane system includes the nuclear envelope, endoplasmic reticulum (both smooth ER and rough ER), Golgi, lysosomes, cell membrane, and vesicles and vacuoles of many different types. The functions of the endomembrane system include the synthesis, modification, and transport of proteins, and the compartmentalization of the cell into function-specific and often materials-specific membrane-bound enclosures. A good example of the compartmentalization function is the formation of lysosomes which contain many different enzymes for degrading biomolecules. Those enzymes are prevented from degrading everything in the cell by being compartmentalized and by having an acidic pH optimum (lysosomes are acidic compartments).

4. Peroxisomes are organelles containing oxidative enzymes for the break-down of many types of biomolecules. These organelles are also involved in the synthesis of bile acids and cholesterol.
5. A vacuole is a large membranous compartment. The main difference between a vacuole and a vesicle is size; vesicles are much smaller. There are many different types of vacuoles (and vesicles), but many are involved in the storage of water, ions, and other molecules.
6. Because it sorts and packages lipids and proteins, like a post office does. The Golgi receives material from the ER on the cis face in vesicles and sends the packaged material in vesicles off the *trans* face.

(Student textbook page 67)

7. Both mitochondria and chloroplasts possess their own DNA molecule(s) which encode some of their own proteins. Both are also surrounded by a double membrane system. Both organelles are also involved in energy transduction and redox reactions. They differ in the form of energy that is initially converted into usable forms for the cell—chloroplasts convert the energy of sunlight into energy-rich organic molecules, whereas mitochondria convert the stored energy in energy-rich organic molecules into other molecules that can act as usable energy. Both organelles are similar and different in terms of their location—mitochondria are found in both plants and animal cells, but chloroplasts are found only in plant cells.
8. The cell wall is a rigid layer surrounding plant (but not animal) cells, composed of proteins and/or carbohydrates. The cell wall provides structural support for a cell as well as a measure of protection against mechanical injury and invasion by other organisms.
9. The cytoskeleton is a network of protein fibres extending throughout the cytosol that provides a cell with internal structural elements that help support the cell and determine its shape. The cytoskeleton also provides for the movement and subsequent placement of organelles in the cytoplasm, the contractile activity of muscle, and the movement of cells through their environment as components (microtubules) of flagella and/or cilia.
10. The three main types of protein filaments of the cytoskeleton are microtubules, microfilaments, and intermediate filaments. All are involved in maintaining cell shape, in addition to other functions. Intermediate filaments provide no motility but do provide a structure and support including the internal framework supporting the nucleus and sites of anchorage for some organelles. Microtubules and microfilaments move organelles, and are involved in the formation and dynamic activities of the spindle used during cell division. They are also used in cilia and flagella for moving cells through their environment or sweeping material across the cell surface. Microfilaments are used during muscle contraction and cleavage furrow formation during cell division.
11. Cilia are relatively short structures composed of an internal shaft made of microtubules, covered with an outer membrane. Cilia move fluid and anything in that fluid, for example, sweeping debris along and out of airways, or helping move an egg through the fallopian tube.
12. Flagella are longer structures but, like cilia, are composed of an internal shaft made of microtubules, covered with an outer membrane. Human sperm possesses a single flagellum used to propel the sperm in the female genital tract in search of an egg (oocyte).

(Student textbook page 70)

13. Answers should include any two of: physically separate the contents of the cell from its aqueous environment; act as a selective barrier for the passage of some molecules through its lipid bilayer; serve as a site for cell recognition events; serve as a site for catalysis by many membrane-bound enzymes; transporting specific molecules into and out of the cell using transport proteins; and serving as a site for communication between cells through signal reception and transduction.
14. Lipids and proteins and their associated carbohydrates. The carbohydrate content is covalently linked to some membrane lipids and proteins, forming glycolipids and glycoproteins, respectively.
15. If membranes were composed only of lipids they would be expected to disallow the passage of non-lipid substances such as polar, hydrophilic molecules. Monosaccharides and amino acids would not be able to cross a lipid-only membrane. It is well known, however, that such substances can cross membranes. This fact is part of what led to the early hypothesis that membranes also possess proteins (many of which were later shown to be involved in the transport of substances across membranes).
16. According to the fluid mosaic model, a membrane is a fluid-like (dynamic) phospholipid bilayer with a variety of proteins. Some of the proteins partially or completely span the lipid bilayer (integral membrane proteins) while others are associated with either face of the lipid bilayer (peripheral membrane proteins).

    Most of the lipid and protein molecules are able to move quite freely in the membrane, with lipids moving locally within their layer of the membrane (but not laterally throughout the bilayer) and exchanging places millions of times per second, and proteins moving laterally as components of the membrane and others moving about at either surface.
17. The molecules (mainly phospholipids) that make up their basic framework are not covalently bound to one another but are instead associated with one another through weak intermolecular interactions that are readily made and broken.
18. When phospholipids are mixed with water, their polar (hydrophilic) head groups orient themselves toward the water molecules and cluster together while their non-polar (hydrophobic) fatty acid "tails" cluster together because of hydrophobic interactions. This results in a bilayer of lipids with no free edge (they are "self-sealing") in which the polar head groups face toward the aqueous environment and their non-polar tails face each other on the inside of the bilayer.

(Student textbook page 74)

19. In a cellular context, a concentration gradient is a difference between the concentration of a substance on either side of a membrane.

20. Diffusion involves the random movements of ions or molecules of a substance in space with a net movement from a region with the higher concentration to regions with lower concentration. Diffusion occurs because all molecules are in motion.

21. Answers should include any three of: temperature, pressure, molecule size, polarity, or molecular charge.
- temperature and pressure both increase (or decrease) the rate of diffusion as they increase (or decrease) and make molecules move faster (or more slowly)
- the rate of diffusion is inversely related to molecule size; the larger a molecule is, the more difficult it is for it to diffuse across a membrane
- although small polar molecules can cross membranes, their rates of diffusion are generally lower than those of non-polar molecules of the same size
- in general, charged molecules and ions cannot diffuse across a cell membrane

22. Both involve the random movement that results in net movement of molecules from a region of higher concentration to a region of lower concentration and both can involve movement from one side of a membrane to the other. Osmosis, however, refers only to the diffusion of water molecules.

23. Isotonic, since our cells are normally not crenated (as they would be if the environment were hypertonic) or swollen (as they would be if the environment were hypotonic).

24. Hypotonic, since this condition fills the central vacuole with water, providing rigidity (turgor) to the plant, which is required to keep it upright.

(Student textbook page 77)

25. Both use membrane proteins and aid in the movement of molecules across a membrane. The processes differ in that facilitated diffusion moves molecules down their gradients ("with" their gradients) in a passive manner that requires no energy input. Active transport, on the other hand, requires a net input of energy to move substances against their gradients.

26. Both are integral membrane proteins used to transport molecules across a membrane. Unlike channel proteins, carrier proteins bind the molecules they transport across the membrane. As a result, they undergo a change in shape during transport and move their solutes at much lower rates than channel proteins do. Since channel proteins do not bind the molecules they transport, very high rates of diffusion are possible through channel proteins.

27. ATP is a nucleotide with three phosphate groups and is used as the main source of energy in cells. In active transport, the hydrolysis of ATP often provides the energy necessary to move a substance against its gradient.

28. One component of an electrochemical gradient is a concentration gradient. It is different because it also includes a charge difference across the membrane (an electrical potential component). Thus, an electrochemical gradient involves two differences across a membrane—a concentration difference and an electric potential difference.

29. In primary active transport, ATP hydrolysis is used directly by a transport protein as a source of energy for transporting a substance across a membrane against its concentration gradient and/or against its electric potential gradient. In secondary active transport the energy released from the hydrolysis of ATP is used indirectly to power the transport of a substance against its gradient. The ATP hydrolysis performed by a different (primary active) transport process sets up an electrochemical gradient that the secondary active transporter uses to power its transport process.

30. It is a primary active transporter that hydrolyzes one molecule of ATP for every three sodium ions it pumps out of a cell and every two potassium ions it pumps in. Both ions are moved against their concentration gradients. The pump works by undergoing a series of shape changes powered initially by ATP hydrolysis and then later by the binding or release of the ions and a phosphate group that is transiently bound to the pump during the pumping cycle. During the pumping cycle the E1 conformation of the protein releases K^+, binds Na^+, and hydrolyzes ATP. In the E2 conformation, the protein is transiently phosphorylated on its cytoplasmic side, releases Na^+ and binds K^+.

Answers to Caption Questions

Figure 2.2 (Student textbook page 59): Chloroplasts and a central vacuole. Chloroplasts use the energy from sunlight to convert carbon dioxide and water into high energy organic molecules. The central vacuole is used to store excess water which contributes to the turgor pressure used by plant cells to maintain their rigidity. It also stores ions, sugars, amino acids, and macromolecules, and has enzymes that can break down a variety of macromolecules and waste products.

Figure 2.14 (Student textbook page 73): The water level in the tube rises because the tube has a solution with a higher solute concentration than the solution in the beaker and the two solutions are separated by a semi-permeable membrane. The difference in solute concentrations results in a difference in the concentration of water between the tube (lower) and the beaker (higher). The situation results in the net diffusion of water (osmosis) from the beaker into the tube, causing the level of water in the tube to rise.

Figure 2.18 (Student textbook page 76): Because the pump moves a dissimilar number of charges across the membrane, with three positive charges being pumped out for very two positive charges pumped in. The activity results in a deficit of positive charge in the cell (a build-up of negative charge).

Answers to Chapter 2 Review Multiple Choice Questions
(Student textbook pages 89–90)

1. e	**2.** d	**3.** b	**4.** c	**5.** a
6. d	**7.** b	**8.** e	**9.** c	**10.** e
11. a	**12.** b	**13.** d	**14.** c	**15.** b

Answers to Chapter 2 Self-Assessment Review Multiple Choice Questions
(Student textbook pages 94-5)

1. c	**2.** e	**3.** d	**4.** b	**5.** d
6. c	**7.** d	**8.** b	**9.** c	**10.** b

Answers to Unit 1 Review Multiple Choice Questions
(Student textbook pages 99–100)

1. a	**2.** e	**3.** c	**4.** e	**5.** e
6. c	**7.** e	**8.** e	**9.** c	**10.** d
11. a	**12.** e	**13.** d	**14.** a	**15.** e

Answers to Unit 1 Self-Assessment Review Multiple Choice Questions
(Student textbook page 104)

1. d	**2.** e	**3.** a	**4.** d	**5.** a
6. e	**7.** b	**8.** a	**9.** b	**10.** c

Unit 2 Metabolic Processes
Chapter 3 Energy and Cellular Respiration
Answers to Learning Check Questions
(Student textbook page 118)

1. Metabolism refers to all of the reactions occurring in an organism. Some of these reactions build new compounds (anabolism) and others break them down (catabolism).
2. Anabolic reactions decrease entropy and require energy to be put into the system. These are endergonic. On the other hand, catabolic reactions increase entropy and release energy. Those are exergonic.
3. Whenever a chemical bond forms between two atoms, energy is released. This causes a decrease in the amount of chemical energy in the molecule each time a bond forms.
4. Energy cannot be created or destroyed, but it can be transformed from one type into another and transferred from one object to another. This means that the total amount of energy in the universe remains constant.
5. *During any process, the universe tends toward disorder.* This relates to the natural tendency of compounds to break apart; increasing entropy. Living organisms use raw materials and input energy to maintain compounds or build new ones. Any decrease in entropy would require an energy input. For any spontaneous process, the overall entropy of the universe would have to increase; which is impossible, based on the first law.
6. Free energy is energy available to do work in an organism. An endergonic reaction requires an input of free energy. An exergonic reaction releases energy. Entropy is related to the constant input and release of energy, which tends to increase disorder in the universe.

(Student textbook page 124)

7. Cellular respiration includes the catabolic pathways that break down energy-rich compounds to produce ATP while aerobic respiration refers to those pathways that require oxygen in order to proceed.
8. $C_6H_{12}O_6(s) + 6O_2(g) \rightarrow 6CO_2(g) + 6H_2O([\ell]) + \text{energy}$
   This chemical equation represents the combustion of sugar; a one-step reaction. Since cellular respiration takes more than two dozen reactions, this must represent a summary of those processes.
9. Glycolysis-cytosol (not in mitochondria); pyruvate oxidation and Krebs cycle (mitochondrial matrix); oxidative phosphorylation (inner mitochondrial membrane).
10. The process in which two molecules of ADP are converted to two molecules of ATP for every molecule of glucose entering glycolysis is called substrate level phosphorylation. This is reaction 7 in the process of glycolysis.
11. Oxidative phosphorylation is the formation of ATP using energy obtained from redox reactions in the electron transport chain (i.e., high-energy electrons carried by NADH and $FADH_2$). In the process, the enzyme ATP synthase adds P_i to ADP molecules to make ATP after the Krebs cycle is complete.
12. Glycolysis is the metabolic pathway that breaks glucose into two pyruvate molecules (ATP).

(Student textbook page 128)

13. NAD^+ is reduced to form NADH, an energy-rich electron carrier that allows for the production of ATP.
14. The Krebs cycle is the cyclic metabolic pathway that acquires acetyl-CoA and ultimately oxidizes it to carbon dioxide while regenerating the compound that picks up more acetyl-CoA; it converts released energy to ATP, NADH, and $FADH_2$.
    Its importance is in the production of large quantities of NADH and FADH2, which generate more ATP than would be formed if substrate level phosphorylation had occurred instead of oxidative phosphorylation.
15. It loses a carbon in the form of CO_2 and is oxidized by NAD^+, resulting in NADH. The remaining two carbons are attached to a co-enzyme called CoA.
16. Each molecule of glucose produces two molecules of pyruvate. Each pyruvate goes through the Krebs cycle independently.
17. Reactions 4, 5, and 7.
18. Reactions 4 and 5 generate CO_2 and NADH, (isocitrate combines with isocitrate dehydrogenase which allows NAD^+ to remove an H atom, oxidizing isocitrate, releasing CO_2 and forming NADH. This repeats in reaction 5 with the enzyme α-Ketoglutarate dehydrogenase).

ATP is produced in reaction 6. ATP is produced by substrate level phosphorylation. This is a complex reaction in which a phosphate group replaces the CoA while the substrate, succinate, is bound to the enzyme. The phosphate group is then added to a molecule of guanosine diphosphate (GDP) forming guanosine triphosphate (GTP). The terminal phosphate group from GTP is then transferred to ADP to produce ATP.

$FADH_2$ is produced when FADH is reduced by the oxidation of succinate in reaction 7.

Answers to Caption Questions

Figure 3.1 (Student textbook page 114): Bricks, boards, and cement have more entropy than the building does. When the building is put together, the entropy decreases. Demolishing the building would be a catabolic reaction.

Figure 3.3 (Student textbook page 115): Answers could include thermal energy, as this is the usual waste product associated with light.

Figure 3.5 (Student textbook page 118): Graph B, since energy is released to form the products. Glucose has more potential energy than carbon dioxide and water does.

Figure 3.15 (Student textbook page 131): Acetyl CoA

Answers to Chapter 3 Multiple Choice Review Questions

(Student textbook pages 147–8)

1. c	**2.** a	**3.** e	**4.** b	**5.** b
6. e	**7.** a	**8.** a	**9.** e	**10.** b
11. d	**12.** b	**13.** d	**14.** c	

Chapter 3 Self-Assessment Multiple Choice Questions

(Student textbook pages 152–3)

1. e	**2.** a	**3.** c	**4.** e	**5.** d
6. b	**7.** d	**8.** c	**9.** a	**10.** d

Chapter 4 Photosynthesis

Answers to Learning Check Questions

(Student textbook page 159)

1. It represents the overall sum of many reactions. It is an oversimplification because water and carbon dioxide don't simply join together to make glucose.
2. Thylakoids are one of many interconnected sac-like membranous disks within the chloroplast. They contain the molecules that absorb energy from the Sun to power photosynthesis.
3. Venn diagrams should reflect the following:

Light Dependent Only	Light Dependent and Light Independent	Light Independent Only
• uses light energy to make ATP and NADPH • water is used	occur in the chloroplast	• use ATP and NADPH to create glucose

4. Green pigments (like chlorophyll) reflect green light. The other colour wavelengths are absorbed by the plant for photosynthesis.
5. Each pigment has an ideal wavelength of light that it absorbs, meaning that a wider variety (range) of wavelengths can be absorbed.
6. Peaks represent greater light absorption. Troughs indicate that more reflection of light (less absorption) has occurred.

(Student textbook page 163)

7. To supply electrons and hydrogens for (ultimately) the production of ATP and NADPH.
8. The movement of hydrogen ions is linked to the synthesis (creation) of ATP by chemiosmosis.
9.

Thylakoid Only	Thylakoid and Cristae	Cristae Only
• found in chloroplast • contains chlorophyll and photosystems • home of electron transport carriers that pump protons across the membrane • result in production of O_2, NADPH, and ATP	• double membrane • electron transport chains • ATP synthase • increase surface area • result in the production of ATP	• found in mitochondria • home of electron transport carriers that pump protons across the membrane • result in production of ATP, NAD^+, and H_2O

10. Sketches should resemble Figure 4.7 on page 160 of the student [illegible]
11. Non-cyclic photophosphorylation produces ATP, NADPH, and oxygen gas. The electron flow is unidirectional from water to $NADP^+$. Cyclic photophosphorylation only produces ATP. Electrons cycle and, therefore, no source of electrons (water) or sink for electrons (NAD^+) is required. Excited electrons leave photosystem I and are passed to an electron acceptor. From the electron acceptor, they pass to the b_6-f complex and back to photosystem I. The proton gradient is generated in the same manner as in non-cyclic photophosphorylation, and ATP synthesis by chemiosmosis occurs.
12. A gradient formed with ions is more effective than one formed with neutral molecules, as both the concentration gradient and a charge gradient across the membrane contributes to the stored energy. So both a concentration gradient and voltage across the membrane build up. Both of these contribute to push a charged solute (ion) back across the membrane through the ATP synthase.

(Student textbook page 167)

13. To fix atmospheric carbon and create a three-carbon sugar called G3P
14. G3P

15. It is a 5C compound that CO_2 immediately binds to. It is unstable and then breaks apart into two 3C compounds. This creates the cyclic nature of the Calvin cycle.
16. For every twelve G3P molecules made in the Calvin cycle, two are used to make glucose. The other ten G3P are used to regenerate six RuBP.
17. PGA is energized by ATP through phosphorylation and reduced by NADPH to produce G3P.
18. Any three of: sucrose, starch, cellulose, RuBP

Answers to Caption Questions

Figure 4.1 (Student textbook page 156): Examples could include other phyla of plantae such as gymnosperms, ferns, and mosses, or other protists like euglena.

Figure 4.8 (Student textbook page 161): Protons would leak across the membrane and ATP could not be produced through chemiosmosis.

Figure 4.11 (Student textbook page 166): glycolysis

Answers to Chapter 4 Review Multiple Choice Questions

(Student textbook pages 179–80)

1. e	**2.** c	**3.** c	**4.** c	**5.** e
6. a	**7.** b	**8.** a	**9.** b	**10.** c
11. b	**12.** d	**13.** c	**14.** c	

Answers to Chapter 4 Self-Assessment Multiple Choice Questions

(Student textbook pages 184–5)

1. d	**2.** c	**3.** e	**4.** d	**5.** e
6. e	**7.** a	**8.** d	**9.** e	**10.** c

Answers to Unit 2 Multiple Choice Review Questions

(Student textbook pages 189–90)

1. b	**2.** a	**3.** c	**4.** d	**5.** c
6. b	**7.** d	**8.** b	**9.** d	**10.** b
11. b	**12.** a	**13.** b1	**4.** c	**15.** a

Answers to Unit 2 Multiple Choice Self-Assessment Questions

(Student textbook pages 194–5)

1. d	**2.** a	**3.** c	**4.** b	**5.** b
6. b	**7.** a	**8.** b	**9.** c	**10.** c

Unit 3 Molecular Genetics

Chapter 5 The Structure and Function of DNA

Answers to Learning Check Questions

(Student textbook page 207)

1. Genetic material must contain information that regulates the production of proteins. It also must be able to accurately replicate itself to maintain continuity in future generations. Genetic material must allow for some mutations so that there is variation within a species.
2. Griffith used two forms of *S. pneumoniae*: a pathogenic S-strain and a non-pathogenic R-strain. After injecting mice with a mixture of heat-killed S-strain and live R-strain, the mice died. Griffith concluded that something from the heat-killed S-strain transferred to the R-strain to transform it into a pathogenic form.
3. Avery, MacLeod, and McCarty conducted a series of experiments and discovered the following:
   - When they treated heat-killed pathogenic bacteria with a protein-destroying enzyme, transformation still occurred.
   - When they treated heat-killed pathogenic bacteria with a DNA-destroying enzyme, transformation did not occur. These results provided strong evidence for DNA's role in transformation.
4. Two different radioactive isotopes were used to trace each type of molecule. One sample of T2 virus was tagged with radioactive phosphorus (^{32}P), since phosphorus is present in DNA and not protein. The other sample of T2 virus was tagged with radioactive sulfur (^{35}S), since sulfur is only found in the protein coat of the capsid.
5. The independent variable in the experiment was the type of radioactive isotope used to tag the virus. The dependent variable in the experiment was the presence of radioactivity inside the infected bacterial cells. Some of the controls include: the usage of the same type of virus in both experiments and the usage of the same protocol for infecting bacterial cells in both experiments.
6. Bacterial cells that are infected by viruses with ^{32}P-labelled DNA would not be radioactive. Bacterial cells infected by viruses with ^{35}S-labelled capsid proteins would be radioactive.

(Student textbook page 212)

7. Answers should be similar to **Figure 5.4** on page 208 of the student textbook, with labels for the following: phosphate group, sugar group, nitrogen-containing base.
8. Nucleotides in DNA have a deoxyribose sugar, while nucleotides in RNA have a ribose sugar with a hydroxyl group at carbon 2. In addition to the sugar group each nucleotide is attached to a phosphate group and a base. The bases are adenine, cytosine, guanine, and thymine in the case of DNA and adenine, cytosine, guanine, and uracil in the case of RNA.
9. Chargaff's rule states that in the DNA nucleotides, the amount of adenine will be more or less equal to the amount of thymine, and the amount of guanine will be equal to the amount of cytosine. The number of A-T nucleotides will not necessarily equal the number of C-G nucleotides. This overturned Levene's earlier hypothesis that the nucleotides occurred in equal amounts and were present in a constant and repeated sequence.
10. Franklin used X-ray photography to analyze the structure of DNA. Her observations provided evidence that DNA has a helical structure with two regularly repeating patterns. She also concluded that the nitrogenous bases were located on the inside of the helical structure, and the sugar-phosphate backbone was located on the outside, facing toward the watery nucleus of the cell. Pauling's methods of assembling three-dimensional models of compounds led to the discovery that many proteins had a helical structure. Watson and Crick also used this information to propose that DNA had a helix shape.

11. Answers should be similar to **Figure 5.7B** on page 213 of the student textbook. Base pairing and directionality of strands should be shown.

12. Nucleic acids are soluble in water. Therefore, the nitrogenous bases, which are somewhat hydrophobic, must be positioned away from the water found in the nucleoplasm, and the polar phosphate groups (which are hydrophilic) must be on the outside of the molecule, interacting with the water.

(Student textbook page 222)

13. The main objective of DNA replication is to produce two identical DNA molecules from a parent DNA molecule.

14. DNA replication occurs during the S-phase of interphase, and prior to cell division. Therefore, DNA replication is important since each new daughter cell must have the same genetic information as the parent cell.

15.
- Conservative model: Two new daughter strands form to create a new double helix, and the original DNA strands re-form into the parent molecule.
- Semi-conservative model: Each new DNA molecule contains one strand of the original DNA and one newly synthesized strand.
- Dispersive model: Parental DNA is broken into fragments. Therefore, the daughter DNA contains a mix of parental and newly synthesized DNA.

16. Nitrogen is a component of DNA and is incorporated into newly synthesized daughter strands. Having a "light" form (^{14}N) and a "heavy" form (^{15}N) allowed the separation of different DNA strands based on the amount of isotope present in the newly synthesized DNA. DNA with more ^{15}N would be denser than DNA with ^{14}N, and therefore could be separated by centrifugation.

17. Meselson and Stahl concluded that DNA replication is semi-conservative. After one round of replication, DNA appeared as a single band, midway between the expected positions of ^{15}N-labelled DNA and 14N-labelled DNA. After the second round of replication, DNA appeared as two bands, with one band corresponding to 14N-labelled DNA and the other band in the position of hybrid DNA (half ^{14}N and half ^{15}N). In additional rounds of replication, the same two bands were observed, therefore supporting the semi-conservative model.

18. Each new cell that is produced must have an exact copy of parental DNA. The daughter strands of DNA are part of a DNA molecule that will be in the daughter cells. This ensures that newly born cells are similar to parents and maintain their genetic identity.

(Student textbook page 227)

19. Initiation: Helicase enzymes unwind DNA to separate it into two strands. A replication bubble is formed when single-strand binding proteins stabilize the separated strands.

Elongation: New DNA strands are synthesized by joining free nucleotides together. This is catalyzed by DNA polymerase, which synthesizes the new strands that are complementary to the parental strand.

Termination: The two new DNA molecules separate from one another.

20. Replication takes place in a slightly different way on each DNA strand because DNA polymerase can only catalyze elongation in the 5′ to 3′ direction. In order for both strands of DNA to be synthesized simultaneously, the method of replication must differ.

21. On the leading strand, DNA synthesis takes place along the DNA molecule in the same direction as the movement of the replication fork. On the lagging strand, DNA synthesis proceeds in the opposite direction to the movement of the replication fork: The lagging strand is synthesized in short fragments called Okazaki fragments.

22. DNA replication requires the use of many enzymes that have specific roles. The presence of numerous specialized enzymes may reflect the importance of having accurate DNA replication, since mutations in DNA can change the genetic makeup of an organism.

23. Answers may include: DNA polymerases have a proofreading function, where they excise incorrect bases and add the correct base. Mismatch repair involves a group of enzymes that identify, remove, and replace incorrect bases.

24. Many tissues and organs require continuous regeneration of new cells. Therefore, DNA replication must be quick and accurate since new daughter cells must receive exact copies of DNA from the parent cell.

Answers to Caption Questions

Figure 5.2 (Student textbook page 205): If live strain had been transferred, the effects would have been due to that strain, not due to the transfer of a substance form it to the R-strain, which makes it pathogenic.

Figure 5.3 (Student textbook page 207): The results would have also shown that protein was not the hereditary material. However, it would not have directly demonstrated the role of DNA as the hereditary material since RNA also contains phosphorus.

Figure 5.10 (Student textbook page 215): Twisting a rubber band around itself is similar to how DNA undergoes supercoiling. The rubber band becomes compacted due to the coils that form from twisting. This model is also useful since it also demonstrates the tension that is created due to supercoiling.

The rubber band model shown in **Figure 5.9** on page 214 of the student textbook shows the rubber band being linearized, where supercoiling in bacterial DNA occurs because it is a circular. The rubber band model also does not reflect the double-stranded nature of DNA

Figure 5.16 (Student textbook page 221): If DNA had not been uniformly labelled with 15N, the banding patterns would not accurately reflect the presence of parental DNA.

Answers to Chapter 5 Review Questions

(Student textbook pages 235-9)

1. d	**2.** b	**3.** e	**4.** b	**5.** c
6. c	**7.** b	**8.** d	**9.** b	**10.** a
11. d	**12.** d	**13.** a	**14.** a	

Answers to Chapter 5 Self-Assessment Questions

(Student textbook pages 240-1)

1. b	**2.** c	**3.** e	**4.** c	**5.** a
6. e	**7.** c	**8.** a	**9.** e	**10.** b

Chapter 6 Gene Expression

Answers to Learning Check Questions

(Student textbook page 246)

1. Garrod's studies on alcaptonuria showed that having the black urine phenotype was due to a recessive inheritance factor that caused that production of a defective enzyme.

2. Answers could use **Figure 6.1** on page 245 of the student textbook as a guideline. The results of the Beadle and Tatum experiment showed that a single gene produces one enzyme (one-gene/one-enzyme hypothesis). This was later modified to the one-gene/one-polypeptide hypothesis since not all proteins are enzymes.

3. The evidence that supported RNA as an intermediate molecule included the following: RNA is found in the nucleus and cytoplasm; the concentration of RNA in the cytoplasm is correlated with protein production; RNA is synthesized in the nucleus and transported to the cytoplasm.

4. Jacob and colleagues saw that bacteria infected with a virus had a newly synthesized virus-specific RNA molecule. This RNA molecule associated with bacterial ribosomes, which are the sites of protein production. Therefore, the RNA molecule carried the genetic information for the production of a viral protein.

5. Having RNA as an intermediate molecule that can transport from the nucleus to the cytoplasm is preferred over DNA having to continually transport itself, which could increase the likelihood of damage to the DNA. Multiple steps in gene expression provide many opportunities for regulation. This allows the cell to have increased control over protein synthesis.

6. 5′-GAUUAACGG-3′

(Student textbook page 254)

7. Answers could include the following: DNA is the genetic material that controls protein synthesis whereas RNA helps DNA and is involved in protein synthesis. DNA has the sugar called deoxyribose whereas RNA has ribose sugar. The bases in DNA are adenine, guanine, thymine, and cytosine; the bases in RNA include adenine, guanine, cytosine, and uracil. DNA is a double-stranded helix whereas RNA is single stranded without a helix.

8. Answers could include information referenced in **Table 6.3** on page 252 of the student textbook: mRNA is the template for translation, while tRNA and rRNA are involved in the translation of mRNA.

9. Initiation: Transcriptional machinery is assembled on the sense strand. RNA polymerase binds to the promoter region of the sense strand.

Elongation: RNA polymerase synthesizes a strand of mRNA that is complementary to the sense strand of DNA.

Termination: RNA polymerase detaches from the DNA strand when it reaches a stop signal. The mRNA strand is release and the DNA double helix re-forms.

10. mRNA would be transported from the nucleus to the cytoplasm for translation.

11. During the elongation phase of transcription, a second RNA polymerase complex can bind to the promoter region immediately after the previous RNA polymerase complex starts moving along the DNA. Therefore, multiple strands of mRNA can be synthesized from one gene at a time. This is advantageous for the cell since increased amounts of mRNA leads to more protein that can be produced in a given amount of time.

12. Errors in transcription would be less damaging than errors in DNA replication. An error in transcription would results in an error in one protein molecule, while an unrepaired error in DNA replication would cause a change in the genetic makeup of an organism.

(Student textbook page 260)

13. Answers may be based on **Table 6.4** on page 257 of the student textbook and should include mRNA, tRNA, ribosomes, and translation factors. The following information may be included:

- mRNA: contains the genetic information which determines a protein's amino acid sequence
- tRNA: a molecule that links mRNA codons with their corresponding amino acid
- ribosomes: a structure composed of rRNA and proteins which provides the site for protein synthesis
- translation factors: accessory proteins which are necessary for each stage of translation

14. A polyribosome is a complex composed of multiple ribosomes along a strand of mRNA. It can produce many copies of a protein at the same time.

15. Translation consumes large amounts of energy since many molecules (i.e., proteins and nuclei acid components) must be synthesized and assembled. Energy is also needed to form peptide bonds that exist between the amino acids in proteins. This energy is provided by mitochondria, which supply the cell with most of its energy.

16. Translation is initiated when initiation factors assemble the translation components. The small ribosomal subunit attaches to the mRNA near the start codon (AUG). The initiator tRNA (with anticodon UAC) binds to the start codon. The large ribosomal subunit then joins to form the complete ribosome. Translation is terminated when a stop codon on the mRNA is reached. This causes the polypeptide and the translation machinery to separate. The polypeptide is then cleaved from the last tRNA by a release factor.

17. Answers could be based on **Figure 6.14** on page 259 of the student textbook.

18. The antibiotic would prevent the initiation of translation of bacterial proteins. Therefore, the bacterial cells would not survive.

(Student textbook page 265)

19. Mutations that occur in reproductive cells can be passed onto future generations. Mutations which occur in somatic cells do not get passed on from one generation to another.
20. No, since the deletion of nucleotides is divisible by three.
21. A silent mutation has no effect on the amino acids of a protein, and therefore would be the least harmful to an organism. Missense mutations cause alterations in the amino acids of a protein, and nonsense mutations results in a shortened protein due to a premature stop codon.
22. Single-gene mutations involve changes to the nucleotide sequence of one gene. Chromosomal mutations involve changes in chromosomes, and therefore may affect many genes.
23. Spontaneous mutations may arise during DNA replication if there is incorrect base pairing by DNA polymerase. DNA transposition can also cause mutations by the movement of specific DNA sequences (transposons, or "jumping genes") moving within and between chromosomes.
24. UV radiation can lead to a mutation by covalently linking thymines to form a thymine dimer. A specific mechanism for repairing thymine dimers is photorepair, where an enzyme specifically recognizes and cleaves thymine dimers. A non-specific mechanism for repairing UV damage is excision repair, where a group of enzyme repair enzymes identify and correct many different types of DNA damage.

(Student textbook page 269)

25. Constitutive genes code for proteins that needed for cell survival. Therefore, they are always active and are expressed at constant levels.
26. The regulation of genes allows for cell specialization. Gene expression is controlled so that only certain genes are active in certain amounts and at certain times in each cell type. Therefore, each cell type can have its own set of proteins.
27. Many genes in bacteria are clustered together in operons and are therefore under transcriptional control of a single promoter. This type of transcriptional control is energy efficient since these genes are transcribed together into a single polycistronic mRNA, rather than individual mRNA molecules. Individual proteins are then synthesized from the polycistronic mRNA.
28. Answer could be based on **Figure 6.23** on page 267 of the student textbook. Diagram should include the regulatory region, promoter region, operator, and the coding region.
29. The lac operon is inducible since the transcription of genes is induced when lactose is present. In the *trp* operon, the genes are actively transcribed until a repressor binds to inhibit transcription.
30. A common analogy used is the regulation of temperature by a thermostat, since the feedback mechanism is similar (i.e., at low temperatures, the thermostat turns on to increase the temperature; at high temperatures, the thermostat turns off to decrease the temperature).

Answers to Caption Questions

Figure 6.1 (Student textbook page 245): Growth would occur after each addition of intermediate since glutamate is the first step of the arginine pathway.

Figure 6.3 (Student textbook page 249): The drugs will also inhibit translation, since it follows transcription.

Figure 6.8 (Student textbook page 253): The increasing length of the mRNA strands indicates the direction of transcription.

Figure 6.11 (Student textbook page 257): The genes that code for tRNA produce tRNA molecules, which are not translated into proteins.

Figure 6.16 (Student textbook page 261): Cysteine.

Figure 6.18 (Student textbook page 263): A silent mutation occurs when a nucleotide substitution has no effect on the polypeptide sequence (i.e., ACU to ACC)

Figure 6.24 (Student textbook page 268): Both operons have the same basic structure. The lac operon is inactive unless lactose is present. The trp operon is normally active, but becomes inactive if enough tryptophan is made.

Answers to Chapter 6 Review Multiple Choice Questions

(Student textbook pages 277-81)

1. b	**2.** d	**3.** c	**4.** a	**5.** b
6. e	**7.** c	**8.** c	**9.** d	**10.** a
11. b	**12.** a	**13.** c	**14.** e	

Answers to Chapter 6 Self-Assessment Multiple Choice Questions

(Student textbook pages 282-3)

1. c	**2.** e	**3.** a	**4.** a	**5.** a
6. c	**7.** a	**8.** d	**9.** c	**10.** b

Chapter 7 Genetic Research And Biotechnology

Answers to Learning Check Questions

(Student textbook page 291)

1. *Specificity:* The cuts made by an endonuclease are specific and predictable. That is, the same enzyme will cut a particular strand of DNA the same way each time, producing an identical set of small DNA fragments called restriction fragments.

   Staggered cuts: Most restriction endonucleases produce a staggered cut that leaves a few unpaired nucleotides on a single strand at each end of the restriction fragment. These short strands, often referred to as "sticky ends," can then form base pairs with other short strands that have complementary strands, creating a recombinant DNA molecule.
2. The construction recombinant DNA molecules require the use of restriction endonucleases. When a gene of interest and a vector have the same restriction endonuclease cutting site, "sticky ends" can be produced when they undergo a restriction endonuclease reaction. The "sticky ends" in the gene of interest can form base pairs with the "sticky ends" of the plasmid to create a recombinant DNA molecule.
3. Gene cloning is a process that produces many identical copies of a gene. Cloning genes is useful for studying gene function and for producing large amounts of mRNA or proteins from a gene for further study.

4. Selectable markers, such as antibiotic resistant genes or lacZ genes, are used to identify and specifically select bacterial colonies which have the recombinant DNA.
5. Answers can use **Figure 7.6** on page 290 of the student textbook as a guide.
6. Answers could include the following: Amplification of "ancient DNA" for evolutionary studies; screening of genetic defects using a DNA sample from a single cell; amplifying DNA from small samples that are found at crime scenes for further analysis.

(Student textbook page 295)

7. DNA fragments are negatively charged. In gel electrophoresis, the negatively charged DNA fragments are attracted to and travel toward the positive terminal.
8. The smaller fragments move more easily through the spaces between the protein molecules of the gel and migrate quicker and further from the loading well than larger fragments.
9. Gel electrophoresis allows the analysis of DNA fragments based on size. When cloning a gene, the size of the plasmid and the size of the gene are known. Therefore, to check if the gene was successfully cloned into a plasmid, researchers will cut the recombinant DNA with the same restriction endonucleases that were originally used to create the recombinant DNA. This sample is then run on a gel to analyze the size of the resulting fragments. If the fragments correspond to the same sizes of the gene and the plasmid, then cloning was successful.
10. DNA fingerprinting involves identifying individuals by analyzing their DNA sequence at certain regions in the genome.
11. Short tandem repeats (STR) are short, repeating sequences of DNA in the genome that vary in length between individuals. In STR profiling, the STRs of an individual are amplified by PCR and analyzed by gel electrophoresis. Since each DNA fragment is fluorescently labelled, a detector can be used to measure the amount of fluorescence emitted from each STR. The resulting printout is a series of peaks that correspond to STRs of varying molecular mass. **Figure 7.10** on page 293 of the student textbook shows an example of a STR profile.
12. Other situations where DNA fingerprinting may be useful are in paternity cases, identifying the remains of murder or accident victims, tracing the movement of wildlife, or in plant and animal breeding programs.

(Student textbook page 304)

13. Genetic engineering is the process of specifically altering the genetic material of an organism. Examples of genetic engineering include the production of transgenic bacteria, plants, and animals.
14. Answers may include the following: Patenting GMOs allows researchers to protect their intellectual property rights, provide financial incentive, and stimulate further research. Patents may also lead to higher costs and exclusivity, therefore shutting out researchers and other individuals who may wish to study or use a GMO. The ethics surrounding the ownership of organisms should also be considered.
15. Expression vectors have sequences which support transcription and translation of an inserted gene. Therefore, if a gene that codes for a therapeutic protein is inserted into an expression vector, it can be transcribed and translated in bacterial cells in large quantities. The protein can then be purified from the bacteria for medicinal use.
16. Bacteria are easy to use and inexpensive to maintain. Using transgenic bacteria to produce human insulin has also removed some of the allergic side effects that were present when insulin was produced in animal sources.
17. Some bacteria can be genetically modified to improve their ability to break down pollutants, such as pesticides, herbicides, and oil spills that are found in water.
18. Answers may include the following: Transgenic bacteria can clean up and reduce environmental pollutants. However, transgenic bacteria can also negatively affect the balance of an ecosystem (i.e., effects on the food supply or habitat of native plants and animals). There is also the potential of mutations which may occur in the transgenic bacteria, which would have unpredictable effects on an ecosystem.

(Student textbook page 307)

19. Transgenic crop plants have increased tolerance to pesticides, herbicides, disease and pests, which allows crops to grow in harsh environments. Transgenic crop plants may also lead to an increase in crop yields, a reduction in spoilage, a reduction in pesticide use, and a reduction in cost.
20. The biolistic method involves bombarding plant cells with particles coated with DNA. Since the bombardment occurs at a high speed, the DNA can penetrate the plant cells and integrate into the plant genome.
21. Answers will be similar to **Figure 7.19** on page 306 of the student textbook.
22. In horizontal gene transfer, transgenic plants may transfer their introduced traits to other organisms, such as other plants, animals, bacteria and fungi. In vertical gene transfer, the transgenic plant could transfer its introduced trait into the genomes of the natural or wild version of the same plant.
23. Health Canada requires 7-10 years of health and safety research for each new GMO intended for human consumption. This research includes investigating how the GMO was developed as well as food safety. The main concern surrounding GMOs is the uncertainty surrounding the long-term effects of GMO consumption.
24. Answers may include: Some transgenic plants express toxins as insecticidal protection, which could affect other unintended organisms. Gene transfer could also occur, where the transgenic plant's insecticide properties are transferred to other organisms or to the genomes of the wild versions of the same plant. Insects which normally feed on the plant may be physically harmed by the insecticide itself, and therefore will need to find a new food source. These insects may then cause harm to another plant which does not have insecticide.

Answers to Caption Questions

Figure 7.1 (Student textbook page 286): AATT

Figure 7.3 (Student textbook page 287): Instead of the two different molecules being ligated together to produce one recombinant molecule, each molecule could react with itself by forming base pairs within the same molecule. Therefore, this would produce at least two different molecules.

Figure 7.4 (Student textbook page 289): The gene is cloned in the last step, where the host divides to produce many cells.

Figure 7.6 (Student textbook page 290): $2^{30} = 1,073,741,824$ copies

Answers to Chapter 7 Review Multiple Choice Questions

(Student textbook pages 319-23)

1. c	**2.** c	**3.** e	**4.** e	**5.** a
6. a	**7.** c	**8.** b	**9.** a	**10.** a
11. c	**12.** a	**13.** b	**14.** c	

Answers to Chapter 7 Self-Assessment Multiple Choice Questions

(Student textbook pages 324-5)

1. a	**2.** e	**3.** a	**4.** b	**5.** d
6. a	**7.** a	**8.** c	**9.** a	**10.** d

Answers to Unit 3 Review Multiple Choice Questions

(Student textbook pages 329-33)

1. c	**2.** a	**3.** e	**4.** b	**5.** b
6. d	**7.** a	**8.** e	**9.** c	**10.** d
11. c	**12.** c	**13.** b	**14.** b	**15.** c

Answers to Unit 3 Self-Assessment Multiple Choice Questions

(Student textbook pages 334-5)

1. c	**2.** b	**3.** c	**4.** a	**5.** a
6. d	**7.** b	**8.** a	**9.** a	**10.** c

Unit 4 Homeostasis

Chapter 8 The Nervous System and Homeostasis

Answers to Learning Check Questions

(Student textbook page 346)

1. cells, tissues, organs, organ systems

2. The tendency of the body to maintain a relatively constant internal environment.

3. A cycle of events in which a variable, such a body temperature, is continually monitored, assessed, and adjusted.

4. • a sensor that detects a change in the internal environment and sends a signal to a control centre;

- a control centre that sets the range of values within which a variable should be maintained, receives information from a sensor, and sends signals to an effector; and
- an effector that receives signals from the control centre and responds, resulting in a change to an internal variable

5. The sensory receptors in the integumentary system communicate with the brain and spinal cord via nerves (the three parts of the nervous system).

6. The stomach is composed of individual cells. Cells of the same type that perform a common function make up tissues, such as those that line the stomach. One or more tissues interact to form more complex structures known as organs, such as the stomach. Several organs, such as the stomach and small intestine, are organized structurally and functionally to form an organ system, such as the digestive system.

(Student textbook page 354)

7. The human nervous system can regulate tens of thousands of activities simultaneously. Its overall function is to collect information about the external conditions in relation to the body's internal state, analyze it, and initiate appropriate responses to maintain homeostasis.

8. Sample answer:

System	Structure	Function
Central nervous system	• brain • spinal cord	• The spinal cord carries messages from the body to the brain. • The brain analyzes and interprets these messages. • The brain then passes response messages through the spinal cord to target structure such as a muscle, gland, or neuron.
Peripheral nervous system	• somatic nervous system • autonomic nervous system	• The sympathetic nervous system controls organs in times of stress (fight or flight). • The parasympathetic nervous system causes a return to a state of rest and controls organs when the body is at rest

9. Neurons are the basic structural and functional units of the nervous system. They are specialized to respond to physical and chemical stimuli, to conduct electrochemical signals, and to release chemicals that regulate various body processes.

Glial cells support neurons. These cells nourish the neurons, remove their wastes, and defend against infection. Glial cells also provide a supporting framework for all of the nervous-system tissue.

10. Diagrams should be similar to the motor neurons shown in **Figure 8.8** on page 351 of the student textbook. Labels should identify the dendrites—receive impulses from other neurons and conduct impulses to the cell body; cell body—site of the cell's metabolic reactions and relay gatekeeper for nerve impulses sent down the axon; axon—conducts impulses away from the cell body; Schwann cell—type of glial cell that wraps around the axon to form the myelin sheath, insulating the axon and speeding up impulses; and the node of Ranvier—spaces between adjacent Schwann cells involved in impulse transmission.

11. • Sensory neurons—gather information from sensory receptors and transmit those impulses to the central nervous system.

- Interneurons—found entirely within the central nervous system act as a link between sensory and motor neurons. They process and integrate incoming sensory information and relay outgoing motor information.
- Motor neurons—transmit information from the central nervous system to effectors (muscles, glands, or other organs).

12. The basic neural pathway used is a reflex arc. Sense organ (eye) detects the ball → initiates an impulse in a sensory neuron → activates spinal cord (interneuron) → activates a motor neuron [arrow] causes muscle to act to move the body out of the way of the ball. This response is much like a withdrawal reflex, that is, an involuntary response triggers a very fast response that moves the body out of harm's way.

(Student textbook page 357)

13. This is the charge difference across the membrane in a resting neuron; usually –70 mV in unstimulated neurons, and is more negative on the inside. This provides energy for the generation of a nerve impulse in response to a stimulus.

14.
- Some negatively charged substances, such as proteins and chloride ions (Cl^-), are trapped inside the cell and unable to diffuse out through the selectively permeable cell membrane.
- Sodium ions (Na^+) and potassium ions (K^+) cannot diffuse unaided from one side of the cell membrane to the other. Special membrane proteins, however, can use the energy of ATP to pump charged particles across the membrane. This sodium-potassium pump pumps out three sodium ions for every two potassium ions pumped into the cell, which results in an unequal distribution of positive charges on either side of the membrane. The build up of positive charges on the outside of the cell creates an electric potential.
- Special transport proteins form ion-specific channels allow potassium ions to diffuse down their concentration gradient and out of the cell. There are sodium ion channels as well, but, in a resting neuron, there are more open channels for potassium ions than for sodium ions. As a result, relatively more potassium ions diffuse out of the cell compared to the number of sodium ions diffusing in. This contributes to the build up of positive charges on the outside of the membrane.

15.
- The shape of a carrier protein in the cell membrane allows it to take up three sodium ions (Na^+) on the cytoplasmic side (inside) of the neuron.
- ATP is split, and a phosphate group is transferred to the carrier protein.
- A change in shape of the carrier protein causes the release of three sodium ions (Na^+) outside the cell. The altered shape permits the uptake of two potassium ions (K^+) from the outside of the extracellular (outside) of the membrane.
- The phosphate group is released from the carrier protein.
- A change in shape of the carrier protein causes the protein in the cell membrane to release the potassium ions (K^+) into the cell. The carrier protein is once again able to take up three sodium ions (Na^+).

16. A nerve impulse is a wave of action potentials that occurs along the length of a neuron, followed by a subsequent repolarization along the neuron. The impulse begins as a movement of ions across a localized area of the cell membrane, which reverses the polarity of the membrane potential in the area. This reversal of polarity results in depolarization in the neighbouring area of the membrane, which has the same effect on the next area of the membrane, and so on. In this way, the impulse propagates in one direction along the length of the neuron, with sodium ion channels and then potassium ion channels opening up in one area after another.

17. Until the neuron goes through the refractory period and is repolarized, it cannot be stimulated again. In other words, it cannot generate another action potential until it has been repolarized.

18. If a threshold potential is reached, a large numbers of sodium ion channels simultaneously open, allowing for an influx of sodium ions and rapid depolarization of the membrane. This rapid change in the membrane potential initiates an action potential. If the sodium ion channels are blocked by tetrodotoxin, then the sodium ions will not flow into the neuron. Without this influx of sodium ions, an action potential will not be generated. Essentially, nervous impulses will not be transmitted and an individual may experience paralysis.

(Student textbook page 367)

19. The two structures of the central nervous system are the brain and spinal cord. The site of neural integration and processing, the central nervous system receives information from the senses, evaluates this information and initiates outgoing responses to the body.

20. The two types of nervous tissue found in the CNS are grey matter and white matter. Grey matter is grey because it contains mostly cell bodies, dendrites, and unmyelinated neurons. White matter is white because it contains myelinated neurons that run together in tracts.

21. The spinal cord is protected by cerebrospinal fluid, soft tissue layers, and the spinal column, a series of backbones (vertebrae).

22. Diagrams should resemble **Figure 8.19** on page 365 of the student textbook.

23. Protects the brain by preventing many toxins and infectious agents from leaving the blood. The blood-brain barrier also supplies the brain with nutrients and oxygen.

24. The hypothalamus helps maintain homeostasis by
- controlling blood pressure
- controlling heart rate
- controlling body temperature
- coordinating actions of the pituitary gland, by producing and regulating the release of many hormones

(Student textbook page 372)

25. Nerves that link the brain and spinal cord to the rest of the body, including the senses, muscles, glands, and internal organs.

26. Twelve pairs of myelinated cranial nerves and 31 pairs of myelinated spinal nerves.
27. Voluntary control that service the head, trunk, and limbs. Its sensory neurons carry information about the external environment inward to the brain, from the receptors in the skin, tendons, and skeletal muscles. Its motor neurons carry information to the skeletal muscles.
28. Sympathetic nervous system and the parasympathetic nervous system, and either stimulate or inhibit the glands, or cardiac or smooth muscle.
29. Stress or danger is likely to trigger a response from the sympathetic nervous system, sometimes called a "fight-or-flight" response.
30. This system prepares the body either to confront a dangerous situation (fight) or to flee from it (flight).

Answers to Caption Questions

Figure 8.1 (Student textbook page 344): cardiac muscle cells [right arrow] cardiac muscle tissue [right arrow] heart [right arrow] circulatory system

Figure 8.3A (Student textbook page 346): The control centre sends messages to several effectors causing constriction of blood vessels in the skin, which decreases heat loss; hormone release that increases metabolism, generating heat; and contracting muscles repeatedly (shivering) which creates heat.

Figure 8.14 (Student textbook page 357): If the sodium channels were unable to close, the neuron would not be able to repolarize. The membrane potential would stay positive instead of returning to its normal resting membrane potential, and the line on the graph would continue to rise.

Figure 8.15 (Student textbook page 358): Increased velocity, 30 times faster than by continuous conduction.

Figure 8.2[illegible] (Student textbook page [illegible]): [illegible] temporal

Figure 8.2[illegible] (Student textbook page 370): Those in muscles of the arms, shoulders, and fingers as well as those that maintain balance and posture.

Answers to Chapter 8 Multiple Choice Review Questions
(Student textbook page 381)

1. d	**2.** a	**3.** e	**4.** a	**5.** c
6. e	**7.** b	**8.** e		

Answers to Chapter 8 Self-Assessment Questions
(Student textbook pages 386–7)

1. b	**2.** d	**3.** e	**4.** c	**5.** e
6. b	**7.** c	**8.** e	**9.** a	**10.** e

Chapter 9 The Endocrine System

Answers to Learning Check Questions
(Student textbook page 392)

1. A hormone is a chemical messenger sent to many parts of the body to produce a specific effect on its target cell(s) or organ(s). Examples include epinephrine and norepinephrine released from the neurons of the adrenal gland.
2. The endocrine glands and the hormones that they secrete make up the endocrine system. The endocrine system works in parallel with the nervous system to maintain homeostasis by releasing hormones from various glands. Whereas the nervous system has a rapid effect on homeostasis, the endocrine system has slower, but longer lasting effects.
3. Sample answer:

System	Structure	Function
Nervous system	Neurons	Cellular communication through neurons; messages transmitted rapidly to precise locations in the body
Endocrine system	Endocrine glands	Cellular communication through chemical messengers called hormones; hormones secreted into bloodstream which transports the hormones throughout the body; typically has slower and longer acting effects, and affects a broader range of cells

4. See **Figure 9.3** on page 391 of the student textbook. Target cells are cells whose activity is affected by a particular hormone. Target cells have receptors for the hormone, and the hormone combines with the receptor as a key fits a lock.
5. The adrenal medulla gland produces epinephrine and norepinephrine. These hormones regulate a short-term stress (fight-or-flight) response. These hormones trigger an increase in breathing rate, heart rate, blood pressure, blood flow to the heart and muscles, and the conversion of glycogen to glucose in the liver.
6. This situation would [illegible] from the endocrine system that would then stimulate the sympathetic nervous system

(Student textbook page 399)

7. The pituitary gland is controlled by the hypothalamus.
8. Answers should include:
   - Neurosecretory cells in the hypothalamus produce releasing and release-inhibiting hormones.
   - These hormones are secreted into a portal system which carries these hormones from the hypothalamus to anterior pituitary.
   - If a stimulating hormone is released, the anterior pituitary would produce and secrete the hormone into the bloodstream.
   - If an inhibiting hormone is produced, the anterior pituitary stops the production and secretion of the hormone.

9. Venn diagrams should show:

Posterior Pituitary Only	Posterior and Anterior Pituitary	Anterior Pituitary Only
- part of nervous system - does not produce hormones; stores and releases hormones ADH and oxytocin - ADH and oxytocin are produced in hypothalamus and transferred to posterior pituitary by neuronal axons	- pituitary gland sits in a bony cavity attached to the hypothalamus - controlled by the hypothalamus via releasing hormones and neurons that run through the connecting stalk - hypothalamus and pituitary gland control many physiological process that maintain homeostasis	- is a true hormone-synthesizing gland - cells produce and release six major hormones: hGH, PRL, TSH, ACTH, FSH, and LH - series of blood vessels (portal system) carries releasing hormones from hypothalamus to the anterior pituitary, and these hormones either stimulate or inhibit the release of hormones from this gland

10. A tropic hormone, since it targets an endocrine gland and stimulates it to release other hormones.
11. Sample answer:
    - increases protein synthesis
    - increases metabolic breakdown and release of fats stored in adipose (fat) tissue
    - increases cell division and growth, especially the growth of cartilage, bone, and muscle (specifically, muscles, connective tissues, and plates at the end of the long bones)
12. Since hGH stimulates the growth of muscles, connective tissue, and the growth plates at the ends of the long bones, which causes elongation of these bones, a lower than normal secretion of hGH during childhood results in these bones not growing as long as normal. As a result, the individual is shorter than normal.

(Student textbook page 402)

13. A butterfly-shaped gland located below the larynx in the neck, which produces thyroxine (T4) hormone to help regulate metabolism and growth.
14. Symptoms include: anxiety, insomnia, heat intolerance, an irregular heartbeat, and weight loss.
15. Sample answer: hypothalamus secretes releasing hormone [arrow] anterior pituitary stimulated [arrow] TSH released into bloodstream [arrow] TSH travels to thyroid gland [arrow] thyroxine released [arrow] cellular respiration stimulated in target cells [arrow] high levels of thyroxine signal shut off of pituitary and hypothalamus releases

    16. Extremely low quantities of thyroxin result in TSH release not being shut off, which results in relentless stimulation of the thyroid gland by TSH, which causes a goitre (enlarged thyroid gland).
17. Venn diagrams should show:
    - Cretinism Only—thyroid does not develop properly in childhood; thyroid produces low quantities of thyroxine (hypothyroidism); individuals are stocky and shorter than average, and have mental developmental delays if not treated with hormone injections early in life
    - Both Cretinism and Graves' Disease—thyroid gland involved
    - Graves' Disease Only—a severe state of hyperthyroidism caused by immune system attacking thyroid; muscles around the eyes swell, making them protrude and interfering with vision
18. A lack of iodine makes it impossible for the body to produce thyroxine, which signals the anterior pituitary gland to stop the secretion of TSH. The endless stimulation of the thyroid gland by TSH causes a goitre (an enlargement of the thyroid gland). This is a visible swelling in the neck accompanied by difficulty breathing and/or swallowing, and coughing.

(Student textbook page 407)

19. A pair of independent hormone-producing organs located on top of the kidneys, each composed of an outer cortex and an inner medulla.
20. They are both formed from nervous system tissue.
21. These hormones trigger an increase in breathing rate, heart rate, blood pressure, and blood flow to the heart and muscles, as well as the conversion of glycogen to glucose in the liver. This moves more oxygen-rich blood to the brain faster and to the muscles needed for fighting or fleeing. Epinephrine causes a rapid release of glucose and fatty acids into the bloodstream, raising blood glucose levels providing the fuel required to make ATP. Senses become keener, memory sharpens, and sensitive to pain lessens.
22. Sample answer:

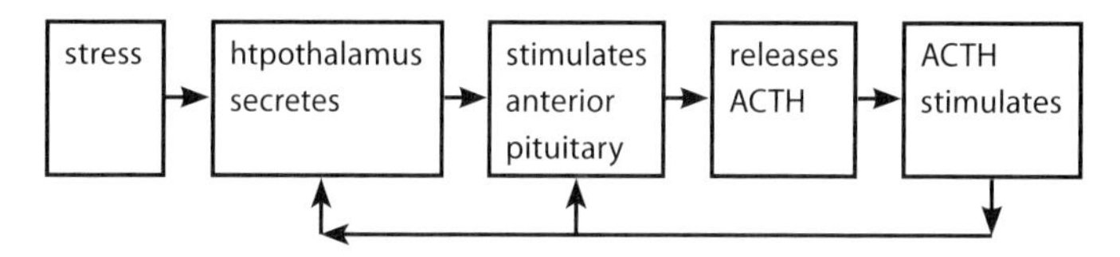

23. Injected epinephrine will open up the air passages to permit breathing.
24. Cortisol inhibits the regeneration of connective tissue such as tendons and ligaments. Too much cortisol would, in effect, weaken the joint, making it more susceptible to injury in the future.

(Student textbook page 410)

25. Beta cells, which secrete insulin, resulting in decreased blood glucose levels, and alpha cells, which secrete glucagon, increasing blood glucose levels.
26. To maintain homeostasis, insulin is released from the pancreas into the blood when blood glucose levels go above the normal range. Insulin decreases blood sugar levels. If blood glucose levels go below normal, the pancreas releases glucagon, which raises the blood glucose levels.
27. Diabetes mellitus is a serious, chronic condition resulting from inadequate production of or response to insulin. Diabetes mellitus results in blood glucose levels rising sharply after meals and remaining at significantly elevated levels.
28. Blindness, kidney failure, nerve damage, and gangrene in the limbs. Diabetes can also be fatal if not properly managed.

29. Diabetes mellitus ensures that blood glucose remains outside of the body's cells. Glucose, along with oxygen, is used by the cells to make energy. If the cells must break down fat or protein to use to make energy, the process is less efficient, and the cells will have less energy to carry out life-sustaining activities.
30. The condition of having levels of blood glucose that are too high.

(Student textbook page 416)

31. To unite a single reproductive cell (egg) from the female parent with a single reproductive cell (sperm) from the male parent and produce offspring.
32. One of several chemical compounds that control the development and function of the reproductive system or secondary sex characteristics.
33. Organizers should indicate that the scrotum contains the testes, which contain the seminiferous tubules (long coiled tubes) where sperm is produced. Sperm are transported to the epididymis, where they mature and become motile. Sperm then move to the storage duct known as the ductus deferens (or vas deferens) awaiting transport to the penis.
34. They are cells that lie between the seminiferous tubules and secrete testosterone.
35. The movement of semen results from a series of interactions between the sympathetic, parasympathetic, and somatic nervous systems. Sensory stimulation, arousal, and coordinated muscular contractions combine to trigger the release of semen from the penis.
36. Enzymes that are responsible for both sperm and hormone (testosterone) production operate most effectively at an optimal temperature. If temperature is elevated by even one degree, sperm and testosterone production are adversely affected.

(Student textbook page 419)

37. The Y chromosome carries a gene called testis-determining factor (TDF) gene that triggers the production of male sex hormones (androgens) which initiate the development of male sex organs and ducts. Without the Y chromosome (an XX combination) the sex organs remain female.
38. The hypothalamus increases production of gonadotropin releasing hormone (GnRH) which acts on the anterior pituitary, causing it to release two different sex hormones: follicle-stimulating hormone (FSH) and luteinizing hormone (LH). In males, these hormones cause the testes to begin producing sperm and to release testosterone. Testosterone acts on various tissues to complete the development of sex organs and sexual characteristics.
39. Inhibin acts on the anterior pituitary to inhibit the production of FSH, creating a negative feedback loop. As the level of FSH drops, the testes release less inhibin. A decrease in the level of inhibin causes the anterior to release more FSH. This feedback loop keeps the level of sperm production relatively constant over time.
40. Andropause is a gradual decline in males' testosterone level beginning around the age of 40. Symptoms include fatigue, depression, loss of muscle and bone mass, and a drop in sperm production.
41. Answers should include:

Structure	Function
Ovaries	Produce eggs and sex hormones
Fimbriae	Sweep an egg into the oviduct
Oviducts (Fallopian tubules)	Conduct an egg from the ovary to the uterus; place where fertilization occurs
Uterus	Houses the developing fetus
Cervix	Opening to the uterus
Vagina	Receives the male penis during sexual intercourse; serves as the birth canal and as the exit for menstrual flow

42. Both a fertilized and an unfertilized ovum travel down the oviduct. Only a fertilized ovum implants itself into the endometrium (lining in the uterus) after leaving the oviduct. An unfertilized ovum exits the body through the cervix and vagina.

(Student textbook page 423)

43. Girls are born with a complete but immature set of reproductive organs. The basic hormones and hormonal processes of female puberty are similar to those of male puberty, beginning with increased production of GnRH which acts on the anterior pituitary to release FSH and LH.

    In girls, FSH and LH act on the ovaries to produce the female sex hormones estrogen and progesterone. These hormones stimulate the development of the female secondary sex characteristic and launch a reproductive cycle.
44. Both are female sex hormones produced in the ovary. Estrogen helps maintain sexual organs and secondary sexual characteristics. Progesterone is produced first by the corpus luteum of the ovary to prepare the uterus for the fertilized ovum, and later by the placenta to maintain pregnancy.
45. Diagrams should look like **Figure 9.27** on page 420 of the student textbook, without the numbered steps.

    Sample caption: A follicle matures by growing layers of follicular cells and a central fluid-filled vesicle. The vesicle contains the maturing ovum. At ovulation, the follicle ruptures and the ovum is released into the oviduct. The follicle develops into a corpus luteum. If pregnancy does not occur, the corpus luteum starts to degenerate after about 10 days.
46.
    - first day of menstruation, corpus luteum has degenerated and blood levels of sex hormones are low
    - new follicle begins to mature and release estrogen
    - after six days, estrogen levels are high enough to cause the endometrium to begin to thickening
    - after ovulation, the release of progesterone by the corpus luteum causes a more rapid thickening of the endometrium
    - if fertilization does not occur, the corpus luteum degenerates
    - levels of the sex hormones drop, the endometrium breaks down, and menstruation begins again

47. Human chorionic gonadotropin (hCG), because it is produced by the embryonic-membranes shortly after the embryo attaches to the uterine lining and builds up rapidly in the woman's body in the first few days of pregnancy.
48. Menopause is a period in a woman's life when a decrease in estrogen and progesterone results in an end of menstrual cycles. It typically occurs around age 50, but it can occur earlier or later. The age-related decrease in the number of functioning follicles leads to a gradual overall decline in the amount of estrogen and progesterone in the blood.

Answers to Caption Questions

Figure 9.3 (Student textbook page 391): Receptor proteins are specifically shaped to bind with a specific hormone, much like a key fits only into a specific lock.
Figure 9.5 (Student textbook page 394): Sensors in the body send signals to the hypothalamus, which secretes releasing hormones that target the pituitary gland. The pituitary gland releases tropic hormones that target various endocrine glands to release other hormones to help the body maintain equilibrium. Once equilibrium has been restored, the signals stop coming from the sensors, and the sequence of hormone release is inhibited.
Figure 9.8 (Student textbook page 398): hGH is considered to be a tropic hormone because it stimulates endocrine glands (liver) to release other hormones (growth factors).
Figure 9.20 (Student textbook page 409): Insulin is released when blood glucose levels rise above the normal range. Insulin decreases blood sugar levels. If blood glucose levels drop below normal, the pancreas releases glucagon, which raises the blood glucose levels.
Figure 9. 21 (Student textbook page 410): In persons with type 1 diabetes, antibodies produced by the immune system destroy beta cells in the pancreas
Figure 9.25 (Student textbook page 417): Negative

Answers to Chapter 9 Multiple Choice Review Questions
(Student textbook pages 435–6)

1. e	**2.** a	**3.** c	**4.** d	**5.** d
6. d	**7.** c	**8.** b	**9.** b	**10.** a
11. b				

Answers to Chapter 9 Self-Assessment Multiple Choice Review Questions
(Student textbook pages 440–1)

1. b	**2.** d	**3.** b	**4.** e	**5.** c
6. b	**7.** e	**8.** b	**9.** e	**10.** d

Chapter 10 Excretion and the Interaction of Systems

Answers to Learning Check Questions
(Student textbook page 446)

1. Respiratory system—excretes carbon dioxide and small amounts of gases, including water
   Skin—excretes water, salts, and some urea in perspiration
   Digestive system—excretes water, salts, lipids, and a variety of pigments and other cellular chemicals
   Excretory system—eliminates most metabolic wastes dissolved or suspended in a solution
2. Excretion reduces the level of toxins in the body that threaten health/ homeostasis.
3. The kidneys assist the endocrine system in hormone secretion. The kidneys secrete the hormones calcitriol (which aids the absorption of calcium) and erythropoietin (which stimulates the production of red blood cells). The kidneys also secrete renin, which leads to the secretion of aldosterone (a hormone) from the adrenal cortex.
4. kidneys → ureters → urinary bladder → urethra
5. It requires control at higher levels of the central nervous system, which develops later.
6. Sample answer: A network of blood vessels associate closely with over one million microscopic structures called nephrons, which are responsible for filtering various substances from blood, transforming it into urine and eliminating toxins from the body to maintain homeostasis.

(Student textbook page 453)

7. Sample answer: Glomerular filtration—moves water and solutes (except proteins) from blood plasma into the nephron
   Tubular reabsorption—removes useful substances such as sodium from the filtrate and returns them to the blood for reuse by body systems
   Tubular secretion—moves additional wastes and excess substances from the blood into the filtrate
   Water reabsorption—removes water from the filtrate in the nephron and returns it to the blood for reuse by body systems
8. Blood pressure and the size of the pores of the capillaries.
9. The glomerular filtrate contains small dissolved molecules in approximately the same concentration as is found in plasma. The large molecules (proteins) are too large to pass through the capillary into the filtrate. The filtrate is identical to blood plasma, minus the proteins and blood cells.
10. The energy-releasing power of ATP is used to drive the active transport of sodium ions and other solutes from the proximal tubule back into the blood.
11. Diagrams should resemble **Figure 10.7** on page 451 of the student textbook. Summary should include that nutrients are actively reabsorbed, negatively charge ions are passively reabsorbed by electrical attraction, and water is reabsorbed by osmosis.
12. The function of the loop of Henle is to reabsorb water and ions from the filtrate.
    - in the descending limb of the loop of Henle, water diffuses from the filtrate to the capillaries by osmosis, increasing the concentration of sodium ions inside the tubule
    - filtrate continues around the bend of the loop of Henle and into the ascending limb; solutes, including sodium ions, diffuse from the filtrate along their concentration gradient and pass into nearby blood vessels
    - at the thick-walled portion of the ascending limb of the loop of Henle, sodium ions are moved out of the filtrate by active transport; at this point, two-third of the sodium and water from the filtrate has been reabsorbed

(Student textbook page 459)

13. They send impulses to release antidiuretic hormone (ADH) from the posterior pituitary gland.

14. Answers should include: osmotic pressure too low → osmoreceptors in hypothalamus → pituitary gland → less ADH released into blood stream → decreased reabsorption of water in kidney tubules and collecting ducts (more water in urine) → osmotic pressure of body fluids increases

15. $OH^- + H_2CO_3 \rightarrow HCO_3^- + H_2O$

When hydroxide ions are added to the blood, they combine with carbonic acid to produce bicarbonate ions and water.

16. If the blood is too acidic, H^+ ions are excreted into the tubules and HCO_3^- ions are reabsorbed. If the blood is too basic, H^+ ions are not excreted and HCO_3^- ions are not reabsorbed.

17. Caffeine is a diuretic and stimulates urine production partly by decreasing ADH release, which decreases the permeability of the tubules and collecting duct to water. As a result, caffeine drinkers may experience an increase in the volume of urine they produce. By contrast, those drinking decaffeinated coffee will reabsorb more water and will produce less urine.

18. Sample answer: I disagree. The relative concentration of water and solutes fluctuate within narrow limits. For example, if a person drinks too much water, the kidneys allow more water to pass into the urine. If water is scarce, the kidneys conserve water by producing more concentrated urine.

(Student textbook page 463)

19. Sample answer: reduced oxygen-carrying capacity of blood → sensors in kidneys → kidneys release erythropoietin → stimulates production of red blood cells in bone marrow [arrow] increases oxygen-carrying capacity of the blood

20. If the level of calcium in the blood falls below normal parathyroid hormone (PTH) is released by the parathyroid gland. PTH stimulates the kidneys to release another hormone (calcitriol, the active form of vitamin D) which promotes calcium (Ca^{2+}) absorption from the digestive tract.

21. Medications can be used to help break down the crystals. If the stones are less than 20 mm in diameter, ultrasound shock waves can be used to disintegrate the crystalline structure so that they can be passed naturally in the urine. Small stones can pass through the urinary tract on their own and do not require treatment. Surgery may be needed to remove larger stones.

22. Answers should include:

Hemodialysis:

- artificial membrane in external device
- connected to artery and vein in arm

Similarities:

- remove wastes and excess fluid from blood when the kidney function is lost due to renal failure
- based on the principle of diffusion of dissolved substances along a concentration gradient through a semipermeable membrane

Peritoneal dialysis:

- uses peritoneum (intestinal lining) as dialysis membrane

23. High blood pressure reduces the kidneys' ability to remove fluids and waste products from the blood, which damages the kidneys.

24. If the kidneys do not operate sufficiently, wastes will not be removed from the blood.

Answers to Caption Questions

Figure 10.2 (Student textbook page 445): The renal artery transports blood to the kidney for filtration of wastes. The filtered blood returns to the circulatory system through the renal veins, which connect to the inferior vena cava.

Figure 10.4 (Student textbook page 447): Bowman's capsule → proximal tubule → loop of Henle (descending loop then ascending loop) → distal tubule → collecting duct.

Figure 10.6 (Student textbook page 450): In the proximal tubule: active reabsorption of all nutrients and positively charged; passive reabsorption of water by osmosis, and of negatively charged ions by electrical attraction to positively charged ions; and active secretion of hydrogen ions. In the distal tubule: active reabsorption of sodium ions; passive reabsorption of negatively charged ions; passive reabsorption of water by osmosis; active tubular secretion of hydrogen ions from the blood into the tubule; and passive secretion of potassium ions by electrical attraction to chloride ions.

Figure 10.9 (Student textbook page 453): Passive reabsorption by osmosis.

Figure 10.11 (Student textbook page 457): A negative feedback system controls the release of ADH, since it is the dilution level of the blood that signals osmoreceptors to stop releasing ADH. As long as the blood is not too dilute, ADH will continue to be released.

Figure 10.13 (Student textbook page 461): So that delicate renal tissues will not be damaged.

Answers to Chapter 10 Review Multiple Choice Questions

(Student textbook page 471)

1. a	**2.** d	**3.** a	**4.** b
5. a	**6.** d	**7.** a	**8.** d

Answers to Chapter 10 Self-Assessment Multiple Choice Questions

(Student textbook page 476)

1. b	**2.** c	**3.** e	**4.** e	**5.** d
6. b	**7.** c	**8.** c	**9.** c	**10.** c

Answers to Unit 4 Review Multiple Choice Questions

(Student textbook pages 481–2)

1. a	**2.** d	**3.** c	**4.** c	**5.** a
6. e	**7.** b	**8.** d	**9.** a	**10.** e
11. b	**12.** a	**13.** c	**14.** d	

Answers to Unit 4 Self-Assessment Multiple Choice Questions

(Student textbook page 488)

1. b	**2.** d	**3.** b	**4.** e	**5.** a
6. e	**7.** a	**8.** b	**9.** d	**10.** a

Unit 5: Population Dynamics

Chapter 11 Describing Populations and Their Growth

Answers to Learning Check Questions

(Student textbook page 505)

1. It is impractical to count individuals in ever-changing populations because of time, money, and human resources; therefore, ecologists estimate populations.
2. Biologists sample subsets of populations using transects, quadrats, and mark-recapture techniques, and then estimate (or extrapolate) to the larger population to estimate the size.
3. In the mark-recapture method, animals are caught, tagged, and released. Then, another set is caught at a later date. The proportion in that set of previously-tagged animals to unmarked animals is used to estimate population size.
4. Aspens have clumped distribution because they can reproduce asexually, sprouting new plants from shoots that grow off their root system. In addition, the trees tend to cluster in areas where soil, moisture, and temperature conditions are optimal for growth.
5. Populations often exhibit a clumped distribution pattern when resources are unevenly distributed. When populations show a uniform distribution pattern, resources tend to be evenly distributed but scarce. Resources are usually plentiful and evenly distributed when population distribution is random.
6. The interactions are positive because they tend to benefit the population. For many species in clumped distributions, interactions give protection from predators, promote hunting efficiency, and allow collective raising of the young.

(Student textbook page 513)

7. Birth, death, immigration, and emigration;
   $\alpha N = (B + I) - (D + E)$
8. In most populations, immigration and emigration are about equal.
9. In order to make sound decisions about managing and protecting species and the environment
10. The growth rate is the change in population size over the specified time period (3 years):

    Change in population size: 3600 – 1500 = 2100

    Growth rate = 2100/3 = 700

    The growth rate is 700 geese per year.
11. Population A shows a greater increase than population B because population A was greater at the start. To make a more meaningful comparison, use the per capita growth rate.
12. A negative per capita growth rate means that the size of the population is declining over time. These changes are likely because deaths outnumber births or emigration exceeds immigration, or both.

(Student textbook page 529)

13. Abiotic factors generally cause a sharp drop in population growth before the habitat reaches carrying capacity. On the other hand, biotic factors slow the growth of the population when it becomes dense.
14. Intraspecific competition for resources occurs among members of the same population, while interspecific competition occurs between two or more populations.
15. The more successful predators thrive, so the traits that make them successful will be passed on. Prey that survive will also pass on any trails that helped them survive. Traits that made prey more susceptible to predation, or that made predators less successful, will be eliminated from the population.
16. As predators feed on prey, the predator population increases and the prey population declines. This leads to competition among predators for food, causing the predator population to decline. With fewer predators, the prey population then increases, resulting in a cycle.
17. There is generally a "ripple effect" in which other populations grow or decline because of reduced competition, predation, or resources.
18. Camouflage protects the butterfly from predators since it appears to be a dead leaf.

Answers to Caption Questions

Figure 11.1 (Student textbook page 498): It is difficult to count every individual in a very large population such as this forest. Counting each individual is limited by human ability, time and money.

Figure 11.5 (Student textbook page 501): The bald eagle benefits from a uniform distribution pattern because it maintains sole access (or nearly) to the limited resources in its territory. The moose benefits from a random distribution pattern because it can move freely to use the abundant resources without the need to defend or share resources. The aspen benefits from a clumped distribution pattern because its population grows where conditions are optimal for survival.

Figure 11.13 (Student textbook page 511): The species may become extinct.

Figure 11.14 (Student textbook page 512): The population size of A was five times the population of B at the start. Since there were more individuals to reproduce in population A its doubling resulted in faster growth.

Figure 11.19 (Student textbook page 517): Spring hatching means that young will experience milder weather and abundant food.

Figure 11.22 (Student textbook page 520): While ice covered everything, access to some food and habitat was eliminated. Damage to trees reduced habitat and resources.

Figure 11.24 (Student textbook page 521): Parasites pass from one host to another, a process that is supported by a high density of hosts.

Figure 11.28 (Student textbook page 523): Organisms with adaptations suited to different ecological niches do not compete with each other even if they are living within the same small area. This assists in each organism satisfying its needs for survival and increases the chance that each will survive.

Figure 11.32 (Student textbook page 526): The survival of the hares would be affected by abiotic factors such as extreme weather, the availability of abundant clean water, or by biotic factors such as predators.

Answers to Chapter 11 Review Multiple Choice Questions
(Student textbook pages 541–2)

1. b **2.** d **3.** d **4.** a **5.** c
6. d **7.** e **8.** e **9.** c **10.** a
11. a **12.** c **13.** e **14.** d

Answers to Chapter 11 Self-Assessment Multiple Choice Questions
(Student textbook pages 546–7)

1. d **2.** c **3.** a **4.** b **5.** b
6. b **7.** d **8.** a **9.** b **10.** e

Chapter 12 Human Populations

Answers to Learning Check Questions
(Student textbook page 554)

1. Throughout most of history, human population growth was fairly stable, showing very little increase. There is a decline around the 1300s due to widespread disease. Since that time, the population has grown exponentially.
2. The amount of time it takes for the population to double.
3. Food supply was increased through agricultural technology and domesticating animals. Predators were staved off with weapons. Medicines controlled diseases. Shelter provided protection from weather and improved storage for food to prolong supply.
4. Sample answer: I would expect Canada's population pyramid to have a rectangular shape because the birth rate is only slightly higher than the death rate, which means that the population grows slowly. The population in each age category tends to be about the same.
5. The population would be shrinking (more deaths than births). Also, a large number of individuals would be past their reproductive years and few would be about to enter their reproductive years. In this type of population, there will be a significant burden on the proportionally small number of working people to meet the obligations of the government and to support number of older people who have retired.
6. The human population was much greater in the 1900s than in the 1300s when the bubonic plague occurred. Because the change in the population in the 1900s was not significant when compared to the size of the population at that time, the decline does not show up on the graph.

(Student textbook page 565)

7. Biomagnification is the increased concentration of a substance (e.g., pesticide, other toxin) at each trophic level of a food chain.
8. The burning of coal releases mercury into the atmosphere. The mercury falls to Earth through deposition (such as precipitation) and either settles directly in lakes or streams or is washed into them after being deposited on land. Within the waterways, the actions of anaerobic microorganisms transform the mercury into methylmercury.
9. Deforestation in impoverished countries is often caused by people trying to survive by selling forest products or clearing land to farm.
10. Sample answer: Improved agricultural practices that discourage deforestation and enhanced forest management practices to promote reforestation.
11. Coal is an inexpensive and readily available source of energy. For many people, including some government leaders and policy makers, affordable energy costs are more important than environmental protection and outweigh the harm that burning coal does to the environment.
12. Plastic items are harmful to marine life because they leach toxic substances and are often ingested, blocking the digestive system. Plastics are still being released into the oceans and will be a problem for a long time because they do not decompose quickly.

Answers to Caption Questions

Figure 12.3 (Student textbook page 552): Differences in the availability of clean water and healthy food. Differences in access to modern medical technology to treat disease and injuries would also affect the survival of members of a population, as would economic disparities that could limit access to suitable shelter.
Figure 12.5 (Student textbook page 555): If humans continue to exceed the biocapacity of Earth the resources will not be sufficient to support the population. Earth's capacity to supply resources will likely be degraded as available land is used for other uses or water is polluted and becomes unusable.

Answers to Chapter 12 Review Multiple Choice Questions
(Student textbook pages 580–1)

1. c **2.** [illegible] **3.** [illegible] **4.** b **5.** a
6. [illegible] **7.** [illegible] **8.** b **9.** c **10.** [illegible]
11. c **12.** e **13.** c **14.** e **15.** b

Answers to Chapter 12 Self-Assessment Multiple Choice Questions
(Student textbook page 584)

1. e **2.** d **3.** e **4.** a **5.** a
6. b **7.** a **8.** b **9.** a **10.** c

Answers to Unit 5 Review Multiple Choice Questions
(Student textbook pages 589–93)

1. a **2.** e **3.** b **4.** d **5.** c
6. a **7.** a **8.** d **9.** b **10.** d
11. e **12.** e **13.** c **14.** c **15.** c
16. e

Answers to Unit 5 Self-Assessment Multiple Choice Questions
(Student textbook page 594)

1. a **2.** d **3.** d **4.** e **5.** d
6. b **7.** e **8.** e **9.** b **10.** a

Glossary

How to Use This Glossary

This Glossary provides the definitions of the key terms that are shown in **boldface** type in the text. Definitions for terms that are *italicized* within the text are included as well. Each glossary entry also shows the number(s) of the sections where you can find the term in its original context.

1.1 = Chapter 1, Section 1
U2P = Unit 2 Project
C3R = Chapter 3 Review
App. A = Appendix A
App. B = Appendix B

A pronunciation guide, using the key below, appears in square brackets after selected words.

a = mask, back
ae = same, day
ah = car, farther
aw = dawn, hot
e = met, less
ee = leaf, clean
i = simple, this
ih = idea, life
oh = home, loan
oo = food, boot
u = wonder, Sun
uh = taken, travel
uhr = insert, turn
s = sit
z = zoo
zh = equation

Emphasis is placed on the syllable(s) in CAPITAL letters.

3′ poly-A tail a series of A nucleotides that are added to the 3′ end of mRNA (6.2)

5′ cap a modified form of a G nucleotide that is added to the 5′ end of mRNA (6.2)

abiotic the term used to describe interactions between organisms and their non-living environment (11.3)

absorbance spectrum a graph that shows the relative amounts of light of different wavelengths that a compound absorbs (4.1)

acceptor stem the 3′ end of a tRNA molecule that is the site of attachment for a particular amino acid, based on the anticodon (6.3)

acetylcholine the primary neurotransmitter of both the somatic nervous system and the parasympathetic nervous system (8.2)

acid a substance that releases hydrogen ions, H^+, when dissolved in water (1.3)

action potential in an axon, the change in charge that occurs when the gates of the K^+ channels close and the gates of the Na^+ channels open after a wave of depolarization is triggered (8.2)

action spectrum a graph that demonstrates the relative effectiveness of different wavelengths of light for promoting photosynthesis by showing the rate at which oxygen is produced by photosynthesis (4.1)

activation energy the energy required to initiate a chemical reaction (1.3)

activator a molecule that binds to the allosteric site of an enzyme and that keeps an enzyme active or causes an increase in the activity of that enzyme (1.3); a protein that binds to a particular DNA sequence to regulate transcription; it increases the rate of transcription of a gene or genes (6.4)

active site the site on an enzyme where the substrate binds andwhere the chemical reaction that is catalyzed by the enzyme takes place (1.3)

adrenal cortex the outer layer of the adrenal glands that produces glucocorticoids and mineralocorticoids, hormones that regulate the long-term stress response; also secretes a small amount of gonadocorticoids, female and male sex hormones that supplement the hormones produced by the gonads (testes and ovaries) (9.3)

adrenal gland one of a pair of organs located on top of the kidneys; composed of two layers: an outer cortex and an inner medulla; each layer produces different hormones and functions as an independent organ (9.3)

adrenal medulla the inner layer of the adrenal glands that produces epinephrine and norepinephrine, hormones that regulate the short-term stress response (9.3)

adrenocorticotropic hormone (ACTH) a hormone synthesized by the anterior pituitary gland to target the adrenal cortex and regulate the production of glucocorticoids (9.3)

aerobic respiration catabolic pathways (a series of chemical reactions) that require oxygen (3.2)

aldosterone a type of mineralocorticoid hormone secreted by the adrenal cortex; stimulates the distal tubule and the collecting duct of the kidneys to increase the absorption of sodium into the bloodstream, which is followed by the passive absorption of water and chloride (9.3)

alien species a species that is accidentally or deliberately introduced to a new location, usually as a result of human activity (Unit 5 Preparation)

all-or-none the principle that governs the response of an axon to a stimulus; if a neuron is stimulated sufficiently, an impulse will travel the length of the axon, but if the stimulus is not sufficient, no impulse will travel down the axon (8.2)

allosteric regulation the regulation of enzyme activity by means of activators and inhibitors binding to allosteric sites (1.3)

allosteric site the site on an enzyme that is not the active site, where other molecules can interact with and regulate the activity of the enzyme by causing a change in the conformation of the enzyme (1.3)

alpha cell a cell of the pancreas that secretes glucagon to increase the level of blood glucose (9.3)

alternative splicing a process that allows one gene to code for more than one protein; as a result, certain cell types are able to produce forms of a protein that are specific for that cell (6.2)

amino acid an organic molecule composed of a central carbon atom bonded to a hydrogen atom, an amino group, a carboxyl group, and a variable R group (1.2)

aminoacyl-tRNA synthetase the enzyme responsible for attaching an amino acid to a tRNA (6.3)

ampicillin an antibiotic, introduced by a British company in 1961, that is used extensively to treat bacterial infections (7.1)

anabolic steroid a synthetic compound that mimics male sex hormones; used to build muscle mass in people who have cancer and AIDS, but also frequently misused by athletes (1.2)

anabolism a metabolic process that uses energy to synthesize a large molecule from smaller molecules (3.1)

anaerobic respiration a form of cellular respiration that occurs in the absence of oxygen; molecules such as carbon dioxide, nitrate, and sulfate are the final electron acceptors (3.3)

andropause in men, a gradual decline in their testosterone level beginning around age 40; symptoms include fatigue, depression, loss of muscle and bone mass, and a drop in sperm production (9.4)

anion the negatively ion that results when an atom or group of atoms gains electrons (1.1)

anneal base pair; in response to lower temperatures, two nucleotide primers anneal, or base pair, with the 3′ ends of the single-stranded DNA to be amplified in a polymerase chain reaction (7.1)

anoxic a term describing an oxygen-free environment (3.3)

antenna complex the numerous pigment molecules that gather the light energy in chloroplasts during photosynthesis (4.1)

anterior pituitary the anterior lobe of the pituitary gland, an endocrine gland that synthesizes and secretes six major hormones: human growth hormone (hGH), prolactin (PRL), thyroid-stimulating hormone (TSH), adrenocorticotropic hormone (ACTH), follicle-stimulating hormone (FSH), and luteinizing hormone (LH) (9.2)

anticodon loop a triplet of bases positioned at one end of a tRNA, which recognizes and base pairs with a codon on mRNA during protein synthesis (6.3)

antidiuresis decreased excretion of urine (10.3)

antidiuretic hormone (ADH) a hormone regulated by the hyphothalamus and released by the pituitary gland that increases the permeability of the distal tubule and the collecting duct in the nephrons of the kidneys, allowing more water to be reabsorbed into the blood from the filtrate (10.3)

antiparallel refers to the directionality of the two strands in a DNA molecule; the strands run in opposite directions, with each end of a DNA molecule containing the 3′ end of one strand and the 5′ end of the other strand (5.1)

antiretroviral agents antiviral drugs that have been developed to inhibit the reverse transcriptase and protease enzymes used by HIV, which can delay the onset of AIDS (Unit 1 Project)

antisense strand (template strand) the DNA strand that is used as a template for RNA synthesis or DNA replication (6.2)

atomic mass the sum of an atom's protons and neutrons (1.1)

ATP (adenosine triphosphate) a nucleotide with three phosphate groups; the breakdown of ATP into ADP and Pi makes energy available for energy-requiring processes in cells (3.1)

autonomic system in vertebrates, the division of the peripheral nervous system that controls involuntary glandular secretions and the functions of smooth and cardiac muscle (8.4)

autoradiography a method of attaching a radioactive tag to dideoxynucleotides in order to visualize the DNA by exposing the gel to X-ray film after electrophoresis (7.1)

available biocapacity Earth's carrying capacity for the human population

axon the long, cylindrical extension of a neuron's cell body that can range from 1 mm to 1 m in length; transmits impulses away from the cell body along its length to the next neuron (8.2)

b6-f complex a protein complex that uses energy released by the photosynthetic electron transport system to pump hydrogen ions from the stroma, across the thylakoid membrane, and into the thylakoid space; this process occurs during the light-dependent reactions of photosynthesis (4.1)

background extinction the death of all the individuals of a species that occurs over long periods of time as ecosystems gradually change (Unit 5 Preparation)

base a substance that releases hydroxide ions, OH^-, when dissolved in water (1.3)

Batesian mimicry q type of mimicry in which a species looks like another species that has an effective defence strategy (11.3)

beta-carotene a member of a very large class of pigments called carotenoids, which absorb blue and green light so they are yellow, orange, and red in colour; beta-carotene is responsible for the orange colour of carrots (4.1)

beta cell a cell of the pancreas that secretes insulin to decrease the level of blood glucose (9.3)

biochemistry the study of the activity and properties of biologically important molecules (1.1)

biodiversity the variety of species on Earth and their range of behavoural, ecological, physiological, and other adaptations, the genetic diversity that exists within each of these species, and the diversity of ecosystems (ecosystem diversity) of which these species are a part (12.2)

biolistic method (gene-gun method) a method used for producing transgenic plants that involves bombarding plant cells with particles coated in DNA that become integrated into the plant genome (7.2)

biomagnification an increase in the concentration of a substance, such as methylmercury or DDT, that occurs in a food chain and that is not broken down by environmental processes (12.2)

bioremediation the use of micro-organisms or other living cells for environmental clean-up (7.2)

biotechnology the technological use of an organism, or a product from an organism, to solve a problem in a way that benefits humans (7.2)

biotic living; refers to interactions among organisms (11.3)

biotic interactions relationships that exist between organisms in ecosystems, including predation, competition, and symbiosis (mutualism and parasitism) (Unit 5 Preparation)

biotic potential the highest possible per capita growth rate for a population (11.2)

blood-brain barrier a protective barrier formed by glial cells and blood vessels that separates the blood from the central nervous system; this barrier selectively controls the entrance of substances into the brain from the blood (8.3)

bond energy the energy required to break (or form) a chemical bond (3.1)

bottleneck effect changes in gene distribution that result from a rapid decrease in population size (Unit 5 Preparation)

Bowman's capsule in the kidney, the cap-like formation at the top of each nephron that serves as a filtration structure; surrounds the glomerulus (10.1)

buffer a substance that minimizes changes in pH by donating or accepting hydrogen ions as needed (1.3)

bycatch aquatic organisms that are caught unintentionally by fishing gear or nets and that are often discarded as waste (12.2)

C3 photosynthesis the process of converting carbon dioxide to glyceraldehyde-3-phosphate using only the Calvin cycle; involves production of a three-carbon intermediate (PGA) (4.2)

calcitonin a hormone that regulates calcium levels in the blood (9.2)

Calvin cycle in photosynthesis, the reactions that convert carbon dioxide to the three-carbon organic molecule glyceraldehyde-3-phosphate (G3P); can occur in the absence or presence of light; also called the dark reactions and the Calvin-Benson cycle (4.2)

carbohydrate a biological macromolecule that contains carbon, hydrogen, and oxygen in a 1:2:1 ratio (1.2)

carbon dioxide fixation the first stage in the synthesis of carbohydrates through the Calvin cycle; a carbon atom in carbon dioxide is chemically bonded to a pre-existing 5-carbon compound (ribulous bisphosphate or RuBP) in the stroma of a chloroplast (4.2)

carotenoids a very large class of pigments of which beta-carotene is a member (4.1)

carrying capacity the maximum population size that a habitat can sustain over an extended period of time (11.2)

catabolism a metabolic process that involves breaking down a molecule into smaller molecules, usually to release energy (3.1)

catalase an enzyme within peroxisomes that breaks down hydrogen peroxide to water and oxygen gas (2.1)

catalyst a substance that speeds up the rate of a chemical reaction by lowering the activation energy for the reaction and thatis not consumed in the reaction (1.3)

catalytic cycle the process in which substrates form an enzyme-substrate complex and are then released from the complex as a product or products, freeing the enzyme for further reactions (1.3)

cation a positively charged ion that results when an atom or group of atoms loses electrons (1.1)

cell body the main portion of a neuron that contains a nucleus and from which dendrites and an axon extend; the site of the cell's metabolic reactions (8.2)

cell cycle the life cycle of a cell (5.2)

cell membrane a thin structure, composed of various macromolecules, that separates the inside of the cell from the extracellular environment; controls the flow of substance into and out of the cell (2.1)

cell wall a rigid layer surrounding plant, algae, fungal, bacterial, and some archaea cells, composed of proteins and/or carbohydrates; gives the cell its shape and structural support (2.1)

central dogma the theory that states that genetic information flows from DNA to RNA to protein (6.1)

central nervous system a network of nerves that includes the brain and spinal cord; integrates and processes information sent by nerves

cerebellum a walnut-shaped structure located below and largely behind the cerebrum; involved in the unconscious coordination of posture, reflexes, and body movements, as well as fine, voluntary motor skills (8.3)

cerebral cortex a thin outer covering of grey matter that covers each cerebral hemisphere of the brain; responsible for language, memory, personality, conscious thought, and other activities that are associated with thinking and feeling (8.3)

cerebrospinal fluid a dense, clear liquid derived from blood plasma, found in the ventricles of the brain, in the central canal of the spinal cord, and in association with the meninges; transports hormones, white blood cells, and nutrients across the blood-brain barrier to the cells of the brain and spinal cord; acts as a shock absorber to cushion the brain (8.3)

cerebrum the part of the brain that is divided into right and left cerebral hemispheres, which contain the centres for intellect, learning and memory, consciousness, and language; it interprets and controls responses to sensory information (8.3)

cgr per capita growth rate (11.2)

Chargaff's rule a rule stating that in DNA, the percent composition of adenine is the same as thymine, and the percent composition of cytosine is the same as guanine (5.1)

chemical energy the potential energy stored in the arrangement of the bonds in a compound (3.1)

chemical mutagen a molecule that can enter the nucleus of a cell and induce mutations by reacting chemically with the DNA; may cause a nucleotide substitution or a frameshift mutation (6.3)

chemiosmosis a process that uses energy in a hydrogen ion gradient across the inner mitochondrial membrane to drive phosphorylation of ADP (3.2)

chloroplast an organelle in the cells of photosynthetic organisms, in which light energy from the Sun is captured and stored in the form of high-energy organic molecules such as glucose (2.1, 4.1)

chromatin the non-condensed form of genetic material that predominates for most of the eukaryotic cell cycle (5.1); chromatin represents the unfolded state of chromosomes

chromosome a strand-like complex of nucleic acids and protein tightly bound together; chromosomes contain the hereditary units known as genes (2.1)

chromosome mutation a mutation that involves changes in chromosomes, and may involve many genes (6.3)

cilia numerous short appendages protruding from a cell; each cilium is composed of a microtubule-based shaft covered in an extension of the cell membrane; allow a cell to propel itself or to propel substances across the surface of the cell

cis face the entry face through which proteins and lipids enter the Golgi apparatus (2.1)

citric acid cycle the cycle that results in the breakdown of carbohydrates to carbon dioxide; also known as the Krebs cycle (3.1)

cloning a process that produces identical copies of genes, cells, or organisms (7.1)

clumped distribution the most common pattern of dispersion within a population, in which individuals are gathered in small groups (11.1)

CO2 assimilation the conversion of carbon dioxide to organic compounds (4.2)

codon in a gene, a sequence of three nucleotide bases that specifies a particular amino acid or a stop codon; codons function during translation (7.1)

coenzyme a non-protein, small, organic molecule that assists an enzyme (1.3)

co-evolution the process by which two or more species of organisms influence each other's evolutionary pathways (11.3)

cofactors inorganic metal ions, such as iron and zinc, that assist enzyme activity (1.3)

cohort a large group of individuals all born at the same time (11.1)

collecting duct in the kidney, the large pipe-like channel arising from the tubule connected to the Bowman's capsule in the nephron; functions as a water-conservation device, reabsorbing water from the filtrate in the nephron (10.1)

commensalism a symbiotic relationship in which one partner benefits and the other partner neither benefits nor is harmed. (11.3)

competition a biotic interaction in which two or more organisms compete for the same resource in the same location at the same time (Unit 5 Preparation)

competitive exclusion principle the principle stating that two species with overlapping niches cannot coexist (11.3)

competitive inhibitor a molecule that binds to the active site of an enzyme and inhibits the ability of the substrate to bind (1.3)

complementary base pairing in DNA, the interaction of bases of nucleotides on opposite strands through hydrogen bond formation (5.1)

concentration the amount of a substance that is dissolved in a solvent (Unit 1 Preparation)

condensation reaction (dehydration reaction) a chemical reaction that results in the formation of a covalent bond between two molecules with the production of a water molecule (1.3)

conservative model a model proposing that DNA replication involves the formation of two new daughter strands from the parent templates, with the two new strands joining to create a new double helix (5.2)

constitutive gene a gene that is constantly being expressed; does not undergo regulation of expression (6.4)

control centre a body structure that sets the range of values within which a variable should be maintained, receives information from the sensor, and sends signals to effectors when needed (8.1)

corpus callosum a bundle of white matter that joins the two cerebral hemispheres of the cerebrum of the brain; sends messages from one cerebral hemisphere to the other, telling each half of the brain what the other half is doing (8.3)

cortisol a type of glucocorticoid hormone released by the adrenal cortex of the adrenal gland in a long-term stress response; triggers an increase in blood glucose levels and reduces inflammation (9.3)

coupled refers to two reactions that can occur as cells use ATP to drive endergonic reactions; if the cleaving of ATP's terminal high-energy bond releases more energy than the other reaction consumes, the two reactions can be coupled so that the energy released by the hydrolysis of ATP can be used to supply the endergonicreaction with energy, resulting in a net release of energy (3.1)

covalent bond the attraction between atoms that results from the sharing of electrons; forms when the electron shells of two non-metal atoms overlap so that valence electrons of each atom are shared between both atoms (1.1)

Crassulaceae CAM plants, which include succulent (water-storing) plants such as cacti and pineapple; use a biochemical pathway identical to the C4 plants, but the reactions take place in the same cell; CAM plants thrive in hot, arid desert conditions (4.2)

cristae folds of the inner membrane of mitochondria (2.1)

cyclic photophosphorylation a pattern of electron flow in the thylakoid membrane of a plant cell that is cyclic and that generates ATP alone (4.1)

cytoplasm the region of the cell that is contained within the cell membrane; includes the cytosol, the organelles, and other life-supporting materials (2.1)

cytoskeleton a network of protein fibres that extends throughout the cytosol providing structure, shape, support, and motility (2.1)

cytosol the fluid part of the cytoplasm in which molecules and ions are dissolved or suspended (2.1)

deforestation the cutting, clearing, or removal of rainforests or similar ecosystems so land can be used as pastureland or cropland (12.2)

deletion a missing region of a chromosome (6.3)

demography the study of statistics related to human populations, such as population size, density, distribution, movement, births, and deaths (12.1)

denaturation the unfolding of proteins; occurs when the normal bonding between R groups is disturbed (1.2)

denatured altered structurally by an external stress (7.1)

dendrites short, branching terminals that receive nerve impulses from other neurons or sensory receptors, and then relay the impulse to the cell body (8.2)

density-dependent factor a biotic interaction that varies in its effect on population growth, depending on the density of the populations involved (11.[illegible])

density-independent factor an abiotic factor, such as fire or flood, that affects population size independent of the population's density (11.3)

deoxyribose a sugar found in DNA nucleotides (1.2)

depolarization a process in which the inside of the cell becomes less negatively charged relative to the outside of the cell (8.2)

diabetes insipidus a condition in which ADH activity is insufficient, thereby causing a person to urinate excessively—perhaps as much as 4 L to 8 L per day; thirst is intense, but water is excreted more quickly than it is consumed, leading to severe dehydration and ion imbalances (10.3)

diabetes mellitus a serious chronic condition that results when the pancreas does not produce enough insulin or the body does not respond properly to insulin; levels of blood glucose tend to rise sharply after meals (hyperglycemia) and remain at significantly elevated levels (9.3)

dialysis a procedure that removes wastes and excess fluid from the blood when kidney function is lost due to renal failure (10.3)

dideoxy sequencing (dideoxy chain termination method) a method for determining the sequence of a DNA fragment using dideoxynucleotides, which cause termination of DNA synthesis during the procedure (7.1)

diploid refers to eukaryotes that contain two copies of each gene (5.1)

disaccharide a carbohydrate composed of two monosaccharides joined by a covalent bond (1.2)

dispersive model a model proposing that the parental DNA molecules were broken into fragments and that both strands of DNA in each of the daughter molecules were made up of an assortment of parental and new DNA (5.2)

distal tubule in the kidney, the tubular portion of the nephron that lies between the loop of Henle and the collecting duct; its main function is reabsorption of water and solutes, and secretion of various substances (10.2)

distribution pattern a pattern in which a population is distributed or spread out in an area; three types are uniform, random, and clumped (11.1)

diuresis increased excretion of urine (10.3)

DNA (deoxyribonucleic acid) a biological macromolecule composed of nucleotides containing the sugar deoxyribose (1.2)

DNA amplification the process of producing large quantities of DNA from a sample (7.1)

DNA fingerprinting (DNA profiling) a technology used to identify individuals by analyzing the DNA sequence of certain regions of their genome (7.1)

DNA ligase an enzyme that catalyzes the joining of Okazaki fragments when the lagging strand is synthesized during DNA replication (5.2); in recombinant DNA technology, an enzyme that seals the breaks in the DNA, forming covalent bonds between the two different fragments; the result is a stable, recombinant DNA molecule (7.1)

DNA polymerase I an enzyme that removes RNA primer and fills gaps between Okazaki fragments on the lagging strand with DNA nucleotides; proofreads newly synthesized DNA (5.2)

DNA polymerase II an enzyme that proofreads newly synthesized DNA (5.2)

DNA polymerase III an enzyme that adds nucleotides to the 3' end of a growing polynucleotide strand (5.2)

DNA replication the process in which two identical DNA molecules are produced from an original, parent DNA molecule (5.2)

DNA sequencing a method for determining the nucleotide sequence, base by base, of a fragment of DNA (7.1)

DNA supercoiling the formation of additional coils in the structure of DNA due to twisting forces on the molecule (5.1)

DNA transposition a process involving the movement of specific DNA sequences within and between chromosomes; can disrupt more extensive regions of genetic information (6.3)

doubling time the time it takes for a population to double in number (12.1)

ductus deferens a storage duct that connects the epididymis to the penis via the ejaculatory duct (9.4)

duplication a section of a chromosome that occurs two or more times (6.3)

ecological footprint the amount of productive land that is required for each person in a defined area, such as a country, for food, water, transport, housing, waste management, and other requirements (12.1)

effector a body structure that responds to signals from a control centre to effect change in a variable (8.1)

electron carriers compounds that pick up electrons from energy-rich compounds and then donate them to low-energy compounds; an electron carrier is recycled, just as ATP is (3.1)

electron transport system a group of protein complexes and small organic molecules embedded in the inner mitochondrial membrane; the components accept and donate electrons to each other in a linear manner (4.1)

electronegativity the property governing how strongly an atom attracts electrons (1.1)

elongation phase an event in DNA replication in which two new strands of DNA are assembled using the parent DNA as a template; the new DNA molecules—each composed of one strand of parent DNA and one strand of daughter DNA—re-form into double helices (5.2)

emigration the movement of individuals out of a population (11.2)

endergonic refers to a chemical reaction that requires energy (3.1)

endocrine gland a ductless gland that secretes hormones directly into the bloodstream (9.1)

endocrine system in vertebrates, a system that works in parallel with the nervous system to maintain homeostasis by secreting hormones, that serve as chemical messengers, from various glands; composed of the hormone-producing glands and tissues of the body (9.1)

endomembrane system the system within the cell that acts to synthesize, modify, and transport proteins and other cell products; includes the endoplasmic reticulum, the Golgi apparatus, vesicles, and the cell membrane, among other structures (2.1)

endoplasmic reticulum (ER) a complex system of channels and sacs composed of double phospholipid membranes enclosing a lumen; made up of two parts, the rough ER and the smooth ER (2.1)

energy the capacity to cause change or to do work (3.1)

enhancer a response element in eukaryotes that increases the rate of transcription (6.4)

enthalpy the total energy of a system, symbolized as H (3.1)

entropy the measure of the disorder in a system; symbolized by S (3.1)

enzyme a biological macromolecule that catalyzes, or speeds up, chemical reactions in biological systems (1.3)

enzyme-substrate complex the combined structure of an enzyme with a substrate that is bound to the enzyme's active site (1.3)

epinephrine a hormone produced by the adrenal cortex that helps regulate the short-term stress response; also known as adrenaline (9.3)

essential amino acids amino acids that cannot be produced by the human body and must be consumed as part of the diet (1.2)

ester linkage a bond between the hydroxyl group on a glycerol molecule and the carboxyl group on a fatty acid (1.2)

estrogen a female sex hormone produced in the ovary; helps maintain sexual organs and secondary sexual characteristics (9.4)

ethanol fermentation a biological process in which many single-celled organisms such as yeasts and some bacteria generate cellular energy in anaerobic conditions by converting glucose to pyruvate and then to ethanol and carbon dioxide (3.3)

ethidium bromide a chemical that associates with DNA and that fluoresces under ultraviolet light (7.1)

euchromatin the less condensed regions of a chromosome; areas that are capable of gene transcription (5.1)

excitatory refers to one of two effects neurotransmitters can have on the postsynaptic membrane; this particular effect causes the receptor proteins to trigger ion channels that open to allow positive ions, such as sodium, to flow into the postsynaptic neuron and the membrane becomes slightly depolarized as a result (8.2)

excretion the process of separating wastes from body fluids and eliminating them from the body; performed by several body systems, including the respiratory, integumentary, digestive, and excretory systems (10.1)

excretory system in animals, the system that regulates the volume and composition of body fluids by excreting metabolic wastes and recycling some substances for reuse; main organs include the kidneys, ureters, bladder, and urethra; also known as the urinary system (10.1)

exergonic refers to a chemical reaction that releases energy (3.1)

exon the coding region in eukaryotic genes along which non-coding regions called introns are interspersed (6.2)

exponential growth the growth pattern exhibited by a population growing at its biotic potential (11.2)

expression vector a plasmid vector that is transformed into a host cell for the purpose of producing a foreign protein (7.2)

extinction the disappearance of all members of a species from Earth (12.2)

extracellular fluid the fluid outside cells (Unit 1 Preparation)

facultative anaerobes organisms that are able to function aerobically as well as anaerobically (3.3)

fatty acid a hydrocarbon chain ending in a carboxyl group (1.2)

fecundity the average number of offspring produced by a female member of a population over her lifetime (11.1)

feedback inhibition a process that allows a cell to ensure that when biochemical reactions occur in pathways, the product of the last reaction of the pathway is a non-competitive inhibitor of the enzyme that catalyses a reaction at the beginning of the pathway; in this way, enough product is available for each reaction and all the reactions are turned off or reduced, as appropriate (1.3)

fermentation an anaerobic metabolic pathway that generates ATP by converting glucose to an alcohol or an organic acid such as lactate (3.3)

fight-or-flight response the body's acute reaction to stress in which the sympathetic nervous system is stimulated; also known as fight-or-flight response; also known as short-term stress response (9.3)

filtrate in the kidney, a filtered fluid that proceeds from the glomerulus into the Bowman's capsule of the nephron (10.1)

flagellum an appendage protruding from a cell;, composed of a microtubule-based shaft covered in an extension of the cell membrane; allows a cell to propel itself; flagella are longer and less numerous compared to cilia (2.1)

follicle-stimulating hormone (FSH) a reproductive hormone that stimulates the development of the sex organs and gamete production in males and females (9.4)

frameshift mutation a mutation caused by the addition or deletion of a number of nucleotides not divisible by three, resulting in a change in the reading frame (6.3)

free energy energy from a chemical reaction that is available for doing work (3.1)

fluid mosaic model a model of the cell membrane based on the changing location and pattern of protein molecules in a fluid phospholipid layer (2.1)

frontal lobe the lobe of the cerebral cortex that integrates information from other parts of the brain and controls reasoning, critical thinking, memory, and personality (8.3)

functional group an atom or group of atoms attached to a molecule, which give the molecule particular chemical and physical properties (1.1)

fusion proteins proteins comprised of the ß-galactosidase enzyme fused to the A chain or the B chain in human insulin manufactured using transgenic bacteria; the presence of the ß-galactosidase enzyme protects the insulin chains from being degraded by the bacteria, as they would be if expressed on their own (7.2)

gel electrophoresis a method that uses an electric field to separate negatively charged DNA fragments according to size as they pass through a gel (7.1)

gene the basic unit of heredity that determines, in whole or part, a genetic trait; a specific sequence of DNA that encodes for proteins and RNA molecules, and can contain sequences that influence production of these molecules (5.1)

gene cloning the process of manipulating DNA to produce many identical copies of a gene or another segment of DNA in foreign cells (7.1)

gene expression the transfer of genetic information from DNA to RNA to protein (6.1)

gene pharming (molecular pharming) the process of using transgenic livestock to produce therapeutic proteins (pharmaceuticals) (7.2)

gene regulation the control of and change in gene expression in response to different conditions in the cell or its environment (6.4)

gene therapy a method for treating genetic disorders by introducing a correct form of the disease-related gene into an individual's genome (7.2)

genetic code a set of rules for determining how genetic information in the form of a nucleotide sequence is converted to an amino acid sequence of a protein; a code specifying the relationship between a nucleotide codon and an amino acid (6.1)

genetic engineering alteration of the genetic material of an organism in a specific manner (7.2)

genetically modified organism (GMO)
an organism whose genetic material has been modified, often through the insertion of a foreign gene into its genome (7.2)

genome the complete genetic makeup of an organism; an organism's total DNA sequence (5.1)

glans penis a variable-length shaft with an enlarged tip, which is surrounded and protected by the foreskin (9.4)

glial cell a support cell of the nervous system that nourishes neurons (nerve-impulse conducting cells), removes their wastes, defends against infection, and provides a supporting framework for all the nervous-system tissue (8.2)

glomerular filtration in the kidney, a process that results in the movement of water and solutes, except proteins, from the blood plasma into the nephron down a pressure gradient (10.2)

glomerulus in the kidney, a fine network of capillaries within the Bowman's capsule of the nephron; arising from the renal artery, the walls of the glomerulus act as a filtration device (10.1)

glucagon a hormone produced by the alpha cells of the islets of Langerhans in the pancreas to stimulate the liver to convert glycogen back into glucose, which is released into the blood (9.3)

glycolysis a metabolic pathway that breaks glucose down to pyruvate (3.2)

glycoprotein a protein that has carbohydrates attached to it (2.1)

glycosidic linkage a covalent bond between monosaccharides (1.2)

goiter the enlargement of the thyroid gland characterized by a large swelling in the throat, often associated with a deficiency of iodine; occurs when the thyroid gland is constantly stimulated by thyroxine-stimulating hormone (TSH), but is unable to synthesize thyroxine to create a negative feedback loop (9.2)

Golgi apparatus a stack of curved membrane sacs that packages, processes, sorts, and distributes proteins, lipids, and other substances within the cell; acts like a "post office" for the cell (2.1)

gonad an organ that produces reproductive cells (gametes); the ovary produces eggs (ova), and the testes produce sperm (9.4)

gonadotropin-releasing hormone (GnRH) a hormone that acts on the anterior pituitary gland to cause it to release two different sex hormones: luteinizing hormone (LH) and follicle-stimulating hormone (FSH) (9.4)

granum in a chloroplast, a structure made up of stacked thylakoids (4.1)

grey matter brain matter that is grey because it contains mostly cell bodies, dendrites, and short, unmyelinated axons (8.3)

heat the transfer of thermal energy from one object to another due to a temperature difference between the objects (3.1)

helicase a group of enzymes that aid in the unwinding of DNA (5.2)

hemodialysis a type of renal (kidney) dialysis that utilizes an artificial membrane in an external device and is connected to an artery and a vein in a person's arm to remove waste and excess fluid from the blood when kidney function is lost due to renal failure (10.3)

heterochromatin highly compacted regions of chromosomes; in general, these regions are transcriptionally inactive because of their tight conformation (5.1)

histone a member of a family of proteins that associate with DNA in eukaryotic cells, which acts to help compact the DNA (5.1)

homeostasis the tendency of the body to maintain a relatively constant internal environment (8.1)

horizontal gene transfer in this type of gene transfer, an introduced trait—such as resistance to an antibiotic or a pesticide—is transferred to other organisms; these organisms could be other plants and animals, or even bacteria or fungi (7.2)

hormone a chemical messenger sent to many parts of the body to produce a specific effect on a target cell or organ (9.1)

hormone replacement therapy (HRT) the administration of low levels of estrogen and/or progesterone to alleviate symptoms of menopause in females (9.4)

host the organism in a symbiotic relationship on which the symbiont feeds (11.3)

housekeeping gene (constitutive gene) a gene that is constantly being expressed; does not undergo regulation of expression (6.4)

Human Genome Project a project that sequenced the human genome and identified all the genes within it (7.1)

human growth hormone (hGH) a hormone that ultimately affects almost every body tissue, by direct stimulation or via tropic effects; stimulates the liver to secrete hormones called growth factors, which, along with hGH, influence many physiological processes (9.2)

hydrocarbon an organic molecule that is made up of only carbon and hydrogen atoms (1.1)

hydrogen bond a weak association between an atom with a partial negative charge and a hydrogen atom with a partial positive charge (1.1)

hydrogenation a food preservation process that involves the chemical addition of hydrogen to unsaturated fatty acids of triglycerides to produce saturated fats (1.2)

hydrolysis reaction a chemical reaction that results in cleavage of a covalent bond whereby a hydrogen atom is attached to one atom in the bond and a hydroxyl group is added to the other atom of the bond with the addition of a water molecule (1.3)

hydrophilic refers to polar molecules that have attractive interactions with water molecules (1.1)

hydrophobic refers to non-polar molecules that do not have attractive interactions with water molecules (1.1)

hydrophobic effect the natural clumping together of non-polar molecules in water (1.1)

hyperglycemia a condition resulting from high levels of blood glucose; occurs in individuals with diabetes mellitus (9.3)

hyperpolarization a change in the membrane potential that occurs when a cell becomes more polarized (8.2)

hyperthyroidism a condition that results when the thyroid produces extremely high levels of thyroxine (9.2)

hypothalamus a region of the forebrain just below the cerebral hemispheres that helps to regulate the body's internal environment, as well as certain aspects of behaviour; the hypothalamus contains neurons that control blood pressure, heart rate, body temperature, and basic drives (such as thirst and hunger) and emotions (8.3)

hypothyroidism a condition that results when the thyroid produces extremely low levels of thyroxine (9.2)

immigration the movement of individuals into a population (11.2)

induced refers to mutations that are caused by agents outside the cell (6.3)

induced fit a change in shape of the active site of an enzyme to accommodate the substrate (the reactant that interacts with the enzyme in an enzyme-catalyzed reaction) (1.3)

inhibin a hormone that acts on the anterior pituitary to inhibit the production of follicle-stimulating hormone (FSH), produces a negative feedback loop that controls the rate of sperm formation (9.4)

inhibitor a molecule that binds to the allosteric or active site of an enzyme and causes a decrease in the activity of that enzyme (1.3)

inhibitory refers to one of two effects that neurotransmitters can have on the postsynaptic membrane; this particular effect causes the receptor to trigger potassium channels to open, allowing potassium ions to flow out, which results in a more negative membrane potential and subsequent hyperpolarization (8.2)

initiation phase an event in DNA replication in which a portion of the DNA double helix is unwound to expose the bases for new base pairing (5.2)

insulin a hormone secreted by the alpha cells of the islets of Langerhans in the pancreas to make target cells more permeable to glucose; enables the body to use sugar and other carbohydrates (9.3)

intermolecular occurring between atoms of different molecules (1.1)

interspecific competition a situation in which two or more populations compete for limited resources (11.3)

intracellular fluid the fluid inside cells (Unit 1 Preparation)

intraspecific refers to individuals of the same species (11.3)

intraspecific competition a situation in which members of the same population compete for resources (11.3)

invasive species a species that can take over the habitat of native species or invade their bodies (Unit 5 Preparation)

inversion a process that occurs when a segment of a chromosome is broken and then is inverted in the opposite direction (6.3)

ion an atom or group of atoms that has gained or lost one or more electrons, giving it a positive or negative charge (1.1)

ionic compound a chemical compound made up of ions that are held together by ionic bonds; an ionic bond is the attractive electrostatic force between a negative ion and a positive ion (1.1)

islets of Langerhans clusters of endocrine cells found throughout the pancreas, consisting of glucagon-producing alpha cells and insulin-producing beta cells (9.3)

isomer one of two or more molecules with the same number and type of atoms, but different structural arrangements (1.2)

isotope atoms of the same element that have different numbers of neutrons (1.1)

juglone a chemical secreted into soil by the black walnut tree to deter other plants from growing near it (11.1)

karoshi the name given to the phenomenon in which many Japanese business people die from heart attacks and strokes; translates as "death from overwork" (9.4)

keystone species a species that can greatly affect population numbers and the health of an ecosystem (Unit 5 Prep)

kidney in vertebrates, one of a pair of organs that filters waste from the blood (which is excreted in urine) and adjusts the concentrations of salts in the blood (10.1)

kinetic energy energy associated with movement (3.1)

Krebs cycle the cyclical metabolic pathway that occurs in the mitochondrial matrix (3.2)

K-selected strategies life strategies used by populations that live close to the carrying capacity of their environment (11.2)

Lac the operon on the *E. coli* chromosome in which are found the genes that encode the enzymes that are needed to break down lactose (6.4)

lactate fermentation a biological process in which some single-celled organisms, as well as some animal muscle cells that are temporarily without oxygen, generate cellular energy by converting glucose to pyruvate that, in turn, is converted into lactate (lactic acid) (3.3)

lacZ the gene that codes for an enzyme that breaks down galactose (7.1)

lag phase in exponential population growth, the brief phase that is followed by a steep increase in the growth curve (11.2)

law of conservation of energy the law stating that the total energy of the universe is constant—energy cannot be created or destroyed, but it can be transformed from one type into another and transferred from one object to another; therefore, the mass of products produced by a chemical reaction is always equal to the mass of the reactants (3.1)

leading strand a new strand from a parent strand synthesized continuously by DNA polymerase (5.2)

life history the survivorship and reproductive patterns shown by individuals in a population (11.1)

life strategies trade-offs that an organism makes to maximize the number of offspring that survive given limited resources; life strategies can be correlated to the type of environment in which the organism lives (11.2)

light-dependent reactions in photosynthesis, the reactions that trap solar energy and use it to generate ATP and NADPH (4.1)

light-independent reactions in photosynthesis, the reactions that assimilate carbon dioxide to produce an organic molecule that can be used to produce biologically important molecules such as carbohydrates (4.1)

limiting factor a factor that limits the growth, distribution, or amount of a population in an ecosystem (Unit 5 Preparation)

linker DNA in the chromosomes of eukaryotic cells, regions that connect nucleosome structures; each neucleosome is composed of double-stranded DNA wrapped around a group of eight histone proteins; linker DNA is often described as "beads on a string" (5.1)

lipid a macromolecule composed of carbon, hydrogen, and oxygen atoms, with a high proportion of non-polar, carbon–hydrogen bonds (1.2)

lipid bilayer a structure that spontaneously forms when phospholipid molecules are in a water-based environment; hydrophilic "heads" are toward the exterior and hydrophobic "tails" are toward the centre of the bilayer (1.2)

loci locations of DNA sequences or genes (7.1)

logistic growth the growth pattern exhibited by a population for which growth is limited by carrying capacity and limited availability of resources (11.2)

long-term stress response a sustained physiological response to stressors, characterized by increases in blood glucose and blood pressure, and a decrease in inflammatory response; regulated by hormones produced by the adrenal cortex (9.3)

loop of Henle in the kidney, the tubular portion of the nephron that lies between the proximal tubule and the distal tubule; its main function is reabsorption of water and ions (10.2)

lumen the internal space within an organelle (2.1)

luteinizing hormone (LH) a reproductive hormone that triggers ovulation, stimulates the formation of the corpus luteum, and (with follicle-stimulating hormone) stimulates estrogen production in the ovaries; also stimulates the release of testosterone in the testes (9.4)

lysosome a membrane-bound vesicle containing enzymes that catalyze hydrolysis reactions, thereby breaking down macromolecules (2.1)

macromolecule a large, complex molecule, usually composed of repeating units of smaller molecules covalently linked together (1.2)

malate a compound into which oxaloacetate is converted before it is transported into the bundle-sheath cells (4.2)

manual sequencing a method in which researchers themselves performed the processes involved, in determining the nucleotide sequence, base by base, of a fragment of DNA without any automation (7.1)

mark recapture a method in which animals are captured, marked with a tag, collar, or band, released, and then recaptured at a later time to determine an estimate of population size (11.1)

mass extinction a large-scale dying out of a large percentage of all living organisms within an area over a short time (Unit 5 Preparation)

matrix the fluid-filled space in the inner membrane of mitochondria (2.1)

mature mRNA mRNA that has undergone processing (6.2)

Maxam-Gilbert sequencing a detailed method of DNA sequencing that relies on radioactive labelling of the single-stranded DNA to be sequenced (7.1)

medulla oblongata a region of the brain that coordinates many reflexes and automatic bodily functions that maintain homeostasis, including heart rate, constriction or dilation of blood vessels, and the rate and depth of breathing, swallowing, and coughing (8.3)

membrane potential an electrical charge separation across a cell membrane; a form of potential energy (8.2)

meninges three layers of tough, elastic tissue within the skull and spinal column that directly enclose the brain and spinal cord (8.3)

menopause the period in a woman's life during which a decrease in estrogen and progesterone results in an end of menstrual cycles, usually occurring around age 50 (9.4)

menstrual cycle in a human female, the period of 20 to45 days during which hormones stimulate the development of the uterine lining, and an egg (ovum) is developed and released from an ovary; if the egg is not fertilized, the uterine lining is shed as the cycle begins again; the menstrual cycle can be divided into the ovarian cycle and the uterine cycle (9.4)

messenger RNA (mRNA) RNA that contains the genetic information of a gene and that carries it to the protein synthesis machinery; it provides the information that determines the amino acid sequence of a protein (6.1)

messenger RNA hypothesis the confirmed postulation that when bacteria were infected by a virus, a virus-specific RNA molecule was synthesized and became associated with pre-existing bacterial ribosomes; this RNA molecule carried the genetic information to produce the viral protein (6.1)

metabolic pathway a sequential series of chemical reactions in living cells; each reaction is catalyzed by an enzyme (3.1)

metabolism the sum of all chemical reactions that occur in the cell (3.1)

midbrain the region of the brain involved in processing information from sensory neurons in the eyes, ears, and nose; it relays visual and auditory information between areas of the hindbrain and forebrain (8.3)

mimicry a structural adaptation in which a harmless species resembles a harmful species in coloration or structure (Unit 5 Preparation)]

minimum viable population size the lowest number of individuals that can sustain a population in the wild for a long period of time (12.2)

mismatch repair a mechanism for repairing errors made during DNA replication, whereby a group of proteins recognize a mispaired nucleotide on the newly synthesized strand and replace it with a correctly paired nucleotide (5.2)

missense mutation a mutation that changes the amino acid sequence of a protein (6.3)

mitochondrion an organelle found in eukaryotic cells that supplies most of the cell's ATP (2.1)

model in the study of population dynamics, a pattern that helps ecologists describe a population (11.1)

molecular biology the study of the structure and functions of nucleic acids and proteins (7.1)

molecular chaperones proteins that interact with the polypeptide chain and, through a series of steps, produce the final properly folded protein (1.2)

molecular formula a representation that shows the number of each type of atom in an element or compound (1.1)

molecule a substance composed of two or more atoms that are covalently bonded together (1.1)

monomer the smallest repeating unit of a polymer (1.2)

monosaccharide a carbohydrate composed of between three and seven carbon atoms (1.2)

monounsaturated refers to an unsaturated fatty acid that has one double bond (1.2)

mortality the proportion of individuals in a population that typically die at a given age over a given period of time (11.2)

Müllerian mimicry a type of mimicry in which many noxious species converge in such as way as to look the same, thus reinforcing a basic distasteful design (11.3)

mutagen an event or substance that increases the rate of changes to the DNA sequence of an organism's genome (6.3)

mutation a change in the nucleotide sequence of a cell's DNA (6.3)

mutualism a type of symbiotic relationship in which both species benefit from the relationship (11.3)

myelin sheath the fatty, insulating layer around the axon of a nerve cell, composed of Schwann cells; protects myelinated neurons and speeds up the rate of nerve impulse transmission (8.2)

natality birth (11.2)

natural selection the process by which characteristics of a population change over many generations as organisms with advantageous heritable traits survive and reproduce, passing their traits to offspring (Unit 5 Prep)

negative feedback system the mechanism of homeostatic response by which the output of a system reverses a change in a variable, bringing the variable back to within normal range (8.2)

nephron a microscopic tube-like filtration unit found in the kidneys that filters and reabsorbs various substances from the blood; produces urine (10.1)

nerve the message pathway of the nervous system; made up of many neurons grouped into bundles and surrounded by protective connective tissue (8.2)

neuromuscular junction the synapse between a motor neuron and a muscle cell (8.2)

neuron a nerve cell; the structural and functional unit of the nervous system, consisting of a nucleus, cell body, dendrites, and axons; specialized to respond to physical and chemical stimuli, to conduct electrochemical signals, and to release chemicals that regulate various body processes (8.2)

neurotransmitter a chemical messenger secreted by neurons to carry a neural signal from one neuron to another, or from a neuron to an effector, such as a gland or muscle fibre (8.2)

neutralization reaction a chemical reaction between an acid and a base, producing water and a salt (1.3)

nodes of Ranvier the exposed areas in the axons of myelinated neurons that contain many voltage-gated Na^1 channels (8.2)

non-competitive inhibitor a molecule that binds to an enzyme at a location that is outside the active site and that inhibits the enzyme's function (1.3)

noncyclic photophosphorylation the production of ATP by the passing of electrons through the Z scheme (4.1)

nonsense mutation a mutation that shortens a protein by introducing a stop codon (6.3)

norepinephrine a neurotransmitter released by sympathetic neurons of the autonomic system to produce an excitatory effect on target muscles; also, a hormone produced by the adrenal medulla (8.4)

nuclear envelope a double membrane surrounding the nucleus of a cell (2.1)

nuclear matrix a filamentous network of proteins that is found inside the nucleus and lines the inner nuclear membrane; the nuclear matrix serves to organize the chromosomes (2.1)

nuclear pore complex a group of proteins forming openings in the nuclear envelope (2.1)

nucleic acid a macromolecule composed of nucleotide monomers (1.2)

nuclein a weakly acidic substance, extracted from nuclei, containing nitrogen and phosphorus (5.1)

nucleoid a structure in bacteria that contains the chromosomal DNA (5.1)

nucleolus a non-membrane-bound structure in the nucleus, which contains RNA and proteins (2.1)

nucleoplasm a thick fluid that fills the nucleus of a cell (2.1)

nucleosome a condensed structure formed when double-stranded DNA wraps around an octamer of histone proteins (5.1)

nucleotide an organic molecule, consisting of a repeating unit of nucleic acids, composed of a sugar group bonded to a phosphate group and a nitrogen-containing base (1.2); (5.1)

occipital lobe a lobe of the cerebral cortex that receives and analyzes visual information, and is needed for recognition of visual stimuli (8.3)

Okazaki fragment a short DNA fragment that is generated during the synthesis of the lagging strand in DNA replication (5.2)

one-gene/one-polypeptide hypothesis the proposal that one gene codes for one polypeptide (or protein) (6.1)

open system a system that can exchange matter and energy with its surroundings, and vice versa (3.1)

operator a DNA sequence element to which a repressor protein binds to regulate transcription (6.4)

operon a cluster of genes grouped together under the control of one promoter; occurs in prokaryotic genomes (6.4)

organic molecule a carbon-containing molecule in which carbon atoms are nearly always bonded to each other and to hydrogen (1.1)

origin of replication the DNA sequence where replication begins (5.2)

orphan disease a disease that does not receive adequate support in terms of research, funding, and public awareness, due to the low incidence of these diseases in the population (1.3)

osmosis the diffusion of water molecules across a membrane, from an area of high concentration of water molecules to an area of lower concentration of water molecules (Unit 1 Preparation)

overexploitation the excessive harvesting or killing of a species until the species no longer exists or has a very small population (12.2)

oxaloacetate a four-carbon compound formed when carbon dioxide is added to phosphoenolpyruvate (PEP) (4.2)

oxic refers to an oxygen-containing environment (3.3)

oxidases enzymes in peroxisomes that catalyze redox reactions (2.1)

oxidation a process involving the loss of electrons; occurs during the breakdown of small organic molecules (1.3)

oxidative phosphorylation the formation of ATP using energy obtained from redox reactions in the electron transport chain (3.2)

oxygen debt in lactate fermentation, the amount of oxygen required to allow lactate to return to the oxidative pathways to be metabolized (3.3)

pancreas a small gland in the abdomen that secretes digestive enzymes into the small intestine; also secretes the hormone insulin (9.3)

parasitism a symbiotic relationship in which a symbiont lives off and harms the host (11.3)

parasympathetic nervous system a division of the autonomic system that regulates involuntary processes in the body; works in opposition to the sympathetic nervous system; typically activated when the body is calm and at rest (8.4)

parathyroid hormone (PTH) a hormone synthesized and released in response to falling concentrations of calcium in the blood; PTH stimulates bone cells to break down bone material (calcium phosphate) and secrete calcium into the blood (9.2)

parietal lobe a lobe of the cerebral cortex that receives and processes sensory information from the skin, and helps to process information about the body's position and orientation (8.3)

patent a government ruling giving an individual or organization the sole title or right to make, use, or sell a particular invention (7.2)

pectins non-cellulose structural polysaccharides found in cell walls; synthesized by the Golgi apparatus in many plants (2.1)

peptide bond a covalent bond formed between two amino acids during protein synthesis (6.3)

per capita a term meaning per individual or per person (11.2)

peripheral nervous system a network of nerves that carry sensory messages to the central nervous system (CNS) and send information from the CNS to the muscles and glands; consists of the autonomic and somatic systems (8.2)

peritoneal dialysis a type of renal (kidney) dialysis that utilizes the lining of the intestines, called the peritoneum, as the dialysis membrane to remove waste and excess fluid from the blood when kidney function is lost due to renal failure (10.3)

peroxisome a membrane-bound sac containing oxidative enzymes that break down excess fatty acids and hydrogen peroxide, and participate in the synthesis of bile acids and cholesterol (2.1)

pH scale a numerical scale ranging from 0 to 14 that is used to classify aqueous solutions as acidic, basic, or neutral (1.3)

phosphodiester bond a covalent bond between adjacent nucleotides; occurs between the phosphate group on one nucleotide and a hydroxyl group on the sugar of the next nucleotide in the strand (1.2)

phosphoenolpyruvate (PEP) a three-carbon compound involved in the uptake of carbon dioxide from the Calvin cycle in the outer layer of mesophyll cells in C4 plants (4.2)

phospholipid a lipid composed of a glycerol molecule bonded to two fatty acids and a phosphate group with an R group; the main component of cell membranes (1.2)

phospholipid bilayer a double layer of phospholipid molecules that line up tail to tail to create a hydrophilic cell membrane (2.1)

photon a packet of energy made up of absorbed light energy (4.1)

photophosphorylation the use of photons of light to drive the phosphorylation of ADP to produce ATP via chemiosmosis (4.1)

photorepair a specific mechanism for repairing the thymine dimer structures (6.3)

photorespiration the reaction of oxygen with ribulose-1,5-bisphosphate in a process that reverses carbon fixation and reduces the efficiency of photosynthesis (4.2)

photosystem in photosynthesis, one of two protein-based complexes composed of clusters of pigments that absorb light energy (4.1)

physical mutagen an agent that physically changes the structure of DNA (6.3)

pigment a compound that absorbs certain wavelengths of visible light while reflecting others (4.1)

plasmid in recombinant DNA technology, a self-replicating, closed circular piece of DNA that can act as a carrier of a gene to be cloned in bacteria (7.1)

point mutation a mutation involving a single base pair substitution, insertion, or deletion (6.3)

polar covalent bond the unequal sharing of electrons in a covalent bond (1.1)

polar molecules molecules such as water, which have regions of partial negative and partial positive charge (1.1)

polarization the process in which the membrane potential of the cell is lowered below its equilibrium value; in nerves, the process of generating a resting membrane potential of −70 mV (8.2)

polycistronic refers to the fact that genes that are involved in the same metabolic pathway are often found in the same operon; since they are all under the control of the same promoter region, these genes are all transcribed together into one continuous mRNA strand (6.4)

polymer a large molecule composed of repeating units of smaller molecules (monomers) (1.2)

polymerase chain reaction (PCR) an automated method for amplifying specific regions of DNA from extremely small quantities (7.1)

polypeptide a polymer composed of many amino acids linked together by covalent bonds between the amino group of one amino acid and the carboxyl group of another amino acid (1.2)

polyribosome a structure composed of multiple ribosomes along a strand of mRNA (6.3)

polysaccharide a carbohydrate polymer composed of many monosaccharides joined by covalent bonds between particular atoms (1.2)

polyunsaturated refers to unsaturated fatty acids with two or more double bonds (1.2)

pons a region that serves as a relay centre between the neurons of the right and left halves of the cerebrum, the cerebellum, and the rest of the brain (8.3)

population a group of individuals of the same species living in the same geographical area (Unit 5 Preparation)

population crash a rapid decrease in a population; the opposite of a population explosion (11.2)

population cycle alternating periods of large and small population sizes (11.3)

population density (D_p) the number of individuals per unit of volume or area (11.1)

population pyramid a type of bar graph that shows the age distribution in a population and that demographers use to study a population (12.1)

population size (*N*) the number of individuals of the same species living within a specific geographical area (11.1)

positive feedback system a mechanism of homeostatic response by which the output of a system strengthens or increases a change in homeostasis (8.1)

posterior pituitary the posterior lobe of the pituitary gland; an endocrine gland that stores and releases antidiuretic hormone (ADH) and oxytocin, which are produced in the hypothalamus and transferred to the posterior pituitary by neuronal axons (9.2)

potential energy energy that is stored (3.1)

precursor mRNA (pre-mRNA) a form of mRNA that has not undergone processing (6.2)

predation a relationship between two different species in which one species feeds on another (Unit 5 Prep)

primase an enzyme that synthesizes an RNA primer (5.2)

primer in DNA replication, a short segment of RNA that is complementary to a part of the DNA strand and serves as a starting point for addition of nucleotides (5.2)

progesterone a female sex hormone produced first by the corpus luteum of the ovary to prepare the uterus for the fertilized egg (ovum), and later by the placenta to maintain pregnancy (9.4)

promoter region a sequence of nucleotides in DNA that indicates where RNA polymerase complex should bind to initiate transcription (6.2)

protective coloration adaptations that help individuals avoid predation; [illegible] camouflage, mimicry, and body coloration used as a warning signal (11.3)

protein a macromolecule composed of amino acid monomers linked by covalent bonds (1.2)

proximal tubule in the kidney, the tubular portion of the nephron that lies between the Bowman's capsule and the loop of Henle; its main function is reabsorption of water and solutes, as well as secretion of hydrogen ions (10.2)

pyruvate a three-carbon compound formed from the breakdown of glucose during glycolysis (4.2)

quadrat an area of specific size used for sampling a population; often used to sample immobile organisms or those that move very little (11.1)

quaternary structure the association of two or more polypeptides to form a protein; one of four levels of protein structure (1.2)

***R* group (side chain)** in the active site of a catalyzed reaction, a group of amino acids that end up close to certain chemical bonds in the substrate, causing these bonds to stretch or bend, which makes the bonds weaker and easier to break (1.2)

radioisotope an unstable isotope that decays over time by emitting radiation (1.1)

radioisotope tracing a method in which doctors inject radioactive material into a patient and trace its movement in the body (1.1)

random distribution a pattern of dispersion within a population, in which individuals do not appear to be specifically positioned relative to another individual (11.1)

reaction centre a combination of the pair of chlorophyll a molecules and proteins (4.1)

reading frame collectively, the codons of mRNA that are read to produce an amino acid sequence; it is set by the start codon (6.3)

recombinant DNA a molecule of DNA composed of genetic material from different sources (7.1)

redox reaction a chemical reaction that involves the transfer of electrons from one substance to another; also called oxidation-reduction reaction (1.3)

reducing power in redox reactions, electrons that pass from one atom to another carry energy with them, so the reduced form of a molecule is always at a higher energy level than the oxidized form; thus, the electrons are said to carry reducing power (3.1)

reduction the chemical process involving the gain of electrons (1.3)

reflex a sudden, involuntary response to certain stimuli (8.2)

refractory period the few milliseconds after an action potential, during which the cell membrane cannot be stimulated to undergo another action potential (8.2)

reflex arc a simple connection of neurons that results in a reflex action in response to a stimulus (8.2)

regenerating RuBP the third stage of the Calvin cycle, in which most of the reduced G3P molecules are used to make more RuBP; energy, supplied by ATP, is required to break and re-form the chemical bonds to make the five-carbon RuBP from G3P (4.2)

regulatory sequence a sequence of DNA that regulates the activity of a gene ([illegible])

relative density an estimate of the number of individuals determined for one area of study compared with an estimate of the number of individuals determined for another area of study (11.1)

releasing hormones chemicals that stimulate the pituitary gland to secrete hormones that act on other endocrine glands (9.1)

renal artery a blood vessel that originates from the aorta and delivers blood to the kidneys; splits into a fine network of capillaries (the glomerulus) within the Bowman's capsule of the nephron (10.1)

renal vein a blood vessel that drains from the kidney; it returns to the body the solutes and water reabsorbed by the kidney (10.1)

replication bubble the unwound, oval-shaped area of a double helix (5.2)

replication fork the Y-shaped area of a double helix (5.2)

replication machine a protein-DNA complex at each replication fork that carries out replication (5.2)

repolarized returned to previous polarization; the effect of the sodium-potassium exchange pump and the small amount of naturally occurring diffusion on the cell membrane of a neuron, which returns to its normal resting potential of -70 mV. (8.2)

repressor a protein that binds to a particular DNA sequence to regulate transcription; inhibits the transcription of a gene or genes (6.4)

r-selected strategies life strategies used by populations that live close to their biotic potential (11.2)

resource partitioning the use of different resources within a habitat (11.3)

resting membrane potential the potential difference across the membrane in a resting neuron (8.2)

restriction endonuclease an enzyme that cleaves (cuts) the interior of double-stranded DNA in a sequence-specific manner (7.1)

restriction enzyme an enzyme that restricts the replication of infecting viruses by cleaving (cutting) viral DNA (7.1)

restriction fragment a small segment of DNA generated by cutting a larger piece of DNA with a restriction enzyme (7.1)

restriction fragment length polymorphism (RFLP) in DNA fingerprinting, a method of analysis in which the bands on the gel used in gel electrophoresis that are unique to each individual are compared to bands with patterns from an individual of known identify (7.1)

restriction site a particular point within the DNA sequence at which the enzymes cut a strand of DNA (7.1)

ribose a sugar found in RNA nucleotides (1.2)

ribosomal RNA (rRNA) RNA that is associated with proteins in the ribosome (6.3)

ribosome a cell structure composed of RNA and proteins, and responsible for synthesis of polypeptides in the cytosol and on the surface of the rough endoplasmic reticulum (2.1); acts as a site of assembly for mRNA, amino acid containing tRNAs, and the enzymes needed for protein synthesis (6.3)

RNA (ribonucleic acid) a macromolecule composed of nucleotides containing the sugar ribose (1.2)

RNA interference the regulation of gene expression by small RNAs; inhibits gene expression by degrading mRNA or inhibiting translation (6.4)

RNA polymerase the main enzyme that catalyzes the formation of RNA from a DNA template (6.2)

rough endoplasmic reticulum (rough ER) a part of the ER that is studded with ribosomes; this region plays a key role in the initial synthesis and sorting of proteins that are destined for the ER, Golgi apparatus, lysosomes, vacuoles, plasma membrane, or the extraceullular environment (2.1)

rubisco short form for ribulose bisphosphate carboxylase, an enzyme that catalyzes the reaction that leads to two identical three-carbon compounds (PGA) in C3 photosynthesis; possibly the most abundant protein on Earth (4.2)

saturated fatty acid a fatty acid that has no double bonds between carbon atoms (1.2)

Schwann cells glial cells that form myelin on axons that travel outside the brain and spinal cord (8.2)

selectable markers resistance genes that researchers can use to specifically select for the bacterial colonies that contain the recombinant DNA of interest (7.1)

semi-conservative model a model proposing that each new molecule of DNA would contain one strand of the original complementary DNA molecule and one new parent strand (5.2)

semi-conservative replication a mechanism of DNA replication in which each newly synthesized DNA molecule is composed of one strand from the original DNA molecule and one new strand (5.2)

sense strand (coding strand) the DNA strand opposite to the anti-sense strand or template strand (6.2)

sensor a body structure that monitors and detects changes in the internal environment (8.1)

sequence specificity in recombinant DNA technology, an important characteristic of restriction endonucleases whereby cuts made by restriction endonucleases are specific and predictable; the same enzyme will cut a particular strand of DNA the same way each time, producing an identical set of DNA fragments, called restriction fragments (7.1)

sex hormone one of several chemical compounds that control the development and function of the reproductive system or secondary sex characteristics (9.4)

short tandem repeat (STR) refers to a technology used to identify individuals based on repeating short sequences of DNA in the genome that vary in length between individuals (7.1)

short-term stress response the body's acute reaction to stress in which the sympathetic nervous system is stimulated; also known as fight-or-flight response (9.3)

sigmoidal refers to an S-shaped curve describing the growth of a population as the population reaches its carrying capacity (11.2)

silent mutation a mutation that does not change the amino acid sequence of a protein (6.3)

single-gene mutation a mutation that involves changes in the nucleotide sequence of one gene (6.3)

single-strand-binding proteins proteins that help to stabilize the newly unwound single strands of DNA, these strands have a tendency, if unchecked, to re-form into a double helix (5.2)

sinusoidal growth a wavelike oscillating growth pattern that is typical of predator-prey interactions (11.3)

site-directed mutagenesis a method of specifically altering the nucleotide sequence of a region of DNA (7.1)

sodium-potassium exchange pump a system involving a carrier protein in the plasma membrane that uses the energy of ATP to transport sodium ions out of and potassium ions into animal cells; important in nerve cells and muscle cells (8.2)

somatic system in vertebrates, the division of the peripheral nervous system that controls voluntary movement of skeletal muscles (8.4)

splicing in mRNA, a process of excising out the introns and combining in the exons (6.2)

spontaneous mutation a mutation resulting from abnormalities in biological processes (6.3)

staggered cuts in recombinant DNA technology, a characteristic of restriction endonucleases whereby a few unpaired nucleotides on a single strand at each end of the restriction fragment are left; these short single-stranded regions are often referred to as sticky ends or overhangs; these sticky ends can form base pairs with other single-stranded regions that have a complementary sequence (7.1)

start codon a three-base sequence that specifies the first amino acid of a protein (6.3)

steroid a lipid composed of four attached carbon-based rings (1.2)

steroid hormone a lipid based hormone, such as testosterone, estrogen, and cortisol (9.1)

sticky ends in DNA, short, single-stranded regions which can form base pairs with other single-stranded regions that have a complementary sequence (7.1)

stomata surface pores on plant surfaces that can be closed to retain water or opened to allow the entry of CO2 needed for photosynthesis and the exit of oxygen and water vapour (4.1)

stop codonvone of three three-base sequences—UAA, UAG, and UGA—that signals the end of translation

strand a polymer of nucleotides (1.2)

stroma the fluid-filled interior surrounding the grana in a chloroplast (4.1)

structural formula a representation that shows how the different atoms of a molecule are bonded together (1.1)

substrate a reactant that interacts with the enzyme in an enzyme-catalyzed reaction (1.3)

substrate level phosphorylation a process in the glytolytic pathway in which a phosphate group is removed from a substrate molecule and combined with an ADP molecule to form ATP (3.2)

surroundings everything in the universe outside of a system (3.1)

survivorship the number or percentage of organisms that typically live to a given age in a given population (11.1)

sustainable refers to the use of resources, such as food, energy, timber, and other items acquired from the environment, at a level that does not exhaust the supply or cause ecological damage (12.2)

symbiont an organism in a symbiotic relationship which lives or feeds in or on the host organism (11.3)

symbiosis an ecological relationship between two species living in direct contact; includes parasitism, mutualism, and commensalism (11.3)

sympathetic nervous system the division of the autonomic system that regulates involuntary processes in the body; works in opposition to the parasympathetic nervous system; typically activated in stress-related situations (8.4)

synapse a junction between two neurons or between a neuron and an effector (muscle or gland) (8.2)

synthesis refers to the chemical reactions that produce a carbohydrate (4.1)

system a whole organism, a group of cells, or a set of substrates and products (3.1)

Taq a certain type of DNA polymerase used for polymerase chain reaction (PCR) (7.1)

target sequence a short sequence of nucleotides within a strand of DNA (7.1)

T-DNA a part of the Ti plasmid that integrates into the plant genome and causes the uncontrolled cell growth that results in a tumour (7.2)

telomere a repetitive section of DNA, near each end of a chromosome; the presence of this sequence helps to prevent loss of important genetic information during replication of the linear DNA in eukaryotic cells (5.2)

temporal lobe a lobe of the cerebral cortex that shares in the processing of visual information but the main function of which is auditory reception (8.3)

termination phase the event in DNA replication in which the replication process is completed and the two new DNA molecules separate from one another; at that point, the replication machine is dismantled (5.2)

tertiary structure a three-dimensional shape of a single polypeptide; one of four levels of protein structure (1.2)

testosterone a reproductive hormone that stimulates the development of the male reproductive tract and secondary sex characteristics; produces only minor effects in females (9.4)

thalamus a region that consists of neurons that provide connections between various parts of the brain (8.3)

thermal energy the kinetic energy of the particles that make up an object moving rapidly in random directions (3.1)

thermodynamics the study of energy changes (3.1)

threshold potential the membrane potential that is sufficient to open voltage-gated Na1 channels and to trigger an action potential, typically around –50 mV (8.2)

thylakoid one of many interconnected sac-like membranous disks within the chloroplast, containing the molecules that absorb energy from the Sun (4.1)

thyroid gland a butterfly-shaped gland located below the larynx in the neck; produces the hormone thyroxine; helps regulate metabolism and growth (9.2)

thyroid-stimulating hormone (TSH) a hormone released by the anterior pituitary that causes the thyroid gland to secrete thyroxine; controlled by a negative feedback mechanism: rising thyroxine levels in the blood detected by the hypothalamus and anterior pituitary suppress the secretion of TSH and, therefore, thyroxine (9.2)

thyroxine (T4) a hormone produced by the thyroid and released into the bloodstream; controls the rate at which the body metabolizes fats, proteins, and carbohydrates for energy (9.2)

Ti plasmid method a method of producing transgenic plants that uses the tumour-inducing plasmid from *Agrobacterium tumefaciens* as the vector for the insertion of a foreign gene into a plant genome (7.2)

T-lymphocytes (T-cells) white blood cells that mature in the thymus (an organ that is part of the body's lymphatic system) and that defend the body against disease (Unit 1 Project)

topoisomerase II enzyme an enzyme that helps to relieve the strain on the double helix sections ahead of the replication forks, which results from the unwinding process in DNA (5.2)

totipotent refers to the capacity of one cell to grow and divide to produce all the different types of cells in a multicellular organism (7.2)

trans face the exit face through which proteins and lipids enter the Golgi apparatus (2.1)

transcription the synthesis of RNA from a DNA template (6.1)

transcription factor one of a set of proteins required for initiation of transcription; required for RNA polymerase complex to bind to the promoter (6.4)

transect a long, relatively narrow rectangular area used for sampling a population (11.1)

transfer RNA (tRNA) an RNA molecule that links the codons on mRNA to the corresponding amino acid for protein synthesis (6.3)

transformation in recombinant DNA technology, a process in which a bacterial host takes up a segment of DNA from the environment under particular experimental conditions (7.1)

transgenic refers to an organism that is produced from the introduction of foreign DNA into its genome, providing it with a new phenotype (7.2)

translation the synthesis of protein from an mRNA template (6.1)

translocation the third stage in transcription; involves a section of one chromosome breaking and fusing to another chromosome (6.3)

transposon a short segment of DNA capable of moving within the genome of an organism; also called a jumping gene (6.3)

triglyceride a lipid molecule composed of a glycerol molecule and three fatty acids linked by ester bonds (1.2)

triplet hypothesis the proposal that the genetic code is read three nucleotide bases at a time; the genetic code consists of a combination of three nucleotides, called a codon (6.1)

tropic hormone a hormone that targets endocrine glands and stimulates them to release other hormones (9.1)

trp an operon that contains five genes that are involved in the synthesis of tryptophan; this operon is normally transcribed, until the cell has sufficient tryptophan (6.4)

tubular reabsorption in the kidney, a process in which water and useful solutes are reabsorbed from the filtrate in the nephron and transported into capillaries for reuse by the body (10.2)

tubular secretion in the kidney, a process that moves additional wastes and excess substances from the blood into the filtrate in the nephron; uses mainly active transport (10.2)

tumour profiling a method that involves determining the DNA sequence of a tumour and, therefore, the type of genetic change has caused a particular cancer (7.1)

turgor pressure the hydrostatic pressure required to stop the net flow of water across a membrane due to osmosis (2.1)

uniform distribution a pattern of dispersion within a population, in which individuals maintain a certain minimum distance between themselves to produce an evenly spaced distribution (11.1)

unsaturated fatty acid a fatty acid that has one or more double bonds between carbon atoms (1.2)

ureter in mammals, a pair of muscular tubes that carry urine from the kidneys to the bladder (10.1)

urethra the tube through which urine exits the bladder and the body (10.1)

urinary bladder the organ where urine is stored before being discharged by way of the urethra (10.1)

urine in the kidneys, a filtrate of the nephron; upon leaving the collecting duct, exits the body through the urethra (10.1)

vacuole a large, membrane-bound sac in plant cells and some other cells that stores water, ions, macromolecules, sugars, and amino acids (2.1)

vector the carrier for a gene to be cloned (7.1)

vertical gene transfer a process that involves the transfer of a gene or trait into the genomes of the natural or wild versions of the same crop (7.2)

vesicle small, membrane-bound sac that stores substances within a cell (2.1)

water reabsorption in the kidney, a process that removes water from the filtrate in the nephron and returns it to the blood for reuse by body systems (10.2)

water-soluble hormone a hormone such as epinephrine, human growth hormone (hGH), thyroxine (T4), and insulin; it cannot diffuse across the cell membrane (9.1)

wax a lipid composed of long carbon-based chains that are solids at room temperature (1.2)

white matter brain matter that is white because it contains myelinated axons that run together in tracts (8.3)

Index

C

D

F

G

H

N

Q

R

S

T

U

Credits

Photo Credits

iv-v (background) ktsimage/iStockphoto; **vi-vii** (background) HYBRID MEDICAL ANIMATION/SCIENCE PHOTO LIBRARY; **vi** (top left) Tan Wei Ming/iStockphoto; **vi** (centre left) konradlew/iStockphoto; **vii** (top right) Richard J. Green/Photo Researchers, Inc.; **vii** (centre right) PROFESSOR OSCAR L. MILLER/SCIENCE PHOTO LIBRARY; **vii** (bottom right) AJ/IRRI/Corbis; **viii-ix** (background) Don Johnston/All Canada Photos/Getty Images;; **viii** (top left) Dennis Kunkel Microscopy, Inc./Phototake/GetStock; **viii** (bottom left) THOMAS DEERINCK, NCMIR/SCIENCE PHOTO LIBRARY; **ix** (top right) Tratong/Dreamstime.com/GetStock; **ix** (centre right) THONY BELIZAIRE/AFP/Getty Images; **x-xi** (background) Photo courtesy of GloFish/via Getty Images; **xii** (background) Angelo Cavalli/Getty Images; **2-3** ktsimage/iStockphoto; **8-9** PAUL D STEWART/SCIENCE PHOTO LIBRARY; **9** (top right) Cathykeifer/Dreamstime.com/GetStock; **10** Scott Camazine/Phototake/GetStock; **15** (bottom right) Wiki commons; **20** (left) JEREMY BURGESS/SCIENCE PHOTO LIBRARY; **20** (right) Don W. Fawcett/Photo Researchers, Inc.; **21** (centre) Science Source/J.D. Litvay/Visuals Unlimited; **24** Wolfgang Polzer/GetStock; **38** () MAXIMILIAN STOCK LTD/SCIENCE PHOTO LIBRARY; **41** Photo courtesy of the Canadian MPS Society; **56-57** DR DAVID FURNESS, KEELE UNIVERSITY/SCIENCE PHOTO LIBRARY; **57** (top centre) Visuals Unlimited/Corbis; **57** (top right) CNRI/SCIENCE PHOTO LIBRARY; **58** © Dr. Dennis Kunkel/Visuals Unlimited; **59** © E.H. Newcomb/W.P. Wergin/Biological Photo Service; **60** (top right) Courtesy Ron Milligan/Scripps Research Institute; **62** PROFESSORS P. MOTTA & T. NAGURO/SCIENCE PHOTO LIBRARY; **63** J.C. REVY, ISM/SCIENCE PHOTO LIBRARY; **64** (top) Courtesy of Herbert W. Israel, Cornell University; **65** (top) Courtesy of Dr. Keith Porter; **66** (bottom left) EYE OF SCIENCE/SCIENCE PHOTO LIBRARY; **66** (bottom right) EYE OF SCIENCE/SCIENCE PHOTO LIBRARY; **67** ROBERT MCNEIL, BAYLOR COLLEGE OF MEDICINE/SCIENCE PHOTO LIBRARY; **71** DR DAVID FURNESS, KEELE UNIVERSITY/SCIENCE PHOTO LIBRARY; **78** (top right) Micrograph Courtesy of the CDC/Dr. Edwin P. Ewing, Jr.; **78** (second from top) © BCC Microimaging, Inc., Reproduced with permission; **78** (bottom two images) © The Company of Biologists Limited ; **79** (right) © Dr. Brigit Satir; **80** Courtesy of Dr. Frances Sharom; **84** (centre) photo by Blaise MacMullin © 2002 Athabasca University, reproduced with permission; **85-86** (background) Purestock/Getty Images; **86** (centre right) LAGUNA DESIGN/SCIENCE PHOTO LIBRARY; **96-97** (background) Ingram Publishing; **96** (bottom left) Ingram Publishing; **98** (centre right) PAUL D STEWART/SCIENCE PHOTO LIBRARY; **98** (bottom right) DR DAVID FURNESS, KEELE UNIVERSITY/SCIENCE PHOTO LIBRARY; **101** (top left) Courtesy Ron Milligan/Scripps Research Institute; **101** (bottom left) Courtesy E.G. Pollock; **106-107** HOGE NOORDEN JACOB VAN ESSEN/AFP/Getty Images; **112-113** Lokinthru/Dreamstime.com/GetStock; **115** (bottom right) Tan Wei Ming/iStockphoto; **117** (top left) Spencer Grant/Photo Edit; **117** (top centre) Spencer Grant/Photo Edit; **141** () Image Source/Getty Images; **143** (left) REUTERS/Hoi-Ying Holman Group/Landov; **143** (right) ROBERT BROOK/SCIENCE PHOTO LIBRARY; **144-145** (background) The McGraw-Hill Companies, Inc./Jill Braaten, photographer; **144** (top centre) JGI/Blend Images LLC; **144** (bottom left) Health Canada; **145** (bottom centre) The McGraw-Hill Companies, Inc./Jill Braaten, photographer; **154-155** Mike Grandmaison/All Canada Photos/Corbis; **155** (top right) Dr. David Newmann/Visuals Unlimited; **156** (centre left) Nancy Nehring/iStockphoto; **156** (centre) Island Effects/iStockphoto.com; **156** (centre right) konradlew/iStockphoto; **163** Pavlo Vakhrushev/iStockphoto; **164** (top left) Courtesy of Dr. Ibrahim Dincer; **164** (top centre) Courtesy of Dr. Ibrahim Dincer; **169** (centre left) HAIBO BI/iStockphoto; **169** (bottom centre) liveostockimages/iStockphoto; **186-187** Science World British Columbia/Photo by Stewart Marshall/Flickr; **186** (centre) Lee W. Wilcox; **186** (centre right) Courtesy Ray F. Evert/University of Wisconsin Madison; **196-197** HYBRID MEDICAL ANIMATION/SCIENCE PHOTO LIBRARY; **198** © Dr. Dennis Kenkel/Visuals Unlimited; **202-203** DR GOPAL MURTI/SCIENCE PHOTO LIBRARY; **204** PR. G GIMENEZ-MARTIN/SCIENCE PHOTO LIBRARY; **210** (left) Science Source/Photo Researchers, Inc.; **210** (right) SCIENCE PHOTO LIBRARY; **211** A. BARRINGTON BROWN/SCIENCE PHOTO LIBRARY; **213** (centre) M. Freeman/PhotoLink/Getty Images; **214** (centre right) Richard J. Green/Photo

Researchers, Inc.; **216** (left side, top to bottom) Dr. Gobal Murti/Visuals Unlimited; Ada L. Olins and Donald E. Olins/Biological Photo Service; Courtesy of Dr. Jerome B. Rattner, Cell Biology and Anatomy, The University of Calgary; Paulson, JR. & Laemmli, UK. The structure of histone-depleted metaphase chromosomes. Cell. 1977 Nov; 12(3):817-28, f. 5; Peter Engelhardt/Department of Department of Pathology, Haartman Institute, University of Helsinki, Finland/Nanomicroscopy Center (NMC), Department of Applied Physics, Aalto University, School of Science and Technology, Finland/Centre of Excellence in Co; Peter Engelhardt/Department of Department of Pathology, Haartman Institute, University of Helsinki, Finland/Nanomicroscopy Center (NMC), Department of Applied Physics, Aalto University, School of Science and Technology, Finland/Centre of Excellence in Co; **217** (bottom left) Ken Lucas/Visuals Unlimited, Inc./Getty Images; **217** (bottom centre) Dex Image/Corbis; **217** (bottom right) SINCLAIR STAMMERS/SCIENCE PHOTO LIBRARY; **223** (bottom) Courtesy of Tsuneko Okazaki; **231** PHANIE/Photo Researchers, Inc.; **242-243** Friday/Dreamstime.com/GetStock; **253** (right) PROFESSOR OSCAR L. MILLER/SCIENCE PHOTO LIBRARY; **258** (bottom right) Courtesy Alexander Rich; **264** Courtesy of the Barbara McClintock Papers, American Philosophical Society; **271** Courtesy of Dr. Sashdev Sidhu; **283** PROFESSOR OSCAR L. MILLER/SCIENCE PHOTO LIBRARY; **284-285** Sean O'Neill/GetStock; **286** (centre right) REUTERS/STR/Landov; **287** (top right) Mark Harmel/GetStock; **289** (bottom centre) VINCE BUCCI/AFP/Getty Images; **293** (top) Patrice Latron/Corbis; **298** Courtesy of Elaine Mardis; **299** © Nic Lehoux; **300** Ada L. Olins and Donald E. Olins/Biological Photo Service; **301** (top) PATRICK LANDMANN/SCIENCE PHOTO LIBRARY; **301** (bottom) Photo courtesy of GloFish/via Getty Images; **304** (left) Courtesy of Dr. Ananda Chakrabarty; **304** (right) Dennis Kunkel Microscopy, Inc./Visuals Unlimited, Inc.; **309** (bottom left) AJ/IRRI/Corbis; **310** REUTERS/Texas A&M University/Landov; **314** CP/Paul Chiasson; **316-317** (background) Photograph by Cecil W. Forsberg/University of Guelph; **316** (bottom left) MCT/Landov; **316** (bottom right) Photograph by Cecil W. Forsberg/University of Guelph; **322** Dennis Kunkel Microscopy, Inc./Visuals Unlimited, Inc.; **323** Najlah Feanny/CORBIS SABA; **326-327** (background) Rosen Karanedev/iStockphoto; **326** (bottom left) CDC/Dr. A. Harrison and Dr. P. Feorino; **326** (bottom centre) Rosen Karanedev/iStockphoto; **328** (top right) DR GOPAL MURTI/SCIENCE PHOTO LIBRARY; **328** (centre right) Friday/Dreamstime.com/GetStock; **328** (bottom right) Sean O'Neill/GetStock; **334** PhotoAlto/PictureQuest; **335** SCIENCE PHOTO LIBRARY; **336-337** Rune Hellestad/Corbis; **339** (bottom left) STEVE GSCHMEISSNER/SCIENCE PHOTO LIBRARY; **342-343** (top centre) Ann C. McKee, MD/Bedford VA Hospital/Boston University School of Medicine; **342** (centre left) Frederick C. Skvara, MD/Visuals Unlimited/Corbis; **342** (centre) Ann C. McKee, MD/Bedford VA Hospital/Boston University School of Medicine; **342** (bottom right) Blair Gable; **343** (centre left) Visuals Unlimited/Corbis; **343** (bottom left) Blair Gable; **344** (top centre) SCIENCE PHOTO LIBRARY; **349** Rob Howard/Corbis; **351** (top right) Dennis Kunkel Microscopy, Inc./Phototake/GetStock; **361** Samantha Craggs/Brock University; **363** (bottom left) MANFRED KAGE/SCIENCE PHOTO LIBRARY; **363** (bottom right) Anatomical Travelogue/Photo Researchers, Inc.; **364** (bottom right) HERVE CONGE, ISM/SCIENCE PHOTO LIBRARY; **367** (centre) VOLKER STEGER/SCIENCE PHOTO LIBRARY; **369** Ralph Hutchings/Visuals Unlimited; **370** Jaboardm/Dreamstime.com/GetStock; **371** (bottom right) PM Images/Getty Images; **373** (bottom left) Otnaydur/Dreamstime.com/GetStock; **376** ASTRID & HANNS-FRIEDER MICHLER/SCIENCE PHOTO LIBRARY; **379** (bottom left) © J. Timothy Cannon, Ph.D.; **379** (bottom left) © J. Timothy Cannon, Ph.D.; **385** © Motor Lab, University of Pittsburgh; **388-389** mayo5/iStockphoto; **389** (top right) Paul Cox/arabianEye/Corbis; **390** Richard Lautens/GetStock; **398** (bottom left) China Daily/Reuters/Corbis; **398** (bottom right) AP Photo/Hays Daily News/Jeff Cooper; **400** (top left) The McGraw-Hill Companies, Inc./Al Telser, photographer; **404** (top) Andrew Stawicki/GetStock; **406** The McGraw-Hill Companies, Inc./Al Telser, photographer; **410** Carolina Biological/Visuals Unlimited/Corbis; **412** BELMONTE/SCIENCE PHOTO LIBRARY; **415** (top right) POWER AND SYRED/SCIENCE PHOTO LIBRARY; **426-427** (background) Brand X Pictures/PunchStock; **426** (bottom right) Damir Spanic/iStockphoto; **427** (top right) David Hoffman/GetStock; **428** Kornilovdream/Dreamstime.com/GetStock; **442-443** Medical Body Scans/Photo Researchers, Inc.; **444** (top left) Getty Images/SW Productions; **444** (top right) Chris Schmidt/iStockphoto; **444** (bottom left) Blend Images/Alamy; **449** () THOMAS DEERINCK, NCMIR/SCIENCE PHOTO LIBRARY; **456** Creatas/PunchStock; **462** (top right) PHOTOTAKE Inc./Alamy; **462** (centre right) © Dr. P. Marazzi/Photo Researchers, Inc.; **463** Frances Roberts/GetStock; **464** SHEILA TERRY/SCIENCE PHOTO LIBRARY; **466** Brand X Pictures/PunchStock; **478-479** (background) Burke/Triolo/Getty Images; **478** (bottom centre) Hrabar/Dreamstime.com/GetStock; **478** (centre right) Darius Ramazani/Corbis; **478** (bottom right) Ingram Publishing/SuperStock; **480** (centre right) Ann C. McKee, MD/Bedford VA Hospital/Boston University School of Medicine; **480** (bottom right) mayo5/iStockphoto; **481** Medical Body Scans/Photo Researchers, Inc.; **490-491** Don Johnston/All Canada Photos/Getty Images; **492** (right) Hugo Willocx/Foto Natura/Minden Pictures; **493** (centre left) Adam Jones/Visuals Unlimited, Inc./Getty Images; **493** (bottom left) © Ian McAllister/Getty Images; **496-497** CDC/James Gathany; **498** All Canada Photos/SuperStock; **499** (bottom) Leslie Garland/LGPL/GetStock; **500** Rick Eglinton/GetStock; **501** (left) Comstock/JupiterImages; **501** (centre) Cindy Creighton/iStockphoto; **501** (right) Socanski/Dreamstime.com/GetStock; **502** Tratong/Dreamstime.com/GetStock; **503** Photodisc Collection/Getty Images; **505** Angelo Cavalli/Getty Images; **506** Robert Koopmans/iStockphoto; **509** Darrell Gulin/Corbis; **510** Gary Meszaros/Visuals Unlimited/Corbis; **511** Mark and Leslie Degner/Wilderness Light; **513** (left) CROWN COPYRIGHT/HEALTH & SAFETY LABORATORY/SCIENCE PHOTO LIBRARY; **513** (right) Andrew McLachlan/All Canada Photos/Corbis; 514 (left) Science Photo Library RF/Getty Images; **515** (bottom right) Sean O'Neill/GetStock; **517** Getty Images; **518** (top left) Patrick J. Lynch/Photo Researchers, Inc.; **518** (bottom) Norma Joseph/GetStock; **520** CP/Ryan Remiorz; **521** (bottom right) Dr. Richard Kessel & Dr. Gene Shih/Visuals Unlimited/Corbis; **522** (top left) Matauw/iStockphoto; **522** (bottom left) MAURICE NIMMO/SCIENCE PHOTO LIBRARY; **522** (centre) Michael DeFreitas North America/Alamy; **524** Chris Stenger/Foto Natura/Minden Pictures; **525** (top left) A & J Visage/GetStock; **525** (top right) Dubults/Dreamstime.com/GetStock; **525** (bottom right) Photodisc Collection/Getty Images; **526** (bottom) Schulz/GetStock; **528** (centre) Creatas Images/PictureQuest; **528** (bottom left) R. Degginger/Photo Researchers, Inc.; **528** (bottom right) Edward R. Degginger/GetStock; **529** (left) Jim Zuckerman/GetStock; **529** (right) Alex Kerstitch/Visuals Unlimited, Inc.; **530** (top left) Peter Garbet/iStockphoto; **530** (centre left) Nigel Cattlin/GetStock; **531** (top right) Anndrey Nekrasov/Alamy; **531** (bottom) Peter Arnold, Inc./Alamy; **532** Jurie Maree/iStockphoto; **533** Murphy_Shewchuk/iStockphoto; **539** (left) Wil Meinderts/Foto Natura/Minden Pictures; **539** (right) Steve Maslowski/Visuals Unlimited/Corbis; **541** Alan Pembleton/GetStock; **542** Andrew Darrington/GetStock; **544** Stephen Durr; **548-549** Jeremy Woodhouse/Getty Images; **559** Jake Lyell/GetStock; **560** THONY BELIZAIRE/AFP/Getty Images; **561** (top centre) ZEPHYR/SCIENCE PHOTO LIBRARY; **561** Yashkru/Dreamstime.com/GetStock; **563** Rosanne Tackaberry/GetStock; **564** bryan konya/iStockphoto; **565** George Steinmetz/Corbis; **566** canadabrian/GetStock; **567** Robert Bird/GetStock; **568** Duncan Usher/GetStock; **569** Harrison Eastwood/Getty Images; **576-577** (background) Visions of America, LLC/Alamy; **576** (centre right) Photothek/Andia/GetStock; **577** (centre) Index Stock/Alamy; **578** Courtesy of Professory Sally Humphries; **580** Scott Bauer/USDA; **581** Alex L. Fradkin/Getty Images; **582** narvikk/iStockphoto; **583** Aleksandar Jaksic/iStockphoto; **585** Frans Lanting Studio/Alamy; **586-587** (background) Clint Farlinger/GetStock; **586** (centre right) Clint Farlinger/GetStock; **586** (bottom right) Rinus Baak/Dreamstime.com; **588** (centre right) CDC/James Gathany; **588** (bottom right) Jeremy Woodhouse/Getty Images; **591** Brand X Pictures/PunchStock; **592** Jared Hobbs/All Canada Photos/Corbis; **593** Royalty-Free/Corbis; **594** Poelzer Wolfgang/GetStock; **604** Tetra Images/Getty Images; **662** © E.H. Newcomb/W.P. Wergin/Biological Photo Service; **663** Courtesy of Herbert W. Israel, Cornell University; **666** (top right) Micrograph Courtesy of the CDC/Dr. Edwin P. Ewing, Jr.; (second from top) © BCC Microimaging, Inc., Reproduced with permission; (bottom two images) © The Company of Biologists Limited.

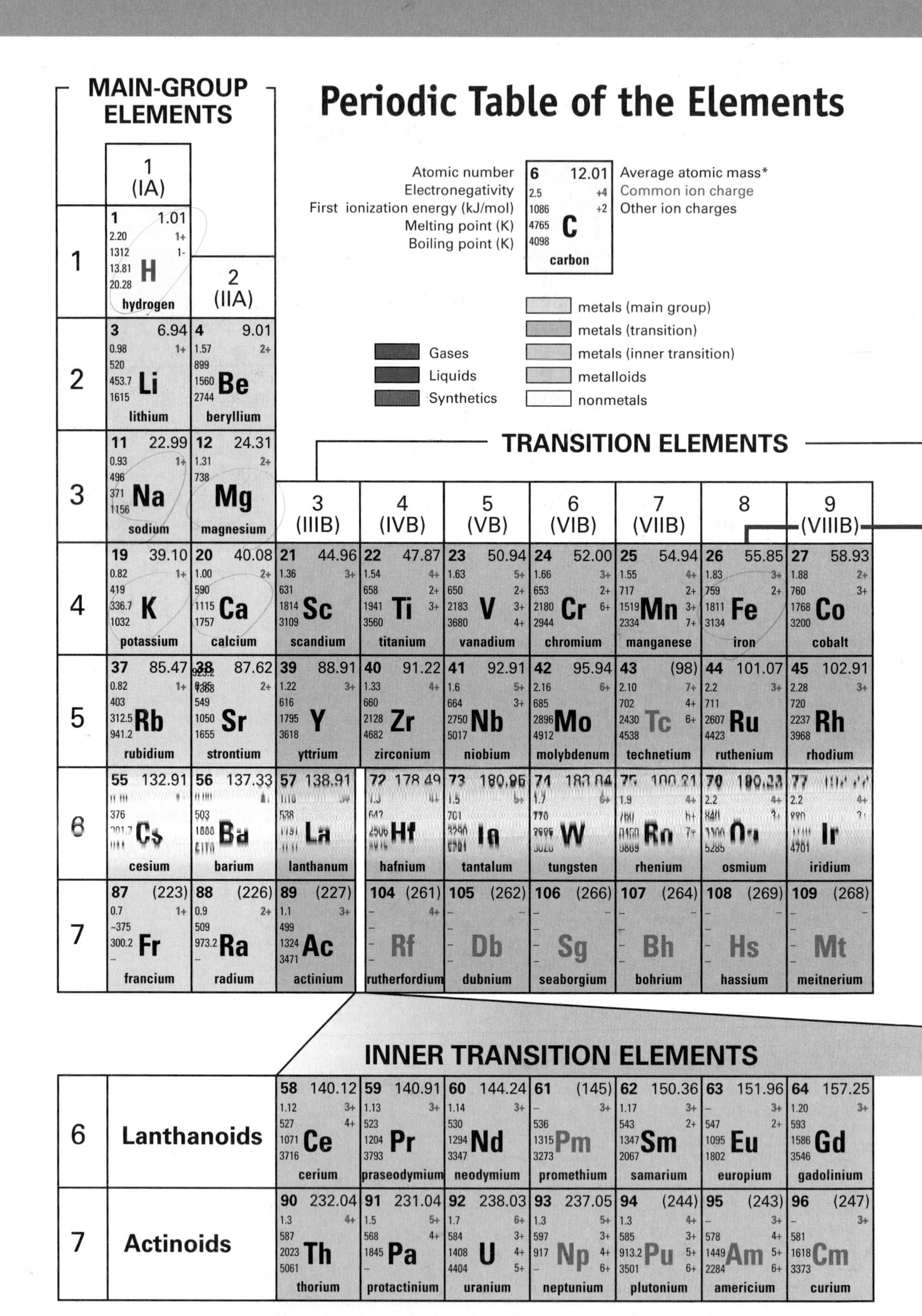

*Average atomic mass data in brackets indicate atomic mass of most stable isotope of the element.

MAIN-GROUP ELEMENTS

10	11 (IB)	12 (IIB)	13 (IIIA)	14 (IVA)	15 (VA)	16 (VIA)	17 (VIIA)	18 (VIIIA)
								2 4.00 – – 2372 5.19 5.02 **He** helium
			5 10.81 2.04 – 800 2348 4273 **B** boron	6 12.01 2.55 – 1086 4765 4098 **C** carbon	7 14.01 3.04 3– 1402 63.15 77.36 **N** nitrogen	8 16.00 3.44 2– 1314 54.36 90.2 **O** oxygen	9 19.00 3.98 1– 1681 53.48 84.88 **F** fluorine	10 20.18 – – 2080 24.56 27.07 **Ne** neon
			13 26.98 1.61 – 577 933.5 2792 **Al** aluminum	14 28.09 1.90 – 786 1687 3538 **Si** silicon	15 30.97 2.19 – 1012 317.3 553.7 **P** phosphorus	16 32.07 2.58 2– 999 392.8 717.8 **S** sulfur	17 35.45 3.16 1– 1256 171.7 239.1 **Cl** chlorine	18 39.95 – – 1520 83.8 87.3 **Ar** argon
28 58.69 1.91 2+ 737 3+ 1728 3186 **Ni** nickel	29 63.55 1.90 2+ 745 1+ 1358 2835 **Cu** copper	30 65.39 1.65 2+ 906 692.7 1180 **Zn** zinc	31 69.72 1.81 3+ 579 302.9 2477 **Ga** gallium	32 72.61 2.01 – 761 1211 3106 **Ge** germanium	33 74.92 2.18 – 947 1090 876.2 **As** arsenic	34 78.96 2.55 2– 941 493.7 958.2 **Se** selenium	35 79.90 2.96 1– 1143 266 332 **Br** bromine	36 83.80 – – 1351 115.8 119.9 **Kr** krypton
46 106.42 2.20 2+ 805 3+ 1828 3236 **Pd** palladium	47 107.87 1.93 1+ 731 1235 2435 **Ag** silver	48 112.41 1.69 2+ 868 594.2 1040 **Cd** cadmium	49 114.82 1.78 3+ 558 429.8 3345 **In** indium	50 118.71 1.96 4+ 708 2+ 505 2875 **Sn** tin	51 121.76 2.05 – 834 903.8 1860 **Sb** antimony	52 127.60 2.1 – 869 722.7 1261 **Te** tellurium	53 126.90 2.66 1– 1009 386.9 457.4 **I** iodine	54 131.29 – – 1170 161.4 165 **Xe** xenon
78 195.08 2.2 4+ 870 2+ 2042 4098 **Pt** platinum	79 196.97 2.4 3+ 890 1+ 1337 3129 **Au** gold	80 200.59 1.9 2+ 1107 1+ 234.3 629.9 **Hg** mercury	81 204.38 1.8 1+ 589 3+ 577.2 1746 **Tl** thallium	82 207.20 1.8 2+ 715 4+ 600.6 2022 **Pb** lead	83 208.98 1.9 3+ 703 5+ 544.6 1837 **Bi** bismuth	84 (209) 2.0 4+ 813 2+ 527.2 1235 **Po** polonium	85 (210) 2.2 1– (926) 575 – **At** astatine	86 (222) – – 1037 202.2 211.5 **Rn** radon
110 (271) – – – – – **Uun** ununnilium	111 (272) – – – – – **Uuu** unununium	112 (277) – – – – – **Uub** ununbium	113	114 (285) – – – – – **Uuq** ununquadium	115	116 (289) – – – – – **Uuh** ununhexium		

65 158.93 – 3+ 565 1629 3503 **Tb** terbium	66 162.50 1.22 3+ 572 1685 2840 **Dy** dysprosium	67 164.93 1.23 3+ 581 1747 2973 **Ho** holmium	68 167.26 1.24 3+ 589 1802 3141 **Er** erbium	69 168.93 1.25 3+ 597 1818 2223 **Tm** thulium	70 173.04 – 3+ 603 2+ 1092 1469 **Yb** ytterbium	71 174.97 1.0 3+ 524 1936 3675 **Lu** lutetium
97 (247) – 3+ 601 4+ 1323 – **Bk** berkelium	98 (251) – 3+ 608 1173 – **Cf** californium	99 (252) – 3+ 619 1133 – **Es** einsteinium	100 (257) – 3+ 627 1800 – **Fm** fermium	101 (258) – 3+ 635 2+ 1100 – **Md** mendelevium	102 (259) – 3+ 642 2+ 1100 – **No** nobelium	103 (262) – 3+ – 1900 – **Lr** lawrencium